T0140072

Studies in Computational Intelligence

Volume 688

Series editor

Janusz Kacprzyk, Polish Academy of Sciences, Warsaw, Poland
e-mail: kacprzyk@ibspan.waw.pl

About this Series

The series "Studies in Computational Intelligence" (SCI) publishes new developments and advances in the various areas of computational intelligence—quickly and with a high quality. The intent is to cover the theory, applications, and design methods of computational intelligence, as embedded in the fields of engineering, computer science, physics and life sciences, as well as the methodologies behind them. The series contains monographs, lecture notes and edited volumes in computational intelligence spanning the areas of neural networks, connectionist systems, genetic algorithms, evolutionary computation, artificial intelligence, cellular automata, self-organizing systems, soft computing, fuzzy systems, and hybrid intelligent systems. Of particular value to both the contributors and the readership are the short publication timeframe and the worldwide distribution, which enable both wide and rapid dissemination of research output.

More information about this series at http://www.springer.com/series/7092

Ahmad Taher Azar · Sundarapandian Vaidyanathan
Adel Ouannas
Editors

Fractional Order Control and Synchronization of Chaotic Systems

Springer

Editors
Ahmad Taher Azar
Faculty of Computers and Information
Benha University
Benha
Egypt

and

Nanoelectronics Integrated Systems Center (NISC)
Nile University
Cairo
Egypt

Sundarapandian Vaidyanathan
Research and Development Centre
Vel Tech University
Chennai, Tamil Nadu
India

Adel Ouannas
Laboratory of Mathematics, Informatics and Systems (LAMIS)
University of Larbi Tebessi
Tebessa
Algeria

ISSN 1860-949X ISSN 1860-9503 (electronic)
Studies in Computational Intelligence
ISBN 978-3-319-84356-8 ISBN 978-3-319-50249-6 (eBook)
DOI 10.1007/978-3-319-50249-6

Softcover reprint of the hardcover 1st edition 2017

Printed on acid-free paper

This Springer imprint is published by Springer Nature
The registered company is Springer International Publishing AG
The registered company address is: Gewerbestrasse 11, 6330 Cham, Switzerland

Preface

About the Subject

Fractional systems and fractional control have received great attention recently, both from an academic and industrial viewpoint, because of their increased flexibility (with respect to integer order systems) which allows a more accurate modeling of complex systems and the achievement of more challenging control requirements.

Chaotic systems have been studied for recent decades after the discovery of the first classical chaotic attractor in 1963. Chaos control and chaos synchronization are especially remarkable and important research fields aiming to affect the dynamics of chaotic systems in order to use them for different kinds of applications that can be examined within many different fields such as computer sciences, mechanics, communication, economics and finance, biology, chemistry, medicine, and geology, among others.

About the Book

The new Springer book, *Fractional Order Control and Synchronization of Chaotic Systems*, consists of 30 contributed chapters by subject experts who are specialized in the various topics addressed in this book. The special chapters have been brought out in this book after a rigorous review process in the broad areas of chaos theory, control systems, computer science, fuzzy logic, neural network, and modeling and engineering applications. Special importance was given to chapters offering practical solutions and novel methods for the recent research problems in the main areas of this book, viz. fractional order control and synchronization of chaotic systems.

Objectives of the Book

This book aims at presenting the latest developments, trends, research solutions, and applications of fractional order control and synchronization of chaotic systems. There are different methods that have been proposed for performing fractional order control, chaos control, and chaos synchronization tasks. But because of some limitations of these methods, newer approaches are designed and proposed by researchers to improve the related works within the field. Also, there are many studies already introduced in order to improve more advanced solutions for problems of fractional order control, chaos control, and chaos synchronization. Most of these studies include the usage of intelligent approaches, optimization methods, and hybrid techniques on the related problems of the control theory. Both novice and expert readers should find this book a useful reference in the field of fractional order control and chaos synchronization.

Organization of the Book

This well-structured book consists of 30 full chapters. They are organized into two parts.

Part I: Fractional Order Control Systems
Part II: Applications of Fractional Order Chaotic Systems

Book Features

- The chapters in this book deal with the recent research problems in the areas of fractional order control, chaos theory, control, synchronization, and engineering applications.
- The chapters in this book contain a good literature survey with a long list of references.
- The chapters in this book are well written with a good exposition of the research problem, methodology, numerical examples, simulation results, and block diagrams.
- The chapters in this book discuss the details of engineering applications and future research areas.

Audience

This book is primarily meant for researchers from academia and industry, who are working in the research areas—Fractional Order Control and Synchronization of Chaotic Systems with applications in engineering, automation, chaos, and control engineering. This book can also be used at the graduate or advanced undergraduate level as a textbook or major reference for courses such as control systems, fractional differential equations, fractional control systems, mathematical modeling, computational science, numerical simulation, fuzzy logic control, and many others.

Acknowledgements

As the editors, we hope that the chapters in this well-structured book will stimulate further research in fractional order control systems, fractional order chaotic systems, and synchronization of chaotic systems and utilize them in the real-world applications.

We hope sincerely that this book, covering so many different topics, will be very useful for all readers.

We would like to thank all the reviewers for their diligence in reviewing the chapters.

Special thanks go to Springer, especially the book Editorial team.

Cairo, Egypt — Ahmad Taher Azar
Chennai, India — Sundarapandian Vaidyanathan
Tebessa, Algeria — Adel Ouannas

Contents

Part I
Fractional Order Control Systems

Comparative Study on Fractional Order PID and PID Controllers on Noise Suppression for Manipulator Trajectory Control

Vineet Kumar and K.P.S. Rana

Abstract The main contribution of this chapter is to demonstrate the sensor and controller noise suppression capabilities of the best tuned Fractional Order-Proportional plus Integral plus Derivative (FO-PID) and classical PID controllers in closed-loop. A complex non-linear and coupled system, a 2-link rigid planar manipulator was considered for the study as it encounters noise in many forms such as sensor and controller noise during the operation in industry. Uniform White Noise (UWN) and Gaussian White Noise (GWN) were considered both for the sensor and the controller in the closed-loop and a comparative study was performed for FO-PID and PID controllers. Both the controllers were tuned using Genetic Algorithm and all the simulations were performed in LabVIEW environment. The simulation results have revealed that FO-PID controller demonstrates superior sensor and controller noise suppression as compared to conventional PID controller in the closed-loop.

Keywords Fractional order PID controller · PID · Sensor noise · Controller noise · Noise suppression and uncertainty

1 Introduction

Noise is generally an inherent part of every measurement and control system. Noise affects the decision-making capability of the controller in the plant as it introduces uncertainty in the process variable and control action. There are several causes of measurement noise in the process industry such as loose wiring, improper soldering, thermal noise, electromagnetic interference, etc. Furthermore, the execution of control action in a closed-loop control system may be seriously affected due to the

V. Kumar (✉) · K.P.S. Rana
Department of Instrumentation and Control Engineering, Netaji Subhas Institute of Technology, Sector-3, Dwarka, New Delhi 110078, India
e-mail: vineetkumar27@gmail.com

K.P.S. Rana
e-mail: kpsrana1@gmail.com

A.T. Azar et al. (eds.), *Fractional Order Control and Synchronization of Chaotic Systems*, Studies in Computational Intelligence 688,
DOI 10.1007/978-3-319-50249-6_1

improper connection of signal line from the controller to the final control element particularly in a vibrating environment. This action may add considerable noise in the control action and may even try to destabilize the process being controlled. Most general nature of sensor and controller noise is likely to be random. Robotic manipulator being a highly non-linear, time-varying and uncertain system noise has adverse effect on the positioning and accuracy of the end-effectors of robotic manipulator. Therefore, it is a challenging task to design such a robust controller which can effectively deal with external noises and uncertainties [5, 7–9, 11, 22, 30, 33, 55, 59].

Conventional PID controller has been the most popular controller in the industries for last sixty years due to its simple structure, low cost, easy design and robust performance. For its implementation it is available in many forms in industry, like pneumatic, hydraulic and electronic etc. Due to these features, it is one of the popular choices of control engineers in robotics as well [1, 3, 6].

The fractional order calculus has a very long mathematical history, but its applications to science and engineering are just recent. One of its popular applications is in fractional order chaotic system [4, 10, 51–53]. Many good works have been reported regarding the control and synchronization of fractional order chaotic system [15, 16, 46–50]. Also, in the last few years, fractional-order calculus has gained extensive attention of researchers and scientists in the area of control engineering to develop fractional order PID (FO-PID) controller [32, 37, 54]. Due to advancements in the electronics now it has become possible to design and realize the FO-PID controllers. It offers additional design Degree of Freedom (DOF) to the control engineers. Advancements also have been observed in the controller tuning methods for customized performance indices. Now-a-days the trend has been the usage of optimization methods such as Genetic Algorithm (GA) and other bio-inspired techniques for tuning the controllers. Therefore, the main goal of this chapter is to demonstrate the capabilities of FO-PID controller to cater the effect of noise in closed loop. In this regard, an intensive simulation studies were performed in closed loop by controlling a complex coupled system i.e. 2-link rigid planar robotic manipulator in presence of sensor and controller noise and using fractional and integer order PID controllers. Uniform White Noise (UWN) and Gaussian White Noise (GWN) were considered for this study. Both the controllers were tuned using GA for minimum Integral of Absolute Error (IAE). The performed comparative study reveals that FO-PID controller with low fractional order derivative term effectively suppress the noise in closed loop. The main contributions of this chapter can be summarized as follows:

- It demonstrates the effect of sensor and controller noise in closed loop for trajectory tracking control of 2-link rigid planar robotic manipulator using fractional and integer order PID controllers.
- It shows that FO-PID controller having low order derivative term effectively suppress the noise effect in closed loop as compared to its counterpart integer order PID controller.

- The performance of fractional and integer order PID controllers tuned with GA were assessed for IAE and it has been found that FO-PID controller outperformed integer order PID controller for sensor as well as controller noise suppression study.

The rest of the chapter is organized as follows: A brief literature review of the related works have been presented in Sect. 2. The implementation of fractional order operator is presented in Sect. 3. In Sect. 4, fractional order PID controller is presented. In Sect. 5, the performance criteria chosen are explained. In Sect. 6, the mathematical model of the 2-link robotic manipulator is described. The detailed description of the performed simulation experiments and obtained results are presented in the Sect. 7. Finally, the conclusions of the present work are drawn in Sect. 8.

2 Literature Survey

Noise in any form may degrade the performance of any controller in closed loop. Therefore, it has drawn attention of various researchers and scientists over the time especially in the field of control system. Many research works have been reported in this regard and presented as follows.

Tsai et al. presented the robustness testing for sensor noise for the fuzzy model-following control applied to a 2-link robotic manipulator [45]. Chaillet et al. investigated the robustness study of PID controller for the external disturbance for a robotic manipulator. A uniform semi-global practical asymptotic stability for PID controller for robotic system with external disturbance was investigated. Simulation was done for the 2-link robotic manipulator with viscous and coulomb friction effects [17]. Song et al. presented a computed torque controller scheme with fuzzy as compensator for the uncertainties in a robotic manipulator. The robustness testing was done with nonlinearities, uncertainties and flexibilities [41]. Tang et al. presented a modified fuzzy PI controller for a flexible–joint robotic manipulator for path tracking performance in handling uncertainty and nonlinearity. The uncertainties used were within 10% tolerance of all nominal system parameter values [42]. Bingul and Karahan investigated the fuzzy logic controller for a 2-link robotic manipulator. The robustness testing was done with model uncertainties, change in used trajectory and white noise addition to the system. White noise with different noise powers were added to each link [12]. Chen presented dynamic structure neural-fuzzy network adaptive controller for robotic manipulator. The robustness was tested with payload variation at second link [18]. Tian and Collins studied the robustness testing of the adaptive neuro-fuzzy control of a flexible manipulator with the tip payload variations [44].

Bingul and Karahan presented a comparative study of fractional PID controllers tuned by Particle Swarm Optimization (PSO) and GA for a 2-link robotic manipulator. The robustness testing included parameter change, trajectory change and addition of white noise. Simulation results, for the robot trajectory experiment, showed that the FO-PID controller tuned by PSO has better performance than the FO-PID controller tuned by the GA [13].

Lin and Huang presented a hierarchical supervisory fuzzy controller for the robot manipulators with oscillatory base. The proposed method had various benefits such as reduced chattering effect, lesser overshoot, faster convergence and lesser online computation time [28]. Peng and Woo investigated a neural-fuzzy controller for the 2-link planar robotic manipulator. In this work, the simulation results suggested that the controller based on online weight adjustment is robust in the presence of uncertainties like friction, unknown disturbance and changing payload [36]. El-Khazali introduces a new design method of FO-PD and FO-PID controllers. A biquadratic approximation of a fractional order differential operator is used to introduce a new structure of finite-order FO-PID controllers. They claimed that using the new FO-PD controller, the controlled system can achieve the desired phase margins without migrating the gain crossover frequency of the uncontrolled system. The proposed FO-PID controller has smaller number of parameters to tune than its existing counterparts. The viability of the design methods is verified using a simple numerical example [21]. Li and Huang presented an adaptive fuzzy terminal sliding mode controller for robotic manipulator. It has several advantages like eliminates chattering, good response with uncertainties and disturbances etc. [27]. Yildirim and Eski presented different neural network implementations as noise analyzer for the robotic manipulator. The performance of Radial Basis Function Neural Network was better than all other networks [56]. Oya et al. investigated a continuous time tracking controller without using velocity measurements of the robotic manipulator. The noise due to quantization error was studied. The proposed controller offered better results than those based on Eular approximation [34]. Zhu and Fang presented a fuzzy neural network algorithm for parallel manipulators. It was designed to cope with external disturbance, payload variation and model uncertainties. The neural network was used to modify the fuzzy rules [58].

Petras presented the hardware implementation of digital FO-PID control for permanent magnet DC motor. The digital and analog realization of proposed controller was done with microprocessor and fractance circuits, respectively [37]. Delavari et al. reported a fractional adaptive PID controller for robotic manipulator in which parameters of PID are updated online and the fractional order parameters are obtained offline [20]. Bingul and Karahan applied a FO-PID controller to a robotic manipulator for trajectory tracking problem using PSO. The simulation results showed that FO-PID controller performed better than that conventional PID [14]. Silva et al. proposed the superiority of fractional order controller over integer order controller for a hexapod robot in which flexibilities and viscous friction were present at the joints of the legs [40]. Sharma et al. presented a comparative performance analysis of fractional order fuzzy PID controller, fuzzy PID, FO-PID and PID controller for a trajectory tracking and disturbance rejection of a 2-link robotic manipulator. Simulation studies revealed that fractional order FPID outperformed rest of the controller [39]. Kumar et al. proposed a robust fractional order Fuzzy

P + Fuzzy I + Fuzzy D controller for nonlinear and uncertain system which offered superior performance as compared to its integer order counterpart fuzzy controller [24].

Pan and Das proposed a FO-PID controller for automatic voltage regulator (AVR) with chaotic multi-objective optimization. In this work, FO-PID controller was not completely superior to PID controller as the simulations results were better for FO-PID for some cases and for conventional PID in other cases [35]. Zamani et al. investigated the performances of FO-PID controllers for AVR systems [57]. Tang et al. developed FO-PID controller for AVR system. The performance of FO-PID was superior to PID controller for the system with or without uncertainties. The Chaotic Ant Swarm algorithm was used for finding optimized parameters [43]. Luo and Chen proposed a systematic tuning procedure for the FO-PD controller for a FO system. The performance of the proposed controller was superior to both FO-PD and integer order PD controller [29]. Monje et al. presented a tuning method of FO-PID controller and ensured the robustness for the noise as well as gain variation. Also, an auto-tuning method using relay test has investigated. Experimental results showed the effectiveness of the proposed tuning methods [32]. Ladaci et al. designed an adaptive internal model controller (AIMC) with a FO system. The robustness of the proposed controller was done against noise. It was superior to conventional AIMC and also, to conventional PID controller [26]. Many other recent applications, such as, binary distillation column control [31] and control of hybrid electric vehicle [25] have also been reported in the literature for making use of fractional order control system.

The literature survey conducted above clearly indicates that several authors have investigated effects of the model uncertainties and external disturbances for the robotic manipulators but the effects of sensor as well as controller noise has not been well explored by the researchers and therefore needs attention.

3 Fractional Order Calculus

In the present work, fractional order operators (differ-integral) are implemented using Grünwald-Letnikov (G-L) method [38, 54]. The definition of GL fractional differ-integral can be expressed as follows:

$$ {}_aD_t^{\gamma} g(t) = \lim_{h \to 0} \frac{1}{h^{\gamma}} \sum_{j=0}^{[(t-a)/h]} (-1)^j \binom{\gamma}{j} g(t-jh) \tag{1}$$

where t and a are the limits, γ is the order of the mathematical operation i.e. μ and $-\lambda$, D is the differ-integral operator, h is the step size considered to be very small and $\binom{\gamma}{j} = \frac{(\gamma)(\gamma-1)(\gamma-2)\ldots\ldots(\gamma-j+1)}{\Gamma(j-1)}$

In the present study, fractional order calculus was realized in z-domain. Backward difference method i.e. $s = (\frac{1-z^{-1}}{T})$, where T is sampling time, is considered to transform the differentiator operator from s-domain to z-domain. Therefore, fractional differentiator operator 's^{γ}' is transformed into z-domain as follows:

$$s^{\gamma} = \left(\frac{1 - z^{-1}}{T}\right)^{\gamma} \tag{2}$$

or

$$s^{\gamma} = T^{-\gamma} \sum_{j=0}^{\infty} (-1)^{j} \frac{(\gamma)(\gamma-1)(\gamma-2)\ldots\ldots(\gamma-j+1)}{\Gamma(j-1)} z^{-j} \tag{3}$$

In terms of discrete time, the differentiator operator 'D' is defined as

$$D^{\gamma} = T^{-\gamma} \sum_{j=0}^{\infty} (-1)^{j} \binom{\gamma}{j} z^{-j} \tag{4}$$

or

$$D^{\gamma} = T^{-\gamma} \sum_{j=0}^{\infty} d_j z^{-j} \tag{5}$$

where, $d_j = (-1)^{j} \binom{\gamma}{j}$ a Binomial coefficient, which can be further arrange in an recursive algorithm.

$$d_j = \left(1 - \frac{1+\gamma}{j}\right) d_{j-1};\ \mathrm{j} = 1, 2, 3, \ldots\ldots\ldots\ldots \tag{6}$$

So, the fractional order operator (differentiator/integrator) of a sequence $g[n]$ can be expressed as follows:

$$D^{\gamma}(g[n]) = T^{-\gamma} \sum_{j=0}^{\infty} d_j g[n-j] \tag{7}$$

Now, it has been clear from the Eq. (8) that in order to realize the fractional order differ-integral operator infinite number of memory is required which seems to unrealistic. Therefore, for the implementation of these fractional order operators a short memory concept was introduced. In this regard, only last few samples have to be stored. In the present case, a memory of 1000 was opted for realization of fractional order mathematical order.

$$D^{\gamma}(g[n]) = T^{-\gamma} \sum_{j=0}^{1000} d_j g[n-j] \tag{8}$$

4 Fractional Order PID Controller ($PI^{\lambda}D^{\mu}$)

The PID controller, in general, can be expressed as

$$U_{PID}(t) = K_C e(t) + K_I \int e(t)dt + K_D \frac{de(t)}{dt} \tag{9}$$

In time domain the Fractional Order PID Controller ($PI^{\lambda}D^{\mu}$) can be expressed as follows:

$$U_{FO-PID}(t) = K_C e(t) + K_I \frac{d^{-\lambda}e(t)}{dt^{-\lambda}} + K_D \frac{d^{\mu}e(t)}{dt^{\mu}} \tag{10}$$

where K_C is proportional constant; K_I is integral constant; K_D is derivative constant; λ is the fractional integral value and μ is the fractional derivative value. $U_{FO-PID}(t)$ is the aggregate output of FO-PID controller and $e(t)$ is the tracking error.

In s-domain, the $PI^{\lambda}D^{\mu}$ controller would become

$$U_{FO-PID}(s) = \left(K_C + K_I \frac{1}{s^{\lambda}} + K_D s^{\mu}\right) E(s) \tag{11}$$

From the Fig. 1, one can clearly understand that the classical PI, PD and PID controllers are unique cases of the FO-PID controller. Particularly, PID can be formed by letting the variables λ and μ as unity. It is due to the selection of values of variables λ and μ, which provides two more DOF to the control engineer in addition to the three controller constants. Hence, for designing a FO-PID controller, five variables need to be tuned in order to get the best desired response.

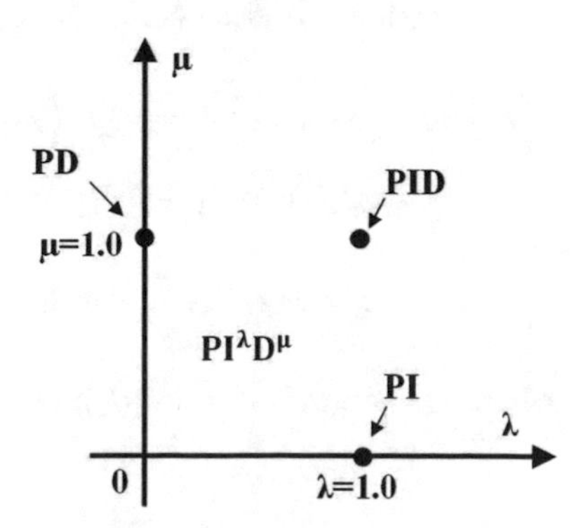

Fig. 1 FO-PID controller

In the present work, the parameters of FO-PID and PID controllers were optimally determined using GA for a defined performance criterion as described in the following section for a 2-link planar rigid manipulator. With the experiments conducted it can be concluded that the overall control behavior of the FO-PID controller is much superior to the classical PID controller.

5 Performance Criteria

For evaluating the set-point tracking performance of the system in closed-loop the IAE and Integral Square of Change in Controller Output (ISCCO) for each link having equal weight were considered as a performance criteria. The IAE and ISCCO are defined as:

$$IAE = \int_0^t |e(t)|dt \tag{12}$$

$$ISCCO = \int_0^t \Delta u^2(t)dt \tag{13}$$

The above performance indices were used for adjusting the various parameters (K_C, K_I, K_D, λ, μ) of FO-PID and PID controllers.

6 Dynamic Model of 2-Link Manipulator

A 2-link planar rigid manipulator having two DOF is shown in Fig. 2. It has two links having length l_1 and l_2, mass m_1 and m_2, respectively. The angular position of link-1 and link-2 are θ_1 and θ_2, respectively and τ_1 and τ_2 are the respective torque for link-1 and link-2. The dynamic model of 2-link planar rigid manipulator described in [2, 19] has been utilized in this work.

The mathematical model of the 2-link planar rigid manipulator is as follows:

$$\begin{aligned}\tau_1 = {} & m_2 l_2^2\left(\ddot{\theta}_1 + \ddot{\theta}_2\right) + m_2 l_1 l_2 c_2\left(2\ddot{\theta}_1 + \ddot{\theta}_2\right) + (m_1 + m_2) l_1^2 \ddot{\theta}_1 - m_2 l_1 l_2 s_2 \dot{\theta}_2^2 \\ & - 2 m_2 l_1 l_2 s_2 \dot{\theta}_1 \dot{\theta}_2 + m_2 l_2 g c_{12} + (m_1 + m_2) l_1 g c_1\end{aligned} \tag{14}$$

$$\tau_2 = m_2 l_1 l_2 c_2 \ddot{\theta}_1 + m_2 l_1 l_2 s_2 \dot{\theta}_1^2 + m_2 l_2 g c_{12} + m_2 l_2^2\left(\ddot{\theta}_1 + \ddot{\theta}_2\right) \tag{15}$$

where $s_2 = \sin(\theta_2)$, $c_1 = \cos(\theta_1)$, $c_2 = \cos(\theta_2)$, and $c_{12} = \cos(\theta_1 + \theta_2)$.

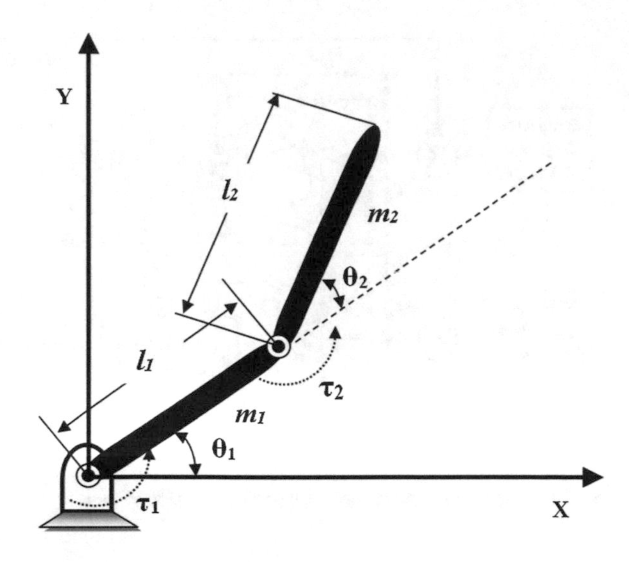

Fig. 2 A 2-link planar rigid manipulator

Table 1 Parameters for a 2-link planar rigid robotic manipulator

Parameters	Link-1	Link-2
Mass (kg)	0.1	0.1
Length (m)	0.8	0.4
Acceleration due to gravity (g) (m/s^2)	9.81	9.81

Equations 14 and 15 give the required torques at the actuators as a function of joint positions, velocities, and accelerations. The parameters of the manipulator are listed in Table 1.

7 Experimental Results

The noise suppression investigations were organized as follows. Firstly, sensor noise suppression was investigated followed by the controller's noise suppression as a second case. Both the experiments were simulated in closed-loop for 2-link rigid manipulator trajectory control using FO-PID and PID controllers. The block diagram of closed-loop control system is shown in Fig. 3. Simulations were performed using National Instrument® software, LabVIEW™ 8.5 and its add-ons Simulation and Control Design toolkit. In the simulation loop, 4th order Runge–Kutta method, an ordinary deferential equation (ODE) with a fixed step size of 10 ms was used.

Ayala and Coelho [2] have considered a reference trajectory based on cubic interpolation polynomial nature to be followed by the proposed 2-link manipulator. The same reference trajectory has been taken in this work. It was defined in [2, 19]:

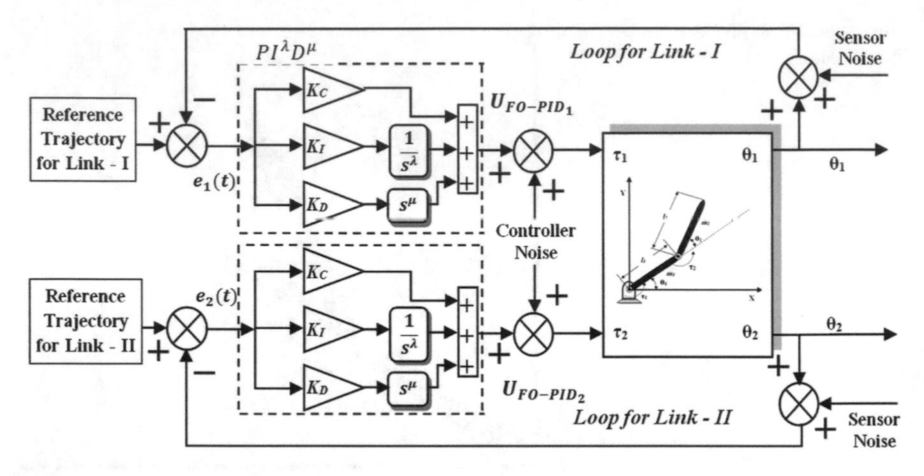

Fig. 3 The block diagram of closed-loop control system

$$\theta_{rt,j}(t) = a_0 + a_1 t + a_2 t^2 + a_0 t^3,\ j = 1, 2 \tag{16}$$

Therefore the joint velocity and acceleration along the reference trajectory becomes

$$\dot{\theta}_{rtf,j}(t) = a_1 + 2a_2 t + 3a_0 t^2,\ j = 1,\ 2 \tag{17}$$

$$\dot{\theta}_{rtf,j}(t) = 2a_2 + 6a_0 t,\ j = 1, 2 \tag{18}$$

where $\theta_{rt,j}(t)$ $[\theta_{rt,1}(t);\ \theta_{rt,2}(t)]$ is the instantaneous desired position for link-1 and link-2, respectively. Also, $\theta_{rtf,j}(t)$ and $\theta_{rtf,j}(t)$ are the final desired values for the position ($\theta_{rtf,1}(t) = 1$ rad and $\theta_{rtf,2}(t) = 2$ rad in $t = 2$ s and $\theta_{rtf,1}(t) = 0.5$ rad and $\theta_{rtf,2}(t) = 4$ rad in final time $t_f = 4$s) and velocity ($\dot{\theta}_{rtf,1}(t) = \dot{\theta}_{rtf,2}(t) = 0$ rad/s in $t = 2$ s and $t_f = 4$ s) of link-1 and link-2 respectively.

The various parameters of FO-PID and PID controllers, in the absence of noise, were tuned using GA, developed in the used LabVIEW environment [23]. The population size was considered to be 20 and the tolerance level was kept as 10^{-6} and the maximum numbers of iterations were kept as 100. The used cost function (*J*), to be minimized, was the weighted sum of the IAE and ISCCO as defined below.

$$J = \int_0^{\infty} [w_1^1 * |e_1(t)| + w_1^2 * |e_2(t)| + w_2^1 * \Delta u_1^2(t) + w_2^2 * \Delta u_2^2(t)]dt \tag{19}$$

In the present work, equal weights i.e., $w_1^1 = w_2^1 = w_1^2 = w_2^2 = 0.25$ were assigned to IAE and ISCCO while optimizing the parameters of FO-PID and PID controllers. Figure 4 shows cost function versus generation plot for both the controllers.

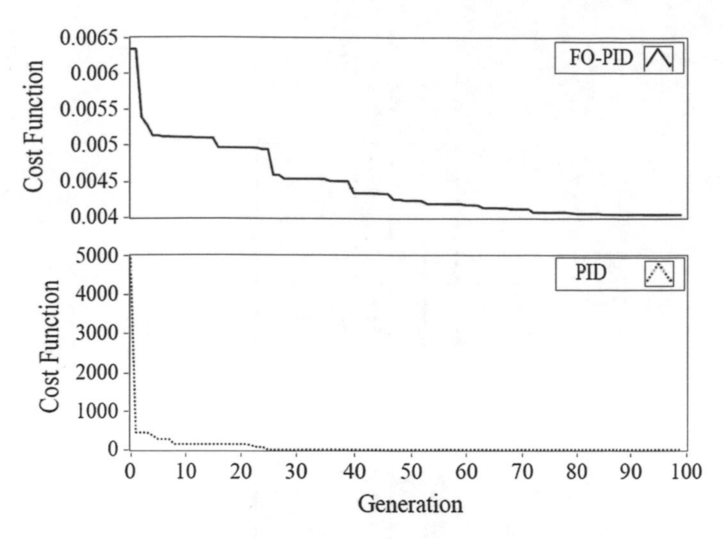

Fig. 4 Cost function versus generation curve for FO-PID and PID controllers

The tuned parameters values for both the controllers are given in Table 2 along with the cost function values. For these gains the obtained values of IAE are tabulated in Table 3. The complete set-point tracking response of 2-link rigid manipulator without sensor and controller noise in closed-loop with FO-PID and PID controllers is shown in Fig. 5. It can be observed that set-point tracking performance of FO-PID controller is better than PID controller. Also, lower variations in the torque of link-1 and link-2 were observed for FO-PID controller as compared to PID controller.

In the present work, point by point UWN and GWN were considered for measurement/sensor and controller noise study. Sensor and controller noise were introduced in the closed-loop control system in both the links of manipulator as shown in the Fig. 3. Noise suppression study, of FO-PID and PID controllers, for sensor and controller noise in closed-loop, is presented in the following section.

7.1 Sensor Noise Suppression

All control systems, in practical cases, are subject to some kind of noise during their operation. Thus, in addition to responding to the input signal, the system should also be able to reject and suppress noise and unwanted signals. There are many forms and sources of noise but the sensor noise play significant impact on the performance of the system. Such noise is typically dominated by high frequencies. Measurement noise usually sets an upper limit on the bandwidth of the loop. Also, it introduces uncertainty in the system. In the subsequent section the sensor noise suppression of FO-PID and PID controllers in the closed-loop is presented.

Table 2 Optimal parameters of FO – PID and PID controller

Controller type	Manipulator link	Performance index	Minimum cost function	Controller parameters				
				K_C	K_I	K_D	λ	μ
$PI^{\lambda}D^{\mu}$	Link-1	IAE and ISCCO	0.0079360	507.102	1035.55	430.616	0.997709	0.17498
	Link-2			10.517	1019.06	312.083	0.894167	0.603051
PID	Link-1	IAE and ISCCO	0.0177892	449.895	0.524	5.550	1	1
	Link-2			238.338	7.755	22.342	1	1

Table 3 Performance index

Controller type	Manipulator link	Performance index
		IAE
$PI^{\lambda}D^{\mu}$	Link-1	0.00215675
	Link-2	0.00196381
PID	Link-1	0.00902466
	Link-2	0.00951855

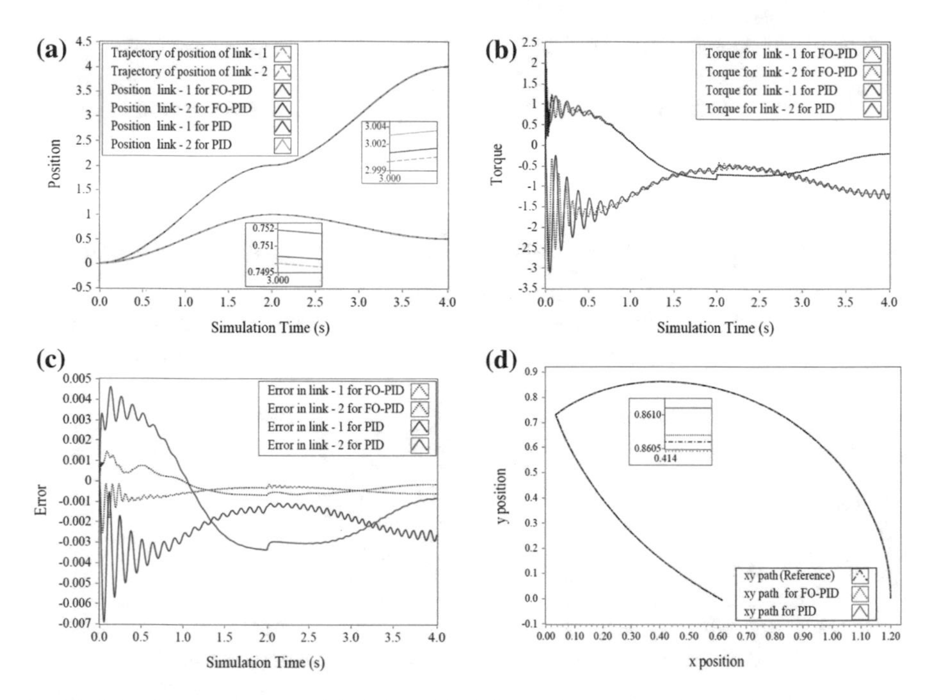

Fig. 5 **Closed**-loop response of link-1 and link-2 of robotic manipulator without sensor and controller noise. **a** Position. **b** Applied torque. **c** Error. **d** xy curve

7.1.1 Uniform White Noise

UWN was introduced as sensor noise in link-1 then in link-2 and finally it was injected in both links together. For this study, in all the cases, amplitude of UWN was varied and the corresponding IAE of FO-PID and PID controllers in closed-loop were recorded. The variation of IAE of FO-PID and PID controllers in closed-loop and amplitude of UWN are plotted in Fig. 6. It can be noted that PID controller fails to suppress the sensor noise even for very small amplitude of UWN while FO-PID controller is able to suppress even the large amplitude UWN sensor noise effectively. This outcome clearly shows the superiority of FO-PID controller over PID controller. To further elaborate a typical case the time response of robotic manipulator in closed-loop with FO-PID and PID controllers and with UWN sensor noise of amplitude 0.002 in both links was shown in Fig. 7. It clearly demonstrates

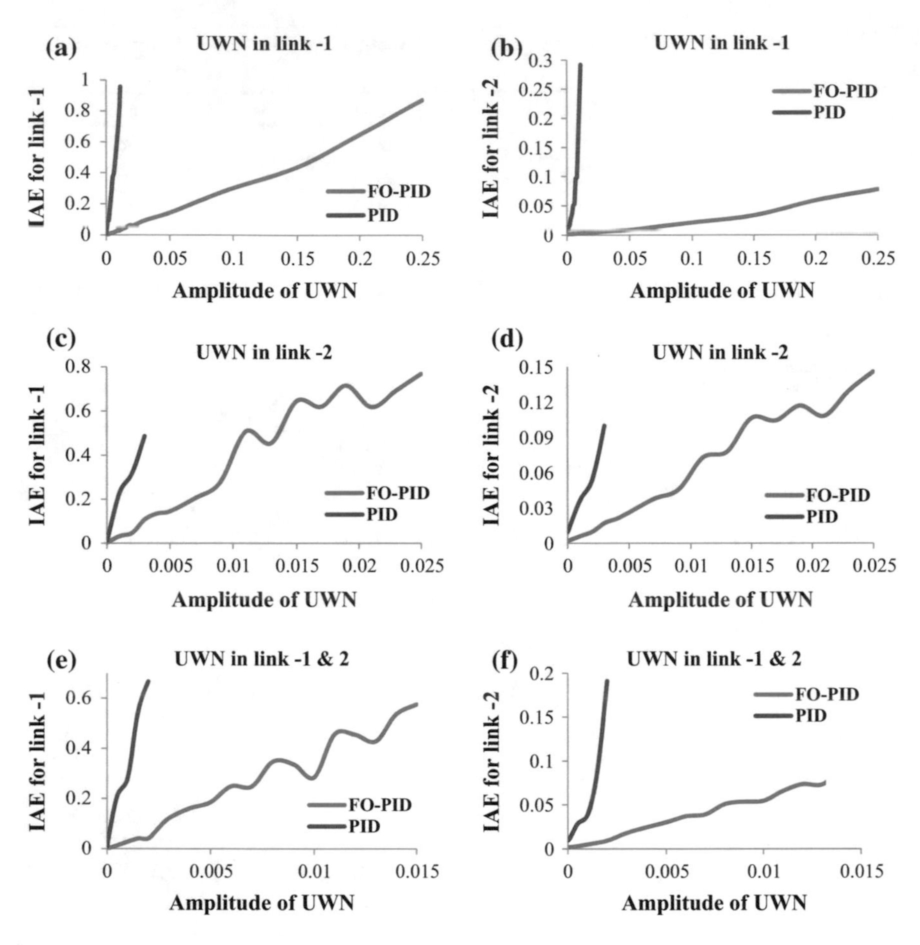

Fig. 6 Effect of sensor noise (UWN) in closed-loop. **a** and **b** in link-1. **c** and **d** in link-2. **e** and **f** in link-1 and link-2

that when sensor noise is present in the both links of robotic manipulator, PID controller fails to track the trajectory and start oscillating around the reference trajectory with increasing amplitude in comparison to FO-PID controller which sticks to the trajectory and follow it without any deviation. In Fig. 7 (d) the resulting xy curve shows it effectively.

7.1.2 Gaussian White Noise

Further, in line with the above, GWN was added as a measurement noise in the link-1, link-2 and link-1 and 2. Figure 8 shows the variation in IAE values for FO-PID and PID controllers as the standard deviation of GWN increases. Again, it can be observed that conventional PID controller fails to suppress this measurement

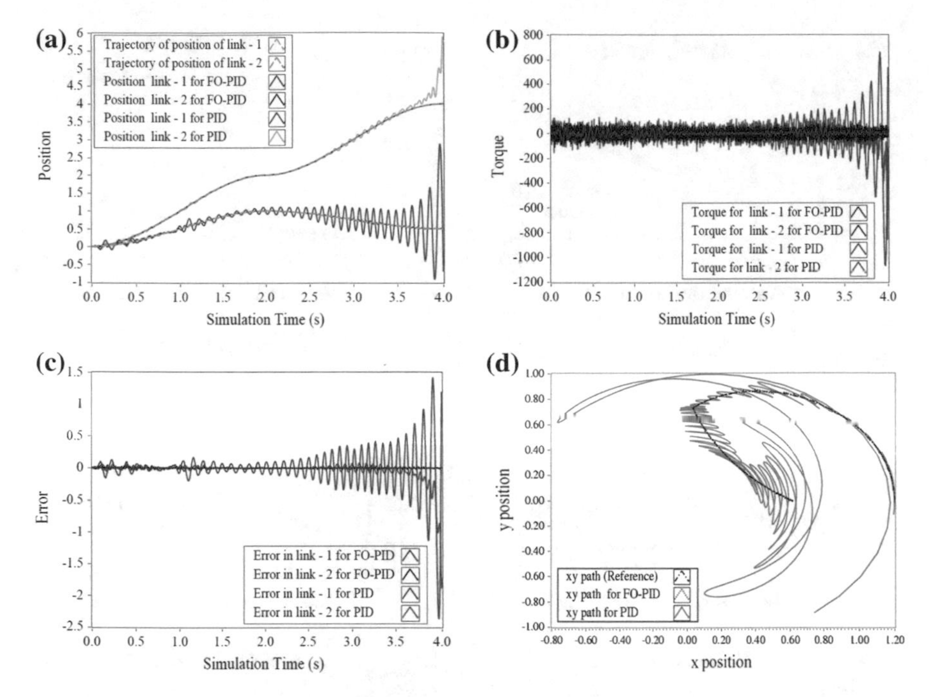

Fig. 7 Closed-loop response of link-1 and link-2 of robotic manipulator with UWN sensor noise of amplitude 0.002 in both links. **a** Position. **b** Applied torque. **c** Error. **d** xy curve

noise as its standard deviation is increased. Also, as the intensity of noise increase manipulator deviates from the desired trajectory and finally become unstable. On the other hand, FO-PID controller effectively handles the sensor noise and eliminates its effect so that robotic manipulator follows the trajectory without any deviation for a sufficient GWN measurement noise as shown in Fig. 8. This investigation clearly demonstrates the superiority of FO-PID controller over PID controller. Figure 9 shows a typical reference trajectory tracking response of manipulator in closed-loop with GWN noise of 0.007 standard deviations, in both links. The time response demonstrates the utility of FO-PID controller over PID controller for sensor noise suppression. From Fig. 9 the trajectory tracking, control action, error and xy curve illustrate it clearly.

7.2 Controller Noise Suppression

Generally, in closed-loop control system the control action is implemented through a manipulate variable. So a control signal is sent from controller to final control element in the plant in order to regulate the manipulate variable to achieve desired set-point. Actually, a physical connection between controller and final control

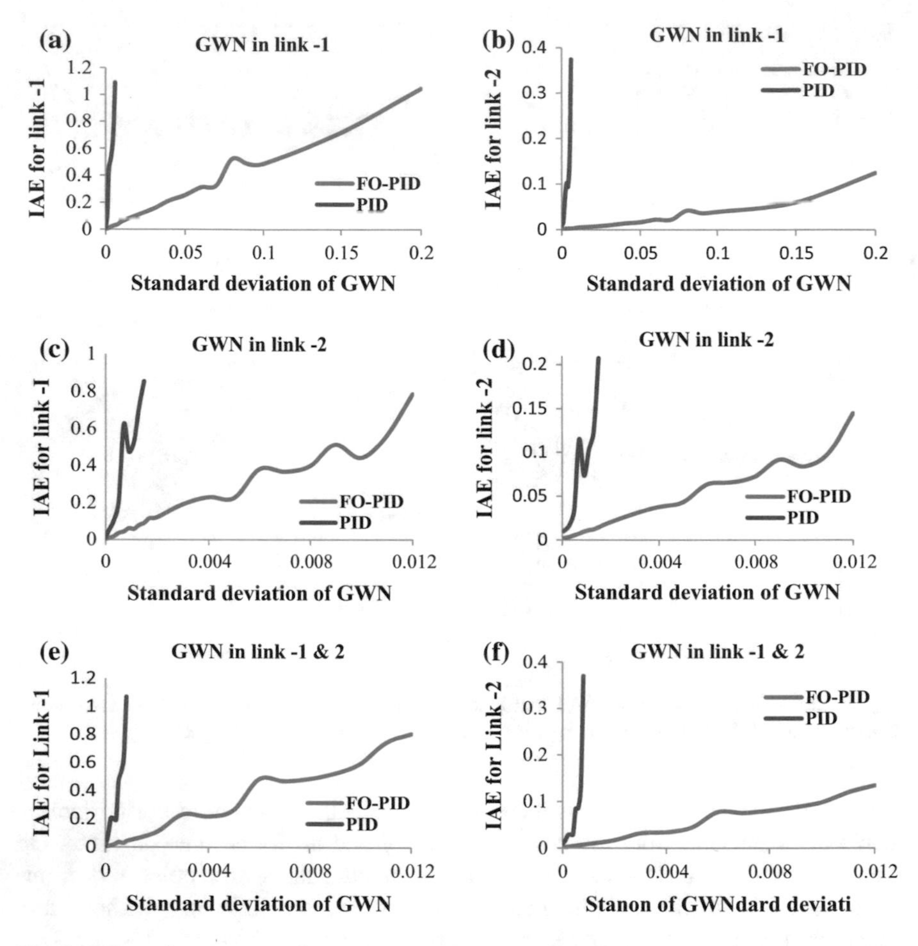

Fig. 8 Effect of sensor noise (GWN) in closed-loop. **a** and **b** in link-1. **c** and **d** in link-2. **e** and **f** in link-1 and link-2

element is required to realize it. In process industry generally the final control elements are placed in the field and due to hazardous area with lots of vibration. Many times connection may become loose and may add additional random noise in the control action. The magnitude of random noise depends upon the environment around the final control element. Also, it has been noted in the industry that many mammals such as mouse cut the wire. In case of partially broken wires, it may pick up the random noise and the magnitude of noise will depend upon the exposure of the conductor to environment and elements around the final control element. Therefore the unexpected noises added in the control signal and corrupt the control action and may try to destabilize the plant or process.

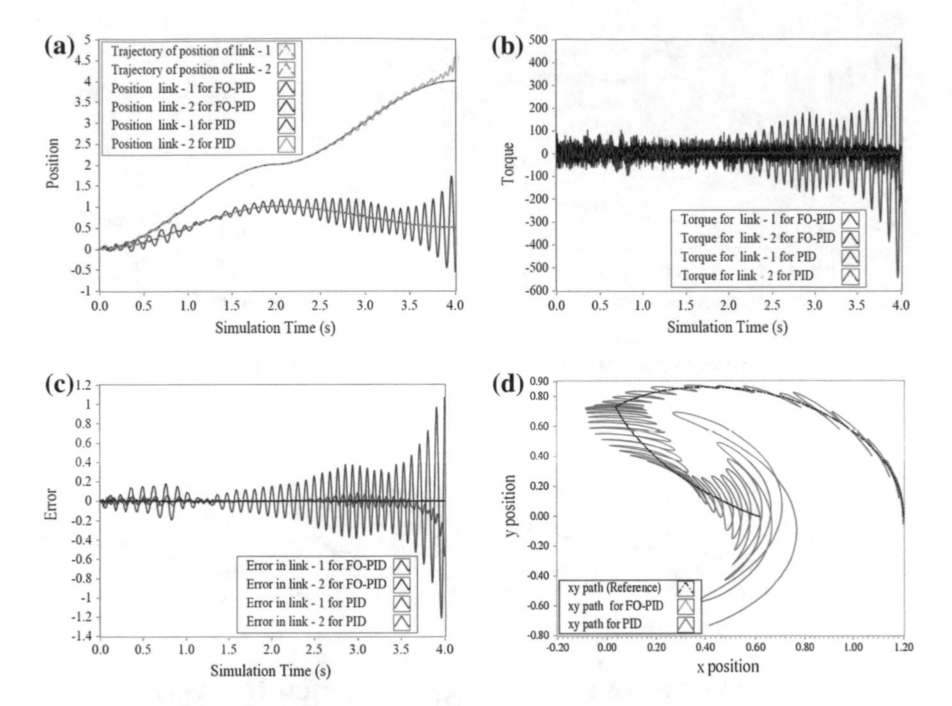

Fig. 9 Closed-loop response of link-1 and link-2 of robotic manipulator with GWN sensor noise of standard deviation of 0.0007 in both links. **a** Position. **b** Applied torque. **c** Error. **d** xy curve

In the present work, UWN and GWN are added in the control action as a controller noise. In the next section, the impacts of both controller noises on the system response are presented.

7.2.1 Uniform White Noise

UWN was added as controller noise in the links independently and then in both links simultaneously. The amplitude of UWN was increased linearly and corresponding IAE for FO-PID and PID controller were recorded. Figure 10 illustrates the variation of the amplitude of UWN and IAE for FO-PID and PID controllers in closed-loop. It can be noted that PID controller acquired considerable IAE and became unstable around the amplitude of UWN has a value 50. While FO-PID controllers have moderate value of IAE and follow trajectory effectively up to the amplitude of UWN having a value 200. The time response curve of the robotic manipulator trajectory control for amplitude of UWN of 24 in both links was shown in Fig. 11. It undoubtedly demonstrates the superiority of FO-PID controller in comparison of PID controller for controller noise suppression.

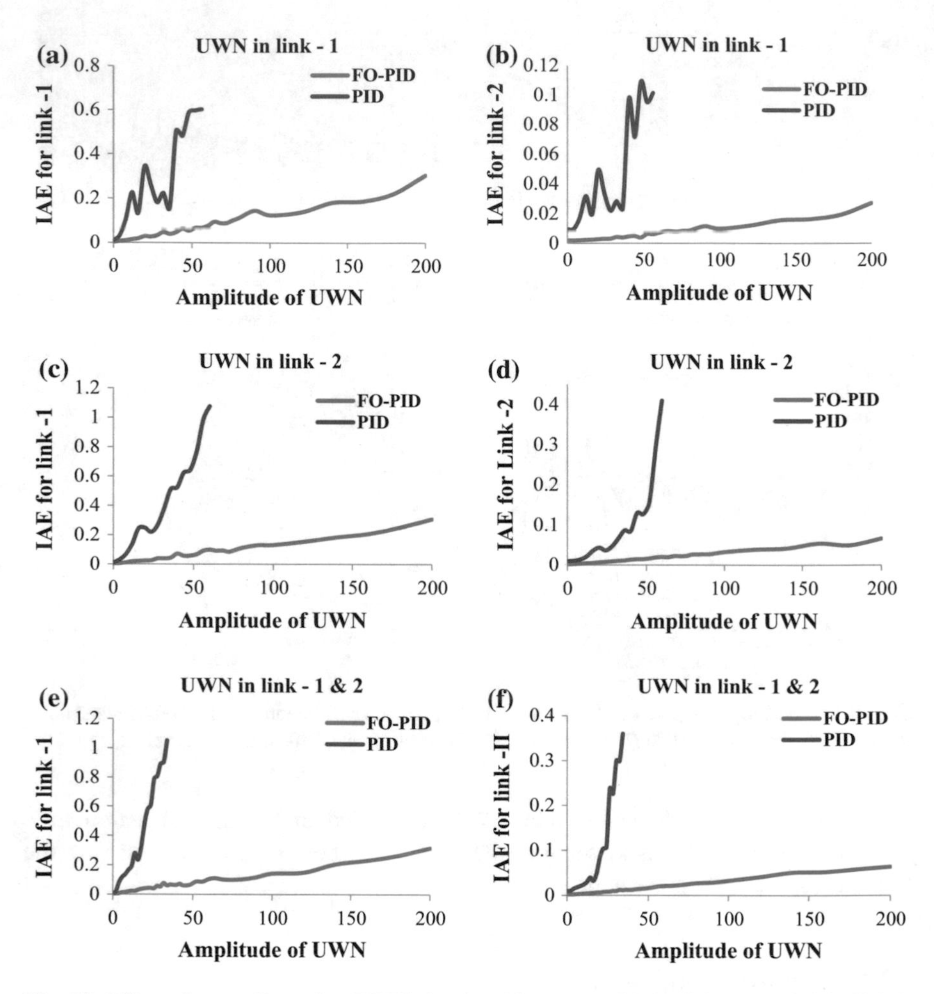

Fig. 10 Effect of controller noise (UWN) in closed-loop. **a** and **b** in link-1. **c** and **d** in link-2. **e** and **f** in link-1 and link-2

7.2.2 Gaussian White Noise

Now, GWN was introduced as a controller noise in closed-loop. The standard deviation of controller noise was varied and the corresponding variation in IAE for FO-PID and PID controller was plotted in Fig. 12. It has been observed that IAE of PID controller increases rapidly and system become unstable around the 40 standard deviation of GWN. In contrast, FO-PID controller keeps its IAE quite small and as the standard deviation of GWN is increased up to 200 and still it has

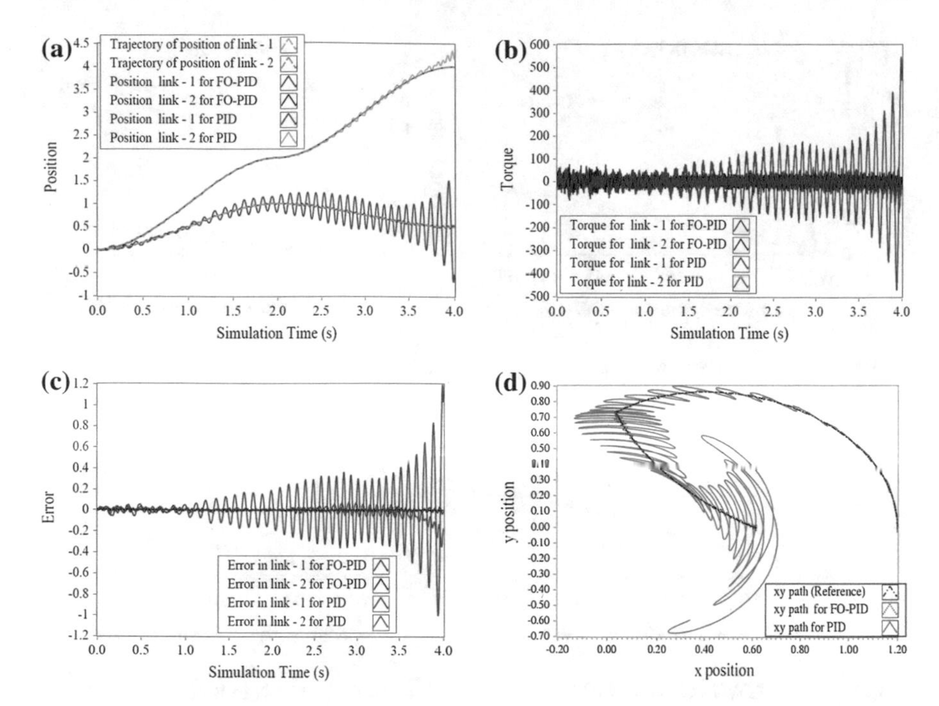

Fig. 11 Closed-loop response of link-1 and link-2 of robotic manipulator with UWN controller noise of amplitude 24 in both links. **a** Position. **b** Applied torque. **c** Error. **d** xy curve

reasonable value of IAE. Figure 13 illustrates the set-point tracking response of FO-PID and PID controller for standard deviation of GWN of 21 in both links. Again, it shows the advantage of FO-PID controller over PID controller.

7.3 Sensor and Controller Noise

In the highly noisy environment, the operation of robotic manipulator can be contaminated by random noises at various areas of closed loop system which would considerably degrade the effectiveness and accuracy of the manipulator. In this section, UWN and GWN are added together in the control action as well as in the sensor in both links. The tracking performance responses of robotic manipulator with UWN sensor and controller noise of the amplitude 0.002 and 24 respectively in both links are shown in Fig. 14. Also, the closed loop responses of the manipulator with GWN sensor and controller noise of the standard deviation of 0.0007 and 21 respectively in both links are shown in Fig. 15.

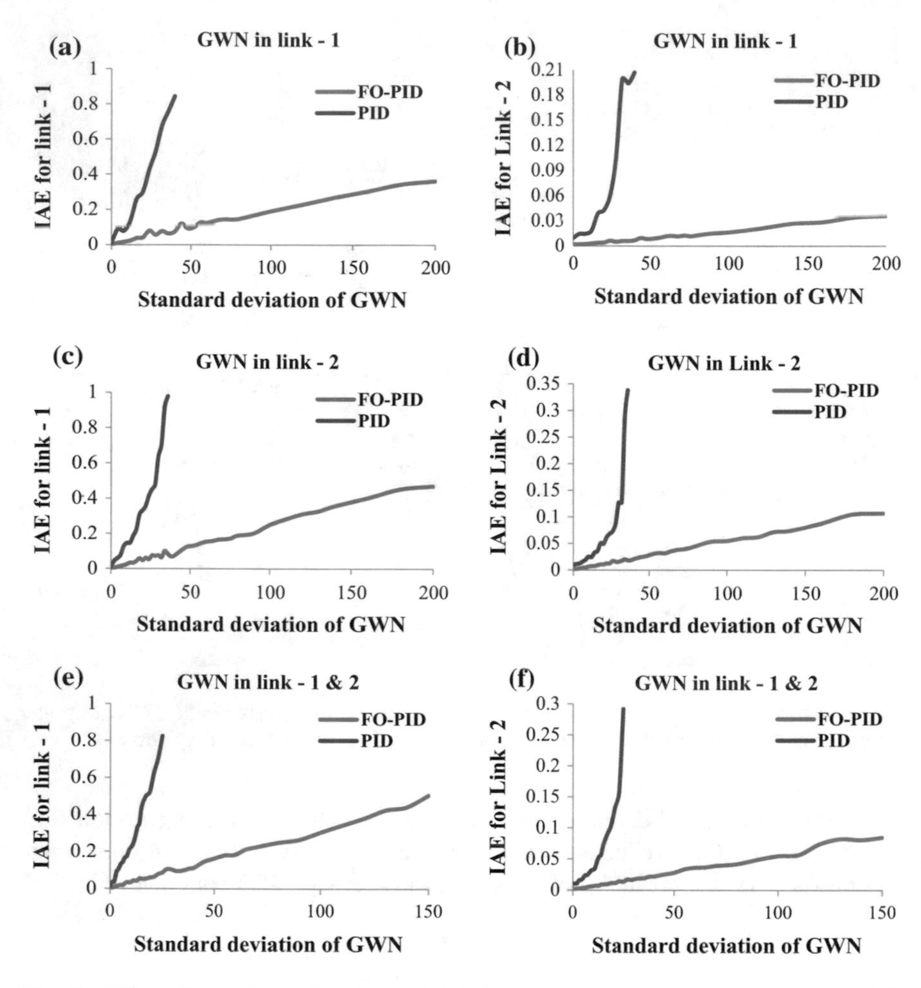

Fig. 12 Effect of controller noise (GWN) in closed-loop. **a** and **b** in link-1. **c** and **d** in link-2. **e** and **f** in link-1 and link-2

The FO-PID controller shows better noise suppression in controller as well as in the sensor due to the fractional order derivative. The frequency response of the fractional and complete derivative is shown in Fig. 16. It is clear from the frequency response that fractional derivative has lower attenuation rate as compared to complete derivative. Specially, in the investigated case of link-1, the order of fractional derivative was around 0.17 and the gain remains below 0 dB till 100 Hz. Further, it is clear from the graph that slope of the fractional derivative is restricted below 20 dB/decade. Therefore, FO-PID is able to successfully suppress the sensor and controller noise for a manipulator.

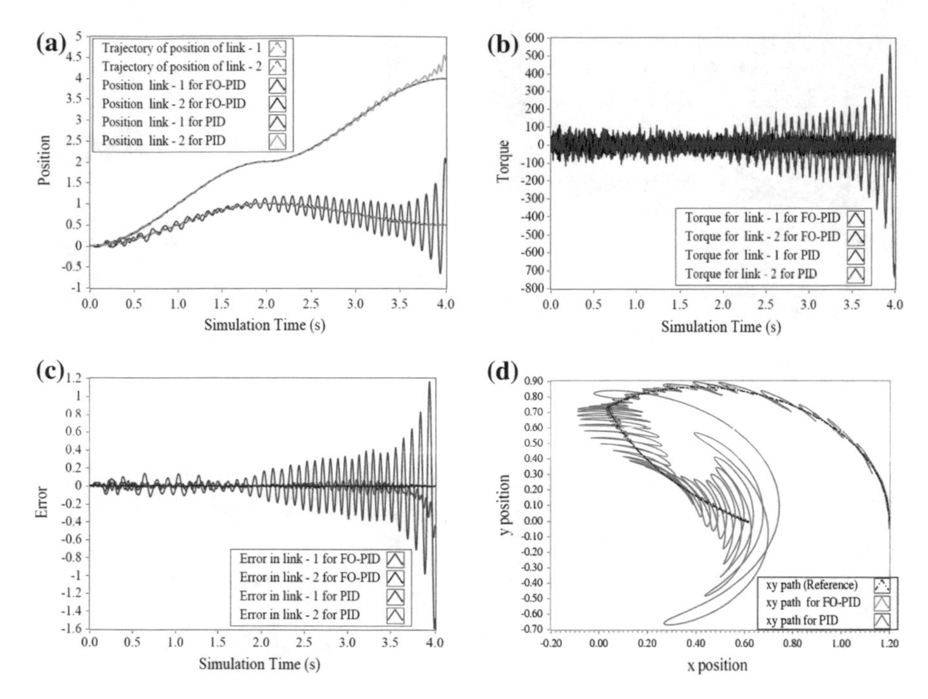

Fig. 13 Closed-loop response of link-1 and link-2 of robotic manipulator with GWN controller noise of amplitude 21 in both links. **a** Position. **b** Applied torque. **c** Error. **d** xy curve

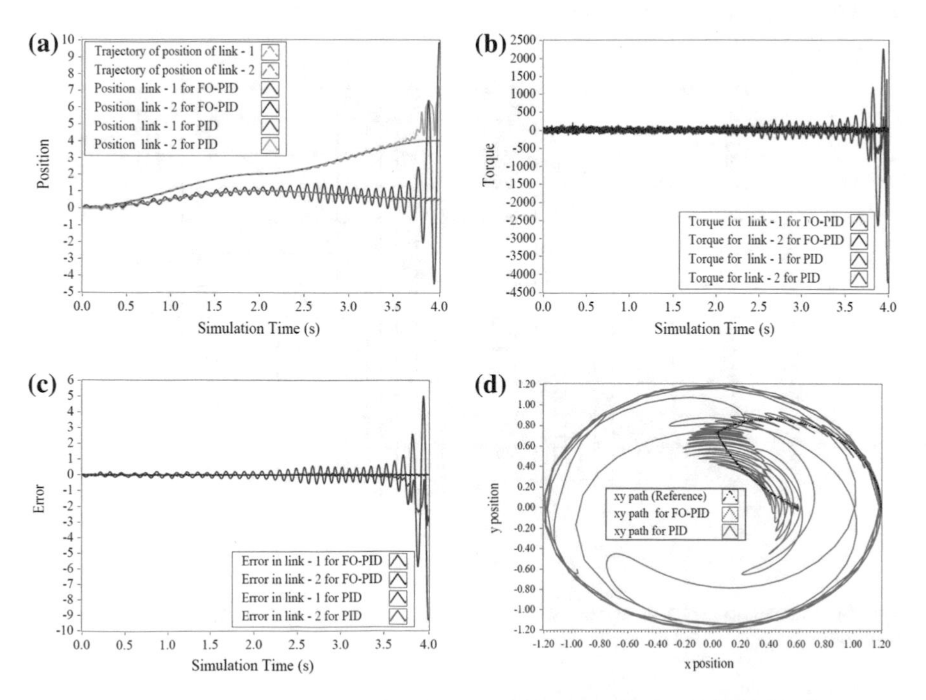

Fig. 14 Closed-loop response of link-1 and link-2 of robotic manipulator with UWN sensor and controller noise of amplitude 0.002 and 24 respectively in both links. **a** Position. **b** Applied torque. **c** Error. **d** xy curve

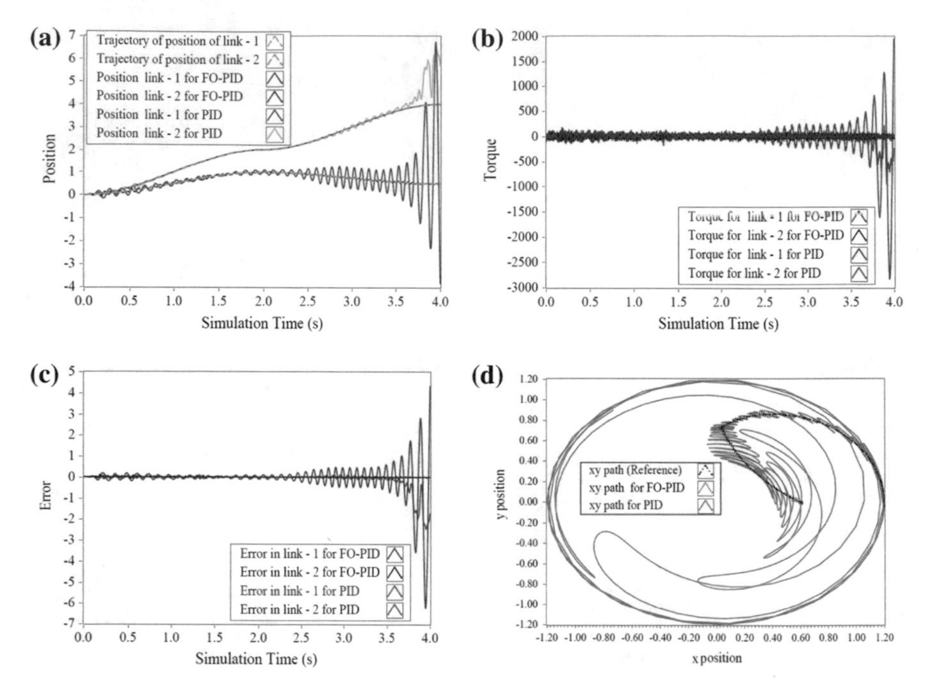

Fig. 15 Closed-loop response of link-1 and link-2 of robotic manipulator with GWN sensor and controller noise of standard deviation of 0.0007 and 21 respectively in both links. **a** Position. **b** Applied torque. **c** Error. **d** xy curve

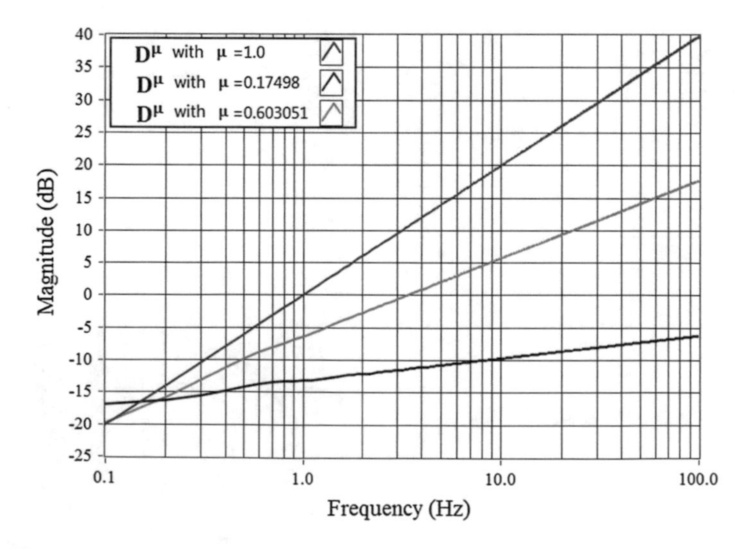

Fig. 16 Frequency response of the fractional derivative controller D^{μ} with $\mu = 1.0, 0.17498$ and 0.603051

8 Conclusion

Noise is an integral part of any real measurement and control experiment. It appears in many forms in control loop in process industry and introduces uncertainty in the system. Uncertain systems become a challenge for a control engineer. Conventional parallel PID controller fails to suppress all kind of random noise due to complete derivative term in it. But the fractional order derivative, employing effectively lower order derivative increases the noise suppression capability of Fractional Order-Proportional plus Integral plus Derivative (FO-PID) controller. The main contribution of this chapter has been to demonstrate the sensor and controller noise suppression of FO-PID over PID controller for 2-link rigid manipulator trajectory control.

In this chapter, FO-PID and PID controller were successfully implemented in closed-loop. The controllers were tuned for minimum weighted sum of Integral of the Absolute value of the Error (IAE) and Integral Square of Change in Controller output (ISCCO), for a non-linear 2-link rigid planner robotic manipulator, using Genetic Algorithm. UWN and GWN were considered for sensor and controller noise in the loop. It has been observed that FO-PID controller outperformed PID controller in both sensor and controller noise suppression in closed-loop.

The present study can be further extended in the future by realizing the fractional order calculus using different implementations techniques reported in the literature. Furthermore, similar investigation can also be performed on intelligent control schemes applied to complex plant. Also, the simulated results can be verified experimentally.

References

1. Aström, K. J., & Hägglund, T. (2006). *Advanced PID controllers* (1st ed.). Research Triangle Park, NC: ISA. 27709.
2. Ayala, H. V. H., & Coelho, L. D. S. (2012). Tuning of PID controller based on a multi-objective genetic algorithm applied to a robotic manipulator. *Expert Systems with Applications, 39*(10), 8968–8974.
3. Azar, A. T., & Serrano, F. E. (2014). Robust IMC-PID tuning for cascade control systems with gain and phase margin specifications. *Neural Computing and Applications, 25*(5), 983–995. Springer. doi:10.1007/s00521-014-1560-x.
4. Azar, A. T., & Serrano, F. E. (2015). Deadbeat control for multivariable systems with time varying delays. In *Chaos modeling and control systems design*. Studies in computational intelligence (Vol. 581, pp. 97–132). Springer-Verlag GmbH: Berlin/Heidelberg. doi:10.1007/978-3-319-13132-0_6.
5. Azar, A. T., & Serrano, F. E. (2015). Adaptive sliding mode control of the furuta pendulum. In: A. T. Azar & Q. Zhu (Eds.), *Advances and applications in sliding mode control systems*. Studies in computational intelligence (Vol. 576, pp. 1–42). Springer-Verlag GmbH: Berlin/Heidelberg. doi:10.1007/978-3-319-11173-5_1.
6. Azar, A. T., & Serrano, F. E. (2015). Design and modeling of anti wind up PID controllers. In: Q. Zhu & A. T. Azar (Eds.), *Complex system modeling and control through intelligent soft*

computations. Studies in fuzziness and soft computing (Vol. 319, pp. 1–44). Springer: Germany. doi:10.1007/9783319128832_1.

7. Azar, A. T., & Vaidyanathan, S. (2015). *Chaos modeling and control systems design*. Studies in computational intelligence (Vol. 581). Springer: Germany. ISBN: 9783319131313.
8. Azar, A. T., & Vaidyanathan, S. (2015). *Computational intelligence applications in modeling and control*. Studies in computational intelligence (Vol. 575) SpringerVerlag: Germany. ISBN: 9783319110165.
9. Azar, A. T., & Vaidyanathan, S. (2015) Handbook of research on advanced intelligent control engineering and automation. Advances in Computational Intelligence and Robotics (ACIR) Book Series, IGI Global: USA.
10. Azar, A. T., & Vaidyanathan, S. (2016). *Advances in chaos theory and intelligent control*. Studies in fuzziness and soft computing (Vol. 337). Springer-Verlag: Germany. ISBN: 978-3-319-30338-3.
11. Azar, A. T., Zhu, Q. (2015). *Advances and applications in sliding mode control systems*. Studies in computational intelligence (Vol. 576). SpringerVerlag: Germany. ISBN: 9783319111728.
12. Bingul, Z., & Karahan, O. (2011). A fuzzy logic controller tuned with PSO for a 2 DOF robot trajectory control. *Expert Systems with Applications, 38*(1), 1017–1031.
13. Bingul, Z., & Karahan, O. (2011). Fractional PID controllers tuned by evolutionary algorithms for robot trajectory control. *Turk Journal of Electrical Engineering and Computer Science, 20*(1), 1123–1136.
14. Bingul, Z., & Karahan, O. (2011, April 13–15). Tuning of fractional PID controllers using PSO algorithm for robot trajectory control. In *Proceedings of IEEE international conference on mechatronics*. Turkey, pp. 955–960.
15. Boulkroune, A., Bouzeriba, A., Bouden, T., Azar, A. T. (2016). Fuzzy adaptive synchronization of uncertain fractional-order chaotic systems. In *Advances in chaos theory and intelligent control*. Studies in fuzziness and soft computing (Vol. 337, pp. 681–697). Springer-Verlag: Germany.
16. Boulkroune, A., Hamel, S., & Azar, A. T. (2016). Fuzzy control-based function synchronization of unknown chaotic systems with dead-zone input. In *Advances in chaos theory and intelligent control*. Studies in fuzziness and soft computing (Vol. 337). Springer-Verlag: Germany.
17. Chaillet, A., Loria, A., Kelly, R. (2006, December 13–15). Robustness of PID controlled manipulators with respect to external disturbance. In *Proceedings of IEEE conference on Decision and Control*. San Diego, USA (pp. 2949–2954).
18. Chen, C. S. (2008). Dynamic structure neural-fuzzy networks for robust adaptive control of robot manipulators. *IEEE Transactions on Industrial Electronics, 55*(9), 3402–3414.
19. Craig, J. J. (1996). *Introduction to robotics: mechanics and control*. New York: Addison-Wesley.
20. Delavari, H., Ghaderi, R., Ranjbar, A., HosseinNia, & S. H., Momani, S. (2010). Adaptive fractional PID controller for robotic manipulator. In*: Proceeding of 4th IFAC Workshop Fractional Differentiation and its Applications*. Badajoz, Spain (pp. 1–7).
21. El-Khazali, R. (2013). Fractional-order $PI^{\lambda}D^{\mu}$ controller design. *Computer and Mathematics with Application, 66*(5), 639646.
22. Franklin, G. F., Powell, J. D., & Workman, M. L. (1998). *Digital control of dynamic systems* (3rd ed.). New York: Addison-Wesley Longman.
23. Kumar, V., Rana, K. P. S., Kumar, A., Sharma, R., Mishra, P., & Nair, S. S. (2013, December 26–28). Development of a genetic algorithm toolkit in LabVIEW. In: *Proceedings of the 3rd International Conference on Soft Computing for Problem Solving (SocProS-13)*. Advances in intelligent systems and computing—Series (Vol. 259, pp. 281–296). Springer: Greater Noida Extension Centre of IIT Roorkee, India. doi:10.1007/978-81-322-1771-8_25.
24. Kumar, V., Rana, K. P. S., Kumar, J., Mishra, P., & Nair, S. S. (2016). A robust fractional order fuzzy p + fuzzy i +fuzzy d controller for nonlinear and uncertain system. *International Journal of Automation and Computing*. Springer publication. doi:10.1007/s11633-016-0981-7.

25. Kumar, V., Rana, K. P. S., & Mishra, P. (2016). Robust speed control of hybrid electric vehicle using fractional order fuzzy pd & pi controllers in cascade control loop. *Journal of the Franklin Institute, 353*(8), 1713–1741.
26. Ladaci, S., Loiseau, J. J., & Charef, A. (2010). Adaptive internal model control with fractional order parameter. *International Journal of Adaptive Control and Signal Processing, 24*(11), 944–960.
27. Li, T. H. S., & Huang, Y. C. (2010). MIMO adaptive fuzzy terminal sliding mode controller for robotic manipulators. *Information Sciences, 180*(23), 4641–4660.
28. Lin, J., & Huang, Z. Z. (2006). A hierarchical supervisory fuzzy controller for robot manipulators with oscillatory bases. In *Proceeding of IEEE International Conference on Fuzzy Systems*. Canada (pp. 2400–2407)
29. Luo, Y., & Chen, Y. Q. (2009). Fractional order [proportional derivative] controller for a class of fractional order systems. *Automatica, 45*(10), 2446–2450.
30. Mekki, H., Boukhetala, D., & Azar, A. T. (2015). Sliding modes for fault tolerant control. In *Advances and applications in sliding mode control systems*. Studies in computational intelligence book series (Vol. 576, pp. 407–433). Springer-Verlag GmbH: Berlin/Heidelberg. doi:10.1007/978-3-319-11173-5_15.
31. Mishra, P., Kumar, V., & Rana, K. P. S. (2015). A fractional order fuzzy PID controller for binary distillation column control. *Expert Systems with Applications, 42*(22), 8533–8549.
32. Monje, C. A., Vinagre, B. M., Feliu, V., & Chen, Y. Q. (2008). Tuning and auto-tuning of fractional order controller for industry applications. *Control Engineering Practice, 16*(7), 798–812.
33. Ogata, K. (2009). *Modern control engineering* (5th ed.). India: Prentice Hall.
34. Oya, M., Wada, M., Honda, H., & Kobayashi, T. (2003). Experimental studies of a robust tracking controller for robot manipulators with position measurements with position measurements contaminated by noises. In *Proceeding of 4th International Conference on Control and Automation*. Montreal, Canada (pp. 664–668)
35. Pan, I., & Das, S. (2012). Chaotic multi-objective optimization based design of fractional order $PI^{\lambda}D^{\mu}$ controller in AVR system. *Electrical Power and Energy Systems, 43*(1), 393–407.
36. Peng, T., & Woo, P. Y. (2002). Neural-fuzzy control system for robotic manipulators. *IEEE Control Systems Magazine, 22*(1), 53–63.
37. Petras, I. (2009). Fractional—order feedback control of a DC motor. *Journal of Electrical Engineering, 60*(3), 117–128.
38. Rana, K. P. S., Kumar, V., Mittra, N., & Pramanik, N. (2016). Implementation of fractional order integrator/differentiator on field programmable gate array. *Alexandria Engineering Journal*. doi:10.1016/j.aej.2016.03.030.
39. Sharma, R., Rana, K. P. S., & Kumar, V. (2014). Performance analysis of fractional order fuzzy PID controllers applied to a robotic manipulator. *Expert Systems with Applications, 41* (9), 4274–4289.
40. Silva, M. F., Machada, J. A. T., & Lopes, A. M. (2004). Fractional order control of a hexapod robot. *Nonlinear Dynamics, 38*(1–4), 417–433.
41. Song, Z., Yi, J., Zhao, D., & Li, X. (2005). A computed torque controller for uncertain robotic manipulator systems: Fuzzy approach. *Fuzzy Sets and Systems, 154*(2), 208–226.
42. Tang, W., Chen, G., & Lee, R. (2001). A modified fuzzy PI controller for a flexible-joint robot arm with uncertainties. *Fuzzy Sets and Systems, 118*(1), 109–119.
43. Tang, Y., Cui, M., Hua, C., Li, L., & Yang, Y. (2012). Optimum design of fractional order $PI^{\lambda}D^{\mu}$ controller for AVR system using chaotic ant swarm. *Expert Systems with Applications, 39*(8), 6887–6896.
44. Tian, L. F., & Collins, C. (2005). Adaptive neuro-fuzzy control of a flexible manipulator. *Mechatronics, 15*(10), 1305–1320.
45. Tsai, C. H., Wang, C. H., & Lin, W. S. (2000). Robust fuzzy model-following control of robot manipulators. *IEEE Transactions on Fuzzy Systems, 8*(4), 462–469.

46. Vaidyanathan, S., & Azar, A. T. (2015). Analysis and control of a 4-D novel hyperchaotic system. In A. T. Azar & S. Vaidyanathan (Eds.), *Chaos modeling and control systems design*. Studies in computational intelligence (Vol. 581, pp. 19–38). Springer-Verlag GmbH: Berlin/Heidelberg. doi:10.1007/978-3-319-13132-0_2.
47. Vaidyanathan, S., & Azar, A. T. (2015). Analysis, control and synchronization of a nine-term 3-D novel chaotic system. In A. T. Azar & S. Vaidyanathan (Eds.), Chaos modeling and control systems design. Studies in Computational Intelligence (Vol. 581, pp. 3–17). Springer-Verlag GmbH: Berlin/Heidelberg. doi:10.1007/978 3 319 13132 0_1.
48. Vaidyanathan, S., & Azar, A. T. (2016). A novel 4-D four-wing chaotic system with four quadratic nonlinearities and its synchronization via adaptive control method. In *Advances in chaos theory and intelligent control*. Studies in fuzziness and soft computing (Vol. 337). Springer-Verlag: Germany.
49. Vaidyanathan, S., & Azar, A. T. (2016). Adaptive backstepping control and synchronization of a novel 3-D Jerk System with an exponential nonlinearity. In *Advances in chaos theory and intelligent control*. Studies in fuzziness and soft computing (Vol. 337). Springer-Verlag: Germany.
50. Vaidyanathan, S., & Azar, A. T. (2016). Adaptive control and synchronization of Halvorsen circulant chaotic systems. In *Advances in chaos theory and intelligent control*. Studies in fuzziness and soft computing (Vol. 337). Springer-Verlag: Germany.
51. Vaidyanathan, S., & Azar, A. T. (2016). Dynamic analysis, adaptive feedback control and synchronization of an eight-term 3-D novel chaotic system with three quadratic nonlinearities. In *Advances in chaos theory and intelligent control*. Studies in fuzziness and soft computing (Vol. 337). Springer-Verlag: Germany.
52. Vaidyanathan, S., & Azar, A. T. (2016). Generalized projective synchronization of a novel hyperchaotic four-wing system via adaptive control method. In *Advances in chaos theory and intelligent control*. Studies in fuzziness and soft computing (Vol. 337). Springer-Verlag: Germany.
53. Vaidyanathan, S., & Azar, A. T. (2016). Qualitative study and adaptive control of a novel 4-D hyperchaotic system with three quadratic nonlinearities. In *Advances in chaos theory and intelligent control*. Studies in fuzziness and soft computing (Vol. 337). Springer-Verlag: Germany.
54. Valério, D., & Costa, J. S. D. (2013). *An introduction to fractional control*. London, United Kingdom: IET.
55. Yi, S. Y., & Chung, M. J. (1997). A robust fuzzy logic controller for robot manipulators with uncertainties. *IEEE Transactions on Systems, Man, and Cybernetics-Part B: Cybernetics, 27* (4), 706–713.
56. Yildirim, S., & Eski, I. (2010). Noise analysis of robot manipulator using neural networks. *Robotics and Computer-Integrated Manufacturing, 26*(4), 282–290.
57. Zamani, M., Ghartemani, M. K., Sadati, N., & Parniani, M. (2009). Design of a fractional order PID controller for an AVR using particle swarm optimization. *Control Engineering Practice, 17*(12), 1380–1387.
58. Zhu, D., & Fang, Y. (2007). Adaptive control of parallel manipulators via fuzzy-neural network algorithm. *Journal of Control Theory and Applications, 5*(3), 295–300.
59. Zhu, Q., & Azar, A. T. (2015). *Complex system modelling and control through intelligent soft computations*. Studies in fuzziness and soft computing. (Vol. 319). Springer-Verlag: Germany. ISBN: 9783319128825.

Control of the Temperature of a Finite Diffusive Interface Medium Using the CRONE Controller

X. Moreau, R. Abi Zeid Daou and F. Christophy

Abstract This chapter deals with the control of the temperature across a finite diffusive interface medium using the CRONE controller (French acronym: *Commande Robuste d'Ordre Non Entier*). In fact, the plant transfer function presents two special properties: a fractional integrator of order 0.5 and a delay factor of a fractional order (when controlling the temperature far from the boundary where the density of flux is applied). The novel approach of this work resides by the use of a fractional controller that would control a fractional order plant. Also note that the choice of the CRONE generation is important as this controller is developed in three generations: the first generation CRONE strategy is particularly appropriate when the desired open-loop gain crossover frequency ω_u is within a frequency range where the plant frequency response is asymptotic (this frequency band will be called a plant asymptotic-behavior band). As for the second generation, it is defined when ω_u is within a frequency range where the plant uncertainties are gain-like along with a constant phase variation. Concerning the third generation, it would be applied when both a gain and a phase variations are observed when dealing with plant's uncertainties. This generation will not be treated in this chapter due to some space constraints. Thus, this chapter will present some case scenarios which will lead to the use of the first two CRONE generations when using three different plants: the first one is constituted of iron, the second of aluminum and the third of copper with variable lengths L and several placements of the temperature sensor x. Simulation results will show the temperature variation across the diffusive interface medium in both time and frequency domains using Matlab and Simulink.

X. Moreau · F. Christophy
IMS Laboratory, Group CRONE, University of Bordeaux, Talence, Bordeaux, France
e-mail: xavier.moreau@u-bordeaux.fr

F. Christophy
e-mail: fadi.christophy@u-bordeaux.fr

R. Abi Zeid Daou (✉)
Faculty of Public Health, Biomedical Technologies Department, Lebanese German University, Sahel Alma, Jounieh, Lebanon
e-mail: r.abizeiddaou@lgu.edu.lb

A.T. Azar et al. (eds.), *Fractional Order Control and Synchronization of Chaotic Systems*, Studies in Computational Intelligence 688,
DOI 10.1007/978-3-319-50249-6_2

These results show how the temperature behaves at different positions for the three materials in use.

Keywords Finite diffusive interface • CRONE controller • Temperature control in homogeneous bars • Robustness • Users specifications • Gain and phase margin variation • Fractional order control

1 Introduction

The fractional calculus is a very old topic that was born following letter exchanges between L'Hopital and Leibtniz in September 30th, 1695 [15, 16]. Most of engineering domains have started the implementation of this topic either in the modelling process or in the control of their dynamic behavior [8]. Hence, the recognition of the fractional order in a system may reside in the identification of the plant transfer function or the control of a whole process using well known fractional controllers as the CRONE controller or the generalized PID controller [9, 17].

The diffusive interface medium is a fractional order system. In fact, the modelling of this medium has shown a semi integration (integration of order half) when considering the density of flux as the input of this medium and the temperature at any given point as the system output [7, 10]. Thus, we will consider in this work the temperature control of a finite diffusive interface after modelling, in previous works, the finite and the semi-infinite diffusive interface media [2, 3].

Hence, the novelty of this work resides in the deployment of a fractional order regulator in order to control a non-integer order plant. For this purpose, the CRONE controller will be used in its first two generations.

As each generation is used for a specific variation in the plant's transfer function, two case studies will be proposed in order to analyze the behavior and the robustness of the first two CRONE generations [1, 13].

However, lots of controllers were already synthesized and applied to several engineering fields and have shown great results. The controllers could be synthesized in both, time or frequency domain. For both cases, the fractional order has been applied. Interested readers can refer to the following papers [4–6] for more controller synthesis methods and applications.

This chapter will be divided as follow: in Sect. 2, the previous works concerning the modelling of the diffusive interface will be presented. The exact model will be proposed as well as the simplified one. Section 3 will present the CRONE controller along with the first two generations. The same example will be applied for both generations with three different gain margins. The main aim is to show the way to synthesis the CRONE controller as well as to compare between these generations. Section 4 will conclude the proposed work and will introduce some future work that may enrich this system.

2 Presentation of the Finite Diffusive Interface Medium

When presenting a finite diffusive interface medium, the heat transfer function is governed by three partial differential equations along with an initial condition regarding the initial time:

$$\begin{cases} \frac{\partial T(x,t)}{\partial t} = \alpha \frac{\partial^2 T(x,t)}{\partial x^2}, \; x > 0, \; t > 0 \\ -\lambda \frac{\partial T(x,t)}{\partial x} = \varphi(t), \; x = 0, \; t > 0 \\ -\lambda \frac{\partial T(x,t)}{\partial x} = 0, \; x = L, \; t > 0 \\ T(x,t) = 0, \; 0 \le x < L, \; t = 0 \end{cases}. \tag{1}$$

As the temperature initial condition is null, the Laplace transform of the first equation of system (1) leads to a differential equation of order 2 with respect to the variable x, as shown in Eq. (2):

$$\frac{\partial^2 \overline{T}(x,s)}{\partial x^2} - \frac{s}{\alpha_d} \overline{T}(x,s) = 0 \text{ where } \overline{T}(x,s) = \mathrm{L}\{T(x,t)\}. \tag{2}$$

The solution of this equation is of the following form [7, 19]:

$$\bar{T}(x,s) = K_1(s)\, \mathrm{e}^{x\sqrt{s/\alpha_d}} + K_2(s)\, \mathrm{e}^{-x\sqrt{s/\alpha_d}}. \tag{3}$$

When taking into consideration the boundary conditions ($x = 0$ and $x = L$), a system of two equations with two unknown values, $K_1(s)$ and $K_2(s)$, is derived as shown in (4):

$$\begin{cases} K_1(s) - K_2(s) = -\frac{1}{\lambda\sqrt{\frac{s}{\alpha_d}}} \bar{\varphi}(s) \\ K_1(s) e^{L\sqrt{\frac{s}{\alpha_d}}} - K_2(s)\, e^{-L\sqrt{\frac{s}{\alpha_d}}} = 0 \end{cases}. \tag{4}$$

The solution of this system, after the introduction of a new parameter $\lambda = \alpha_d\, \rho\, C_p$, leads to the expressions of $K_1(s)$ and $K_2(s)$, as shown below:

$$\begin{cases} K_1(s) = \frac{1}{\sqrt{\lambda \rho\, C_p\, s}} \frac{\mathrm{e}^{-L\sqrt{s/\alpha_d}}}{\mathrm{e}^{L\sqrt{s/\alpha_d}} - \mathrm{e}^{-L\sqrt{s/\alpha_d}}} \overline{\varphi}(s) \\ K_2(s) = \frac{1}{\sqrt{\lambda \rho\, C_p\, s}} \frac{\mathrm{e}^{L\sqrt{s/\alpha_d}}}{\mathrm{e}^{L\sqrt{s/\alpha_d}} - \mathrm{e}^{-L\sqrt{s/\alpha_d}}} \overline{\varphi}(s) \end{cases}. \tag{5}$$

The introduction of system Eq. (5) in Eq. (3) and the replacement of the flux density $\overline{\varphi}(s)$ by the flux $\overline{\phi}(s)$ $\left(\bar{\varphi}(s) = \overline{\phi}(s)/S\right)$ lead to the below transfer function of the whole system:

$$H(x,s,L) = \frac{\overline{T}(x,s,L)}{\overline{\phi}(s)} = \frac{1}{S\sqrt{\lambda \rho\, C_p\, s}} \frac{\mathrm{e}^{-(L-x)\sqrt{s/\alpha_d}} + \mathrm{e}^{(L-x)\sqrt{s/\alpha_d}}}{\mathrm{e}^{L\sqrt{s/\alpha_d}} - \mathrm{e}^{-L\sqrt{s/\alpha_d}}}, \tag{6}$$

which can be also presented as follow after introducing the hyperbolical functions [3]

$$H(x,s,L) = \frac{\bar{T}(x,s,L)}{\bar{\phi}(s)} = \frac{1}{S\sqrt{\lambda \rho\, C_p\, s}} \frac{1}{\tanh\left(L\sqrt{\frac{s}{\alpha_d}}\right)} \frac{\cosh\left((L-x)\sqrt{\frac{s}{\alpha_d}}\right)}{\cosh\left(L\sqrt{\frac{s}{\alpha_d}}\right)}. \tag{7}$$

Hence, to sum up, the system transfer function can be partitioned in several blocks as shown in Fig. 1 and as mentioned in systems (8)–(10).

$$H(x,s,L) = H_0\, I^{0.5}(s)\, F(0,s,L)\, G(x,s,L), \tag{8}$$

where

$$\begin{cases} H_0 = \frac{s^{0.5}\,\overline{T}(0,s,\infty)}{\overline{\phi}(s)} = \frac{1}{S\,\eta_d} \\ I^{0.5}(s) = \frac{\overline{T}(0,s,\infty)}{s^{0.5}\overline{T}(0,s,\infty)} = \frac{1}{s^{0.5}} \\ F(0,s,L) = \frac{\overline{T}(0,s,L)}{\overline{T}(0,s,\infty)} = \frac{1}{\tanh\left(\sqrt{\frac{s}{\omega_L}}\right)} \\ G(x,s,L) = \frac{\overline{T}(x,s,L)}{\overline{T}(0,s,L)} = \frac{\cosh\left(\sqrt{\frac{s}{\omega_{Lx}}}\right)}{\cosh\left(\sqrt{\frac{s}{\omega_L}}\right)} \end{cases}, \tag{9}$$

knowing that

$$\begin{cases} \eta_d = \sqrt{\lambda \rho\, C_p} \\ \omega_L = \frac{\alpha_d}{L^2} \\ \omega_{Lx} = \frac{\alpha_d}{(L-x)^2} \end{cases}, \tag{10}$$

where η_d represents the thermal effusivity.

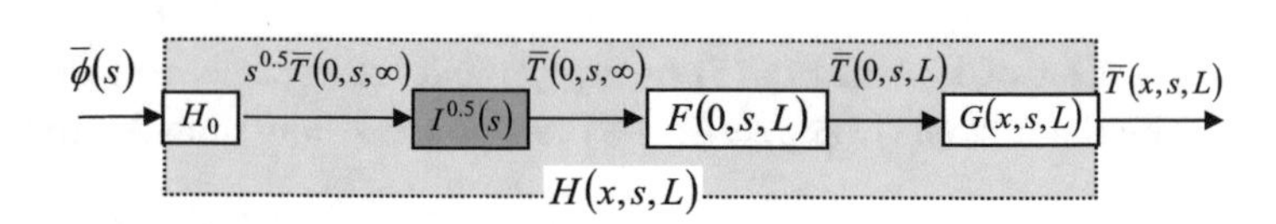

Fig. 1 Block diagram of the finite medium transfer function

Table 1 Physical characteristics of the used materials

Material	α_d	η_d	H_0	ω_L(rad/s)			ω_x (rad/s)		
	m^2/s	$W.K^{-1}.m^{-2}.s^{0.5}$	$K.s^{0.5}.W^{-1}$	$L = 0.25$ m	$L = 0.5$ m	$L = 1$ m	$x = 0$	$x = 0.5$ cm	$x = 1$ cm
Copper	117×10^{-6}	3.72×10^{4}	0.269	19×10^{-4}	4.68×10^{-4}	1.17×10^{-4}	Inf.	4.68	1.17
Aluminum	97×10^{-6}	2.41×10^{4}	0.416	16×10^{-4}	3.88×10^{-4}	0.97×10^{-4}	Inf.	3.88	0.97
Iron	23×10^{-6}	1.67×10^{4}	0.596	3.68×10^{-4}	0.92×10^{-4}	0.23×10^{-4}	Inf.	0.92	0.23

Note that, for analysis purposes, a relation between ω_L, ω_{Lx} and the diffusive time constant, τ_L, were introduced:

$$\begin{cases} \omega_L = \left(1 - \frac{x}{L}\right)^2 \omega_{Lx} \\ \tau_L = \frac{1}{\omega_L} = \frac{L^2}{\alpha_d} \end{cases}. \quad (11)$$

The approximation of this system was already presented in several previous works. Interested authors can refer to the following references [3, 10, 12]. Just need to know that Oustaloup approximation and Maclaurin series were at the core of this approximation. As a conclusion, the finite diffusive interface medium can be approximated by the following transfer function:

$$H(x,s,L) = \frac{\overline{T}(x,s,L)}{\overline{\phi}(s)} = H_0 \frac{1}{s^{0.5}} \frac{1}{\tanh\left(\sqrt{\frac{s}{\omega_L}}\right)} e^{-\sqrt{\frac{s}{\omega_x}}}, \quad (12)$$

where $\omega_x = \alpha_d / x^2$.

At the end of this section, let us define the different materials to be used for the control process. In fact, the aluminum, the copper and the iron were used for the simulations later in this chapter. All physical values of these three materials will be presented in Table 1.

3 CRONE Controller

CRONE is the acronym for *Commande Robuste d'Ordre Non Entier* (non-integer order robust control). While the first two approaches use the real fractional integration or differentiation operator, the third uses the complex differentiation operator. In the frequency domain, they enable to synthesize simply and methodologically, linear robust control laws. The control schematic used is based on the classic unity-feedback configuration. Thus, Fig. 2 shows a general scheme used for the control-system design.

The equations associated to this scheme are given by:

- Output: $Y(s) = S(s)\,D_m(s) + SP(s)\,D_u(s) + T(s)\,Y_{ref}(s)$ (13)

- Errorsignal: $\varepsilon(s) = -S(s)\,D_m(s) - SP(s)\,D_u(s) + S(s)\,Y_{ref}(s)$ (14)

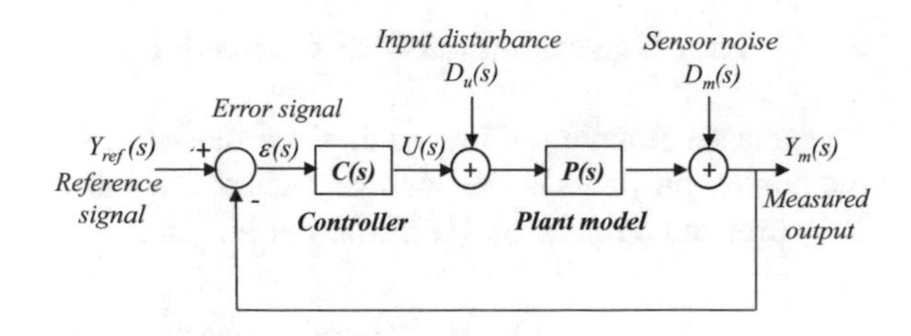

Fig. 2 Scheme used for the control-system design

- Controlsignal: $U(s) = -SC(s)\,D_m(s) + T(s)\,D_u(s) + SC(s)\,Y_{ref}(s)$ (15)

with

$$\begin{cases} S(s) = \frac{1}{1+\beta(s)}: & \text{sensitivity function} \\ T(s) = 1 - S(s): & \text{complementary sensitivity function} \\ SP(s) = S(s)\,P(s) & \\ SC(s) = S(s)\,C(s) & \\ \beta(s) = C(s)\,G(s): & \text{open} - \text{loop transfer function} \end{cases} . \quad (16)$$

As the main purpose is not to present the CRONE controller but to provide the tools used to fit the user specifications concerning the stability degree, the rapidity, the precision in the steady state mode, the saturation as well as the sensibility of the system towards the disturbances, we will present hereafter the general transfer functions of the first two CRONE generations along with the conditions that must be filled in order to apply each of these generations.

The user specifications lead us to set the following parameters:

- Concerning the ***stability degree***, the phase margin M_Φ varies between $90° \geq M_\Phi \geq 45°$;
- Concerning the ***speed***, desired open-loop gain crossover frequency ω_u (or ω_{cg}) is equal to 1 rad/s;
- Concerning the ***precision in the steady state response***, a null static error;
- Concerning the ***saturation***, a maximum input value of 12 W is allowed.

3.1 Synthesis with Gain Variations Only

In this first part, we will treat the gain variations only while considering a constant phase. Thus, the first CRONE generation could be used as well as the second one.

3.1.1 First Generation CRONE Controller

The transfer function of the synthesized model of the plant that would be used for the control purposes at the boundary where the flux is applied (e.g., for $x = 0$ mm) is Poinot and Trigeassou [18], Malti et al. [14]:

$$P_1(s) = H(0, s, L) = H_0 \frac{1}{s^{0.5}} \frac{1}{\tanh\left(\sqrt{\frac{s}{\omega_L}}\right)}, \tag{17}$$

where the nominal state parameters (for the aluminum case) are:

$$\begin{cases} H_0 = H_0(\text{Alu}) = 0.416 \text{ K s}^{0.5} \text{ W}^{-1} \\ \omega_L = \omega_L(\text{Alu}, L = 1\text{m}) = 0.97\ 10^{-4} \text{ rad/s} \end{cases}, \tag{18}$$

and their variation ranges taking into account the two other materials already shown in Table 1:

$$\begin{cases} H_0 \in \left[\underline{H_0} = H_0(\text{Cop}) = 0.269 \quad ; \quad \overline{H_0} = H_0(\text{Iron}) = 0.596\right] \text{K s}^{0.5} \text{ W}^{-1} \\ \omega_L \in \left[\underline{\omega_L} = \omega_L(\text{Iron}, L = 1\text{ m}) = 0.23\ 10^{-4} \text{ rad/s} \quad ; \quad \overline{\omega_L} = \omega_L(\text{Cop}, L = 0.25\text{ m}) = 19\ 10^{-4} \text{ rad/s}\right] \end{cases} \tag{19}$$

As for the first generation CRONE controller, its transfer function is of the following form [11]:

$$C_F(s) = C_0 \left(\frac{1 + s/\omega_I}{s/\omega_I}\right)^{m_I} \left(\frac{1 + s/\omega_l}{1 + s/\omega_h}\right)^{m} \frac{1}{(1 + s/\omega_F)^{m_F}}, \tag{20}$$

where m_I, $m_F \in \mathbf{N}$, $\omega_I < \omega_F \in \mathbb{R}$ and $\omega_l < \omega_h \in \mathbb{R}$.

If we choose $\omega_I = \omega_l$ and $\omega_h = \omega_F$ in order to simplify the transfer function of the controller while taking into consideration the user specifications already defined at the beginning of this section, we can get:

$$C_F(s) = C_0 \left(\frac{\omega_l}{s}\right)^{m_I} \frac{(1 + s/\omega_l)^{m_I + m}}{(1 + s/\omega_h)^{m_F + m}}. \tag{21}$$

Referring to the synthesis characteristics, we choose:

- $m_I = 1$, in order to get a null static error as the plant contains an integration of order 1 at low frequencies;
- $m = (M_\Phi - 180° - \arg P_1(j\omega_u))/90°$, which is based on the definition of the phase margin M_Φ, knowing that $\arg P_1(j\omega_u) = -45°$, thus $M_\Phi \in [45°; 90°]$, hence $m \in [-1;\ -0.5]$;
- $m_F = 1$, in order to limit the input sensitivity;
- $\omega_{unom} = 1$ rad/s, value defined by the authors;

Based on these values, the expression of $C_F(s)$ can be rewritten as follow:

$$C_F(s) = C_0\left(\frac{\omega_l}{s}\right)\left(\frac{1+s/\omega_l}{1+s/\omega_h}\right)^{1+m}. \tag{22}$$

Thus, knowing that $m \in [-1; -0.5]$ and $\omega_l < \omega_h$, the fractional form of the controller $C_F(s)$ can be expressed by an integrator of order 1 in series with a lead compensator of order $1 + m$, m being a non-integer value. The expression of $C_F(s)$ is thus characterized by four parameters (m, ω_l, ω_h et C_0) that could be defined based on three below steps:

Step 1 m is determined based on the phase margin M_Φ;

Step 2 ω_l and ω_h are defined in such a way that the fractional asymptotic behavior of the controller should vary in a frequency range between $[\omega_A, \omega_B]$ around the nominal gain cutoff frequency ω_{unom}. In order to keep the stability degree robustness, it is necessary to set the following:

$$\forall \omega_u \in [\omega_{u\min}; \omega_{u\max}],\ \omega_A \le \omega_u \le \omega_B \quad \Rightarrow \quad \begin{cases} \omega_A \le \omega_{u\min} \\ \omega_B \ge \omega_{u\max} \end{cases}, \tag{23}$$

and

$$\begin{cases} \omega_l = \omega_A/b \\ \omega_h = b\omega_B \end{cases}, \quad \text{where } b > 1. \tag{24}$$

If we consider that ω_l and ω_h are geometrically distributed around the cutoff frequency ω_{unom} and if we suppose that $r = \omega_B/\omega_A$, ω_l and ω_h would be calculated as follow:

$$\begin{cases} \sqrt{\omega_l\,\omega_h} = \omega_{unom} \\ \frac{\omega_h}{\omega_l} = b^2\,r \end{cases} \quad \Rightarrow \quad \begin{cases} \omega_l = \omega_{unom}/(b\sqrt{r}) \\ \omega_h = \omega_{unom} b\sqrt{r} \end{cases}. \tag{25}$$

The value of the ratio r is deduced from the slope $-n20$ dB/dec (n being defined based on the open loop transfer function order around ω_{unom}) and from the gain variation $\Delta\beta$ due to the parametric uncertainties, thus:

$$r = \Delta\beta^{1/n}. \tag{26}$$

Step 3 C_0 is calculated in order to respond to the speed specifications. Hence, C_0 can be calculated based on the following relation:

$$|\beta(j\omega_u)| = 1 \quad \Leftrightarrow \quad |C_F(j\omega_u)||P_1(j\omega_u)| = 1, \tag{27}$$

which can be also expressed as follows:

$$C_0\left(\frac{\omega_l}{\omega_u}\right)\left(\frac{1+(\omega_u/\omega_l)^2}{1+(\omega_u/\omega_h)^2}\right)^{\left(\frac{1+m}{2}\right)}|P_1(j\omega_u)|=1, \quad (28)$$

thus,

$$C_0=\left[\left(\frac{\omega_l}{\omega_u}\right)\left(\frac{1+(\omega_u/\omega_l)^2}{1+(\omega_u/\omega_h)^2}\right)^{\left(\frac{1+m}{2}\right)}|P_1(j\omega_u)|\right]^{-1}. \quad (29)$$

Finally, the last step consists on presenting the controller fractional order transfer function $C_F(s)$ in a rational form $C_R(s)$. Different approaches were proposed but Oustaloup approximation remains one of the best. Hence, applying it will lead to the following general form of the controller:

$$C_R(s)=C_0\left(\frac{\omega_l}{s}\right)\prod_{i=1}^{N}\left(\frac{1+s/\omega_i'}{1+s/\omega_i}\right), \quad (30)$$

where

$$\begin{cases}\frac{\omega_{i+1}'}{\omega_i'}=\frac{\omega_{i+1}}{\omega_i}=\alpha\eta>1\\ \frac{\omega_i}{\omega_i'}=\alpha \text{ et } \frac{\omega_{i+1}'}{\omega_i}=\eta\\ \alpha\eta=(\omega_h/\omega_l)^{1/N}\\ \alpha=(\alpha\eta)^{1+m} \text{ et } \eta=(\alpha\eta)^{-m}\\ \omega_1'=\omega_1\eta^{1/2} \text{ et } \omega_N=\omega_h\eta^{-1/2}\end{cases}. \quad (31)$$

The remaining part of this paragraph will show the 1st generation CRONE controller computation for three phase margin values $M_\Phi = 45°$, 67.5° and 90° and its robustness when applying it to the three materials (aluminum, copper and iron).

Example 1: Phase Margin $M_\Phi = 45°$

If $M_\Phi = 45°$, then $m=(M_\Phi-180°-\arg P_1(j\omega_u))/90°=-1$. The controller transfer function can be expressed as shown in Eq. (32):

$$C_F(s)=C_0\left(\frac{\omega_l}{s}\right). \quad (32)$$

This is a particular case as the controller is expressed as an integrator of order 1 characterized by only one parameter $C_0^* = C_0\,\omega_l$ whose main purpose is to take in consideration the speed specification, thus:

$$|\beta(j\omega_u)| = 1 \quad \Leftrightarrow \quad |C_F(j\omega_u)||P_1(j\omega_u)| = 1, \tag{33}$$

which can also be expressed as follow:

$$\frac{C_0^*}{\omega_u}|P_1(j\omega_u)| = 1, \tag{34}$$

thus,

$$C_0^* = \frac{\omega_u}{|P_1(j\omega_u)|} \quad \Rightarrow \quad C_0^* = 2.405 V/^\circ. \tag{35}$$

As the controller is of an integer form in this case, the transfer functions of $C_F(s)$ and $C_R(s)$ are similar.

Example 2: Phase Margin $\boldsymbol{M_\Phi = 67.5^\circ}$

If $M_\Phi = 67.5^\circ$, then $m = \left(M_\Phi - 180^\circ - \arg P_1(j\omega_u)\right)/90^\circ = -0.75$. The controller transfer function can be expressed as shown in Eq. (36):

$$C_F(s) = C_0\left(\frac{\omega_l}{s}\right)\left(\frac{1 + s/\omega_l}{1 + s/\omega_h}\right)^{0.25}. \tag{36}$$

In this second example, the controller could be expressed as an integrator of order 1 in series with a lead compensator of order 0.25. The gain C_0 and the open-loop gain crossover frequency are defined as follow:

$$\begin{cases} \left.\begin{array}{l} \omega_A = \omega_{u\min} = 72.77\ 10^{-2}\text{rad/s} \\ \omega_B = \omega_{u\max} = 137.43\ 10^{-2}\ \text{rad/s} \end{array}\right\} \quad \Rightarrow \quad r = 1.89 \\ b = 25 \quad \Rightarrow \quad \begin{cases} \omega_l = 2.91\ 10^{-2}\text{rad/s} \\ \omega_h = 34.36\ \text{rad/s} \end{cases} \\ C_0 = 34.125\text{V s}/^\circ \end{cases}. \tag{37}$$

The rational form of this controller (who is expressed in Eq. (30)) will have the below values:

$$\begin{cases} N = 6 \\ \omega_1' = 8.17\ 10^{-2}\ \text{rad/s} & \omega_1 = 3.37\ 10^{-2}\ \text{rad/s} \\ \omega_2' = 26.55\ 10^{-2}\ \text{rad/s} & \omega_2 = 10.96\ 10^{-2}\ \text{rad/s} \\ \omega_3' = 86.3\ 10^{-2}\ \text{rad/s} & \omega_3 = 35.64\ 10^{-2}\ \text{rad/s}. \\ \omega_4' = 2.81\ \text{rad/s} & \omega_4 = 1.159\ \text{rad/s} \\ \omega_5' = 9.12\ \text{rad/s} & \omega_6 = 3.77\ \text{rad/s} \\ \omega_6' = 29.65\ \text{rad/s} & \omega_7 = 12.25\ \text{rad/s} \end{cases} \quad (38)$$

Example 3: Phase Margin $\boldsymbol{M_\Phi = 90°}$

If $M_\Phi = 90°$, then $m = \left(M_\Phi - 180^\circ - \arg P_1(j\omega_u)\right)/90^\circ = -0.5$. The controller transfer function can be expressed as shown in Eq. (39):

$$C_F(s) = C_0\left(\frac{\omega_l}{s}\right)\left(\frac{1 + s/\omega_l}{1 + s/\omega_h}\right)^{0.5}. \quad (39)$$

In this example, the controller could be expressed as an integrator of order 1 in series with a lead compensator of order 0.5. The gain C_0 and the open-loop gain crossover frequency are defined based on system (40):

$$\begin{cases} \left.\begin{aligned} \omega_A = \omega_{u\min} = 67.21\ 10^{-2}\ \text{rad/s} \\ \omega_B = \omega_{u\max} = 148.8\ 10^{-2}\ \text{rad/s} \end{aligned}\right\} \Rightarrow \quad r = 2.214 \\ b = 25 \Rightarrow \quad \begin{cases} \omega_l = 2.69\ 10^{-2}\ \text{rad/s} \\ \omega_h = 37.2\ \text{rad/s} \end{cases} \\ C_0 = 14.67\text{V s}/° \end{cases}, \quad (40)$$

As for the parameters of the rational transfer function $C_R(s)$, they are as follow:

$$\begin{cases} N = 6 \\ \omega_1' = 6.64\ 10^{-2}\ \text{rad/s} & \omega_1 = 3.63\ 10^{-2}\ \text{rad/s} \\ \omega_2' = 22.16\ 10^{-2}\ \text{rad/s} & \omega_2 = 12.13\ 10^{-2}\ \text{rad/s} \\ \omega_3' = 73.98\ 10^{-2}\ \text{rad/s} & \omega_3 = 40.49\ 10^{-2}\ \text{rad/s}. \\ \omega_4' = 2.47\ \text{rad/s} & \omega_4 = 1.352\ \text{rad/s} \\ \omega_5' = 8.24\ \text{rad/s} & \omega_6 = 4.51\ \text{rad/s} \\ \omega_6' = 27.52\ \text{rad/s} & \omega_7 = 15.06\ \text{rad/s} \end{cases} \quad (41)$$

Performances

Figures 3, 4, 5, 6, 7, 8 show the frequency domain and time domain performances for the three examples when applying the three different materials. In fact, Fig. 3 shows the Bode diagrams of $C_F(s)$ for the three listed examples ($M_\Phi = 45°$ (a), $M_\Phi = 67.5°$ (b) and $M_\Phi = 90°$ (c)).

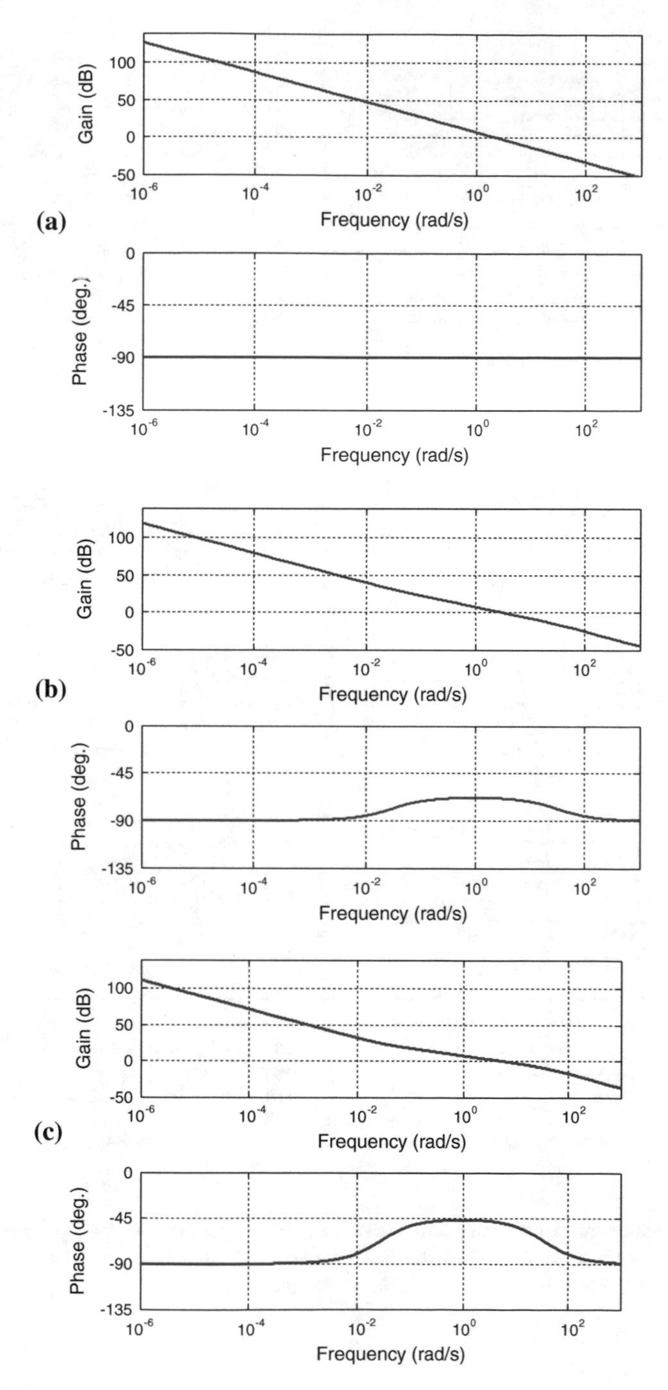

Fig. 3 Bode diagrams for CRONE controller $C_F(s)$ for $M_\Phi = 45°$ (**a**), $M_\Phi = 67.5°$ (**b**) and $M_\Phi = 90°$ (**c**)

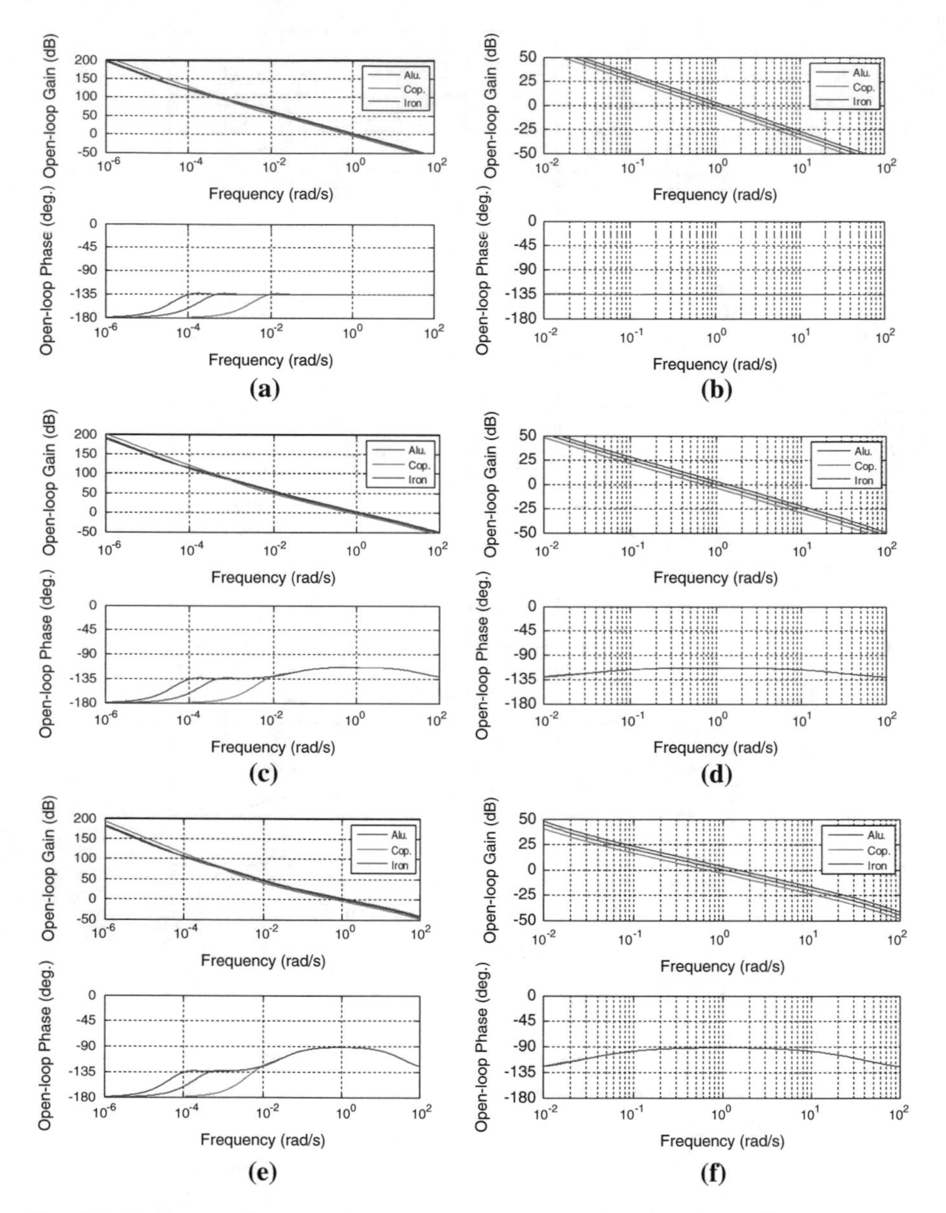

Fig. 4 Bode diagrams for the open loop transfer function over the interval [10^{-6} rad/s; 10 rad/s] (**a**, **c**, **e**), and with a zoom around the gain crossover frequency $\omega_u = 1$ rad/s (**b**, **d**, **f**), for $M_\Phi = 45°$ (**a**, **b**), $M_\Phi = 67.5°$ (**c**, **d**) and $M_\Phi = 90°$ (**e**, **f**)

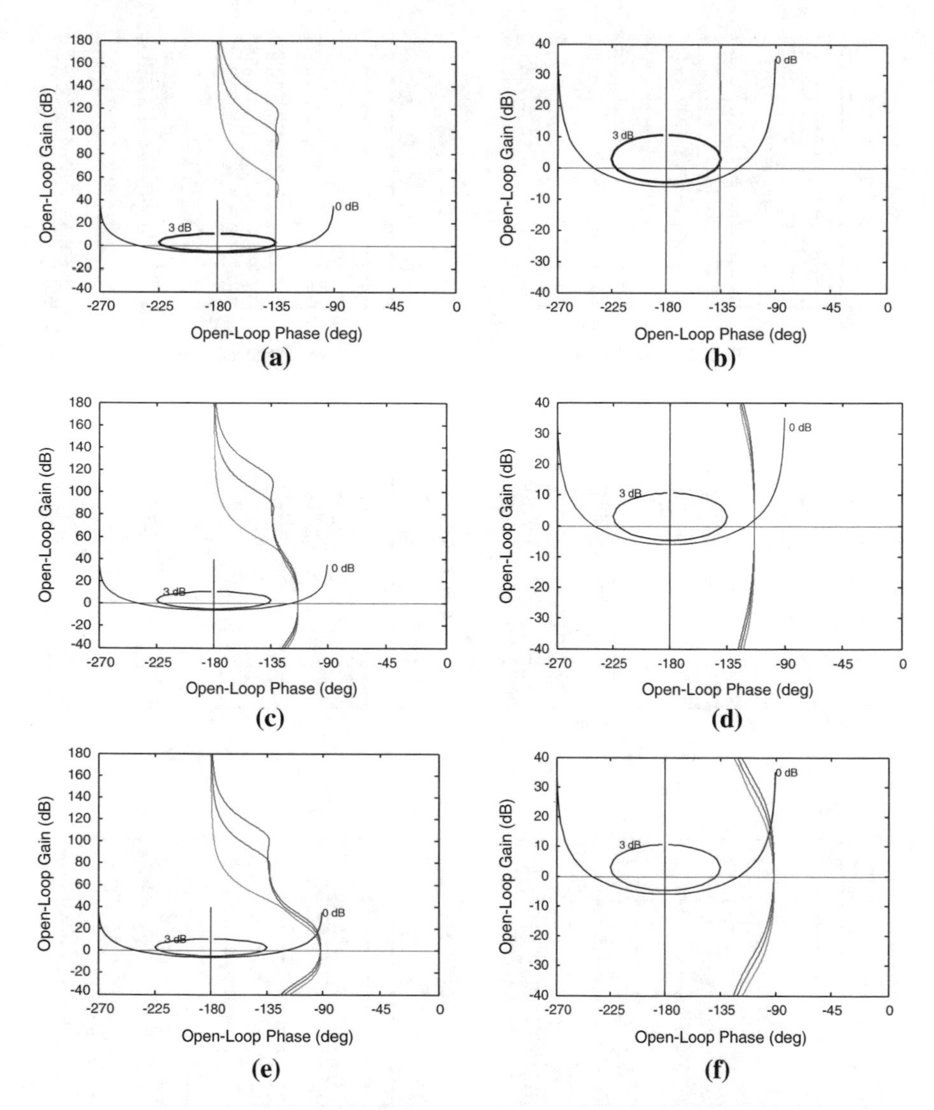

Fig. 5 Black-Nichols plots for the open loop transfer function over the interval [−40 dB; 180 dB] (**a**, **c**, **e**) and with a zoom around the gain crossover frequency $\omega_u = 1$ rad/s (**b**, **d**, **f**), for $M_\Phi = 45°$ (**a**, **b**), $M_\Phi = 67.5°$ (**c**, **d**) and $M_\Phi = 90°$ (**e**, **f**)

Figure 4 shows the open loop Bode diagrams over the frequency bandwidth [10^{-6} rad/s; 10 rad/s] in (a) (c) (e), with a particular zoom around the open-loop gain crossover frequency $\omega_u = 1$ rad/s in (b) (d) (f), for $M_\Phi = 45°$ (a) (b), $M_\Phi = 67.5°$ (c) (d) and $M_\Phi = 90°$ (e) (f).

Figure 5 presents the Black-Nichols plots of the open loop transfer function over the interval [−40 dB; 180 dB] in (a) (c) (e) with a particular zoom around the

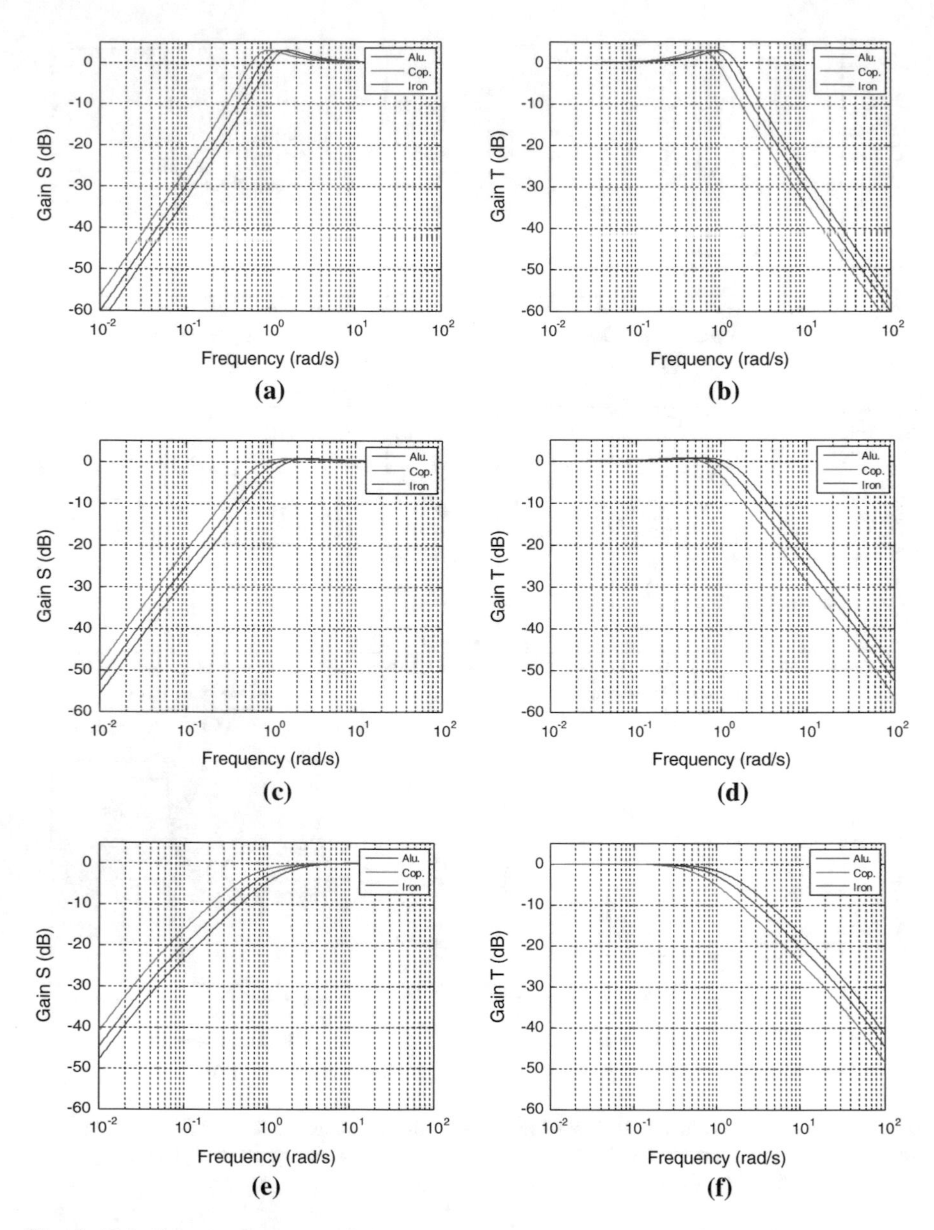

Fig. 6 Gain diagrams for the sensitivity functions: *S(jω)* (**a**, **c**, **e**) and *T(jω)* (**b**, **d**, **f**), for $M_{\Phi} = 45°$ (**a**, **b**), $M_{\Phi} = 67.5°$ (**c**, **d**) and $M_{\Phi} = 90°$ (**e**, **f**)

open-loop gain crossover frequency $\omega_u = 1$ rad/s in (b) (d) (f), for $M_{\Phi} = 45°$ (a) (b), $M_{\Phi} = 67.5°$ (c) (d) and $M_{\Phi} = 90°$ (e) (f).

Figure 6 shows the gain diagrams for the sensitivity functions ***S(jω)*** in (a) (c) (e) and ***T(jω)*** in (b) (d) (f), for $M_{\Phi} = 45°$ (a) (b), $M_{\Phi} = 67.5°$ (c) (d) and $M_{\Phi} = 90°$ (e) (f).

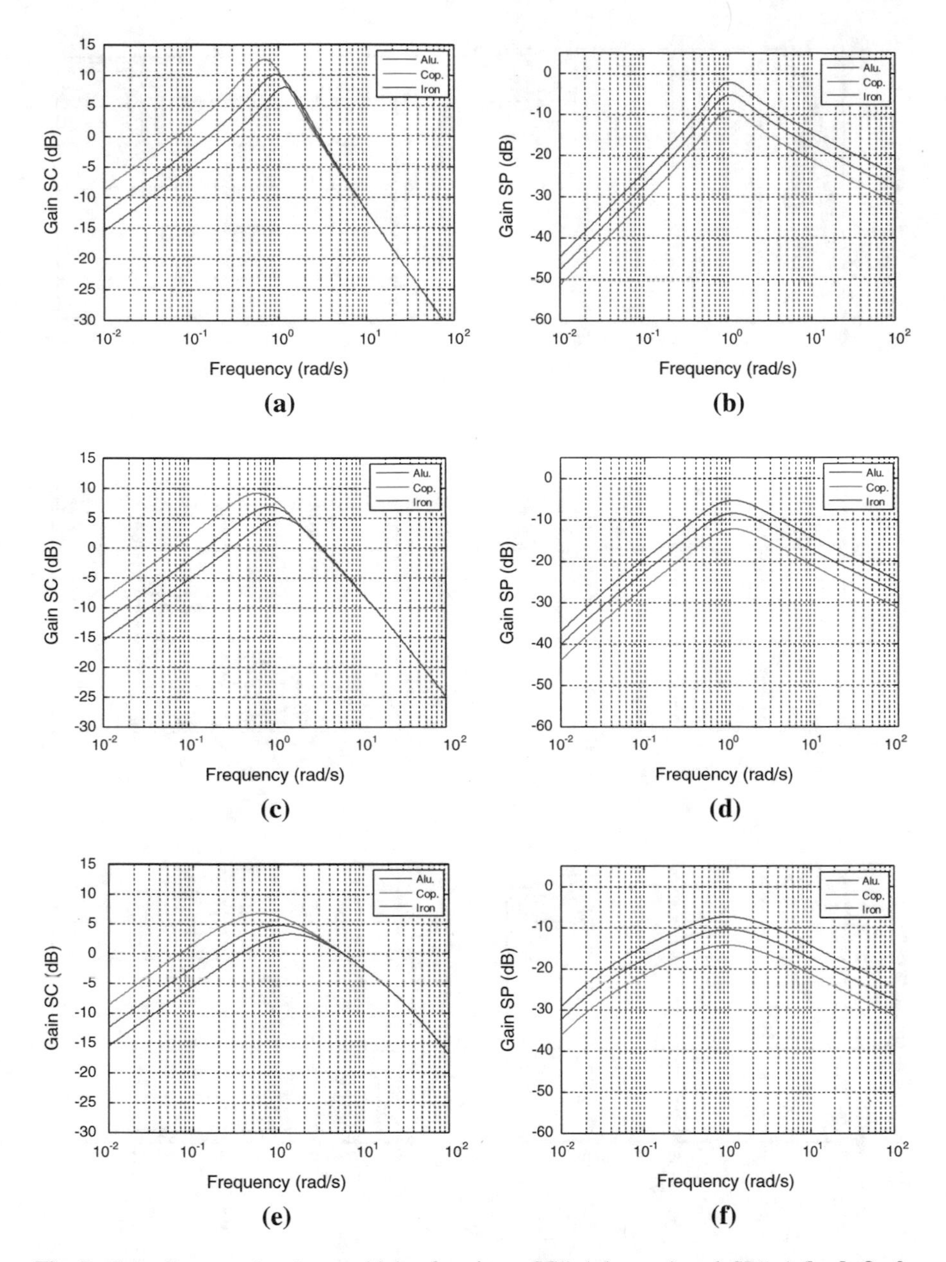

Fig. 7 Gain diagrams for the sensitivity functions: *SC(jω)* (**a**, **c**, **e**) and *SP(jω)* (**b**, **d**, **f**), for $M_\Phi = 45°$ (**a**, **b**), $M_\Phi = 67.5°$ (**c**, **d**) and $M_\Phi = 90°$ (**e**, **f**)

Figure 7 presents the gain diagrams of the sensitivity functions ***SC(jω)*** in (a) (c) (e) et ***SP(jω)*** in (b) (d) (f), for $M_\Phi = 45°$ (a) (b), $M_\Phi = 67.5°$ (c) (d) and $M_\Phi = 90°$ (e) (f).

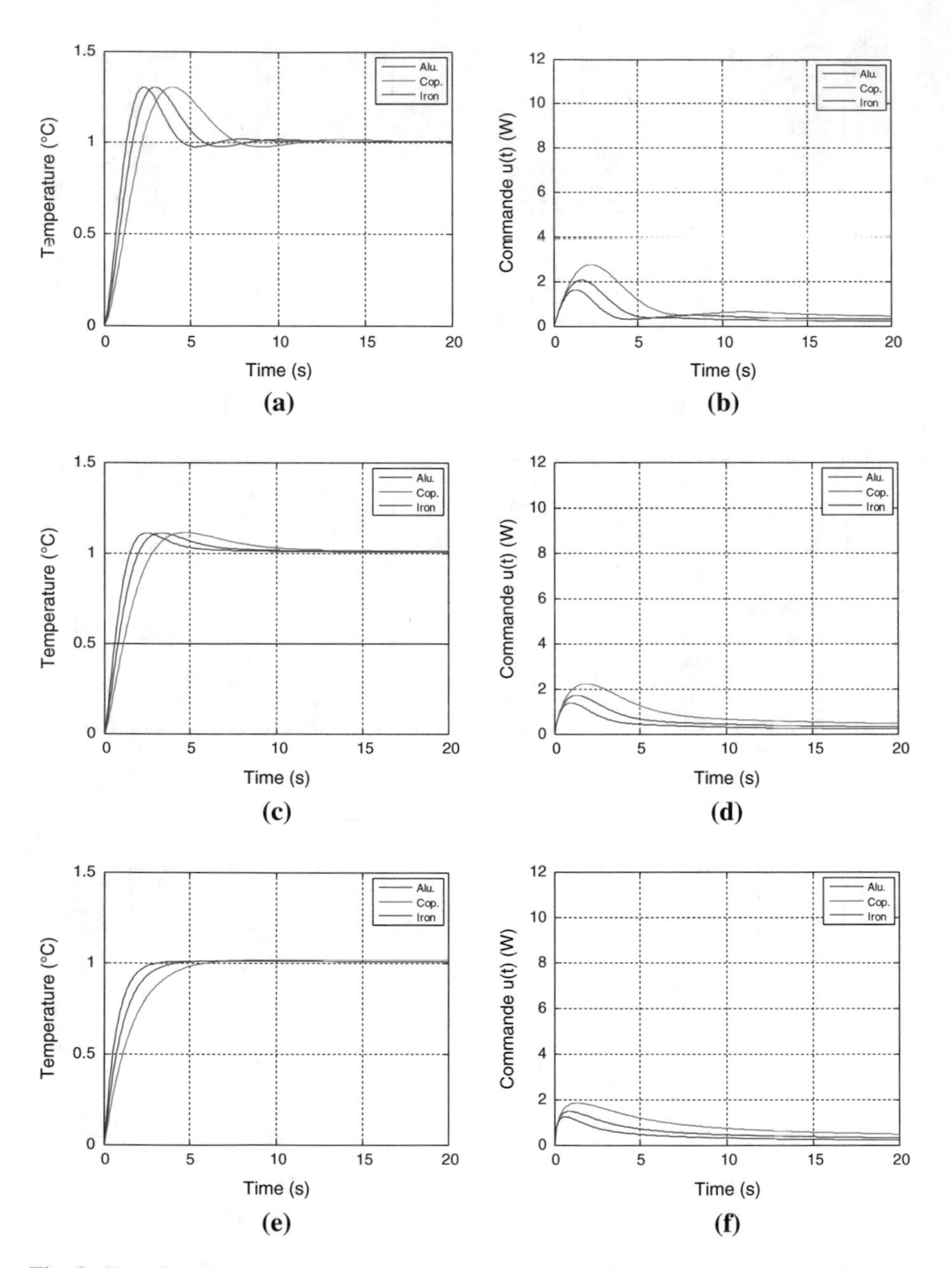

Fig. 8 Time domain responses for a step input of 1 °C: output temperature (**a**, **c**, **e**) and control signal (**b**, **d**, **f**), for $M_{\Phi} = 45°$ (**a**, **b**), $M_{\Phi} = 67.5°$ (**c**, **d**) and $M_{\Phi} = 90°$ (**e**, **f**)

Finally, Fig. 8 shows the output temperature variations for step input of amplitude 1 °C in (a) (c) (e) and for the control signal in (b) (d) (f), for $M_{\Phi} = 45°$ (a) (b), $M_{\Phi} = 67.5°$ (c) (d) and $M_{\Phi} = 90°$ (e) (f).

The robustness study is also presented in these figures as the plots contain the behavior of the three different materials.

3.1.2 Second Generation CRONE Controller

Let's start by a review of the plant simplified transfer function. In fact, the model $P_{2.1}(s)$ will be used to synthesize the second generation CRONE controller. Its transfer function is as follow:

$$P_{2.1}(s) = H_0^* \frac{(1 + s/\omega_L)^{0.5}}{s/\omega_L}, \tag{42}$$

where

$$H_0^* = \frac{H_0}{\omega_L^{0.5}}, \tag{43}$$

and where the values of the variables H_0 and ω_L were already presented in Eq. (14) along with their intervals (system (19)).

As for the CRONE controller synthesis, it is done a posteriori when applying the second generation. Hence, the open loop transfer function is expressed as follow:

$$\beta(s) = \beta_0 \left(\frac{1 + s/\omega_l}{s/\omega_l}\right)^{n_l} \left(\frac{1 + s/\omega_h}{1 + s/\omega_l}\right)^{n} \frac{1}{(1 + s/\omega_h)^{n_h}}, \tag{44}$$

where ω_l and ω_h represent the transitional low and high frequencies, n a real non-integer order between 1 and 2 near the frequency ω_u, n_l and n_h are the orders of the asymptotic behavior at low and high frequencies and β_0 a constant which ensures unity gain at frequency ω_u.

Thus, after defining the open loop transfer function, the computation of the fractional order CRONE controller is defined as follow:

$$C_F(s) = \beta(s)\, P_{2.1}^{-1}(s). \tag{45}$$

Hence, $C_F(s)$ could be written as follow:

$$C_F(s) = \beta_0 \left(\frac{1 + s/\omega_l}{s/\omega_l}\right)^{n_l} \left(\frac{1 + s/\omega_h}{1 + s/\omega_l}\right)^{n} \frac{1}{(1 + s/\omega_h)^{n_h}} \frac{s/\omega_L}{H_0^*(1 + s/\omega_L)^{0.5}}. \tag{46}$$

Referring to the user specifications already shown at the start of this section, the parameters values of Eq. (46) are set as follow:

- $n_l = 2$, in order to get a null static error;

- $n = (180° - M_\Phi)/90°$, in order to have a phase margin $M_\Phi \in [45°; 90°]$; thus $n \in [1; 1.5]$;
- $n_h = 1.5$, in order to limit the input sensitivity;
- $\omega_{unom} = 1$ rad/s, value defined by the user and applied for all examples for a comparative study.

Hence, expression (46) can be rewritten as follow:

$$C_F(s) = C_0\left(\frac{\omega_l}{s}\right)\left(\frac{1+s/\omega_l}{1+s/\omega_h}\right)^{2-n}\left(\frac{1+s/\omega_h}{1+s/\omega_L}\right)^{0.5}, \tag{47}$$

where

$$C_0 = \frac{\beta_0\ \omega_l}{H_0^*\ \omega_L}. \tag{48}$$

As for the first generation, this paragraph will also study the system behavior while applying the second generation CRONE controller for three different cases when the phase margin M_Φ is equal to 45°, 67.5° and 90°.

Example 4: Phase Margin $M_\Phi = 45°$

If $M_\Phi = 45°$, then $n = (180° - M_\Phi)/90° = 1.5$. The open loop transfer function can be expressed as follow:

$$\beta(s) = \beta_0 \frac{(1+s/\omega_l)^{0.5}}{(s/\omega_l)^2}, \tag{49}$$

and the controller transfer function would be:

$$C_F(s) = C_0\left(\frac{\omega_l}{s}\right)\left(\frac{1+s/\omega_l}{1+s/\omega_L}\right)^{0.5}. \tag{50}$$

If we suppose that $\omega_l = \omega_L$, $C_F(s)$ can be rewritten as shown in Eq. (51):

$$C_F(s) = C_0\left(\frac{\omega_l}{s}\right). \tag{51}$$

This expression is identical to the one of the first generation CRONE controller obtained in example 1 of Sect. 3.1.1.

Example 5: Phase Margin $\boldsymbol{M_\Phi = 67.5°}$

If $M_\Phi = 67.5°$, then $n = (180° - M_\Phi)/90° = 1.25$. The open loop transfer function can be expressed as shown in Eq. 52:

$$\beta(s) = \beta_0 \left(\frac{\omega_l}{s}\right)^2 \frac{(1 + s/\omega_l)^{0.75}}{(1 + s/\omega_h)^{0.25}}. \tag{52}$$

Thus, the controller transfer function would be as follow:

$$C_F(s) = C_0 \left(\frac{\omega_l}{s}\right) \frac{(1 + s/\omega_l)^{0.75}}{(1 + s/\omega_L)^{0.5}} \frac{1}{(1 + s/\omega_h)^{0.25}}. \tag{53}$$

As for the first example, if we choose $\omega_l = \omega_L$, the form of $C_F(s)$ will be similar to the first generation CRONE controller as it appears in Eq. (54)

$$C_F(s) = C_0 \left(\frac{\omega_l}{s}\right) \left(\frac{1 + s/\omega_l}{1 + s/\omega_h}\right)^{0.25}. \tag{54}$$

Example 6: Phase Margin $\boldsymbol{M_\Phi = 90°}$

If $M_\Phi = 90°$, then $n = (180° - M_\Phi)/90° = 1$. The open loop transfer function can be expressed as shown in Eq. (55):

$$C_F(s) = C_0 \left(\frac{\omega_l}{s}\right) \left(\frac{1 + s/\omega_l}{1 + s/\omega_h}\right)^{0.5}. \tag{55}$$

Hence, the controller $C_F(s)$ transfer function would be:

$$C_F(s) = C_0 \left(\frac{\omega_l}{s}\right) \frac{(1 + s/\omega_l)}{(1 + s/\omega_L)^{0.5}} \frac{1}{(1 + s/\omega_h)^{0.5}}. \tag{56}$$

As the two previous cases, when choosing $\omega_l = \omega_L$, the form of $C_F(s)$ of the second generation will be similar to the first generation CRONE controller as presented in relation (57):

$$C_F(s) = C_0 \left(\frac{\omega_l}{s}\right) \left(\frac{1 + s/\omega_l}{1 + s/\omega_h}\right)^{0.5}. \tag{57}$$

Performances

In this last section, the study will be divided into two parts referring to value of x.

For x = 0

To sum up, for the three cases studied regarding the value of the phase margin ($M_\Phi = 45°$, $67.5°$ and $90°$), the form of the synthesized controller when using the first generation and the second generation CRONE controllers is the same. Thus, the transfer function of this controller is the same for examples 1 and 4.

However, some small differences exist between examples 5 and 6 on one hand and examples 2 and 3 on the other hand because of the choice of $\omega_l = \omega_L$. This assumption leads to $\omega_l = 0.97\ 10^{-4}$ rad/s (nominal case for the Aluminum for $L = 1$ m). In fact, in the open loop, this difference can be expressed by the absence of an asymptotical behavior around the phase $-135°$ over the interval [10^{-4} rad/s; 10^{-2} rad/s] as it was the case for examples 2 and 3.

Nevertheless, for the low frequencies ($\omega < 10^{-4}$ rad/s), around the open-loop gain crossover frequency $\omega_u = 1$ rad/s and at high frequencies, the open loop behavior is identical when comparing examples 2 and 5 or examples 3 and 6. This can explain the fact that the closed loop dynamics are similar for both CRONE generations.

For X > 0

It is important to analyze the sensitivity of the stability degree at position of x through the phase margin a posteriori when controlling the temperature $T(0,t)$ for $x = 0$ and studying the influence of the temperature sensor when this latter is not placed at $x = 0$ exactly.

Hence, for $x > 0$, the open loop transfer function $\beta(s, x)$ can be expressed as follow:

$$\beta(s,x) = C(s)P_{2.1}(s)\, e^{-\left(\frac{s}{\omega_x}\right)^{0.5}}, \tag{58}$$

which can be simplified when introduction the nominal open loop transfer function $\beta_{nom}(s)$,

$$\beta(s,x) = \beta_{nom}(s)\, e^{-\left(\frac{s}{\omega_x}\right)^{0.5}}, \tag{59}$$

whose frequency response $\beta(j\omega, x)$ is of the following form

$$\beta(j\omega,x) = \beta_{nom}(j\omega)\, e^{-\left(j\frac{\omega}{\omega_x}\right)^{0.5}}. \tag{60}$$

Knowing that

$$e^{-\left(j\frac{\omega}{\omega_x}\right)^{0.5}} = m(x,\omega)e^{-j\theta(x,\omega)} \quad \text{where} \quad \begin{cases} m(x,\omega) = e^{-\left(\frac{\omega}{2\omega_x}\right)^{0.5}} = e^{-x\left(\frac{\omega}{2\alpha_d}\right)^{0.5}} \\ \theta(x,\omega) = -\left(\frac{\omega}{2\omega_x}\right)^{0.5} = -x\left(\frac{\omega}{2\alpha_d}\right)^{0.5} \end{cases}, \tag{61}$$

the open loop gain and phase can be expressed as follow:

$$\begin{cases} |\beta(j\omega,x)| = |\beta_{nom}(j\omega)| m(x,\omega) \\ \arg\beta(j\omega,x) = \arg\beta_{nom}(j\omega) + \theta(x,\omega) \end{cases}. \tag{62}$$

The expression of the phase margin $M_\Phi(x)$ at the gain crossover frequency, ω_u, can be represented as follow:

$$\begin{aligned} M_\Phi(x) &= \pi + \arg\beta(x, j\omega_u) \\ &= \pi + \arg\beta_{nom}(j\omega_u) + \theta(x,\omega_u) \\ &= \left(\pi - n\frac{\pi}{2}\right) - \sqrt{\frac{\omega_u}{2\alpha_d}}x. \end{aligned} \tag{63}$$

Referring to Eq. (63), one can conclude the following:

- Concerning well defined values of n, ω_u and α_d, $M_\Phi(x)$ is a decreasing linear function depending on x;
- The negative slope is proportional to ω_u and inversely proportional to the thermal diffusivity of the material α_d;
- The value of the order n selected based on the phase margin value does not affect the slope; thus, it does not alter the sensitivity of the stability degree of the controller for position x.

However, a special sensor placement, noted x_{crit}, allows to get a null phase margin (which yields to an oscillatory system on closed loop). Above this value, the system becomes unstable. Thus, the value x_{crit} is the following:

$$M_\Phi(x) = 0 \quad \Rightarrow \quad x_{crit} = \left(\pi - n\frac{\pi}{2}\right)\sqrt{\frac{2\alpha_d}{\omega_u}}. \tag{64}$$

When x varies between 0 and x_{crit}, another particular position exists. It will be known as the *limit position*, x_{lim}, which corresponds to a minimal phase margin, $M_{\Phi min}$, that will be set depending on the user specifications as shown in Eq. (65):

$$x_{\lim} = \sqrt{\frac{2\alpha_d}{\omega_u}}\left(\pi - n\frac{\pi}{2} - M_{\Phi\min}\right) \quad \in \quad [0; x_{crit}]. \tag{65}$$

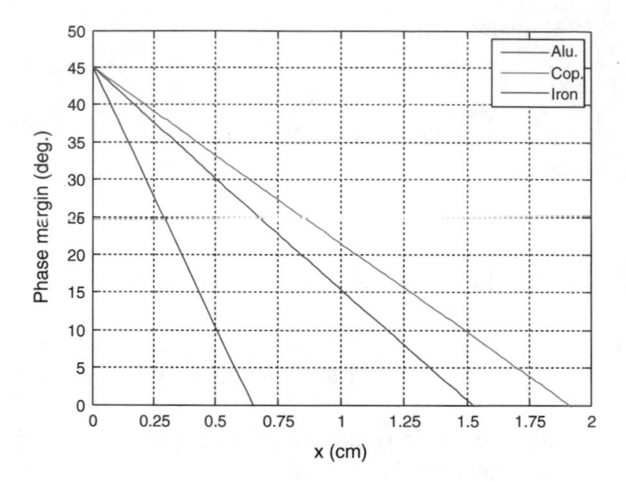

Fig. 9 Phase margin variation $M_{\Phi}(x)$ for $n = 1.5$ with respect to the position x (in cm) and of the used material

Thus, Fig. 9 shows the variation of the position x for the three materials when the phase margin $M_{\Phi}(x)$ varies between 45° and 0°. For this example, the critical values are: 0.65 cm for the iron, 1.52 cm for the aluminum and 1.91 cm for the copper. Another example resides by the determination of the phase margin when x is equal to 0.5 cm: for the iron, $M_{\Phi} = 10.5°$, for the aluminum, $M_{\Phi} = 30.2°$ and $M_{\Phi} = 33.2°$ for the copper.

As an example, two simulations were realized when using the controller obtained through example 4 ($M_{\Phi}(0) = 45°$). In the first simulation (Fig. 10a), the feedback is realize based on the temperature value $T(t,0)$ measured at $x = 0$ cm. The temperature variation $T(t,0)$ at $x = 0$ and $T(t,x)$ at $x = 5$ mm are shown in Fig. 10a, c, e). In the second simulation (Fig. 10b), the feedback is realized using temperature value $T(t,x)$ measured at $x = 5$ mm. The temperature variation $T(t,0)$ at $x = 0$ and $T(t,x)$ at $x = 5$ mm are shown in Fig. 10b, d, f. Through this second case study, the influence of the sensor position uncertainties (by a value of 5 mm) is shown clearly.

Thus, Fig. 11 shows the responses for the temperature $T(t,0)$ at $x = 0$ (figures a et b), $T(t,x)$ at $x = 5$ mm (figures c et d) at the corresponding control signal $u(t)$

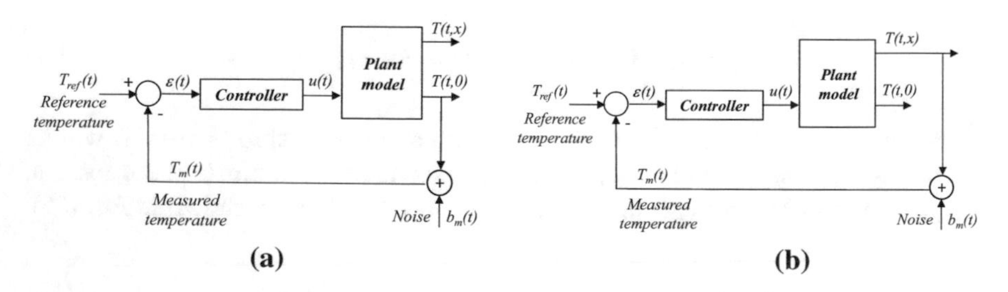

Fig. 10 Control blocks of the two study cases based on the synthesized controller of example 4 ($M_{\Phi}(0) = 45°$): **a** feedback realized based on the temperature at $x = 0$; **b** feedback realized based on the temperature at $x = 5$ mm

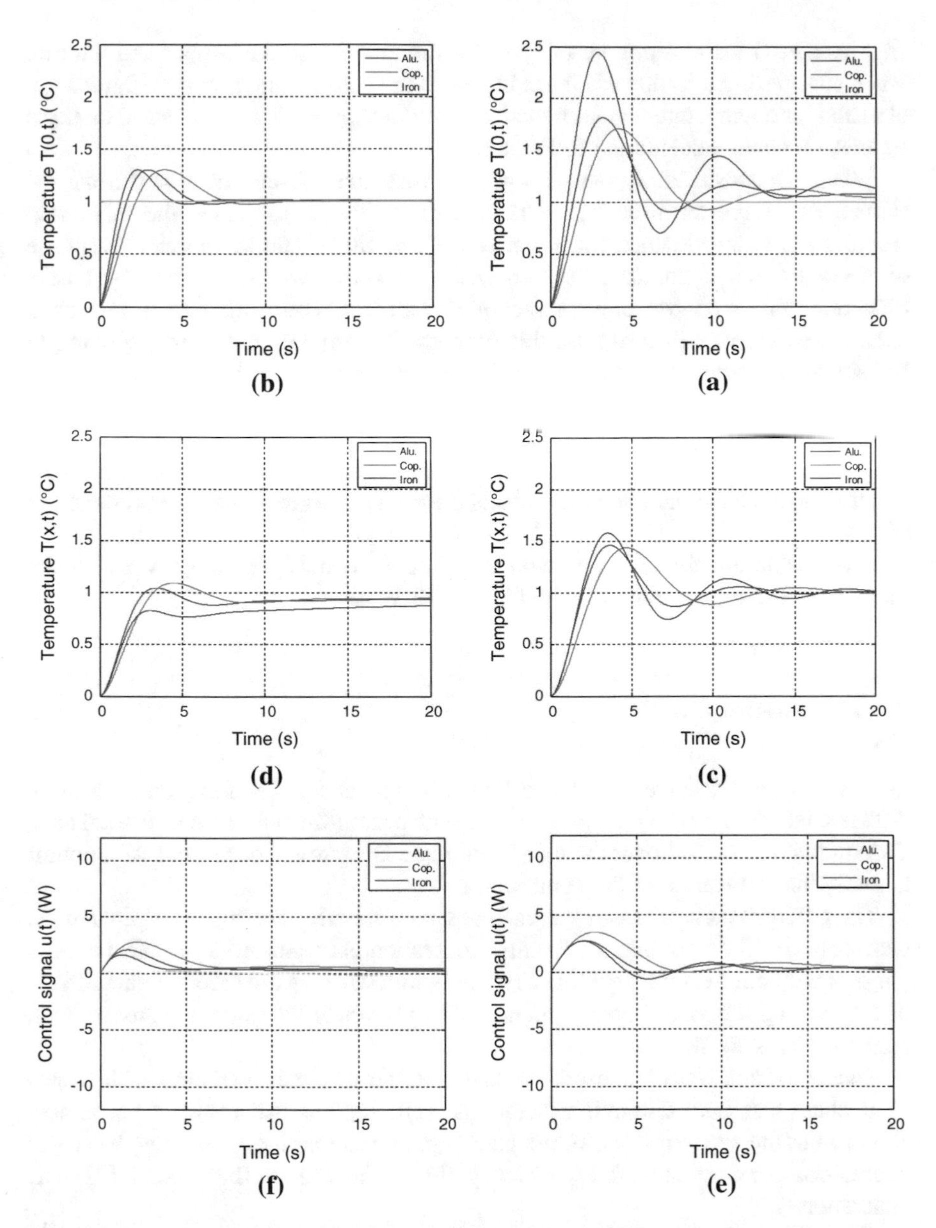

Fig. 11 Temperature response of $T(t,0)$ at $x = 0$ (1st line: **a**, **b** $T(t,x)$ at $x = 5$ mm (2nd line: **c**, **d**) and for the control signal u(t) (3rd line: **e**, **f**) for a step input of 1 °C for the aluminum, the copper and the iron, when considering a feedback control based on the temperature at $x = 0$ (figures **a**, **c**, **e**) and for a temperature measured at $x = 5$ mm (figures: **b**, **d**, **f**)

(figures e et f) for a step input of 1 °C for the aluminum, the copper and the iron when considering a feedback control based on the temperature at $x = 0$ (figures a, c et e) and for a temperature measured at $x = 5$ mm (figures: b, d et f). Based on these figures, one can conclude the following:

When the **positioning sensor uncertainties are absent at $x = 0$ mm**, the robustness of the stability degree is confirmed (Fig. 11a). Thus, the three step responses are almost equal. So, when comparing them with the nominal response obtained for the aluminum, the two other responses would be expressed as a dilatation (for the copper) or a contraction (for the iron) concerning the time domain axis. However, when $x > 0$ mm, this property is no more conserved (referring to the simulation where the sensor is positioned at 5 mm—Fig. 11c).

However, when the **positioning uncertainties are present at $x = 5$ mm**, the robustness of the stability degree is no more conserved whatever the material in use is and for any positioning of the temperature sensor (for $x = 0$, Fig. 11b). This result is logic as the variation of x affects the phase margin as already presented in Fig. 8.

A particular attention should be point out on the control signal $u(t)$ whose value remains below the saturation limit ($U_{max} = 12$ W).

4 Conclusions

In this chapter, we have introduced first the general transfer function of a finite diffusive interface medium in order to study the heat diffusion across its central axis. The study is conducted over three different materials (Iron, Copper and Aluminum) to study the robustness of the controller.

The CRONE controller was the one used in this study. The first two generations were applied. The controller of the first generation is calculated a priori where the phase is constant over all the frequency bandwidth whereas the second generation is deduced using the loop shaping and it applies whenever the phase is constant with gain variations for the plant.

Two scenarios were proposed: the first one shows a gain variation with a constant phase and both CRONE generations were applied. All results were almost similar. For the second scenario, the plant's gain was varying while the phase was maintained constant around ω_u which yield in the use of the second CRONE generation.

Concerning the future works, lot of adjustments could be made to enrich this work. First, some new scenarios could be proposed in order to analyze the behavior of the third generation CRONE controllers and to compare this controller to other ones. Added to that, the implementation of some observers could be interesting as we will not be able to measure the temperature at any point of the bar due to some physical/technical limitations. The last and most interesting point is to implement physically this system and to be able to compare the real measurements to the simulated ones. Whenever the test bench is realized, the identification of the equation that would model this system will be made and a comparison between the real one and the approximated one will be proposed. Then, the control of this plant using the LabView software along with the data acquisition board will be performed. The performance analysis of the observers will be a novel study applied in a fractional order environment.

References

1. Abi Zeid Daou, R., & Moreau, X. (2015). *Fractional calculus: Applications*. New York: Nova.
2. Abi Zeid Daou, R., Moreau, X., Assaf, R., & Christohpy, F. (2012). Analysis of HTE fractional order system in the thermal diffusive interface—part 1: Application to a semi-infinite plane medium. In *International Conference on Advances in Computational Tools for Engineering Applications*, Lebanon.
3. Assaf, R., Moreau, X., Abi Zeid Daou, R., & Christohpy, F. (2012). Analysis of HTE fractional order system in HTE thermal diffusive interface—part 2: Application to a finite medium. In *International Conference on Advances in Computational Tools for Engineering Applications*, Lebanon.
4. Azar, A. T., & Serrano, F. (2014). Robust IMC-PID tuning for cascade control systems with gain and phase margin specifications. *Neural Computing and Applications, 25*(5).
5. Azar, A. T., & Serrano, F. (2015). Design and modeling of anti wind up PID controllers. *Complex System Modelling and Control Through Intelligent Soft Computations, 319*.
6. Azar, A. T., & Zhu, Q. (2015). Advances and applications in sliding mode control systems. Study in computational intelligence (Vol. 576). Springer.
7. Battaglia, J., Cois, O., Puigsegur, L., & Oustaloup, A. (2001). Solving an inverse heat conduction problem using a non-integer identified model. *International Journal of Heat and Mass Transfer, 44*(14), 2671–2680.
8. Boulkroune, A., Bouzeriba, A., Bouden, T., & Azar, A. T. (2016). *Fuzzy adaptative synchronization of uncertain fractional-order chaotic systems*. Control: Advances in Chaos Theory and Intelligent. 337.
9. Charef, A., & Fergani, N. (2010). PIλDμ controller tuning for desired closed-loop response using impulse response. In *Workshop on Fractional Derivation and Applications,* Spain.
10. Cois, O. (2002). *Systèmes linéaires non entiers et identification par modèle non entier: application en thermique*. Bordeaux: Université Bordeaux I.
11. CRONE Group. (2005). *CRONE control design module*. Bordeaux: Bordeaux University.
12. Lin, J. (2001). *Modélisation et identification de systèmes d'ordre non entier*. Poitiers: Université de Poitiers.
13. Magin, R., Ortigueira, M., Podlubny, I., & Trujillo, J. (2011). On the fractional signals and systems. *Signal Processing, 91*(3), 350–371.
14. Malti, R., Sabatier, J., & Akçay, H. (2009). Thermal modeling and identification of an aluminium rod using fractional calculus. In *15th IFAC Symposium on System Identification*, France.
15. Miller, K., & Ross, B. (1993). *An introduction to the fractional calculus and fractional differential equations*. New York: Wiley.
16. Oldham, K., & Spanier, J. (1974). *The fractional calculus*. New York: Academic Press.
17. Oustaloup, A. (1975). *Etude et Réalisation d'un systme d'asservissement d'ordre 3/2 de la fréquence d'un laser à colorant continu*. Bordeaux: Universitu of Bordeaux.

18. Poinot, T., & Trigeassou, J. (2004). Identification of fractional systems using an output-error technique. *Nonlinear Dynamics, 38*(1), 133–154.
19. Trigeassou, J.-C., Poinot, T., Lin, J., Oustaloup, A., & Levron, F. (1999). Modeling and identification of a non integer order system. In *European Control Conference.* Karlsruhe: IFAC.

Grey Predictor Assisted Fuzzy and Fractional Order Fuzzy Control of a Moving Cart Inverted Pendulum

Amanvir Singh Sidana, Akarsh Kumar, Akshit Kanda, Vineet Kumar and K.P.S. Rana

Abstract In this chapter, a fractional order fuzzy PD controller with grey predictor (FOFPD-GP) is presented for effective control of a moving cart inverted pendulum. FOFPD-GP was tuned with the help of Genetic Algorithm for minimum settling time and its performance has been assessed using Integral of Absolute Error (IAE) and Integral of Square Error (ISE). Further, a comparative study of FOFPD-GP with its potential counterparts such as fuzzy PD with grey predictor (FPD-GP) controller, a fractional order fuzzy PD (FOFPD) controller and fuzzy PD (FPD) controller has also been carried out to assess its relative performance. Additionally, the pendulum was subjected to the impulse and sinusoidal disturbances and the disturbance rejection capabilities of the investigated controllers were analyzed and have been presented in this chapter. The simulation results revealed that FOFPD-GP controller outperformed all the other controllers under study by offering least IAE and ISE values.

Keywords Inverted pendulum · Fuzzy logic controller · Fractional order controller · Grey predictor

A.S. Sidana · A. Kumar · A. Kanda · V. Kumar (✉) · K.P.S. Rana
Division of Instrumentation and Control Engineering,
Netaji Subhas Institute of Technology, Sector – 3, Dwarka,
New Delhi 110078, India
e-mail: vineetkumar27@gmail.com
URL: http://www.nsit.ac.in

A.S. Sidana
e-mail: aman.sidana@gmail.com
URL: http://www.nsit.ac.in

A. Kumar
e-mail: akarshkumar072@gmail.com

A. Kanda
e-mail: akshitkanda@gmail.com

K.P.S. Rana
e-mail: kpsrana1@gmail.com

A.T. Azar et al. (eds.), *Fractional Order Control and Synchronization of Chaotic Systems*, Studies in Computational Intelligence 688,
DOI 10.1007/978-3-319-50249-6_3

1 Introduction

Inverted pendulum, an inherently non-linear and unstable system, has always been a topic of interest for control engineers since many decades. It has been a classical benchmark problem for designing, testing and evaluating contemporary control techniques. Inverted pendulum finds uses in military and space application, such as space shuttles and missiles, where there is requirement to maintain a precise vertical orientation.

Conventional PID controllers have been in use for a very long time. They have proved to be efficient controllers, providing satisfactory response at a very moderate cost. The evidence of their popularity lies in the fact that even today, 90% of the industry employs PID controller, one of the most popular conventional controller. PID controllers have been able to provide efficient output when tuned appropriately. However, conventional controllers fail to serve the purpose when the plant is non-linear and uncertain. This has led the scholars to search for alternative solutions.

For the past three decades lots of research has been reported on in the intelligent controllers. One of the most important outcomes of this research has been fuzzy logic control. It tries to mimic the process of human decision making based on 'if-else' logic. Fuzzy logic controller (FLC) is seen as the most suitable option to replace the conventional PID as it provides an easier option of implementing rules that resemble instructions given by a human operator. However, fuzzy logic lacks the capability to predict future data and take necessary actions. This aspect can be incorporated into the plant by using a grey predictor (GP).

Grey system theory was first introduced by Professor Deng Julong [17]. A system can be defined with a color that represents the amount of clear information about that system. For instance, a system can be called as "black box", if its internal characteristics or mathematical equations that describe its dynamics are completely unknown. On the other hand, if the description of the system is completely known, it is named as a white system. Similarly, a system that has both known and unknown information is defined as a grey system. In real life, every system can be considered as a grey system because there are always some uncertainties associated with the physical systems.

The use of fractional order calculus in the field of control engineering is another interesting development that has taken place over the last several years. Fractional order calculus has been used to define chaotic systems accurately. It allows description and modeling of a real system more accurately than the classical integer order calculus methods. When fractional order calculus is used as a part of the controller, its action resembles that of adding more tuning knobs to the controller, which helps in the generation of the desired response.

Good works have been reported on intelligent fractional controllers and grey prediction, but their combined potential appears to be underexplored. This has been the main motivation for this chapter, which aims to investigate GP based fractional order fuzzy PD (FOFPD) controller. In this chapter, a fractional order fuzzy PD controller with grey predictor (FOFPD-GP) has been implemented on a moving cart inverted pendulum. The performance of the FOFPD-GP has been compared with fuzzy PD (FPD), fuzzy PD with grey predictor (FPD-GP) and FOFPD for settling time when an impulse disturbance is given at the controller output. The gains of the controllers were tuned with the help of inbuilt optimization tool genetic algorithm (GA). Later, the pendulum was subjected to impulse and sinusoidal disturbances, and the disturbance rejection capabilities of the controllers were investigated by comparing the Integral of Absolute Error (IAE) and Integral of Square Error (ISE) values.

This chapter is organized as follows: Following the introduction in Sect. 1, a brief literature survey in Sect. 2 related to the proposed study has been presented. In Sect. 3, a complete description of the moving cart inverted pendulum is given. Fractional order calculus and its implementation on a controller as Oustaloup's recursive approximation (ORA) are described in Sect. 4. In Sect. 5, GP and its mathematical model are described. Subsequently, mathematical model of FPD, design and implementation of FPD, FOFPD, FPD-GP and FOFPD-GP controllers are described with the help of block diagrams and their corresponding Simulink diagrams in Sect. 6. Finally, results for settling time of investigated controllers and their comparison on robustness have been presented in Sect. 7 followed by the conclusion and future scope in Sect. 8.

2 Literature Survey

As already has been mentioned above, the popularity of PID is due to its ease of implementation, cost effectiveness and its ability to provide a satisfactory response. Large numbers of PID variants have been developed to suit the needs of verities of the plants. Azar and Serrano presented an internal model control plus PID tuning procedure for cascade control systems based on the gain and phase margin specifications of the inner and outer loop [2]. Azar and Serrano also developed PI loop shaping control design implementing a describing function to find the limit cycle oscillations and the appropriate control gains, thus showing the stabilization of cart-pendulum system with the proposed control scheme [6]. However, conventional controllers do not give successful results when used with non-linear plants. A survey on classical PID as well as fuzzy PID (FPID) controllers has been presented by Kumar et al. where it was realized that classical PID controllers are effective for linear systems but not suitable for nonlinear systems [21]. Consequently, the focus has been shifted from conventional to intelligent control.

For the last four decades, with the advent of soft computing, it has become possible to implement relatively complex control structures with ease. There have been numerous successful attempts for control using intelligent control systems [9]. Among the different intelligent techniques, fuzzy logic was proposed by Zadeh [45–51] and FLC was initially proposed by Mamdani [27, 28]. Meghni et al. presented a second-order sliding mode and FLC for optimizing energy management

[29]. Giove et al. used fuzzy logic to prevent dialysis hypotensive episodes [15]. Wang et al. presented an observer-based adaptive fuzzy neural network controller with supervisory mode for a certain class of higher order unknown non-linear dynamic systems [44]. An implementation in industries, for the first time, was presented by King and Mamdani for a complex dynamic and poorly defined system [20]. Nour et al. modeled a non-linear inverted pendulum on Simulink and implemented PID controller and FLC on it [32]. It was observed that FLC provided a better control action than PID controller. A similar trend was noticed in the works of Prasad et al. and Tyagi et al. [23, 40]. Kumar et al. proposed a new formula-based fuzzy PI controller and the effectiveness of the controller was assessed by controlling outlet flow concentration of a nonlinear non-thermic catalytic continuous stirred-tank reactor [22]. Boulkroune et al. presented an interesting work dealing with adaptive fuzzy control-based function vector synchronization between two chaotic systems with both, unknown dynamic disturbances and input nonlinearities [12]. In another stimulation work, Boulkroune et al. used a fuzzy adaptive controller for a fractional order chaotic system with uncertain dynamics to realize a practical projective synchronization [11]. However, fuzzy wasn't predictive in nature as mentioned above, thus, it gave way to GP.

GP theory distinguished with its ability to deal with systems that have partially unknown parameters. With the use of grey system mathematics (for instance, grey equations and grey matrixes) it is possible to generate meaningful information using little poor data. GP has ability to predict the future outputs of a system by using recently obtained data [19]. Over the last two decades, grey system theory has been developed rapidly and caught the attention of researchers with its successful real-time practical applications. It has been applied to analysis, modeling, prediction, decision making and control of various systems such as social, economic, financial, scientific and technological, agricultural, industrial, transportation, mechanical, meteorological, ecological, geological, medical, military, etc. [18]. GP has been used with sliding mode control of higher order non-linear systems and non-linear liquid level systems [42]. In both these cases, it was observed that using a GP along with the sliding mode and fuzzy controller independently, the response of the system improved. This showed superior performance of the GP.

Some of the areas of application of advanced intelligent systems in modeling and control of multi-disciplinary complex processes are electronic, chemical, mechanical, and aerospace, as explained by Azar and Vaidyanathan [7, 8]. Azar and Zhu presented quality works on control of non-linear, uncertain and coupled systems like robot arms, internal combustion engines etc. using sliding mode control tuned by GA [10]. Zhu and Azar also presented different soft computing methods for management of waste, wind-up control and application in biomedical systems [52]. Azar and Serrano presented soft computing method for wind-up control [5]. Azar and Serrano also devised an adaptive sliding mode consisting of a sliding mode control law with an adaptive gain, making the controller more flexible and reliable than other sliding mode control algorithms and nonlinear control strategies, for a furuta pendulum [3]. Mekki et al. highlighted the benefits of sliding modes when applied to the field of fault tolerant control [30]. Azar and Serrano proposed a novel approach for the dead beat control of multivariable discrete time systems [4]. Azar presented an adaptive neuro-fuzzy inference system as a novel approach for post-dialysis urea rebound prediction [1].

Another development in control engineering has been that of the fractional order control systems which make use of fractional order calculus. Fractional order calculus was described as a paradox from which useful results can be obtained. It has been used to describe systems, especially chaotic systems, and provide an effective control structure [16, 38]. Fractional controller helps in providing intermediate options to the plant. Ghoudelbourk et al. implemented a fractional pitch angle controller in a wind turbine to tap maximum energy in wind power generation [14]. FOFPD and fractional order fuzzy PI have (FOFPI) been used in cascaded loops for speed control of highly non-linear hybrid electric vehicle [24]. Recently, Sharma et al. investigated a fractional order fuzzy PID (FOFPID) on a two-link planar rigid robotic manipulator. The resulting response was seen to outperform fuzzy PID (FPID), fractional order PID (FOPID) and conventional PID [41].

The survey presented above shows that lots of work has been done on intelligent fractional controllers and grey prediction, but their combined utility appears to be underexplored. Thus the aim of the chapter is to investigate the GP based FOFPD controller and check its effectiveness against the potential counterparts.

3 Problem Formulation and Plant Model

This section presents the problem formulation and mathematical model of the considered moving cart inverted pendulum system along with its respective initial conditions and system parameters. Before delving into designing a controller for the plant, one needs to have an accurate mathematical model that can be replicated.

3.1 Inverted Pendulum

An inverted pendulum is a pendulum that has its centre of mass above its pivot point. Whereas a pendulum is stable when hanging downwards, an inverted pendulum is in its unstable equilibrium position when upright. Even a slight disturbance from its upright position can bring the pendulum down, so it requires active control. There are various kinds of inverted pendulums that are used in the research field by scholars, such as moving cart inverted pendulum and multiple segmented inverted pendulums on a cart. Another platform is a two wheeled balancing inverted pendulum having the ability to spin on the spot offering a great deal of manoeuvrability.

A moving cart inverted pendulum has been used in this chapter. One of the main reasons for its choice is its wide practical applications such as rocket launchers, hover boards, etc. The pendulum is maintained at a desired reference angle by changing the position of the cart, so effectively it is a form of stabilization control. In this chapter, a reference angle of 1° has been considered for positioning the pendulum. For the purpose of effective control, it is necessary to understand the dynamics of the moving cart pendulum system, which can be done by deriving the mathematical model of the pendulum.

3.2 Mathematical Modeling

The moving cart inverted pendulum under investigation is shown in Fig. 1 [13, 33]. It consists of a rod free to move about the pivot in the x-axis. The mass of the cart is *M*, the mass of the rod is *m*, the length of the rod is *2l, x* is the displacement of the cart from the origin, the angle of the rod with the perpendicular at the pivot is θ and *u* is the control force acting on the cart so as to the reference angle. The surface is taken to be frictionless and the mechanical joint is assumed to be smooth.

Figure 2 shows the free body diagram of the plant from which the non-linear dynamics of the system are derived. Both the forces *H* and *V* are internal forces which the pivot and rod exert on each other whenever the rod is subjected to any disturbance. *H* is the horizontal force acting on both the rod and the pivot but in opposite directions. Similarly, *V* is the vertical force acting on rod and pivot but in the opposite directions.

Net torque about the end of the rod **not** pivoted is,

$$I^*\dot{\theta} = V^*l^*sin\theta - H^*l^*cos\theta \quad (1)$$

where *I* is the moment of inertia of the rod about the rod's end.

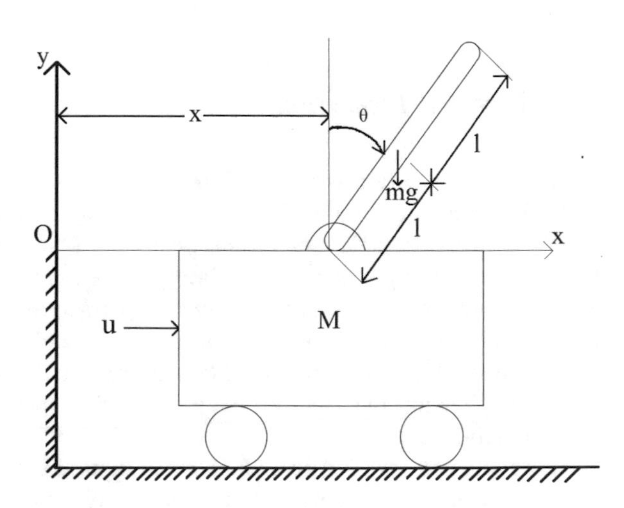

Fig. 1 A moving cart inverted pendulum

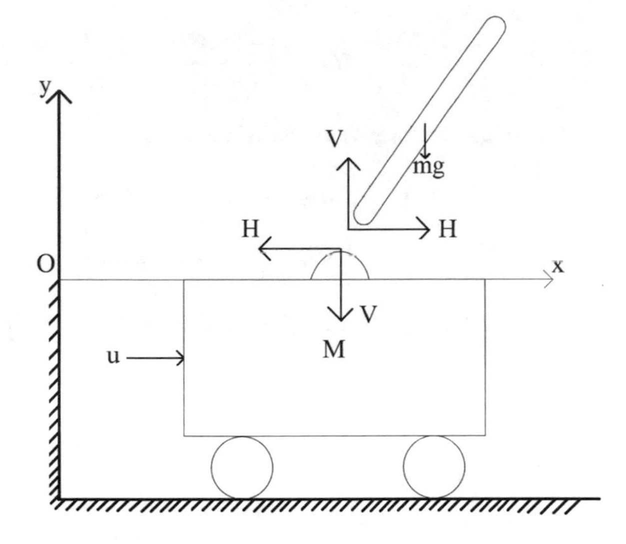

Fig. 2 Free body diagram of an inverted pendulum

Force balancing in the *x*-direction for the rod,

$$m\frac{d^2(x + l * sin\theta)}{dt^2} = H \tag{2}$$

Force balancing in the *y*-direction for the rod,

$$m\frac{d^2(l * cos\theta)}{dt^2} = V - mg \tag{3}$$

Force balancing in the *x*-direction for the cart,

$$M\frac{d^2x}{dt^2} = u - H \tag{4}$$

Using Eqs. 2 and 4,

$$M*\ddot{x} = u - \{m\ddot{x} + m * l * (\dot{\theta}cos\theta - \theta^2 sin\theta)\} \tag{5}$$

$$\ddot{x}(M + m) = u - m*l*(\dot{\theta}cos\theta - \theta^2 sin\theta) \tag{6}$$

Putting value of $\ddot{x}$ from Eq. 6 in Eq. 4,

$$H = u - M(\frac{u + ml(\theta^2 \sin\theta - \dot{\theta}\cos\theta)}{M + m}) \tag{7}$$

$$H = \frac{mu - Mm^{*}l^{*}\left(\theta^{2}\sin\theta - \dot{\theta}\cos\theta\right)}{M+m} \tag{8}$$

Force balance in y-direction of the rod,

$$V - mg = -m * l * \left(\theta^{2} sin\theta + \dot{\theta} cos\theta\right) \tag{9}$$

$$V = m\left\{g - l * \left(\theta^{2} sin\theta + \dot{\theta} cos\theta\right)\right\} \tag{10}$$

Using values of H and V from Eqs. 8 and 10 respectively and subsequently putting in Eq. 1,

$$I\dot{\theta} = m\left\{g - l\left(\theta^{2} sin\theta + \dot{\theta} cos\theta\right)\right\} l * sin\theta - l * cos\theta\left\{\frac{mu - Mm * l\left(\theta^{2} sin\theta - \dot{\theta} cos\theta\right)}{M+m}\right\} \tag{11}$$

Using the value of moment of inertia,

$$I = \frac{m(2 * l)^{2}}{3} = \frac{4m * l^{2}}{3} \tag{12}$$

$$\frac{\dot{\theta}}{ml}\left(\frac{4ml^{2}}{3} + ml^{2}\sin^{2}\theta - \frac{Ml^{2}\cos^{2}\theta}{M+m}\right) = \left(\begin{array}{c} g\sin\theta - \frac{u\cos\theta}{M+m} - l\theta^{2}\sin\theta\cos\theta \\ + \frac{M^{*}l^{*}\cos\theta\sin\theta\left(\theta^{2}\right)}{M+m} \end{array}\right) \tag{13}$$

$$\dot{\theta} * l * \left(\frac{4}{3} - \frac{mcos^{2}\theta}{M+m}\right) = gsin\theta + cos\theta\left(\frac{-u - m * l * \theta^{2} sin\theta}{M+m}\right) \tag{14}$$

$$\dot{\theta} = \frac{gsin\theta + cos\theta\left(\frac{-u - m * l * \theta^{2} sin\theta}{M+m}\right)}{l * \left(\frac{4}{3} - \frac{mcos^{2}\theta}{M+m}\right)} \tag{15}$$

As can be clearly seen from Eq. 15, the dynamics of the plant is non-linear and a suitable controller is required to maintain the pendulum at a certain position. In this study, the value of m is 1 kg, M is 2 kg and l is 1 m.

4 Fractional Order Calculus

The mention of fractional calculus can be dated back to 1695, when L' Hôpital commented on the 'meaning of derivatives with non-integer order' as "*It will be an apparent paradox from which one day useful consequences will be derived*". For a few centuries, the development of fractional calculus has been in theory, but with the advent of high computational devices, it can be utilized in effective control of complex plants. Fractional derivatives describe memory and hereditary properties in an extremely appropriate manner. This is the main advantage of fractional order derivatives when compared with integer-order models, in which such effect is neglected.

The use of fractional order calculus for the purpose of control emerged with Bode [31]. Bode presented an elegant solution to robust design problem where it was desired to have the closed loop performance invariant to changes in the amplifier gain. He came up with fractional order integrator with transfer function $G(s) = \left(\frac{\omega_{cg}}{s}\right)^{\alpha}$, known as Bode's ideal transfer function where ω_{cg} is the gain crossover frequency. Fractional calculus is a generalization of integration and differentiation to a non-integer order fundamental operator ${}_aD_t^r$, where a and t are the limits of the operation. The continuous integro-differential operator is defined as:

$$ {}_aD_t^r = \begin{cases} \frac{d^r}{dt^r}, & R(r) > 0 \\ 1, & R(r) = 0 \\ \int\limits_a^t (d\tau)^{-r}, & R(r) < 0 \end{cases} \tag{16} $$

where r is the order of the operation, generally $r \in R$, but r could also be a complex number.

The three equivalent forms of the fractional integro-differential most commonly used are the Grunwald-Letnikov (GL) definition, the Riemann-Liouville (RL) and the Caputo definition. The GL definition is given as:

$$ {}_aD_t^r f(t) = \lim_{h \to 0} h^{-r} \sum_{j=0}^{\left[\frac{t-a}{h}\right]} (-1)^j \binom{r}{j} f(t - jh), \tag{17} $$

where [.] means the integer part.

The RL definition is given as:

$$ {}_aD_t^r f(t) = \frac{1}{\Gamma(n-r)} \frac{d^n}{dt^n} \int\limits_a^t \frac{f(\tau)}{(t-\tau)^{r-n+1}} d\tau \tag{18} $$

where $(n - 1 < r < n)$ and $\Gamma(.)$ is the *Gamma* function.

The Caputo definition can be written as:

$$ \omega_{z,1} = \omega_L \sqrt{\eta} \tag{19} $$

for $(n - 1 < r < n)$. The initial conditions for the fractional order differential equations with the Caputo derivatives are in the same form as for the integer-order differential equations.

The most usual way of making use, both in simulations and hardware implementations, of transfer functions involving fractional powers of ‘s’ is to approximate them with usual (integer order) transfer functions with a similar behaviour [35–37]. So as to perfectly mimic a fractional transfer function, an integer transfer function would have to include an infinite number of poles and zeroes. Oustaloup’s approximation makes use of recursive distribution of poles and zeros to obtain series of rational functions whose frequency response fit the frequency response of the irrational function within specific frequency band. This method, also known as ORA is defined as follows for frequency band of $[\omega_l; \omega_h]$.

$$s^{\lambda}_{[\omega_l,\, \omega_h]} = k \prod_{n=1}^{N} \frac{1 + \frac{s}{\omega_{z,n}}}{1 + \frac{s}{\omega_{p,n}}} \tag{20}$$

$$\omega_{z,1} = \omega_L \sqrt{\eta}, \tag{21}$$

$$\begin{aligned} \omega_{p,n} &= \omega_{z,n}\alpha, \qquad n = 1 \ldots N, \\ \omega_{z,n+1} &= \omega_{p,n}\eta, n = 1 \ldots N-1, \end{aligned} \tag{22}$$

$$\alpha = \left(\frac{\omega_h}{\omega_l}\right)^{\frac{\lambda}{N}} \tag{23}$$

$$\eta = \left(\frac{\omega_h}{\omega_l}\right)^{(1-\lambda)/N} \tag{24}$$

where k is a constant that should be chosen such that the magnitude of the approximate shall have unity gain (0 dB) at 1 rad/s. N represents the number of poles and zeros which should be chosen beforehand. Large value of N permits good approximation but increases the computational complexity. On the other hand, small value results in simpler approximation but could cause appearance of ripple in gain and phase behavior. Low and high frequencies band limitations could avoid the use of infinite numbers of rational transfer function besides limiting the high frequency gain of the derivative effect [44].

5 Grey Predictor

GP is employed when there is a lack of information about the system model. It extracts present and past information from a plant to generate future values in order to minimize the error. In real life, due to noise that arises from both the inside and outside of the system (and the limitations of our cognitive abilities), the information one perceives about that system is always uncertain and limited in scope [19]. One of the characteristics of grey system is the construction of model with small amount of data. Grey prediction models can be used to predict the future values of the system with high accuracy.

5.1 GP Model

The main task of grey system theory is to extract realistic governing laws of the system using available data. This process is known as the generation of the grey sequence. It is argued that even though the available data of the system, which generally consists of white numbers, is too complex or chaotic, it always contains some governing laws. If the randomness of the data obtained from a system is somehow smoothed, it is easier to derive any special characteristics of that system. For instance, the sequence that represents the speed values of a motor might be given as:

$$Q(0) = (820, 840, 835, 850, 890) \tag{25}$$

It is obvious that the sequence does not have a clear regularity. If accumulating generation is applied to original sequence, $Q(1)$ is obtained which has a clear growing tendency [25].

$$Q(1) = (820, 1660, 2495, 3345, 4235) \tag{26}$$

As one plots the data points from Eq. 25 which represents the speed values of a motor, a random graph as shown in Fig. 3 with no definite pattern or future growing tendency, is obtained.

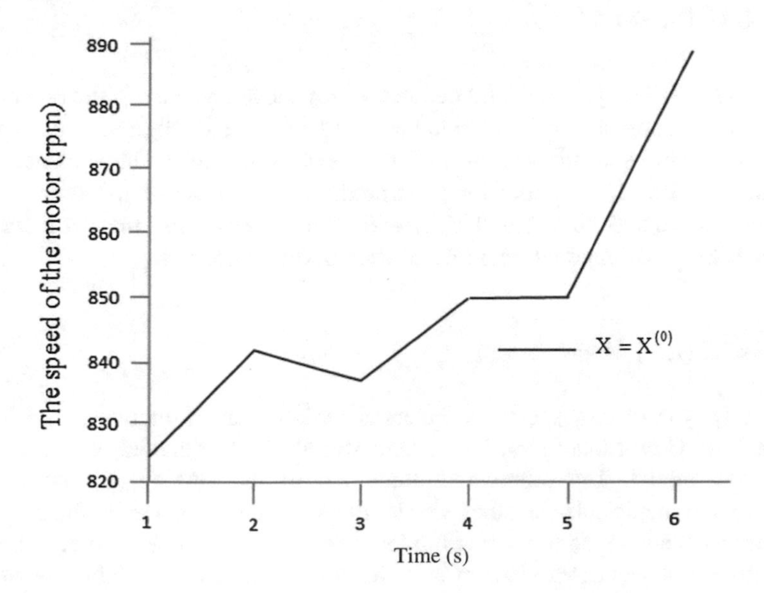

Fig. 3 The original data set

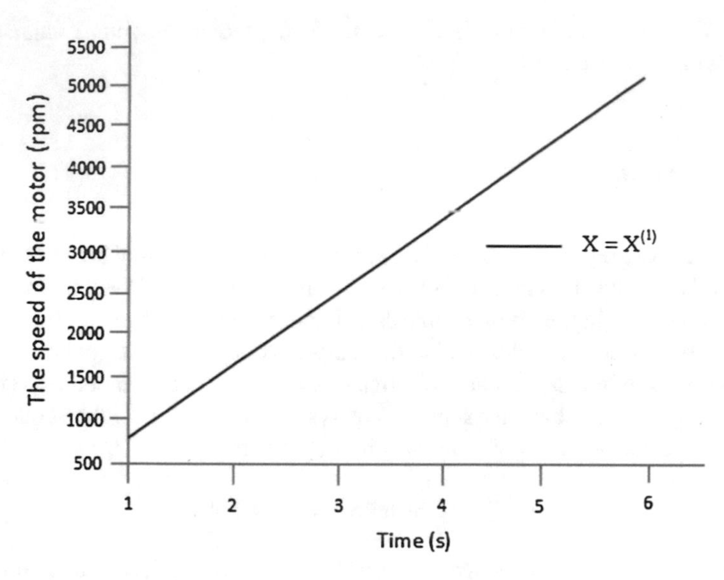

Fig. 4 The accumulated data set

As data points of accumulated data are plotted from Eq. 26, a graph as shown in Fig. 4 is obtained. This graph has a clear growing tendency and gives more information about the system than the original data set.

5.2 *GM (n, m) Model*

In grey systems theory, GM (n, m) denotes a grey model, where n is the order of the difference equation and m is the number of variables. Although various types of grey models can be mentioned, most of the previous researchers have focused their attention on GM (1, 1) model for their predictions because of its computational efficiency. It should be noted that in real time applications, the computational burden is the most important parameter after performance [18].

5.2.1 GM (1, 1) Model

GM (1, 1) type of grey model is the most widely used in literature and is pronounced as "Grey Model First Order One Variable". This model is a time series forecasting model. The differential equations of the GM (1, 1) model have time-varying coefficients. In other words, the model is renewed as the new data becomes available to the prediction model. The GM (1, 1) model can only be used in positive data sequences [19]. In this chapter, an inverted pendulum plant with reference angle more than 0° is used, so that the value of θ is never negative. Hence, the GM (1, 1) model can be used to forecast the position of the pendulum.

In order to smooth the randomness, the primitive data obtained from the system to form a GM (1, 1) it is subjected to an operator named accumulating generation operation (AGO) [19]. The differential equation (i.e. GM (1, 1)) thus evolved is solved to obtain the n-step ahead predicted value of the system. Finally, using the predicted value, the inverse accumulating operation (IAGO) is applied to find the predicted values of original data [18].

5.2.2 Mathematical Modeling of GP

Consider a single input and single output system. Assuming that the time sequence $Q^{(0)}$ represents the output of the system

$$Q^{(0)} = (q^{(0)}(1), q^{(0)}(2), \ldots q^{(0)}(m)),\ m \geq 4 \tag{27}$$

where $Q^{(0)}$ is a non-negative sequence and m is the sample size of the data. This sequence is then subjected to AGO to obtain the sequence, $Q^{(1)}$. Through Eq. 28, it can be observed that $Q^{(1)}$ is monotone increasing

$$Q^{(1)} = (q^{(1)}(1), q^{(1)}(2), \ldots q^{(1)}(m)),\ m \geq 4 \tag{28}$$

where $q^{(1)}(j) = \sum_{i=1}^{j} q^{(0)}(i),\ j = 1, 2, 3, \ldots m$

This shows that each term in the $Q^{(1)}$ sequence is actually a cumulated sum of all the terms from the beginning till that term.

The mean sequence $W^{(1)}$ of $Q^{(1)}$, is defined as:

$$W^{(1)} = (w^{(1)}(1), w^{(1)}(2), \ldots w^{(1)}(m)) \tag{29}$$

where $w^{(1)}(j)$ terms in Eq. 29 are actually the mean values of adjacent data terms, which can be obtained as shown in Eq. 30,

$$w^{(1)}(j) = 0.5q^{(1)}(j) + 0.5q^{(1)}(j-1), j = 2, 3, \ldots, m \tag{30}$$

The least square estimate sequence of the grey difference equation of GM (1, 1) is defined as follows:

$$q^{(0)}(j) + cw^{(1)}(j) = v \tag{31}$$

The whitening equation is therefore written as follows:

$$\frac{dq^{(1)}(t)}{dq} + cq^{(1)}(t) = v \tag{32}$$

In Eq. 32, $[c, v]^T$ is a sequence of parameters that can be found from Eq. 33:

$$[c, v]^T = (O^T O)^{-1} O^T P \tag{33}$$

where,

$$P = [q^{(0)}(2), q^{(0)}(3), \ldots q^{(0)}(m)]^T \tag{34}$$

$$O = \begin{pmatrix} -w^{(1)}(2) & 1 \\ -w^{(1)}(3) & 1 \\ \vdots & \vdots \\ -w^{(1)}(n) & 1 \end{pmatrix} \tag{35}$$

The solution of $q^{(1)}(t)$ at time k is obtained as follows:

$$q_z^{(1)}(j+1) = [q^{(0)}(1) - \frac{v}{c}]e^{-cj} + \frac{v}{c} \tag{36}$$

To obtain the predicted value of the primitive data at time $(j + 1)$, IAGO is used to establish the grey model as shown in Eq. 37:

$$q_z^{(0)}(j+1) = [q^{(0)}(1) - \frac{v}{c}]e^{-cj}(1 - e^c) \tag{37}$$

The parameter '$-c$' in the GM (1, 1) model is called 'development coefficient' and gives information about the development of states. The parameter 'v' is called the 'grey action quantity' which reflects changes in the data that have arisen because of being derived from the background values [18].

GM (1,1) Rolling Model: GM (1,1) rolling model is based on the forward data sequence to build the GM (1,1). For instance, using $q(0)(j)$, $q(0)(j+1)$, $q(0)(j+2)$ and $q(0)(j+3)$, model predicts the value of the next point $q(0)(j+4)$. In the next step, the first point is always shifted to the second. It means that $q(0)(j+1)$, $q(0)(j+2)$, $q(0)(j+3)$ and $q(0)(j+4)$ are used to predict the value of $q(0)(j+5)$. This procedure is repeated till the end of the sequence and the method is called rolling check. GM (1,1) rolling model is used to predict the long continuous data sequences such as the output of a system, price of a specific product, trend analysis for finance statements, social parameters, etc. In this chapter, GM (1,1) rolling model is used to predict the future outputs of the moving cart inverted pendulum system [10].

6 Integer/Fractional Order FPD Controller with/without GP

This section introduces the integer and fractional order FLC design and its implementation. For this task, initially the FPD controller is described along with its mathematical equations and block diagram in Sect. 6.1. Following this, in Sect. 6.2, FOFPD controller is introduced by replacing the integer order derivative with a fractional order derivative and its application model is explained. Sections 6.3 and 6.4 describe the way GP is combined with FPD and FOFPD controllers in order to transform them into FPD-GP and FOFPD-GP controllers respectively. The membership functions, rule base, inference mechanism and defuzzification technique which have been used in the operation of all the four controllers are elaborated in Sect. 6.5.

6.1 FPD Controller

The standard equation of a conventional PD controller in time domain is defined as:

$$u_{PD}(t) = K_p^{'}[e(t) + T_d\dot{e}(t)] \tag{38}$$

or

$$u_{PD}(t) = \left[K_p^{'}e(t) + K_d^{'}\dot{e}(t)\right] \tag{39}$$

where $K_p^{'}$ is the proportional constant, T_d is the derivative time constant, $K_d^{'}$ is the derivative gain, $u_{PD}(t)$ is the output of PD controller, $e(t)$ is error.

In discrete form, Eq. 39 can be written as:

$$u_{PD}(k) = \left[K_{\mathrm{p}}^{'}e(k) + K_{\mathrm{d}}^{'}r(k)\right] \tag{40}$$

where $r(k) = (e(k) - e(k-1))\ /T$ is the rate of change of error and T is the sampling period.

Now, the FPD controller can be designed based on the discrete form of the conventional PD controller as given in Eq. 40. The inputs to the FPD controller are error $e(k)$ and rate of change of error $r(k)$ and the output is $u_{FPD}(k)$ [39].

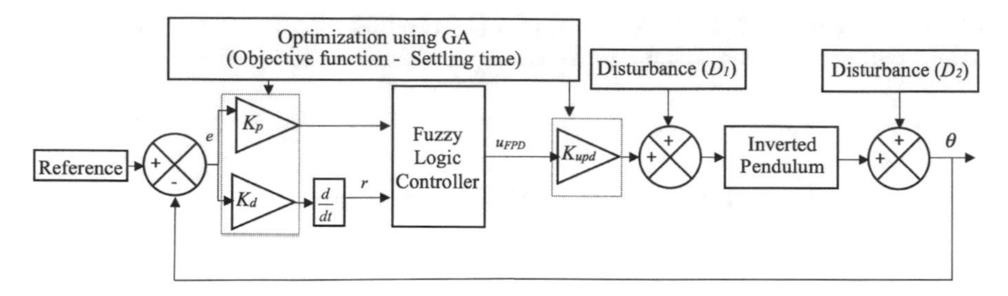

Fig. 5 Block diagram of FPD controller

$$u_{FPD}(k) = K_{upd}D_F\{F_F[K_pe(k), K_dr(k)]\} \quad (41)$$

Equation 41 gives the control action of a FPD controller, where K_p and K_d are the scaling factors of the inputs and K_{upd} is the scaling factor of the defuzzified output. F_F refers to the fuzzification of inputs and D_F refers to the defuzzification of the fuzzified output. The implementation of the FPD controller can be shown through a block diagram in Fig. 5.

6.2 FOFPD Controller

When fractional calculus is implemented in conjunction with FPD controller, a better response can be expected than a conventional FPD. This can be attributed to the fact that the FOFPD controller would be less sensitive to the parametric variations of the system due to an extra degree of freedom. In terms of mathematical variations from the FPD controller, the $\frac{d}{dt}$ term in Eq. 39 would be replaced by $\frac{d^\lambda}{dt^\lambda}$ where $\lambda \in (0, 1)$. The block diagram in Fig. 6 shows the design of the FOFPD controller.

As can be observed from Fig. 6, putting back $\lambda = 1$ would result in FPD controller. The actual FOFPD controller has been realized using a transfer function

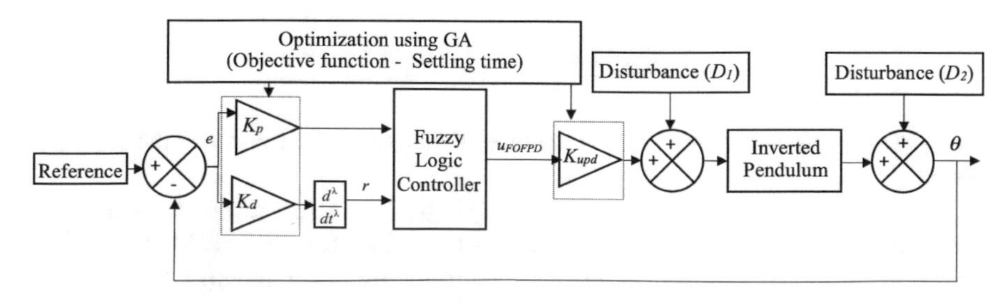

Fig. 6 Block diagram of FOFPD controller

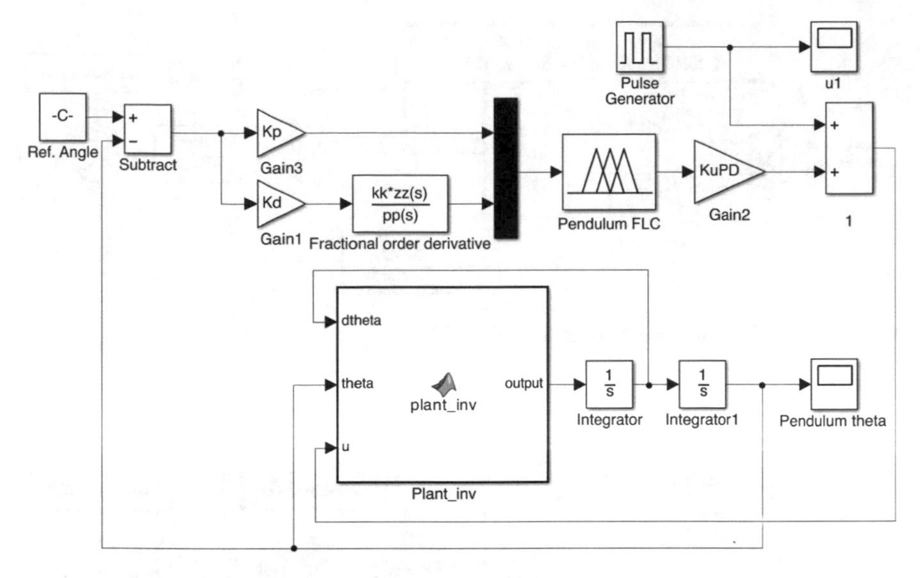

Fig. 7 Simulink model of FOFPD controller on inverted pendulum

which employs ORA as discussed earlier in Sect. 3. The Simulink model implementing the same is shown in Fig. 7.

6.3 FPD-GP Controller

The GP designed for the inverted pendulum uses data terms obtained from previous four samples to predict the future value. This helps in minimization of error and consequently, in a better transient response. GP is positioned in the feedback path of the FPD controller. Hence, the error received by FPD controller is more or less similar to the error which would have resulted in the next sample time in absence of GP. Thus, in a way, GP can be said to have accelerated the error detection process and help in the error reduction a step ahead. GP was implemented with the help of 'Interpreted MATLAB function'. The above working can be demonstrated using a block diagram shown in Fig. 8.

6.4 FOFPD-GP Controller

FOFPD-GP controller is constructed by juxtaposition of FOFPD controller and GP. It combines the advantages of fractional calculus and GP, i.e. the fractional part compensates for the dynamical instabilities of the plant and makes the response more robust, while the GP helps in prediction of future values which the controller

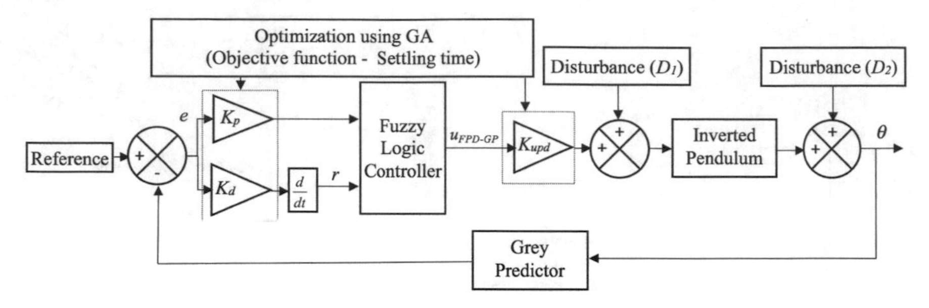

Fig. 8 Block diagram of FPD-GP controller

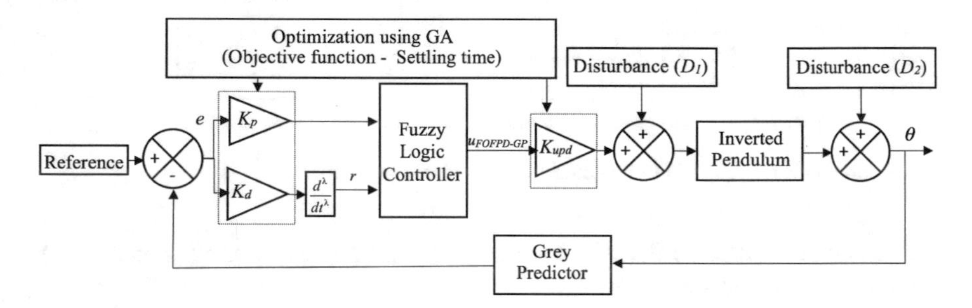

Fig. 9 Block diagram of a FOFPD-GP controller

can act upon and reduce the error. The block diagram in Fig. 9 clearly depicts the entire setup.

The practical realization of the FOFPD-GP controller using a Simulink model can be shown in Fig. 10.

6.5 Framework of Fuzzy Controllers

A typical FPD controller consists of the arrangement shown in Fig. 11.

The membership functions and their universes of discourse need to be designed depending upon the plant model. Following this, the rule base must be defined for all the possible combination of inputs. Fuzzy inference would then map the given inputs into output according to the rule base. Finally, a suitable defuzzification technique is required to convert the fuzzy output signal into a crisp control signal, which is made available to the plant input. The fuzzy logic used for all the controllers in this study is identical.

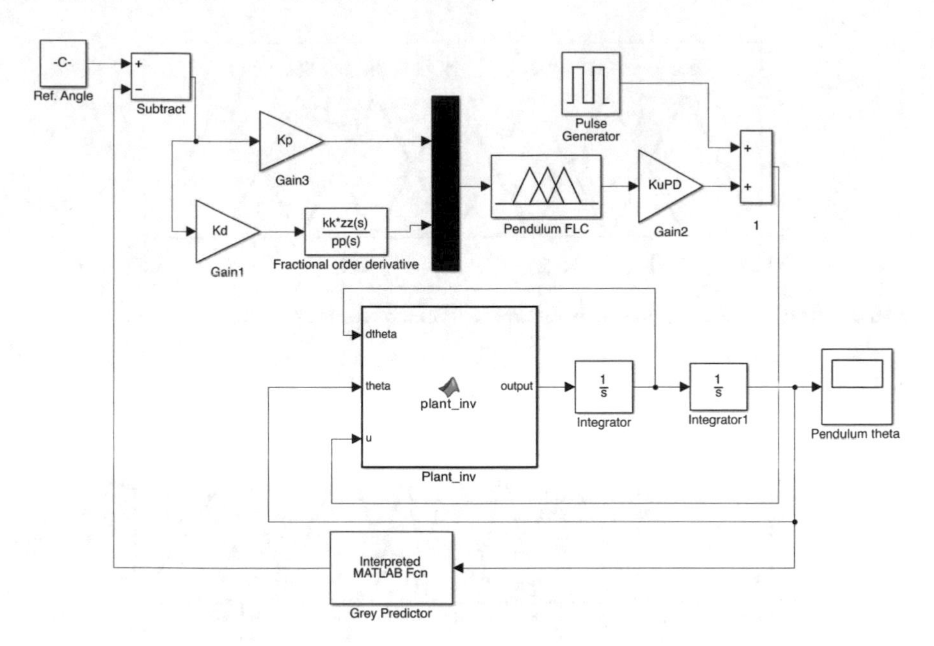

Fig. 10 Simulink model of FOFPD-GP controller on inverted pendulum

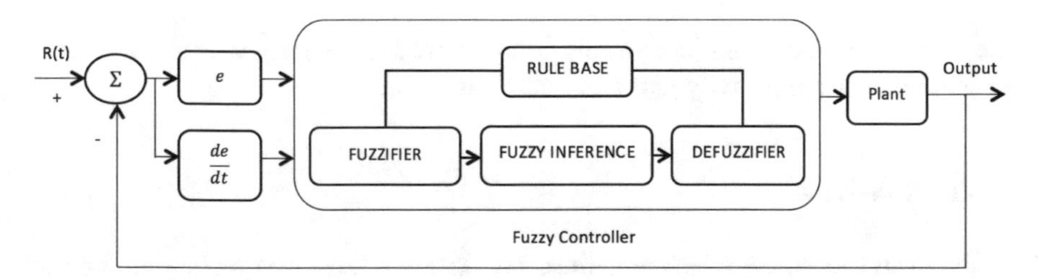

Fig. 11 Block diagram of a FPD controller

6.5.1 Membership Functions

Seven membership functions have been used for both the inputs and the output. Membership functions for error and rate of change of error are same and defined in the range $-\pi/2 < e(k)/r(k) < \pi/2$ as shown in Fig. 12. The reason for choice of this universe of discourse is accredited to the fact that outside this range, the pendulum rod falls down and needs different control techniques like swing up control to get back to its unstable equilibrium position.

Membership function for output is defined in the range $-15 < u < 15$ as shown in Fig. 13. Asymmetric membership functions are selected instead of the usual symmetric ones because of the fact that the practical range of angle θ never exceeds 5° and for error larger than this value, the system becomes less dynamic. Therefore,

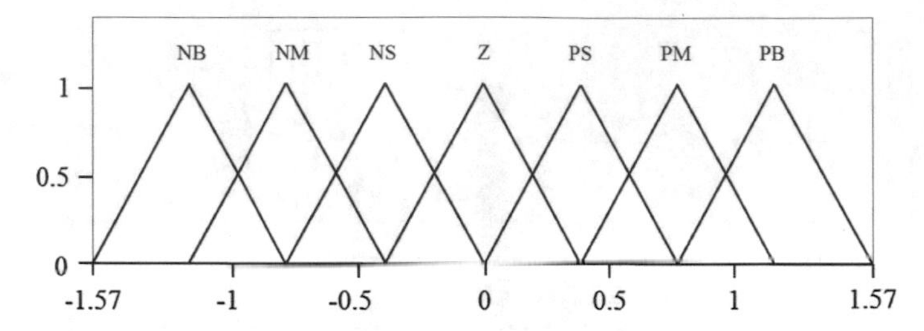

Fig. 12 Input membership functions for error and rate of change of error

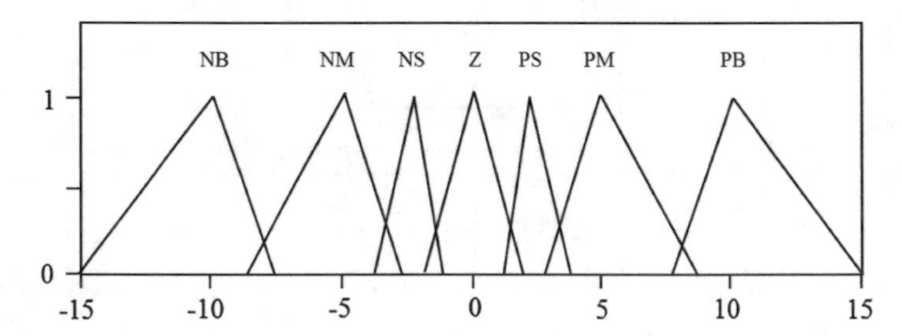

Fig. 13 Output membership functions

the membership functions close to the 0° mark are spaced close to each other so as to provide better sensitivity and more amplification [26].

6.5.2 Rule Base

In the matrix form, the rules are defined in Table 1. The reasoning behind the rule base can be explained by taking a particular case as an example, if the error and rate of change of error are NB and NM respectively, this suggests that the current output

Table 1 Rule table

$\dot{e}$	e						
	NB	NM	NS	Z	PS	PM	PB
NB	PB	PB	PB	PB	PM	PS	Z
NM	PB	PB	PB	PM	PS	Z	NS
NS	PB	PB	PM	PS	Z	NS	NM
Z	PB	PM	PS	Z	NS	NM	NB
PS	PM	PS	Z	NS	NM	NB	NB
PM	PS	Z	NS	NM	NB	NB	NB
PB	Z	NS	NM	NB	NB	NB	NB

where NB, NM, NS, Z, PS, PM and PB represent Negative Big, Negative Medium, Negative Small, Zero, Positive Small, Positive Medium and Positive Big respectively

is at a very large distance from the reference and moving away from it at a considerable speed. Therefore, PB is required to bring the output back close to reference position [37].

6.5.3 Inference Engine and Defuzzification Method

The inference method used in the above analysis for all the controllers is Mamdani's max-min inference process.

Defuzzification refers to the process of converting the fuzzy output signal to a crisp non-fuzzy value. It is done because fuzzy values cannot be directly used for actuators applications. The defuzzification technique employed in this study is centroid method. Mathematically, it's given as

$$z^* = \frac{\int \mu_C(z) \cdot z\,dz}{\int \mu_C(z)\,dz} \tag{42}$$

7 Results and Discussions

Application of FPD and FPD-GP controller on an inverted pendulum aims to control the rod at the reference position when an impulse disturbance is given. However, in order to further reduce the settling time, a transfer function employing ORA is used to convert FPD and FPD-GP into FOFPD and FOFPD-GP controllers respectively. A comparative study for settling time is demonstrated for the four abovementioned controllers and their controller gains tuned through GA are presented in Sect. 6.1. The error and control signal comparisons for all the controllers have also been shown. Further, all the controllers are subjected to sinusoidal disturbances at plant input and plant output individually for their robust testing. This study has been presented in Sects. 6.2 and 6.3. The performance of FOFPD and FOFPD-GP controller is found to be more efficient than FPD and FPD-GP controller respectively with significant reduction in settling time.

7.1 Optimization of Controllers

The gains and the fractional order exponent (for FOFPD and FOFPD-GP), used in the controllers are tuned for their optimum performance in the given bounds for all the controllers used in the chapter. The tuning is done using GA which is available as an inbuilt optimization tool in MATLAB. Three gains (K_p, K_d and K_{upd}) and the fractional order exponent λ (in case of FOFPD and FOFPD-GP) are tuned. The parameters' settings used for tuning are shown in Table 2.

Table 2 Various parameters values for GA

Parameter	Values
Population size	50
Generation	30
Function tolerance	1E-6
Lower bound gains	1E-6
Upper bound gains	100
Solver	Runga-Kutta (ode4)
Step size	1 ms

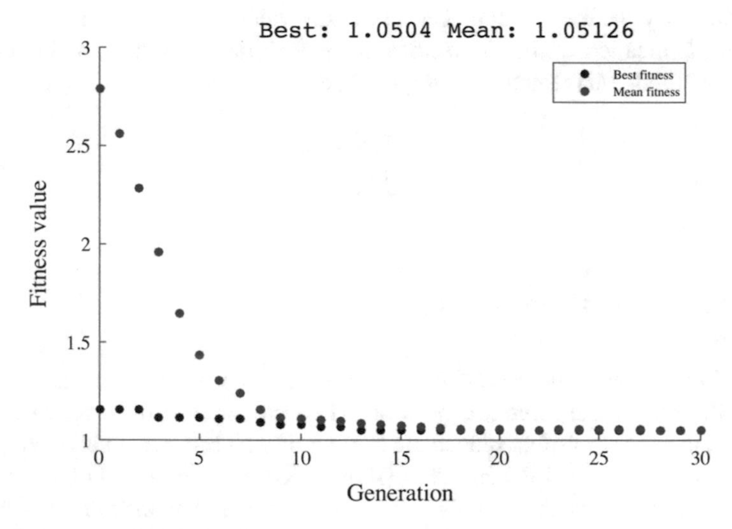

Fig. 14 Convergence plot for FPD

The gains are tuned so as to produce minimum settling time. A pulse of width 10 ms and amplitude 100 N is applied at 1 s to the plant at the controller output, which disturbs the plant from its initial position. The best fitness plots for all the four types of controllers are shown below.

The convergence plot in Fig. 14 comes out to be smooth and becomes almost constant after 15 generations.

From Fig. 15, it can be observed that there are only minor variations after 10 generations. However, the graph settles completely only towards the latter part.

Figure 16 depicts the plot that is monotonically decreasing and more or less settles around 20th generation.

The graph in Fig. 17 shows that the plot for mean fitness settles very late to the best fitness plot (around 25th generation). However, the major fall occurs by the 10th generation only.

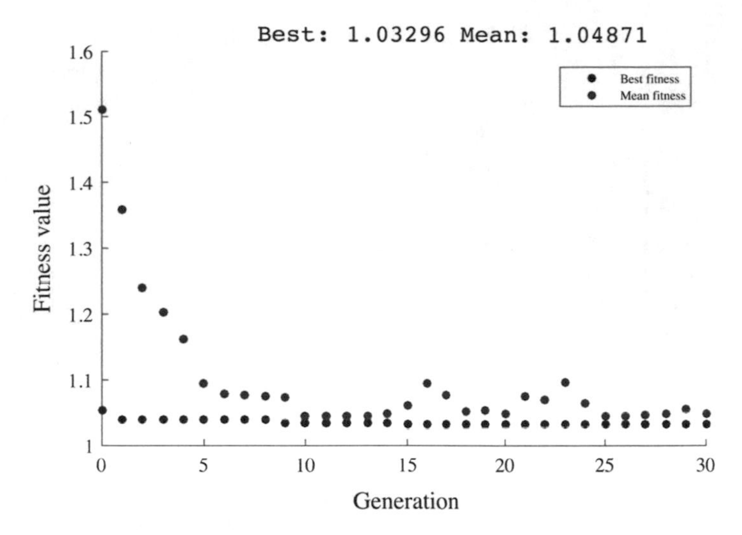

Fig. 15 Convergence plot for FOFPD

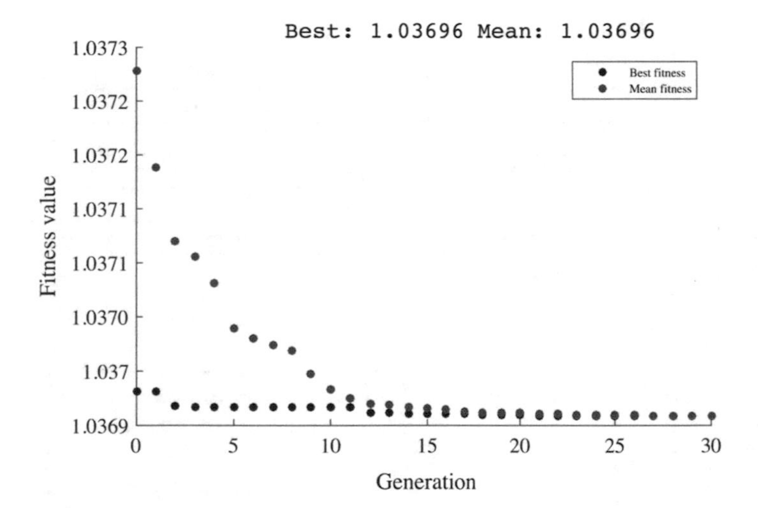

Fig. 16 Convergence plot for FPD-GP

From the above convergence plots, it can be observed that the mean fitness value settles very close to the best value and continues to be stable till the last generation. The exact settling time and tuned gains' values for various controllers are listed in Table 3.

The reference set point tracking responses for all the investigated controllers are shown in Fig. 18 for the case when an impulse was given at the controller output.

As it can be observed from Fig. 18, the settling time obtained from FOFPD-GP is minimum as compared to all the other controllers. Also, the undershoot is least

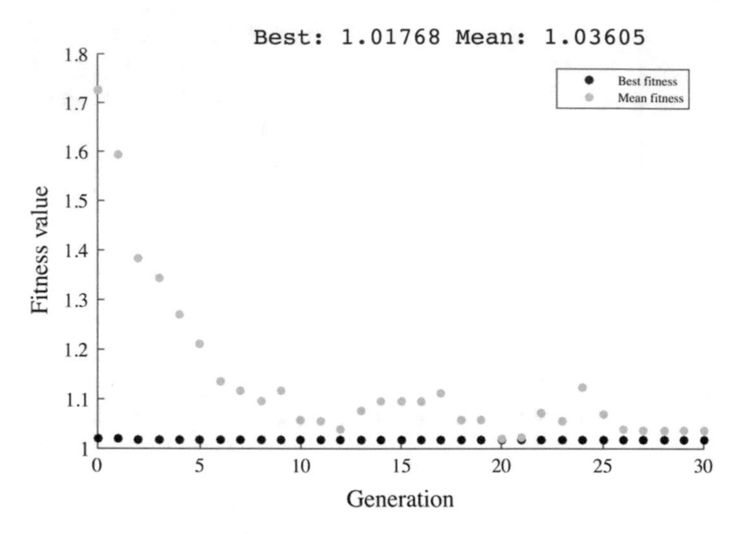

Fig. 17 Convergence plot for FOFPD-GP

Table 3 Tuned gains and obtained settling time values for the controllers

Parameters	FPD	FPD-GP	FOFPD	FOFPD-GP
K_p	99.976	93.7628	98.106	40.2241
K_d	1.449	1E-6	1.7	35.0111
K_{upd}	22.908	99.7668	32	86.9323
λ	1	1	0.91	0.254
Settling time (sec)	0.0504	0.0370	0.0351	0.0175

where λ refers to the order of derivative

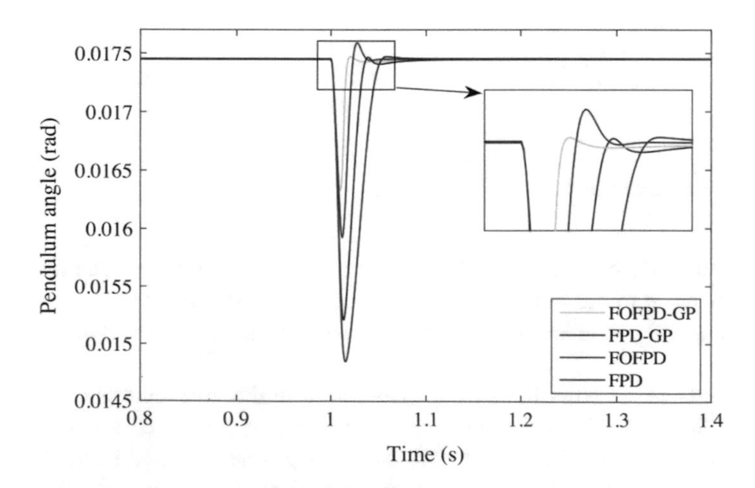

Fig. 18 Set point tracking responses of controllers

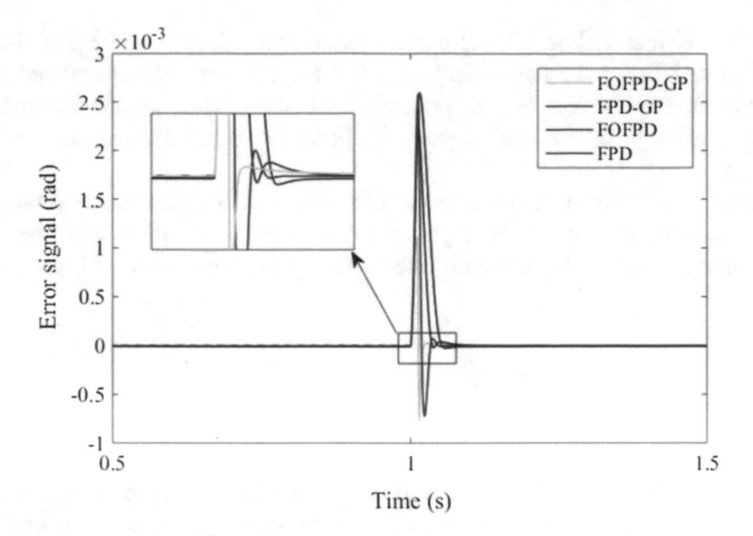

Fig. 19 Error comparison for all the controllers

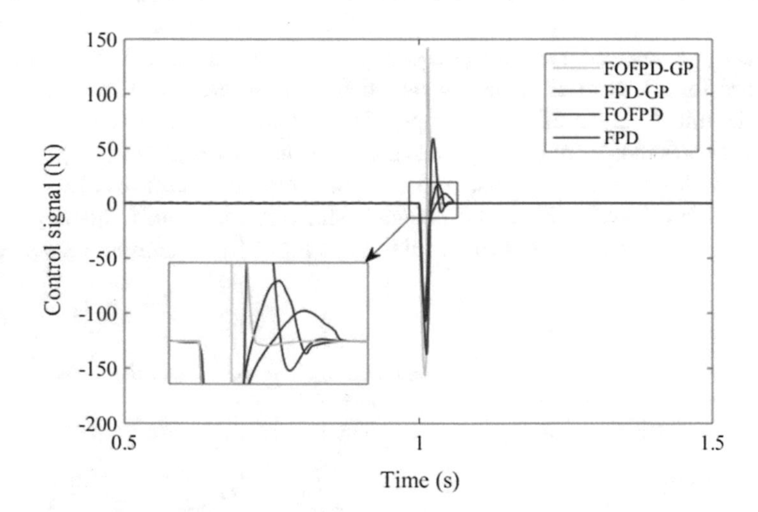

Fig. 20 Control signal comparison for all the controllers

for FOFPD-GP controller, thereby demonstrating its effectiveness in maintaining the pendulum angle at reference position. The corresponding error and the control signal comparisons for all the controllers are shown in Figs. 19 and 20.

It can be observed from Fig. 19, that the error signals of FOFPD and FOFPD-GP controllers come out to be smaller in amplitude and tend to settle earlier than the error signal of their integral counterparts, FPD and FPD-GP. Also, it can be clearly seen that the error for FOFPD-GP comes out to be the least.

As it is visible in Fig. 20, the control signals for FOFPD and FOFPD-GP are smoother and faster than FPD and FPD-GP respectively. This dominant performance of the control signals is responsible for better overall settling time response of the fractional controllers. Again, FOFPD-GP outperformed all the other controllers.

Further, the controllers developed, with their tuned gains, were subjected to various disturbances at different points to test their robustness and disturbance rejection capabilities. The following subsections present the relevant investigations.

7.2 Disturbance at Controller Output

This disturbance is analogous to a vibrating cart which has inverted pendulum mounted on cart. For the same, a sinusoidal disturbance was given at the controller output, D_1 at time 0 s. Now, additionally; an impulse signal at 1 s was introduced as shown in Fig. 5. The impulse given at time t = 1 s has a pulse width of 10 ms and amplitude 100 N. The sinusoidal disturbance was varied in two different ways. First, the frequency was kept constant and the amplitude was varied, then the amplitude was kept constant and frequency was varied. IAE and ISE values were calculated for all the four responses obtained from different controllers. In all the cases, the tuned values of gains computed earlier were used. Firstly, a sinusoidal signal with constant amplitude and varying frequencies was given as a disturbance. The amplitude of the signal is taken to be 5 N. A plot for responses of FOFPD-GP, FPD-GP, FOFPD and FPD controllers when sine disturbance of frequency 50 rad/s and an impulse at t = 1 s is given is shown in Fig. 21 as a sample case study.

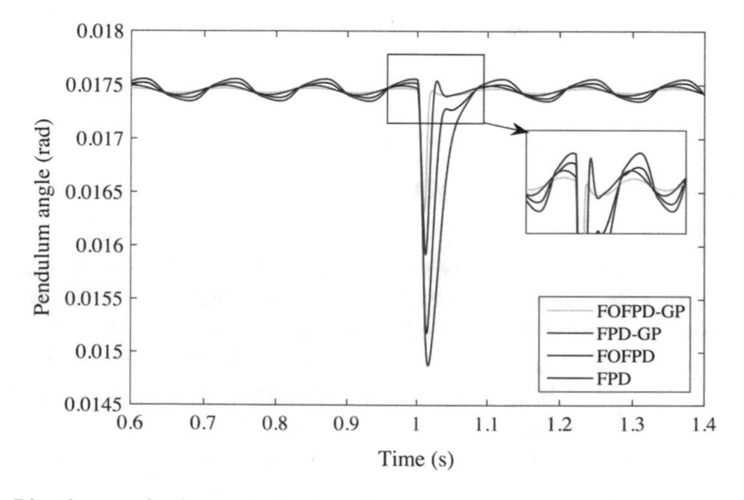

Fig. 21 Disturbance rejection study for sinusoidal disturbance of 5 N and 50 rad/s

Table 4 Performance analysis: Disturbance of constant amplitude and varying frequency at D_1

Frequency (rad/s)	FOFPD-GP		FPD-GP	
	IAE	ISE	IAE	ISE
50	0.0400	9.94E-06	0.1095	3.45E-05
75	0.0399	9.88E-06	0.1127	3.48E-05
100	0.0392	9.81E-06	0.1170	3.52E-05
125	0.0388	9.35E-06	0.1222	3.56E-05
150	0.0385	9.72E-06	0.1278	3.61E-05
175	0.0390	9.72E-06	0.1330	3.65E-05
Frequency (rad/s)	FOFPD		FPD	
	IAE	ISE	IAE	ISE
50	0.1348	7.03E-05	0.2773	1.47E-04
75	0.1236	7.01E-05	0.2507	1.47E-04
100	0.1162	6.98E-05	0.2297	1.43E-04
125	0.1112	6.93E-05	0.2105	1.39E-04
150	0.1037	6.89E-05	0.1926	1.34E-04
175	0.1029	6.81E-05	0.1765	1.29E-04

It can be observed from Fig. 21 that the response obtained for FOFPD-GP is the most robust as it doesn't oscillate from its mean position as much as responses of other controllers. Therefore, the IAE and ISE values are also found to be lower for FOFPD-GP controller. The IAE and ISE values for constant amplitude and varying frequencies are shown in Table 4.

Again, it can be confirmed from Table 4 that the addition of fractional components improves the response by a significant factor. FOFPD-GP outperformed FPD-GP, FOFPD and FPD controllers as its IAE and ISE values for all the considered frequencies came out to be lowest.

Next, a sinusoidal signal with constant frequency and varying amplitude was applied. The frequency of the sinusoidal disturbance was taken to be 100 rad/s. Comparative plot depicting responses of FOFPD-GP, FPD-GP, FOFPD and FPD controllers for amplitude 6 N and frequency 100 rad/s is shown in Fig. 22 as a typical study.

It can be observed from Fig. 22 that the response obtained for FOFPD-GP is most robust, as its amplitude variations are very less in response to sinusoidal signal. Therefore, the IAE and ISE values are also lowest for FOFPD-GP. Table 5 shows the IAE and ISE values for constant frequency and varying amplitudes.

As can be clearly noted from Table 5, the controllers with fractional derivatives performed considerably better than those without fractional components. FOFPD-GP outperformed FPD-GP, FOFPD and FPD controllers as its IAE and ISE values for all the frequencies came out to be lowest.

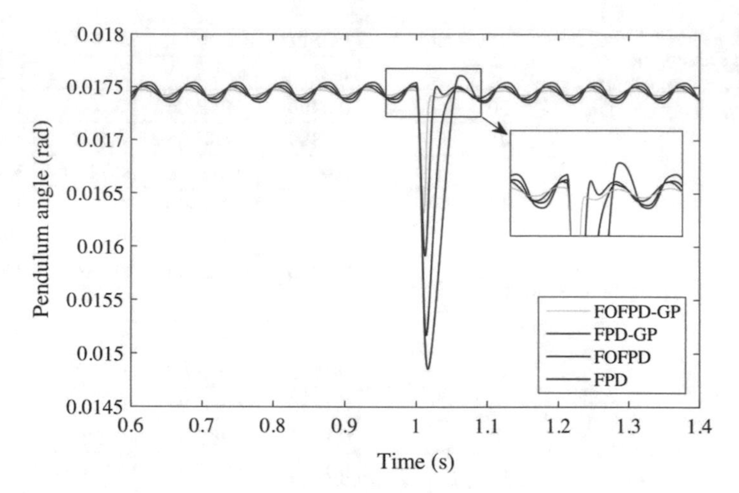

Fig. 22 Disturbance rejection study for sinusoidal disturbance of 6 N and 100 rad/s

Table 5 Performance analysis: disturbance of constant frequency and varying amplitude at D_1

Amplitude	FOFPD-GP		FPD-GP	
	IAE	ISE	IAE	ISE
1	0.0186	9.58E-06	0.0426	3.15E-05
2	0.0235	9.57E-06	0.0609	3.20E-05
3	0.0287	9.62E-06	0.0794	3.27E-05
4	0.0336	9.68E-06	0.0981	3.38E-05
5	0.0392	9.81E-06	0.1170	3.52E-05
6	0.0445	9.99E-06	0.1361	3.69E-05
Amplitude	FOFPD		FPD	
	IAE	ISE	IAE	ISE
1	0.0542	6.17E-05	0.0977	1.24E-04
2	0.0689	6.31E-05	0.1290	1.28E-04
3	0.0840	6.57E-05	0.1615	1.32E-04
4	0.1002	6.76E-05	0.1950	1.37E-04
5	0.1168	6.98E-05	0.2297	1.43E-04
6	0.1327	7.24E-05	0.2653	1.51E-04

7.3 Disturbance at Plant Output

Referring to Fig. 5, a sinusoidal disturbance was added to the plant output at D_2 to further analyze the relative performances of the investigated controllers. The sinusoidal disturbance was given at $t = 1.5$ s for 1 time period and it was varied in two ways in line with the previous case i.e. frequency and amplitude one at a time. In all the cases, the tuned values of gains computed earlier were used.

First, a sinusoidal signal with varying amplitudes and constant frequency was given as a disturbance. The frequency of the signal was taken to be 2πrad/s. Controller outputs for amplitude 5π/180 rad and frequency 2πrad/s are shown in Fig. 23 as a sample case study.

It can be observed from Fig. 23 that the response obtained from FOFPD-GP is the least unwavering, as it doesn't oscillate from its mean position as much as the response of other controllers. Therefore, the IAE and ISE values also come out to be lowest for FOFPD-GP. Table 6 shows the IAE and ISE values of the controllers for the case of constant frequency and varying amplitudes.

Table 6 again confirms that controllers with fractional components outperformed the controllers with integer derivatives and FOFPD-GP offers the best response among all the controllers.

Next, amplitude of the sinusoidal disturbance is taken to be π/180 rad/s. Controller outputs for amplitude π/180 rad and frequency π rad/s are shown in Fig. 24.

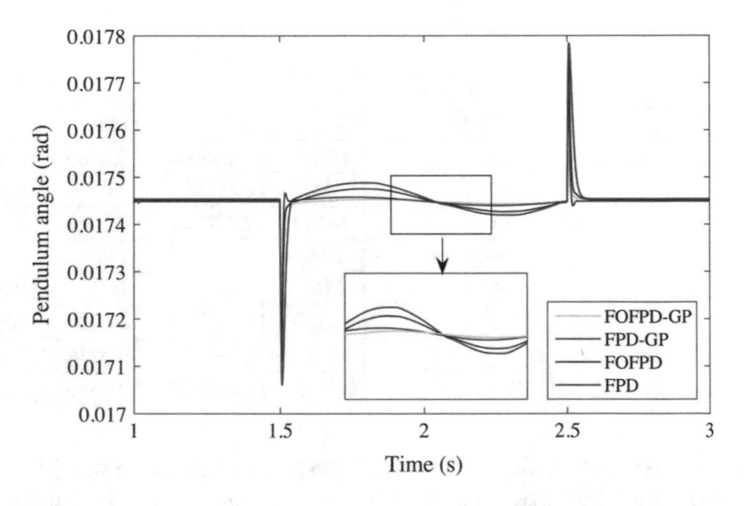

Fig. 23 Disturbance rejection study for sinusoidal disturbance of 5π/180 rad and 2π rad/s

Table 6 Performance analysis: disturbance of constant frequency and varying amplitude at D_2

Amplitude (rad)	FOFPD-GP		FPD-GP	
	IAE	ISE	IAE	ISE
π/180	0.0113	2.39E-07	0.0147	1.03E-06
3π/180	0.0340	1.97E-05	0.0466	3.19E-05
5π/180	0.0570	6.20E-05	0.0850	1.1E-04
Amplitude (rad)	FOFPD		FPD	
	IAE	ISE	IAE	ISE
π/180	0.0346	1.88E-06	0.0437	4.1E-06
3π/180	0.0928	3.59E-05	0.1368	7.24E-05
5π/180	0.1551	1.36E-04	0.2342	2.19E-04

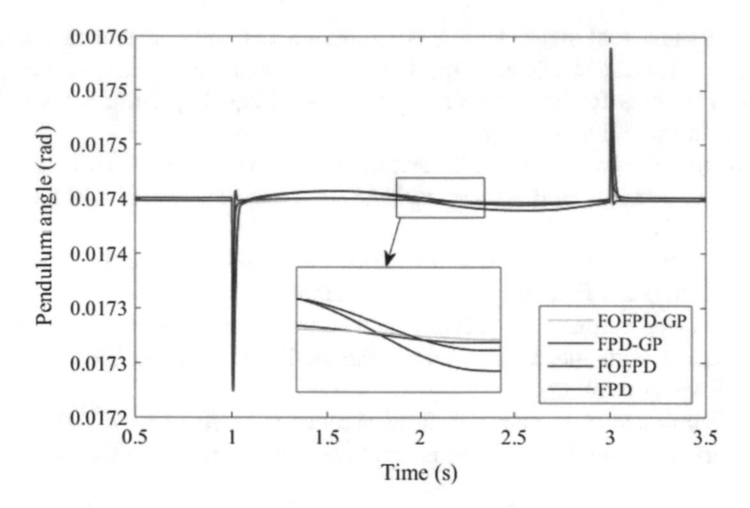

Fig. 24 Disturbance rejection study for sinusoidal disturbance of π/180 rad and π rad/s

Table 7 Performance analysis: disturbance of constant amplitude and varying frequency at D_2

Frequency (rad/s)	FOFPD-GP		FPD-GP	
	IAE	ISE	IAE	ISE
π/2	0.0052	5.10E-08	0.0209	2.23E-07
π	0.0093	4.49E-07	0.0268	8.65E-07
2π	0.0147	2.39E-07	0.0437	4.10E-06
Frequency (rad/s)	FOFPD		FPD	
	IAE	ISE	IAE	ISE
π/2	0.0160	8.50E-08	0.0429	3.32E-07
π	0.0086	2.26E-07	0.0244	6.16E-07
2π	0.0113	1.00E-06	0.0346	1.88E-06

Again, it is revealed from Fig. 24 that the response obtained from FOFPD-GP is more robust, as it doesn't oscillate from its mean position as much as the response of FPD-GP, FOFPD and FPD controllers. Therefore, the IAE and ISE values are also lowest for FPD-GP. Table 7 compares the IAE and ISE values for constant amplitude and varying frequencies:

As can be proved from Table 7, FOFPD-GP and FOFPD controllers perform considerably better than their integer counterparts, FPD-GP and FPD respectively. Further, it is to be noted from Table 7, that FOFPD-GP outperformed FPD-GP, FOFPD and FPD controllers.

8 Conclusions and Future Scope

Fractional Order Fuzzy PD with Grey Predictor (FOFPD-GP), Fuzzy PD with Grey Predictor (FPD-GP), Fractional Order Fuzzy PD (FOFPD) and Fuzzy PD (FPD) controllers were successfully implemented and compared on a moving cart inverted pendulum. The simulations were done using Simulink on MATLAB. Initially, an impulse disturbance was given to the plant model for which the plant was tuned for minimum settling time using inbuilt optimization tool Genetic Algorithm. It was observed that the settling time of FOFPD-GP was improved from FPD by 2.88 times, from FPD-GP by 2.11 times and FOFPD by 2 times. Based on these facts, it is deduced that FOFPD-GP controller outperformed the other controllers under investigation.

Further, when sinusoidal disturbances were given at plant input and plant output, it was noted that the Integral of Absolute Error (IAE) and Integral of Square Error (ISE) values for FOFPD-GP came out to be better than FOFPD, FPD-GP and FPD for both the disturbances when the amplitude was kept constant and frequency was varied or when frequency was kept constant and amplitude was varied. Overall FOFPD-GP offered better responses than FOFPD, FPD-GP and FPD controllers in all the investigated cases for the moving cart inverted pendulum.

As a future scope, effectiveness of other inference mechanisms, different membership functions, defuzzification methods and other control varieties of fuzzy control schemes may be explored for moving cart inverted pendulum or such complex plants. Additionally, new and advanced and optimization techniques can also be applied to further effectively tune the control gains.

References

1. Azar, A. T. (2010). Adaptive neuro-fuzzy systems. In: A. T Azar (Ed.), *Fuzzy systems*. Vienna, Austria: IN-TECH. ISBN 978-953-7619-92-3.
2. Azar, A. T., & Serrano, F. E. (2014). Robust IMC-PID tuning for cascade control systems with gain and phase margin specifications. *Neural Computing and Applications*, *25*(5), 983–995. doi:10.1007/s00521-014-1560-x.
3. Azar, A. T., & Serrano, F. E. (2015). Adaptive sliding mode control of the Furuta pendulum. In: A. T. Azar, & Q. Zhu (Eds.), *Advances and applications in sliding mode control systems*. Studies in computational intelligence (Vol. 576, pp. 1–42). Berlin/Heidelberg: Springer-Verlag GmbH.
4. Azar, A. T., & Serrano, F. E. (2015). Deadbeat control for multivariable systems with time varying delays. In: A. T. Azar, & S. Vaidyanathan (Eds.), *Chaos modeling and control systems design*. Studies in computational intelligence (Vol. 581, pp. 97–132). Berlin/Heidelberg: Springer-Verlag GmbH. doi:10.1007/978-3-319-13132-0_6.
5. Azar, A. T., & Serrano, F. E. (2015). Design and modeling of anti wind up PID controllers. In: Q. Zhu, & A. T Azar (Eds.), *Complex system modelling and control through intelligent soft computations*. Studies in fuzziness and soft computing (Vol. 319, pp. 1–44). Germany: Springer. doi:10.1007/978-3-319-12883-2_1.

6. Azar, A. T., & Serrano, F. E. (2016). Stabilization of mechanical systems with backlash by PI loop shaping. *International Journal of System Dynamics Applications (IJSDA), 5*(3), 20–47.
7. Azar, A. T., & Vaidyanathan, S. (2015) *Computational intelligence applications in modeling and control*. Studies in computational intelligence (Vol. 575). Germany: Springer. ISBN: 978-3-319-11016-5.
8. Azar, A T., & Vaidyanathan, S. (2015). *Handbook of research on advanced intelligent control engineering and automation*. Advances in Computational Intelligence and Robotics (ACIR) Book Series. USA: IGI Global.
9. Azar, A. T., & Vaidyanathan, S. (2016). *Advances in chaos theory and intelligent control*. Studies in fuzziness and soft computing (Vol. 337). Germany: Springer. ISBN 978-3-319-30338-3.
10. Azar, A. T., & Zhu, Q. (2015). *Advances and applications in sliding mode control systems*. Studies in computational intelligence (Vol. 576). Germany: Springer. ISBN: 978-3-319-11172-8 A1:A42.
11. Boulkroune, A., Bouzeriba, A., Bouden, T., & Azar, A. T. (2016). *Fuzzy adaptive synchronization of uncertain fractional-order chaotic systems*. Advances in chaos theory and intelligent control. Studies in fuzziness and soft computing (Vol. 337). Germany: Springer.
12. Boulkroune, A., Hamel, S., & Azar, A. T. (2016). Fuzzy control-based function synchronization of unknown chaotic systems with dead-zone input. In: *advances in chaos theory and intelligent control*. Studies in fuzziness and soft computing (Vol. 337). Germany: Springer.
13. Burns, R. S. (2001). *Advanced control engineering*. Oxford: Butterworth-Heinemann. ISBN 0-7506-5100-8.
14. Ghoudelbourk, S., Dib, D., Omeiri, A., & Azar, A. T. (2016). MPPT control in wind energy conversion systems and the application of fractional control (PIα) in pitch wind turbine. *International Journal of Modelling, Identification and Control (IJMIC)*, in press.
15. Giove, S., Azar, A. T., & Nordio, M. (2013). Fuzzy logic control for dialysis application. In: A. T Azar (Ed.), *Biofeedback systems and soft computing techniques of dialysis* (Vol. 405, pp. 1181–1222). Berlin/Heidelberg: Springer-Verlag GmbH. doi:10.1007/978-3-642-27558-6_9.
16. Grigorenko, I., & Grigorenko, E. (2003). Chaotic dynamics of the fractional Lorenz system. *Physical Review Letters, 91*(3), 034101.
17. Julong, D. (1989). Introduction to grey system theory. *The Journal of grey system, 1*(1), 1–24.
18. Kayacan, E., & Kaynak, O. (2006). An adaptive grey fuzzy PID controller with variable prediction horizon. In: SCIS & ISIS (Vol. 2006, pp. 760–765).
19. Kayacan, E., & Kaynak, O. (2006). Grey prediction based control of a non-linear liquid level system using PID type fuzzy controller. In: *2006 IEEE International Conference on Mechatronics* (pp. 292–296).
20. King, P. J., & Mamdani, E. H. (1977). The application of fuzzy control systems to industrial processes. *Automatica, 13*(3), 235–242.
21. Kumar, V., Nakra, B. C., & Mittal, A. P. (2011). A review on classical and fuzzy PID controllers. *International Journal of Intelligent Control and Systems, 16*(3), 170–181.
22. Kumar, V., Rana, K. P. S., & Sinha, A. K. (2011). Design, performance, and stability analysis of a formula-based fuzzy PI controller. *International Journal of Innovative Computing, Information and Control, 7*(7), 4291–4308.
23. Kumar, P., Nema, S., & Padhy, P. K. (2014). Design of fuzzy logic based PD controller using cuckoo optimization for inverted pendulum. In: 2014 International Conference on Advanced Communication Control and Computing Technologies (ICACCCT) (pp. 141–146).
24. Kumar, V., Rana, K. P. S., & Mishra, P. (2016). Robust speed control of hybrid electric vehicle using fractional order fuzzy PD & PI controllers in cascade control loop. Journal of the Franklin Institute, http://dx.doi.org/10.1016/j.jfranklin.2016.02.018, in press.
25. Liu, S., Forrest, J., & Yang, Y. (2013). A summary of the progress in grey system research. In: Proceedings of 2013 IEEE International Conference on Grey systems and Intelligent Services (GSIS) (pp. 1–10). IEEE.

26. Magana, M. E., & Holzapfel, F. (1998). Fuzzy-logic control of an inverted pendulum with vision feedback. *IEEE Transactions on Education, 41*(2), 165–170.
27. Mamdani, E. H. (1974). Application of fuzzy algorithms for control of simple dynamics plant. *The Proceedings of Institute of Electrical, Control and Science, 121*, 1585–1588.
28. Mamdani, E. H., & Baaklini, N. (1975). Prescriptive method for deriving control policy in a fuzzy-logic controller. *Electronics Letters, 25*(11), 625–626.
29. Meghni, B., Dib, D., & Azar, A. T. (2016). A second-order sliding mode and fuzzy logic control to optimal energy management in wind turbine with battery storage. *Neural Computing and Applications*, 1–18.
30. Mekki, H., Boukhetala, D., & Azar, A. T. (2015). Sliding modes for fault tolerant control. In: A. T. Azar, & Q. Zhu, (Eds.), *Advances and applications in sliding mode control systems*. Studies in computational intelligence book series (Vol. 576, pp. 407–433). Berlin/Heidelberg: Springer GmbH.
31. Monje, C. A., Chen, Y. Q., Vinagre, B. M., Xue, D., & Vicente, F. (2010) *Fractional-order systems and controls*. Springer-Verlag London Limited. ISBN: 978–1-84996-334-3.
32. Nour, M. I. H,. Ooi, J., & Chan, K. Y. (2007). Fuzzy logic control vs. conventional PID control of an inverted pendulum robot. In: *International Conference on Intelligent and Advanced Systems, 2007, ICIAS 2007* (pp. 209–214).
33. Ogata, K. (2010). *Modern control engineering*. New Jersey: PHI Learning Pvt. Ltd. ISBN 0-13-615673-8.
34. Oustaloup, A. (1983). *Systémes Asservis Linéaires d'Ordre Fractionnaire: Théorie et Pratique*. Paris: Editions Masson.
35. Oustaloup, A. (1991). *La commande CRONE*. Paris: Editions Hermès.
36. Oustaloup, A., Levron, F., Mathieu, B., & Nanot, F. M. (2000). Frequency-band complex noninteger differentiator: characterization and synthesis. *IEEE Transactions on Circuits and Systems I: Fundamental Theory and Applications, 47*(1), 25–39.
37. Passino, K. M., Yurkovich, S., & Reinfrank, M. (1998). *Fuzzy control* (Vol. 20). Menlo Park, CA: Addison-wesley.
38. Petráš, I. (2009). Chaos in the fractional-order Volta's system: modeling and simulation. *Nonlinear Dynamics, 57*(1–2), 157–170.
39. Pivoňka, P. (2002). Comparative analysis of fuzzy PI/PD/PID controller based on classical PID controller approach. In: *Proceedings of the 2002 IEEE International Conference on Fuzzy systems* (pp. 541–546).
40. Prasad, L. B., Gupta, H. O., & Tyagi, B. (2011). Intelligent control of nonlinear inverted pendulum dynamical system with disturbance input using fuzzy logic systems. In: *2011 International Conference on Recent Advancements in Electrical, Electronics and Control Engineering* (pp. 136–141).
41. Sharma, R., Rana, K. P. S., & Kumar, V. (2014). Performance analysis of fractional order fuzzy PID controllers applied to a robotic manipulator. *Expert Systems with Applications, 41* (9), 4274–4289.
42. Shuhong, Z., Mianyun, C., & Yexin, S. (2001). The application of grey system theory combined with fuzzy reasoning in high order nonlinear control system. In: *Proceedings OF 2001 International Conferences on Info-tech and Info-net, 2001* (ICII 2001-Beijing) (Vol. 4, pp. 291–296). IEEE.
43. Valério, D., & da Costa, J. S. (2006). Tuning of fractional PID controllers with Ziegler–Nichols-type rules. *Signal Processing, 86*(10), 2771–2784.
44. Wang, C. H., Liu, H. L., & Lin, T. C. (2002). Direct adaptive fuzzy-neural control with state observer and supervisory controller for unknown nonlinear dynamical systems. *IEEE Transactions on Fuzzy Systems, 10*(1), 39–49.
45. Zadeh, L. A. (1965). Fuzzy sets. *Information and Control, 8*(3), 338–353.
46. Zadeh, L. A. (1970). Towards a theory of fuzzy systems. In R. E. Kalman & N. D. Claris (Eds.), *Aspects of networks and system theory* (pp. 469–490). New York: Rinehart and Winston.

47. Zadeh, L. A. (1972). A rationale for fuzzy control. *Journal of Dynamic Systems, Measurement, and Control, 94*(1), 3–4.
48. Zadeh, L. A. (1973). Outline of a new approach to the analysis of complex systems and decision processes. *IEEE Transactions on systems, Man, and Cybernetics, 1*, 28–44.
49. Zadeh, L. A. (1975). The concept of a linguistic variable and its application to approximate reasoning—I. *Information Sciences, 8*(3), 199–249.
50. Zadeh, L. A. (1975). The concept of a linguistic variable and its application to approximate reasoning—II. *Information Sciences, 8*(4), 301–357.
51. Zadeh, L. A. (1975). The concept of a linguistic variable and its application to approximate reasoning—III. *Information Sciences, 9*(1), 43–80.
52. Zhu, Q., & Azar, A. T. (2015). Complex system modelling and control through intelligent soft computations. In: *Studies in fuzziness and soft computing* (Vol 319). Germany: Springer. ISBN: 978-3- 319-12882-5.

H_∞ Design with Fractional-Order $PI^\lambda D^\mu$ Type Controllers

De-Jin Wang

Abstract This chapter focuses on H_∞ performance design for fractional-delay systems with fractional-order $PI^\lambda D^\mu$ type controllers, including $PI^\lambda D^\mu$ controller, PI^λ controller and PD^μ controller. The method adopted here is based on parameter plane approach. Firstly, the stabilizing region boundary lines in the plain of the two controller's gains are drawn for other parameters of the controller to be fixed, and the stabilizing region is identified using a graphical stability criterion applicable to fractional-delay systems. Secondly, in the stabilizing region, the modern H_∞-norm constraint of sensitivity function or complementary sensitivity function is mapped into the stabilizing region by means of the explicit algebraic equations obtained according to the definition of H_∞-norm. Thirdly, in the stabilizing region, the classical phase-margin and gain-margin curves are drawn using the technique of the gain and phase margin tester. Thus, the co-design of the modern and classical performances is realized. Finally, in time-domain, the dynamic behaviors and the robustness to the plant uncertainties are simulated via Matlab toolbox and compared with integer-order *PID* controller. Also, the influence of varying the fractional orders (λ and μ) on the step responses is simulated.

Keywords Fractional-order $PI^\lambda D^\mu$ controllers • H_∞-norm • Gain and phase margins • Sensitivity functions • Complementary sensitivity functions • Parameter plane approach

1 Introduction

This chapter mainly concerns with the H_∞ performance design of control systems with fractional-order $PI^\lambda D^\mu$ type controllers, including fractional-order PD^μ, PI^λ

D.-J. Wang (✉)
School of Electronic Information and Automation, Tianjin University of Science and Technology, Tianjin 300222, People's Republic of China
e-mail: wdejin56@sina.com

A.T. Azar et al. (eds.), *Fractional Order Control and Synchronization of Chaotic Systems*, Studies in Computational Intelligence 688,
DOI 10.1007/978-3-319-50249-6_4

and $PI^{\lambda}D^{\mu}$ controllers, based on the parameter plane approach. From the point of view of application, the parameters tuning of $PI^{\lambda}D^{\mu}$ type controllers is an important issue in order to achieve better control effects, just as in the case of conventional (integer-order) PID controllers. It is well-known that H_{∞} performance is a robustness measure of control systems. The H_{∞} design with $PI^{\lambda}D^{\mu}$ type controllers discussed in this chapter gives a new parameters tuning of fractional-order controllers, and this is the motivation behind this work. The main contribution of the chapter lies in the following three aspects: (1) A graphical stability criterion for time-delay systems is applied to fractional-delay systems, giving a new approach to the stability analysis of fractional-order systems and the design of fractional-order controllers. The stabilizing region in the two gains' plane of $PI^{\lambda\lambda}D^{\mu}$ type controller, for other parameters to be fixed, is drawn and identified directly, avoiding the complicated stability testing of other methods, such as D-partition technique used in Hamamci [8]. (2) In the stabilizing region, the gain and phase margins design is considered via the gain and phase margin tester (GPMT) technique [8]. (3) H_{∞} design with fractional-order controllers is proposed by calculating the H_{∞}-norm of sensitivity or complementary sensitivity function of the closed-loop in the stabilizing region using an algebraic method, and the relationship between the H_{∞} region and the fractional orders of $PI^{\lambda}D^{\mu}$ controller is discussed via examples. The design procedure given in this chapter is simple and flexible in the parameters tuning of fractional-order controllers. The content of this chapter is mainly based on the authors work published in the literature [26, 28, 29].

The rest of the chapter is organized as follows. The next section is the description of the related work in the field. Section 3 presents the basic knowledge used in the following sections. Sections 4, 5 and 6 are the main results of the chapter, discussing, in the parameter plane of the controller, the stabilizing regions, the phase and gain margins regions and the H_{∞} constraint regions using fractional-order PD^{μ}, PI^{λ} and $PI^{\lambda}D^{\mu}$ controllers, respectively. Finally, in Sect. 7 concluding remarks are given.

2 Related Work

In recent years, the investigation of fractional-order systems has attracted considerable attention. In this field, the parameter's tuning of fractional-order controllers is an important issue. The existing simple fractional-order controllers include the following four types: TID (tilt-integral-derivative) controller [16], CRONE Controller [19, 20], fractional lead-lag compensator [17, 24] and $PI^{\lambda}D^{\mu}$ controller [22, 23] The $PI^{\lambda}D^{\mu}$ controller is the extension of conventional PID controller, which introduces two more tunable parameters, i.e., the order of the integrator and the order of the differentiator. Two special cases of $PI^{\lambda}D^{\mu}$ controller are fractional-order PI^{λ} [15, 18], and PD^{μ} [10, 26] controllers. The parameter's tuning of $PI^{\lambda}D^{\mu}$ type controllers can be classified into two groups, i.e., the analytical method [5, 15, 18] and the graphical approach [8–10, 26]. The former considers calculating the controller parameters based on some frequency domain design specifications, such as phase margin, gain cross-over frequency and robustness to the gain variation of the plant (ISO-damping). The later focuses on plotting the stabilizing boundary curves in the parameter space of the controller. These boundary curves are described by real root boundary (RRB), infinite root boundary (IRB) and complex root boundary (CRB). Then, in the stabilizing region, the performance specifications are taken into account. The gain-margin and the phase-margin design was given in [8, 9] using GPMT technique. Other investigations on the tuning of $PI^{\lambda}D^{\mu}$ controllers can be found in [1, 14, 21, 31]. In Akbari [1], the problem related to robust BIBO stability analysis of the fractional *PID*-based control systems was studied for a class of uncertain plants modeled by a four-parameter model structure. Another type of fractional-order *PD* controller, called *FO*-[*PD*] controller, was proposed in Lou [14] and the fairness issue in comparing with other controllers was addressed. The authors in Padula [21], presented a set of tuning rules for *PID* and fractional-order *PID* controllers for integral and unstable processes in order to minimize the integrated absolute error criterion, and the results can be used to quantify the performance improvement that can be obtained by using the fractional-order controller instead of the integer one. In Yeroglu and Tan [31], two design techniques for tuning the parameters of fractional-order *PID* controller were given. The first one used the idea of the Ziegler-Nichols and the Astrom-Hagglund methods. The second one was related with the robust fractional-order *PID* controllers, and a design procedure was given using the Bode envelopes of the control systems with parametric uncertainty. It is worthy of mentioning that the fractional-order chaotic systems were also studied in the recent literature [4].

On the other hand, it is well-known that H_∞-norm is a robustness measure of closed-loop systems. For integer-order processes, using integer-order *PID* controller or other types of low-order controllers, sensitivity function, complementary sensitivity function and robust performance of the closed-loop systems were considered [3, 12, 13, 25, 30], respectively, and the corresponding controllers were designed. To the best of our knowledge, no fractional-order *PID* type controller synthesis exists for achieving H_∞ specification in the literature. This chaptor attempts to fill this gap.

3 Preliminaries

In this section, we give some elementary concepts and definitions related in the following sections, including the form of fractional-order $PI^{\lambda}D^{\mu}$ controllers, the parameter plane approach for the stability analysis of fractional-delay systems controlled by $PI^{\lambda}D^{\mu}$ controllers, and the definition of H_∞-norm of a transfer function.

3.1 Fractional-Order $PI^{\lambda}D^{\mu}$ Controllers

The conventional *PID* controller has the following form in time domain

$$u(t) = K\left(e(t) + \frac{1}{T_i}\int_0^t e(\tau)d\tau + T_d\frac{de(t)}{dt}\right) \tag{1}$$

where u is the control variable and e the control error. The control variable is thus a sum of three terms: the P-term, which is proportional to the error, the I-term, which is proportional to the integral of the error, and the D-term, which is proportional to the derivative of the error. The controller parameters are proportional gain K, the integral time T_i and the derivative time T_d. A higher proportional action (higher proportional gain K) gives a faster response speed and a lower steady-state error, but a bigger overshoot (more oscillatory response behavior). A stronger integral action (smaller integral time T_i) can reduce and eliminate the steady-state error. The derivative action (the derivative time T_d) increases the damping of the system and improves the stability property.

The *PID* algorithm given by (1) can be represented by the transfer function

$$C(s) = K(1 + \frac{1}{T_i s} + T_d s) \tag{2}$$

where s is the Laplace variable. Another representation of the *PID* algorithm is of the form

$$C(s) = k_p + \frac{k_i}{s} + k_d s \tag{3}$$

The parameters between (2) and (3) have the following relations

$$k_p = K$$

$$k_i = \frac{K}{T_i}$$

$$k_d = KT_d$$

where k_p is called the proportional-gain, k_i the integral-gain and k_d the derivative-gain, respectively.

By contrast with the conventional *PID* controller, the output of the fractional-order $PI^{\lambda}D^{\mu}$ controller, in time domain, is

$$u(t) = k_p e(t) + k_i D^{-\lambda} e(t) + k_d D^{\mu} e(t) \tag{4}$$

where D^{μ} is Caputo's fractional derivative of order μ with respect to t and with the starting point at $t = 0$

$$D^{\mu} e(t) = \frac{1}{\Gamma(1-\delta)} \int_0^t \frac{e^{(m+1)}(\tau)}{(t-\tau)^{\delta}} d\tau (\mu = m + \delta, m \in Z, 0 < \delta < 1) \tag{5}$$

$D^{-\lambda}$ is Caputo's fractional integral of order λ

$$D^{-\lambda} e(t) = \frac{1}{\Gamma(\lambda)} \int_0^t \frac{e(\tau)}{(t-\tau)^{1-\lambda}} d\tau (\lambda > 0) \tag{6}$$

where $\Gamma(\bullet)$ is the gamma function. The transfer function of such a controller has the form in frequency domain

$$C(s) = k_p + \frac{k_i}{s^{\lambda}} + k_d s^{\mu}, (0 < \lambda, \mu < 2) \tag{7}$$

Setting $\mu = 0$ gives a PI^{λ} controller, and letting $\lambda = 0$ corresponds to a PD^{μ} controller. They are two special cases of fractional $PI^{\lambda}D^{\mu}$ controller.

In comparison with the conventional *PID* controller, the fractional $PI^{\lambda}D^{\mu}$ controller has two more tunable parameters, i.e., the integral order λ and the derivative order μ. Taking $\lambda = \mu = 1$, one recovers a conventional *PID* controller. So the $PI^{\lambda}D^{\mu}$ controller is a generalization of *PID* controller. It is shown in Podlubny [22] that a suitable way to the more efficient control of fractional-order systems is to use the fractional $PI^{\lambda}D^{\mu}$ type controllers. Thus, the parameter's tuning of the $PI^{\lambda}D^{\mu}$ controller is an important topic from the point of view of application.

3.2 Parameter Plane Approach

From (7) in the previous subsection, one observes that there exist five tunable parameters in $PI^{\lambda}D^{\mu}$ controller, i.e., three gains k_p, k_i and k_d, and two fractional orders λ and μ (in the case of PI^{λ} controller and PD^{μ} controller, only three parameters remained, i.e., k_p, k_i and λ, and k_p, k_d and μ, respectively). Select three parameters, e.g., λ, μ and k_p, to be fixed, one can study the stability region in the plane of the other two parameters, e.g., in (k_d, k_i)-plane. The general principle of the stability analysis of a control system in the parameter plane is stated as follows [6].

Let the characteristic equation of a control system be

$$F(\alpha,\beta,s)=0 \tag{8}$$

where s is the Laplace variable, $\alpha,\beta\in R$ denote a two-dimensional parameter plane, i.e., (α,β)-plane. With the help of the Implicit Function Theorem, solving the equation

$$F(\alpha,\beta,j\omega)=0 \tag{9}$$

for (α, β) as a function of $\omega\in R$ finds the stability boundary in (α, β)-plane such that for $(\alpha,\beta)(\alpha,\beta)$ on such a boundary, the characteristic equation has a root exactly on the imaginary axis in s-plane. These boundary curves divide the parameter plane into regions, and by additional arguments, when α and β appear linearly, it is possible to conclude which side of the boundary curve corresponds to stability region or instability region. In practice, it may require a computer to draw the corresponding boundary curves in(α,β)-plane parameterized by ω.

To conclude this subsection, one gives a graphical stability criterion applicable to fractional-delay systems. Given a characteristic equation as defined in (9), one partitions $F(\alpha,\beta,j\omega)$ into its real and imaginary parts

$$F(\alpha,\beta,j\omega)=F_1(\alpha,\beta,\omega)+jF_2(\alpha,\beta,\omega) \tag{10}$$

where

$$\begin{aligned} F_1(\alpha,\beta,\omega)&=ReF(\alpha,\beta,j\omega)\\ F_2(\alpha,\beta,\omega)&=ImF(\alpha,\beta,j\omega) \end{aligned} \tag{11}$$

Suppose one has found, in one way or another, a point (α^0,β^0) in the (α, β)-plane such that

$$\begin{aligned} F_1\left(\alpha^0,\beta^0,\omega\right)&=0\\ F_2\left(\alpha^0,\beta^0,\omega\right)&=0 \end{aligned} \tag{12}$$

i.e., the characteristic equation has a root at position ω on the imaginary axis in s-plane. Let J denote the Jacobian Matrix of partial derivatives with respect to (α,β) evaluated at that point, i.e.,

$$J=\begin{bmatrix} \frac{\partial F_1}{\partial\alpha} & \frac{\partial F_1}{\partial\beta} \\ \frac{\partial F_2}{\partial\alpha} & \frac{\partial F_2}{\partial\beta} \end{bmatrix}_{(\alpha^0,\beta^0,\omega)} \tag{13}$$

Then, according to the Explicit Function Theorem, if the Jacobian (13) is non-singular, the Eq. (12) has a unique local solution curve $(\alpha(\omega),\beta(\omega))$. Moreover, the following proposition holds.

Proposition 1 *The critical roots are in the right half-plane if the point in the parameter plane, relative to the selected values of α and β, lies at the left side of the curve ($\alpha(\omega)$, $\beta(\omega)$) when one follows this curve in the direction of increasing ω, whenever $\det J < 0$ and at the right side when $\det J > 0$. Here J is the Jacobian Matrix defined in (13).*

As the coefficients in the characteristic Eq. (9) are real, one concludes that if $j\omega$ is a root of (9), so is the complex conjugate of it. Therefore, it is sufficient to consider $\omega \in [0, \infty)$. Correspondingly, the boundary curves between the stability and instability regions in the parameter plane are defined by the following three parts: (1) Real root boundary (RRB) for $\omega = 0$ (A real root crosses over the imaginary axis at $s = 0$). (2) Complex root boundary (CRB) for $\omega \in (0, \infty)$ (A pair of complex roots crosses over the imaginary axis at $s = j\omega$). (3) Infinite root boundary (IRB) for $\omega = \infty$ (A real root crosses over the imaginary axis at $s = \infty$).

3.3 H_∞-Norm of Sensitivity and Complementary Sensitivity

Consider SISO LTI unity feedback system as shown in Fig. 1, which is composed of a plant $G(s)$ and a controller $C(s)$. The sensitivity transfer function is defined as

$$S(s) \equiv \frac{1}{1 + C(s)G(s)} \tag{14}$$

i.e., the transfer function from the reference input to the tracking error. On the other hand, the transfer function from the reference input to the output of the closed-loop system is

$$T(s) \equiv 1 - S(s) = \frac{C(s)G(s)}{1 + C(s)G(s)} \tag{15}$$

referred to as the complementary sensitivity transfer function. From (15), it is clear that the two functions T and S are co-complementary. The name sensitivity function comes from the fact that S is the sensitivity of the closed-loop transfer function T to an infinitesimal perturbation in G [7].

H_∞-norm of a transfer function $G(s)$ is defined as

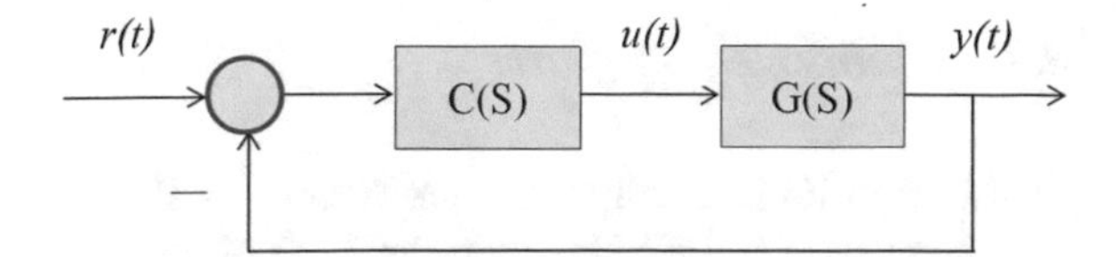

Fig. 1 Unity feedback system

$$\|G(s)\|_\infty \equiv \sup_\omega |G(j\omega)| \tag{16}$$

Note that the H_∞-norm of G equals the distance in the complex plane from the origin to the farthest point on the Nyquist plot of G. It also appears as the peak value on the Bode magnitude plot of G.

Now, recall the sensitivity function defined in (14). one wants the tracking error to have amplitude less than ε, a measure of goodness of tracking. Then the performance specification can be expressed as

$$\|S(s)\|_\infty < \varepsilon \tag{17}$$

or

$$\|W_1(s)S(s)\|_\infty < 1 \tag{18}$$

where $W_1(s) = 1/\varepsilon$, called the weighting function. Usually, $W_1(s)$ is a stable frequency-dependent function, and is chosen as a low pass filter to get a good tracking.

A typical robust stability test, one for the multiplicative uncertain model, is the following H_∞-norm constraint of the weighted complementary sensitivity function defined in (15)

$$\|W_2(s)T(s)\|_\infty < 1 \tag{19}$$

where $W_2(s)$ is the weighting function. Typically, $|W_2(j\omega)|$ is an increasing function of ω, representing that uncertainty increases with increasing frequency.

4 H_∞ Design with Fractional PD^μ Controllers

We first begin with the H_∞ design with simple fractional PD^μ controllers. The parameter plane approach discussed in Sect. 3.2 is adopted to draw and identify the stabilizing region in the two gains' plane of the controller. Then, in the stabilizing region, the H_∞-norm constraint of complementary sensitivity function is considered.

4.1 *Stabilizing Region*

Consider SISO LTI unity feedback system as shown in Fig. 1, which is composed of a plant $G(s)$ and a PD^μ controller $C(s)$ given by, respectively

$$G(s) = \frac{N(s)}{D(s)} e^{-Ls} \tag{20}$$

$$C(s) = k_p + k_d s^\mu \tag{21}$$

where N(s) and D(s) are integer-order or fractional-order polynomials, $L > 0$ demotes the time-delay, $\mu \in (0, 2)$ is the derivative-order of the controller, k_p and k_d are the proportional-gain and derivative-gain, respectively, of PD^μ controller. The objective of this subsection is to determine the stabilizing parameters region of PD^μ controller in (k_d, k_p)-plane, for a fixed $\mu \in (0, 2)$, using the parameter plane approach given in Sect. 3.2. To this end, the closed-loop characteristic quasi-polynomial is first given

$$\Delta(s) = D(s) + N(s)(k_p + k_d s^\mu)e^{-Ls}$$

Multiplying both sides of the above equality by e^{Ls} yields

$$\Delta^*(s) = D(s)e^{Ls} + N(s)(k_p + k_d s^\mu) \tag{22}$$

Substituting $s = j\omega$ into (22) gives rise to

$$\Delta^*(j\omega) = D(j\omega)e^{jL\omega} + N(j\omega)[k_p + k_d(j\omega)^\mu] \tag{23}$$

Let

$$\begin{aligned} D(j\omega) &= D_r(\omega) + jD_i(\omega) \\ N(j\omega) &= N_r(\omega) + jN_i(\omega) \end{aligned} \tag{24}$$

and notice that

$$j^\alpha = \cos\frac{\alpha\pi}{2} + j\sin\frac{\alpha\pi}{2}, (0 < \alpha < 2) \tag{25}$$

where $D_r(\omega)$ (or $N_r(\omega)$) and $D_i(\omega)$ (or $N_i(\omega)$) stand for the real and imaginary components of $D(j\omega)$ (or $N(j\omega)$), respectively, one partitions $\Delta^*(j\omega)$ into its real and imaginary parts

$$\Delta^*(j\omega) = \Delta_r(\omega) + j\Delta_i(\omega)$$

where

$$\begin{aligned} \Delta_r(\omega) = |D(j\omega)|\cos[L\omega + \alpha(\omega)] + k_d\omega^\mu \\ |N(j\omega)|\cos[\beta(\omega) + \mu\pi/2] + k_p N_r(\omega) \end{aligned} \tag{26}$$

$$\Delta_i(\omega) = |D(j\omega)|\sin[L\omega + \alpha(\omega)] + k_d\omega^{\mu} \\ |N(j\omega)|\sin[\beta(\omega) + \mu\pi/2] + k_p N_i(\omega) \tag{27}$$

With $|D(j\omega)|$ (or $|N(j\omega)|$) being the modulus of the complex variable $D(j\omega)$ (or $N(j\omega)$), and $\alpha(\omega)$ (or $\beta(\omega)$) being the phase function of $D(j\omega)$ (or $N(j\omega)$).

From (26) and (27), it is clear that both Δ_r and Δ_i depends on the parameters (k_d, k_p, μ, ω). Suppose one has found, in one way or another, a point $(k_d^0, k_p^0, \mu, \omega)$ in (k_d, k_p)-plane, for a fixed μ, such that

$$\begin{cases} \Delta_r = \Delta_r(k_d^0, k_p^0, \mu, \omega) = 0 \\ \Delta_i = \Delta_i(k_d^0, k_p^0, \mu, \omega) = 0 \end{cases} \tag{28}$$

i.e., there exists a root on the imaginary axis. According to the Explicit Function Theorem, if the Jacobian

$$J = \begin{bmatrix} \frac{\partial \Delta_r}{\partial k_d} & \frac{\partial \Delta_r}{\partial k_p} \\ \frac{\partial \Delta_i}{\partial k_d} & \frac{\partial \Delta_i}{\partial k_p} \end{bmatrix}_{(k_d^0, k_p^0, \mu, \omega)} \tag{29}$$

is nonsingular, then the Eq. (28) has a unique local solution curve $(k_d(\omega), k_p(\omega))$. Moreover, the proposition stated in the Sect. 3.2 holds, where the Jacobian is defined by (29).

As the coefficients in the characteristic Eq. (23) and L are real, if $j\omega$ is a root of (23), so is the complex conjugate of it. Therefore, it's sufficient to consider $\omega \in [0, \infty)$. For $\omega = 0$, let $\Delta^*(j\omega) = 0$, one obtains a part of the marginal stability curve (RRB)

$$k_p = -D(j0)/N(j0) \tag{30}$$

For $\omega \in (0, \infty)$ one first solves Eq. (28) for k_d and k_p in terms of μ and ω

$$k_d = -\frac{|D(j\omega)|\sin[L\omega + \alpha(\omega) - \beta(\omega)]}{\omega^{\mu}|N(j\omega)|\sin(\frac{\mu\pi}{2})} \tag{31}$$

$$k_p = \frac{|D(j\omega)|\sin[L\omega + \alpha(\omega) - \beta(\omega) - \frac{\mu\pi}{2}]}{|N(j\omega)|\sin(\frac{\mu\pi}{2})} \tag{32}$$

then, (31) and (32) gives another part of the marginal stability curve (CRB) in (k_d, k_p)-plane, and from Proposition 1, one can determine which side of the curve belongs to stability region, according to the sign of det J given by

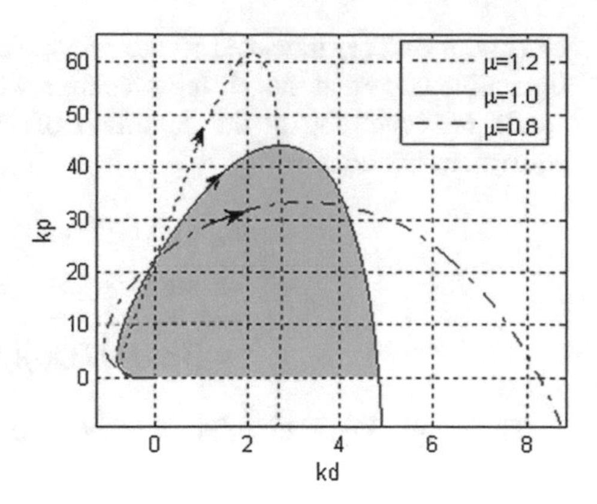

Fig. 2 Stabilizing regions for different μ

$$\det J = -\omega^{\mu}|N(j\omega)|^2 \sin\left(\frac{\mu\pi}{2}\right) < 0, \forall\omega > 0, \mu \in (0,2) \tag{33}$$

Note that in this case the IRB does not exist.

Example 1 Consider the following fractional-order integrating process with time delay

$$G(s) = \frac{1}{s^{1.5}} e^{-0.1s} \tag{34}$$

which is investigated in Hamamci [8]. Apply PD^{μ} controller (21) to this process, one plots the stabilizing boundary curves in (k_d, k_p)-plane, using (31) and (32), and (30) $(k_p = 0)$, for different fractional-order μ as shown in Fig. 2. The arrows along the curves denote the direction of increasing ω. Then, according to Proposition 1 and the sign of the determinant of Jacobian (33), the stabilizing regions can be identified as the closed areas above the k_d-axis for each μ (for $\mu = 1$, the area filled). It is clear that different μ corresponds to different shapes and areas of stabilizing regions.

4.2 H_∞ *Design of Complementary Sensitivity*

In this subsection, the H_∞-norm constraint of complementary sensitivity function

$$T(s) = \frac{C(s)G(s)}{1 + C(s)G(s)} = \frac{L(s)}{1 + L(s)} \tag{35}$$

where $L(s) = C(s)G(s)$ stands for the open-loop transfer function, is considered in the stabilizing region of PD^{μ} controller. As mentioned in Sect. 3.3, the H_∞-norm constraint of $T(s)$ is related to the robust stability of the closed-loop for the multiplicative uncertain model. Introducing a weighting function $W_2(s)$ and a positive scalar M, according to the definition of H_∞-norm (16), one has the following equivalent relations

$$\begin{aligned}&\|W_2(s)T(s)\|_\infty \le M\\ \Leftrightarrow &\sup_\omega |W_2(j\omega)T(j\omega)| \le M\\ \Leftrightarrow &|W_2(j\omega)T(j\omega)| \le M, \forall \omega \in R\end{aligned} \tag{36}$$

The weight $W_2(s)$ describes the frequency characteristic of the H_∞ specification. To determine the (k_p, k_d) values in the stabilizing region for which the H_∞-norm constraint (36) of the weighted complementary sensitivity is satisfied, consider the following controller transformation

$$C(s) = k_p + k_d s^\mu = k_p(1 + \frac{k_d}{k_p} s^\mu) = x(1 + ys^\mu) \tag{37}$$

where $x = k_p$, $y = k_d/k_p$, and the open-loop transfer function can be written as

$$L(s) = C(s)G(s) = x[G(s) + ys^\mu G(s)]$$

Letting $s = j\omega$ yields

$$L(j\omega) = x[G(j\omega) + y(j\omega)^\mu G(j\omega)]$$

Partition $G(j\omega)$ into its real and imaginary components,

$$G(j\omega) = A(\omega) + jB(\omega)$$

and notice (25), one has

$$L(j\omega) = x[A(\omega) + yA_1(\omega)] + jx[B(\omega) + yB_1(\omega)] \tag{38}$$

where

$$\begin{aligned}A_1(\omega) &= \omega^\mu[A(\omega)\cos\frac{\mu\pi}{2} - B(\omega)\sin\frac{\mu\pi}{2}]\\ B_1(\omega) &= \omega^\mu[B(\omega)\cos\frac{\mu\pi}{2} + A(\omega)\sin\frac{\mu\pi}{2}]\end{aligned}$$

From (35) and (36), the following holds

$$\left|\frac{W_2(j\omega)L(j\omega)}{1 + L(j\omega)}\right| \le M, \quad \forall \omega \in R \tag{39}$$

Define

$$f_1 = |1 + L(j\omega)|^2 - M_1|L(j\omega)|^2$$

where

$$M_1 = \left|\frac{W_2(j\omega)}{M}\right|^2$$

and from (38), one gets

$$f_1 = 1 + 2x(A + yA_1) + x^2[(A + yA_1)^2 + (B + yB_1)^2](1 - M_1) \geq 0 \tag{40}$$

Inequality (40) defines an optimization problem, and when the equality holds for some frequency ω, the minimum value of f_1 is reached. One wants to find the pair (x, y), accordingly the pair (k_p, k_d), for a fixed parameter μ, such that the equality in (40) is satisfied for some frequency ω. To this end, differentiating f_1 with respect to ω and letting the corresponding derivative at that frequency to be zero gives

$$\begin{aligned} f_2 = df_1/d\omega = 2x(A + yA_1) + x^2[2(A + yA_1) \\ (A + yA_1) + 2(B + yB_1)(B + yB_1)] \\ (1 - M_1) - x^2[(A + yA_1)^2 + (B + yB_1)]M_1 \end{aligned}$$

Eliminate x^2 in equations

$$\begin{cases} f_1 = 0 \\ f_2 = 0 \end{cases}$$

one solves

$$x = \frac{2(\dot{A} + y\dot{A}_1)}{C_0 + C_1 y + C_2 y^2} \tag{41}$$

where

$$\begin{aligned} C_0 &= 2(AA + BB)(\gamma_1 - 1) + \dot{\gamma}_1(A^2 + B^2) \\ C_1 &= 2(AA_1 + AA_1 + BB_1 + BB_1)(\gamma_1 - 1) + 2\dot{\gamma}_1(AA_1 + BB_1) \\ C_2 &= 2(A_1A_1 + B_1B_1)(\gamma_1 - 1) + \dot{\gamma}_1(A_1^2 + B_1^2) \end{aligned}$$

Substitute (41) into $f_1 = 0$ in (40), one obtains a fourth-order equation about y for each ω

$$D_0 + D_1 y + D_2 y^2 + D_3 y^3 + D_4 y^4 = 0 \tag{42}$$

where

$$\begin{aligned}
D_0 &= C_0^2 + 4A\dot{A}C_0 + 4(1-\gamma_1)e_{00}\\
D_1 &= 2C_0C_1 + 4e_{10} + 8(1-\gamma_1)e_{11}\\
D_2 &= 2C_0C_2 + C_1^2 + 4e_{20} + 4(1-\gamma_1)e_{21}\\
D_3 &= 2C_1C_2 + 4e_{30} + 8(1-\gamma_1)e_{31}\\
D_4 &= C_2^2 + 4A_1\dot{A}_1C_2 + 4(1-\gamma_1)e_{40}
\end{aligned}$$

$$\begin{aligned}
e_{00} &= (A^2+B^2)\dot{A}^2\\
e_{10} &= A\dot{A}C_1 + (\dot{A}A_1 + A\dot{A}_1)C_0\\
e_{11} &= (AA_1 + BB_1)\dot{A}^2 + (A^2+B^2)\dot{A}\dot{A}_1\\
e_{20} &= A\dot{A}C_2 + (\dot{A}A_1 + A\dot{A}_1)C_1 + A_1\dot{A}_1C_0
\end{aligned}$$

$$\begin{aligned}
e_{21} &= (A_1^2+B_1^2)\dot{A}^2 + 4(AA_1+BB_1)\dot{A}\dot{A}_1 + (A^2+B^2)\dot{A}_1^2\\
e_{30} &= (A\dot{A}_1 + \dot{A}A_1)C_2 + A_1\dot{A}_1C_1\\
e_{31} &= (A_1^2+B_1^2)\dot{A}\dot{A}_1 + (AA_1+BB_1)\dot{A}_1^2\\
e_{40} &= (A_1^2+B_1^2)\dot{A}_1^2
\end{aligned}$$

The pair (x, y), accordingly the pair (k_p, k_d), which defines the H_∞ boundary curve for a range of frequencies, can be found in the following manner. Solve the fourth-order Eq. (42) for y, for an appropriately selected frequency ω, one gets 4 roots, and substitute these 4 roots into (41), respectively, one obtains the corresponding 4 parameters x. Then, recover the original controller parameters by relations $k_p = x$ and $k_d = xy$, and the solution pair, if any, is the real pair (k_p, k_d) which is located in the stabilizing region. By changing the frequency ω, the real pairs $(k_p(\omega), k_d(\omega))$ plot the H_∞ boundary curve in the stabilizing region.

Remark 1 When the sensitivity function is considered, the above H_∞ design procedure for complementary sensitivity function can be applied similarly.

Example 2 Example 1 revisited. In this case, we consider the problem of determining the admissible parameters of PD^μ controllers in the stabilizing region for which $\|W_2(s)T(s)\|_\infty \le M$, where the weighting function $W_2(s)$ is chosen as a high pass filter

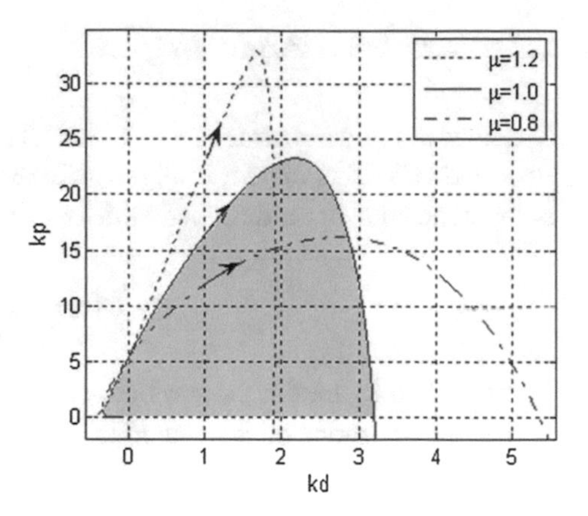

Fig. 3 H_∞ regions for different μ

$$W_2(s) = \frac{s+0.1}{s+1}$$

and $M = 2$. For a fixed μ, solving Eqs. (42) and then (41) for different ω gives rise to the pair $(k_p(\omega), k_d(\omega))$, which defines the H_∞ boundary curves in the stabilizing region in (k_d, k_p)-plane as shown in Fig. 3. The arrows along the curves indicate the direction of increasing ω. The filled area corresponds to the H_∞ region $\|W_2(s)T(s)\|_\infty < M$ in the case of $\mu = 1$. It is observed that different μ gives different shapes of H_∞ region. In fact, further observation shows that for $\mu > 1$, a larger admissible k_p value is allowed. On the contrary, when $\mu < 1$, as μ decreases, a wider range of k_d value is expected, showing that with fractional-order PD^μ controller, the ranges of k_p and k_d values are wider than those with conventional PD controllers, which means that when one considers other specifications in the H_∞ region, a better performance can be achieved using fractional-order controllers.

5 H_∞ Design with Fractional PI^λ Controllers

The classical relative stability tests are the gain-margin and the phase-margin. In this section, we utilize the gain and phase margin tester technique to handle this problem, and consider the stability margins and the H_∞ co-design with fractional-order PI^λ controllers.

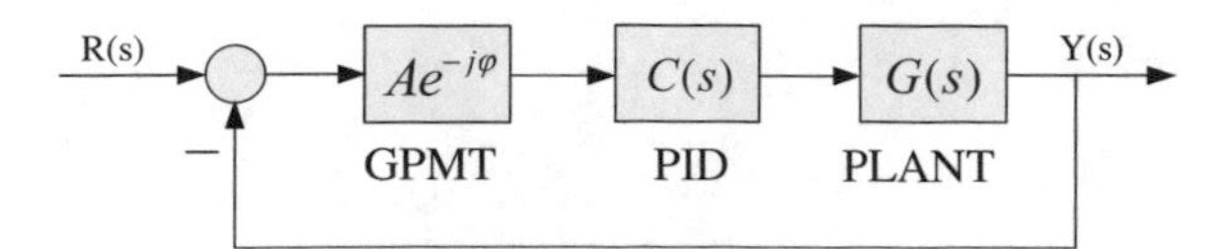

Fig. 4 General structure of unity feedback

5.1 Stability Margin Region

Consider the general structure of SISO LTI unity feedback with a compensator of so-called GPMT as shown in Fig. 4, where $G(s)$ is the plant given in (20) and $C(_s)$ is the controller to be designed which is of the fractional-order PI^{λ} form

$$C(s) = k_p + \frac{k_i}{s^{\lambda}} \tag{43}$$

where k_p and k_i are the proportional-gain and integral-gain, respectively, $\lambda \in (0, 2)$ is the integral-order of the integrator. $Ae^{-\phi}$ in Fig. 4 denotes the GPMT, which provides information on plotting the boundary lines of constant gain-margin and phase-margin in the parameter plane of PI^{λ} controller, corresponding to the following three cases: (a) setting $A = 1$, one obtains the boundary for a given phase-margin φ. (b) setting $\varphi = 0$, the boundary lines for a desired gain-margin A. And (c) to find the stability boundary lines, one needs to set $A = 1$ and $\varphi = 0$, simultaneously. In practical control systems, the block GPMT does not exist, it is only employed to design control systems satisfying desired gain-margin and/or phase-margin.

The objective of this subsection is to determine the stabilizing region in (k_p, k_i)-plane for a fixed $\lambda \in (0, 2)$, and then, the regions, in the stability region, satisfying the gain-margin and phase-margin, utilizing the parameter plane approach discussed in Sect. 3.2. To this end, the closed-loop characteristic quasi-polynomial in Fig. 4 is first computed as

$$\Delta(s) = s^{\lambda}D(s)e^{(Ls + j\varphi)} + A(k_i + k_p s^{\lambda})N(s) = 0 \tag{44}$$

Substituting $s = j\omega$ into (44) gives rise to

$$\Delta(j\omega) = (j\omega)^{\lambda}D(j\omega)e^{j(L\omega + \varphi)} + A[k_i + k_p(j\omega)^{\lambda}]N(j\omega) = 0 \tag{45}$$

Let

$$D(j\omega) = D_r(\omega) + jD_i(\omega) = |D(j\omega)|e^{j\alpha(\omega)}$$
$$N(j\omega) = N_r(\omega) + jN_i(\omega) = |N(j\omega)|e^{j\beta(\omega)}$$

where $D_r(\omega)(orN_r(\omega))$ and $D_i(\omega)(orN_i(\omega))$ represent the real and imaginary parts, respectively, of $D(j\omega)(orN(j\omega))$, and $|D(j\omega)|(or|N(j\omega)|)$ and $\alpha(\omega)(or\beta(\omega))$ stand for the modulus and the phase, respectively, of the complex variable $D(j\omega)(orN(j\omega))$. Decomposing $\Delta(j\omega)$ into its real and imaginary components yields

$$\Delta(j\omega) = \Delta_r(\omega) + j\Delta_i(\omega)$$

where

$$\begin{aligned}\Delta_r(\omega) &= \omega^{\lambda}|D(j\omega)|\cos(L\omega + \alpha(\omega) - \beta(\omega) + \varphi + \frac{\lambda\pi}{2}) \\ &+ A|N(j\omega)|(k_i + k_p\omega^{\lambda}\cos\frac{\lambda\pi}{2})\end{aligned} \tag{46}$$

$$\begin{aligned}\Delta_i(\omega) &= \omega^{\lambda}|D(j\omega)|\sin(L\omega + \alpha(\omega) - \beta(\omega) + \varphi + \frac{\lambda\pi}{2}) \\ &+ A|N(j\omega)|k_p\omega^{\lambda}\sin\frac{\lambda\pi}{2}\end{aligned} \tag{47}$$

Note that in the deriving of (46) and (47), one has utilized the expression (25).

Similar to the analysis in the previous section, only the frequency interval $\omega \in [0, \infty)$ needs to be considered. For $\omega = 0$, letting $\Delta(j\omega) = 0$ leads to a piece of stabilizing boundary line (RRB)

$$k_i = 0, \ if \ N(j0) \neq 0 \tag{48}$$

and for $\omega \in (0, \infty)$, solving the following equations

$$\begin{cases} \Delta_r(\omega) = B_1(\omega)k_p + C_1(\omega)k_i + D_1(\omega) = 0 \\ \Delta_i(\omega) = B_2(\omega)k_p + C_2(\omega)k_i + D_2(\omega) = 0 \end{cases} \tag{49}$$

for k_p and k_i, using Gramer's rule, yields

$$k_p(\omega, \lambda, A, \varphi) = \frac{C_1(\omega)D_2(\omega) - C_2(\omega)D_1(\omega)}{J(A, \varphi)}, \quad \forall\omega > 0 \tag{50}$$

$$k_i(\omega, \lambda, A, \varphi) = \frac{-B_1(\omega)D_2(\omega) + B_2(\omega)D_1(\omega)}{J(A, \varphi)}, \quad \forall\omega > 0 \tag{51}$$

where

$$B_1(\omega) = A|N(j\omega)|\omega^{\lambda}\cos\frac{\lambda\pi}{2}$$
$$C_1(\omega) = A|N(j\omega)|$$
$$D_1(\omega) = |D(j\omega)|\omega^{\lambda}\cos(L\omega + \alpha(\omega) - \beta(\omega) + \varphi + \frac{\lambda\pi}{2})$$
$$B_2(\omega) = A|N(j\omega)|\omega^{\lambda}\sin\frac{\lambda\pi}{2}$$
$$C_2(\omega) = 0$$
$$D_2(\omega) = |D(j\omega)|\omega^{\lambda}\sin(L\omega + \alpha(\omega) - \beta(\omega) + \varphi + \frac{\lambda\pi}{2})$$

And

$$J(A,\varphi) \underline{\underline{\Delta}} \begin{vmatrix} B_1 & C_1 \\ B_2 & C_2 \end{vmatrix} = -B_2C_1 = -A^2|N(j\omega)|^2\omega^{\lambda}\sin\frac{\lambda\pi}{2} < 0, \quad \forall\omega > 0,\ \lambda \in (0,2) \tag{52}$$

denotes the Jacobian of Eq. (49). By setting $A = 1$ and $\varphi = 0^0$ in (50) and (51), one has

$$k_p(\omega,\lambda,A=1,\varphi=0^0) = -\frac{|D(j\omega)|\sin(L\omega + \alpha(\omega) - \beta(\omega) + \frac{\lambda\pi}{2})}{|N(j\omega)|\sin\frac{\lambda\pi}{2}}, \forall\omega > 0 \tag{53}$$

$$k_i(\omega,\lambda,A=1,\varphi=0^0) = \frac{\omega^{\lambda}|D(j\omega)|\sin(L\omega + \alpha(\omega) - \beta(\omega))}{|N(j\omega)|\sin\frac{\lambda\pi}{2}}, \forall\omega > 0 \tag{54}$$

Using (53) and (54), another piece of stabilizing boundary curve (CRB) in (k_p, k_i)-plane, for a fixed $\lambda \in (0,2)$, can be drawn. This curve separates the (k_p, k_i)-plane into stable parameter region (SPR) and unstable parameter region (*UPR*). Using the arguments as in Proposition 1, if $J(A=1,\varphi=0^0) > 0, \forall\omega > 0, \lambda \in (0,2)$, the SPR is to the left of the stabilizing boundary curve following the direction of increasing ω. Similarly, the SPR is to the right of the stabilizing boundary curve following the direction of increasing ω, while $J(A=1,\varphi=0^0) < 0, \forall\omega > 0$, $\forall\lambda \in (0,2)$.

Next, for a desired gain-margin A or a phase-margin φ, from (50) and (51), it follows that

$$k_p(\omega,\lambda,A,\varphi) = -\frac{|D(j\omega)|\sin(L\omega + \alpha(\omega) - \beta(\omega) + \varphi + \frac{\lambda\pi}{2})}{A|N(j\omega)|\sin\frac{\lambda\pi}{2}}, \forall\omega > 0 \tag{55}$$

$$k_i(\omega,\lambda,A,\varphi) = \frac{\omega^{\lambda}|D(j\omega)|\sin(L\omega + \alpha(\omega) - \beta() + \varphi)}{A|N(j\omega)|\sin\frac{\lambda\pi}{2}}, \forall\omega > 0 \tag{56}$$

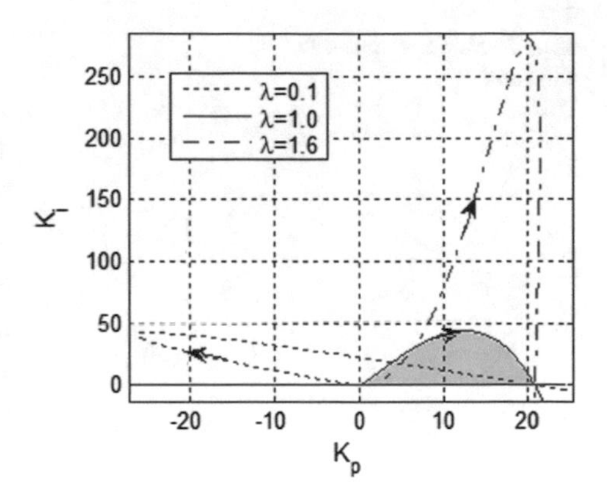

Fig. 5 Stabilizing regions for different λ

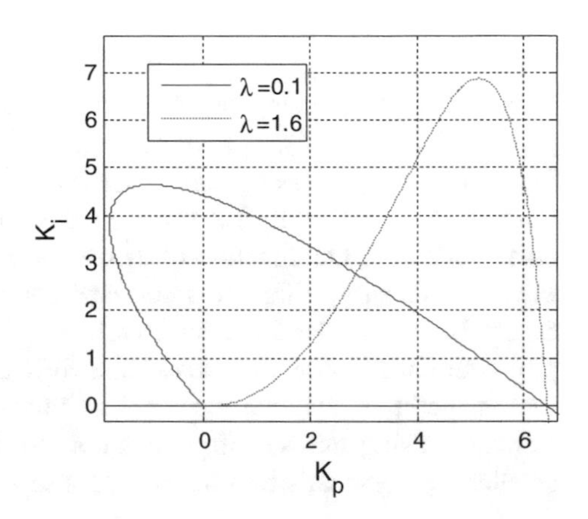

Fig. 6 Phase-margin curves for different λ

It is clear that for a fixed λ the pairs $(k_p(\omega,\lambda,A,\varphi=0^0),k_i(\omega,\lambda,A,\varphi=0^0)),\forall\omega>0$, draw the curve, in the stabilizing region, satisfying the desired gain-margin A, and the pairs $(k_p(\omega,\lambda,A=1,\varphi),k_i(\omega,\lambda,A=1,\varphi)),\forall\omega>0$, plot the curve, in the stabilizing region, satisfying the given phase-margin φ. The area, in the stabilizing region, which meets that the gain-margin or the phase margin is greater than the desired one, can be identified as to the right of the curve following the direction of increasing ω, according to the sign of $J(A,\varphi)$ given in (52).

Example 3 Consider the following fractional-order integrating process with time-delay

$$G(s)=\frac{1}{s^{1.2}}e^{-0.1s} \tag{57}$$

which is studied in Hamamci [8]. Applying PI^{λ} controller (43) to the process (57), one plots the stabilizing boundary curves in (k_p,k_i)-plane for different

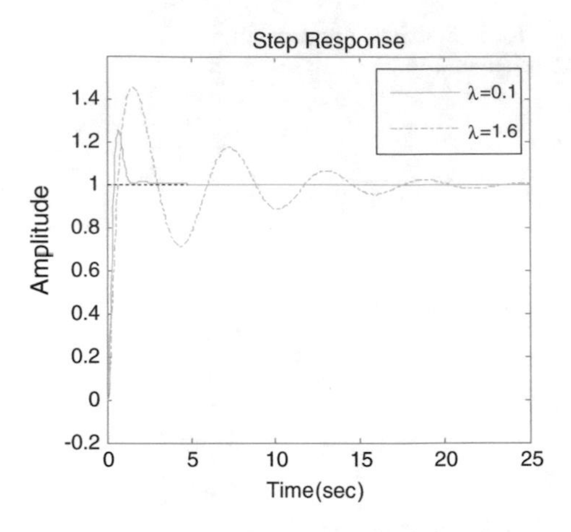

Fig. 7 Step responses for different λ

fractional-order λ, using the pairs $(k_p(\omega,\lambda,A=1,\varphi=0^0),\ k_i(\omega,\lambda,A=1,\varphi=0^0))$ given in (53) and (54), as shown in Fig. 5. The arrows along each curve denote the direction of increasing ω. Because, in our case, $J(A=1,\varphi=0^0)<0$, $\forall\omega>0, \forall\lambda\in(0,2)$, the SPR for each λ is to the right of the stabilizing boundary curve following the direction of increasing ω as shown in Fig. 5 by the filled area for $\lambda=1$. Clearly, different λ corresponds to different shapes and areas of SPR. Select $\lambda=0.1$ and $\lambda=1.6$, respectively, corresponding to relatively larger SPR, one plots the phase margin curves in the SPR as shown in Fig. 6. Choose the intersection point of the two curves, one reads $k_p=2.868$ and $k_i=2.785$. The step responses using this set of parameters are depicted in Fig. 7. It is observed that smaller λ corresponds to a better response behavior in time domain.

5.2 H_∞ *Design of Sensitivity*

This subsection considers the H_∞-norm performance design of the sensitivity function of the closed-loop system shown in Fig. 4 without the GPMT block. The open-loop transfer function is described by

$$L(s)=C(s)G(s)$$

Then, the sensitivity function is written as

$$S(s)=\frac{1}{1+L(s)} \tag{58}$$

Introducing a weighting function $W_1(s)$ and a positive scalar M, according to the definition of H_∞-norm in (16), one has the following equivalent relations

$$\begin{aligned} &\|W_1(s)S(s)\|_\infty \leq M \\ &\Leftrightarrow \sup_\omega |W_1(j\omega)S(j\omega)| \leq M \\ &\Leftrightarrow |W_1(j\omega)S(j\omega)| \leq M,\ \forall \omega \in R \end{aligned} \tag{59}$$

In order to determine the real pairs (k_p, k_i) in the stabilizing region for which the H_∞-norm constraint (59) of the weighted sensitivity is satisfied, consider the following controller transformation

$$C(s) = k_p + \frac{k_i}{s^\lambda} = k_p(1 + \frac{k_i/k_p}{s^\lambda}) = x(1 + \frac{y}{s^\lambda}) \tag{60}$$

where $x = k_p, y = k_i/k_p$. Then, the open-loop transfer function can be expressed as

$$L(s) = C(s)G(s) = x\left[G(s) + \frac{yG(s)}{s^\lambda}\right]$$

Let $s = j\omega$, one gets

$$L(j\omega) = C(j\omega)G(j\omega) = x[G(j\omega) + \frac{yG(j\omega)}{(j\omega)^\lambda}] \tag{61}$$

Decomposing $G(j\omega)$ into its real and imaginary parts

$$G(j\omega) = A(\omega) + jB(\omega)$$

and noting (25) and the following

$$\frac{G(j\omega)}{(j\omega)^\lambda} = A_1(\omega) + jB_1(\omega)$$

where

$$A_1(\omega) = \frac{1}{\omega^\lambda}[A(\omega)\cos\frac{\lambda\pi}{2} + B(\omega)\sin\frac{\lambda\pi}{2}]$$

$$B_1(\omega) = \frac{1}{\omega^\lambda}[B(\omega)\cos\frac{\lambda\pi}{2} - A(\omega)\sin\frac{\lambda\pi}{2}]$$

one obtains

$$L(j\omega) = x[A(\omega) + yA_1(\omega) + j(B(\omega) + yB_1(\omega))] \tag{62}$$

From (58) and (59), the following holds

$$\left|\frac{W_1(j\omega)}{1 + L(j\omega)}\right| \leq M, \quad \forall\omega \in R \tag{63}$$

Define the following function of ω

$$f_1(\omega) = |1 + L(j\omega)|^2 - M_1 \geq 0$$

where

$$M_1 = \left|\frac{W_1(j\omega)}{M}\right|^2$$

and from (63), one has

$$f_1(\omega) = 1 + 2x(A + yA_1) + x^2(A + yA_1)^2 + x^2(B + yB_1)^2 - M_1 \geq 0 \tag{64}$$

Inequality (64) defines an optimization problem, and when the equality holds for some frequency ω, one gets the minimum value of $f_1(\omega)$. For a prescribed positive scalar $M > 0$, one wants to find the pair (x, y), accordingly, the pair (k_p, k_i), for a fixed fractional-order $\lambda \in (0, 2)$, such that the equality in (64) is obtained for some frequency ω. To this end, differentiating $f_1(\omega)$ with respect to ω and letting the corresponding derivative at that frequency to be zero yields

$$\begin{aligned} f_2(\omega) = \frac{df_1(\omega)}{d\omega} &= 2x(A + yA_1) + 2x^2(A + yA_1)(A + yA_1) \\ &+ 2x^2(B + yB_1)(B + + yB_1) - M_1 = 0 \end{aligned}$$

Eliminating x^2 in equations

$$\begin{cases} f_1(\omega) = 0 \\ f_2(\omega) = 0 \end{cases}$$

One solves

$$x = \frac{g_0 + g_1 y + g_2 y^2}{h_0 + h_1 y + h_2 y^2 + h_3 y^3} \tag{65}$$

where

$$
\begin{aligned}
g_0 &= 2(1-M_1)(AA+BB)+M_1(A^2+B^2)\\
g_1 &= 2(1-M_1)(AA_1+AA_1+BB_1+BB_1)+2M_1(AA_1+BB_1)\\
g_2 &= 2(1-M_1)(A_1A_1+B_1B_1)+M_1(A_1^2+B_1^2)
\end{aligned}
$$

$$
\begin{aligned}
h_0 &= 2\dot{A}(A^2+B^2)-4A(A\dot{A}+B\dot{B})\\
h_1 &= 2A_1(\dot{A}^2+B^2)+4A(AA_1+BB_1)-4A_1(A\dot{A}+B\dot{B})\\
&\quad -4A(A\dot{A}_1+\dot{A}A_1+BB_1+BB_1)\\
h_2 &= 2A(A_1^2+B_1^2)+4A_1(AA_1+BB_1)-4A(A_1A_1+B_1B_1)\\
&\quad -4A_1(AA_1+AA_1+BB_1+BB_1)\\
h_3 &= 2\dot{A}_1(A_1^2+B_1^2)-4A_1(A_1A_1+B_1B_1)
\end{aligned}
$$

Substituting x in (65) into $f_1(\omega)=0$ in (64) gives rise to a sixth-order equation about y for each ω

$$
v_0+v_1y+v_2y^2+v_3y^3+v_4y^4+v_5y^5+v_6y^6=0 \tag{66}
$$

where

$$
v_0=(1-M_1)h_0^2+(A^2+B^2)g_0^2+2g_0h_0A
$$

$$
\begin{aligned}
v_1 &= 2(1-M_1)h_0h_1+2(AA_1+BB_1)g_0^2+2(A^2+B^2)g_0g_1\\
&\quad +2[(g_0h_1+g_1h_0)A+g_0h_0A_1]
\end{aligned}
$$

$$
\begin{aligned}
v_2 &= (1-M_1)(2h_0h_2+h_1^2)+(A_1^2+B_1^2)g_0^2+4(AA_1+BB_1)g_0g_1\\
&\quad +(A^2+B^2)(2g_0g_2+g_1^2)+2[(g_0h_2+g_1h_1+g_2h_0)A\\
&\quad +(g_0h_1+g_1h_0)A_1]
\end{aligned}
$$

$$
\begin{aligned}
v_3 &= 2(1-M_1)(h_0h_3+h_1h_2)+2(A_1^2+B_1^2)g_0g_1\\
&\quad +2(AA_1+BB_1)(2g_0g_2+g_1^2)+2(A^2+B^2)g_1g_2\\
&\quad +2[(g_0h_3+g_1h_2+g_2h_1)A+(g_0h_2+g_1h_1+g_2h_0)A_1]
\end{aligned}
$$

$$
\begin{aligned}
v_4 &= (1-M_1)(2h_1h_3+h_2^2)+(A_1^2+B_1^2)(2g_0g_2+g_1^2)\\
&\quad +4(AA_1+BB_1)g_1g_2+(A^2+B^2)g_2^2\\
&\quad +2[(g_1h_3+g_2h_2)A+(g_0h_3+g_1h_2+g_2h_1)A_1]
\end{aligned}
$$

$$
\begin{aligned}
v_5 &= 2(1-M_1)h_2h_3+2(A_1^2+B_1^2)g_1g_2+2(AA_1+BB_1)g_2^2\\
&\quad +2[g_2h_3A+(g_1h_3+g_2h_2)A_1]
\end{aligned}
$$

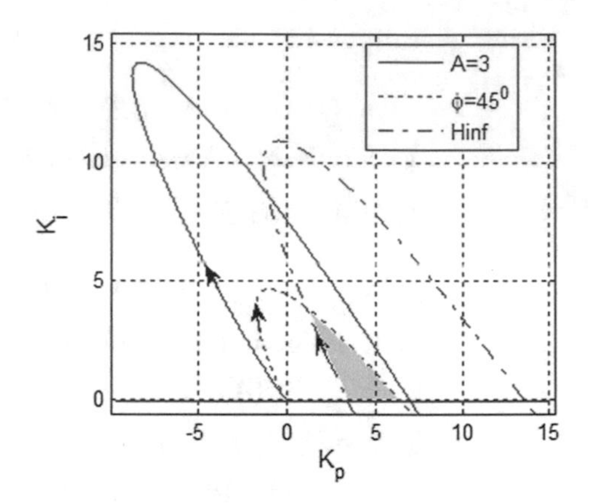

Fig. 8 GM, PM and H_∞ co-design

$$v_6 = (1 - M_1)h_3^2 + (A_1^2 + B_1^2)g_2^2 + 2g_2h_3A_1$$

The pair (x, y), accordingly, the pair (k_p, k_i), which define the H_∞ boundary curve for a range of frequencies, can be found in the following manner. Solve the sixth-order Eq. (66) for y, for an appropriately selected frequency ω, one gets 6 roots, and substitute these 6 roots into (65), respectively, one obtains the corresponding 6 parameters x. Then, recover the original controller parameters by relations $k_p = x$ and $k_i = xy$, and the solution pair, if any, is the real pair (k_p, k_i) which is located in the stabilizing region. By changing the frequency ω, the real pairs $(k_p(\omega), k_i(\omega))$ plot the H_∞ boundary curve in the stabilizing region.

Example 4 Example 3 revisited. In this example, we first consider the H_∞ design for the integrating process using PI^λ controller, then, give the stability margins and H_∞ co-design. In order to obtain a better performance in time domain, one fixes the fractional-order $\lambda = 0.1$, see Fig. 7. The corresponding stabilizing region in (k_p, k_i)-plane is shown in Fig. 5. The weighted H_∞-norm constraint of sensitivity function is given by $\|W_1(s)S(s)\|_\infty \le M$, where the weighting function $W_1(s)$ is chosen as a low pass filter

$$W_1(s) = \frac{5}{s+1}$$

and the scalar $M = 1.8$. Solving the sixth-order Eq. (66) and then Eq. (65) at an appropriately selected frequency ω gives rise to the six pairs $(k_p(\omega), k_i(\omega))$. One selects the pair from the six pairs, which is real and located in the stabilizing region, and this pair defines the H_∞ boundary curve, in the stabilizing region, as ω changes, as shown in Fig. 8 by the dash-dotted line. Next, one draws the phase-margin curve of $\varphi = 45^0$, as shown by the dashed line in Fig. 8, and the gain-margin curve of $A = 3$, as shown by the solid line in Fig. 8. To the right along each curve are the regions satisfying $\varphi > 45^0$, $A > 3$ and $M < 1.8$, respectively. Finally, the union of the three regions, shown by the filled area in Fig. 8, gives the solution of phase-margin, gain-margin and H_∞ co-design.

6 H_∞ Design with Fractional $PI^\lambda D^\mu$ Controllers

In this section, the general form of fractional-order $PI^\lambda D^\mu$ controller is used to conduct the H_∞ design in the stabilizing region. First, two fractional orders of the controller is optimized to expect a larger stabilizing region. Then, in the stabilizing region, the curves of H_∞-norm constraint are drawn along the similar line as in the previous two sections. Finally, the influence of the two fractional orders on the step responses, in time domain, is discussed by simulation.

6.1 Stabilizing Region

Consider SISO unity feedback system as shown in Fig. 1, where $G(s)$ represents the plant given in (20) and $C(s)$ is the $PI^\lambda D^\mu$ controller of the following form

$$C(s) = k_P + \frac{k_i}{s^\lambda} + k_d s^\mu \tag{67}$$

with k_p, k_i and k_d being the proportional-gain, integral-gain and derivative-gain, respectively, of the controller, λ and μ, $0 < \lambda, \mu < 2$, being the integral-order and the derivative-order, respectively, of the controller.

The objective of this subsection is to determine the stabilizing regions in (k_d, k_i)-plane for fixed k_p, λ and μ values, using the parameter plane approach stated in Sect. 3.2 To this end, the closed-loop characteristic quasi-polynomial in Fig. 1 is first computed as

$$\Delta(s) = s^\lambda D(s) + (k_p s^\lambda + k_i + k_d s^{\lambda+\mu})N(s)e^{-Ls} \tag{68}$$

Since the delay term e^{-Ls} has no finite roots in s-plane, the stability property of $\Delta(s)$ is equivalent to that of $\Delta^*(s)$, where

$$\Delta^*(s) = s^\lambda D(s)e^{Ls} + (k_p s^\lambda + k_i + k_d s^{\lambda+\mu})N(s) \tag{69}$$

Next, substituting $s = j\omega$ into (69), one has

$$\Delta^*(j\omega) = (j\omega)^\lambda D(j\omega)e^{jL\omega} + [k_p(j\omega)^\lambda + k_i + k_d(j\omega)^{\lambda+\mu}]N(j\omega) \tag{70}$$

Using the relation (25) and the followings

$$D(j\omega) = D_r(\omega) + jD_i(\omega) \tag{71}$$

$$N(j\omega) = N_r(\omega) + jN_i(\omega) \tag{72}$$

and partitioning $\Delta^*(j\omega)$ into its real and imaginary components yields

$$\Delta^*(j\omega) = \Delta_r(\omega) + j\Delta_i(\omega)$$

where

$$\begin{aligned}\Delta_r(\omega) &= \omega^\lambda |D(j\omega)| \cos[L\omega + \alpha(\omega) - \beta(\omega) + \frac{\lambda\pi}{2}] \\ &\quad + |N(j\omega)|(k_p\omega^\lambda \cos\frac{\lambda\pi}{2} + k_i + k_d\omega^{\lambda+\mu} \cos\frac{\lambda+\mu}{2}\pi)\end{aligned} \tag{73}$$

$$\begin{aligned}\Delta_i(\omega) &= \omega^\lambda |D(j\omega)| \sin[L\omega + \alpha(\omega) - \beta(\omega) + \frac{\lambda\pi}{2}] \\ &\quad + \omega^\lambda |N(j\omega)|(k_p \sin\frac{\lambda\pi}{2} + k_d\omega^{\mu} \sin\frac{\lambda+\mu}{2}\pi)\end{aligned} \tag{74}$$

with $|D(j\omega)|$ representing the modular of the complex function $D(j\omega)$ in (71), $|N(j\omega)|$ the modular of $N(j\omega)$ in (72), and $\alpha(\omega)$ denoting the phase of $D(j\omega)$, $\beta(\omega)$ the phase of $N(j\omega)$, respectively.

Along the similar line as the discussion in Sect. 4.1, only the frequency interval $\omega \in [0, \infty)$ is considered. For $\omega = 0$, letting $\Delta^*(j\omega) = 0$, one arrives at

$$k_i = 0, \ \textit{if} \ N(j\omega) \neq 0 \tag{75}$$

Condition (75) defines a straight-line in (k_d, k_i)-plane, which gives a piece of stabilizing boundary line (RRB). For $\omega \in (0, \infty)$, considering the following equations

$$\begin{cases} \Delta_r(\omega) = 0 \\ \Delta_i(\omega) = 0 \end{cases} \tag{76}$$

one solves this equations for k_d and k_i in terms of k_p, λ and μ. Then, according to the Implicit Function Theorem, if the Jacobian

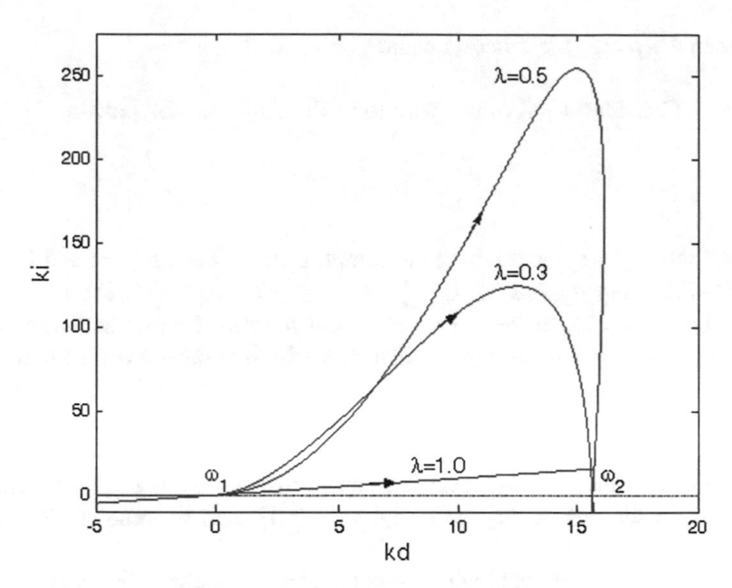

Fig. 9 Stabilizing regions for $\mu = 1$ and different λ

$$J = \begin{vmatrix} \frac{\partial \Delta_r}{\partial k_d} & \frac{\partial \Delta_r}{\partial k_i} \\ \frac{\partial \Delta_i}{\partial k_d} & \frac{\partial \Delta_i}{\partial k_i} \end{vmatrix} = -\omega^{\lambda+\mu}|N(j\omega)| \sin\frac{\lambda+\mu}{2}\pi \tag{77}$$

is not equal to zero, the Eq. (76) has a unique local solution curve $(k_d(\omega), k_i(\omega))$ given by

$$k_d(\omega) = -\frac{|D(j\omega)| \sin[L\omega + \alpha(\omega) - \beta(\omega) + \frac{\lambda\pi}{2}] + k_p|N(j\omega)| \sin\frac{\lambda\pi}{2}}{\omega^{\mu}|N(j\omega)| \sin\frac{\lambda+\mu}{2}\pi} \tag{78}$$

$$k_i(\omega) = \frac{|D(j\omega)| \sin[L\omega + \alpha(\omega) - \beta(\omega) - \frac{\mu\pi}{2}] - k_p|N(j\omega)| \sin\frac{\mu\pi}{2}}{|N(j\omega)| \sin\frac{\lambda+\mu}{2}\pi} \tag{79}$$

The pairs $(k_d(\omega), k_i(\omega)), \omega > 0$, plot another piece of stabilizing boundary curve (CRB) in (k_d, k_i)-plane. This curve separates the (k_d, k_i)-plane into stable parameter region and unstable parameter region. Using the Proposition 1, where J is the Jacobian defined in (77). one can identify to which side of the curve is the stable parameter region.

In our case, from (77), the followings are true.

(1) When $\lambda + \mu < 2$, to the right of the curve $(k_d(\omega), k_i(\omega))$ is the stabilizing region.
(2) When $\lambda + \mu > 2$, to the left is the stabilizing region.

(3) When $\lambda+\mu=2$, the criterion cannot be applied.

Example 5 Consider a DC motor with the following transfer function

$$G(s)=\frac{k}{s^2}e^{-0.1s} \tag{80}$$

which has two poles at the origin of the complex plane. Taking $k=1$ and $k_p=1$, one plots the stabilizing regions in (k_d, k_i)-plane for $\mu=1$ and different λ as shown in Fig. 9. The arrows along the curves represent the direction of increasing ω. The stabilizing regions are to the right of each curve. In the figure, $\lambda=1$ is approximated by taking $\lambda=0.9999$. It is observed that the stabilizing regions are of sector forms, and for $\lambda=\mu=1$ (integer-order *PID*), the stabilizing region corresponds to a triangle. When μ takes other values, the change of the stabilizing regions with λ is similar to that in Fig. 9. It is worthy of noting that when $\lambda+\mu>2$, the stabilizing regions are to the left of the curve $(k_d(\omega), k_i(\omega))$, and become smaller as $\lambda+\mu$ increases.

From Fig. 9, bigger stabilizing regions can be realized by setting smaller fractional-order λ. This means that a better system performance can be achieved by using fractional-order *PID* instead of integer-order one.

6.2 Fractional Order Optimization

In control system synthesis, bigger stabilizing regions are expected to provide wider room for system performances. From Example 5 in the previous subsection, it is shown that the shapes and areas of the stabilizing regions change with the fractional orders λ and μ. In this subsection, optimal λ and μ are computed in the sense of achieving bigger stabilizing regions.

It is observed, from Fig. 9, that when $k_i(\omega)=0$, the stabilizing boundary curves intersect with k_d-axis, and the corresponding intersection frequencies ω_1 and ω_2 are the first and second solutions of the following equation

$$|D(j\omega)|\sin[L\omega+\alpha(\omega)-\beta(\omega)-\frac{\mu\pi}{2}]-k_p|N(j\omega)|\sin\frac{\mu\pi}{2}=0 \tag{81}$$

which is obtained by letting $k_i(\omega)=0$ in (79). For instance, in Fig. 9, the two intersection frequencies ω_1 and ω_2, the first and second solutions of the Eq. (81), read $\omega_1=1.0025$ and $\omega_2=15.6672$, respectively. It is clear, from (81), that the solutions ω_1 and ω_2 are independent of λ for fixed μ and k_p. This fact leads one to utilize the Leibniz Sector Formula [12] to calculate the areas of the stabilizing regions

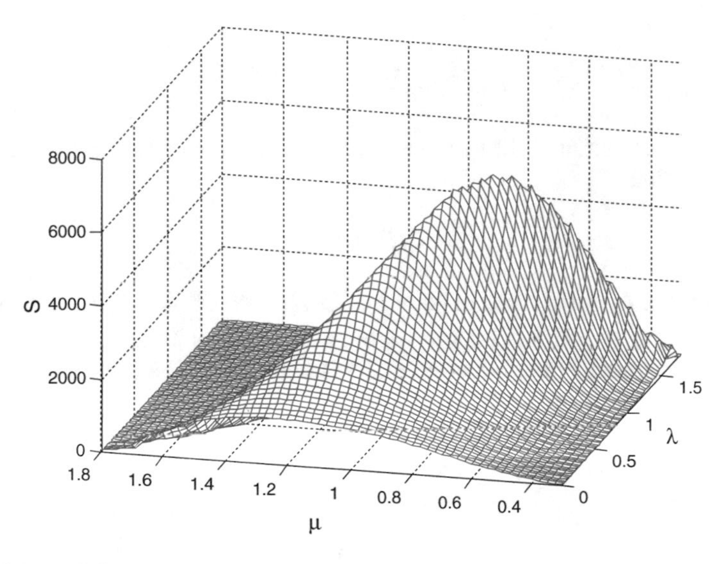

Fig. 10 3-D mesh lines

$$S(\lambda)=\frac{1}{2}\int_{\omega_1}^{\omega_2}[k_d(\omega)\dot{k}_i(\omega)-\dot{k}_d(\omega)k_i(\omega)]d\omega \tag{82}$$

where $\dot{x}(\omega)$ denotes the first-derivative of x with respect to ω. From (82), the relation curve between $S(\lambda)$ and λ can be drawn for fixed μ and k_p, and by gridding μ, 3-dimensinal mesh lines of $S(\lambda)$ with respect to λ and μ can be plotted as shown in the following example.

Example 6 Example 5 revisited. For the same k and k_p values as in Example 5, by using Formula (82), the 3-D mesh lines of $S(\lambda)$ with respect to λ and μ are plotted in Fig. 10, from which the optimal values of λ and μ can be selected corresponding to bigger stabilizing regions. Hence, the 3-D mesh lines provide a guard line for the selection of the fractional orders in the tuning of $PI^\lambda D^\mu$ parameters.

6.3 H_∞ Design of Sensitivity

In Fig. 1, one considers the H_∞-norm constraint of sensitivity function

$$\|W_1(s)S(s)\|_\infty < M \tag{83}$$

For simplicity, the weighting function is taken to be $W_1(s)=1$. Then according to the definition of H_∞-norm in (16), the sensitivity function is bounded for all frequencies

$$\left|\frac{1}{1+C(j\omega)G(j\omega)}\right| \leq M, \ \forall \omega \geq 0 \tag{84}$$

From this condition, when $M \to \infty$, Eq. (84) becomes the closed-loop characteristic equation

$$1 + C(j\omega)G(j\omega) = 0$$

If controller $C(j\omega)$ is taken to be of the form of $PI^{\lambda}D^{\mu}$ as in (67), then, the same equation $\Delta^{*}(j\omega) = 0$ as in (70) can be obtained, i.e., as M approaches to ∞, the stabilizing boundary curve is recovered. This point can be verified in the following example.

In order to determine the (k_d, k_i) values in the stabilizing region for which the sensitivity constraint (84) is satisfied, one performs the following controller transformation

$$\begin{aligned} C(s) &= k_p + k_i \frac{1}{s^{\lambda}} + k_d s^{\mu} \\ &= k_p + k_d \left(s^{\mu} + \frac{k_i/k_d}{s^{\lambda}}\right) \\ &= k_p + x\left(s^{\mu} + y\frac{1}{s^{\lambda}}\right) \end{aligned} \tag{85}$$

where $x = k_d$, $y = k_i/k_d$. Under this transformation, the open-loop transfer function is written as

$$L(s) = C(s)G(s) = k_p G(s) + x[G_1(s) + yG_2(s)] \tag{86}$$

where $G_1(s) = s^{\mu}G(s)$, $G_2(s) = G(s)/s^{\lambda}$. Substituting $s = j\omega$ into $L(s)$ and partitioning $G(j\omega)$, $G_1(j\omega)$ and $G_2(j\omega)$ into their real and imaginary parts

$$\begin{aligned} G(j\omega) &= A(\omega) + jB(\omega) \\ G_1(j\omega) &= A_1(\omega) + jB_1(\omega) \\ G_2(j\omega) &= A_2(\omega) + jB_2(\omega) \end{aligned}$$

one gets

$$L(j\omega) = k_p(A + jB) + x[A_1 + jB_1 + y(A_2 + jB_2)] \tag{87}$$

where the frequency dependency of A, B, A_1, B_1, A_2 and B_2 has been omitted. From (84), it follows that

$$|1+L(j\omega)|^2 \geq \frac{1}{M^2}$$

Define

$$f_1(\omega) = |1+L(j\omega)|^2 - M_1$$

where $M_1 = 1/M^2$, and from (87), one has

$$f_1(\omega) = C_1 + 2xD_1 + x^2E_1 \geq 0 \tag{88}$$

where

$$\begin{aligned}
C_1 &= 1 - M_1 + 2k_P A + k_p^2(A^2 + B^2) \\
D_1 &= A_1 + yA_2 + k_P[A(A_1 + yA_2) + B(B_1 + yB_2)] \\
E_1 &= (A_1 + yA_2)^2 + (B_1 + yB_2)^2
\end{aligned}$$

Inequality (88) defines an optimization problem, and when the equality is solved for some frequency ω, the minimum value of $f_1(\omega)$ is reached. The design objective is to find the pair $(x,\ y)$, accordingly the pair (k_d, k_i), for fixed parameters λ, μ and k_p, such that the equality in (88) holds for some ω. To this end, one differentiates $f_1(\omega)$ with respect to ω and let the corresponding derivative at that frequency to be zero

$$f_2(\omega) = \frac{df_1(\omega)}{d\omega} = C_2 + xD_2 + x^2E_2 = 0 \tag{89}$$

where

$$\begin{aligned}
C_2 &= k_P\dot{A} + k_p^2(A\dot{A} + B\dot{B}) \\
D_2 &= \dot{A}_1 + y\dot{A}_2 + k_P[\dot{A}(A_1 + yA_2) + A(\dot{A}_1 + y\dot{A}_2) + \dot{B}(B_1 + yB_2) + B(\dot{B}_1 + y\dot{B}_2)] \\
E_2 &= (A_1 + yA_2)(\dot{A}_1 + y\dot{A}_2) + (B_1 + yB_2)(\dot{B}_1 + y\dot{B}_2)
\end{aligned}$$

Eliminating x^2 in equations

$$\begin{cases} f_1(\omega) = 0 \\ f_2(\omega) = 0 \end{cases}$$

one solves

$$x = \frac{C_2E_1 - C_1E_2}{2D_1E_2 - D_2E_1} = \frac{e_{00} + e_{01}y + e_{02}y^2}{e_{10} + e_{11}y + e_{12}y^2 + e_{13}y^3} \tag{90}$$

where

$$
\begin{aligned}
e_{00} &= C_2 a_{10} - C_1 b_{10} \\
e_{01} &= C_2 a_{11} - C_1 b_{11} \\
e_{02} &= C_2 a_{12} - C_1 b_{12} \\
e_{10} &= 2d_{10} b_{10} - d_{20} a_{10} \\
e_{11} &= 2(d_{10} b_{11} + d_{11} b_{10}) - (d_{21} a_{10} + d_{20} a_{11})
\end{aligned}
$$

$$
\begin{aligned}
e_{12} &= 2(d_{11} b_{11} + d_{10} b_{12}) - (d_{20} a_{12} + d_{21} a_{11}) \\
e_{13} &= 2d_{11} b_{12} - d_{21} a_{12} \\
a_{10} &= A_1^2 + B_1^2 \\
a_{11} &= 2(A_1 A_2 + B_1 B_2) \\
a_{12} &= A_2^2 + B_2^2
\end{aligned}
$$

$$
\begin{aligned}
b_{10} &= A_1 \dot{A}_1 + B_1 \dot{B}_1 \\
b_{11} &= A_1 \dot{A}_2 + \dot{A}_1 A_2 + B_1 \dot{B}_2 + \dot{B}_1 B_2 \\
b_{12} &= A_2 \dot{A}_2 + B_2 \dot{B}_2 \\
d_{10} &= A_1 + k_p (AA_1 + BB_1) \\
d_{11} &= A_2 + k_p (AA_2 + BB_2) \\
d_{20} &= \dot{A}_1 + k_p (\dot{A}A_1 + A\dot{A}_1 + \dot{B}B_1 + B\dot{B}_1) \\
d_{21} &= \dot{A}_2 + k_p (\dot{A}A_2 + A\dot{A}_2 + \dot{B}B_2 + B\dot{B}_2)
\end{aligned}
$$

Substituting (90) into $f_1(\omega) = 0$ in (88) yields the following sixth-order equation about y for each ω.

$$
f_0 + f_1 y + f_2 y^2 + f_3 y^3 + f_4 y^4 + f_5 y^5 + f_6 y^6 = 0 \tag{91}
$$

where

$$
\begin{aligned}
f_0 &= C_1 e_{10}^2 + 2d_{10} e_{00} e_{10} + a_{10} e_{00}^2 \\
f_1 &= 2C_1 e_{10} e_{11} + 2(d_{10} e_{00} e_{11} + d_{10} e_{01} e_{10} + d_{11} e_{00} e_{10} + a_{10} e_{00} e_{01}) + a_{11} e_{00}^2 \\
f_2 &= C_1 (e_{11}^2 + 2e_{10} e_{12}) + 2(d_{10} e_{00} e_{12} + d_{10} e_{02} e_{10} + d_{11} e_{01} e_{10} + d_{10} e_{01} e_{11} + d_{11} e_{00} e_{11} \\
&\quad + a_{10} e_{00} e_{02} + a_{11} e_{00} e_{01}) + a_{10} e_{01}^2 + a_{12} e_{00}^2 \\
f_3 &= 2C_1 (e_{10} e_{13} + e_{11} e_{12}) + 2(d_{10} e_{00} e_{13} + d_{11} e_{02} e_{10} + d_{10} e_{01} e_{12} + d_{11} e_{00} e_{12} \\
&\quad + d_{10} e_{02} e_{11} + d_{11} e_{01} e_{11} + a_{10} e_{01} e_{02} + a_{11} e_{00} e_{02} + a_{12} e_{00} e_{01}) + a_{11} e_{01}^2
\end{aligned}
$$

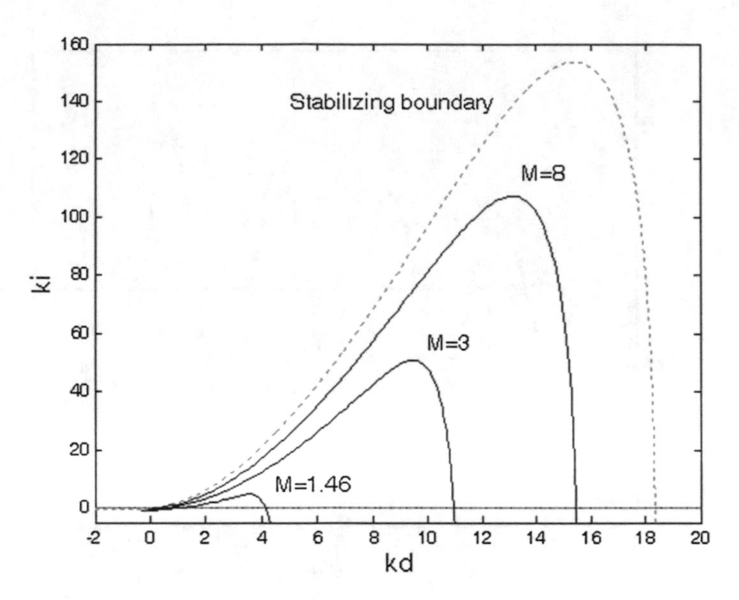

Fig. 11 Sensitivity constraints for different M

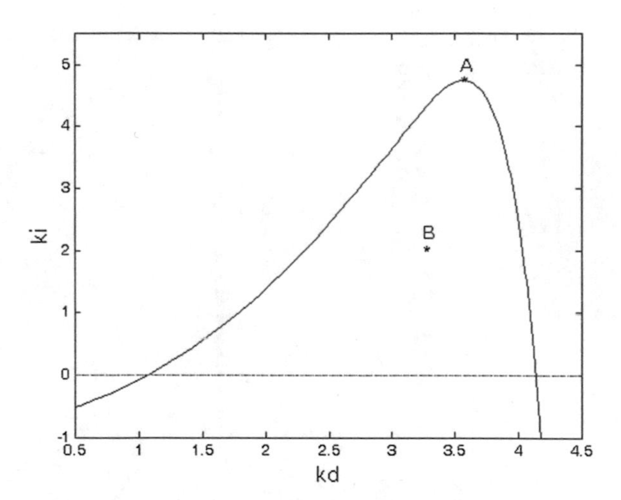

Fig. 12 Sensitivity constraint for $M = 1.46$

$$f_4 = C_1(e_{12}^2 + 2e_{11}e_{13}) + 2(d_{10}e_{01}e_{13} + d_{11}e_{00}e_{13} + d_{11}e_{02}e_{11} + d_{10}e_{02}e_{12} + d_{11}e_{01}e_{12} + a_{11}e_{01}e_{02} + a_{12}e_{00}e_{02}) + a_{10}e_{02}^2 + a_{12}e_{01}^2$$

$$f_5 = 2C_1e_{12}e_{13} + 2(d_{10}e_{02}e_{13} + d_{11}e_{01}e_{13} + d_{11}e_{02}e_{12} + a_{12}e_{01}e_{02}) + a_{11}e_{02}^2$$

$$f_6 = C_1e_{13}^2 + 2d_{11}e_{02}e_{13} + a_{12}e_{02}^2$$

The pair (x, y) can be found in the following way. Solve the sixth-order linear Eq. (91) for y, for an appropriately selected frequency ω, one gets 6 roots, and

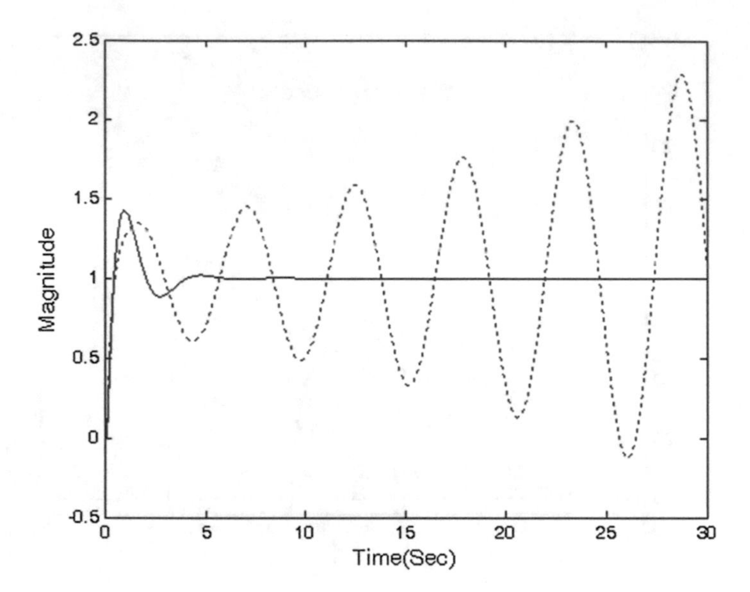

Fig. 13 Step responses corresponding to point A

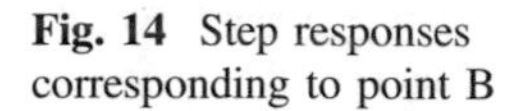

Fig. 14 Step responses corresponding to point B

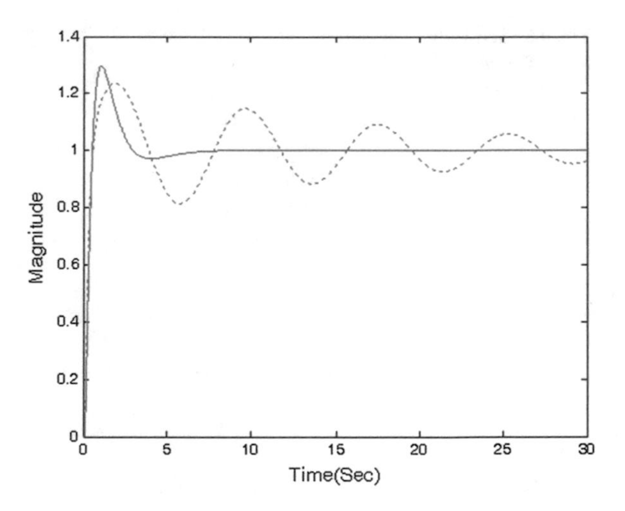

substitute these roots into (90), respectively, one obtains the corresponding 6 parameters x. Then, using the relations $k_d = x$ and $k_i = xy$, one recovers the original controller gains. The solution pair, if any, is the real pair $(k_d,\ k_i)$ which is located in the stabilizing region. For an appropriately chosen frequency interval, the real pairs $(k_d(\omega),\ k_i(\omega))$ draw the sensitivity boundary curve for the given M in the stabilizing region.

Example 7 Examples 5 and 6 revisited. In this example, for fixed fractional orders $\lambda = 0.5$ and $\mu = 0.9$, corresponding to a relatively large stabilizing region, see

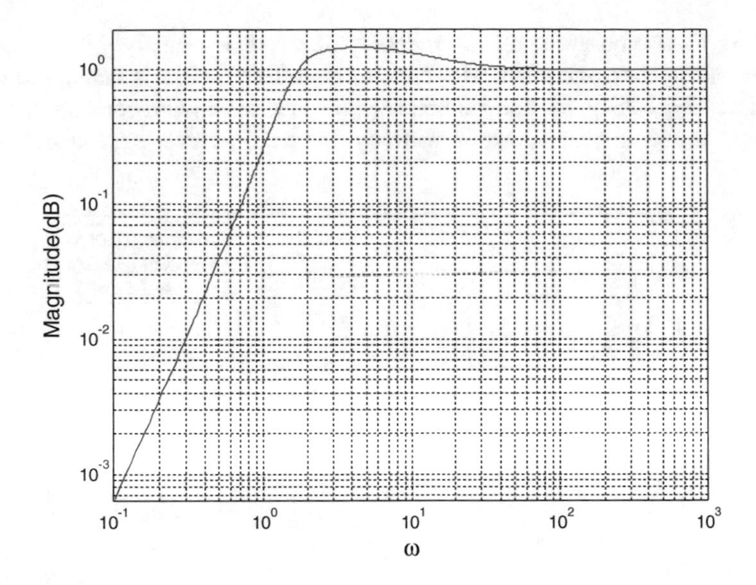

Fig. 15 Sensitivity curve corresponding to point A

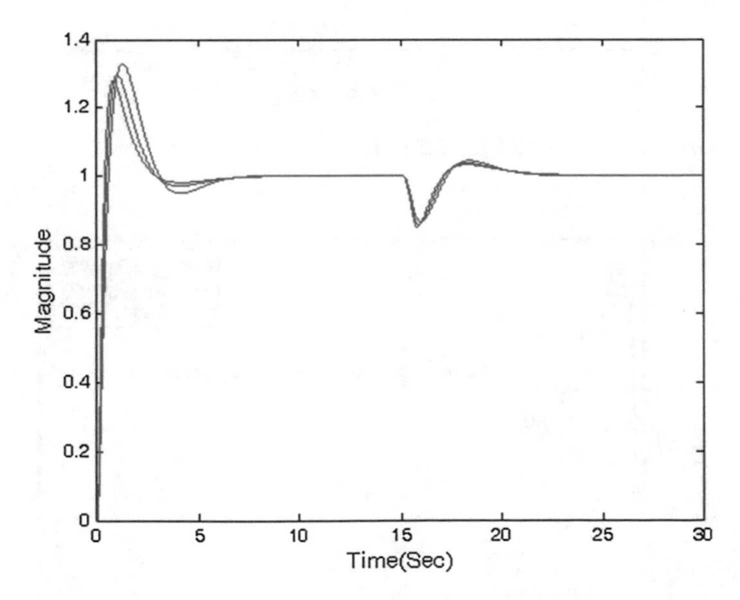

Fig. 16 Robustness to the change of open-loop gain

Fig. 10, one wants to find the pairs $(k_d(\omega),\ k_i(\omega))$ in the stabilizing region such that the sensitivity meets $M \leq 1.46$.

For different sensitivity bound M, the sensitivity boundary curves in the stabilizing region are depicted in Fig. 11. Figure 12 is the zoomed-in version of Fig. 11

for $M = 1.46$. It is observed that as M decreases, the region satisfying (84) becomes smaller and smaller, and when $M = 1.26$, the sensitivity curve is tangent to the k_d-axis, i.e., $M = 1.26$ gives the minimal achievable sensitivity constraint. On the contrary, as M increases, the sensitivity boundary curve approaches to the

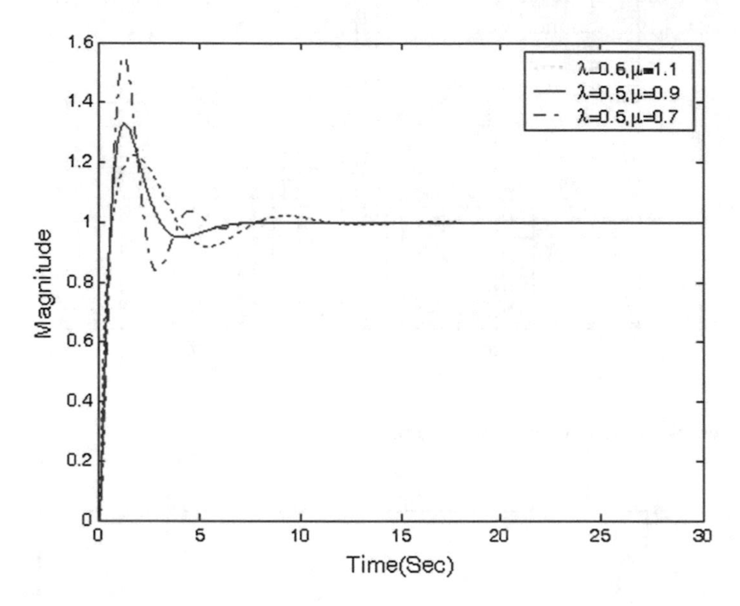

Fig. 17 Step responses for $\lambda = 0.5$ and different μ

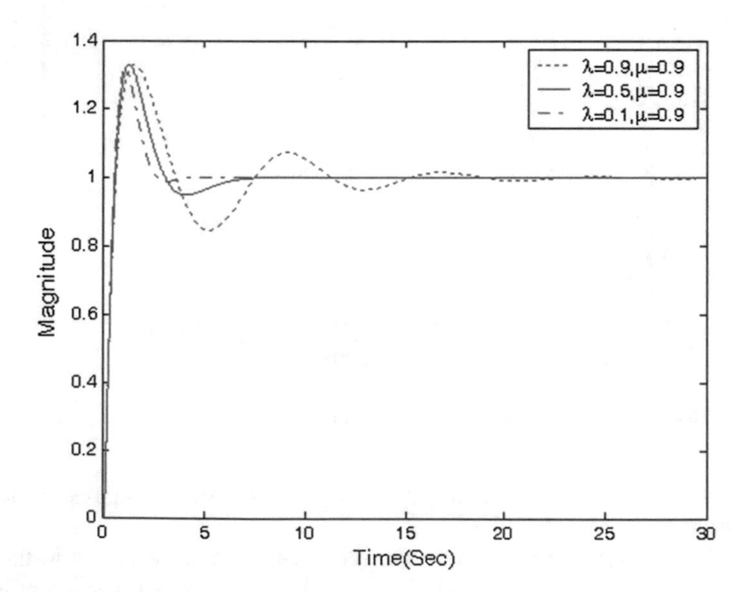

Fig. 18 Step responses for $\mu = 0.9$ and different λ

stabilizing boundary, and for sufficiently large M, the two curves coincide with each other, i.e., the stabilizing boundary curve is recovered.

Next, one considers the dynamic behaviors in time domain and the robustness to the plant uncertainties and compares these properties with integer-order *PID* controller. First, the step responses via Matlab simulation are studied. Select a point A on the sensitivity boundary curve for $M = 1.46$ (corresponding to a phase margin 40° approximately [31]), see Fig. 12, one reads the gain parameters of the controller as $k_d = 3.57$ and $k_i = 4.75$. The step responses of $PI^\lambda D^\mu$ controller $(\lambda = 0.5, \mu = 0.9)$ and the integer-order *PID* controller $(\lambda = \mu = 1)$ are plotted in Fig. 13. It is seen that the integer *PID* (dotted line) gives an unstable response, because the gain parameter pair $(k_d,\ k_i) = (3.57,\ 4.75)$ is outside the stabilizing region of integer *PID*. Choose another point B inside the constraint curve for $M = 1.46$, see Fig. 12, corresponding to $k_d = 3.25$ and $k_i = 2.00$. The step responses of the two controllers are plotted in Fig. 14. The overshoot of the response of the $PI^\lambda D^\mu$ controller at point B is less than that at Point A shown in Fig. 13, since point B is inside the constraint for $M = 1.46$ and gives a phase margin greater than 40°. In this case, the response of integer *PID* (dotted line) is stable, but has severe oscillatory behavior. Second, one investigates the robustness of the design. For the point A along the curve of $M = 1.46$, the sensitivity curve given by the left-hand side of (84) is drawn, see Fig. 15. It is observed that sensitivity meets the constraint $M \leq 1.46\ (7.57\,dB)$ for all frequencies. To verify the robustness to the change of the plant gain, one adds a load impulse to the plant at 15 s, see Fig. 16, and observes the responses for $k = 1.2$ and $k = 0.8$ ($\pm 20\%$ variations around the nominal value of $k = 1.0$). Clearly, the robustness to the change of the gain within its interval is satisfactory. To compare the robustness with integer *PID* controller, one considers point B in Fig. 12, corresponding to stable responses for both fractional and integer *PID* controllers, see Fig. 14. Similar to Fig. 15, the sensitivity curves can be drawn, and the maximum achievable sensitivities read $M = 1.37$ at $\omega = 5.65\,rad/s$ for $\lambda = 0.5$ and $\mu = 0.9$, and $M = 1.32$ at $\omega = 9.50\,rad/s$ for $\lambda = \mu = 1$. Clearly, the robustness measure achieved by fractional *PID* controller is better than that by integer *PID* controller.

Finally, the influence of varying λ and μ on the step responses is simulated. Considering point B in Fig. 12, for fixed $\lambda = 0.5$, as μ increases, one observes that the overshoot of the step response decreases, see Fig. 17, showing that the damping of the system is increased and the stability is improved with stronger derivative action. For fixed $\mu = 0.9$, as λ increases, the integral action becomes stronger, and more oscillatory behavior of the response happens, see Fig. 18. Similarly, the influence of varying the gains k_d and k_i around point B and the gain k_p on the response can be discussed, and the results are similar to that of integer *PID*. Details can be found in [2, 27].

7 Conclusions

In this chapter, two problems have been discussed. One is the determination of stabilizing parameters set of fractional-order $PI^{\lambda}D^{\mu}$ type controllers, based on a graphical stability criterion fit for fractional-order systems with time-delay, exhibiting simple and direct characteristics in identifying the stabilizing regions for fractional-order systems. The other is the computation of both of the classical phase and gain margins and the modern H_{∞} constraints in the stabilizing region. In the case of two margins, the GPMT technique has been used, and for the computation of H_{∞} boundary lines, an algebraic approach to the design of $PI^{\lambda}D^{\mu}$ type controllers has been developed. Further design along this line is to consider other system performances in the stabilizing regions of the fractional order controllers, such as the H_2-norm calculation.

References

1. Akbari, M. K., & Haeri, M. (2010). Robustness in fractional proportional-integral-derivative-based closed-loop systems. *IET Control Theory and Applications, 4*(10), 1933–1944.
2. Astrom, K. J., & Hagglund, T. (1995). *PID* controllers: Theory, design and tuning. Instrument Society of American, Research Triangle Park, NC.
3. Blanchini, F., Lepschy, A., Miani, S., & Viaro, U. (2004). Characterization of *PID* and lead/lag compensators satisfying given H_{∞} specifications. *IEEE Transactions on Automatic Control, 49*(5), 736–740.
4. Boulkroune, A., Bouzeriba, A., Bouden, T., & Azar, A. T. (2016). Fuzzy adaptive synchronization of uncertain fractional-order chaotic systems. In *Advances in Chaos theory and intelligent control*. Studies in fuzziness and soft computing (Vol. 337). Germany: Springer.
5. Das, S., Saha, S., Das, S., & Gupta, A. (2011). On the selection of tuning methodology of *FOPID* controllers for the control of higher order processes. *ISA Transactions, 50*, 376–388.
6. Diekmann, O., van Gils, S. A., verduyn Lunel, S. M., & Walther, H.-O. (1995). Delay equations: Functional-, complex- and nonlinear analysis. In *Applied mathematical sciences*. Springer.
7. Doyle, J., Francis, B., & Tennenbaum, A. (1990). *Feedback control theory*. Macmilan Publishing Co.
8. Hamamci, S. E. (2007). An algorithm for stabilization of fractional-order time-delay systems using fractional-order *PID* controllers. *IEEE Transactions on Automatic Control, 52*(10), 1964–1969.
9. Hamamci, S. E. (2008). Stabilization using fractional-order *PI* and *PID* controllers. *Nonlinear Dynamics, 51*, 329–343.
10. Hamamci, S. E., & Kolsal, M. (2010). Calculation of all stabilizing fractional-order *PD* controllers for integrating time delay systems. *Computers & Mathematics with Applications, 59*(5), 1630–1636.
11. Hildebrandt, S. (2003). *Analysis 2*. Springer.

12. Ho, M. T. (2003). Synthesis of H_∞ *PID* controllers: A parametric approach. *Automatica, 39*, 1069–1075.
13. Ho, M. T., & Lin, C. Y. (2003). *PID* controller design for robust performance. *IEEE Transactions on Automatic Control, 48*(8), 1404–1409.
14. Lou, Y., & Chen, Y. Q. (2009). Fractional-order (proportional derivative) controller for a class of fractional order systems. *Automatica, 45*, 2446–2450.
15. Lou, Y., Chen, Y. Q., Wang, C. Y., & Pi, Y. G. (2010). Tuning fractional order proportional integral controllers for fractional order systems. *Journal of Process Control, 20*, 823–832.
16. Lure, J. (1994). Three parameter tunable tilt-integral-derivative (TID) controller. US Patent US53711670.
17. Monje, C. A., Calderon, A. J., Vigagre, B. M., & Feliu, V. (2004). The fractional-order lead compensator. In *Second IEEE Conference on Computational cybernetics*, Vienna (pp. 347–352).
18. Monje, C. A., Calderon, A. J., Vigagre, B. M., Chen, Y., & Feliu, V. (2004). On fractional PI^λ controllers; Some tuning rules for robustness to plant uncertainties. *Nonlinear Dynamics, 38*, 369–381.
19. Oustaloup, A., Mathieu, B., & Lanusse, P. (1995). The CRONE control of resonant plants: Application to a flexible transmission. *European Journal of Control, 1*(2), 113–121.
20. Oustaloup, A., Moreau, X., & Nouillant, M. (1996). The CRONE suspension. *Control Enginering Practice, 4*(8), 1101–1108.
21. Padula, F., & Visioli, A. (2012). Optimal tuning rules for proportional-integral-derivative and fractional-order proportional-integral-derivative controllers for integrating and unstable processes. *IET Control Theory and Applications, 6*(6), 776–786.
22. Podlubny, L. (1999). *Fractional differential equations*. London, UK: Academic Press.
23. Podlubny, L. (1999). Fractional-order systems and $PI^\lambda D^\mu$ controllers. *IEEE Transactions on Automatic Control, 44*(1), 208–214.
24. Raynaud, H.-P., & Zergalnoh, A. (2000). State-space representation for fractional order controllers. *Automatica, 36*, 1017–1021.
25. Tantaris, R. N., Keel, L. H., & Bhattacharyya, S. P. (2006). H_∞ design with first-order controllers. *IEEE Transactions on Automatic Control, 51*(8), 1343–1347.
26. Wang. D.-J. (2012). H_∞ design with fractional-order PD^μ controller. *Automatica, 48*(5), 974–977.
27. Wang, D.-J. (2012). A *PID* controller set of guaranteeing stability and gain and phase margins for time-delay systems. *Journal of Process Control, 22*(7), 1298–1306.
28. Wang, D.-J., & Gao, X.-L. (2013). Stability margins and H_∞ co-design with fractional-order PI^λ controllers. *Asian Journal of Control, 15*(3), 691–697.
29. Wang, D.-L., Li, W., & Guo, M.-L. (2013). Tuning of $PI^\lambda D^\mu$ controllers based on sensitivity constraint. *Journal of Process Control, 23*(6), 861–867.
30. Yaniv, O., & Nagurka, M. (2004). Design of *PID* controllers satisfying gain margin and sensitivity constraint on a set of plants. *Automatica, 40*, 111–116.
31. Yeroglu, C., & Tan, N. (2011). Note on fractional-order proportional-integral-differential controller design. *IET Control Theory and Applications, 5*(17), 1978–1989.

On the Electronic Realizations of Fractional-Order Phase-Lead-Lag Compensators with OpAmps and FPAAs

Carlos Muñiz-Montero, Luis A. Sánchez-Gaspariano, Carlos Sánchez-López, Víctor R. González-Díaz and Esteban Tlelo-Cuautle

Abstract It is well known that the fractional-order phase-lead-lag compensators can achieve control objectives that are not always possible by using their integer-order counterparts. However, up to now one can find only a few of publications discussing the strategies for parameters' tuning of these compensators, with only simulation results reported. This is due in part to the implicit difficulties on the implementation of circuit elements with frequency responses of the form $s^{\pm\lambda}$ that are named "fractances". In this regard, there exist approximations with rational functions, but the drawback is the difficulty to approximate the required values with the ones of the commercially-available resistances and capacitors. Consequently, fractional compensators have not been appreciated by the industry as it is in the academia. Therefore, motivated by the lack of reported implementations, this chapter is structured as a tutorial that deals with the key factors to perform, with the frequency-domain approach, the design, simulation and implementation of integer-order and fractional-order phase-lead-lag compensators. The circuit implementations are performed with

C. Muñiz-Montero
Electronics and Telecommunications Department,
Universidad Politécnica de Puebla, 72640 Cuanalá, Puebla, Mexico
e-mail: carlos.muniz@uppuebla.edu.mx

L.A. Sánchez-Gaspariano · V.R. González-Díaz
Faculty of Electronics, Benemérita Universidad Autónoma de Puebla,
Av. San Claudio y 18 Sur Jardines de San Manuel, 72592 Puebla, Mexico
e-mail: luis.sanchez@uppuebla.edu.mx

V.R. González-Díaz
e-mail: vicrodolfo.gonzalez@correo.buap.mx

C. Sánchez-López
Department of Electronics, Universidad Autónoma de Tlaxcala,
90300 Apizaco, Tlaxcala, Mexico
e-mail: carlsan@ieee.org

E. Tlelo-Cuautle (✉)
Department of Computer Science, CINVESTAV,
Av. Instituto Politécnico Nacional No. 2508 Col. San Pedro Zacatenco,
07360 Mexico City, Mexico
e-mail: etlelo@cs.cinvestav.mx

A.T. Azar et al. (eds.), *Fractional Order Control and Synchronization of Chaotic Systems*, Studies in Computational Intelligence 688,
DOI 10.1007/978-3-319-50249-6_5

Operational Amplifiers (OpAmps) and with Field Programmable Analog Arrays (FPAA). Emphasis is focused in the obtaining of commercially-available values of resistances and capacitors. Therefore, the design procedure starts with the use of equations that provide the exact and unique solution for each parameter of the compensator, avoiding conventional trial-and-error procedures. Then, five OpAmp-based configurations for integer-order and fractional-order realizations are described in terms of basic analog building blocks, such as integrators or differential amplifiers, among others. The corresponding design equations are also provided. Then, six examples are presented for both, OpAmp-based and FPAA-based implementations with the simulation and experimental results discussed regarding other results reported in the literature.

Keywords Fractional calculus · Fractional-order lead/lag compensators · Field programmable analog array

1 Introduction

Proportional-Integral-Derivative (*PID*) controllers [4–6], and lead/lag compensators are the control strategies most used in today's industry. The phase-lag compensator reduces the static error by increasing the low-frequency gain without any resulting instability, and increases the phase margin of the system to yield the desired overshoot [30]. Meanwhile, the phase-lead compensator change the phase diagram to reduce the percent overshoot and to reduce the peak time [30]. The design of lead/lag compensators may require four-step and twelve-step trial-and-error approaches, respectively [30]. Typically, during the design stage the plant is modeled by its transfer function with integer orders q on the Laplace frequency s^q. However, experimental evidences show that physical systems can be modeled with higher accuracy using fractional-order transfer functions [16, 20, 46]. On this direction, it is known that the fractional-order PID controllers and phase-lead-lag compensators have better performance than their integer counterparts [16, 22, 27, 28, 35]. That is due to the addition of degrees of freedom, which can be used to incorporate additional control objectives. For instance, in the case of lead/lag compensators it can be established a constrain in the initial value of the error signal (actuator's constraint) [35].

Although fractional calculus has been studied from Leibniz in 1965, its practical use has been restricted. It was not until the development of new computing environments and numerical calculus (e.g. MATLAB) when researchers introduced this theory to the modeling and control of systems [16, 24, 36, 38]. In fact, those computing environments allowed other complex control strategies such as the reported in [7–10, 12, 13, 47]. That way, in 1999 Podlubny proposed the first fractional PID controller [32]. Up to now one can find several realizations for this kind of controller [22, 24, 27, 28, 32, 38, 41, 46]. In addition, researchers have developed the corresponding rules for parameters' tuning, some of them considering Ziegler-Nichols rules [17, 25, 40], optimization methods [29, 36], techniques in the frequency domain [18, 25, 39], which offer robustness to the controller facing parametric uncertainties of the process and presence of external perturbations [2]; or also in techniques for intelligent computing, such as: neural networks [31], genetic algorithms [14], or fuzzy logic [43, 45]. In general, these techniques can be classified as analytical, numeric or rules-based ones. A summary of them is given in [29, 41, 42]. Unfortunately, the fractional-order lead/lag compensators have not been reported as abundantly as the fractional PID controllers. In [29, 34] there have been studied the following fractional-order lead/lag compensator (and rational-order approximations for such a compensator)

$$C(s) = K\left(\frac{1+\alpha\tau s}{1+\tau s}\right)^{\lambda}, \qquad \lambda \in (0,\infty) \tag{1}$$

In [35] it was reported a method for the unique solution of the parameters α, τ and q of the compensator

$$C(s) = K\left(\frac{1+\alpha\tau s^{q}}{1+\tau s^{q}}\right), \qquad q \in (0,2) \tag{2}$$

Unfortunately, to the best of the authors' knowledge, analog implementations of this compensator have not been reported. As in the case of PID controllers, it is due to the difficulties to accomplish the design of circuit elements with frequency responses of the form $s^{-\lambda}$ or s^{μ} that are named "fractances". The fractances are circuit elements with constant phase response at all frequencies [23]. For instance, very few physical realizations have been reported related to "fractal capacitances" [11, 21]. Unfortunately, those elements are bulky, require chemical compounds with difficult manipulation and the order λ cannot be modified easily. As alternatives, there exist approximations with rational functions in s for the operators $s^{-\lambda}$ or s^{μ}, that are obtained from Carlson methods, Oustaloup, continuous fractions expansion (CFE) [16, 33], among others. The resulting functions are implemented with arrays of resistances, capacitors and inductors in ladder networks [33]. The drawback of these realizations is the difficulty to approximate the required values with the ones of the commercially-available resistances and capacitors [19], in addition they can require negative impedance converters [3, 33], or inductors [15].

From the difficulties on the implementations mentioned above and motivated by the lack of reported implementations of (2), this chapter is structured as a tutorial that deals with the key factors to perform the design, simulation and implementation of integer and fractional-order phase-lead-lag compensators. It is proposed the use of first-order analog approximations for fractional derivatives and integrals, with the main advantage of using integrators of integer order, differential amplifiers, two-inputs adder amplifiers, and conventional lead-lag networks, all of them realized with OpAmps. Most important is that the resulting circuits can be implemented with commercially-available resistances and capacitors, avoiding the use of negative impedance converters or inductors. Each design is realized obtaining the parameters of (2) with the procedure reported in [35]. Five configurations for integer and fractional-order compensators are verified experimentally from realizations using OpAmps uA741 and using an Application Specific Integrated Circuit (ASIC) that is known as Field-programmable Analog Array (FPAA) AN231E04 from Anadigm [1]. Six design examples of both integer and fractional-order phase-lead-lag compensators are presented.

2 Theoretical Background

This section describes the calculus of derivatives and integrals of fractional order, the corresponding Laplace transforms, and the fractional order transfer function. From these concepts fractional order phase-lead-lag compensators are described in Sect. 2.3.

2.1 *Derivative and Integral of Fractional Order*

The Riemann-Lieuville definition for calculation of fractional derivatives and integrals establishes [20]

$$\mathscr{D}_t^{\alpha} f(t) = \frac{1}{\Gamma(m-\alpha)} \left(\frac{d}{dt}\right)^m \int_0^t \frac{f(\tau)}{(t-\tau)^{\alpha-m+1} d\tau} \tag{3}$$

where $\alpha \in \mathbb{R}, m-1 < \alpha < m, m \in \mathbb{N}$ and $\Gamma(\cdot)$ is Gamma function. For $\alpha > 0$, $\alpha < 0$ and $\alpha = 0$ one gets the fractional derivative, integral and identity function.

2.2 *Laplace Fractional Operator and Fractional-Order Transfer Function*

Laplace Transform with initial conditions equal to zero of (3) is given by [20]

$$\mathscr{L}\left\{\mathscr{D}_t^{\alpha} f(t)\right\} = s^{\alpha} F(s) \tag{4}$$

where $F(s)$ denotes Laplace transform of $f(t)$, and s^{α} is the Laplace operator of fractional order expressed as

$$s^{\alpha} = (j\omega)^{\alpha} = \omega^{\alpha}\left[\cos\left(\frac{\alpha\pi}{2}\right) + j\sin\left(\frac{\alpha\pi}{2}\right)\right] \tag{5}$$

Since Laplace transform is a lineal operator, (4) can be applied to a differential equation of fractional order with coefficients $a_k, b_k \in \mathbb{R}$ and input and output signals $u(t)$ and $e(t)$ to obtain the transfer function [2]

$$H(s) = \frac{U(s)}{E(s)} = \frac{b_m s^{\beta_m} + b_{m-1} s^{\beta_{m-1}} + \cdots + \beta_0 s^{\beta_0}}{a_n s^{\alpha_n} + a_{n-1} s^{\alpha_{n-1}} + \cdots + \alpha_0 s^{\alpha_0}} \quad (6)$$

where $U(s) = \mathscr{L}\{u(t)\}$ and $E(s) = \mathscr{L}\{e(t)\}$.

2.3 *Integer and Fractional-Order Phase-Lead-Lag Compensators*

The phase-lag compensator reduces the static error by increasing the low-frequency gain without any resulting instability. This compensator also increases the phase margin of the system to yield the desired overshoot in the transient response [30]. In most cases reported in the literature, that design process is a four-step trial-and-error approach based on graphic approximation (Bode plots) [30]. On the other hand, the phase-lead compensator is designed, via Bode plots, to change the phase diagram in order to increase the phase margin, reduce the percent overshoot, and increase the bandwidth (by increasing the gain crossover frequency) to obtain a faster transient response with a reduced peak time [30]. Typically, the design procedure of this compensator requires a twelve-step trial-and-error approach.

In 2003 and 2013 Wang and Tavazoei reported, respectively, exact and unique solutions for integer-order and fractional-order phase-lead-lag compensators when the gain and phase that the compensator must provide are known for a given frequency. The advantage of those methods is that no trial-and-error or other guesswork is needed. Considering this advantage, in this work are employed the procedures described by Wang and Tavazoei. This way, this section summarizes the design equations developed in [35, 44]. Examples of the use of these equations are provided in Sect. 5.

2.3.1 Exact Solution for Integer-Order Phase-Lead-Lag Compensation [44]

Consider M dB and p rad ($-\pi/2 \le p \le \pi/2$) as the required magnitude and phase which should be provided by the integer-order phase-lead-lag compensator at a frequency $\omega = \omega_c$ to yield the desired transient response. This goal is obtainable by means of the compensator

$$C(s) = \frac{1 + \alpha\tau s}{1 + \tau s} \quad (7)$$

if and only if

$$c > \sqrt{1 + \delta^2} \quad \textit{and} \quad 0 < p < \pi/2 \quad \textit{(phase-lead compensation)} \quad (8)$$

$$c < \frac{1}{\sqrt{1 + \delta^2}} \quad \textit{and} \quad -\pi/2 < p < 0 \quad \textit{(phase-lag compensation)} \quad (9)$$

where $c = 10^{M/20}$ and $\delta = \tan(p)$. If (8) or (9) are satisfied, the compensator parameters α and τ can be calculated as

$$\alpha = \frac{c(c\sqrt{1+\delta^2}-1)}{c-\sqrt{1+\delta^2}} \quad and \quad \tau = \frac{c-\sqrt{1+\delta^2}}{c\delta\omega_c} \tag{10}$$

2.3.2 Exact Solution for Fractional-Order Phase-Lead-Lag Compensation [35]

Consider M dB and p rad ($-\pi/2 \leq p \leq \pi/2$) as the required magnitude and phase which should be provided by a fractional-order compensator at the frequency $\omega = \omega_c^q$. This objective is obtainable by means of

$$C_f(s) = K\left(\frac{1+\alpha\tau s^q}{1+\tau s^q}\right), \quad q \in (0,2) \tag{11}$$

if and only if

$$\cot\left(\frac{q\pi}{2}\right) < \frac{c\cos(p)-1}{c\sin(p)}, \quad 0 < p < \frac{\pi}{2} \quad (\textit{phase-lead compensation}) \tag{12}$$

$$\cot\left(\frac{q\pi}{2}\right) < \frac{c-\cos(p)}{\sin(p)}, \quad -\frac{\pi}{2} < p < 0 \quad (\textit{phase-lag compensation}) \tag{13}$$

If (12) or (13) are satisfied, the parameters α and τ can be calculated as

$$\alpha = \frac{uv\tan\left(\frac{q\pi}{2}\right)-1}{v\tan\left(\frac{q\pi}{2}\right)-1}, \qquad \tau = \frac{1}{\omega_c^q}\left[v\sin\left(\frac{q\pi}{2}\right)-\cos\left(\frac{q\pi}{2}\right)\right] \tag{14}$$

where

$$u = c\frac{c-\cos(p)}{c\cos(p)-1}, \qquad v = \frac{c\cos(p)-1}{c\sin(p)} \tag{15}$$

In the case of the fractional-order phase-lead compensator ($0 < p < \pi/2$), the value of q is selectable in the range $(q^*, 2)$, where

$$q^* = \frac{2}{\pi}\tan^{-1}\left(\frac{c\sin(p)}{c\cos(p)-1}\right), \quad \textit{for} \quad c > \frac{1}{\cos(p)} \tag{16}$$

$$q^* = 2 + \frac{2}{\pi}\tan^{-1}\left(\frac{\sin(p)}{c-\cos(p)}\right), \quad \textit{for} \quad c < \frac{1}{\cos(p)} \tag{17}$$

Similarly, for the fractional-order phase-lag compensator ($-\pi/2 < p < 0$), the value of q is selectable in the range $(q^*, 2)$, where

$$q^* = \frac{2}{\pi}\tan^{-1}\left(\frac{\sin(p)}{c-\cos(p)}\right), \quad for \quad c < \cos(p) \tag{18}$$

$$q^* = 2 + \frac{2}{\pi}\tan^{-1}\left(\frac{c\sin(p)}{c\cos(p)-1}\right), \quad for \quad c > \cos(p) \tag{19}$$

2.3.3 Exact Solution for Integer-Order Phase-Lead-Lag Compensation with Actuator's Constrains [35]

One advantage of the fractional-order phase-lead and phase-lag compensators regarding their integer-order counterparts is the fact that the order q, selectable in the range $(q^*, 2)$, represents and additional degree of freedom, which can be used to satisfy another control objective. This way, the exact value of q can be chosen based on an acceptable value for the initial peak of the control signal, i.e., establishing an actuator constraint.

Supposing that it is desired that the initial peak of the control signal be equal to u_0 when the input is a unit step reference, for lead compensation ($u_0 \in (K, \infty)$) the exact value of q must be selected as

$$q = \begin{cases} \frac{2}{\pi}\tan^{-1}\left(\frac{u_0 - K}{v(u_0 - Ku)}\right) & \text{if } v(u_0 - Ku) > 0 \\ 1 & \text{if } v(u_0 - Ku) = 0 \\ 2 + \frac{2}{\pi}\tan^{-1}\left(\frac{u_0 - K}{v(u_0 - Ku)}\right) & \text{if } v(u_0 - Ku) < 0 \end{cases} \tag{20}$$

and in the case of lag compensation ($u_0 \in (0, K)$) the order q must be calculated as

$$q = \begin{cases} 2 + \frac{2}{\pi}\tan^{-1}\left(\frac{u_0 - K}{v(u_0 - Ku)}\right) & \text{if } v(u_0 - Ku) > 0 \\ 1 & \text{if } v(u_0 - Ku) = 0 \\ \frac{2}{\pi}\tan^{-1}\left(\frac{u_0 - K}{v(u_0 - Ku)}\right) & \text{if } v(u_0 - Ku) < 0 \end{cases} \tag{21}$$

2.4 *Realization of Analog Fractances*

The challenge in implementing fractional-order transfer functions and, consequently, fractional-order phase-lead-lag compensators is related to the non-existence of circuit elements that reproduce the operator (5). Those elements are called fractances [23], which are characterized by a magnitude response with roll-off $\pm 20\alpha$ deci-

$$Z = z_1 + \cfrac{1}{y_2 + \cfrac{1}{z_3 + \cfrac{1}{y_4 + \ldots}}}$$

$$\begin{aligned}
Z &= \frac{(s+2)(s+4)}{(s+1)(s+3)} = \frac{s^2+6s+8}{s^2+4s+3} = 1 + \frac{2s+5}{s^2+4s+3} \\
&= 1 + \cfrac{1}{\cfrac{s^2+4s+3}{2s+5}} = 1 + \cfrac{1}{s/2 + \cfrac{3s/2+3}{2s+5}} \\
&= 1 + \cfrac{1}{s/2 + \cfrac{1}{\cfrac{2s+5}{3s/2+3}}} = 1 + \cfrac{1}{s/2 + \cfrac{1}{4/3 + \cfrac{1}{3s/2+3}}} \\
&= \boxed{1} + \cfrac{1}{\boxed{s/2} + \cfrac{1}{\boxed{4/3} + \cfrac{1}{\boxed{3s/2} + \boxed{\frac{1}{1/3}}}}}
\end{aligned}$$

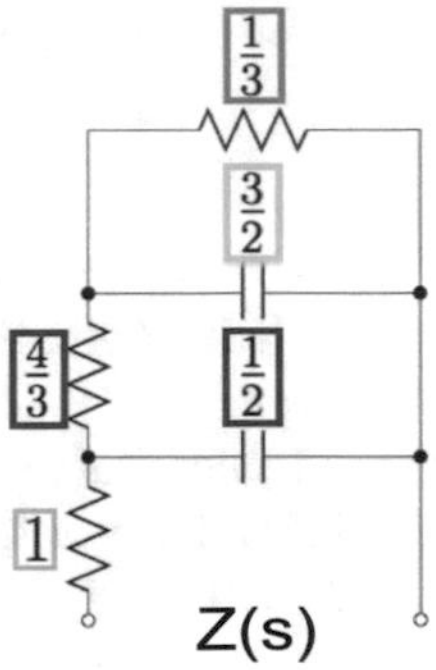

Fig. 1 Method of Cauer for circuit synthesis

bels by decade, and a constant-phase response at all frequencies of $\pm 90\alpha$ degrees. For instance, very few physical realizations have been reported related to "fractal capacitances", capacitors with impedance $Z = 1/(s^{\alpha}C)$ [11, 21]. Unfortunately, those elements are bulky, require chemical compounds with difficult manipulation and the order α cannot be modified easily. Alternatively, the fractances can be approached in a desired bandwidth with rational functions from the methods of Newton, Carlson, Muir, Oustaloup, Matsuda, power series expansion (PSE), continuous fractions expansion (CFE), among others [16, 23, 33]. Once a rational function is obtained it can be synthesized with ladder networks of Cauer, or Foster [33], tree structure, or transmission lines [16, 33]. The circuit components can be resistors, inductors [15], capacitors and sometimes negative impedance converters [3, 33]. One example for synthesis by Cauer method is given in Fig. 1. The drawback of these realizations is the difficulty to approximate the required values with the ones of the commercially-available resistances and capacitors [19].

Once the procedures to design integer-order and fractional-order phase-lead-lag compensators have been described, in the following sections will be focused the problem of circuit implementation.

3 Basic Building Blocks

This section presents the OpAmp-based basic building blocks that will be employed in Sect. 4 to perform the synthesis of integer-order and fractional-order phase-lead-lag compensators.

3.1 Inverting Integrator (IInt)

The OpAmp-based Inverting Integrator of Fig. 2a uses capacitive feedback to integrate the input signal V_{in1}. The transfer function of this circuit is given by

$$\frac{V_{out_1}}{V_{in_1}} = -\frac{1}{R_g C_g s} = -\frac{1}{s} \quad (22)$$

where C_g can be used as degree of freedom and $R_g = 1/C_g$. Then, magnitude (Z_m) and frequency (Ω_f) denormalizations can be used to obtain commercially available values of the passive elements and the desired frequency response.

3.2 Non-inverterting and Inverting Amplifiers (NIA, IA)

The OpAmp-based Non-inverting Amplifier (NIA) and Inverting Amplifier (IA) depicted in Fig. 2b and c use resistive feedback to amplify the input signals V_{in2} and V_{in3}. The corresponding transfer functions are

$$\frac{V_{out_2}}{V_{in_2}} = 1 + \frac{R_{e2}}{R_{e1}}, \qquad \frac{V_{out_3}}{V_{in_3}} = -\frac{R_{f2}}{R_{f1}} \quad (23)$$

3.3 Weighted Differential Amplifier and Differential Amplifier

The Weighted Differential Amplifier (WDA) amplifies the weighted difference between two voltages. A particular case is the Differential Amplifier (DA), which amplifies the difference between the two voltages but does not amplify the particular voltages. Fig. 2d shows an implementation of a WDA with R_{g1} and R_{g2} used to control the gains and with R_g as degree of freedom. By nodal analysis the output voltage V_{out_4} results

$$V_{out_4} = \frac{R_g}{R_{g1}} V_{1A} - \frac{R_g}{R_{g2}} V_{2A} \quad (24)$$

Alternatively, by omitting in Fig. 2d the landed resistors R_{g1} and R_{g2} and by choosing $R_{g1} = R_{g2} = R_{g3}$ is obtained the DA of Fig. 2e with output voltage

$$V_{out_5} = \frac{R_g}{R_{g3}} \left(V_{1B} - V_{2B}\right) \quad (25)$$

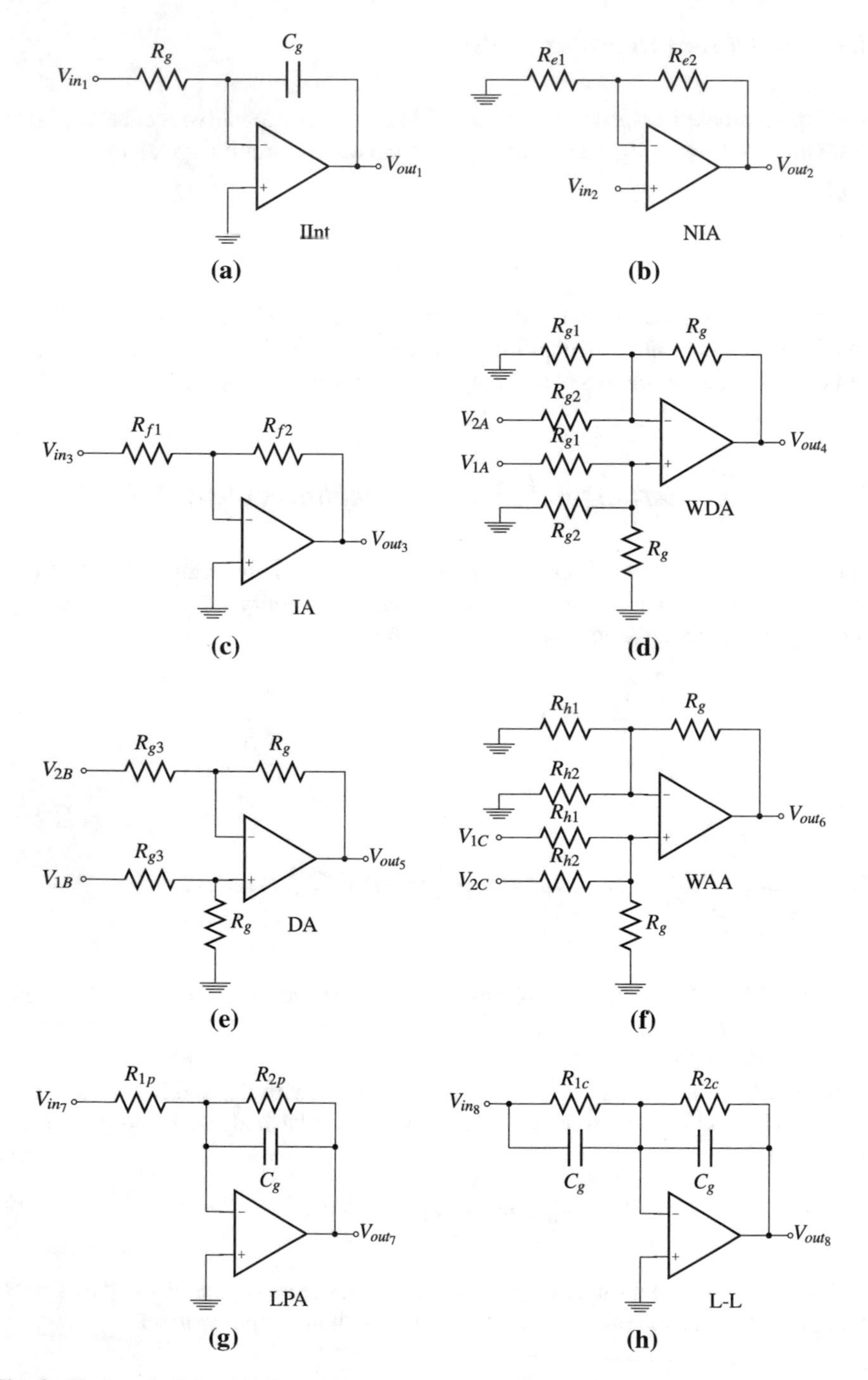

Fig. 2 OpAmp building blocks: **a** Inverting Integrator (IInt). **b** Non-inverting Amplifier (NIA). **c** Inverting Amplifier (IA). **d** Two-input weighted differential amplifier (WDA). **e** Differential Amplifier (DA). **f** Two-input weighted adder amplifier (WAA). **g** Low-Pass Amplifier (LPA). **h** Lead/lag network

3.4 Two-Input Weighted Adder Amplifier (WAA)

It produces an output V_{out_6} equal to the weighted sum of the two inputs V_{1C} and V_{2C}. The realization of Fig. 2f uses R_g as degree of freedom and R_{h1} and R_{h2} to control the weighted factors by means of

$$V_{out_6} = \frac{R_g}{R_{h1}} V_{1C} + \frac{R_g}{R_{h2}} V_{2C} \tag{26}$$

3.5 Lowpass Amplifier (LPA)

Figure 2g shows a first-order inverting Low Pass Filter Amplifier (LPA) ("bilinear filter"). The DC gain and corner frequency of this circuit are $|H|_{s=0} = R_{2p}/R_{1p}$ and $\omega_c = 1/(R_{2p}C_g)$ (C_g can be used as degree of freedom), R_{2p} determines the corner frequency and R_{1p} the DC gain. The transfer function V_{out_7}/V_{in_7} results

$$\frac{V_{out_7}}{V_{in_7}} = -\frac{\frac{1}{R_{1p}C_g}}{s + \frac{1}{R_{2p}C_g}} \tag{27}$$

3.6 Lead/Lag Network (L-L)

The transfer function of the circuit of Fig. 2h, its pole, its zero and its DC gain can be calculated as

$$\frac{V_{out_8}}{V_{in_8}} = -\frac{R_{2c}}{R_{1c}} \left(\frac{1 + sR_{1c}C_g}{1 + sR_{2c}C_g} \right) \tag{28}$$

$$\omega_z = \frac{1}{R_{1c}C_g}, \quad \omega_p = \frac{1}{R_{2c}C_g}, \quad |H|_{(s=0)} = \frac{R_{2c}}{R_{1c}} \tag{29}$$

Clearly, this network provides positive (leading) phase shift if the zero of the transfer function is closer to the origin of the s-plane than the pole, which occurs if $R_{1c} > R_{2c}$. Conversely, with $R_{1c} < R_{2c}$ the pole is closer than the zero to the origin of the s-plane, and the network provides negative (lagging) phase shift of the output signal relative to the input signal at all frequencies.

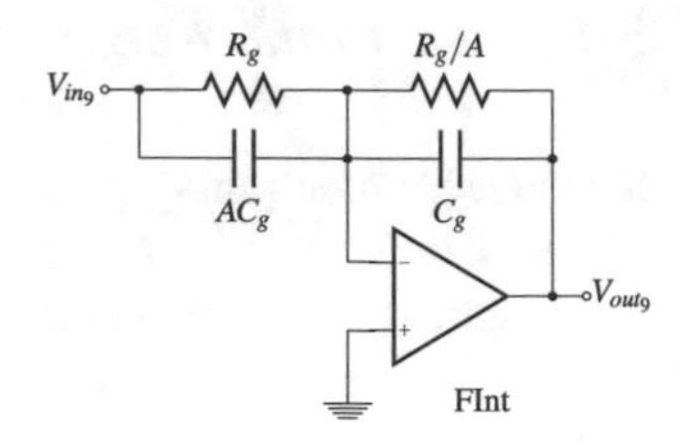

Fig. 3 First-order approximation of a Fractional-order Integrator (FInt)

3.7 *Fractional Integrator (FInt)*

As mentioned before, the operator s^{-q} cannot be implemented directly, it is required to perform an approximation. Consider the transfer function $V_{out_9}/V_{in_9} = -s^{-q}$. Then, an approximation of order one of a Fractional-order Integrator given by

$$\frac{V_{out_9}}{V_{in_9}} = -\frac{1}{s^q} \approx -\frac{(1-q)s+(1+q)}{(1+q)s+(1-q)} = -\frac{As+1}{s+A} = -\frac{1}{A}\left(\frac{1+As}{1+\frac{s}{A}}\right), \quad A = \frac{1-q}{1+q} \tag{30}$$

can be implemented with an adequate selection of the capacitors and resistances of the Lead-Lag network in Fig. 2h, resulting the circuit of Fig. 3.

4 OPAMP-Based Realization of Integer-Order and Fractional-Order Phase-Lead-Lag Compensators

This section presents OPAMP-based realization of integer-order and fractional-order phase-lead-lag compensators performed with the basic building blocks of Figs. 2 and 3.

4.1 *Integer-Order Phase-Lead-Lag Compensator*

A phase-lead-lag compensator with DC unity-gain can be realized by means of the OpAmp-based network shown in Fig. 4, which consists of a L-L network connected in series with an IA block (with $R_{f1} = R_{1c}$ and $R_{f2} = R_{2c}$). The transfer function of this network is expressed by

$$\frac{V_{oc}}{V_{ic}} = \frac{1+sR_{1c}C_g}{1+sR_{2c}C_g} = \frac{1+\alpha\tau s}{1+\tau s} \tag{31}$$

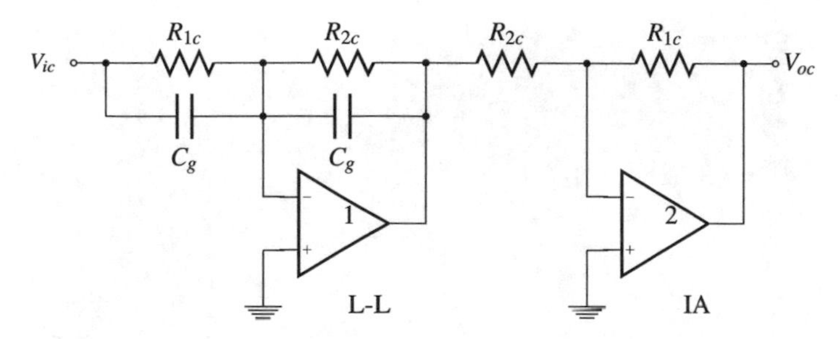

Fig. 4 Integer-order phase-lead-lag compensator

where

$$\tau = R_{2c}C_g, \quad \alpha = \frac{R_{1c}}{R_{2c}} \tag{32}$$

Therefore, for $\alpha > 1$ (i.e. with $R_{1c} > R_{2c}$) and $\alpha < 1$ (i.e. with $R_{2c} > R_{1c}$) are obtained, respectively, phase-lead and phase-lag responses.

4.2 *Fractional-Order Phase-Lead-Lag Compensator (1 < q < 2)*

Consider the fractional-order phase-lead-lag transfer function given by

$$\frac{V_{oc}(s)}{V_{ic}(s)} = \frac{1 + \alpha\tau s^q}{1 + \tau s^q} \tag{33}$$

where 1< q <2 is assumed. Algebraic manipulation on (33) leads to

$$V_{oc}(s) + V_{oc}(s)\tau s^q = V_{ic}(s) + V_{ic}(s)\alpha\tau s s^{q-1} \tag{34}$$

and after dividing both sides of (34) by τs^q and regrouping similar terms, it results in

$$V_{oc}(s) = \frac{V_{ic}(s) - V_{oc}(s)}{\tau s s^{q-1}} + \alpha V_{ic}(s) \tag{35}$$

This way, the fractional-order phase-lead-lag compensator with 1< q <2 can be realized starting from the block diagram shown in Fig. 5a. The corresponding implementation using OpAmps is shown in Fig. 5b. Here, the algebraic operation $(V_{ic}(s) - V_{oc}(s))/\tau$ is performed by the block DA of Fig. 2e, with $R_{g3} = \tau R_g$. The operation $1/s^q = (1/s)(1/s^{q-1})$ is realized by means of the series array of an integer-order Inverter Integrator IInt (with $R_g = 1/C_g$) and a Fractional-order Integrator FInt whit $\tilde{A}$ calculated from (30) as

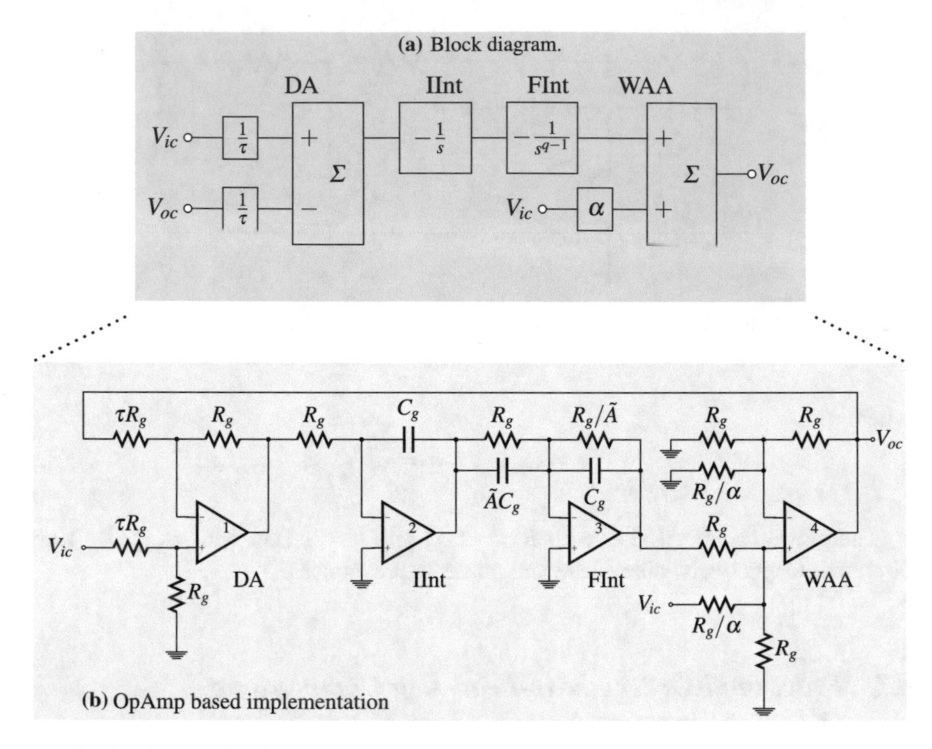

Fig. 5 Fractional-order phase-lead-lag compensator (case $1 < q < 2$)

$$\tilde{A} = \frac{1-(q-1)}{1+(q-1)} = \frac{2}{q} - 1 \tag{36}$$

Finally, to complete (35), the output of FInt is added to αV_{ic} by means of the block WAA, using $R_g/R_{h1} = 1$ and $R_g/R_{h2} = \alpha$.

4.3 *Fractional-Order Phase-Lead-Lag Compensator ($0 < q < 1$)*

For the case of fractional-order phase-lead-lag compensators with $0 < q < 1$, the block IInt of Fig. 5 must be omitted. Consequently, the block WAA must be changed by a block WDA to avoid a positive feedback, with the output of the block FInt connected to the inverting input of the block WDA, resulting the implementation depicted at Fig. 6. In this circuit, the block DA is designed with $R_{g3} = \tau R_g$, the block WDA is realized with $R_{g1} = R_g/\alpha$ and $R_{g2} = R_g$, and A is calculated with (30).

Two alternative implementations for fractional-order phase-lead-lag networks with $0 < q < 1$ and with fewer active but more passive elements are presented below.

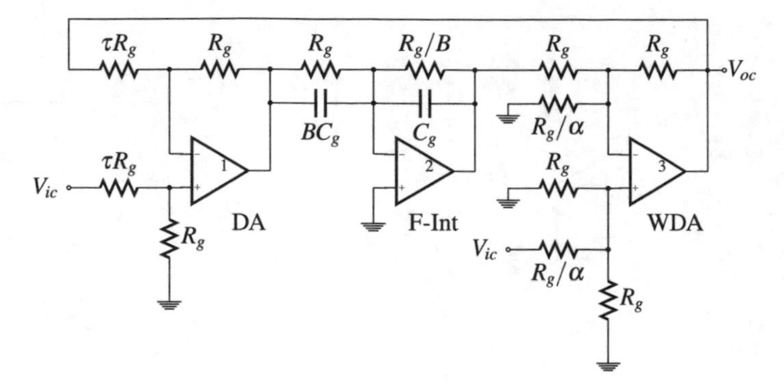

Fig. 6 Fractional-order phase-lead-lag compensator (case $0 < q < 1$)

4.4 *Fractional-Order Phase-Lead-Lag Compensators $0 < q < 1$ (Cauer's Approximation)*

The circuit of Fig. 7a is a well known phase-lead configuration, but with a capacitor C substituted by a fractal capacitor with fractance $1/(s^q C)$. It can be easily demonstrated with nodal analysis that the transfer function of this circuit becomes

$$\frac{V_{oc}(s)}{V_{ic}(s)} = \frac{1 + R_{1ca}Cs^q}{1 + \left(R_{1ca}||R_{2ca}\right)Cs^q} = \frac{1 + \alpha\tau s^q}{1 + \tau s^q} \tag{37}$$

$$\tau = \left(R_{1ca}||R_{2ca}\right)C, \quad \alpha = 1 + \frac{R_{1ca}}{R_{2ca}} \tag{38}$$

and the fractal capacitor can be approximated with any of the methods mentioned in Sect. 2.4 (for a given n-th order of approximation) and, subsequently, implemented by Cauer networks by means of Continuous Fraction Expansion method.

Analogously, the circuit of Fig. 7b is a well known phase-lag configuration with the capacitor C substituted by a fractal capacitor with fractance $1/(s^q C)$. In this case the transfer function takes the form

$$\frac{V_{oc}(s)}{V_{ic}(s)} = \frac{1 + R_{2ca}Cs^q}{1 + \left(R_{1ca} + R_{2ca}\right)Cs^q} = \frac{1 + \alpha\tau s^q}{1 + \tau s^q} \tag{39}$$

$$\tau = \left(R_{1ca} + R_{2ca}\right)C, \quad \alpha = \frac{R_{2ca}}{R_{1ca} + R_{2ca}} \tag{40}$$

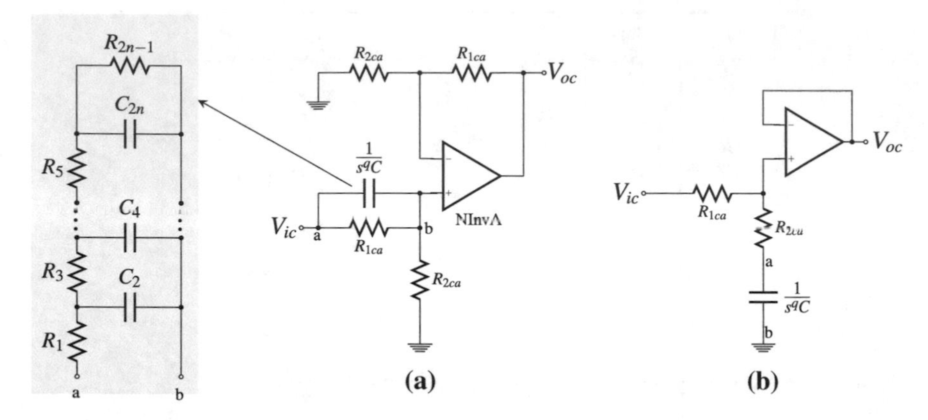

Fig. 7 Fractional-order compensators with $0 < q < 1$ and Cauer networks. **a** Phase-lead compensator. **b** Phase-lag compensator

5 Examples of Phase-Lead-Lag Compensated Systems Implemented with OpAmps

To validate the proposals of implementation of Sect. 4, this Section presents simulation or experimental results of systems that use the circuits in Figs. 4, 5, and 7 as compensators connected in series with an integer-order plant in unity-gain negative feedback configuration. Additionally, to explain the procedures described in Sect. 2.3 to design integer-order and fractional-order phase-lead-lag compensators, will be employed the system modeled by the following transfer function

$$G(s) = \frac{100K}{s(s+36)(s+100)} \tag{41}$$

This system has been considered as an academic example in [30, 44] for integer-order compensation, and in [35] for fractional-order compensation. Gains $K = 5839$ and $K = 1440$ have been employed in the lag and lead compensators, respectively, to satisfy steady-state error specifications. Figure 8 shows an OpAmp-based implementation of $G(s)$ by means of blocks IInv, LPA and IA. By equating the transfer function of this circuit with (41), results

$$\frac{\left(\frac{1}{R_{1p}C_g}\right)\left(\frac{1}{R_{1p}C_g}\right)}{s\left(s+\frac{1}{R_{2p}C_g}\right)\left(s+\frac{1}{R_{3p}C_g}\right)} = \frac{100K}{s(s+36)(s+100)} \tag{42}$$

and by choosing $1/(R_{2p}C_g) = 36$, $1/(R_{3p}C_g) = 100$, $1/(R_{1p}C_g) = 10\sqrt{K}$, $C_g = 1$ F and $R_g = 1\,\Omega$ are obtained: $R_{1p} = 1/\sqrt{100K}\,\Omega$, $R_{2p} = 0.0277\,\Omega$ and $R_{3p} = 0.01\,\Omega$.

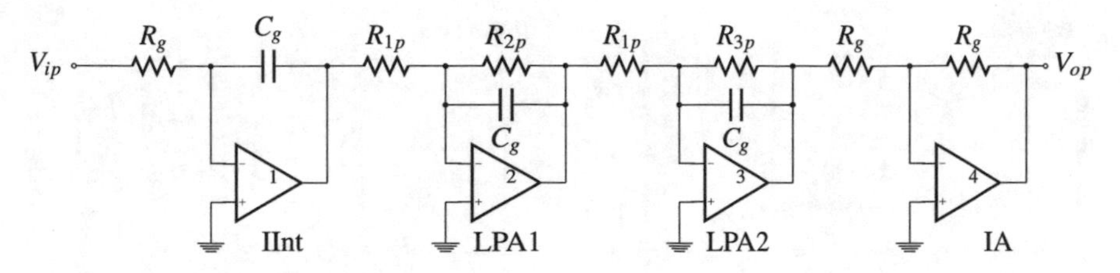

Fig. 8 Implementation of the plant $G(s)$ with OpAmps

Table 1 Design details of the plant $G(s)$

Block	Element	Theoretical value	Employed value (commercially available)
IInt	R_g	100 KΩ	100 KΩ
	C_g	10 nF	10 nF
LPA1	R_{1p}	130 Ω for $K = 5839$	120 Ω
		260 Ω for $K = 1440$	270 Ω
	R_{2p}	2.77 KΩ	2.7 KΩ
	C_g	10 nF	10 nF
LPA2	R_{1p}	130 Ω for $K = 5839$	120 Ω
		260 Ω for $K = 1440$	270 Ω
	R_{3p}	1 KΩ	1 KΩ
	C_g	10 nF	10 nF
IA	R_g	100 KΩ	100 KΩ

Then, impedance ($Z_m = 1E5$) and frequency ($\Omega_f = 1000$) denormalizations are carried out over this elements to obtain the values detailed in Table 1.

5.1 *Example 1: Integer Order Phase-Lag Compensator ($K = 5839$)*

Figure 9a shows the Bode diagram of $G(s)$ with $K = 5839$. As can be observed, the system presents a phase margin of 67° and a gain of 22.9 dB at the desired crossover frequency $\omega_c = 11$ rad/s. Assuming a required phase margin $PM = 62°$ to yield the desired transient response, a phase $p = -5°$ and a magnitude $M = -22.9$ dB must be provided by the phase-lag compensator to obtain a composite Bode diagram that goes through 0 dB at $\omega_c = 11$ rad/s. Therefore, using the procedure presented in Sect. 2.3 for integer-order compensators are calculated $c = 10^M/20 = 0.0711$ and $\delta = \tan(p) = -0.087$. Substituting these values of c and δ in (10) are obtained $\alpha = 0.0711$ and $\tau = 14.3472$, resulting the integer order phase-lag compensator

$$C(s) = \frac{1 + \alpha\tau s}{1 + \tau s} = \frac{1 + (0.0711)(14.3472)s}{1 + 14.3472s} \tag{43}$$

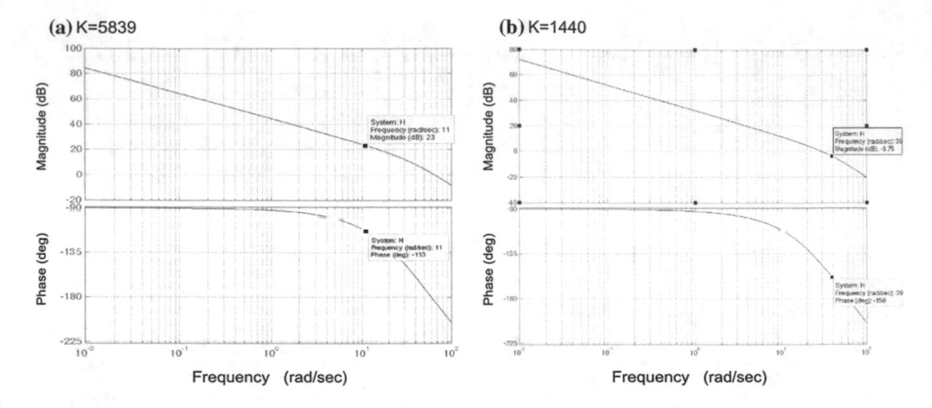

Fig. 9 Bode magnitude and phase plots of $G(s) = 100K/s(s + 36)(s + 100)$ with **a** $K = 5839$; **b** $K = 1440$

Table 2 Design details of the integer-order phase-lag compensator of Fig. 4 and Example 1 with $K = 5839$

Block	Element	Theoretical value	Employed value (commercially available)
L-L	R_{1c}	100 KΩ	100 KΩ
	R_{2c}	1.43 MΩ	1.5 MΩ
	C_g	10 nF	10 nF
InvA	R_{1c}	100 KΩ	100 KΩ
	R_{2c}	1.43 MΩ	1.5 MΩ

Additionally, it can be verified that with $\delta = -0.087$ and $c = 0.0711$ the condition (9) is satisfied, guarantying the existence of the compensator, which is implemented with the circuit of Fig. 4. Therefore, by choosing $C_g = 1$ and $\alpha = 0.0711$ in (32) are obtained $R_{2c} = 14.34\,\Omega$ and $R_{1c} = 1.02\,\Omega$. Then, impedance ($Z_m = 1E5$) and frequency ($\Omega_f = 1000$) denormalizations are carried out to obtain the values of elements detailed in Table 2.

The system in Fig. 8 in the unity negative feedback structure with the lag-phase compensator of Fig. 4 was simulated using HSPICE and the model of the OpAmp uA741. The details of design are listed in Tables 1 and 2. Figure 10 shows the results for an step-input of 1 V and frequency 166.6 Hz. The resulted overshoot was 9.7% with a peak time 258.4 μs. This overshoot corresponds to a second order system with phase margin $PM = 58.93°$ and damping factor $\zeta = 0.596$. These results show a good agreement with the results given in [44] (PM = 62°, $\zeta = 0.591$, overshoot = 10%) and [30] (overshoot = 9.8%, denormalized peak time = 260 μs), validating the proposal of implementation.

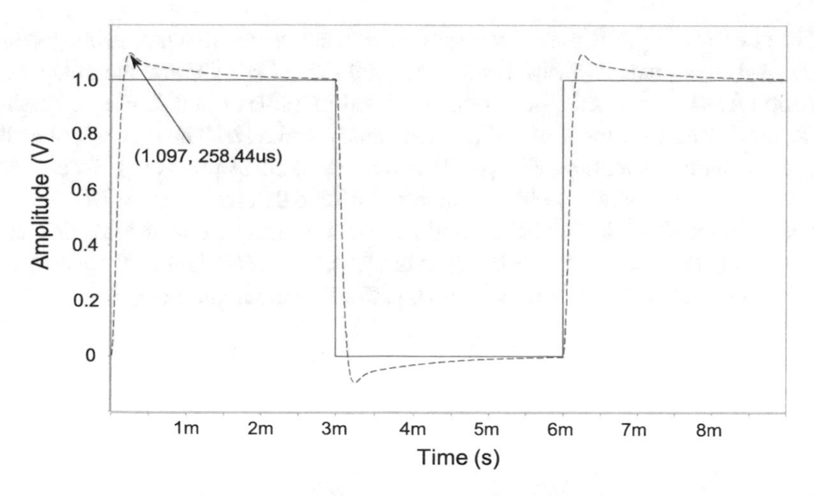

Fig. 10 Time-domain simulation results of the integer-order phase-lag compensator $C(s) = [1 + (0.0711)(14.3472)s]/[1 + 14.3472s]$ realized with the circuit of Fig. 4 with design details of Table 2, in unity negative feedback configuration, in which the plant $G(s) = 583900/[s(s+36)(s+100)]$ is implemented with the circuit of Fig. 8 with the design details of Table 1. Denormalizations in impedance and frequency $Z_m = 1E5$ and $\Omega_f = 1000$ were also performed

5.2 *Example 2: Integer Order Phase-Lead Compensator (K = 1440)*

Figure 9b shows the Bode diagram of $G(s)$ with $K = 1440$. The phase margin and gain of the system at the expected crossover frequency $\omega_c = 39$ rad/s are $PM = 180°–159° = 21°$ and $M = -3.7668$ dB, respectively. Consider a required phase margin $PM = 45.5°$ to yield the desired transient response. Consequently, the phase-lead compensator must provide a phase of $p = 24.5°$ and a gain of $M = 3.7668$ dB to yield the desired transient response with a Bode diagram that goes through 0 dB at $\omega_c = 39$ rad/s. By means of the procedure presented in Sect. 2.3, for integer-order compensators, are calculated: $c = 10^M/20 = 1.5429$ and $\delta = \tan(p) = 0.4473$. Substituting these values of c and δ in (10) results $\alpha = 2.3799$ and $\tau = 0.0166$. The corresponding integer-order phase-lead compensator becomes

$$C(s) = \frac{1+\alpha\tau s}{1+\tau s} = \frac{1+(2.3799)(0.0166)s}{1+0.0166s} = 2.3795\left(\frac{s+25.31}{s+60.24}\right) \tag{44}$$

and the existence of this phase-lead compensator is guaranteed because of with $\delta = 0.4473$ and $c = 1.54$ the condition (8) is satisfied. To proceed with the implementation it is employed the circuit of Fig. 4. By selecting $C_g = 1$ F and $\alpha = 2.3799$ in (32) are obtained $R_{2c} = 0.0166\,\Omega$ and $R_{1c} = 0.0395\,\Omega$. Then, impedance ($Z_m = 1E5$) and frequency ($\Omega_f = 1000$) denormalizations are carried out to obtain the details of design summarized in Table 3.

The system in Fig. 8 in the unity negative feedback structure with the integer-order phase-lead compensator of Fig. 4 was simulated using HSPICE and the model of the OpAmp uA741. The details of design are listed in Tables 1 and 3. Figure 11 shows the results for an step-input of 1 V and frequency 166.6 Hz. The resulted overshoot was 18.2% with a peak time 77.5 μs. This overshoot corresponds to a second order system with phase margin 49.9° (compared with the theoretical value of 45.5°) and damping factor $\zeta = 0.476$. These results show a good agreement with the results given in [44] (PM = 45.5°, $\zeta = 0.427$, overshoot = 22.6%) and [30] (overshoot = 22.6%, denormalized peak time = 72 μs), validating the implementation.

Table 3 Design details of the integer-order phase-lead compensator of Fig. 4 and Example 2 with $K = 1440$

Block	Element	Theoretical value	Employed value (commercially available)
L-L	R_{1c}	3.95 KΩ	3.9 KΩ
	R_{2c}	1.66 KΩ	1.2 KΩ + 470 Ω
	C_g	10 nF	10 nF
InvA	R_{1c}	3.95 KΩ	3.9 KΩ
	R_{2c}	1.66 KΩ	1.2 KΩ + 470 Ω

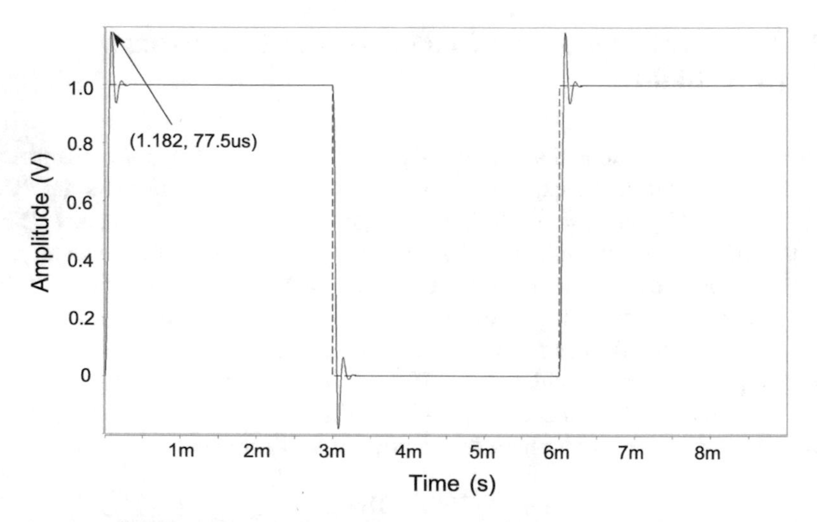

Fig. 11 Time-domain simulation results of the phase-lead compensator $C(s) = [1 + (2.3799)(0.0166)s]/[1 + 0.0166s]$ realized with the circuit of Fig. 4 with design details of Table 3, in unity negative feedback configuration, in which the plant $G(s) = 144000/[s(s + 36)(s + 100)]$ is implemented with the circuit of Fig. 8 with the design details of Table 1. Denormalizations in impedance and frequency $Z_m = 1E5$ and $\Omega_f = 1000$ were also performed

5.3 Example 3: Fractional-Order Phase-Lead Compensator ($1 < q < 2$)

Consider the system $G(s)$ given by (41) with $K = 1440$. As was explained in Example 2, a desired phase margin 45.5° with a gain cross-frequency of 39 rad/s correspond to $M = 3.76$ dB, $c = 1.5429, p = 24.1$ and $\delta = 0.4473$. Using the procedure described in Sect. 2.3 for fractional-order compensators and according to (16), this goal is achievable by means of a fractional-order phase-lead compensator with $q \in (q^*, 2) = (0.6338, 2)$. By taking advantage of the additional grade of liberty it can be obtained the exact value of q by establishing a desired value of the initial peak of the control signal, i.e., by using an actuator constraint. For instance, assuming a desired initial peak of the control signal of $u_0 = 2500$ and according to (20) and (15) it is obtained $q = 1.33$. Substituting this value in (14) and (15) the resulting fractional-order phase-lead compensator is expressed by

$$C(s) = \frac{1 + \alpha\tau s^q}{1 + \tau s^q} = \frac{1 + (1.736)\left(8.1395 \times 10^{-3}\right) s^{1.33}}{1 + 8.1395 \times 10^{-3} s^{1.33}} \tag{45}$$

This fractional-order phase-lead compensator satisfies the condition (12) with $\delta = 0.4557$ and $c = 1.5429$. The implementation is performed with the circuit of Fig. 5 selecting $R_g = 1\,\Omega, \alpha = 1.736, \tau = 8.1395 \times 10^{-3}$ and $\tilde{q} = q - 1 = 0.33$. This way, the following results are obtained: $\tilde{A} = (1 - \tilde{q})/(1 + \tilde{q}) = 0.5037, \tau R_g = 8.1395 \times 10^{-3}\,\Omega, R_g/\tilde{A} = 1.9853\,\Omega, \tilde{A}/R_g = 0.5037$ F and $R_g/\alpha = 0.576\,\Omega$. Then, impedance ($Z_m = 1\text{E}5$) and frequency ($\Omega_f = 1000$) denormalizations are carried out to obtain the details of design summarized in Table 4.

The system in Fig. 8 in the unity negative feedback structure with the fractional-order lead-phase compensator of Fig. 5 was implemented on protoboard with OpAmps uA741 and the details of design listed in Tables 1 and 4. The experimental

Table 4 Design details of the fractional-order phase-lead compensator of Fig. 4 and Example 3 with $K = 1440$

Block	Element	Theoretical value	Employed value (commercially available)
DA	R_g	100 KΩ	100 KΩ
	τR_g	813.9 Ω	820 Ω
IInt	R_g	100 KΩ	100 KΩ
	C_g	10 nF	10 nF
L-L	R_g	100 KΩ	100 KΩ
	$R_g/\tilde{A}$	200 KΩ	200 KΩ
	$\tilde{A}C_g$	5 nF	5 nF
	C_g	10 nF	10 nF
WAA	R_g	100 KΩ	100 KΩ
	R_g/α	57.6 KΩ	47 KΩ + 10 KΩ

ELVIS II National Instruments (Vin=(1V, 100Hz). VCC=+15V, VEE= =-15V)

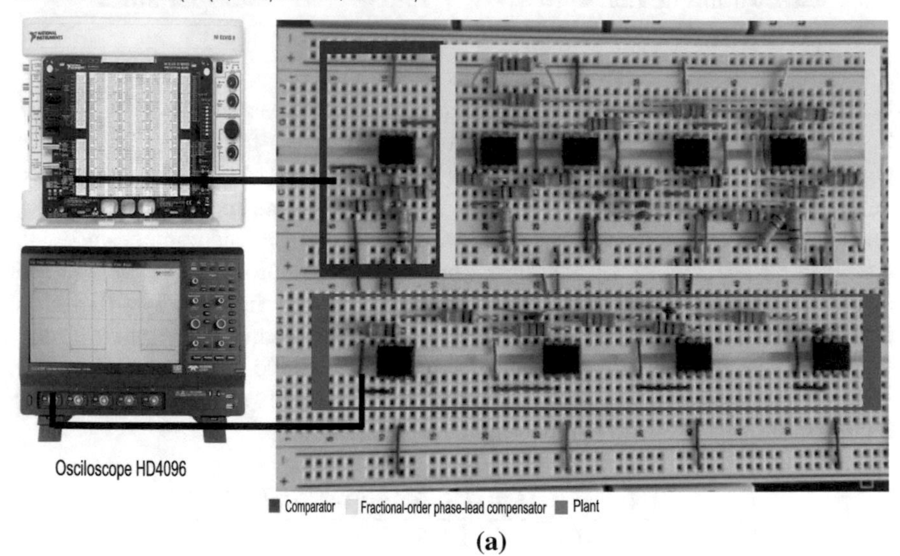

(a)

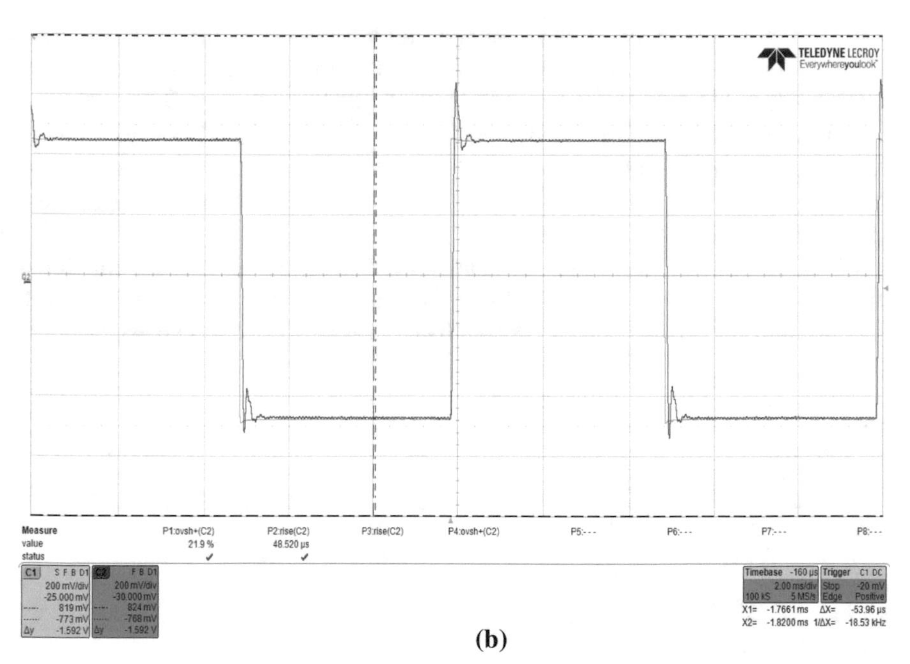

(b)

Fig. 12 **a** Experimental setup of the fractional-order phase-lead compensator $C(s) = [1 + (1.736)(8.1395 \times 10^{-3})s^{1.33}]/[1 + 8.1395 \times 10^{-3}s^{1.33}]$ realized with the circuit of Fig. 5 in unity negative feedback configuration with plant $G(s) = 144000/[s(s + 36)(s + 100)]$ implemented with the circuit of Fig. 8. **b** Time-domain experimental results

setup is shown in Fig. 12a. It consists of an input square signal of 1 V in amplitude, offset = 0.5 V, and frequency of 100 Hz, supplied from the experimental platform ELVIS II from National Instruments. This device also provides bias voltages of ±15 V to the OpAmps. Figure 12b shows the measured time response. The output was measured with an Oscilloscope HD4096 Teledyne Lecroy. The resulted overshoot was 21.9%. This overshoot corresponds to a second order system with phase margin 46.32° and damping factor $\zeta = 0.435$. It can be noted that the theoretical (overshoot = 22.7%, $\zeta = 0.426$, PM = 45.5°), the simulation (overshoot = 21.13%, $\zeta = 0.4434$, PM = 47.05°), the experimental results (overshoot = 21.9%, $\zeta = 0.435$, PM = 46.32°) and the results reported in Fig. 7 of [35] are in good agreement, thus validating the proposal of implementation.

5.4 Example 4: Fractional-Order Phase-Lead Compensator with $0 < q < 1$ and Cauer Approximation

With an unconstrained initial peak of the control signal, the value of q in Example 3 may be any in the range $(q^*, 2) = (0.6338, 2)$. Consider $q = 0.7$. Substituting this value in (14) and (15) the fractional-order phase-lead compensator becomes

$$C(s) = \frac{1 + \alpha\tau s^q}{1 + \tau s^q} = \frac{1 + (7.442)\left(9.5243 \times 10^{-3}\right) s^{0.7}}{1 + 9.5243 \times 10^{-3} s^{0.7}} \tag{46}$$

and this compensator satisfies the condition (12) with $\delta = 0.4557$ and $c = 1.5429$. In this case, the circuit implementation can be performed with the circuit of Fig. 6. However, this implementation is similar to the presented in Example 3. Instead, in this example will be explored the realization with network approximation by means of the Cauer method. Then, consider the circuit implementation of Fig. 7. With $q = 0.7, C = 1$ and a fourth-order approximation of the impedance $1/s^q$ is obtained [23]

$$\begin{aligned} \frac{1}{s^q} &= \frac{Q_0 s^4 + Q_1 s^3 + Q_2 s^2 + Q_3 s + Q_4}{Q_4 s^4 + Q_3 s^3 + Q_2 s^2 + Q_1 s + Q_0} \\ &= \frac{0.037 s^4 + 2.324 * s^3 + 9.921 s^2 + 7.8 s + 1}{s^4 + 7.8 s^3 + 9.92 s^2 + 2.324 s + 0.037} \end{aligned} \tag{47}$$

where

$$\begin{aligned} Q_0 &= q^4 - 10q^3 + 35q^2 - 50q + 24 \\ Q_1 &= -4q^4 + 20q^3 + 40q^2 - 320q + 384 \\ Q_2 &= 6q^4 - 150q^2 + 864 \\ Q_3 &= 4q^4 - 10q^3 + 40q^2 + 320q + 384 \\ Q_4 &= q^4 + 10q^3 + 35q^2 + 50q + 24 \end{aligned} \tag{48}$$

Alternatively, the approximation of $1/s^q$ can be obtained by other methods of approximation, such as Crone, Carlson or Matsuda, employing the Ninteger tool box of MATLAB [26]. Moreover, the synthesis of the fractance (47) can be obtained with the repeated division process of terms described in Sect. 2.4 (Continued Fraction Expansion) by means of the tool box FOMCON of MATLAB and the following code [26, 37]

$$\gg a1 = [1\ \ 7.807\ \ 9.921\ \ 2.323\ \ 0.037];$$
$$\gg b1 = [0.037\ \ 2.3239\ \ 9.921\ \ 7.807\ \ 1];$$
$$\gg [q, expr] = polycfe(b1, a1)$$

resulting

$$\frac{1}{s^{0.7}} = R_1 + \cfrac{1}{sC_2 + \cfrac{1}{R_3 + \cfrac{1}{sC_4 + \cfrac{1}{R_5 + \cfrac{1}{sC_6 + \cfrac{1}{R_7 + \cfrac{1}{sC_8 + \cfrac{1}{R_9}}}}}}}} \tag{49}$$

The values of $R_1, C_2, R_3, C_4, R_5, C_6, R_7, C_8$ and R_9 are shown in Table 5. Impedance ($Z = 1E5$) and frequency ($\Omega = 1000$) denormalizations were carried out. Finally, using $\alpha = 7.442, C = 1$ and $\tau = 9.5243 \times 10^{-3}$ are solved simultaneously both equations in (38) to obtain $R_{1ca} = 0.07088\,\Omega$ and $R_{2ca} = 0.011\,\Omega$. Again by using the impedance denormalization $Z = 1E5$ over these elements are obtained the values indicated in Table 5.

The system in Fig. 8 in the unity negative feedback structure with the fractional-order lead-phase compensator with Cauer network approximation of Fig. 7 was simulated using HSPICE and the model of the OpAmp uA741. The details of design

Table 5 Design details of the fractional-order phase-lead compensator of Fig. 7 and Example 4 with $q < 1$ and $K = 1440$

Element	Value	Element	Value
R_1	3.7 KΩ	R_7	534 KΩ
C_2	4.9 nF	C_8	9.8 nF
R_3	65.38 KΩ	R_9	1.87 MΩ
C_4	5.6 nF	R_{1ca}	7 KΩ
R_5	224.8 KΩ	R_{2ca}	1.12 KΩ
C_6	6.8 nF		

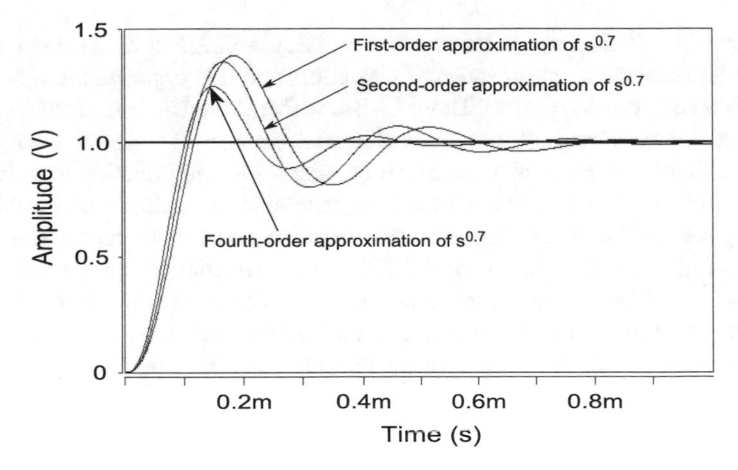

Fig. 13 Time-domain simulation results of the fractional-order phase-lead compensator $C(s) = [1 + (7.442)(9.5243 \times 10^{-3})s^{0.7}]/[1 + 9.5243 \times 10^{-3}s^{0.7}]$ realized with the circuit of Fig. 7 with design details of Table 5, in unity negative feedback configuration, in which the plant $G(s) = 144000/[s(s + 36)(s + 100)]$ is implemented with the circuit of Fig. 8 with the design details of Table 1. Denormalizations in impedance and frequency $Z = 1E5$ and $\Omega = 1000$ were also performed

are listed in Tables 1 and 5. Figure 13 shows the results for an step-input of 1 V, offset 0.5 V and frequency 100 Hz. The resulted overshoot was 25% with a peak time 75.3 μs. This overshoot corresponds to a second order system with phase margin 43.46° and damping factor $\zeta = 0.4037$. These results show a good agreement with the results given in Example 3 and in Fig. 6 of [35], validating the implementation. However, in Fig. 13 it can also be observed that the network approximation is not appropriate in this case when the order of the approximation is less than 4. It is due to the limited bandwidth where the approximation is valid.

6 Phase-Lead Compensated Systems Implemented with FPAA

In this section is illustrated, starting from the designs of Examples 2 and 3, the implementation and experimental validation of integer-order and fractional-order phase-lead compensators by means of FPAAs, which are processors for analog signals, equivalents to the digital processors FPGAs (Field Programmable Gate Arrays). FPAAs are devices of specific purpose with the characteristics of being reconfigurable electrically. They are used to implement a variety of analog functions, such as: integration, derivation, weighted sum/subtraction, filtering, rectification, comparator, multiplication, division, analog-digital conversion, voltage references, signal conditioning, amplification, nonlinear functions, generation of arbitrary signals, among others. Since FPAAs are reconfigurable, one can implement complex prototypes in a short time. In this work the FPAA AN231E04 from Anadigm [1], is used. It uses technology of switched capacitors and it is organized into four configurable analog blocks (CABs). Those CABs are distributed in a matrix of size 2×2, supported by resources of programmable interconnections, seven configurable analog cells of input-output with active elements for amplification and dynamic reduction of offset and noise, an on-chip generator of multiple non-overlapped clock-signals and internal voltage references to eliminate temperature effects. It also includes a look-up table (LTU) of 8×256 bits for function synthesis and nonlinear signals, and for analog-digital conversion. The configuration data is saved into an internal SRAM, which allows reprogramming the device without interrupting its operation. The circuits are designed using the software Anadigmdesigner2, in which the user has access to a library of functional circuits CAMs (Configurable Analog Modules). Those CAMs are mapped in a portion for each CAB. The CABs have matrices of switches and capacitors, two OpAmps, a comparator, and digital logic for programming.

6.1 Example 5: FPAA Implementation of the Integer-Order Phase-Lead Compensator of Example 2

Figure 14a shows an implementation, using the FPAA AN231E04, equivalent to the closed-loop system designed in Example 2. The corresponding experimental setup is illustrated in the same figure. Details of the design and the corresponding transfer functions of every employed building block are listed in Table 6. The comparator producing the signal error $e(t) = V_{in}(t) - V_{out}(t)$ is realized using a CAM "SumDiff" (adder-subtractor) with gains 1 and -1 at each input. The integer order lead controller (44) is implemented by CAM "FilterBilinear 1" (Bilinear filter), designed to produce a transfer function with DC gain 2.3795, and one pole ($\omega_p = 60.24$ rad/s) and one zero ($\omega_z = 25.31$ rad/s), both denormalized by a factor $\Omega_f = 1000$ ($f_p = 9.57$ kHz, $f_z = 4.02$ kHz).

On the other hand, the plant (41) with $K = 1440$ is modeled by two low-pass filters $H_2(s) = -36/(s + 36)$ and $H_3(s) = -100/(s + 100)$ (CAM "Bilinear Filter 2" and CAM "Bilinear Filter 3", respectively), and by an "integrator" CAM, $H_1(s) = -40/s$, in series with a block of gain -1 (CAM "GainInv1"). The frequency denormalization $\Omega_f = 1000$ was realized by substituting in each block s by s/Ω_f.

The experimental configuration of Fig. 14a has a differential input of 0.5 V in amplitude and frequency 500 Hz, and a common-mode component of 1.5 V. It is provided by the array of three OpAmps and the target ELVIS II from National Instruments. This device also provides bias voltages of ± 15 V to the OpAmps. The output is converted from differential mode to simple mode with a differential amplifier. The output is measured with an Oscilloscope HD4096 Teledyne Lecroy (see Fig. 14b). The resulted overshoot was 21.1%. This overshoot corre-

Table 6 Details of the design in FPAA of the closed-loop system conformed by the plant (41) and the integer-order phase-lead controller (44) with $\Omega_f = 1000$

Operation	CAM	Transfer function	Characteristics
Comparator	SumDiff (Sum/difference)	$V_o = V_{in} - V_{out}$	$G_1 = 1$ $G_2 = -1$
Lead controller	Filter bilinear 1 (pole-zero filter)	$-G\left(\frac{1+s/\omega_z}{1+s/\omega_p}\right)$	$G = 1$ $f_p = \frac{\omega_p \Omega_f}{2\pi} = \frac{(60.24)(1000)}{2\pi} =$ 9.57 kHz $f_z = \frac{\omega_z \Omega_f}{2\pi} = \frac{(25.31)(1000)}{2\pi} =$ 4.02 kHz
Plant	Filter bilinear 2 (lowpass filter)	$-G\left(\frac{\omega_p}{s+\omega_p}\right)$	$G = 1$ $f_p = \frac{\omega_p \Omega_f}{2\pi} = \frac{(36)(1000)}{2\pi} =$ 5.74 kHz
	Filter bilinear 3 (lowpass filter)	$-G\left(\frac{\omega_p}{s+\omega_p}\right)$	$G = 1$ $f_p = \frac{\omega_p \Omega_f}{2\pi} = \frac{(100)(1000)}{2\pi} =$ 15.9 kHz
	Integrator	$\frac{K_{int}}{s}$	$K_{int} * \Omega_f = 40 * 1000 =$ $40000 = 0.04/\mu s$
	GainHold (Inverter amplifier)	$-G$	$G = 1$

G_i: gain of the CAM. f: corner frequency of the CAM. K_{int}: integration constant of the CAM

sponds to a second order system with phase margin 47° (compared with the theoretical value of 45.5°) and damping factor $\zeta = 0.443$. The simulation results of Example 2 and the experimental results of the system of Fig. 14a are in good agreement, thus validating this proposal of implementation.

6.2 Example 6: FPAA Implementation of the Fractional-Order Phase-Lead Compensator of Example 3

Consider the fractional-order phase-lead compensator with $q = 1.33$ presented in Example 3 (see (45)). This controller can be reformulated by using a first-order approximation of the fractional derivator $s^{\tilde{q}} = (Bs + 1)/(s + B)$, with $0 < \tilde{q} < 1$, resulting

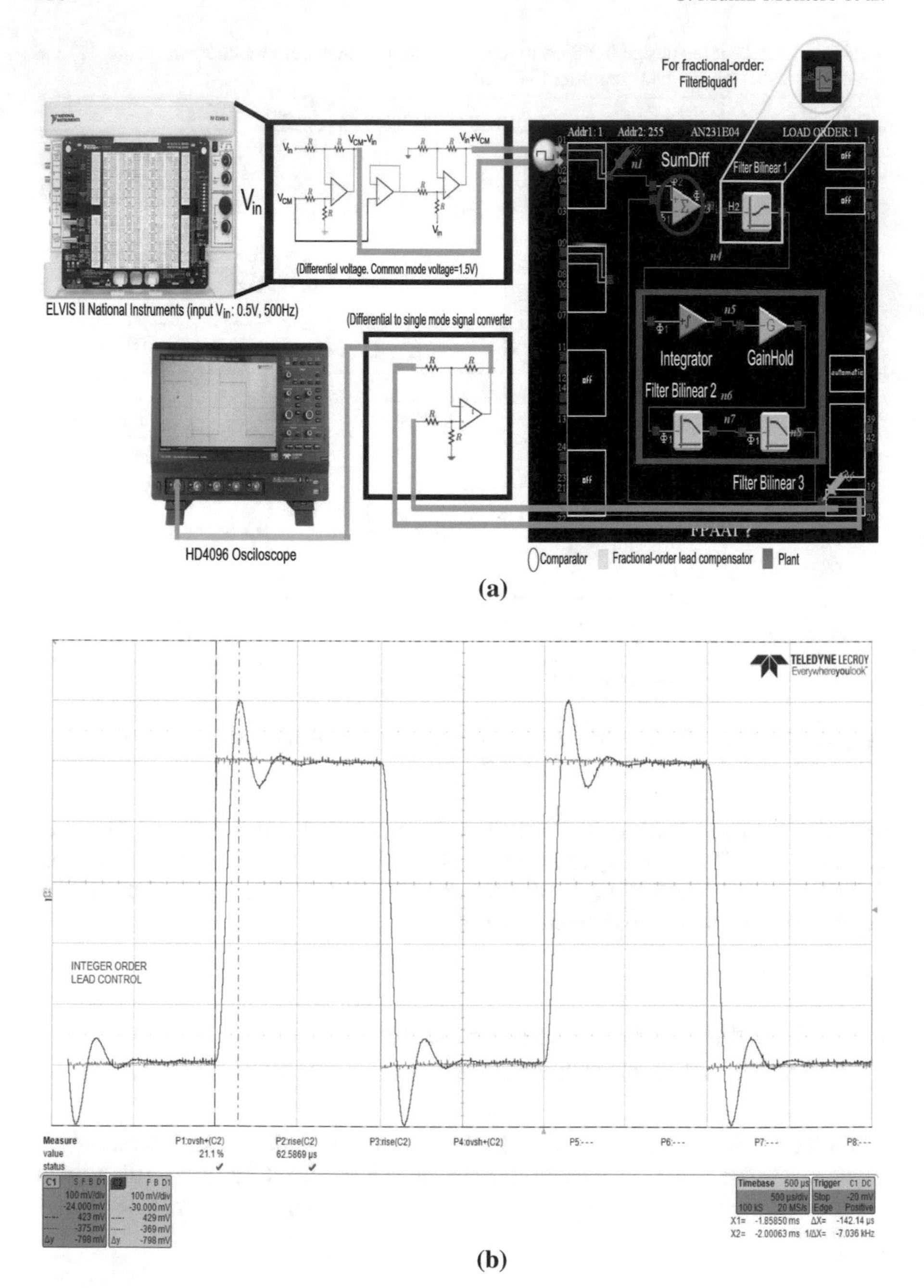

Fig. 14 **a** Experimental setup of the integer-order phase-lead compensator $C(s) = [1 + (2.3799)(0.0166)s]/[1 + 0.0166s]$ realized with an FPAA, in unity negative feedback configuration, with plant $G(s) = 144000/[s(s + 36)(s + 100)]$ also implemented in the FPAA. A denormalization in frequency $\Omega_f = 1000$ was performed. **b** Time-domain experimental results

$$C(s) = \frac{1+\alpha\tau s^q}{1+\tau s^q} = \frac{1+\alpha\tau s^{1+\tilde{q}}}{1+\tau s^{1+\tilde{q}}} = \frac{1+\alpha\tau s\left(\frac{Bs+1}{s+B}\right)}{1+\tau s\left(\frac{Bs+1}{s+B}\right)}$$

$$= \alpha\left(\frac{s^2+\frac{\omega_z}{Q_z}s+\omega_z^2}{s^2+\frac{\omega_p}{Q_p}s+\omega_p^2}\right) \tag{50}$$

where

$$\tilde{q} = q-1, \qquad s^{\tilde{q}} \approx \frac{Bs+1}{s+B}, \qquad B = \frac{1+\tilde{q}}{1-\tilde{q}} \tag{51}$$

and

Table 7 Details of the design in FPAA of the closed-loop system conformed by the plant (41) and the fractional-order lead controller (50) with $\alpha = 1.736, \tau = 8.1395 \times 10^{-3}, q = 1.33$ and $\Omega_f = 3000$

Operation	CAM	Transfer function	Characteristics
Comparator	SumDiff (Sum/Difference)	$V_o = V_{in} - V_{out}$	$G_1 = 1$ $G_2 = -1$
Lead controller	Filter biquad (pole-zero filter)	$G\left(\frac{s^2+\frac{\omega_z}{Q_z}s+\omega_z^2}{s^2+\frac{\omega_p}{Q_p}s+\omega_p^2}\right)$	$G = 1.736$ $f_p = \frac{\omega_p\Omega_f}{2\pi} = \frac{(11.084)(3000)}{2\pi} =$ 5.29 kHz $Q_p = 0.1776$ $f_z = \frac{\omega_z\Omega_f}{2\pi} = \frac{(8.4122)(3000)}{2\pi} =$ 4.016 kHz $Q_z = 0.2326$
Plant	Filter bilinear 2 (lowpass filter)	$-G\left(\frac{\omega_p}{s+\omega_p}\right)$	$G = 1$ $f_p = \frac{\omega_p\Omega_f}{2\pi} = \frac{(36)(3000)}{2\pi} =$ 17.22 kHz
	Filter bilinear 3 (lowpass filter)	$-G\left(\frac{\omega_p}{s+\omega_p}\right)$	$G = 1$ $f_p = \frac{\omega_p\Omega_f}{2\pi} = \frac{(100)(3000)}{2\pi} =$ 47.7 kHz
	Integrator	$\frac{K_{int}}{s}$	$K_{int} * \Omega_f = 40 * 3000 =$ $120000 = 0.12/\mu s$
	GainHold (Inverter amplifier)	$-G$	$G = 1$

G_i: gain of the CAM. f: corner frequency of the CAM. K_{int}: integration constant of the CAM

$$\omega_p = \frac{1}{\sqrt{\tau}}, \quad Q_p = \frac{\sqrt{\tau}B}{\tau + 1}, \quad \omega_z = \frac{1}{\sqrt{\alpha\tau}}, \quad Q_z = \frac{\sqrt{\alpha\tau}B}{\alpha\tau + 1} \tag{52}$$

This way, the fractional-order phase-lead compensator with $1 < q < 2$ can be realized by means of a biquad filter. Based on this idea, Fig. 14a shows the implementation in FPAA AN231E04 of the closed-loop controlled system designed in Example 3. Details of the design and the transfer functions of every building block are listed in Table 7. The comparator producing the signal error $e(t) = V_{in}(t) - V_{out}(t)$ is realized using a CAM "SumDiff" (adder-subtractor). The fractional-order lead controller (50) is implemented by CAM "FilterBiquad1" (Biquad filter, see Fig. 14a). The parameters of this block are calculated using (51) and (52) with $\alpha = 1.736$, $\alpha\tau = 1.4131 \times 10^{-2}$, $\tau = 8.1395 \times 10^{-3}$ and $\tilde{q} = 0.33$, been obtained: $\omega_p = 11.084$ rad/s, $\omega_z = 8.412$ rad/s, $Q_p = 0.1776$ and $Q_z = 0.2326$. Then, a frequency denormalization $\Omega_f = 3000$ is realized, resulting $f_p = 5.29$ kHz and $f_z = 4.02$ kHz. The same denormalization is performed in the plant, which is designed as in the Example 5 (see Fig. 14a). The design details are summarized in Table 7.

The experimental configuration is equal to the reported in Fig. 14a, but with the CAM "FilterBiquad1" instead of the CAM "FilterBilinear1". It has a differential input of 0.5 V in amplitude, offset 0.25 V and frequency 500 Hz, and a common-mode component of 1.5 V from the array of three OpAmps and target ELVIS II from National Instruments. This device also provides bias voltages of $\pm$15V to the OpAmps. The output is converted from differential mode to simple mode with a

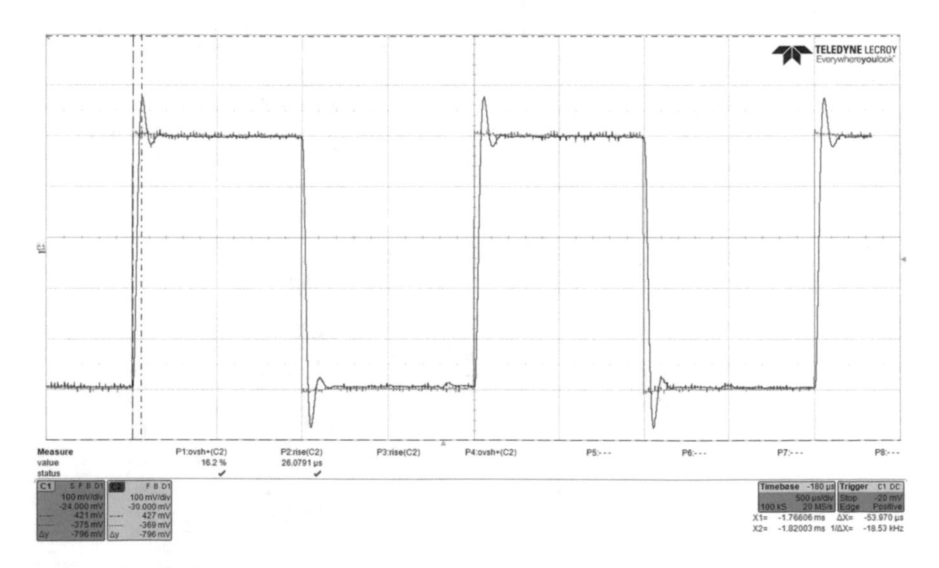

Fig. 15 Time-domain experimental result of the fractional-order phase-lead compensator $C(s) = [1 + (1.736)(8.1395 \times 10^{-3})s^{1.33}]/[1 + (8.1395 \times 10^{-3})s^{1.33}]$ realized with an FPAA, in unity negative feedback configuration, with plant $G(s) = 144000/[s(s + 36)(s + 100)]$ also implemented in the FPAA. A denormalization in frequency $\Omega_f = 3000$ was performed

differential amplifier. The output is measured with an Oscilloscope HD4096 Teledyne Lecroy (see Fig. 15). The resulted overshoot was 16.2%, compared with the 21.9% obtained with the implementation of Example 3. This overshoot corresponds to a second order system with phase margin 51.9° and damping factor $\zeta = 0.5$. The simulation results of Example 3 and the experimental results of Example 6 are similar, but with a difference of 5% regarding the overshoot. This difference is attributed to the different amplitudes and frequencies of the input signals (the FPAA has swing limits), the resolution of the programmable gains that can be implemented with the FPAA and the different denormalization frequencies.

7 Conclusion

The OPAMP-based and FPAA-based design of integer-order and fractional-order phase-lead-lag compensators for the case of $q \in (0, 1)$ and $q \in (1, 2)$, have been introduced. Each design was realized considering: (i) the parameters obtained with the procedures reported in [44] (integer-order) and [35] (fractional-order) when the gain and the phase required at a particular frequency are known for a desired time-domain response; (ii) Five configurations of compensators realized with basic OPAMP building blocks and (iii) Two more configurations (fractional and integer orders) with FPAAs. The OPAMP building blocks employed include inverter integrators, inverter and non-inverter amplifiers, differential amplifiers, weighted adders, first order low-pass filters, fractional-order phase-lead-lag ladders and fractional integrators. With all those blocks, the design equations were established taking care that some resistor and all capacitor values were chosen like degrees of freedom. Six design examples of both integer and fractional-order phase-lead-lag compensators were presented. In order to compare the performance of the different compensators, an integer-order plant was used. The compensation was made in series with the plant in a unity feedback loop. Simulation and experimental results agree with theory. An interesting result is the fact that the proposed solutions are good enough for the case of $q \in (1, 2)$ when the order of the approximation of s^q is one; however, those are not adequate when $q \in (0, 1)$. In those cases is necessary employ an approximation of higher order. It was evident with the third example that when the Cauer ladders are not employed, it is possible to synthesize realizations with commercial values of capacitors and resistors based on frequency and impedance transformations. Nevertheless, according with the fourth example, when $q \in (0, 1)$, the number of active elements can rise significantly since the approximation of the operator $s^{\pm\alpha}$ is required to be of high-order. In this case, the Cauer ladder solution might be an option despite of its inherent complexity for the computation of the commercial element values. To the author's knowledge, there very few (two) fractional-order phase lead-lag compensators reported in the literature. Therefore, the circuit solutions presented in this chapter offer useful alternatives that can be occupied for diverse controllers. Furthermore, some other applications such as memristors, filters and chaotic systems might benefit from the proposed strategies of implementation.

Acknowledgements This work was supported in part by the National Council for Science and Technology (CONACyT), Mexico, under Grant 181201, 222843 and 237991; in part by the Universidad Autónoma de Tlaxcala (UATx), Tlaxcala de Xicothencatl, TL, Mexico, under Grant CACyPI-UATx-2015; and in part by the Program to Strengthen Quality in Educational under Grant P/PROFOCIE-2015-29MSU0013Y-02

References

1. AN231E04 Datasheet Rev 1.2. (2012). http://www.anadigm.com/an231e04.asp
2. Angel, L., & Viola, J. (2015). Design and statistical robustness analysis of FOPID, IOPID and SIMC PID controllers applied to a motor-generator system. *IEEE Latin America Transactions, 13*(12), 3724–3734. doi:10.1109/TLA.2015.7404900.
3. Antoniou, A. (1972). Floating negative-impedance converters. *IEEE Transactions on Circuit Theory, 19*(2), 209–212.
4. Azar, A. T., & Serrano, F. E. (2014). Robust IMC-PID tuning for cascade control systems with gain and phase margin specifications. *Neural Computing and Applications, 25*(5), 983–995. doi:10.1007/s00521-014-1560-x.
5. Azar, A. T., & Serrano, F. E. (2015). Design and modeling of anti wind up PID controllers. In Q. Zhu, A. T. Azar (Eds.), *Complex system modelling and control through intelligent soft computations*. Germany: Springer. doi:10.1007/978-3-319-12883-21.
6. Azar, A. T., & Serrano, F. E. (2016). Stabilization of mechanical systems with backlash by PI loop shaping. *International Journal of System Dynamics Applications, 5*(3), 20–47.
7. Azar, A. T., & Vaidyanathan, S. (2015). *Chaos modeling and control systems design* (Vol. 581). Studies in computational intelligence. Germany: Springer.
8. Azar, A. T., & Vaidyanathan, S. (2015). *Computational intelligence applications in modeling and control* (Vol. 575). Studies in computational intelligence. Germany: Springer. ISBN 978-3-319-11016-5.
9. Azar, A. T., & Vaidyanathan, S. (2015c). *Handbook of research on advanced intelligent control engineering and automation*. Advances in computational intelligence and robotics (ACIR) book series. USA: IGI Global. ISBN 978-1-466-67248-2.
10. Azar, A. T., & Vaidyanathan, S. (2016). *Advances in chaos theory and intelligent control*. Studies in fuzziness and soft computing. Germany: Springer.
11. Biswas, K., Sen, S., & Dutta, P. K. (2006). Realization of a constant phase element and its performance study in a differentiator circuit. *IEEE Transactions on Circuits and Systems II: Express Briefs, 53*(9), 802–806. doi:10.1109/TCSII.2006.879102.
12. Boulkroune, A., Bouzeriba, A., Bouden, T., & Azar, A. T. (2016). Fuzzy adaptive synchronization of uncertain fractional-order chaotic systems. *Advances in chaos theory and intelligent control* (Vol. 337). Studies in fuzziness and soft computing. Germany: Springer.
13. Boulkroune, A., Hamel, S., & Azar, A. T. (2016). Fuzzy control-based function synchronization of unknown chaotic systems with dead-zone input. In *Advances in chaos theory and intelligent control* (Vol. 337). Studies in fuzziness and soft computing. Germany: Springer.
14. Cao, J., Liang, J., & Cao, B. (2005). Optimization of fractional order PID controllers based on genetic algorithms. In *2005 International Conference on Machine Learning and Cybernetics* (Vol. 9, pp. 5686–5689). doi:10.1109/ICMLC.2005.1527950.
15. Charef, A. (2006). Analogue realization of fractional-order integrator, differentiator and fractional $PI^{\lambda}D^{\mu}$ controller. *IEE Proceedings—Control Theory and Applications, 153*(6), 714–720.
16. Chen, Y., Petras, I., & Xue, D. (2009). Fractional order control—a tutorial. In *2009 American Control Conference* (pp. 1397–1411). doi:10.1109/ACC.2009.5160719.
17. Deepak, V. D., & Kumari, U. S. (2014). Modified method of tuning for fractional PID controllers. In *2014 International Conference on Power Signals Control and Computations, EPSCICON 2014* (pp. 8–10). doi:10.1109/EPSCICON.2014.6887481.

18. Dobra, P., Trusca, M., & Duma, R. (2012). Embedded application of fractional order control. *Electronics Letters*, *48*(24), 1526–1528. doi:10.1049/el.2012.1829.
19. Dorcák, L., Terpák, J., Petrá, I., Valsa, J., & González, E. (2012). Comparison of the electronic realization of the fractional-order system and its model. In *Carpathian Control Conference (ICCC), 2012 13th International* (pp. 119–124). doi:10.1109/CarpathianCC.2012.6228627.
20. Herrmann, R. (2011). *Fractional calculus* (1st ed.). World Scientific publishing Co.
21. Jesus, I. S., & Tenreiro, J. A. (2009). Development of fractional order capacitors based on electrolyte processes. *Nonlinear Dynamics*, *56*(1–2), 45–55. doi:10.1007/s11071-008-9377-8.
22. Khubalkar, S., Chopade, A., Junghare, A., Aware, M., & Das, S. (2016). Design and realization of stand-alone digital fractional order PID controller for buck converter fed DC motor. *Circuits, Systems, and Signal Processing*, *35*(6), 2189–2211. doi:10.1007/s00034-016-0262-2.
23. Krishna, B. T. (2011). Studies on fractional order differentiators and integrators: A survey. *Signal Processing*, *91*(3), 386–426. doi:10.1016/j.sigpro.2010.06.022.
24. Lachhab, N., Svaricek, F., Wobbe, F., & Rabba, H. (2013). Fractional order PID controller (FOPID)-toolbox. In *Control Conference (ECC), 2013 European* (pp. 3694–3699).
25. Ltifi, A., Ghariani, M., & Neji, R. (2013). Performance comparison on three parameter determination method of fractional PID controllers. In *2013 14th International Conference on Sciences and Techniques of Automatic Control and Computer Engineering (STA)* (pp. 453–460). doi:10.1109/STA.2013.6783171.
26. Mahmood, A. K., & Saleh, S. A. R. (2015). Realization of fractional order differentiator by analogue electronic circuit. *International Journal of Advances in Engineering and Technology*, *1*, 1939–1951.
27. Marzaki, M. H., Rahiman, M. H. F., Adnan, R., & Tajjudin, M. (2015). Real time performance comparison between PID and fractional order PID controller in SMISD plant. In *2015 IEEE 6th Control and System Graduate Research Colloquium (ICSGRC)* (pp. 141–145). doi:10.1109/ICSGRC.2015.7412481.
28. Mehta, S., & Jain, M. (2015). Comparative analysis of different fractional PID tuning methods for the first order system. In *2015 International Conference on Futuristic Trends on Computational Analysis and Knowledge Management (ABLAZE)* (pp. 640–646). doi:10.1109/ABLAZE.2015.7154942.
29. Monje, C. A., Chen, Y., Vinagre, B. M., Xue, D., & Feliu-Batlle, V. (2010). *Fractional-order systems and controls, fundamentals and applications* (1st ed.). Advances in industrial control. Springer.
30. Nise, N. S. (2011). *Control systems engineering* (6th ed.). Wiley. ISBN 978-0-470-91373-4.
31. Ou, B., Song, L., & Chang, C. (2010). Tuning of fractional PID controllers by using radial basis function neural networks. In *2010 8th IEEE International Conference on Control and Automation (ICCA)* (pp. 1239–1244). doi:10.1109/ICCA.2010.5524367.
32. Podlubny, I. (1999). Fractional-order systems and $PI^{\lambda}D^{\mu}$-controllers. *IEEE Transactions on Automatic Control*, *44*(1), 208–214. doi:10.1109/9.739144.
33. Podlubny, I., Petras, I., Vinagre, B. M., O'Leary, P., & Dorcak, L. (2002). Analogue realizations of fractional-order controllers. *Nonlinear Dynamics*, *29*(1), 281–296.
34. Raynaud, H. F., & Zergainoh, A. (2000). Brief state-space representation for fractional order controllers. *Automatica*, *36*(7), 1017–1021. doi:10.1016/S0005-1098(00)00011-X.
35. Tavazoei, M. S., & Tavakoli-Kakhki, M. (2014). Compensation by fractional-order phase-lead/lag compensators. *IET Control Theory and Applications*, *8*(5), 319–329. doi:10.1049/iet-cta.2013.0138.
36. Tepljakov, A., Petlenkov, E., & Belikov, J. (2011). FOMCON: Fractional-order modeling and control toolbox for MATLAB. In *Mixed Design of Integrated Circuits and Systems (MIXDES), 2011 Proceedings of the 18th International Conference* (pp. 684–689).
37. Tepljakov, A., Petlenkov, E., & Belikov, J. (2011). FOMCON: Fractional-order modeling and control toolbox for MATLAB. In *Proceedings of the 18th International Conference Mixed Design of Integrated Circuits and Systems—MIXDES 2011* (Vol. 4, pp. 684–689).

38. Tepljakov, A., Petlenkov, E., & Belikov, J. (2012). A flexible MATLAB tool for optimal fractional-order PID controller design subject to specifications. In *Control Conference (CCC), 2012 31st Chinese* (pp. 4698–4703).
39. Truong, V., & Moonyong, L. (2013). Analytical design of fractional-order proportional-integral controllers for time-delay processes. *ISA Transactions*, *52*(5), 583–591. doi:10.1016/j.isatra.2013.06.003.
40. Valerio, D., & da Costa, J. (2006). Tuning of fractional PID controllers with Ziegler Nichols-type rules. *Signal Processing*, *86*(10), 2771–2784. doi:10.1016/j.sigpro.2006.02.020.
41. Valerio, D., & Da Costa, J. S. (2006). Tuning-rules for fractional PID controllers. In *Proceedings of the 2nd IFAC Workshop on Fractional Differentiation and its Applications FDA06, Porto, Portugal.*
42. Valerio, D., & Da Costa, J. S. (2010). A review of tuning methods for fractional PIDs. In *Preprints, IFAC Workshop on Fractional Differentiation and its Applications, Badajoz, Spain.*
43. Varshney, P., & Gupta, S. K. (2014). Implementation of fractional Fuzzy PID controllers for control of fractional-order systems. In *Proceedings of the 2014 International Conference on Advances in Computing, Communications and Informatics, ICACCI 2014* (pp. 1322–1328). doi:10.1109/ICACCI.2014.6968376.
44. Wang, F. Y. (2003). The exact and unique solution for phase-lead and phase-lag compensation. *IEEE Transactions on Education*, *46*(2), 258–262. doi:10.1109/TE.2002.808279.
45. Xue, D., Liu, L., & Pan, F. (2015). Variable-order fuzzy fractional PID controllers for networked control systems. In *2015 IEEE 10th Conference on Industrial Electronics and Applications (ICIEA)* (pp. 1438–1442). doi:10.1109/ICIEA.2015.7334333.
46. Zhong, J., & Li, L. (2015). Tuning fractional-order $PI^{\lambda}D^{\mu}$ controllers for a solid-core magnetic bearing system. *IEEE Transactions on Control Systems Technology*, *23*(4), 1648–1656.
47. Zhu, Q., & Azar, A. T. (2015). Complex system modelling and control through intelligent soft computations. In *Studies in fuzziness and soft computing* (Vol. 319). Germany: Springer. ISBN 978-3-319-12882-5.

Robust Adaptive Supervisory Fractional Order Controller for Optimal Energy Management in Wind Turbine with Battery Storage

B. Meghni, D. Dib, Ahmad Taher Azar, S. Ghoudelbourk and A. Saadoun

Abstract To address the challenges of poor grid stability, intermittency of wind speed, lack of decision-making, and low economic benefits, many countries have set strict grid codes that wind power generators must accomplish. One of the major factors that can increase the efficiency of wind turbines (WTs) is the simultaneous control of the different parts in several operating area. A high performance controller can significantly increase the amount and quality of energy that can be captured from wind. The main problem associated with control design in wind generator is the presence of asymmetric in the dynamic model of the system, which makes a generic supervisory control scheme for the power management of WT complicated. Consequently, supervisory controller can be utilized as the main building block of a wind farm controller (offshore), which meets the grid code requirements and can increased the efficiency of WTs, the stability and intermit-

B. Meghni
Faculty of Applied Science, Department of Electrical Engineering,
University of KasdiMerbah, 3000 Ouargla, Algeria
e-mail: maghni_1990@yahoo.fr

D. Dib · S. Ghoudelbourk
Department of Electrical Engineering, University Larbi Tebessi,
Tébessa, Algeria
e-mail: dibdjalel@gmail.com

S. Ghoudelbourk
e-mail: sghoudelbourk@yahoo.fr

A.T. Azar (✉)
Faculty of Computers and Information, Benha University,
AlQalyubiyah, Banha, Egypt
e-mail: ahmad_t_azar@ieee.org; ahmad.azar@fci.bu.edu.eg

A.T. Azar
Nanoelectronics Integrated Systems Center (NISC),
Nile University, Cairo, Egypt

A. Saadoun
Department of Electronics, University of Badji Mokhtar,
Annaba, Algeria
e-mail: saadoun_a@yahoo.fr

A.T. Azar et al. (eds.), *Fractional Order Control and Synchronization of Chaotic Systems*, Studies in Computational Intelligence 688,
DOI 10.1007/978-3-319-50249-6_6

tency problems of wind power generation. This Chapter proposes a new robust adaptive supervisory controller for the optimal management of a variable speed turbines (VST) and a battery energy storage system (BESS) in both regions (II and III) simultaneously under wind speed variation and grid demand changes. To this end, the second order sliding mode (SOSMC) with the adaptive gain super-twisting control law and fuzzy logic control (FLC) are used in the machine side, BESS side and grid side converters. The control objectives are fourfold:

(i) Control of the rotor speed to track the optimal value;
(ii) Maximum Power Point Tracking (MPPT) mode or power limit mode for adaptive control;
(iii) Maintain the DC bus voltage close to its nominal value;
(iv) Ensure: a smooth regulation of grid active and reactive power quantity, a satisfactory power factor correction and a high harmonic performance in relation to the AC source and eliminating the chattering effect.

Results of extensive simulation studies prove that the proposed supervisory control system guarantees to track reference signals with a high harmonic performance despite external disturbance uncertainties.

Keywords Power management • A high performance • Supervisory control • Wind turbine • Fuzzy logic • Second order sliding mode control, power limit

1 Introduction

Because of the environmental problems, the oil crisis and the growing demand for energy, wind energy is one of the most mature of the different renewable energy technologies which received a lot of concern and attention in perceptible many parts of the world [1].

Nowadays, specifically related to offshore wind turbines, wind conversion technology showed new aspects of its construction and operation. Current WT operate at variable speed based PMSG and without speed multiplier, this type of wind turbines increase energy efficiency, reduce mechanical stress, can work with a high power factor and improve the quality of the electrical energy produced by WT compared to fixed speed [2].

For this type (VSWT), doubly fed induction generator (DFIG) and permanent magnet synchronous generators PMSGs are the most used technologies [3]. Today, due to its simple structure with characteristic self-excitation that can work with: a good performance, high reliability, good performance control and a great capacity for maximizing the power extracting by the MPPT, the PMSG topology is required, it is recommended to be connected to the variable speed wind turbines (VSWTs) [4]. Moreover, this technology is better in the case of offshore wind, as maintenance is simpler and less expensive compared to a technology using a gearbox.

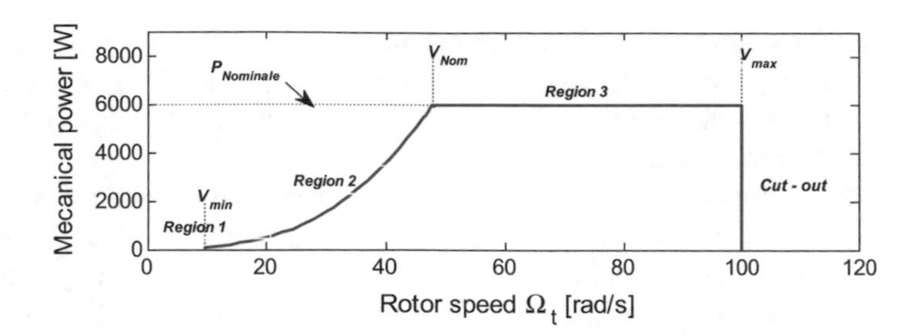

Fig. 1 Operation regions for VSWT

Practically and for safety reasons turbines and uniform stability between supply and demand of energy, wind turbines work only in a specified range of wind speeds limited by $(v\,cut - in)$ and $(v\,cut - out)$ as is shown in Fig. 1, where the possibility of three different operating zones [5].

- Region 1: when the wind speed is below the speed $(V\,cut - in)$ wind, no maximize efficiency that occurs in this region.
- Region 2: when the wind speed exceeds $(V\,cut - in)$ but under the rated $(Vnom)$. In this area the main controller is to increase the efficiency of the power extracted from the WTs, so it operates at its maximum power point (MPP).
- Region 3: when the wind speed is greater than (Vnom) but under the cut – out wind speed (Vmax); the task of the controller is to keep the power captured at a fixed or nominal value instead of trying to maximize it. Another important controller objective in this last region, is to keep the electrical and structural conditions in a safety region [5, 6].

Despite this characteristic, the utmost challenge of wind power generation is the inherently sporadic nature of the wind which can deviate quickly. Its intermittent availability is the main impediment to power quality and flow control. Therefore, the stability and power quality of the grid operation is affected. Consequently, the fluctuations of wind power should be reduced to prevent a degradation of the grid's performance [7].

In the new universal grid code for wind power generation, the power oscillation damping by wind energy is included. For instance, the energy storage system (ESS) is integrated with the renewable sources which are connected into the power grid to maintain the safe operation of the power grid, balance the supply and demand sides, and enhance fault ride-through ability and damp short-term power oscillation [8]. There are different types of ESSs in the power systems such as batteries ESS, superconducting magnetic ESS, compressed air, hydrogen ESS, gravitational potential energy with water reservoirs, electric double layer capacitor, and flywheels [9]. The BESS is one of the most rapid growing storage technologies. The BESS installation cost and generating noise are relatively lower than the other storage technologies.

However, these features remain restricted with respect to the practical domain of variation of wind speed. Therefore, effective architecture of control systems are one of the major issues in hybrid VSWT (managing variations in load demand and to extract maximum power) to prevent possible degradation on the quality of electrical energy delivered into the electric grid, where variations of the loads and generations are significant in the system.

This goal has been and is a motivation for researchers and investors to search on robust and effective control strategies for VSWT to overcome various constraints such as the optimal tracking point controller has been developed in many literatures especially for wind energy systems [5], to track the maximum wind power available at each instant. They include namely, tip speed ratio (TSR) [5], sliding mode control (SMC) [10], the hill-climb searching (HCS) [5] and fuzzy logic control (FLC) [11] techniques. Several methods of power limitation control and pitch power control have been advanced in some studies [12–14]. Effective techniques have been developed in many papers to control the optimal rotational speed of the wind turbine in order to determine the maximum power coefficient for a given wind speed. Some of these variable speed techniques are FOSMC [15], SOSMC [16] and FLC [17], robust control [18]. Numerous recent studies [9] advanced the benefit of the energy storage techniques to mitigate the unpredictable character of wind energies, ensuring more efficient management of the available resources and provide operation freedom to wind generation that allows time-shifting between generation and demand. More robust and efficient strategies based on modern control techniques such as fuzzy logic control [19], robust control [20] and sliding mode control [21] have been widely developed and implemented to smooth regulation of the grid active and reactive powers exchanges between the PMSG and the grid.

1.1 Contribution of the Paper

Based on our previous research experience on WT/battery hybrid power system and their supervisory control system [22], a key contribution of this study is to present a comprehensive model of the proposed structure based on WT/battery hybrid power system and to implement **the novel supervisory control system in order to optimize the power output and protecting all the system.** This proposed management strategy is achieved through the combination of the efficient and robust control methods described in recent publications in order to develop a perfect wind system.

Furthermore, contrary to the works found in the literature, this work presents a new approach to control and management for extended Control & of VSWT that includes simultaneously the two **operating regions (II and III)** whatever the speed variation the wind. This adaptation **(commutation of control system in both regions)** may provide better performance in all possible operational scenarios of the wind: extract the maximum power from the wind, storage of excess power, compensation of power between supply and demand, limiting the upper power at nominal generator or demand.

For this reason (To reach this goal), the control objectives are three in number; the first on the generator side converter, the second on bidirectional DC/DC converter and the last one on the grid side converter.

1. The main function of the generator side controller is to track the maximum power via speed loop control based on SOSMC. An MPPT control algorithm, based on an FLC, has been used to regulate the rotational speed in order to tack accurately the PMSG working point to its maximum power point (MPP) and to derive the rotational speed reference. (Force the working point of the PMSG to its MPP and to provide the rotational speed reference). In the case studied in control subsystem, two adaptive and commutative operation modes are distinguished:
   - When the aerodynamic power is not enough to reach the synchronous speed, the system operates at mode 1(tracking mode MPPT): maximum power extraction, whereas, if the wind speed exceeds the rated value, the system switches to mode 2 (Power Limitation mode): power limitation, which leads the PMSG to provide its rated power below nominal speed of the rotor.
2. A bidirectional DC/DC converter is connected between the battery and the DC bus. This converter is controlled to maintain the DC bus voltage at its rated value, allowing the active power, bi-directional active power flow from the battery, through the charge/discharge of the device in response to the variations of the operating conditions (regardless of the variations of the operating conditions).
3. In the grid side converter, a SOSMC controller has been used to achieve smooth regulation of active and reactive power quantities exchange between the BESS and the grid according to grid demand under real fluctuating wind speed (region II and III).

1.2 Organization of the Research Work

The paper is organized as follows. In Sect. 2, are view of the previous related work is presented. In Sect. 3, the modeling of the PMSG wind turbine and ESS is described. Section 4 presents in detail the proposed supervisory control applied in this work. The simulations results and robustness test are illustrated in Sect. 5. Finally, the conclusions of the obtained results and future work are shown in Sect. 6.

2 Related Work

Much research has been do in recent years to improve system performance monitoring and management of the energy produced by the WTs where several techniques have been proposed various structures of WT, such as artificial neural networks (ANN) [23–25], FLC [26–28], the first order sliding mode [15, 29, 30] and the SOSMC [31–33]. The main reason for this interest is that these techniques and structures were used to achieve a new and robust optimization that actually works for VSWT problems that cannot be solved by conventional techniques. Similarly, recent contributions remained within certain limits due to multiple ignored or neglected issues such as MPPT, power control, power limit, speed control and energy storage system. These problems contribute to the degradation of the performance of WTs.

In Ref. [34], a PSIM software integration study was performed in more than one MPPT used FLC technique where the results are limited, because the study is applied to as simple structure without a storage system, security system in rated wind speed and without power control for the grid side.

Moreover, in the second part of the study by [35], the authors proposed a technique based on the strategy FOSMC to control the active and reactive power injected to the grid. The simulation results show a poor power quality produced in the presence of the chattering phenomenon.

Furthermore, the reference [3] shows an original method for the sensorless MPPT of a small power wind turbine using a permanent magnet synchronous generator (PMSG). On the other hand the operating range of this system is limited in the region 2, because it not contains any limit power control or pitch control in rated wind speed.

On [9, 36–38] the authors treated with details, modeling and control of a hybrid system composed of DFIG/WT and ESS. The results are convincing, but the authors have not applied effective and robust control techniques in key points such as MPPT, speed control, power control of the grid, which makes these reliable studies only normal and stable working conditions (wind speed constant, no default in the wind channel and no fault in the grid).Also In these studies the authors adopted only on the operating area 2 (This is what makes activity limited by WT).

The advanced control algorithms FLC and SOMSC were proposed by [11] to simulate VSWT-based DFIG. Although the results were attractive, choosing the type of machine used remains not determinant. Further- more, the authors assessed a method of limiting the power extracted from a gust of wind, while they have ignored the use of storage means to earn excess wind power. In the Refs. [39–42], the authors have developed MPPT techniques to optimize the performance of the power extracted from the wind in (region II). These contributions have given a great success in this field. These studies are not considered to be a perfect solution unless we can successfully integrate advanced techniques for power management to consumption (power control). Other hand, when the demand for power exceeds the power extracted, the MPPT technology will not be able to satisfy the demand; hence, the need for a compensation system ESS.

In reference [22] the authors generally succeed in the choice of the supervisory control system which applied in hybrid VSWT. But they are neglect the protection of the wind system (electric and mechanic) under the rated wind speed in addition they are limiting the operating range of wind turbine.

Despite the amazing results presented in the previous studies, these remains limited and insufficient to face various restrictions (because these control strategies are usually used separately for each type of study to VSWT). The new supervisory control system proposed in WTs has become a solution necessary for the optimal management of the electric and mechanical power in any part of the wind channel. This new system is based on the combination in the same type WT of new algorithms (robust, effective and flexible) to work in various areas simultaneously (Region II and III) with high precision, which makes it universal for different types of WT.

3 Dynamic Model of PMSG-WT

In this section the studied system is presented. In order to achieve the system control and a first study by simulation before implementation, its modelling is required. Figure 2 shows a representative topology of the investigated wind energy system. As illustrated in this figure the offshore VSWT a horizontal axis turbine with a three-bladed rotor design directly transmits the aerodynamic torque and power to PMSG (without a gearbox). The generator power is then fed to the utility grid by means of power electronic devices (two back-to-back IGBT bridges AC/DC/AC) interconnected by a common DC bus. This WT is supported by an ESS associated with DC bus system composed of a lead acid battery with a bidirectional DC/DC converter.

3.1 Wind Turbine Aerodynamic

For a variable speed wind turbine, the mechanical power and torque extracts from the wind turbine is proportional to the wind speed and can be calculated by the following formulas [10]:

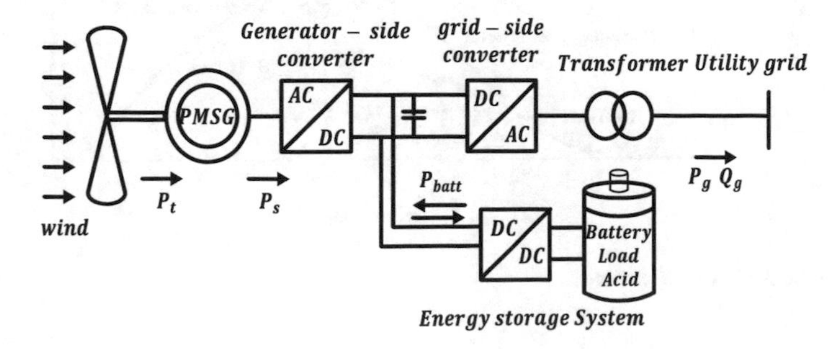

Fig. 2 Wind generation system configuration

$$P_t = \frac{1}{2}\rho\pi R_t^2 v^3 C_p(\lambda, \beta) \tag{1}$$

$$T_t = \frac{P_t}{\Omega_t} = \frac{1}{2\lambda}\rho\pi R_t^3 v^2 C_p(\lambda, \beta) \tag{2}$$

Which λ present the ratio between the wind speed and the turbine angular speed. This ratio is called tip speed ration:

$$\lambda = \frac{R_t \Omega_t}{v} \tag{3}$$

where: λ is the tip speed ratio, C_p is the power coefficient, β is the pitch angle, Ω_t is the rotor speed (rad/s), R_t is the rotor-planeradius (m), ρ is the air density (Kg/m^3).

The coefficient C_p is a variable magnitude as a function of λ, the theoretically possible maximum value of the power is β. The C_p is different for each wind turbine, as shown in [43]. Theoretically, the Betz limit is ≈0.5926 further and practically, friction and the force dragged reduce this value to 0.5 for large wind turbines [44]. Calculating another analytic expression $C_p(\lambda)$ for different values of β is also possible.

For a pitch angle β given, the analytical expression commonly used is a polynomial regression as follows [45]:

$$C_p = 0.073\left(\frac{151}{\lambda_i} - 0.058\beta - 0.002\beta^{2.14} - 13.2\right)e^{\frac{-18.4}{\lambda_i}} \tag{4}$$

Where $\lambda_i = \frac{1}{\frac{1}{\lambda - 0.02,\beta} - \frac{0.003}{\beta^2 + 1}}$

The aerodynamic characteristics of a variable speed turbine are usually represented by the relationship $C_p(\lambda)$ as illustrated in Fig. 3.

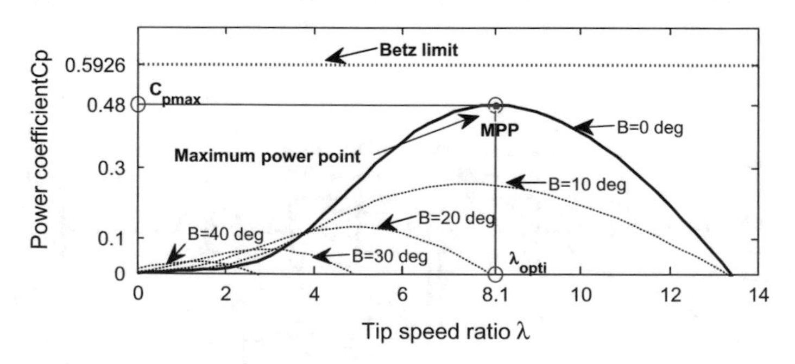

Fig. 3 Typical C_p versus λ curve

From this figure and according to Eqs. (1, 3), we can conclude that for a fixed value of $\beta = 0$, C_p only becomes a nonlinear function of λ. According to Eq. (3), there is a relationship between λ and Ω_t and at some, power is maximized at some Ω_t optimal speed Ω_{topti}. This rate corresponds to a λ_{opti}. The value of λ is constant for all the maximum power point (MPP) [5].

Thus, to extract maximum power at wind speeds of variable λ must be adjusted to its optimum value λ_{opti} followed a maximal power coefficient value C_p, to follow the optimum operating point. From Eqs. (1, 3), we get [46]:

$$P_{tmax} = \frac{1}{2}\rho\,\pi\,R_t^5\left(\frac{C_{p.max}}{\lambda_{opti}^3}\right)\,\Omega_{topti}^3 \tag{5}$$

$$\Omega_{topti} = \frac{\lambda_{opti}v}{R_t} \tag{6}$$

3.2 Model of PMSG

The simple dynamic model of three-phase PMSG in d, q reference frame can be represented by the following voltages equations [45]:

$$V_{sd} = R_s I_{sd} + L_d\frac{dI_{sd}}{dt} - \omega_e L_q I_{sq} \tag{7}$$

$$V_{sq} = R_s I_{sq} + L_q\frac{dI_{sq}}{dt} + \omega_e(L_d I_{sd} + \psi_m) \tag{8}$$

Where $V_{sd}, V_{sq}(V)$ are the direct and quadrature components of the PMSG voltages, R_s, L_d and L_q respectively are the resistance, the direct and the quadrature inductance of the PMSG winding, $\psi_m(wb)$ represents the magnet flux, $\omega_e(rad/s)$ is the electrical rotational speed of PMSG $I_{sd}, I_{sq}(A)$ are the direct and quadrature components of the PMSG currents respectively.

Now the mechanical dynamic equation of a PMSG is given by [47]:

$$\frac{d\Omega_t}{dt} = \frac{1}{J_T}T_e + \frac{D_T}{J_T}\Omega_t - \frac{1}{J_T}T_t \tag{9}$$

The electromagnetic torque of a p-pole machine is obtained as [47]:

$$T_e = \frac{3}{2}n_p(\psi_m I_{sq} + (L_d - L_q)I_{sd}I_{sq}) \tag{10}$$

Where $(N.m)$ is the electromagnetic torque,n_p is the number of pole pairs, D_T is the damping coefficient, J_T is the moment of inertia.

3.3 Model of the Grid

The dynamic model of the grid connection in reference frame rotating synchronously with the grid voltage is given as follows [10].

$$V_{dg} = V_{di} - R_g I_{dg} \quad L_{dg} \frac{dI_{dg}}{dt} + L_{qg} w_g I_{qg} \tag{11}$$

$$V_{qg} = V_{qi} - R_g I_{qg} - L_{qg} \frac{dI_{qg}}{dt} - L_{dg} w_g I_{dg} \tag{12}$$

The DC-link system equation can be given by:

$$C \frac{dV_{DC}}{dt} = \frac{3}{2} \frac{V_{dg}}{V_{DC}} I_{dg} - I_{DC} \tag{13}$$

Where: $V_{dg}, V_{qg}(V)$ are the direct and quadrature components of the grid voltages, $V_{di}, V_{qi}(V)$ are the inverter voltages components, (R_g, L_{dg}, L_{qg}) are resistance, the direct and quadrature grid inductance respectively, $I_{dg}, I_{qg}(A)$ are the direct and quadrature components of the grid currents respectively, V_{DC} is the DC-link voltage, I_{DC} is the grid side transmission line current and C is the DC-link capacitor.

The power equations in the synchronous reference frame are given by [34]:

$$P_g = \frac{3}{2}(V_{dg} I_{dg} + V_{qg} I_{qg}) \tag{14}$$

$$Q_g = \frac{3}{2}(V_{dg} I_{qg} - V_{qg} I_{dg}) \tag{15}$$

After orienting the reference frame along the grid voltage, V_{qg} equals to zero by aligning the *d*-axis. Then, the active and reactive power can be obtained in this new reference from the following equations [34]:

$$P_g = \frac{3}{2} V_{dg} I_{dg} \tag{16}$$

$$Q_g = \frac{3}{2} V_{dg} I_{qg} \tag{17}$$

3.4 Model of the ESS

The lead-acid battery used in this work is modeled by the battery model included in Sim Power Systems [48] where it is modeled as a variable voltage source in series

Fig. 4 Lead-acid battery simplest model [49]

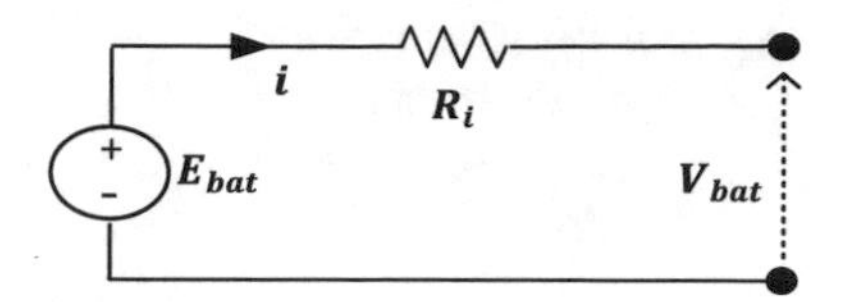

with an equivalent internal resistance (see Fig. 4). The battery voltage is given by Eq. (18).

$$V_{bat} = E_{bat} - R_i i \tag{18}$$

The voltage of the rated load for the period of the charging or discharging of the battery depends on the internal battery parameters such as: the battery current, the hysteresis phenomenon of the battery during the charging and discharging cycles and the capacity extracted [36].

$$\begin{cases} E_{batdisch} = E_0 - K\frac{Q}{Q-i_t}i^* - K\frac{Q}{Q-i_t}i_t + f_{hyst-disch}(i) \\ E_{batcharg} = E_0 - K\frac{Q}{0.1Q+|i_t|}i^* - K\frac{Q}{Q-i_t}i_t + f_{hyst-char}(i) \end{cases} \tag{19}$$

Where V_{bat} is the battery rated voltage, $E_{batdisch}$ is the discharge voltage,$E_{batcharg}$ is the charge voltage, E_0 is internal EMF, R_i is internal resistance, K is the polarisation constant (V/Ah), Q is battery capacity (Ah), i_t is the actual battery charge and i^* is the filtered current.

In most electrochemical batteries, it is important to maintain the SOC within limits recommended to prevent internal damage. The instantaneous value of the load condition is calculated by [36]:

$$SOC = 1 - \frac{Q_e}{C(0,\theta)}, \quad DOC = 1 - \frac{Q_e}{C(I_{avg},\theta)} \tag{20}$$

Where:SOC is battery state of charge, $Q_e(A.s)$ is the battery's charge, DOC is battery depth of charge, $I_{avg}(A)$ is the mean discharge current, $C(A.s)$ is the battery's capacity.

3.5 *Battery Converter Modeling*

Different converters based PWM DC/DC are used to connect the various energy sources to the DC bus, these converters are used to control the flow of energy between sources to maintain the DC bus at a constant value. In this present work, the electro-chemical battery is connected to the DC bus of PMSG through a bidirectional converter (buck-boost) DC/DC power. The structure of this converter is shown in Fig. 5. It consists of a high-frequency inductor, an output filtering capacitor, and two IGBT-diodes switches.

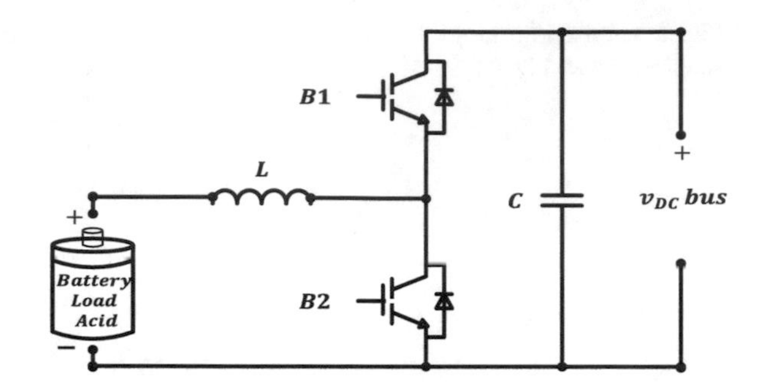

Fig. 5 Bidirectional DC/DC battery power converter

This makes it possible to charging and discharging the battery in both directions to keep the DC bus voltage to a reference value independent of variations the battery voltage. During the charging phase, the power flows from the DC link bus to the BESS through the B_1 switch and B_2 diode. Therefore, the converter can acts like a unidirectional buck converter. On another side, the battery discharges through the B_2 switch and B_1 diode, furnishing energy to the DC bus. In this period, the converter acts like a unidirectional boost converter [7].

4 Proposed Control System

The supervisory control system of hybrid WT has the responsibility to provide appropriate regulation, stability, protection, optimization and tracking objectives for the WT rotor speed in various constraints (supply and demand changing, sporadic nature of the wind, different operating region (region II or III) and the unexpected faults in the grid).

In order to achieve this goal, we have optimized our concept of classical supervisory control [22], for makes the WT operate in a wide range of wind speeds (region II and III).

The major objective most of the control systems used in this paper are:

- In Region II: Capture of maximum energy from the wind, through the combination advanced control based on FLC-SOSMC applied to machine side converter (MSC).
- In Region III: Power limitation, above the rated wind speed, this control must limit the extracted power by adaptive control (FLC-SOSMC-MPPT and the sensed extracted power as a feedback).
- In both regions (II and III): Managing energy between generated and consumed energies of the hybrid system components using the supervisory controller (through the battery side converter BSC).
- In both regions (II and III): Power quality improvement, through the robust and efficiency strategy control (SOSMC) applied in the grid side converter (GSC).

4.1 Control of the Machine Side Converter (MSC)

VSWTs are designed to achieve maximum aerodynamic effective on a wide range of wind speeds, which sometimes include several areas of operation. However, this degree of freedom requires a system of speed/sophisticated and robust power control to overcome various constraints to monitor the point of maximum available power (Region II) and to limit the power captured when the wind speed exceeds a certain par value (region III). In this section, we showed the design of the machine side controller as shown in Fig. 6. (Control side of the converter machine (MSC)) that includes two additional operating modes (adaptive).

1. **Tracking mode (Region II)**

In this area and to meet the total demand for power, the turbine operates at variable speed under a nominal wind speed between $(vcut - in)$ and $(vcut - out)$. For this reason the cascade control structure with two control loop have been created. In the outer loop, a (MPPT) algorithm based on advanced technology FLC-SOSMC is designed to permanently extract the optimal aerodynamic energy in order to generate the electromagnetic torque reference. Whereas, in the inner loop we controlled the *dq*-axis current of the generator according to the Eq. (8) and by field oriented control strategy (FOC) of the PMSG to ensure that the system works around the optimal point, which corresponds to the maximum power extracted by the turbine [50].

For a fixed value of $\beta = 0$ and for each wind speed, the MPPT algorithm uses fuzzy logic controller generates the reference speed that maximizes the extracted power from the turbine as shown in Fig. 7.

To ensure a quick and smooth tracking of the maximum power without the knowledge of the characteristic of the turbine and the wind speed measurement,

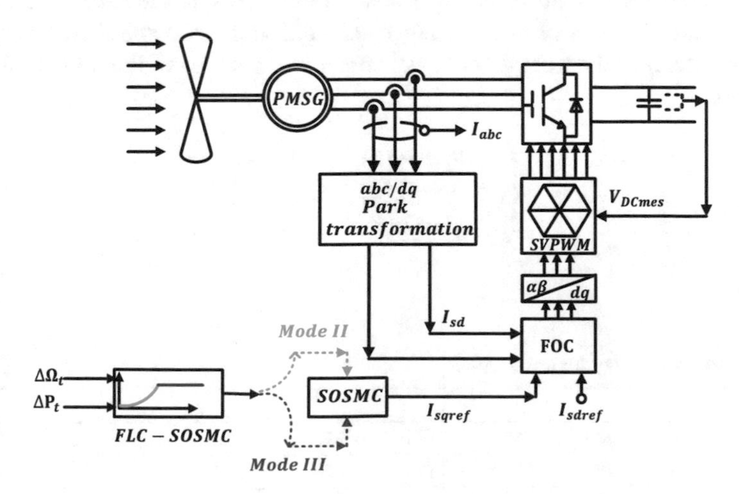

Fig. 6 Control block diagram of machine side converter

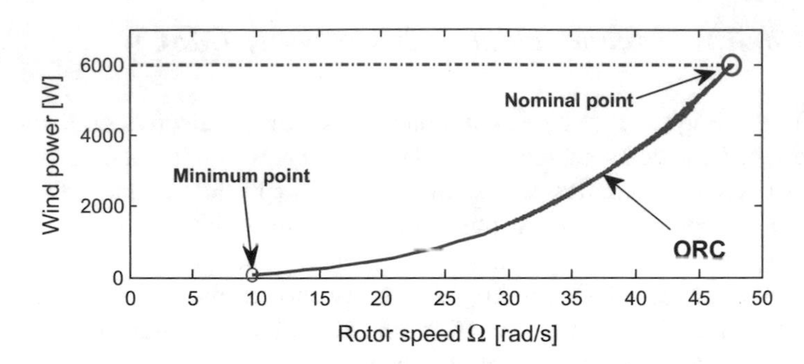

Fig. 7 Real tracking the optimal rotational speed ORC in "region II"

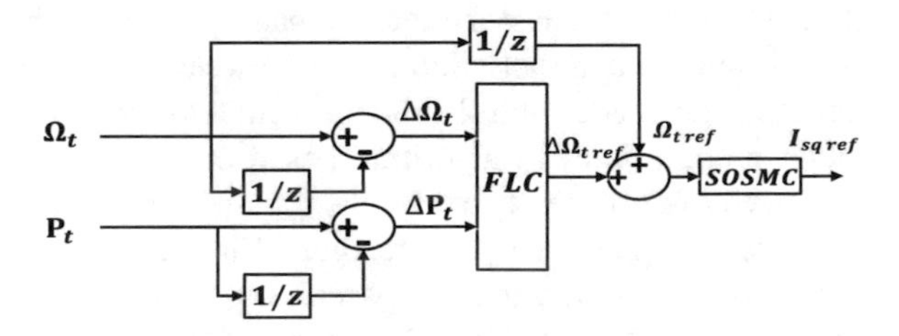

Fig. 8 Input and output of fuzzy controller

FLC technique is used for generating a reference speed allowing the WT run around the maximum points at varying wind speeds as shown in Fig. 8. This technique has become universal for the different types of WT [11]. The proposed fuzzy controller has two inputs and an output. The base rule of the system is given in Table 1, and the variation step in the speed reference and power is indicated in Fig. 9.

The Eq. (21) show that the relationship between the optimum speed rotation, the extracted power and wind speed are linear. For this reason, the MPPT-FLC device based on measurement of power change ΔP_t and rotational speed $\Delta \Omega_t$ propose a variation $\Delta \Omega_{tref}$ of the turbine rotational speed reference according to the following equations:

$$\begin{cases} \Delta P_t = P_t(k) - P_t(k-1) \\ \Delta \Omega_t = \Omega_t(k) - \Omega_t(k-1) \\ \Omega_{tref} = \Omega_t(k-1) + \Delta \Omega_{tref} \end{cases} \tag{21}$$

Table 1 Rules of fuzzy logic controller

$\Delta \Omega_t$	ΔP_t								
	NBB	NB	NM	NS	ZE	PS	PM	PB	PBB
N	PBB	PB	PM	PS	ZE	NS	NM	NB	NBB
ZE	NB	NB	NM	NS	ZE	PS	PM	PB	PB
P	NBB	NM	NS	NB	ZE	PM	PM	PM	PBB

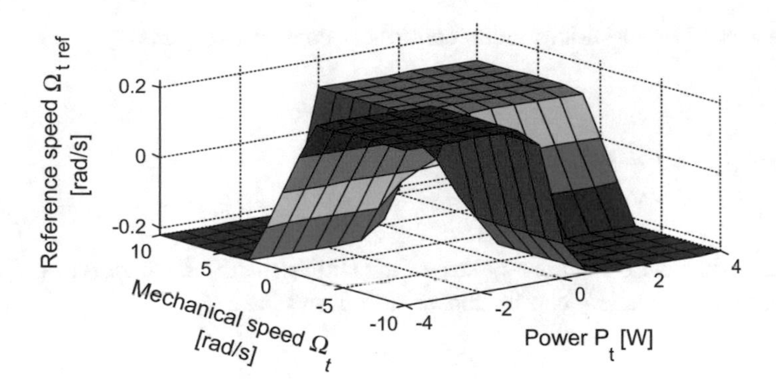

Fig. 9 Control surface of 2D fuzzy controller

Fig. 10 Detailed block diagram of the power limit

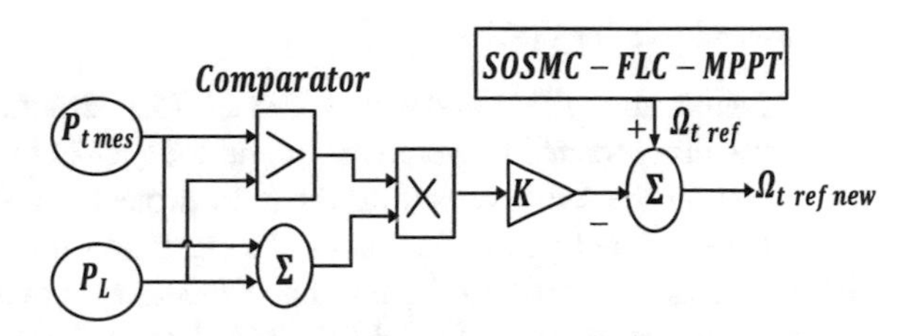

In the same part, the reference speed (along which tracks the MPPs) is used in the speed regulator input to generate the q-axis current component as shown in Fig. 8. Therefore, a novel SOSMC algorithm is proposed to achieve the speed control of PMSG for each wind speed in order to maximize the extracted power at the turbine output. In this algorithm, the chattering phenomenon can be limited so as to improve the PMSG performance when compared to the classical FOSMC (Fig. 10).

Let us introduce the following sliding surface for the speed Ω.

$$s_{\Omega_t} = \Omega_{tref} - \Omega_t \tag{22}$$

Then we have:

$$\dot{s}_{\Omega_t} = \dot{\Omega}_{tref} - \dot{\Omega}_t = \dot{\Omega}_{tref} - \frac{1}{J}(T_t + T_e - F\,\Omega_t)$$

If we define the functions G_{Ω_t} as follows:

$$G_{\Omega_t} = \dot{\Omega}_{tref} - \frac{1}{J}(T_t - F\,\Omega_t)$$

Thus: $\ddot{s}_{\Omega_t} = \dot{G}_{\Omega_t} - \frac{\dot{T}_e}{J}$

The control algorithm proposed which is based on super twisting algorithm (ST) has been introduced by Levant [51].

The second order sliding mode controllers contain two parts:

$$I_{sqref} = I_{sqeq} + I_{sqN} \tag{23}$$

Where: $I_{sqN} = I_1 + I_2$

With: $\begin{cases} \dot{I}_1 = -N_1 sign(s_{\Omega_t}) \\ I_2 = N_2 \sqrt{|s_{\Omega_t}|} sign(s_{\Omega_t}) \end{cases}$

In order to ensure the convergence of the sliding manifolds to zero in finite time, the constants N_1 and N_2 can be chosen as follows [52]:

$$\begin{cases} N_1 > \frac{\mu_i}{L_g} \\ N_2 \geq \frac{\mu_i(k_i + \mu_i)}{L_g^2(k_i - \mu_i)} \end{cases} \tag{24}$$

2. **Power limit (region III)**

In this region, above the rated wind speed 11.75 m/s as shown in Fig. 11. The control must limit the extracted power in the tolerable beach between $P_l - 1.2P_l$ and the Ω_t of the turbine in the stable operation mode to protect the wind turbine and PMSG, so when the extracted power increases during the nominal value the control circuit shown in Fig. 10, lowered the reference speed, to prevent steady-state high power amounts. A speed controller circuit added to the previous SOSMC-FLC-MPPT design, the sensed stator power as a feedback. When the power exceeds the maximum value, the reference speed must be reduced by the amount, K [53]. The new reference speed $\Omega_{t\,ref\,new}$ is given by:

$$\Omega_{t\,ref\,new} = \Omega_{t\,ref} - \Delta\Omega_t \tag{25}$$

4.2 Control of the Battery (ESS) Side Converter (BSC)

This converter is controlled in order to maintain the DC bus voltage close to its nominal value (800 v) at different operating conditions (wind speed). Since

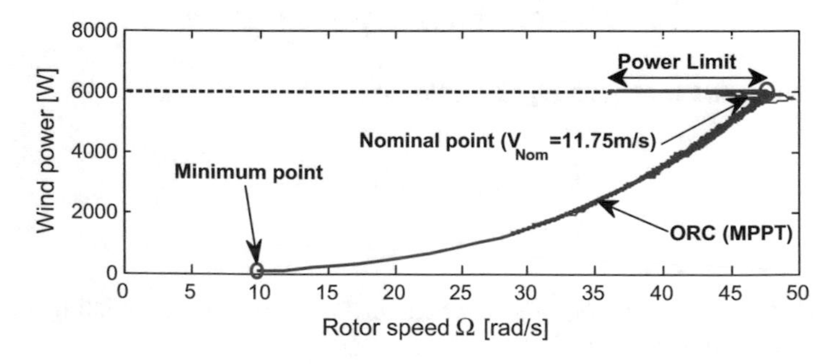

Fig. 11 Real tracking the optimal rotational speed ORC in "region II and III"

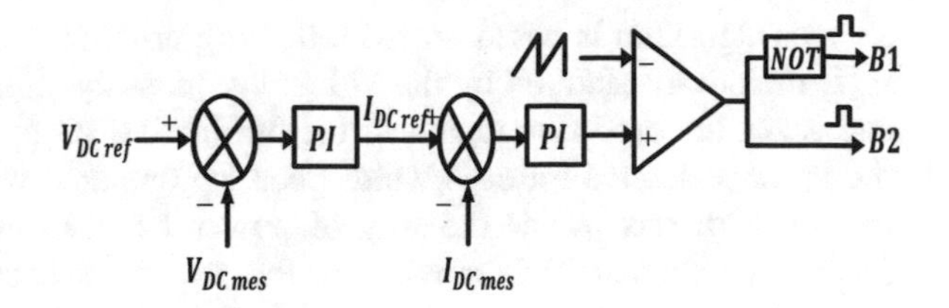

Fig. 12 Control block diagram of the Battery ESS

increasing the output power (grid side) rather than the input power to DC-link capacitor (PMSG side) causes a decrease of the ESS voltage and vice versa.

More detailed, Fig. 12, describes the control strategy for the bidirectional DC/DC converter; this controller uses contains two cascaded control loops. The outer control loop compares the measured DC link voltage V_{DC} link to the desired $V_{DC\,ref}$ DC link in order to generate the reference battery current $I_{DC\,ref}$ for the inner control loop. The current $I_{DC\,ref}$ is compared to the measured battery current $I_{DC\,mes}$ in order to generate the gating signals for the IGBT switches. The DC/DC converter charges or discharges the battery according to the duty ratio of the two IGBT switches [54].

The algorithm presented below in Fig. 13 describes excessively different operating modes (charging and discharging of the battery) in nominal and instantaneous variations of the following variables (wind speed (T, N), rated power the turbine (N), the supply and demand side of the network power and the power extracted).

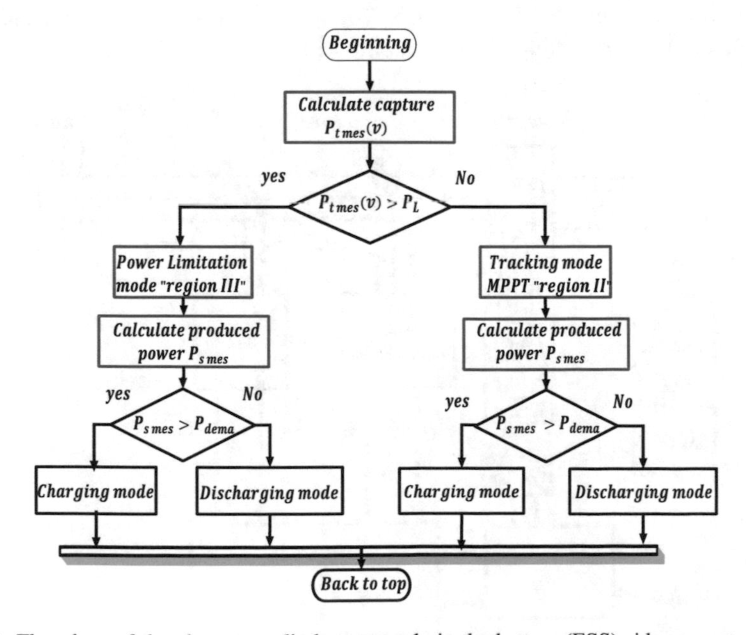

Fig. 13 Flowchart of the charges or discharges cycle in the battery (ESS) side converter

This algorithm is based on the following criteria:

If the power captured by the WT is less than the nominal power of the turbine can select the operating region of the turbine/PMSG by MPPT. In the same case, the power extracted for each wind speed is compared with the power specified by the demand: this means the way of power between the battery and the PMSG (charge or discharge).Otherwise, if the speed exceeds the nominal wind, the operating area of the turbine is moved to the region III, mode 2 (power limitation) and therefore the power produced to be always constant where the load and the system of discharge V_{DC} are imposed by the amount of power required at the electrical grid.

4.3 Control of the Grid Side Converter (GSC)

The main purpose of the GSC is to provide and organized the power required by the user, regardless of the operating conditions. In this part, a new DPC using SOSMC approach and space vector modulation SVM is proposed and realized for the control of the both active and reactive power. Figure 14 shows the schematic diagram of the control of GSC. In the first one, the external loop for controlling the DC link uses a voltage BESS with a bidirectional DC/DC converter is designed.In the task of the second one, internal loop contains an active and reactive power controller based on nonlinear controller SOSMC. The approach of the DPC-SOSMC-SVM strategy directly generates the reference voltage references for the grid side converter unlike the conventional vector method [35].

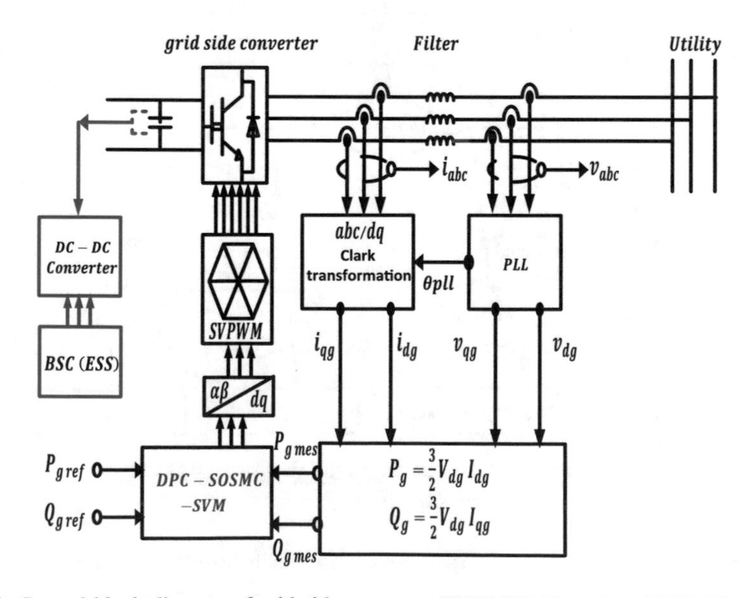

Fig. 14 Control block diagram of grid side converter "DPC-SVM based on SOSMC"

Improved control performance and quality of the power fed to the grid requires advanced and robust control techniques to overcome various constraints. For this reason, a common practice in the treatment of problems of control of the flow of active and reactive power of the grid is to use a conventional linearization approach [55, 56]. However, due to the invisibility of the system uncertainties and external disturbances marring the process control such methods come at the price of poor system performance and low reliability.

To consider these problems, a nonlinear and robust control is required [10, 57–68]. Many methods can be used for this purpose, the SMC control is shown to be particularly suitable for non-linear systems, offering effective structures [11, 21, 22, 30–32].

- **High order sliding mode controller design**

The phenomenon of chattering is the major disadvantage to the practical implementation of a control by sliding mode of order 1. An effective method to deal with this problem is to use a higher order control by sliding mode that generalizes the idea of simple order sliding mode. A command of nth order is the nth derivatives to mitigate the effect chattering keeping the main properties of the original approach as the robustness [69].

The active and reactive grid powers are derived as follows:

$$\begin{cases} P_g = \frac{3}{2} V_{dg} I_{dg} \\ Q_g = \frac{3}{2} V_{dg} I_{qg} \end{cases} \tag{26}$$

The optimal reactive power is set to zero to ensure a unity power factor operation of this system: $Q_{gref} = 0$ whereas the optimal active power P_{gref} can be written depending on the needs of the grid. The block diagram of the SOSMC applied to the grid side converter is illustrated in Fig. 14.

Let us introduce the following sliding surface for the active and reactive powers p_g, Q_g.

$$\begin{cases} s_P = P_{gref} - P_g \\ s_Q = Q_{gref} - Q_g \end{cases} \tag{27}$$

After the first derivation of the both surfaces:
Then we will have

$$\begin{cases} \dot{s}_P = \dot{P}_{gref} - \frac{1.5V_{dg}}{L_g}\left(-V_{dg} - R_g I_{dg} + L_g w_g I_{qg}\right) - \frac{V_{id}}{L_g} \\ \dot{s}_Q = \dot{Q}_{gref} - \frac{1.5V_{qg}}{L_g}\left(-V_{qg} - R_g I_{qg} - L_g w_g I_{dg}\right) - \frac{V_{iq}}{L_g} \end{cases} \tag{28}$$

If we define the functions and G_P and G_Q as follows:

$$\begin{cases} G_P = \dot{P}_{gref} - \frac{1.5V_{dg}}{L_g}\left(-V_{dg} - R_g I_{dg} + L_g w_g I_{qg}\right) \\ G_Q = \dot{Q}_{gref} - \frac{1.5V_{qg}}{L_g}\left(-V_{qg} - R_g I_{qg} - L_g w_g I_{dg}\right) \end{cases} \tag{29}$$

After the second derivation of the both surfaces:
Thus we have

$$\begin{cases} \ddot{s}_P = \dot{G}_p - \frac{\dot{V}_{id}}{L_g} \\ \ddot{s}_Q = \dot{G}_Q - \frac{\dot{V}_{iq}}{L_g} \end{cases} \tag{30}$$

The control algorithm proposed which is based on super twisting algorithm (ST) has been introduced by Levant [31]. The second order sliding mode controllers contain two parts:
where

$$\begin{cases} V_P^{ref} = V_p^N + V_p^{eq} \\ V_p^N = w_1 + w_2 \end{cases} \tag{31}$$

with

$$\begin{cases} \dot{w}_1 = -k_1 sign(s_P) \\ w_2 = -M_1\sqrt{|s_P|}sign(s_P) \end{cases} \tag{32}$$

and

$$\begin{cases} V_Q^{ref} = V_Q^N + V_Q^{eq} \\ V_Q^N = w_1 + w_2 \end{cases} \tag{33}$$

$$\begin{cases} \dot{w}_1 = -k_2 sign(s_Q) \\ w_2 = -M_2\sqrt{|s_Q|}sign(s_Q) \end{cases} \tag{34}$$

In order to ensure the convergence of the sliding manifolds to zero in finite time, the constants k_i and M_i can be chosen as follows [70].

$$\begin{cases} k_i > \frac{\mu_i}{L_g} \\ M_i \geq \frac{\mu_i(k_i+\mu_i)}{L_g^2(k_i-\mu_i)} \\ |G_i| < \mu_i; i = 1,2 \end{cases}$$

5 Results and Discussion

The performance of the proposed supervisory control system has been evaluated by numerous simulations (in several areas and in different conditions) using Matlab–Simulink package, under a wind speed profile of (11.75 m/s) mean value as depicted in Fig. 15. The system parameters are given in the Appendix A.

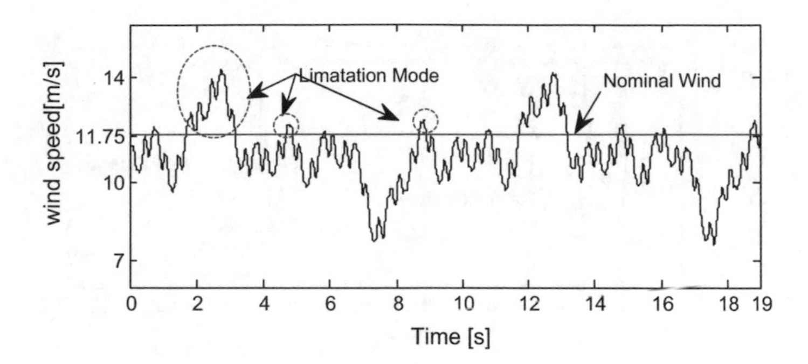

Fig. 15 Wind speed variation in (m/s)

It is noted that in Fig. 15 at nominal wind speed of 11.75 m/s, the WT operates in MPPT Mode (Region II) and the MPPT controller proposed in this paper (FLC-SOSMC-MPPT) ensures the optimum monitoring point of maximum power with high reliability while maintaining the power coefficient to maximum $C_{pmax} = 0.48$ with an optimum value of $\lambda_{opti} = 8.1$, as shown in Figs. 16a, b.

Figure 16c, d, describes the performance of the control law (FLC-SOSMC-MPPT), i.e., the quality of tracking the maximum power point, and one can see the corresponding distribution of the operating point around ORC. From this Figure: a smooth tracking with a high efficiency of the power extracted, with minimal mechanical stress on the turbine shaft.

To protect the wind power generation system (turbine/PMSG) above the rated wind speed (region III), a control mode 2nd was applied (power limitation). Consequently the C_p and λ are reduced. The curve shown in Fig. 16c, demonstrating the reliability and the ability to adapt (switching) of the Control & System in (parts II and III).

The fuzzy logic controller is used to find the optimum speed Ω_{topti} that follows the maximum power point to variable wind speedsin mode 1(tracking mode), also this Ω_{topti} is used as the input to generate a new reference speed $\Omega_{t\,ref\,new}$ in controlmode 2 (power limitation mode).

On the other hand, a SOSMC algorithm is then applied to control the speed of a PMSG, the robustness of which is investigated and finally the performance of the SOSMC is compared with that obtained by a FOSMC as illustrated in Fig. 17a. In both regions, two controllers are able to track the desired slip trajectory precisely which is an inherent advantage of the sliding mode controller. However, the result using conventional FOSMC shows some chattering in all the response as illustrated in Fig. 17a, this phenomenon is highly undesirable as it may lead to vibration on the mechanical part (high-frequency mechanical efforts on the turbine shaft). The SOSMC, on the other hand, get rid of the chattering phenomenon, giving a smooth tracking trajectory and lower slip error and lower control effort as it can be seen in Fig. 17b, where the value for the speed error is limited by a maximum value [0.1, −0.1] rad/s (negligible error).

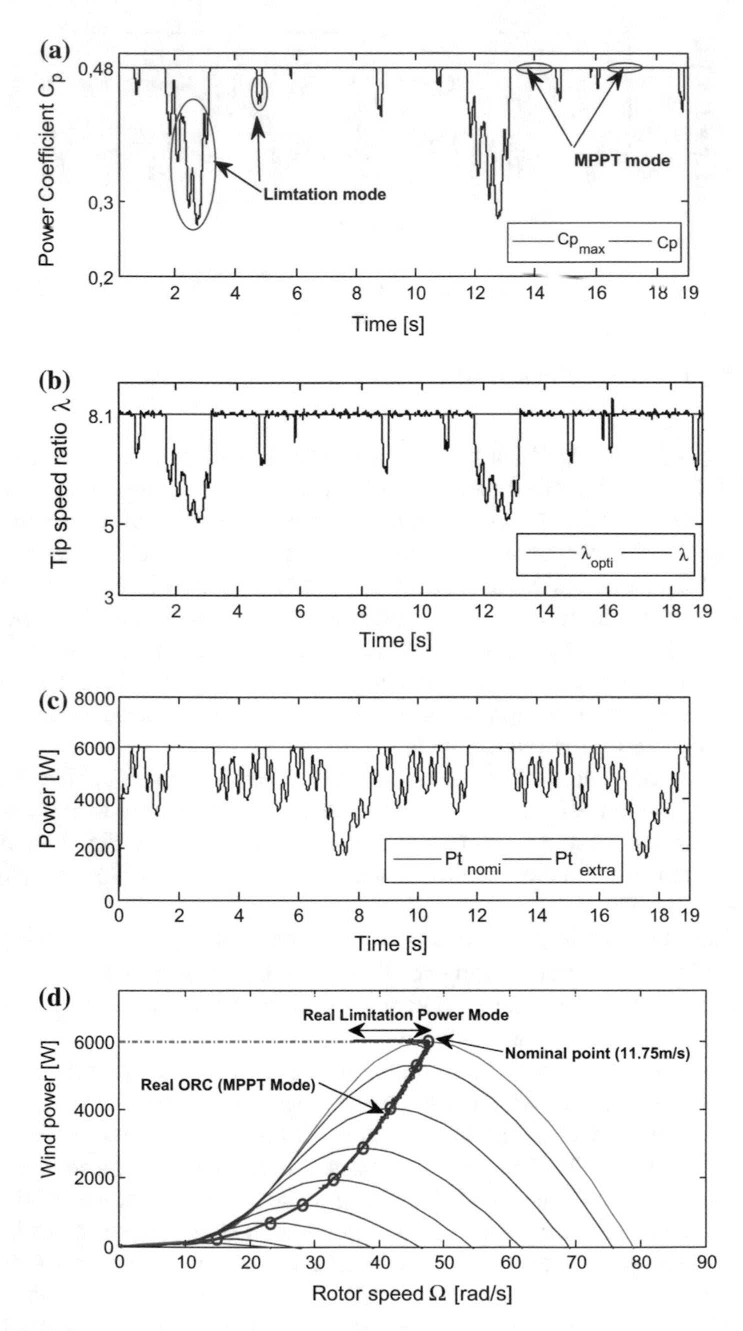

Fig. 16 MPPT FLC-SOSMC, **a** Power coefficient C_p, **b** The tip speed ratio λ, **c** The power generation in both region (II, III) and **d** The real power characteristic of the WT used in this study (II and III)

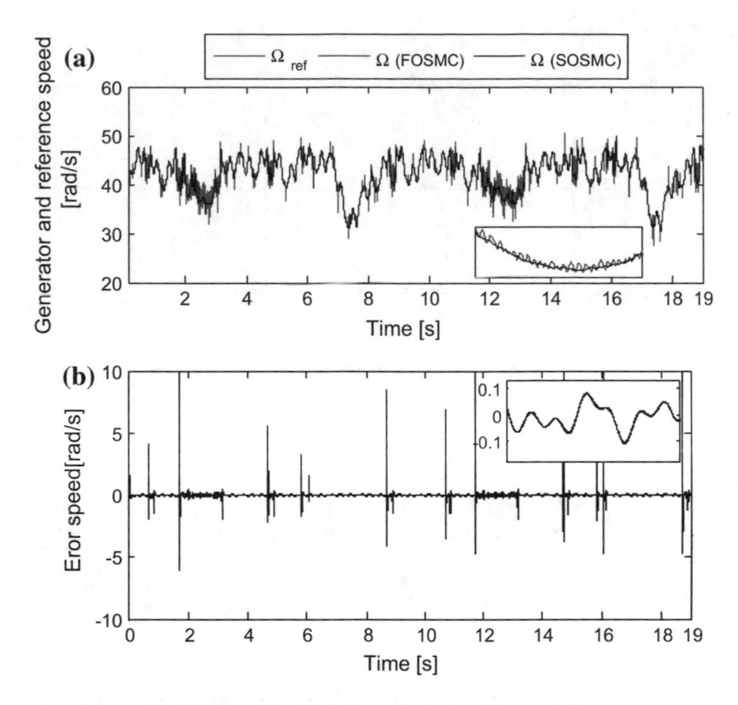

Fig. 17 **a** Generator speed used (FOSMC/SOSMC) and **b** Error speed

Under instantaneous variation for the (generated power, power demand), the battery is used to stabilize the voltage at the PMSG DC bus, the second main task of the control system is to maintain the DC Link voltage close to its nominal value 800 V, as it can be seen in Fig. 18a, by designing a battery DC/DC power converter for this purpose. The active power response, measured at different points of the hybrid system (in region II and III, is shown in Fig. 18b, c.

The requested power (demand power 'reference') is initially set to 3000 W in the first (5 s). A t = 5 s, the demand is increased to a value up to 4000 W t = 10 s and after she continues to increase to 7000 W during the period of [10–14 s], finally an decreased of 5000 W has been produced in the rest of the simulation time [14–19 s] as shown in Fig. 18b, c. In all cases we take into consideration the operating area of WT (2 or 3).

1. From [0 s to 5 s] as shown in Fig. 18b, c. The active power demanded by the grid is $P_{deman} = 3000W$. During this period, the reference power is lower than the available wind power [$P_{deman} < P_{extr}$]. At that moment, a battery recharging cycle begins and lasts until the SOC achieved 50.048% or higher. During the recharging time, the battery is charged by the power surplus provided by the PMSG. The advantage of storing excess wind power has become an important option in the field of WTs.

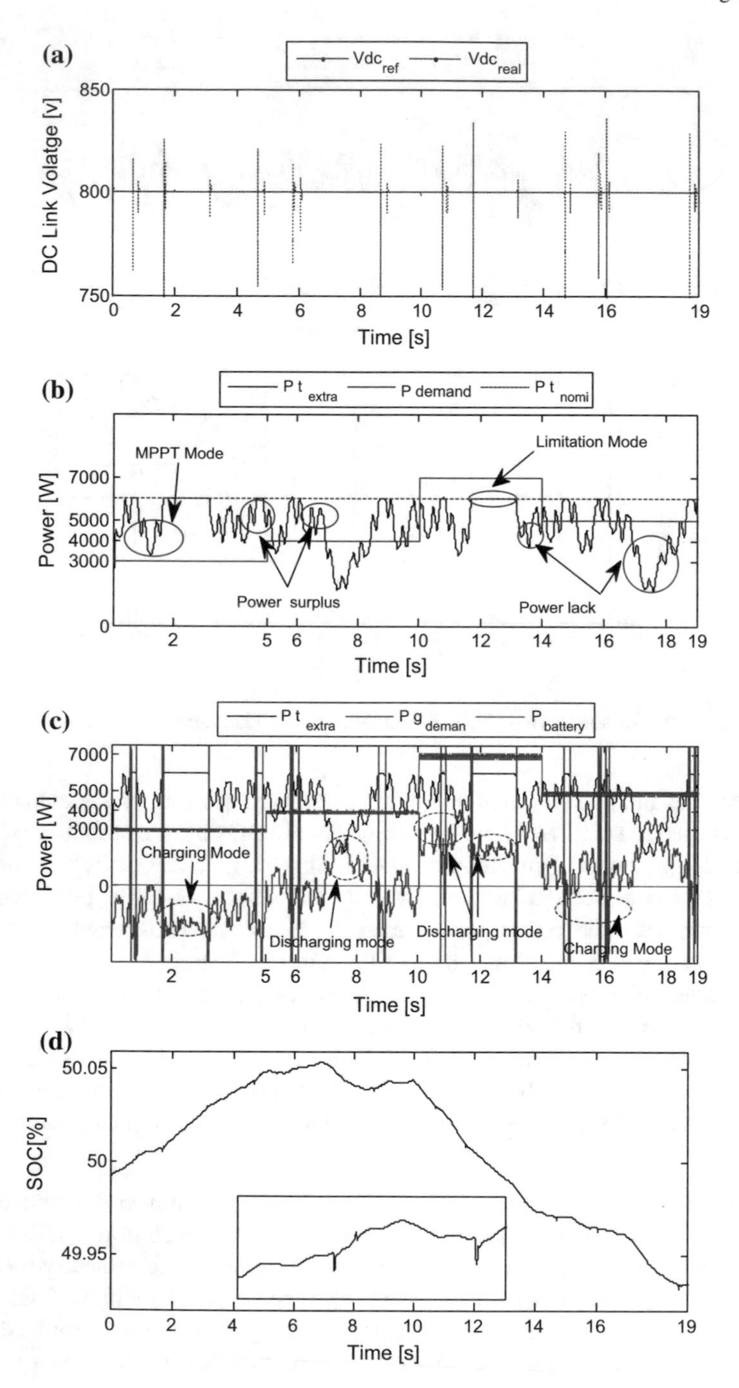

Fig. 18 ESS control **a** DC link voltage, **b** 'extrcted, nominal, demand' active power, **c** 'grid, battery, extracted' active power and **d** SOC of Battery

2. For instance we find, from [5 s s to 10 s], the active power demanded by the grid is $P_{deman} = 4000W$(a predetermined value) which is “lower or higher” the available wind power as presented in Fig. 18b, c. For this reason: First, when the active power demanded by the grid is higher [$P_{deman} > P_{extr}$], than the available wind power(extracted power), the battery supplements (discharging mode) the output of PMSG in order to provide the demanded power until it reaches its lowest recommended SOC of 50.03%. Then, when the power demanded by the grid is lower [$P_{deman} < P_{extr}$], below the available wind power, this power surplus provided by the PMSG is also stored in the battery (charging mode), enabling SOC to increase (50.044%) as observed in Fig. 18d.
3. During the period [10–14 s] where the demand for power is 7000 W maximum, the latter is greater than the power extracted by the turbine $[P_{deman} > P_{extr}]$ as observed in Fig. 18b, c. In this case, the ESS provides electrical power (discharging mode) to compensate for lack of the power supplied to the power grid (the gap between demand and production). During this period, a battery discharge cycle begins and lasts until 14 s, and a remarkable decrease in SOC continues up to 49.97% as illustrated in Fig. 18d. This hybrid system is able to work in unpredictable conditions and overcoming various constraints. According to Fig. 18b, c.
4. The same phenomena of the first interval [5–10 s] is similar to the during the time interval [14–19 s]. During the period [14–19 s] where the demand for power is 5000 W, the latter or lower is greater than the power extracted by the turbine $P_{deman} >< P_{extr}$ as observed in Fig. 18b, c. In this case, the ESS (provides/absorbed) electrical power (discharging/charging) to compensate or store for (lack/surplus) of the power (supplied to the grid/provided by the PMSG).

This hybrid system is able to work in unpredictable conditions (region II or III) and overcoming various constraints. According to Fig. 18d. In the previous four steps, SOS rapidly varied to achieve every moment the cycles of charge/discharge of the battery.

In the last one: active and reactive power control has been achieved by direct power control DPC-SVM.

To assess the merit of our choice SOSMC, a comparison was made between four types of controllers (PI conventional, fractional PI, FOSMC and SOSMC), these controllers are compared with a separately [(PI/$PI_{fractional}$) (FOSMC/SOSMC) and ($PI_{fractional}$/SOSMC)] for objective to clarify their efficacy compared to our selection. Other shares, these results are supported by a harmonic analysis of each regulator as shown in Table 2.

Table 2 A comparison of the four Controller types

Controller types	PI (conventional)	PI (fractional)	FOSMC	SOSMC
Current THD	3.96%	0.93%	3.86%	**0.59%**

Figures 19 and 20 display the (active and reactive powers) on the grid side, controlled via the proposed DPC-SVM and used four different types of regulators.

From Figs. 19a and 20a, both types of controllers (SOSMC) and (FOSMC) are able to track the desired slip trajectory precisely but, under the super twister control algorithm, the (direct, quadrature currents and active, reactive powers grid) track their reference values with chattering-free smooth profiles. By comparison, the result of the conventional FOSMC shows some chattering in all the responses. This phenomenon is highly undesirable as it may produce current distortion. The reactive power is set at zero for unity power factor is shown in Fig. 20a.

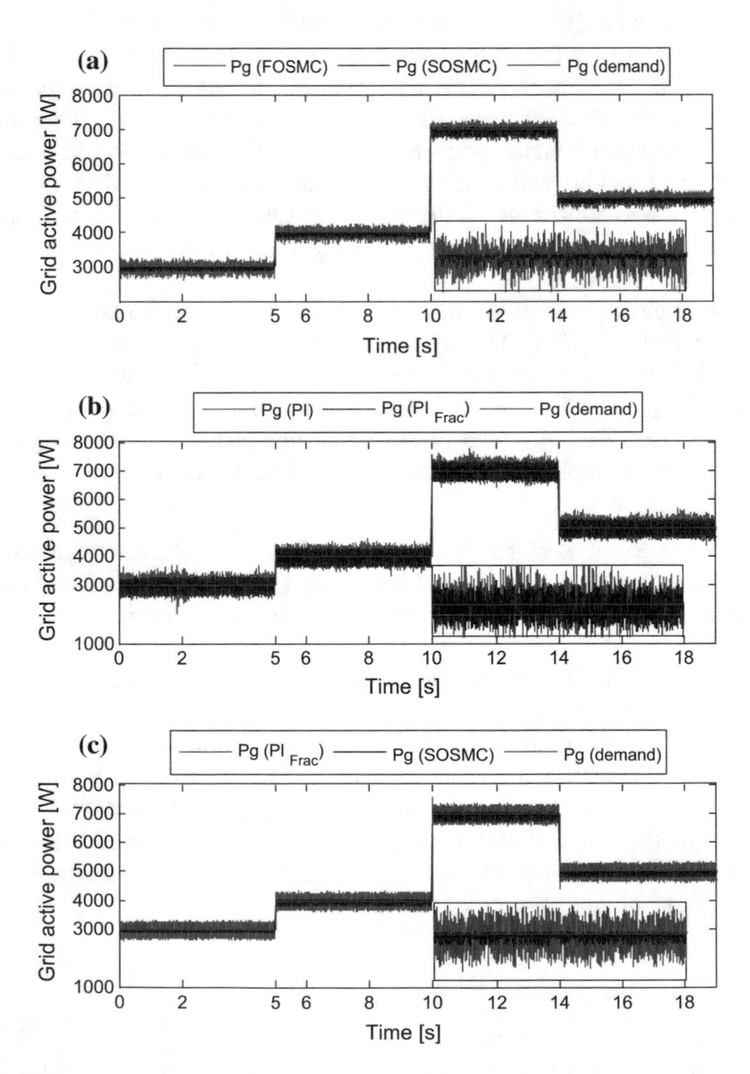

Fig. 19 Grid active power used DPC-SVM, **a** "FOSMC/SOSMC", **b** "PI/PI frac'' and **c** "SOSMC/PI frac"

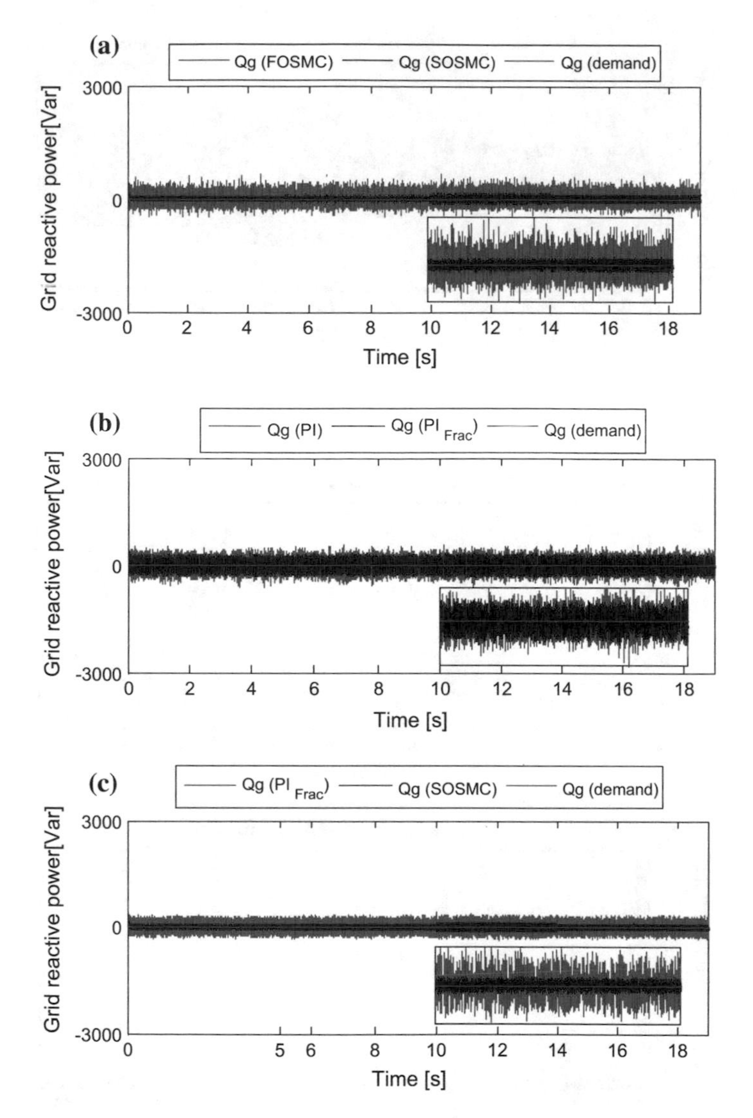

Fig. 20 Grid active power used DPC-SVM, **a** "FOSMC/SOSMC", **b** "PI/PI frac'' and **c** "SOSMC/PI frac"

While comparatively, the Figs. 19b and 20b describes these quantities (grid active and reactive power) under the conventional PI and PI fractional. From these figures we have observed that the two controllers are able to track the desired reference precisely with a clear priority to the PI fractional controller.

Figures 19c and 20c assembly the two best resulting regulators (SOSMC, $PI_{fractional}$) in "Figs. 19a, b and 20a, b", this comparison are given a general way the reason and contentment of our choice of the (SOMSC) controller. After these

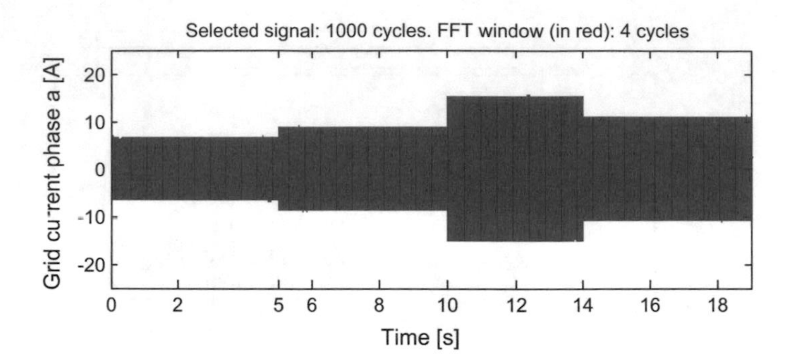

Fig. 21 Grid current phase “A”

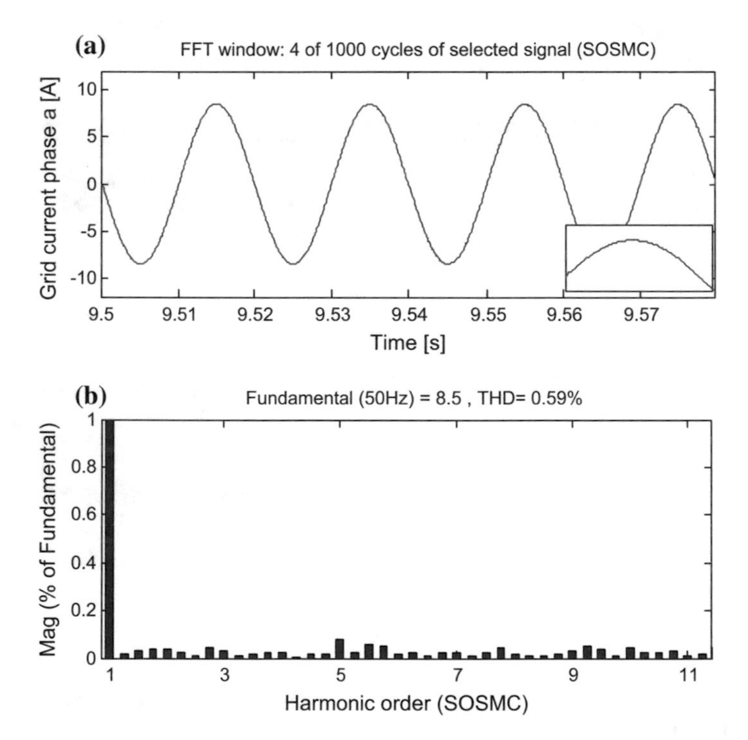

Fig. 22 **a** Grid current for SOSMC, **b** THD for SOSMC

Figures we have observed a high efficiency and smooth to track the desired slip trajectory precisely unlike the “$PI_{fractional}$”.

The results obtained in this section show a good performance to follow the desired path (active and reactive power) compared to conventional regulators such as (FOSMC, PI and $PI_{fractional}$).

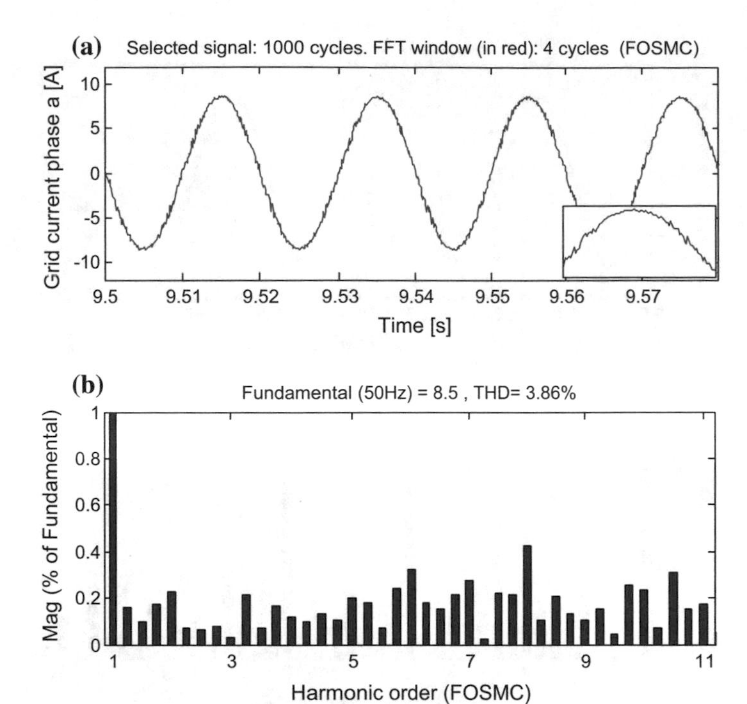

Fig. 23 **a** Grid current for FOSMC, **b** THD for FOSMC

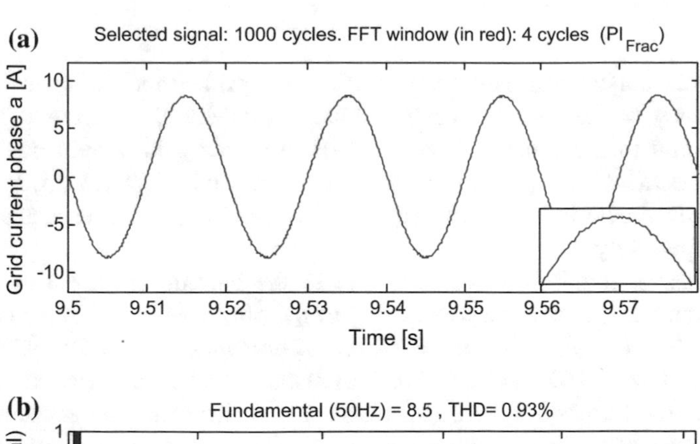

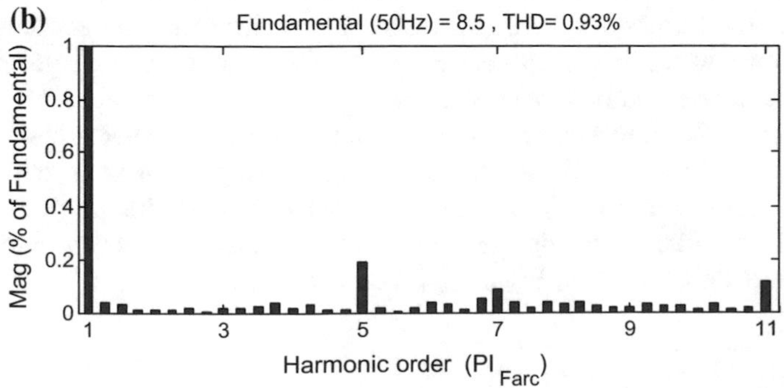

Fig. 24 **a** Grid current for PI frac, **b** THD for PI frac

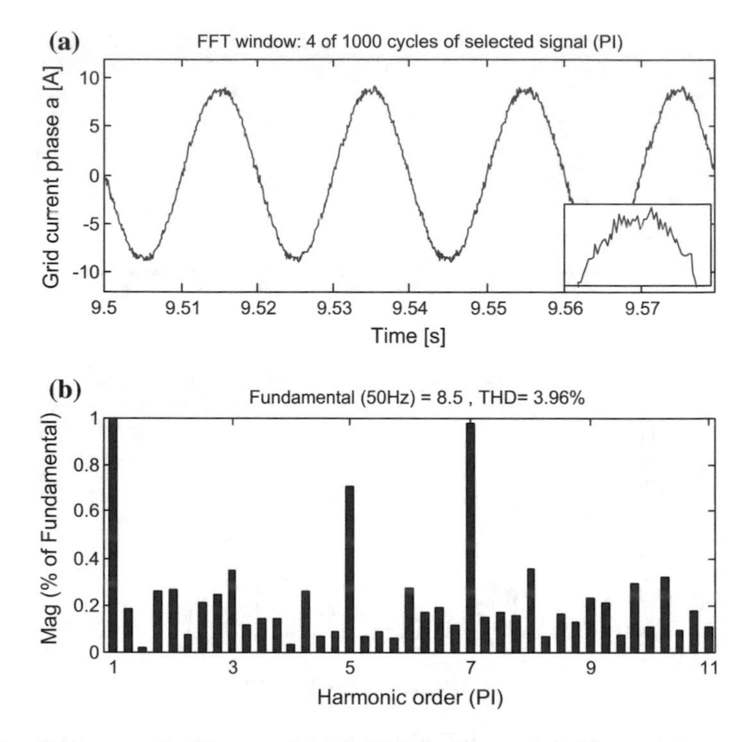

Fig. 25 **a** Grid current for PI conventional, **b** THD for PI conventional

Figure 21 illustrates a sample waveform of the grid current for phase A. To see the efficiency of the proposed control strategy (SOMSC), a Harmonic frequency spectrum and total harmonic distortion (THD) of the grid current controlled by (SOSMC, FOSMC, PI_{frac} and PI) controllers are shown in Figs. 22, 23, 24, and 25a, b. The THD shown in Figs. 23(b) and Fig. 25b are bigger and reaches 3.86% and 3.96% respectively.

That means, some unwanted distortion in the current waveform as shown in Figs. 23a and 25a, it means a poor quality of the power delivered to the power grid when we use ($PI_{conventional}$ and FOSMC). Compared with (SOSMC/$PI_{fractional}$) Where is reduced THD (0.59% and 0.93%) respectively as summarized in Table 2. In the proposed control system (SOMSC), we noticed that the distortion of the electrical current no longer appears Fig. 22a. This efficiency is due to the elimination by filtering odd harmonics Fig. 22b.

In Table 2, A comparison of the four strategies of controller types is summarized. The current THD shows that an important improvement in terms of odd harmonic mitigation (3–11). It can be concluded that the proposed method (SOMSC) can filter more than 80 and 40% of single harmonics compared to classical methods (FOSMC, $PI_{conventional}$) and ($PI_{fractional}$).

6 Conclusion and Future Work

Wind speed is often considered as one of the most difficult parameters to estimate because of its intermittent, which can deviate quickly. Although much effort has been dedicated to make WTs in wide operating range (to include more than one region at the same time).

In this chapter, a novel supervisory control scheme has been proposed to optimize the power management of a hybrid renewable energy system in the both regions (II and III) and to evaluate the coordinate operation of a grid connected of PMSG wind turbine and BESS.

In comparison with the existing works, our proposed architecture of control systems (supervisory control) provide to the WTs a freedom to work in a wide range with high protection in rated wind speed. The effectiveness of the control architecture has been checked by simulation study and compared to conventional control techniques, the proposed configuration considered was tested under varied operating conditions.

The significance of this work is indicated below:

a. *System Design*

- The WT proposed in this study is (Horizontal-axis/variable speed wind turbine/three-bladed).
- Due to its simple structure with characteristic self-excitation that can work with. The PMSG is nowadays a popular choice for variable speed WTs.
- Due to the inherently sporadic nature of the wind which can deviate quickly. The utilization of the BESS in WT increases the safe operation of the power grid, balance the supply and demand sides.
- The three converters are controlled precisely to mitigate grid power disturbances and ensure the maximum power extraction for each wind speed.

b. *control system Design*

- The generator side controller used to track and limited the maximum power generated from WT by controlling the rotational speed of the PMSG in the (region II and III) for this reason a FLC-SOSMC were designed to extract the maximum aerodynamic power up to the rated power (tracking mode), regardless of the turbine power-speed slope. If the wind power exceeds the rated value, the system switches to power regulation mode, via the power limitation circuit (limitation power mode).
- The battery DC/DC converter is controlled to maintain the DC bus voltage at its rated value. The ESS provides or stores the power mismatching between the actual wind power and grid demand. As a result, the hybrid system is

able to provide constant power to the grid as the incoming wind varies and increases the power transferred to the grid when required.

- In the grid side converter, the described DPC-SVM based on high order SMC has been designed to control the active and reactive powers exchanged between the PMSG and the grid. On the other hand, the grid power quantities provided by the SOSMC strategy shows smooth waveforms with good tracking indices and small THD compared with (FOSMC, $PI_{conventional}$ and $PI_{fractional}$).

c. *Results judged (obtained contribution)*

- It can be judged that the proposed structure (VSWT, PMSG, BESS) along with the integrating of the supervisory control strategy based on robust nonlinear techniques (FLC, SOSMC) turns the classical system into an enhanced stable, reliable and effective hybrid system with uniform protection, ensuring the power quality and management in the system under various operating conditions (region II and III).

d. *Future work*

- In future research, the authors wish to explore a methodology to store excess wind power when the BESS is in a state of saturation, there is more wind power and a decrease in demand.
- In addition, the present work focused on the implementation of efficient and robust control system for power management (limitation, tracking and delivered) by the wind hybrid system. To evaluates this work, the experimental results are necessary (test bench).
- We also plan to implement this method on several proposed wind turbines (small wind farm with high power), with optimization in untreated Points by others as: intelligent control for limiting the power in rated wind speeds or above par and controlling the AC/DC/AC.

Appendix A

Tables 3, 4 and 5.

Table 3 Nomenclatures and Abbreviations

Nomenclature	
V_{sd}, V_{sq}, I_{sd}, I_{sq}	The direct and quadrature components of the PMSG voltages and currents respectively
R_s, L_d, L_q	The resistance, the direct and the quadrature inductance of the PMSG respectively
ψ_m	The magneticflux
T_e	The electromagnetic torque
ω_e	The electrical rotational speed of PMSG
n_p	The number of pole pairs
$V_{dg}, V_{qg}, I_{dg}, I_{qg}$	The direct and quadrature components of the grid voltages and currents respectively
V_{di}, V_{qi}	The inverter voltages components
R_g, L_{dg}, L_{qg}	The resistance, the direct and quadrature grid inductance respectively
V_{bat}, E_0	The battery rated voltage and the internal EMF respectively
$E_{batdisch}, E_{batcharg}$	The discharge and charge voltage respectively
R_i, Q	The internal resistance and the battery capacity respectively
i_t, i^*	The actual battery charge and the filtered current respectively
BESS	The battery energy storage system
WTG	The wind turbine generator
MPP	The maximum power point
FLC	The fuzzylogiccontroller
SOSMC	The second order sliding mode control
FOSMC	The first order sliding controller
WECS	The wind energy conversion system
PMSG	The permanent magnetsynchronousgenerators
MPPT	The Maximum Power Point Tracking
ESS	The energystorage system
TSR	The tip speed ratio
HCS	The hill-climbsearching
MSC	The machine sideconverter
GSC	The gridsideconverter
BSC	The batterysideconverter
WT	The wind turbine

Table 4 PMSG parameters

Nominal power	$P = 6$ kw
Stator resistance	$R_s = 0.4\ \Omega$
Direct stator inductance	$L_d = 8.4$ mH
Stator inductance quadrature	$L_q = 8.4$ mH
Field flux	$\psi_{fl} = 0.4$ wb
Number of pole pairs	$n_p = 12$
Inertia	$J_t = 0.089$ kg.m^2
Friction	$f = 0.0016$ N.m

Table 5 Wind turbine parameters

Radius of the turbine	$R_t = 3.2$ m
Volume density of the air	$\rho = 1.225$ kg.m^3
The pitch angle	$\beta = 0^\circ$
specific optimal speed	$\lambda_{opti} = 8.1$
Coefficient of maximum power	$C_{pmax} = 0.48$

References

1. Nikolova, S., Causevski, A., & Al-Salaymeh, A. (2013). Optimal operation of conventional power plants in power system with integrated renewable energy sources. *Energy Conversion and Management, 65*(2013), 697–703.
2. Zou, Y., Elbuluk, M. E., & Sozer, Y. (2013). Simulation comparisons and implementation of induction generator wind power systems. *IEEE Transactions on Industry Applications, 49*(3), 1119–1128.
3. Carranza, O., Figueres, E., Garcerá, G., & Gonzalez-Medina, R. (2013). Analysis of the control structure of wind energy generation systems based on a permanent magnet synchronous generator. *Applied Energy, 103*(2013), 522–538.
4. Aissaoui, A. G., Tahour, A., Essounbouli, N., Nollet, F., Abid, M., & Chergui, M. I. (2013). A Fuzzy-PI control to extract an optimal power from wind turbine. *Energy Conversion and Management, 65*(2013), 688–696.
5. Abdullah, M. A., Yatim, A. H. M., Tan, C. W., et al. (2012). A review of maximum power point tracking algorithms for wind energy systems. *Renewable and Sustainable Energy Reviews, 16*(5), 3220–3227.
6. Jaramillo-Lopez, F., Kenne, G., & Lamnabhi-Lagarrigue, F. (2016). A novel online training neural network-based algorithm for wind speed estimation and adaptive control of PMSG wind turbine system for maximum power extraction. *Renewable Energy, 86*(2016), 38–48.
7. Syed, I. M., Venkatesh, B., Wu, B., & Nassif, A. B. (2012). Two-layer control scheme for a supercapacitor energy storage system coupled to a Doubly fed induction generator. *Electric Power Systems Research, 86*(2012), 76–83.
8. Domínguez-García, J. L., Gomis-Bellmunt, O., Bianchi, F. D., & Sumper, A. (2012). Power oscillation damping supported by wind power: a review. *Renewable and Sustainable Energy Reviews, 16*(7), 4994–5006.
9. Zhao, H., Wu, Q., Hu, S., Xu, H., & Rasmussen, C. N. (2015). Review of energy storage system for wind power integration support. *Applied Energy, 137*(2015), 545–553.
10. Azar, A. T., & Zhu, Q. (2015). *Advances and Applications In Sliding Mode Control Systems. Studies in computational intelligence* (vol. 576). Germany: Springer. ISBN: 978-3-319-11172-8.
11. Abdeddaim, S., & Betka, A. (2013). Optimal tracking and robust power control of the DFIG wind turbine. *International Journal of Electrical Power & Energy Systems, 49*(2013), 234–242.
12. Gao, R., & Gao, Z. (2016). Pitch control for wind turbine systems using optimization, estimation and compensation. *Renewable Energy, 91*(2016), 501–515.
13. Kumar, A., & Verma, V. (2016). Photovoltaic-grid hybrid power fed pump drive operation for curbing the intermittency in PV power generation with grid side limited power conditioning. *International Journal of Electrical Power & Energy Systems, 82*(2016), 409–419.
14. Yin, X. X., Lin, Y. G., Li, W., Liu, H. W., & Gu, Y. J. (2015). Adaptive sliding mode back-stepping pitch angle control of a variable-displacement pump controlled pitch system for wind turbines. *ISA Transactions, 58*(2015), 629–634.

15. Saravanakumar, R., & Jena, D. (2015). Validation of an integral sliding mode control for optimal control of a three blade variable speed variable pitch wind turbine. *International Journal of Electrical Power & Energy Systems, 69*(2015), 421–429.
16. Kim, H., Son, J., & Lee, J. (2011). A high-speed sliding-mode observer for the sensorless speed control of a PMSM. *IEEE Transactions on Industrial Electronics, 58*(9), 4069–4077.
17. Ramesh, T., Panda, A. K., & Kumar, S. S. (2015). Type-2 fuzzy logic control based MRAS speed estimator for speed sensorless direct torque and flux control of an induction motor drive. *ISA Transactions, 57*(2915), 262–275.
18. Thirusakthimurugan, P., & Dananjayan, P. (2007). A novel robust speed controller scheme for PMBLDC motor. *ISA Transactions, 46*(4), 471–477.
19. Pichan, M., Rastegar, H., & Monfared, M. (2013). Two fuzzy-based direct power control strategies for doubly-fed induction generators in wind energy conversion systems. *Energy, 51* (2013), 154–162.
20. Uhlen, K., Foss, B. A., & Gjøsæter, O. B. (1994). Robust control and analysis of a wind-diesel hybrid power plant. *IEEE Transactions on Energy Conversion, 9*(4), 701–708.
21. Evangelista, C., Valenciaga, F., & Puleston, P. (2013). Active and reactive power control for wind turbine based on a MIMO 2-sliding mode algorithm with variable gains. *IEEE Transactions on Energy Conversion, 28*(3), 682–689.
22. Billela, M., Dib, D., & Azar, A. T. (2016). A Second order sliding mode and fuzzy logic control to Optimal Energy Management in PMSG Wind Turbine with Battery Storage. In *Neural Computing and Applications*. Springer. doi:10.1007/s00521-015-2161-z.
23. Assareh, E., & Biglari, M. (2015). A novel approach to capture the maximum power from variable speed wind turbines using PI controller, RBF neural network and GSA evolutionary algorithm. *Renewable and Sustainable Energy Reviews, 51*(2015), 1023–1037.
24. Witczak, P., Patan, K., Witczak, M., Puig, V., & Korbicz, J. (2015). A neural network-based robust unknown input observer design: Application to wind turbine. *IFAC-PapersOnLine, 48* (21), 263–270.
25. Ata, R. (2015). Artificial neural networks applications in wind energy systems: a review. *Renewable and Sustainable Energy Reviews, 49*(2015), 534–562.
26. Suganthi, L., Iniyan, S., & Samuel, A. A. (2015). Applications of fuzzy logic in renewable energy systems–A review. *Renewable and Sustainable Energy Reviews, 48*(2015), 585–607.
27. Banerjee, A., Mukherjee, V., & Ghoshal, S. P. (2014). Intelligent fuzzy-based reactive power compensation of an isolated hybrid power system. *International Journal of Electrical Power & Energy Systems, 57*(2014), 164–177.
28. Castillo, O., & Melin, P. (2014). A review on interval type-2 fuzzy logic applications in intelligent control. *Information Sciences, 279*(2014), 615–631.
29. Mérida, J., Aguilar, L. T., & Dávila, J. (2014). Analysis and synthesis of sliding mode control for large scale variable speed wind turbine for power optimization. *Renewable Energy, 71* (2014), 715–728.
30. Hong, C.-M., Huang, C.-H., & Cheng, F.-S. (2014). Sliding Mode Control for Variable-speed Wind Turbine Generation Systems Using Artificial Neural Network. *Energy Procedia, 61* (2014), 1626–1629.
31. Benbouzid, M., Beltran, B., Amirat, Y., Yao, G., Han, J., & Mangel, H. (2014). Second-order sliding mode control for DFIG-based wind turbines fault ride-through capability enhancement. *ISA Transactions, 53*(3), 827–833.
32. Liu, J., Lin, W., Alsaadi, F., & Hayat, T. (2015). Nonlinear observer design for PEM fuel cell power systems via second order sliding mode technique. *Neurocomputing, 168*(2015), 145–151.
33. Evangelista, C. A., Valenciaga, F., & Puleston, P. (2012). Multivariable 2-sliding mode control for a wind energy system based on a double fed induction generator. *International Journal of Hydrogen Energy, 37*(13), 10070–10075.

34. Eltamaly, A. M., & Farh, H. M. (2013). Maximum power extraction from wind energy system based on fuzzy logic control. *Electric Power Systems Research, 97*(2013), 144–150.
35. Meghni, B., Saadoun, A., Dib, D., & Amirat, Y. (2015). Effective MPPT technique and robust power control of the PMSG wind turbine. *IEEJ Transactions on Electrical and Electronic Engineering, 10*(6), 619–627.
36. Sarrias, R., Fernández, L. M., García, C. A., & Jurado, F. (2012). Coordinate operation of power sources in a doubly-fed induction generator wind turbine/battery hybrid power system. *Journal of Power Sources, 205*(2012), 354–366.
37. Sarrias-Mena, R., Fernández-Ramírez, L. M., García-Vázquez, C. A., & Jurado, F. (2014). Improving grid integration of wind turbines by using secondary batteries. *Renewable and Sustainable Energy Reviews, 34*(2014), 194–207.
38. Sharma, P., Sulkowski, W., & Hoff, B. (2013). Dynamic stability study of an isolated wind-diesel hybrid power system with wind power generation using IG, PMIG and PMSG: A comparison. *International Journal of Electrical Power & Energy Systems, 53*(2013), 857–866.
39. Liu, J., Meng, H., Hu, Y., Lin, Z., & Wang, W. (2015). A novel MPPT method for enhancing energy conversion efficiency taking power smoothing into account. *Energy Conversion and Management, 10*(2015), 738–748.
40. Nasiri, M., Milimonfared, J., & Fathi, S. H. (2014). Modeling, analysis and comparison of TSR and OTC methods for MPPT and power smoothing in permanent magnet synchronous generator-based wind turbines. *Energy Conversion and Management, 86*(2014), 892–900.
41. Daili, Y., Gaubert, J.-P., & Rahmani, L. (2015). Implementation of a new maximum power point tracking control strategy for small wind energy conversion systems without mechanical sensors. *Energy Conversion and Management, 97*(2015), 298–306.
42. Kortabarria, I., Andreu, J., de Alegría, I. M., Jiménez, J., Gárate, J. I., & Robles, E. (2014). A novel adaptative maximum power point tracking algorithm for small wind turbines. *Renewable Energy, 63*(2014), 785–796.
43. Ghedamsi, K., & Aouzellag, D. (2010). Improvement of the performances for wind energy conversions systems. *International Journal of Electrical Power & Energy Systems, 32*(9), 936–945.
44. Poitiers, F., Bouaouiche, T., & Machmoum, M. (2009). Advanced control of a doubly-fed induction generator for wind energy conversion. *Electric Power Systems Research, 79*(7), 1085–1096.
45. Hong, C.-M., Chen, C.-H., & Tu, C.-S. (2013). Maximum power point tracking-based control algorithm for PMSG wind generation system without mechanical sensors. *Energy Conversion and Management, 69*(2013), 58–67.
46. Zou, Y., Elbuluk, M., & Sozer, Y. (2013). Stability analysis of maximum power point tracking (MPPT) method in wind power systems. *IEEE Transactions on Industry Applications, 49*(3), 1129–1136.
47. Narayana, M., Putrus, G. A., Jovanovic, M., Leung, P. S., & McDonald, S. (2012). Generic maximum power point tracking controller for small-scale wind turbines. *Renewable Energy, 44*(2012), 72–79.
48. Yin, M., Li, G., Zhou, M., & Zhao, C. (2007). Modeling of the wind turbine with a permanent magnet synchronous generator for integration. In *Power Engineering Society General Meeting, IEEE 2007*, June 24-28, 2007, Tampa, FL, (pp. 1–6). doi:10.1109/PES.2007.385982.
49. SimPowerSystems, T. M. (2010). Reference, Hydro-Qu{é}bec and the MathWorks. Inc., Natick, MA.
50. Jain, B., Jain, S., & Nema, R. K. (2015). Control strategies of grid interfaced wind energy conversion system: An overview. *Renewable and Sustainable Energy Reviews, 47*(2015), 983–996.

51. Benelghali, S., El Hachemi Benbouzid, M., Charpentier, J. F., Ahmed-Ali, T., & Munteanu, I. (2011). Experimental validation of a marine current turbine simulator: Application to a permanent magnet synchronous generator-based system second-order sliding mode control. *IEEE Transactions on Industrial Electronics, 58*(1), 118–126.
52. Rafiq, M., Rehman, S., Rehman, F., Butt, Q. R., & Awan, I. (2012). A second order sliding mode control design of a switched reluctance motor using super twisting algorithm. *Simulation Modelling Practice and Theory, 25*(2012), 106–117.
53. Soler, J., Daroqui, E., Gimeno, F.J., Seguí-Chilet, S., & Orts, S. (2005). Analog low cost maximum power point tracking PWM circuit for DC loads. In *Proceedings of the Fifth IASTED International Conference on Power and Energy Sysemst, Benalmadena,* Spain, June 15–17, 2005.
54. Gkavanoudis, S. I., & Demoulias, C. S. (2014). A combined fault ride-through and power smoothing control method for full-converter wind turbines employing Supercapacitor Energy Storage System. *Electric Power Systems Research, 106*(2014), 62–72.
55. Pena, R., Cardenas, R., Proboste, J., Asher, G., & Clare, J. (2008). Sensorless control of doubly-fed induction generators using a rotor-current-based MRAS observer. *IEEE Transactions on Industrial Electronics, 55*(1), 330–339.
56. Tapia, G., Tapia, A., & Ostolaza, J. X. (2007). Proportional–integral regulator-based approach to wind farm reactive power management for secondary voltage control. *IEEE Transactions on Energy Conversion, 22*(2), 488–498.
57. Azar, A. T. (2012). Overview of type-2 fuzzy logic systems. *International Journal of Fuzzy System Applications, 2*(4), 1–28.
58. Azar, A. T. (2010). *Fuzzy systems*. Vienna: IN-TECH. ISBN 978-953-7619-92-3.
59. Azar, A.T., & Vaidyanathan, S. (2015). Handbook of research on advanced intelligent control engineering and automation. In *Advances in Computational Intelligence and Robotics (ACIR) Book Series*, IGI Global, USA.
60. Azar, A. T., & Vaidyanathan, S. (2015). *Computational intelligence applications in modeling and control*. Studies in computational intelligence (vol. 575). Germany: Springer. ISBN 978-3-31911016-5.
61. Azar, A. T., & Vaidyanathan, S. (2015). *Chaos modeling and control systems design, studies in computational intelligence* (Vol. 581). Germany: Springer. ISBN 978-3-319-13131-3.
62. Zhu, Q., & Azar, A. T. (2015). Complex system modelling and control through intelligent soft computations. *Studies in fuzziness and soft computing* (vol. 319). Germany: Springer. ISBN: 978-3-31912882-5 123.
63. Azar, A.T., & Serrano, F.E. (2015). Design and modeling of anti wind up PID controllers. In Q. Zhu & A. T. Azar (Eds.), *Complex system modelling and control through intelligent soft computations, Studies in Fuzziness and Soft Computing* (vol. 319, pp. 1–44). Germany: Springer, Germany. doi:10.1007/978-3-319-12883-2_1.
64. Azar, A. T., & Serrano, F. E. (2015). Adaptive sliding mode control of the furuta pendulum. In A. T. Azar & Q. Zhu, (Eds.) *Advances and Applications in Sliding Mode Control systems, Studies in Computational Intelligence*, (vol. 576, pp. 1–42). Berlin/Heidelberg: Springer GmbH. doi:10.1007/978-3-319-11173-5_1.
65. Azar, A. T., & Serrano, F. E. (2015). Deadbeat control for multivariable systems with time varying delays. In A. T. Azar & S. Vaidyanathan (Eds.), *Chaos modeling and control systems design, studies in computational intelligence* (vol 581, pp 97–132). Berlin: Springer GmbH. doi:10.1007/978-3-319-13132-0_6.
66. Mekki, H., Boukhetala, D., & Azar, A. T. (2015). Sliding modes for fault tolerant control. In A.T. Azar & Q Zhu (Eds.) *Advances and applications in sliding mode control systems, studies in computational intelligence book Series* (vol. 576, pp 407–433). Berlin: Springer GmbH. doi:10.1007/978-3-319-11173-5_15.
67. Luo, Y., & Chen, Y. (2012). Stabilizing and robust fractional order PI controller synthesis for first order plus time delay systems. *Automatica, 48*(9), 2159–2167.

68. Ebrahimkhani, S. (2016). *Robust fractional order sliding mode control of doubly-fed induction generator (dfig)-based wind turbines*. *ISA transactions*, 2016. In Press.
69. Munteanu, I., Bacha, S., Bratcu, A. I., Guiraud, J., & Roye, D. (2008). Energy-reliability optimization of wind energy conversion systems by sliding mode control. *IEEE Transactions on Energy Conversion, 23*(3), 975–985.
70. Beltran, B., Ahmed-Ali, T., & Benbouzid, M. E. H. (2008). Sliding mode power control of variable-speed wind energy conversion systems. *IEEE Transactions on Energy Conversion, 23*(2), 551–558.

Robust Adaptive Interval Type-2 Fuzzy Synchronization for a Class of Fractional Order Chaotic Systems

Khatir Khettab, Yassine Bensafia and Samir Ladaci

Abstract This chapter presents a novel Robust Adaptive Interval Type-2 Fuzzy Logic Controller (RAIT2FLC) equipped with an adaptive algorithm to achieve synchronization performance for fractional order chaotic systems. In this work, by incorporating the H^{∞} tracking design technique and Lyapunov stability criterion, a new adaptive fuzzy control algorithm is proposed so that not only the stability of the adaptive type-2 fuzzy control system is guaranteed but also the influence of the approximation error and external disturbance on the tracking error can be attenuated to an arbitrarily prescribed level via the H^{∞} tracking design technique. The main contribution in this work is the use of the interval type-2 fuzzy logic controller and the numerical approximation method of Grünwald-Letnikov in order to improve the control and synchronization performance comparatively to existing results. By introducing the type-2 fuzzy control design and robustness tracking approach, the synchronization error can be attenuated to a prescribed level, even in the presence of high level uncertainties and noisy training data. A simulation example on chaos synchronization of two fractional order Duffing systems is given to verify the robustness of the proposed AIT2FLC approach in the presence of uncertainties and bounded external disturbances.

Keywords Robust adaptive control · Interval type-2 fuzzy · Upper and lower membership functions fractional systems · Chaos synchronization · Stability

---

K. Khettab (✉)
Department of Electrical Engineering, Mohamed Boudiaf University of M'sila, 28000 Sétif, Algeria
e-mail: zoubirhh@yahoo.fr

Y. Bensafia
Department of Electrical Engineering, Bouira University, 10000 Béjaia, Algeria
e-mail: bensafiay@yahoo.fr

S. Ladaci
EEA Department, National Polytechnic School of Constantine, BP 75A RP Ali Mendjeli, 25000 Constantine, Algeria
e-mail: samir_ladaci@yahoo.fr

A.T. Azar et al. (eds.), *Fractional Order Control and Synchronization of Chaotic Systems*, Studies in Computational Intelligence 688,
DOI 10.1007/978-3-319-50249-6_7

1 Introduction

Fractional calculus deals with derivatives and integrations of arbitrary order [41, 53] and has found many applications in many fields of physics, applied mathematics and engineering. Moreover, many real-world physical systems are well characterized by fractional order differential equations, i.e., equations involving both integer and non integer order derivatives [25]. It is observed that the description of some systems is more accurate when the fractional derivative is used. For instance, electrochemical processes and flexible structures are modeled by fractional order models [38, 41]. Nowadays, many fractional-order differential systems behave chaotically, such as the fractional-order Chua system [52], the fractional-order *Duffing* system [2], the fractional-order Lu system, the fractional order Chen system [51].

Recently, due to its potential applications in secure communication and control processing, the study of chaos synchronization in fractional order dynamical systems and related phenomena is receiving growing attention.

The synchronization problem of fractional order chaotic systems was first investigated by *Deng* and *Li* who carried out synchronization in the case of the fractional *Lü* system. Afterwards, they studied chaos synchronization of the *Chen* system with a fractional order in a different manner [18, 19].

Fuzzy logic controllers are generally considered applicable to plants that are mathematically poorly understood and where experienced human operators are available for providing a qualitative "rule of thumb".

Based on the universal approximation theorem [11, 60] (fuzzy logic controllers are general enough to perform any nonlinear control actions) there is rapidly growing interest in systematic design methodologies for a class of nonlinear systems using fuzzy adaptive control schemes. An adaptive fuzzy system is a fuzzy logic system equipped with a training algorithm in which an adaptive controller is synthesized from a collection of fuzzy IF–THEN rules and the parameters of the membership functions characterizing the linguistic terms in the IF–THEN rules change according to some adaptive law for the purpose of controlling a plant to track a reference trajectory.

In this work we consider Type-2 fuzzy sets which are extension of type-1 fuzzy sets introduced in the first time by Zadeh [66]. Basic concepts of type-2 fuzzy sets and systems were advanced and well established in [9, 20, 45, 54]. In 1998, Mendel and Karnik [20] introduced five different kinds of type reduction methods which are extended versions of type-1 defuzzification methods. Qilian and Mendel [54] proposed an efficient and simplified method for computing the input and antecedent operations for interval type-2 fuzzy logic controller (IT2FLC) using the concept of upper and lower Membership functions. Karnik and Mendel developed the centroid of an interval type-2 fuzzy set (IT2FS), not only for an IT2FS and IT2FLCs but also for general type-2 FSs and introduced an algorithm for its computation. Mendel [17, 44] described important advances for both general and interval type-2 fuzzy sets and systems in 2007. Because of the calculation complexity especially in the type reduction, use of IT2FLC is still controversial. Seplveda et al. showed that using adequate hardware implementation, IT2FLC can be efficiently utilized in applications that require high speed processing. Thus, the type-2 FLS has been successfully applied to several fuzzy controller designs [12, 17, 36, 42, 62].

In this paper, by incorporating the H^{∞} tracking design technique [35, 38, 43] and Lyapunov stability criterion, a new adaptive fuzzy control algorithm is proposed so that not only the stability of the adaptive type-2 fuzzy control system is guaranteed but also the influence of the approximation error and external disturbance on the tracking error can be attenuated to an arbitrarily prescribed level via the H^{∞} tracking design technique. The proposed design method attempts to combine the attenuation technique, type-2 fuzzy logic approximation method, and adaptive control algorithm for the robust tracking control design of the nonlinear fractional order systems with a large uncertainty or unknown variation in plant parameters and structures.

This chapter is organized as follows: Sect. 2 presents a brief review on the state of the art for the addressed problem. In Sect. 3, an introduction to fractional derivatives and its relation to the approximation solution will be addressed and the basic definition and preliminaries for fractional order systems. A description of the interval type-2 fuzzy logic is presented in Sect. 4. Section 5 and 6 generally propose adaptive type-2 fuzzy robust H^{∞} control of uncertain fractional order systems in the presence of uncertainty and its stability analysis. In Sect. 7, application of the proposed method on fractional order expression chaotic systems (Duffing oscillator) is investigated. Finally, the simulation results and conclusion will be presented in Sect. 8.

2 Related Work: A Brief Review

Fractional adaptive control is a growing research topic gathering the interest of a great number of researchers and control engineers [32]. The main argument of this community is the significant enhancement obtained with these new real-time controllers comparatively to integer order ones [53].

Since the pioneering works of Vinagre et al. [59] and Ladaci and Charef [26, 27], an increasing number of works are published focusing on various fractional order adaptive schemes such as: fractional order model reference adaptive control [13, 27, 63], fractional order adaptive pole placement control [34], fractional high-gain adaptive control [29], fractional multi-model adaptive control [33], robust fractional adaptive control [30], fractional extremum seeking control [48], Fractional IMC-based adaptive control [31], fractional adaptive sliding mode control [15], fractional adaptive PID control [28, 47] … etc.

The study and design of fractional adaptive control laws for nonlinear systems is also an actual leading research direction [5, 6, 50, 57]. Many control strategies have been proposed in literature to deal with the control and synchronization problems of various nonlinear and chaotic fractional order systems [1, 55]. Nonlinear fractional adaptive control is wide meaning concept with many different control approaches such as: fractional order adaptive backstepping output feedback control scheme [64], adaptive feedback control scheme based on the stability results of linear fractional order systems [49], Adaptive Sliding Control [36, 65], Adaptive synchronization of fractional-order chaotic systems via a single driving variable [67], H∞ robust adaptive control [22, 37], etc. Whereas, in order to deal with nonlinear systems presenting uncertainties or unknown model parameters, many authors have used fuzzy systems [3, 7, 8, 58]. In this work, we use Type-2 Fuzzy logic systems [4, 40].

3 Basic Preliminaries for Fractional Order Systems

Fractional calculus (integration and differentiation of arbitrary 'fractional' order) is an old concept which dates back to Cauchy, Riemann Liouville and Leitnikov in the 19th century. It has been used in mechanics since at least the 1930s and in electrochemistry since the 1960s. In control field, several theoretical physicists and mathematicians have studied fractional differential operators and systems [14, 53].

Fractional order operator is a generalization of integration and differentiation to non integer order fundamental operators, denoted by ${}_aD_t^{\alpha}$, where a and t are the limits of the operator. This operator is a notation for taking both the fractional integral and functional derivative in a single expression defined as [22–24, 51]:

$$ {}_aD_t^q = \begin{cases} \frac{d^q}{dt^q} & q>0 \\ 1 & q=0 \\ \int\limits_a^t (d\tau)^{-q} & q<0 \end{cases} \tag{1} $$

There are some basic definitions of the general fractional integration and differentiation. The commonly used definitions are those of *Riemann–Liouville* and *Grünwald-Letnikov* [29, 30, 56].

The Riemann-Liouville (R-L) integral of order $\lambda > 0$ is defined as:

$$ I_{RL}^{\lambda} f(t) = D^{-\lambda} f(t) \quad = \frac{1}{\Gamma(\lambda)} \int\limits_0^t (t-\tau)^{\lambda-1} f(\tau) d\tau \tag{2} $$

and the expression of the R-L fractional order derivative of order $\mu > 0$ is:

$$ D_{RL}^{\mu} f(t) = \frac{1}{\Gamma(n-\mu)} \frac{d^n}{dt^n} \int\limits_0^t (t-\tau)^{n-\mu-1} f(\tau) d\tau \tag{3} $$

with Γ(.) is the Euler's gamma function and the integer n is such that $(n-1) < \mu < n$. This fractional order derivative of Eq. (3) can also be defined from Eq. (2) as:

$$D^{\mu}_{RL}f(t) = \frac{d^n}{dt^n}\left\{ I_{RL}^{(n-\mu)} f(t) \right\} \tag{4}$$

The *Grünwald–Letnikov* definition of the fractional derivative, is expressed as:

$${}^{GL}_{t_0}D^q_t f(t) = \lim_{n\to 0}\frac{1}{h^n}\sum_{j=0}^{\left[\frac{t-q}{h}\right]}(-1)^j\binom{q}{j}f(t-jh) \tag{5}$$

where $\left[\frac{t-q}{h}\right]$ indicates the integer part and $(-1)^j\binom{q}{j}$ are binomial coefficients $c_j^{(q)}\ (j=0,1,\ldots)$.

The calculation of these coefficients is done by formula of following recurrence:

$$c_0^{(q)} = 1, \quad c_j^{(q)} = \left(1 - \frac{1+q}{j}\right)c_{j-1}^{(q)}$$

The general numerical solution of the fractional differential equation:

$${}^{GL}_{a}D^q_t y(t) = f(y(t), t),$$

can be expressed as follows:

$$y(t_k) = f(y(t_k), t_k)h^q - \sum_{j=v}^{k} c_j^{(q)} y(t_{k-j}). \tag{6}$$

The Fundamental Predictor-Corrector Algorithm

The fractional Adams-Bashforth-Moulton method used to approximate the fractional order integral operator was introduced in [14]. In fact it is more practical to use a numerical fractional integration method to compute fractional order integration or derivation as the approximating transfer functions are of relatively high orders.

Consider the differential equation

$$D^{\alpha}y(x) = f(x, y(x)) \tag{7}$$

with initial conditions:

$$y^{(k)}(0) = y_0^{(k)}, \qquad k = 0, 1, \ldots, m-1, \tag{8}$$

where $m = [\alpha]$ and the real numbers $y^{(k)}(0) = y_0^{(k)},\ k = 0, 1, \ldots, m-1,$ are assumed to be given.

The basics of this technique take profit of an interesting analytical property: the initial value problem (4), (5) is equivalent to the Volterra integral equation

$$y(x) = \sum_{k=0}^{[\alpha]-1} y^{(k)}(0)\frac{x^k}{k!} + \frac{1}{\Gamma(\alpha)}\int_0^{\infty} (x-t)^{\alpha-1} f(t, y(t))dt \tag{9}$$

Introducing the equispaced nodes $t_j = jh$ with some given $h > 0$ and by applying the trapezoidal integral technique to compute (6), the corrector formula becomes

$$\begin{aligned} y_h(t_{n+1}) = & \sum_{k=0}^{[\alpha]-1} \frac{t_{n+1}}{k!} y^{(k)}(0) + \frac{h^\alpha}{\Gamma(\alpha+2)} f\left(t_{n+1}, y_h^P(t_{n+1})\right) \\ & + \frac{h^\alpha}{\Gamma(\alpha+2)} \sum_{j=0}^{n} a_{j,n+1} f\left(t_j, y_h(t_j)\right) \end{aligned} \tag{10}$$

where

$$\begin{aligned} a_{0,n+1} = & \, n^{\alpha+1} - (n-\alpha)(n+1)^\alpha \\ a_{j,n+1} = & \, (n-j+2)^{\alpha+1} + (n-j)^{\alpha+1} - 2(n-j+1)^{\alpha+1}, \\ & (1 \le j \le n) \end{aligned} \tag{11}$$

and $y_h^P(t_{n+1})$ is given by,

$$y_h^P(t_{n+1}) = \sum_{k=0}^{[\alpha]-1} \frac{t_{n+1}}{k!} y^{(k)}(0) + \frac{1}{\Gamma(\alpha)} \sum_{j=0}^{n} b_{j,n+1} f\left(t_n, y_h(t_j)\right) \tag{12}$$

where now

$$b_{j,n+1} = \frac{h^\alpha}{\alpha}\left((n+1-j)^\alpha - (n-j)^\alpha\right) \tag{13}$$

This approximation of the fractional derivative within the meaning of *Grünwald-Letnikov* is on the one hand equivalent to the definition of *Riemann-Liouville* for a broad class of functions [46], on the other hand, it is well adapted to the definition of *Caputo* (Adams method) because it requires only the initial conditions and has a physical direction clearly. In this work, the *Grünwald–Letnikov method* is used for numerical evaluation of the fractional derivative.

4 Interval Type-2 Fuzzy Systems

A brief overview of the basic concepts of Interval type-2 fuzzy systems is presented in the following [21, 37, 40]. If we consider a type-1 membership function, as in Fig. 1, then a type-2 membership function can be produced. In this case, for a

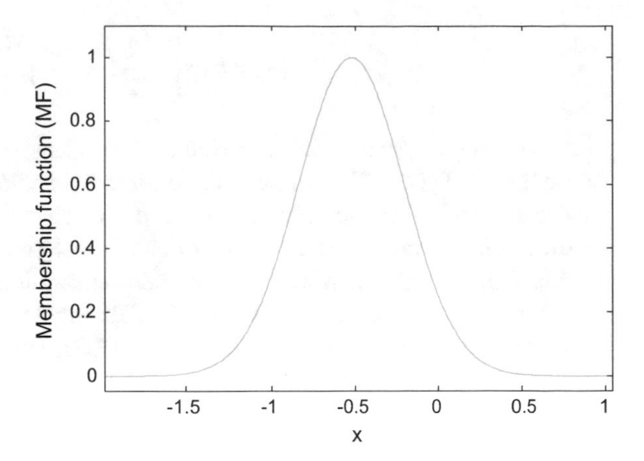

Fig. 1 Example of a type-1 membership function

specific value x' the membership function (u'), takes on different values, which are not all weighted the same, so we can assign membership grades to all of those points.

A type-2 fuzzy set in a universal set X is denoted as $\tilde{A}$ and can be characterized in the following form:

$$\tilde{A} = \int_{x \in X} \mu_{\tilde{A}}(x, v) / x, \forall v \in J_x \subseteq [0, 1] \tag{14}$$

$$\mu_{\tilde{A}}(x) = \int_{v \in J_x} f_x(v) / v,$$

in which $0 \leq \mu_{\tilde{A}}(x) \leq 1$.

The 2-D interval type-2 Gaussian membership function with uncertain mean $m \in [m_1, m_2]$ and a fixed deviation σ is shown in Fig. 2.

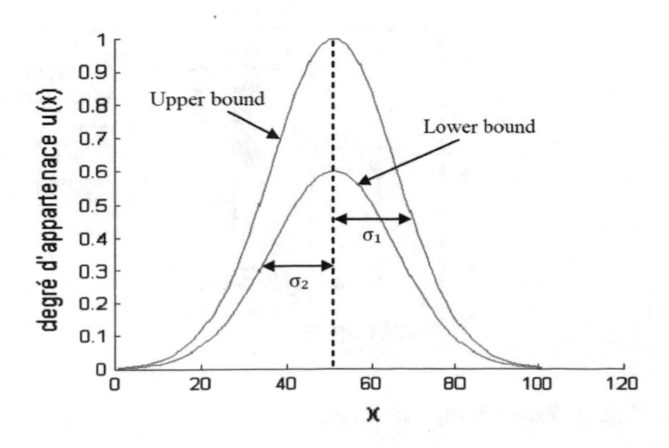

Fig. 2 Interval type-2 membership function

$$\mu_{\tilde{A}}(x) = \exp\left[-\frac{1}{2}\left(\frac{x-m}{\sigma}\right)^2\right]$$

A fuzzy logic system (FLS) described using at least one type-2 fuzzy set is called a type-2 FLS. Type-1 FLSs are unable to directly handle rule uncertainties, because they use type-1 fuzzy sets that are certain. On the other hand, type-2 FLSs, are useful in circumstances where it is difficult to determine an exact numeric membership function, and there are measurement uncertainties [40].

A type-2 FLS is characterized by IF–THEN rules, where their antecedent or consequent sets are now of type-2. Type-2 FLSs, can be used when the circumstances are too uncertain to determine exact membership grades such as when the training data is affected by noise. Similarly, to the type-1 FLS, a type-2 FLS includes a fuzzifier, a rule base, fuzzy inference engine, and an output processor, as we can see in Fig. 3 for a Mamdani-model.

An IT2FS is described by its Lower $\underline{\mu}_{\tilde{A}}(x)$ and Upper $\bar{\mu}_{\tilde{A}}(x)$ membership functions. For an IT2FS, the footprint of uncertainty (FOU) is described in terms of lower and upper MFs as:

$$FOU(\tilde{A}) = \bigcup_{x \in X}\left[\underline{\mu}_{\tilde{A}}(x), \bar{\mu}_{\tilde{A}}(x)\right]$$

The type-reducer generates a type-1 fuzzy set output, which is then converted in a numeric output through running the defuzzifier. This type-1 fuzzy set is also an interval set, for the case of our FLS we used center of sets type reduction, y(X) which is expressed as [10] :

$$y(X) = [y_l, y_r]$$

where $y_l = \frac{\sum_{i=1}^{M} f_l^i y_l^i}{\sum_{i=1}^{M} f_l^i}$ and $y_r = \frac{\sum_{i=1}^{M} f_r^i y_r^i}{\sum_{i=1}^{M} f_r^i}$

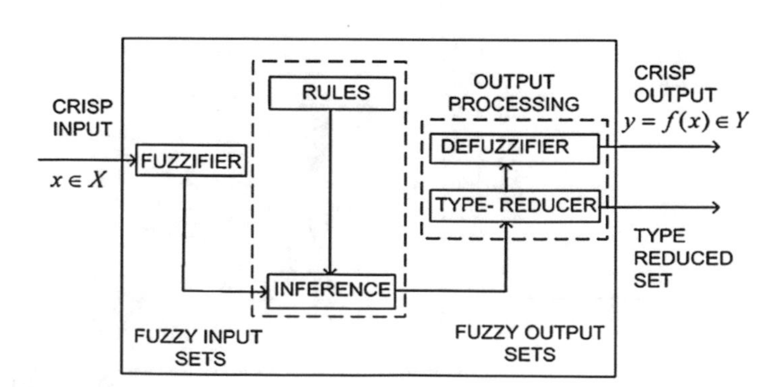

Fig. 3 Type-2 fuzzy logic system

The values of y_l and y_r define the output interval of the type-2 fuzzy system, which can be used to verify if training or testing data are contained in the output of the fuzzy system. This measure of covering the data is considered as one of the design criteria in finding an optimal interval type-2 fuzzy system. The other optimization criteria, is that the length of this output interval should be as small as possible.

From the type-reducer, we obtain an interval set y(X), to defuzzify it we use the average of y_l and y_r, so the defuzzified output of an interval singleton type-2 FLS is [10]:

$$y(X) = \left(\frac{1}{2}\right)(y_l + y_r) \tag{15}$$

where y_l and y_r are the left most and right most points of the Interval type-1 set: $y_l = \sum_{i=1}^{M} y_l^i \xi_l^i = \underline{\xi_l^T \theta_l}$ and $y_r = \sum_{i=1}^{M} y_r^i \xi_r^i = \underline{\xi_r^T \theta_r}$

$$y(X) = (y_l + y_r)/2 = \underline{\xi^T \theta} \tag{16}$$

where $\underline{\xi^T} = (1/2)\underline{[\xi_r^T \xi_l^T]}$ and $\underline{\theta} = [\underline{\theta_r} \quad \underline{\theta_l}]$

5 H^∞ Adaptive Interval Type-2 Fuzzy Control Scheme

Consider an incommensurate fractional order SISO nonlinear dynamic system of the form [22, 37, 40]

$$\begin{cases} x_1^{(q_1)} = x_2 \\ \vdots \\ x_{n-1}^{(q_{n-1})} = x_n \\ x_n^{(q_n)} = f(X,t) + g(X,t)u + d(t) \\ y = x_1 \end{cases}$$

if $q_1 = q_2 = \ldots = q_n = q$ the above system is called a commensurate order system, then equivalent form of the above system is described as:

$$\begin{cases} x^{(nq)} = f(X,t) + g(X,t)u + d(t) \\ y = x_1 \end{cases} \tag{17}$$

where $X = [x_1, x_2, \ldots, x_n]^T = \left[x, x^{(q)}, x^{(2q)}, \ldots, x^{((n-1)q)}\right]^T$ is the state vector, $f(X,t)$ *and* $g(X,t)$ are unknown but bounded nonlinear functions which express system dynamics, $d(t)$ is the external bounded disturbance and $u(t)$ is the control input. The control objective is to force the system output y to follow a given bounded reference signal y_d, under the constraint that all signals involved must be bounded. For simplicity, in this paper adaptive IT2FLC for a commensurate order system is proposed only, since the stability condition for the incommensurate order system can be converted to that for the commensurate order system [21, 24, 40].

To begin with, the reference signal vector y_d and the tracking error vector e will be defined as

$$\underline{y}_d = \left[y_d, y_d^{(q)}, \ldots, y_d^{((n-1)q)}\right]^T \in R^n$$
$$\underline{e} = \underline{y}_d - \underline{y} = \left[e, e^{(q)}, \ldots, e^{((n-1)q)}\right]^T \in R^n,$$
$$e^{(iq)} = y_d^{(iq)} - y^{(iq)}.$$

Let $\underline{k} = [k_1, k_2, \ldots, k_n]^T \in R^n$ to be chosen such that the stable condition$|arg(eig(A))| > q\pi/2$ is met, where $0 < q < 1$ and $eig(A)$ represents the eigenvalues of the matrix A given in (23).

If the functions $f(X,t)$ *and* $g(X,t)$ are known and the system is free of external disturbance d, then the control law of the certainty equivalent controller is obtained as [38, 61].

$$u_{eq} = \frac{1}{g(X,t)}\left[-f(X,t) + y_d^{(nq)} + \underline{k}^T \underline{e}\right]. \tag{18}$$

Substituting (19) into (18), we have:

$$e^{(nq)} + k_n e^{(n-1)q} + \ldots + k_1 e = 0$$

which is the main objective of control, $\lim_{t \to \infty} e(t) = 0$.

However, $f(X,t)$ *and* $g(X,t)$ are unknown and external disturbance $d(t) \neq 0$, the ideal control effort (18) cannot implemented. We replace $f(X,t)$ *and* $g(X,t)$ by the interval type-2 fuzzy logic system $f\left(X|\underline{\theta}_f\right)$ *and* $g\left(X|\underline{\theta}_g\right)$ in a specified form as (16, 17), i.e.,

$$\begin{aligned} f\left(X|\underline{\theta}_f\right) &= \frac{f_l + f_r}{2} = \frac{1}{2}\left(\xi_{fl}^T \underline{\theta}_{fl} + \xi_{fr}^T \underline{\theta}_{fr}\right) = \xi_f^T \underline{\theta}_f \\ g\left(X|\underline{\theta}_g\right) &= \frac{g_l + g_r}{2} = \frac{1}{2}\left(\xi_{gl}^T \underline{\theta}_{gl} + \xi_{gr}^T \underline{\theta}_{gr}\right) = \xi_g^T \underline{\theta}_g \end{aligned} \tag{19}$$

where $\underline{\theta}_f = \left[\underline{\theta}_{fr} \quad \underline{\theta}_{fl}\right]$ and $\underline{\theta}_g = \left[\underline{\theta}_{gr} \quad \underline{\theta}_{gl}\right]$.

Here the fuzzy basis function $\xi(X) = \left(\frac{1}{2}\right)\left[\underline{\xi}_r \quad \underline{\xi}_l\right] = \xi_f(X) = \xi_g(X)$ depends on the type-2 fuzzy membership functions and is supposed to be fixed, while $\underline{\theta}_f$ *and* $\underline{\theta}_g$ are adjusted by adaptive laws based on a Lyapunov stability criterion.

Therefore, the resulting control effort can be obtained as [24, 39],

$$u = \frac{1}{g\left(X|\underline{\theta}_g\right)}\left[-f\left(X|\underline{\theta}_f\right) + y_d^{(nq)} + \underline{k}^T\underline{e} - u_a\right]$$

so

$$u = \frac{1}{\frac{1}{2}\left(\xi_{gl}^T\underline{\theta}_{gl} + \xi_{gr}^T\underline{\theta}_{gr}\right)}\left[-\frac{1}{2}\left(\xi_{fl}^T\underline{\theta}_{fl} + \xi_{fr}^T\underline{\theta}_{fr}\right) + y_d^{(nq)} + \underline{k}^T\underline{e} - u_a\right] \tag{20}$$

where the robust compensator u_a is employed to attenuate the external disturbance and the fuzzy approximation errors.

By substituting (20) into (17), we have

$$\begin{aligned} x^{(nq)} &= f(X,t) + g(X,t)u + d(t) + g\left(X|\underline{\theta}_g\right)u - g\left(X|\underline{\theta}_g\right)u \\ &= \left[f(X,t) - \frac{1}{2}\left(\xi_{fl}^T\underline{\theta}_{fl} + \xi_{fr}^T\underline{\theta}_{fr}\right)\right] + y_d^{(nq)} + \underline{k}^T\underline{e} - u_a \\ &\quad + \left[g(X,t) - \frac{1}{2}\left(\xi_{gl}^T\underline{\theta}_{gl} + \xi_{gr}^T\underline{\theta}_{gr}\right)\right]u + d(t) \end{aligned} \tag{21}$$

then

$$\begin{aligned} e^{(nq)} = &\left[f(X,t) - \frac{1}{2}\left(\xi_{fl}^T\underline{\theta}_{fl} + \xi_{fr}^T\underline{\theta}_{fr}\right)\right] + \underline{k}^T\underline{e} - u_a + d(t) \\ &+ \left[g(X,t) - \frac{1}{2}\left(\xi_{gl}^T\underline{\theta}_{gl} + \xi_{gr}^T\underline{\theta}_{gr}\right)\right]u = 0 \end{aligned} \tag{22}$$

Equation (22) can be rewritten in state space representation as:

$$\begin{aligned} \underline{e}^{(q)} = A\underline{e} + B\Bigg[&\frac{1}{2}\left(\xi_{fl}^T\underline{\theta}_{fl} + \xi_{fr}^T\underline{\theta}_{fr}\right) - f(X,t) + u_a + \left(\frac{1}{2}\left(\xi_{gl}^T\underline{\theta}_{gl} + \xi_{gr}^T\underline{\theta}_{gr}\right)\right. \\ &\left. - g(X,t)u - d(t)\right)\Bigg] \end{aligned} \tag{23}$$

where $A = \begin{bmatrix} 0 & 1 & 0 & 0 & \cdots & 0 & 0 \\ 0 & 0 & 1 & 0 & \cdots & 0 & 0 \\ \vdots & \vdots & \vdots & \vdots & \ddots & \vdots & \vdots \\ 0 & 0 & 0 & 0 & \cdots & \vdots & 1 \\ -k_1 & -k_2 & -k_3 & -k_4 & \cdots & -k_{(n-1)} & -k_n \end{bmatrix}$, $B = \begin{bmatrix} 0 \\ 0 \\ \vdots \\ 0 \\ 1 \end{bmatrix}$

The optimal parameter estimations $\underline{\theta}_f^*$ and $\underline{\theta}_g^*$ are defined:

$$\underline{\theta}_f^* = \arg \min_{\underline{\theta}_f \varepsilon \Omega_f} [\sup_{x \varepsilon \Omega_x} \left| f\left(X|\underline{\theta}_f\right) - f(X,t) \right|] \tag{24}$$

$$\underline{\theta}_g^* = \arg \min_{\underline{\theta}_g \varepsilon \Omega_g} [\sup_{x \varepsilon \Omega_x} \left| g\left(X|\underline{\theta}_g\right) - g(X,t) \right|] \tag{25}$$

where Ω_f, Ω_g *and* $\Omega_{\underline{x}}$ are constraint sets of suitable bounds on $\underline{\theta}_f, \underline{\theta}_g$ and x respectively and they are defined as $\Omega_f = \left\{\underline{\theta}_f \middle| |\underline{\theta}_f| \leq M_f\right\}$, $\Omega_g = \left\{\underline{\theta}_g \middle| |\underline{\theta}_g| \leq M_g\right\}$ et $\Omega_x = \{x||x| \leq M_x\}$ where M_f, M_g et M_x are positive constants.

By using (24)–(25), an error dynamic Eq. (23) can be expressed as:

$$\underline{e}^{(q)} = A\underline{e} + B\left[f\left(X|\underline{\theta}_f\right) - f\left(X|\underline{\theta}_f^*\right) + u_a + \left(g\left(\underline{x}|\underline{\theta}_g\right) - g\left(X|\underline{\theta}_g^*\right)\right)u - d(t)\right] \tag{26}$$

Also, the minimum approximation error is defined as:

$$\omega_1 = g\left(X|\underline{\theta}_g^*\right) - g(X,t) + f\left(X|\underline{\theta}_f^*\right) - f(X,t) - d(t) \tag{27}$$

If $\underline{\theta}_f = \underline{\theta}_f - \underline{\theta}_f^*$ and $\underline{\theta}_g = \underline{\theta}_g - \underline{\theta}_g^*$, (27) can be rewritten as:

$$\underline{e}^{(q)} = A\underline{e} + B\left[\xi(X)^T \underline{\theta}_f + \xi(X)^T \underline{\theta}_g u + u_a + \omega_1\right] \tag{28}$$

Following the preceding consideration, the following theorem can be obtained [35].

6 Stability Analysis

Theorem 1 *Consider the commensurate fractional order SISO nonlinear dynamic system* (17) *with control input* (20)*, if the robust compensator* u_a *and the type-2 fuzzy-based adaptive laws are chosen as*

$$u_a = -\frac{1}{r} B^T P \underline{e} \tag{29}$$

$$\underline{\theta}_{fr}^{(q)} = -r_1 \xi_{fr} B^T P \underline{e} \tag{30}$$

$$\underline{\theta}_{fl}^{(q)} = -r_2 \xi_{fl} B^T P \underline{e} \tag{31}$$

$$\underline{\theta}_{gr}^{(q)} = -r_3 \xi_{gr} B^T P \underline{e} u \tag{32}$$

$$\underline{\theta}_{gl}^{(q)} = -r_4 \xi_{gl} B^T P \underline{e} u \tag{33}$$

where $\underline{\theta}_f^{(q)} = \left[\underline{\theta}_{fr}^{(q)} \quad \underline{\theta}_{fl}^{(q)} \right]$, $\underline{\theta}_g^{(q)} = \left[\underline{\theta}_{gr}^{(q)} \quad \underline{\theta}_{gl}^{(q)} \right]$, $\underline{\xi}_f^T = \left(\frac{1}{2}\right) \left[\underline{\xi}_{fr}^T \quad \underline{\xi}_{fl}^T \right]$ and $\underline{\xi}_g^T = \left(\frac{1}{2}\right) \left[\underline{\xi}_{gr}^T \quad \underline{\xi}_{gl}^T \right]$

where $r > 0, r_i > 0, i = 1 \sim 4,$ *and* $P = P^T > 0$ *is the solution of the following Riccati-like equation :*

$$\mathrm{PA} + \mathrm{A^T P} + \mathrm{Q} - \mathrm{PB}\left(\frac{2}{\mathrm{r}} - \frac{1}{\rho^2}\right)\mathrm{B^T P} = 0 \tag{34}$$

where $Q = Q^T > 0$ *is a prescribed weighting matrix. Therefore, the* H^∞ *tracking performance can be achieved for a prescribed attenuation level* ρ *which satisfies* $2\rho^2 \geq r$ *and all the variables of the closed-loop system are bounded.*

In order to analyze the closed-loop stability, the Lyapunov function candidate is chosen as

$$\begin{aligned} V(t) = {} & \frac{1}{2}\underline{e}^T(t) P \underline{e}(t) + \frac{1}{2r_1}\left(\underline{\theta}_{fr}^T\right)\left(\underline{\theta}_{fr}\right) + \frac{1}{2r_2}\left(\underline{\theta}_{fl}^T\right)\left(\underline{\theta}_{fl}\right) \\ & + \frac{1}{2r_3}\left(\underline{\theta}_{gr}^T\right)\left(\underline{\theta}_{gr}\right) + \frac{1}{2r_4}\left(\underline{\theta}_{gl}^T\right)\left(\underline{\theta}_{gl}\right) \end{aligned} \tag{35}$$

Taking the derivative of (36) with respect to time, we get

$$\begin{aligned} V^{(q)}(t) = {} & \frac{1}{2}\left(\underline{e}^{(q)}(t)\right)^T P \underline{e}(t) + \frac{1}{2}\underline{e}^T(t) P \underline{e}(t) \\ & + \frac{1}{r_1}\left(\underline{\theta}_{fr}^T\right)\left(\underline{\theta}_{fr}^{(q)}\right) + \frac{1}{r_3}\left(\underline{\theta}_{gr}^T\right)\left(\underline{\theta}_{gr}^{(q)}\right) + \frac{1}{r_2}\left(\underline{\theta}_{fl}^T\right)\left(\underline{\theta}_{fl}^{(q)}\right) + \frac{1}{r_4}\left(\underline{\theta}_{gl}^T\right)\left(\underline{\theta}_{gl}^{(q)}\right) \\ = {} & \frac{1}{2}\left\{A\underline{e} + B\left[\xi^T \underline{\theta}_f + \xi^T \underline{\theta}_g u + u_a + \omega_1\right]\right\}^T P \underline{e} + \frac{1}{2}\underline{e}^T(t) P \left\{A\underline{e} + B\left[\xi^T \underline{\theta}_f\right.\right. \\ & \left.\left. + \xi^T \underline{\theta}_g u + u_a + \omega_1\right]\right\} + \frac{1}{r_1}\left(\underline{\theta}_{fr}^T\right)\left(\underline{\theta}_{fr}^{(q)}\right) + \frac{1}{r_3}\left(\underline{\theta}_{gr}^T\right)\left(\underline{\theta}_{gr}^{(q)}\right) + \frac{1}{r_2}\left(\underline{\theta}_{fl}^T\right)\left(\underline{\theta}_{fl}^{(q)}\right) \\ & + \frac{1}{r_4}\left(\underline{\theta}_{gl}^T\right)\left(\underline{\theta}_{gl}^{(q)}\right) \end{aligned} \tag{36}$$

obtained after a simple manipulation

$$
\begin{aligned}
V^{(q)}(t) = {} & \frac{1}{2}\underline{e}^T(A^T + PA)\underline{e} + \underline{e}^T PBu_a + \underline{e}^T PB\omega_1 + \{\underline{\theta}_{fr}^T[\xi_{fr}B^T P\underline{e} \\
& + \frac{1}{r_1}(\underline{\theta}_{fr}^{(q)})]\} + \left\{\underline{\theta}_{fl}^T\left[\xi_{fl}B^T P\underline{e} + \frac{1}{r_2}(\underline{\theta}_{fl}^{(q)})\right]\right\} + \{\underline{\theta}_{gr}^T[\xi_{gr}B^T P\underline{e}u \\
& + \frac{1}{r_3}(\underline{\theta}_{gr}^{(q)})]\} + \left\{\underline{\theta}_{gl}^T\left[\xi_{gl}B^T P\underline{e}u + \frac{1}{r_4}(\underline{\theta}_{gl}^{(q)})\right]\right\}
\end{aligned}
\tag{37}
$$

From (29) the robust compensator u_a, and the fuzzy-based adaptive laws are given (30)–(33), $V^{(q)}(t)$ in (37) can be rewritten as:

$$
\begin{aligned}
V^{(q)}(t) &= -\frac{1}{2}\underline{e}^T Q\underline{e} - \frac{1}{2\rho^2}\underline{e}^T PBB^T\underline{e} + \underline{e}^T PB\omega_1 \\
&= -\frac{1}{2}\underline{e}^T Q\underline{e} - \frac{1}{2}\left(\frac{1}{\rho}B^T P\underline{e} - \rho\omega_1\right)^T\left(\frac{1}{\rho}B^T P\underline{e} - \rho\omega_1\right) + \frac{1}{2}\rho^2\omega_1^T\omega_1 \\
&\leq -\frac{1}{2}\underline{e}^T Q\underline{e} + \frac{1}{2}\rho^2\omega_1^T\omega_1
\end{aligned}
\tag{38}
$$

Integrating (38) from $t = 0$ to $t = T$, we have

$$
V(T) - V(0) \leq -\frac{1}{2}\int_0^T\left(\underline{e}^T Q\underline{e}dt + \frac{1}{2}\rho^2\omega_1^T\omega_1\right)dt \tag{39}
$$

Since $V(T) \geq 0$, (39) can be rewritten as follows:

$$
\begin{aligned}
\int_0^T \underline{e}^T Q\underline{e}dt \leq {} & e^T(0)Pe(0) + \theta^T(0)\theta(0) \\
& + \rho^2\int_0^T \omega_1^T\omega_1 dt
\end{aligned}
\tag{40}
$$

Therefore, the H^∞ tracking performance can be achieved. The proof is completed.

7 Simulation Results

The chaotic behaviors in a fractional order modified Duffing system studied numerically by phase portraits are given by [16, 22]. In this section, we will apply our adaptive fuzzy H^∞ controller to synchronize two different fractional order chaotic *Duffing systems.*

Consider the following two fractional order chaotic Duffing systems [23]:

- Drive system:

$$\begin{cases} D^q y_1 = y_2 \\ D^q y_2 = y_1 - 0.25 y_2 - y_1^3 + 0.3\cos(t) \end{cases} \quad (41)$$

- Response system:

$$\begin{cases} D^q x_1 = x_2 \\ D^q x_2 = x_1 - 0.3 x_2 - x_1^3 + 0.35\cos(t) + u(t) + d(t) \end{cases} \quad (42)$$

where the external disturbance $d(t) = 0.1\,sin(t)$. The main objective is to control the trajectories of the response system to track the reference trajectories obtained from the drive system. The initial conditions of the drive and response systems are chosen as:

$$\begin{bmatrix} x_1(0) \\ x_2(0) \end{bmatrix} = \begin{bmatrix} 0 \\ 0 \end{bmatrix} \; and \; \begin{bmatrix} y_1(0) \\ y_2(0) \end{bmatrix} = \begin{bmatrix} 0.2 \\ -0.2 \end{bmatrix}, \text{(respectively)}.$$

The simulations results for fractional order $q = 0.98$ are illustrated as follows:

The Fig. 4 represents the 3D phase portrait of the drive and response systems without control input. It is obvious that the synchronization performance is bad without a control effort supplied to the response system.

The different values of $0 < q < 1$ are considered in order to show the robustness of the proposed adaptive fuzzy H^{∞} control with our law.

According to the two state output ranges, the membership functions of x_i, for $f\left(X|\underline{\theta}_f\right)$ and $g\left(X|\underline{\theta}_g\right)$ are selected as follows:

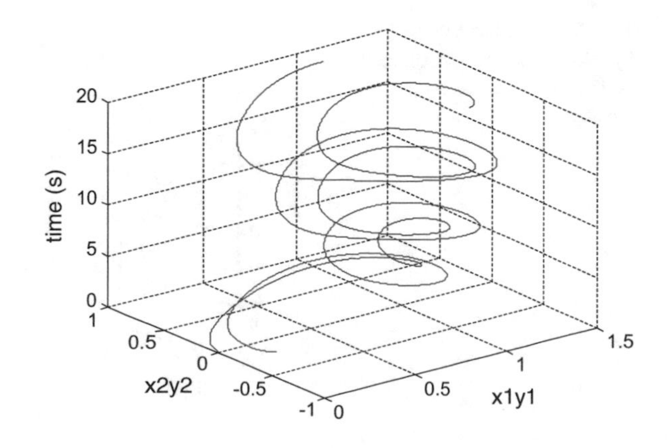

Fig. 4 3D phase portrait of the drive and response systems without control input (*Before* the control input)

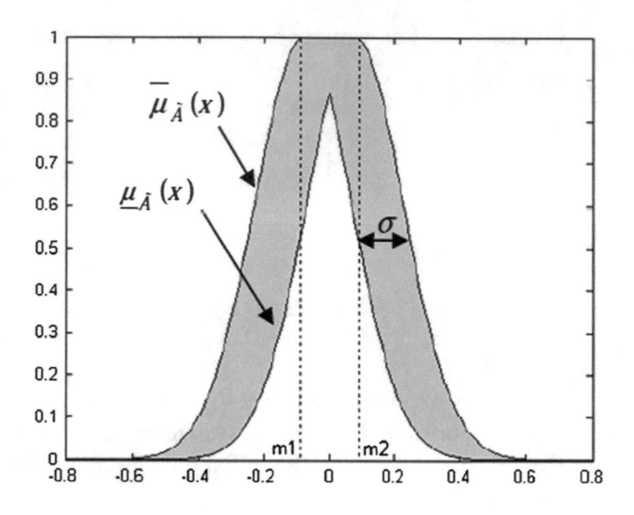

Fig. 5 Interval Type-2 Fuzzy sets Gaussian with uncertain standard deviation σ

$\mu_{F_i^l}(x_i) = \exp\left[-0.5\left(\frac{x_i - \overline{x}}{0.8}\right)^2\right]$ $i = 1, 2$ and $l = 1, \ldots, 7$ where $\bar{x}$ is selected from the interval $[-1, 2]$. (Figure 5)

From the adaptive laws (30)–(33) and the robust compensator (29), the control law of the response system can be obtained as:

$$u = \frac{1}{\xi^T(X)\underline{\theta}_g}\left[-\xi^T(X)\underline{\theta}_f + y_d^{(nq)} + \underline{k}^T\underline{e} - u_a\right] \tag{43}$$

According to Theorem 1, the controlled error system can be stabilized, i.e., the master system (41) can synchronize the slave system (42) with the control law (20).

The Figs. 6, 7, 8, 9, 10 represent the different simulation results of the drive and response systems with control input (43) for the fractional order q = 0.98.

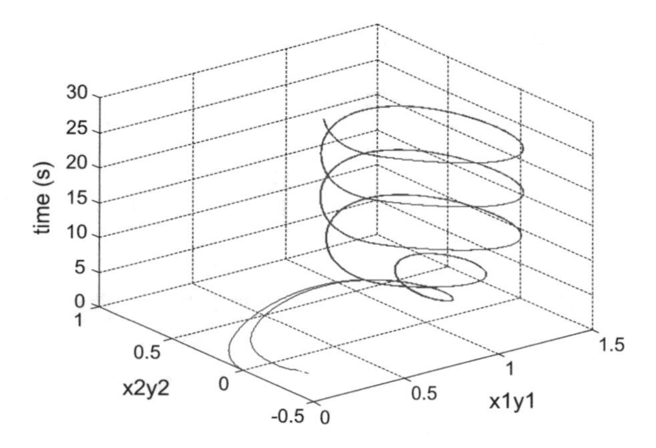

Fig. 6 3D phase portrait, synchronization performance, of the drive and response systems

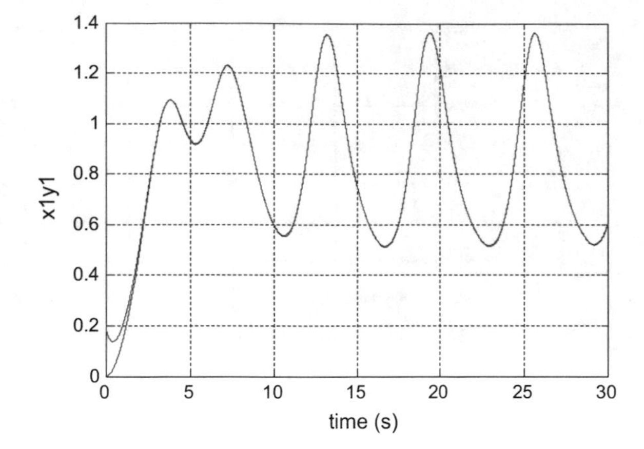

Fig. 7 State trajectories: x_1y_1

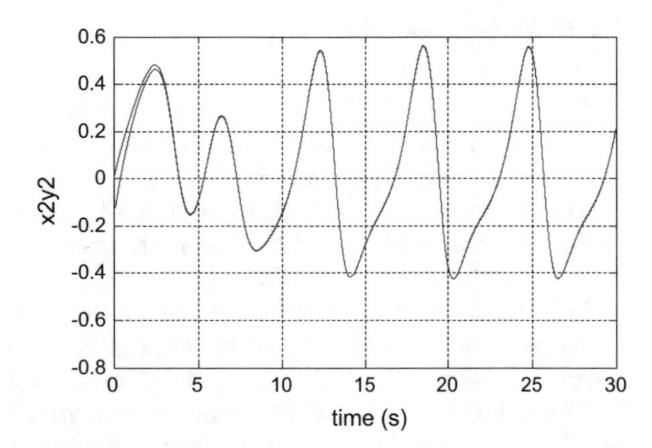

Fig. 8 State trajectories: x_2y_2

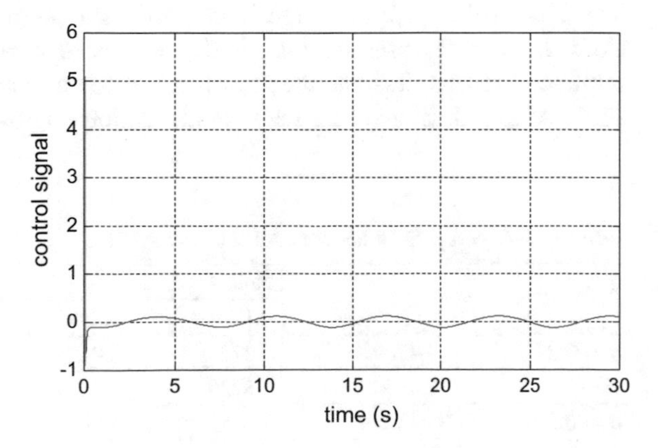

Fig. 9 Control effort: $u(t)$

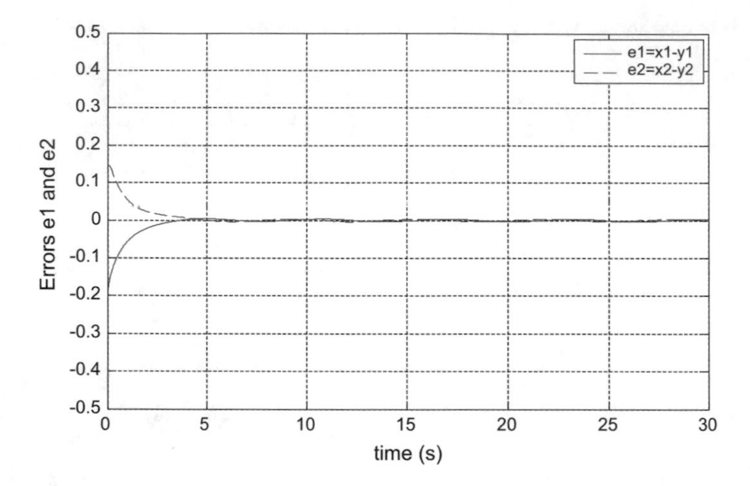

Fig. 10 Errors signals: $e_1\&e_2$

It is clearly seen from Fig. 10 that the tracking errors $e_1(t)$ and $e_2(t)$ converge both to zero in less than 5 s. Synchronization is perfectly achieved as shown by the state trajectories in Figs. 7 and 8.

The control signal can be observed in the Fig. 9. It indicates that the obtained results are comparable with the solution presented in [23], but fluctuations of the control function are much smaller.

In order to have a quantitative comparison between both methods, a white Gaussian noise is applied to the measured signal with various signals to Noise and Integral of Absolute Error (IAE) is selected as the criterion.

The results in Table 1 clearly indicate that the performance of our proposed type-2 fuzzy controller surpasses the type-1fuzzy method [22]. As can be seen in high SNRs both of the methods have similar performance, however in low SNRs type-1 controller [22] has large IAEs while our proposed controller has still low IAEs. Despite the presence of additive noises in measured signals, the obtained simulation results illustrate the robustness of the proposed control strategy and the utility of introducing type-2 fuzzy modelization approach.

Table 1 IAE comparison between AT2FC and AT1FC

q		Noises			
		0.1sin$t$	0.3sin$t$	0.5sin$t$	0.7sin$t$
q = 0.89	T1-Fuzzy	3.3787	3.4135	3.4481	3.4725
	T2-Fuzzy	3.3720	3.4044	3.4348	3.4607
q = 0.98	T1-Fuzzy	3.9251	4.0016	4.0731	4.1320
	T2-Fuzzy	3.8879	3.9163	3.9418	3.9607

8 Conclusion

In this paper a novel adaptive interval type-2 fuzzy using H^{∞} control is proposed to deal with chaos synchronization between two different uncertain fractional order chaotic systems. The use of interval type-2 helps to minimize the added computational burden and hence renders the overall system to be more practically applicable.

Based on the Lyapunov synthesis approach, free parameters of the adaptive fuzzy controller can be tuned on line by the output feedback control law and adaptive laws. The simulation example, chaos synchronization of two fractional order Duffing systems, is given to demonstrate the effectiveness of the proposed methodology. The significance of the proposed control scheme in the simulation for different values of q is manifest. Simulation results show that a fast synchronization of drive and response can be achieved and as q is reduced the chaos is seen reduced, i.e., the synchronization error is reduced, accordingly.

Future research efforts will concern observer-based nonlinear adaptive control of uncertain or unknown fractional order systems. The problem of online identification and parameters estimation for such systems is also a good challenge. Another topic of interest is the design of new robust adaptive control laws for the class of fractional nonlinear systems based on various control configurations such as: (Internal model control) IMC, (Model reference Adaptive Systems) MRAS and the Strictly Positive realness property.

References

1. Aguila-Camacho, N., Duarte-Mermoud, M. A., & Delgado-Aguilera, E. (2016). Adaptive synchronization of fractional Lorenz systems using a re duce d number of control signals and parameters. *Chaos, Solitons and Fractals, 87*, 1–11.
2. Arena, P., Caponetto, R., Fortuna, L., & Porto, D. (1997). Chaos in a fractional order Duffing system. In: *Proceedings ECCTD* (pp. 1259–1262).
3. Azar, A. T. (2010). Adaptive neuro-fuzzy systems. In: A. T. Azar (ed.), Fuzzy Systems. IN-TECH, Vienna, Austria. ISBN 978-953-7619-92-3.
4. Azar, A. T. (2012). Overview of type-2 fuzzy logic systems. *International Journal of Fuzzy System Applications (IJFSA), 2*(4), 1–28.
5. Azar, A. T., Vaidyanathan, S. (2015). Chaos modeling and control systems design. Studies in computational intelligence (Vol. 581). Germany: Springer. ISBN 978-3-319-13131-3.
6. Azar, A. T., Vaidyanathan, S. (2016). Advances in chaos theory and intelligent control. Studies in fuzziness and soft computing (Vol. 337). Germany: Springer. ISBN 978-3-319-30338-3.
7. Boulkroune, A., Bouzeriba, A., Bouden, T., & Azar, A. T. (2016). Fuzzy adaptive synchronization of uncertain fractional-order chaotic systems. Advances in chaos theory and intelligent control. Studies in fuzziness and soft computing (Vol. 337). Germany: Springer.
8. Boulkroune, A., Hamel, S., & Azar, A. T. (2016). Fuzzy control-based function synchronization of unknown chaotic systems with dead-zone input. Advances in chaos theory and intelligent control. Studies in fuzziness and soft computing (Vol. 337). Germany: Springer.

9. Castillo, O., & Melin, P. (2008). *Type-2 fuzzy logic: Theory and applications*. Heidelberg: Springer.
10. Castillo, O., & Melin, P. (2012). Recent advances in interval type-2 fuzzy systems. *Springer Briefs in Computational Intelligence*. doi:10.1007/978-3-642-28956-9_2.
11. Castro, J. L. (1995). Fuzzy logical controllers are universal approximators. *IEEE Transactions on Systems, Man and Cybernetics, 25*, 629–635.
12. Cazarez-Castro, N. R., Aguilar, L. T., & Castillo, O. (2012). Designing type-1 and type-2 fuzzy logic controllers via fuzzy Lyapunov synthesis for non-smooth mechanical systems. *Engineering Applications of Artificial Intelligence, 25*(5), 971–979.
13. Chen, Y., Wei, Y., Liang, S., & Wang, Y. (2016). Indirect model reference adaptive control for a class of fractional order systems. *Communications in Nonlinear Science and Numerical Simulation*.
14. Diethlem, K. (2003). Efficient solution of multi-term fractional differential equations using P (EC)mE methods. *Computing, 71*, 305–319.
15. Efe, M. O. (2008). Fractional fuzzy adaptive sliding-mode control of a 2-DOF direct-drive robot arm. *IEEE Transactions on Systems, Man, and Cybernetics—Part B: Cybernetics, 38* (6), 1561–1570.
16. Ge, Z.-M., & Ou, C.-Y. (2008). Chaos synchronization of fractional order modified Duffing systems with parameters excited by a chaotic signal. *Chaos, Solitons and Fractals, 35*(4), 705–717.
17. Ghaemi, M., & Akbarzadeh T. M. R. (2011). Optimal design of adaptive interval type-2 fuzzy sliding mode control using genetic algorithm. In *Proceedings of the 2nd International Conference on Control, Instrumentation, and Automation, Shiraz, Iran* (pp. 626–631).
18. Hartley, T. T., Lorenzo, C. F., & Qammer, H. K. (1995). Chaos on a fractional Chua's system. *IEEE Transactions on Circuits and Systems Theory and Applications, 42*(8), 485–490.
19. Hilfer, R. (2001). *Applications of fractional calculus in physics*. New Jersey: World Scientific.
20. Karnik, N. N., & Mendel, J. M. (1998). Type-2 fuzzy logic systems: Type-reduction. In *Proceedings of the IEEE International Conference on Systems, Man, and Cybernetics, San Diego, CA* (pp. 2046–2051).
21. Karnik, N. N., Mendel, J. M., & Liang, Q. (1999). Type-2 fuzzy logic systems. *IEEE Transactions on Fuzzy Systems, 7*, 643–658.
22. Khettab, K., Bensafia, Y., & Ladaci, S. (2014). Robust adaptive fuzzy control for a class of uncertain nonlinear fractional systems. In *Proceedings of the Second International Conference on Electrical Engineering and Control Applications ICEECA'2014, Constantine, Algeria*.
23. Khettab, K., Bensafia, Y., & Ladaci, S. (2015). Fuzzy adaptive control enhancement for non-affine systems with unknown control gain sign. In *Proceedings of the International Conference on Sciences and Techniques of Automatic Control and Computer Engineering, STA'2015, Monastir, Tunisia* (pp. 616–621).
24. Khettab, K., Ladaci, S., & Bensafia, Y. (2016). Fuzzy adaptive control of fractional order chaotic systems with unknown control gain sign using a fractional order Nussbaum gain. *IEEE/CAA Journal of Automatica Sinica*.
25. Kilbas, A. A., Srivastava, H. M., & Trujillo, J. J. (2006). *Theory and applications of fractional differential equations*. Amsterdam: Elsevier.
26. Ladaci, S., & Charef, A. (2002). Commande adaptative à modèle de référence d'ordre fractionnaire d'un bras de robot (in French) Communication Sciences & Technologie, ENSET Oran, Algeria (Vol. 1, pp. 50–52).
27. Ladaci, S., & Charef, A. (2006). On fractional adaptive control. *Nonlinear Dynamics, 43*(4), 365–378.
28. Ladaci, S., & Charef, A. (2006). An adaptive fractional $PI^{\lambda}D^{\mu}$ controller. In *Proceedings of the Sixth Int. Symposium on Tools and Methods of Competitive Engineering, TMCE 2006, Ljubljana, Slovenia* (1533–1540).

29. Ladaci, S., Loiseau, J. J., & Charef, A. (2008). Fractional order adaptive high-gain controllers for a class of linear systems. *Communications in Nonlinear Science and Numerical Simulation, 13*(4), 707–714.
30. Ladaci, S., Charef, A., & Loiseau, J. J. (2009). Robust fractional adaptive control based on the strictly positive realness condition. *International Journal of Applied Mathematics and Computer Science, 19*(1), 69–76.
31. Ladaci, S., Loiseau, J. J., & Charef, A. (2010). Adaptive internal model control with fractional order parameter. *International Journal of Adaptive Control and Signal Processing, 24*, 944–960.
32. Ladaci, S., & Charef, A. (2012). Fractional order adaptive control systems: A survey. In E.W. Mitchell & S.R. Murray (Eds.), *Classification and application offractals* (pp. 261–275). Nova Science Publishers Inc.
33. Ladaci, S., & Khettab, K. (2012). Fractional order multiple model adaptive control. *International Journal of Automation and Systems Engineering, 6*(2), 110–122.
34. Ladaci, S., & Bensafia, Y. (2016). Indirect fractional order pole assignment based adaptive control. *Engineering Science and Technology, an International Journal, 19*, 518–530.
35. Lee, C.-H., & Chang, Y.-C. (1996). H^{∞} Tracking design of uncertain nonlinear SISO systems: Adaptive fuzzy approach. *IEEE Transactions on Systems, 4*(1).
36. Lin, T.-C., Chen, M.-C., Roopaei, M., & Sahraei, B. R. (2010). Adaptive type-2 fuzzy sliding mode control for chaos synchronization of uncertain chaotic systems. In *Proceedings of the IEEE International Conference on Fuzzy Systems (FUZZ), Barcelona* (pp. 1–8).
37. Lin, T. C., Kuo, M. J., & Hsu, C. H. (2010). Robust adaptive tracking control of multivariable nonlinear systems based on interval type-2 fuzzy approach. *International Journal of Innovative Computing, Information and Control, 6*(1), 941–961.
38. Lin, T.-C., & Kuo, C.-H. (2011). H^{∞} synchronization of uncertain fractional order chaotic systems: Adaptive fuzzy approach. *ISA Transactions, 50*, 548–556.
39. Lin, T. C., Kuo, C.-H., Lee, T.-Y., & Balas, V. E. (2012). Adaptive fuzzy H^{∞} tracking design of SISO uncertainnonlinear fractional order time-delay systems. *Nonlinear Dynamics, 69*, 1639–1650.
40. Lin, T. C., Lee, T.-Y., Balas, & V. E. (2011). Synchronization of uncertain fractional order chaotic systems via adaptive interval type-2 fuzzy sliding mode control. In *Proceedings of the IEEE International Conference on Fuzzy Systems, Taipei, Taiwan.*
41. Lin, T.-C., Lee, T.-Y., & Balas, V. E. (2011). Adaptive fuzzy sliding mode control for synchronization of uncertain fractional order chaotic systems. *Chaos, Solitons and Fractals, 44*, 791–801.
42. Lin, T. C., Liu, H. L., & Kuo, M. J. (2009). Direct adaptive interval type-2 fuzzy control of multivariable nonlinear systems. *Engineering Applications of Artificial Intelligence, 22*(3), 420–430.
43. Lin, T. C., Wang, C.-H., & Liu, H.-L. (2004). Observer-based indirect adaptive fuzzy-neural tracking control for nonlinear SISO systems using VSS and H^{∞} approaches. *Fuzzy Sets and Systems, 143*(2), 211–232.
44. Mendel, J. M. (2007). Advances in type-2 fuzzy sets and systems. *Information Sciences, 177* (1), 84–110.
45. Mendel, J. M., & John, R. I. B. (2002). Type-2 fuzzy sets made simple. *IEEE Transactions on Fuzzy Systems, 10*(2), 117–127.
46. N'Doye, I. (2011). Généralisation du lemme de Gronwall-Bellman pour la stabilisation des systèmes fractionnaires. Ph.D. Thesis, Ecole doctorale IAEM Lorraine, Morocco.
47. Neçaibia, A., & Ladaci, S. (2014). Self-tuning fractional order $PI^{\lambda}D^{\mu}$ controller based on extremum seeking approach. *International Journal of Automation and Control, Inderscience, 8*(2), 99–121.
48. Neçaibia, A., Ladaci, S., Charef, A., & Loiseau, J. J. Fractional order extremum seeking control. In *Proceedings of the 22nd Mediterranean Conference on Control and Automation* (pp. 459–462) (MED'14, Palermo, Italy on June 16–19).

49. Odibat, Z. M. (2010). Adaptive feedback control and synchronization of non-identical chaotic fractional order systems. *Nonlinear Dynamics, 60*, 479–487.
50. Ouannas, A., Azar, A. T., & Vaidyanathan, S. (2016). A robust method for new fractional hybrid chaos synchronization. *Mathematical Methods in the Applied Sciences*. doi:10.1002/mma.4099.
51. Petráš, I. (2006). A note on the fractional-order cellular neural networks. In *Proceedings of the IEEE International World Congress on Computational Intelligence, International Joint Conference on Neural Networks* (pp. 16–21).
52. Petráš, I. (2008). A note on the fractional-order Chua's system. *Chaos, Solitons and Fractals, 38*(1), 140–147.
53. Podlubny, I. (1999). *Fractional differential equations*. San Diego: Academic Press.
54. Qilian, L., & Mendel, J. M. (2000). Interval type-2 fuzzy logic systems: Theory and design. *IEEE Transactions on Fuzzy Systems, 8*(5), 535–550.
55. Rabah, K., Ladaci, S., & Lashab, M. (2016). Stabilization of a Genesio-Tesi chaotic system using a fractional order $PI^{\lambda}D^{\mu}$ regulator. *International Journal of Sciences and Techniques of Automatic Control and Computer Engineering, 10*(1), 2085–2090.
56. Srivastava, H. M., & Saxena, R. K. (2001). Operators of fractional integration and their applications. *Applied Mathematics and Computation, 118*, 1–52.
57. Tian, X., Fei, S., & Chai, L. (2014). Adaptive control of a class of fractional-order nonlinear complex systems with dead-zone nonlinear inputs. In *Proceedings of the 33rd Chinese Control Conference, Nanjing, China* (pp. 1899–1904).
58. Vaidyanathan, S., & Azar, A. T. (2016). Takagi-Sugeno fuzzy logic controller for Liu-Chen four-scroll chaotic system. *International Journal of Intelligent Engineering Informatics, 4*(2), 135–150.
59. Vinagre, B. M., Petras, I., Podlubny, I., & Chen, Y.-Q. (2002). Using fractional order adjustment rules and fractional order reference models in model-reference adaptive control. *Nonlinear Dynamics, 29*, 269–279.
60. Wang, L. X. (1992). Fuzzy systems are universal approximators. In *Proceedings of the IEEE International Conference on Fuzzy Systems, San Diego* (pp. 1163–1170).
61. Wang, C.-H., Liu, H.-L., & Lin, T.-C. (2002). Direct adaptive fuzzy-neural control with state observer and supervisory controller for unknown nonlinear dynamical systems. *IEEE Transactions on Fuzzy Systems, 10*(1), 39–49.
62. Wang, C.-H., Cheng, C.-S., & Lee, T.-T. (2004). Dynamical optimal training for interval type-2 fuzzy neural network (T2FNN). *IEEE Transactions on Systems, Man, and Cybernetics. Part B, Cybernetics, 34*(3), 1462–1477.
63. Wei, Y., Sun, Z., Hu, Y., & Wang, Y. (2015). On fractional order composite model reference adaptive control. *International Journal of Systems Science*. doi:10.1080/00207721.2014.998749.
64. Wei, Y., Tse, P. W., Yao, Z., & Wang, Y. (2016). Adaptive backstepping output feedback control for a class of nonlinear fractional order systems. *Nonlinear Dynamics*. doi:10.1007/s11071-016-2945-4.
65. Yuan, J., Shi, B., & Yu, Z. (2014). Adaptive sliding control for a class of fractional commensurate order chaotic systems. mathematical problems in engineering. Article ID 972914.
66. Zadeh, L. A. (1975). The concept of a linguistic variable and its application to approximate reasoning. *Information Sciences, 8*(3), 199–249.
67. Zhang, R., & Yang, S. (2011). Adaptive synchronization of fractional-order chaotic systems via a single driving variable. *Nonlinear Dynamics, 66*, 831–837.

Optimal Fractional Order Proportional—Integral—Differential Controller for Inverted Pendulum with Reduced Order Linear Quadratic Regulator

M.E. Mousa, M.A. Ebrahim and M.A. Moustafa Hassan

Abstract The objective of this chapter is to present an optimal Fractional Order Proportional—Integral-Differential (FOPID) controller based upon Reduced Linear Quadratic Regulator (RLQR) using Particle Swarm Optimization (PSO) algorithm and compared with PID controller. The controllers are applied to Inverted Pendulum (IP) system which is one of the most exciting problems in dynamics and control theory. The FOPID or PID controller with a feed-forward gain is responsible for stabilizing the cart position and the RLQR controller is responsible for swinging up the pendulum angle. FOPID controller is the recent advances improvement controller of a conventional classical PID controller. Fractional-order calculus deals with non-integer order systems. It is the same as the traditional calculus but with a much wider applicability. Fractional Calculus is used widely in the last two decades and applied in different fields in the control area. FOPID controller achieves great success because of its effectiveness on the dynamic of the systems. Designing FOPID controller is more flexible than the standard PID controller because they have five parameters with two parameters over the standard PID controller. The Linear Quadratic Regulator (LQR) is an important approach in the optimal control theory. The optimal LQR needs tedious tuning effort in the context of good results. Moreover, LQR has many coefficients matrices which are designer

M.E. Mousa (✉)
Ministry of Civil Aviation, Cairo, Egypt
e-mail: mohamedmousa181@gmail.com

M.A. Ebrahim
Faculty of Engineering at Shoubra, Department of Electrical Engineering,
Benha University, Cairo, Egypt
e-mail: mohamed.mohamed@feng.bu.edu.eg; mohamedahmed_en@yahoo.com

M.A. Ebrahim
FCLab FR CNRS 3539, FEMTO-ST UMR CNRS 6174, UTBM,
90010 Belfort Cedex, France

M.A. Moustafa Hassan
Faculty of Engineering, Department of Electrical Engineering,
Cairo University, Giza, Egypt
e-mail: mmustafa@eng.cu.edu.eg; project_1426@hotmail.com

A.T. Azar et al. (eds.), *Fractional Order Control and Synchronization of Chaotic Systems*, Studies in Computational Intelligence 688,
DOI 10.1007/978-3-319-50249-6_8

dependent. These difficulties are talked by introducing RLQR. RLQR has an advantage which allows for the optimization technique to tune fewer parameters than classical LQR controller. Moreover, all coefficients matrices that are designer dependent are reformulated to be included into the optimization process. Tuning the controllers' gains is one of the most crucial challenges that face FOPID application. Thanks to the Metaheuristic Optimization Techniques (MOTs) which solves this dilemma. PSO technique is one of the most widely used MOTs. PSO is used for the optimal tuning of the FOPID controller and RLQR parameters. The control problem is formulated to attain the combined FOPID controllers' gains with a feed forward gain and RLQR into a multi-dimensions control problem. The objective function is designed to be multi-objective by considering the minimum settling time, rise time, undershoot and overshoot for both the cart position and the pendulum angle. It is evident from the simulation results, the effectiveness of the proposed design approach. The obtained results are very promising. The design procedures are presented step by step. The robustness of the proposed controllers is tested for internal and external large and fast disturbances.

Keywords Fractional order Proportional-Integral-Derivative · Inverted pendulum · Linear quadratic regulator · Particle swarm optimization technique · Proportional-Integral-Derivative · Reduced linear quadratic regulator · Robustness verification

1 Introduction

The Inverted Pendulum (IP) System is a classical benchmark for the control designers. The IP system is a physical system consists of pendulum carried by a cart and swinging around the fixed pivot [1]. The concept of the IP system is used in many modern technological applications like the landing of aircraft, space satellites, launching and guidance of the missile operations, spacecraft, statistics applications, and biomechanics [2]. Also, IP system is used on a large scale in many areas and applications including medical, transportation, robotics, aerospace, and military [3]. The IP system is considered the heart of many industrial applications. Some of these industrial applications are control of our ankle joint during quiet standing up, Segway's, quad rotor helicopters and walking robots [4]. IP system is the subject of an interesting from the standpoint of control because of their intrinsic nonlinearity [5]. It is used to illustrate the ideas in the nonlinear control and control of the chaotic system. The evaluation of various control theories is based on the inverted pendulum system.

IP system is a physical system consists of a bar which is usually made of aluminum and swinging around the fixed pivot. This fixed pivot will be installed on the vehicle which moves in the horizontal direction only. The center of gravity of the normal pendulum is under the axis of rotation and therefore, his condition is stable when it is directed to the bottom while the center of gravity of the inverted pendulum is over its axis of rotation. In the inverted pendulum problem, the pendulum tried to be in a vertical position to be heading up. The swinging up of the pendulum makes the situation of the system abnormal. So, the permanent controller should be applied to the system to keep the pendulum vertically upright. The nonlinearity and inherently instability of the system adding complexity to the problem especially when the proposed controllers will apply to the nonlinear system without any linearization [6]. Fast swinging up of the pendulum angle and stabilizing the cart at a certain position is required. The proposed controllers should be robust against the various system disturbances. Different control techniques are applied on the inverted pendulum to show the performance and effectiveness of the techniques [7].

In the IP problem, a lot of control techniques are proposed to make the pendulum balance in inverted to be heading up. Since this situation is abnormal, the status of "unbalanced" Basically, a permanent effort is needed to keep it this way, at any moment this effort is stopped, the system will collapse but return again to put the natural stability beyond. Normally be inaugurated with a turnover point centered on a moving vehicle accidentally. If the pendulum starts from a vertical position without applying any control strategy, it will begin to fall off and the cart will move in the opposite direction which means that any change in pendulum moving will effect on the cart and vice versa. The desired objectives of inverted pendulum control are:

(a) Maintaining the pendulum vertically upright.
(b) Stabilizing the cart.

The major objectives of the chapter are:

1 Modeling the IP system and presenting the linearized model at the certain operating point.
2 Design Reduced Linear Quadratic Regulator (RLQR) with minimum tuning parameters.
3 Design HybridPID controller in conjunction with feed forward and RLQR.
4 Design Hybrid Fractional Order PID controller with RLQR.
5 Develop multi-objective function which guarantees overall system stability in terms of minimum overshoot, settling time and steady state error for the both outputs of the IPsystem.
6 Propose Particle Swarm Optimization (PSO) technique for tuning the IP control system parameters.
7 Verify the robustness of the proposed controllers on changing the IP system parameters.
8 Validate the effectiveness of the proposed controllers using various types of disturbances.

The rest of the chapter is organized as follows: Sect. 2 illustrates a comprehensive literature survey of all related works. Section 3 formulates the dynamic model of a simple IP system. In Sect. 4, different classical and metaheuristic optimization techniques are used for optimality regions classification and verification. Additionally, the selection criteria of the most robust controller is reported. Simulation results are considered in Sect. 5. The conclusions and the perspectives are drawn in Sect. 6. Finally, the future work is illustrated in Sect. 7.

2 Related Work

There are many types of control techniques that are applied on the IP which has two outputs: position and angle. The presented methods for IP control are classified into seven groups:

- Classical methods such as PID controllers [8].
- Adaptation methods [9].
- Artificial methods such as fuzzy logic control [10], neural network [11], Genetic Algorithm [12] and PSO [13].
- Hybrid control [14].
- Sliding mode control [15].
- Time optimal control [16].
- Predictive control [17].

Some of these techniques are applied for tuning the angle controller gains while the position controller gains are constant. These strategies try to find the best gains to achieve the desired angle response. After that, the same procedures are carried out for tuning the position controller gains while maintaining the angle controller gains constant. The old control strategies deal with the IP as a single input single output system (SISO). In recent years, there are many control strategies deal with the IP as Single Input Multi Output System (SIMO). Some of control techniques are applied to the Inverted Pendulum system as follow:

Fuzzy Logic Controller

Fuzzy logic controller is used based on the single input rule modules. The input terms of the fuzzy controller are: the angle, angular velocity of the pendulum, the position and velocity of the cart and the output term is driving force. The authors in [18] represented a nonlinear plant with a Takagi-Sugeno fuzzy model. Each control rule is derived by using "parallel distributed compensation" in the controller design. To solve linear matrix inequality problems, Convex programming techniques are used as the control design problems can be reduced to LMI.

Lyapunov Approach

Lyapunov approach is used in PID adaptive control for self-tuning method for a class of nonlinear control systems [19]. There are three PID control gains parameters are adjustable and updated online with a stable adaptation mechanism. By introducing a supervisory control and a modified adaptation law with projection, the stability of closed-loop nonlinear PID control system is analyzed. Finally; a tracking control of an inverted pendulum system is used to demonstrate the control performance. Properties of simple strategies for swinging up an inverted pendulum are discussed in [20]. It turns out that the inverted pendulum swing behavior depends mainly on, the ratio of the maximum acceleration of the pivot to the acceleration of gravity. There are great ideas to minimum time solutions by make a comparison of energy-based strategies with minimum time strategy.

Energy Control Methods

In [21] generalized energy control methods are used to swing-up and stabilization of a cart–pendulum system with some restriction such, cart track length and control force. By using energy control principles, the pendulum is swung up to the upright unstable equilibrium configuration with Starting from a pendant position. In order to prevent the cart from going outside the limited length, an “energy well” must be built within the cart track. When getting adequate amount of energy by the pendulum and maintained it, it goes into a “cruise” mode. Finally, the stabilizing controller is activated around a linear zone about the upright configuration when the pendulum is closed to the upright configuration. This way has worked well both in simulation and a practical setup and derived the conditions for stability by using the multiple Lyapunov functions approach. The feedback of an inverted pendulum is not linear although inverted pendulum is one of the typical examples of nonlinear control systems. A new method to design back stepping-like controller is proposed by Saeki in [22]. By combining Saeki’s method with the energy function method, produces a swing-up controller. Firstly, to prevent the effect of the pendulum, the control input is given of the cart. Secondly, design the input that guarantees the convergence of the acceleration of the cart to the desired value. Thirdly, an energy function was used to design the swing-up control law. The energy function-based controller is used to swing up the pendulum and the potential function-based controllers used to stabilize the inverted pendulum.

Sliding Mode Control

In [15], the authors developed a Second-order sliding mode control synthesis for under-actuated mechanical systems and operates it under uncertainty conditions. The output is specified in such a way that the corresponding zero dynamics is locally asymptotically stable in order to locally stabilize an under-actuated system around an unstable equilibrium. And then, provide the desired stability property of the closed-loop system by applying a quasi-homogeneous second-order sliding mode controller, driving the system to the zero dynamics manifold in finite time [23]. It does not rely on the generation of first-order sliding modes, although the present synthesis exhibits an infinite number of switches on a finite time interval, while providing robustness features similar to those possessed by their standard sliding mode counterparts. The performance issues of the proposed method are illustrated in numerical and experimental studies of a cart–pendulum system.

Optimal Control

Optimal control with time invariant nonlinear controller is presented in [24] for the inverted pendulum, which is defined for all pendulum angles. The external field is calculated by solving the Euler–Lagrange equations backward in time. The time-optimal feedback control that brings a pendulum to the upper unstable equilibrium position is obtained in [25]. The technique is based on the maximum principle and analytical investigations and numerical computations.

The nonlinear model predictive control is applied in [26] to an inverted pendulum apparatus. A standard sequential quadratic programming approach is used to solve non-convex constrained optimization problem involves 61-variables with 241-constraints.

In this chapter, a new objective function with new artificial intelligent based technique for tuning the controllers' gains of SIMO inverted pendulum system is proposed. The difficulty of the proposed strategy comes from that any change in angle will effect on the position and vice versa. The tuning process of FOPID controller and RLQR is not aneasy task as there are five parameters for the FOPID and two weighting matrices for RLQR. These gains directly affect the angle and cart response so it is a complicated problem.

3 Mathematical Modeling

Modeling and control of the IP are the prerequisites of autonomous walking. The primary approach to derive the model is the Euler-Lagrange approach. The IP is one of the most difficult systems in control theory due to the non-linearity [27]. It is inherently unstable system with single input and multi-outputs so applying classical control methods did not lead to good results. If there is a stick on hand and the objective is to make it always in a vertical position, it is needed to move the hand to keep the stick in a vertical position. On the other hand, a force is applied to keep the stick in a vertical position. Similarly in the control of inverted pendulum, a force is applied to make the pendulum always upright vertical without any deviation about zero. The IP system has two degrees of freedom of motions as shown in Fig. 1. The first degree of freedom is the motion of the cart along the x-axis and the second

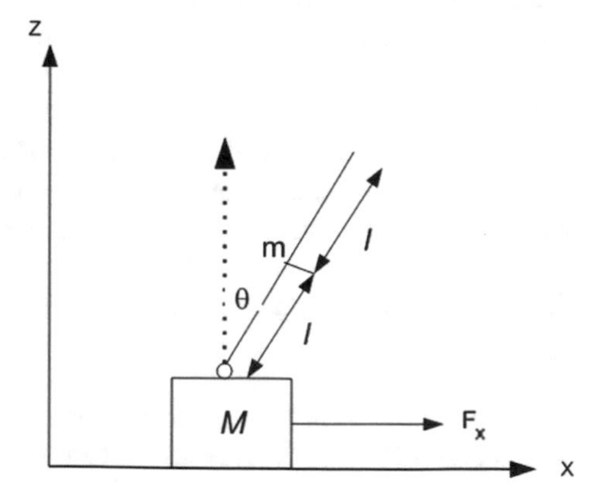

Fig. 1 Inverted pendulum system modeling

Table 1 Inverted pendulum system parameters

Symbol	Parameter	Value	Unit
M	Mass of the cart	0.455	Kg
m	Mass of the pendulum	0.21	Kg
l	The distance from the pivot to the mass center of the pendulum	0.61/2	m
g	The acceleration of gravity	9.8	m/s^2

degree of freedom is the rotation of the pendulum aboutthex-zplane. The mathematical model can be defined as, a set of mathematical equations representing some of the phenomena in a way that gives insight into the origins and the consequences of the behavior of the system. The more accurate the mathematical model is, the more complex the equations will be. The mathematical model should be easy to understand. So accuracy and the simplicity are the two main parameters that should take into consideration while modeling. It can be seen that the equations describing the system are non-linear. Taylor series expansion is used in order to obtain a linear model to convert the non-linear equations to linear ones; finally, produce a linear model that will be helpful in linear control design. The system has two equilibrium points: one point is stable such as the pendant position and the other point is the unstable equilibrium point such as the inverted position. For our purpose, the second is required to make linearization to the model about it. So, a very small deviation from the vertical is assumed.

The parameters of the IP system are illustrated in Table 1.

The IP System is nonlinear and inherently unstable system. The modeling equations of the IP system are very important which allow the controller to stabilize the cart position and swinging up the pendulum angle. The dynamic differential equation of the system is derived according to the Euler-Lagrange equations. The IP system is a highly coupling system as it not allowable to derive each output equation individually. The Euler-Lagrange formula can derive the system as a multi-outputs system and get the state-space representation of the system states.

3.1 *Applying Lagrangian to Inverted Pendulum System*

The following steps should be followed to put the IP system in Lagrangian formula [28]:

(a) Obtain the kinetic and potential energies.
(b) Substitute in the Lagrangian formula $L = K - P$
(c) Find $\partial L/\partial q$
(d) Find $\partial L/\partial \dot{q}$ then find $\frac{d}{dt}\left(\frac{\partial L}{\partial \dot{q}}\right)$.
(e) Solve the Euler-Lagrange equation with the generalized force $\frac{d}{dt}\left(\frac{\partial L}{\partial \dot{q}}\right) - \frac{\partial L}{\partial q} = Q_q$

where:

Q_q The generalized forces
q The generalized coordinates

The nonlinear differential equations of the motion are as follow:

$$X = \frac{F_x - mg\cos\theta\sin\theta + ml\theta^2\sin\theta}{M + m\sin^2\theta} \tag{1}$$

$$\dot{\theta} = \frac{(M+m)g\sin\theta - ml\theta^2\sin\theta\cos\theta - F_x\cos\theta}{Ml + ml\sin^2\theta} \tag{2}$$

Here, the states of the system are defined as the following to represent the state space on the inverted pendulum:

$$X_1 = X/X_2 = \dot{X}/X_3 = \theta/X_4 = \theta,$$

where:

X The position of the cart that move along the x-axis
$\dot{X}$ The velocity of the cart that moves along x-axis
θ The angle position from the vertical position
θ The velocity of the pendulum that swings along Z-axis

When the system is linearized, the state space representation could easily be obtained as:

$$\begin{bmatrix} \dot{x}_1 \\ \dot{x}_2 \\ \dot{x}_3 \\ \dot{x}_4 \end{bmatrix} = \begin{bmatrix} 0 & 1 & 0 & 0 \\ 0 & 0 & -4.5231 & 0 \\ 0 & 0 & 0 & 1 \\ 0 & 0 & 46.9609 & 0 \end{bmatrix} \begin{bmatrix} x_1 \\ x_2 \\ x_3 \\ x_4 \end{bmatrix} + \begin{bmatrix} 0 \\ 2.1978 \\ 0 \\ -7.2059 \end{bmatrix} F \tag{3}$$

$$y = \begin{bmatrix} 1 & 0 & 0 & 0 \\ 0 & 1 & 0 & 0 \\ 0 & 0 & 1 & 0 \\ 0 & 0 & 0 & 1 \end{bmatrix} \begin{bmatrix} X_1 \\ X_2 \\ X_3 \\ X_4 \end{bmatrix} + \begin{bmatrix} 0 \\ 0 \\ 0 \\ 0 \end{bmatrix} F_X \tag{4}$$

4 The Proposed Control Techniques:

The proposed control techniques for the IP system in this chapter are:

- *Linear Quadratic Regulator.*
- *Proportional Integral Derivative Controller.*
- *Fractional Order Proportional Integral Derivative Controller.*
- *Particle Swarm Optimization based PI/RLQR and FOPID/RLQR.*

(a) ***Linear Quadratic Regulator***

Given a linear time-invariant state-space model of the system:

$$\dot{x} = Ax + B \tag{5}$$

$$y = Cx + Du \tag{6}$$

The LQR is used to minimize the following cost function [29]:

$$J = \frac{1}{2}\int_{0}^{\infty} [x^T Qx + u^T Ru]dt \tag{7}$$

where:

Q and R are weighting matrices which are selected by the designer.

This selection process depends on the experience of the designer which in turn is a tedious effort in multi-dimensions problems. In LQR, the following Riccati equation should be solved [29]:

$$PA + A^T P - PBR^T B^T P + Q = 0 \tag{8}$$

The matrix P is obtained by solving the above equation so the controller gain can be calculated according to Eq. (9) as explained in [14]:

$$K = R^{-1} B^T P \tag{9}$$

The Q and R matrices are the main design parameters which greatly affect on the controller gain. In this chapter; PSO technique is used to get Q and R matrices according to specific constraints.

(b) ***Proportional Integral Derivative Controller***

The proportional-integral-derivative (PID) controller is used in most control systems. It consists of three gains: proportional gain (K_p), integral gain (K_i) and derivative gain (K_d). Each of the PID controller gains has an action on the error. The error is the difference between a setpoint designed by the user and some measured process variables. The continuous form of a PID controller, with input e and output U, is presented in Eq. (10).

$$U_{PID} = K_P e(t) + K_i \int_{0}^{t} e(t)dt + K_d \frac{d}{dt} e(t) \tag{10}$$

(c) ***Fractional Order PID Controller with Feed Forward Gain (FOPID)***

Fractional Order Proportional-Integral-Derivative (FOPID) controller is the recent advances improvement controller of a conventional classical PID controller [30]. The earliest studies concerning fractional calculus presented in the 19th century made by some researchers such as Liouville (1832), Holmgren (1864), and Riemann (1953) as introduced in [31], and others made some contributions in this field in the past. Fractional-order calculus deals with non-integer order systems. It is the same as the traditional calculus but with a much wider applicability. Fractional Calculus is used widely in the last two decades and applied in different fields in the control area.

Fractional order Proportional-Integral-Derivative controller achieves great success because of its effectiveness on the dynamic of the systems. Designing FOPID Controller is more flexible than the standard PID Controller because they have five parameters with two parameters over the standard PID controller. The operator ${}_aD_t^q$ is commonly used in fractional calculus which is defined as the differentiation integration operator and discussed as presented in Eq. (11):

$$
{}_aD_i^q = \left\{ \begin{array}{cc} \frac{d^q}{dt^q} & q > 0 \\ 1 & q = 0 \\ \int_a^t (d\tau)^{-q} & q < 0 \end{array} \right\} \tag{11}
$$

where:

q Fractional order (can be complex)
a and t The limits of operation

There are different definitions for fractional derivatives. The widely used definitions are as following:

(a) Grunwald–Letnikov definition.
(b) Riemann–Liouville definition.
(c) Caputo definition.

These definitions will be discussed below:

(a) Grunwald–Letnikov definition

The Grunwald–Letnikov definition is given by Eq. (12):

$$
{}_aD_t^q f(t) = \frac{d^q f(t)}{d(t-a)^q} = \lim_{N \to \infty} \left[\frac{t-a}{N}\right] \sum_{j=0}^{N-1} (-1)^j \binom{q}{j} f(t - j\left[\frac{t-a}{N}\right]) \tag{12}
$$

(b) Riemann–Liouville definition

The Riemann–Liouville definition is the easiest definition and defined as presented in Eq. (13):

$$ {}_aD_t^q f(t) = \frac{d^q f(t)}{d(t-a)^q} = \frac{1}{\Gamma(n-q)} \frac{d^n}{dt^n} \int_0^t (t-\tau)^{n-q-1} f(\tau) d\tau \tag{13} $$

where:

n The first integer $(n-1 \leq q < n)$

Γ The Gamma function $(\Gamma(z) = \int_0^\infty t^{z-1} e^{-t} dt)$

(c) Caputo definition

The Caputo definition is given by Eq. (14):

$$ {}_aD_t^q f(t) = \begin{cases} \frac{1}{\Gamma(m-q)} \int_0^t \frac{f^{(m)}}{(t-\tau)^{q+1-m}} d\tau & m-1 < q < m \\ \frac{d^m}{dt^m} f(t) & q = m \end{cases} \tag{14} $$

where:
m: The first integer larger than q

Fractional differential equation simulation is not easy as compared with the ordinary differential ones. Approximation and numerical methods are used for solving fractional order differential equations Fractional order control calculus presented by Tustin for the position control of massive objects a half century ago. Provided some of the other researches were presented by Manabe around (1960). However, the fractional-order control was not included in the control engineering because of the major limitations of the possibilities and a lack of adequate amount of mathematical knowledge and computational power at this. The researchers have concluded in the past decades that the (fractional order differential equations) could model diverse systems fuller than integer-order ones and provide an excellent instrument for describing dynamic processes. In fractional order controllers, in addition to parameters of the classical proportional-integral-derivative constants, there are two extra parameters (λ and μ) as discussed in [32]. The parameters λ and μ are the order of s in integral and derivative respectively so a specific algorithm is required to make tuning for the parameters of the FOPID Controller. This will improve the system performance in terms of flexibility and durability better than the classical PID controller.

The differential equation of the FOPID controller is described as:

$$ U(t) = K_P e(t) + K_i D^{-\lambda} e(t) + K_d D^{\mu} e(t) \tag{15} $$

After the introduction of this definition, it became easy to see that the classical types of PID controller such as integral order PID, PI, or PD become special cases of the most general fractional order PID controller. In other words, the FOPID controller expands the integer-order PID controller from point to plane, as shown in

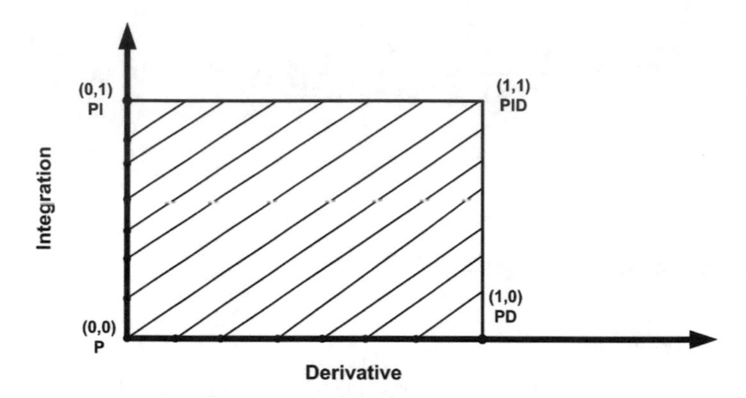

Fig. 2 Schematic view of PID for all probabilities

Fig. 2. Taking Laplace Transform of Eq. (15), the controller expression in s-domain is obtained as:

$$C(s) = \frac{U(s)}{E(s)} = K_P + \frac{K_i}{s^{\lambda}} + K_d s^{\mu} \tag{16}$$

(d) ***Particle Swarm Optimization***

Particle Swarm Optimization (PSO) is a stochastic optimization technique developed by Eberhart and Kennedy in 1995 as given in [33]. The PSO algorithm is inspired by social behavior of bird flocking, animal hoarding, or fish schooling. In PSO, the potential solutions, called particles, fly through the problem space by following the current optimum particles as explained in [34]. PSO has been successfully applied in many areas [35, 36], [37] and [38, 39].

PSO simulates the behavior of bird flocking. When a group of birds flying in the sky searching for the food. The food is located at the specific place through the searching area but not all the birds know where the food is. Each bird estimates a position of the food and the bird which have the least distance from the food position, will follow the group. By iterations, the birds can reach to the food easily. PSO started with random values for the particles and searching for the optimal solution that achieves the minimum values of the objective functions. During each iteration, the best value of the objective functions obtained in each iteration is called local best (*pbest*). The best value of the local best values obtained through the iterations is called global best (*gbest*) as explained in [40]. After finding the local best and global best values in each iteration, the particles update its velocity and position according to Eqs. (17)–(18) respectively as introduced in [33].

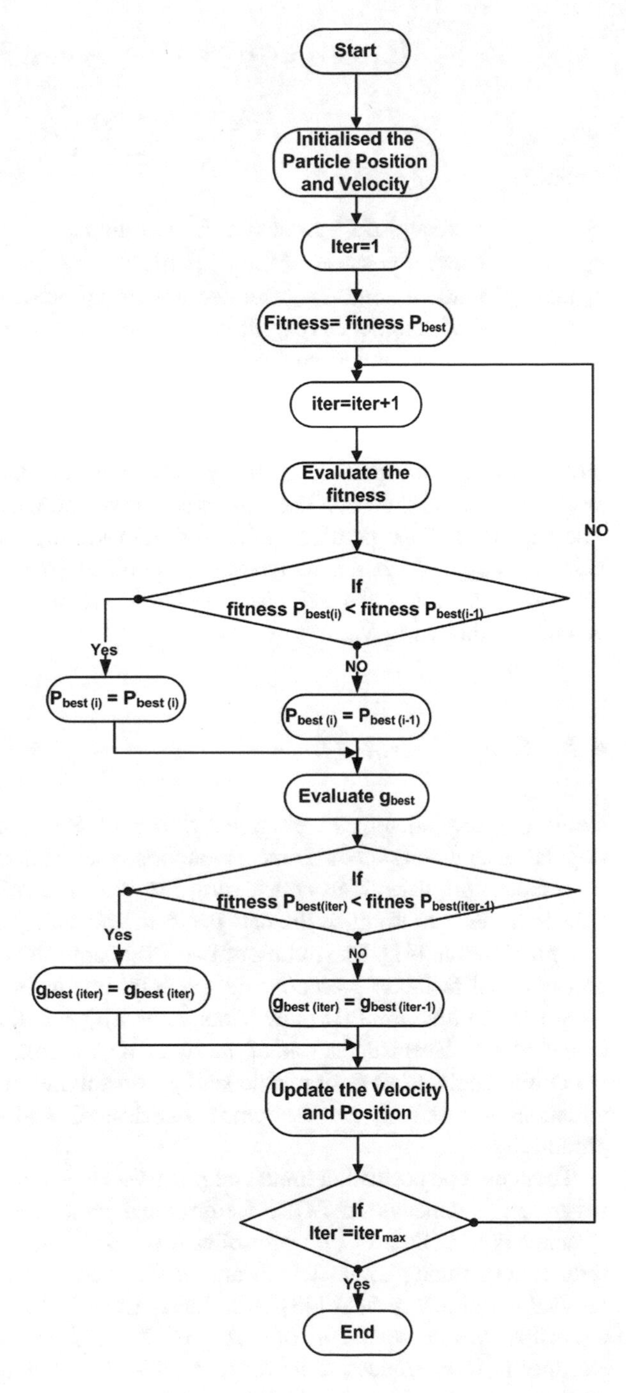

Fig. 3 Flowchart of the PSO

$$V_{k+1}^{i} = wv_{k}^{i} + c_1 r_1 (pbest^{i} - x_{k}^{i}) + c_2 r_2 (gbest - x_{k}^{i}) \quad (17)$$

$$x_{k+1}^{i} = x_{k}^{i} + v_{k+1}^{i} \quad (18)$$

where:

v_k^i	Velocity of ith particle at k^{th} iteration
x_k^i	Current position of the ith particle,
r_1 and r_2	random numbers generated uniformly between 0 and 1
c_1	Self-confidence (cognitive) factor
c_2	swarm confidence (social) factor
w	is the inertia weight factor

The 1st term in Eq. (18) represents the effect of the inertia of the particle, the 2nd term represents the particle memory influence, and the 3rd term represents the swarm (society) influence. The flow chart of the procedure is illustrated in Fig. 3. The velocities of the particles on each dimension may be clamped to a maximum velocity V_{max}, which the parameter is specified by the user. If the sum of the accelerations causes the velocity on that dimension to exceed V_{max}, then the velocity is limited to V_{max}.

4.1 Control Strategy:

Various control strategies are applied to the IP. PID controller is one of the most popular ones among them. Some researches concentrate on swinging up the angle in vertical upright without considering the dynamics of the cart. Employing two PID controllers to stabilize the cart position and swinging up the pendulum angle was presented in [41]. The tuning of two PID controllers is atedious effort by using conventional methods. Recently, the artificial intelligent computational techniques were used to tune the PID parameters. Linear Quadratic Regulator (LQR) is suggested as a replacement for one of the two PID controllers [42] for swinging up the pendulum angle. LQR design is depending on solving the Riccati equation. Riccati equation is based on two designing matrices Q and R which have seventeen parameters.

They must be positive definite and positive semi-definite respectively. Tuning of seventeen parameters of LQR, feedforward gain, and in addition to the three parameters of FOPID or PID controller is time-consuming and more complex. The reduction of tuning parameters is one of the most important topics in the computational evolutionary field [43]. A reduced order LQR with FOPID and with PID controller was presented in this chapter. In this proposed technique the tuning parameters were reduced to be thirteen instead of twenty-one in the case of using PID and reduced from twenty-three to fifteen in the case of using FOPID. It simplifies the optimization problem and has a great effect on the computation time.

Designing the gains of LQR is mainly depending on the choice of the Q and R matrices which selected by the designer. This may take a long time to obtain the best values of the two matrices parameters. Therefore, trial and error method is time-consuming. The process of selecting the matrices becomes more difficult when the system has a large dimension of system state space matrices. In this chapter, an evolutionary optimization based RLQR controller, FOPID controller and compensating gain (K_f) design for an inverted pendulum system is introduced and compared to another one with PID controller. The weighting matrices Q and R are positive semi-definite and positive-definite respectively. This means that the term $x^T Qx$ in Eq. (7) is always positive or zero at each time t for all functions x(t). Furthermore, the second term in Eq. (7) is always positive at each time t for all values of u(t). Therefore J is always positive at each time. To ensure that the weighting matrices Q and R are positive semi-definite and positive-definite respectively and to reduce the dimension of Q and R matrices as explained in [43], it is assumed that:

$$Q = W^T * W \tag{19}$$

$$R = V^T * V \tag{20}$$

where:

W amatrixof m * n dimension
V matrix of k * l dimension

In this chapter, It is assumed that: m = 2, n = 4, k = 1, l = 1.

So the modified Riccati equation is given by Eq. (21):

$$PA + A^T P - PB(V^T * V)^T B^T P + (W^T * W) = 0 \tag{21}$$

The proposed optimization technique is used to tune the W and V matrices to guarantee that Q and R will bc positive semi-definite and positive-definite respectively. After that, the modified Riccati equation is solved to find the reduced Linear Quadratic Regulator gains according to the following equation:

$$K = (V^T * V)^{-1} B^T P \tag{22}$$

Hint: The modified Riccati equation can be solved in Matlab by using the command lqr (A, B, Q, R)
where:
A and B: System State-space matrices.
Q and R: weighting matrices of RLQR gains.
K: RLQR gain ($K = [K_1 K_2 K_3 K_4]$)

The four gains (K_1, K_2, K_3, K_4) of the RLQR will be calculated. The states that affect the pendulum angle are cart velocity, pendulum position, and the pendulum velocity so the feedback from the angle output having these states. Hence, there is no necessity to use the controller gain K_1 which controls the cart position state.

4.2 Problem Formulation

The optimization problem has 15 variables $(K_P, K_i, K_d, \lambda, \mu, K_F, Wmatrix$ and $Vmatrix)$ in the case of using FOPID controller. Also, it has 13 variables $(K_P, K_i, K_d, K_F, Wmatrix$ and $Vmatrix)$ in the case of using PID controller. PSO run to find the best values for all variables that achieve the minimum Overshoot, Steady state error and Settling Time. The difficulty of the algorithm is to achieve the minimum Overshoot, Steady state error and settling time for both the cart position and the pendulum angle at the same time as presented in Eq. (23). In this chapter, the algorithm runs according to a Multi-objective function that has the constraints which give an acceptable response to the two outputs.

The global _best_ Fitness is determined according to:

$$Global - best - fitness = min(e_{ss},\ o.s,\ T_s) \tag{23}$$

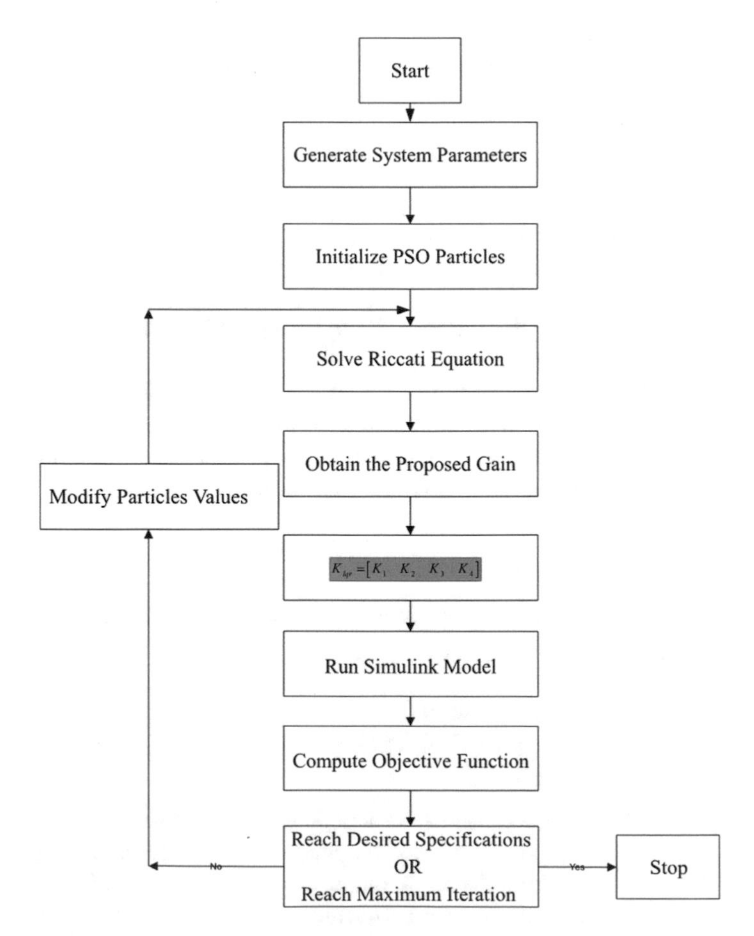

Fig. 4 The flow chart of the proposed control strategy

The steps of the proposed algorithm are illustrated in Fig. 4 as follow:

Step1: Generating the inverted pendulum system parameters.
Step2: Initialize values of PSO particles.
Step3: Solve Riccati equation.
Step4: Obtain the proposed RLQR gain.
Step5: Consider last three values only of the RLQR gain.
Step6: Run the Simulink model.
Step7: Computing the objective function of the algorithm.
Step8: Check achieving minimum value for steady state error, settling time and overshoot.
Step9: Stop the algorithm when achieving minimization for the objective function or exceeding the maximum iteration number.

5 Simulation and Results:

The IP is among the most difficult systems to control in the field of control engineering. It consists of two control loops as presented in Fig. 5. For the purpose of effective comparison, the system is equipped with FOPID with RLQR and PID with RLQR. The first one is FOPID with RLQR (FOPID/RLQR) controller. The (FOPID/RLQR) gains are responsible for stabilizing the cart and swinging up the pendulum to be in a vertical position. The second one is PID with RLQR (PID/RLQR) controller. In the first control loop there is a feedback signal from the cart position output to the summing point with the input signal (unit step) then a feed forward controller is applied. The feedback in the second control loop is extremely different as the factors that effect on swinging up the angle are the speed of the cart, the pendulum angle and the angular velocity of the pendulum. A feed-forward estimator is used to estimate the speed of the cart and the angular velocity of the pendulum as presented in the model as illustrated in Fig. 5. The estimations for the speed and angular velocity are collected and then introduced into the inverted pendulum. A feed forward amplifier (K_f) affects the dynamic response of the inverted pendulum.

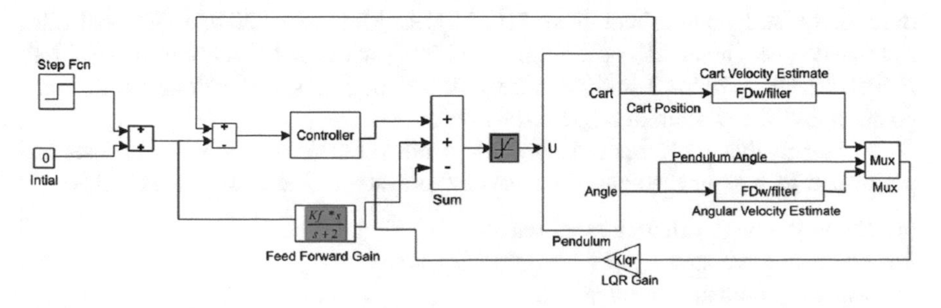

Fig. 5 Block diagram of inverted pendulum system

Table 2 The obtained gains from PSO

	K_P	K_i	K_d	λ	μ	K_f
FOPID/RLQR	10	19.4	−2.1	−1	1.6	55.7
PID/RLQR	10	0.009	−29.2320	1	1	−50

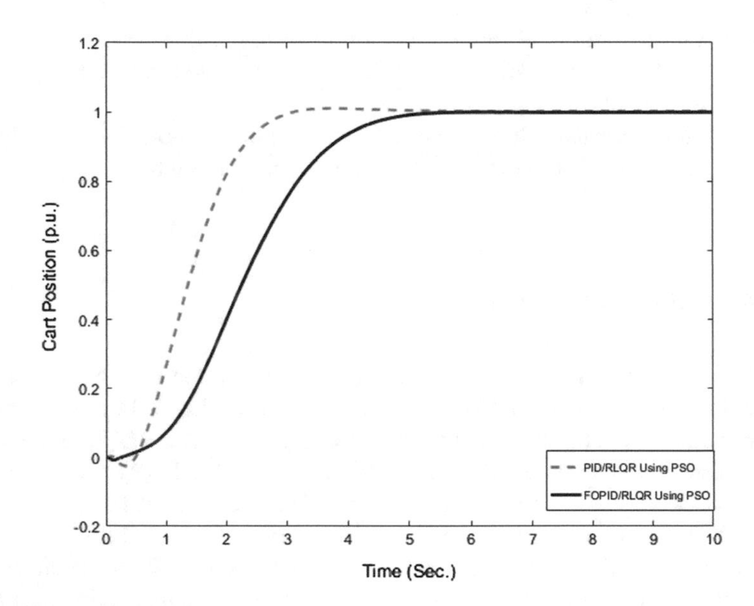

Fig. 6 System response of cart position with different controllers

The obtained gains from PSO are given in Table 2. The results are presented in Figs. 6 and 7.

The RLQR gains for both FOPID/RLQR and PID/RLQR are K_{lqr} = [−50.03 −353.79 −85.9] and K_{lqr}=[−62.03 −251.45 −72.69] respectively.

The simulation results illustrate that both controllers had succeeded in stabilizing the cart position and swinging up the pendulum angle effectively. Although, FOPID/RLQR controller stabilized the angle of the inverted pendulum with less over shoot and under shoot than PID/RLQR. Moreover, PID/RLQR controller stabilized the inverted pendulum position with less settling time than FOPID/RLQR. Tables 3 and 4 presented the output specifications of the cart position and the pendulum angle respectively.

To ensure that the proposed controllers are robust, three robustness tests are performed to measure the effectiveness of the system. The cases are as follows:

(a) Set points with different amplitudes.
(b) Increasing the step input with different ranges.
(c) System parametersperturbation.

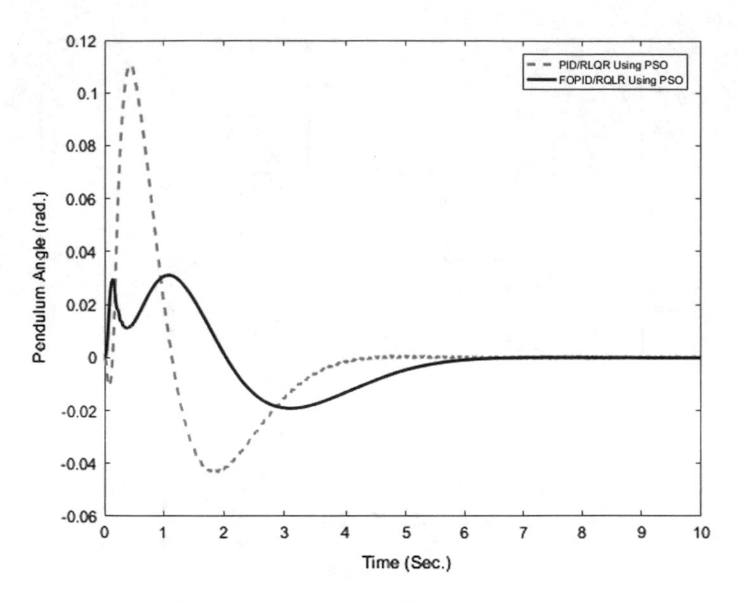

Fig. 7 System response of pendulum angle with different controllers

Table 3 The output specifications of the cart position

Time response specifications	PID/RLQR using PSO	FOPID/RLQR using PSO
Rise time (s)	1.5638	2.58
Settling time (s)	2.7873	4.6234
Overshoot (%)	1.2%	0.00072206%
Undershoot (%)	2.4988%	0.8306%

Table 4 The output specifications of the pendulum angle

Time response specifications	PID/RLQR using PSO	FOPID/RLQR using PSO
Maximum value (rad.)	0.1109	0.0312
Minimum value (rad.)	−0.0432	−0.0191
Settling time (s)	3.5758	5.6728

(a) **Set points with different amplitudes**

A series of set points as shown in Fig. 8 are applied to the IP system to validate the effectiveness of the proposed controllers. It is evident from the results the effectiveness of the controllers in stabilizing the cart position and swinging up the pendulum angle. Both controllers are succeeded in stabilizing the cart position and swinging up the pendulum angle as shown in Figs. 9 and 10.

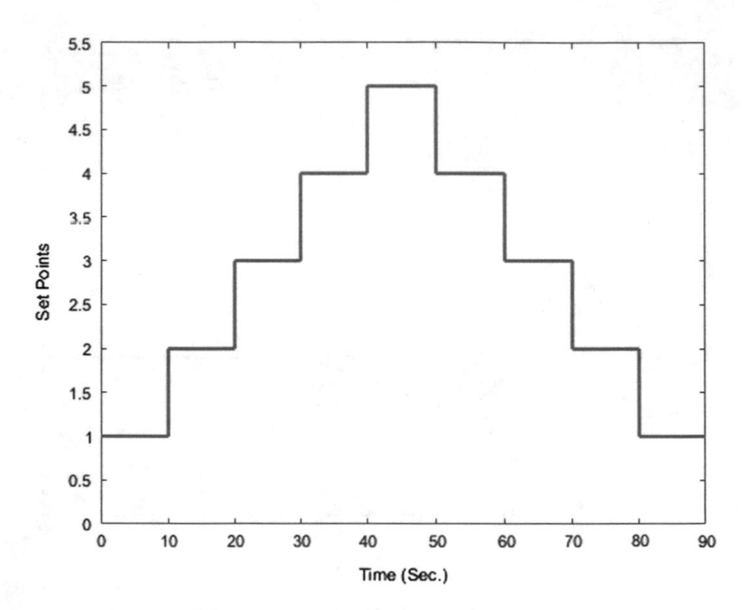

Fig. 8 Series of set points with different amplitudes

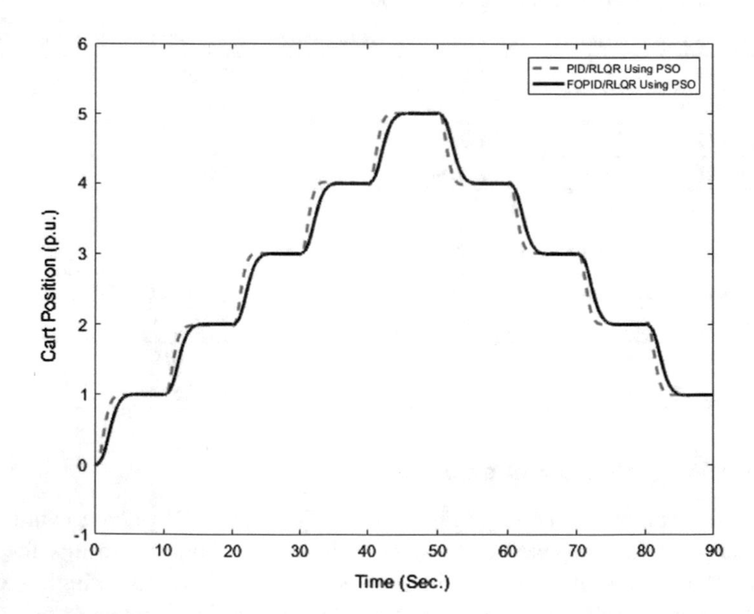

Fig. 9 System cart position response of different controllers with series of set points

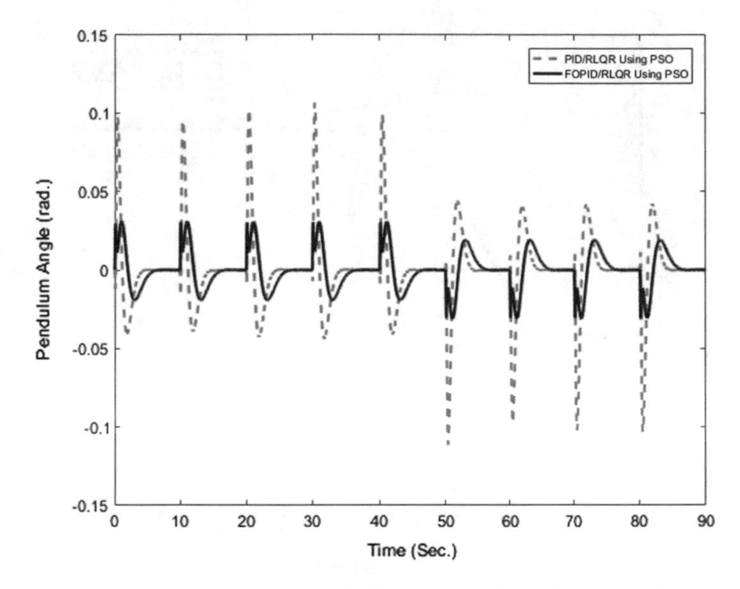

Fig. 10 System pendulum angle response of different controllers with series of set points

Robustness Verification

(a) **Increasing the step input with different ranges.**

To measure the effectiveness of the system, the step input is increased with different ranges. The controller which cannot withstand the increasing of the step input will be not a robust controller design. Firstly, step input is increased with 10% then it will be increased by 20%. Figures 11 and 12 illustrate dynamic responses of the cart position and the pendulum angle respectively when the step input increased with 10%. It is noted that the two proposed controllers succeeded in keeping balance to the inverted pendulum system when the step input increased with 10%.

The step input increased with 20% from its value. Although PID/RLQR controller using PSO can stabilize the inverted pendulum system with 10% increasing of the step input, it cannot keep the system in balance with 20% increase. FOPID/RLQR controllers using PSO only the controller that succeeded in the robustness test related to the increasing of the step input. Figure 13 presented the response of the cart position. Figures 14 and 15 illustrated the pendulum angle response in case of PID/RLQR and FOPID/RLQR using PSO.

(b) **System parameters perturbation.**

This test is one of the most important tests in checking the robustness of the inverted pendulum system. In this test, inverted pendulum parameters are increased with 10% from their values. Each controller will be applied to the nonlinear system with the new parameters. If the FOPID/RLQR controller is succeeded in controlling the same system with the new parameters, the controller will be very robust. Table 5 Illustrates

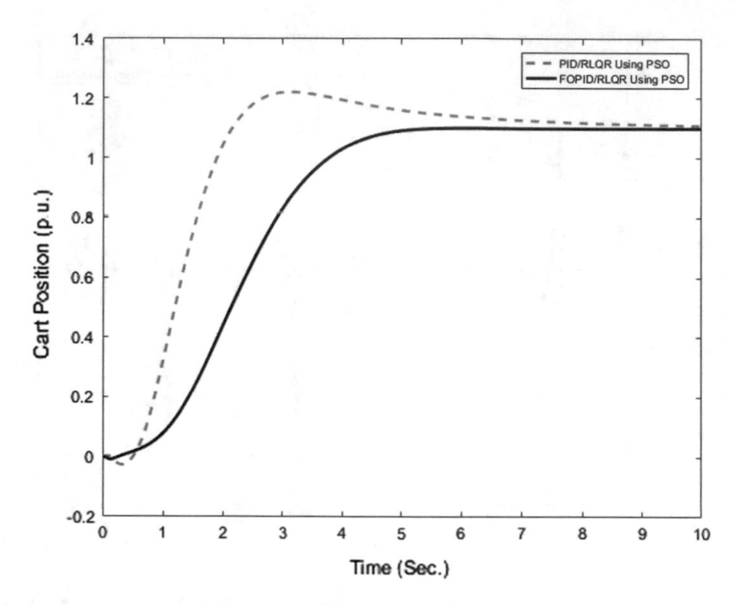

Fig. 11 Dynamic response of cart position with 10% increase of step input

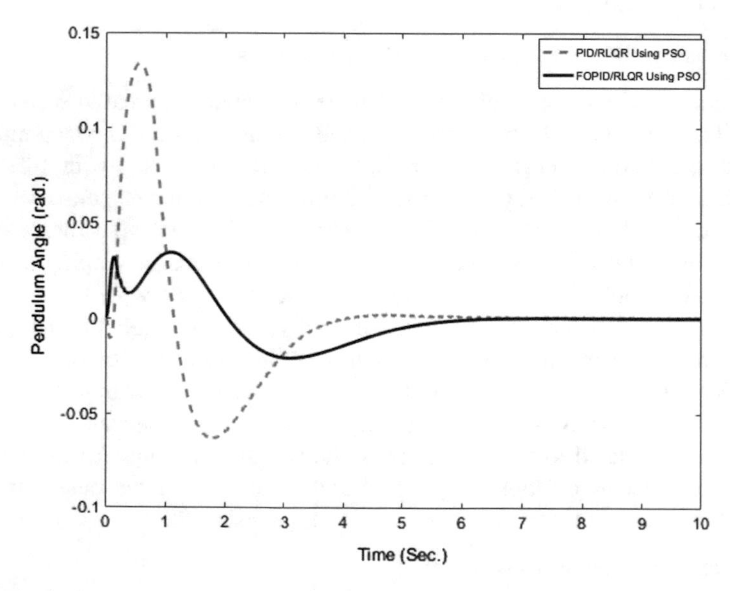

Fig. 12 Dynamic response of pendulum angel with 10% increase of step input

the IP system parameters with 10% increasing. It is noted that the FOPID/RLQR using PSO can stabilize the cart position and swing up the pendulum angle while PID/RLQR using PSO failed in balancing the system (Figs. 16, 17 and 18)

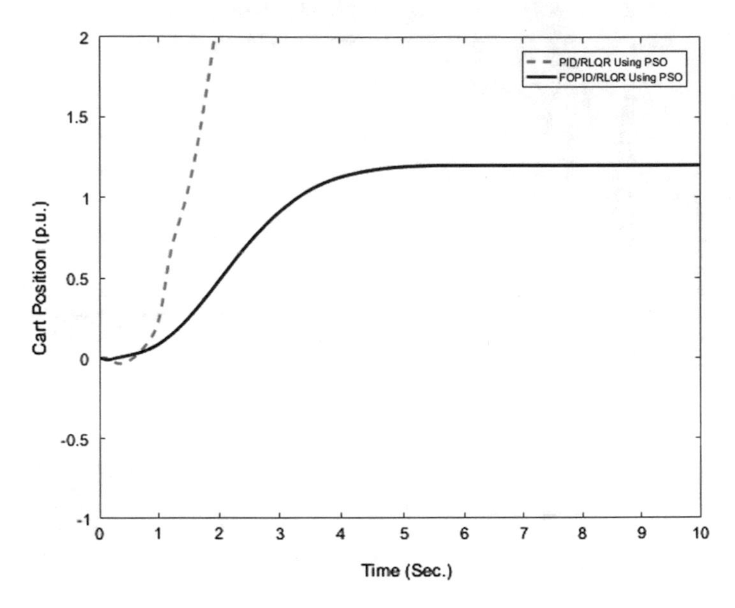

Fig. 13 Dynamic response of cart position with 20% increase of step input

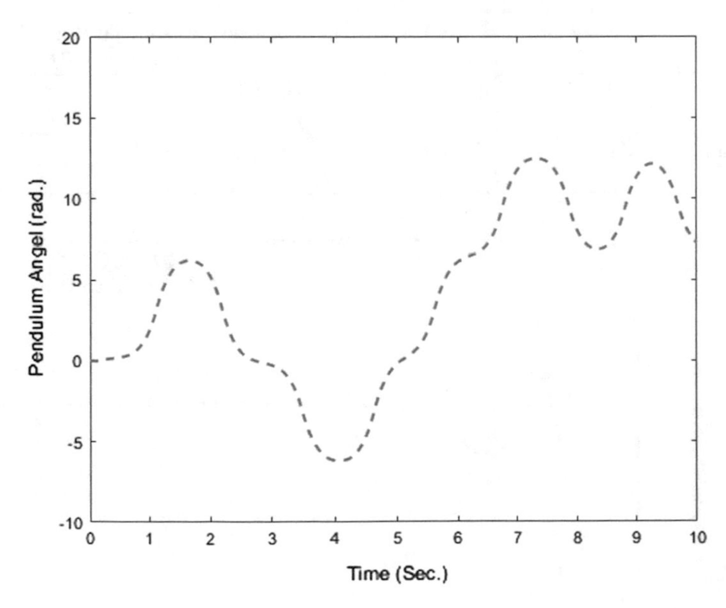

Fig. 14 Dynamic response of pendulum angel with 20% increase of step input using PID/RLQR controller

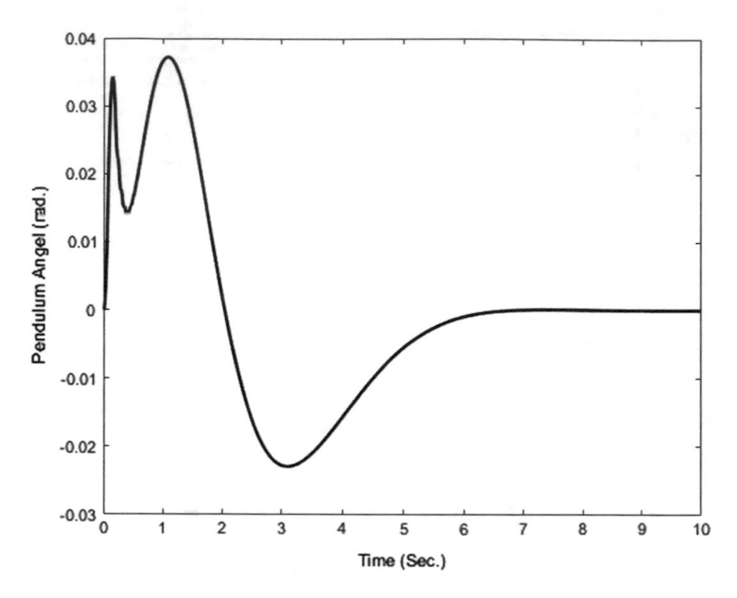

Fig. 15 Dynamic response of pendulum angel with 20% increase of step input using FOPID/RLQR controller

Table 5 Parameters of the inverted pendulum system with 10% increasing

System parameters	Mass of the cart (M)	Mass of the pendulum (m)	Distance from the pivot to the mass center of the pendulum (l)
Parameters values	0.455	0.21	0.61/2
Increasing with 10%	0.5005	0.2310	0.3355

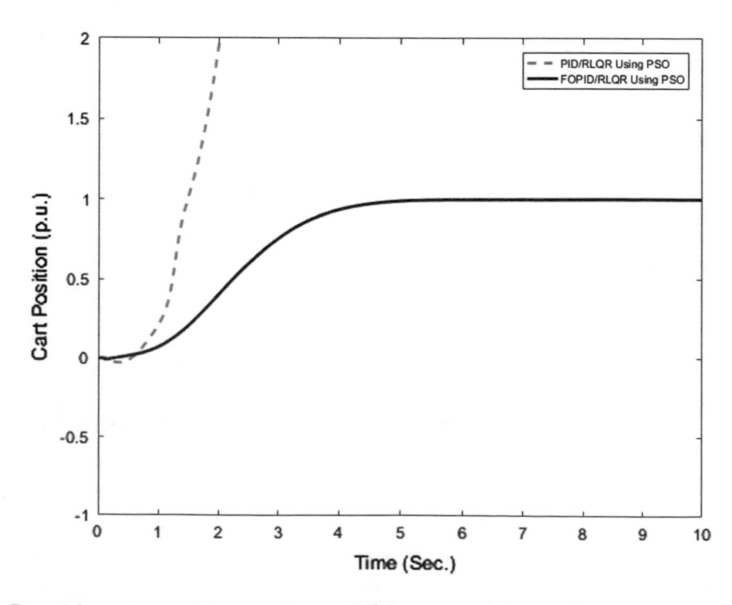

Fig. 16 Dynamic response of cart position (10% Increase of System Parameters)

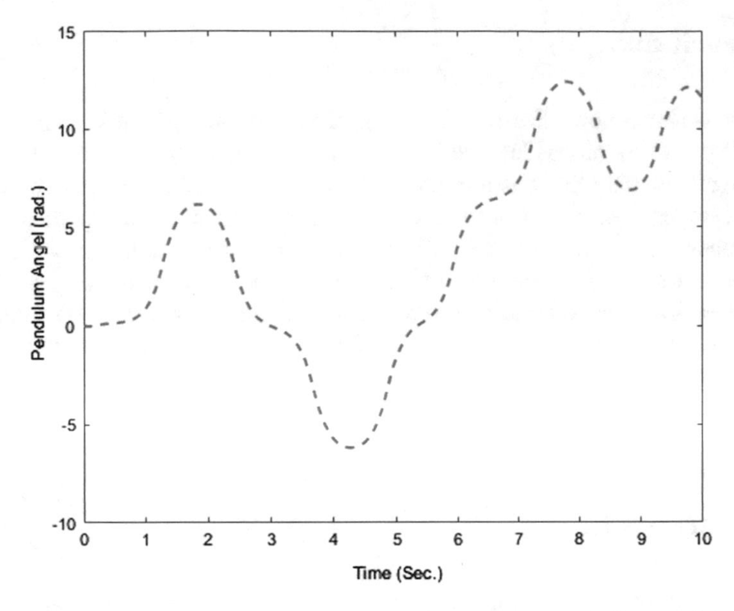

Fig. 17 Dynamic response of pendulum angle using PID/RLQR controller (10% increase of system parameters)

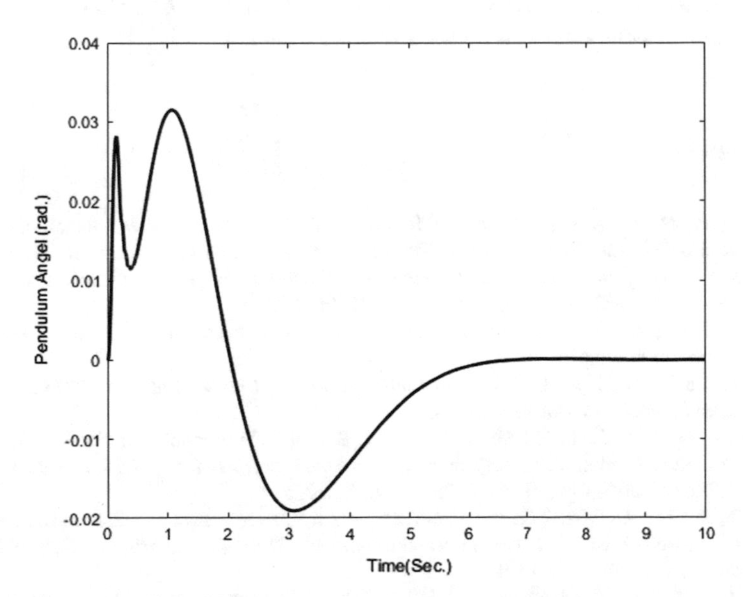

Fig. 18 Dynamic response of pendulum angle using FOPID/RLQR Controller (10% increase of system parameters)

6 Conclusion

In this chapter, a new effective control method integrating both PSO-based Fractional Order Proportional Integral Derivative (FOPID) and Reduced Linear Quadratic Regulator (RLQR) was introduced. This chapter demonstrates that PSO can solve searching and tune the controller parameters more efficiently than conventional ones. The inherited instabilities in the inverted pendulum were treated effectively. Modeling of the inverted pendulum was performed using MATLAB. The simulation was conducted in order to cover the full range of operating conditions and severe disturbances. The application of the proposed control method showed its ability to stabilize the inverted pendulum. The obtained results are very promising.

7 Future Work

In control system engineering there are a lot of techniques for balancing the inverted pendulum system. PSO is one of the evolutionary computational techniques which proposed in this chapter. Balancing the inverted pendulum system with other techniques is left as a future work. Adaptive control and using other computational techniques can be used for this purpose.

References

1. Nasir, A. N. K., Ahmad, M. A., & Rahmat, M. F. A. (2008). Performance comparison between LQR and PID controllers for an inverted pendulum system. In *International conference on power control and optimization: Innovation in Power control for optimal industry* (Vol. 1052, No. 1, pp. 124–128). AIP Publishing.
2. Chakraborty, S. (2014). An experimental study for stabilization of inverted pendulum. (Doctoral dissertation).
3. Li, Z., Yang, C., & Fan, L. (2012). *Advanced control of wheeled inverted pendulum systems*. Springer Science and Business Media.
4. Noh, K. K., Kim, J. G., & Huh, U. Y. (2004). Stability experiment of a biped walking robot with inverted pendulum. In *30th annual conference of IEEE industrial electronics society, 2004. IECON 2004* (Vol. 3, pp. 2475–2479). IEEE.
5. Li, Z., & Yang, C. (2012). Neural-adaptive output feedback control of a class of transportation vehicles based on wheeled inverted pendulum models. *IEEE Transactions on Control Systems Technology, 20*(6), 1583–1591.
6. Azar, A. T., & Vaidyanathan, S. (Eds.). (2016). *Advances in chaos theory and intelligent control* (Vol. 337). Springer.
7. Azar, A. T., & Serrano, F. E. (2015a). Stabilizatoin and control of mechanical systems with backlash. *Handbook of research on advanced intelligent control engineering and automation*. Advances in computational intelligence and robotics (ACIR), IGI-Global, USA (pp. 1–60).

8. Altinoz, O. T., Yilmaz, A. E., & Weber, G. W. (2010). Chaos particle swarm optimized PID controller for the inverted pendulum system. In *2nd international conference on engineering optimization*, Lisbon, Portugal.
9. Azar, A. T., & Serrano, F. E. (2015b). Adaptive sliding mode control of the Furuta pendulum. In *Advances and Applications in Sliding Mode Control systems* (pp. 1–42). Springer International Publishing.
10. Azar, A. T., & Vaidyanathan, S. (2014). Handbook of research on advanced intelligent control engineering and automation. IGI Global.
11. Pengpeng, Z., Lei, Z., & Yanhai, H. (2013). BP neural network control of single inverted pendulum. In *2013 3rd international conference on computer science and network technology (ICCSNT)* (pp. 1259–1262). IEEE.
12. Jaleel, J. A., & Francis, R. M. (2013). Simulated annealing based control of an inverted pendulum system. In *2013 International conference on control communication and computing (ICCC)* (pp. 204–209). IEEE.
13. Hanafy, T. O. (2012). Stabilization of inverted pendulum system using particle swarm optimization. In *8th international conference on informatics and systems (INFOS)*, 2012 (pp. BIO-207). IEEE.
14. Zhao, J., & Spong, M. W. (2001). Hybrid control for global stabilization of the cart–pendulum system. *Automatica, 37*(12), 1941–1951.
15. Park, M. S., & Chwa, D. (2009). Swing-up and stabilization control of inverted-pendulum systems via coupled sliding-mode control method. *IEEE Transactions on Industrial Electronics, 56*(9), 3541–3555.
16. Mason, P., Broucke, M., & Piccoli, B. (2008). Time optimal swing-up of the planar pendulum. *IEEE Transactions on Automatic Control, 53*(8), 1876–1886.
17. Mills, A., Wills, A., & Ninness, B. (2009). Nonlinear model predictive control of an inverted pendulum. In *2009 American control conference* (pp. 2335–2340). IEEE.
18. Wang, H. O., Tanaka, K., & Griffin, M. F. (1996). An approach to fuzzy control of nonlinear systems: stability and design issues. *IEEE Transactions on Fuzzy Systems, 4*(1), 14–23.
19. Chang, W. D., Hwang, R. C., & Hsieh, J. G. (2002). A self-tuning PID control for a class of nonlinear systems based on the Lyapunov approach. *Journal of Process Control, 12*(2), 233–242.
20. Åström, K. J., & Furuta, K. (2000). Swinging up a pendulum by energy control. *Automatica, 36*(2), 287–295.
21. Åström, K. J. (1999). Hybrid control of inverted pendulums. In *Learning, control and hybrid systems* (pp. 150–163). London: Springer.
22. Mazenc, F., & Bowong, S. (2003). Tracking trajectories of the cart-pendulum system. *Automatica, 39*(4), 677–684.
23. Azar, A. T., & Zhu, Q. (Eds.). (2015). *Advances and applications in sliding mode control systems*. Berlin: Springer.
24. Holzhüter, T. (2004). Optimal regulator for the inverted pendulum via Euler-Lagrange backward integration. *Automatica, 40*(9), 1613–1620.
25. Chernousko, F. L., & Reshmin, S. A. (2007). Time-optimal swing-up feedback control of a pendulum. *Nonlinear Dynamics, 47*(1–3), 65–73.
26. Srinivasan, B., Huguenin, P., & Bonvin, D. (2009). Global stabilization of an inverted pendulum–control strategy and experimental verification. *Automatica, 45*(1), 265–269.
27. Azar, A. T., & Vaidyanathan, S. (Eds.). (2015). *Chaos modeling and control systems design*. Germany: Springer.
28. Karnopp, D. C., Margolis, D. L., & Rosenberg, R. C. (2012). *System dynamics: modeling, simulation, and control of mechatronic systems*. Wiley.
29. Ogata, K. (2010). Modern control engineering. книга.
30. Box, G. E. (1957). Evolutionary operation: A method for increasing industrial productivity. *Applied statistics* (pp. 81–101).
31. Friedman, G. J. (1959). Digital simulation of an evolutionary process. *General systems yearbook* (Vol. 4, pp. 171–184).

32. Dianati, M., Song, I., & Treiber, M. (2002). *An introduction to genetic algorithms and evolution strategies*. Technical report, University of Waterloo, Ontario, N2L 3G1, Canada.
33. James, K., & Russell, E. (1995). Particle swarm optimization. In *Proceedings of 1995 IEEE international conference on neural networks* (pp. 1942–1948).
34. Mehdi, H., & Boubaker, O. (2011). Position/force control optimized by particle swarm intelligence for constrained robotic manipulators. In *11th international conference on intelligent systems design and applications (ISDA)* (pp. 190–195). IEEE.
35. Ebrahim, M. A., El-Metwally, A., Bendary, F. M., Mansour, W. M., Ramadan, H. S., Ortega, R., & Romero, J. (2011). Optimization of proportional-integral-differential controller for wind power plant using particle swarm optimization technique. *International Journal of Emerging Technologies in Science and Engineering*.
36. Ebrahim, M. A., Mostafa, H. E., Gawish, S. A., & Bendary, F. M. (2009). Design of decentralized load frequency based-PID controller using stochastic particle swarm optimization technique. In International conference on electric power and energy conversion systems, 2009. EPECS'09 (pp. 1–6). IEEE.
37. Jagatheesan, K., Anand, B., & Ebrahim, M. A. (2014). Stochastic particle swarm optimization for tuning of PID controller in load frequency control of single area reheat thermal power system. *International Journal of Electrical Power and Energy, 8*(2), 33–40.
38. Mousa, M. E., Ebrahim, M. A., & Hassan, M. M. (2015). Stabilizing and swinging-up the inverted pendulum using pi and pid controllers based on reduced linear quadratic regulator tuned by PSO. *International Journal of System Dynamics Applications (IJSDA), 4*(4), 52–69.
39. Ali, A. M., Ebrahim, M. A., & Hassan, M. M. (2016). Automatic voltage generation control for two area power system based on particle swarm optimization. *Indonesian Journal of Electrical Engineering and Computer Science, 2*(1), 132–144.
40. Kennedy, J. (1997). The particle swarm: social adaptation of knowledge. In *IEEE international conference on evolutionary computation* (pp. 303–308). IEEE.
41. Wang, J. J. (2011). Simulation studies of inverted pendulum based on PID controllers. *Simulation Modelling Practice and Theory, 19*(1), 440–449.
42. Ciprian, P. P., Luminita, D., & Lucia, P. (2011). Control optimization using MATLAB. INTECH Open Access Publisher.
43. Branch, S. T. (2011). Optimal design of LQR weighting matrices based on intelligent optimization methods.

Towards a Robust Fractional Order PID Stabilizer for Electric Power Systems

Magdy A.S. Aboelela and Hisham M. Soliman

Abstract This chapter deals with the design and application of a robust Fractional Order PID (FOPID) power system stabilizer tuned by Genetic Algorithm (GA). The system's robustness is assured through the application of *Kharitonov's* theorem to overcome the effect of system parameter's changes within upper and lower pounds. The FOPID stabilizer has been simplified during the optimization using the Oustaloup's approximation for fractional calculus and the "nipid" toolbox of Matlab during simulation. The objective is to keep robust stabilization with maximum attained degree of stability against system's uncertainty. This optimization will be achieved with the proper choice of the FOPID stabilizer's coefficients (k_p, k_i, k_d, λ, and δ) as discussed later in this chapter. The optimization has been done using the GA which limits the boundaries of the tuned parameters within the allowable domain. The calculations have been applied to a single machine infinite bus (SMIB) power system using Matlab and Simulink. The results show superior behavior of the proposed stabilizer over the traditional PID.

Keywords Power system · Power system stabilizer (PSS) · Genetic algorithm · Robust control · Single machine infinite bus (SMIB) · Kharitonov's theorem · Matlab/simulink

M.A.S. Aboelela (✉)
Faculty of Engineering, Electric Power and Machines Department,
Cairo University, Giza, Egypt
e-mail: aboelelamagdy@gmail.com

H.M. Soliman
Department of Electrical and Computer Engineering, College of Engineering,
Sultan Qaboos University, Muscat, Oman

A.T. Azar et al. (eds.), *Fractional Order Control and Synchronization of Chaotic Systems*, Studies in Computational Intelligence 688,
DOI 10.1007/978-3-319-50249-6_9

1 Intoduction

Low or negative damping in a power system can lead to spontaneous appearance of large power oscillations. Several methods for increasing the damping in a power system are available such as static voltage condenser (SVC), high voltage direct current (HVDC) and power system stabilizer (PSS). Operating conditions of a power system are continually changing due to load patterns, electric generation variations, disturbances, transmission topology and line switching [18].

To enhance system damping; the generators are equipped with power system stabilizers that provide supplementary feedback stabilizing signals in the excitation systems [23]. The control strategy should be capable of manipulating the PSS effectively. The PSS should provide robust stability over a wide range of operating conditions, easy to implement, improves transient stability, low developing time and least cost [18]. Various topologies and many control methods have been proposed for PSS design, such as adaptive controller [33], robust controller [3, 8, 9], extended integral controller [41], state feedback controller [19], fuzzy logic controller [4] and variable structure controller [11]. In Kothari et al. [15] an adaptive fuzzy PSS that behaves like a PID controller that provides faster stabilization of the frequency error signal with less dependency on expert knowledge is proposed. In Malik et al. [20], an indirect adaptive PSS is designed using two input signals, the speed deviation and the power deviation to a neural network controller.

The robust PSS has the ability to maintain stability and achieves desired performance while being insensitive to the perturbations. Among the various robustness techniques, H_∞ optimal control [5] and the structured singular value (SSV or μ) technique [31] have received considerable attention. But, the application of μ technique for controller design is complicated due to the computational requirements of μ design. Besides the high order of the resulting controller, also introduces difficulties with regard to implementation [14, 34].

The H_∞ optimal controller design is relatively simpler than the μ synthesis in terms of the computational burden [5, 35, 36].

Since power systems are highly non-linear, conventional fixed-parameter PSSs cannot cope with wide changes of the operating conditions. There are two main approaches to stabilize a power system over a wide range of operating conditions; namely adaptive control [1, 10, 37] and robust control [3, 19, 35]. However, adaptive controllers have generally poor performance during the learning phase; unless they are properly initialized. Successful operation of adaptive controllers requires the measurements to satisfy strict persistent excitation conditions; otherwise the adjustment of the controller's parameters fail [2, 5, 13].

This chapter is organized as follow: In Sect. 2, we present a brief introduction to fractional calculus and its approximation. Section 3 presents the GA. Section 4 illustrates the system under investigation. Section 5 presents the problem formulation and the problem solution is discussed in Sect. 6. The design procedure of

FOPID PSS is introduced in Sect. 7 with different loading and working conditions. Section 8 and some references are given in Sect. 9. The chapter has three Appendices A, B, and C.

2 Fractional Order PID Controller ($PI^{\lambda}D^{\delta}$) Design

The PSS proposed in this chapter belongs to the class of robust controllers. It relies on the Kharitonov's theorem and GA optimization. The use of the Kharitonov's theorem enables us to consider a finite number of plants to be stabilized. The resulting controller will be able to stabilize the original system at any operating point within the design range. We propose to tune the controller's parameters using the genetic algorithm optimization technique [11, 12, 16].

Proportional-Integral-Derivative (PID) controllers are widely being used in industries for process control applications. The merit of using PID controllers lie in its simplicity of design and good performance including low percentage overshoot and small settling time for slow industrial processes. The performance of PID controllers can be further improved by appropriate settings of fractional-I and fractional-D actions [24, 25, 28, 29].

In a fractional PID controller, the I- and D-actions being fractional have wider scope of design. Naturally, besides setting the proportional, derivative and integral constants Kp,Td and Ti respectively, we have two more parameters: the power of s in integral and derivative actions-λ and δ respectively. Finding [k_p, k_i, k_d, λ, and δ] as an optimal solution to a given process thus calls for optimization on the five-dimensional space. Classical optimization techniques cannot be used here because of the roughness of the objective function surface. We, therefore, use a derivative-free optimization, guided by the collective behavior of social swarm and determine optimal settings of k_p, k_i, k_d, λ, and δ [1].

The performance of the optimal fractional PID controller is better than its integer counterpart. Thus the proposed design will find extensive applications in real industrial processes. Traces of work on fractional PID are available in the current literature [1, 7, 22, 24–29, 32] on control engineering. A frequency domain approach based on the expected crossover frequency and phase margin is mentioned in Vinagre et al. [39]. A method based on pole distribution of the characteristic equation in the complex plane was proposed in Petras [24]. A state-space design method based on feedback poles placement can be viewed in Dorcak et al. [7]. The fractional controller can also be designed by cascading a proper fractional unit to an integer-order controller [26].

Moreover, researchers reported that controllers making use of factional order derivatives and integrals could achieve performance and robustness results superior to those obtained with conventional (integer order) controllers. The Fractional-order PID controller (FOPID) controller is the expansion of the conventional PID controller based on fractional calculus [1].

The differential equation of the $PI^{\lambda}D^{\delta}$ controller is described in time domain by

$$u(t) = k_p e(t) + k_i D_t^{-\lambda} e(t) + k_d D_t^{\delta} e(t) \tag{1}$$

The continuous transfer function of the $PI^{\lambda}D^{\delta}$ controller is obtained through Laplace transform as

$$G_c(s) = k_p + k_i s^{-\lambda} + k_d s^{\delta} \tag{2}$$

It is obvious that, the FOPID controller does not only need the design three parameters k_p, k_i and k_d, but also the design of two orders λ, δ of integral and derivative controllers. The orders λ, δ are not necessarily integer, but any real numbers [25].

3 Genetic Algorithm Operation

To illustrate the working process of genetic algorithm, the steps to realize a basic GA are listed below [11, 12, 16]:

Step 1: Represent the problem variable domain as a chromosome of fixed length; choose the size of the chromosome population N, the crossover probability Pc and the mutation probability Pm.

Step 2: Define a fitness function to measure the performance of an individual chromosome in the problem domain. The fitness function establishes the basis for selecting chromosomes that will be mated during reproduction.

Step 3: Randomly generate an initial population of size N: $x_1, x_2, \ldots, x_N$.

Step 4: Calculate the fitness of each individual chromosome: $f(x_1), f(x_2), \ldots, f(x_N)$.

Step 5: Select a pair of chromosomes for mating from the current population. Parent chromosomes are selected with a probability related to their fitness. High fit chromosomes have a higher probability of being selected for mating than less fit chromosomes.

Step 6: Create a pair of offspring chromosomes by applying the genetic operators.

Step 7: Place the created offspring chromosomes in the new population.

Step 8: Repeat Step 5 until the size of the new population equals that of initial population, N.

Step 9: Replace the initial (parent) chromosome population with the new (offspring) population.

Step 10: Go to Step 4, and repeat the process until the termination criterion is satisfied.

A GA is an iterative process. Each iteration is called a generation. A typical number of generations for a simple GA can range from 50 to over 500. A common practice is to terminate a GA after a specified number of generations and then examine the best chromosomes in the population. If no satisfactory solution is found, then the GA is restarted [21, 31].

The GA moves from generation to generation until a stopping criterion is met. The stopping criterion could be maximum number of generations, population convergence criteria, lack of improvement in the best solution over a specified number of generations or target value for the objective function.

Evaluation functions or objective functions of many forms can be used in a GA so that the function can map the population into a partially ordered set. The computational flowchart of the GA optimization process employed in the present study is given in Fig. 1.

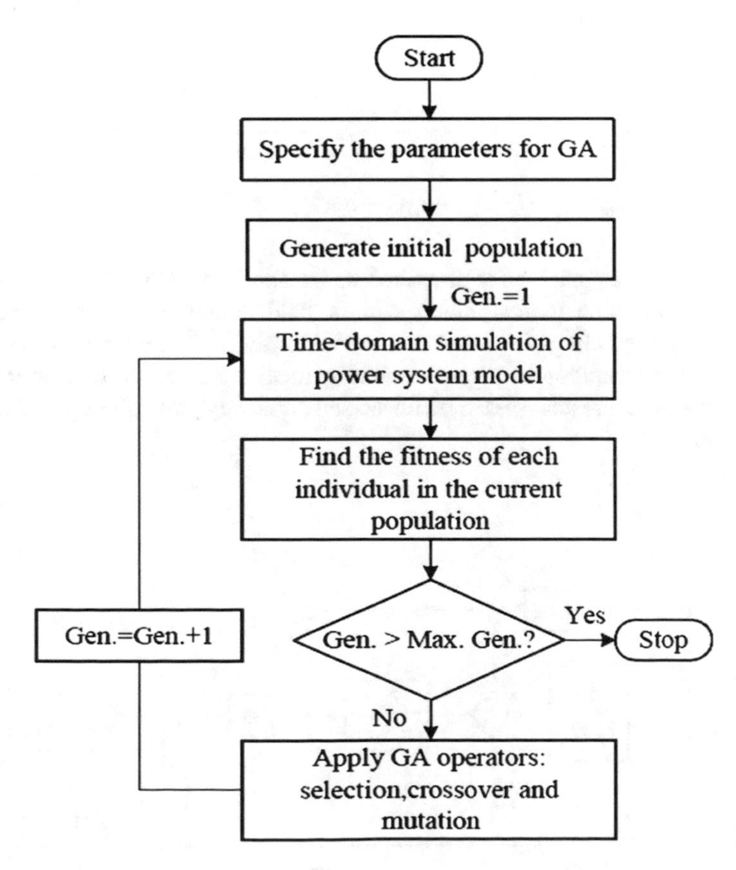

Fig. 1 The computational flowchart of the GA

4 System Investigated

A single machine-infinite bus (SMIB) system is considered for the present investigations. A machine connected to a large system through a transmission line may be reduced to a SMIB system, by using Thevenin's equivalent of the transmission network external to the machine. Because of the relative size of the system to which the machine is supplying power, the dynamics associated with machine will cause virtually no change in the voltage and frequency of the Thevenin's voltage (infinite bus voltage). The Thevenin's equivalent impedance shall henceforth be referred to as equivalent impedance (i.e. Re + jXe) [6].

Figure 2 shows the system under study which consists of a single machine connected to an infinite bus through a tie-line. The machine is equipped with a static exciter. The non-linear equations of the system are

$$\begin{aligned}\dot{\omega} &= \frac{T_m - T_e}{M}\\ \dot{\delta} &= \omega_0 \omega\\ \dot{E}'_q &= \frac{1}{T'_{do}}\left(E_{fd} - \frac{x_d + x_e}{x_d' + x_e}E'_q + \frac{x_d + x'_d}{x_d' + x_e}V\cos\delta\right)\\ \dot{E}_{fd} &= \frac{1}{T_E}\left(k_E E_{ref} - k_E V_t - E_{fd}\right)\end{aligned} \tag{3}$$

The synchronous machine is described as the fourth order model. The two-axis synchronous machine representation with a field circuit in the direct axis but without damper windings is considered for the analysis. The equations describing the steady state operation of a synchronous generator connected to an infinite bus through an external reactance can be linearized about any particular operating point as follows:

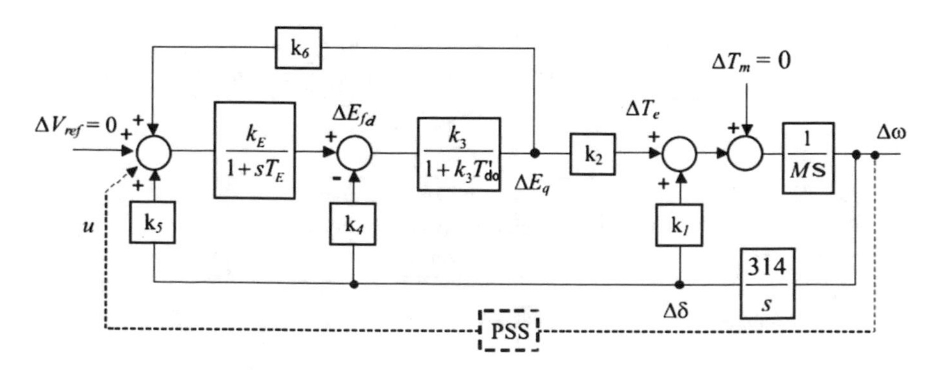

Fig. 2 The block diagram for closed loop SMIB System

$$\Delta T_m - \Delta P = M\frac{d^2\Delta\delta}{dt^2} \tag{4}$$

$$\Delta P = K_1\Delta\delta + K_2\Delta E'_q \tag{5}$$

$$\Delta E'_q = \frac{K_3}{1 + sT'_{d0}K_3}\Delta E_{fd} - \frac{K_3K_4}{1 + sT'_{d0}K_3}\Delta\delta \tag{6}$$

$$\Delta V_t = K_5\Delta\delta + K_6\Delta E'_q \tag{7}$$

The synchronous machine is described by Heffron-Philips model as described in Fig. 2. The K-constants are given in Appendix A. The data definitions are given in Appendix B. The system data are illustrated in Appendix C.

The interaction between the speed and voltage control equations of the machine is expressed in terms of six constants k_1–k_6. These constants with the exception of k_3, which is only a function of the ratio of impedance, are dependent upon the actual real and reactive power loading as well as the excitation levels in the machine [6].

The system equation can be expressed in the following state variable form:

$$\begin{aligned} \dot{X}(t) &= AX(t) + Bu(t) \\ y(t) &= Cx(t) \end{aligned} \tag{8}$$

$$X(t) = [\Delta\delta \quad \Delta\omega \quad \Delta E'_q \quad \Delta E_{fd}]^T,$$

$$A = \begin{bmatrix} 0 & \omega_0 & 0 & 0 \\ \frac{-k_1}{M} & 0 & -\frac{k_2}{M} & 0 \\ -\frac{k_4}{TT'_{do}} & 0 & -\frac{1}{T} & -\frac{1}{T'_{do}} \\ -\frac{k_5\,k_E}{T_E} & 0 & -\frac{k_6\,k_E}{T_E} & -\frac{1}{T_E} \end{bmatrix}, \tag{9}$$

$$B = \begin{bmatrix} 0 & 0 & 0 & \frac{k_E}{T_E} \end{bmatrix}', C = [0 \quad 1 \quad 0 \quad 0].$$

5 Problem Formulation

The system can be represented by the block diagram proposed by deMello and Concordia [40] which can be cast as shown in Fig. 2. The parameters of the model are load dependent, thus, they have to be calculated at each operating point. Analytical expressions for the parameters k1–k6, as derived in Soliman et al. [35], Soliman and Sakr [36], are listed in Appendix A. The parameters, k1–k6, are functions of the loading condition (*P* and *Q*). By varying P and/or Q to cover a wide range of system loading, the parameters K_1–K_6 are computed.

The use of the high-gain voltage regulators usually destabilizes the system. This effect is usually complemented compensated by the inclusion of a stabilizing signal generated by the PSS to provide the required damping. In most cases, the speed deviation signal $\Delta\omega$ is used as an input to the PSS.

To design the PSS, it is convenient to represent the system in the transfer function form as shown in Fig. 3. An analytical expression for the transfer function is derived based on the obtained parameters by using Mason's rule. The resulting transfer function is

$$\frac{\Delta\omega}{U}(s) = \frac{bs}{a_4 s^4 + a_3 s^3 + a_2 s^2 + a_1 s + a_0} \tag{10}$$

The transfer-function coefficients expressed in terms of the k-parameters are:

$$\begin{aligned} a_4 &= MTT_E \\ a_3 &= M(T + T_E) \\ a_2 &= M = 314k_1 TT_E + k_E k_3 k_6 M \\ a_1 &= 314k_1(T + T_E) - 314k_2 k_3 k_4 T_E \\ a_0 &= 314(k_1 - k_2 k_3 k_4 + k_E k_1 k_3 k_6) \\ b &= k_E k_2 k_3 \end{aligned} \tag{11}$$

The coefficients of the transfer function are load-dependent. So, the PSS has to be adjusted at different loads. To scan the whole range of operation, the load dependency may require the analysis of a large number of points with a new model generated at each operating condition.

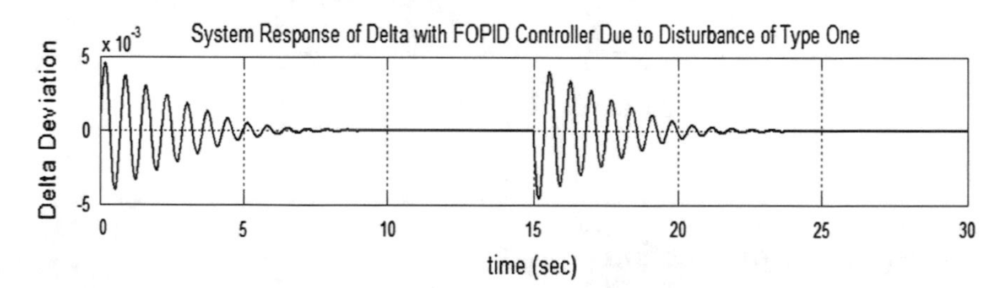

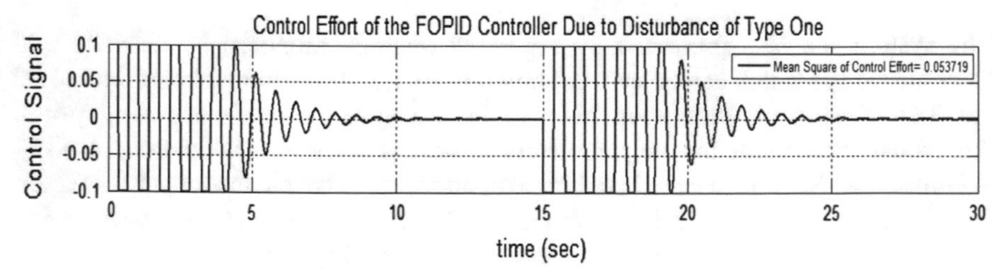

Fig. 3 System response to 0.2 pu torque disturbance at (P = 0.8, Q = 0.3)

A proposed technique, based on the Kharitonov's theorem and GA, is used to design a fixed parameters robust FOPID controller to stabilize the non-linear system over the specified range of operating conditions [P_{min}, P_{max}] and [Q_{min}, Q_{max}]. In this technique, the problem is transformed to simultaneous stabilization of a finite number of extreme plants. We will show in the next section that we need to stabilize exactly eight characteristic polynomials.

5.1 Mathematical Tools and Problem Solution

5.1.1 Kharitonov's Theorem

The Kharitonov's theorem studies the robust stability of an interval polynomial family [40]. A polynomial

$$p = a_n s^n + a_{n-1} s^{n-1} + \cdots + a_0 \tag{12}$$

is said to be an interval polynomial if each coefficient a_i is independent of the others and varies within an interval having lower and upper bounds that is,

$$a_i = [a_i^-, a_i^+] \tag{13}$$

The Kharitonov's theorem states that "*An interval polynomial*

$$p = \sum_{i=0}^{n} [a_i^-, a_i^+] s^i \tag{14}$$

is robustly stable if and only if the following four Kharitonov's polynomials

$$\begin{aligned} p_1 &= a_0^- + a_1^- s + a_2^+ s^2 + a_3^+ s^3 + a_4^- s^4 + \cdots \\ p_2 &= a_0^+ + a_1^+ s + a_2^- s^2 + a_3^- s^3 + a_4^+ s^4 + \cdots \\ p_3 &= a_0^+ + a_1^- s + a_2^- s^2 + a_3^+ s^3 + a_4^+ s^4 + \cdots \\ p_4 &= a_0^- + a_1^+ s + a_2^+ s^2 + a_3^- s^3 + a_4^- s^4 + \cdots \end{aligned} \tag{15}$$

are stable".

Assuming that the coefficient function *ai* depends continuously on the vector = $[P\ Q]^T$(machine loading *P* and *Q*), we define the bounds

$$\begin{aligned} a_i^{*-} &= \min_r(a_i) \\ a_i^{*+} &= \min_r(a_i) \end{aligned} \tag{16}$$

and simply construct the polynomial described by

$$p^{*}(s) = \sum_{i=0}^{n} [a_i^{*-}, a_i^{*+}] s^i \tag{17}$$

Then the robust stability of polynomial (17) implies the robust stability of (12), El-Metwally et al. [10].

5.1.2 Oustaloup's Recursive Filter to Approximate FOPID

Some continuous filters have been summarized in [36]. Among the filters, the well-established Oustaloup recursive filter has a very good fitting to the fractional-order differentiators. Assume that the expected fitting range is (ω_b, ω_h). The filter can be written as

$$G_f(s) = K \prod_{K=-N}^{N} \frac{s + \omega'_k}{s + \omega_k} \tag{18}$$

where the poles, zeros, and gain of the filter can be evaluated such that

$$\begin{aligned} \omega'_k &= \omega_b \left(\frac{\omega_h}{\omega_b}\right)^{\frac{k+N+\frac{1}{2}(1-\gamma)}{2N+1}} \\ \omega_k &= \omega_b \left(\frac{\omega_h}{\omega_b}\right)^{\frac{k+N+\frac{1}{2}(1+\gamma)}{2N+1}} \end{aligned} \tag{19}$$

and

$$K = \omega_h^{\gamma}$$

Thus, the any signal $y(t)$ signal can be filtered through this filter and the output of the filter can be regarded as an approximation for the derivative term of the FOPID with $\gamma = \delta$ or the integral counterpart with $\gamma = -\lambda$. The resulted transfer function of the FOPID is the sum of the proportional term k_p plus the filter approximation of the integral term $(k_i s^{-\lambda})$ plus the derivative term $(k_D s^{\delta})$. The result will be the approximated transfer function of the FOPID controller $G_c(s)$ as given by Eq. (2).

In general $G_c(s)$ can be assumed to be in the form:

$$G_c(s) = \frac{N(s)}{D(s)} \tag{20}$$

As shown in Fig. 3, the closed loop characteristic equation can be written as

$$1 + G_c(s)G_p(s) = 0 \tag{21}$$

where $G_p(s) = \frac{\Delta\omega}{U}(s)$ is the plant transfer function [10].

5.1.3 The 16 Kharitonov's Polynomials

Given the plant family with Kharitonov's polynomials N_1, …, N_4 and D_1, …, D_4 for the numerator and denominator, respectively, we define the 16 Kharitonov's plants as El-Metwally et al. [10].

$$G_c^i(s) = \frac{N_{i_1(s)}}{D_{i_2}(s)}, i_1 = 1, 2, \ldots, 4\, and\, i_2 = 1, 2, \ldots, 4 \tag{22}$$

where i = 1, 2, …, 16. If the controller can stabilize all the 16 closed loop polynomials given as

$$1 + G_c^i(s)G_p(s) = 0 \tag{23}$$

Then the closed loop system (23) is robustly stable, where $i_1 = 1, 2, \ldots, 4\, and\, i_2 = 1, 2, \ldots, 4$, El-Metwally et al. [10].

Applying the above mathematical tools to the single machine–infinite bus system (Fig. 1), we have the vector r which is composed of two independent components.

$$r = [P \quad Q]^T \tag{24}$$

In the system under study, the numerator of the transfer function is a first order polynomial (bs). Thus, the coefficient b has two extreme values b^+ and b^-; that is, the 16 plants corresponding to (23) are reduced to 8 plants only.

6 Problem Solution

To stabilize the system over the required ranges of P and Q, the following eight polynomials must be stable.

We will use the genetic algorithm to find the values of k_p, k_i, k_d, λ, and δ that correspond to the following optimization problem

$$\min_{\text{kp, ki, kd}, \lambda, \text{and}\delta} (\max(\lambda_e)) \tag{25}$$

Subject to

$$\begin{aligned} k_p^{min} &\le k_p \le k_p^{max} \\ k_i^{min} &\le k_i \le k_i^{max} \\ k_D^{min} &\le k_D \le k_D^{max} \\ \lambda^{min} &\le \lambda \le \lambda^{max} \\ \gamma^{min} &\le \gamma \le \gamma^{max} \end{aligned} \tag{26}$$

where λ_e is a vector containing the real parts of the roots of the eight equations resulting from (25). This means that the parameters k, z and p must stabilize the eight polynomials in Eq. (25). On the other hand, the swarm optimization algorithm attempts to push the closed loop poles to the left as far as possible by minimizing the maximum real part of the roots resulting from (25). The problem can be tackled using a different approach. If we divide the range of P and Q into small steps, the resulting grid will represent the possible operating points.

For each point on the grid, a model can be derived. Applying the genetic algorithm optimization technique to stabilize such systems is possible. However, there is no guarantee that stability is preserved for intermediate points inside the grid. The proposed technique eliminates this shortcoming via the Kharitonov's theorem.

7 PSS Design for Different Machine Loadability

The design objective, in this chapter, is to implement the machine loadability, of the system under study, over the range $Q \in [-0.4, 0.4]$ and $P \in [0.2, 1.2]$. The design procedure can be summarized as follows:

- Develop the linearized model as shown in Fig. 2. The machine parameters and the k-parameter calculations are given in the Appendices A and C.
- Based on the analytical expressions for $a_0, a_1, \ldots, a_4$ and b in (11), calculate the maximum and minimum values of the aforementioned parameters using any standard optimization technique. Note that a_3 and a_4 do not depend on the values of P and Q.
- Using (29) and replacing a_i by a_i^*, construct the four Kharitonov's polynomials as in (15). Compute the roots of the 8 extreme polynomials and take the largest real part of the roots as the objective function to be minimized.
- Use the GA to find a solution for the optimization problem (26) such that the roots of (25) lie in the left hand side of the s-plane away from the imaginary axis as much as possible. Thus the shortest settling time of oscillations is achieved

The above procedure is applied to the system under study as follows: Consider the system transfer function (10). The extreme values of its coefficients are calculated as

$$\begin{aligned} a_i^{*-} &= \min_{P,Q}(a_i) \text{ and } a_i^{*+} = \max_{P,Q}(a_i^*) \\ b_i^{*-} &= \min_{P,Q}(b_i) \text{ and } b_i^{*+} = \max_{P,Q}(b_i^*) \end{aligned} \tag{27}$$

The results of the above calculations are

$$a_4^* = 1,\ a_3^* = 22,\ a_2^* \in [64\ 106],\ a_1^* \in [388\ 1002],\ a_0^*[392\ 2624]\ and\ b^* \in [2.7\ 12.4]$$

Then, the four Kharitonov's polynomials are:

$$\begin{aligned} p_1 &= 392 + 388s + 106s^2 + 22s^3 + s^4 \\ p_2 &= 2624 + 1003s + 64s^2 + 22s^3 + s^4 \\ p_3 &= 2624 + 388s + 64s^2 + 22s^3 + s^4 \\ p_4 &= 392 + 1003s + 106s^2 + 22s^3 + s^4 \end{aligned} \tag{28}$$

7.1 Design of a Robust PSS Using GA

The plant transfer function (10) is analyzed using eight extreme plants given by

$$\begin{aligned} G_p(s) &= \frac{\Delta\omega}{U}(s) = \frac{b^- s}{p_i}, i = 1, 2, \ldots, 4 \\ G_p(s) &= \frac{\Delta\omega}{U}(s) = \frac{b^+ s}{p_i}, i = 1, 2, \ldots, 4 \end{aligned} \tag{29}$$

To reach the optimization goal, proper adjustment of the GA parameters are needed. This requires the determination of population size (N = 100 is sufficient), the bit size for each binary parameter (16 is reasonable size), and the upper and lower bounds of the optimization of FOPID PSS (for k_p, k_i, and k_d, [0 100] is an acceptable range but for λ and δ [0 1.5] is found to be a proper choice in our case [38].

The results obtained using the GA on FOPID PSS design procedure mentioned in this chapter are delineated in Table 1. The same procedure can be successfully applied to the case of PID PSS considering the limits of λ and δ of the FOPID PSS as [1 1]. Results of this case are also shown in Table 1.

The proposed PSS is tested over three operating condition.

Table 1 GA estimated parameters for PID and FOPID PSS

Controller	k_p	k_i	k_d	δ	λ
PID (minimum = −1.3961)	45.36	45.452	62.2	N/A	N/A
FOPID (minimum = −1.3849)	48.50	93.666	79.8	0.61	1.3

7.2 The Normal Loading Test

The first operating point is $P = 0.8$ pu and $Q = 0.3$ pu represents the normal loading conditions. The system was exposed to a 0.20 p.u step increase in the input torque reference at 0.5 s. The disturbance was removed at 15 s, .e. the signal duration is 14.5 s, and the system returned to the original operating point by the end of disturbance. The regulated system without a stabilizer is stable at this point [10]. However, the mechanical disturbance pushes the system close to the stability bound. Figure 3 shows the machine speed deviation and the machine power angle (δ). It is clear that if the power system stabilizer is not employed, the rotor angle oscillation will have a very slow damping behavior. On the other hand, the proposed FOPID stabilizer successfully suppresses and damps the oscillations in almost three seconds. The controller signal is shown in Fig. 3. It is clear that the controller is utilizing the full control range limited by the maximum standard power system stabilizer signal ±0.1 pu.

The Simulink models for the FOPID PSS applications are illustrated in Figs. 4 and 5. The FOPID PSS block is represented by "NIPID" block of "ninteger" blockset of Matlab [14, 38].

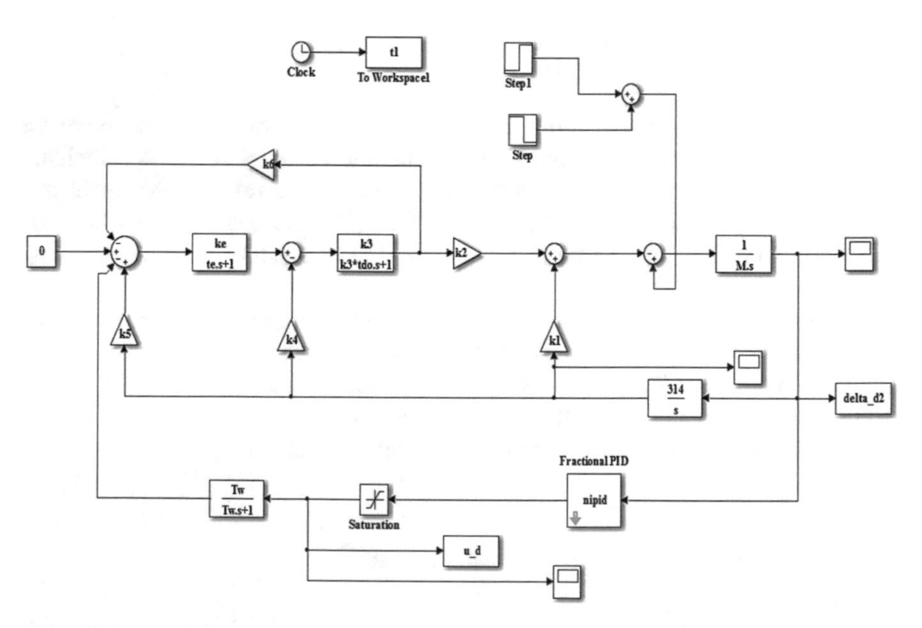

Fig. 4 Matlab/Simulink Model with FOPID PSS and Torque Disturbance Signal

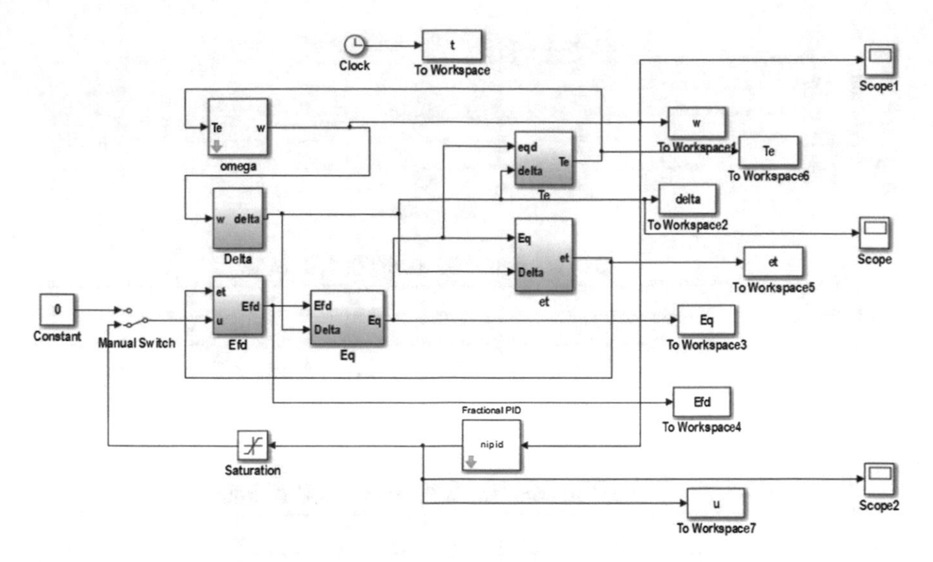

Fig. 5 Matlab/Simulink Model of SMIB with FOPID PSS

7.3 Overload Test

In this test, the machine was operating at $P = 1.2$ pu and $Q = 0.2$ pu. The machine speed deviation is unstable at this operating point [10]. Figure 6 shows the effectiveness of the proposed FOPID PSS to stabilize the system during over loading conditions [17].

7.4 Full Load with Leading Power Factor Test

The second operating point is $P = 1$ pu and $Q = -0.4$ pu. This point lies in the unstable region for the regulated system without a stabilizer as illustrated in Fig. 7a. The system at this operating point was exposed to a three phase to ground short circuit at 3 s and this will stay only for 100 m s and then cured. Figure 7b illustrates that the proposed FOPID stabilizer can damp the power angle and angular frequency oscillations within a short period of time with the same value of tuned parameters given in Table 1.

Finally, for the more illustration, the effect of the PID and FOPID PSSs on the stabilization of the SMIB power system described herein is shown in Figs. 8 and 9 for only the case of normal operation with $P = 0.8$ pu and $Q = 0.3$ pu without disturbance. It is clear that The damping effect of the FOPID PSS is noticeable compared with that of the PID PSS. The control effort in both PID and FOPID PSSs are shown in Fig. 10a, b. Obviously, the control effort of the FOPID PSS is much less than that of the PID in both magnitude and mean square error.

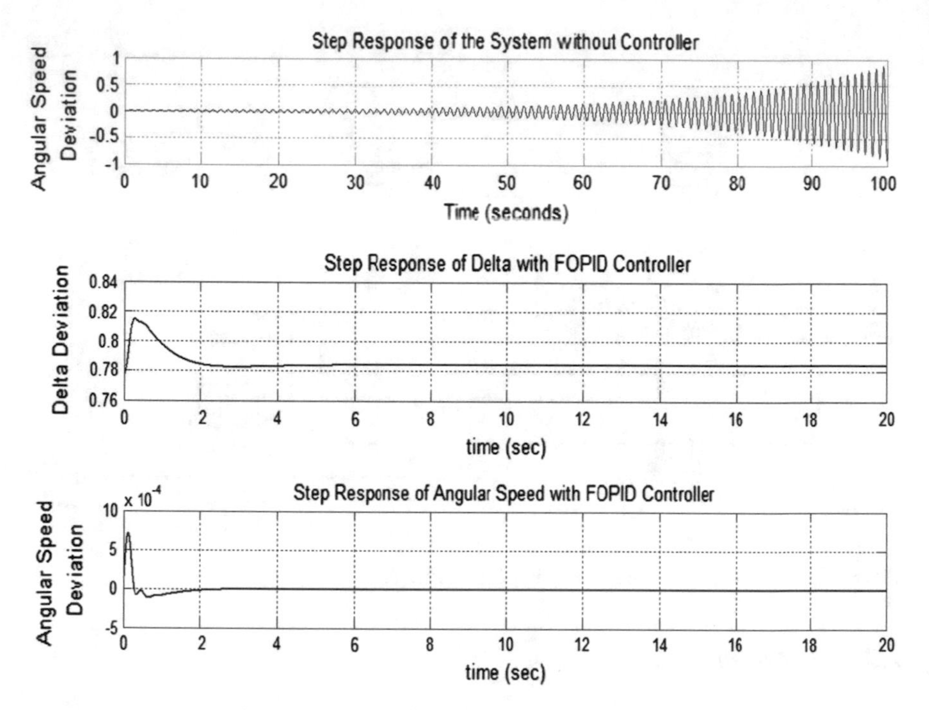

Fig. 6 **$\Delta\delta$ and** $\Delta\omega$ after adding FOPID PSS type in normal operation at ($P = 1.2$, $Q = 0.2$)

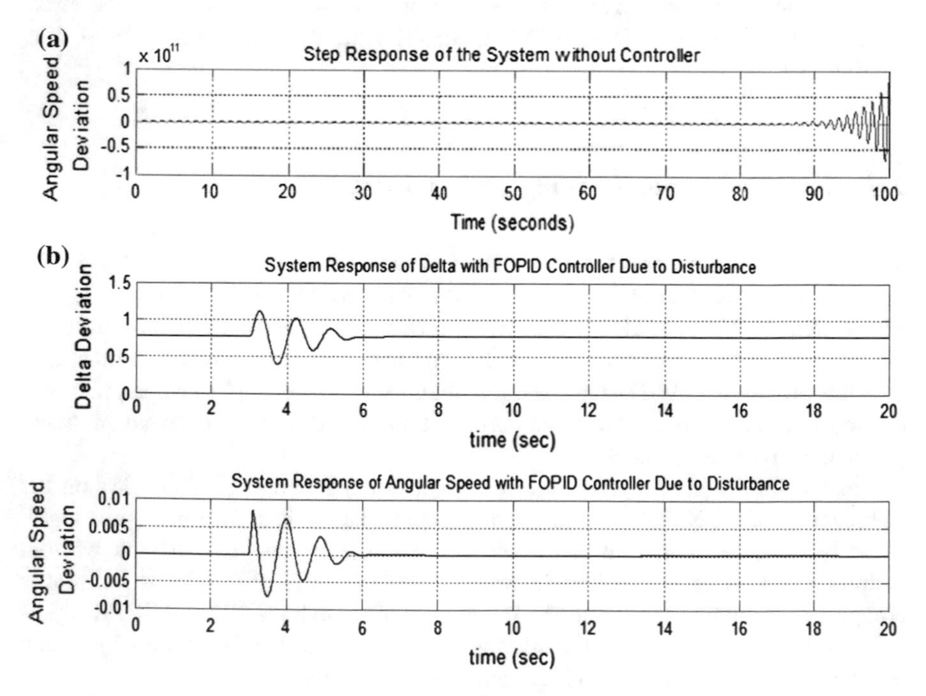

Fig. 7 **a** $\Delta\delta$ without Controller ($P = 1$, $Q = -0.4$). **b** $\Delta\delta$ and $\Delta\omega$ due to a three line to ground fault at 3 s staying for 100 ms after adding FOPID PSS type ($P = 1$, $Q = -0.4$)

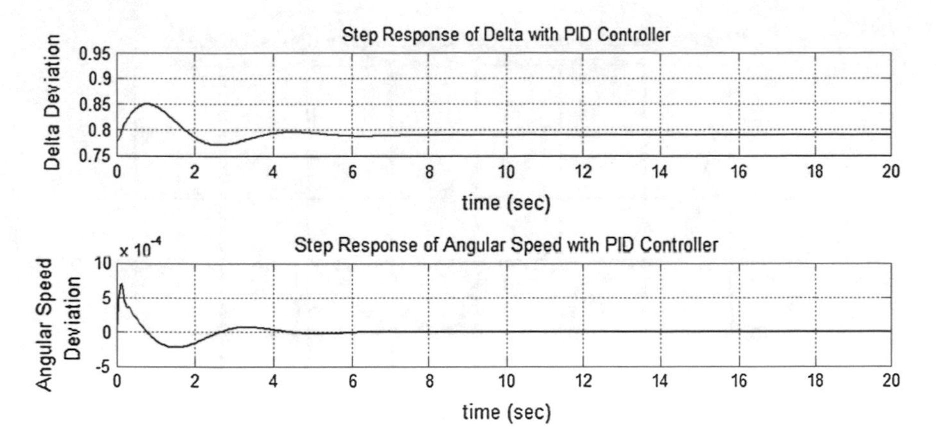

Fig. 8 $\Delta\delta$ and $\Delta\omega$ after adding PID PSS type in normal operation ($P = 0.8$, $Q = 0.3$)

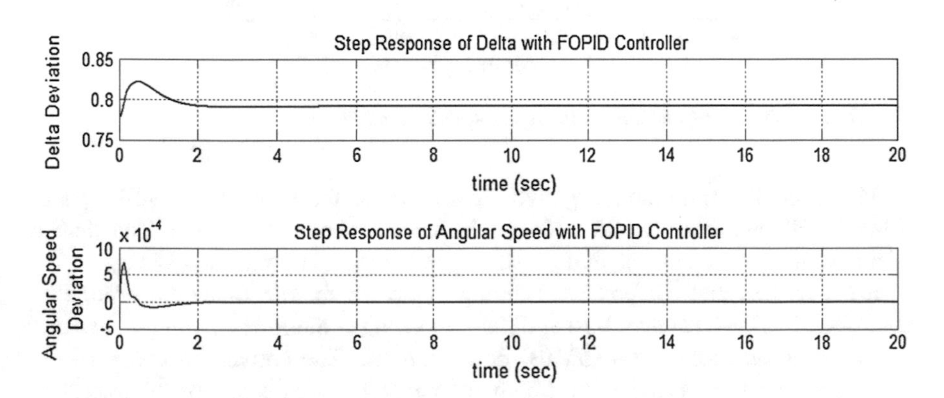

Fig. 9 $\Delta\delta$ and $\Delta\omega$ after adding FOPID PSS type in normal operation ($P = 0.8$, $Q = 0.3$)

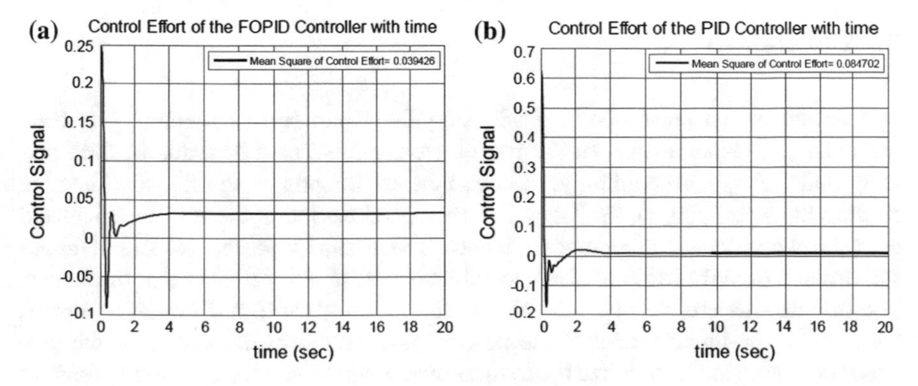

Fig. 10 **a** Control Effort of the FOPID PSS in normal operation ($P = 0.8$, $Q = 0.3$) **b** Control Effort of the PID PSS in normal operation ($P = 0.8$, $Q = 0.3$)

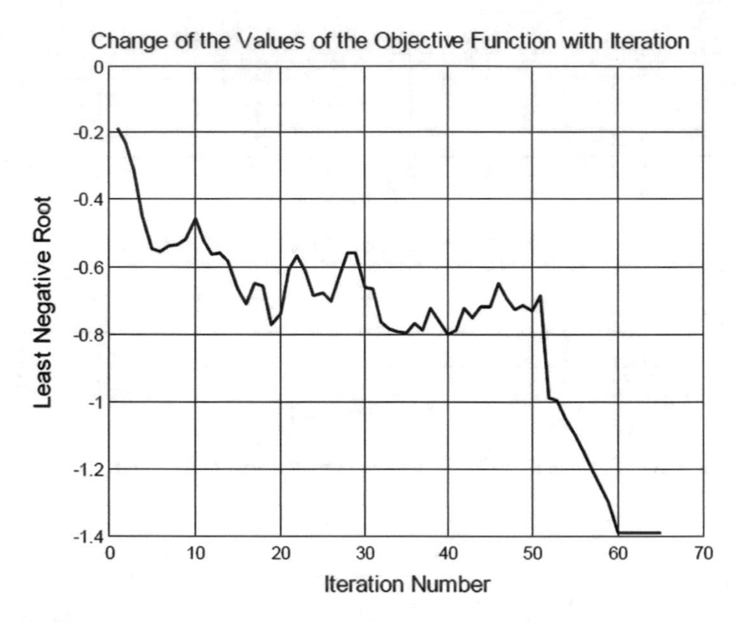

Fig. 11 The Objective Function vs. Iterations in case of FOPID PSS

Moreover, the minimum negative eigenvalue of the stabilized SMIB system using the PID and FOPID PSS is almost the same as shown in Table 1. The change of this value for the case of FOPID PSS with iteration is delineated if Fig. 11.

It is apparent that the presented tuning algorithm for the fractional order PID controllers has been found robust at different loading conditions of a single machine connected to an infinite bus (SMIB) power system. The convergence rate of the presented algorithm is noticeable which encourage the application of the fractional order PID (FOPID) controllers on some other industrial applications.

8 Conclusion

The design of a robust FOPID PSS using the Kharitonov's theorem has been proposed. The *k*-parameters of the model are parameterized in terms of the operating point (*P*, *Q*). Accordingly, the coefficients 'bounds of the transfer function relating the stabilizing control signal to the speed deviation have been calculated over the whole range of operating points. The design is based on simultaneous stabilization of eight extreme plants to achieve a satisfactory dynamic performance. The calculations are based on the GA optimization algorithm. Simulation results based on a non-linear model of the power system confirm the ability of the proposed compensator to stabilize the system over a wide range of operating points as illustrated with various examples.

The performance of the conventional PID PSS, designed with the same procedure, as compared with the FOPID PSS shows less oscillation damping of both the changes in angle δ and the angular speed ω.

For future work, authors recommend the extension of the method to the case of multi machines power systems. Also some other evolutionary techniques such as bat inspiration, gravitational techniques and imperialist colony may be tried to determine the best tuning of the fractional order PID controllers.

Appendix A: Derivation of k-Constants

All the variables with subscript 0 are values of variables evaluated at their pre-disturbance steady-state operating point from the known values of P_0, Q_0 and V_{t0}.

$$i_{q0} = \frac{P_0 V_{to}}{\sqrt{(P_0 x_q)^2 + (V_{t0}^2 + Q_0 x_q)^2}} \tag{A1}$$

$$v_{d0} = i_{q0} x_q \tag{A2}$$

$$v_{qo} = \sqrt{V_{t0}^2 - v_{t0}^2} \tag{A3}$$

$$i_{d0} = \frac{Q_0 + x_q i_{q0}^2}{v_{q0}} \tag{A4}$$

$$E_{q0} = v_{q0} + i_{d0} x_q \tag{A5}$$

$$E_0 = \sqrt{(v_{d0} + x_e i_{q0})^2 + (v_{q0} - x_e i_{d0})^2} \tag{A6}$$

$$\delta_0 = \tan^{-1}\frac{(v_{d0} + x_e i_{q0})}{(v_{q0} - x_e i_{d0})} \tag{A7}$$

$$\mathrm{K}_1 = \frac{x_q - x'_d}{x_e + x'_d} i_{q0} E_0 \sin\delta_0 + \frac{E_{q0} E_0 \cos\delta_0}{x_e + x_q} \tag{A8}$$

$$\mathrm{K}_2 = \frac{E_0 \sin\delta_0}{x_e + x'_d} \tag{A9}$$

$$\mathrm{K}_3 = \frac{x'_d + x_e}{x_d + x_e} \tag{A10}$$

$$K_4 = \frac{x_q - x'_d}{x_e + x'_d} E_0 \sin \delta_0 \tag{A11}$$

$$K_5 = \frac{x_q}{x_e + x_q} \frac{v_{d0}}{V_{t0}} E_0 \cos \delta_0 - \frac{x'd}{x_e + x'_d} \frac{v_{q0}}{V_{t0}} E_0 \sin \delta_0 \tag{A12}$$

$$K_6 = \frac{x_e}{x_e + x'_d} \frac{v_{q0}}{V_{t0}} \tag{A13}$$

Appendix B

Nomenclature

All quantities are per unit on machine base.

D	Damping Torque Coefficient
M	Inertia constant
ω	Angular speed
δ	Rotor angle
I_d, I_q	Direct and quadrature components of armature current
x_d and x_q	Synchronous reactance in d and q axis
x'_d and x'_q	Direct axis and Quadrature axis transient reactance
E_{fd}	Equivalent excitation voltage
K_E	Exciter gain
T_E	Exciter time constant
T_m and T_e	Mechanical and Electrical torque
T'_{do}	Field open circuit time constant
V_d and V_q	Direct and quadrature components of terminal voltage
K_1	Change in T_e for a change in δ with constant flux linkages in the d axis
K_2	Change in T_e for a change in d axis flux linkages with constant δ
K_3	Impedance factor
K_4	Demagnetizing effect of a change in rotor angle
K_5	Change in V_t with change in rotor angle for constant E'_q
K_6	Change in V_t with change in E'_q constant rotor angle

Appendix C

The system data are as follows:

Machine (p.u):

$$\begin{aligned} x_d &= 1.6 \quad x'_d = 0.32 \\ x_q &= 1.55 \quad T'_{d0} = 6\,\text{s} \\ \text{D} &= 0.0 \quad \text{M} = 10\,\text{s} \end{aligned} \tag{C1}$$

Transmission line (p.u):

$$\text{r}_\text{e} = 0.0 \quad \text{x}_\text{e} = 0.4 \tag{C2}$$

Exciter:

$$\text{K}_\text{E} = 25.00 \quad \text{T}_\text{E} = 0.05\,\text{s} \tag{C3}$$

Nominal Operating point:

$$\begin{aligned} V_{t0} &= 1.0 \quad P_0 = 0.8 \\ Q_0 &= 0.3 \quad \delta_0 = 45^\circ \\ &\qquad \omega_0 = 314 \end{aligned} \tag{C4}$$

Others

$$\begin{aligned} \text{k}_3 &= 1/2.78 \\ \text{v} &= 1.0 \\ \text{T}_\text{w} &= 5 \end{aligned} \tag{C5}$$

References

1. Aboelela, M. A. S., Ahmed, M. F., & Dorrah, H. T. (2012). Design of aerospace control systems using fractional PID controller. *Journal of Advanced Research, 3*(2), 185–192.
2. Azar, A. T., & Serrano F. E. (2015). Adaptive sliding mode control of the furuta pendulum. In: A. T. Azar & Q. Zhu (eds.), *Advances and applications in sliding mode control systems*, Studies in computational intelligence (Vol. 576, pp. 1–42). Berlin/Heidelberg: Springer-Verlag GmbH.
3. Barmish, B. R. (1994). New tools for robustness of linear systems. Macmillan Publisher.
4. Chen, G., & Malik, O. (1995). Tracking constrained adaptive power system stabilizer. *IEE Proceedings, Generation, Transmission and Distribution, 142*, 149–156.
5. Chen, S., & Malik, O. P. (1995). H∞ optimization-based power system stabilizer design. *IEE Proceedings, Generation, Transmission and Distribution, 142*, 179–184.
6. deMello, F. P., & Concordia, C. (1969). Concepts of synchronous machine stability as affected by excitation control. IEEE Transactions on Power Apparatus and Systems, PAS-88, 316–327.

7. Dorcak, L., Petras, I., Kostial, I., & Terpak, J. (2001). State space controller design for the fractional-order regulated system. In *Proceedings of the International Carpathian Control Conference* (pp. 15–20).
8. Doyle, J. C., Francis, B. A., & Tannenbaum, A. R. (1992). *Feedback control theory*. New York: Macmillan Press.
9. Duc, G., & Font, S. (1999). *H theory and μ- analyse, tools for robustness*. Paris: HERMES Science Publications.
10. El-Metwally, K. A., Elshafei, A. L., Soliman, H. M. (2006). A robust power-system stabilizer design using swarm optimization. *International Journal Of Modeling, Identification and Control, 1*(4).
11. Goldberg, D. E. (1989). *Algorithms in search, optimization, and machine learning*. Addison-Wiley Publishing Company, Inc.
12. Goldberg, D. E. (1991). *Genetic algorithms in search optimization and machine learning*. Reading, MA: Addison-Wesley Publishing Company, Inc.
13. Gosh, A., Ledwich, M. O., & Hope, G. (1989). Power system stabilizers based on adaptive control techniques. *IEEE Transactions on Power Apparatus and Systems, PAS-103, 8*, 1983–1989.
14. Klein, M., Rogers, G. J., Moorty, S., & Kundur, P. (1992). Analytical investigation of factors influencing PSS performance. *IEEE Transaction on EC, 7*(3), 382–390.
15. Kothari, M. L., Bhattacharya, K., & Nada, J. (1996). Adaptive power system stabilizer based on pole shifting technique. *IEE Proceedings, 143*, Pt. C, No. 1, 96–98.
16. Koza, J. R. (1991). Genetic evolution and co-evolution of computer programs. In C. G. Langton, C. Taylor, J. D. Farmer, & S. Rasmussen (Eds.), *Artificial life II: SFI studies in the sciences of complexity* (Vol. 10). Addison-Wesley.
17. Kundur, P. (1994). Power system stability and control. McGraw-Hill.
18. Lee, S. S., & Park, J. K. (1998). Design of reduced order observer based variable structure power system stabilizer for immeasurable state variables. *IEE Proceedings Conference Transmission and Distribution, 145*(5), 525–530.
19. MacFarlane, D. C., & Glover, K. (1992). A loop shaping design procedure using H∞ synthesis. *IEEE Transactions on Automatic Control, AC-37*, 759–769.
20. Malik, O. P., Chen, G., Hope, G., Qin, Y., & Xu, G. (1992). An adaptive self-optimizing pole shifting control algorithm. *IEE Proceedings of D, 139*, 429–438.
21. Mehran, R., Farzan, R., & Hamid, M. (2003). Tuning of power system stabilizers via genetic algorithm for stabilization of power systems. In *Proceedings of the IEEE International Conference on Systems, Man & Cybernetics* (pp. 649–654). Washington, D.C., USA, 5–8 October.
22. Milos, S., & Martin, C. (2006). The fractional order PID controller outperforms the classical one. In *7th International Scientific-Technical Conference–Process Control*, June 13–16, Kouty nad Desnou, Czech Republic.
23. Mrad, F., Karaki, S., & Copti, B. (2000). An adaptive fuzzy synchronous machine stabilizer. *IEEE Transactions on Systems, Man, and Cybernetics, 30*(1), 131–137.
24. Petras, I. (1999). The fractional order controllers: Methods for their synthesis and application. *Journal of Electrical Engineering, 50*(9), 284–288.
25. Petras, I., Lubomir, D., & Imrich, K. (1998). Control quality enhancement by fractional order. In *2nd National Conference on Recent Trends in Information Systems (ReTIS-08) Controllers. Acta Montanistica Slovaca, 3*(2), 143–148.
26. Petras, I., & Vinagre, B. M. (2002). Practical applications of digital fractional order controller to temperature control. In *Acta Montanistica, Slovaca Rocnik, 2*, 131–137.
27. Podlubny, I. (1999). Fractional-order systems and $PI^{\lambda}D^{\delta}$ controllers. *IEEE Trans. on Automatic Control, 44*(1), 208–213.
28. Podlubny, I. P., Petras, I., Blas, M. V., Yang-Quan, C., O' Leary, P., & Lubomir, D. (2003). Realization of fractional order controllers. *Acta Montanistica Slovaca, 8*.
29. Podlubny, I. P., Vinagre, B. M., O' Leary, P., & Dorcak L. (2002). Analogue realizations of fractional-order controllers. *Nonlinear Dynamics, 29*, 281–296.

30. Rashidi M., Rashidi F., & Monavar, H. (2003). Tuning of power system stabilizers via genetic algorithm for stabilization of power systems. In *Proceedings of the IEEE International Conference on Systems, Man & Cybernetics* (pp. 649–654). Washington, D.C., USA, 5–8 October.
31. Samarasinghe, V. G., & Pahalawaththa, N. C. (1997). Damping of multimodal oscillations in power systems using variable structure control techniques. *IEE Proceedings of Generation, Transmission and Distribution, 144*(3), 323–331.
32. Schlegel, M., & Cech, M. (2006). The fractional order PID controller outperforms the classical one. In *7th International Scientific-Technical Conference*, June 13–16, Kouty nad Desnou, Czech Republic.
33. Shamsollahi, P., & Malik O. P. (1999). Adaptive control applied to synchronous generator. *IEEE Transactions on Energy Conversion, I4*(4), l341–1346.
34. Shu, H., & Chen, T. (1997). Robust digital design of power system stabilizers. In *Proceedings of the American Control Conference*, Albuquerque, 1953–1957.
35. Soliman, H., Elshafei, A. L., Shaltout, A. A., & Morsi, M. F. (2000). Robust power system stabilizer. *IEE Proceedings, Generation, Transmission and Distribution, 147*(5), 285–291.
36. Soliman, H. M., & Sakr, M. M. F. (1988). Wide-range power system pole placer. *IEE Proceedings*, *135*, Pt. C, No. 3, 195–201.
37. Sun, C., Zhao, Z., Sun, Y., & Lu, Q. (1996). Design of non-linear robust excitation control for multi-machine power system. *IEE Proceedings, Generation, Transmission and Distribution, 143*, 253–257.
38. Valério, D., & Sá Da Costa, J. (2004). Ninteger: A fractional control toolbox for Mat lab. In *First IFAC Workshop on Fractional Differentiation and Its Applications*. Bordeaux: IFAC.
39. Vinagre, B. M., Podlubny, I., Dorcak, L., & Feliu, V. (2000). On fractional PID controllers: A frequency domain approach. In *Proceedings of IFAC Workshop on Digital Control—Past, Present and Future of PID Control* (pp. 53–58).
40. Xue, D., Chen, Y., & Atherton, D. P. (2007). *Linear Feedback Control. Society for Industrial and Applied Mathematics*. Philadelphia.
41. Young-Hyun, M., Heon-Su, R., Jong-Gi, L., Kyung-Bin, S., & Myong-Chul, S. (2002). Extended integral control for load frequency control with the consideration of generation-rate constraints. *International Journal of Electrical Power & Energy Systems, 24*(4), 263–269.

Application of Fractional Order Controllers on Experimental and Simulation Model of Hydraulic Servo System

M. El-Sayed M. Essa, Magdy A.S. Aboelela and M.A.M. Hassan

Abstract Hydraulic Servo System (HSS) plays an important role in industrial applications and other fields such as plastic injection machine, material testing machines, flight simulator and landing gear system of the aircraft. The main reason of using hydraulic systems in many applications is that, it can provide a high torque and high force. The hydraulic control problems can be classified into force, position, acceleration and velocity problems. This chapter presents a study of using fractional order controllers for a simulation model and experimental position control of hydraulic servo system. It also presents an implementation of a non-linear simulation model of Hydraulic Servo System (HSS) using MATLAB/SIMULINK based on the physical laws that govern the studied system. A simulation model and experimental hardware of hydraulic servo system have been implemented to give an acceptable closed loop control system. This control system needs; for example, a conventional controller or fractional order controller to make a hydraulic system stable with acceptable steady state error. The utilized optimization techniques for tuning the proposed fractional controller are Genetic Algorithm (GA). The utilized simulation model in this chapter describes the behavior of BOSCH REXROTH of Hydraulic Servo System (HSS). Furthermore the fractional controllers and conventional controllers will be tuned by Genetic Algorithm. In addition, the hydraulic system has a nonlinear effect due to the friction between cylinders and pistons, fluid compressibility and valve dynamics. Due to these effects, the simulation and experimental results show that using fractional order controllers will give better response, minimum performance indices values, better disturbance rejection, and better sinusoidal trajectory than the conventional PID/PI controllers. It also shows

M.E.-S.M. Essa (✉)
IAET, Imbaba Airport, Giza, Egypt
e-mail: mohamed.essa@iaet.edu.eg

M.A.S. Aboelela · M.A.M. Hassan
Faculty of Engineering, Electric Power and Machines Department,
Cairo University, Giza, Egypt
e-mail: aboelelamagdy@gmail.com

M.A.M. Hassan
e-mail: mmustafa_98@hotmail.com; mmustafa@eng.cu.edu.eg

A.T. Azar et al. (eds.), *Fractional Order Control and Synchronization of Chaotic Systems*, Studies in Computational Intelligence 688,
DOI 10.1007/978-3-319-50249-6_10

that the fractional controller based on Genetic Algorithm has the desired robustness to system uncertainties such as the perturbation of the viscous friction, Coulomb friction, and supply pressure.

Keywords Hydraulic servo system · Genetic Algorithm · PID · Fractional order controllers · MATLAB/SIMULINK · BOSCH REXROTH

1 Introduction

Hydraulic control systems are widely used in many industrial fields due to their small size-to-power ratio and the ability to apply very large force and torque. The Hydraulic Servo System (HSS) applications include: manufacturing systems, material test machines, active suspension systems, mining machinery, fatigue testing, paper machines, injection molding machines, robotics, and aircraft fields.

In hydraulic control system, the main purpose of control is to achieve a desirable response from the system. In light of this requirement; the development of the controller has been established for adjusting measured response to be as close as possible to the desired response. The control signal errors are generally compared with velocity, position, force, pressure, and other system parameters. An HSS is a system consisting of motor, servo, controller, power supply, and other system accessories [19]. In HSS, the system controls the cylinder position to track a certain position trajectory values enforced by the operator. The cylinder movement must precisely follow position, speed, and acceleration profiles. Controller tuning was one of the difficulties that has been faced during implementation of the controllers. Many approaches have been developed for tuning the controller response optimally. This ranges from trial and error, root locus, Zeigler Nicholas (ZN) method and evolutionary techniques [33]. Evolutionary techniques have been evolved from observing complex behaviors of human and other animals, event happening in nature and arrive at a mathematical model representing criteria under study [40]. One of the evolutionary techniques is Swarm Intelligence (SI), which models social behavior of organisms living in swarms.

In the field of hydraulic system control, a wide selection of control design techniques and applications have been figured out. Electro-hydraulic problems are classified into many control problems such as:

(a) Position control problems.
(b) Velocity control problems.
(c) Force control problems.

Due to the importance of hydraulic systems in industrial applications, so many researchers have studied HSS. The dynamics of hydraulic systems are highly nonlinear as stated in [39] and the system may be subjected to non-smooth and discontinuous nonlinearities due to directional change of valve opening, friction…

etc. There have been some studies on analysis and implementation of the nonlinear tracking control law for HSS. This provided exponential stability for force tracking and position tracking to furnish an accurate friction model [39].

In this chapter, the experimental setup of cylinder load has different connection with the recent research. There are different types of automatic controllers that have been applied to HSS to give accurate tracking of position, acceleration, pressure and force. A mathematical modeling and simulation of HSS have been implemented to obtain the observed system response with sinusoidal input. It is then used to design a PID controller based on GA as introduced in [20]. The mathematical modeling of HSS and experimental setup have been developed for force tracking control using the nonlinear fuzzy controller as given in [2]. Whilst the using of Particle Swarm Optimization (PSO) to design an optimal robust PI-controller for HSS that achieves both the robustness and performance measures has been explained in [30]. The using of PSO technique has been extended to identify controller's gains for the Scott Russell mechanism as investigated in [16]. The objective of the HSS controller is to give almost a zero steady state error in motion of the actuator and force output. Thus, these requirements have been satisfied by using PSO based on H_{∞} loop shaping control for MIMO HSS, as stated in [31]. The enhancement of stability and robustness of HSS by utilizing the fuzzy strategy approximation for antibodies inhibit adjustment function with immune algorithm based on PSO for PID tuning has been presented in [45]. While the study of external torque of hydraulic actuator and then design a controller using modern control theory have been introduced in [41]. The using of Fuzzy Logic Controller (FLC) for position control of electro hydraulic actuator and ant colony optimization technique that is used to attain the best value for parameters of fuzzy neural network has been stated in [25]. The improvement of position tracking performance based on invariance principle and feed-forward compensation is developed by pole-zero placement theory of the system as described in [44]. A hydraulic position servo system control is implemented by utilizing a Particle Swarm Optimization (PSO) algorithm for control PID loops is presented in [35]. The force control of hydraulic servo system is implemented by designing fuzzy controllers to minimize the force overshoot and preserve the load from failure as illustrated in [8]. The two most common approaches that have been developed to compensate the nonlinear behavior of HSS are adaptive control and variable structure control. The acceleration feedback control by using the variable structure controller in the presence of important friction nonlinearities is introduced and described in [7]. A nonlinear controller based on Lyapunov stability theories that considers the valve's dynamics is used for position control of HSS, as stated in [37].

The dynamic characteristics of HSS are usually very complex and highly nonlinear, so a self organizing and self learning fuzzy algorithm for position control of hydraulic servo drive is represented and discussed in [10]. A sliding mode control, enhanced by the fuzzy PI controller to a typical position control of electro-hydraulic system is confirmed in [27]. Whereas the optimization of PID controller parameters and overcomes of the nonlinearities of HSS based on GA are explained in [3].

A simulation Model of position control of HSS with MATLAB/SIMULINK program is performed and the model is verified experimentally using the Data Acquisition card.

The HSS real time consists of the following hydraulic elements:

- Oil tank with capacity of 100 L.
- Pressure, Temperature and Flow Displays.
- Filter for oil return.
- An axial piston pump swash plate type with variable flow rate pump.
- Servo valve with electrical position feedback (−10 to 10 V), Type 4WRSE, both are made by Rexroth Bosch.
- Pressure relief valve.
- Two hydraulic cylinders with face-to-face connection.
- External length measurement (Position transducer).
- Pressure, Temperature and Flow sensors.

The two cylinders are connected in such away to simulate hydraulic symmetric linear actuator. In addition, the nominal oil pressure is 10 Map. The oil pump is driven by three phase electrical motor 5.5 kW at 1500 rpm. The measuring system consists of one length transducer (measurement range 0–500 mm) which is connected to the piston rod. It is supplied from 24 VDC to generate an electrical signal from −10 to 10 V. The measuring signals are acquired by Data Acquisition card (PCI-NI 6014) from National Instruments with sampling rate 200 ski/s, and then sent to the PC-HP with 1 GB RAM, Windows XP operating system on 2.72 GHz processor.

1.1 Objectives of the Chapter

The objectives of The Study can be summarized as:

- Investigate a simulation model for HSS.
- Prepare the HSS for laboratory testing.
- Develop a Genetic Algorithm (GA) based PID/PI and FOPID/FOPI controllers tuning methodology for optimizing the control of simulated HSS and real time HSS.

1.2 Organization of the Chapter

Section 1 presents an introduction to the chapter. While Sect. 2 displays a description of Hydraulic servo system. Section 3 explains the mechanical and experimental setup. Whilst, system controllers design is given in Sect. 4 but the tuning method of the proposed controllers is displayed in Sect. 5. Section 6 presents application of GA to hydraulic servo system. While conclusion and future work are given in Sects. 7 and 8 respectively.

2 Hydraulic Servo Systems (HSS)

Electro-hydraulic problems are classified into position control problems, velocity control problems and force control problems. The common types of electro-hydraulic servos are [13]:

- Position servo (linear or angular)
- Velocity or speed servo (linear or angular)
- Force or torque servo.

2.1 Modeling and Simulation of HSS

A mathematical model of a HSS is presented, which includes the most non-linear effects that are involved in the hydraulic system. The problem that has been studied is illustrated in Figs. 1 and 2. The objective of the modeling and simulation of the electro-hydraulic servo system is to design a suitable controller for piston position control. In this section a nonlinear model of a HSS is developed by simulation using SIMULINK/MATLAB program. For more details about the same model but for force tracking control that is illustrated and discussed in [2]. The model describes the behavior of a servo system BOSCH REXROTH [22] servo valve and includes the nonlinearities due to friction forces, valve dynamics, oil compressibility and load influence.

Figure 3 illustrates a focus view of hydraulic cylinder connection and real photo of valve and cylinders connection inside the laboratory. An electro hydraulic servo valve under regulated supply pressure P_s drives the double rod cylinder. Two cylinders can achieve a double rod cylinder configuration, which are mounted into

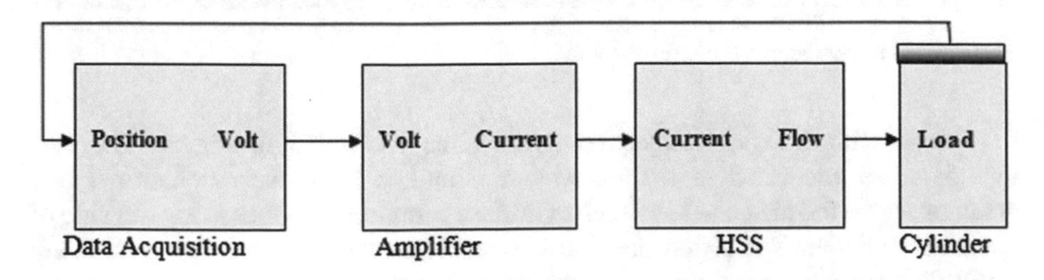

Fig. 1 Block diagram of hydraulic servo system

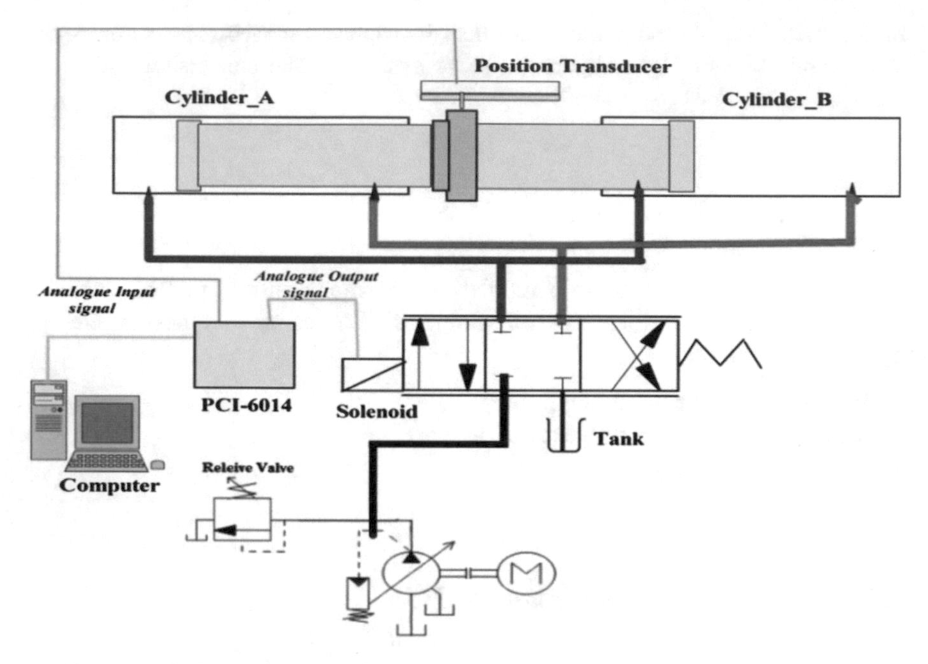

Fig. 2 Schematic diagram of the experimental system

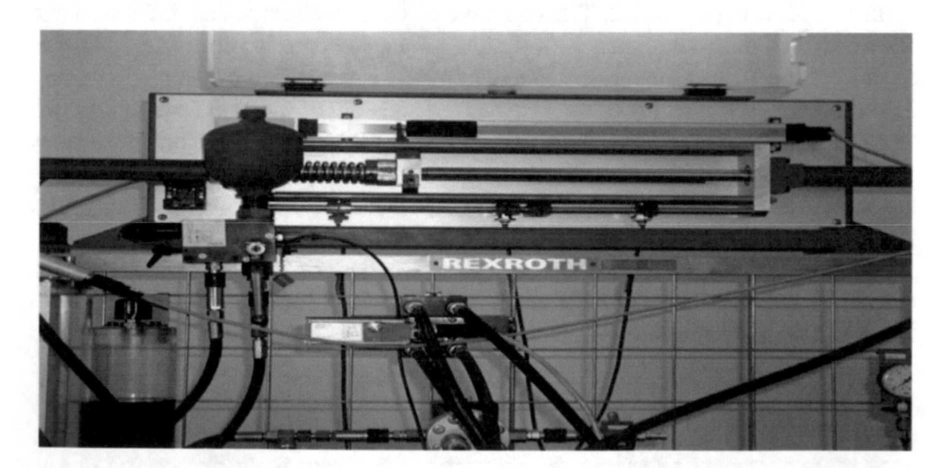

Fig. 3 Real hydraulic servo system

the frame with the face-to-face connection, as illustrated in Fig. 3. The two cylinders are connected in such a way to simulate hydraulic symmetric linear actuator where the piston side of each cylinder is connected to the piston rod side of the other cylinder. The piston position is considered the feedback signal by using linear displacement transducer. The amount of flow rate Q_A in chamber (A) and the

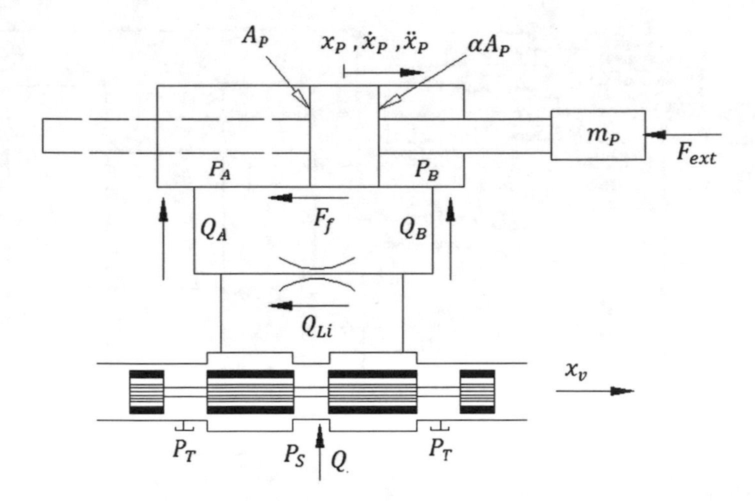

Fig. 4 Valve-cylinder combinations with variables definitions

other amount flow rate Q_B in chamber (B) of the cylinder are function of both the valve spool position x_v and the cylinder pressures P_A and P_B. The objective of this system is to control the piston position of a hydraulic cylinder to track a desired position as closely as possible.

The simplification of nonlinear HSS modeling based on standard assumptions in practical are summarized as:

(1) Low frequency operation.
(2) Pipeline effects do not play a role in the input-output behavior.
(3) Ideal oil supply, constant pressure supply and constant tank pressure.
(4) The possible dynamic behavior of the pressure in the pipelines between valve and actuator is assumed negligible.

Due to the previous assumption, the model of a HSS is composed of two subsystems (valve and cylinder) as shown in Fig. 4 and explained in [12, 15]. The Complete block diagram of HSS is illustrated in Fig. 5.

Where

A_P	Piston area (m^2)
α	Ratio of ring side area to piston side area
m_P	Piston mass (kg)
P_A	Pressure in chamber A (Pa)
P_B	Pressure in chamber B (Pa)
P_S	Supply pressure (Pa)
P_T	Tank pressure (Pa)
Q_A	Flow rate in chamber A (m^3/s)

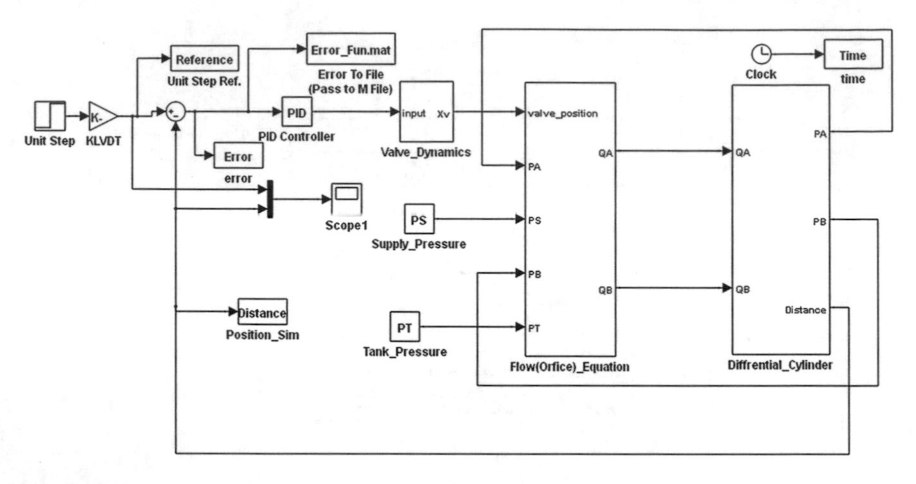

Fig. 5 Complete hydraulic servo system model

Q_B	Flow rate in chamber B (m^3/s)
Q_{Li}, Q_{Le}	Internal and external leakage flow (m^3/s)
$x_P, \dot{x}_P, \ddot{x}_P$	Piston position, velocity, acceleration, respectively (m)
x_v	Valve spool position (m)
F_{ext}	External force (N).

2.1.1 Hydraulic Cylinder Modeling

The hydraulic cylinder includes the pressure, dynamic modeling, the load equally and the piston friction with the cylinder. The differential equations governing the dynamics of the actuator are given in [15, 26]. More details about the hydraulic cylinder modeling can be found in [2, 13]. The total overview of the differential cylinder SIMULINK model is displayed in Fig. 6.

2.1.2 Pressure Dynamics of Hydraulic Chamber

The pressure dynamics equations for the chamber (A) and chamber (B) are displayed in Eqs. (1) and (2).

$$\dot{P}_A = \frac{1}{C_{hA}}(Q_A - A_P\dot{x}_P + Q_{Li} - Q_{LeA}) \tag{1}$$

$$\dot{P}_B = \frac{1}{C_{hB}}(Q_B + \alpha A_P\dot{x}_P - Q_{Li} - Q_{LeB}) \tag{2}$$

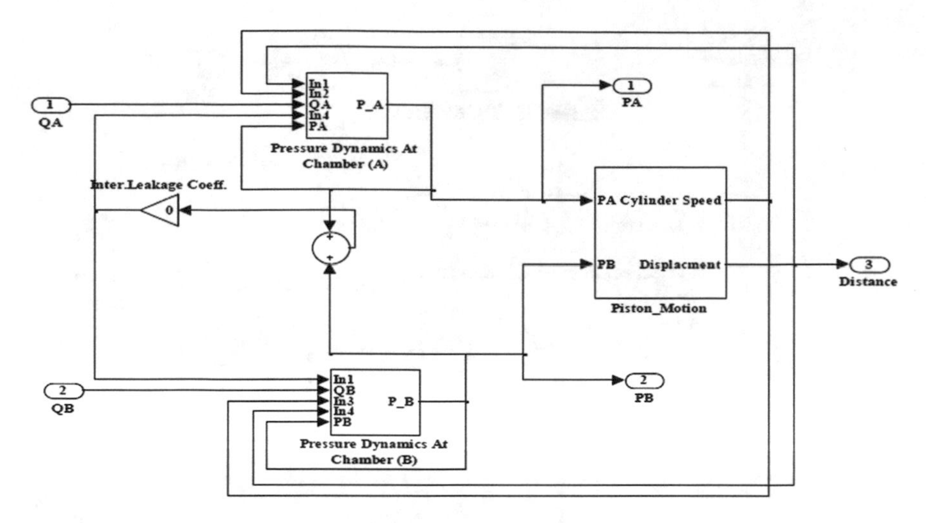

Fig. 6 Block diagram of differential cylinder

where: Q_{Li} is the internal leakage flow. Assume that external leakage flow Q_{LeA} and Q_{LeB} are negligible may be due to the high performance of REXROTH Equipment Company. The hydraulic capacitance of chamber A, C_{hA}, and chamber B, C_{hB}, are given by Eqs. (3) and (4).

$$C_{hA} = C_h(P_A, x_P) = \frac{V_A(x_p)}{\overline{\beta}(P_A)} = \frac{V_{P1,A}(x_{P0} + x_P)A_P}{\overline{\beta}(P_A)} \tag{3}$$

$$C_{hB} = C_h(P_B, x_P) = \frac{V_B(x_p)}{\overline{\beta}(P_B)} = \frac{V_{P1,B}(x_{P0} - x_P)\alpha A_P}{\overline{\beta}(P_B)} \tag{4}$$

$$V_A = V_{P1,A} + \left(\frac{S}{2} + x_P\right)A_P = V_{A0} + x_P A_P \tag{5}$$

$$V_B = V_{P1,B} + \left(\frac{S}{2} - x_P\right)\alpha A_P = V_{B0} - x_P \alpha A_P \tag{6}$$

where: S is the cylinder stroke. $V_{P1,A}$ and $V_{P1,B}$ are the pipeline volumes at A-side and B-side respectively. The initial chamber volumes are assumed that the piston is centered such that these are equal. That is:

$$V_{A0} = V_{B0} = V_0 \tag{7}$$

The commonly used equation for calculation the effective bulk modulus β for hydraulic cylinders is given by Eq. (8) as given in [26].

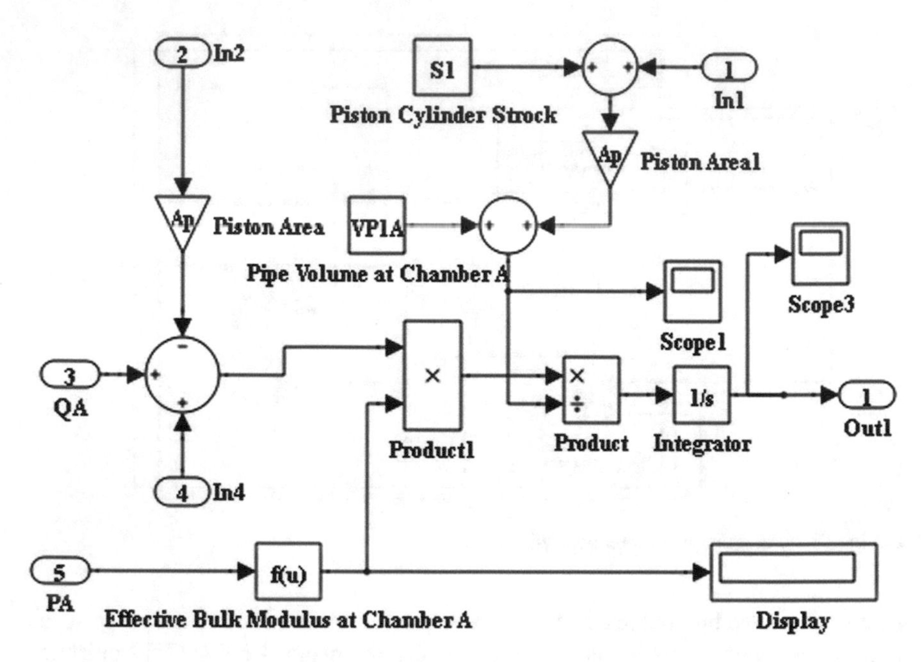

Fig. 7 Pressure dynamics of Side A model

$$\overline{\beta}(p) = a_{1\beta_{max}} \log(a_2 \frac{p}{p_{max}} + a_3) \tag{8}$$

where $a_1 = 50$, $a_2 = 90$, $a_3 = 3$, $\beta_{max} = 1800$ MPa, and $p_{max} = 28$ MPa. The simulation model of dynamic pressure in chamber A and B are illustrated in Figs. 7 and 8.

2.1.3 Load Equation

Equation (9) illustrates the equation of piston motion which governing the load motion. After applying the Newton's second law to the forces that applied to the piston, the resultant equation is given as follows [15].

$$m_t \ddot{x}_P + K_S x_p + F_f(\dot{x}_P) = (P_A - \alpha P_B) A_P \tag{9}$$

where

K_S Spring stiffness
F_f Friction force
m_t Total mass.

In Eq. (9), there is an external force (F_{ext}) equal to $K_S x_p$ which has been applied as an input force or a disturbing force on the piston. It is achieved by connecting a

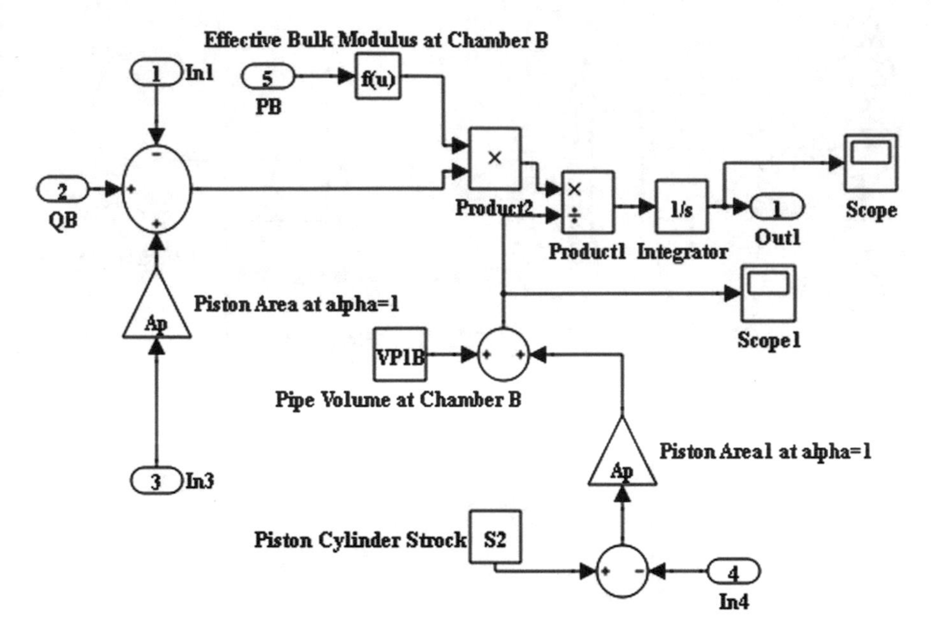

Fig. 8 Pressure dynamics of Side B model

spring at the outer end of the piston. For simplicity, we assume this external force to be zero. This means that, the external force is neglected.

The total mass m_t consists of the piston mass m_P, the mass of hydraulic fluid in the cylinder chambers and in the pipelines, m_{Afl} and m_{Bfl} respectively. Assume the mass of fluid is neglected compared to the piston mass.

$$m_t = m_P + m_{Afl} + m_{Bfl} \tag{10}$$

From Eq. (9) The SIMULINK model of piston, motion is presented in Fig. 9.

2.1.4 Piston Friction

The asymmetry of the friction forces that occurs in differential cylinders can be represented by using one experimental function with referred to stribeck curve as illustrated in Eq. (11). The friction model is shown in Fig. 10 and explained by Jelali and Kroll [15], Merritt [26].

$$F_f(\dot{x}_P) = \left\{ \begin{array}{l} \sigma^+ \dot{x}_P + Sign(\dot{x}_P)\left[F_{CO}^+ + F_{SO}^+ exp\left(-\frac{|\dot{x}_P|}{C_S^+}\right)\right] \forall \dot{x}_P \geq 0 \\ \sigma^- \dot{x}_P + Sign(\dot{x}_P)\left[F_{CO}^- + F_{SO}^- exp\left(-\frac{|\dot{x}_P|}{C_S^-}\right)\right] \forall \dot{x}_P < 0 \end{array} \right\} \tag{11}$$

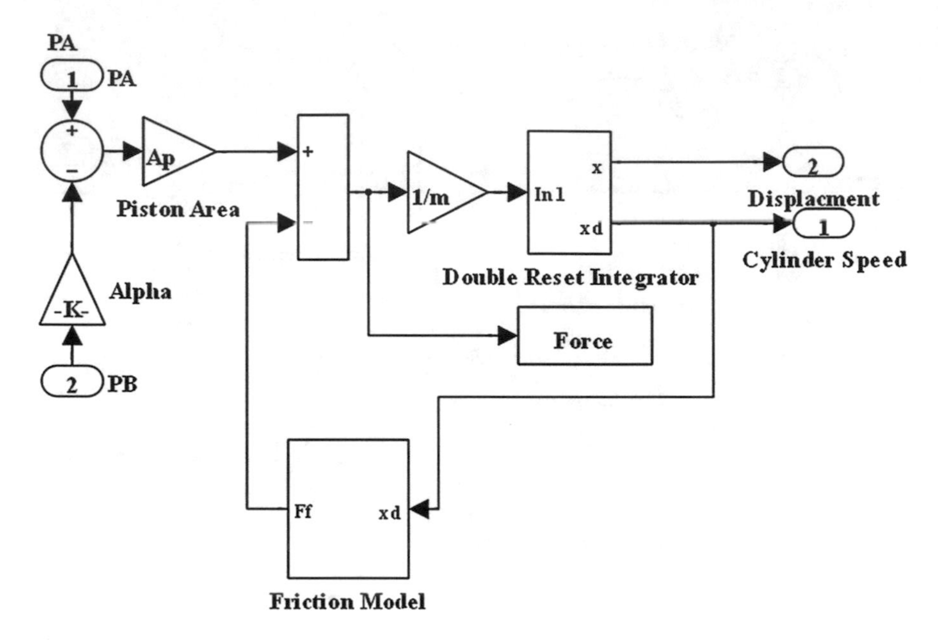

Fig. 9 Piston motion model

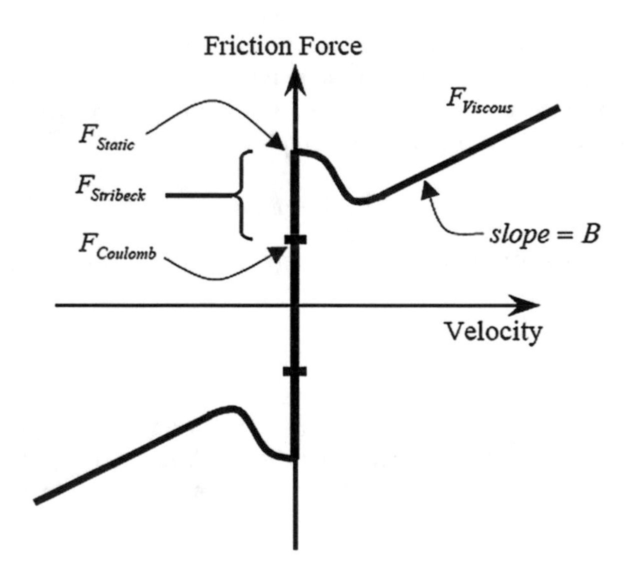

Fig. 10 Friction model

where

σ	Viscous friction parameter
F_{CO}^{+}, F_{CO}^{-}	Differential cylinder Coulomb friction parameter
F_{SO}^{+}, F_{SO}^{-}	Differential cylinder Stribeck friction parameter
C_{S}^{+}, C_{S}^{-}	Stribeck velocity range.

An auxiliary force is required to be added to the friction function to prevent the non unique relation between $\dot{x}_P$ and F_f at $\dot{x}_P = 0$ and between x_p and F_c at $x_p = 0$ and then capable to calculate F_f and F_c. More details about auxiliary force are explained in [15]. The friction model SIMULINK block diagram is presented in Fig. 11.

The Figs. 7, 8, 9 and 11 can be arranged to form the total block diagram of a hydraulic cylinder as illustrated in Fig. 6.

2.1.5 Servo Valve Model

The type of the utilized servo valve type (4WRSE) is a four-way spool valve with a critical center, which it is illustrated in Fig. 12.

The classical continuity equation, which governs flow direction in the servo valve, is presented in Eqs. (12), (13), (14) and (15).

$$Q_A = Q_1 - Q_2 \tag{12}$$

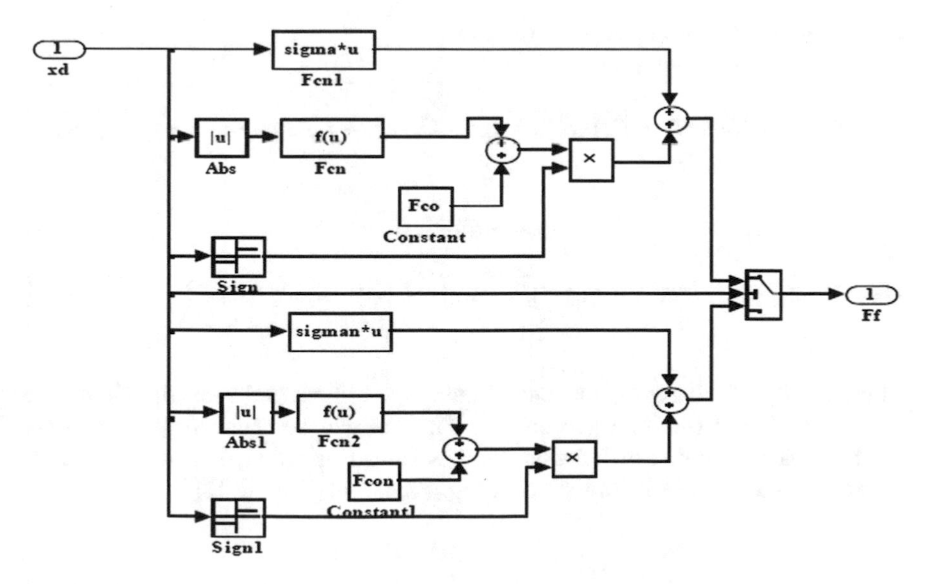

Fig. 11 Friction model diagram

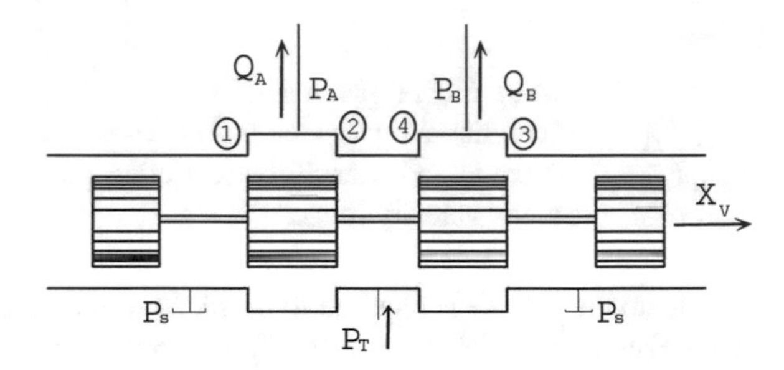

Fig. 12 Zero lapped four ports spool valve

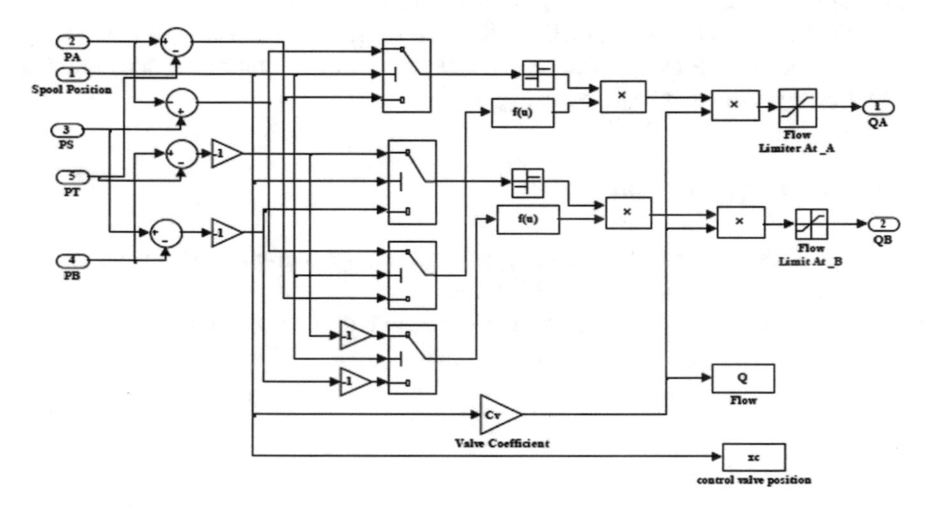

Fig. 13 Simplified block diagram of servo valve

$$Q_A = C_{V1} sg(x_v) Sign(P_S - P_A)\sqrt{|P_S - P_A|} - C_{V2} sg(-x_v) Sign(P_A - P_T)\sqrt{|P_A - P_T|} \tag{13}$$

$$Q_B = Q_3 - Q_4 \tag{14}$$

$$Q_B = C_{V3} sg(-x_v) Sign(P_S - P_B)\sqrt{|P_S - P_B|} - C_{V4} sg(x_v) Sign(P_B - P_T)\sqrt{|P_B - P_T|} \tag{15}$$

where C_{Vi} is a discharge coefficients of the valve orifices and subscript (i) takes a value from 1 to 4 (no. of valve orifice). The C_{Vi} will be equal if all orifices are identical. The definition of function $sg(x)$ is shown in Eq. [16]. Simplified block diagram of servo valve is illustrated in Fig. 13 and is given in [2].

$$sg(x) = \begin{Bmatrix} x & for & x \geq 0 \\ 0 & for & x < 0 \end{Bmatrix} \tag{16}$$

For simplicity, a first order model for the servo valve is development using system identification toolbox in MATLAB to capture most of dynamic behavior that includes large number of parameters as given in [2, 20]. Equation (17) shows the general form of first order model as given in [2]. The development model of valve dynamics is introduced in Eq. (18).

$$\dot{x}_v = \frac{1}{\tau}x_v + \frac{K_v}{\tau}u \tag{17}$$

where τ is the time constant, K_v is the valve gain and u is the valve input signal. Considering the valve dynamics in Eq. (17) with a time constant 0.0033 s and valve gain 0.98 that yields the resulting transfer function in Eq. (18).

$$\frac{x_v(s)}{u(s)} = K_{Conv.}\frac{300.7}{s + 306.8} \tag{18}$$

where the input u(s) is the command voltage input for the valve and the conversion factor, $K_{Conv.}$ converts the voltage reading out of the valve Linear Variable Differential Transducer (LVDT) to actual spool displacement in meters. The type of valve center is defined by the width of the land compared to the width of the port in the valve sleeve when the spool is in neutral position. The utilized type is a critical-center or zero-lapped valve which has a land width identical to the port width.

At this end, from Eq. (1) to Eq. (17) can be combined to form a total simulated model of HSS. Finally, the complete block diagram of HSS consists of the main following block diagrams.

- Differential Cylinder Block Diagram.
- Valve Dynamics Block Diagram.
- Flow Orifice Block Diagram.

3 Mechanical and Experimental Setup

The hydraulic power unit is illustrated in Fig. 14 and a real time picture of the experimental HSS is illustrated in Fig. 15 and the system components are shown in Fig. 16.

The experimental hydraulic system is mainly consists of the following components as described by [13]:

- Oil tank with capacity of 100 L.
- Pressure, Temperature and Flow Displays.
- Filter for return oil.
- An axial piston pump swash plate type with variable flow rate pump A10VSO.

Fig. 14 Hydraulic power unit

Fig. 15 Real time picture of the experiment HSS

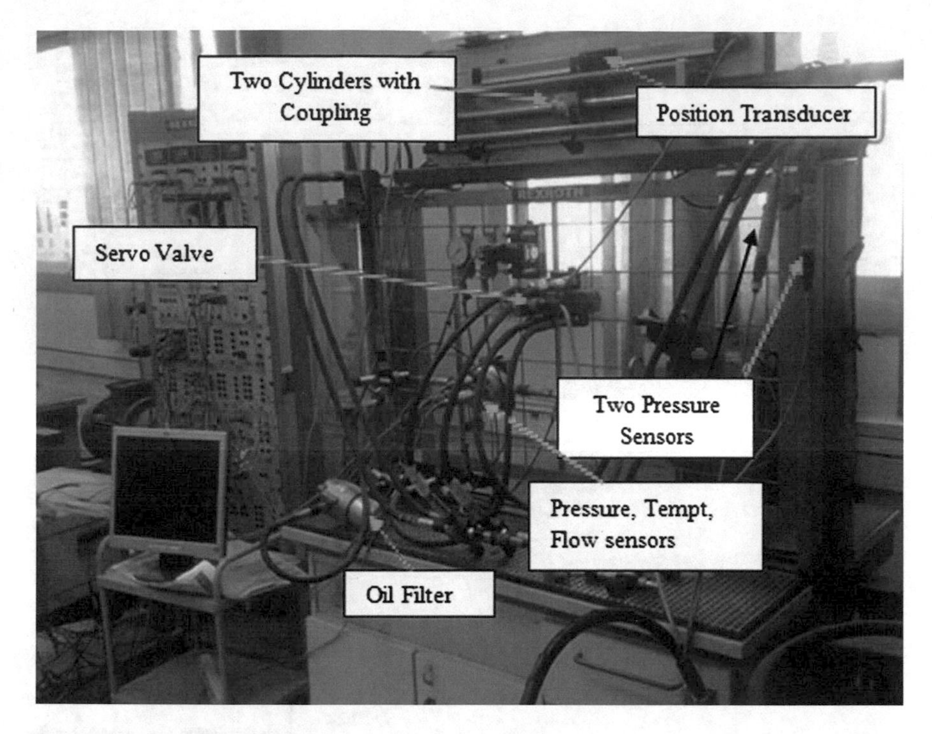

Fig. 16 Real time picture of the experiment components

- Servo valve with electrical position feedback (−10 to 10 V), Type 4WRSE, both are made by Rexroth Bosch.
- Pressure relief valve.
- Two hydraulic cylinders with face-to-face connection.
- External length measurement (Position transducer).
- Pressure, Temperature and Flow sensors.

The main purpose of experimental setup is to verify the simulation model for piston position of HSS and applying the controller design on practical system. Real time photos of the experimental HSS are illustrated in Figs. 14, 15 and 16. In the Experimental system, the two cylinders are connected to simulate hydraulic symmetric linear actuator. The utilized nominal oil pressure is 10 MPa and the oil pump is driven by a three phase electrical motor 5.5 kW at 1500 rpm. The measured system consists of one length transducer has the range of 0–500 mm, which connected to the piston rod, as illustrated in Fig. 15. When a 24 V supplies the transducer, it generates a signal from −10 to 10 V. Data Acquisition Card (PCI-NI 6014) from National Instruments [28] acquires the measuring signals, and then they are sent to the hp-PC with 2.71 GHz processor, 2 GB RAM, and operating system Windows XpSP3. It has a sampling rate (200 kS/s), number of channels (16 single-ended or 8 differential) and 16 bit resolution. The final SIMULINK/ MATLAB model of HSS is illustrated in Fig. 17.

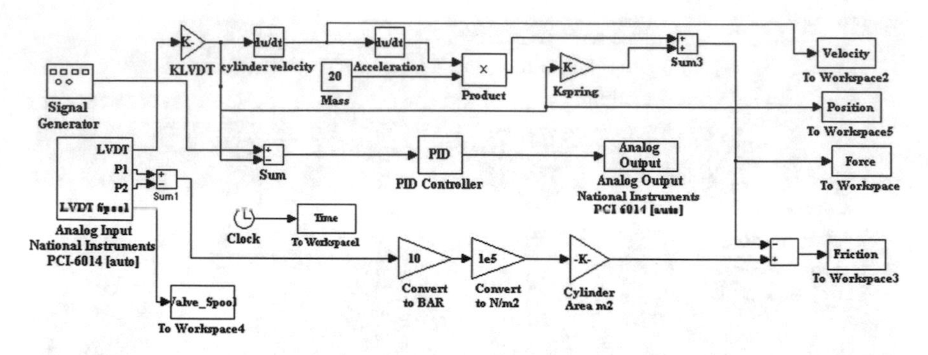

Fig. 17 SIMULINK model of experimental HSS

Fig. 18 System identification

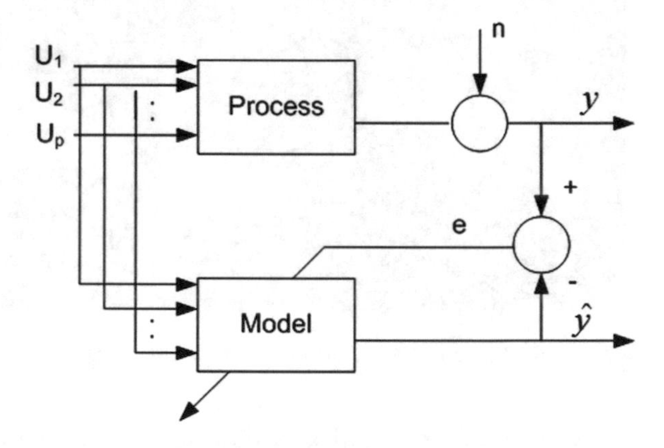

3.1 System Identification

The System Identification allows to build a mathematical models of a dynamic system based on measured data. The model quality is typically measured in terms of the error between the (disturbed) process output and the model output. This error is utilized to adjust the parameters of the model. Schematic diagram of system identification definition is illustrated in Fig. 18 [13, 29].

The main steps that have to be performed for successful identification of a system are illustrated in Fig. 19 and explained in details by Ljung [18].

The purpose of this step is to collect a set of input/output data that describes how the system acts over its entire range of operation. The idea is to motivate the system with a random input u, and observe the impact on the output y.

3.2 Model Representations for System Identification

System identification can be classified into two approaches based on model representation. The first one is input-output model form which is identical to the

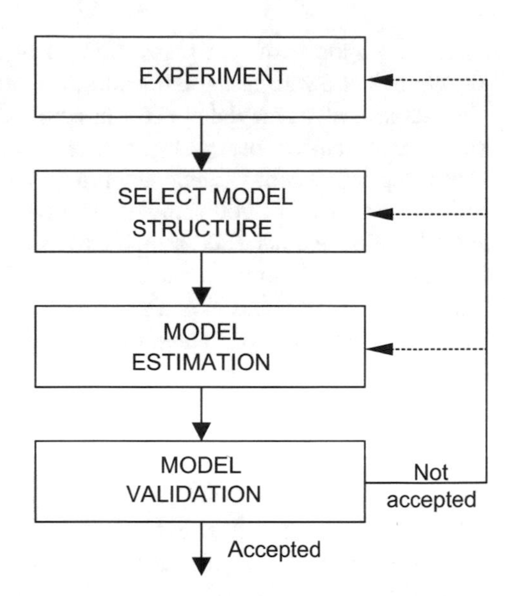

Fig. 19 System identification steps [13]

transfer function representation. The second approach produces models in state space form. Models with state space representation allow identification of multi input multi output (MIMO) systems. The first approach has been utilized in position control of HSS. System Identification Toolbox constructs mathematical models of dynamic systems from measured input-output data. It provides MATLAB functions, SIMULINK blocks, and an interactive tool for creating and using models of dynamic systems not easily modeled from first principles or specifications. Time domain and frequency domain input-output data can be used to identify continuous time, discrete time transfer functions, process models, and state-space models.

3.3 HSS Identification for Position Control

The process models in the system identification toolbox [20] are used to build a continuous time model. It has been used to build and estimate a continuous transfer function for the position control of HSS. Process models consist of the basic type static gain, time constant and time delay as presented in [21]. The mathematical representation of process model is illustrated in Eq. (19). The process model with integrator is described in Eq. (20).

$$P_1(s) = \mathrm{K} \cdot \mathrm{e}^{-\mathrm{T_d} * \mathrm{s}} \frac{1 + \mathrm{T_Z} * \mathrm{s}}{(1 + \mathrm{T_{P1}} * \mathrm{s})(1 + \mathrm{T_{P2}} * \mathrm{s})} \tag{19}$$

$$P_2(s) = \mathrm{K} \cdot \mathrm{e}^{-\mathrm{T_d} * \mathrm{s}} \frac{1 + \mathrm{T_Z} * \mathrm{s}}{\mathrm{s} \cdot (1 + \mathrm{T_{P1}} * \mathrm{s})(1 + \mathrm{T_{P2}} * \mathrm{s})} \tag{20}$$

where the number of real poles (0, 1, 2 or 3) can be determined, as well as the occurrence of a zero in the numerator, the presence of an integrator term (1/s) and the presence of a time delay (Td). In addition, an under damped (complex) pair of poles may replace the real poles. The excitation signal for identification is multi-step signal with variable amplitudes (−2.5 to 2.5 V) and variable frequencies and over arrange of 3000 samples. The first 1500 sample are used to estimate the model while the other 1500 sample are used to the validation step. The experiment is done in closed loop. To calculate the estimated model, the percentage Best Fit (BF) criterion is used as explained in [18]. It measures how much better the model describes the process compared to the mean of the output. The Best Fit description is illustrated in Eq. (21).

$$\text{Best Fit} = \left(1 - \frac{|y - \hat{y}|}{|y - \bar{y}|}\right) \times 100 \tag{21}$$

where y is the measured output, $\hat{y}$ is the simulated or predicted model output, and $\bar{y}$ is the mean of y. A part of measured and simulated outputs is illustrated in Fig. 20 for the identified 3rd order model with integrator. It shows that the model perfectly captures most of the dynamics of the system. The measured and simulated output is illustrated in Fig. 21. The identified continuous-time model here gives Best Fit of 91.88%, which it is an acceptable result.

After the above identification, Eq. (22) introduces the identified continuous time transfer function model and then this equation is discretized to be in z-domain as shown in Eq. (23).

$$\frac{x_P(s)}{u_v(s)} = \frac{1520\,s + 100}{s\,(s^3 + 93.2\,s^2 + 1122\,s + 45.32)} \tag{22}$$

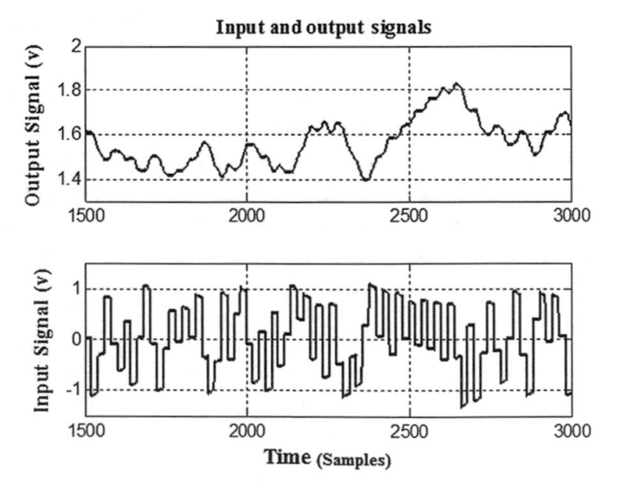

Fig. 20 Input and output signals for HSS identification

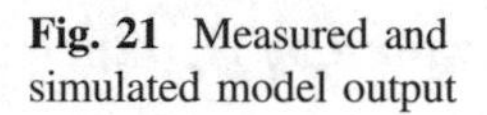

Fig. 21 Measured and simulated model output

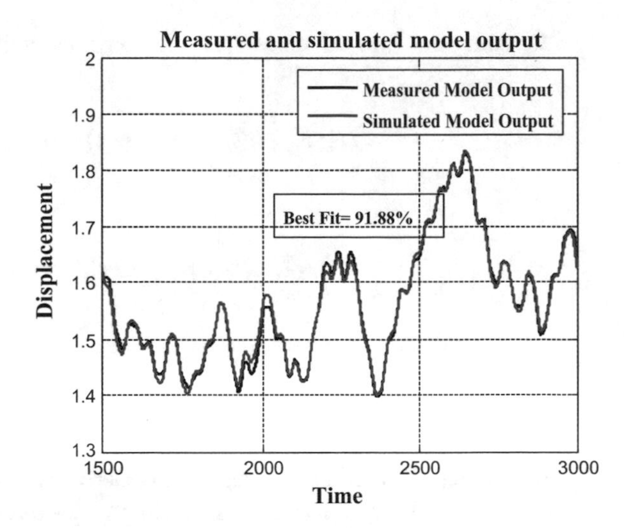

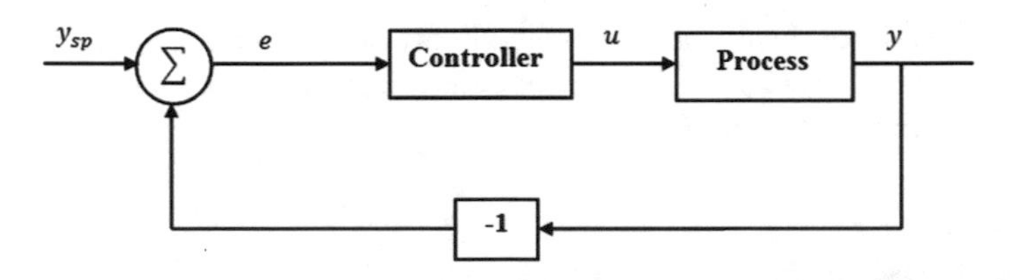

Fig. 22 Block diagram of a process with a feedback controller [12]

$$\frac{x_P(z)}{u_v(z)} = \frac{0.000202\,\mathrm{z}^{-1} + 0.0004\,\mathrm{z}^{-2} - 0.0005\,\mathrm{z}^{-3} - 0.0001\,\mathrm{z}^{-4}}{1 - 3.321\,\mathrm{z}^{-1} + 4.037\,\mathrm{z}^{-2} - 2.109\,\mathrm{z}^{-3} + 0.3938\,\mathrm{z}^{-4}} \tag{23}$$

4 System Controllers Design

4.1 PID Controller

The PID controller abbreviation is a proportional–Integral–Derivative controller. PID controller is the most common controller form of feedback control system, which is widely used in industrial control systems [43]. The objective of using PID controller is to minimize the difference between a measured process variable and a desired set point by adjusting the process control inputs [5]. Block diagram of a process with a feedback controller is illustrated in Fig. 22 and depicted in [1].

PID controller form is represented mathematically and described by [1, 42]:

$$u(t) = \mathrm{K_p}\left(\mathrm{e(t)} + \frac{1}{\mathrm{T_i}}\int_0^t \mathrm{e}(\tau)\mathrm{d}\tau + \mathrm{T_d}\frac{\mathrm{de(t)}}{\mathrm{dt}}\right) \tag{24}$$

$$u(t) = \mathrm{K_p e(t)} + \mathrm{K_i}\int_0^t \mathrm{e}(\tau)\mathrm{d}\tau + \mathrm{K_d}\frac{\mathrm{de(t)}}{\mathrm{dt}} \tag{25}$$

where

$u(t)$	Controller output
y(t)	System output
K_p, K_i and K_d	Proportional, Integral and Derivative coefficients respectively
T_i, T_d	Integral and derivative time respectively
e(t)	The system error.

The system error (e(t)) is the difference between the output y(t) and the desired set point as shown in Fig. 22. The mathematical representation of PID controller is displayed in Eqs. (24) and (25).

4.2 *Fractional Order PID Controller*

Fractional order controller is one of the elegant way that enhance the performance of conventional PID controllers,, where integral and derivative actions have, in general, non-integer orders.

In a fractional order controller, besides the proportional, integral and derivative constants, denoted by K_p, K_i and K_d respectively, there are two more adjustable parameters such that the powers of 's' in integral and derivative actions are λ and δ respectively. The values of λ and δ lies between 0 and 1. This provides more flexibility and opportunity to better adjust the dynamical properties of the control system. The fractional order controller revels good robustness. The robustness of fractional controller is more highlighted in presence of a non-linear actuator. The concept of a fractional order PID control system is explained by Das [9], Machado [24], Podlubny [32] and is illustrated in Fig. 23. The fractional order controller is considered as a special case of the classical controller, so that when putting the values of λ and δ equal to 1, it will give the conventional PID controller and when put the values of $\lambda = 1$ and $\delta = 0$, it will give the PI controller.

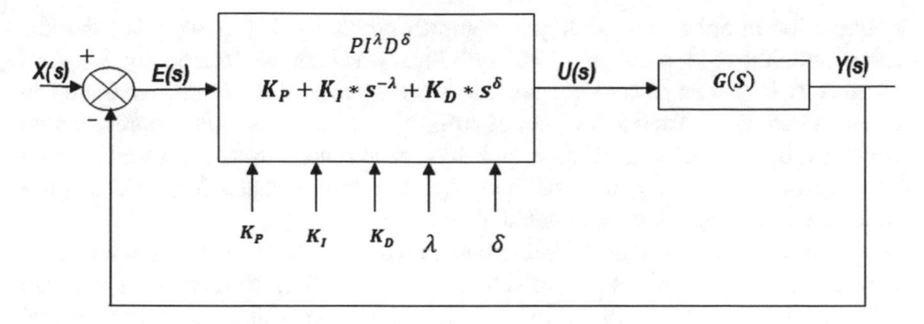

Fig. 23 Fractional order PID control system [13]

Where:

$x(t)$ Input Signal
$e(t)$ Error Signal
$G(s)$ System or Plant Transfer Function
$y(t)$ Output Signal
$u(t)$ Controller Signal.

Fractional order PID controller form is represented mathematically as follows [9]:

$$u(t)_{fc} = K_p \cdot e(t) + K_i \cdot s^{-\lambda} \cdot e(t) + K_d \cdot s^{\delta} \cdot e(t) \tag{26}$$

where: $u(t)_{fc}$ is the controller output and $e(t)$ is the system error.

5 PID and FOPID Controller Tuning

PID controllers and FOPID controller tuning for position control of HSS is designed in this chapter by incorporating Genetic Algorithm (GA). They have been designed and implemented in simulation model of HSS and experimental hardware.

5.1 Genetic Algorithm (GA)

Genetic Algorithm (GA) is an important tool to search and optimize many engineering and scientific problems. These applications includes different fields such that airlines management revenue, artificial creative and automated design for computers and mechatronics. The basic principles of GA were first proposed by Holland [38]. GA is considered as a stochastic optimization algorithm that was originally motivated by the mechanisms of natural selection and evolutionary genetics [6, 38]. It uses a direct analogy of such natural evolution to do global optimization in order to solve highly complex problems [14]. It supposes that the scope solution of a problem is an individual and can be formed by a set of parameters. These parameters are regarded as the genes of a chromosome and can be structured by concatenated values of string. The form of variables representation is defined by the encoding scheme. But these representations of the variables may be represented by binary, real numbers, or other forms, depending on the application data. Its range is usually defined by the problem.

GA includes a population of individuals, referred to as chromosomes, and each chromosome consists of a string of cells called genes [38]. Chromosomes undergo selection in the presence of variation inducing operators such as crossover and mutation. The crossover in GA occurs with a user specified probability called the "crossover probability" and is problem dependant. The mutation operator is considered to be a background operator that is mainly used to explore new areas within the search space and to add diversity to the population of chromosomes in order to prevent them from being trapped within a local optimum. But the mutation is applied to the offspring chromosomes after crossover is performed. A selection operator selects chromosomes for mating in order to generate offspring. The selection process is usually biased toward fitter chromosomes. A fitness function is used to evaluate chromosomes and reproductive success varies with fitness. The Genetic Algorithm (GA) works on a population using a set of operators that are applied on the population. This population is a set of points in the design space and the initial population is generated randomly by default. Where the next generation of the population is computed using the fitness of the individuals in the current generation.

The genetic algorithm involves a population of individuals called chromosomes where each on represents the solution of the studied problem (parameters of PID/PI and FOPID/FOPI controllers) which its performance is evaluated based on fitness function [11]. A group of chromosomes is selected to undergo to selection, crossover and mutation stages based on the fitness of each individual. The application of selection, crossover and mutation operations yields to create new individuals that give better solutions then the parents leading to optimal solution. The steps of tuning the proposed controllers by GA as follow [36]:

i. Setting of the GA parameters and generate initial, random population of individuals.
ii. Evaluate the fitness function for each chromosome.
iii. Perform selection, crossover and mutation.
iv. Repeat the fitness evaluation until end of generation.

In general, genetic algorithms use some variation of the following procedure to search for an optimal solution [11, 36]:

(a) Initialization
(b) Selection
(c) Crossover
(d) Mutation
(e) Repeat

In the first step (Initialization), an initial population of solutions is randomly generated, and the objective function is estimated for each member of this initial generation as described in [3]. While in the selection step, the individual members are chosen stochastically either to parent the next generation or to be passed on to it. The parent or the passing will occur in the members whose fitness is higher. The solution of fitness based on its objective value which the better objective value means higher fitness. Whereas the cross over means that some of the selected solutions are passed to a crossover operator. The crossover operator combines two or more parents to produce new offspring solutions for the next generation. The crossover operator tends to produce new offspring that keep the common characteristics of the parent solutions, while combining the other behavior in new ways. In this way new areas of the search space are explored, hopefully while retaining

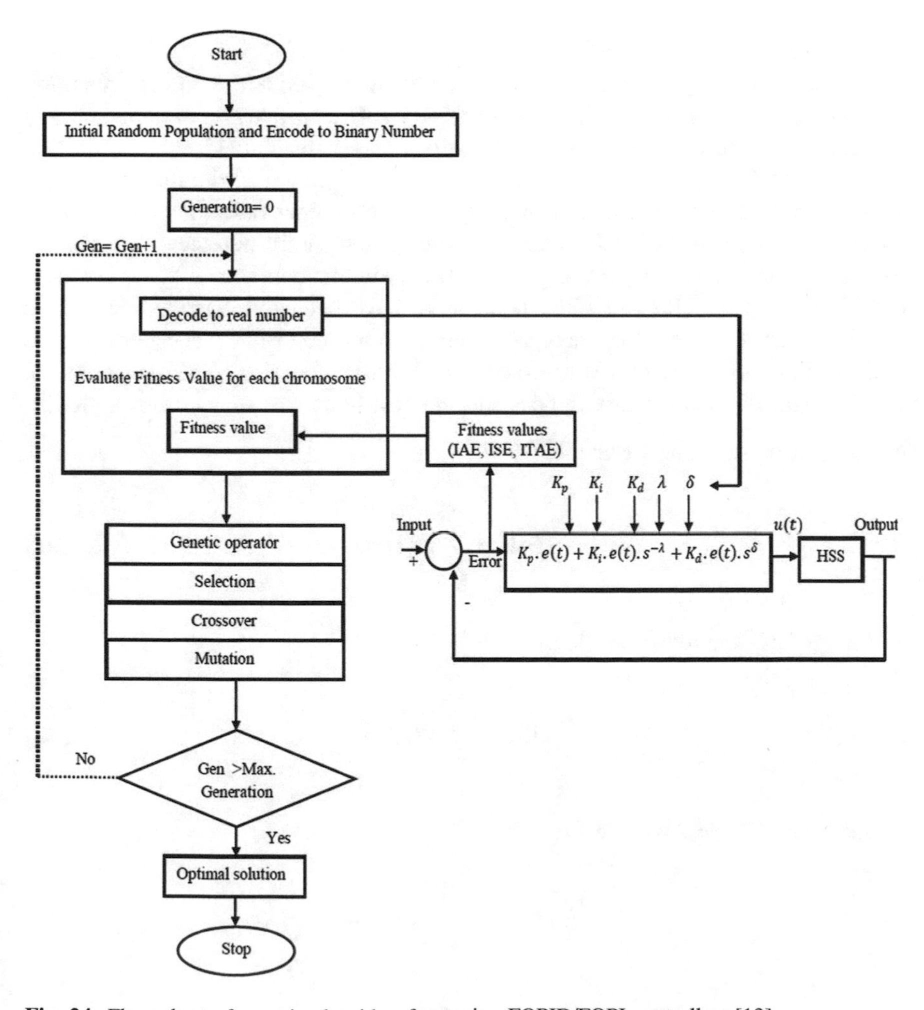

Fig. 24 Flow chart of genetic algorithm for tuning FOPID/FOPI controllers [13]

optimal solution characteristics. In mutation step some of the next-generation solutions are passed to a mutation operator, which introduces random variations in the solutions. The purpose of the mutation operator is to ensure that the solution space is adequately searched to prevent premature convergence to a local optimum. Finally, the current generation of solutions is replaced by the new generation. If the stopping criterion is not satisfied, the process returns to the selection phase. Figure 24 presents the flowchart of GA for tuning PID/PI and FOPID/FOPI controllers as adopted from [36].

6 Application of GA to HSS

6.1 Position Control of HSS

Hydraulic control systems are widely used in many industrial fields, including manufacturing systems; materials test machines, active suspension systems, mining machinery, fatigue testing, flight simulation, paper machines, ships, injection moulding machines, robotics, and aluminum mill equipment. Hydraulic systems are also common in aircraft, where their high power-to-weight ratio [34] and accurate control makes them an ideal choice for actuation of flight surfaces. The control objective is to control the piston position for hydraulic actuator, a PSO, GA, and AWPSO based on PID and FPID controllers have been implemented for piston position control. Error signal acts as an input to the controller. The performance indices (IAE, ISE and ITAE) are used as objective function. The mathematical equations for the performance indices and the cost functions are as follows [4, 23]:

- Integral of Absolute Error (IAE)

$$\mathrm{IAE} = \int_{0}^{\infty} |e(t)|dt \tag{27}$$

- Integral of Squared Error (ISE)

$$\mathrm{ISE} = \int_{0}^{\infty} e(t)^2 dt \tag{28}$$

- Integral of Time Absolute Error (ITAE)

$$\mathrm{ITAE} = \int_{0}^{\infty} t|e(t)|dt \tag{29}$$

For GA the objective function is defined as follows [12, 13, 17]:

$$f = \frac{1}{(performance\ index)} \tag{30}$$

6.2 Parameters of HSS Model

The descriptions and values for position control parameters of HSS model are illustrated in Table 1 [12].

6.3 Simulation Results

The step time of the utilized unit step in the simulation model is 1 s. The settling time, overshoot, undershoot, cross correlation between the reference sinusoidal and output signals of the model, conventional and fractional order gains values for the three performance indices (IAE, ISE and ITAE) are shown in Table 2. The step response and the error of fractional controller and conventional that based on GA with IAE, PID/FOPID and PI/FOPI controllers are shown in Fig. 25. While the

Table 1 Parameters values and description for HSS model

Parameter	Description	Value
A_P	Piston area	0.0012 m^2
C_S^+, C_S^-	Stribeck velocity range	0.015 m/s
C_v	Discharge coefficient of the valve orifices	$9.4281 \times 10^{-5} m^3/s\sqrt{N}$
F_{CO}^+, F_{CO}^-	Cylinder Coulomb friction parameters	300 N, 250 N
K_S	Spring stiffness coefficient	0 N/m
m_t	Total moving mass	20 kg
P_S	Working supply pressure	20 MPa
P_T	Tank pressure	0.1 MPa
P_n	Nominal pressure	10 MPa
Q_n	Nominal flow rate	1.333×10^{-4} m^3/s
S_1, S_2	Cylinder Stroke	0.25 m
$V_{P1,A}, V_{P1,A}$	Pipeline volumes at A-side and B-side	0.000001 m^3
σ^+, σ^-	Cylinder viscous friction parameters	20 N s/m
α	Ratio of ring side area to piston side area	1
F_{SO}^+, F_{SO}^-	Cylinder Stribeck friction parameters	50 N, 120 N
x_{max}, x_{min}	Cylinder stroke limit	±0.28 m
$x_{v,max}$	Maximum valve stroke	$\|2 \times 10^{-3}\|$ m

Table 2 Simulation results values using GA

Tuning Method	Performance Index	Controller Type	Kp	Integral term		Derivative term		Settling time (sec.)	Over Shoot (%)	XCF(*) Values
				Ki	λ	Kd	δ			
GA	IAE	PID	52.7271	21.97	1	1.21	1	6.39	11.51	0.932
		FOPID	50.3399	23.82	0.05	0.52	0.1	1.65	1.4	0.998
		PI	30.82	13.56	1	0	---	7.20	18.7	0.923
		FOPI	33.0016	12.10	0.14	0	---	2.43	No O.S	0.997
	ISE	PID	53.84	20.74	1	1.47	1	6.64	11.5	0.912
		FOPID	53.5242	20.36	0.01	0.37	0.7	1.64	1.4	0.997
		PI	47.05	17.38	1	0	---	6.92	11.87	0.909
		FOPI	30.6712	12.91	0.27	0	---	2.68	1.4	0.993
	ITAE	PID	51.3985	24.26	1	0.59	1	5.98	14.03	0.895
		FOPID	50.6874	20.51	0.13	0.54	0.3	1.70	0.71	0.998
		PI	42.4784	17.34	1	0	---	6.82	13.66	0.926
		FOPI	30.84	12.66	0.34	0	---	2.79	No O.S	0.996

Where: XCF is the cross correlation coefficient between sinusoidal reference signal and output signal for different techniques.

piston position and the error with ISE based on GA are displayed in Fig. 26. Whilst the step response and the error based on GA with ITAE are illustrated in Fig. 27.

From Table 2, it is visible that the GA for different performance indexes IAE, ISE and ITAE gives different values for the control gains. There are two main reasons for this difference in gains by using different methods and different performance indices. The first one is the different setting of the gains' range for GA in the Matlab code and different setting of the algorithm initial parameters. The user has to consider only positive values of the optimization parameters and consequently a constrained optimization algorithm will be invoked. The way of interaction of this constrained optimization with the initial conditions of each algorithm may also lead to different results. The second reason is the different objective function for the technique, where it may be IAE, or ISE or ITAE.

6.4 Experimental Results

The step time of the utilized unit step in the experimental system is 1 s. The settling time, overshoot, undershoot, cross correlation between the reference sinusoidal and output signals of the model, conventional and fractional order gains values for the three performance indices (IAE, ISE and ITAE) for GA is shown in Table 3. In addition, for the comparison between fractional controller and conventional

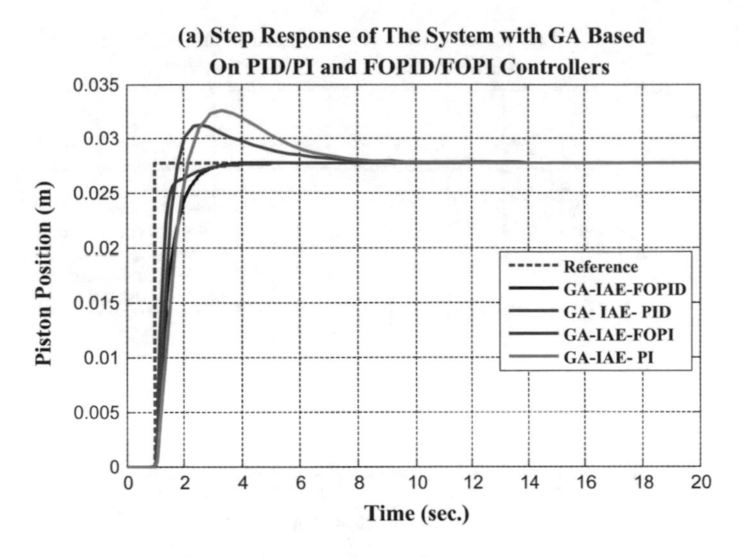

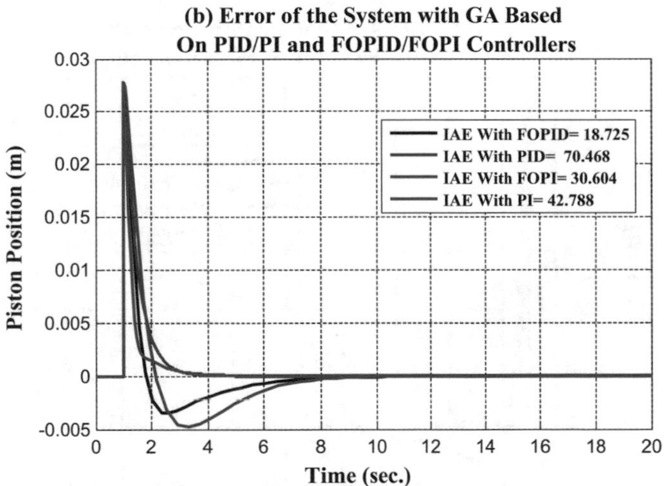

Fig. 25 Piston position and error of HSS simulation model with FOPID/FOPI and PID/PI based on GA and IAE

controller, the step response and the error of system based on GA with IAE, PID/FOPID and PI/FOPI controllers are illustrated in Fig. 28. Whilst the piston position and the error with ISE based on GA are shown in Fig. 29. While the step response and the error based on GA with ITAE are displayed in Fig. 30. The same reasons for parameters different that has been discussed in Sect. 6.3 are the same in the experimental work.

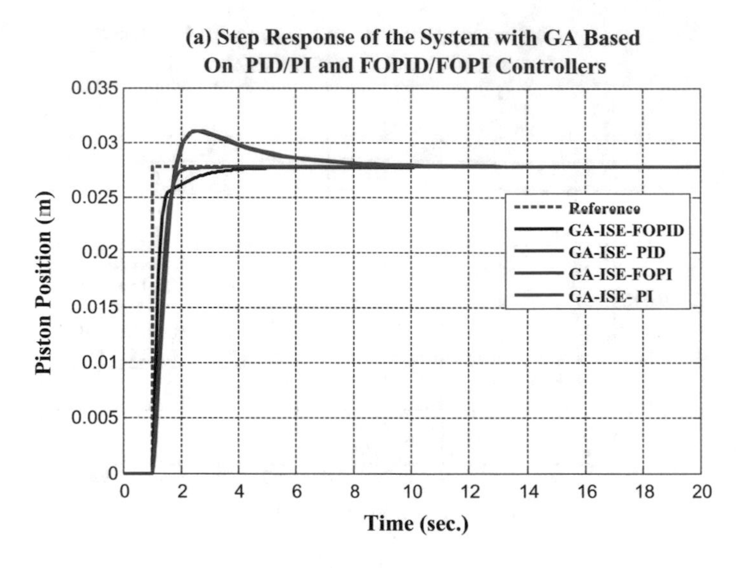

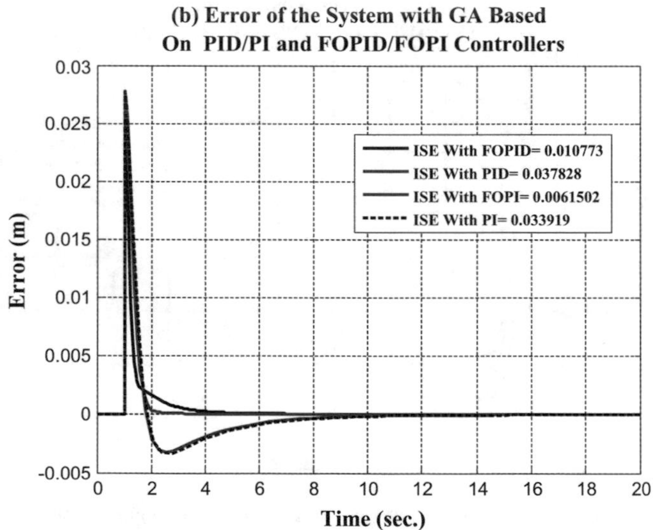

Fig. 26 Piston position and error of HSS simulation model with FOPID/FOPI and PID/PI based on GA and ISE

6.5 Discussion of Simulation Results

The argument of the simulation model depends on the results for GA that have been illustrated in Table 2. In case of IAE, the GA for conventional and fractional controllers gave an acceptable settling time in seconds which are within the permissible range (0–30 s). But due to the nonlinearities of the HSS, the settling time

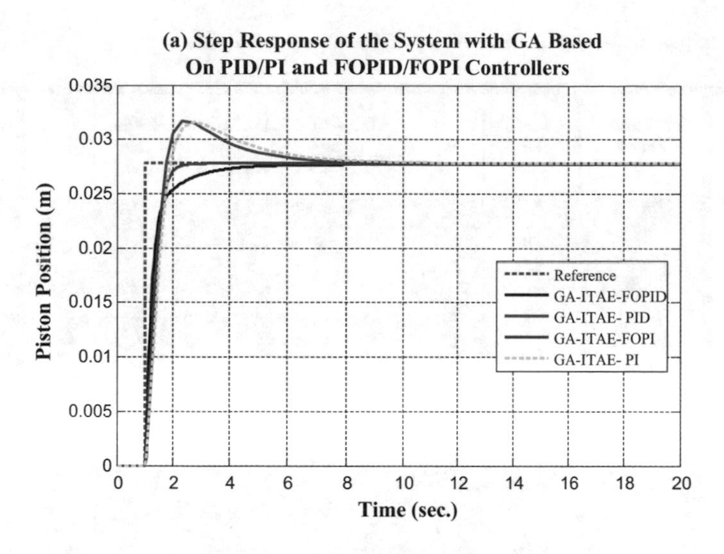

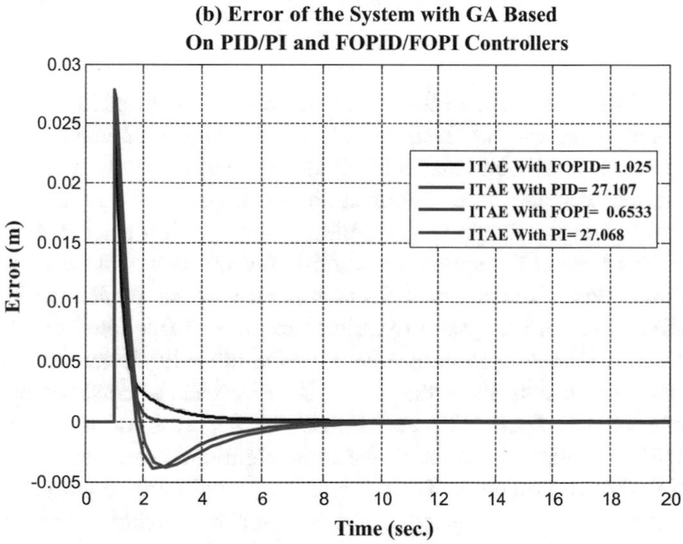

Fig. 27 Piston position and error of HSS simulation model with FOPID/FOPI and PID/PI based on GA and ITAE

of the system response based on fractional order controller is the minimum value which is around 1.5 s compared with the other results. Furthermore the percentage of system overshoot in case of fractional controller is 1% which is the minimum value compared with the other results. While in case of ISE, the GA for conventional and fractional controllers also gave an adequate settling time in seconds which are inside the permissible range (0–30 s). But the settling time of the system response anchored in fractional order controller is the minimum value which is

Table 3 Experimental results values using PSO, AWPSO and GA

Tuning Method	Performance Index	Controller Type	Kp	Integral term		Derivative term		Settling time (sec.)	Over Shoot (%)	XCF(*) Values
				Ki	λ	Kd	δ			
GA	IAE	PID	52.7271	21.97	1	1.21	1	6.7327	11.51	0.89
		FOPID	50.3399	23.82	0.05	0.52	0.1	2.22	No O.S	0.92
		PI	30.82	13.56	1	0	---	7.5516	18.7	0.93
		FOPI	33.0016	12.10	0.14	0	---	3.17	1	0.985
	ISE	PID	53.84	20.74	1	1.47	1	7.0548	11.87	0.94
		FOPID	53.5242	20.36	0.01	0.37	0.7	2.07	No O.S	0.98
		PI	47.05	17.38	1	0	---	8.0585	11.87	0.89
		FOPI	30.6712	12.91	0.27	0	---	3.79	0.71	0.91
	ITAE	PID	51.3985	24.26	1	0.59	1	6.4047	14.39	0.89
		FOPID	50.6874	20.51	0.13	0.54	0.3	2.54	No O.S	0.98
		PI	42.4784	17.34	1	0	---	7.336	16.18	0.92
		FOPI	30.84	12.66	0.34	0	---	3.93	0.71	0.95

around 1.5 s in relation to the other types of controllers. In addition, there isn't a system overshoot in case of fractional order controller compared with available system overshoot in conventional controllers. In addition to the results, in case of ITAE, it shows that the used optimization technique 'GA' gave an acceptable settling time values for fractional controllers, which within the permissible range. But the settling time of the system response based on fractional order controller is the minimum value which around 1.5 s compared with the other technique results. Additionally, there isn't a system overshoot in case of fractional order controller compared with available system overshoot in the other Evolutionary techniques.

The simulation results show that, there isn't systems undershoot for the three performance indices (IAE, ISE and ITAE) in the case of using PID/PI and FOPID/FOPI controllers. When using the same mentioned parameters of the PID/PI and FOPID/FOPI controllers in Table 2, the Fractional Order controller that based on GA technique give an efficient sinusoidal wave tracking, where it gives an acceptable cross correlation coefficients. On a global view to the responses, it is found that the nonlinear controller or the fractional order controller based on GA is the better controller than classical controller in determination the optimal parameters of the proposed controller. Moreover, the settling time and system overshoot of the three performance indices in case of Fractional Order PID (FOPID) controller is the minimum value compared with the other results. In fact the fractional controller shows its good performance in reducing the settling time and overshoot from available overshoot value to non overshoot. It is also found that there isn't system undershoot for all the optimization techniques. Furthermore the used FOPID gives a better system response and results compared with FOPI controller results.

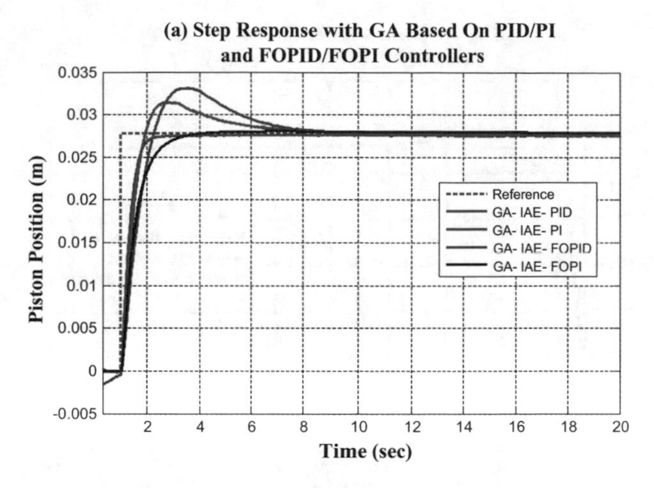

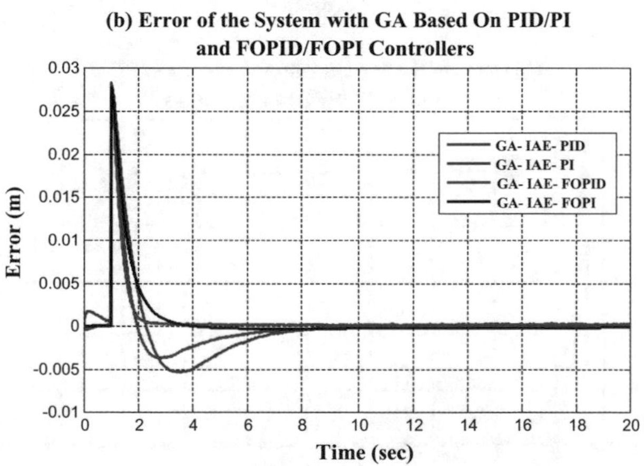

Fig. 28 Piston position and error of experimental HSS with FOPID/FOPI and PID/PI based on GA and IAE

The resultant performance indices that are displayed in the figures must be multiplied by 10^{-3} to get the actual values for the performance indices.

6.6 *Effects of Changing Reference Profile for the Simulation HSS*

A changing of reference profile with 50% of the set point value is added to the control signal (unit step input) at the process input and drive the system away from its desired operating point from (t = 20) seconds to (t = 40) seconds during the stability condition of the system. The changing in profile based on GA with

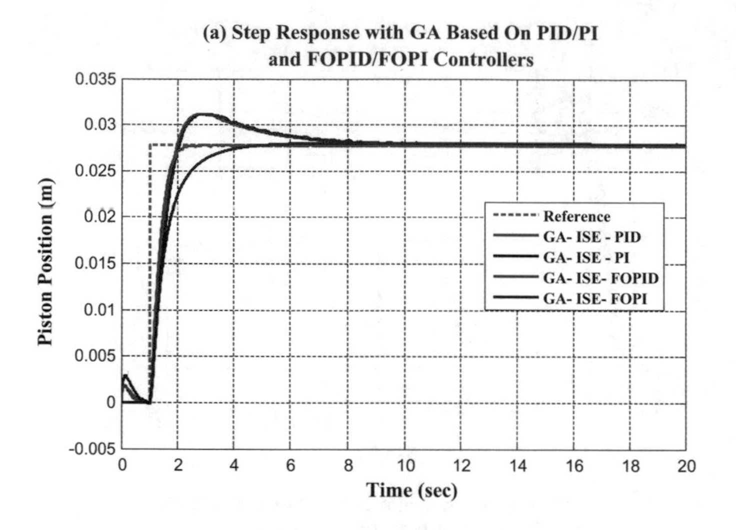

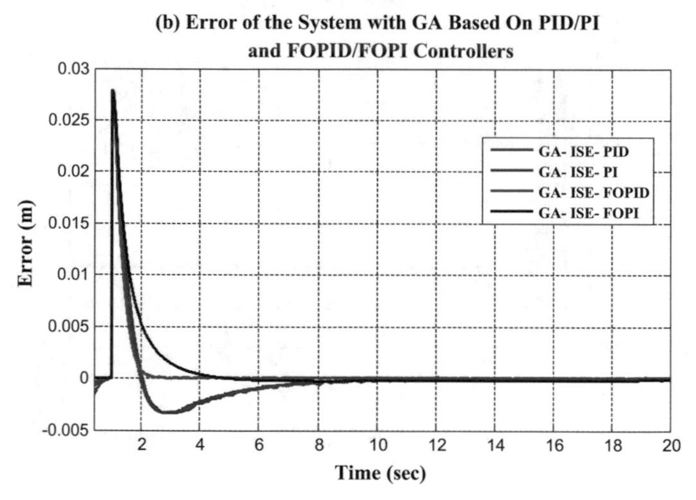

Fig. 29 Piston position and error of experimental HSS with FOPID/FOPI and PID/PI based on GA and ISE

classical controller is shown in Fig. 31. Whereas the profile's changing of the HSS model based on GA with nonlinear controller is displayed in Fig. 32. The figures show that the fractional order controller based on GA has better results in the case of profile changing in relation to other techniques and the system behaves stronger ant changing profile ability.

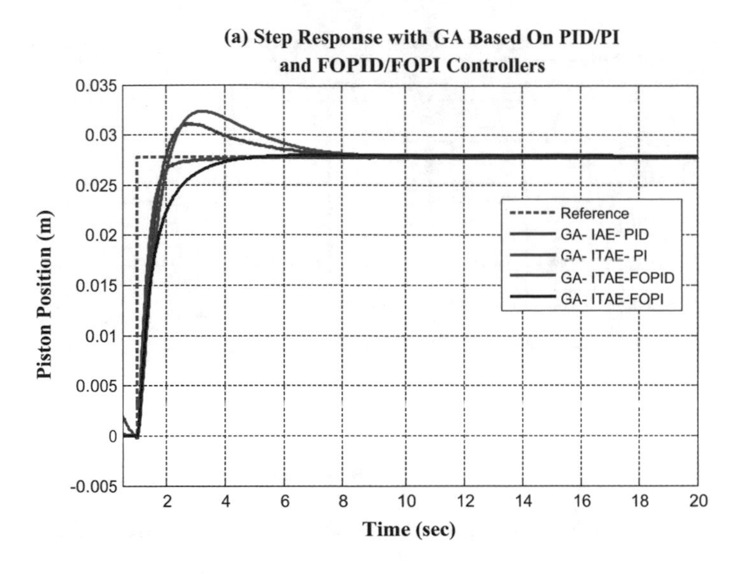

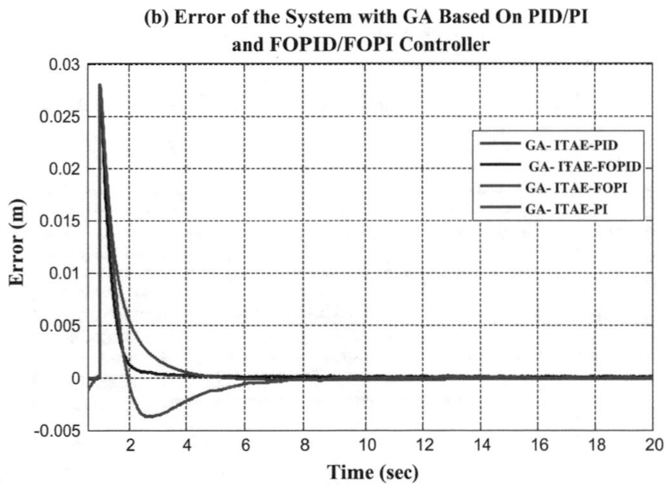

Fig. 30 Piston position and error of experimental HSS with FOPID/FOPI and PID/PI based on GA and ITAE

6.7 Discussion of Experimental Results

The cases of the experimental system depend on the results Table for PSO, AWPSO, GA that illustrated in Table 3. The settling time of the system response based on fractional order controller using GA is the minimum value which around 2.5 s in relation to the other controller results. Moreover there isn't a system overshoot in case of fractional controller in compared with the available overshoot values in conventional controllers. In addition, the settling time of the system

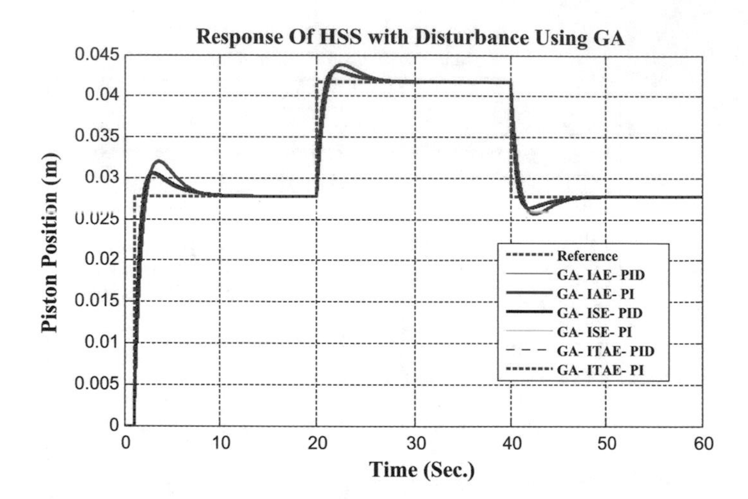

Fig. 31 Response of HSS simulation model with 50% changing in profile based on GA

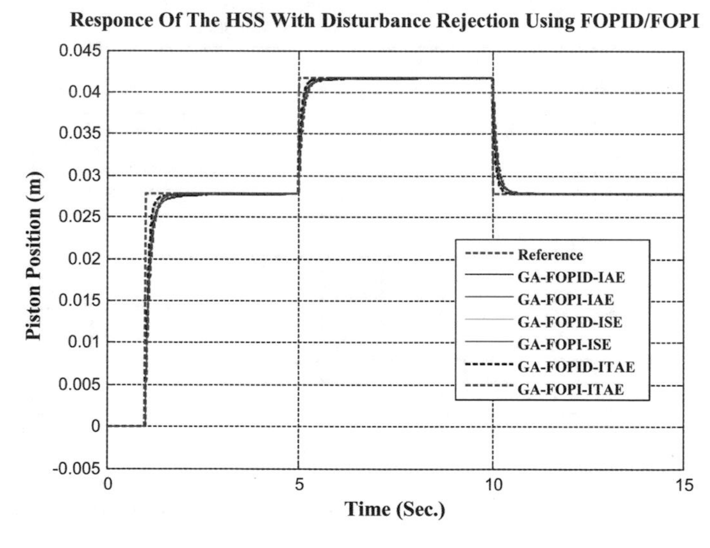

Fig. 32 Response of HSS simulation model with 50% changing in profile based on GA and fractional order controller

response anchored in fractional order controller is the minimum value which approximately 2.5 s with regard to the other results. Additionally, there isn't a system overshoot in case of fractional order controller compared with available system overshoot in the conventional controllers. The settling time of the system response based on fractional order controller is the minimum value which around 2.5 s corresponding to the other results. Moreover, there isn't a system overshoot in case of fractional order controller compared with available system overshoot in the other controllers.

The experimental results illustrate that, there isn't systems undershoot for the three performance indices (IAE, ISE and ITAE) in the case of using PID/PI and FOPID/FOPI controllers. When using the same mentioned parameters of the PID/PI and FOPID/FOPI controllers in Table 3, the Fractional Order controller that based on GA technique give an efficient sinusoidal wave tracking, where it gives an acceptable cross correlation coefficients. On a global analysis to thc rcsponscs, it is found that the nonlinear controller or the fractional order controller based on GA is the better controller than classical controller in determination the best parameters of the projected controller. On the way, the settling time and system overshoot of the three performance indices in case of Fractional Order PID (FOPID) controller is the minimum value compared with the other results. In fact the fractional controller provides its robustness in reducing the settling time and overshoot from available overshoot value to non overshoot. It is also found that there isn't system undershoot for all the optimization techniques. In addition, the used FOPID gives a better system response and results compared with FOPI controller results.

6.8 *Effects of Changing Reference Profile for the Experimental HSS*

The same changing in profile signal that has been added in the previous signal is also has been added in experimental system. The changing in profile reference based on GA with classical controller is shown in Fig. 33. Whereas the changing in

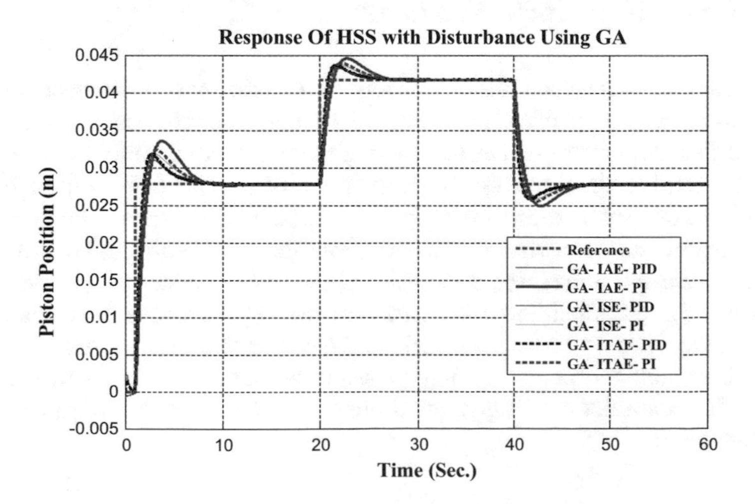

Fig. 33 Response of experimental HSS with 50% changing in profile based on GA

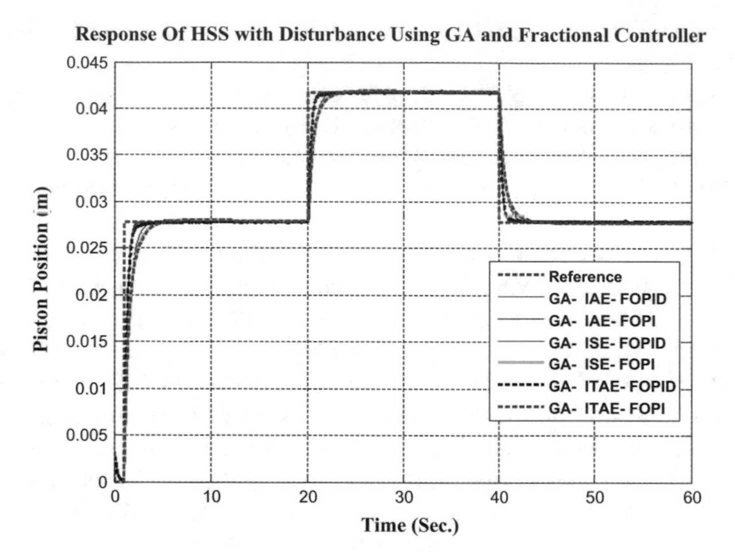

Fig. 34 Response of experimental HSS with 50% changing in profile based on GA and fractional order controller

profile of the HSS model based on GA with nonlinear controller is displayed in Fig. 34. The figures illustrate that the fractional order controller based on GA has better resistance the changing in profile with regard to conventional controllers. It also shows that the system behaves stronger ant changing in profile ability.

6.9 *Sensitivity of HSS Parameters*

The hydraulic systems have many numbers of perturbations in parameters such as perturbation in supply pressure, Coulomb friction and viscous friction. It was assumed that the system parameters have a perturbation of 20%. Tables 4 and 5 show the settling time and system overshoots of the position control of HSS according to the perturbation of the supply pressure, viscous and coulomb frictions. The simulation and experimental results show that the settling time and system overshoots are still around the same values in case of nominal parameters. It also shows that the proposed controller based on the GA technique have the desired robustness to system uncertainties such as the perturbation of the viscous friction, Coulomb friction and pump's supply pressure. In addition it shows that the fractional order controller is still give good time performance with compared to classical controllers.

Table 4 Time performance of HSS due to supply pressure sensitivity

Tuning Method	Performance Index	Controller Type	Simulation Model of HSS				Experimental HSS			
			Increasing in Supply Pressure by 20 %		Decreasing in Supply Pressure by 20 %		Increasing in Supply Pressure by 20 %		Decreasing in Supply Pressure by 20 %	
			Settling time (sec.)	Over Shoot (%)	Settling time (sec.)	Over Shoot (%)	Settlin g time (sec.)	Over Shoot (%)	Settling time (sec.)	Over Shoot (%)
GA	IAE	PID	6.192	11.21	6.99	12.81	6.62	11.01	6.93	11.91
		FOPID	1.553	1.1	1.95	1.98	2.02	No O.S	2.52	No O.S
		PI	7.12	17.9	7.80	19.87	7.13	18.1	7.8516	19.7
		FOPI	2.23	No O.S	2.873	No O.S	3.17	1	3.17	1
	ISE	PID	6.446	10.2	6.864	12.85	6.96	10.89	7.6548	12.97
		FOPID	1.445	1.34	1.984	1.94	2.07	No O.S	2.578	No O.S
		PI	6.724	10.97	7.45	12.47	8.05	10.89	8.659	12.87
		FOPI	2.482	1.24	3.28	1.84	3.71	0.5	3.95	1.75
	ITAE	PID	5.88	13.53	6.88	15.63	6.24	13.12	6.98	15.39
		FOPID	1.60	0.61	2.10	1.61	2.54	No O.S	2.98	No O.S
		PI	6.62	13.16	7.92	14.96	7.336	16.18	7.336	16.18
		FOPI	2.595	No O.S	3.69	No O.S	3.87	0.62	4.56	1.71

Table 5 Time performance of HSS due to friction parameters sensitivity

Tuning Method	Performance Index	Controller Type	Simulation Model of HSS				Simulation Model of HSS			
			Increasing in Viscosity Friction by 20 %		Decreasing in Viscosity Friction by 20 %		Increasing in Coulomb Friction by 20 %		Decreasing in Coulomb Friction by 20 %	
			Settling time (sec.)	Over Shoot (%)	Settling time (sec.)	Over Shoot (%)	Settlin g time (sec.)	Over Shoot (%)	Settling time (sec.)	Over Shoot (%)
GA	IAE	PID	6.611	10.43	6.611	10.43	6.614	10.432	6.613	10.432
		FOPID	1.775	1.6	1.769	1.69	1.776	1.68	1.775	1.6
		PI	6.952	15.64	6.955	15.66	6.953	15.63	6.952	15.64
		FOPI	2.735	No O.S	2.736	No O.S	2.736	No O.S	2.736	No O.S
	ISE	PID	6.95	11.96	6.955	11.97	6.955	11.97	6.955	11.97
		FOPID	1.775	1.5	1.775	1.5	1.775	1.5	1.775	1.5
		PI	6.916	11.96	6.95	11.96	6.946	11.96	6.916	11.96
		FOPI	2.665	No O.S	2.665	No O.S	2.665	No O.S	2.665	No O.S
	ITAE	PID	5.984	15.04	5.984	15.04	5.984	15.04	5.984	15.04
		FOPID	1.915	1	1.918	1	1.916	1	1.918	1
		PI	6.843	7.58	6.845	7.59	6.844	7.582	6.846	7.584
		FOPI	2.987	No O.S	2.988	No O.S	2.979	No O.S	2.9799	No O.S

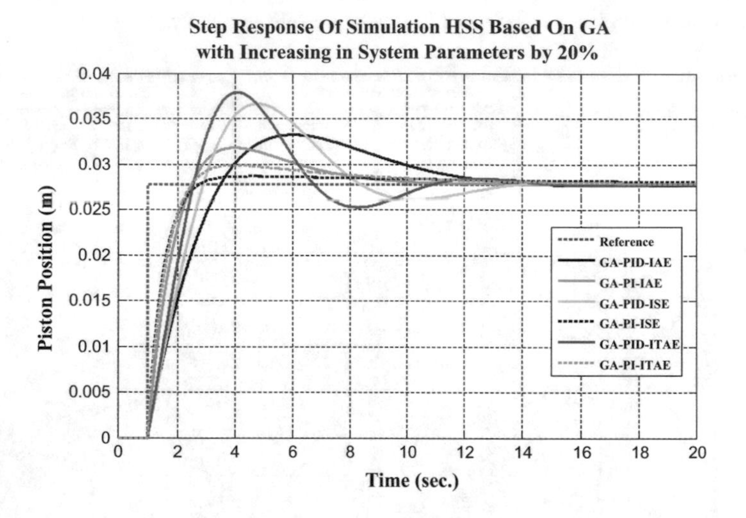

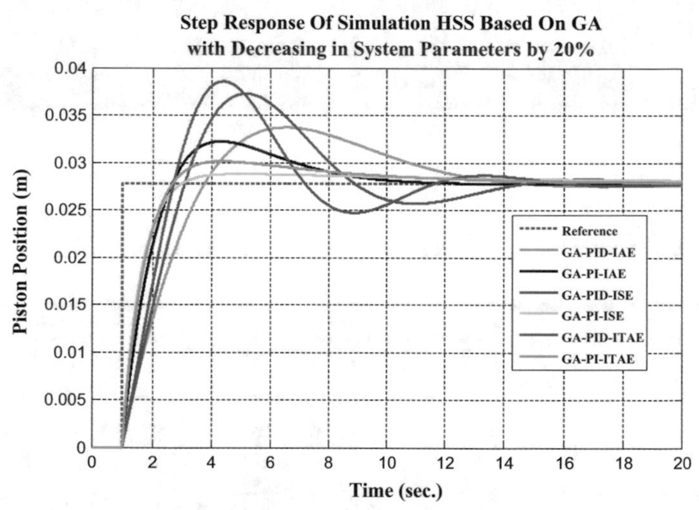

Fig. 35 Step response of simulation HSS based on GA with increasing and decreasing in system parameters

After a deep study of position control of HSS, The recommended controller that gives good time performance, tracking the change in reference profile and robust controller for parameters sensitivity is the fractional order controllers. Figures 35 and 37 present the system response based on GA in case of increasing and decreasing in HSS parameters for conventional controllers. While Figs. 36 and 38 illustrate the system response based on GA and fractional order controllers in case of parameters sensitivity in HSS for the simulation HSS model and experimental hardware respectively.

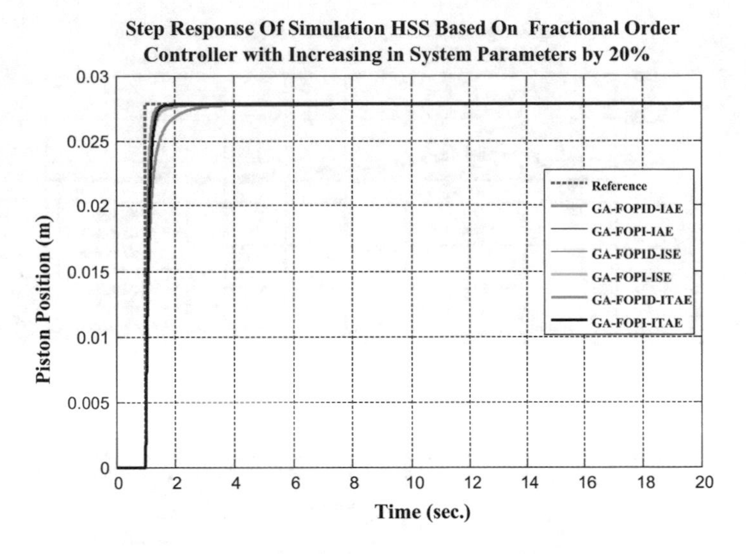

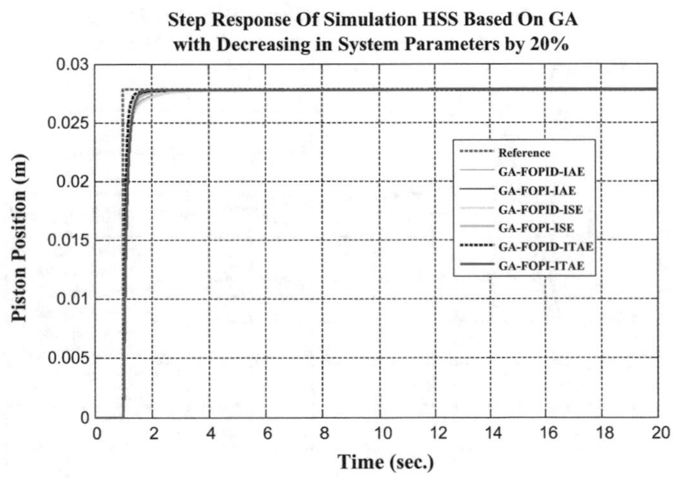

Fig. 36 Step response of simulation HSS based on FOPID/FOPI with increasing and decreasing in system parameters

6.10 Validation Between Simulation and Experimental Results

The main objective of this Section is to illustrate the convergence and validation of results and graphics between the methods, which represents the Hydraulic Servo System (HSS). The results show that a good validation between the following method.

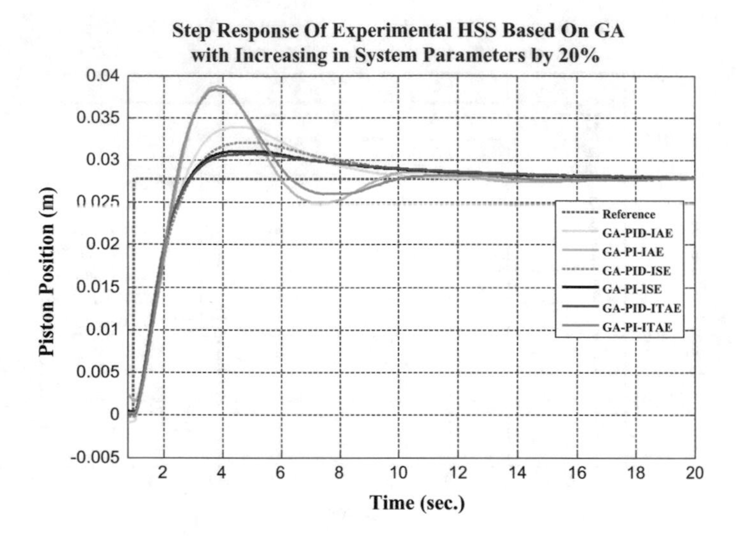

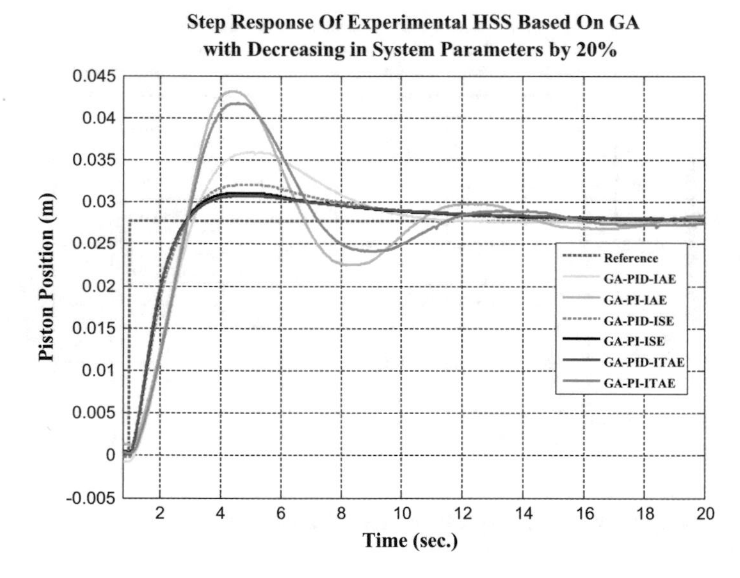

Fig. 37 Step response of experimental HSS based on GA with increasing and decreasing in system parameters

I. Simulation model based on physical laws.
II. Experimental system.
III. Identified model based on input–output data.

The decision of good validation between the above mentioned models is based on that there are a small deviation between the settling time, overshoot and graphs. Figures 39 and 40 show the validation of the results using PID/FOPID controllers based on GA.

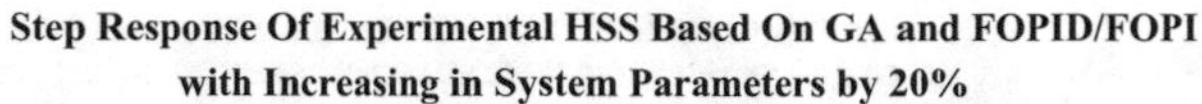

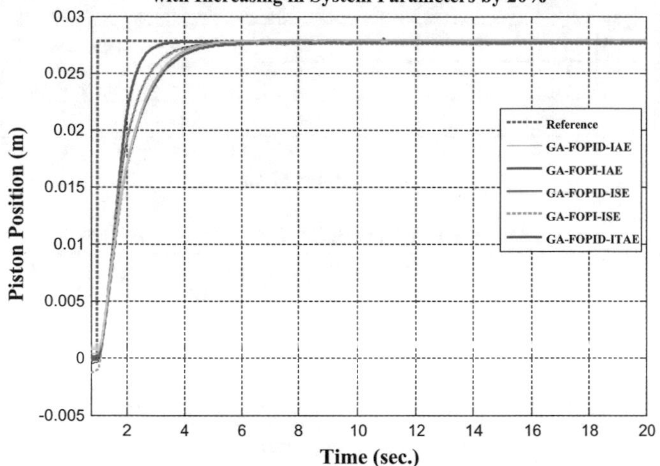

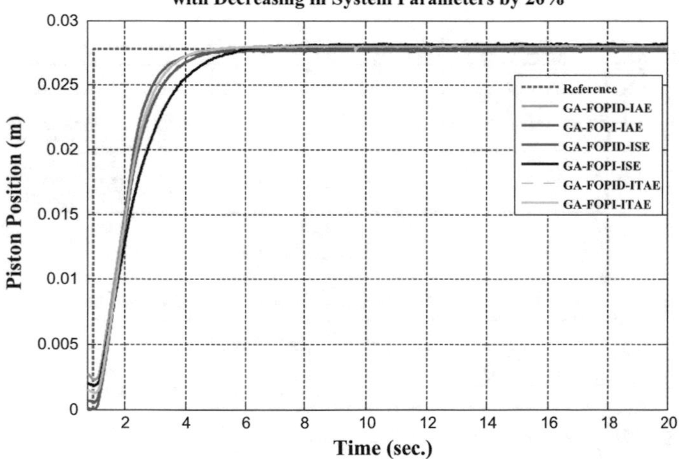

Fig. 38 Step response of experimental HSS based on FOPID/FOPI with increasing and decreasing in system parameters

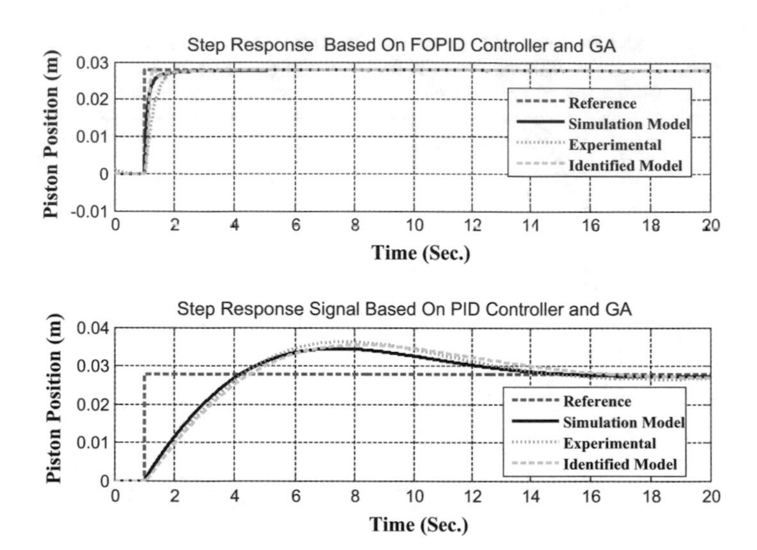

Fig. 39 Validation of step response results using GA

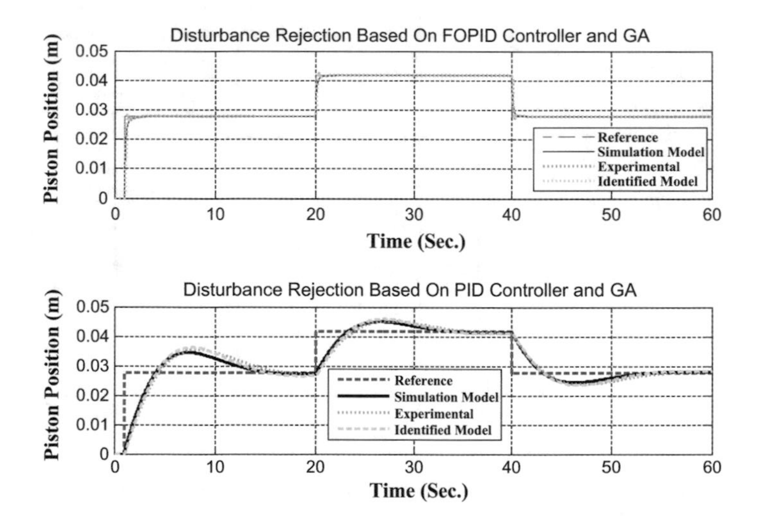

Fig. 40 Validation of disturbance results using GA

7 Conclusion

This chapter presents application of GA to design the following controllers;

(a) PID/PI controllers.
(b) Fractional Order PID/PI controllers.

This design is implemented on simulation model and real time of Position Control for Hydraulic Servo System. The utilized optimization technique and tuning method in this research is Genetic Algorithm (GA).

A SIMULINK model for a typical electro-hydraulic servo system was implemented and modified which included major nonlinearities and was verified on an experimental system. The hardware components are related to Bosch REXROTH German Company. The HSS plays an important role in the industrial applications such as the machine tool industry, material handling, mobile equipment, plastics, steel plants, oil exploration and automotive testing, so it is important to design a robust control system in this field. The experimental and simulation model of HSS are considered as a single optimization problems. The three performance indices (IAE, ISE and ITAE) have been used as the objective functions in GA. Defining the objective function for the system depends on the dynamics of the system and the desired performance for the system. The results demonstrated that the minimum settling time in case of GA based on conventional and fractional order controllers are 5.98 s and 1.64 s respectively. In addition, in case of GA based on classical and fractional controller, the minimum settling times are 6.404 and 2.017 s respectively. A changing in profile signal with 50% from the set point signal are is applied to HSS model and real time system with GA, it showed that there are a spikes and dips in GA based on conventional controller. But in case of fractional order controller, it showed that better achievement of changing in profile in relation to other techniques and the system behaves stronger ant changing in profile ability.

The simulation and experimental results showed that the nonlinear controller or fractional order controller achieved a desired performance for position control of HSS by reducing settling time and overshoots with measurable values. There are no spikes or dips appeared in the output response, while the system reached steady state smoothly compared to the other utilized techniques. It also displayed that the system responses of simulation model and experimental system with the fractional order controller based on GA are reliable and robust system with disturbance rejection. Due to the nonlinearities of HSS because of the frictions forces, valve dynamics, oil compressibility and load influence, it is recommended to utilize a non linear controller such Fractional Order Controller to avoid these effects of HSS nonlinearities.

8 Future Work

The simulation and experimental results showed that the fractional order controller achieved better dynamic response of the HSS system accurately tracks the trajectory and remains robust to disturbances. More desirable performance and future work of the HSS system can be achieved by utilizing the following considerations:

- Using conventional controller with Fuzzy Logic Controller (FLC) based on PSO, AWPSO and GA.
- Utilizing fractional order controllers with nonlinear controllers like Fuzzy Logic Controller (FLC) to adequate with the nonlinearities of HSS.
- Implementing a two degree of freedom controller.
- Implementing HSS model for tracking the higher frequency signals.
- Implementing force and pressure trajectory for HSS to be familiar with the operation of HSS.
- Controller design to assurance stability and performance for change from position tracking to pressure/force tracking.
- Design a fractional controller based on GA to achieve better and more desirable performance of the system in terms of pressure and force tracking.
- Design a controller based on PSO, AWPSO and GA with multi objective functions of settling time and overshoots.

References

1. Astrom, K. J., & Hägglund, T. (1995). *PID controllers: Theory, design and tuning* (2nd ed.). Research Triangle Park: International Society of Automation. ISBN 1-55617-516-7.
2. Ahmed, O. A. (2009). Control of fluid power system-nonlinear fuzzy control for force tracking of electro-hydraulic servo system. Master Thesis, Faculty of Engineering (at Shoubra), Benha University, Egypt.
3. Aly, A. A. (2011). PID parameters optimization using genetic algorithm technique for electro hydraulic servo control system. *Intelligent Control and Automation*, *2*(02), 69–76.
4. Azar, A. T., & Serrano, F. E. (2014). Robust IMC-PID tuning for cascade control systems with gain and phase margin specifications. *Neural Computing and Applications*, *25*(5), 983–995. doi:10.1007/s00521-014-1560-x.
5. Azar, A. T., & Serrano, F. E. (2015). Design and modeling of anti wind up PID controllers. In Q. Zhu & A. T Azar (Eds.), *Complex system modelling and control through intelligent soft computations*. Studies in Fuzziness and Soft Computing (Vol. 319, pp. 1–44). Germany: Springer. doi:10.1007/978-3-319-12883-2_1.
6. Azar, A. T., & Vaidyanathan, S. (2015). *Chaos modeling and control systems design*. Studies in Computational Intelligence (Vol. 581). Germany: Springer. ISBN 978-3-319-13131-3.
7. Bonchis, A., Corke, P. I., Rye, D. C., & Ha, Q. P. (2001). Variable structure methods in hydraulic servo systems control. *Automatica*, *37*(4), 589–595.
8. Chen, H. M., Shen, C. S., & Lee, T. E. (2013). Implementation of precision force control for an electro-hydraulic servo press system. In *Proceeding of Third International Conference on Intelligent System Design and Engineering Applications (ISDEA)* (pp. 854–857). IEEE.

9. Das, S. (2008). *Functional fractional calculus for system identification and controls*. Berlin, Heidelberg: Springer.
10. Edvard, D., & Uroš, Ž. (2011). An intelligent electro-hydraulic servo drive positioning. *Strojniškivestnik-Journal of Mechanical Engineering*, *57*(5), 394–404.
11. Elbayomy, K. M., Zongxia, J., & Huaqing, Z. (2008). PID controller optimization by GA and its performances on the electro-hydraulic servo control system. *Chinese Journal of Aeronautics*, *21*(4), 378–384.
12. Essa, M., Aboelela M. A. S., & Hassan, M. A. M. (2014a). Position control of hydraulic servo system using proportional-integral-derivative controller tuned by some evolutionary techniques. *Journal of Vibration and Control*, 1077546314551445.
13. Essa, M., Aboelela M. A. S., & Hassan, M. A. M. (2014b). Position control of hydraulic servo systems using evolutionary techniques. Master thesis, Faculty of Engineering (Cairo University), Egypt.
14. Goldberg, D. E., & Holland, J. H. (1988). Genetic algorithms and machine learning. *Machine learning*, *3*(2), 95–99. doi:10.1023/A:1022602019183.
15. Jelali, M., & Kroll, A. (2012). *Hydraulic servo-systems: Modeling, identification and control*. Springer, London Ltd. ISBN 978-1-4471-1123-8.
16. Kao, C. C. (2009). Applications of particle swarm optimization in mechanical design. *Journal of Gaoyuan University*, *15*, 93–116.
17. Kim, J.-S., Kim, J.-H., Park, J-M., Park, S.-M., Choe, W.-Y., & Heo, H. (2008). Auto tuning PID controller based on improved genetic algorithm for reverse osmosis plant. *World Academy of Science, Engineering and Technology*, *47*(2), 384–389.
18. Ljung, L. (2010). System Identification Toolbox 7 Getting started guide. http://www.mathworks.com.
19. Maneetham, D., & Afzulpurkar, N. (2010). Modeling, simulation and control of high speed nonlinear hydraulic servo system. *Journal of Automation Mobile Robotics and Intelligent Systems, 4*, 94–103.
20. Mahdi, S. M. (2012). Controlling a nonlinear servo hydraulic system using PID controller with a genetic algorithm tool. *IJCCCE, 12*(1), 42–52.
21. Mathwork. (2014). Simulink Design Optimization Toolbox. http://www.mathworks.com.
22. Mannesmann Rexroth. (2005) 4/3-way high response directional valve direct actuated, with electrical position feedback Type 4WRSE. http://www.boschrexroth.com.
23. Mahony, T. O., Downing, C. J., & Fatla, K. (2000). Genetic algorithm for PID parameter optimization: Minimizing error criteria. *Process Control and Instrumentation*, 148–153.
24. Machado, J. T. (1997). Analysis and design of fractional-order digital control systems. *Systems Analysis Modeling Simulation*, *27*(2–3), 107–122.
25. Md Rozali, S., Rahmat, M. F. A., Husain, A. R., & Kamarudin, M. N. (2014). Design of adaptive back stepping with gravitational search algorithm for nonlinear system. *Journal of Theoretical and Applied Information Technology (JATIT)*, *59*(2), 460–468.
26. Merritt, H. E. (1967). *Hydraulic control systems*. New York: Wiley. ISBN 0-471-59617-5.
27. Mihajlov, M., Nikolić, V., & Antić, D. (2002). Position control of an electro-hydraulic servo system using sliding mode control enhanced by fuzzy PI controller. *Facta Universitatis-Series: Mechanical Engineering*, *1*(9), 1217–1230.
28. National Instrument. (2002). DAQ NI 6013/6014 User Manual. http://www.ni.com.
29. Nelles, O. (2013). *Nonlinear system identification: From classical approaches to neural networks and fuzzy models*. Berlin, Heidelberg: Springer. ISBN 978-3-642-08674-8.
30. Olranthichachat, P., & Kaitwanidvilai, S. (2011). Design of optimal robust PI controller for electro-hydraulic servo system. *Engineering Letters, 19*(3), 197–203.
31. Olranthichachat, P., & Kaitwanidvilai, S. (2011b). Structure specified robust H∞ loop shaping control of a MIMO electro-hydraulic servo system using particle swarm optimization. In *Proceedings of the International Multi Conference of Engineers and Computer Scientists* (Vol. 2).
32. Podlubny, I. (1999). Fractional-order systems and $PI^{\lambda}D^{\delta}$ controllers. *IEEE Transactions on Automatic Control, 44*(1), 208–214.

33. Puangdownreong, D., & Sukulin, A. (2012). Obtaining an optimum PID controllers for unstable systems using current search. *International Journal of Systems Engineering, Applications & Development 6*(2), 188–195.
34. Rydberg, K. E. (2008). Hydraulic servo systems. TMHP51 fluid and mechanical engineering systems. *Linköping University*, 1–45.
35. Roozbahani, H., Wu, H., & Handroos, H. (2011). Real-time simulation based robust adaptive control of hydraulic servo system. In *Proceeding of IEEE International Conference on Mechatronics (ICM)*, 779–784.
36. Saad, M. S., Jamaluddin, H., & Darus, I. Z. (2012). PID controller tuning using evolutionary algorithms. *WSEAS Transactions on Systems and Control*, *7*(4), 139–149.
37. Sirouspour, M. R., & Salcudean, S. E. (2000). On the nonlinear control of hydraulic servo-systems. In *IEEE International Conference on Robotics and Automation Proceedings. ICRA'00* (Vol. 2, pp. 1276–1282). IEEE.
38. Sivanandam, S. N., & Deepa, S. N. (2007). *Introduction to genetic algorithms*. Berlin, Heideberg, New York: Springer. ISBN 978-3-540-73189-4.
39. Sohl, G. A., & Bobrow J. E. (1999). Experiments and simulations on the nonlinear control of a hydraulic servo system. *IEEE Transactions on Control Systems Technology*, *7*(2), 238–247.
40. Tooby, J., & Cosmides, L. (1989). Adaption versus phylogeny: The role of animal psychology in the study of human behavior. *International Journal of Comparative Psychology*, *2*(3), 175–188.
41. Tian, J. (2013). Study on control strategy of electro-hydraulic servo loading system. *Indonesian Journal of Electrical Engineering*, *11*(9), 5044–5047.
42. Truong, D. Q., Kwan, A. K., Hung, H. T., & Yoon II, J. (2008). A study on hydraulic load simulator using self tuning grey predictor-fuzzy PID. In *Proceedings of the 17th World Congress the International Federation of Automatic Control (IFAC)* (Vol. 41, No. 2, pp. 13791–13796).
43. Xue, D., Atherton, D. P., & Chen, Y. (2007). *Linear feedback control: Analysis and design with MATLAB* (Vol. 14). Siam.
44. Yao, J. J., Di, D., Jiang, G., & Liu, S. (2012). High precision position control of electro-hydraulic servo system based on feed-forward compensation. *Research Journal of Applied Sciences, Engineering and Technology.*, *4*(4), 289–298.
45. Yu, M., & Lichen, G. (2013). Fuzzy immune PID control of hydraulic system based on PSO algorithm. *Indonesian Journal of Electrical Engineering*, *11*(2), 890–895.

Control and Synchronization of Fractional-Order Chaotic Systems

Ahmed G. Radwan, Wafaa S. Sayed and Salwa K. Abd-El-Hafiz

Abstract The chaotic dynamics of fractional-order systems and their applications in secure communication have gained the attention of many recent researches. Fractional-order systems provide extra degrees of freedom and control capability with integer-order differential equations as special cases. Synchronization is a necessary function in any communication system and is rather hard to be achieved for chaotic signals that are ideally aperiodic. This chapter provides a general scheme of control, switching and generalized synchronization of fractional-order chaotic systems. Several systems are used as examples for demonstrating the required mathematical analysis and simulation results validating it. The non-standard finite difference method, which is suitable for fractional-order chaotic systems, is used to solve each system and get the responses. Effect of the fractional-order parameter on the responses of the systems extended to fractional-order domain is considered. A control and switching synchronization technique is proposed that uses switching parameters to decide the role of each system as a master or slave. A generalized scheme for synchronizing a fractional-order chaotic system with another one or with a linear combination of two other fractional-order chaotic systems is presented. Static (time-independent) and dynamic (time-dependent) synchronization, which could generate multiple scaled versions of the response, are discussed.

Keywords Active nonlinear control · Amplitude modulation · Hidden attractors · Lyapunov stability · Non-standard finite difference schemes · Static and dynamic synchronization

A.G. Radwan (✉) · W.S. Sayed · S.K. Abd-El-Hafiz
Faculty of Engineering, Engineering Mathematics and Physics Department,
Cairo University, Giza 12613, Egypt
e-mail: agradwan@ieee.org

W.S. Sayed
e-mail: wafaa.s.sayed@eng.cu.edu.eg

S.K. Abd-El-Hafiz
e-mail: salwa@computer.org

A.G. Radwan
Nanoelectronics Integrated Systems Center, Nile University, Cairo 12588, Egypt

A.T. Azar et al. (eds.), *Fractional Order Control and Synchronization of Chaotic Systems*, Studies in Computational Intelligence 688,
DOI 10.1007/978-3-319-50249-6_11

1 Introduction

Chaotic systems and their implementations have been studied heavily during the last four decades [41, 42, 45, 50, 80]. The sensitivity of chaotic systems to parameters and initial conditions is required for many applications such as chemical reactions [18], biological systems [33, 67], circuit theory [40, 46, 51, 52], electronics [62], control [4, 5], secure communication [14, 16, 27] and cryptography [1, 2, 6, 26, 47, 48, 56, 58–60]. Much attention has been devoted to the search for better and more efficient methods for obtaining the analytical or numerical solutions or controlling the responses of chaotic systems. During the last few decades, fractional calculus has also become a powerful tool in describing the dynamics of complex systems which appear frequently in several branches of science and engineering. Therefore, fractional differential equations and their numerical techniques find numerous applications in the field of viscoelasticity, robotics, feedback amplifiers, electrical circuits, control theory, electro analytical chemistry, fractional multi-poles, electromagnetics, bioengineering, and image encryption [10, 17, 30, 39, 49, 54, 55, 57, 63, 64].

The chaotic dynamics of fractional-order systems began to attract the interest of the scientific community in recent years associated with the advances in numerical methods for solving fractional-order systems and their electronic implementations [10]. In addition, fractional calculus is more suitable for modeling the continuous non-standard behaviors of nature due to the flexibility offered by the extra degrees of freedom. Recently, most of the chaotic dynamical systems based on integer-order calculus have been extended into the fractional-order domain to fit the experimental data much precisely than the integer-order modeling.

The coupling of two or more chaotic systems is referred to as synchronization. Control and synchronization of fractional-order chaotic systems have found their way to many applications such as biological and physical systems, structural engineering, ecological models, secure communication and cryptography [3, 7–9, 11, 12, 15, 22, 35, 36, 66, 76, 79]. Since the introduction of the concept of synchronization of two chaotic signals starting at different initial conditions [38], there has been a lot of work on chaos control and synchronization. Chaotic synchronization represents a challenge due to the sensitivity to initial conditions characteristic of chaotic systems. Two trajectories starting at slightly different initial conditions exponentially diverge from each other in the long-term evolution. Several papers handled conventional synchronization of two identical chaotic systems, their anti-synchronization [21], as well as synchronization of two different systems [77]. More recent researches extended the concept to fractional-order domain [33], introduced generalized scaled dynamic (time-dependent) synchronization [43, 61], and provided the capability of control and switching for exchanging roles between master and slave systems [19, 44].

The aim of this chapter is to introduce several methods for control and synchronization of fractional-order chaotic systems using active nonlinear control technique. Several chaotic systems are extended to fractional-order domain and the effect of the fractional-order parameter on the output responses is studied. The concept of active control using two on/off switches for the synchronization between two fractional-order chaotic systems is proposed. A generalized synchronization scheme is applied to synchronize two identical or different fractional-order chaotic systems. In addition, a new chaotic system is formed as a linear combination of two systems where the generalized synchronization scheme is applied to synchronize a system with the linear combination of two other systems. A block diagram of the generalized synchronization scheme and the associated mathematical analysis are presented. The control signals are obtained in terms of the responses, parameters and scaling factors. Generalizations permit conventional synchronization, anti-synchronization, static scaling, as well as dynamic scaling where the type of synchronization and the scaling factor vary as time advances. Mathematical analysis and various examples are presented at different values of fractional-orders using Grünwald-Letnikov method of approximation and Non-Standard Finite Difference (NSFD) discretization technique [24]. Simulation results, including time series and strange attractors, are consistent with the performed analysis.

Section 2 of this chapter provides a brief introduction to relevant concepts and a survey of previous works on control and synchronization of chaotic dynamics. Section 3 provides the preliminaries of numerical solution of fractional-order differential equations and reviews the properties of the systems chosen for numerical simulations. Section 4 illustrates the effect of parameters and fractional-orders on the responses of the utilized chaotic systems. Section 5 presents active nonlinear control and synchronization using two on/off switches for the synchronization between two different chaotic systems. Section 6 discusses the analysis required to get the control signals, which is suitable for achieving any required synchronization case. This analysis is validated through simulation results for two identical or different fractional-order chaotic systems. Section 7 proposes analysis and simulation results in case of synchronizing a fractional-order chaotic system with a linear combination of two other systems. In addition, results show that the linear combination provides another way of controlling the obtained attractor diagram. Finally, Sect. 8 summarizes the main contributions of the chapter.

2 Control and Synchronization of Chaotic Dynamics

System parameters and fractional-orders represent a way of controlling the type of obtained response with no external control procedure. Chaos control requirements differ according to the given specifications and application. It is sometimes required to stabilize the system and force it to follow a certain periodic solution, while other cases require conservative systems with quasi-periodic solutions. Other modeling applications as well as pseudo-random number generation and utilization in cryptography require chaotic responses.

Continuous flows expressed in terms of ordinary differential equations can have numerous types of post transient solution(s). Reporting when these systems of differential equations exhibit chaos represents a rich research field. Research efforts have been exerted (e.g., [25, 65]) to come up with simple novel chaotic flows other than the well-known conventional systems (Lorenz, Rössler, …). These researches depend on Poincaré-Bendixson theorem, [20] which states that for any autonomous first-order ordinary differential equations with continuous functions to have chaotic solutions it requires at least three dimensions with at least one nonlinear term. Some systematic numerical search methods have been developed for detecting the presence of chaotic solutions for new equations that contain multiple parameters. These parameters mainly appear as the coefficients of each term in the system of differential equations. Methods aim at setting many coefficients to zero with the others set to ± 1 if possible or otherwise to a small integer or decimal fraction with the fewest possible digits. These systems, with the least number of existing coefficients and nonlinear terms, should exhibit chaotic properties of aperiodic bounded long-time evolution and sensitive dependence on initial conditions for some ranges of parameters.

Continuous chaotic systems can be classified into two wide categories. Dissipative systems, to which most of the studied systems belong, usually exhibit chaos for most initial conditions in a specified range of parameters. On the other hand, a conservative system exhibits periodic and quasi-periodic solutions for most values of parameters and initial conditions, and can exhibit chaos for special values only. Consequently, dissipative systems usually appear in most applications of chaos theory such as chaos-based communication, physical and financial modeling. It should be noted that conservative systems have another different set of applications where they are useful to study the development of chaos in some kinds of systems.

Another important classification of chaotic or strange attractors is either self-excited or hidden attractors. A self-excited attractor has a basin of attraction that is associated with or excited from unstable equilibria. For example, the well-known Lorenz and Rössler attractors are self-excited. From a computational point of view, this allows one to use a numerical method in which a trajectory started from a point, on the unstable manifold in the neighborhood of an unstable equilibrium, reaches an attractor and identifies it. On the other hand, a hidden attractor has a basin of attraction that does not intersect with small neighborhoods of any equilibrium points. Hidden attractors cannot be found by the previous method and are important in engineering applications because they allow unexpected and potentially disastrous responses to perturbations in a structure like a bridge or an airplane wing.

As for external control methods, Pecora and Carroll [38] were the first to introduce the concept of synchronization of two systems with different initial conditions. Many chaotic synchronization schemes have also been introduced during the last decade such as adaptive control [68–73], time delay feedback approach [13, 37], sliding mode control [11, 23], nonlinear feedback synchronization, and active control [22]. However, most of these methods have been tested for two identical chaotic systems. When Ho and Hung [22] presented and applied the concept of active control method on the synchronization of chaotic systems, many recent papers investigated this technique for different systems and in different applications [28, 74].

The synchronization of three chaotic fractional-order Lorenz systems with bidirectional coupling in addition to the chaos synchronization of two identical systems via linear control were investigated in [34, 78]. Moreover, two different fractional-order chaotic systems can be synchronized using active control as in [7]. The hyper-chaotic synchronization of the fractional-order Rössler system, which exists when its order is as low as 3.8, was shown by Yua and Lib [78].

Anti-synchronization is a phenomenon in which the state vectors of the synchronized systems have the same amplitude but opposite signs to those of the driving system. Therefore, the sum of two signals is expected to converge to zero when anti-synchronization appears. Since the discovery of anti-synchronization experimentally in the context of self-synchronization, it has been applied in many different fields, such as biological and physical systems, structural engineering, and ecological models [75]. Liu et al. [29] shows that either synchronization or anti-synchronization can appear depending on the initial conditions of the coupled pendula. Active control method is used to study the anti-synchronization for two identical and nonidentical systems [7, 22].

Before we proceed to presenting our work on control and synchronization of fractional-order chaotic systems, the numerical methods associated with fractional-order differential equations are briefly reviewed in the next section.

3 Fractional-Order Chaotic Systems and Their Numerical Solution

Finding robust and stable numerical and analytical methods for solving the fractional differential equations has recently been an active research topic. These methods include the fractional difference method, the Adomian decomposition method, the homotopy-perturbation method, the variational iteration method, and the Adams-Bashforth-Moulton method. Recently, the non-standard finite difference (NSFD) scheme [31, 32] has been applied for the numerical solutions of fractional differential equations [24]. The scheme has been developed as an alternative method for solving a wide range of problems whose mathematical models involve algebraic, differential, biological models, and chaotic systems. The definition of Grünwald-Letnikov derivative has been used in numerical analysis to discretize the fractional differential equations. The technique has many advantages over the classical techniques, and provides an efficient numerical solution.

The Caputo fractional derivative [17] of order α is defined as:

$$
\begin{aligned}
D^{\alpha} f(t) &= \frac{d^{\alpha} f(t)}{dt^{\alpha}} \\
&= \begin{cases} \frac{1}{\Gamma(m-\alpha)} \int_0^t \frac{f^m(\tau)}{(t-\tau)^{\alpha-m+1}} d\tau & m-1 < \alpha < m \\ \frac{d^m}{dt^m} f(t) & \alpha = m \end{cases},
\end{aligned} \tag{1}
$$

where m is the first integer greater than α and $\Gamma(.)$ is the gamma function defined by:

$$\Gamma(z) = \int_0^\infty e^{-t} t^{z-1} dt, \qquad \Gamma(z+1) = z\Gamma(z). \tag{2}$$

Consider the fractional-order differential equation

$$D^\alpha x(t) = f(t, x). \tag{3}$$

Grünwald-Letnikov method of approximation [24] is defined as follows:

$$D^\alpha x(t) = \lim_{h \to 0} h^{-\alpha} \sum_{j=0}^{t/h} (-1)^j \binom{\alpha}{j} x(t - jh), \tag{4}$$

where h is the step size. This equation can be discretized as follows:

$$\sum_{j=0}^{n+1} c_j{}^\alpha x(t - jh) = f(t_n, x(t_n)), \qquad j = 1, 2, 3, \ldots \tag{5}$$

where $t_n = nh$ and $c_j{}^\alpha$ are the Grünwald-Letnikov coefficients defined as:

$$c_j{}^\alpha = \left(1 - \frac{1+\alpha}{j}\right) c_{j-1}{}^\alpha, \quad j = 1, 2, 3, \ldots, \quad c_0{}^\alpha = h^{-\alpha}. \tag{6}$$

The NSFD discretization technique is based on replacing the step size h by a function $\phi(h)$ [24, 33] and applying it with (5) to solve (3). In the rest of this paper, NSFD with $\phi(h) = 1 - e^{-h}$ is used to solve the systems of differential equations. In addition, a time step of 0.005 is employed according to the system properties and a total simulation time of 200 points is used except where stated otherwise.

Same algebraic manipulation can be applied to a system of three fractional-order differential equations

$$D^\alpha x = f_1(x, y, z), \tag{7a}$$

$$D^\beta y = f_2(x, y, z), \tag{7b}$$

$$D^\gamma z = f_3(x, y, z), \tag{7c}$$

where $0 < \alpha, \beta, \gamma \leq 1$, to obtain the corresponding solutions. All state variables (x, y, z, …), scaling factors (s_x, s_y, s_z, …), and control functions (u_x, u_y, u_z, …) that will appear later on are in general functions of time, i.e., their values may change at every time instant.

3.1 Systems Utilized for Synchronization Purposes

The first three systems are Lü, Newton-Leipnik and Chua's circuit, which have appeared before in fractional-order form in [39, 53] and others. The rest of the utilized systems appeared before in integer-order [25, 65], yet, in this section, they are extended to fractional-order domain. One of the systems is the slave, while the master may be one of the other two systems or a linear combination of them as detailed later on in Sect. 7. Table 1 shows the equations of the selected systems in fractional-order domain and their strange attractors in the integer-order case. They are a dissipative

Table 1 Equations of the utilized systems, their properties, discretized solutions and attractor diagrams

System	Lü system [39]	Newton-Leipnik [39]	Chua's circuit [53]
Equations	$D^{\alpha}x = a(y-x)$ $D^{\beta}y = -xz + by$ $D^{\gamma}z = xy - cz$	$D^{\alpha}x = -ax + y + 10yz$ $D^{\beta}y = -x - 0.4y + 5xz$ $D^{\gamma}z = -5xy + bz$	$D^{\alpha}x = a(y - f(w)x)$ $D^{\beta}y = z - x$ $D^{\gamma}z = -by + cz$ $Dw = x$, where $f(w) = \begin{cases} d, & \lvert w\rvert < 1 \\ e, & \lvert w\rvert \geq 1 \end{cases}$
Parameters	$(a,b,c) = (36,20,3)$	$(a,b) = (0.4,0.175)$	$(a,b,c,d,e) = (4,1,0.65,0.2,5)$
Initial Point	$(x_0,y_0,z_0) = (0.2,0.5,0.3)$	$(x_0,y_0,z_0) = (0.19,0,-0.18)$	$(x_0,y_0,z_0,w_0) = (0.01,0.02,0.01,0.05)$
Discretized Solution	$x_{n+1} = \frac{-\sum_{j=1}^{n+1} c_j{}^{\alpha} x_{n+1-j} + ay_n - ax_n}{c_0{}^{\alpha}}$ $y_{n+1} = \frac{-\sum_{j=1}^{n+1} c_j{}^{\beta} y_{n+1-j} - x_{n+1}z_n + by_n}{c_0{}^{\beta}}$ $z_{n+1} = \frac{-\sum_{j=1}^{n+1} c_j{}^{\gamma} z_{n+1-j} + x_{n+1}y_{n+1} - cz_n}{c_0{}^{\gamma}}$	$x_{n+1} = \frac{-\sum_{j=1}^{n+1} c_j{}^{\alpha} x_{n+1-j} - ax_n + y_n + 10y_nz_n}{c_0{}^{\alpha}}$ $y_{n+1} = \frac{-\sum_{j=1}^{n+1} c_j{}^{\beta} y_{n+1-j} - x_{n+1} - 0.4y_n + 5x_{n+1}z_n}{c_0{}^{\beta}}$ $z_{n+1} = \frac{-\sum_{j=1}^{n+1} c_j{}^{\gamma} z_{n+1-j} - 5x_{n+1}y_{n+1} + bz_n}{c_0{}^{\gamma}}$	$x_{n+1} = \frac{-\sum_{j=1}^{n+1} c_j{}^{\alpha} x_{n+1-j} + ay_n - af(w)x_n}{c_0{}^{\alpha}}$ $y_{n+1} = \frac{-\sum_{j=1}^{n+1} c_j{}^{\beta} y_{n+1-j} + z_n - x_{n+1}}{c_0{}^{\beta}}$ $z_{n+1} = \frac{-\sum_{j=1}^{n+1} c_j{}^{\gamma} z_{n+1-j} - by_{n+1} + cz_n}{c_0{}^{\gamma}}$ $w_{n+1} = w_n + hx_n$
Attractor $\alpha = 1$ $\beta = 1$ $\gamma = 1$	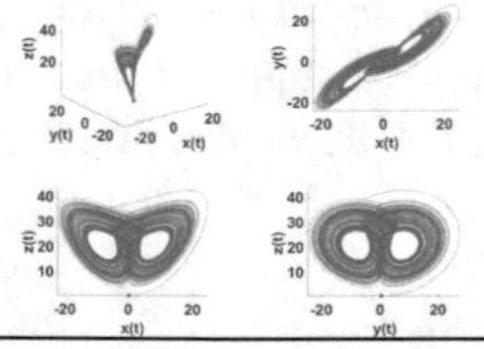	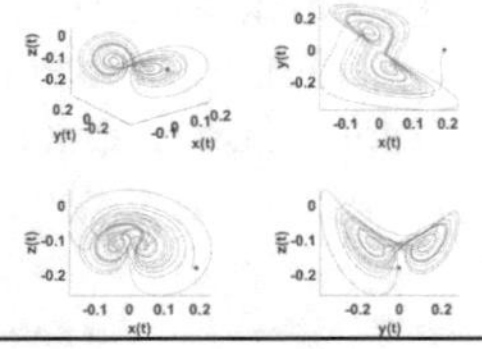	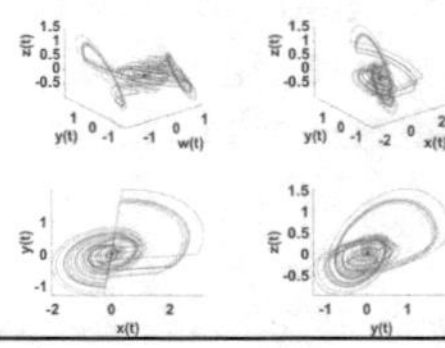

System	A dissipative hidden attractor [25]	A dissipative self-excited attractor [65]	A conservative system [25]
Equations	$D^{\alpha}x = -y$ $D^{\beta}y = x + z$ $D^{\gamma}z = 2y^2 + xz - a$	$D^{\alpha}x = y$ $D^{\beta}y = z$ $D^{\gamma}z = -y - az + b\left(\frac{x^2}{c} - c\right)$	$D^{\alpha}x = y$ $D^{\beta}y = -x - zy$ $D^{\gamma}z = y^2 - a$
Parameters	$a = 0.35$	$(a,b,c) = (0.6,0.58,1)$	$a = 1$
Initial Point	$(x_0,y_0,z_0) = (0,0.4,1)$	$(x_0,y_0,z_0) = (0.25,0.2,0.25)$	$(x_0,y_0,z_0) = (0,5,0)$
Discretized Solution	$x_{n+1} = \frac{-\sum_{j=1}^{n+1} c_j{}^{\alpha} x_{n+1-j} - y_n}{c_0{}^{\alpha}}$ $y_{n+1} = \frac{-\sum_{j=1}^{n+1} c_j{}^{\beta} y_{n+1-j} + x_n + z_n}{c_0{}^{\beta}}$ $z_{n+1} = \frac{-\sum_{j=1}^{n+1} c_j{}^{\gamma} z_{n+1-j} + 2y_ny_{n+1} + 2x_{n+1}z_n - a}{c_0{}^{\gamma} + x_{n+1}}$	$x_{n+1} = \frac{-\sum_{j=1}^{n+1} c_j{}^{\alpha} x_{n+1-j} + y_n}{c_0{}^{\alpha}}$ $y_{n+1} = \frac{-\sum_{j=1}^{n+1} c_j{}^{\beta} y_{n+1-j} + z_n}{c_0{}^{\beta}}$ $z_{n+1} = \frac{-\sum_{j=1}^{n+1} c_j{}^{\gamma} z_{n+1-j} - y_{n+1} - az_n + b\left(\frac{x_{n+1}{}^2}{c} - c\right)}{c_0{}^{\gamma}}$	$x_{n+1} = \frac{-\sum_{j=1}^{n+1} c_j{}^{\alpha} x_{n+1-j} + y_n}{c_0{}^{\alpha}}$ $y_{n+1} = \frac{-\sum_{j=1}^{n+1} c_j{}^{\beta} y_{n+1-j} - x_{n+1} - 2z_ny_n}{c_0{}^{\beta} - z_n}$ $z_{n+1} = \frac{-\sum_{j=1}^{n+1} c_j{}^{\gamma} z_{n+1-j} + y_{n+1}{}^2 - a}{c_0{}^{\gamma}}$
Attractor $\alpha = 1$ $\beta = 1$ $\gamma = 1$	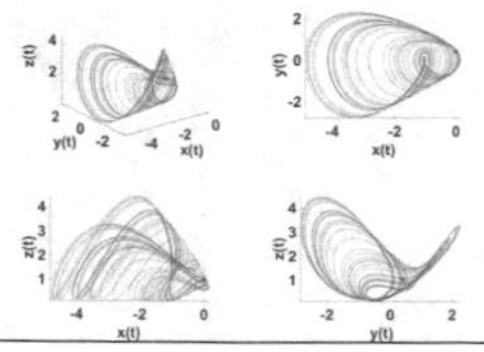	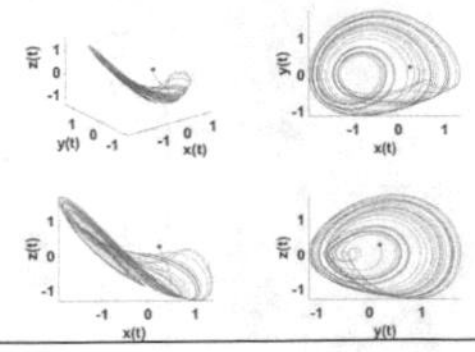	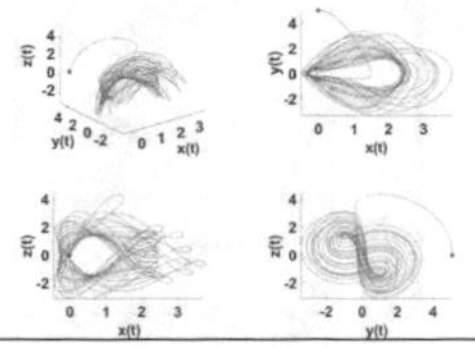

hidden attractor with no equilibria and quadratic non-linearity [25], a dissipative self-excited attractor with quadratic non-linearity [65], and a conservative one [25] with the equations shown in Table 1.

Discretized solutions to the systems could be obtained using (5) and NSFD. Non-linear terms including the same state variable that is being calculated are replaced with the aid of the nonlocal discrete representations. For example, in the equation of $D^{\beta}y$, the following rules are used for replacement:

$$y \approx y_n \quad y^2 \approx y_n y_{n+1}, \quad xy \approx 2x_{n+1}y_n - x_{n+1}y_{n+1}, \text{ and } zy \approx 2z_n y_n - z_n y_{n+1}. \qquad (8)$$

The relations used for solving the systems are given in Table 1.

Subscripts will be used later on to characterize different roles that a system could act as a master or slave. There are various possible values for the fractional-orders where the effect of fractional-orders and criteria of choosing them are studied in the next section.

4 Sensitivity to Fractional-Orders and Parameters Variation

In this section, we discuss the sensitivity of the six presented systems to parameters and fractional-orders. Numerical simulations are used to identify when they generate periodic or chaotic responses. In addition, we compare the shape of their attractors in integer-order and fractional-order. For simplicity, the three fractional-orders in

Table 2 Lü system responses versus fractional-order α and parameter a

	$\alpha = 0.92$	$\alpha = 0.95$	$\alpha = 0.98$	$\alpha = 1$
$a = 30$	Chaotic	Chaotic	Chaotic	Chaotic
$a = 36$	Stable	Chaotic	Chaotic	Chaotic

Table 3 Newton-Leipnik system responses versus fractional-order α and parameters

	$\alpha = 0.92$	$\alpha = 0.95$	$\alpha = 0.98$	$\alpha = 1$
$a = 0.4$ $b = 0.175$	Stable	Chaotic	Chaotic	Chaotic
$a = 0.6$ $b = 0.175$	Stable	Stable	Chaotic	Chaotic

Table 4 Chua's circuit system responses versus fractional-order α and parameter a

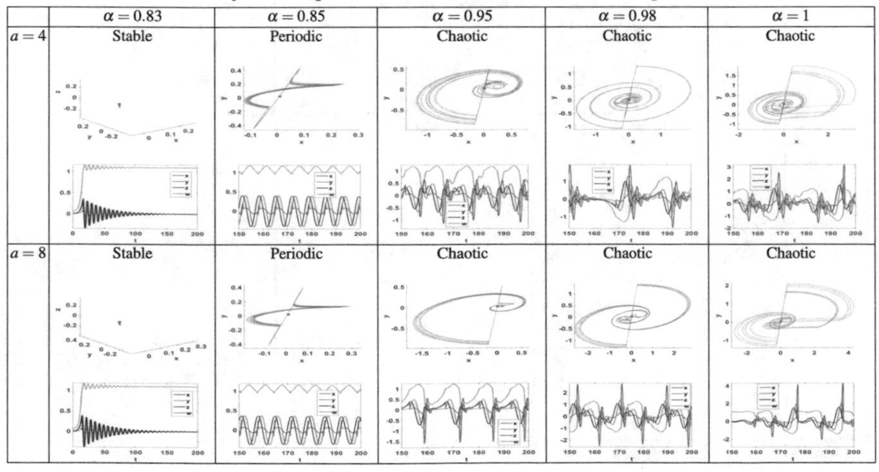

	$\alpha = 0.83$	$\alpha = 0.85$	$\alpha = 0.95$	$\alpha = 0.98$	$\alpha = 1$
$a = 4$	Stable	Periodic	Chaotic	Chaotic	Chaotic
$a = 8$	Stable	Periodic	Chaotic	Chaotic	Chaotic

the system of fractional differential equations are assumed to be equal, i.e., in this section $\alpha = \beta = \gamma$ and the unified fractional-order is denoted by α.

Tables 2, 3, 4, 5, 6 and 7 show the post-transient time series of the 3 phase space dimensions x, y and z as well as the post-transient attractor diagram with the initial point marked in red illustrating the obtained type of solution (periodic or chaotic) for different values of the fractional-order. It should be noted that in the upcoming

Table 5 Response type of the dissipative system with hidden attractor at various values of fractional-order α and parameter a

	$\alpha = 0.96$	$\alpha = 0.98$	$\alpha = 1$
$a = 0.15$	Periodic 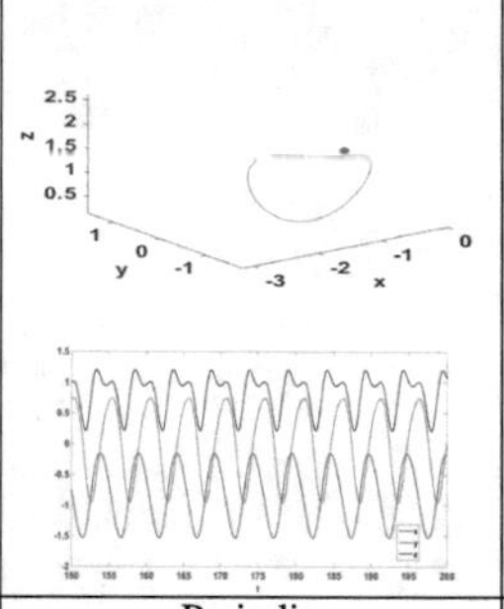	Periodic 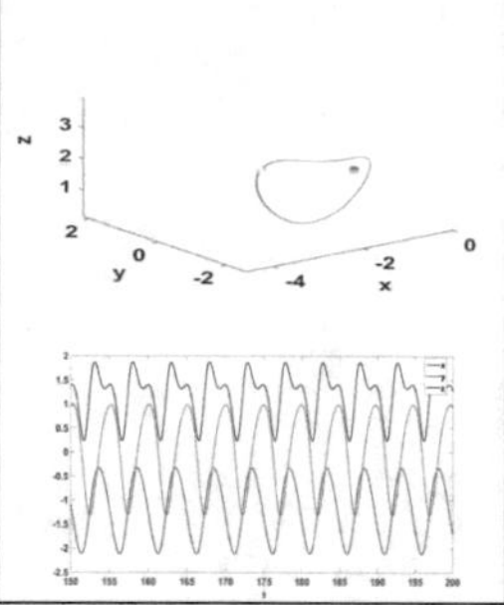	Chaotic 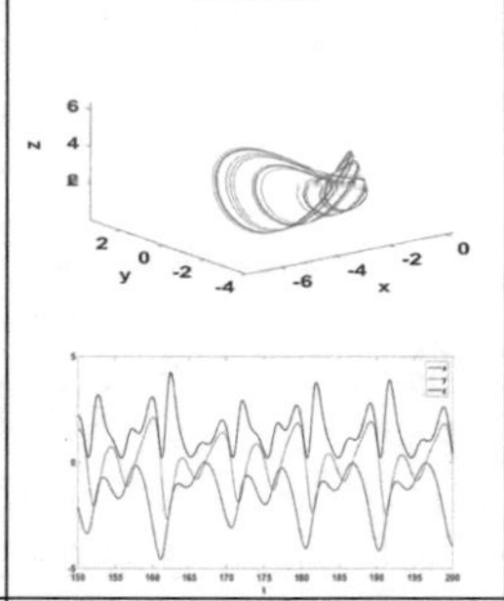
$a = 0.25$	Periodic 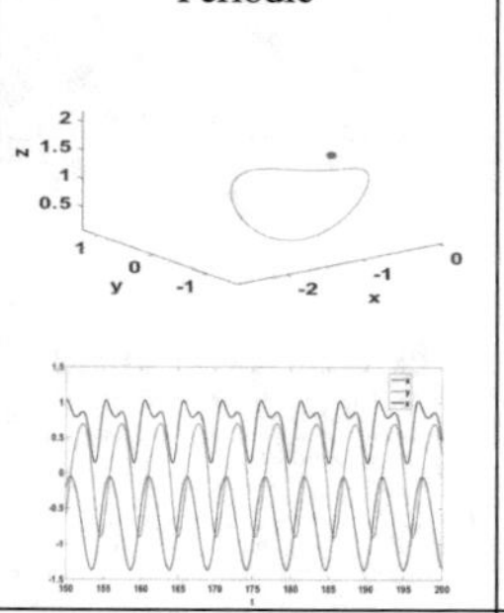	Periodic 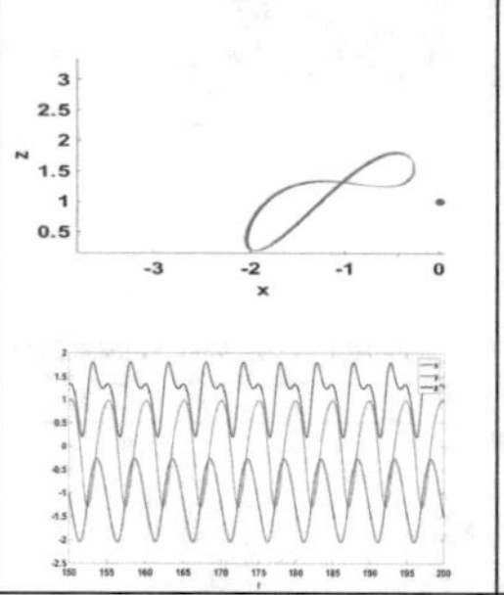	Chaotic 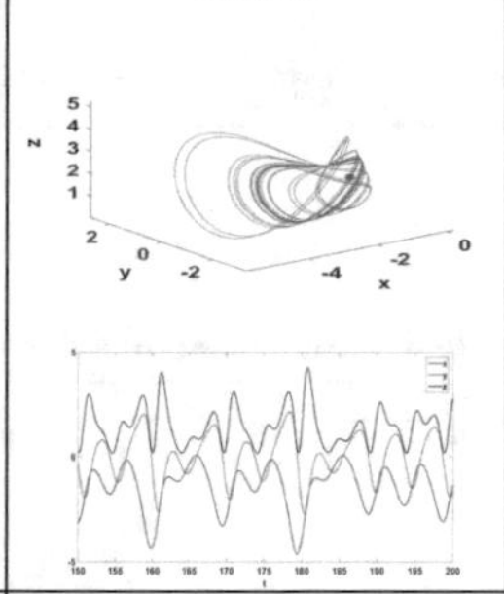
$a = 0.35$	Divergence 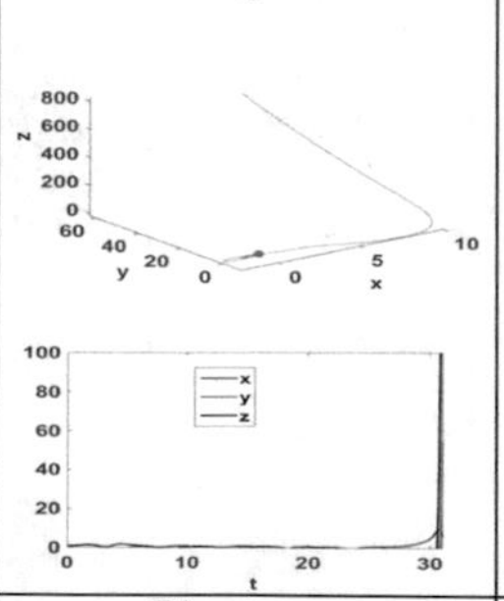	Periodic 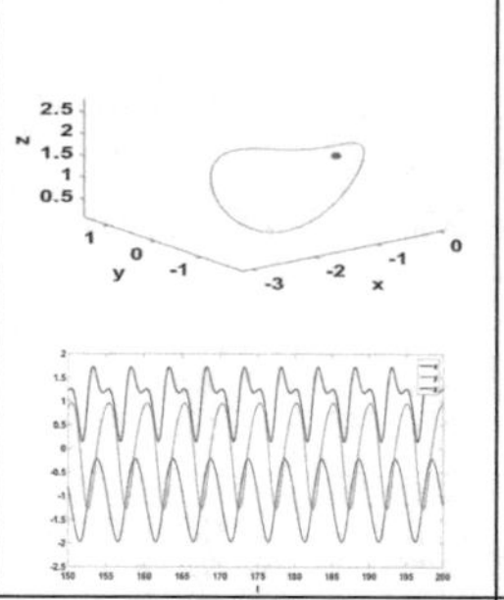	Chaotic 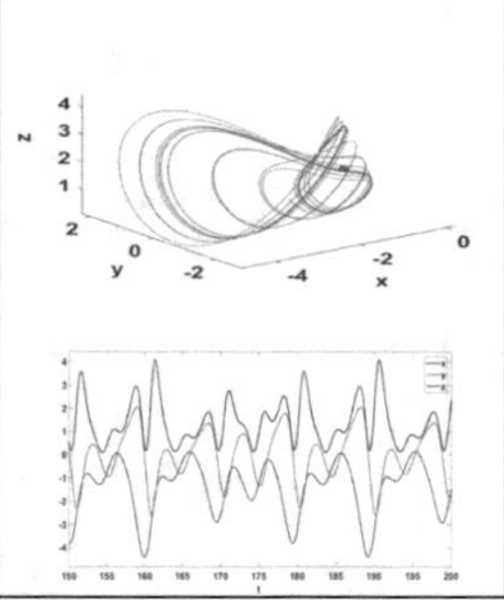
$a = 0.45$	Divergence 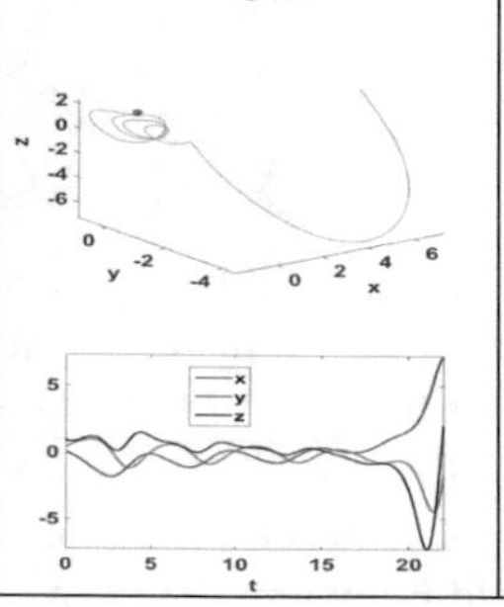	Periodic 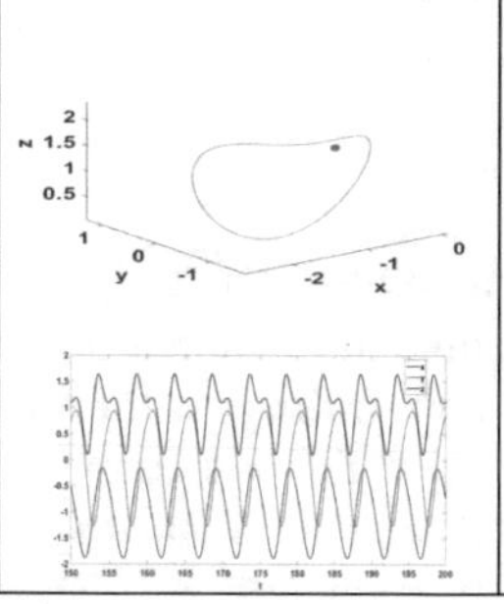	Chaotic 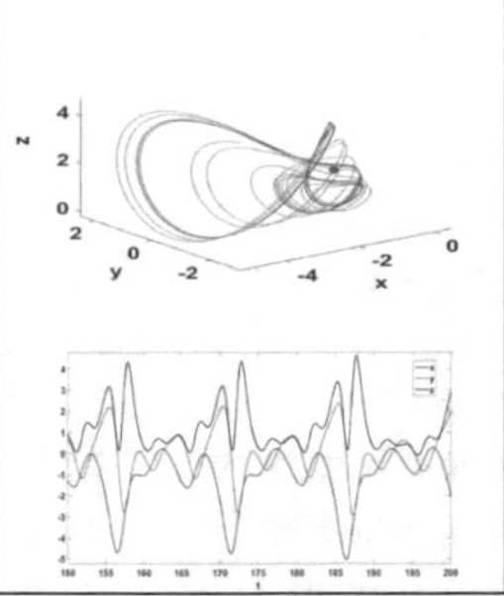

Table 6 Response type of the dissipative system with self-excited attractor at various values of fractional-order α and parameters a and b

	$\alpha = 0.95$	$\alpha = 0.97$	$\alpha = 0.98$	$\alpha = 0.99$	$\alpha = 1$
$a = 0.6$ $b = 0.58$	Periodic	Periodic	Periodic	Chaotic	Chaotic
$a = 1.2$ $b = 1.2$	Periodic	Periodic	Periodic	Periodic	Divergence

Table 7 Conservative system responses versus fractional-order α and parameter a

	$\alpha = 0.95$	$\alpha = 0.97$	$\alpha = 0.99$	$\alpha = 1$
$a = 1$	Periodic	Periodic	Chaotic	Chaotic
$a = 1.5$	Periodic	Periodic	Quasi-periodic	Chaotic

sections, transient regions of the time series and attractors are shown to illustrate that synchronization takes place early at the beginning of simulation time.

The dissipative system with hidden attractor exhibits a narrow range of fractional-orders that yield chaotic behavior and may exhibit divergent responses. Consequently, it is utilized in Sects. 6 and 7 as a slave system to control its response. For the dissipative system with self-excited attractor, the parameter c is just a scaling parameter [65], so the effect of a, b as well as α is considered.

5 Control and Switching Synchronization

In this section, a control and switching technique for synchronizing the response of any chaotic system to follow another pattern is presented. This can be achieved through two switches that control the role of each system whether it acts as a master or a slave. Figure 1 shows the general block diagram that describes the proposed technique for two chaotic systems. Conventional synchronization is defined as changing the response of the slave system to synchronize with the master chaotic system and exactly follow its pattern. This purpose is achieved using active control functions which affect only the slave response without any loading on the master system [7, 43].

The switching synchronization technique is applied to the Lü system and the Newton-Leipnik system. Hence, their equations with the switches and control functions effect being considered are given by:

$$D^{\alpha}x_1 = a_1(y_1 - x_1) - S_1u_x, \tag{9a}$$

$$D^{\beta}y_1 = b_1y_1 - x_1z_1 - S_1u_y, \tag{9b}$$

$$D^{\gamma}z_1 = x_1y_1 - c_1z_1 - S_1u_z, \tag{9c}$$

and

$$D^{\alpha}x_2 = -a_2x_2 + y_2 + 10y_2z_2 + S_2u_x, \tag{10a}$$

$$D^{\beta}y_2 = -x_2 - 0.4y_2 + 5x_2z_2 + S_2u_y, \tag{10b}$$

$$D^{\gamma}z_2 = b_2z_2 - 5x_2y_2 + S_2u_z, \tag{10c}$$

where S_1 and S_2 are on-off parameters (digital bit), which either have the values "1" or "0" according to the required dependence between both systems as shown in Fig. 1. The unknown terms (u_x, u_y, u_z) in (9) and (10) are active control functions to be determined, and the error functions can be defined as:

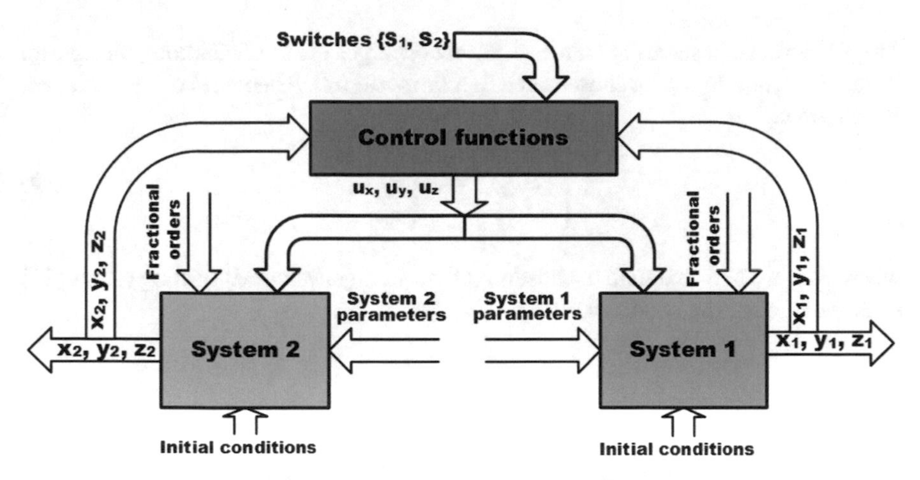

Fig. 1 Block diagram of the switched synchronization scheme between two different fractional-order chaotic systems

$$e_x = x_2 - x_1,\ e_y = y_2 - y_1,\ e_z = z_2 - z_1. \tag{11}$$

Equation (11) together with (9) and (10) yield the error system

$$D^{\alpha} e_x = -a_2(e_x + x_1) + (1 + 10(e_z + z_1))(e_y + y_1) - a_1(y_1 - x_1) + (S_1 + S_2)u_x, \tag{12a}$$

$$D^{\beta} e_y = -(e_x + x_1) - 0.4(e_y + y_1) + 5(e_x + x_1)(e_z + z_1) - b_1 y_1 + x_1 z_1 + (S_1 + S_2)u_y, \tag{12b}$$

$$D^{\gamma} e_z = b_2(e_z + z_1) - 5(e_x + x_1)(e_y + y_1) - x_1 y_1 + c_1 z_1 + (S_1 + S_2)u_z. \tag{12c}$$

The active control functions (u_x, u_y, u_z) are defined as follows

$$(S_1 + S_2)u_x = V_x(e_x) - (1 + 10(e_z + z_1))(e_y + y_1) + a_1(y_1 - x_1) + a_2 x_1, \tag{13a}$$

$$(S_1 + S_2)u_y = V_y(e_y) + e_x + x_1 + (0.4 + b_1)y_1 - 5(e_x + x_1)(e_z + z_1) - x_1 z_1, \tag{13b}$$

$$(S_1 + S_2)u_z = V_z(e_z) - (b_2 + c_1)z_1 + 5(e_x + x_1)(e_y + y_1) + x_1 y_1. \tag{13c}$$

The terms V_x, V_y and V_z are linear functions of the error terms e_x, e_y and e_z. With the choice of u_x, u_y and u_z given by (13) the error system between the two chaotic systems (12) becomes

$$D^{\alpha} e_x = -a_2 e_x + V_x(e_x),\ D^{\beta} e_y = -0.4 e_y + V_y(e_y),\ D^{\gamma} e_z = b_2 e_z + V_z(e_z). \tag{14}$$

There is no need to solve (14) if the solution converges to zero. Therefore, the control terms V_x, V_y and V_z can be chosen such that the system (15) becomes stable with zero steady state.

$$\begin{pmatrix} V_x(e_x) \\ V_y(e_y) \\ V_z(e_z) \end{pmatrix} = A \begin{pmatrix} e_x \\ e_y \\ e_z \end{pmatrix}, \tag{15}$$

where A is a 3×3 real matrix chosen so that all eigenvalues λ_i of the system (15) satisfy the following condition:

$$|arg(\lambda_i)| > \frac{\alpha\pi}{2}. \tag{16}$$

Hence, the matrix A is chosen as follows

$$A = \begin{pmatrix} a_2 - k_x & 0 & 0 \\ 0 & 0.4 - k_y & 0 \\ 0 & 0 & -b_2 - k_z \end{pmatrix}. \tag{17}$$

Then the eigenvalues of the linear system (15) satisfy the necessary and sufficient condition (16) for all fractional-orders $\alpha < 2$ [53]. In this specific case, $k_x = k_y = k_z = 100$ is chosen to overcome the large difference between ranges of x, y and z between the two chosen systems.

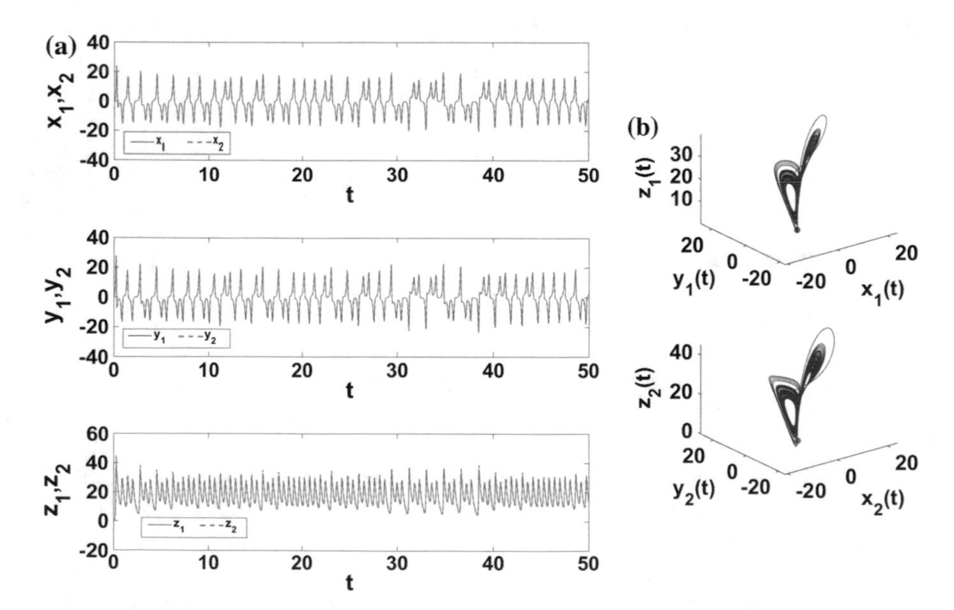

Fig. 2 Static switching setting Lü as a master and Newton-Leipnik as a slave **a** Time series, **b** Attractor diagrams

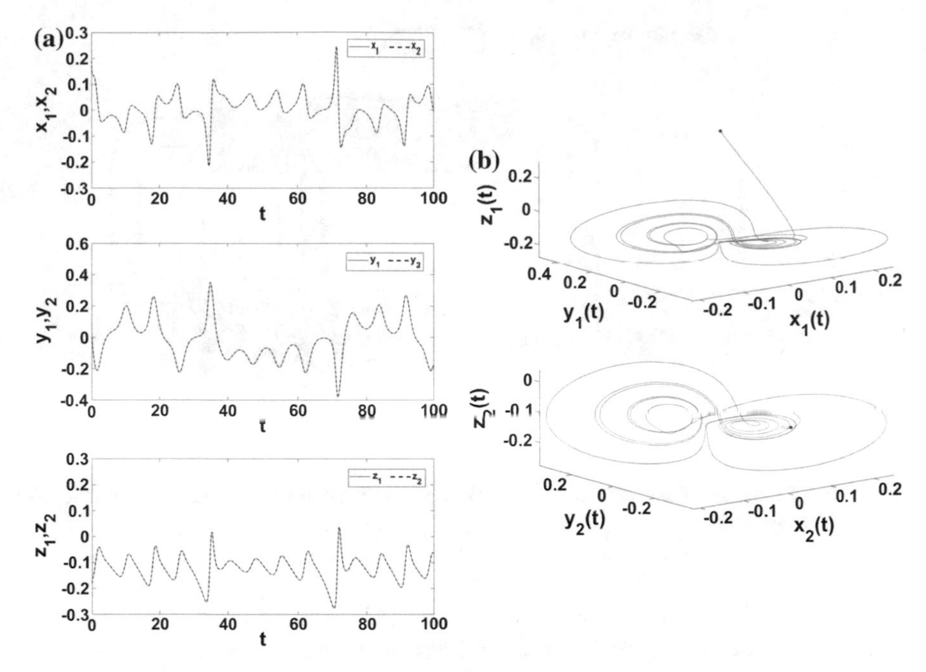

Fig. 3 Static switching setting Newton-Leipnik as a master and Lü as a slave **a** Time series, **b** Attractor diagrams

Simulation results validate the previous analysis as shown in Fig. 2. Time series and attractor diagrams are shown for the case $(S_1, S_2) = (0, 1)$ at $(\alpha, \beta, \gamma) = (0.95, 0.96, 0.97)$. Lü system works normally and the Newton-Leipnik system adapts its response to follow the Lü system. Time series of Lü are represented by the solid lines while the dotted lines correspond to those of Newton-Leipnik. Similarly, Fig. 3 shows time series and attractor diagrams in the reverse case when $(S_1, S_2) = (1, 0)$. Dynamic or mixed synchronization could also be achieved in which the switches become functions of time. In this case, the role of each system is not fixed throughout the simulation time, i.e., both systems can exchange their roles at any time instant.

6 Synchronization of Two Fractional-Order Chaotic Systems

6.1 Generalized Synchronization

Generalized synchronization aims at changing the response of the slave system to follow a given relation with the master system. Based on the active control method for synchronization and anti-synchronization [21] and the generalized synchroniza-

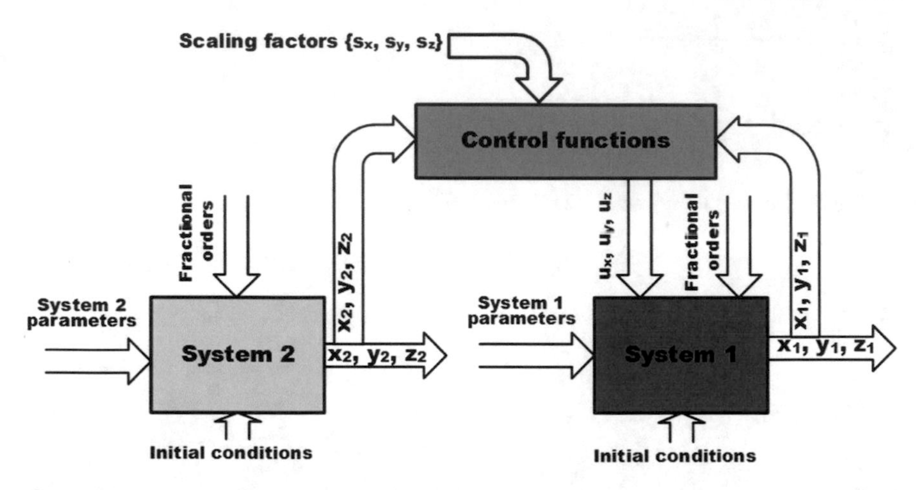

Fig. 4 Block diagram of the generalized synchronization scheme between two different fractional-order chaotic systems

tion that has been applied to identical systems [43], a more general synchronization scheme is adapted as shown in Fig. 4 such that it becomes suitable for different systems too. For 3D phase space systems, the error vector between the generated responses of the two synchronized systems is given by

$$e = \begin{pmatrix} e_x \\ e_y \\ e_z \end{pmatrix} = \begin{pmatrix} x_m + s_x x_s \\ y_m + s_y y_s \\ z_m + s_z z_s \end{pmatrix}, \tag{18}$$

where (x_m, y_m, z_m) and (x_s, y_s, z_s) are the responses of the master and slave systems, respectively. Hence,

$$x_s(t) = -\frac{x_m(t)}{s_x(t)}, \quad y_s(t) = -\frac{y_m(t)}{s_y(t)}, \quad z_s(t) = -\frac{z_m(t)}{s_z(t)}. \tag{19}$$

This generalized synchronization permits various special cases to appear at different values of the scaling factors. The cases listed below are used in this section to validate the proposed generalized synchronization technique.

6.1.1 Case 1: Scaled Synchronization and Anti-synchronization

In this case, each s_i, $i \in \{x, y, z\}$, is a constant value which is time-independent. All responses (x, y and z for 3-D) could be scaled with the same factor or each with a different factor. For s_i negative, the slave response is in phase (synchronized) with the master response, while for positive values of s_i they have an opposite phase (anti-

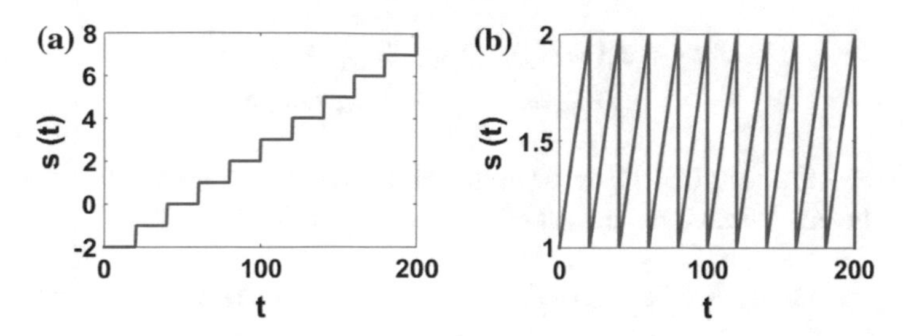

Fig. 5 Examples on scaling factors **a** $s(t) = -2 + int(t/20)$ and **b** $s(t) = 1 + (mod(t/20))/20$

synchronized). Moreover, when $|s_i| < 1$, the slave response has a higher amplitude than the master response, whereas it has a lower amplitude when $|s_i| > 1$ according to (19).

6.1.2 Case 2: Scaling Factors $s_i(t)$ Are Functions of Time

Multiple cases in which s_i is a function of time could be described. For example, $s_i(t) = c + int(t/m)$, where c, m are constants and $int(.)$ returns the quotient of integer division. This is a stair-case function which performs scaling in a variable manner as time advances. The type of synchronization (anti-synchronization) and/or its scale change every m time units as the example shown in Fig. 5a. Another example is $s_i(t) = c + (mod(t/m))/m$, where c, m are constants and $mod(.)$ returns the remainder of integer division. This is a periodic ramp function which is time-dependent too. The value of the scaling factor increases within every interval of m time units and resets at the end of each interval as the example shown in Fig. 5b.

6.2 Simulation Results for Two Fractional-Order Systems

First, we consider generalized synchronization of two identical systems in which only parameters and/or initial conditions differ. For this purpose, fractional-order Chua's circuit is used which is a 4-D system. The slave (response) and master (drive) systems are described, respectively, by the following equations. However, the initial condition of the drive system is different from that of the response system.

$$\begin{aligned} D^{\alpha}x_1 &= a_1(y_1 - f(w_1)x_1) + u_x,\ D^{\beta}y_1 = z_1 - x_1 + u_y, \\ D^{\gamma}z_1 &= -b_1y_1 + c_1z_1 + u_z,\ Dw_1 = x_1 + u_w. \end{aligned} \tag{20}$$

$$D^{\alpha}x_2 = a_2(y_2 - f(w_2)x_2),\ D^{\beta}y_2 = z_2 - x_2,$$
$$D^{\gamma}z_2 = -b_2y_2 + c_2z_2,\ Dw_2 = x_2. \tag{21}$$

Extending Eq. (18) to 4-D, substituting in it and calculating the fractional derivatives of the error functions, the set of Eq. (22) is obtained.

$$D^{\alpha}e_x = a_2(y_2 - f(w_2)x_2) + s_x(a_1(y_1 - f(w_1)x_1) + u_x), \tag{22a}$$

$$D^{\alpha}e_y = z_2 - x_2 + s_y(z_1 - x_1 + u_y), \tag{22b}$$

$$D^{\alpha}e_z = -b_2y_2 + c_2z_2 + s_z(-b_1y_1 + c_1z_1 + u_z), \tag{22c}$$

$$D^{\alpha}e_w = x_2 + s_w(x_1 + u_w). \tag{22d}$$

For the purpose of synchronization, all terms except those which are function of the corresponding error term should be canceled. For example, in the equation of $D^{\alpha}e_x$ only e_x should appear. Hence, the vector of control functions u is given by:

$$u_x = \frac{1}{s_x}\left(-a_2(y_2 - f(w_2)x_2) - s_x(-a_1(y_1 - f(w_1)x_1)) - k_xe_x\right), \tag{23a}$$

$$u_y = \frac{1}{s_y}\left(-z_2 + x_2 - s_y(z_1 - x_1) - k_ye_y\right), \tag{23b}$$

$$u_z = \frac{1}{s_z}\left(b_2y_2 - c_2z_2 - s_z(-b_1y_1 + c_1z_1) - k_ze_z\right), \tag{23c}$$

$$u_w = \frac{1}{s_w}\left(-x_2 - s_wx_1 - k_we_w\right), \tag{23d}$$

which result in decaying error functions as the values of k_x, k_y, k_z and k_w are positive. The procedure is simple for this case, however, a more detailed analysis for the general case is provided in Sect. 7.

Figure 6 shows samples of successfully achieved generalized synchronization between the slave and master system at $(\alpha, \beta, \gamma) = (0.93, 0.95, 0.97)$ for different parameter values $(a_1, b_1, c_1) = (4.5, 0.9, 0.6)$ and $(a_2, b_2, c_2) = (4, 1, 0.65)$ and starting at different initial conditions $(x_{10}, y_{10}, z_{10}, w_{10}) = (0.02, 0.03, 0.02, 0.06)$ and $(x_{20}, y_{20}, z_{20}, w_{20}) = (0.01, 0.02, 0.01, 0.05)$.

Figure 6a shows static synchronization with $s_x = -3$ where the x-time series of the slave synchronizes with that of the master system at a scaling factor of (1/3). Figure 6b shows static synchronization with $s_y = 0.25$ where the y-time series of

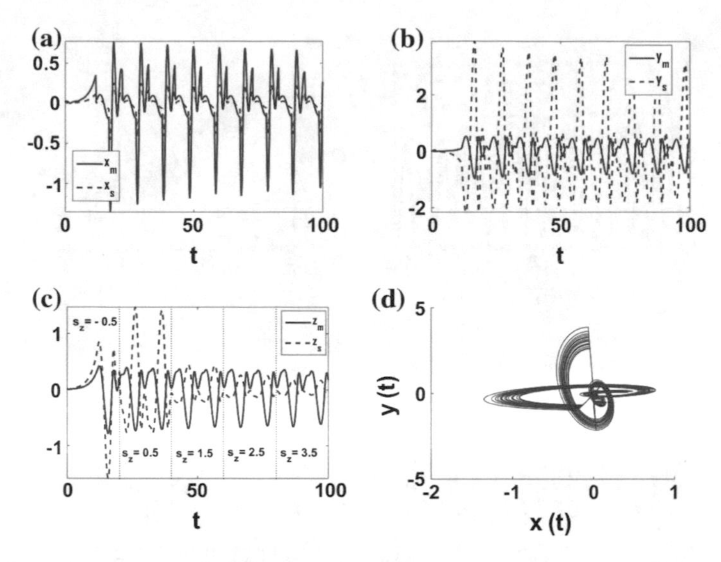

Fig. 6 Generalized synchronization of Chua's circuit at **a** $s_x = -3$, **b** $s_y = 0.25$, **c** $s_z = -0.5 + int(t/20)$ and **d** Attractor diagrams of the slave system (in *red*) different from the original attractor (in *blue*)

the slave is anti-synchronized with that of the master system at a scaling factor of 4. Figure 6c shows dynamic synchronization at $s_z(t) = -0.5 + int(t/20)$ where the scaling factor starts with a value equals -0.5 and increases by 1 every 20 time units. Figure 6d shows the resulting attractor diagram with new shape after applying these scaling functions.

Further simulation results are also presented to illustrate the generalized synchronization of two different fractional-order systems whether they generate periodic or chaotic responses. The following equations represent a dissipative hidden attractor [25] as the slave system (system 1) as well as a dissipative self-excited attractor [65] (system 2) and a conservative system [25] (system 3) as the master systems alternatively.

$$D^{\alpha}x_1 = -y_1 + u_x, \qquad D^{\beta}y_1 = x_1 + z_1 + u_y, \qquad D^{\gamma}z_1 = 2{y_1}^2 + x_1z_1 - a_1 + u_z. \tag{24}$$

$$D^{\alpha}x_2 = y_2, \qquad D^{\beta}y_2 = z_2, \qquad D^{\gamma}z_2 = -y_2 - a_2z_2 + b_2\left(\frac{{x_2}^2}{c_2} - c_2\right). \tag{25}$$

$$D^{\alpha}x_3 = y_3, \qquad D^{\beta}y_3 = -x_3 - z_3y_3, \qquad D^{\gamma}z_3 = {y_3}^2 - a_3. \tag{26}$$

The methodology of obtaining the control signals using active nonlinear control method is performed similar to the previous section and as discussed in [44].

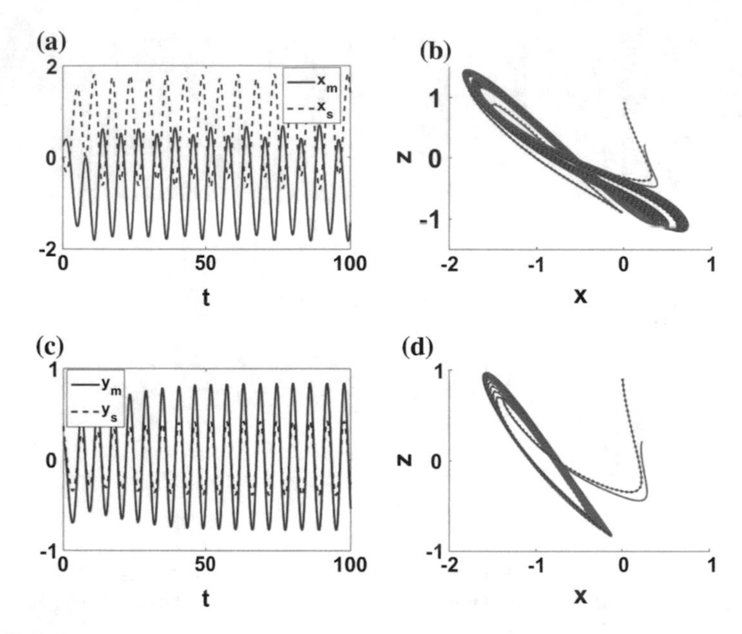

Fig. 7 Static synchronization (Case 1) with system 2 as master and system 1 as slave at **a** $\{\alpha, \beta, \gamma\} = \{0.97, 1, 0.95\}$ and $s_x = 1$, **b** $s_x = s_y = s_z = -1$, **c** $\{\alpha, \beta, \gamma\} = \{0.93, 0.93, 0.93\}$ and $s_y = -2$, and **d** $s_x = s_y = s_z = -1$

Consider the synchronization of the slave system with system 2 as the master system. The plots in Fig. 7 match the expected behavior where at $s_x = 1$, the slave response is the exact anti-synchronization of the master response, and at $s_y = -2$ it is the halved-synchronization. The master and slave attractor diagrams are shown to be co-incident in the case of full-synchronization $s_x = s_y = s_z = -1$. It is worth mentioning that the system exhibits different attractors when varying the values of fractional-orders as illustrated by the two *y-z* projections plotted at $\{\alpha, \beta, \gamma\} = \{0.97, 1, 0.95\}$ and $\{\alpha, \beta, \gamma\} = \{0.93, 0.93, 0.93\}$.

Figure 8 shows the synchronization of the slave system with system 3 as the master system. The integer-order case, or autonomous system of three first order ordinary differential equations, is shown to follow the same expected behavior as a special case of generalized fractional-order.

Figure 9 shows dynamic synchronization at $s_y(t) = -2.5 + int(t/20)$ where the scaling factor starts with a value of -2.5 and increases by 1 every 20 time units. It also shows dynamic synchronization at $s_x = -0.5 + int(t/40)$ where the scaling factor starts with a value of -0.5 and increases by 1 every 40 time units.

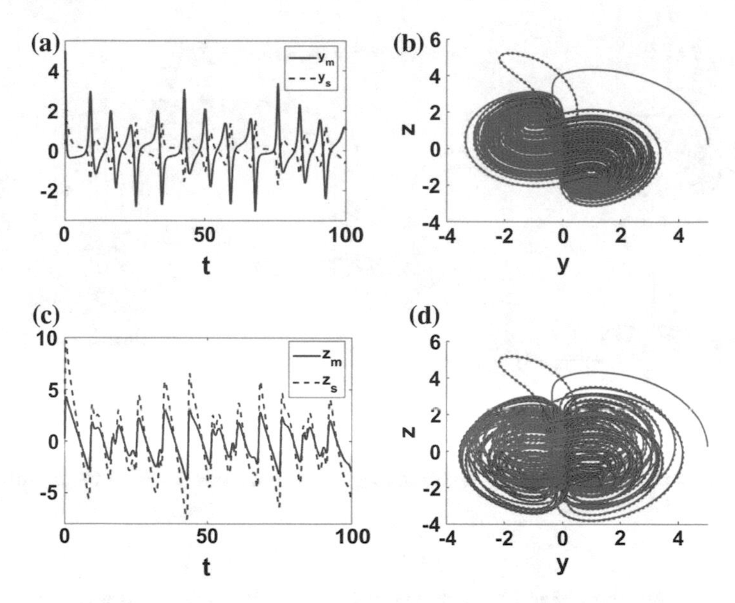

Fig. 8 Static synchronization (Case 1) with system 3 as master and system 1 as slave at **a** $\{\alpha, \beta, \gamma\} = \{1, 1, 1\}$ and $s_y = 2$, **b** $s_x = s_y = s_z = -1$, **c** $\{\alpha, \beta, \gamma\} = \{0.97, 1, 0.99\}$ and $s_z = -0.5$, and **d** $s_x = s_y = s_z = -1$

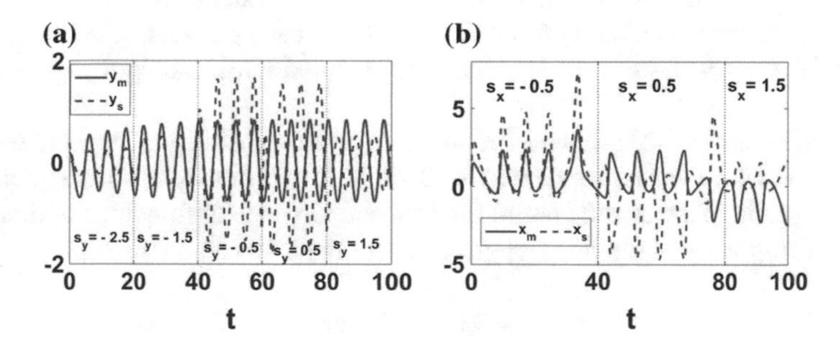

Fig. 9 Dynamic synchronization (Case 2) with system 2 as master and system 1 as slave at **a** $\{\alpha, \beta, \gamma\} = \{0.93, 0.93, 0.93\}$ and $s_y = -2.5 + int(t/20)$ and **b** $\{\alpha, \beta, \gamma\} = \{0.97, 1, 0.99\}$ and $s_x = -0.5 + int(t/40)$

7 Synchronization of a Fractional-Order Chaotic System and a Linear Combination of Two Other Systems

In this section, a novel fractional-order chaotic system is formed as a linear combination of two fractional-order systems and another system is synchronized with this linear combination. This linear combination represents another means of controlling the system response and forcing it to yield chaos. The block diagram of the generalized synchronization scheme is shown in Fig. 10 where the linear combination of systems 2 and 3 is the master and system 1 is the slave.

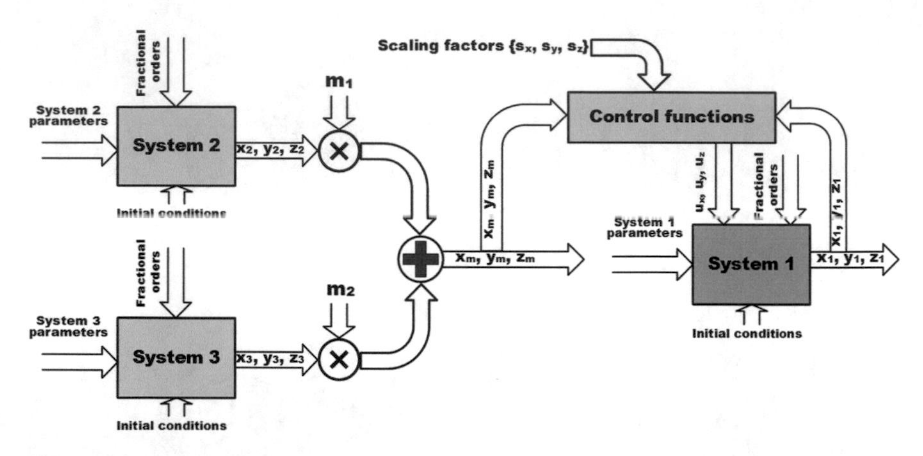

Fig. 10 Block diagram of generally synchronizing a fractional-order chaotic system with a linear combination of two systems

Table 8 shows the attractor diagrams of different linear combinations of systems 2 and 3 at various values of fractional-orders where the post-transient part is colored in blue. Several examples show that the linear combination can yield more chaotic responses or sequences with long periods in fractional-order in comparison with the single systems shown in Tables 6 and 7. These cases can be proven to exhibit chaotic behavior through well-known techniques such as maximum Lyapunov exponent calculation.

The procedure in [44] is applied but with an added capability of generalized synchronization with the cases explained in Sect. 6. As previously mentioned, all state variables, scaling factors and control functions are in general functions of time. The combined responses of the two systems shown in Fig. 10 can be written as:

$$x_m = m_1 x_2 + m_2 x_3, \tag{27a}$$

$$y_m = m_1 y_2 + m_2 y_3, \tag{27b}$$

$$z_m = m_1 z_2 + m_2 z_3, \tag{27c}$$

substituting in (18) and calculating the fractional derivatives, the set of equations (28) is obtained. For the purpose of synchronization, all terms except those which are function of the corresponding error term should be canceled. For example, in the equation of $D^\alpha e_x$ only e_x should appear. Hence, the vector of control functions u is given by (29).

Table 8 Attractor diagrams of different linear combinations of systems 2 and 3 at various values of fractional-orders

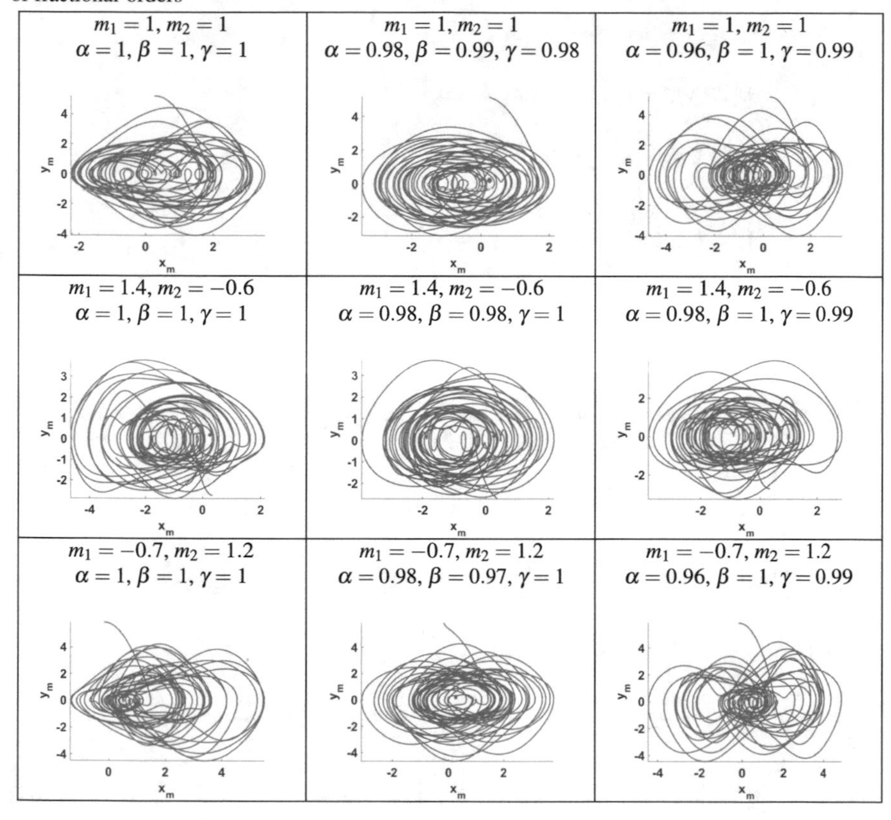

$$
\begin{aligned}
D^{\alpha} e_x = D^{\alpha}\left(x_m + s_x x_s\right) &= D^{\alpha}\left(m_1 x_2 + m_2 x_3 + s_x x_1\right) = m_1 y_2 + m_2 y_3 - s_x y_1 + s_x u_x \\
&= m_1 y_2 + m_2 y_3 - s_x\left(\frac{e_y - y_m}{s_y}\right) + s_x u_x, \qquad (28a)
\end{aligned}
$$

$$
\begin{aligned}
D^{\beta} e_y &= D^{\beta}\left(y_m + s_y y_s\right) = D^{\beta}\left(m_1 y_2 + m_2 y_3 + s_y y_1\right) \\
&= m_1 z_2 - m_2 x_3 - m_2 z_3 y_3 + s_y x_1 + s_y z_1 + s_y u_y \\
&= m_1 z_2 - m_2 x_3 - m_2 z_3 y_3 + s_y\left(\frac{e_x - x_m}{s_x}\right) + s_y\left(\frac{e_z - z_m}{s_z}\right) + s_y u_y, \qquad (28b)
\end{aligned}
$$

$$
\begin{aligned}
D^{\gamma} e_z &= D^{\gamma}\left(z_m + s_z z_s\right) \\
&= -m_1 y_2 - m_1 a_2 z_2 + m_1 b_2 \left(\frac{{x_2}^2}{c_2} - c_2\right) + m_2 {y_3}^2 - m_2 a_3 \\
&\quad + 2 s_z {y_1}^2 + s_z x_1 z_1 - s_z a_1 + s_z u_z \\
&= -m_1 \left(y_2 + a_2 z_2 - b_2 \left(\frac{{x_2}^2}{c_2} - c_2\right)\right) + m_2 \left({y_3}^2 - a_3\right) \\
&\quad + s_z \left(2\left(\frac{e_y - y_m}{s_y}\right)^2 + \left(\frac{e_x - x_m}{s_x}\right)\left(\frac{e_z - z_m}{s_z}\right) - a_1 + u_z\right).
\end{aligned} \tag{28c}
$$

Therefore, the control functions can be obtained by using (15) as follows:

$$
u_x = V_x(e_x) - \frac{m_1}{s_x} y_2 - \frac{m_2}{s_x} y_3 - \frac{1}{s_y} y_m + \frac{1}{s_y} e_y = V_x(e_x) - \frac{1}{s_x} y_m - \frac{1}{s_y} y_m + \frac{1}{s_y} e_y, \tag{29a}
$$

$$
u_y = V_y(e_y) - \frac{m_1}{s_y} z_2 + \frac{m_2}{s_y} x_3 + \frac{m_2}{s_y} z_3 y_3 + \frac{1}{s_x} x_m + \frac{1}{s_z} z_m - \frac{1}{s_x} e_x - \frac{1}{s_z} e_z, \tag{29b}
$$

$$
\begin{aligned}
u_z = V_z(e_z) &+ \frac{m_1}{s_z} y_2 + \frac{m_1}{s_z} a_2 z_2 - \frac{m_1}{s_z} b_2 \left(\frac{{x_2}^2}{c_2} - c_2\right) - \frac{m_2}{s_z} {y_3}^2 + \frac{m_2}{s_z} a_3 \\
&- 2\left(\frac{e_y - y_m}{s_y}\right)^2 - \left(\frac{e_x - x_m}{s_x}\right)\left(\frac{-z_m}{s_z}\right) + a_1.
\end{aligned} \tag{29c}
$$

Recalling that $(e_x - x_m)/s_x = x_1$ from (18), the following equations for fractional derivatives of error are, thus, obtained:

$$
D^{\alpha} e_x = s_x V_x(e_x), \tag{30a}
$$

$$
D^{\beta} e_y = s_y V_y(e_y), \tag{30b}
$$

$$
D^{\gamma} e_z = s_z V_z(e_z) + x_1 e_z. \tag{30c}
$$

Based on the nonlinear control theory and Lyapunov stability theory [53], these derivatives should be decaying functions of the error. The terms $V_x(e_x)$, $V_y(e_y)$, and $V_z(e_z)$ form a system of linear equations in the errors e_x, e_y, and e_z. They should be chosen carefully to form a stable system with zero steady state [44]. Consequently, they should force negative eigen values for the synchronization system:

$$\begin{pmatrix} V_x(e_x) \\ V_y(e_y) \\ V_z(e_z) \end{pmatrix} = \begin{pmatrix} -\frac{k_x}{s_x} & 0 & 0 \\ 0 & -\frac{k_y}{s_y} & 0 \\ 0 & 0 & -\left(\frac{k_z+x_1}{s_z}\right) \end{pmatrix} \begin{pmatrix} e_x \\ e_y \\ e_z \end{pmatrix}. \quad (31)$$

Here, the coefficients k_x, k_y, k_z are simply chosen as ones. It could be proved, similar to [43, 44], that the designed controller achieves the general required synchronization function.

Simulation results are presented, which validate the synchronization of a chaotic system with a linear combination of two other systems. The plots shown in Figs. 11, 12 and 13 match the explanation of different cases of generalized synchronization explained in Sect. 6 with the same parameters and initial values given in Table 1. Various generalized static and dynamic synchronization cases for different values of the linear combination's coefficients m_1 and m_2 are demonstrated.

The case $m_1 = m_2 = 1$ represents synchronizing the slave system with the sum of the two other systems. In addition, the response of the slave system follows the selected synchronization case among the cases illustrated in Sect. 6 and according to (19). For example, when $s_x = 0.5$, x-time series of the slave system is the doubled anti-synchronization of that of the linear combination (master system). When $s_z(t) = 1 + (mod(t/50))/50$, the z-time series of the slave starts as the exact anti-synchronized version of that corresponding to the master. Then, the scaling factor

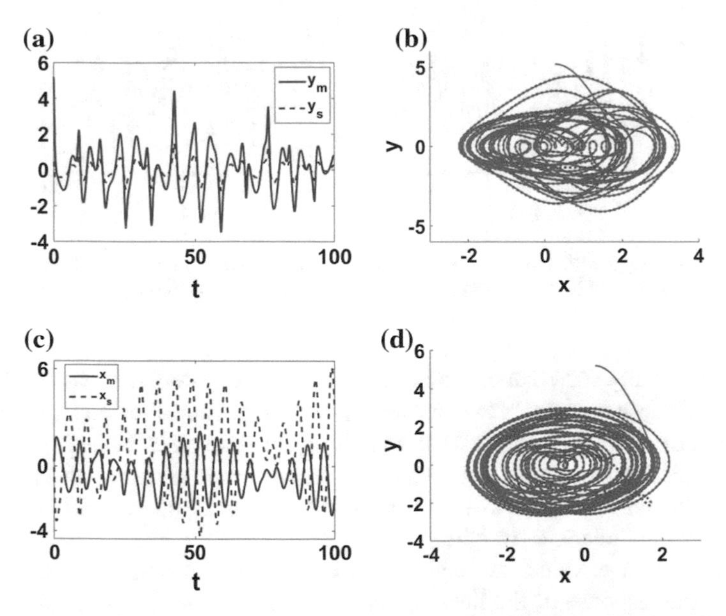

Fig. 11 Static synchronization (Case 1) of system 1 with a linear combination of systems 2 and 3 at $m_1 = m_2 = 1$ and **a** $\{\alpha, \beta, \gamma\} = \{1, 1, 1\}$ and $s_y = -3$, **b** $s_x = s_y = s_z = -1$, **c** $\{\alpha, \beta, \gamma\} = \{0.97, 1, 0.95\}$ and $s_x = 0.5$, and **d** $s_x = s_y = s_z = -1$

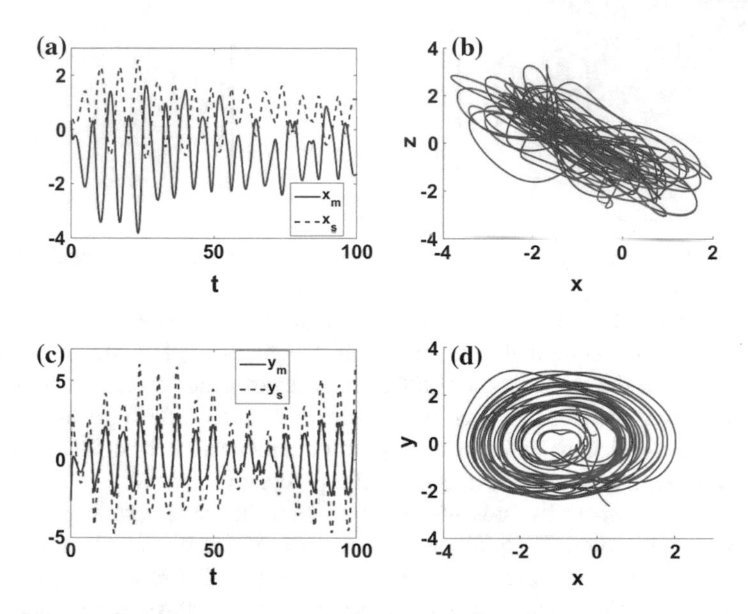

Fig. 12 Static synchronization (Case 1) of system 1 with a linear combination of systems 2 and 3 at $m_1 = 1.4$ and $m_2 = -0.6$ and **a** $\{\alpha, \beta, \gamma\} = \{0.97, 0.99, 0.99\}$ and $s_x = 1.5$, **b** $s_x = s_y = s_z = -1$, **c** $\{\alpha, \beta, \gamma\} = \{0.98, 0.98, 0.98\}$ and $s_y = -0.5$, and **d** $s_x = s_y = s_z = -1$

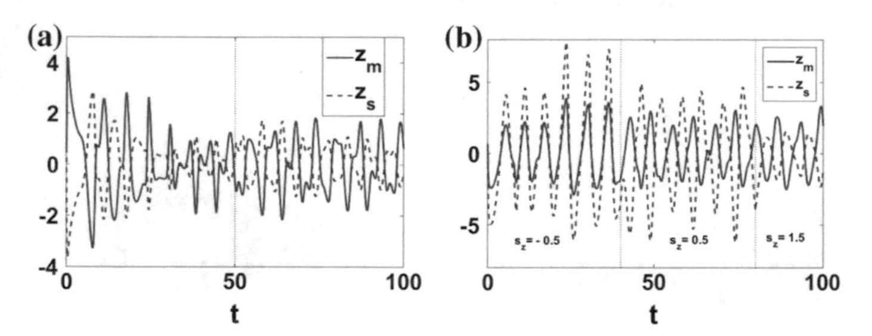

Fig. 13 Dynamic synchronization (Case 2) **a** $m_1 = m_2 = 1$, $\{\alpha, \beta, \gamma\} = \{0.97, 1, 0.95\}$ and $s_z = 1 + (mod(t/50))/50$ and **b** $m_1 = 2$ and $m_2 = -0.5$, $\{\alpha, \beta, \gamma\} = \{0.98, 0.98, 0.98\}$ and $s_z = -0.5 + int(t/40)$

increases gradually such that the amplitude of the slave system decreases till $t = 50$. At $t = 50$, the system returns to exact anti-synchronization followed by the gradual decrease in the amplitude of the slave. Synchronization at other values of m_1 and m_2 could be described similarly, e.g., $m_1 = 1.4$ and $m_2 = -0.6$ shown in Fig. 12. The resulting attractor diagram is usually similar to that of the system which has a higher value for the coefficient, or weight.

The time series at values of fractional-orders around those in Figs. 11, 12 and 13 show that the response of the linear combination is chaotic, i.e., the values do not repeat. Setting $m_1 = 1$ and $m_2 = 0$, or alternatively $m_1 = 0$ and $m_2 = 1$, yields the same results as those in Sect. 6. Other values for the coefficients are possible and yield consistent results too.

8 Conclusions

Six chaotic systems were selected and utilized in their fractional-order form to analyze and validate three proposed block diagrams of synchronization systems. Discretized solutions to the systems were obtained using the Grünwald-Letnikov method of approximation and the nonstandard finite difference method for discretization. Synchronization techniques were based on active nonlinear control and Lyapunov stability. The nonlinear controller is designed to ensure the stability and convergence of the proposed synchronization scheme.

The first block diagram presents a switching synchronization scheme between two different chaotic systems or one chaotic system with different parameters using the active control method. By using the proposed technique, it is possible to perform static synchronization (switching control independent of time), mono-dynamic synchronization (one of the control switches depends on time) or bi-dynamic synchronization (the two switches are time dependent). The concepts introduced in this block diagram have been verified by using the fractional-order version of two different known chaotic systems, which are the Lü and the Newton-Leipnik chaotic systems. Moreover, the switching parameters can be a function of time to introduce a new concept of static and dynamic switching of synchronizations, which makes the system more flexible as shown from the results.

The second block diagram presents a generalized synchronization scheme that has been validated to work for different chaotic systems as well as identical systems. This generalized synchronization permits both static and dynamic synchronization or anti-synchronization with various scaling factors. Hence, conventional synchronization is considered a very narrow subset from the proposed technique where the scale between the output response and the input response can be controlled via control functions and this scale may be either constant (positive, negative) or time dependent. Many examples including synchronization and anti-synchronization, between identical or different systems with the same or different system parameters and initial conditions, are discussed. The scaling functions are chosen to be positive/negative and constant/dynamic, which covers all possible cases.

The proposed technique utilizing dynamic scaling functions can be useful in amplitude modulation applications in which the amplitude of the output signal should be a function of the input signal. The scaling factors in this case play the role of information signal, which is modulated by the chaotic dynamics of the system to give the modulated signal. Demodulation can be done similarly by reversing the operation.

Finally, a new chaotic system, constructed as a linear combination of two different systems, was introduced with two extra parameters that correspond to more degrees of freedom and response controlling capability. The generalized synchronization method was shown to successfully synchronize a third system with the system formed by the linear combination through both mathematical analysis and simulation results. All cases of the generalized synchronization scheme were validated in generalized fractional-order domain, with integer-order as a special case, for different choices of the linear combination's coefficients and values of the scaling factors. Time series for various system responses and attractor diagrams were plotted to demonstrate different cases of generalized synchronization.

References

1. Abd-El-Hafiz, S. K., Radwan, A. G., & AbdElHaleem, S. H. (2015). Encryption applications of a generalized chaotic map. *Applied Mathematics & Information Sciences*, *9*(6), 1–19.
2. Abd-El-Hafiz, S. K., AbdElHaleem, S. H., & Radwan, A. G. (2016). Novel permutation measures for image encryption algorithms. *Optics and Lasers in Engineering*, *85*, 72–83.
3. Agrawal, S., Srivastava, M., & Das, S. (2012). Synchronization of fractional order chaotic systems using active control method. *Chaos, Solitons & Fractals*, *45*(6), 737–752.
4. Azar, A. T., & Vaidyanathan, S. (2015). *Chaos modeling and control systems design*. Springer.
5. Azar, A. T., & Vaidyanathan, S. (2016). *Advances in chaos theory and intelligent control* (Vol. 337). Springer.
6. Barakat, M. L., Mansingka, A. S., Radwan, A. G., & Salama, K. N. (2013). Generalized hardware post-processing technique for chaos-based pseudorandom number generators. *ETRI Journal*, *35*(3), 448–458.
7. Bhalekar, S., & Daftardar-Gejji, V. (2010). Synchronization of different fractional order chaotic systems using active control. *Communications in Nonlinear Science and Numerical Simulation*, *15*(11), 3536–3546.
8. Boulkroune, A., Bouzeriba, A., Bouden, T., & Azar, A. T. (2016a). Fuzzy adaptive synchronization of uncertain fractional-order chaotic systems. In *Advances in chaos theory and intelligent control* (pp. 681–697). Springer.
9. Boulkroune, A., Hamel, S., Azar, A. T., & Vaidyanathan, S. (2016b). Fuzzy control-based function synchronization of unknown chaotic systems with dead-zone input. In *Advances in chaos theory and intelligent control* (pp. 699–718). Springer.
10. Caponetto, R. (2010). *Fractional order systems: Modeling and control applications* (Vol. 72). World Scientific.
11. Chen, D., Zhang, R., Ma, X., & Liu, S. (2012). Chaotic synchronization and anti-synchronization for a novel class of multiple chaotic systems via a sliding mode control scheme. *Nonlinear Dynamics*, *69*(1–2), 35–55.
12. Chen, D., Wu, C., Iu, H. H., & Ma, X. (2013). Circuit simulation for synchronization of a fractional-order and integer-order chaotic system. *Nonlinear Dynamics*, *73*(3), 1671–1686.
13. Chen, S., & Lü, J. (2002). Synchronization of an uncertain unified chaotic system via adaptive control. *Chaos, Solitons & Fractals*, *14*(4), 643–647.
14. Chien, T. I., & Liao, T. L. (2005). Design of secure digital communication systems using chaotic modulation, cryptography and chaotic synchronization. *Chaos, Solitons & Fractals*, *24*(1), 241–255.
15. Faieghi, M. R., & Delavari, H. (2012). Chaos in fractional-order Genesio-Tesi system and its synchronization. *Communications in Nonlinear Science and Numerical Simulation*, *17*(2), 731–741.
16. Frey, D. R. (1993). Chaotic digital encoding: An approach to secure communication. *IEEE Transactions on Circuits and Systems II: Analog and Digital Signal Processing*, *40*(10), 660–666.
17. Gorenflo, R., & Mainardi, F. (1997). *Fractional calculus*. Springer.
18. Han, S. K., Kurrer, C., & Kuramoto, Y. (1995). Dephasing and bursting in coupled neural oscillators. *Physical Review Letters*, *75*(17), 3190.
19. Henein, M. M. R., Sayed, W. S., Radwan, A. G., & Abd-El-Hafiez, S. K. (2016). Switched active control synchronization of three fractional order chaotic systems. In *13th International*

Conference on Electrical Engineering/Electronics, Computer, Telecommunications and Information Technology.
20. Hirsch, M. W., Smale, S., & Devaney, R. L. (2012). *Differential equations, dynamical systems, and an introduction to chaos*. Academic Press.
21. Ho, M., Hung, Y., & Chou, C. (2002). Phase and anti-phase synchronization of two chaotic systems by using active control. *Physics Letters A*, *296*(1), 43–48.
22. Mc, Ho, & Hung, Y. C. (2002). Synchronization of two different systems by using generalized active control. *Physics Letters A*, *301*(5), 424–428.
23. Hosseinnia, S., Ghaderi, R., Mahmoudian, M., Momani, S., et al. (2010). Sliding mode synchronization of an uncertain fractional order chaotic system. *Computers & Mathematics with Applications*, *59*(5), 1637–1643.
24. Hussian, G., Alnaser, M., & Momani, S. (2008). Non-standard discretization of fractional differential equations. In: *Proceeding of 8th Seminar of Differential Equations and Dynamical Systems in, Isfahan, Iran*.
25. Jafari, S., Sprott, J. C., & Golpayegani, S. M. R. H. (2013). Elementary quadratic chaotic flows with no equilibria. *Physics Letters A*, *377*(9), 699–702.
26. Kocarev, L., & Lian, S. (2011). *Chaos-based cryptography: Theory, algorithms and applications* (vol. 354). Springer.
27. Lau, F., & Tse, C. K. (2003). *Chaos-based digital communication systems*. Springer.
28. Li, C., & Yan, J. (2007). The synchronization of three fractional differential systems. *Chaos, Solitons & Fractals*, *32*(2), 751–757.
29. Liu, J., Ye, C., Zhang, S., & Song, W. (2000). Anti-phase synchronization in coupled map lattices. *Physics Letters A*, *274*(1), 27–29.
30. Magin, R. L. (2006). *Fractional calculus in bioengineering*. Begell House Redding.
31. Mickens, R. E. (2000). *Applications of nonstandard finite difference schemes*. World Scientific.
32. Mickens, R. E. (2005). *Advances in the applications of nonstandard finite difference schemes*. World Scientific.
33. Moaddy, K., Radwan, A. G., Salama, K. N., Momani, S., & Hashim, I. (2012). The fractional-order modeling and synchronization of electrically coupled neuron systems. *Computers & Mathematics with Applications*, *64*(10), 3329–3339.
34. Odibat, Z. M., Corson, N., Aziz-Alaoui, M., & Bertelle, C. (2010). Synchronization of chaotic fractional-order systems via linear control. *International Journal of Bifurcation and Chaos*, *20*(01), 81–97.
35. Ouannas, A., Azar, A. T., & Abu-Saris, R. (2016a). A new type of hybrid synchronization between arbitrary hyperchaotic maps. *International Journal of Machine Learning and Cybernetics*, 1–8.
36. Ouannas, A., Azar, A. T., & Vaidyanathan, S. (2016b). A robust method for new fractional hybrid chaos synchronization. *Mathematical Methods in the Applied Sciences*.
37. Park, J., & Kwon, O. (2005). A novel criterion for delayed feedback control of time-delay chaotic systems. *Chaos, Solitons & Fractals*, *23*(2), 495–501.
38. Pecora, L. M., & Carroll, T. L. (1990). Synchronization in chaotic systems. *Physical Review Letters*, *64*(8), 821.
39. Petras, I. (2011). *Fractional-order nonlinear systems: Modeling, analysis and simulation*. Springer Science & Business Media.
40. Radwan, A. (2012). Stability analysis of the fractional-order $RL_\beta C_\alpha$ circuit. *Journal of Fractional Calculus and Applications*, *3*(1), 1–15.
41. Radwan, A., Soliman, A., & El-Sedeek, A. (2004). MOS realization of the modified Lorenz chaotic system. *Chaos, Solitons & Fractals*, *21*(3), 553–561.
42. Radwan, A., Soliman, A. M., & Elwakil, A. S. (2007). 1-D digitally-controlled multiscroll chaos generator. *International Journal of Bifurcation and Chaos*, *17*(01), 227–242.
43. Radwan, A., Moaddy, K., & Hashim, I. (2013). Amplitude modulation and synchronization of fractional-order memristor-based Chua's circuit. In *Abstract and applied analysis* (Vol. 2013). Hindawi Publishing Corporation.

44. Radwan, A., Moaddy, K., Salama, K. N., Momani, S., & Hashim, I. (2014a). Control and switching synchronization of fractional order chaotic systems using active control technique. *Journal of advanced research*, *5*(1), 125–132.
45. Radwan, A. G. (2013a). On some generalized discrete logistic maps. *Journal of advanced research*, *4*(2), 163–171.
46. Radwan, A. G. (2013b). Resonance and quality factor of the fractional circuit. *IEEE Journal on Emerging and Selected Topics in Circuits and Systems*, *3*(3), 377–385.
47. Radwan, A. G., & Abd-El-Hafiz, S. K. (2013). Image encryption using generalized tent map. In *IEEE 20th International Conference on Electronics, Circuits, and Systems (ICECS)* (pp. 653–656). IEEE.
48. Radwan, A. G., & Abd-El-Hafiz, S. K. (2014). The effect of multi-scrolls distribution on image encryption. In *21st IEEE International Conference on Electronics, Circuits and Systems (ICECS), 2014* (pp. 435–438). IEEE.
49. Radwan, A. G., & Fouda, M. E. (2013). Optimization of fractional-order RLC filters. *Circuits, Systems, and Signal Processing*, *32*(5), 2097–2118.
50. Radwan, A. G., Soliman, A. M., & El-Sedeek, A. L. (2003). An inductorless CMOS realization of Chua's circuit. *Chaos, Solitons & Fractals*, *18*(1), 149–158.
51. Radwan, A. G., Elwakil, A. S., & Soliman, A. M. (2008a). Fractional-order sinusoidal oscillators: design procedure and practical examples. *IEEE Transactions on Circuits and Systems I: Regular Papers*, *55*(7), 2051–2063.
52. Radwan, A. G., Soliman, A. M., & Elwakil, A. S. (2008b). First-order filters generalized to the fractional domain. *Journal of Circuits, Systems, and Computers*, *17*(01), 55–66.
53. Radwan, A. G., Moaddy, K., & Momani, S. (2011a). Stability and non-standard finite difference method of the generalized Chua's circuit. *Computers & Mathematics with Applications*, *62*(3), 961–970.
54. Radwan, A. G., Shamim, A., & Salama, K. N. (2011b). Theory of fractional order elements based impedance matching networks. *IEEE Microwave and Wireless Components Letters*, *21*(3), 120–122.
55. Radwan, A. G., Abd-El-Hafiz, S. K., & AbdElHaleem, S. H. (2012). Image encryption in the fractional-order domain. In *International Conference on Engineering and Technology (ICET), 2012* (pp. 1–6). IEEE.
56. Radwan, A. G., Abd-El-Hafiz, S. K., & AbdElHaleem, S. H. (2014b). An image encryption system based on generalized discrete maps. In *21st IEEE International Conference on Electronics, Circuits and Systems (ICECS)* (pp. 283–286). IEEE.
57. Radwan, A. G., Abd-El-Hafiz, S. K., & AbdElHaleem, S. H. (2015a). Image encryption based on fractional-order chaotic generators. In *2015 International Symposium on Nonlinear Theory and its Applications NOLTA2015*, Kowloon, Hong Kong, China, 1–4 December 2015 (pp. 688–691). IEEE.
58. Radwan, A. G., AbdElHaleem, S. H., & Abd-El-Hafiz, S. K. (2015b). Symmetric encryption algorithms using chaotic and non-chaotic generators: A review. *Journal of Advanced Research*.
59. Sayed, W. S., Radwan, A. G., & Fahmy, H. A. (2015a). Design of a generalized bidirectional tent map suitable for encryption applications. In *2015 11th International Computer Engineering Conference (ICENCO)* (pp. 207–211). IEEE.
60. Sayed, W. S., Radwan, A. G., & Fahmy, H. A. H. (2015b). Design of positive, negative, and alternating sign generalized logistic maps. *Discrete Dynamics in Nature and Society*, *2015*, Article ID 586783, 2015.
61. Sayed, W. S., Radwan, A. G., & Abd-El-Hafiez, S. K. (2016). Generalized synchronization involving a linear combination of fractional-order chaotic systems. In *13th International Conference on Electrical Engineering/Electronics, Computer, Telecommunications and Information Technology*.
62. Schöll, E. (2001). *Nonlinear spatio-temporal dynamics and chaos in semiconductors* (Vol. 10). Cambridge University Press.
63. Shamim, A., Radwan, A. G., & Salama, K. N. (2011). Fractional Smith chart theory. *IEEE Microwave and Wireless Components Letters*, *21*(3), 117–119.

64. Soltan, A., Radwan, A. G., & Soliman, A. M. (2012). Fractional order filter with two fractional elements of dependant orders. *Microelectronics Journal, 43*(11), 818–827.
65. Sprott, J. C. (2000). A new class of chaotic circuit. *Physics Letters A, 266*(1), 19–23.
66. Srivastava, M., Ansari, S., Agrawal, S., Das, S., & Leung, A. (2014). Anti-synchronization between identical and non-identical fractional-order chaotic systems using active control method. *Nonlinear Dynamics, 76*(2), 905–914.
67. Strogatz, S. H. (2014). *Nonlinear dynamics and chaos: With applications to physics, biology, chemistry, and engineering*. Westview Press.
68. Vaidyanathan, S., & Azar, A. T. (2016a). Adaptive backstepping control and synchronization of a novel 3-D jerk system with an exponential nonlinearity. In *Advances in chaos theory and intelligent control* (pp. 249–274). Springer.
69. Vaidyanathan S, & Azar AT (2016b) Adaptive control and synchronization of Halvorsen circulant chaotic systems. In *Advances in chaos theory and intelligent control* (pp. 225–247). Springer.
70. Vaidyanathan, S., & Azar, A. T. (2016c). Dynamic analysis, adaptive feedback control and synchronization of an eight-term 3-D novel chaotic system with three quadratic nonlinearities. In *Advances in chaos theory and intelligent control* (pp. 155–178). Springer.
71. Vaidyanathan, S., & Azar, A. T. (2016d). Generalized projective synchronization of a novel hyperchaotic four-wing system via adaptive control method. In *Advances in chaos theory and intelligent control* (pp. 275–296). Springer.
72. Vaidyanathan, S., & Azar, A. T. (2016e). A novel 4-D four-wing chaotic system with four quadratic nonlinearities and its synchronization via adaptive control method. In *Advances in chaos theory and intelligent control* (pp. 203–224). Springer.
73. Vaidyanathan, S., & Azar, A. T. (2016f). Qualitative study and adaptive control of a novel 4-D hyperchaotic system with three quadratic nonlinearities. In *Advances in chaos theory and intelligent control* (pp. 179–202). Springer.
74. Vincent, U. (2008). Chaos synchronization using active control and backstepping control: A comparative analysis. *Nonlinear Analysis, 13*(2), 253–261.
75. Wedekind, I., & Parlitz, U. (2001). Experimental observation of synchronization and anti-synchronization of chaotic low-frequency-fluctuations in external cavity semiconductor lasers. *International Journal of Bifurcation and Chaos, 11*(04), 1141–1147.
76. Wu, X., Wang, H., & Lu, H. (2012). Modified generalized projective synchronization of a new fractional-order hyperchaotic system and its application to secure communication. *Nonlinear Analysis: Real World Applications, 13*(3), 1441–1450.
77. Yassen, M. T. (2005). Chaos synchronization between two different chaotic systems using active control. *Chaos, Solitons & Fractals, 23*(1), 131–140.
78. Yu, Y., & Li, H. X. (2008). The synchronization of fractional-order Rössler hyperchaotic systems. *Physica A: Statistical Mechanics and its Applications, 387*(5), 1393–1403.
79. Yuan, L. G., & Yang, Q. G. (2012). Parameter identification and synchronization of fractional-order chaotic systems. *Communications in Nonlinear Science and Numerical Simulation, 17*(1), 305–316.
80. Zidan, M. A., Radwan, A. G., & Salama, K. N. (2012). Controllable V-shape multiscroll butterfly attractor: System and circuit implementation. *International Journal of Bifurcation and Chaos, 22*(06), 1250,143.

Adaptive Control of a Novel Nonlinear Double Convection Chaotic System

Sundarapandian Vaidyanathan, Quanmin Zhu and Ahmad Taher Azar

Abstract This research work describes a six-term novel nonlinear double convection chaotic system with two nonlinearities. First, this work presents the 3-D dynamics of the novel nonlinear double convection chaotic system and depicts the phase portraits of the system. Our novel nonlinear double convection chaotic system is obtained by modifying the dynamics of the Rucklidge chaotic system (1992). Next, the qualitative properties of the novel chaotic system are discussed in detail. The novel chaotic system has three equilibrium points. We show that the equilibrium point at the origin is a saddle point, while the other two equilibrium points are saddle-foci. The Lyapunov exponents of the novel nonlinear double convection chaotic system are obtained as $L_1 = 0.2089$, $L_2 = 0$ and $L_3 = -3.2123$. The Lyapunov dimension of the novel chaotic system is obtained as $D_L = 2.0650$. Next, we present the design of adaptive feedback controller for globally stabilizing the trajectories of the novel nonlinear double convection chaotic system with unknown parameters. Furthermore, we present the design of adaptive feedback controller for achieving complete synchronization of the identical novel nonlinear double convection chaotic systems with unknown parameters. The main adaptive control results are proved using Lyapunov stability theory. MATLAB simulations are depicted to illustrate all the main results derived in this research work for the novel nonlinear double convection system.

S. Vaidyanathan (✉)
Research and Development Centre, Vel Tech University,
Avadi, Chennai 600062, Tamil Nadu, India
e-mail: sundarcontrol@gmail.com; sundarvtu@gmail.com

Q. Zhu
Department of Engineering Design and Mathematics,
University of the West of England, Bristol, UK
e-mail: quan.zhu@uwe.ac.uk

A.T. Azar
Faculty of Computers and Information, Benha University, Banha, Egypt
e-mail: ahmad_t_azar@ieee.org; ahmad.azar@fci.bu.edu.eg

A.T. Azar
Nanoelectronics Integrated Systems Center (NISC), Nile University, Cairo, Egypt

A.T. Azar et al. (eds.), *Fractional Order Control and Synchronization of Chaotic Systems*, Studies in Computational Intelligence 688,
DOI 10.1007/978-3-319-50249-6_12

Keywords Chaos · Chaotic systems · Rucklidge system · Double convection · Adaptive control · Feedback control · Synchronization

1 Introduction

Chaotic systems are defined as nonlinear dynamical systems which are sensitive to initial conditions, topologically mixing and with dense periodic orbits [1–3]. Sensitivity to initial conditions of chaotic systems is popularly known as the *butterfly effect*. Small changes in an initial state will make a very large difference in the behavior of the system at future states.

Poincaré [4] suspected chaotic behaviour in the study of three bodies problem at the end of the 19th century, but chaos was experimentally established by Lorenz [5] only a few decades ago in the study of 3-D weather models.

In the chaos literature, there is great interest in the modelling and applications of chaotic systems in several fields such as mechanics, electrical systems, memristors, neurology, economics, lasers, chemical reactions, population dynamics, gyroscopes, communication devices, cryptosystems, neural networks, etc. [1–3].

In the last five decades, there is significant interest in the literature in discovering new chaotic systems [6]. Some popular chaotic systems are Lorenz system [5], Rössler system [7], Arneodo system [8], Henon-Heiles system [9], Genesio-Tesi system [10], Sprott systems [11], Chen system [12], Lü system [13], Rikitake dynamo system [14], Liu system [15], Shimizu system [16], Rucklidge system [17], etc.

In the recent years, many new chaotic systems have been found such as Pandey system [18], Qi system [19], Li system [20], Wei-Yang system [21], Zhou system [22], Zhu system [23], Sundarapandian systems [24, 25], Dadras system [26], Tacha system [27], Vaidyanathan systems [28–55], Vaidyanathan-Azar systems [56–60], Pehlivan system [61], Sampath system [62], Akgul system [63], Pham system [64–67], etc.

Chaos theory and control systems have many important applications in science and engineering [1, 2, 68–71]. Some commonly known applications are oscillators [72–78], lasers [79, 80], chemical reactions [81–91], biology [92–99], ecology [100, 101], encryption [102, 103], cryptosystems [104, 105], mechanical systems [106–110], secure communications [111–113], robotics [114–116], cardiology [117, 118], intelligent control [119, 120], neural networks [121–123], memristors [124, 125], etc.

Synchronization of chaotic systems is a phenomenon that occurs when two or more chaotic systems are coupled or when a chaotic system drives another chaotic system. Because of the butterfly effect which causes exponential divergence of the trajectories of two identical chaotic systems started with nearly the same initial conditions, the synchronization of chaotic systems is a challenging research problem in the chaos literature.

Major works on synchronization of chaotic systems deal with the complete synchronization of a pair of chaotic systems called the *master* and *slave* systems. The design goal of the complete synchronization is to apply the output of the master system to control the slave system so that the output of the slave system tracks the output of the master system asymptotically with time. Active feedback control is used when the system parameters are available for measurement. Adaptive feedback control is used when the system parameters are unknown.

Pecora and Carroll pioneered thc research on synchronization of chaotic systems with their seminal papers [126, 127]. The active control method [128–137] is typically used when the system parameters are available for measurement.

Adaptive control method [138–154] is typically used when some or all the system parameters are not available for measurement and estimates for the uncertain parameters of the systems. Adaptive control method has more relevant for many practical situations for systems with unknown parameters. In the literature, adaptive control method is preferred over active control method due to the wide applicability of the adaptive control method.

Intelligent control methods like fuzzy control method [155, 156] are also used for the synchronization of chaotic systems. Intelligent control methods have advantages like robustness, insensitive to small variations in the parameters, etc.

Sampled-data feedback control method [157–160] and time-delay feedback control method [161–163] are also used for synchronization of chaotic systems. Backstepping control method [164–171] is also used for the synchronization of chaotic systems, which is a recursive method for stabilizing the origin of a control system in strict-feedback form.

Another popular method for the synchronization of chaotic systems is the sliding mode control method [172–180], which is a nonlinear control method that alters the dynamics of a nonlinear system by application of a discontinuous control signal that forces the system to "slide" along a cross-section of the system's normal behavior.

In fluid dynamics modelling, cases of two-dimensional convection in a horizontal layer of Boussinesq fluid with lateral constraints were studied by Rucklidge [17]. When the convection takes place in a fluid layer rotating uniformly about a vertical axis and in the limit of tall thin rolls, convection in an imposed vertical magnetic field and convection in a rotating fluid layer are both modeled by the Rucklidge chaotic system, which exhibits chaotic attractor similar to the Lorenz system [5].

In this research work, by modifying the Rucklidge dynamics [17], we obtain a six-term novel nonlinear double convection chaotic system. Section 2 describes the 3-D dynamical model and phase portraits of the novel chaotic system.

Section 3 describes the dynamic analysis of the novel double convection chaotic system. We show that the novel chaotic system has three equilibrium points. In addition, we show that the equilibrium point at the origin is a saddle point, while the other two equilibrium points are saddle-foci. Thus, it follows that all the three equilibrium points of the novel double convection chaotic system are unstable.

The Lyapunov exponents of the novel nonlinear double convection chaotic system are obtained as $L_1 = 0.2089$, $L_2 = 0$ and $L_3 = -3.2123$. Since the sum of the Lyapunov exponents of the novel chaotic system is negative, this chaotic system is dissipative. Also, the Lyapunov dimension of the novel chaotic system is obtained as $D_L = 2.0650$.

Section 4 describes the adaptive feedback control of the novel nonlinear double convection chaotic system with unknown parameters. Section 5 describes the adaptive feedback synchronization of the identical novel nonlinear double convection chaotic systems with unknown parameters. The adaptive feedback control and synchronization results are proved using Lyapunov stability theory [181].

MATLAB simulations are depicted to illustrate all the main results for the 3-D novel nonlinear double convection chaotic system. Section 6 concludes this work with a summary of the main results.

2 A Novel Nonlinear Double Convection Chaotic System

In fluid dynamics, Rucklidge chaotic system [17] for nonlinear double convection is described by the 3-D dynamics

$$\begin{aligned} \dot{x}_1 &= -ax_1 + bx_2 - x_2x_3 \\ \dot{x}_2 &= x_1 \\ \dot{x}_3 &= -x_3 + x_2^2 \end{aligned} \tag{1}$$

where x_1, x_2, x_3 are the states and a, b are constant, positive parameters.

Rucklidge system (1) is *chaotic* when the parameter values are taken as

$$a = 2, \quad b = 6.7 \tag{2}$$

For numerical simulations, we take the initial state of the Rucklidge chaotic system (1) as

$$x_1(0) = 0.2, \quad x_2(0) = 0.2, \quad x_3(0) = 0.2 \tag{3}$$

The Lyapunov exponents of the Rucklidge chaotic system (1) for the parameter values (2) and the initial values (3) are numerically determined as

$$L_1 = 0.1868, \quad L_2 = 0, \quad L_3 = -3.1890 \tag{4}$$

The Lyapunov dimension of the Rucklidge chaotic system (1) is calculated as

$$D_L = 2 + \frac{L_1 + L_2}{|L_3|} = 2.0586, \tag{5}$$

which is fractional.

In this work, we propose a novel nonlinear double convection system by modifying the third differential equation of the Rucklidge system (1).

Our novel nonlinear double convection system is described as follows.

$$\begin{aligned} \dot{x}_1 &= -ax_1 + bx_2 - x_2x_3 \\ \dot{x}_2 &= x_1 \\ \dot{x}_3 &= -x_3 + |x_2| \end{aligned} \tag{6}$$

where x_1, x_2, x_3 are the states and a, b are constant, positive parameters.

In this work, we show that novel chaotic system (6) exhibits a strange chaotic attractor, when the parameter values are taken as

$$a = 2, \quad b = 12 \tag{7}$$

For numerical simulations, we take the initial state of the Rucklidge chaotic system (6) as

$$x_1(0) = 0.2, \quad x_2(0) = 0.2, \quad x_3(0) = 0.2 \tag{8}$$

The Lyapunov exponents of the novel chaotic system (6) for the parameter values (7) and the initial values (8) are numerically determined as

$$L_1 = 0.2089, \quad L_2 = 0, \quad L_3 = -3.2123 \tag{9}$$

The Lyapunov dimension of the novel double convection system (6) is calculated as

$$D_L = 2 + \frac{L_1 + L_2}{|L_3|} = 2.0650, \tag{10}$$

which is fractional.

We note that the Maximal Lyapunov Exponent (MLE) of the novel chaotic system (6) is $L_1 = 0.2089$, which is greater than the Maximal Lyapunov Exponent (MLE) of the Rucklidge chaotic system (1), *viz.* $L_1 = 0.1868$. Also, we find that the Lyapunov dimension of the novel chaotic system (6) is $D_L = 2.0650$, which is greater than the Lyapunov dimension of the Rucklidge chaotic system (1), *viz.* $D_L = 2.0586$.

Figure 1 describes the phase portrait of the strange chaotic attractor of the novel double-convection chaotic system (6). Thus, the strange attractor of the novel double-convection chaotic system (6) is a *two-scroll* attractor.

Figure 2 describes the 2-D projection of the strange chaotic attractor of the novel double-convection chaotic system (6) on (x_1, x_2)-plane.

Figure 3 describes the 2-D projection of the strange chaotic attractor of the novel double-convection chaotic system (6) on (x_2, x_3)-plane.

Figure 4 describes the 2-D projection of the strange chaotic attractor of the novel double-convection chaotic system (6) on (x_1, x_3)-plane.

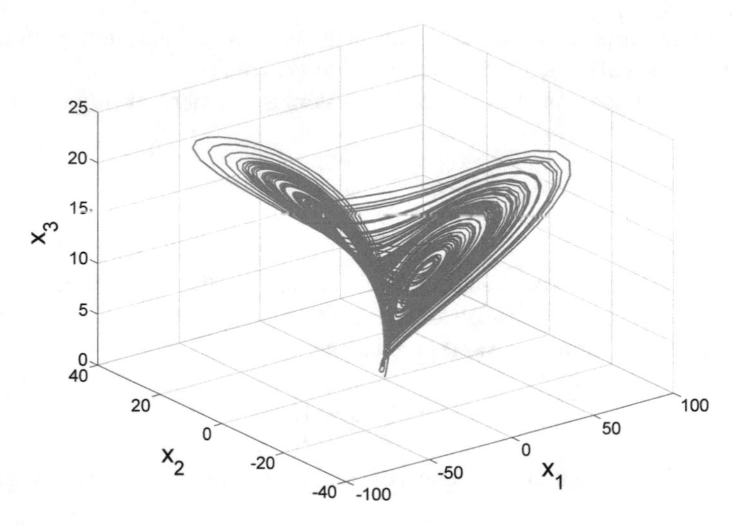

Fig. 1 Strange attractor of the novel double convection chaotic system in $\mathbb{R}^3$

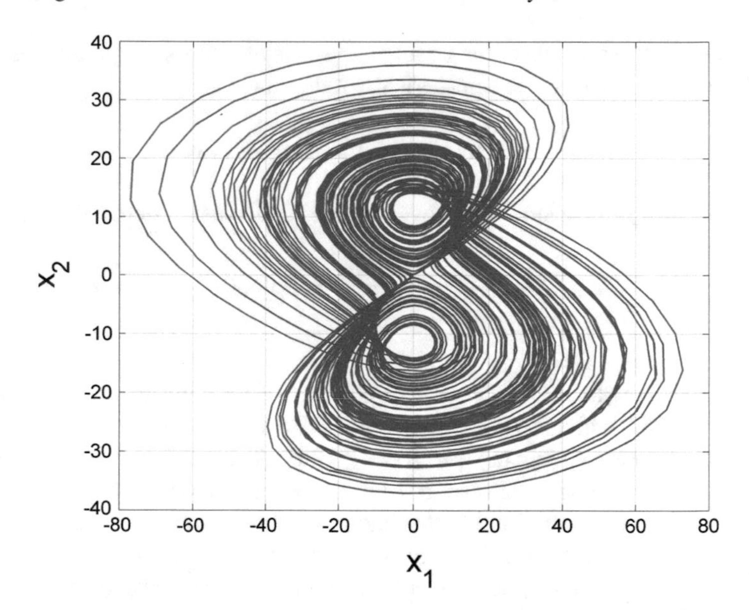

Fig. 2 2-D projection of the novel double convection chaotic system on (x_1, x_2)-plane

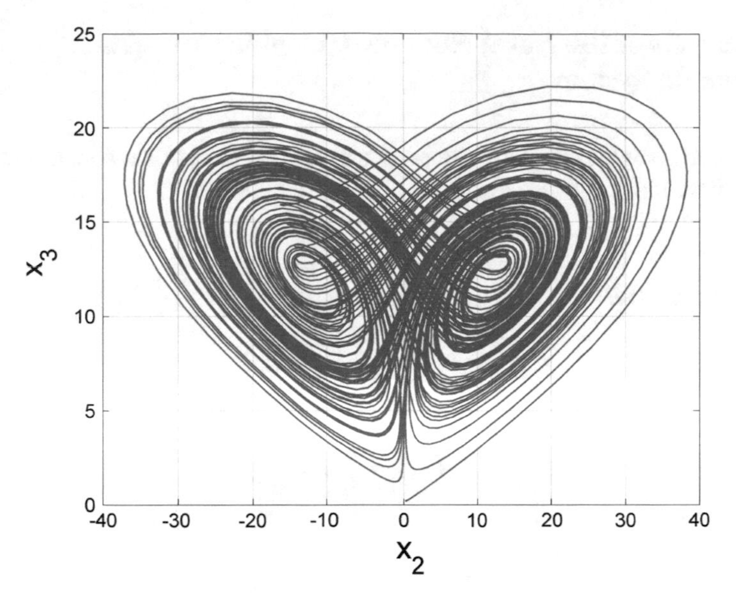

Fig. 3 2-D projection of the novel double convection chaotic system on (x_2, x_3)-plane

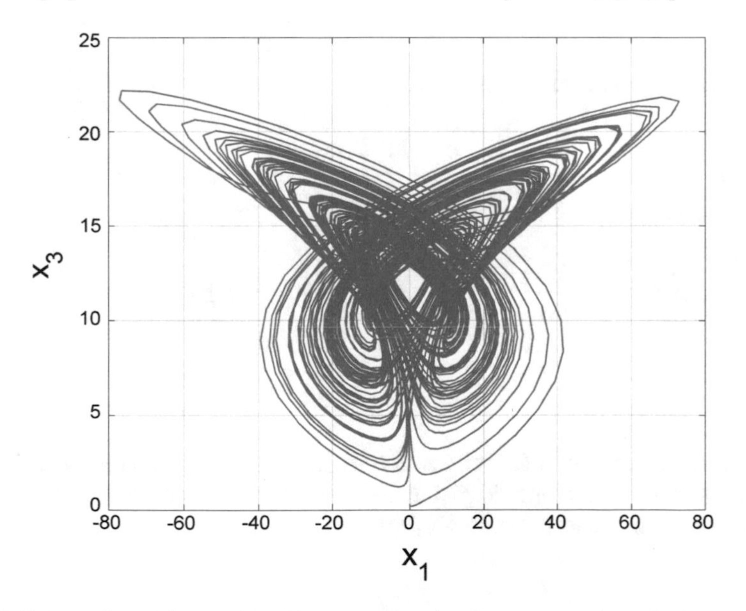

Fig. 4 2-D projection of the novel double convection chaotic system on (x_1, x_3)-plane

3 Analysis of the Novel Nonlinear Double Convection Chaotic System

This section gives the qualitative properties of the novel nonlinear double convection chaotic system (6).

3.1 *Dissipativity*

In vector notation, the novel system (6) can be expressed as

$$\dot{x} = f(x) = \begin{bmatrix} f_1(x) \\ f_2(x) \\ f_3(x) \end{bmatrix}, \tag{11}$$

where

$$\begin{aligned} f_1(x) &= -ax_1 + bx_2 - x_2x_3 \\ f_2(x) &= x_1 \\ f_3(x) &= -x_3 + |x_2| \end{aligned} \tag{12}$$

We take the parameter values as

$$a = 2, \quad b = 12 \tag{13}$$

The divergence of the vector field f on $\mathbb{R}^3$ is obtained as

$$\text{div } f = \frac{\partial f_1(x)}{\partial x_1} + \frac{\partial f_2(x)}{\partial x_2} + \frac{\partial f_3(x)}{\partial x_3} = -a + 0 - 1 = -\mu \tag{14}$$

where

$$\mu = a + 1 = 3 > 0 \tag{15}$$

Let Ω be any region in $\mathbb{R}^3$ with a smooth boundary. Let $\Omega(t) = \Phi_t(\Omega)$, where Φ_t is the flow of the vector field f. Let $V(t)$ denote the volume of $\Omega(t)$.

By Liouville's theorem, it follows that

$$\frac{dV(t)}{dt} = \int\limits_{\Omega(t)} (div\, f) dx_1\, dx_2\, dx_3 \tag{16}$$

Substituting the value of *div f* in (16) leads to

$$\frac{dV(t)}{dt} = -\mu \int\limits_{\Omega(t)} dx_1\, dx_2\, dx_3 = -\mu V(t) \tag{17}$$

Integrating the linear differential equation (17), $V(t)$ is obtained as

$$V(t) = V(0)\exp(-\mu t) \tag{18}$$

From Eq. (18), it follows that the volume $V(t)$ shrinks to zero exponentially as $t \to \infty$.

Thus, the novel nonlinear double convection chaotic system (6) is dissipative. Hence, any asymptotic motion of the system (6) settles onto a set of measure zero, *i.e.* a strange attractor.

3.2 Symmetry

It is easy to verify that the nonlinear double convection chaotic system (6) is invariant under the coordinates transformation

$$(x_1, x_2, x_3) \mapsto (-x_1, -x_2, -x_3) \tag{19}$$

Since the transformation (19) persists for all values of the system parameters, it follows that the nonlinear double convection chaotic system (6) has rotation symmetry about the x_3-axis and that any non-trivial trajectory of the system (6) must have a twin trajectory.

3.3 Invariance

It is easily seen that the x_3-axis is invariant for the flow of the novel chaotic system (6). The invariant motion along the x_3-axis is characterized by the scalar dynamics

$$\dot{x}_3 = -x_3, \tag{20}$$

which is globally exponentially stable.

3.4 Equilibria

The equilibrium points of the novel chaotic system (6) are obtained by solving the nonlinear equations

$$\begin{aligned} f_1(x) &= -ax_1 + bx_2 - x_2x_3 = 0 \\ f_2(x) &= x_1 \qquad\qquad\qquad\quad\; = 0 \\ f_3(x) &= -x_3 + |x_2| \qquad\qquad\; = 0 \end{aligned} \tag{21}$$

We take the parameter values as in the chaotic case, *viz.* $a = 2$ and $b = 12$.

Solving the nonlinear system (21), we obtain three equilibrium points of the novel chaotic system (6), *viz.*

$$E_0 = \begin{bmatrix} 0 \\ 0 \\ 0 \end{bmatrix}, \quad E_1 = \begin{bmatrix} 0 \\ 12 \\ 12 \end{bmatrix}, E_2 = \begin{bmatrix} 0 \\ -12 \\ 12 \end{bmatrix} \tag{22}$$

The Jacobian matrix of the novel chaotic system (6) at (x_1, x_2, x_3) is obtained as

$$J(x) = \begin{bmatrix} -2 & 12 - x_3 & -x_2 \\ 1 & 0 & 0 \\ 0 & \text{sign}(x_2) & -1 \end{bmatrix} \tag{23}$$

The matrix $J_0 = J(E_0)$ has the eigenvalues

$$\lambda_1 = -1, \quad \lambda_2 = -4.6056, \quad \lambda_3 = 2.6056 \tag{24}$$

This shows that the equilibrium point E_0 is a saddle-point, which is unstable.

The matrix $J_1 = J(E_1)$ has the eigenvalues

$$\lambda_1 = -3.4348, \quad \lambda_{2,3} = 0.2174 \pm 1.8564i \tag{25}$$

This shows that the equilibrium point E_1 is a saddle-focus, which is unstable.

The matrix $J_2 = J(E_2)$ has the eigenvalues

$$\lambda_1 = -3.4348, \quad \lambda_{2,3} = 0.2174 \pm 1.8564i \tag{26}$$

This shows that the equilibrium point E_2 is a saddle-focus, which is unstable.

The matrix $J_3 = J(E_3)$ has the eigenvalues

$$\lambda_1 = -5.4615, \quad \lambda_{2,3} = 1.7307 \pm 2.0469i \tag{27}$$

Hence, E_0, E_1, E_2 are all unstable equilibrium points of the novel double-convection chaotic system (6), where E_0 is a saddle point and E_2, E_3 are saddle-foci.

3.5 Lyapunov Exponents and Lyapunov Dimension

We take the initial values of the novel chaotic system (6) as in (8) and the parameter values of the novel chaotic system (6) as in (7).

Then the Lyapunov exponents of the novel chaotic system (6) are numerically obtained as

$$L_1 = 0.2089, \quad L_2 = 0, \quad L_3 = -3.2123 \tag{28}$$

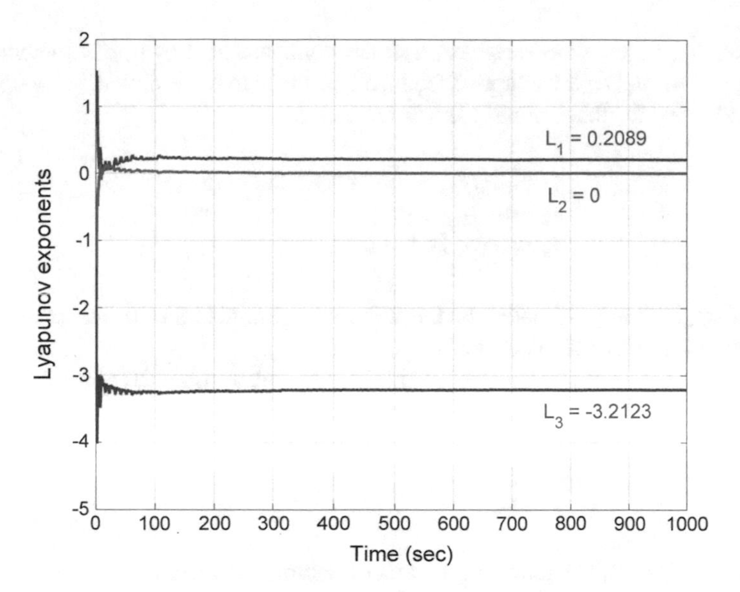

Fig. 5 Lyapunov exponents of the novel double convection chaotic system

Since $L_1 + L_2 + L_3 = -3.0034 < 0$, the system (6) is dissipative.
Also, the Lyapunov dimension of the system (6) is obtained as

$$D_L = 2 + \frac{L_1 + L_2}{|L_3|} = 2.0650, \tag{29}$$

which is fractional.

Figure 5 depicts the Lyapunov exponents of the novel double convection chaotic system (6). From this figure, it is seen that the Maximal Lyapunov Exponent (MLE) of the novel chaotic system (6) is $L_1 = 0.2089$.

4 Adaptive Control of the Novel Double Convection Chaotic System

This section derives new results for the adaptive controller design in order to stabilize the unstable novel double convection chaotic system with unknown parameters for all initial conditions.

The controlled novel 3-D chaotic system is given by

$$\begin{aligned} \dot{x}_1 &= -ax_1 + bx_2 - x_2x_3 + u_1 \\ \dot{x}_2 &= x_1 + u_2 \\ \dot{x}_3 &= -x_3 + |x_2| + u_3 \end{aligned} \tag{30}$$

where x_1, x_2, x_3 are state variables, a, b are constant, unknown, parameters of the system and u_1, u_2, u_3 are adaptive feedback controls to be designed.

An adaptive feedback control law is taken as

$$\begin{aligned} u_1 &= \hat{a}(t)x_1 - \hat{b}(t)x_2 + x_2x_3 - k_1x_1 \\ u_2 &= -x_1 - k_2x_2 \\ u_3 &= x_3 - |x_2| - k_3x_3 \end{aligned} \tag{31}$$

where $\hat{a}(t), \hat{b}(t)$ are estimates for the unknown parameters a, b, respectively, and k_1, k_2, k_3 are positive gain constants.

The closed-loop control system is obtained by substituting (31) into (30) as

$$\begin{aligned} \dot{x}_1 &= -[a - \hat{a}(t)]x_1 + [b - \hat{b}(t)]x_2 - k_1x_1 \\ \dot{x}_2 &= -k_2x_2 \\ \dot{x}_3 &= -k_3x_3 \end{aligned} \tag{32}$$

To simplify (32), we define the parameter estimation error as

$$\begin{aligned} e_a(t) &= a - \hat{a}(t) \\ e_b(t) &= b - \hat{b}(t) \end{aligned} \tag{33}$$

Using (33), the closed-loop system (32) can be simplified as

$$\begin{aligned} \dot{x}_1 &= -e_ax_1 + e_bx_2 - k_1x_1 \\ \dot{x}_2 &= -k_2x_2 \\ \dot{x}_3 &= -k_3x_3 \end{aligned} \tag{34}$$

Differentiating the parameter estimation error (33) with respect to t, we get

$$\begin{aligned} \dot{e}_a &= -\dot{\hat{a}} \\ \dot{e}_b &= -\dot{\hat{b}} \end{aligned} \tag{35}$$

Next, we consider the quadratic Lyapunov function defined by

$$V(x_1, x_2, x_3, e_a, e_b, e_c, e_p) = \frac{1}{2}\left(x_1^2 + x_2^2 + x_3^2 + e_a^2 + e_b^2\right), \tag{36}$$

which is positive definite on $\mathbb{R}^5$.

Differentiating V along the trajectories of (34) and (35), we get

$$\dot{V} = -k_1x_1^2 - k_2x_2^2 - k_3x_3^2 + e_a\left[-x_1^2 - \dot{\hat{a}}\right] + e_b\left[x_1x_2 - \dot{\hat{b}}\right] \tag{37}$$

In view of Eq. (37), an update law for the parameter estimates is taken as

$$\begin{aligned} \dot{\hat{a}} &= -x_1^2 \\ \dot{\hat{b}} &= x_1 x_2 \end{aligned} \tag{38}$$

Theorem 1 *The novel double convection chaotic system (30) with unknown system parameters is globally and exponentially stabilized for all initial conditions* $x(0) \in \mathbb{R}^3$ *by the adaptive control law (31) and the parameter update law (38), where* $k_i, (i = 1, 2, 3)$ *are positive constants.*

Proof The result is proved using Lyapunov stability theory [181]. We consider the quadratic Lyapunov function V defined by (36), which is positive definite on $\mathbb{R}^5$.

Substitution of the parameter update law (38) into (37) yields

$$\dot{V} = -k_1 x_1^2 - k_2 x_2^2 - k_3 x_3^2, \tag{39}$$

which is a negative semi-definite function on $\mathbb{R}^5$.

Therefore, it can be concluded that the state vector $x(t)$ and the parameter estimation error are globally bounded, *i.e.*

$$\left[x_1(t)\ x_2(t)\ x_3(t)\ e_a(t)\ e_b(t)\ e_c(t)\ e_p(t) \right]^T \in \mathbf{L}_\infty. \tag{40}$$

We define

$$k = \min\{k_1, k_2, k_3\} \tag{41}$$

Then it follows from (39) that

$$\dot{V} \le -k\|\mathbf{x}\|^2 \ \text{ or } \ k\|\mathbf{x}\|^2 \le -\dot{V} \tag{42}$$

Integrating the inequality (42) from 0 to t, we get

$$k \int_0^t \|\mathbf{x}(\tau)\|^2 \, d\tau \le - \int_0^t \dot{V}(\tau)\, d\tau = V(0) - V(t) \tag{43}$$

From (43), it follows that $\mathbf{x}(t) \in \mathbf{L}_2$.

Using (34), it can be deduced that $\dot{x}(t) \in \mathbf{L}_\infty$.

Hence, using Barbalat's lemma, we can conclude that $\mathbf{x}(t) \to 0$ exponentially as $t \to \infty$ for all initial conditions $\mathbf{x}(0) \in \mathbb{R}^3$.

This completes the proof. □

For numerical simulations, the parameter values of the novel system (30) are taken as in the chaotic case, *viz.* $a = 2$ and $b = 12$. The gain constants are taken as $k_i = 6, (i = 1, 2, 3)$.

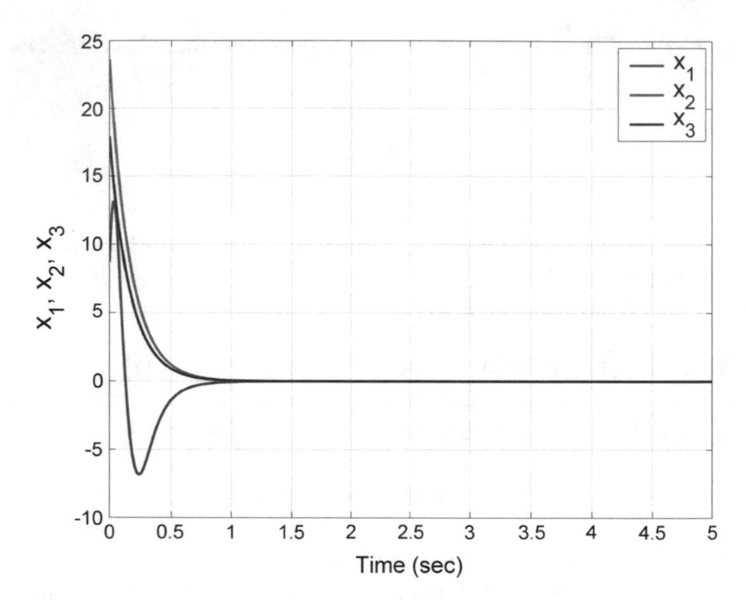

Fig. 6 Time-history of the states $x_1(t), x_2(t), x_3(t)$

The initial values of the parameter estimates are taken as

$$\hat{a}(0) = 15.4, \quad \hat{b}(0) = 4.7 \tag{44}$$

The initial values of the novel double convection chaotic system (30) are taken as

$$x_1(0) = 8.7, \quad x_2(0) = 23.6, \quad x_3(0) = 17.9 \tag{45}$$

Figure 6 shows the time-history of the controlled states $x_1(t), x_2(t), x_3(t)$.

Figure 6 depicts the exponential convergence of the controlled states and the efficiency of the adaptive controller defined by (31).

5 Adaptive Synchronization of the Identical 3-D Novel Chaotic Systems

This section derives new results for the adaptive synchronization of the identical novel chaotic systems with unknown parameters.

The master system is given by the novel chaotic system

$$\begin{aligned} \dot{x}_1 &= -ax_1 + bx_2 - x_2x_3 \\ \dot{x}_2 &= x_1 \\ \dot{x}_3 &= -x_3 + |x_2| \end{aligned} \tag{46}$$

where x_1, x_2, x_3 are state variables and a, b, c, p are constant, unknown, parameters of the system.

The slave system is given by the controlled novel chaotic system

$$\begin{aligned}\dot{y}_1 &= -ay_1 + by_2 - y_2y_3 + u_1\\ \dot{y}_2 &= y_1 + u_2\\ \dot{y}_3 &= -y_3 + |y_2| + u_3\end{aligned} \tag{47}$$

where y_1, y_2, y_3 are state variables and u_1, u_2, u_3 are adaptive controls to be designed.

The synchronization error is defined as

$$e_i = y_i - x_i, \quad (i = 1, 2, 3) \tag{48}$$

The error dynamics is easily obtained as

$$\begin{aligned}\dot{e}_1 &= -ae_1 + be_2 - y_2y_3 + x_2x_3 + u_1\\ \dot{e}_2 &= e_1 + u_2\\ \dot{e}_3 &= -e_3 + |y_2| - |x_2| + u_3\end{aligned} \tag{49}$$

An adaptive control law is taken as

$$\begin{aligned}u_1 &= \hat{a}(t)e_1 - \hat{b}(t)e_2 + y_2y_3 - x_2x_3 - k_1e_1\\ u_2 &= -e_1 - k_2e_2\\ u_3 &= e_3 - |y_2| + |x_2| - k_3e_3\end{aligned} \tag{50}$$

where $\hat{a}(t), \hat{b}(t)$ are estimates for the unknown parameters a, b, respectively, and k_1, k_2, k_3 are positive gain constants.

The closed-loop control system is obtained by substituting (50) into (49) as

$$\begin{aligned}\dot{e}_1 &= -\left[a - \hat{a}(t)\right]e_1 + \left[b - \hat{b}(t)\right]e_2 - k_1e_1\\ \dot{e}_2 &= -k_2e_2\\ \dot{e}_3 &= -k_3e_3\end{aligned} \tag{51}$$

To simplify (51), we define the parameter estimation error as

$$\begin{aligned}e_a(t) &= a - \hat{a}(t)\\ e_b(t) &= b - \hat{b}(t)\end{aligned} \tag{52}$$

Using (52), the closed-loop system (51) can be simplified as

$$\begin{aligned}\dot{e}_1 &= -e_ae_1 + e_be_2 - k_1e_1\\ \dot{e}_2 &= -k_2e_2\\ \dot{e}_3 &= -k_3e_3\end{aligned} \tag{53}$$

Differentiating the parameter estimation error (52) with respect to t, we get

$$\begin{aligned} \dot{e}_a &= -\dot{\hat{a}} \\ \dot{e}_b &= -\dot{\hat{b}} \end{aligned} \tag{54}$$

Next, we find an update law for parameter estimates using Lyapunov stability theory.

Consider the quadratic Lyapunov function defined by

$$V(e_1, e_2, e_3, e_a, e_b) = \frac{1}{2} \left(e_1^2 + e_2^2 + e_3^2 + e_a^2 + e_b^2 \right), \tag{55}$$

which is positive definite on $\mathbb{R}^5$.

Differentiating V along the trajectories of (53) and (54), we get

$$\dot{V} = -k_1 e_1^2 - k_2 e_2^2 - k_3 e_3^2 + e_a \left[-e_1^2 - \dot{\hat{a}} \right] + e_b \left[e_1 e_2 - \dot{\hat{b}} \right] \tag{56}$$

In view of Eq. (56), an update law for the parameter estimates is taken as

$$\begin{aligned} \dot{\hat{a}} &= -e_1^2 \\ \dot{\hat{b}} &= e_1 e_2 \end{aligned} \tag{57}$$

Theorem 2 *The identical novel chaotic systems (46) and (47) with unknown system parameters are globally and exponentially synchronized for all initial conditions* $x(0), y(0) \in \mathbb{R}^3$ *by the adaptive control law (50) and the parameter update law (57), where* $k_i, (i = 1, 2, 3)$ *are positive constants.*

Proof The result is proved using Lyapunov stability theory [181].

We consider the quadratic Lyapunov function V defined by (55), which is positive definite on $\mathbb{R}^7$.

Substitution of the parameter update law (57) into (56) yields

$$\dot{V} = -k_1 e_1^2 - k_2 e_2^2 - k_3 e_3^2, \tag{58}$$

which is a negative semi-definite function on $\mathbb{R}^7$.

Therefore, it can be concluded that the synchronization error vector $e(t)$ and the parameter estimation error are globally bounded, *i.e.*

$$\left[e_1(t)\ e_2(t)\ e_3(t)\ e_a(t)\ e_b(t)\ e_c(t)\ e_p(t) \right]^T \in \mathbf{L}_\infty. \tag{59}$$

Define

$$k = \min \{k_1, k_2, k_3\} \tag{60}$$

Then it follows from (58) that

$$\dot{V} \leq -k\|e\|^2 \;\; \text{or} \;\; k\|\mathbf{e}\|^2 \leq -\dot{V} \tag{61}$$

Integrating the inequality (61) from 0 to t, we get

$$k \int_0^t \|\mathbf{e}(\tau)\|^2 \, d\tau \leq -\int_0^t \dot{V}(\tau) \, d\tau = V(0) - V(t) \tag{62}$$

From (62), it follows that $\mathbf{e}(t) \in \mathbf{L}_2$.

Using (53), it can be deduced that $\dot{\mathbf{e}}(t) \in \mathbf{L}_\infty$.

Hence, using Barbalat's lemma, we can conclude that $\mathbf{e}(t) \to 0$ exponentially as $t \to \infty$ for all initial conditions $\mathbf{e}(0) \in \mathbb{R}^3$.

This completes the proof. □

For numerical simulations, the parameter values of the novel systems (46) and (47) are taken as in the chaotic case, *viz.* $a = 2$ and $b = 12$.

The gain constants are taken as $k_i = 6$ for $i = 1, 2, 3$.

The initial values of the parameter estimates are taken as

$$\hat{a}(0) = 9.2, \quad \hat{b}(0) = 4.9 \tag{63}$$

The initial values of the master system (46) are taken as

$$x_1(0) = 15.2, \quad x_2(0) = 7.3, \quad x_3(0) = -5.4 \tag{64}$$

The initial values of the slave system (47) are taken as

$$y_1(0) = 3.4, \quad y_2(0) = 12.5, \quad y_3(0) = 9.8 \tag{65}$$

Figures 7, 8 and 9 show the complete synchronization of the identical chaotic systems (46) and (47).

Figure 7 shows that the states $x_1(t)$ and $y_1(t)$ are synchronized in 1 s (MATLAB).

Figure 8 shows that the states $x_2(t)$ and $y_2(t)$ are synchronized in 1 s (MATLAB).

Figure 9 shows that the states $x_3(t)$ and $y_3(t)$ are synchronized in 1 s (MATLAB).

Figure 10 shows the time-history of the synchronization errors $e_1(t), e_2(t), e_3(t)$. From Fig. 10, it is seen that the errors $e_1(t), e_2(t)$ and $e_3(t)$ are stabilized in 1 s (MATLAB).

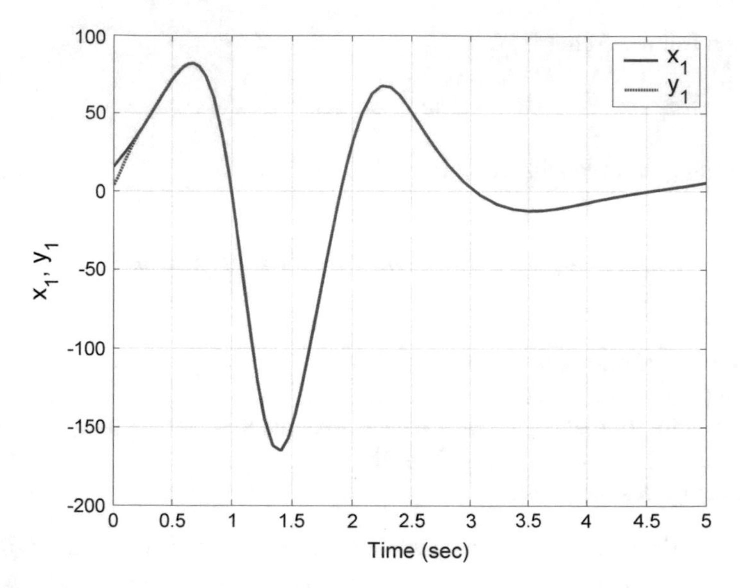

Fig. 7 Synchronization of the states x_1 and y_1

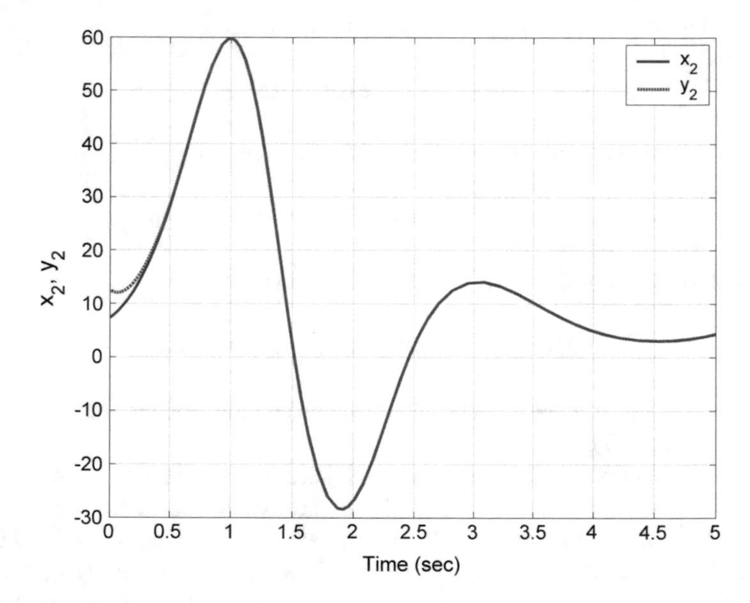

Fig. 8 Synchronization of the states x_2 and y_2

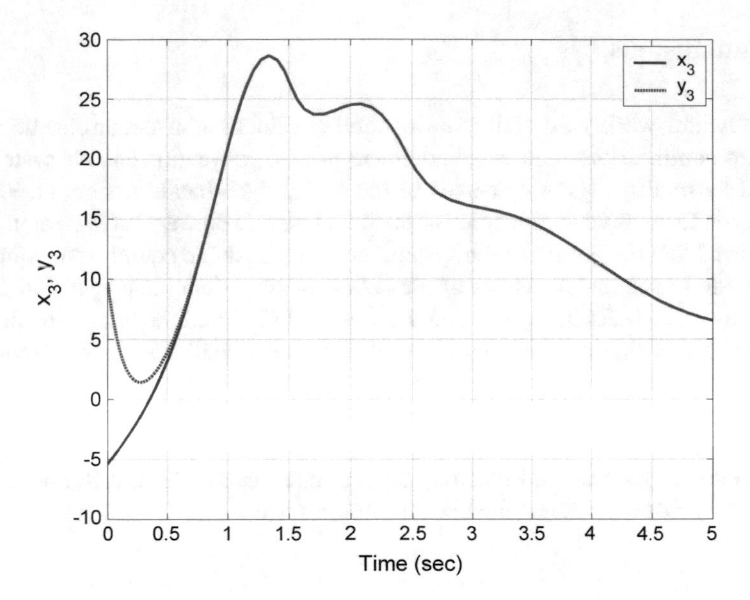

Fig. 9 Synchronization of the states x_3 and y_3

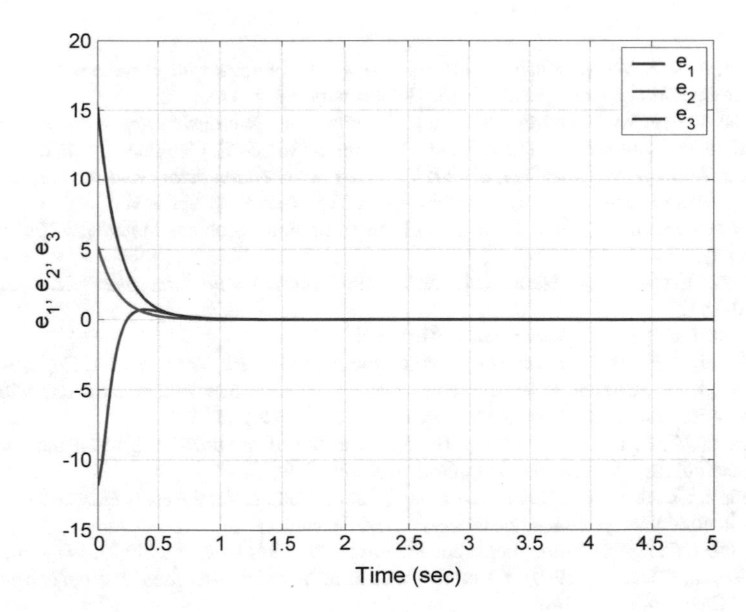

Fig. 10 Time-history of the synchronization errors e_1, e_2, e_3

6 Conclusions

In this research work, we described novel nonlinear double convection chaotic system with two nonlinearities. Our novel nonlinear double convection chaotic system was derived by modifying the dynamics of the Rucklidge chaotic system (1992). We discussed the qualitative properties of the novel double convection system in detail. We showed that the novel chaotic system has three unstable equilibrium points. We derived the Lyapunov exponents of the novel nonlinear double convection chaotic system as $L_1 = 0.2089$, $L_2 = 0$ and $L_3 = -3.2123$. The Lyapunov dimension of the novel chaotic system has been deduced as $D_L = 2.0650$. Next, we presented the designs of adaptive feedback controller and adaptive synchronizer for the nonlinear double convection chaotic system with unknown parameters. The main adaptive control results were proved using Lyapunov stability theory. MATLAB simulations were shown to validate and illustrate all the main results derived in this research work for the novel nonlinear double convection system.

References

1. Azar, A. T., & Vaidyanathan, S. (2015a). *Chaos modeling and control systems design*. Studies in computational intelligence (Vol. 581). Germany: Springer.
2. Azar, A. T., & Vaidyanathan, S. (2015b). *Computational intelligence applications in modeling and control*. Studies in computational intelligence (Vol. 575). Germany: Springer.
3. Azar, A. T., & Vaidyanathan, S. (2016). *Advances in chaos theory and intelligent control*. Studies in fuzziness and soft computing (Vol. 337). Germany: Springer.
4. Barrow-Green, J. (1997). Poincaré and the three body problem. *American Mathematical Society*.
5. Lorenz, E. N. (1963). Deterministic periodic flow. *Journal of the Atmospheric Sciences*, *20*(2), 130–141.
6. Sprott, J. C. (2010). *Elegant chaos*. World Scientific.
7. Rössler, O. E. (1976). An equation for continuous chaos. *Physics Letters A*, *57*(5), 397–398.
8. Arneodo, A., Coullet, P., & Tresser, C. (1981). Possible new strange attractors with spiral structure. *Communications in Mathematical Physics*, *79*, 573–579.
9. Henon, M., & Heiles, C. (1964). The applicability of the third integral Of motion: Some numerical experiments. *The Astrophysical Journal*, *69*, 73–79.
10. Genesio, R., & Tesi, A. (1992). Harmonic balance methods for the analysis of chaotic dynamics in nonlinear systems. *Automatica*, *28*(3), 531–548.
11. Sprott, J. C. (1994). Some simple chaotic flows. *Physical Review E*, *50*(2), 647–650.
12. Chen, G., & Ueta, T. (1999). Yet another chaotic attractor. *International Journal of Bifurcation and Chaos*, *9*(7), 1465–1466.
13. Lü, J., & Chen, G. (2002). A new chaotic attractor coined. *International Journal of Bifurcation and Chaos*, *12*(3), 659–661.
14. Rikitake, T. (1958). Oscillations of a system of disk dynamos. *Mathematical Proceedings of the Cambridge Philosophical Society*, *54*(1), 89–105.
15. Liu, C., Liu, T., Liu, L., & Liu, K. (2004). A new chaotic attractor. *Chaos, Solitions and Fractals*, *22*(5), 1031–1038.
16. Shimizu, T., & Morioka, N. (1980). On the bifurcation of a symmetric limit cycle to an asymmetric one in a simple model. *Physics Letters A*, *76*(3–4), 201–204.

17. Rucklidge, A. M. (1992). Chaos in models of double convection. *Journal of Fluid Mechanics*, *237*, 209–229.
18. Pandey, A., Baghel, R. K., & Singh, R. P. (2012). Synchronization analysis of a new autonomous chaotic system with its application in signal masking. *IOSR Journal of Electronics and Communication Engineering*, *1*(5), 16–22.
19. Qi, G., & Chen, G. (2006). Analysis and circuit implementation of a new 4D chaotic system. *Physics Letters A*, *352*, 386–397.
20. Li, D. (2008). A three-scroll chaotic attractor. *Physics Letters A*, *372*(4), 387–393.
21. Wei, Z., & Yang, Q. (2010). Anti-control of Hopf bifurcation in the new chaotic system with two stable node-foci. *Applied Mathematics and Computation*, *217*(1), 422–429.
22. Zhou, W., Xu, Y., Lu, H., & Pan, L. (2008). On dynamics analysis of a new chaotic attractor. *Physics Letters A*, *372*(36), 5773–5777.
23. Zhu, C., Liu, Y., & Guo, Y. (2010). Theoretic and numerical study of a new chaotic system. *Intelligent Information Management*, *2*, 104–109.
24. Sundarapandian, V. (2013). Analysis and anti-synchronization of a novel chaotic system via active and adaptive controllers. *Journal of Engineering Science and Technology Review*, *6*(4), 45–52.
25. Sundarapandian, V., & Pehlivan, I. (2012). Analysis, control, synchronization, and circuit design of a novel chaotic system. *Mathematical and Computer Modelling*, *55*(7–8), 1904–1915.
26. Dadras, S., & Momeni, H. R. (2009). A novel three-dimensional autonomous chaotic system generating two, three and four-scroll attractors. *Physics Letters A*, *373*, 3637–3642.
27. Tacha, O. I., Volos, C. K., Kyprianidis, I. M., Stouboulos, I. N., Vaidyanathan, S., & Pham, V. T. (2016). Analysis, adaptive control and circuit simulation of a novel nonlinear finance system. *Applied Mathematics and Computation*, *276*, 200–217.
28. Vaidyanathan, S. (2013a). A new six-term 3-D chaotic system with an exponential nonlinearity. *Far East Journal of Mathematical Sciences*, *79*(1), 135–143.
29. Vaidyanathan, S. (2013b). Analysis and adaptive synchronization of two novel chaotic systems with hyperbolic sinusoidal and cosinusoidal nonlinearity and unknown parameters. *Journal of Engineering Science and Technology Review*, *6*(4), 53–65.
30. Vaidyanathan, S. (2014a). A new eight-term 3-D polynomial chaotic system with three quadratic nonlinearities. *Far East Journal of Mathematical Sciences*, *84*(2), 219–226.
31. Vaidyanathan, S. (2014b). Analysis and adaptive synchronization of eight-term 3-D polynomial chaotic systems with three quadratic nonlinearities. *European Physical Journal: Special Topics*, *223*(8), 1519–1529.
32. Vaidyanathan, S. (2014c). Analysis, control and synchronisation of a six-term novel chaotic system with three quadratic nonlinearities. *International Journal of Modelling, Identification and Control*, *22*(1), 41–53.
33. Vaidyanathan, S. (2014d). Generalized projective synchronisation of novel 3-D chaotic systems with an exponential non-linearity via active and adaptive control. *International Journal of Modelling, Identification and Control*, *22*(3), 207–217.
34. Vaidyanathan, S. (2015b). A 3-D novel highly chaotic system with four quadratic nonlinearities, its adaptive control and anti-synchronization with unknown parameters. *Journal of Engineering Science and Technology Review*, *8*(2), 106–115.
35. Vaidyanathan, S. (2015m). Analysis, properties and control of an eight-term 3-D chaotic system with an exponential nonlinearity. *International Journal of Modelling, Identification and Control*, *23*(2), 164–172.
36. Vaidyanathan, S. (2016a). A novel 2-D chaotic enzymes-substrates reaction system and its adaptive backstepping control. In A. T. Azar & S. Vaidyanathan (Eds.), *Advances in chaos theory and intelligent control*. Studies in fuzziness and soft computing (Vol. 337, pp. 507–528). Germany: Springer.
37. Vaidyanathan, S. (2016b). A novel 3-D conservative jerk chaotic system with two quadratic nonlinearities and its adaptive control. In A. T. Azar & S. Vaidyanathan (Eds.), *Advances in chaos theory and intelligent control*. Studies in fuzziness and soft computing (Vol. 337, pp. 349–376). Germany: Springer.

38. Vaidyanathan, S. (2016c). A novel 3-D jerk chaotic system with three quadratic nonlinearities and its adaptive control. *Archives of Control Sciences*, *26*(1), 19–47.
39. Vaidyanathan, S. (2016d). A novel 4-D hyperchaotic thermal convection system and its adaptive control. In A. T. Azar & S. Vaidyanathan (Eds.), *Advances in chaos theory and intelligent control*. Studies in fuzziness and soft computing (Vol. 337, pp. 75–100). Germany: Springer.
40. Vaidyanathan, S. (2016e). A novel double convecton system, its analysis, adaptive control and synchronization. In A. T. Azar & S. Vaidyanathan (Eds.), *Advances in chaos theory and intelligent control*. Studies in fuzziness and soft computing (Vol. 337, pp. 553–579). Germany: Springer.
41. Vaidyanathan, S. (2016f). A seven-term novel 3-D jerk chaotic system with two quadratic nonlinearities and its adaptive backstepping control. In A. T. Azar & S. Vaidyanathan (Eds.), *Advances in chaos theory and intelligent control*. Studies in fuzziness and soft computing (Vol. 337, pp. 581–607). Germany: Springer.
42. Vaidyanathan, S. (2016g). Analysis, adaptive control and synchronization of a novel 3-D chaotic system with a quartic nonlinearity and two quadratic nonlinearities. In A. T. Azar & S. Vaidyanathan (Eds.), *Advances in chaos theory and intelligent control*. Studies in fuzziness and soft computing (Vol. 337, pp. 429–453). Germany: Springer.
43. Vaidyanathan, S. (2016h). Analysis, control and synchronization of a novel 4-D highly hyperchaotic system with hidden attractors. In A. T. Azar & S. Vaidyanathan (Eds.), *Advances in chaos theory and intelligent control*. Studies in fuzziness and soft computing (Vol. 337, pp. 529–552). Germany: Springer.
44. Vaidyanathan, S. (2016j). Dynamic analysis, adaptive control and synchronization of a novel highly chaotic system with four quadratic nonlinearities. In A. T. Azar & S. Vaidyanathan (Eds.), *Advances in chaos theory and intelligent control*. Studies in fuzziness and soft computing (Vol. 337, pp. 405–428). Germany: Springer.
45. Vaidyanathan, S. (2016k). Global chaos synchronization of a novel 3-D chaotic system with two quadratic nonlinearities via active and adaptive control. In A. T. Azar & S. Vaidyanathan (Eds.), *Advances in chaos theory and intelligent control*. Studies in fuzziness and soft computing (Vol. 337, pp. 481–506). Germany: Springer.
46. Vaidyanathan, S. (2016l). Qualitative analysis and properties of a novel 4-D hyperchaotic system with two quadratic nonlinearities and its adaptive control. In A. T. Azar & S. Vaidyanathan (Eds.), *Advances in chaos theory and intelligent control*. Studies in fuzziness and soft computing (Vol. 337, pp. 455–480). Germany: Springer.
47. Vaidyanathan, S., & Azar, A. T. (2015b). Analysis, control and synchronization of a nine-term 3-D novel chaotic system. In A. T. Azar & S. Vaidyanathan (Eds.), *Chaos modelling and control systems design*. Studies in computational intelligence (Vol. 581, pp. 19–38). Germany: Springer.
48. Vaidyanathan, S., & Madhavan, K. (2013). Analysis, adaptive control and synchronization of a seven-term novel 3-D chaotic system. *International Journal of Control Theory and Applications*, *6*(2), 121–137.
49. Vaidyanathan, S., & Pakiriswamy, S. (2015). A 3-D novel conservative chaotic system and its generalized projective synchronization via adaptive control. *Journal of Engineering Science and Technology Review*, *8*(2), 52–60.
50. Vaidyanathan, S., & Volos, C. (2015). Analysis and adaptive control of a novel 3-D conservative no-equilibrium chaotic system. *Archives of Control Sciences*, *25*(3), 333–353.
51. Vaidyanathan, S., Volos, C., Pham, V. T., Madhavan, K., & Idowu, B. A. (2014b). Adaptive backstepping control, synchronization and circuit simulation of a 3-D novel jerk chaotic system with two hyperbolic sinusoidal nonlinearities. *Archives of Control Sciences*, *24*(3), 375–403.
52. Vaidyanathan, S., Rajagopal, K., Volos, C. K., Kyprianidis, I. M., & Stouboulos, I. N. (2015b). Analysis, adaptive control and synchronization of a seven-term novel 3-D chaotic system with three quadratic nonlinearities and its digital implementation in LabVIEW. *Journal of Engineering Science and Technology Review*, *8*(2), 130–141.

53. Vaidyanathan, S., Volos, C. K., Kyprianidis, I. M., Stouboulos, I. N., & Pham, V. T. (2015d). Analysis, adaptive control and anti-synchronization of a six-term novel jerk chaotic system with two exponential nonlinearities and its circuit simulation. *Journal of Engineering Science and Technology Review*, *8*(2), 24–36.
54. Vaidyanathan, S., Volos, C. K., & Pham, V. T. (2015e). Analysis, adaptive control and adaptive synchronization of a nine-term novel 3-D chaotic system with four quadratic nonlinearities and its circuit simulation. *Journal of Engineering Science and Technology Review*, *8*(2), 174–184.
55. Vaidyanathan, S., Volos, C. K., & Pham, V. T. (2015f). Global chaos control of a novel nine-term chaotic system via sliding mode control. In A. T. Azar & Q. Zhu (Eds.), *Advances and applications in sliding mode control systems*. Studies in computational intelligence (Vol. 576, pp. 571–590). Germany: Springer.
56. Vaidyanathan, S., & Azar, A. T. (2016a). A novel 4-D four-wing chaotic system with four quadratic nonlinearities and its synchronization via adaptive control method. In A. T. Azar & S. Vaidyanathan (Eds.), *Advances in chaos theory and intelligent control*. Studies in fuzziness and soft computing (Vol. 337, pp. 203–224). Germany: Springer.
57. Vaidyanathan, S., & Azar, A. T. (2016b). Adaptive backstepping control and synchronization of a novel 3-D jerk system with an exponential nonlinearity. In A. T. Azar & S. Vaidyanathan (Eds.), *Advances in chaos theory and intelligent control*. Studies in fuzziness and soft computing (Vol. 337, pp. 249–274). Germany: Springer.
58. Vaidyanathan, S., & Azar, A. T. (2016d). Dynamic analysis, adaptive feedback control and synchronization of an eight-term 3-D novel chaotic system with three quadratic nonlinearities. In A. T. Azar & S. Vaidyanathan (Eds.), *Advances in chaos theory and intelligent control*. Studies in fuzziness and soft computing (Vol. 337, pp. 155–178). Germany: Springer.
59. Vaidyanathan, S., & Azar, A. T. (2016e). Generalized projective synchronization of a novel hyperchaotic four-wing system via adaptive control method. In A. T. Azar & S. Vaidyanathan (Eds.), *Advances in chaos theory and intelligent control*. Studies in fuzziness and soft computing (Vol. 337, pp. 275–296). Germany: Springer.
60. Vaidyanathan, S., & Azar, A. T. (2016f). Qualitative study and adaptive control of a novel 4-D hyperchaotic system with three quadratic nonlinearities. In A. T. Azar & S. Vaidyanathan (Eds.), *Advances in chaos theory and intelligent control*. Studies in fuzziness and soft computing (Vol. 337, pp. 179–202). Germany: Springer.
61. Pehlivan, I., Moroz, I. M., & Vaidyanathan, S. (2014). Analysis, synchronization and circuit design of a novel butterfly attractor. *Journal of Sound and Vibration*, *333*(20), 5077–5096.
62. Sampath, S., Vaidyanathan, S., Volos, C. K., & Pham, V. T. (2015). An eight-term novel four-scroll chaotic system with cubic nonlinearity and its circuit simulation. *Journal of Engineering Science and Technology Review*, *8*(2), 1–6.
63. Akgul, A., Moroz, I., Pehlivan, I., & Vaidyanathan, S. (2016). A new four-scroll chaotic attractor and its engineering applications. *Optik*, *127*, 5491–5499.
64. Pham, V. T., Vaidyanathan, S., Volos, C. K., & Jafari, S. (2015). Hidden attractors in a chaotic system with an exponential nonlinear term. *European Physical Journal—Special Topics*, *224*(8), 1507–1517.
65. Pham, V. T., Jafari, S., Vaidyanathan, S., Volos, C., & Wang, X. (2016a). A novel memristive neural network with hidden attractors and its circuitry implementation. *Science China Technological Sciences*, *59*(3), 358–363.
66. Pham, V. T., Vaidyanathan, S., Volos, C., Jafari, S., & Kingni, S. T. (2016b). A no-equilibrium hyperchaotic system with a cubic nonlinear term. *Optik*, *127*(6), 3259–3265.
67. Pham, V. T., Vaidyanathan, S., Volos, C. K., Jafari, S., Kuznetsov, N. V., & Hoang, T. M. (2016c). A novel memristive time-delay chaotic system without equilibrium points. *European Physical Journal: Special Topics*, *225*(1), 127–136.
68. Azar, A. T. (2010). *Fuzzy systems*. Vienna, Austria: IN-TECH.
69. Azar, A. T., & Vaidyanathan, S. (2015c). *Handbook of research on advanced intelligent control engineering and automation*. Advances in Computational Intelligence and Robotics (ACIR). USA: IGI-Global.

70. Azar, A. T., & Zhu, Q. (2015). *Advances and applications in sliding mode control systems*. Studies in computational intelligence (Vol. 576). Germany: Springer.
71. Zhu, Q., & Azar, A. T. (2015). *Complex system modelling and control through intelligent soft computations*. Studies in fuzzines and soft computing (Vol. 319). Germany: Springer.
72. Vaidyanathan, S. (2015p). Anti-synchronization of Mathieu-Van der Pol chaotic systems via adaptive control method. *International Journal of ChemTech Research*, *8*(11), 638–653.
73. Vaidyanathan, S. (2015t). Global chaos control of Mathieu-Van der pol system via adaptive control method. *International Journal of ChemTech Research*, *8*(9), 406–417.
74. Vaidyanathan, S. (2015v). Global chaos synchronization of Duffing double-well chaotic oscillators via integral sliding mode control. *International Journal of ChemTech Research*, *8*(11), 141–151.
75. Vaidyanathan, S. (2015w). Global chaos synchronization of Mathieu-Van der Pol chaotic systems via adaptive control method. *International Journal of ChemTech Research*, *8*(10), 148–162.
76. Vaidyanathan, S. (2015x). Global chaos synchronization of novel coupled Van der Pol conservative chaotic systems via adaptive control method. *International Journal of ChemTech Research*, *8*(8), 95–111.
77. Vaidyanathan, S. (2015y). Global chaos synchronization of the forced Van der Pol chaotic oscillators via adaptive control method. *International Journal of PharmTech Research*, *8*(6), 156–166.
78. Vaidyanathan, S. (2016i). Anti-synchronization of Duffing double-well chaotic oscillators via integral sliding mode control. *International Journal of ChemTech Research*, *9*(2), 297–304.
79. Li, N., Pan, W., Yan, L., Luo, B., & Zou, X. (2014). Enhanced chaos synchronization and communication in cascade-coupled semiconductor ring lasers. *Communications in Nonlinear Science and Numerical Simulation*, *19*(6), 1874–1883.
80. Yuan, G., Zhang, X., & Wang, Z. (2014). Generation and synchronization of feedback-induced chaos in semiconductor ring lasers by injection-locking. *Optik—International Journal for Light and Electron Optics*, *125*(8), 1950–1953.
81. Vaidyanathan, S. (2015c). A novel chemical chaotic reactor system and its adaptive control. *International Journal of ChemTech Research*, *8*(7), 146–158.
82. Vaidyanathan, S. (2015d). A novel chemical chaotic reactor system and its output regulation via integral sliding mode control. *International Journal of ChemTech Research*, *8*(11), 669–683.
83. Vaidyanathan, S. (2015h). Adaptive control design for the anti-synchronization of novel 3-D chemical chaotic reactor systems. *International Journal of ChemTech Research*, *8*(11), 654–668.
84. Vaidyanathan, S. (2015i). Adaptive control of a chemical chaotic reactor. *International Journal of PharmTech Research*, *8*(3), 377–382.
85. Vaidyanathan, S. (2015j). Adaptive synchronization of chemical chaotic reactors. *International Journal of ChemTech Research*, *8*(2), 612–621.
86. Vaidyanathan, S. (2015l). Adaptive synchronization of novel 3-D chemical chaotic reactor systems. *International Journal of ChemTech Research*, *8*(7), 159–171.
87. Vaidyanathan, S. (2015n). Anti-synchronization of Brusselator chemical reaction systems via adaptive control. *International Journal of ChemTech Research*, *8*(6), 759–768.
88. Vaidyanathan, S. (2015o). Anti-synchronization of chemical chaotic reactors via adaptive control method. *International Journal of ChemTech Research*, *8*(8), 73–85.
89. Vaidyanathan, S. (2015r). Dynamics and control of Brusselator chemical reaction. *International Journal of ChemTech Research*, *8*(6), 740–749.
90. Vaidyanathan, S. (2015s). Dynamics and control of Tokamak system with symmetric and magnetically confined plasma. *International Journal of ChemTech Research*, *8*(6), 795–803.
91. Vaidyanathan, S. (2015u). Global chaos synchronization of chemical chaotic reactors via novel sliding mode control method. *International Journal of ChemTech Research*, *8*(7), 209–221.
92. Das, S., Goswami, D., Chatterjee, S., & Mukherjee, S. (2014). Stability and chaos analysis of a novel swarm dynamics with applications to multi-agent systems. *Engineering Applications of Artificial Intelligence*, *30*, 189–198.

93. Kyriazis, M. (1991). Applications of chaos theory to the molecular biology of aging. *Experimental Gerontology*, *26*(6), 569–572.
94. Vaidyanathan, S. (2015a). 3-cells cellular neural network (CNN) attractor and its adaptive biological control. *International Journal of PharmTech Research*, *8*(4), 632–640.
95. Vaidyanathan, S. (2015e). Adaptive backstepping control of enzymes-substrates system with ferroelectric behaviour in brain waves. *International Journal of PharmTech Research*, *8*(2), 256–261.
96. Vaidyanathan, S. (2015f). Adaptive biological control of generalized Lotka-Volterra three-species biological system. *International Journal of PharmTech Research*, *8*(4), 622–631.
97. Vaidyanathan, S. (2015g). Adaptive chaotic synchronization of enzymes-substrates system with ferroelectric behaviour in brain waves. *International Journal of PharmTech Research*, *8*(5), 964–973.
98. Vaidyanathan, S. (2015k). Adaptive synchronization of generalized Lotka-Volterra three-species biological systems. *International Journal of PharmTech Research*, *8*(5), 928–937.
99. Vaidyanathan, S. (2015q). Chaos in neurons and adaptive control of Birkhoff-Shaw strange chaotic attractor. *International Journal of PharmTech Research*, *8*(5), 956–963.
100. Gibson, W. T., & Wilson, W. G. (2013). Individual-based chaos: Extensions of the discrete logistic model. *Journal of Theoretical Biology*, *339*, 84–92.
101. Suérez, I. (1999). Mastering chaos in ecology. *Ecological Modelling*, *117*(2–3), 305–314.
102. Lang, J. (2015). Color image encryption based on color blend and chaos permutation in the reality-preserving multiple-parameter fractional Fourier transform domain. *Optics Communications*, *338*, 181–192.
103. Zhang, X., Zhao, Z., & Wang, J. (2014). Chaotic image encryption based on circular substitution box and key stream buffer. *Signal Processing: Image Communication*, *29*(8), 902–913.
104. Rhouma, R., & Belghith, S. (2011). Cryptoanalysis of a chaos based cryptosystem on DSP. *Communications in Nonlinear Science and Numerical Simulation*, *16*(2), 876–884.
105. Usama, M., Khan, M. K., Alghatbar, K., & Lee, C. (2010). Chaos-based secure satellite imagery cryptosystem. *Computers and Mathematics with Applications*, *60*(2), 326–337.
106. Azar, A. T., & Serrano, F. E. (2014). Robust IMC-PID tuning for cascade control systems with gain and phase margin specifications. *Neural Computing and Applications*, *25*(5), 983–995.
107. Azar, A. T., & Serrano, F. E. (2015a). Adaptive sliding mode control of the Furuta pendulum. In A. T. Azar & Q. Zhu (Eds.), *Advances and applications in sliding mode control systems*. Studies in computational intelligence (Vol. 576, pp. 1–42). Germany: Springer.
108. Azar, A. T., & Serrano, F. E. (2015b). Deadbeat control for multivariable systems with time varying delays. In A. T. Azar & S. Vaidyanathan (Eds.), *Chaos modeling and control systems design*. Studies in computational intelligence (Vol. 581, pp. 97–132). Germany: Springer.
109. Azar, A. T., & Serrano, F. E. (2015c). Design and modeling of anti wind up PID controllers. In Q. Zhu & A. T. Azar (Eds.), *Complex system modelling and control through intelligent soft computations*. Studies in fuzziness and soft computing (Vol. 319, pp. 1–44). Germany: Springer.
110. Azar, A. T., & Serrano, F. E. (2015d). Stabilizatoin and control of mechanical systems with backlash. In A. T. Azar & S. Vaidyanathan (Eds.), *Handbook of research on advanced intelligent control engineering and automation*. Advances in Computational Intelligence and Robotics (ACIR) (pp. 1–60). USA: IGI-Global.
111. Feki, M. (2003). An adaptive chaos synchronization scheme applied to secure communication. *Chaos, Solitons and Fractals*, *18*(1), 141–148.
112. Murali, K., & Lakshmanan, M. (1998). Secure communication using a compound signal from generalized chaotic systems. *Physics Letters A*, *241*(6), 303–310.
113. Zaher, A. A., & Abu-Rezq, A. (2011). On the design of chaos-based secure communication systems. *Communications in Nonlinear Systems and Numerical Simulation*, *16*(9), 3721–3727.
114. Mondal, S., & Mahanta, C. (2014). Adaptive second order terminal sliding mode controller for robotic manipulators. *Journal of the Franklin Institute*, *351*(4), 2356–2377.

115. Nehmzow, U., & Walker, K. (2005). Quantitative description of robot-environment interaction using chaos theory. *Robotics and Autonomous Systems*, *53*(3–4), 177–193.
116. Volos, C. K., Kyprianidis, I. M., & Stouboulos, I. N. (2013). Experimental investigation on coverage performance of a chaotic autonomous mobile robot. *Robotics and Autonomous Systems*, *61*(12), 1314–1322.
117. Qu, Z. (2011). Chaos in the genesis and maintenance of cardiac arrhythmias. *Progress in Biophysics and Molecular Biology*, *105*(3), 247–257.
118. Witte, C. L., & Witte, M. H. (1991). Chaos and predicting varix hemorrhage. *Medical Hypotheses*, *36*(4), 312–317.
119. Azar, A. T. (2012). Overview of type-2 fuzzy logic systems. *International Journal of Fuzzy System Applications*, *2*(4), 1–28.
120. Li, Z., & Chen, G. (2006). *Integration of fuzzy logic and chaos theory, studies in fuzziness and soft computing* (Vol. 187). Germany: Springer.
121. Huang, X., Zhao, Z., Wang, Z., & Li, Y. (2012). Chaos and hyperchaos in fractional-order cellular neural networks. *Neurocomputing*, *94*, 13–21.
122. Kaslik, E., & Sivasundaram, S. (2012). Nonlinear dynamics and chaos in fractional-order neural networks. *Neural Networks*, *32*, 245–256.
123. Lian, S., & Chen, X. (2011). Traceable content protection based on chaos and neural networks. *Applied Soft Computing*, *11*(7), 4293–4301.
124. Pham, V. T., Volos, C. K., Vaidyanathan, S., Le, T. P., & Vu, V. Y. (2015b). A memristor-based hyperchaotic system with hidden attractors: Dynamics, synchronization and circuital emulating. *Journal of Engineering Science and Technology Review*, *8*(2), 205–214.
125. Volos, C. K., Kyprianidis, I. M., Stouboulos, I. N., Tlelo-Cuautle, E., & Vaidyanathan, S. (2015). Memristor: A new concept in synchronization of coupled neuromorphic circuits. *Journal of Engineering Science and Technology Review*, *8*(2), 157–173.
126. Carroll, T. L., & Pecora, L. M. (1991). Synchronizing chaotic circuits. *IEEE Transactions on Circuits and Systems*, *38*(4), 453–456.
127. Pecora, L. M., & Carroll, T. L. (1990). Synchronization in chaotic systems. *Physical Review Letters*, *64*(8), 821–824.
128. Karthikeyan, R., & Sundarapandian, V. (2014). Hybrid chaos synchronization of four-scroll systems via active control. *Journal of Electrical Engineering*, *65*(2), 97–103.
129. Sarasu, P., & Sundarapandian, V. (2011a). Active controller design for the generalized projective synchronization of four-scroll chaotic systems. *International Journal of Systems Signal Control and Engineering Application*, *4*(2), 26–33.
130. Sarasu, P., & Sundarapandian, V. (2011b). The generalized projective synchronization of hyperchaotic Lorenz and hyperchaotic Qi systems via active control. *International Journal of Soft Computing*, *6*(5), 216–223.
131. Sundarapandian, V. (2010). Output regulation of the Lorenz attractor. *Far East Journal of Mathematical Sciences*, *42*(2), 289–299.
132. Sundarapandian, V., & Karthikeyan, R. (2012b). Hybrid synchronization of hyperchaotic Lorenz and hyperchaotic Chen systems via active control. *Journal of Engineering and Applied Sciences*, *7*(3), 254–264.
133. Vaidyanathan, S. (2011). Hybrid chaos synchronization of Liu and Lu systems by active nonlinear control. *Communications in Computer and Information Science*, *204*, 1–10.
134. Vaidyanathan, S. (2012d). Output regulation of the Liu chaotic system. *Applied Mechanics and Materials*, *110–116*, 3982–3989.
135. Vaidyanathan, S., & Rajagopal, K. (2011a). Anti-synchronization of Li and T chaotic systems by active nonlinear control. *Communications in Computer and Information Science*, *198*, 175–184.
136. Vaidyanathan, S., & Rajagopal, K. (2011b). Global chaos synchronization of hyperchaotic Pang and Wang systems by active nonlinear control. *Communications in Computer and Information Science*, *204*, 84–93.
137. Vaidyanathan, S., & Rasappan, S. (2011). Global chaos synchronization of hyperchaotic Bao and Xu systems by active nonlinear control. *Communications in Computer and Information Science*, *198*, 10–17.

138. Vaidyanathan, S., Pham, V. T., & Volos, C. K. (2015). A 5-D hyperchaotic Rikitake dynamo system with hidden attractors. *European Physical Journal: Special Topics*, *224*(8), 1575–1592.
139. Sarasu, P., & Sundarapandian, V. (2012a). Adaptive controller design for the generalized projective synchronization of 4-scroll systems. *International Journal of Systems Signal Control and Engineering Application*, *5*(2), 21–30.
140. Sarasu, P., & Sundarapandian, V. (2012b). Generalized projective synchronization of three-scroll chaotic systems via adaptive control. *European Journal of Scientific Research*, *72*(4), 504–522.
141. Sarasu, P., & Sundarapandian, V. (2012c). Generalized projective synchronization of two-scroll systems via adaptive control. *International Journal of Soft Computing*, *7*(4), 146–156.
142. Sundarapandian, V., & Karthikeyan, R. (2011a). Anti-synchronization of hyperchaotic Lorenz and hyperchaotic Chen systems by adaptive control. *International Journal of Systmes Signal Control and Engineering Application*, *4*(2), 18–25.
143. Sundarapandian, V., & Karthikeyan, R. (2011b). Anti-synchronization of Lü and Pan chaotic systems by adaptive nonlinear control. *European Journal of Scientific Research*, *64*(1), 94–106.
144. Sundarapandian, V., & Karthikeyan, R. (2012a). Adaptive anti-synchronization of uncertain Tigan and Li systems. *Journal of Engineering and Applied Sciences*, *7*(1), 45–52.
145. Vaidyanathan, S. (2012b). Anti-synchronization of Sprott-L and Sprott-M chaotic systems via adaptive control. *International Journal of Control Theory and Applications*, *5*(1), 41–59.
146. Vaidyanathan, S. (2013c). Analysis, control and synchronization of hyperchaotic Zhou system via adaptive control. *Advances in Intelligent Systems and Computing*, *177*, 1–10.
147. Vaidyanathan, S. (2015z). Hyperchaos, qualitative analysis, control and synchronisation of a ten-term 4-D hyperchaotic system with an exponential nonlinearity and three quadratic nonlinearities. *International Journal of Modelling, Identification and Control*, *23*(4), 380–392.
148. Vaidyanathan, S., & Azar, A. T. (2015a). Analysis and control of a 4-D novel hyperchaotic system. In A. T. Azar & S. Vaidyanathan (Eds.), *Chaos modeling and control systems design*. Studies in computational intelligence (Vol. 581, pp. 19–38). Germany: Springer.
149. Vaidyanathan, S., & Azar, A. T. (2016c). Adaptive control and synchronization of Halvorsen circulant chaotic systems. In A. T. Azar & S. Vaidyanathan (Eds.), *Advances in chaos theory and intelligent control*. Studies in fuzziness and soft computing (Vol. 337, pp. 225–247). Germany: Springer.
150. Vaidyanathan, S., & Pakiriswamy, S. (2013). Generalized projective synchronization of six-term Sundarapandian chaotic systems by adaptive control. *International Journal of Control Theory and Applications*, *6*(2), 153–163.
151. Vaidyanathan, S., & Rajagopal, K. (2011c). Global chaos synchronization of Lü and Pan systems by adaptive nonlinear control. *Communications in Computer and Information Science*, *205*, 193–202.
152. Vaidyanathan, S., & Rajagopal, K. (2012). Global chaos synchronization of hyperchaotic Pang and hyperchaotic Wang systems via adaptive control. *International Journal of Soft Computing*, *7*(1), 28–37.
153. Vaidyanathan, S., Volos, C., & Pham, V. T. (2014a). Hyperchaos, adaptive control and synchronization of a novel 5-D hyperchaotic system with three positive Lyapunov exponents and its SPICE implementation. *Archives of Control Sciences*, *24*(4), 409–446.
154. Vaidyanathan, S., Volos, C., Pham, V. T., & Madhavan, K. (2015c). Analysis, adaptive control and synchronization of a novel 4-D hyperchaotic hyperjerk system and its SPICE implementation. *Archives of Control Sciences*, *25*(1), 5–28.
155. Boulkroune, A., Bouzeriba, A., Bouden, T., & Azar, A. T. (2016a). Fuzzy adaptive synchronization of uncertain fractional-order chaotic systems. In A. T. Azar & S. Vaidyanathan (Eds.), *Advances in chaos theory and intelligent control*. Studies in fuzziness and soft computing (Vol. 337, pp. 681–697). Germany: Springer.

156. Boulkroune, A., Hamel, S., Azar, A. T., & Vaidyanathan, S. (2016b). Fuzzy control-based function synchronization of unknown chaotic systems with dead-zone input. In A. T. Azar & S. Vaidyanathan (Eds.), *Advances in chaos theory and intelligent control*. Studies in fuzziness and soft computing (Vol. 337, pp. 699–718). Germany: Springer.
157. Gan, Q., & Liang, Y. (2012). Synchronization of chaotic neural networks with time delay in the leakage term and parametric uncertainties based on sampled-data control. *Journal of the Franklin Institute*, *349*(6), 1955–1971.
158. Li, N., Zhang, Y., & Nie, Z. (2011). Synchronization for general complex dynamical networks with sampled-data. *Neurocomputing*, *74*(5), 805–811.
159. Xiao, X., Zhou, L., & Zhang, Z. (2014). Synchronization of chaotic Lur'e systems with quantized sampled-data controller. *Communications in Nonlinear Science and Numerical Simulation*, *19*(6), 2039–2047.
160. Zhang, H., & Zhou, J. (2012). Synchronization of sampled-data coupled harmonic oscillators with control inputs missing. *Systems and Control Letters*, *61*(12), 1277–1285.
161. Chen, W. H., Wei, D., & Lu, X. (2014). Global exponential synchronization of nonlinear time-delay Lur'e systems via delayed impulsive control. *Communications in Nonlinear Science and Numerical Simulation*, *19*(9), 3298–3312.
162. Jiang, G. P., Zheng, W. X., & Chen, G. (2004). Global chaos synchronization with channel time-delay. *Chaos, Solitons and Fractals*, *20*(2), 267–275.
163. Shahverdiev, E. M., & Shore, K. A. (2009). Impact of modulated multiple optical feedback time delays on laser diode chaos synchronization. *Optics Communications*, *282*(17), 3568–2572.
164. Rasappan, S., & Vaidyanathan, S. (2012a). Global chaos synchronization of WINDMI and Coullet chaotic systems by backstepping control. *Far East Journal of Mathematical Sciences*, *67*(2), 265–287.
165. Rasappan, S., & Vaidyanathan, S. (2012b). Hybrid synchronization of n-scroll Chua and Lur'e chaotic systems via backstepping control with novel feedback. *Archives of Control Sciences*, *22*(3), 343–365.
166. Rasappan, S., & Vaidyanathan, S. (2012c). Synchronization of hyperchaotic Liu system via backstepping control with recursive feedback. *Communications in Computer and Information Science*, *305*, 212–221.
167. Rasappan, S., & Vaidyanathan, S. (2013). Hybrid synchronization of n-scroll chaotic Chua circuits using adaptive backstepping control design with recursive feedback. *Malaysian Journal of Mathematical Sciences*, *7*(2), 219–246.
168. Rasappan, S., & Vaidyanathan, S. (2014). Global chaos synchronization of WINDMI and Coullet chaotic systems using adaptive backstepping control design. *Kyungpook Mathematical Journal*, *54*(1), 293–320.
169. Suresh, R., & Sundarapandian, V. (2013). Global chaos synchronization of a family of n-scroll hyperchaotic Chua circuits using backstepping control with recursive feedback. *Far East Journal of Mathematical Sciences*, *73*(1), 73–95.
170. Vaidyanathan, S., & Rasappan, S. (2014). Global chaos synchronization of n-scroll Chua circuit and Lur'e system using backstepping control design with recursive feedback. *Arabian Journal for Science and Engineering*, *39*(4), 3351–3364.
171. Vaidyanathan, S., Idowu, B. A., & Azar, A. T. (2015a). Backstepping controller design for the global chaos synchronization of Sprott's jerk systems. *Studies in Computational Intelligence*, *581*, 39–58.
172. Sundarapandian, V., & Sivaperumal, S. (2011). Sliding controller design of hybrid synchronization of four-wing chaotic systems. *International Journal of Soft Computing*, *6*(5), 224–231.
173. Vaidyanathan, S. (2012a). Analysis and synchronization of the hyperchaotic Yujun systems via sliding mode control. *Advances in Intelligent Systems and Computing*, *176*, 329–337.
174. Vaidyanathan, S. (2012c). Global chaos control of hyperchaotic Liu system via sliding control method. *International Journal of Control Theory and Applications*, *5*(2), 117–123.

175. Vaidyanathan, S. (2012e). Sliding mode control based global chaos control of Liu-Liu-Liu-Su chaotic system. *International Journal of Control Theory and Applications*, *5*(1), 15–20.
176. Vaidyanathan, S. (2014e). Global chaos synchronization of identical Li-Wu chaotic systems via sliding mode control. *International Journal of Modelling, Identification and Control*, *22*(2), 170–177.
177. Vaidyanathan, S., & Azar, A. T. (2015c). Anti-synchronization of identical chaotic systems using sliding mode control and an application to Vaidhyanathan-Madhavan chaotic systems. *Studies in Computational Intelligence*, *576*, 527–547.
178. Vaidyanathan, S., & Azar, A. T. (2015d). Hybrid synchronization of identical chaotic systems using sliding mode control and an application to Vaidhyanathan chaotic systems. *Studies in Computational Intelligence*, *576*, 549–569.
179. Vaidyanathan, S., & Sampath, S. (2011). Global chaos synchronization of hyperchaotic Lorenz systems by sliding mode control. *Communications in Computer and Information Science*, *205*, 156–164.
180. Vaidyanathan, S., & Sampath, S. (2012). Anti-synchronization of four-wing chaotic systems via sliding mode control. *International Journal of Automation and Computing*, *9*(3), 274–279.
181. Khalil, H. K. (2001). *Nonlinear systems*. New Jersey, USA: Prentice Hall.

On the Terminal Full Order Sliding Mode Control of Uncertain Chaotic Systems

Anchan Saxena, Apeksha Tandon, Awadhi Saxena, K.P.S. Rana and Vineet Kumar

Abstract Over the years, several forms of sliding mode control (SMC), such as conventional SMC, terminal SMC (TSMC) and fuzzy SMC (FSMC) have been developed to cater to the control needs of complex, non-linear and uncertain systems. However, the chattering phenomenon in conventional SMC and the singularity errors in TSMC make the application of these schemes relatively impractical. In this chapter, terminal full order SMC (TFOSMC), the recent development in this line, has been explored for efficient control of the uncertain chaotic systems. Two important chaotic systems, Genesio and Arneodo-Coullet have been considered in fractional order as well as integer order dynamics. The investigated fractional and integer order chaotic systems are controlled using fractional order TFOSMC and integer order TFOSMC, respectively and the control performance has been assessed for settling time, amount of chattering, integral absolute error (IAE) and integral time absolute error (ITAE). To gauge the relative performance of TFOSMC, a comparative study with FSMC, tuned by Cuckoo Search Algorithm for the minimum IAE and amount of chattering has also been performed using settling time, amount of chattering, IAE and ITAE performances. The intensive simulation studies presented in this chapter clearly demonstrate that the settling time, amount

A. Saxena · A. Tandon · A. Saxena · K.P.S. Rana (✉) · V. Kumar
Division of Instrumentation and Control Engineering,
Netaji Subhas Institute of Technology, Sector-3,
Dwarka, New Delhi 110078, India
e-mail: kpsrana1@gmail.com
URL: http://www.nsit.ac.in

A. Saxena
e-mail: anchansaxena@gmail.com

A. Tandon
e-mail: apeksha94@live.com

A. Saxena
e-mail: awadhi.saxena@gmail.com

V. Kumar
e-mail: vineetkumar27@gmail.com

A.T. Azar et al. (eds.), *Fractional Order Control and Synchronization of Chaotic Systems*, Studies in Computational Intelligence 688,
DOI 10.1007/978-3-319-50249-6_13

of chattering and steady-state tracking errors offered by TFOSMC are significantly lower than that of FSMC; therefore, making TFOSMC a superior scheme.

Keywords Sliding mode control • Chaotic system • Fractional order • Genesio • Arneodo-Coullet

1 Introduction

Chaos is a non-linear complex phenomenon characterized by a high sensitivity to initial conditions which implies that two chaotic trajectories starting infinitesimally close to each other will diverge exponentially with time, giving rise to an infinite number of unstable periodic orbits. Chaotic dynamics result in a trajectory wherein the system states move in the neighborhood of one of these periodic orbits for a while, then erratically move to a different unstable, periodic orbit where it remains for a limited time, and so forth [23]. Coupled with the fact that experimental conditions are never known perfectly, these systems are inherently unpredictable even while being mathematically deterministic [10, 31].

Chaos has been found to occur in a wide variety of disciplines such as the Raleigh-Bernard convection in fluid dynamics, the Belousov-Zhaobitinsky reaction in chemistry [47], multimode solid state lasers in optics [51], the Chua-Matsumoto oscillator in electronics [30], population models [49], meteorology, in physiological models such as certain heart and respiratory rhythms [32] and so on. Dynamics of chaotic systems can be described using integer as well as non-integer (fractional) order calculus. Novel methods of modeling and control system designing of chaotic systems has always been a sought after area of research [11, 12, 54, 55]. Fractional order calculus allows us to describe and model a real system more accurately than the classical integer order calculus methods. Consequently, it has been reported that the dynamics of several chaotic systems can also be elegantly described by fractional order dynamical equations making use of fractional order operators [29, 35, 46, 66, 67, 76]. In the light of aforementioned potential applications and related issues, stabilization and control of fractional order chaotic systems can be considered to be of fundamental importance [9, 14, 28, 66, 70].

Over the course of time, several schemes have been proposed for control of non-linear complex systems; one of the most recent one has been the terminal full order sliding mode control (TFOSMC) proposed by Feng et al. [26]. It has been claimed to be more efficient over its counter parts. Claimed superiority of TFOSMC has motivated the authors to explore its applications on the Genesio and Arneodo-Coullet chaotic systems for their effective control. Therefore, the objective of this chapter is to demonstrate the application of TFOSMC scheme to effectively control both fractional as well as integer order Genesio and Arneodo-Coullet chaotic systems in the presence of system uncertainties and external disturbances. The numerical simulations, as demonstrated later, clearly indicate that the output of TFOSMC is smooth and chatter-free to a good extent while simultaneously it is able to address the problems of singularity and finite-time convergence. Further, to gauge the relative performance of TFOSMC controller, a comparative study has also been performed with fuzzy sliding mode control (FSMC), whose gains have been tuned by cuckoo search algorithm (CSA) [27] for minimum integral absolute error (IAE) and amount of chattering [25]. Extensive simulation studies have been presented which demonstrate that the settling time, amount of chattering, IAE and integral time absolute error (ITAE) offered by TFOSMC are significantly lower than that of FSMC; therefore, making TFOSMC a superior scheme. Several contributions of this chapter can be listed as follows:

1. For the first time, implementation of TFOSMC for effective control of chaotic systems has been demonstrated in this chapter.
2. Genesio and Arneodo-Coullet chaotic systems have been successfully controlled in the presence of system uncertainties and external disturbances.
3. Control performance of TFOSMC, assessed in terms of settling time, amount of chattering, IAE and ITAE has been found to be superior over its potential counterpart, FSMC (tuned using CSA).

Rest of the chapter is organized as follows. Section 2 provides a brief survey of the related works carried out in the domain of chaotic system control. In Sect. 3, some requisite preliminaries of fractional calculus are presented. Section 4 provides the dynamical models of the two investigated systems (fractional and integer order Genesio and Arneodo-Coullet) along with their 3D chaotic attractors and the

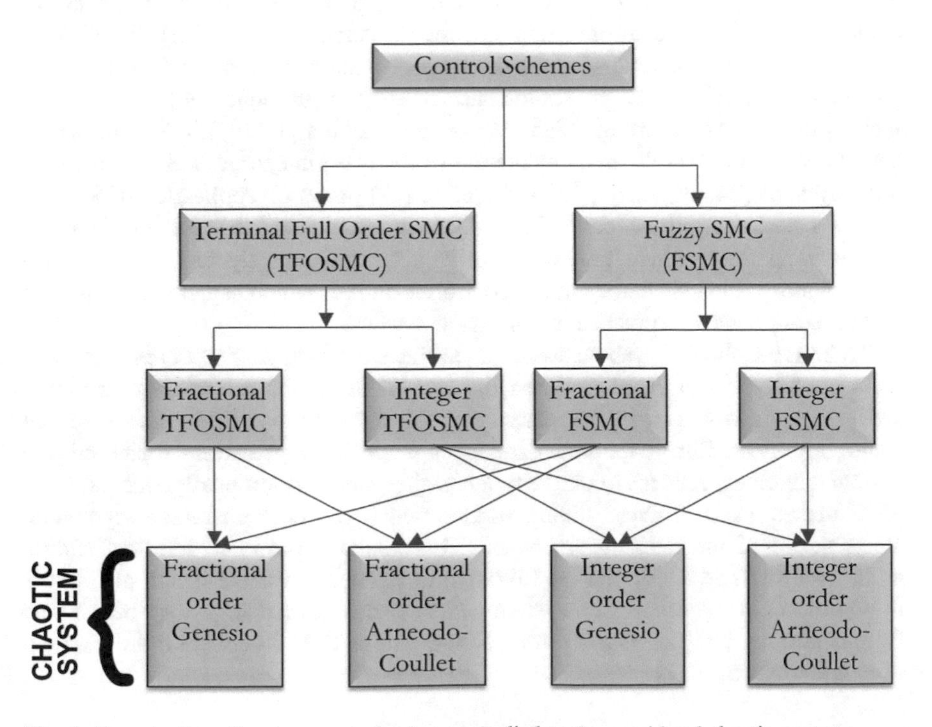

Fig. 1 Organization of various control schemes applied to the considered chaotic systems

problem formulation. In Sect. 5, description and design of the various SMC variants namely conventional SMC, FSMC and TFOSMC are presented followed by MATLAB simulation results illustrating their performances. Finally, Sect. 6 provides a comparative study between FSMC and TFOSMC and Sect. 7 concludes the findings with some future directions.

To summarize the presented work, flowchart in Fig. 1 depicts various combinations of control schemes and the considered chaotic systems in this chapter. As seen in Fig. 1, eight case studies have been investigated in this chapter resulting from the two chaotic systems of integer as well as fractional order dynamics and two control schemes.

2 Literature Survey

The stabilization and control of complex systems with characteristic non-linearities and uncertainities has been one of the prime topics of research inviting works on several control methodologies based on classical, modern and robust control [5–8]. For chaotic systems, in the initial phase, two approaches used for their control were the OGY (Ott, Grebogi and Yorke) method [42] and the Pyragas continuous control method [48] both of which require a preliminary determination of the unstable orbits of the system before the control law can be designed. Over the time, several forms of sliding mode control (SMC) have emerged to cater to the control needs of complex, non-linear and uncertain chaotic systems [13, 56, 63]. SMC is a non-linear control method wherein the system state trajectories are driven to a predefined manifold called the sliding surface and are subsequently kept in a close vicinity of the surface through high frequency switching [1, 37, 45]. Several works have been reported on the implementation of SMC on integer as well as fractional order systems [24, 39, 52, 64, 73]. Chen et al. [21] proposed application of SMC to control a class of fractional order chaotic systems. However, the finite time delays in conventional SMC where the switching is not infinitesimally fast, resulted in a phenomenon called chattering in the controller output, which can cause damage to system components in practical engineering systems.

To limit the chattering about switching surface, the boundary layer approach was introduced by Liu et al. [37]. When the system uncertainties are large, a higher switching gain with a wider boundary layer was required to eliminate the increased chattering effect. But, if the boundary layer width is progressively increased, the system effectively reduces to one without sliding mode. Additionally, conventional SMC makes use of a linear sliding surface which can only guarantee asymptotic convergence of the tracking errors. In [3], Aghababa proposed terminal sliding mode control (TSMC) of chaotic Lorenz and Arneodo systems which guarantees finite-time convergence of the system states to the desired trajectory but suffers from chattering and singularity errors. Non-singular TSMC addressed the problem of singularity errors [2].

Since the introduction of fuzzy set theory by Zadeh [75], fuzzy logic based schemes have been successfully applied to a variety of applications over the past four decades [18–20, 52, 74]. Yau and Chen proposed to control the chaotic Genesio system by replacing the discontinuous signum function in the reaching law by fuzzy logic control (FLC) [72]. Several key breakthroughs have been brought about in the control and synchronization of chaotic systems using adaptive techniques. Vaidyanathan and Azar have led the way in this regard with numerous successful applications of different adaptive control methods such as feedback and backstepping on many complex chaotic systems with unknown parameters [57–62]. Additionally, adaptive fuzzy controllers have also been used to control and synchronize chaotic systems [9, 15, 16, 34, 36, 40].

Recently, Feng et al. [26] proposed TFOSMC for two general non-linear systems in which a full order sliding manifold is utilized. During the full order sliding mode, the system had desirable full-order dynamics, rather than reduced-order dynamics. Furthermore, the derivatives of the terms with fractional powers do not appear in the control law, avoiding the control singularities. However, being relatively new TFOSMC has not yet been implemented on chaotic systems. Thus, the aim of this chapter is to effectively control the aforementioned two chaotic systems by means of TFOSMC and to prepare a performance analysis between the results obtained by TFOSMC and those obtained by FSMC on the basis of settling time, amount of chattering, IAE and ITAE.

3 Some Preliminaries of Fractional Calculus

For the past three decades, significant progress has been witnessed in fractional order calculus (FOC) as it finds extensive applications in modeling phenomena such as diffusion, turbulence, electromagnetism, signal processing, and quantum evolution of complex systems [4].

Several types of fractional order sliding mode controllers have been proposed in literature as they offer greater robustness and lower chattering in comparison to integer order controllers, though at the cost of higher computational requirements [22]. The design idea of the fractional order controller was first proposed by Oustaloup [43, 44]. To obtain a finite approximation of fractional order systems in a desired range of frequencies, he gave the approximation algorithm that is widely used wherein a frequency band of interest is considered within which the following approximation holds.

Suppose that the desired frequency range is given by $[\omega_l, \omega_h]$. The function considered for fractional order integrator/differentiator [17, 55] approximation is of the form:

$$H(s) = s^{\gamma}, \quad \gamma \in R, \quad \gamma \in [-1; 1] \tag{1}$$

The Oustaloup's approximation function to this fractional order differentiator s^{γ} can be written as,

$$\widehat{H}(s) = \left(\frac{\omega_u}{\omega_h}\right)^{\gamma} \prod_{k=-N}^{N} \frac{1 + s/\omega_k^{'}}{1 + s/\omega_k} \tag{2}$$

where,

$$\omega_k^{'} = \omega_l \left(\frac{\omega_h}{\omega_l}\right)^{\frac{k+N+1/2-\gamma/2}{2N+1}} \tag{3}$$

and

$$\omega_k = \omega_l \left(\frac{\omega_h}{\omega_l}\right)^{\frac{k+N+1/2+\gamma/2}{2N+1}} \tag{4}$$

are respective zeros and poles of rank k. The total number of zeros or poles is given by $2N+1$. Frequency $\omega_u = \sqrt{\omega_l \omega_h}$ is the geometric mean of lower and upper bounds of the frequencies.

4 Chaotic System Descriptions and Problem Formulation

This section presents the mathematical models of the considered chaotic systems along with their respective initial conditions and system parameters. The systems are graphically introduced with the help of their resulting chaotic attractor and uncontrolled state trajectories. It may be noted that the considered systems are (i) Fractional order $(\gamma = 0.993)$ and Integer order $(\gamma = 1)$ Genesio system and (ii) Fractional order $(\gamma = 0.993)$ and Integer order $(\gamma = 1)$ Arneodo-Coullet system. These systems are described in brief in the following sub-sections with the help of their uncontrolled chaotic attractor and their uncontrolled system states.

4.1 Genesio Chaotic System

The Genesio chaotic system arises from a jerk equation and represents jerky dynamics, which is the third derivative of position. Genesio chaotic system is described as [72]:

$$\begin{aligned} D^{\gamma}x_1 &= x_2 \\ D^{\gamma}x_2 &= x_3 \\ D^{\gamma}x_3 &= -cx_1 - bx_2 - ax_3 + x_1^2 \end{aligned} \tag{5}$$

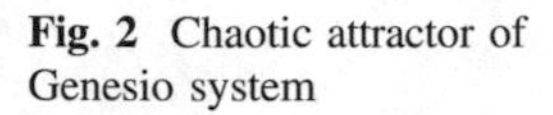

Fig. 2 Chaotic attractor of Genesio system

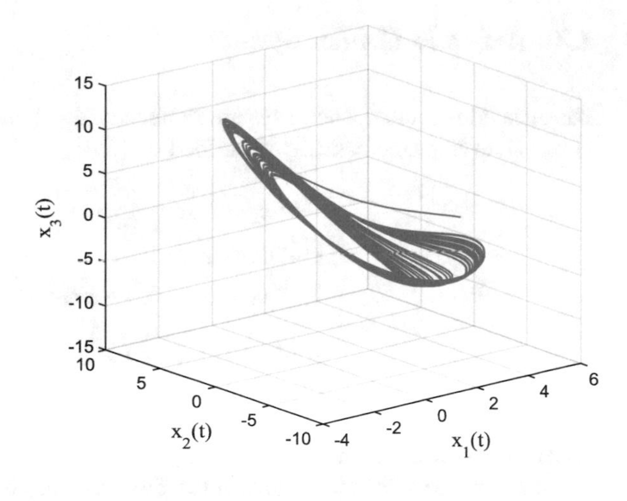

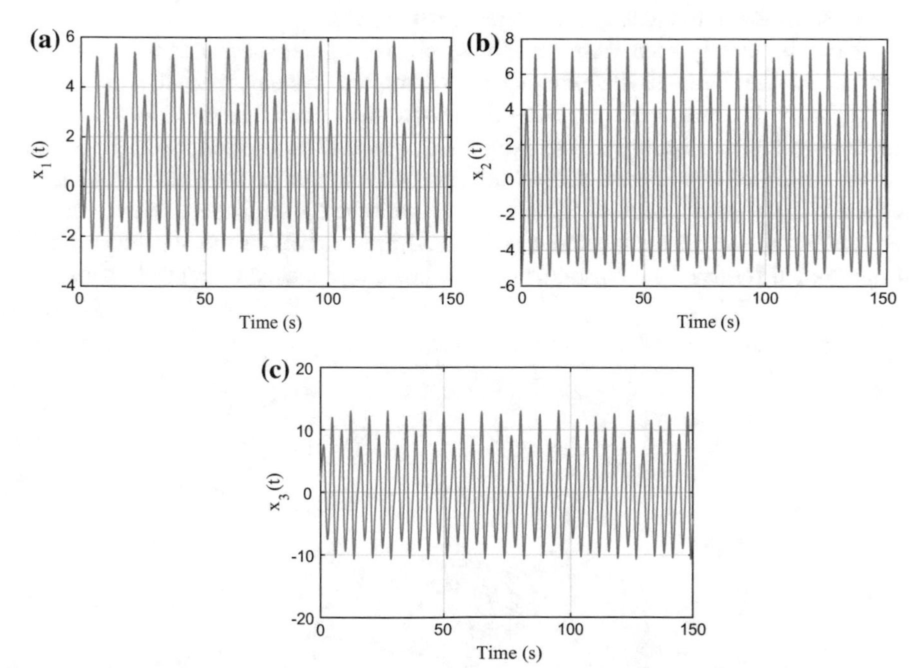

Fig. 3 Uncontrolled trajectories of Genesio system: **a** state x_1; **b** state x_2; **c** state x_3

where x_1, x_2 and x_3 are state variables and a, b and c are the positive real constants. For instance, the Genesio system is chaotic for the parameter values of $a = 1.2$, $b = 2.92$, $c = 6$ and $\gamma = 0.993$. In this work, initial conditions of this system are considered as $x_1 = 3$, $x_2 = -4$ and $x_3 = 2$.

Figures 2 and 3 show the uncontrolled Genesio three-dimensional chaotic attractor and the time responses of the states, respectively.

4.2 Arneodo-Coullet System

The Arneodo-Coullet chaotic system represents the dynamics of a forced oscillator. The system is mathematically described as [53]:

$$\begin{aligned} D^{\gamma} x_1 &- x_2 \\ D^{\gamma} x_2 &= x_3 \\ D^{\gamma} x_3 &= c x_1 - b x_2 - a x_3 - x_1^3 \end{aligned} \tag{6}$$

where x_1, x_2 and x_3 are state variables and a, b and c are the positive real constants. For instance, the Arneodo-Coullet system is chaotic for the parameter values of $a = 0.45$, $b = 1.1$, $c = 0.8$ and $\gamma = 0.993$. In this work, initial conditions of this system are considered as $x_1 = -1.2$, $x_2 = 1.2$ and $x_3 = 0.4$. Figures 4 and 5 show the uncontrolled Arneodo-Coullet three-dimensional chaotic attractor and the time response of the individual system states, respectively.

It can be clearly seen from the above chaotic dynamics that it consists of a motion in the three-dimensional space where the system state moves in the neighborhood of one of the periodic orbits for a while, then falls close to a different unstable, periodic orbit where it remains for a limited time, and so forth. This results in a complicated and unpredictable wandering over longer periods of time. Control of chaos is the stabilization of these unstable periodic orbits. The result is to render an otherwise chaotic motion more stable and predictable. In the subsequent sections, varied forms of sliding mode control law are proposed to control chaos in a

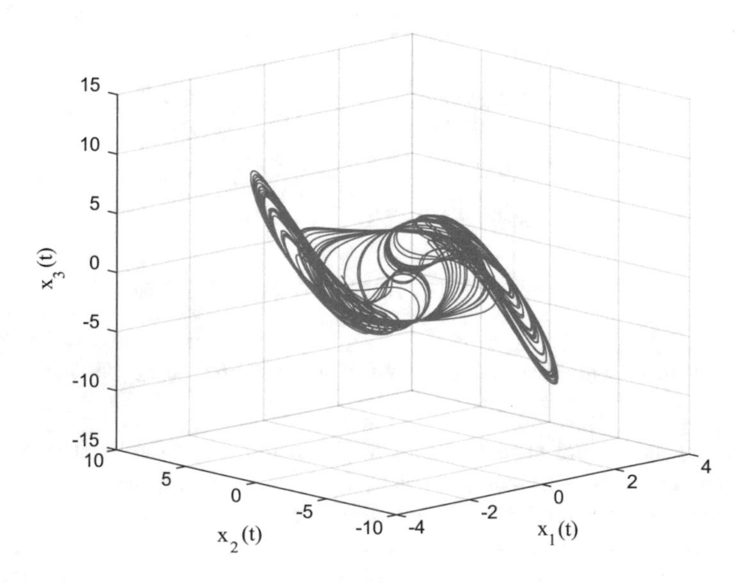

Fig. 4 Chaotic attractor of Arneodo-Coullet system

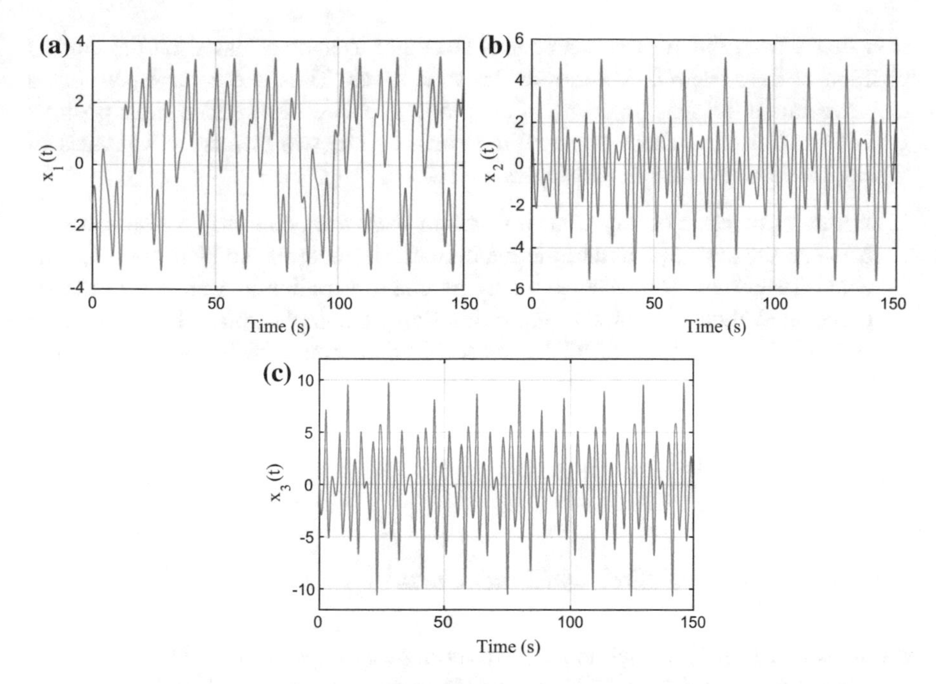

Fig. 5 Uncontrolled trajectories of Arneodo-Coullet system: **a** state x_1; **b** state x_2; **c** state x_3

class of fractional and integer order chaotic systems [47, 48]. The controllers are derived to stabilize the states of these chaotic systems, even if the systems with uncertainty are in the presence of external disturbance.

5 Sliding Mode Control

This section is organized as follows. Section 5.1 presents the description and design of conventional SMC along with numerical simulations to demonstrate the chattering phenomenon. In Sect. 5.2, the FSMC design for each of the considered systems is presented. The simulations are graphically illustrated and the resulting performance indices are given in tabular form. A brief overview of the CSA is also provided. Section 5.3 presents the design of TFOSMC for each of the considered systems along with the requisite finite-time stability analysis. The simulations are graphically illustrated and the resulting performance indices are given in tabular form.

SMC is a variable structure, non-linear control technique that possesses desirable features of accuracy and robustness. SMC is designed to drive the system states onto a particular manifold called the sliding surface. Once the sliding surface is reached, SMC keeps the states in a close neighborhood of the sliding surface. In SMC, the feedback control law is not a continuous function of time. Instead, it switches from one continuous control structure to another based on the current position of the system trajectories in the state space. Thus, the control path has a negative gain if the state trajectory of the plant is "above" the surface and a positive gain if the trajectory drops "below" the surface. The two primary advantages of sliding mode control are elucidated as follows:

1. In the formulation of any control problem there are bound to be discrepancies between the actual plant and the mathematical model of the plant used for the controller design. This mismatch may be due to variation in system parameters, unmodeled dynamics or the approximation of complex plant behavior by a simplified model. With SMC, the closed loop response of the system becomes relatively insensitive to parametric uncertainties.
2. The dynamic behavior of the system may be controlled by an appropriate choice of the switching function.

5.1 Conventional Sliding Mode Control

As shown in Fig. 6, conventional SMC comprises typically of a sliding surface described by $s = 0$ and the sliding motion along the surface [41]. The sliding motion comprises of a reaching phase and a sliding phase. The SMC controller design involves the design of the switching surface and a second phase derives the control law required to drive the system state trajectories onto the sliding surface.

5.1.1 Sliding Surface Design

A typical nth order non-linear chaotic dynamical system with the relative degree n may be directly described as follows [72]:

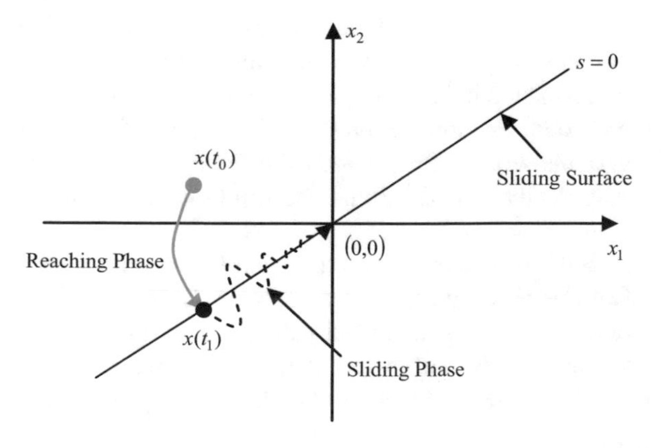

Fig. 6 Phase plane plot of a system with sliding mode control

$$\begin{aligned} \dot{x}_i &= x_{i+1}, & 1 \le i \le n-1 \\ \dot{x}_n &= b_0(X,t) + \Delta b(X,t) + d(t) + u(t), & X = [x_1 \quad x_2 \quad \cdots \quad x_n] \end{aligned} \tag{7}$$

where $X(t) = [x_1(t) \quad x_2(t) \quad \ldots \quad x_n(t)] = [x(t) \quad \dot{x}(t) \quad \cdots \quad x^{(n-1)}(t)] \in R^n$ is the state vector, $\Delta b(X,t)$ and $b_0(X,t)$ are uncertain part and known part of chaotic systems, respectively, $u(t) \in R$ is the controller output, and $d(t)$ is the external disturbance of system. In general, the uncertain term $\Delta b(X,t)$ and disturbance term $d(t)$ are assumed to be bounded i.e.

$$|\Delta b(X,t)| \le \alpha \text{ and } |d(t)| \le \beta \tag{8}$$

where α and β are positive.

From a geometrical point of view, $s = 0$ defines a surface in the error space. The control problem is to get the system to track an n-dimensional desired vector $X_d(t)$, such that,

$$X_d(t) = [x_{d1}(t) \quad x_{d2}(t) \quad \ldots \quad x_{dn}(t)] = \left[x_d(t) \quad \dot{x}_d(t) \quad \cdots \quad x_d^{(n-1)}(t)\right] \tag{9}$$

The tracking error can be then defined as,

$$\begin{aligned} E(t) &= X(t) - X_d(t) \\ &= \left[x(t) - x_d(t) \quad \dot{x}(t) - \dot{x}_d(t) \quad \cdots \quad x^{(n-1)}(t) - x_d^{(n-1)}(t)\right] \\ &= \left[e(t) \quad \dot{e}(t) \quad \cdots \quad e^{(n-1)}(t)\right] = [e_1(t) \quad e_2(t) \quad \cdots \quad e_n(t)] \end{aligned} \tag{10}$$

The resulting state response of tracking error vector should satisfy,

$$\lim_{t \to \infty} \|E(t)\| = \lim_{t \to \infty} \|X(t) - X_d(t)\| \to 0 \tag{11}$$

where, $\| \bullet \|$ is the Euclidean norm of a vector.

The sliding surface depends on the tracking error e and its derivatives, and is usually of the Proportional-Derivative (PD) form given as follows [72],

$$s = e_n + \sum_{i=1}^{n-1} c_i e_i \tag{12}$$

When the closed-loop system is in the sliding mode, it satisfies $\dot{s} = 0$ and then the equivalent control law is obtained by,

$$u_{eq} = -b_0(X,t) - \Delta b(X,t) - d(t) - \sum_{i=1}^{n-1} c_i e_{i+1} + x_d^{(n)}(t) \tag{13}$$

If the reaching law is u_r, then the overall control u is determined by,

$$u = u_{eq} + u_r \tag{14}$$

5.1.2 Reaching Laws

Generally three reaching laws as described below are used.

Constant Rate Reaching Law This reaching law is normally used in conventional SMC and is given by:

$$\begin{gathered} \dot{s} = -K\,\mathrm{sgn}(s), \quad K > 0 \\ \dot{s} = \begin{cases} K, & s < 0 \\ -K, & s > 0 \end{cases} \end{gathered} \tag{15}$$

It constraints the switching variable to reach the switching surface s at a constant rate K. If K is too small, the reaching time will be too long and on the other hand if K is too large, there will be severe chattering.

Exponential Reaching Law It is given by the following expression:

$$\dot{s} = -K\mathrm{sgn}(s) - \beta s, K > 0, \ \beta > 0 \tag{16}$$

where, $-\beta s$ is the exponential term, and its solution is $s = s(0)e^{-\beta t}$. Clearly, by adding the proportional term $-\beta s$, the state is forced to approach the switching manifolds faster when s is large.

Power Rate Reaching Law This law, as stated below, offers a fast and low chattering reaching mode.

$$\dot{s} = -K|s|^{\alpha}\mathrm{sgn}(s), \ K > 0.1 > \alpha > 0 \tag{17}$$

This reaching law increases the reaching speed when the state is far away from the switching manifold. However, it reduces the rate when the state is near the manifold.

5.1.3 SMC Implementation on Chaotic Systems

The design and implementation of conventional SMC for Genesio and Arneodo-Coullet chaotic systems has been described in this section. The resulting state trajectories and controller output have been graphically depicted in order to demonstrate the phenomenon of chattering. It may be noted that this specific study is notional with a purpose to demonstrate in closed loop while the control was manually tuned.

Conventional SMC of Genesio Chaotic System

The dynamical model of the Genesio system is as follows,

$$\begin{aligned} D^{\gamma}x_1 &= x_2 \\ D^{\gamma}x_2 &= x_3 \\ D^{\gamma}x_3 &= -cx_1 - bx_2 - ax_3 + x_1^2 + \Delta b(X,t) + d(t) + u(t) \end{aligned} \tag{18}$$

The initial conditions of the system are $x_1 = 3$, $x_2 = -4, x_3 = 2$ and $\gamma = 1$. The system is perturbed by an uncertainty term $\Delta b(X,t)$ and excited by a disturbance term $d(t)$. Here, $\Delta b(X,t) = 0.1\sin 4\pi x_1 \sin 2\pi x_2 \sin \pi x_3$ and $d(t) = 0.1\sin(t)$ satisfy, respectively, $|\Delta b(X,t)| \leq \alpha = 0.1$ and $|d(t)| \leq \beta = 0.1$. Control objective is to drive the uncertain chaotic system to the desired trajectory $x_d(t)$. Selecting $c_1 = 10$ and $c_2 = 6$ to result in a stable sliding mode. Therefore, the switching surface is,

$$s(t) = e_3(t) + c_1e_1(t) + c_2e_2(t) \tag{19}$$

The equivalent control law is obtained as,

$$u_{eq} = 1.2x_1 + 2.92x_2 + 6x_3 - x_1^2 - c_1e_2(t) - c_2e_3(t) \tag{20}$$

Taking $K = 1$ the constant rate reaching law becomes,

$$u_r = -K\,\mathrm{sgn}(s) = -\mathrm{sgn}(s) \tag{21}$$

Thus, the overall control law becomes,

$$u = 1.2x_1 + 2.92x_2 + 6x_3 - x_1^2 - c_1e_2(t) - c_2e_3(t) - \mathrm{sgn}(s) \tag{22}$$

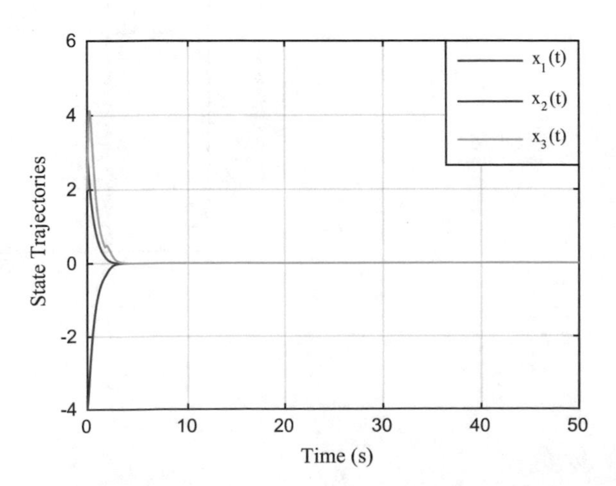

Fig. 7 State trajectories of Genesio system controlled using conventional SMC

The simulations were run for $t = 50\,\text{s}$ using the 4th order Runge-Kutta method with a step time of 0.001 s. The obtained simulation results are shown in Figs. 7, 8 and 9 representing the states' time responses, controller output and sliding surface dynamics, respectively.

As indicated by the resulting plots, the system states settle at $t \approx 3.8\,\text{s}$ (computed for the worst trajectory) and as expected, the chattering behaviour of conventional SMC is clearly demonstrated in the controller output. It may be noted that though the chattering can be reduced but cannot be removed in this scheme.

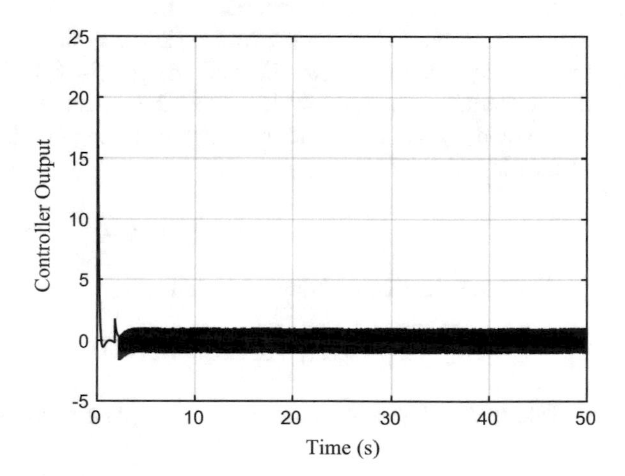

Fig. 8 Control action versus time plot for conventional SMC controlled Genesio system

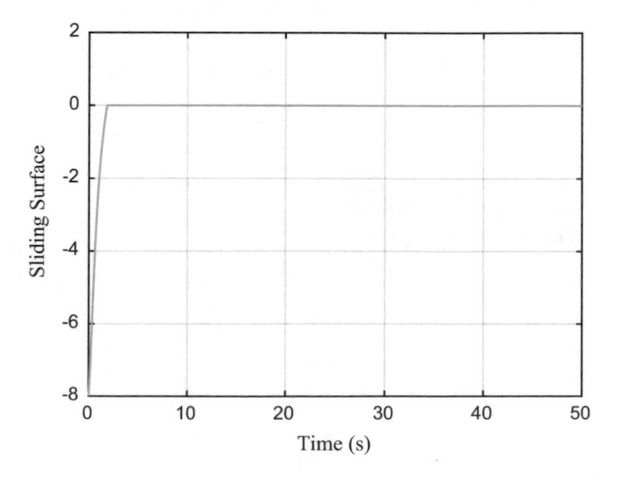

Fig. 9 Time response of sliding surface for conventional SMC controlled Genesio system

Conventional SMC of Arneodo-Coullet Chaotic System

The dynamical model of the Arneodo-Coullet system is as follows,

$$\begin{aligned} D^{\gamma}x_1 &= x_2 \\ D^{\gamma}x_2 &= x_3 \\ D^{\gamma}x_3 &= cx_1 - bx_2 - ax_3 - x_1^3 + \Delta b(X,t) + d(t) + u(t) \end{aligned} \tag{23}$$

The initial conditions of the system are $x_1 = -1.2$, $x_2 = 1.2, x_3 = 0.4$ and $\gamma = 1$. The system is perturbed by an uncertainty term $\Delta b(X,t)$ and excited by a disturbance term $d(t)$. Here, $\Delta b(X,t) = 0.1\sin 4\pi x_1 \sin 2\pi x_2 \sin \pi x_3$ and $d(t) = 0.1\sin(t)$ satisfy, respectively, $|\Delta b(X,t)| \le \alpha = 0.1$ and $|d(t)| \le \beta = 0.1$. Control objective is to drive the uncertain chaotic system to the desired trajectory $x_d(t)$. Selecting $c_1 = 10$ and $c_2 = 6$ to result in a stable sliding mode. Now, the switching surface is,

$$s(t) = e_3(t) + c_1e_1(t) + c_2e_2(t) \tag{24}$$

The equivalent control law is obtained as,

$$u_{eq} = -0.8x_1 + 1.1x_2 + 0.45x_3 + x_1^3 - c_1e_2(t) - c_2e_3(t) \tag{25}$$

Taking $K = 12$, the constant rate reaching law becomes,

$$u_r = -K\text{sgn}(s) = -12\text{sgn}(s) \tag{26}$$

Thus, the overall control law becomes,

$$u = -0.8x_1 + 1.1x_2 + 0.45x_3 + x_1^3 - c_1e_2(t) - c_2e_3(t) - 12\text{sgn}(s) \tag{27}$$

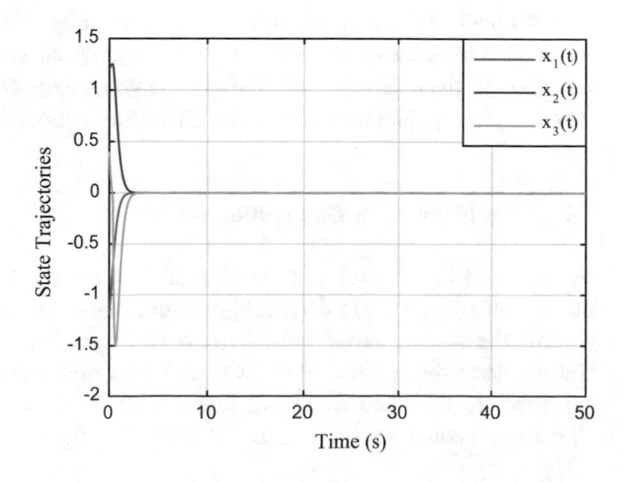

Fig. 10 State trajectories of Arneodo-Coullet system controlled using conventional SMC

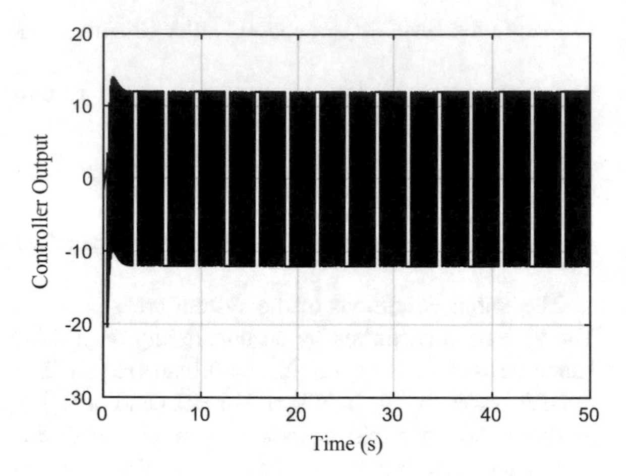

Fig. 11 Control action versus time plot for conventional SMC controlled Arneodo-Coullet system

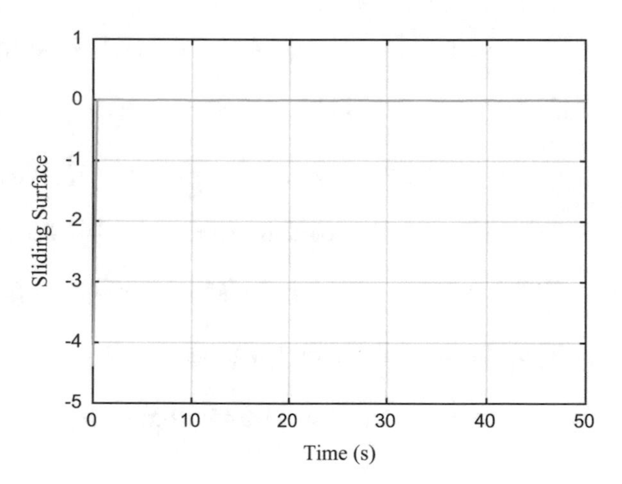

Fig. 12 Time response of sliding surface for conventional SMC controlled Arneodo-Coullet system

The obtained simulation results are shown in Figs. 10, 11 and 12 representing the states' time responses, controller output and sliding surface dynamics, respectively. As indicated by the resulting plots, the system states settle at $t \approx 5.8$ s (computed for the worst trajectory) and the problem of chattering is obtained in the controller output.

5.1.4 Problems with Conventional SMC

As observed in the results of the simulations, conventional SMC suffers from two main problems [50]. (1) Chattering: In the theoretical description of sliding mode control, the system stays confined to the sliding surface and need only be viewed as sliding along the surface. However, real implementations of sliding mode control approximate this theoretical behaviour with a high-frequency switching control signal that causes the system to "chatter" in a tight neighbourhood of the sliding surface. This phenomenon is called chattering and it may damage the actuators in a practical system. (2) Asymptotic Convergence: The sliding surface adopted in conventional sliding mode control is a linear dynamical equation $s(t) = e_3(t) + c_1 e_1(t) + c_2 e_2(t)$. The linear sliding surface can only guarantee asymptotic error convergence in the sliding motion, i.e., the output error cannot converge to zero in finite time. This is practically undesired.

5.2 Fuzzy Sliding Mode Control

Fuzzy logic formalizes the human ability to reason and judge under uncertainty [50, 75]. In traditional SMC, the reaching law is selected as $u_r = k_w u_w$ and the overall control u is determined by [55]:

$$u = u_{eq} + u_r = u_{eq} + k_w u_w \tag{28}$$

where k_w is the switching gain (positive) and u_w is obtained by

$$u_w = -\operatorname{sgn}(s) \tag{29}$$

where,

$$\operatorname{sgn}(s) = \begin{cases} 1, & \text{for} \quad s > 0, \\ 0, & \text{for} \quad s = 0, \\ -1, & \text{for} \quad s < 0, \end{cases} \tag{30}$$

However, the signum function in the overall control law u will cause chattering in the controller output due to finite time delays in the switching. This problem can be tackled by using FLC [65]. A set of rules derived from expert knowledge determine the dynamic behavior of the FLC. On the basis of these rules, the Takagi-Sugeno-Kang fuzzy inference mechanism provides the necessary control action. Since the rules of the fuzzy controller are based on SMC, it is called fuzzy SMC (FSMC) [24, 68, 69].

The employed control scheme is depicted in Fig. 13 [72]; the overall control law is the algebraic sum of the equivalent control part and the FSMC output.

The equivalent control part is obtained from the system equations and the reaching law is selected as:

$$u_r = k_{fs} u_{fs} \tag{31}$$

where, k_{fs} is the normalization factor (fuzzy gain) of the output variable, and u_{fs} is the output obtained from FSMC, which is determined by s and $\dot{s}$. The overall control law u is then obtained as:

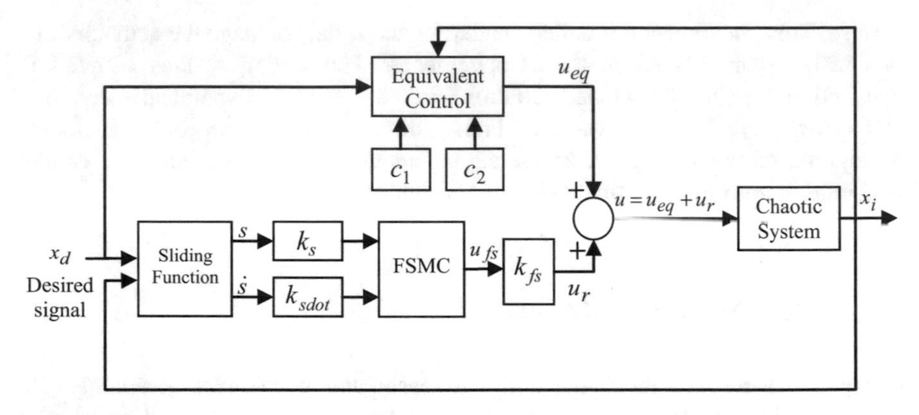

Fig. 13 FSMC implementation scheme

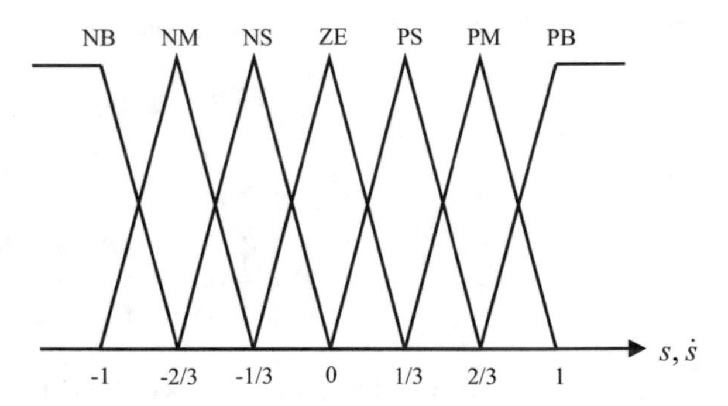

Fig. 14 Membership functions for FSMC

$$u = u_{eq} + u_r = u_{eq} + k_{fs} u_{fs} \quad (32)$$

The fuzzy control rules depend on the sliding surface s and the rate of change of the sliding surface $\dot{s}$.

$$u_{fs} = FSMC(s, \dot{s}) \quad (33)$$

The membership functions of input linguistic variables s and $\dot{s}$, and the membership functions of output linguistic variable u_{fs} are shown in Figs. 14 and 15, respectively. They are partitioned into seven fuzzy membership functions expressed as negative big (NB), negative medium (NM), negative small (NS), zero (ZE), positive small (PS), positive medium (PB) and positive big (PB) in order to cover the entire sample space. The fuzzy rule table is given in Table 1 [72].

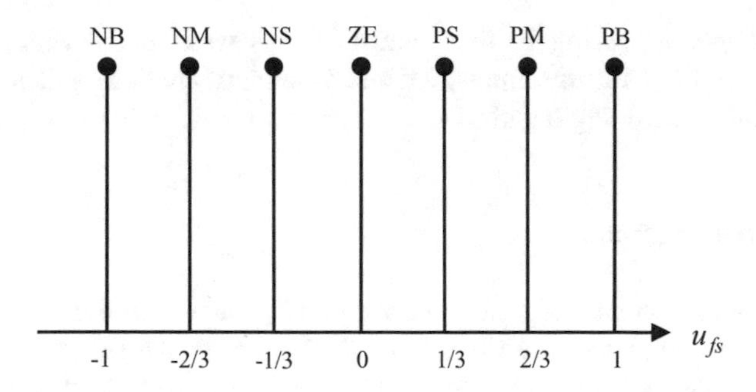

Fig. 15 Fuzzy output singletons

Table 1 Fuzzy rules for FSMC

u_{fs}		s						
		NB	NM	NS	ZE	PS	PM	PB
$\dot{s}$	NB	PB	PB	PB	PB	PM	PS	ZE
	NM	PB	PB	PB	PM	PS	ZE	NS
	NS	PB	PB	PM	PS	ZE	NS	NM
	ZE	PB	PM	PS	ZE	NS	NM	NB
	PS	PM	PS	ZE	NS	NM	NB	NB
	PM	PS	ZE	NS	NM	NB	NB	NB
	PB	ZE	NS	NM	NB	NB	NB	NB

5.2.1 Design and Implementation of FSMC

As determined by Eq. (32), the overall control action u is obtained as

$$u = u_{eq} + u_r = u_{eq} + k_{fs}u_{fs} \tag{34}$$

In the practical system, the system uncertainty $\Delta b(X, t)$ and external disturbance are not known and the equivalent controller output reduces to,

$$u_{eq} = -b_0(X, t) - \sum_{i=1}^{n-1} c_i e_{i+1} + x_d^{(n)}(t) \tag{35}$$

Now the overall control becomes,

$$u = u_{eq} + k_{fs}u_{fs} = -b_0(X, t) - \sum_{i=1}^{n-1} c_i e_{i+1} + x_d^{(n)}(t) + k_{fs}u_{fs} \tag{36}$$

The constants appearing in the control law as well as the fuzzy gains were optimised using CSA for minimum IAE and amount of chattering. The algorithm is described in the following subsection:

Cuckoo Search Algorithm

CSA is a new meta-heuristic search algorithm of global optimization based on the behaviour of cuckoos proposed by Yang & Deb. This algorithm is based on the parasitic behaviour of some cuckoo species in combination with the Lévy flight behaviour of some birds and fruit flies [71].

It is based on the following natural operations:

1. How cuckoos lay their eggs in the host nests.
2. How, if undetected, the eggs are hatched to chicks by the hosts.

Before applying CSA over various structural engineering problems, CSA was benchmarked using standard problems such as the Travelling Salesman's problem. The theoretical analysis of CSA deals with how the cuckoo eggs banish the host eggs, thus allowing for an environment where their survival rate is improved. It is based on the following three idealized rules [27]:

1. Each cuckoo lays one egg at a time, and dumps it in a randomly chosen nest.
2. The best nests with high quality of eggs (solutions) will carry over to the next generations.
3. The number of available host nests is fixed, and a host can discover an alien egg with a probability $P_a \in [0, 1]$. In this case, the host bird can either throw the egg away or abandon the nest so as to build a completely new nest in a new location. Each egg in a nest represents a solution and a cuckoo egg represents new solution.

When generating new solutions $x(t+1)$ for, say cuckoo *i,* a Lévy flight is performed and the formula used is:

$$x_i(t+1) = x_i(t) + \alpha \oplus L\acute{e}vy(\lambda) \tag{37}$$

Where α is the step size and λ is the Lévy coefficient

The following is a pseudo-code for CSA [33]:

Begin

Objective function $f(x), x = [x_1, x_2, \ldots, x_d]^T$;

Initialize a population of n head nests $x_i\ (i = 1,2,\ldots,n)$;

While $(t < \max\ generations)$ or $(stop\ criterion)$

Get a cuckoo (say i) randomly and generate a new solution;

Evaluate its quality/fitness F_i;

Choose a nest among n (say j) randomly;

If $(F_i > F_j)$

Replace j by the new solution;

End

Abandon a fraction (F_i) of worse nests;

Keep the best solutions (or nests with quality solutions);

Rank the solutions and find the current best;

End while

Post process results and visualization;

End

In this work, for CSA, following parameter settings were used:

1. Discovery rate of alien eggs = 0.25
2. Number of nests = 20
3. Total iterations = 30
4. $\lambda = 1.5$

The fitness/cost function to be minimized using CSA was taken as $y = 0.2 \times IAE(e_1 + e_2 + e_3) + 0.4 \times Amount\ of\ chattering$.

FSMC of Genesio Chaotic System

In this section, results of simulations for both the integer and fractional order Genesio chaotic system [63, 64] are presented. Considering the fractional order Genesio chaotic system, the control objective is to drive the uncertain chaotic system to the desired trajectory $x_d(t)$.

Therefore, the fractional order switching surface is proposed as,

$$s(t) = D^{\gamma-1}e_3(t) + c_1 D^{\gamma-1}e_1(t) + c_2 D^{\gamma-1}e_2(t) \tag{38}$$

The overall control law is obtained as,

$$u(t) = 1.2x_1 + 2.92x_2 + 6x_3 - x_1^2 + x_d^{(3)}(t) - c_1e_2 - c_2e_3 + k_{fs}u_{fs} \quad (39)$$

For $\gamma = 0.993$ and initial conditions $x_1 = 3$, $x_2 = -4$ and $x_3 = 2$, the CSA optimised values of the aforementioned fuzzy gains are given in Table 2.

All subsequent simulations were run for $t = 10s$ using the 4th order Runge-Kutta method with a step time of 0.001 s.

The simulation results obtained with the CSA optimised values of the aforementioned fuzzy gains are shown in Figs. 16 and 17. Table 3 presents the assessment of the system for the performances indices viz. settling time, amount of chattering, IAE, ITAE and the cost function.

For $\gamma = 1$, the integer order Genesio system exhibits the following results, as shown in Figs. 18 and 19, using the CSA optimised gains given in Table 4.

The depicted figures clearly show that the tracking errors and state trajectories converge to zero with a settling time $t \approx 4.5$ s for fractional order Genesio chaotic system and $t \approx 3.4$ s for integer order Genesio chaotic system, indicating that stabilisation is indeed realised and the controller output is almost chatter-free. The performance parameters for fractional and integer FSMC are recorded in Tables 3 and 5, respectively.

Table 2 CSA tuned fuzzy gains for fractional order Genesio System

Parameter	CSA optimized values
k_{fs}	17.5333
k_s	7.8159
k_{sdot}	14.7451
c_1	10.9875
c_2	5.9221

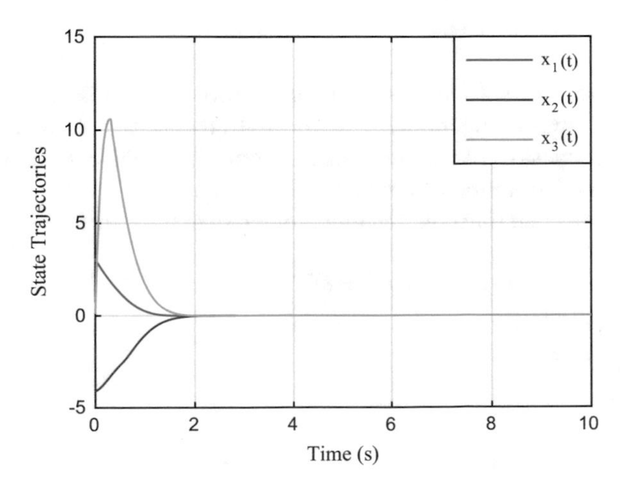

Fig. 16 State trajectories of fractional order Genesio system controlled using FSMC

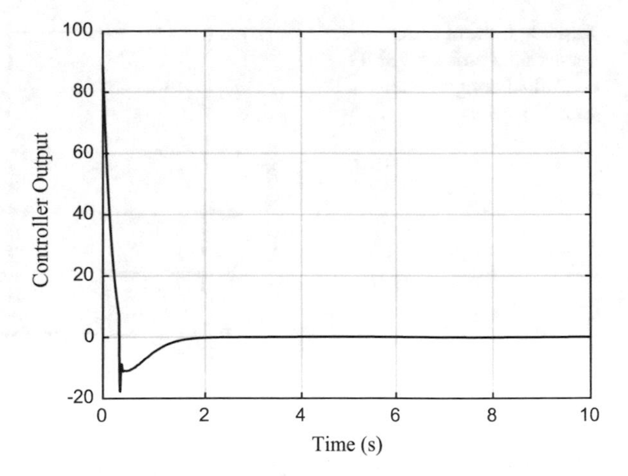

Fig. 17 Control action versus time plot for FSMC controlled fractional order Genesio system

Table 3 Performance parameters of FSMC for fractional order Genesio system

Performance index	Value
Settling time (s)	4.5
$\sum \lvert \Delta u \rvert$	181.95
IAE	e_1: 1454.5 e_2: 3001.8 e_3: 4082.4
ITAE	e_1: 586.06 e_2: 1483.4 e_3: 3071.8
Cost Function	1703.980

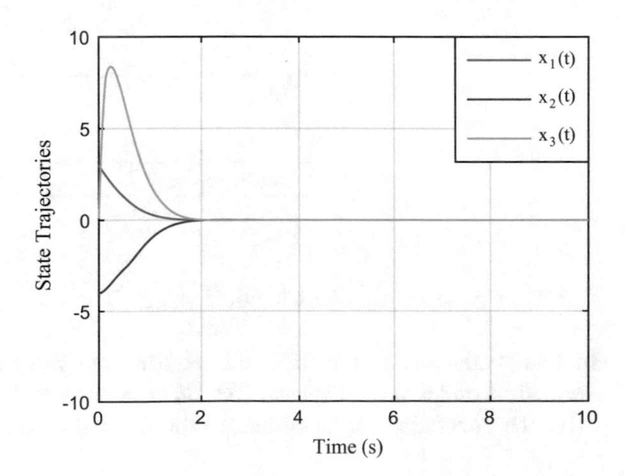

Fig. 18 State trajectories of integer order Genesio system controlled using FSMC

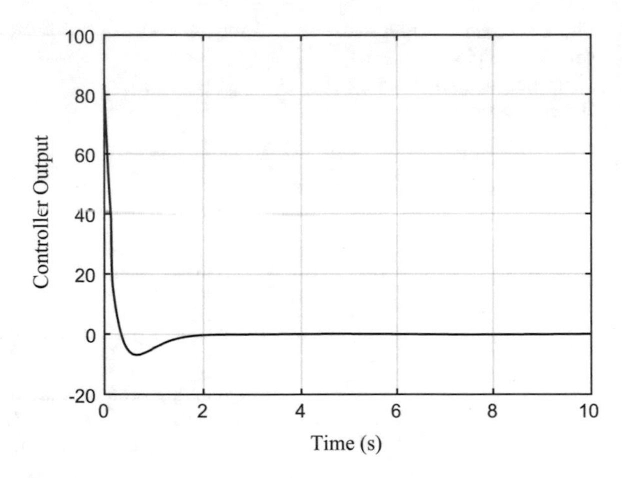

Fig. 19 Control action versus time plot for FSMC controlled integer order Genesio system

Table 4 CSA tuned fuzzy gains for integer order Genesio system

Parameter	CSA optimized values
k_{fs}	19.1568
k_s	4.8771
k_{sdot}	25.4487
c_1	18.6205
c_2	2.7342

Table 5 Performance parameters for FSMC of integer order Genesio system

Performance index	Value
Settling time (s)	3.4
$\sum \lvert\Delta u\rvert$	223.19
IAE	e_1: 1429.3 e_2: 3005.6 e_3: 4416.2
ITAE	e_1: 552.2 e_2: 1432.0 e_3: 3192.4
Cost Function	1617.692

FSMC of Arneodo-Coullet Chaotic System

In this section, results of simulations for both the integer and fractional order Arneodo-Coullet chaotic system [63, 64] are presented. The control objective is to drive the uncertain chaotic system to the desired trajectory $x_d(t)$.

Therefore, the fractional order switching surface is proposed as,

$$s(t) = D^{\gamma-1}e_3(t) + c_1 D^{\gamma-1}e_1(t) + c_2 D^{\gamma-1}e_2(t) \quad (40)$$

The overall control law is obtained as,

$$u(t) = -0.8x_1 + 1.1x_2 + 0.45x_3 + x_1^3 + x_d^{(3)}(t) - c_1 e_2 - c_2 e_3 + k_{fs}u_{fs} \quad (41)$$

The CSA optimised values of the aforementioned fuzzy gains are given in Table 6. For $\gamma = 0.993$ and initial conditions $x_1 = -1.2$, $x_2 = 1.2$ and $x_3 = 0.4$, the obtained simulation results are shown in Figs. 20 and 21 and are summarized in (Table 7).

For $\gamma = 1$, the Arneodo-Coullet system has an integer order and the CSA optimised values of the fuzzy gains are given in Table 8. The obtained simulation results are shown in Figs. 22 and 23.

The depicted figures show that the tracking errors and state trajectories converge to zero with a settling time $t \approx 2.2$ s for fractional order Arneodo-Coullet chaotic system and $t \approx 2.5$ s for integer order Arneodo-Coullet chaotic system, indicating that stabilisation is indeed realised. Further, the controller output as shown in Figs. 21 and 23 is smooth and chatter-free. The performance parameters for fractional and integer FSMC are recorded in Tables 7 and 9, respectively.

Table 6 CSA tuned fuzzy gains for fractional order Arneodo-Coullet system

Parameter	CSA optimized values
k_{fs}	85.7679
k_s	0.0500
k_{sdot}	0.1000
c_1	55.3645
c_2	9.7040

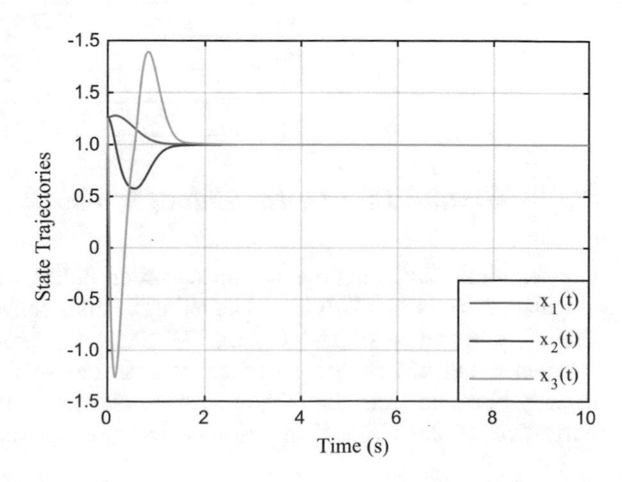

Fig. 20 State trajectories of fractional order Arneodo-Coullet system controlled using FSMC

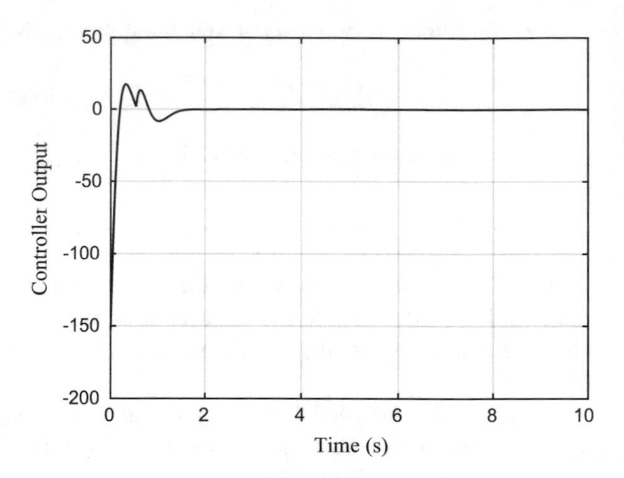

Fig. 21 Control action versus time plot for FSMC controlled fractional order Arneodo-Coullet system

Table 7 Performance parameters of FSMC for fractional order Arneodo-Coullet system

Performance index	Value
Settling time (s)	2.2
$\sum \lvert \Delta u \rvert$	380.62
IAE	e_1: 700.18 e_2: 1208.0 e_3: 4391.8
ITAE	e_1: 275.03 e_2: 725.91 e_3: 2171.8
Cost function	1084.272

Table 8 CSA tuned fuzzy gains for integer order Arneodo-Coullet system

Parameter	CSA optimized values
k_{fs}	50.9436
k_s	22.0565
k_{sdot}	12.1092
c_1	39.5731
c_2	9.3487

5.3 Terminal Full Order Sliding Mode Control

Conventional SMC employs a reduced order sliding surface which results in the singularity errors in TSMC and chattering in both conventional SMC and TSMC. Therefore, a variant of TSMC, i.e., TFOSMC was proposed by Feng et al. [26], wherein a full order sliding surface was chosen so that the control law can be directly obtained from the sliding surface. Consequently, the need for taking the derivative of the sliding surface containing terms having fractional powers is

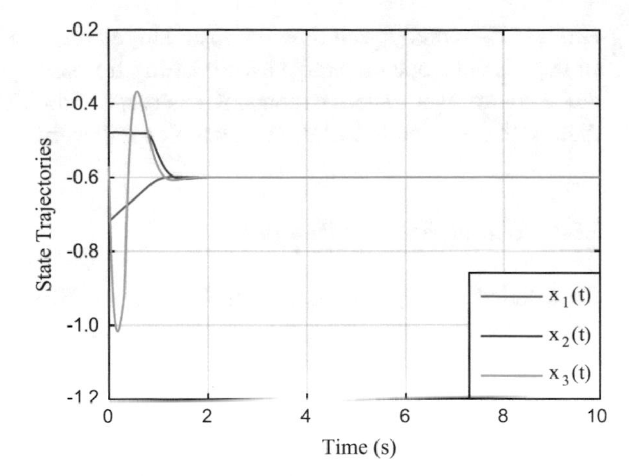

Fig. 22 State trajectories of integer order Arneodo-Coullet system controlled using FSMC

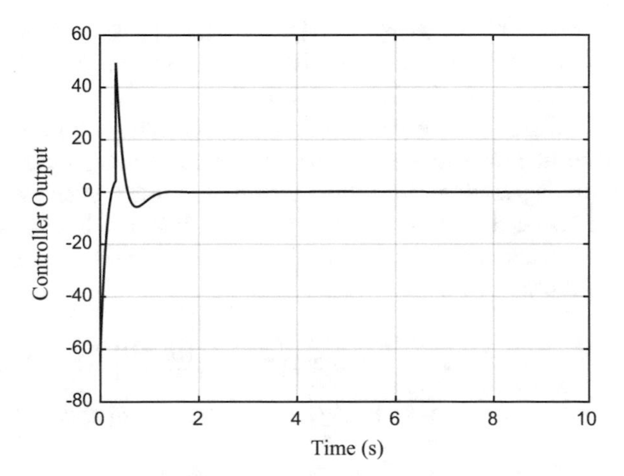

Fig. 23 Control action versus time plot for FSMC controlled integer order Arneodo-Coullet system

Table 9 Performance parameters for FSMC of integer order Arneodo-Coullet system

Performance index	Value
Settling time (s)	2.5
$\sum \lvert \Delta u \rvert$	240.94
IAE	e_1: 940.07 e_2: 1309.3 e_3: 3602.5
ITAE	e_1: 447.95 e_2: 1004.8 e_3: 2376.4
Cost Function	1180.697

eliminated, thereby avoiding control singularities. Here, a continuous control strategy is developed to achieve a chattering free sliding mode control. Since sliding mode control is a system dependent scheme, during ideal sliding mode motion, the systems has desirable full order dynamics rather than reduced-order dynamics.

5.3.1 Sliding Surface Design

Considering a non-linear system with order n [26]:

$$\begin{aligned} \dot{x}_1 &= x_2 \\ \dot{x}_2 &= x_3 \\ &\vdots \\ \dot{x}_{n-1} &= x_n \\ \dot{x}_n &= f(X,t) + d(t) + b(X,t)u \end{aligned} \tag{42}$$

where, $X(t) = [\,x_1(t) \quad x_2(t) \quad \ldots \quad x_n(t)\,] \in R^n$ represents the system state vector, $f(X,t)$ and $b(X,t) \neq 0$ are two non-linear functions and u is the controller output. The function $d(t)$ represents the external disturbance.

SMC is implemented for non-linear systems to force the system states onto the desired trajectory along the pre-defined sliding surface, through an induced ideal sliding motion along the surface. A control strategy is developed to realize the illustrated technique utilizing a finite-time reaching phase.

A terminal sliding mode (TSM) manifold for the above system can be selected as follows:

$$\begin{aligned} s &= x_1^{(n)} + c_n \mathrm{sgn}\left(x_1^{(n-1)}\right)\left|x_1^{(n-1)}\right|^{\alpha_n} + \cdots + c_1 \mathrm{sgn}(x_1)|x_1|^{\alpha_1} \\ &= \dot{x}_n + c_n \mathrm{sgn}(x_n)|x_n|^{\alpha_n} + \cdots + c_1 \mathrm{sgn}(x_1)|x_1|^{\alpha_1} \end{aligned} \tag{43}$$

Where c_i and $\alpha_i (i = 1, 2, \ldots, n)$ are constants. Parameter c_i can be selected such that the polynomial $p^n + c_n p^{n-1} + \cdots + c_2 p^2 + c_1$, which corresponds to TSM manifold, satisfies Hurwitz criterion. α_i can be determined according to the following relation:

$$\begin{aligned} &\alpha_1 = \alpha, && \mathrm{n} = 1 \\ &\alpha_{i-1} = \tfrac{\alpha_i \alpha_{i+1}}{2\alpha_{i+1} - \alpha_i}, && \mathrm{i} = 2, 3, \ldots, \mathrm{n} \ \forall \mathrm{n} \geq 2 \end{aligned} \tag{44}$$

where, $\alpha_{n+1} = 1$, $\alpha_n = \alpha$, $\alpha \in (1 - \zeta, 1)$, $\zeta \in (0, 1)$.

Establishing ideal sliding mode satisfied by $s = 0$, the system dynamics follow

$$\dot{x}_n + c_n \mathrm{sgn}(x_n)|x_n|^{\alpha_n} + \cdots + c_1 \mathrm{sgn}(x_1)|x_1|^{\alpha_1} = 0 \tag{45}$$

or

$$\begin{aligned} \dot{x}_1 &= x_2 \\ \dot{x}_2 &= x_3 \\ &\vdots \\ \dot{x}_{n-1} &= x_n \\ \dot{x}_n &= -c_n \text{sgn}(x_n)|x_n|^{\alpha_n} - \cdots - c_1 \text{sgn}(x_1)|x_1|^{\alpha_1} \end{aligned} \tag{46}$$

The above non-linear system will reach $s = 0$ in finite time and then converge to zero, the equilibrium point, along $s = 0$ within finite-time, if the sliding mode surface s is selected as (43) and the control is designed as follows:

$$u = b^{-1}(X, t)(u_{eq} + u_n) \tag{47}$$

$$u_{eq} = -f(X, t) - c_n \text{sgn}(x_n)|x_n|^{\alpha_n} - \cdots - c_1 \text{sgn}(x_1)|x_1|^{\alpha_1} \tag{48}$$

$$\dot{u}_n + T u_n = v \tag{49}$$

$$v = -(k_d + k_T + \eta)\text{sgn}(s) \tag{50}$$

where $u_n(0) = 0$; c_i and $\alpha_i (i = 1, 2, \ldots, n)$ are all constants, as defined in (44); η is a positive constant; k_d is a constant defined as follows:

The derivative of $d(t)$ in system (43) is bounded—$|\dot{d}(t)| \leq k_d$ where $k_d > 0$ is a constant. Two constants $T \geq 0$ and k_T are selected to satisfy the following condition:

$$k_T \geq T l_d \tag{51}$$

In the above condition, the control signal is equivalent to a low-pass filter, where $v(t)$ is the input and $u_n(t)$ is the output of the filter. The Laplace transfer function of the filter (49) is:

$$\frac{u_n(s)}{v(s)} = \frac{1}{s + T} \tag{52}$$

where $\omega = T$ is the bandwidth of the low-pass filter, $v(t)$ is the virtual control and is non-smooth because of the switching function and $u_n(t)$ is the output of the low-pass filter, softened to be a smooth signal. It may be noted that a pure integrator is more difficult for hardware implementation in practical applications than the low-pass filter that is why it has been replaced with the above low-pass filter. Differentiating terms $c_i \text{sgn}(x_i)|x_i|^{\alpha_i}$ are prevented in the TSM manifold from deriving the control laws. Therefore, singularity is avoided, and the ideal TSM, $s = 0$ is nonsingular.

The Lyapunov stability [38] is shown to be satisfied by taking the Lyapunov function as $V = \frac{1}{2}s^2$. For the considered TSM manifold,

$$s = d(t) + u_n \tag{53}$$

Taking the time derivative,

$$\dot{s} = \dot{d}(t) + \dot{u}_n = \dot{d}(t) + Tu_n - Tu_n = \dot{d}(t) + v - Tu_n \tag{54}$$

Substituting (50) into the above equation,

$$\dot{s} = \dot{d}(t) - (k_d + k_T + \eta)\text{sgn}(s) - Tu_n \tag{55}$$

$$s\dot{s} = \dot{d}(t)s - (k_d + k_T + \eta)|s| - Tu_n s = \left(\dot{d}(t)s - k_d|s|\right) + (-Tu_n s - k_T|s|) - \eta|s| \tag{56}$$

From above equations,

$$\dot{V} = s\dot{s} \leq -\eta|s| < 0 \text{ for } |s| \neq 0 \tag{57}$$

which implies that the system takes finite time to reach $s = 0$.

5.3.2 Design and Implementation of TFOSMC

The design and implementation of TFOSMC for both fractional and integer order Genesio and Arneodo-Coullet chaotic systems has been described in this section.

TFOSMC of Genesio Chaotic System

Considering the fractional order Genesio chaotic system, a TSM manifold is designed as follows:

$$s = D^{\gamma}x_3 + 15\text{sgn}x_3|x_3|^{7/10} + 66\text{sgn}x_2|x_2|^{7/13} + 80\text{sgn}x_1|x_1|^{7/16} \tag{58}$$

where, the parameters α_1, α_2 and α_3 are kept as 7/10, 7/13 and 7/16, respectively. The polynomial is selected as $p^3 + 15p^2 + 66p + 80 = (p+2)(p+5)(p+8)$ satisfying Hurwitz criterion. It may be noted that the considered sliding surface designed is free from the system dynamics.

Based on Eq. (47), $u = u_{eq} + u_n$ is designed as:

$$\begin{aligned} u_{eq} = {} & 1.2x_1 + 2.92x_2 + 6x_3 - x_1^2 - 15\text{sgn}x_3|x_3|^{7/10} \\ & - 66\text{sgn}x_2|x_2|^{7/13} - 80\text{sgn}x_1|x_1|^{7/16} \end{aligned} \tag{59}$$

$$\dot{u}_n + 0.1u_n = v \tag{60}$$

$$v = -10\mathrm{sgn}(s) \tag{61}$$

For $\gamma = 0.993$ and initial conditions $x_1 = 3$, $x_2 = -4$ and $x_3 = 2$, the results obtained are illustrated in Figs. 24 and 25.

For $\gamma = 1$, the Genesio system has an integer order and the results obtained are illustrated in Figs. 26 and 27.

The depicted Figs. 24 and 26 show that the state trajectories converge to zero with a settling time $t \approx 1.81$ s for fractional order Genesio chaotic system and $t \approx 2.35$ s for integer order Genesio chaotic system. Further, the controller outputs shown in Figs. 25 and 27 are smooth and chatter-free. The performance parameters for fractional and integer order TFOSMC are recorded in Tables 10 and 11, respectively.

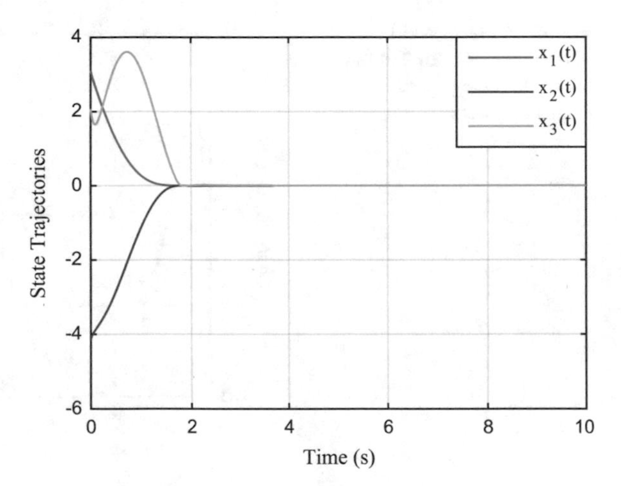

Fig. 24 State trajectories of fractional order Genesio system controlled using TFOSMC

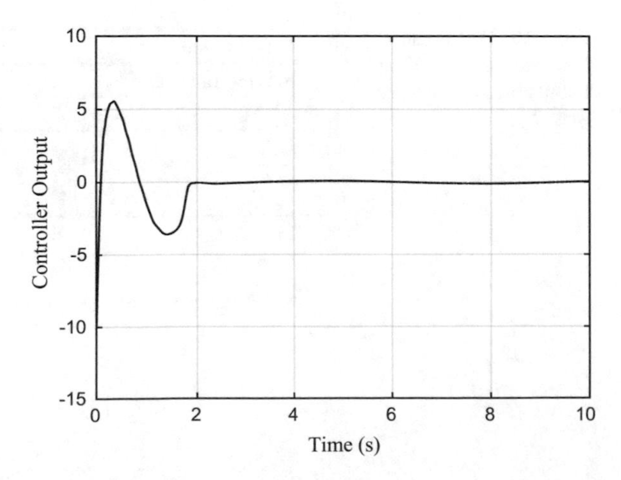

Fig. 25 Control action versus time plot for TFOSMC controlled fractional order Genesio system

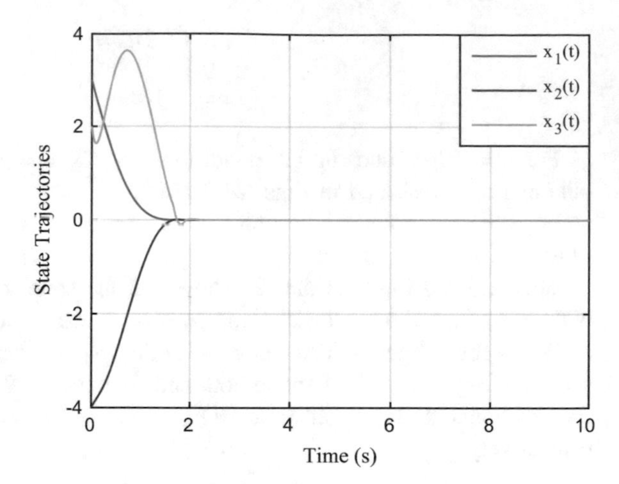

Fig. 26 State trajectories of integer order Genesio system controlled using TFOSMC

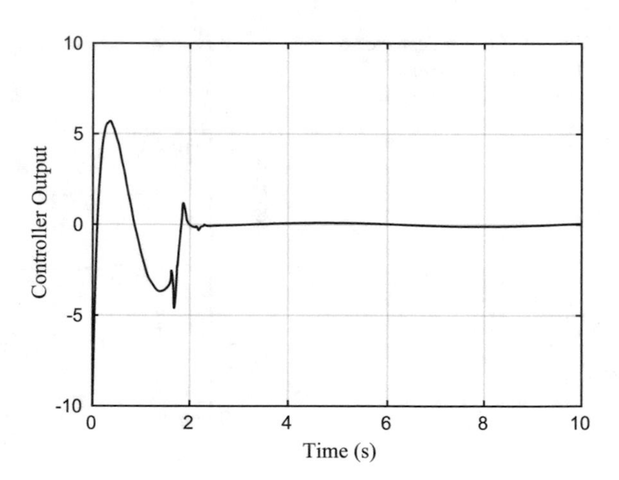

Fig. 27 Control action versus time plot for TFOSMC controlled integer order Genesio system

Table 10 Performance parameters for TFOSMC of fractional-order Genesio system

Performance index	Value
Settling time (s)	1.81
$\sum \lvert \Delta u \rvert$	79.61
IAE	e_1: 1422.2 e_2: 3001.1 e_3: 3999.3
ITAE	e_1: 507.68 e_2: 1438.5 e_3: 3036.2

Table 11 Performance parameters for TFOSMC of integer order Genesio system

Performance index	Value
Settling time (s)	2.35
$\sum \lvert\Delta u\rvert$	51.507
IAE	e_1: 1426.1 e_2: 3003.9 e_3: 4023.7
ITAE	e_1: 509.62 e_2: 1427.9 e_3: 3041.6

TFOSMC of Arneodo-Coullet Chaotic System

Considering the fractional order Arneodo-Coullet chaotic system, a TSM manifold is designed as follows:

$$s = D^{\gamma} x_3 + 15\mathrm{sgn}x_3 |x_3|^{7/10} + 66\mathrm{sgn}x_2 |x_2|^{7/13} + 80\mathrm{sgn}x_1 |x_1|^{7/16} \tag{62}$$

where, the parameters α_1, α_2 and α_3 are kept as 7/10, 7/13 and 7/16, respectively. The polynomial is selected as $p^3 + 15p^2 + 66p + 80 = (p+2)(p+5)(p+8)$ satisfying Hurwitz criterion. It may be noted that the considered sliding surface designed is free from the system dynamics.

Based on Eq. (47), $u = u_{eq} + u_n$ is designed as:

$$\begin{aligned} u_{eq} = & -0.8x_1 + 1.1x_2 + 0.45x_3 + x_1^3 - 15\mathrm{sgn}x_3 |x_3|^{7/10} \\ & - 66\mathrm{sgn}x_2 |x_2|^{7/13} - 80\mathrm{sgn}x_1 |x_1|^{7/16} \end{aligned} \tag{63}$$

$$\dot{u}_n + 0.1u_n = v \tag{64}$$

$$v = -10\mathrm{sgn}(s) \tag{65}$$

For $\gamma = 0.993$ and initial conditions $x_1 = -1.2$, $x_2 = 1.2$ and $x_3 = 0.4$, the results for the state trajectories and the controller output are shown in Figs. 28 and 29, respectively.

For $\gamma = 1$, the Arneodo-Coullet system has an integer order and the results obtained are shown in Figs. 30 and 31.

The depicted Figs. 28 and 30 show that the state trajectories converge to zero with a settling time $t \approx 1.8$ s for fractional order Arneodo-Coullet chaotic system and $t \approx 1.9$ s for integer order Arneodo-Coullet chaotic system, indicating that stabilisation is indeed realised. Further, the controller output as shown in Figs. 29 and 31 is smooth and chatter-free. The performance parameters for fractional and integer FSMC are recorded in Tables 12 and 13, respectively.

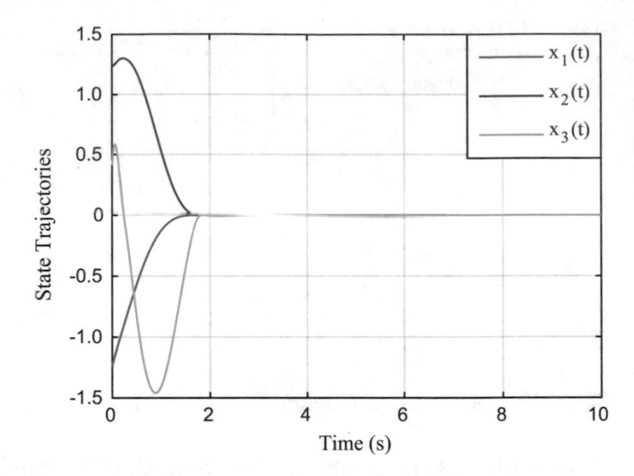

Fig. 28 State trajectories of fractional order Arneodo-Coullet system controlled using TFOSMC

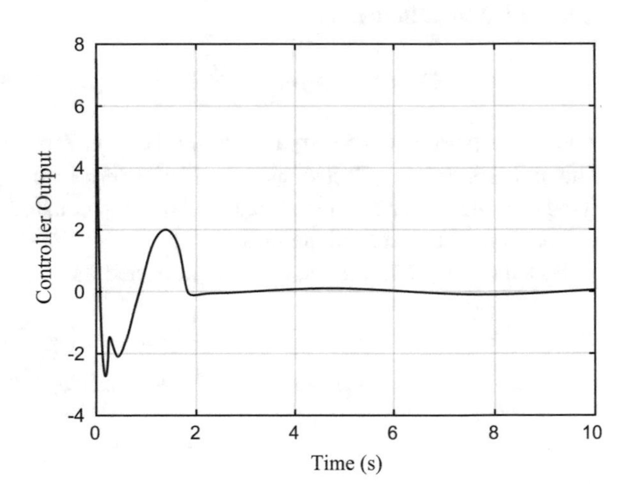

Fig. 29 Control action versus time plot for TFOSMC controlled fractional order Arneodo-Coullet system

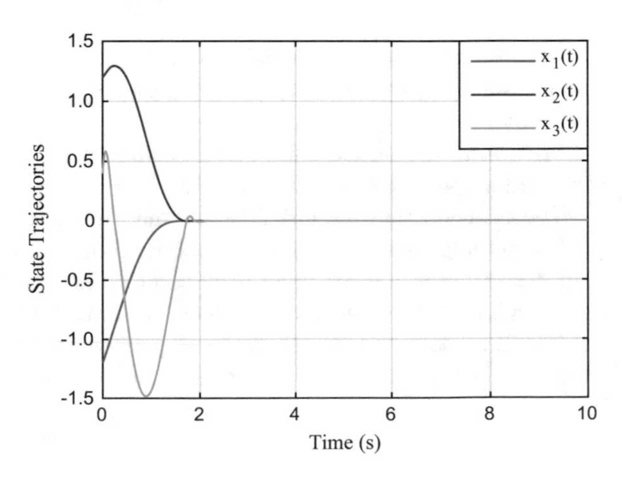

Fig. 30 State trajectories of integer order Arneodo-Coullet system controlled using TFOSMC

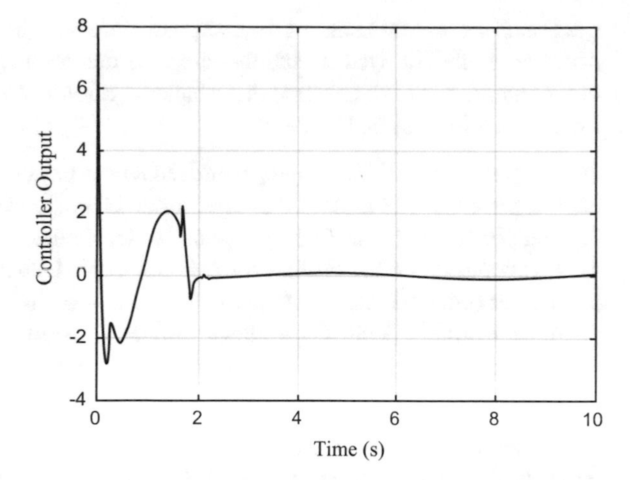

Fig. 31 Control action versus time plot for TFOSMC controlled integer order Arneodo-Coullet system

Table 12 Performance parameters for TFOSMC of fractional-order Arneodo-Coullet system

Performance index	Value
Settling time (s)	1.8
$\sum \lvert \Delta u \rvert$	29.1851
IAE	e_1: 624.21 e_2: 1200.2 e_3: 1381.6
ITAE	e_1: 232.66 e_2: 631.49 e_3: 1230.3

Table 13 Performance parameters for TFOSMC of integer order Arneodo-Coullet system

Performance index	Value
Settling time (s)	1.9
$\sum \lvert \Delta u \rvert$	31.832
IAE	e_1: 625.54 e_2: 1201.2 e_3: 1396.6
ITAE	e_1: 233.50 e_2: 629.914 e_3: 1231.8

6 Results and Discussions

In this chapter, TFOSMC and FSMC control schemes are successfully implemented for the considered two chaotic systems namely Genesio and Arneodo-Coullet. The chaotic systems are considered in integer as well as in fractional order dynamics and both the control schemes are therefore implemented in the form of integer as well as fractional order controllers. The fractional order controllers were applied to fractional order chaotic systems and the integer order controllers were applied to integer order chaotic systems. External disturbances and uncertainties were also considered for all the resulting eight cases:

1. Integer order FSMC on integer order Genesio system
2. Integer order TFOSMC on integer order Genesio system
3. Fractional order FSMC on fractional order Genesio system
4. Fractional order TFOSMC on fractional order Genesio system
5. Integer order FSMC on integer order Arneodo-Coullet system
6. Integer order TFOSMC on integer order Arneodo-Coullet system
7. Fractional order FSMC on fractional order Arneodo-Coullet system
8. Fractional order TFOSMC on fractional order Arneodo-Coullet system

For arriving at the final results, following comparative studies are performed between the performances of TFOSMC and FSMC. For this purpose, the performance of fractional order TFOSMC, applied to the fractional order chaotic systems, was compared with the fractional order FSMC, applied to the same plant. Similarly, the performance of integer order TFOSMC, applied to the integer order chaotic systems, was compared with the integer order FSMC, applied to the same plant. For each case, the state trajectories and controller output were compared graphically in addition to the other performance indices like settling time, amount of chattering, IAE and ITAE organized and presented in a tabular form. Resulting percentage improvements were also calculated and have been presented for each of these performance indices which clearly demonstrate the efficiency of TFOSMC over FSMC.

6.1 Comparison Between FSMC and TFOSMC

The state trajectories of the fractional order Genesio system when controlled by fractional order FSMC and TFOSMC are as shown along with the controller outputs for the same in Fig. 32. Figure 32a–c show the comparative time response of the individual state trajectories and Fig. 32d depicts the controller output for each of the two control schemes.

The state trajectories of the integer order Genesio system when controlled by integer order FSMC and TFOSMC are as shown along with the controller outputs for the same in Fig. 33. Figure 33a–c show the comparative time response of the individual state trajectories and Fig. 33d depicts the controller output for each of the two control schemes.

The data comparing the performance indices of FSMC and TFOSMC for Genesio chaotic system is recorded in Table 14. As tabulated, the settling time shows an improvement of 59.77% and 30.88% for fractional and integer TFOSMC, respectively. Further, the chattering also reduces by 56.24% and 76.5% in the case of fractional and integer order TFOSMC, respectively. It can be inferred all the performance indices show a positive percentage improvement.

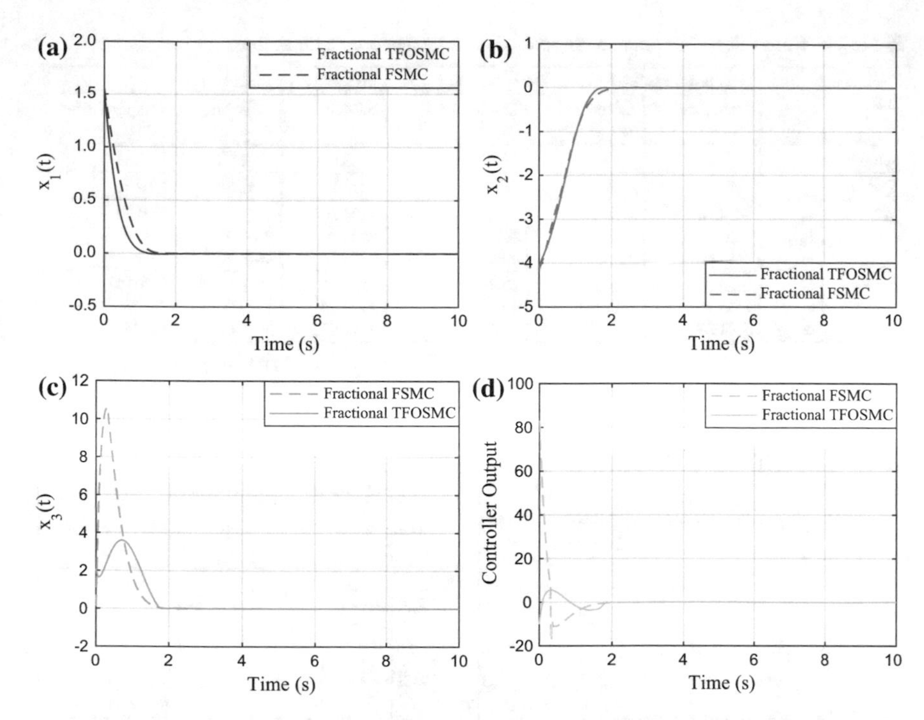

Fig. 32 Comparative performance of Fractional order FSMC and fractional order TFOSMC on fractional order Genesio System: **a** state x_1; **b** state x_2; **c** state x_3; **d** controller output

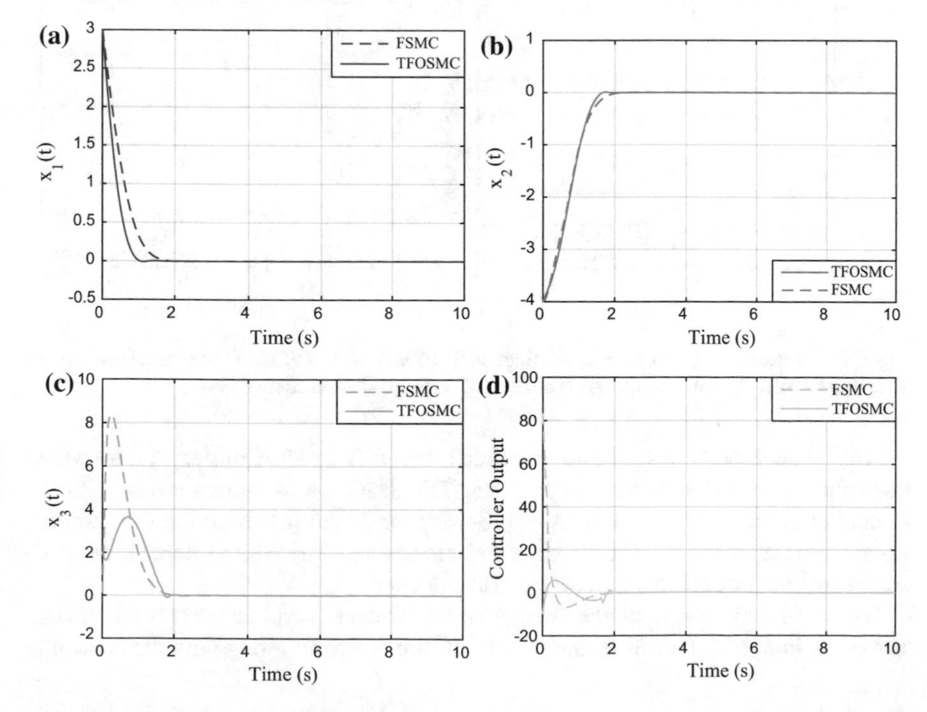

Fig. 33 Comparative performance of integer order FSMC and TFOSMC on integer order Genesio System: **a** state x_1; **b** state x_2; **c** state x_3; **d** controller output

Table 14 Controller performance comparison for Genesio chaotic system

Performance index		Fractional order system ($\gamma = 0.993$)			Integer order system ($\gamma = 1$)		
		Fractional FSMC	Fractional TFOSMC	Improvement (%)	FSMC	TFOSMC	Improvement (%)
Settling time		4.5	1.81	59.77	3.4	2.35	30.88
$\sum\lvert\Delta u\rvert$		181.95	79.61	56.24	223.19	51.50	76.50
IAE	e_1	1454.50	1422.20	2.20	1429.30	1426.10	0.20
	e_2	3001.80	3001.10	0.02	3005.60	3003.90	0.08
	e_3	4082.40	3999.30	2.03	4416.20	4023.70	8.89
ITAE	e_1	586.06	507.68	13.40	552.27	509.62	7.78
	e_2	1483.40	1438.50	3.03	1432.00	1427.90	0.34
	e_3	3071.80	3036.20	1.13	3192.40	3041.60	4.73

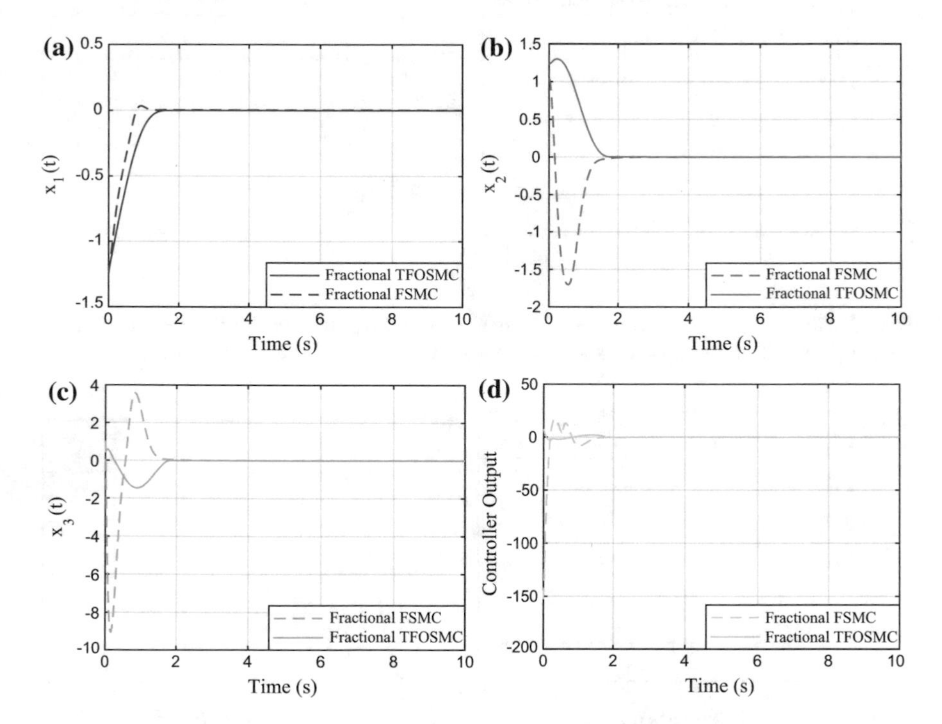

Fig. 34 Comparative performance of fractional FSMC and TFOSMC on fractional order Arneodo-Coullet System: **a** state x_1; **b** state x_2; **c** state x_3; **d** controller output

The state trajectories of the fractional order Arneodo-Coullet system when controlled by fractional order FSMC and TFOSMC are as shown along with the controller outputs for the same in Fig. 34. Figure 34 (a)-(c) show the comparative time response of the individual state trajectories and Fig. 34 (d) depicts the controller output for each of the two control schemes.

The state trajectories of the integer order Arneodo-Coullet system when controlled by integer order FSMC and TFOSMC are as shown along with the controller

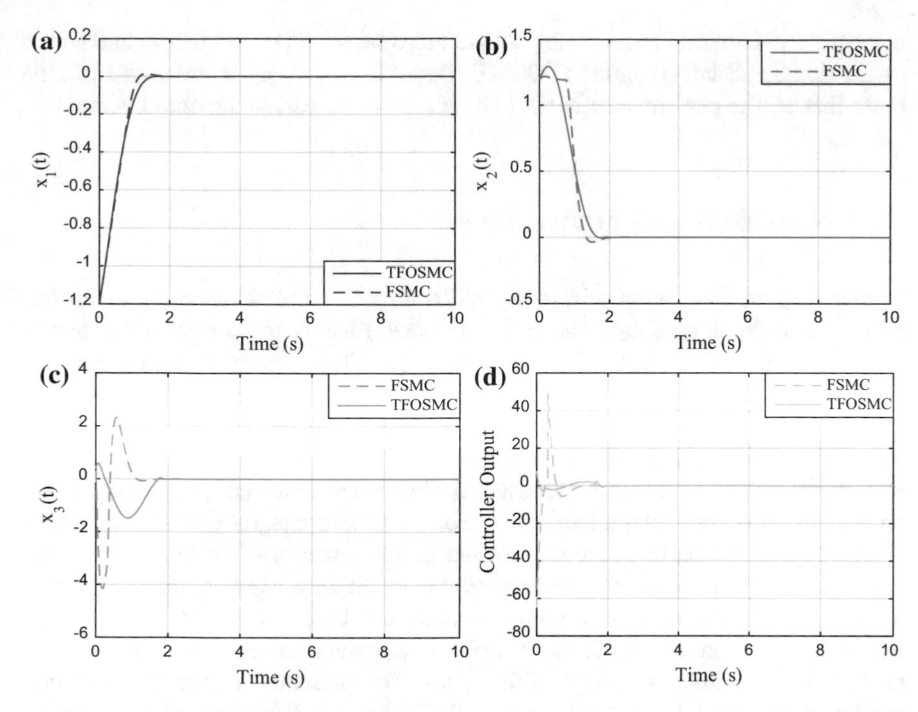

Fig. 35 Comparative performance of integer order FSMC and TFOSMC on integer order Arneodo-Coullet System: **a** state x_1; **b** state x_2; **c** state x_3; **d** controller output

Table 15 Controller performance comparison for Arneodo-Coullet Chaotic system

Performance index		Fractional order system ($\gamma = 0.993$)			Integer order system ($\gamma = 1$)		
		Fractional FSMC	Fractional TFOSMC	Improvement (%)	FSMC	TFOSMC	Improvement (%)
Settling time		2.2	1.8	29.42	2.5	1.9	24.00
$\sum\|\Delta u\|$		380.62	29.19	92.30	240.94	31.83	86.70
IAE	e_1	700.18	624.21	10.80	940.07	625.54	33.52
	e_2	1208.00	1200.20	0.66	1309.30	1201.20	8.23
	e_3	4391.80	1381.60	68.40	3602.50	1396.60	61.20
ITAE	e_1	275.03	232.66	15.63	447.95	233.50	47.80
	e_2	725.91	631.49	15.63	1004.80	629.91	37.30
	e_3	2171.80	1230.30	43.30	2376.40	1231.80	48.10

outputs for the same in Fig. 35. Figure 35 (a)-(c) show the comparative time response of the individual state trajectories and Fig. 35 (d) depicts the controller output for each of the two control schemes.

The data comparing the performance parameters of FSMC and TFOSMC for Genesio chaotic system is recorded in Table 15. As tabulated, the settling time shows an improvement of 29.42% and 24% for fractional and integer TFOSMC, respectively. Further, the chattering also reduces by 92.30% and 86.70% in the case of fractional and integer order TFOSMC, respectively. It can be inferred from the table that all the performance indices show a positive percentage improvement.

7 Conclusions and Future Scope

In this chapter, application of a recently developed control scheme known as terminal full order sliding mode control (TFOSMC) has been successfully explored for efficient control of uncertain chaotic systems. Two important chaotic systems, Genesio and Arneodo-Coullet have been considered in fractional order as well as integer order dynamics. The investigated fractional and integer order chaotic systems are controlled using fractional order TFOSMC and integer order TFOSMC, respectively and the control performance has been assessed for settling time, amount of chattering, integral absolute error (IAE) and integral time absolute error (ITAE). Furthermore, to gauge the relative performance of TFOSMC, a comparative study with its potential counterpart, Fuzzy Sliding Mode Control (FSMC), tuned by Cuckoo Search Algorithm for minimum IAE and amount of chattering was also carried out and the relative performance was assessed using settling time, amount of chattering, IAE and ITAE. From the presented intensive simulation studies on integer order and fractional order Genesio and Arneodo-Coullet chaotic systems, it was clearly observed that all the above mentioned performance indices exhibited significant improvements when TFOSMC was employed instead of FSMC. Another notable outcome of this study has been the significantly lower and smoother controller output and reduced chattering in case of TFOSMC. Based on these detailed investigations and presented results it is concluded that TFOSMC is a better control scheme over FSMC to control the chaotic systems.

Future work, in this line, can be pursued with the performance investigations of cross implementations of the controllers and systems i.e. application of fractional-order controllers on integer order plants and vice versa. Furthermore, applications on the other chaotic systems can also be taken up. Apart from controlling the sys-tem trajectories, investigations on the chaotic systems' synchronization can also be considered.

References

1. Ablay, G. (2009). Sliding mode control of uncertain unified chaotic systems. *Nonlinear Analysis: Hybrid Systems, 3*(4), 531–535.
2. Aghababa, M. P. (2012). Finite-time chaos control and synchronization of fractional-order nonautonomous chaotic (hyperchaotic) systems using fractional nonsingular terminal sliding mode technique. *Nonlinear Dynamics, 69*(1–2), 247–261.

3. Aghababa, M. P. (2013). Design of a chatter-free terminal sliding mode controller for nonlinear fractional-order dynamical systems. *International Journal of Control, 86*(10), 1744–1756.
4. Axtell, M., & Bise, M. E. (1990). Fractional calculus application in control systems. In *Proceedings of the IEEE 1990 National Aerospace and Electronics Conference, 1990. NAECON 1990* (pp. 563–566) IEEE.
5. Azar, A. T., Serrano, F. E. (2014). Robust IMC-PID tuning for cascade control systems with gain and phase margin specifications. *Neural Computing and Applications, 25*(5), 983–995. Springer. doi:10.1007/s00521-014-1560-x.
6. Azar, A. T., Serrano, F. E. (2015). Design and modeling of anti wind up PID controllers. In Q. Zhu & A. T. Azar (Eds.), *Complex system modelling and control through intelligent soft computations*. Studies in Fuzziness and Soft Computing (Vol. 319, pp 1–44). Germany: Springer. doi:10.1007/978-3-319-12883-2_1.
7. Azar, A. T., & Serrano, F. E. (2015). Stabilization and control of mechanical systems with backlash. In: A. T. Azar & S. Vaidyanathan (Eds.), *Advanced intelligent control engineering and automation*. Advances in Computational Intelligence and Robotics (ACIR) Book Series. USA: IGI-Global.
8. Azar, A. T., & Serrano, F. E. (2016). Stabilization of mechanical systems with backlash by PI loop shaping. *International Journal of System Dynamics Applications (IJSDA), 5*(3), 20–47.
9. Azar, A. T., & Vaidyanathan, S. (2016). *Advances in Chaos theory and intelligent control*. Studies in Fuzziness and Soft Computing (Vol. 337). Germany: Springer. ISBN 978–3-319-30338-3.
10. Azar, A. T., & Vaidyanathan, S. (2014c). *Handbook of research on advanced intelligent control engineering and automation* (1st ed.). Advances in Computational Intelligence and Robotics (ACIR) Book Series Hershey, PA: IGI Global.
11. Azar, A. T., & Vaidyanathan, S. (2015). *Chaos modeling and control systems design*. Studies in Computational Intelligence (vol. 581). Germany: Springer.
12. Azar, A. T., & Vaidyanathan, S. (Eds). (2015b). *Computational intelligence applications in modeling and control*. Studies in Computational Intelligence (Vol. 575). Germany: Springer.
13. Azar, A. T., & Zhu, Q. (2015). *Advances and applications in sliding mode control systems*. Studies in Computational Intelligence (Vol. 576). Germany: Springer.
14. Bai, J., & Yu, Y. (2010). Sliding mode control of fractional-order hyperchaotic systems. In *2010 International Workshop on Chaos-Fractal Theories and Applications*.
15. Boulkroune, A., Bouzeriba, A., Bouden, T., Azar, A. T. (2016). Fuzzy adaptive synchronization of uncertain fractional-order chaotic systems. *Advances in Chaos Theory and Intelligent Control*. Studies in Fuzziness and Soft Computing (Vol. 337). Germany: Springer.
16. Boulkroune, A., Hamel, S., & Azar, A. T. (2016). Fuzzy control-based function synchronization of unknown chaotic systems with dead-zone input. *Advances in Chaos theory and intelligent control*. Studies in Fuzziness and Soft Computing (Vol. 337). Germany: Springer.
17. Carlson, G. E., & Halijak, C. A. (1961). Simulation of the fractional derivative operator and the fractional integral operator. *Kansas State University Bulletin, 45*(7), 1–22.
18. Chang, W., Park, J. B., Joo, Y. H., & Chen, G. (2002). Design of robust fuzzy-model-based controller with sliding mode control for SISO nonlinear systems. *Fuzzy Sets and Systems, 125* (1), 1–22.
19. Chen, C. S., & Chen, W. L. (1998). Robust adaptive sliding-mode control using fuzzy modeling for an inverted-pendulum system. *IEEE Transactions on Industrial Electronics, 45* (2), 297–306.
20. Chen, D., Zhang, R., Sprott, J. C., & Ma, X. (2012). Synchronization between integer-order chaotic systems and a class of fractional-order chaotic system based on fuzzy sliding mode control. *Nonlinear Dynamics, 70*(2), 1549–1561.
21. Chen, D. Y., Liu, Y. X., Ma, X. Y., & Zhang, R. F. (2012). Control of a class of fractional-order chaotic systems via sliding mode. *Nonlinear Dynamics, 67*(1), 893–901.
22. Chen, Y., Petras, I., & Xue, D. (2009). Fractional order control-a tutorial. In *2009 American Control Conference* (pp. 1397–1411). IEEE.

23. Cuautle, E. T. (2011). *Chaotic systems*. InTech.
24. Delavari, H., Ghaderi, R., Ranjbar, A., & Momani, S. (2010). Fuzzy fractional order sliding mode controller for nonlinear systems. *Communications in Nonlinear Science and Numerical Simulation, 15*(4), 963–978.
25. Dotoli, M., Lino, P., Maione, B., Naso, D., & Turchiano, B. (2002). A tutorial on genetic optimization of fuzzy sliding mode controllers: Swinging up an inverted pendulum with restricted travel. In *Proceedings of EUNITE 2002*.
26. Feng, Y., Han, F., & Yu, X. (2014). Chattering free full-order sliding-mode control. *Automatica, 50*(4), 1310–1314.
27. Gandomi, A. H., Yang, X. S., & Alavi, A. H. (2013). Cuckoo search algorithm: A metaheuristic approach to solve structural optimization problems. *Engineering with Computers, 29*(1), 17–35.
28. Gang-Quan, S., Hui, C., & Yan-Bin, Z. (2011). A new four-dimensional hyperchaotic Lorenz system and its adaptive control. *Chinese Physics B, 20*(1), 010509.
29. Grigorenko, I., & Grigorenko, E. (2003). Chaotic dynamics of the fractional Lorenz system. *Physical Review Letters, 91*(3), 034101.
30. Hartley, T. T., Lorenzo, C. F., & Qammer, H. K. (1995). Chaos in a fractional order Chua's system. *IEEE Transactions on Circuits and Systems I: Fundamental Theory and Applications, 42*(8), 485–490.
31. Hoppensteadt, F. C. (2013). *Analysis and simulation of chaotic systems* (Vol. 94). Springer Science & Business Media.
32. Ionescu, C., Machado, J. T., & De Keyser, R. (2011). Fractional-order impulse response of the respiratory system. *Computers & Mathematics with Applications, 62*(3), 845–854.
33. Kamat, S., & Karegowda, A. G. (2014). A brief survey on cuckoo search applications. *International Journal of Innovative Research in Computer and Communication Engineering, 2*.
34. Khari, S., Rahmani, Z., & Rezaie, B. (2016). Designing fuzzy logic controller based on combination of terminal sliding mode and state feedback controllers for stabilizing chaotic behaviour in rod-type plasma torch system. *Transactions of the Institute of Measurement and Control, 38*(2), 150–164.
35. Laskin, N. (2000). Fractional market dynamics. *Physica A: Statistical Mechanics and its Applications, 287*(3), 482–492.
36. Lin, T. C., Chen, M. C., & Roopaei, M. (2011). Synchronization of uncertain chaotic systems based on adaptive type-2 fuzzy sliding mode control. *Engineering Applications of Artificial Intelligence, 24*(1), 39–49.
37. Liu, J., & Wang, X. (2012). *Advanced sliding mode control for mechanical systems* (pp. 41–80). Germany: Springer.
38. Lyapunov Exponent. http://en.wikipedia.org/wiki/Lyapunov_exponent. Accessed 20 February 2016.
39. Mekki, H., Boukhetala, D., & Azar, A. T. (2015). Sliding modes for fault tolerant control. *Advances and applications in Sliding Mode Control systems*. Studies in Computational Intelligence Book Series (Vol. 576, pp. 407–433). Berlin/Heidelberg: Springer GmbH. doi:10.1007/978-3-319-11173-5_15.
40. Mohadeszadeh, M., & Delavari, H. (2015). Synchronization of uncertain fractional-order hyper-chaotic systems via a novel adaptive interval type-2 fuzzy active sliding mode controller. *International Journal of Dynamics and Control*, 1–10.
41. Nazzal, J. M., & Natsheh, A. N. (2007). Chaos control using sliding-mode theory. *Chaos, Solitons & Fractals, 33*(2), 695–702.
42. Ott, E., Grebogi, C., & Yorke, J. A. (1990). Controlling chaos. *Physical Review Letters, 64*(11), 1196.
43. Oustaloup, A. (Eds). (2006). *Proceedings of The Second IFAC Symposium on Fractional Differentiation and its Applications (FDA06)*, Porto, Portugal, July 19–21 2006. IFAC. Oxford, UK: Elsevier Science Ltd.

44. Oustaloup, A., Mathieu, B., & Lanusse, P. (1995). The CRONE control of resonant plants: Application to a flexible transmission. *European Journal of control, 1*(2), 113–121.
45. Perruquetti, W., & Barbot, J. P. (2002). *Sliding mode control in engineering*. CRC Press.
46. Petráš, I. (2009). Chaos in the fractional-order Volta's system: Modeling and simulation. *Nonlinear Dynamics, 57*(1–2), 157–170.
47. Petrov, V., Gaspar, V., Masere, J., & Showalter, K. (1993). Controlling chaos in the Belousov—Zhabotinsky reaction. *Nature, 361*(6409), 240–243.
48. Pyragas, K. (1992). Continuous control of chaos by self-controlling feedback. *Physics Letters A, 170*(6), 421–428.
49. Rivero, M., Trujillo, J. J., Vázquez, L., & Velasco, M. P. (2011). Fractional dynamics of populations. *Applied Mathematics and Computation, 218*(3), 1089–1095.
50. Ross, T. J. (2009). *Fuzzy logic with engineering applications*. New York: Wiley.
51. Roy, R., Murphy, T. W., Jr., Maier, T. D., Gills, Z., & Hunt, E. R. (1992). Dynamical control of a chaotic laser: Experimental stabilization of a globally coupled system. *Physical Review Letters, 68*(9), 1259.
52. Ullah, N., Khattak, M. I., & Khan, W. (2015). Fractional order fuzzy terminal sliding mode control of aerodynamics load simulator. In *Proceedings of the Institution of Mechanical Engineers, Part G: Journal of Aerospace Engineering* (pp. 0954410015580804).
53. Vaidyanathan, S. (2011). Output regulation of Arneodo-Coullet chaotic system. In *International Conference on Computer Science and Information Technology* (pp. 98–107). Germany: Springer.
54. Vaidyanathan, S., & Azar, A. T. (2015a). Analysis and control of a 4-D novel hyperchaotic system. In A. T. Azar & S. Vaidyanathan (Eds.), *Chaos modeling and control systems design*. Studies in Computational Intelligence (Vol. 581, pp. 19–38). Berlin/Heidelberg: Springer GmbH.
55. Vaidyanathan, S., & Azar, A. T. (2015b). Anti-synchronization of identical chaotic systems using sliding mode control and an application to Vaidyanathan-Madhavan chaotic systems. In A. T. Azar & Q. Zhu (Eds.), *Advances and applications in sliding mode control systems*. Studies in Computational Intelligence Book Series (Vol. 576, pp. 527–547). Berlin/Heidelberg: Springer GmbH.
56. Vaidyanathan, S., & Azar, A. T. (2015c). Hybrid synchronization of identical chaotic systems using sliding mode control and an application to Vaidyanathan chaotic systems. In A. T. Azar & Q. Zhu (Eds.), *Advances and applications in sliding mode control systems*. Studies in Computational Intelligence Book Series (Vol. 576, pp. 549–569). Berlin/Heidelberg: Springer GmbH.
57. Vaidyanathan, S., & Azar, A. T. (2016). A Novel 4-D four-wing chaotic system with four quadratic nonlinearities and its synchronization via adaptive control method. *Advances in Chaos theory and intelligent control*. Studies in Fuzziness and Soft Computing (Vol. 337). Germany: Springer.
58. Vaidyanathan, S., & Azar, A. T. (2016). Adaptive backstepping control and synchronization of a novel 3-D jerk system with an exponential nonlinearity. *Advances in Chaos theory and intelligent control*. Studies in Fuzziness and Soft Computing (Vol. 337). Germany: Springer-Verlag.
59. Vaidyanathan, S., & Azar, A. T. (2016). Adaptive control and synchronization of Halvorsen circulant chaotic systems. *Advances in Chaos theory and intelligent control*. Studies in Fuzziness and Soft Computing (Vol. 337). Germany: Springer.
60. Vaidyanathan, S., & Azar, A. T. (2016). Dynamic analysis, adaptive feedback control and synchronization of an eight-term 3-D novel chaotic system with three quadratic nonlinearities. *Advances in Chaos theory and intelligent control*. Studies in Fuzziness and Soft Computing, Vol. 337, Springer-Verlag, Germany.
61. Vaidyanathan, S., & Azar, A. T. (2016). Generalized Projective Synchronization of a Novel Hyperchaotic Four-Wing System via Adaptive Control Method. *Advances in Chaos theory*

and intelligent control. Studies in Fuzziness and Soft Computing (Vol. 337). Germany: Springer-Verlag.

62. Vaidyanathan, S., & Azar, A. T. (2016). Qualitative study and adaptive control of a novel 4-D hyperchaotic system with three quadratic nonlinearities. *Advances in Chaos theory and intelligent control*. Studies in Fuzziness and Soft Computing (Vol. 337). Germany: Springer-Verlag.
63. Vaidyanathan, S., Sampath, S., & Azar, A. T. (2015). Global chaos synchronisation of identical chaotic systems via novel sliding mode control method and its application to Zhu system. *International Journal of Modelling, Identification and Control (IJMIC), 23*(1), 92–100.
64. Valrio, D. (2012). Introducing fractional sliding mode control. In *Proceedings of The Encontro de Jovens Investigadores do LAETA FEUP*.
65. Vinagre, B. M., Chen, Y. Q., & Petráš, I. (2003). Two direct Tustin discretization methods for fractional-order differentiator/integrator. *Journal of the Franklin Institute, 340*(5), 349–362.
66. Westerlund, S., & Ekstam, L. (1994). Capacitor theory. *IEEE Transactions on Dielectrics and Electrical Insulation, 1*(5), 826–839.
67. Wilkie, K. P., Drapaca, C. S., & Sivaloganathan, S. (2011). A nonlinear viscoelastic fractional derivative model of infant hydrocephalus. *Applied Mathematics and Computation, 217*(21), 8693–8704.
68. Wong, L. K., Leungn, F. H. F., & Tam, P. K. S. (2001). A fuzzy sliding controller for nonlinear systems. *IEEE Transactions on Industrial Electronics, 48*(1), 32–37.
69. Wu, J. C., & Liu, T. S. (1996). A sliding-mode approach to fuzzy control design. *IEEE Transactions on Control Systems Technology, 4*(2), 141–151.
70. Wu, X., Lu, H., & Shen, S. (2009). Synchronization of a new fractional-order hyperchaotic system. *Physics Letters A, 373*(27), 2329–2337.
71. Yang, X. S., & Deb, S. (2009). Cuckoo search via Lévy flights. In *Nature & Biologically Inspired Computing, 2009. NaBIC 2009. World Congress* (pp. 210–214). IEEE.
72. Yau, H. T., & Chen, C. L. (2006). Chattering-free fuzzy sliding-mode control strategy for uncertain chaotic systems. *Chaos, Solitons & Fractals, 30*(3), 709–718.
73. Yin, C., Zhong, S. M., & Chen, W. F. (2012). Design of sliding mode controller for a class of fractional-order chaotic systems. *Communications in Nonlinear Science and Numerical Simulation, 17*(1), 356–366.
74. Yuan, L., & Wu, H. S. (2010). Terminal sliding mode fuzzy control based on multiple sliding surfaces for nonlinear ship autopilot systems. *Journal of Marine Science and Application, 9* (4), 425–430.
75. Zadeh, L. A. (1988). *Fuzzy logic*.
76. Zaslavsky, G. M. (2002). Chaos, fractional kinetics, and anomalous transport. *Physics Reports, 371*(6), 461–580.

Stabilization of Fractional Order Discrete Chaotic Systems

M.K. Shukla and B.B. Sharma

Abstract Chaos is almost ubiquitous in field of science and engineering. The insurgent and typically unpredictable behavior exhibited by nonlinear systems is seen as chaos. In recent decades, fractional (non-integer) order chaotic systems have also been developed and their applications have invited a lot of interest of research community. Along with fractional order continuous time chaotic systems, researchers have also explored fractional order discrete chaotic systems to some extent. These systems can also be exploited for the same application for which continuous versions are used, thus providing increased flexibility and reliability. Although the mathematics of fractional discrete calculus is still in development phase, still with the help of available knowledge, research community has started giving attention to this emerging field. A number of contributions are available in literature in the area of fractional discrete calculus and its applications in control systems. One can represent linear systems using fractional difference equations in state space domain. Similarly, fractional difference equations can be used to represent nonlinear dynamical systems, especially chaotic systems. Fractional order discrete chaotic systems offer a new domain of exploration to research fraternity. As the work reported is limited, so the need arises to review and consolidate it. The analysis, control and synchronization of fractional order chaotic systems is the aim of this chapter. This chapter initially, gives a brief overview of fractional difference equations and their solution. Thereafter, the results obtained so far in this area are discussed and presented. Chaotic behavior of discrete fractional versions of the famous Logistic map and Henon map is studied first. Further, control of the same class of systems is tackled. The main contribution of the work is to present analysis of control of fractional Henon map using backstepping control which is a well-known technique to researchers in area of nonlinear control. Simulation results are obtained using MATLAB and are presented at the end to validate the results.

M.K. Shukla (✉) · B.B. Sharma
National Institute of Technology, Hamirpur, HP, India
e-mail: acdcmks@gmail.com

B.B. Sharma
e-mail: bharat.nit@gmail.com

A.T. Azar et al. (eds.), *Fractional Order Control and Synchronization of Chaotic Systems*, Studies in Computational Intelligence 688,
DOI 10.1007/978-3-319-50249-6_14

Keywords Chaotic systems · Fractional order · Backstepping · Synchronization · Henon map

1 Introduction

In recent decades, fractional (non-integer) order chaotic systems have also been developed and their applications have invited a lot of interest of research community. Fractional calculus which has been considered as the extension of the integer-order calculus to non-integer order calculus, came into picture in the 17th century. Due to unavailability of solution techniques, fractional calculus has not been explored much for almost three hundred years. In recent years, various methods for approximation of the fractional derivative and integral came into existence, which further enabled us to use it in wide areas such as, bioengineering [1], diffusion of heat [2], signal processing [3], robotics [4], electrical engineering [5], etc. Fractional calculus finds wide applicability in control systems too. The application revolves around two pillars: one is of fractional order systems and the other is fractional order control. It has been established that fractional calculus gives a more realistic modelling of systems and the best performance is achieved when a fractional order controller is employed for a fractional order system [6].

Chaos is one of the ubiquitous phenomenon found in nature. A large amount of literature is available covering their behavior, control and synchronization. Different control and synchronization strategies are presented in [7–13]. Some of the recent contributions in this area is given in [14–23]. The study of chaotic behavior of various systems have also been a great topic of interest. Various researchers have shown interest in this area also [24–30].

Chaotic systems can also be modelled more accurately by non-integer order differential equations and hence are called fractional order chaotic systems. It is known that chaotic behavior is exhibited by the systems, for system order more than three, but recently, researchers have established that some fractional-order differential systems exhibit chaotic behavior for total order less than three as explained in the pioneering work by Hartley et al. [31]. Chaotic behavior and analysis of continuous fractional order chaotic systems has now become a well-studied field with a number of research papers and applications oriented results. In last ten years, researchers have employed various control strategies for stabilization and synchronization of fractional order chaotic systems and applied the results for real time applications also. Several fractional order chaotic systems have been studied such as fractional-order Chen system [32, 33], fractional-order Lorenz system [34, 35]. Various other fractional order chaotic systems have also been studied in recent years such as, Rössler's system [36], Coullet system [37], modified Van der Pol Duffing system [38] and Liu's system [39]. Recent developments in the area of fractional order chaos can be found in [40–43].

Recently, the prominent definitions of continuous fractional derivatives and integrals have been discretized and lead to a new area of research and study i.e. fractional difference equations. It is developing at a vast pace and is being applied to various fields. Atici and Eloe [44] described the discrete initial value problem and presented the existence results. Laplace transform for solving discrete fractional equations in the nabla's sense, was given by Holm [45]. Abedel [46] systemically discussed the Caputo and the Riemann–Liouville fractional differences as well as their properties. More details on discrete fractional calculus can be found in [47–52]. Although, enough work has been done in the area of fractional order discrete linear systems, nonlinear systems version has not been studied much. As we know that discrete time systems are represented by difference equations, one may think of using fractional order difference equations for modelling discrete time systems. In this course, fractional difference may be proved to be powerful tool for the more realistic modeling of dynamics of discrete complex systems. Various definitions of discrete derivatives have been put forward by researchers [53]. In the view of this development, it is required to see the effect of fractional difference in nonlinear dynamical systems. To study the effect of this fractional order in the chaotic behavior of the nonlinear systems is the motivation behind this work. Also, once the chaotic behavior is ascertained, the need arises to design controllers for the stabilization and synchronization of this category of systems. With these developments in this area, the real time problems like secure communication, cryptography etc. can be solved efficiently and satisfactorily.

Very few research papers have been published till now, in this area. Analysis of fractional order logistic map and its synchronization has been carried out by Wu et al. [54–56]. Fractional order Henon map has been introduced by Hu [57]. Further, extending the work, in this chapter, we will investigate the chaotic behaviors of some fractional order discrete chaotic systems (logistic map and Henon map). Bifurcation diagrams are given which validate the chaotic behaviors of these systems. Further, we shall apply backstepping based strategy for control of fractional order discrete chaotic systems which belong to strict-feedback class of systems. The fractional order Henon map which belong to this class is studied and a backstepping controller is designed for its stabilization. Backstepping is a well-known recursive technique, based on Lyapunov stability criterion and is used for controller design for nonlinear systems. The simulation results confirm the efficacy of the proposed controller.

The organization of the chapter is as follows: Sect. 2 gives a brief introduction of discrete fractional calculus. A brief mathematical background is discussed here. Section 3 deals with study of fractional order logistic map and Henon system. Chaotic behavior is proved by bifurcation diagram. Backstepping based control of fractional order Henon map has been described in Sect. 3.4 along with simulation results for validation. The chapter is concluded in Sect. 4.

2 Fundamentals of Discrete Fractional Calculus

Discrete fractional calculus is still in development phase and is topic of research in the field of mathematics. Keeping this point in mind, we will just give a brief introduction which will be sufficient for the reader and will help in understanding the analysis of fractional order discrete chaotic systems. In order to define fractional sum, we can express first order difference as

$$\Delta f(t) = f(t+1) - f(t) \tag{1}$$

On the similar grounds, second order difference will be,

$$\Delta^2 f(t) = \Delta f(t+1) - \Delta f(t)$$

$$\Delta^2 f(t) = f(t+2) - 2f(t+1) + f(t) \tag{2}$$

The general nth order difference is given as

$$\Delta^n f(t) = \Delta^{n-1} f(t+1) - \Delta^{n-1} f(t) = \sum_{k=0}^{n} \binom{n}{k} (-1)^k f(t+n-k) \tag{3}$$

On the similar grounds we can define fractional sum of order α.

Definition 1 For $f : \mathbb{N}_a \to \mathbb{R}$ and $0 < \alpha$, fractional order sum of order α is defined below as in [44],

$$\Delta_a^{-\alpha} f(t) = \frac{1}{\Gamma(\alpha)} \sum_{s=a}^{t-\alpha} (t - \sigma(s))^{(\alpha-1)} f(s) \tag{4}$$

where, $t \in N_{a+\alpha}$, a is the starting point, $a \in \mathbb{R}$, $\sigma(s) = s+1$, and the generalized falling factorial, $t^{(\alpha)} = \frac{\Gamma(t+1)}{\Gamma(t+1-\alpha)}$. Some important properties of factorial function are given below as in [58].

(i) $\Delta t^{(\alpha)} = \alpha t^{(\alpha-1)}$
(ii) $(t-\mu)t^{(\mu)} = t^{(\mu+1)}$, where $\mu \in \mathbb{R}$
(iii) $\mu^{(\mu)} = \Gamma(\mu+1)$
(iv) If $t \le r$, then $t^{(\alpha)} \le r^{(\alpha)}$ for any $\alpha > r$
(v) If $0 < \alpha < 1$, then $t^{(\alpha\nu)} \ge \left(t^{(\nu)}\right)^{(\alpha)}$
(vi) $t^{(\alpha+\beta)} = (t-\beta)^{(\alpha)} t^{(\beta)}$

Caputo has given the two basic definitions of fractional difference which are derived from the continuous definition of fractional derivative given by Caputo.

Definition 2 Caputo like delta fractional difference of $f(t)$ on $\mathbb{N}_a$ and $\mathbb{N}_b$ is defined as, [46]

$$\Delta_C^{\alpha} f(t) = \frac{1}{\Gamma(n-\alpha)} \sum_{s=a}^{t-(n-\alpha)} (t-\sigma(s))^{(n-\alpha-1)} \Delta_s^n f(s) \tag{5}$$

$$\nabla_C^{\alpha} f(t) = \frac{1}{\Gamma(n-\alpha)} \sum_{s=t+(n-\alpha)}^{b} (\rho(s)-t)^{(n-\alpha-1)} \nabla_b^n f(s) \tag{6}$$

where, for two real numbers a and b,

$\mathbb{N}_a = \{a, a+1, a+2, \ldots\}$ and $\mathbb{N}_b = \{b, b-1, b-2, \ldots\}$.

Also, $n = \lceil \alpha \rceil + 1$, $\rho(s) = s-1$.

It is clear from the above definitions, that fractional difference on time scales gives a potential approach for discrete fractional modeling.

2.1 Fractional Order Difference Equation

With the above definitions in mind we can define fractional order nonlinear difference equation as

$$\begin{aligned} &\Delta_a^{\alpha} u(t) = f(t+\alpha-1, u(t+\alpha-1)) \\ &\Delta^k u(a) = u_k, m = \lceil \alpha \rceil + 1, k = 0, 1, \ldots, m-1 \end{aligned} \tag{7}$$

The above equation has a finite fractional difference form but the output is dependent on the difference results of all the past states. This feature depicts the discrete systems' long memory effects or long interactions. Solution of the above equation is not obvious and has to be dealt in a different manner. The solution as given in [51] can be expressed as,

$$u(t) = u_0(t) + \frac{1}{\Gamma(\alpha)} \sum_{s=a+m-\alpha}^{t-\alpha} (t-\sigma(s))^{(\alpha-1)} f(s) f(s+\alpha-1, u(s+\alpha-1)) \tag{8}$$

where the initial condition, $u_0(t) = \sum_{k=0}^{m-1} \frac{(t-a)^{(k)}}{k!} \Delta^k u(a)$.

3 Fractional Order Discrete Chaotic Systems

Chaos and chaos synchronization find a variety of applications. Discrete maps are easy to use and can readily generate chaos and hence, are being used in the different areas of engineering and science. In this section we shall first study the existing fractional discrete chaotic systems and the effect of fractional difference on their chaotic behavior. We shall describe fractional order logistic map and Henon map, one by one.

3.1 Fractional Order Logistic Map

The classical logistic map, which gives the basic representation of demographical model, is given by

$$u(n+1) = \mu u(n)(1 - u(n)), u(0) = c \tag{9}$$

The fractional logistic map with fractional order α, is expressed as,

$$\Delta_a^\alpha u(t) = \mu u(t + \alpha - 1)(1 - u(t + \alpha - 1))$$

$t \in \mathbb{N}_a = a, a+1, a+2, \ \ldots \ a \in \mathbb{R}, \ u(a) = c, \ 0 < \alpha \leq 1.$

The solution of the above equation is expressed as,

$$u(t) = u(0) + \frac{\mu}{\Gamma(\alpha)} \sum_{s=1-\alpha}^{t-\alpha} (t-s-1)^{(\alpha-1)} u(s+\alpha-1)(1 - u(s+\alpha-1)) \tag{10}$$

For numerical solution, the above expression can be written as,

$$u(n) = u(0) + \frac{\mu}{\Gamma(\alpha)} \sum_{j=1}^{n} \frac{\Gamma(n-j+\alpha)}{\Gamma(n-j+1)} u(j-1)(1 - u(j-1)) \tag{11}$$

As explained in [56], the fractional version of logistic map has a discrete kernel function. The memory effect is exhibited here as the present state of evolution depends on all previous states. For $\alpha = 0.1$, $u(0) = 0.3$ and $n = 100$, the behavior of fractional order logistic map is depicted in Figs. 1 and 2. Figure 1 depicts bifurcation diagram of fractional order logistic map. As it is evident from the Fig. 2, the system exhibits chaotic behavior for $\mu = 3.0$. Fractional order logistic map can exhibit chaotic behavior for different values of values of fractional order α. The parameter μ may be varied to obtain the chaotic behavior.

3.2 Fractional Order Henon Map

Henon map is one of the common example of discrete maps which exhibits chaotic behavior. Classical Henon map has been well studied and analyzed. A number of control and synchronization techniques have been developed for Henon map. Integer order Henon map can be expressed as,

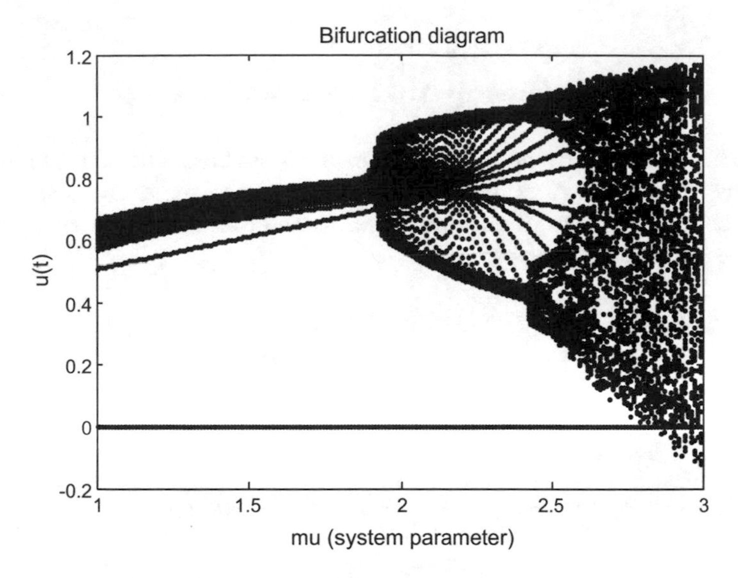

Fig. 1 Bifurcation diagram of fractional order logistic map for order $\alpha = 0.1$ when system parameter μ is varied from 1 to 3

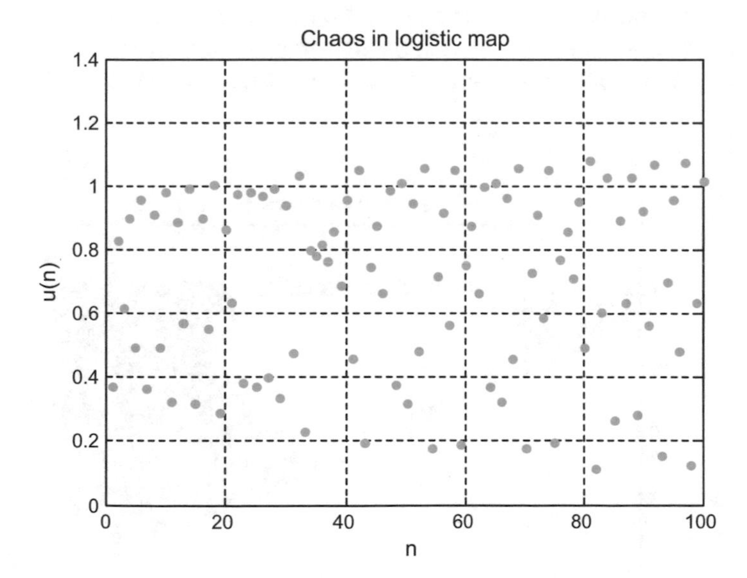

Fig. 2 Chaotic behavior of fractional order logistic map for order $\alpha = 0.1$ when system parameter $\mu = 3.0$

$$\begin{aligned} x(n) &= by(n-1) \\ y(n) &= x(n-1) + 1 - a(y(n-1))^2 + u \end{aligned} \tag{12}$$

The fractional order version of Henon map has not been studied much. A recent paper by Hu [57] describes fractional order Henon map and shows that it exhibits chaos for various values of fractional order α. The fractional order Henon map as expressed in [57], is given as:

$$\begin{aligned} \Delta_a^\alpha x(t) &= by(t+\alpha-1) - x(t+\alpha-1) \\ \Delta_a^\alpha y(t) &= x(t+\alpha-1) + 1 - a(y(t+\alpha-1))^2 - y(t+\alpha-1) + u \end{aligned} \tag{13}$$

For simulation purpose the numerical formula can be written as,

$$\begin{aligned} x(n+1) &= x(a) + \frac{1}{\Gamma(\alpha)} \sum_{j=1}^{n} \frac{\Gamma(n-j+\alpha)}{\Gamma(n-j+1)} (by(n) - x(n)) \\ y(n+1) &= y(a) + \frac{1}{\Gamma(\alpha)} \sum_{j=1}^{n} \frac{\Gamma(n-j+\alpha)}{\Gamma(n-j+1)} \left(x(n) + 1 - a(x(n))^2 - y(n) \right) \end{aligned} \tag{14}$$

The fractional order Henon map exhibits chaotic behavior for different values of α and different parameter values. The bifurcation diagram is presented in Fig. 3 for fractional order $\alpha = 0.95$ and system parameter a, varying from 0.2 to 1.8. The phase portrait is shown in Fig. 4 for system parameters $a = 1.5$, $b = 0.2$.

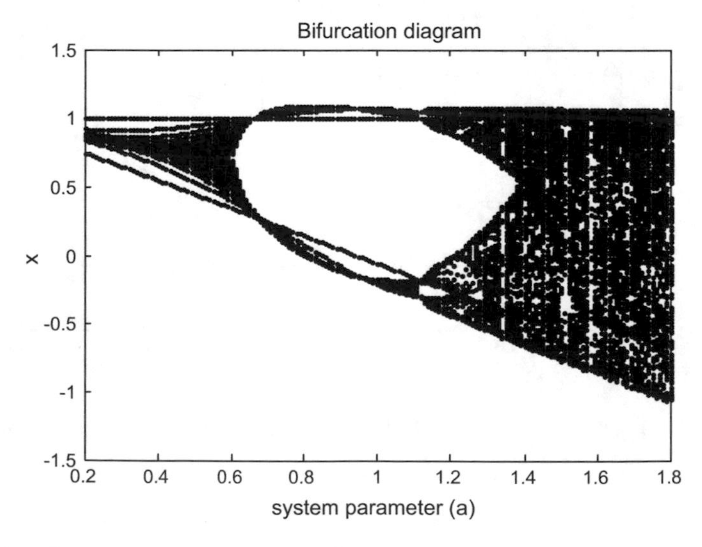

Fig. 3 Bifurcation diagram of fractional order Henon map for order $\alpha = 0.95$ when system parameter a is varied from 0.2 to 1.8

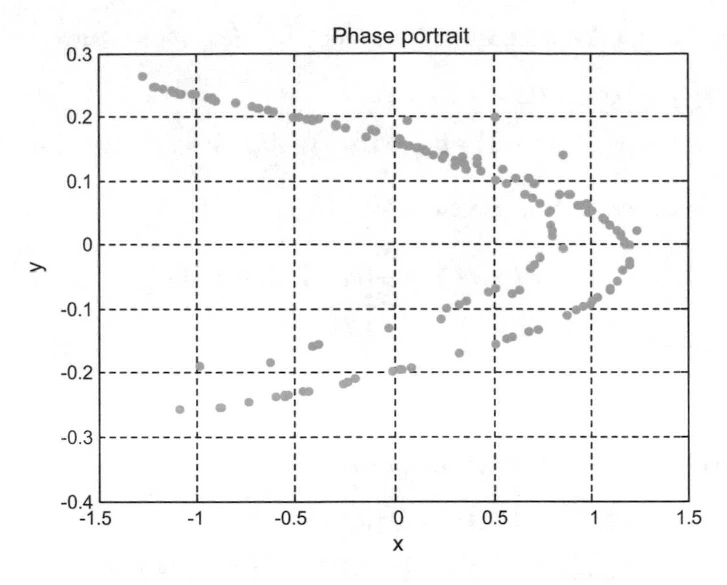

Fig. 4 Phase portrait of fractional order Henon map for order $\alpha = 0.95$ and system parameters $a = 1.5, b = 0.2$

3.3 *Stabilization of Fractional Order Henon Map*

In this section we design a Lyapunov based stabilizing controller using backstepping technique in order to achieve stability of Henon map. Backstepping technique is applied on the systems which are in strict-feedback form. In each step a virtual controller is obtained satisfying the Lyapunov stability criterion, and further we step back towards the final state equation which contains the final control. In the final step, controller for whole system is obtained and the time derivative of the overall Lyapunov function is made negative semi-definite. Here, Lyapunov function for discrete system has been chosen as a modulus function.

For the first equation in (13), let $z_1(t+\alpha-1) = x(t+\alpha-1)$ and $z_2(t+\alpha-1) = y(t+\alpha-1) - \beta_1(t+\alpha-1)$, where β_1 is the virtual controller.

We can have,

$$\begin{aligned} &\Delta_a^\alpha z_1(t) = \Delta_a^\alpha x(t) \\ &\Rightarrow \Delta_a^\alpha z_1(t) = b(z_2(t+\alpha-1) + \beta_1(t+\alpha-1)) - x(t+\alpha-1) \end{aligned} \tag{15}$$

The Lyapunov function for the above subsystem can be chosen as,

$$V_1(t) = |z_1(t)|$$

Taking the fraction difference of Lyapunov function, we can have,

$$\begin{aligned}\Delta_a^\alpha V_1(t) &\le |\Delta_a^\alpha z_1(t)| - |z_1(t+\alpha-1)| \\ &\le |b(z_2(t+\alpha-1)+\beta_1(t+\alpha-1)) - x(t+\alpha-1)| - |z_1(t+\alpha-1)|\end{aligned}$$

The virtual controller is chosen as

$$\beta_1(t+\alpha-1) = \frac{1}{b}(c_1+1)x(t+\alpha-1) \tag{16}$$

which results in,

$$\Delta_a^\alpha V_1(t) \le (c_1-1)|z_1(t+\alpha-1)| + |bz_2(t+\alpha-1)|$$

Similarly, for the second subsystem we have,

$$\begin{aligned}\Delta_a^\alpha z_2(t) &= \Delta_a^\alpha y(t) - \Delta_a^\alpha \beta_1(t) \\ \Delta_a^\alpha z_2(t) &= x(n-1)+1-a(y(n-1))^2+u-\Delta_a^\alpha \beta_1(t)\end{aligned} \tag{17}$$

Overall Lyapunov function will be

$$\begin{aligned}V_2(t) &= V_1(t) + |z_2(t)| \\ &\Rightarrow \Delta_a^\alpha V_2(t) \le (c_1-1)|z_1(t+\alpha-1)| + |bz_2(t+\alpha-1)| \\ &\quad + \left|\left(x(n-1)+1-a(y(n-1))^2+u-\Delta_a^\alpha \beta_1(t)\right)\right| \\ &\quad - |bz_2(t+\alpha-1)|\end{aligned}$$

The final controller can be obtained as,

$$\begin{aligned}u = c_2&\left\{y(t+\alpha-1) - \frac{1}{b}(c_1+1)x(t+\alpha-1)\right\} - x(t+\alpha-1) \\ &+ a(y(t+\alpha-1))^2 + y(t+\alpha-1) + (c_1+1)y(t+\alpha-1)\end{aligned} \tag{18}$$

And the final Lyapunov function will satisfy the following condition

$$\Delta_a^\alpha V_2(t) \le (c_1-1)|z_1(t+\alpha-1)| + (c_2-b-1)|z_2(t+\alpha-1)| \tag{19}$$

Here c_1 and c_2 are the design parameters which can be chosen arbitrarily.

3.4 Results and Discussions

The simulations have been performed on MATLAB software. For the parameters $a = 1.2$, $b = 0.15$ and design parameters $c_1 = c_2 = 0.1$, the stabilization of states is

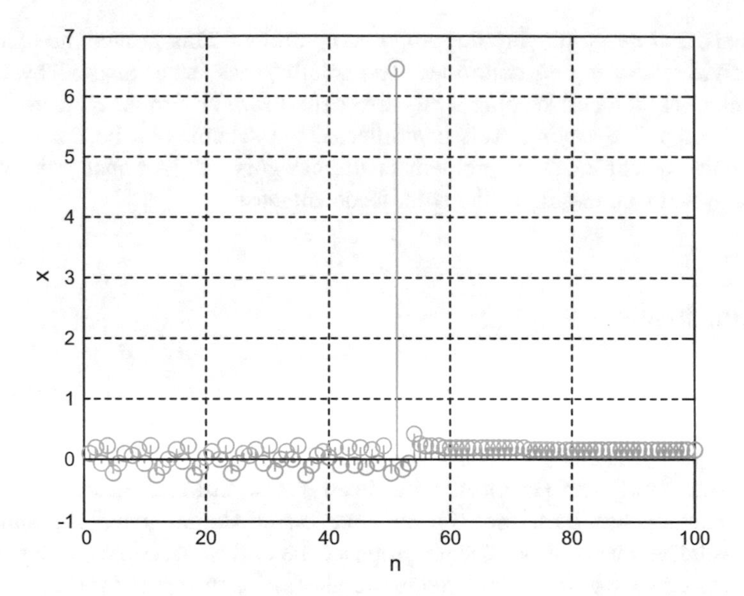

Fig. 5 Stabilization of state x of fractional order Henon map

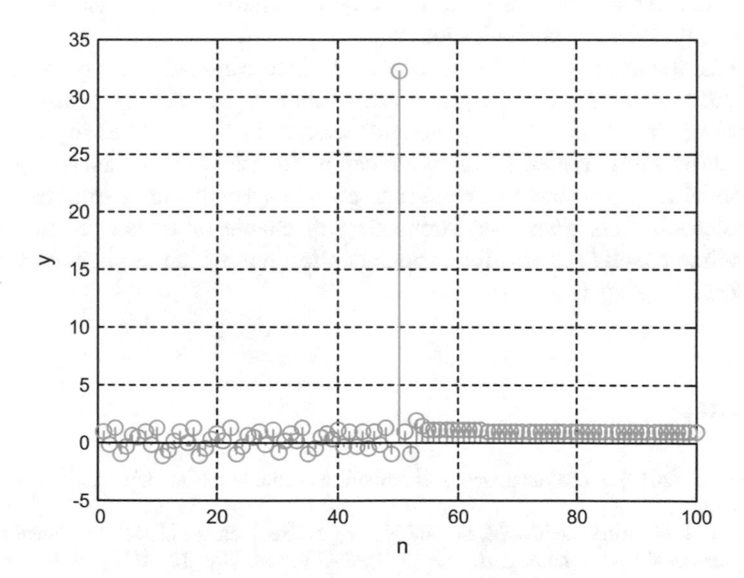

Fig. 6 Stabilization of state y of fractional order Henon map

depicted in Figs. 5 and 6. The initial conditions are chosen as, $x(0) = 0.5$, $y(0) = 0.6$. The control action has been employed at $n = 50$.

It can be seen in the figures that states x and y of the system are bounded after $n = 50$, as controller is applied at this instant. We can conclude that the chaotic behavior of the fractional order Henon map has been suppressed and the system has been stabilized by employing the proposed controller. This proves the efficacy of the derived backstepping controller. The stability has been ensured by using a Lyapunov function and keeping it negative definite which further indicates that the system energy is decaying. This controller design scheme can be further used to address the synchronization problem of the category of fractional order discrete chaotic systems belonging to the strict-feedback class.

4 Conclusion

Study of fractional order discrete time chaotic maps is a new area of study. In this paper, the recent developments in this area have been discussed and a contribution has been made regarding stabilization of such systems. Chaotic behavior of fractional logistic map and Henon map has been discussed here. Backstepping based control strategy has been used for stabilization of Henon map. The simulation results validate the control scheme proposed here. The fractional order discrete chaotic systems are going to find various application as they show variety of chaotic behaviors for various values of system parameters and fractional order α. This special characteristic enables us to use chaotic systems in the area of chemistry, cryptography, secure communication etc.

For this, the need is to design a controller which can lead to synchronization of two chaotic systems in master-slave configuration. In case of continuous-time fractional order chaotic systems, enough literature in there addressing the control and synchronization issues. Future work can be focused towards analyzing chaotic behaviors of fractional order discrete-time chaotic systems along with their control and synchronization. These developments will enable us to use fractional order discrete-time chaotic systems for various applications which are beneficial for the upliftment of the society.

References

1. Yang, Q. (2010). Novel analytical and numerical methods for solving fractional dynamical systems.
2. Angulo, J. M., Ruiz-Medina, M. D., Anh, V. V., & Grecksch W. (2000). Fractional diffusion and fractional heat equation. *Advances in Applied Probability, 32*, 1077–1099.
3. Sun, R. (2007). *Fractional order signal processing: Techniques and applications.*
4. Singh, A. P., Srivastava, T., & Srivastava, P. (2011). Fractional equations of motion via fractional modeling of underactuated robotic systems.
5. Jesus, I. S., & Machado, J. A. T. (2009). Development of fractional order capacitors based on electrolyte processes. *Nonlinear Dynamics, 56*, 45–55.
6. Podlubny, I. (1994). Fractional-order systems and fractional-order controllers. *Proceedings of the Conference Internationale Francophone d'Automatique*, 2002.

7. Science, N., Phenomena, C., Handa, H., & Sharma, B. B. (2016). Chaos, solitons and fractals novel adaptive feedback synchronization scheme for a class of chaotic systems with and without parametric uncertainty. *Chaos, Solitons and Fractals, 86*, 50–63.
8. Shukla, M., & Sharma, B. B. (2016). Control of uncertain fractional order lorenz system via adaptive backstepping control, 0–5.
9. Chauhan, V., Kumar, V., & Shukla, M. (2015). Dynamic surface control of, 10–14.
10. Ramsingh, G., & Sharma, B. B. (2012). Synchronization of chaotic system using u-model based adaptive inverse control.
11. Sharma, B. B., & Kar, I. N. (2008). Adaptive control of wing rock system in uncertain environment using contraction theory. In *2008 American Control Conference*, 2963–2968.
12. Sharma, B. B., Kar, I. N., & Ieee, M. (2009). Contraction based adaptive control of a class of nonlinear systems, 808–813.
13. Sharma, V., Agrawal, V., Sharma, B. B., & Nath, R. (2016). Unknown input nonlinear observer design for continuous and discrete time systems with input recovery scheme. *Nonlinear Dynamics*, 1–14.
14. Boulkroune, A., Hamel, S., Azar, A. T., & Vaidyanathan, S. (2016). Fuzzy control-based function synchronization of unknown chaotic systems with dead-zone input. In *Advances in Chaos Theory and Intelligent Control* (pp. 699–718). Springer.
15. Vaidyanathan, S., & Azar, A. T. (2015). Anti-synchronization of identical chaotic systems using sliding mode control and an application to vaidyanathan–madhavan chaotic systems. In *Advances and Applications in Sliding Mode Control Systems* (pp. 527–547). Springer.
16. Vaidyanathan, S., Sampath, S., & Azar, A. T. (2015). Global chaos synchronisation of identical chaotic systems via novel sliding mode control method and its application to Zhu system. *International Journal of Modelling, Identification and Control, 23*, 92–100.
17. Azar, A. T., & Vaidyanathan, S. (2014). *Handbook of research on advanced intelligent control engineering and automation*. IGI Global.
18. Vaidyanathan, S., & Azar, A. T. (2016). Dynamic analysis, adaptive feedback control and synchronization of an eight-term 3-D novel chaotic system with three quadratic nonlinearities. In *Advances in Chaos Theory and Intelligent Control* (pp. 155–178). Springer.
19. Vaidyanathan, S., & Azar, A. T. (2016). A novel 4-D four-wing chaotic system with four quadratic nonlinearities and its synchronization via adaptive control method. In *Advances in Chaos Theory and Intelligent Control* (pp. 203–224). Springer.
20. Vaidyanathan, S., & Azar, A. T. (2016). Adaptive control and synchronization of halvorsen circulant chaotic systems. In *Advances in Chaos Theory and Intelligent Control*. pp. 225–247. Springer.
21. Vaidyanathan, S., & Azar, A. T. (2016). Adaptive backstepping control and synchronization of a novel 3-D jerk system with an exponential nonlinearity. In *Advances in Chaos Theory and Intelligent Control* (pp. 249–274). Springer.
22. Vaidyanathan, S., & Azar, A. T. (2016). Generalized projective synchronization of a novel hyperchaotic four-wing system via adaptive control method. In *Advances in Chaos Theory and Intelligent Control* (pp. 275–296). Springer.
23. Vaidyanathan, S., Idowu, B. A., & Azar, A. T. (2015). Backstepping controller design for the global chaos synchronization of Sprott's jerk systems. In *Chaos Modeling and Control Systems Design* (pp. 39–58). Springer.
24. Vaidyanathan, S., & Azar, A. T. (2016). Qualitative study and adaptive control of a novel 4-d hyperchaotic system with three quadratic nonlinearities. In *Advances in Chaos Theory and Intelligent Control* (pp. 179–202). Springer.
25. Vaidyanathan, S., Azar, A. T., Rajagopal, K., & Alexander, P. (2015). Design and SPICE implementation of a 12-term novel hyperchaotic system and its synchronisation via active control. *International Journal of Modelling, Identification and Control, 23*, 267–277.
26. Azar, A. T., & Vaidyanathan, S. (2015). *Chaos modeling and control systems design*. Springer.
27. Zhu, Q., Azar, & A. T. (2015). *Complex system modelling and control through intelligent soft computations*. Springer.

28. Vaidyanathan, S., & Azar, A. T. (2015). Hybrid synchronization of identical chaotic systems using sliding mode control and an application to Vaidyanathan chaotic systems. In *Advances and Applications in Sliding Mode Control systems* (pp. 549–569). Springer.
29. Vaidyanathan, S., & Azar, A. T. (2015). Analysis, control and synchronization of a nine-term 3-D novel chaotic system. In *Chaos Modeling and Control Systems Design* (pp. 19–38). Springer.
30. Vaidyanathan, S., & Azar, A. T. (2015). Analysis and control of a 4-D novel hyperchaotic system. In *Chaos Modeling and Control Systems Design* (pp. 3–17). Springer.
31. Hartley, T. T., Lorenzo, C. F., & Qammer, H. K. (1995). Chaos in a fractional order Chua's system.
32. Asheghan, M. M., Beheshti, M. T. H., & Tavazoei, M. S. (2011). Robust synchronization of perturbed Chen's fractional-order chaotic systems. *Communications in Nonlinear Science and Numerical Simulation, 16*, 1044–1051.
33. Wang, J., Xiong, X., & Zhang, Y. (2006). Extending synchronization scheme to chaotic fractional-order Chen systems. *Physica A: Statistical Mechanics and its Applications, 370*, 279–285.
34. Grigorenko, I., & Grigorenko, E. (2003). Chaotic dynamics of the fractional Lorenz system. *Physical Review Letters, 91*, 34101.
35. Xi, H., Li, Y., & Huang, X. (2014). Generation and nonlinear dynamical analyses of fractional-order memristor-based Lorenz systems. *Entropy, 16*, 6240–6253.
36. Li, C., & Chen, G. (2004). Chaos and hyperchaos in the fractional-order Rössler equations. *Physica A: Statistical Mechanics and its Applications, 341*, 55–61.
37. Shahiri, M. T., Ranjbar, A., Ghaderi, R., Karami, M., & Hoseinnia, S. H. (2010). *Adaptive Backstepping Chaos Synchronization of Fractional order Coullet Systems with Mismatched Parameters., 2010*, 1–6.
38. Barbosa, R. S., Machado, J. A. T., Vinagre, B. M., & Calderon, A. J. (2007). Analysis of the van der pol oscillator containing derivatives of fractional order. *Journal of Vibration and Control, 13*, 1291–1301.
39. Hegazi, A. S., Ahmed, E., & Matouk, A. E. (2013). On chaos control and synchronization of the commensurate fractional order Liu system. *Communications in Nonlinear Science and Numerical Simulation 18*, 1193–1202.
40. Azar, A. T., & Vaidyanathan, S. (2016). *Advances in chaos theory and intelligent control*. Springer.
41. Boulkroune, A., Bouzeriba, A., Bouden, T., & Azar, A. T. (2016). Fuzzy adaptive synchronization of uncertain fractional-order chaotic systems. In *Advances in Chaos Theory and Intelligent Control* (pp. 681–697). Springer.
42. Azar, A. T., & Vaidyanathan, S. (2015). *Computational intelligence applications in modeling and control*. Springer.
43. Shukla, M., Sharma, B. B. (2015). Hybrid projective synchronization of fractional order volta' s system via active control.
44. Atici, F. M., & Eloe, P. W. (2009). Initial value problems in discrete fractional calculus. 137, 981–989.
45. Holm, M. T. (2011). The Laplace transform in discrete fractional calculus. *Computers and Mathematics with Applications, 62*, 1591–1601.
46. Abdeljawad, T. (2011). On Riemann and Caputo fractional differences. *Computers and Mathematics with Applications, 62*, 1602–1611.
47. Eloe, P. W., & Science, C. (2009). Discrete fractional calculus with the nabla operator, 1–12.
48. Guermah, S.: Discrete-Time Fractional-Order Systems : Modeling and Fractional-Order Stability Issues Systems: Modeling and Stability Issues.
49. Sierociuk, D. (2015). Stability of discrete fractional order state-space, *14*, 1543–1556.
50. Wu, G.-C., Baleanu, D., Zeng, S.-D., & Luo, W.-H. (2016). Mittag-Leffler function for discrete fractional modelling. *Journal of King Saud University-Science, 28*, 99–102.
51. Chen, F., Luo, X., & Zhou, Y. (2011). Existence results for nonlinear fractional difference equation.

52. Mohan, J. J., Deekshitulu, G. V. S. R. (2012). Fractional order difference equations, 2012.
53. Holm, M. T., & Holm, M. (2011). The theory of discrete fractional calculus: Development and application.
54. Wu, G.-C., & Baleanu, D. (2014). Chaos synchronization of the discrete fractional logistic map. *Signal Processing, 102*, 96–99.
55. Wu, G. -C., Baleanu, D., Xie, H. -P., & Chen, F. -L. (2016). Chaos synchronization of fractional chaotic maps based on the stability condition. *Physica A: Statistical Mechanics and its Applications*.
56. Wu, G.-C., & Baleanu, D. (2014). Discrete fractional logistic map and its chaos. *Nonlinear Dynamics, 75*, 283–287.
57. Hu, T. (2014). Discrete chaos in fractional Henon map. *Applied Mathematics, 5*, 2243.
58. Atici, F. M., & Eloe, P. W. (2007). A transform method in discrete fractional calculus. *International Journal of Difference Equations, 2*, 165–176.

Part II
Applications of Fractional Order Chaotic Systems

A Three-Dimensional No-Equilibrium Chaotic System: Analysis, Synchronization and Its Fractional Order Form

Viet-Thanh Pham, Sundarapandian Vaidyanathan, Christos K. Volos, Ahmad Taher Azar, Thang Manh Hoang and Vu Van Yem

Abstract Recently, a new classification of nonlinear dynamics has been introduced by Leonov and Kuznetsov, in which two kinds of attractors are concentrated, i.e. self-excited and hidden ones. Self-excited attractor has a basin of attraction excited from unstable equilibria. So, from that point of view, most known systems, like Lorenz's system, Rössler's system, Chen's system, or Sprott's system, belong to chaotic systems with self-excited attractors. In contrast, a few unusual systems such as those with a line equilibrium, with stable equilibria, or without equilibrium, are classified into chaotic systems with hidden attractor. Studying chaotic system with hidden attractors has become an attractive research direction because hidden attractors play an important role in theoretical problems and engineering applications. This chapter presents a three-dimensional autonomous system without any equilibrium point which can generate hidden chaotic attractor. The fundamental dynamics properties of such no-equilibrium system are discovered by using phase portraits,

V.-T. Pham (✉) · T.M. Hoang · V. Van Yem
School of Electronics and Telecommunications,
Hanoi University of Science and Technology, Hanoi, Vietnam
e-mail: pvt3010@gmail.com

T.M. Hoang
e-mail: thang.hoangmanh@hust.edu.vn

V. Van Yem
e-mail: yem.vuvan@hust.edu.vn

S. Vaidyanathan
Research and Development Centre, Vel Tech University, Tamil Nadu, India

C.K. Volos
Physics Department, Aristotle University of Thessaloniki, Thessaloniki, Greece
e-mail: volos@physics.auth.gr

A.T. Azar
Faculty of Computers and Information, Benha University, Benha, Egypt
e-mail: ahmad.azar@fci.bu.edu.eg; ahmad_t_azar@ieee.org

A.T. Azar
Nanoelectronics Integrated Systems Center (NISC),
Nile University, Cairo, Egypt

A.T. Azar et al. (eds.), *Fractional Order Control and Synchronization of Chaotic Systems*, Studies in Computational Intelligence 688,
DOI 10.1007/978-3-319-50249-6_15

Lyapunov exponents, bifurcation diagram, and Kaplan–Yorke dimension. Chaos synchronization of proposed systems is achieved and confirmed by numerical simulation. In addition, an electronic circuit is implemented to evaluate the theoretical model. Finally, fractional-order form of the system with no equilibrium is also investigated.

Keywords Chaos · Hidden attractor · No-Equilibrium · Lyapunov exponents · Bifurcation · Synchronization · Circuit · SPICE

1 Introduction

In 1963, Lorenz found a chaotic system when studying a model for atmospheric convection [50]. The most well-known feature of a chaotic system is the sensitivity on initial conditions, named "butterfly effect". This means that a small variation on initial conditions of a system will generate a totally different chaotic trajectory. After the invention of Lorenz, there has been significant interest in chaos theory, chaotic systems, and chaos-based applications [5–8, 18, 19, 67, 72, 104]. Especially, various new chaotic systems have been discovered such as Rössler's system [61], Arneodo's system [4], Chen's system [18], Lü's system [51], Vaidyanathan's systems [79, 85, 87, 88, 90], time-delay systems [11], nonlinear finance system [75], four-scroll chaotic attractor [2] and so on [58, 82]. Complex behaviors of chaotic system were used in different applications. True random bits were generated by using a double-scroll chaotic attractor [103]. Volos et al. controlled autonomous mobile robots via chaotic path planning [96]. Han et al. implemented a fingerprint images encryption scheme based on chaotic attractors [27]. Hoang and Nakagawa proposed applications of time delay systems in secure communication due to their complex dynamics [31]. Application of synchronization of Chua's circuits with multi-scroll attractors in communications was introduced in [24]. In addition, Akgul et al. presented engineering applications of a new four-scroll chaotic attractor [2].

When studying chaotic systems, their equilibrium points play important role [69, 98]. As have been known, most reported chaotic systems have a countable number of equilibrium points [68]. Therefore, chaos in these systems can be proved by using Shilnikov criteria where at least one unstable equilibrium for emergence of chaos is required [66]. However, a few chaotic systems without equilibrium have been proposed recently [34]. We cannot apply the Shinikov method for verifying chaos in such systems because they have neither homoclinic nor heteroclinic orbits. Chaotic systems without equilibrium are categorized as systems with "hidden attractor" and have been received significant attention [44, 46].

In this chapter, a novel system is introduced and its chaotic attractors are displayed. The special is that such new system does not have equilibrium points. This chapter is organized as follows. The related works are summarized in the next section. The model of the new system is proposed in Sect. 3. Dynamics and properties of the new system are investigated in Sect. 4. The adaptive anti-synchronization scheme is studied in Sect. 5. Section 6 presents a circuital implementation of the theoretical model. Fractional-order form of the new no-equilibrium system is proposed in Sect. 7. Finally, conclusions are drawn in the last section.

2 Related Work

The terminology "hidden attractor" has been proposed recently when Leonov and Kuznetsov introduced types of attractors: self-excited attractors and hidden attractors [42, 44, 46, 47]. A self-excited attractor has a basin of attraction that is excited from unstable equilibria. In contrast, hidden attractor cannot be found by using a numerical method in which a trajectory started from a point on the unstable manifold in the neighbourhood of an unstable equilibrium [33]. "Hidden attractor" plays a vital role in nonlinear theory and practical problems [41, 46, 54, 59, 65]. Therefore, various noticeable results relating to this topic has been reported in recent years. The presence of hidden attractors has witnessed in a smooth Chua's system [48], in mathematical model of drilling system [45], in nonlinear control systems [43], or in a multilevel DC/DC converter [107]. Hidden attractors appear in a 4-D Rikitake dynamo system [94], in 5-D hyperchaotic Rikitake dynamo system [92], in a 5D Sprott B system [52] or in a chaotic system with an exponential nonlinear term [56]. Other works on hidden attractors were introduced in [16, 35, 64, 69] and references therein.

Interesting that chaotic systems without equilibrium belong to a class of nonlinear systems with "hidden attractor" [34]. A few three-dimensional chaotic systems without equilibrium points have been discovered. Wei applied a tiny perturbation into the Sprott D system to create a new system with no equilibia [99]. Wang and Chen proposed a no-equilibrium system when constructing a chaotic system with any number of equilibria [98]. Especially, Jafari et al. found catalog of chaotic flows with no equilibria [34].

Moreover, four-dimensional chaotic systems without equilibrium points have been investigated recently. Based on a memristive device, a novel four-dimensional system has been proposed [57]. The peculiarity of the memristive system is that it does not display any equilibria and exhibits periodic, chaotic, and also hyperchaotic dynamics. Vaidyanathan has presented analysis, control and synchronization of a ten-term novel 4-D highly hyperchaotic system with three quadratic nonlinearities [81]. The author have been shown that it is a novel hyperchaotic system does not have any equilibrium point. Dynamics, synchronization and SPICE implementation of a memristive system with hidden hyperchaotic attractor have been reported in [55]. Investigation of new systems without equilibrium is still an attractive topic and should receive further attention.

3 Model of the No-Equilibrium System

Jafari et al. introduced a list of simple chaotic flows without equilibrium (denoted NE_1-NE_{14}) [34]. Interestingly, the system NE_8 can display coexisting hidden attractors [64]. The system NE_8 is described as

$$\begin{cases} \dot{x} = y \\ \dot{y} = -x - yz \\ \dot{z} = xy + 0.5x^2 - a, \end{cases} \tag{1}$$

where x, y, z are state variables and a is a positive parameter. The Lyapunov exponents of system NE_8 in (1) are $\lambda_1 = 0.0314$, $\lambda_2 = 0$, $\lambda_3 = -10.2108$ and the Kaplan–Yorke dimension is $D_{KY} = 2.0031$ (for $a = 1.3$) [34].

Based on system NE_8 in (1), in this work we study a new system in the following form

$$\begin{cases} \dot{x} = y \\ \dot{y} = -x - yz \\ \dot{z} = xy + ax^2 + by^2 - c, \end{cases} \tag{2}$$

where a, b, c are three positive parameters and $c \neq 0$. A detailed study of dynamics and properties of no-equilibrium system in (2) is presented in the next section.

4 Dynamics and Properties of the No-Equilibrium System

The equilibrium points of the system in (2) are found by solving $\dot{x} = 0$, $\dot{y} = 0$, and $\dot{z} = 0$, that is

$$y = 0, \tag{3}$$

$$-x - yz = 0, \tag{4}$$

$$xy + ax^2 + by^2 - c = 0, \tag{5}$$

From (3), (4), we have $x = y = 0$. Therefore Eq. (5) is inconsistent. In the other words, there is no real equilibrium in the system (2).

We consider the system (2) for the selected parameters $a = 0.5$, $b = 0.1$, $c = 1.3$ and the initial conditions are

$$(x(0), y(0), z(0)) = (0, 0.1, 0). \tag{6}$$

Lyapunov exponents, which measure the exponential rates of the divergence and convergence of nearby trajectories in the phase space of the chaotic system [68, 72], are calculated by using the algorithm in [102]. As a result, the Lyapunov exponents of the system (2) are

$$\lambda_1 = 0.0453, \lambda_2 = 0, \lambda_3 = -3.2903. \tag{7}$$

The non-equilibrium chaotic system is dissipative because the sum of the Lyapunov exponents is negative. It is worth noting that this non-equilibrium system can be classified as a nonlinear system with hidden strange attractor because its basin of attractor does not contain neighbourhoods of equilibria [44, 46]. The 2-D and 3-D projections of the chaotic attractors without equilibrium in this case are presented in Figs. 1, 2, 3 and 4.

It has been known that the Kaplan–Yorke dimension, which presents the complexity of attractor [23], is defined by

$$D_{KY} = j + \frac{1}{\left|\lambda_{j+1}\right|} \sum_{i=1}^{j} \lambda_i, \tag{8}$$

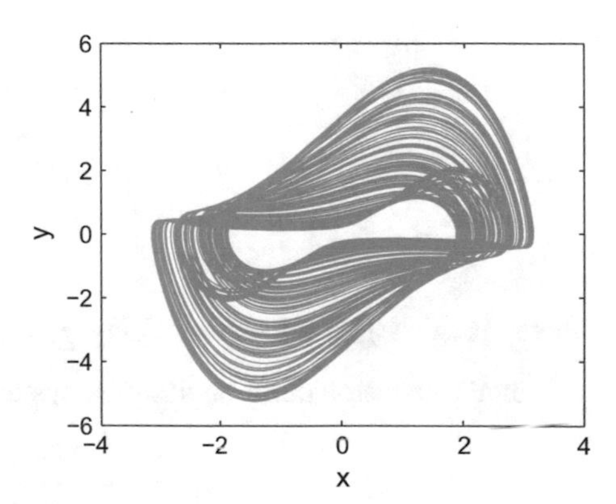

Fig. 1 2-D projection of the chaotic system without equilibrium in (2) in the (x, y)-plane

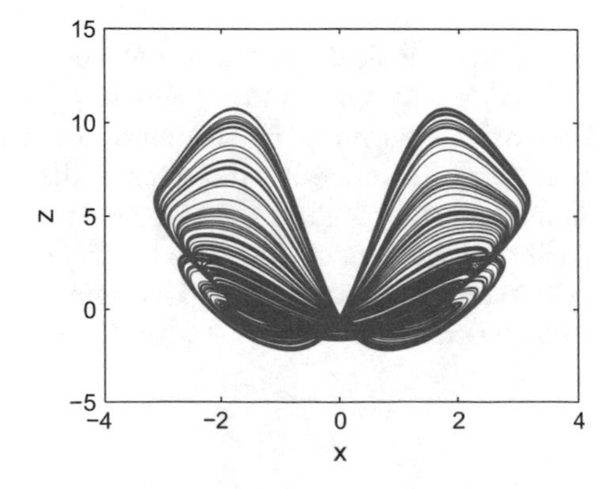

Fig. 2 2-D projection of the chaotic system without equilibrium in (2) in the (x, z)-plane

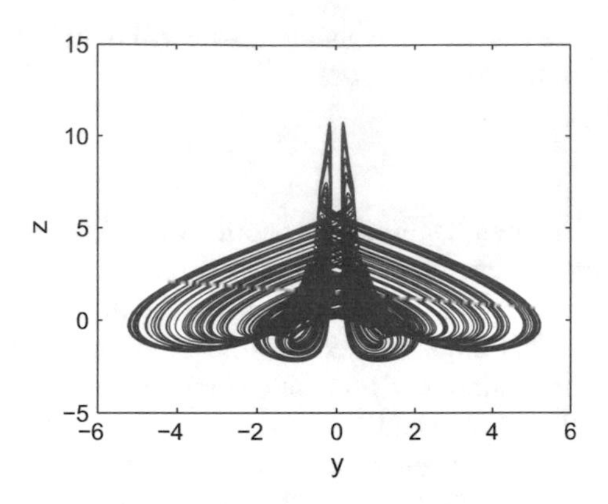

Fig. 3 2-D projection of the chaotic system without equilibrium in (2) in the (y, z)-plane

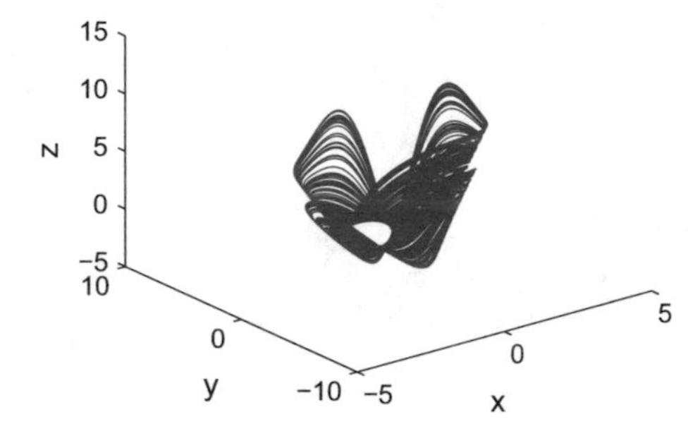

Fig. 4 3-D projection of the chaotic system without equilibrium in (2) in the (x, y, z)-space

where j is the largest integer satisfying $\sum_{i=1}^{j} \lambda_i \geq 0$ and $\sum_{i=1}^{j+1} \lambda_i < 0$. Thus, the calculated fractional dimension of no-equilibrium system in (2) when $a = 0.5, b = 0.1, c = 1.3$ is

$$D_{KY} = 2 + \frac{\lambda_1 + \lambda_2}{|\lambda_3|} = 2.0138. \tag{9}$$

Equation (9) indicates a strange attractor.

It is easy to see that the system in (2) has rotational symmetry with respect to the z-axis as evidenced by their invariance under the transformation from (x, y, z) to $(-x, -y, z)$. Therefore, any projection of the attractor has symmetry around the origin. In the other words there is a symmetric coexisting attractor as shown in Fig. 2. In addition, the attractor of the system is displayed in Fig. 4. The bifurcation diagrams of system in (2) illustrated in Fig. 5 indicate the presence of muti-stability. For example, there are coexisting attractors when $b = 0.15$ as shown in Figs. 6, 7, 8, and 9.

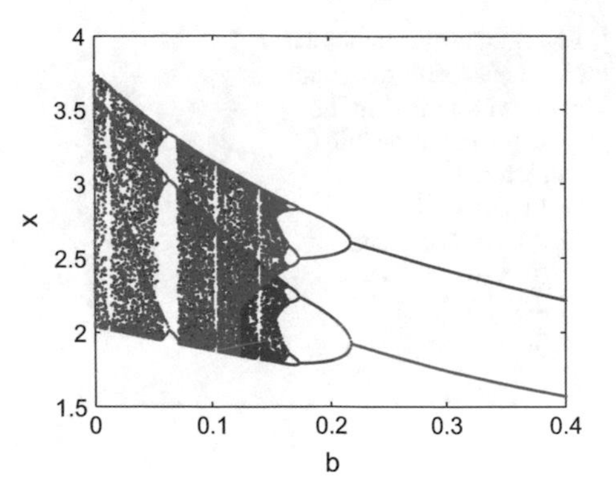

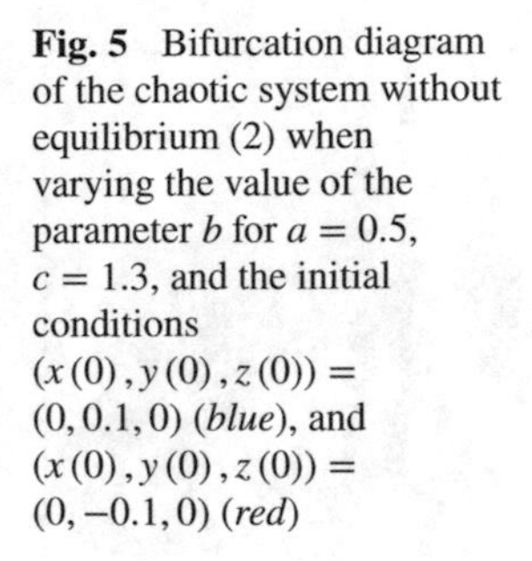

Fig. 5 Bifurcation diagram of the chaotic system without equilibrium (2) when varying the value of the parameter b for $a = 0.5$, $c = 1.3$, and the initial conditions $(x(0), y(0), z(0)) = (0, 0.1, 0)$ (*blue*), and $(x(0), y(0), z(0)) = (0, -0.1, 0)$ (*red*)

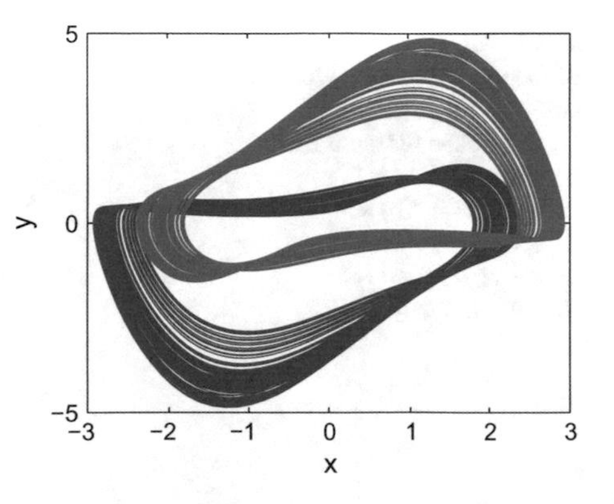

Fig. 6 Coexisting attractors of the chaotic system without equilibrium in (2) in the (x, y)-plane

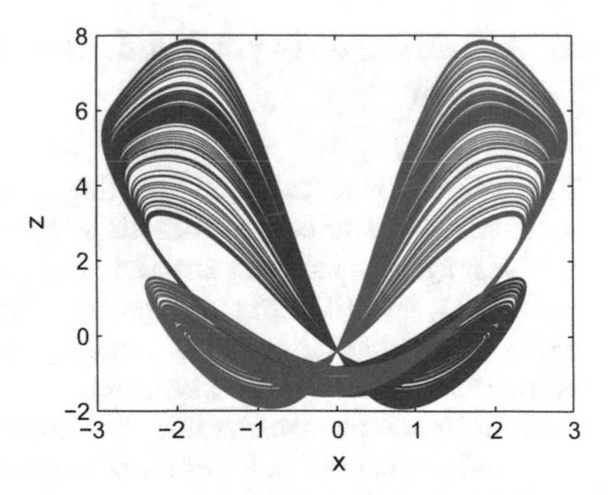

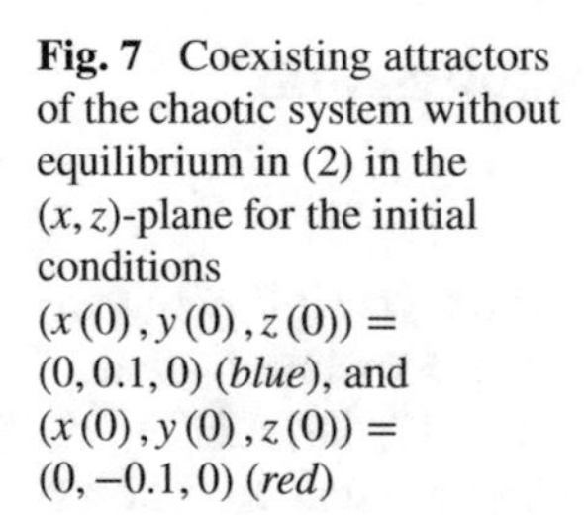

Fig. 7 Coexisting attractors of the chaotic system without equilibrium in (2) in the (x, z)-plane for the initial conditions $(x(0), y(0), z(0)) = (0, 0.1, 0)$ (*blue*), and $(x(0), y(0), z(0)) = (0, -0.1, 0)$ (*red*)

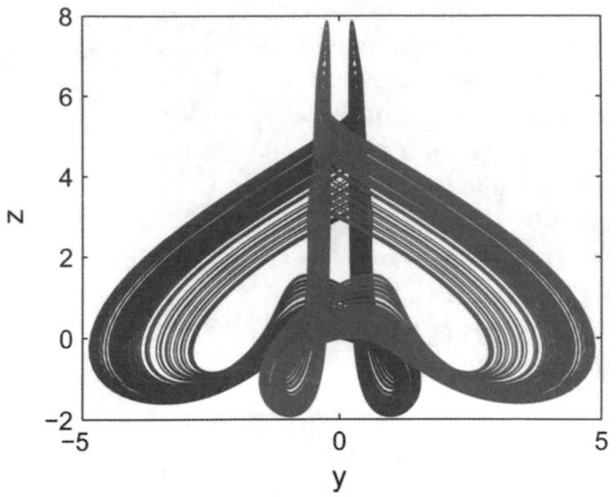

Fig. 8 Coexisting attractors of the chaotic system without equilibrium in (2) in the (y, z)-plane for the initial conditions $(x(0), y(0), z(0)) = (0, 0.1, 0)$ (*blue*), and $(x(0), y(0), z(0)) = (0, -0.1, 0)$ (*red*)

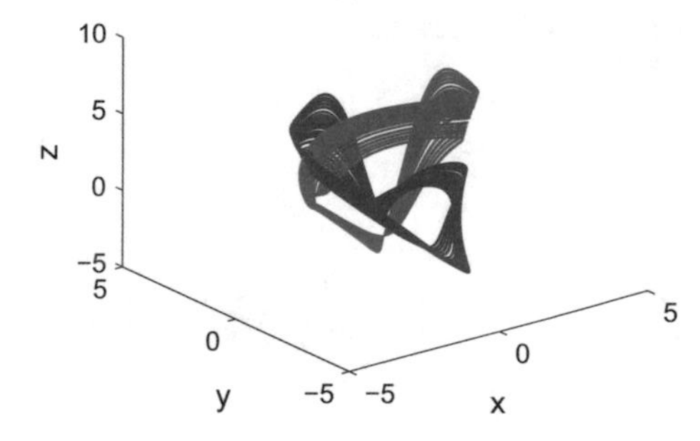

Fig. 9 Coexisting attractors of the chaotic system without equilibrium in (2) in the (x, y, z)-space for the initial conditions $(x(0), y(0), z(0)) = (0, 0.1, 0)$ (*blue*), and $(x(0), y(0), z(0)) = (0, -0.1, 0)$ (*red*)

5 Adaptive Anti-synchronization of the No-Equilibrium System

The most vital practical feature relating to chaotic systems is the possibility of synchronization of two coupled chaotic systems [12, 21, 38, 53]. Synchronization of nonlinear systems has been discovered extensively in literature [17, 24, 39, 59, 70, 80, 83, 84, 91, 106]. Some important obtained results can be listed as follows: synchronized states in a ring of mutually coupled self-sustained nonlinear electrical oscillators [101], ragged synchronizability of coupled oscillators [71], various synchronization phenomena in bidirectionally coupled double-scroll circuits [95], observer for synchronization of chaotic systems with application to secure data transmission was studied in [1], or shape synchronization control [32]. Moreover, various kinds of synchronizations have been reported, for example lag synchronization [60], frequency synchronization [3], projective-anticipating synchronization [30], anti-synchronization [78], adaptive synchronization [86, 93], hybrid chaos synchronization [39], generalized projective synchronization [89], fuzzy control-based function synchronization [15] or fast synchronization [37] etc. It is interesting that fuzzy adaptive synchronization of uncertain fractional-order chaotic systems has been introduced in [14].

In this Section, we study the adaptive anti-synchronization of identical no-equilibrium systems with three unknown parameters. Here the no-equilibrium system in (2) is considered as the master system as

$$\begin{cases} \dot{x}_1 = y_1 \\ \dot{y}_1 = -x_1 - y_1 z_1 \\ \dot{z}_1 = x_1 y_1 + a{x_1}^2 + b{y_1}^2 - c, \end{cases} \tag{10}$$

in which x_1, y_1, z_1 are state variables. The slave system is considered as the controlled no-equilibrium system and its dynamics is described as

$$\begin{cases} \dot{x}_2 = y_2 + u_x \\ \dot{y}_2 = -x_2 - y_2 z_2 + u_y \\ \dot{z}_2 = x_2 y_2 + a{x_2}^2 + b{y_2}^2 - c + u_z, \end{cases} \tag{11}$$

where x_2, y_2, z_2 are the states of the slave system. Here the adaptive controls are u_x, u_y, and u_z. These controls will be designed for the anti-synchronization of the master and slave systems. $A(t)$, $B(t)$ and $C(t)$ are used in order to estimate unknown parameters a, b and c.

The anti-synchronization error between no-equilibrium systems (10) and (11) is given by the following relation

$$\begin{cases} e_x = x_1 + x_2 \\ e_y = y_1 + y_2 \\ e_z = z_1 + z_2. \end{cases} \tag{12}$$

As a result, the anti-synchronization error dynamics is described by

$$\begin{cases} \dot{e}_x = e_y + u_x \\ \dot{e}_y = -e_x - \left(y_1 z_1 + y_2 z_2\right) + u_y \\ \dot{e}_z = \left(x_1 y_1 + x_2 y_2\right) + a\left({x_1}^2 + {x_2}^2\right) + b\left({y_1}^2 + {y_2}^2\right) - 2c + u_z. \end{cases} \tag{13}$$

Our aim is to construct the appropriate controllers u_x, u_y, u_z to stabilize system (13). Therefore, we propose the following controllers for system (13):

$$\begin{cases} u_x = -e_y - k_x e_x \\ u_y = e_x + \left(y_1 z_1 + y_2 z_2\right) - k_y e_y \\ u_z = -\left(x_1 y_1 + x_2 y_2\right) - A(t)\left({x_1}^2 + {x_2}^2\right) \\ \qquad -B(t)\left({y_1}^2 + {y_2}^2\right) + 2C(t) - k_z e_z. \end{cases} \tag{14}$$

in which k_x, k_y, k_z are positive gain constants for each controllers and the estimate values for unknown system parameters are $A(t)$, $B(t)$, and $C(t)$. The update laws for the unknown parameters are determined as

$$\begin{cases} \dot{A} = e_z\left(x_1{}^2 + x_2{}^2\right) \\ \dot{B} = e_z\left(y_1{}^2 + y_2{}^2\right) \\ \dot{C} = -2e_z. \end{cases} \tag{15}$$

Then, the main result of this section will be introduced and proved.

Theorem 15.1 *If the adaptive controller (14) and the updating laws of parameter (15) are chosen, the anti-sychronization between the master system (10) and the slave system (11) is achieved.*

Proof It is noting that the parameter estimation errors $e_a(t)$, $e_b(t)$ and $e_c(t)$ are given as

$$\begin{cases} e_a(t) = a - A(t) \\ e_b(t) = b - B(t) \\ e_c(t) = c - C(t). \end{cases} \tag{16}$$

Differentiating (16) with respect to t, we have

$$\begin{cases} \dot{e}_a(t) = -\dot{A}(t) \\ \dot{e}_b(t) = -\dot{B}(t) \\ \dot{e}_c(t) = -\dot{C}(t). \end{cases} \tag{17}$$

Substituting adaptive control law (14) into (13), the closed-loop error dynamics is defined as

$$\begin{cases} \dot{e}_x = -k_x e_x \\ \dot{e}_y = -k_y e_y \\ \dot{e}_z = (a - A(t))\left(x_1{}^2 + x_2{}^2\right) \\ \qquad + (b - B(t))\left(y_1{}^2 + y_2{}^2\right) - 2(c - C(t)) - k_y e_y \end{cases} \tag{18}$$

Then substituting (16) into (18), we have

$$\begin{cases} \dot{e}_x = -k_x e_x \\ \dot{e}_y = -k_y e_y \\ \dot{e}_z = e_a(t)\left(x_1{}^2 + x_2{}^2\right) + e_b(t)\left(y_1{}^2 + y_2{}^2\right) - 2e_c(t) - k_z e_z. \end{cases} \tag{19}$$

We consider the Lyapunov function given as

$$V(t) = V(e_x, e_y, e_z, e_a, e_b, e_c) = \frac{1}{2}\left(e_x^2 + e_y^2 + e_z^2 + e_a^2 + e_b^2 + e_c^2\right). \tag{20}$$

The Lyapunov function (20) is clearly definite positive.

Taking time derivative of (20) along the trajectories of (12) and (16) we have

$$\dot{V}(t) = e_x\dot{e}_x + e_y\dot{e}_y + e_z\dot{e}_z + e_a\dot{e}_a + e_b\dot{e}_b + e_c\dot{e}_c. \tag{21}$$

From (17), (19), and (21) we get

$$\dot{V}(t) = -k_x e_x^2 - k_y e_y^2 - k_z e_z^2 + e_a\left[e_z\left({x_1}^2 + {x_2}^2\right) - \dot{A}\right] + e_b\left[e_z\left({y_1}^2 + {y_2}^2\right) - \dot{B}\right] - e_c\left(2e_z + \dot{C}\right). \tag{22}$$

Then by applying the parameter update law (15), Eq. (22) become

$$\dot{V}(t) = -k_x e_x^2 - k_y e_y^2 - k_z e_z^2. \tag{23}$$

Obviously, derivative of the Lyapunov function is negative semi-define. According to Barbalat's Lemma in Lyapunov stability theory [40, 63], it follows that $e_x(t) \to 0$, $e_y(t) \to 0$, and $e_z(t) \to 0$ exponentially when $t \to 0$, i.e. anti-synchronization between master and slave system is achieved. This completes the proof. □

We illustrate the proposed anti-synchronization scheme with a numerical example. The parameters of the no-equilibrium systems are selected as $a = 0.5$, $b = 0.1$, $c = 1.3$ and the positive gain constant as $k = 4$. The initial conditions of the master system in (10) and the slave system in (11) have been chosen as $x_1(0) = 0.0$, $y_1(0) = 0.1$, $z_1(0) = 0$, and $x_2(0) = 0.5$, $y_2(0) = 1$, $z_2(0) = 0.8$, respectively. We assumed that the initial values of the parameter estimates are $A(0) = 1$, $B(0) = 0.4$, and $C(0) = 1.5$.

We see that when adaptive control law in (14) and the update law for the parameter estimates in (15) are applied, the anti-synchronization of the master in (10) and slave

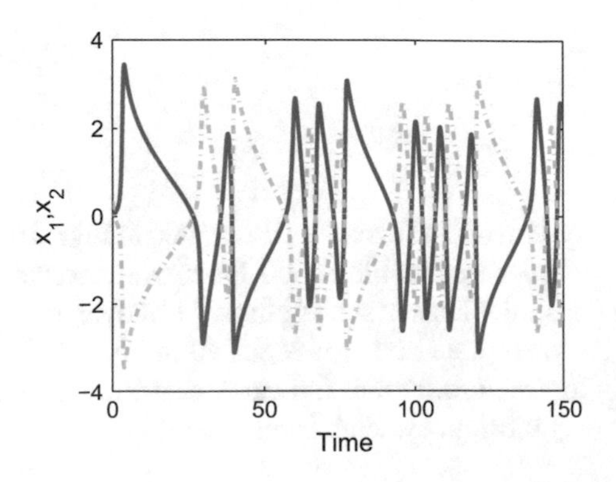

Fig. 10 Anti-synchronization of the states $x_1(t)$ and $x_2(t)$

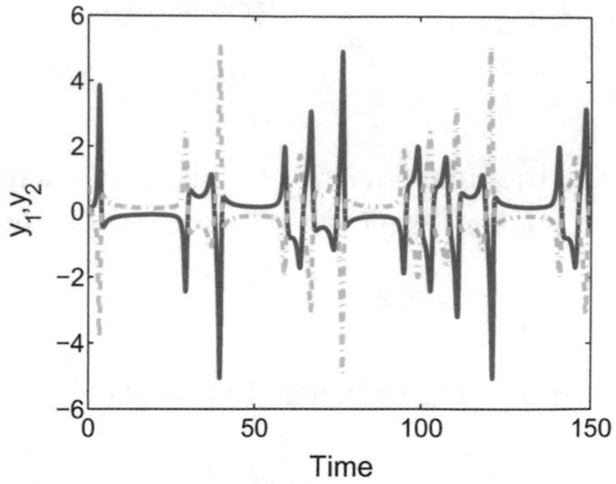

Fig. 11 Anti-synchronization of the states $y_1(t)$ and $y_2(t)$

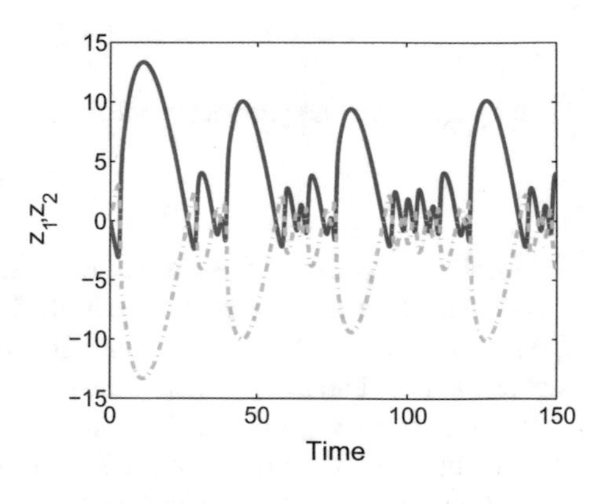

Fig. 12 Anti-synchronization of the states $z_1(t)$ and $z_2(t)$

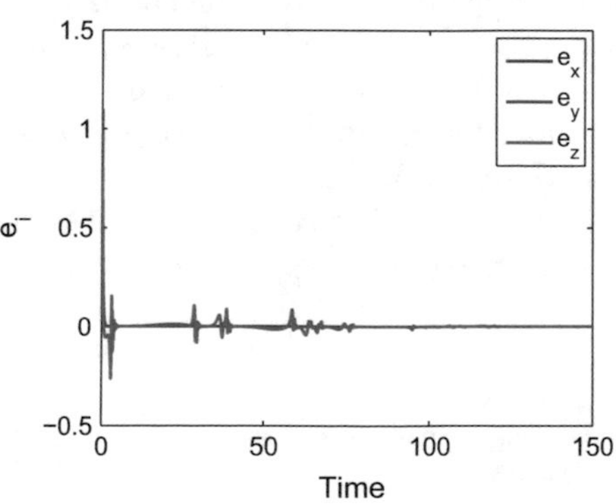

Fig. 13 Time series of the anti-synchronization errors e_x, e_y, and e_z

system in (11) occurs as illustrated in Figs. 10, 11 and 12. Time series of master states are denoted as blue solid lines while corresponding slave states are plotted as red dash-dot lines in such figures. Moreover, the time-history of the anti-synchronization errors e_x, e_y, and e_z is reported in Fig. 13. The anti-synchronization errors converge to the zero, which indicates that the chaos anti-synchronization between the no-equilibrium systems is realized.

6 Circuit Implementation of the No-Equilibrium System

Electronic circuits have been used for emulating theoretical chaotic models [13, 17, 22, 74]. In addition, circuit implementation of chaotic models plays an important role from the point of application view. Circuital realization of chaotic systems has been applied in various engineering fields such as secure communication, signal processing, random bit generator, or path planning for autonomous mobile robot etc. [10, 24, 62, 96, 97, 103].

Therefore, in this section, we introduce an electronic circuit which emulates the theoretical model in (2). By using the operational amplifiers approach [22], the circuit is designed and presented in Fig. 14. The state variables x, y, z of no-equilibrium system in (2) are the voltages across the capacitor C_1, C_2, and C_3, respectively. As seen in Fig. 14, theoretical model in (2) is realized by using only common electronic components such as resistors, capacitors, operational amplifiers and analog multipliers. By applying Kirchhoff's laws to the electronic circuit in Fig. 14, its corresponding circuital equations are derived in the following form

$$\begin{cases} \dfrac{dv_{C_1}}{dt} = \dfrac{1}{R_1C_1}v_{C_2} \\ \dfrac{dv_{C_2}}{dt} = -\dfrac{1}{R_2C_2}v_{C_1} - \dfrac{1}{10R_3C_2}v_{C_2}v_{C_3} \\ \dfrac{dv_{C_3}}{dt} = \dfrac{1}{10R_4C_3}v_{C_1}v_{C_2} + \dfrac{1}{10R_5C_3}v_{C_1}^2 + \dfrac{1}{10R_6C_3}v_{C_2}^2 - \dfrac{1}{R_7C_3}V_c, \end{cases} \tag{24}$$

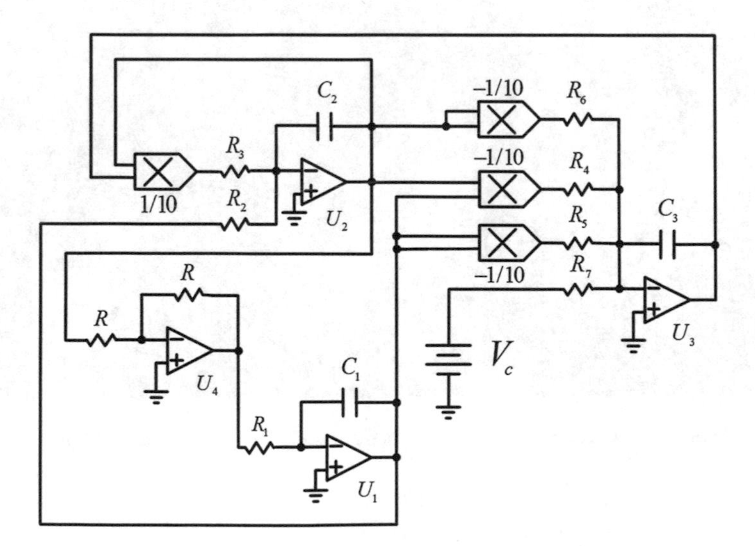

Fig. 14 Schematic of the designed circuit which modelling system without equilibrium in (2)

in which v_{C_1}, v_{C_2}, and v_{C_3} are the voltages across the capacitors C_1, C_2, and C_3, respectively.

In this work, the power supply to all active devices are $\pm 15V_{DC}$ and we use the operational amplifiers TL084. The values of components in Fig. 14 are chosen as follows: $R_1 = R_2 = R_6 = R_7 = R = 10k\Omega, R_3 = R_4 = 1k\Omega, R_5 = 2k\Omega, V_c = 1.3V_{DC}$, and $C_1 = C_2 = C_3 = 10nF$. For the chosen set of components, the values of parameters in system (2) are: $a = 0.5$, $b = 0.1$, and $c = 1.3$.

The designed circuit has implemented in SPICE. The obtained results are reported in Figs. 15 and 16 which display the attractors of the circuit in different phase planes (v_{C_1}, v_{C_2}), (v_{C_1}, v_{C_3}), and (v_{C_2}, v_{C_3}) respectively (Fig. 17). It is easy to see that there is a good agreement between the theoretical attractors (Figs. 1–2) and the circuital ones (Figs. 15 and 16). It can be concluded that the circuit simulations are consistent with the numerical simulations. Moreover, the designed circuit, which is built by using off-the-shelf electronic components, can be applied in practical applications.

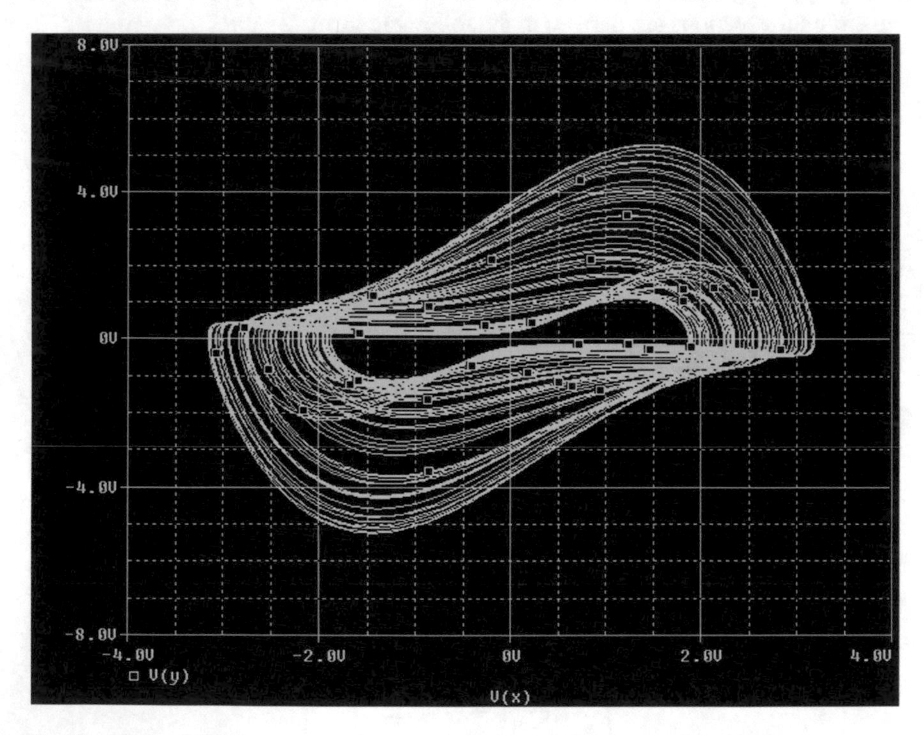

Fig. 15 Obtained SPICE attractor of the designed circuit in the (v_{C_1}, v_{C_2}) phase plane

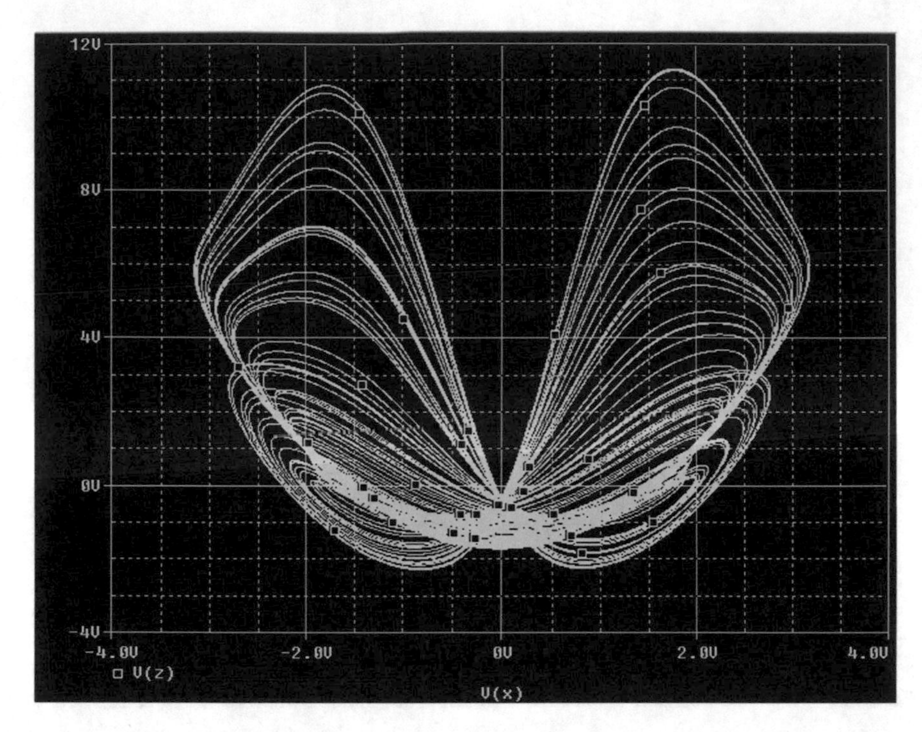

Fig. 16 Obtained SPICE attractor of the designed circuit in the (v_{C_1}, v_{C_3}) phase plane

7 Fractional Order Form of the No-Equilibrium System

As have been known that, practical models such as heat conduction, electrode-electrolyte polarization, electronic capacitors, dielectric polarization, viso-elastic systems are more adequately described by the fractional-order different equations [9, 29, 36, 73, 77, 100]. Existence of chaos in fractional-order systems are investigated [26, 28, 49, 105]. In this section, we consider the fractional-order from of the no-equilibrium system which is described as

$$\begin{cases} \frac{d^q x(t)}{dt^q} = y \\ \frac{d^q y(t)}{dt^q} = -x - yz \\ \frac{d^q z(t)}{dt^q} = xy + ax^2 + by^2 - c, \end{cases} \tag{25}$$

where a, b, c are three positive parameters and $c \neq 0$ for the commensurate order $0 < q \leq 1$. Fractional-order system (25) has been studied by applying Adams–Bashforth-Mounlton numerical algorithm [20, 25, 76]. It is interesting that chaos exists in fractional-order system (25). Figures 18, 19, and 20 display chaotic attractors generated from fractional-order system (25) for the commensurate order $q = 0.99$, the parameters $a = 0.5$, $b = 0.1$, $c = 1.3$ and the initial conditions

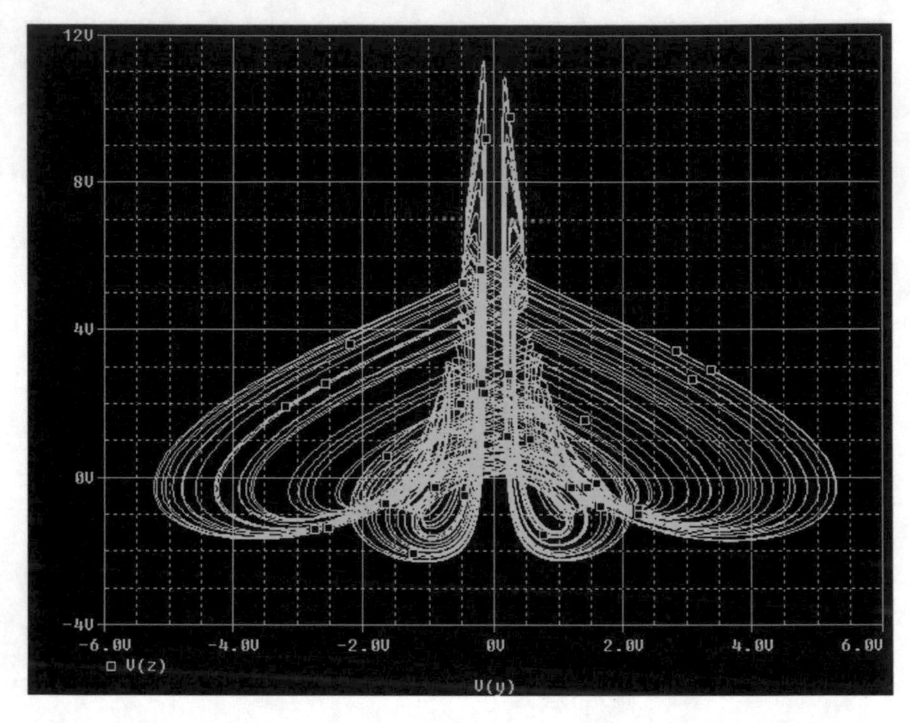

Fig. 17 Obtained SPICE attractor of the designed circuit in the (v_{C_2}, v_{C_3}) phase plane

Fig. 18 2-D projection of the fractional-order system without equilibrium (2) in the (x, y)-plane

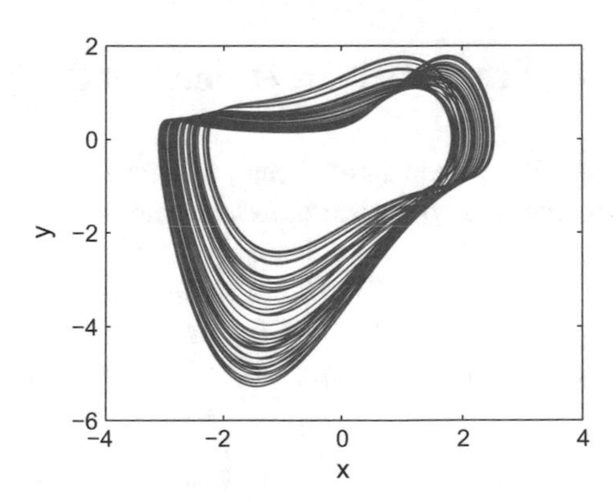

$$(x(0), y(0), z(0)) = (0, 0.1, 0). \tag{26}$$

This research would enable future engineering applications by considering the advantages of the system without equilibrium and the fractional order theory.

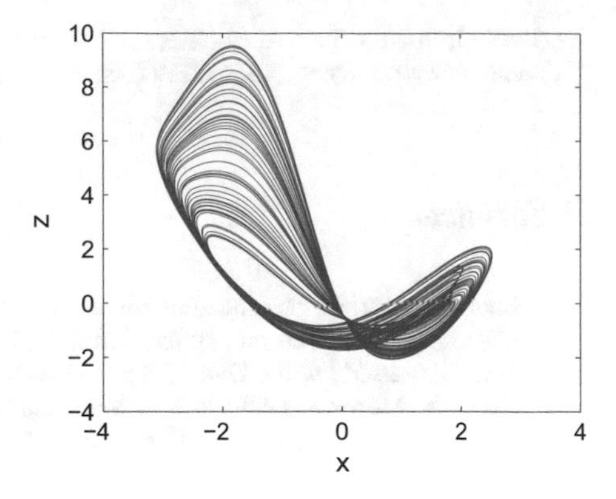

Fig. 19 2-D projection of the fractional-order system without equilibrium (2) in the (x, z)-plane

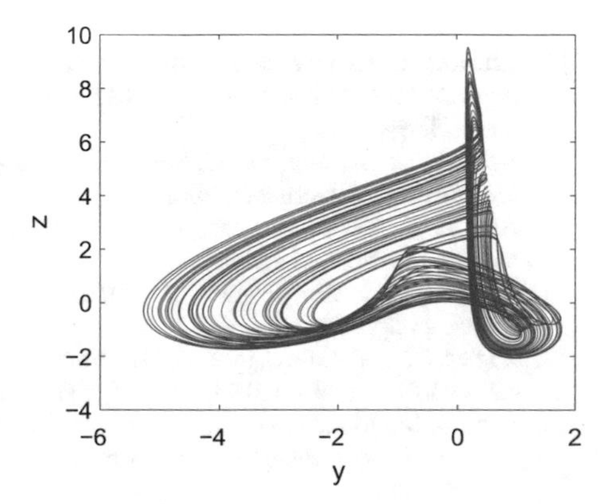

Fig. 20 2-D projection of the fractional-order system without equilibrium (2) in the (y, z)-plane

8 Conclusion

This work introduces a new autonomous chaotic system with special features. There is no any equilibrium points in the proposed system, therefore it is classified as a system with hidden attractor. There is a coexistence of different attractors in the system when changing the values of initial conditions. We have discovered the dynamical properties of such system without equilibrium by using phase portraits, bifurcation diagram, Lyapunov exponents and Kaplan–Yorke dimension. The possibility of synchronization of no-equilibrium systems is studied through an anti-synchronization scheme. The proposed no-equilibrium system are suitable for chaos-based engineering applications because of its complex behavior as well as its feasibility, which has been confirmed by designing an electronic circuit. Fractional order of the proposed system has been given and the result showed that the attractor has no equilibrium.

Potential applications of the proposed system should be investigated. Further studies about fractional-order chaotic systems without equilibrium will be presented in our future works.

Acknowledgements This research is funded by Vietnam National Foundation for Science and Technology Development (NAFOSTED) under grant number 102.02–2012.27

References

1. Aguilar-Lopez, R., Martinez-Guerra, R., & Perez-Pinacho, C. (2014). Nonlinear observer for synchronization of chaotic systems with application to secure data transmission. *European Physics Journal Special Topics*, *223*, 1541–1548.
2. Akgul, A., Moroz, I., Pehlivan, I., & Vaidyanathan, S. (2016). A new four-scroll chaotic attractor and its enginearing applications. *Optik*, *127*, 5491–5499.
3. Akopov, A., Astakhov, V., Vadiasova, T., Shabunin, A., & Kapitaniak, T. (2005). Frequency synchronization in clusters in coupled extended systems. *Physics Letters A*, *334*, 169–172.
4. Arneodo, A., Coullet, P., & Tresser, C. (1981). Possible new strange attractors with spiral structure. *Communications in Mathematical Physics*, *79*, 573–579.
5. Azar, A. T., & Vaidyanathan, S. (2015a). *Chaos modeling and control systems design*. Germany: Springer.
6. Azar, A. T., & Vaidyanathan, S. (2015b). *Computational intelligence applications in modeling and control*. Germany: Springer.
7. Azar, A. T., & Vaidyanathan, S. (2015c). *Handbook of research on advanced intelligent control engineering and automation*. USA: IGI Global.
8. Azar, A. T., & Vaidyanathan, S. (2016). *Advances in chaos theory and intelligent control*. Germany: Springer.
9. Bagley, R. L., & Calico, R. A. (1991). Fractional-order state equations for the control of viscoelastically damped structers. *Journal of Guidance, Control, and Dyanmics*, *14*, 304–311.
10. Barakat, M., Mansingka, A., Radwan, A. G., & Salama, K. N. (2013). Generalized hardware post processing technique for chaos-based pseudorandom number generators. *ETRI Journal*, *35*, 448–458.
11. Barnerjee, T., Biswas, D., & Sarkar, B. C. (2012). Design and analysis of a first order time-delayed chaotic system. *Nonlinear Dynamics*, *70*, 721–734.
12. Boccaletti, S., Kurths, J., Osipov, G., Valladares, D. L., & Zhou, C. S. (2002). The synchronization of chaotic systems. *Physics Reports*, *366*, 1–101.
13. Bouali, S., Buscarino, A., Fortuna, L., Frasca, M., & Gambuzza, L. V. (2012). Emulating complex business cycles by using an electronic analogue. *Nonlinear Analysis: Real World Applications*, *13*, 2459–2465.
14. Boulkroune, A., Bouzeriba, A., Bouden, T., & Azar, A. T. (2016a). Fuzzy adaptive synchronization of uncertain fractional-order chaotic systems. In A. T. Azar & S. Vaidyanathan (Eds.), *Advances in Chaos Theory and Intelligent Control* Studies in Fuzziness and Soft Computing (Vol. 337, pp. 681–697). Germany: Springer.
15. Boulkroune, A., Hamel, S., & Azar, A. T. (2016b). Fuzzy control-based function synchronization of unknown chaotic systems with dead-zone input. In A. T. Azar & S. Vaidyanathan (Eds.), *Advances in chaos theory and intelligent control* Studies in fuzziness and soft computing (Vol. 337, pp. 699–718). Germany: Springer.
16. Brezetskyi, S., Dudkowski, D., & Kapitaniak, T. (2015). Rare and hidden attractors in van der pol-duffing oscillators. *European Physics Journal Special Topics*, *224*, 1459–1467.
17. Buscarino, A., Fortuna, L., & Frasca, M. (2009). Experimental robust synchronization of hyperchaotic circuits. *Physica D*, *238*, 1917–1922.
18. Chen, G., & Ueta, T. (1999). Yet another chaotic attractor. *International Journal of Bifurcation and Chaos*, *9*, 1465–1466.
19. Chen, G., & Yu, X. (2003). *Chaos control: theory and applications*. Berlin: Springer.
20. Diethelm, K., Ford, N. J., & Freed, A. D. (2002). A predictor-corrector approach for the numerical solution of fractional differential equations. *Nonlinear Dynamics*, *29*, 3–22.

21. Fortuna, L., & Frasca, M. (2007). Experimental synchronization of single-transistor-based chaotic circuits. *Chaos, 17*, 043118-1–5.
22. Fortuna, L., Frasca, M., & Xibilia, M. G. (2009). *Chua's circuit implementation: Yesterday*. World Scientific, Singapore: Today and Tomorrow.
23. Frederickson, P., Kaplan, J. L., Yorke, E. D., & York, J. (1983). The lyapunov dimension of strange attractors. *Journal of Differential Equations, 49*, 185–207.
24. Gamez-Guzman, L., Cruz-Hernandez, C., Lopez-Gutierrez, R., & Garcia-Guerrero, E. E. (2009). Synchronization of chua's circuits with multi-scroll attractors: Application to communication. *Communications in Nonlinear Science and Numerical Simulation, 14*, 2765–2775.
25. Gejji, D., & Jafari, H. (2005). Adomian decomposition: A tool for solving a system of fractional differential equations. *Journal of Mathematical Analysis and Applications, 301*, 508–518.
26. Grigorenko, I., & Grigorenko, E. (2003). Chaotic dynamics of the fractional-order lorenz system. *Physics Review Letters, 91*, 034101.
27. Han, F., Hu, J., Yu, X., & Wang, Y. (2007). Fingerprint images encryption via multi-scroll chaotic attractors. *Applied Mathematics and Computing, 185*, 931–939.
28. Hartley, T. T., Lorenzo, C. F., & Qammer, H. K. (1995). Chaos on a fractional Chua's system. *IEEE Transactions on Circuits System I: Fundamental Theory and Applications, 42*, 485–490.
29. Heaviside, O. (1971). *Electromagnetic theory*. New York, USA: Academic Press.
30. Hoang, T. M., & Nakagawa, M. (2007). Anticipating and projective–anticipating synchronization of coupled multidelay feedback systems. *Physics Letters A, 365*, 407–411.
31. Hoang, T. M., & Nakagawa, M. (2008). A secure communication system using projective-lag and/or projective-anticipating synchronizations of coupled multidelay feedback systems. *Chaos, Solitons & Fractals, 38*, 1423–1438.
32. Huang, Y., Wang, Y., Chen, H., & Zhang, S. (2016). Shape synchronization control for three-dimensional chaotic systems. *Chaos, Solitons & Fractals, 87*, 136–145.
33. Jafari, S., & Sprott, J. C. (2013). Simple chaotic flows with a line equilibrium. *Chaos, Solitons & Fractals, 57*, 79–84.
34. Jafari, S., Sprott, J. C., & Golpayegani, S. M. R. H. (2013). Elementary quadratic chaotic flows with no equilibria. *Physics Letters A, 377*, 699–702.
35. Jafari, S., Sprott, J. C., & Nazarimehr, F. (2015). Recent new examples of hidden attractors. *European Physics Journal Special Topics, 224*, 1469–1476.
36. Jenson, V. G., & Jeffreys, G. V. (1997). *Mathematical methods in chemical enginerring*. New York, USA: Academic Press.
37. Kajbaf, A., Akhaee, M. A., & Sheikhan, M. (2016). Fast synchronization of non-identical chaotic modulation-based secure systems using a modified sliding mode controller. *Chaos, Solitons & Fractals, 84*, 49–57.
38. Kapitaniak, T. (1994). Synchronization of chaos using continuous control. *Physical Review E, 50*, 1642–1644.
39. Karthikeyan, R., & Vaidyanathan, S. (2014). Hybrid chaos synchronization of four-scroll systems via active control. *Journal of Electrical Engineering, 65*, 97–103.
40. Khalil, H. (2002). *Nonlinear systems*. New Jersey, USA: Prentice Hall.
41. Kuznetsov, N. V., Leonov, G. A., & Seledzhi, S. M. (2011). Hidden oscillations in nonlinear control systems. *IFAC Proceedings, 18*, 2506–2510.
42. Leonov, G. A., & Kuznetsov, N. V. (2011a). Algorithms for searching for hidden oscillations in the Aizerman and Kalman problems. *Doklady Mathematics, 84*, 475–481.
43. Leonov, G. A., & Kuznetsov, N. V. (2011b). Analytical–numerical methods for investigation of hidden oscillations in nonlinear control systems. *IFAC Proceedings, 18*, 2494–2505.
44. Leonov, G. A., & Kuznetsov, N. V. (2013). Hidden attractors in dynamical systems: From hidden oscillation in Hilbert-Kolmogorov, Aizerman and Kalman problems to hidden chaotic attractor in Chua circuits. *International Journal of Bifurcation and Chaos, 23*, 1330002.
45. Leonov, G. A., Kuznetsov, N. V., Kiseleva, M. A., Solovyeva, E. P., & Zaretskiy, A. M. (2014). Hidden oscillations in mathematical model of drilling system actuated by induction motor with a wound rotor. *Nonlinear Dynamics, 77*, 277–288.

46. Leonov, G. A., Kuznetsov, N. V., Kuznetsova, O. A., Seldedzhi, S. M., & Vagaitsev, V. I. (2011a). Hidden oscillations in dynamical systems. *Transactions on Systems and Control*, *6*, 54–67.
47. Leonov, G. A., Kuznetsov, N. V., & Vagaitsev, V. I. (2011b). Localization of hidden Chua's attractors. *Physics Lett. A*, *375*, 2230–2233.
48. Leonov, G. A., Kuznetsov, N. V., & Vagaitsev, V. I. (2012). Hidden attractor in smooth Chua system. *Physica D*, *241*, 1482–1486.
49. Li, C. P., & Peng, G. J. (2004). Chaos in Chen's system with a fractional-order. *Chaos, Solitons & Fractals*, *20*, 443–450.
50. Lorenz, E. N. (1963). Deterministic non-periodic flow. *Journal of Atmospheric Science*, *20*, 130–141.
51. Lü, J., & Chen, G. (2002). A new chaotic attractor coined. *International Journal of Bifurcation and Chaos*, *12*, 659–661.
52. Ojoniyi, O. S., & Njah, A. N. (2016). A 5D hyperchaotic Sprott B system with coexisting hidden attractor. *Chaos, Solitons & Fractals*, *87*, 172–181.
53. Pecora, L. M., & Carroll, T. L. (1990). Synchronization in chaotic signals. *Physics Review A*, *64*, 821–824.
54. Pham, V.-T., Jafari, S., Volos, C., Wang, X., & Golpayegani, S. M. R. H. (2014a). Is that really hidden? The presence of complex fixed-points in chaotic flows with no equilibria. *International Journal of Bifurcation and Chaos*, *24*, 1450146.
55. Pham, V.-T., Vaidyanathan, S., Volos, C. K., Hoang, T. M., & Yem, V. V. (2016). Dynamics, synchronization and SPICE implementation of a memristive system with hidden hyperchaotic attractor. In A. T. Azar & S. Vaidyanathan (Eds.), *Advances in chaos theory and intelligent control* Studies in Fuzziness and Soft Computing (Vol. 337, pp. 35–52). Germany: Springer.
56. Pham, V. T., Vaidyanathan, S., Volos, C. K., & Jafari, S. (2015a). Hidden attractors in a chaotic system with an exponential nonlinear term. *European Physics Journal Special Topics*, *224*, 1507–1517.
57. Pham, V.-T., Volos, C., & Gambuzza, L. V. (2014). A memristive hyperchaotic system without equilibrium. *Scientific World Journal*, *2014*, 368986.
58. Pham, V.-T., Volos, C., & Vaidyanathan, S. (2015b). Multi-scroll chaotic oscillator based on a first-order delay differential equation. In A. T. Azar & S. Vaidyanathan (Eds.), *Chaos modelling and control systems design* Studies in Computational Intelligence (Vol. 581, pp. 59–72). Germany: Springer.
59. Pham, V.-T., Volos, C. K., Jafari, S., Wei, Z., & Wang, X. (2014c). Constructing a novel no-equilibrium chaotic system. *International Journal of Bifurcation and Chaos*, *24*, 1450073.
60. Rosenblum, M. G., Pikovsky, A. S., & Kurths, J. (1997). From phase to lag synchronization in coupled chaotic oscillators. *Physics Review Letters*, *78*, 4193–4196.
61. Rössler, O. E. (1976). An equation for continuous chaos. *Physics Letters A*, *57*, 397–398.
62. Sadoudi, S., Tanougast, C., Azzaz, M. S., & Dandache, A. (2013). Design and FPGA implementation of a wireless hyperchaotic communication system for secure realtime image transmission. *EURASIP Journal of Image and Video Processing*, *943*, 1–18.
63. Sastry, S. (1999). *Nonlinear systems: Analysis, stability, and control*. USA: Springer.
64. Shahzad, M., Pham, V. T., Ahmad, M. A., Jafari, S., & Hadaeghi, F. (2015). Synchronization and circuit design of a chaotic system with coexisting hidden attractors. *European Physics Journal Special Topics*, *224*, 1637–1652.
65. Sharma, P. R., Shrimali, M. D., Prasad, A., Kuznetsov, N. V., & Leonov, G. A. (2015). Control of multistability in hidden attractors. *European Physics Journal Special Topics*, *224*, 1485–1491.
66. Shilnikov, L. P., Shilnikov, A. L., Turaev, D. V., & Chua, L. O. (1998). *Methods of qualitative theory in nonlinear dynamics*. Singapore: World Scientific.
67. Sprott, J. C. (2003). *Chaos and times-series analysis*. Oxford: Oxford University Press.
68. Sprott, J. C. (2010). *Elegant chaos: Algebraically simple chaotic flows*. Singapore: World Scientific.

69. Sprott, J. C. (2015). Strange attractors with various equilibrium types. *European Physics Journal Special Topics*, *224*, 1409–1419.
70. Srinivasan, K., Senthilkumar, D. V., Murali, K., Lakshmanan, M., & Kurths, J. (2011). Synchronization transitions in coupled time-delay electronic circuits with a threshold nonlinearity. *Chaos*, *21*, 023119.
71. Stefanski, A., Perlikowski, P., & Kapitaniak, T. (2007). Ragged synchronizability of coupled oscillators. *Physics Review E*, *75*, 016210.
72. Strogatz, S. H. (1994). *Nonlinear Dynamics and chaos: With applications to physics, biology, chemistry, and engineering*. Massachusetts: Perseus Books.
73. Sun, H. H., Abdelwahad, A. A., & Onaral, B. (1894). Linear approximation of transfer function with a pole of fractional-order. *IEEE Transactions on Automatic Control*, *29*, 441–444.
74. Sundarapandian, V., & Pehlivan, I. (2012). Analysis, control, synchronization, and circuit design of a novel chaotic system. *Mathematical Computational Modelling*, *55*, 1904–1915.
75. Tacha, O. I., Volos, C. K., Kyprianidis, I. M., Stouboulos, I. N., Vaidyanathan, S., & Pham, V. T. (2016). Analysis, adaptive control and circuit simulatio of a novel nonlineaar finance system. *Applied Mathematics and Computation*, *276*, 200–217.
76. Tavazoei, M. S., & Haeri, M. (2008). Limitations of frequency domain approximation for detecting chaos in fractional-order systems. *Nonlinear Analysis*, *69*, 1299–1320.
77. Tavazoei, M. S., & Haeri, M. (2009). A proof for non existence of periodic solutions in time invariant fractional-order systems. *Automatica*, *45*, 1886–1890.
78. Vaidyanathan, S. (2012). Anti-synchronization of four-wing chaotic systems via sliding mode control. *International Journal of Automation and Computing*, *9*, 274–279.
79. Vaidyanathan, S. (2013). A new six-term 3-D chaotic system with an exponential nonlineariry. *Far East Journal of Mathematical Sciences*, *79*, 135–143.
80. Vaidyanathan, S. (2014). Analysis and adaptive synchronization of eight-term novel 3-D chaotic system with three quadratic nonlinearities. *The European Physical Journal Special Topics*, *223*, 1519–1529.
81. Vaidyanathan, S. (2016). Analysis, control and synchronization of a novel 4-D highly hyperchaotic system with hidden attractors. In A. T. Azar & S. Vaidyanathan (Eds.), *Advances in chaos theory and intelligent control* Studies in Fuzziness and Soft Computing (Vol. 337, pp. 529–552). Germany: Springer.
82. Vaidyanathan, S., & Azar, A. T. (2015a). Analysis, control and synchronization of a nine-term 3-D novel chaotic system. In A. T. Azar & S. Vaidyanathan (Eds.), *Chaos Modelling and Control Systems Design* Studies in Computational Intelligence (Vol. 581, pp. 19–38). Germany: Springer.
83. Vaidyanathan, S., & Azar, A. T. (2015b). Anti-synchronization of identical chaotic systems using sliding mode control and an application to Vaidhyanathan-Madhavan chaotic systems. *Studies in Computational Intelligence*, *576*, 527–547.
84. Vaidyanathan, S., & Azar, A. T. (2015c). Hybrid synchronization of identical chaotic systems using sliding mode control and an application to Vaidhyanathan chaotic systems. *Studies in Computational Intelligence*, *576*, 549–569.
85. Vaidyanathan, S., & Azar, A. T. (2016a). A novel 4-D four-wing chaotic system with four quadratic nonlinearities and its synchronization via adaptive control method. In A. T. Azar & S. Vaidyanathan (Eds.), *Advances in Chaos Theory and Intelligent Control* Studies in Fuzziness and Soft Computing (Vol. 337, pp. 203–224). Germany: Springer.
86. Vaidyanathan, S., & Azar, A. T. (2016b). Adaptive backstepping control and synchronization of a novel 3-D jerk system with an exponential nonlinearity. In A. T. Azar & S. Vaidyanathan (Eds.), *Advances in Chaos Theory and Intelligent Control* Studies in Fuzziness and Soft Computing (Vol. 337, pp. 249–274). Germany: Springer.
87. Vaidyanathan, S., & Azar, A. T. (2016c). Adaptive control and synchronization of Halvorsen circulant chaotic systems. In A. T. Azar & S. Vaidyanathan (Eds.), *Advances in Chaos Theory and Intelligent Control* Studies in Fuzziness and Soft Computing (Vol. 337, pp. 225–247). Germany: Springer.

88. Vaidyanathan, S., & Azar, A. T. (2016d). Dynamic analysis, adaptive feedback control and synchronization of an eight-term 3-D novel chaotic system with three quadratic nonlinearities. In A. T. Azar & S. Vaidyanathan (Eds.), *Advances in Chaos Theory and Intelligent Control* Studies in Fuzziness and Soft Computing (Vol. 337, pp. 155–178). Germany: Springer.
89. Vaidyanathan, S., & Azar, A. T. (2016e). Generalized projective synchronization of a novel hyperchaotic four-wing system via adaptive control method. In A. T. Azar & S. Vaidyanathan (Eds.), *Advances in Chaos Theory and Intelligent Control* Studies in Fuzziness and Soft Computing (Vol. 337, pp. 275–296). Germany: Springer.
90. Vaidyanathan, S., & Azar, A. T. (2016f). Qualitative study and adaptive control of a novel 4-D hyperchaotic system with three quadratic nonlinearities. In A. T. Azar & S. Vaidyanathan (Eds.), *Advances in Chaos Theory and Intelligent Control* Studies in Fuzziness and Soft Computing (Vol. 337, pp. 179–202). Germany: Germany.
91. Vaidyanathan, S., Idowu, B. A., & Azar, A. T. (2015a). Backstepping controller design for the global chaos synchronization of Sprott's jerk systems. *Studies in Computational Intelligence, 581*, 39–58.
92. Vaidyanathan, S., Pham, V. T., & Volos, C. K. (2015b). A 5-d hyperchaotic rikitake dynamo system with hidden attractors. *The European Physical Journal Special Topics, 224*, 1575–1592.
93. Vaidyanathan, S., Volos, C., Pham, V. T., Madhavan, K., & Idowo, B. A. (2014). Adaptive backstepping control, synchronization and circuit simualtion of a 3-D novel jerk chaotic system with two hyperbolic sinusoidal nonlinearities. *Archives of Control Sciences, 33*, 257–285.
94. Vaidyanathan, S., Volos, C. K., & Pham, V. T. (2015c). Analysis, control, synchronization and spice implementation of a novel 4-d hyperchaotic rikitake dynamo system without equilibrium. *Journal of Engineering Science and Technology Review, 8*, 232–244.
95. Volos, C. K., Kyprianidis, I. M., & Stouboulos, I. N. (2011). Various synchronization phenomena in bidirectionally coupled double scroll circuits. *Communications in Nonlinear Science and Numerical Simulation, 71*, 3356–3366.
96. Volos, C. K., Kyprianidis, I. M., & Stouboulos, I. N. (2012). A chaotic path planning generator for autonomous mobile robots. *Robotics and Automation Systems, 60*, 651–656.
97. Volos, C. K., Kyprianidis, I. M., & Stouboulos, I. N. (2013). Image encryption process based on chaotic synchronization phenomena. *Signal Processing, 93*, 1328–1340.
98. Wang, X., & Chen, G. (2013). Constructing a chaotic system with any number of equilibria. *Nonlinear Dynamics, 71*, 429–436.
99. Wei, Z. (2011). Dynamical behaviors of a chaotic system with no equilibria. *Physics Letters A, 376*, 102–108.
100. Westerlund, S., & Ekstam, L. (1994). Capacitor theory. *IEEE Transactions on Dielectrics and Electrical Insulation, 1*, 826–839.
101. Woafo, P., & Kadji, H. G. E. (2004). Synchronized states in a ring of mutually coupled self-sustained electrical oscillators. *Physical Review E, 69*, 046206.
102. Wolf, A., Swift, J. B., Swinney, H. L., & Vastano, J. A. (1985). Determining Lyapunov exponents from a time series. *Physica D, 16*, 285–317.
103. Yalcin, M. E., Suykens, J. A. K., & Vandewalle, J. (2004). True random bit generation from a double-scroll attractor. *IEEE Transactions on Circuits Systems I, Regular Papers, 51*, 1395–1404.
104. Yalcin, M. E., Suykens, J. A. K., & Vandewalle, J. (2005). *Cellular neural networks*. World Scientific, Singapore: Multi-Scroll Chaos and Synchronization.
105. Yang, Q. G., & Zeng, C. B. (2010). Chaos in fractional conjugate lorenz system and its scaling attractor. *Communications in Nonlinear Science and Numerical Simulation, 15*, 4041–4051.
106. Zhu, Q., & Azar, A. T. (2015). *Complex system modelling and control through intelligent soft computations*. Germany: Springer.
107. Zhusubaliyev, Z. T., & Mosekilde, E. (2015). Multistability and hidden attractors in a multilevel DC/DC converter. *Mathematics and Computers in Simulation, 109*, 32–45.

Comparison of Three Different Synchronization Schemes for Fractional Chaotic Systems

S.T. Ogunjo, K.S. Ojo and I.A. Fuwape

Abstract The importance of synchronization schemes in natural and physical systems including communication modes has made chaotic synchronization an important tool for scientist. Synchronization of chaotic systems are usually conducted without considering the efficiency and robustness of the scheme used. In this work, performance evaluation of three different synchronization schemes: Direct Method, Open Plus Closed Loop (OPCL) and Active control is investigated. The active control technique was found to have the best stability and error convergence. Numerical simulations have been conducted to assert the effectiveness of the proposed analytical results.

1 Introduction

Strogatz [40] defined chaos as the aperiodic long term behaviour in a deterministic system that exhibit sensitive dependence on initial conditions. Using Lyapunov exponents, a chaotic system is one with at least one positive Lyapunov exponent. A system with more than one positive Lyapunov exponent is referred to as an hyperchaotic system. Since the proposition of the first chaotic system by Lorenz [19], the study of chaos has evolved due to development of high computing resources and mathematical procedures for analysis [39]. Chaotic systems has been developed in the form of maps, ordinary differential equations, partial differential equations and fractional order differential equations and presence of chaos investigated. Due to complex nature of natural systems, the study of chaos has been extended to time

S.T. Ogunjo (✉) · I.A. Fuwape
Federal University of Technology, Akure, Ondo State, Nigeria
e-mail: stogunjo@futa.edu.ng

I.A. Fuwape
e-mail: iafuwape@futa.edu.ng

K.S. Ojo
University of Lagos, Lagos, Nigeria
e-mail: kaojo@unilag.edu.ng

A.T. Azar et al. (eds.), *Fractional Order Control and Synchronization of Chaotic Systems*, Studies in Computational Intelligence 688,
DOI 10.1007/978-3-319-50249-6_16

series analysis in natural systems with the development of appropriate tools [25]. The sensitivity of chaotic system to initial conditions implies that two more systems with different initial conditions will exhibit different dynamics. However, with the addition of appropriate functions, trajectories of similar or different chaotic systems can be made to coincide [24]. This is referred to as synchronization.

Synchronization of chaos refers to a process wherein two (or many) chaotic systems (either equivalent or nonequivalent) adjust a given property of their motion to a common behavior due to a coupling or to a forcing (periodical or noisy) [5]. The first evidence of synchronization phenomenon was given by Huygen's pendulum clocks [17] while the first synchronization of chaotic system was proposed by Pecora and Carroll [37]. Since the pioneering work of Pecora and Carroll [37], the study of chaos synchronization has gained a lot of interest because of its applications. Several methods of secure communication and encryption has been proposed based on chaos synchronization [30]. The principle assumes that communication between two persons X and Y embedded in a chaotic signal can only be retrieved if the right system parameters (keys) are known. A practical demonstration of secure communication is presented in Strogatz [40].

As the study of chaos synchronization evolves, several types of synchronization such as generalized synchronization [35], lag synchronization [20], complete synchronization, phase synchronization and projective synchronization [32], modified and function projective synchronization [18], etc. have been discovered. In order to achieve any of these type of synchronization, different synchronization techniques such as backstepping [31], active control, direct method [42], Open Plus Close Loop (OPCL) [16] etc. have been developed and implemented. Early studies of different types of synchronization using any of the mentioned techniques on dynamical systems usually involves two systems. Over the years, real life applications of synchronization requires the synchronization of different systems and a given number of systems higher than the traditional two systems. This has given rise to reduced and increased order synchronization [24, 35], combination synchronization [26, 33, 34], combination-combination synchronization [27, 29] and compound combination synchronization [28].

The Caputo's definition of fractional order differ-integral equations are given as

$$ {}_{a}^{C}D_{t}^{\alpha} f(t) = \frac{1}{\Gamma(\alpha - m)} \int_{a}^{t} \frac{f^{(m)}\tau}{(t-\tau)^{\alpha+1-m}} d\tau \tag{1} $$

where $m - 1 < \alpha \leq m \; \varepsilon \; \mathbb{N}$ and $\alpha \; \varepsilon \; \mathbb{R}$ is a fractional order of the differ-integral of the function $f(t)$ [10]. Applications of fractional order are found in transmission line theory, chemical analysis of aqueous solutions, design of heat-flux meters, rheology of soils, growth of intergranular grooves on metal surfaces, quantum mechanical calculations, and dissemination of atmospheric pollutants [7]. Analysis of football player's motion has been analysed using fractional calculus [9]. Several chaotic fractional order systems have been proposed, these include: fractional order Lorenz system, fractional order Chua system, fractional order memristor based system, fractional order Duffing system, fractional order Chen system etc. There is a growing interest in fractional order systems due to its many applications in control and natural systems.

The use of Grunwald-Letnikov's definition for solving fractional order differential equations is described by Concepcion et al. [8] and stated here.

Using the approximation

$$\mathfrak{D}^{\alpha} f(t) \approx \Delta_h^{\alpha} f(t) \tag{2}$$

$$\Delta_h^{\alpha} f(t)|_{t=kh} = h^{-\alpha} \sum_{j=0}^{k} (-1)^j \binom{\alpha}{j} f(kh - jh). \tag{3}$$

For a system given by $a\mathfrak{D}^{\alpha}u(t) + bu(t) = q(t)$, with $a = 1$ and zero initial conditions

$$h^{-\alpha} \sum_{j=0}^{k} w_j^{(\alpha)} y_{k-j} + b y_k = q_k \tag{4}$$

where $t_k = kh,\ y_k = y(t_k),\ y_0 = 0,\ q_k = q(t_k),\ k = 0, 1, 2, \cdots$, and

$$w_j^{(\alpha)} = (-1)^j \binom{\alpha}{j} \tag{5}$$

the numerical solution is then obtained using

$$y_k = -bh^{\alpha} y_{k-1} - \sum_{j=1}^{k} w_j^{(\alpha)} y_{k-j} + h^{\alpha} q_k \tag{6}$$

Synchronization of fractional order systems have been conducted by many researchers. Synchronization of a system consisting of multiple drive and one response was carried out in Zhou et al. [44]. Design, realization, control and synchronization of a novel 4D hyperchaotic fractional order system was carried out using time-delayed feedback control [11]. Generalized synchronization of a novel fractional order chaotic system in different order and dimension has been investigated with success using nonlinear feedback control [43].

In realization of chaos synchronization for real life application such as communication systems, it is intuitive to choose a method and technique which will minimize cost and error while giving the desired robust outputs. Ojo et al. [31] compared the backstepping and active control technique for complete synchronization of chaotic systems. From their results, active control transient error dynamics convergence and synchronization time are achieved faster via the backstepping than that of the active control technique but the control function obtained via the active control is simpler with a more stable synchronization time and hence, it is more suitable for practical implementation. There is the need to investigate an efficient and robust method of synchronization in light of growing interest in fractional order chaotic systems. The aim of this chapter is to compare the performance of three different techniques for complete synchronization of an hyperchaotic fractional order chaotic system. System performance will be investigated using both linear and nonlinear tools.

2 Related Work

Comparison of two different synchronization scheme was carried out on integer order chaotic systems [31]. Recent advances in synchronization of fractional chaotic systems has seen results such as hybrid synchronization [36], exponential synchronization with mixed uncertainties [22], combination-combination synchronization [21], synchronization of nonidentical systems using modified active control [13], synchronization of fractional order switching chaotic system [15], synchronization of fractional order hyperchaotic systems using a new adaptive sliding mode control [23], combination synchronization using nonlinear feedback control method [3], reduced order synchronization of fractional order systems using adaptive control [2], fuzzy adaptive synchronization [6] and robust methods [36] have been reported. Circuit realization of a fractional order chaotic systems has also been implemented [11].

3 Synchronization Methods

A mathematical definition of synchronization was proposed by Wu and Chua [42]. Two systems $\dot{x} = f(x, y, t)$ and $\dot{y} = g(x, y, t)$ are uniform-synchronized with error bound ε if there exist $\delta > 0$ and $T \geq 0$ such that

$$\|x_{i,j}(t_0) - y_{k,l}(t_0)\| \leq \delta \tag{7}$$

In order to achieve this, several techniques have been proposed. In the following subsections, three of the popular techniques are discussed.

3.1 Direct Method

The mathematical definition of Lyapunov Direct Method was given by Wu and Chen [42] and is stated here. Consider the systems $\dot{x} = f(x, t)$ and $\dot{y} = f(y, t)$. Supposed that there exist a Lyapunov function $V(t, x, y)$ such that for all $t \geq t_0$

$$a(\|x - y\|) \leq V(t, x, y) \leq b(\|x - y\|)$$

where $a(\cdot)$ and $b(\cdot)$ are functions. Supposed that there exist $\mu > 0$ such that for all $t > t_0$ and $\|x - y\| \geq \mu$

$$\dot{V}(t, x, y) \leq -c$$

for some constant $c > 0$ where $\dot{V}(t, x, y)$ is the generalized derivative of V along the trajectories of the systems Wu and Chua [42]. The Lyapunov direct method has been used successfully for the synchronization of ...., anti-synchronization etc.

3.2 *Open Plus Closed Loop (OPCL)*

The method of Open Plus Closed Loop was proposed by Grosu [16]. The method has been used for robust synchronization [14].

Consider a drive system $\dot{y} = F(y)$ and a response system given by $\dot{x} = F(x) + D(x, g)$ where $x, y \varepsilon \mathfrak{R}^n$ and $g = \alpha y$, α is a constant. The goal is to satisfy the condition

$$\lim_{t \to \infty} (x(t) - g(t)) = 0$$

From the OPCL theory, there exist an open-loop action, H given by

$$H(g, dg/dt, t) = \frac{dg}{dt} - F(g, t)$$

and a linear feedback (closed-loop), K given as

$$K(g, x, t) = \left(\frac{dF}{dg} - A \right) [g(t) - x(t)]$$

where $g(t) \varepsilon \mathfrak{R}^n$ is an arbitrary smooth function and A is an arbitrary constant Hurwitx matrix with negative real part [14]. The driving term D, can be written as the sum of the open-loop and closed-loop as

$$D = \frac{dg}{dt} - F(g, t) + \left(\frac{dF}{dg} - A \right) [g(t) - x(t)] \tag{8}$$

3.3 *Active Control*

The Active Control method of synchronization was proposed by Bai and Lonngren [4]. Considering a drive system $\dot{x} = f(x)$ and a response defined as $\dot{y} = f(y) + u(t)$, where $u(t)$ are the control functions. Defining the error function as

$$\lim_{t \to \infty} \|e(t)\| = \lim_{t \to \infty} \|f(x) - f(y)\| = 0 \tag{9}$$

we define a subcontroller $v(t) = -\mathbb{K}e$, where $\mathbb{K}$ is a linear controller gain for control of response feedback strength. The error term can be written as

$$\dot{e}(t) = A_i e(t) + v(t)$$

where A_i are residuals of the system parameters. Substituting the values of $v(t)$m, we obtain

$$\dot{e}(t) = Ze(t) \tag{10}$$

where $Z = (A_i - K)$. If all the eigenvalues of the matrix Z have negative real parts, it is an Hurwitz matrix, which implies that the zero solution of the closed loop system is globally asymptotically stable [1].

4 System Description

The Lorenz system was proposed by Lorenz [19] and can be regarded as the first deterministic system. It is a 3D autonomous system given by

$$\begin{aligned} \dot{x}_1 &= a(x_2 - x_1) \\ \dot{x}_2 &= cx_1 - x_2 - x_1 x_3 \\ \dot{x}_3 &= x_1 x_2 - bx_3 \end{aligned} \tag{11}$$

The system has been found to be chaotic when $a = 10$, $b = 8/3$, $c = 28$ with Lyapunov exponents 1.49, 0, −22.46 indicating a strange attractor. Chaotic synchronization of the Lorenz system has been done using different techniques such as increased and reduced order using Active control [24], complete synchronization using OPCL [14].

Gao et al. [12] introduced the 3D fractional order chaotic Lorenz system with order 0.98.

$$\begin{aligned} \frac{d^{q_1} x_1}{dt^{q_1}} &= a(x_2 - x_1) \\ \frac{d^{q_2} x_2}{dt^{q_2}} &= cx_1 - x_2 - x_1 x_3 \\ \frac{d^{q_3} x_3}{dt^{q_3}} &= x_1 x_2 - bx_3 \end{aligned} \tag{12}$$

By adding a nonlinear term $\dot{x}_4 = -x_2 x_3 + rx_4$ to Eq. 11, a new system given by

$$\begin{aligned}
\dot{x}_1 &= a(x_2 - x_1) + x_4 \\
\dot{x}_2 &= cx_1 - x_2 - x_1x_3 \\
\dot{x}_3 &= x_1x_2 - bx_3 \\
\dot{x}_4 &= -x_2x_3 + rx_4
\end{aligned} \tag{13}$$

was obtained. The system was found to be hyperchaotic when $r = -1$.

A 4D hyperchaotic fractional system developed based on Eqs. (12) and (13) system will be used in this paper

$$\begin{aligned}
\frac{d^{q_1}x_1}{dt^{q_1}} &= a(x_2 - x_1) + x_4 \\
\frac{d^{q_2}x_2}{dt^{q_2}} &= cx_1 - x_2 - x_1x_3 \\
\frac{d^{q_3}x_3}{dt^{q_3}} &= x_1x_2 - bx_3 \\
\frac{d^{q_4}x_4}{dt^{q_4}} &= -x_2x_3 + rx_4
\end{aligned} \tag{14}$$

where the parameters are chosen as $a = 10$, $b = 8/3$, $c = 28$, $r = -1$. The system has Lyapunov exponents $\lambda_1 = 0.3362$, $\lambda_2 = 0.1568$, $\lambda_3 = 0$, $\lambda_4 = -15.172$ when the order of the system is 0.98 [38, 41]. The attractor of the fractional order Lorenz system is shown in Fig. 1 and the uncontrolled time series in Fig. 2.

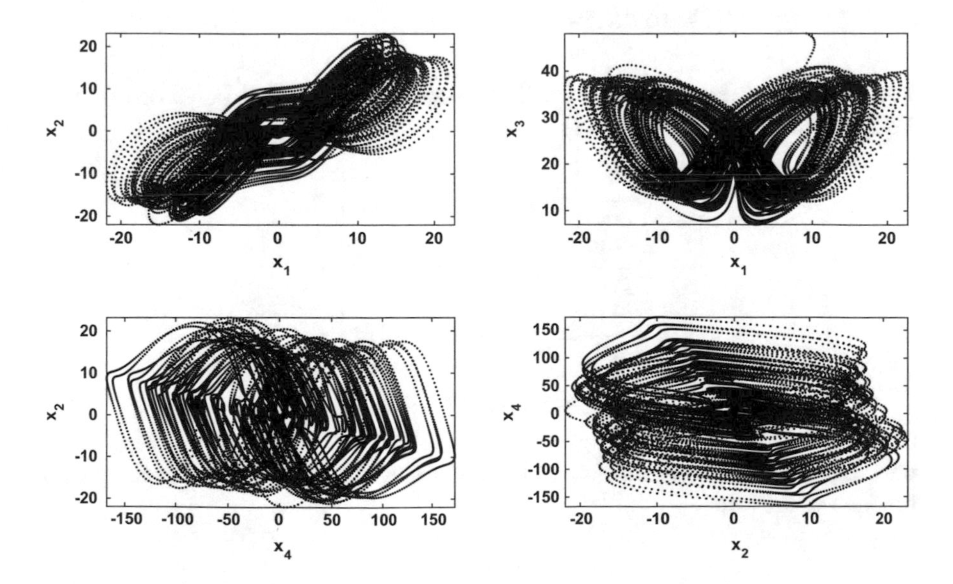

Fig. 1 Phase space of the system

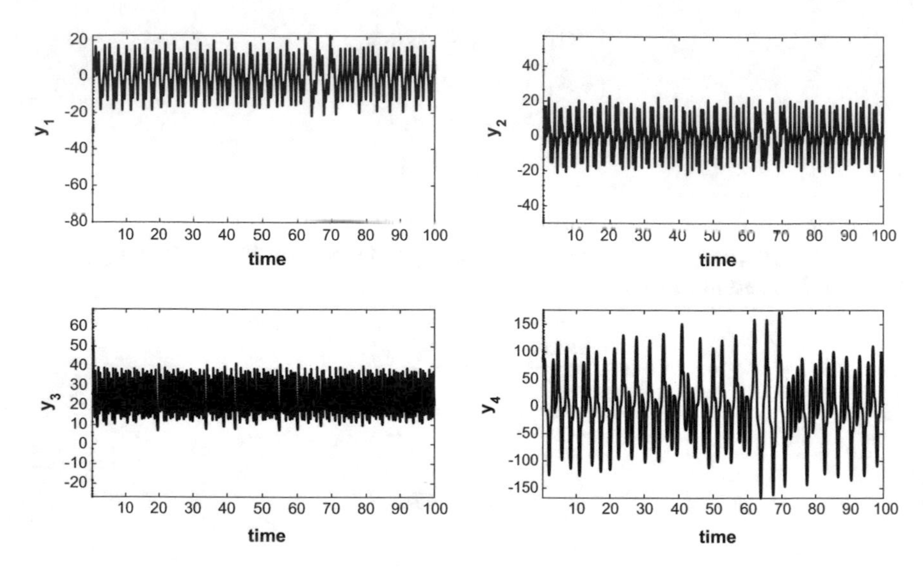

Fig. 2 (2) Time series of each component of the system (x_1, x_2, x_3, x_4) against time

5 Synchronization of Chaos in Fractional Order Lorenz System Using Direct Method

5.1 *Design of Controllers*

Let the drive system of the 4D fractional order Lorenz system be as described in Eq. 14 and the response system as

$$\begin{aligned} \frac{d^{q_1}y_1}{dt^{q_1}} &= a(y_2 - y_1) + y_4 + u_1(t) \\ \frac{d^{q_2}y_2}{dt^{q_2}} &= cy_1 - y_2 - y_1y_3 + u_2(t) \\ \frac{d^{q_3}y_3}{dt^{q_3}} &= y_1y_2 - by_3 + u_3(t) \\ \frac{d^{q_4}y_4}{dt^{q_4}} &= -y_2y_3 + ry_4 + u_4(t) \end{aligned} \tag{15}$$

where $u_i(t)(i = 1, 2, 3, 4)$ is the control function to be determined. We define the error function of the form

$$\begin{aligned} e_1 &= y_1 - \alpha_1 x_1 \\ e_2 &= y_2 - \alpha_2 x_2 \\ e_3 &= y_3 - \alpha_3 x_3 \\ e_4 &= y_4 - \alpha_4 x_4 \end{aligned} \tag{16}$$

where $\alpha_i (i = 1, 2, 3, 4)$ are scaling parameters.

Definition 1 If the scaling factors $\alpha_i (i = 1, 2, 3, 4)$ are chosen such that

$$\alpha_1 = \alpha_2 = \alpha_3 = \alpha_4 = 1$$

complete synchronization of the drive-response system is achieved.

Definition 2 If the scaling factors $\alpha_i (i = 1, 2, 3, 4)$ are chosen such that

$$\alpha_1 = \alpha_2 = \alpha_3 = \alpha_4 = -1$$

complete anti-synchronization of the drive-response system is achieved.

Definition 3 If the scaling factors $\alpha_i (i = 1, 2, 3, 4)$ are chosen such that

$$\alpha_1 = \alpha_2 = \alpha_3 = \alpha_4 = p$$

where $p \neq 0$ or 1 projective synchronization of the drive-response system is achieved.

Definition 4 The drive-response system is said to experience projective antisynchronization if the scaling factors $\alpha_i (i = 1, 2, 3, 4)$ are chosen such that

$$\alpha_1 = \alpha_2 = \alpha_3 = \alpha_4 = -p$$

where $p \neq 0$ or 1.

Definition 5 The drive-response system is said to experience modified projective antisynchronization if the scaling factors $\alpha_i (i = 1, 2, 3, 4)$ are chosen such that

$$\alpha_1 \neq \alpha_2 \neq \alpha_3 \neq \alpha_4 \neq -p$$

where $p \neq 0$ or 1.

Definition 6 If $\alpha_1 = \alpha_3 = 1$ and $\alpha_2 = \alpha_4 = -1$, hybrid synchronization is achieved.

Definition 7 If $\alpha_1 = \alpha_3$ and $\alpha_2 = \alpha_4 = p$, where $p \neq 0, \pm 1$ projective hybrid synchronization is achieved.

Definitions (1)–(7) can be obtained from any of the synchronization method described in this chapter.

The error system is then obtained as

$$\begin{aligned}
D^q e_1 &= a(e_2 - e_1) + e_4 + u_1 \\
D^q e_2 &= ce_1 - e_2 - (y_1 y_3 - \alpha_2 x_1 x_3) + u_2 \\
D^q e_3 &= -be_3 + y_1 y_2 - \alpha_3 x_1 x_2 + u_3 \\
D^q e_4 &= re_4 - (y_2 y_3 - \alpha_4 x_2 x_3) + u_4
\end{aligned} \tag{17}$$

Choosing a quadratic Lyapunov function of the form

$$V = \frac{1}{2}\sum_{i=1}^{4} e_i^2 \tag{18}$$

Its derivative is obtained as

$$\dot{V} = \sum_{i=1}^{4} e_i D^q e_i \tag{19}$$

Applying this to the system,

$$\begin{aligned}\dot{V} = e_1[a(e_2 - e_1) + e_4 + u_1] + e_2[ce_1 - e_2 - (y_1y_3 - \alpha_2x_1x_3) + u_2] \\ + e_3[-be_3 + y_1y_2 - \alpha x_1x_2 + u_3] + [re_4 - (y_2y_3 - \alpha_4x_2x_3) + u_4)]\end{aligned} \tag{20}$$

If,

$$\begin{aligned} u_1 &= -ae_2 - e_4 \\ u_2 &= -ce_1 + (y_1y_3 - \alpha x_1x_3) \\ u_3 &= -y_1y_2 + \alpha x_2x_3 \\ u_4 &= -re_4 + (y_2y_3 - \alpha_4x_2x_3) - ke_4 \end{aligned} \tag{21}$$

Then,

$$\dot{V} = -ae_1^2 - e_2^2 - be_3^2 - ke_4^2 < 0 \tag{22}$$

since a, b, k are positive numbers. We assume $k = 1$.

5.2 *Numerical Simulation Results*

To verify the effectiveness of the synchronization between the drive and response systems using the Lyapunov Direct Method, we used Eq. (6) with the initial conditions $x_i(i = 1, 2, 3, 4)$ and $y_i(i = 1, 2, 3, 4)$ taken as $(-80\ \ 50\ \ 50\ \ 100)$ and $(0.08\ \ -0.5\ \ 1\ \ 0)$ respectively. The order of the system is taken to be 0.98. A time step of 0.005 was used the systems parameters used are $a = 10$, $b = 8/3$, $c = 28$, $r = -0.99$ to ensure chaotic dynamics of the state variables. Solving the drive (Eq. 14) and response (Eq. 15) with the control defined in Eq. (21). The results are shown in Figs. 3, 4, 5, 6 and 7 for the different scaling parameters ($\alpha = 1, -1, 2, -2$). The drive and response systems could be seen to achieve synchronization as indicated by the convergence of the error state variables to zero (i.e. $e_i(1, 2, 3, 4) \to 0$). From the results obtained, the effectiveness of the controller was confirmed.

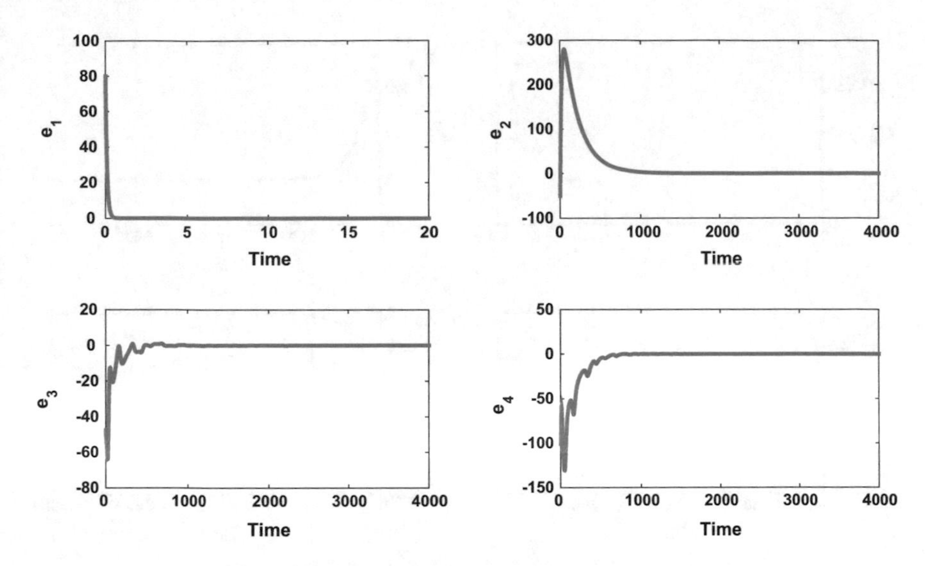

Fig. 3 Error dynamics between Slave and Master system using direct method for $\alpha_1 = \alpha_2 = \alpha_3 = \alpha_4 = 1$

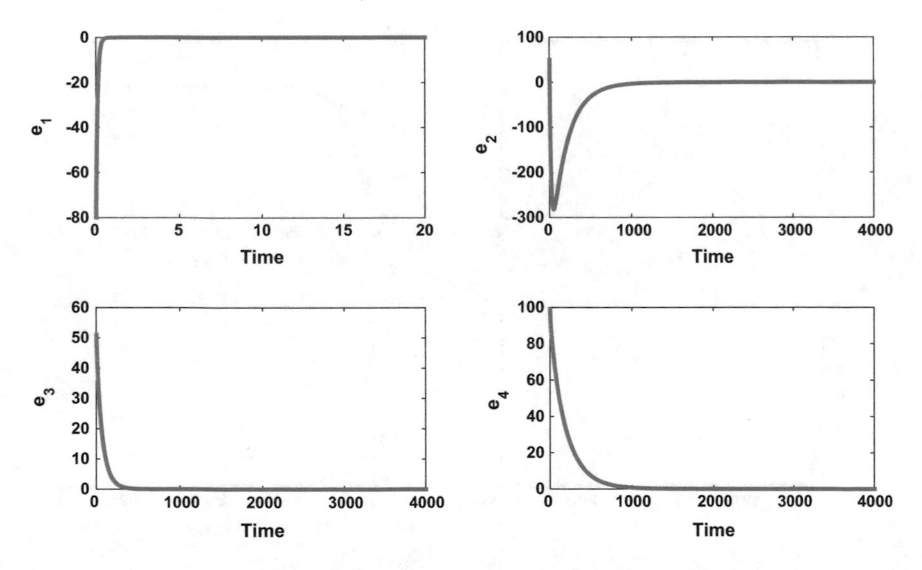

Fig. 4 Error dynamics between Slave and Master system using direct method for $\alpha_1 = \alpha_2 = \alpha_3 = \alpha_4 = -1$

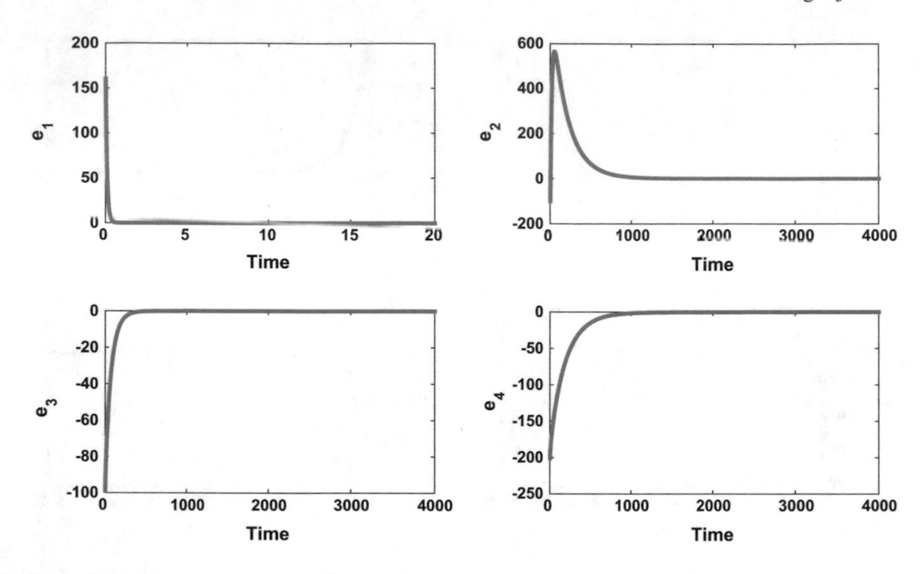

Fig. 5 Error dynamics between Slave and Master system using direct method for $\alpha_1 = \alpha_2 = \alpha_3 = \alpha_4 = 2$

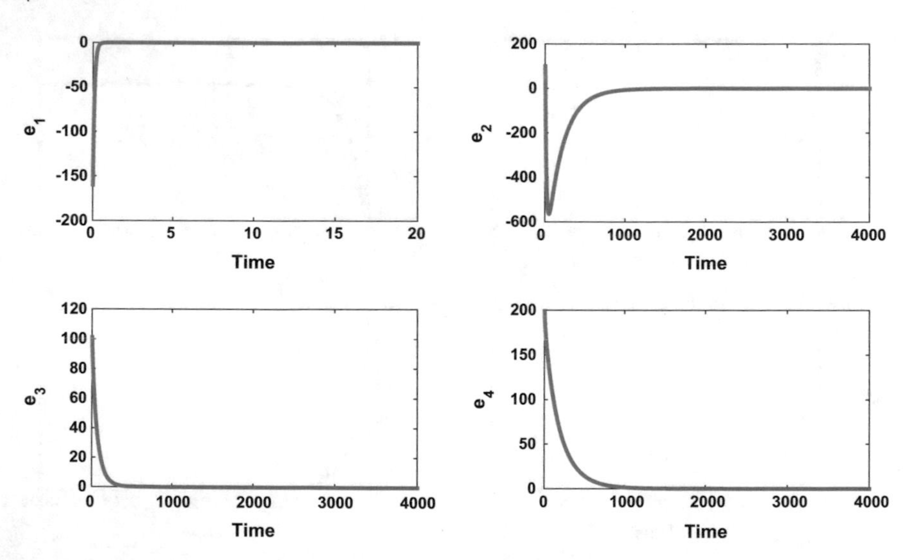

Fig. 6 Error dynamics between Slave and Master system using direct method for $\alpha_1 = \alpha_2 = \alpha_3 = \alpha_4 = -2$

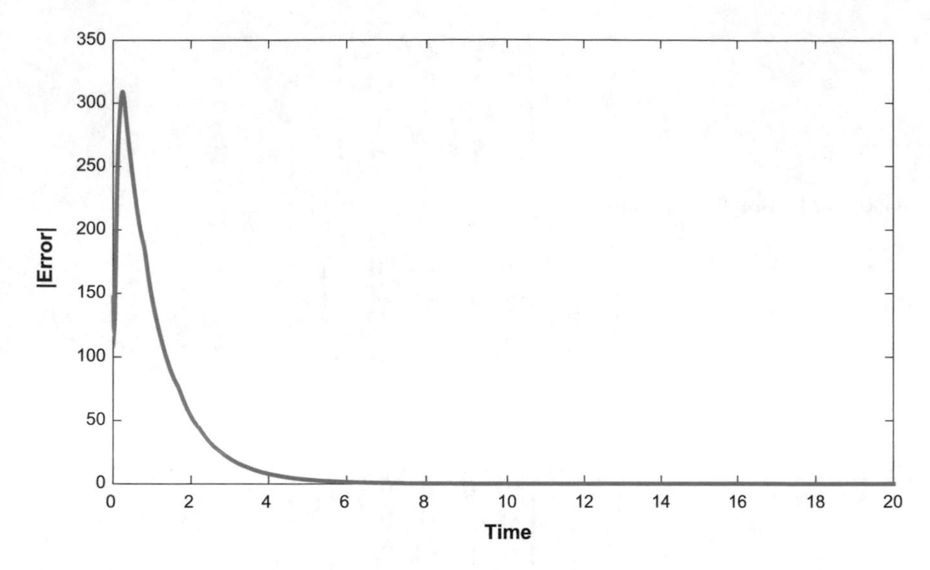

Fig. 7 Error dynamics between Slave and Master system using direct method for $\alpha_1 = \alpha_2 = \alpha_3 = \alpha_4 = 1$

6 Synchronization of Chaos in Fractional Order Lorenz System Using OPCL

6.1 Design of Controllers

If the drive system is taken as Eq. 14 and the response as Eq. 15, then the error state of the system can be written as

$$\begin{aligned} e_1 &= y_1 - g_1 \\ e_2 &= y_2 - g_2 \\ e_3 &= y_3 - g_3 \\ e_4 &= y_4 - g_4 \end{aligned} \tag{23}$$

where

$$\begin{bmatrix} g_1 \\ g_2 \\ g_3 \\ g_4 \end{bmatrix} = \begin{bmatrix} \alpha_1 x_1 \\ \alpha_2 x_2 \\ \alpha_3 x_3 \\ \alpha_4 x_4 \end{bmatrix} \text{ and } \begin{bmatrix} \dot{g_1} \\ \dot{g_2} \\ \dot{g_3} \\ \dot{g_4} \end{bmatrix} = \begin{bmatrix} \alpha_1 \dot{x_1} \\ \alpha_2 \dot{x_2} \\ \alpha_3 \dot{x_3} \\ \alpha_4 \dot{x_4} \end{bmatrix}$$

Defining the drive system as a function of g

$$f(g) = \begin{bmatrix} a(g_2 - g_1) + g_4 \\ cg_1 - g_2 - g_1g_3 \\ g_1g_2 - bg_3 \\ -g_2g_3 + rg_4 \end{bmatrix} \tag{24}$$

Also, the Jacobian is obtained as

$$\frac{\partial f_g}{\partial g} = \begin{pmatrix} -a & a & 0 & 1 \\ (c - g_3) & -1 & -g & 0 \\ g_2 & g_1 & -b & 0 \\ 0 & -g_3 & -g_2 & r \end{pmatrix} \tag{25}$$

To ensure the stability of the system, we choose as Hurwitz matrix

$$H = \begin{pmatrix} -1 & 0 & 0 & 0 \\ 0 & -1 & 0 & 0 \\ 0 & 0 & -1 & 0 \\ 0 & 0 & 0 & -1 \end{pmatrix} \tag{26}$$

Using the OPCL theory, the controller is defined as

$$U = \dot{g} - f(g) + \left[H - \frac{\partial f_g}{\partial g}\right] e \tag{27}$$

The control $u_i (i = 1, 2, 3, 4)$ is then obtained as

$$\begin{aligned} u_1 &= \alpha_1\dot{x}_1 - a(g_2 - g_1) - g_4 + (a - 1)e_1 - a_2e_2 - e_4 \\ u_2 &= \alpha_2\dot{x}_2 - [cg_1 - g_2 - g_1g_3] - (c - g_3)e_1 + g_1e_3 \\ u_3 &= \alpha_3\dot{x}_3 - [g_1g_2 - bg_3] - g_2e_1 - g_1e_2 + (b - 1)e_3 \\ u_4 &= \alpha_4\dot{x}_4 - [rg_4 - g_2g_3] + g_3e_2 + g_2e_3 - (1 + r)e_4 \end{aligned} \tag{28}$$

6.2 *Numerical Simulation Results*

To verify the effectiveness of the synchronization between the drive and response systems using the OPCL method, we used Eq. (6) with the initial conditions $x_i(i = 1, 2, 3, 4)$ and $y_i(i = 1, 2, 3, 4)$ taken as $(-80\ \ 50\ \ 50\ \ 100)$ and $(0.08\ \ -0.5\ \ 1\ \ 0)$ respectively. The order of the system is taken to be 0.98. A time step of 0.005 was used the systems parameters used are $a = 10$, $b = 8/3$, $c = 28$, $r = -0.99$ to ensure chaotic dynamics of the state variables. Solving the drive (Eq. 14) and response (Eq. 15) with the control defined in Eq. (28). The results are shown in Figs. 8, 9, 10, 11 and 12 for the different scaling parameters ($\alpha = 1, -1, 2, -2$). The drive and

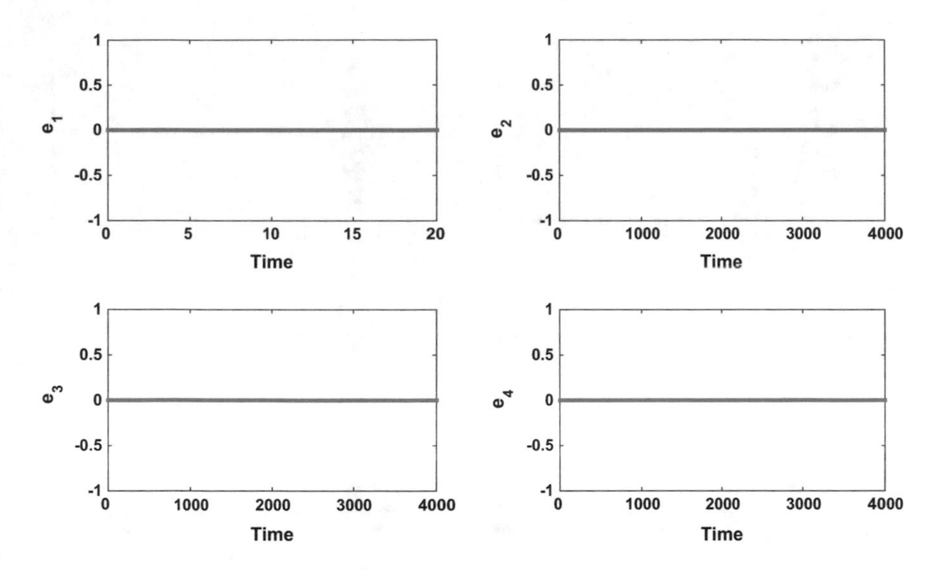

Fig. 8 Error dynamics between Slave and Master system using OPCL method for $\alpha_1 = \alpha_2 = \alpha_3 = \alpha_4 = 1$

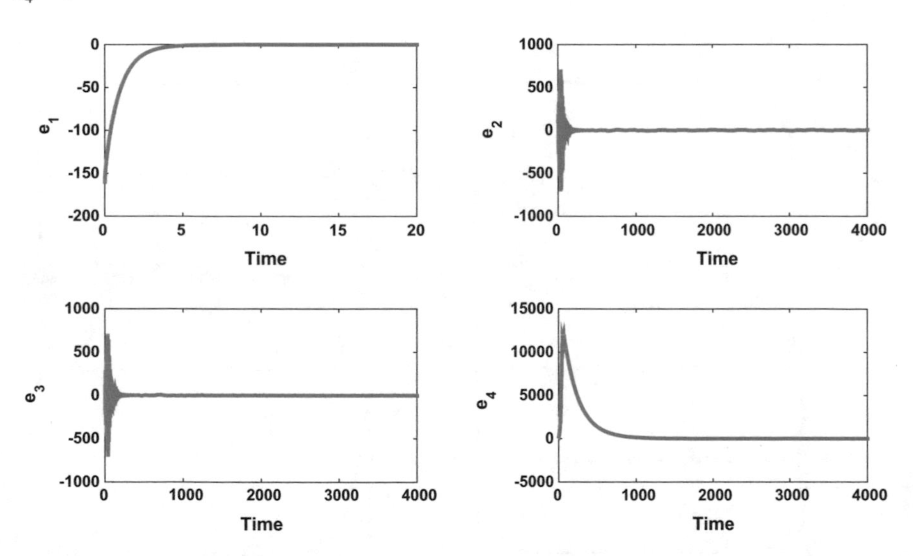

Fig. 9 Error dynamics between Slave and Master system using OPCL method for $\alpha_1 = \alpha_2 = \alpha_3 = \alpha_4 = -1$

response systems could be seen to achieve synchronization as indicated by the convergence of the error state variables to zero (i.e. $e_i(1, 2, 3, 4) \to 0$). From the results obtained, the effectiveness of the controller was confirmed.

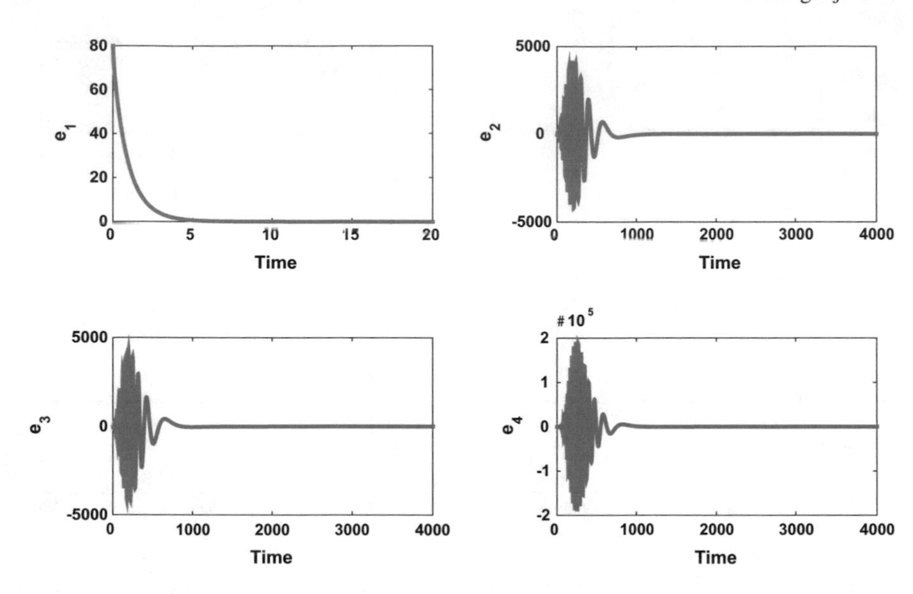

Fig. 10 Error dynamics between Slave and Master system using OPCL method for $\alpha_1 = \alpha_2 = \alpha_3 = \alpha_4 = 2$

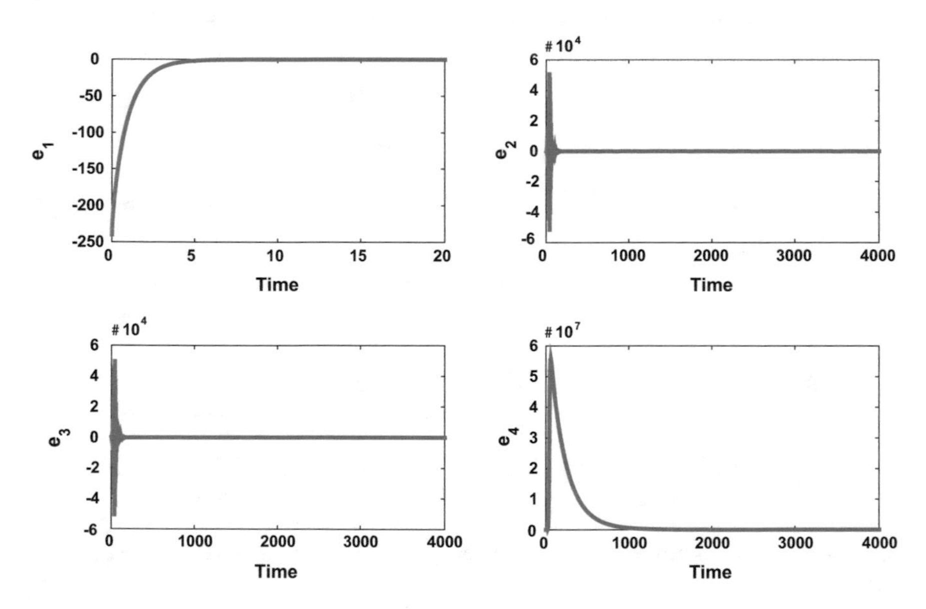

Fig. 11 Error dynamics between Slave and Master system using OPCL method for $\alpha_1 = \alpha_2 = \alpha_3 = \alpha_4 = -2$

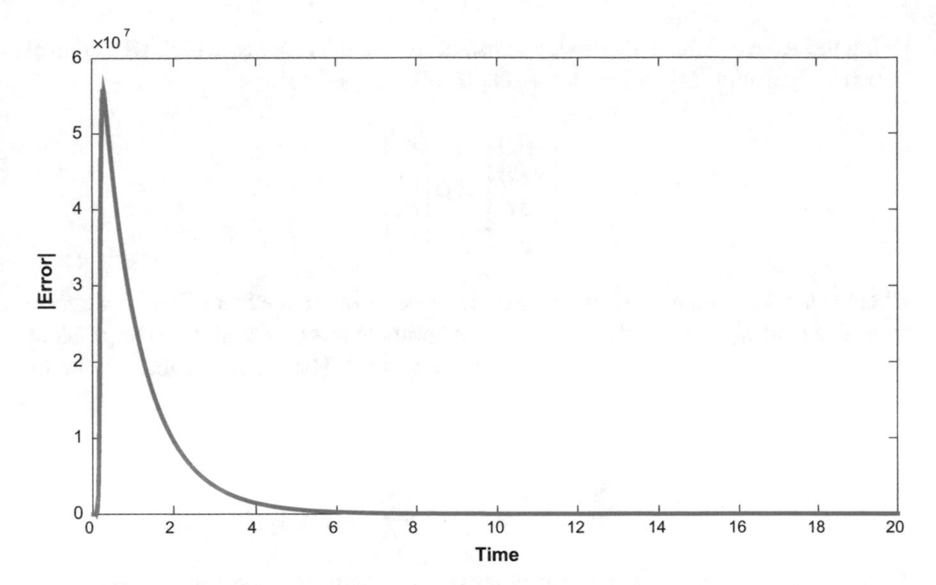

Fig. 12 Error dynamics between Slave and Master system using OPCL method for $\alpha_1 = \alpha_2 = \alpha_3 = \alpha_4 = 1$

7 Synchronization of Chaos in Fractional Order Lorenz System Using Active Control

7.1 Design of Controllers

If the drive system is taken as Eq. 14, the response as Eq. 15, then the error state of the system as Eq. 16, in line with the method of Active Control, we can eliminate terms which cannot be expressed as linear terms in e_1, e_2, e_3, e_4 as

$$\begin{aligned} u_1 &= v_1(t) \\ u_2 &= y_1 y_3 - \alpha_2 x_1 x_3 + v_2(t) \\ u_3 &= -y_1 y_2 + \alpha_3 x_1 x_2 + v_3(t) \\ u_4 &= y_2 y_3 - \alpha_4 x_2 x_3 + v_4(t) \end{aligned} \tag{29}$$

the parameter $v_i(t)(i = 1, 2, 3, 4)$ will be obtained later. Substituting Eq. (29) into Eq. (17) yields

$$\begin{aligned} D^q e_1 &= a(e_2 - e_1) + e_4 + v_1(t) \\ D^q e_2 &= c e_1 - e_2 + v_2(t) \\ D^q e_3 &= -b e_3 + v_3(t) \\ D^q e_4 &= r e_4 + v_4(t) \end{aligned} \tag{30}$$

Using the Active Control method, a constant matrix D is chosen which will control the error dynamics (Eq. 30) such that the feedback matrix is

$$\begin{pmatrix} v_1(t) \\ v_2(t) \\ v_3(t) \\ v_4(t) \end{pmatrix} = D \begin{pmatrix} e_1 \\ e_2 \\ e_3 \\ e_4 \end{pmatrix}$$

where D is a 4×4 matrix. There are various choices of the feedback D which can be chosen to control the error dynamics but we optimize this choice so that the problem of controller complexity is significantly reduced [31]. Hence, D is chosen to be of the form

$$D = \begin{pmatrix} (a - \lambda_1) & -a & 0 & -1 \\ -c & (1 - \lambda_2) & 0 & 0 \\ 0 & 0 & (b - \lambda_3) & 0 \\ 0 & 0 & 0 & -(r + \lambda_4) \end{pmatrix} \tag{31}$$

The eigenvalues $\lambda_i (i = 1, 2, 3, 4)$ are chosen to be negative in order to achieve a stable synchronization between the drive and response system.

7.2 Numerical Simulation Results

To verify the effectiveness of the synchronization between the drive and response systems using the Active control method, we used Eq. (6) with the initial

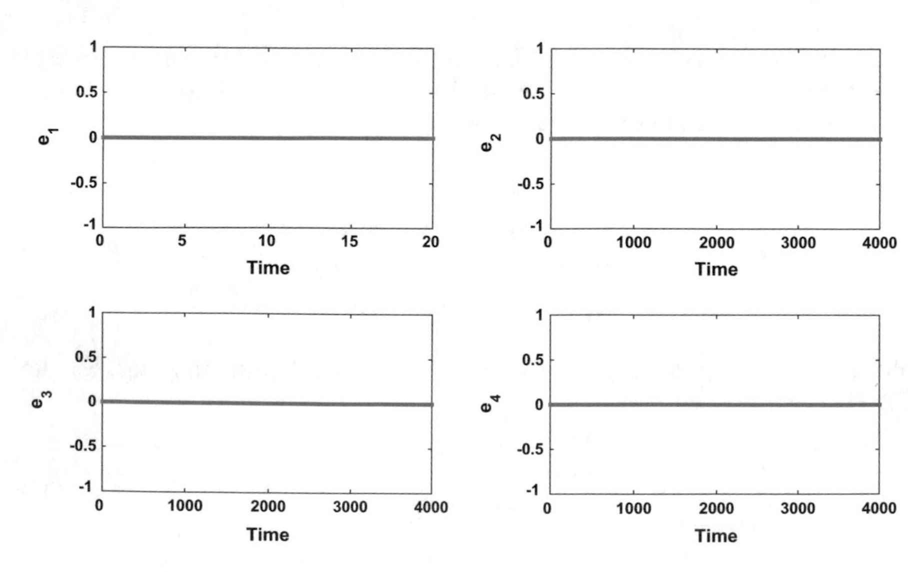

Fig. 13 Error dynamics between Slave and Master system using active control method for $\alpha_1 = \alpha_2 = \alpha_3 = \alpha_4 = 1$

conditions $x_i(i = 1, 2, 3, 4)$ and $y_i(i = 1, 2, 3, 4)$ taken as $(-80\ \ 50\ \ 50\ \ 100)$ and $(0.08\ \ -0.5\ \ 1\ \ 0)$ respectively. The order of the system is taken to be 0.98. A time step of 0.005 was used the systems parameters used are $a = 10$, $b = 8/3$,

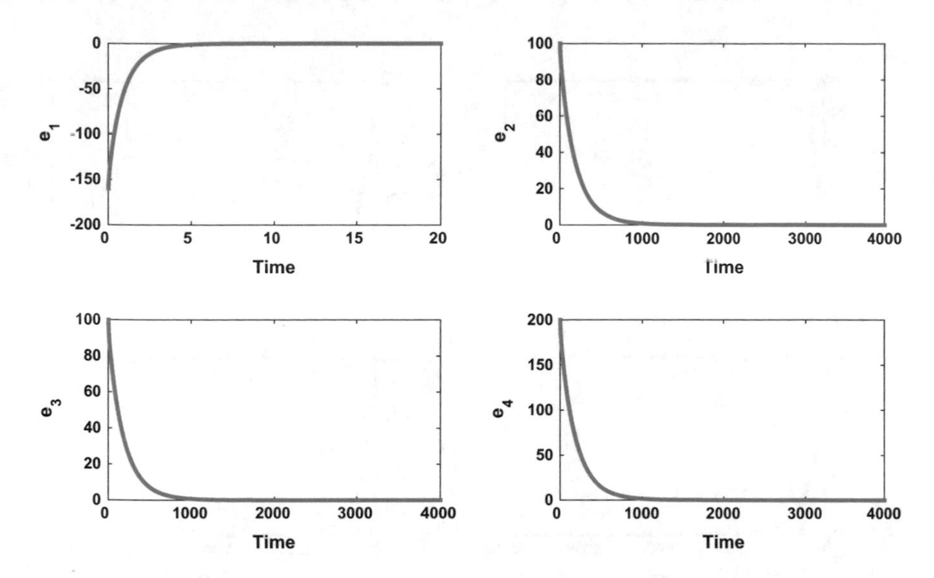

Fig. 14 Error dynamics between Slave and Master system using active control method for $\alpha_1 = \alpha_2 = \alpha_3 = \alpha_4 = -1$

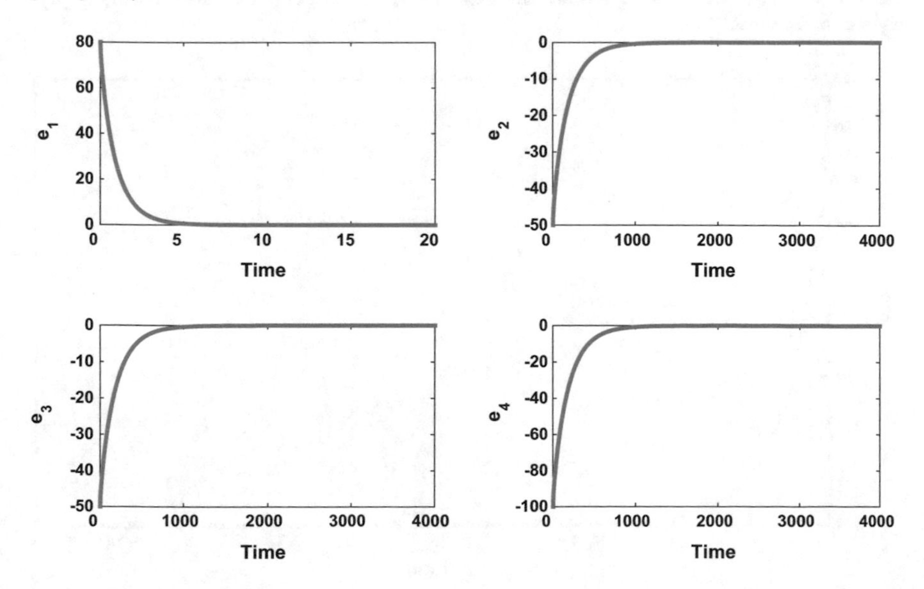

Fig. 15 Error dynamics between Slave and Master system using active control method for $\alpha_1 = \alpha_2 = \alpha_3 = \alpha_4 = 2$

$c = 28$, $r = -0.99$ to ensure chaotic dynamics of the state variables. Solving the drive (Eq. 14) and response (Eq. 15) with the control defined in Eq. (31). The results are shown in Figs. 13, 14, 15, 16 and 17 for the different scaling parameters ($\alpha = 1, -1, 2, -2$). The drive and response systems could be seen to achieve synchro-

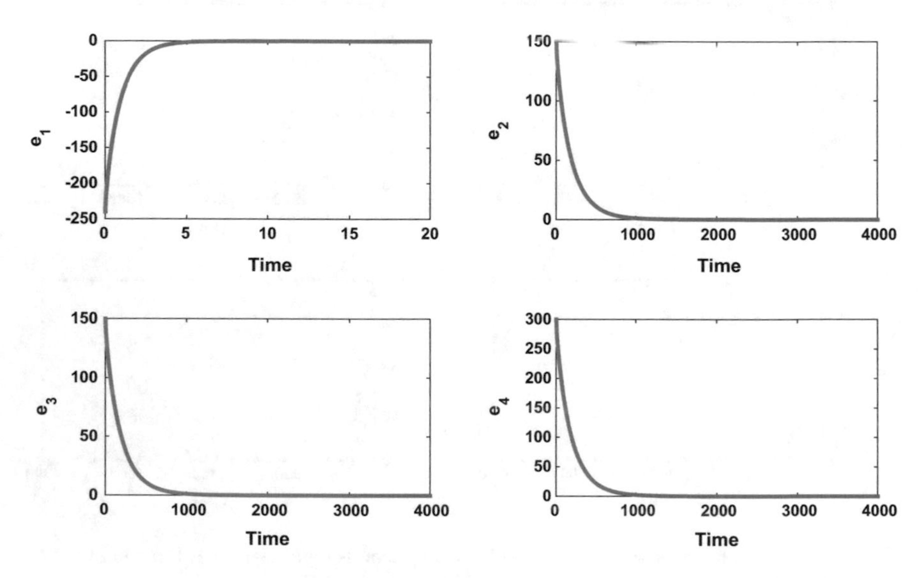

Fig. 16 Error dynamics between Slave and Master system using active control method for $\alpha_1 = \alpha_2 = \alpha_3 = \alpha_4 = -2$

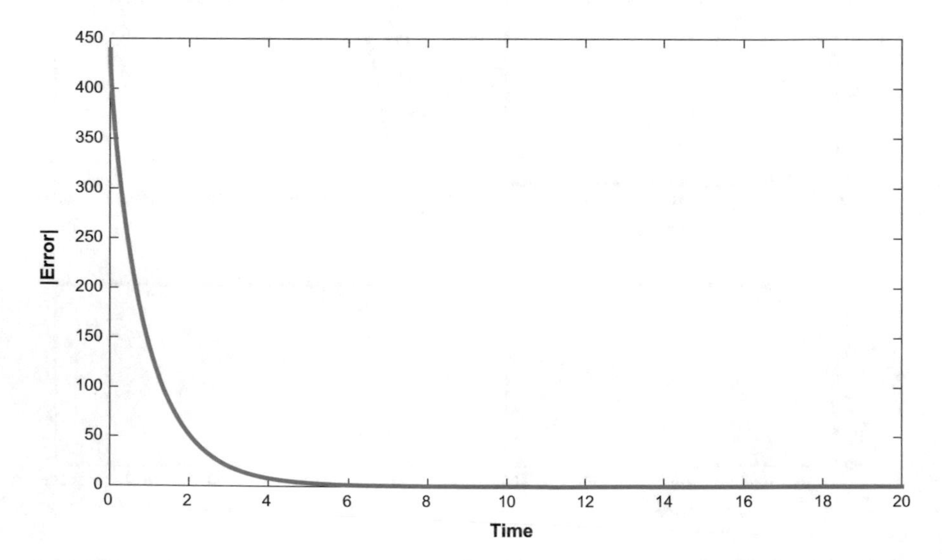

Fig. 17 Error dynamics between Slave and Master system using active control method for $\alpha_1 = \alpha_2 = \alpha_3 = \alpha_4 = 1$

nization as indicated by the convergence of the error state variables to zero (i.e. $e_i(1,2,3,4) \rightarrow 0$). From the results obtained, the effectiveness of the controller was confirmed.

8 Comparison of Direct Method, OPCL and Active Control Techniques

The performance of the three different synchronization scheme is to be compared. The error components for the system and error magnitude are presented in Figs. 18 and 19 respectively. From Fig. 18, apart from the top-left figure, the convergence of the synchronization technique in order of increasing speed is: active control, Lyapunov Direct Method and OPCL. The same trend and order could be observed in the error magnitude as shown in Fig. 19. The behaviour of the error dynamics before achieving convergence is important. The speed of convergence is referred to as synchronization time [24, 31]. From Fig. 18, the active control technique was found to have minimal variations before attaining convergence while the two other techniques show different behaviours in fluctuation. From the dynamics of the error dynamics, it could be observed that the OPCL method showed the highest variation in error amplitude before convergence while the active control has the lowest amplitude variation.

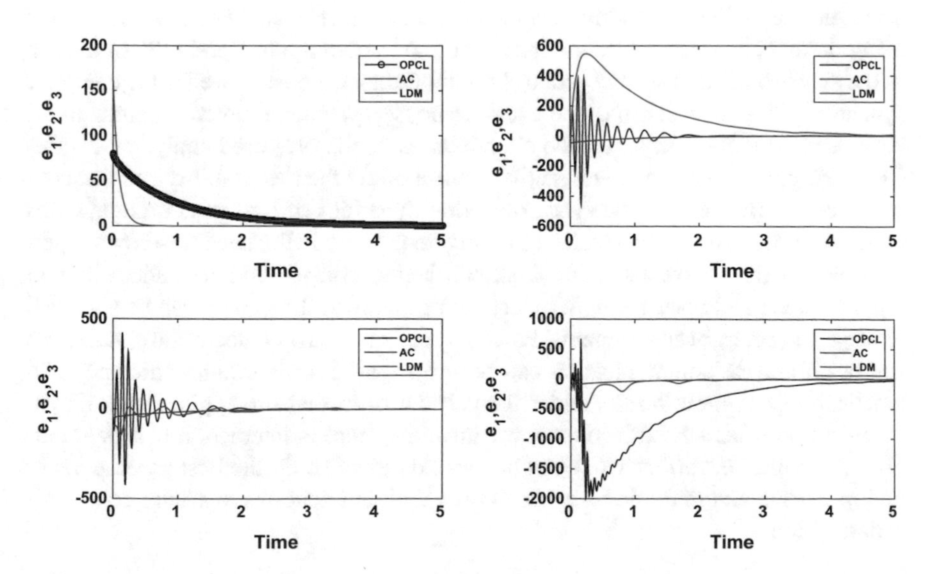

Fig. 18 Error dynamics between Slave and Master system for each component of the system for each of the synchronization techniques under consideration when $\alpha_1 = \alpha_2 = \alpha_3 = \alpha_4 = 2$

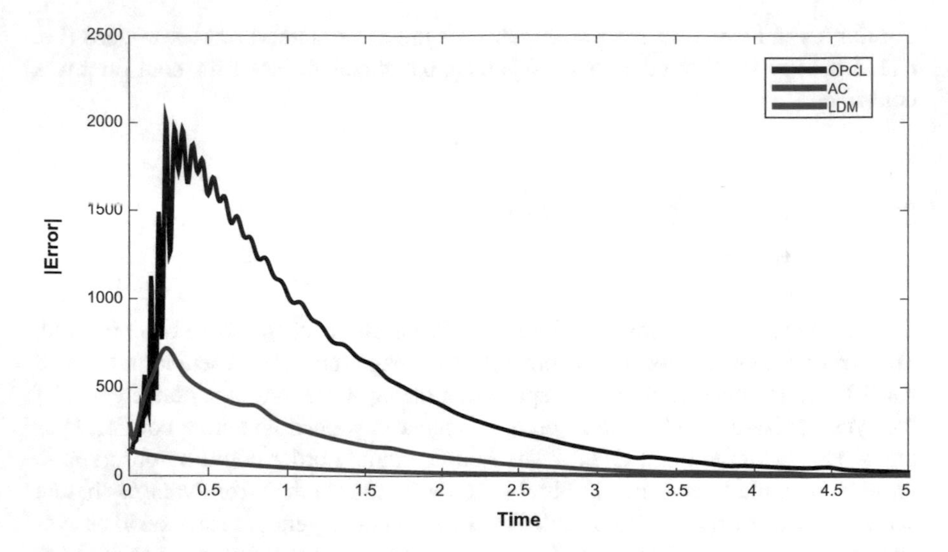

Fig. 19 Error dynamics between Slave and Master system for the three synchronization techniques when $\alpha_1 = \alpha_2 = \alpha_3 = \alpha_4 = 1$

9 Conclusion

The synchronization of chaotic fractional order Lorenz system has been investigated using three techniques: active control, Lyapunov direct method and OPCL. In each of the synchronization scheme, control functions have been achieve for the complete synchronization between the drive and response systems. Numerical simulations have been conducted to assert the effectiveness of the proposed analytical results. Comparing the three techniques, active control offers the best stability and fast convergence of error terms. The synchronization dynamics of fractional order systems under periodic driving force can be investigated. There is the need to study the performance of the different synchronization schemes considered here under different types of noise and noise strength to test their reliability. This study can be extended to maps and integer order systems. Practical realization using electronic simulations and/or circuit for communication can be investigated to determine efficiency and practicability of these results under field scenario. In real life applications of synchronization schemes for secure communication, there is interaction between multiple users, hence, further work can be carried out to study the best scheme under multiple drive and multiple response system with applications to secure communication systems.

References

1. Ahmad, I., Saaban, A. B., & Shahzad, M. (2015). A research on active control to synchronize a new 3D chaotic system. *Systems*, *4*(2), 1–14. doi:10.3390/systems4010002.
2. Al-Sawalha, M. M., & Shoaib, M. (2016). Reduced-order synchronization of fractional order chaotic systems with fully unknown parameters using modified adaptive control. *Journal of Nonlinear Science and Applications*, *9*, 1815–1825.
3. Alam, Z., Yuan, L., & Yang, Q. (2016). Chaos and combination synchronization of a new fractional-order system with two stable node-foci. *IEEE/CAA Journal of Automatica Sinica*, *3*(2), 157–164. doi:10.1109/JAS.2016.7451103.
4. Bai, E. W., & Lonngren, K. E. (1997). Synchronization of two lorenz systems using active control. *Chaos, Solitons & Fractals*, *8*(1), 51–58.
5. Boccaletti, S., Kurths, J., Osipov, G., Valladares, D. L., & Zhou, C. S. (2002). The synchronization of chaotic systems. *Physics Reports*, *366*, 1–101.
6. Bouzeriba, A., Boulkroune, A., & Bouden, T. (2015). Fuzzy adaptive synchronization of uncertain fractional-order chaotic systems. *International Journal of Machine Learning and Cybernetics*, 1–16. http://dx.doi.org/10.1007/s13042-015-0425-7.
7. Caponetto, R., Dongola, G., Fortuna, L., & Petras, I. (2010). *Fractional order systems: Modeling and control applications, A* (Vol. 72). World Scientific Publishing Cp. Pte. Ltd.
8. Concepcion, A. M., Chen, Y. Q., Vinagre, B. M., Xue, D., & Feliu, V. (2010). *Fractional-order systems and control: Fundamentals and applications*. London: springer.
9. Couceiro, M. S., Clemente, F. M., & Martins, F. M. L. (2013). Analysis of football player' s motion in view of fractional calculus. *Central European Journal of Physics*, *11*(6), 714–723. doi:10.2478/s11534-013-0258-5.
10. Dzieliński, A., Sierociuk, D., & Sarwas, G. (2011). Some applications of fractional order calculus. *Bulletin of the Polish Academy of Sciences: Technical Sciences*, *58*(4), 583–592. doi:10.2478/v10175-010-0059-6.
11. El-Sayed, A., Nour, H., Elsaid, A., Matouk, A., & Elsonbaty, A. (2016). Dynamical behaviors, circuit realization, chaos control, and synchronization of a new fractional order hyperchaotic system. *Applied Mathematical Modelling*, *40*(5), 3516–3534. doi:10.1016/j.apm.2015.10.010.
12. Gao, Y. B., Sun, B. H., & Lu, G. P. (2013). Modified function projective lag synchronization of chaotic systems with disturbance estimations. *Applied Mathematical Modelling*, *37*(7), 4993–5000.
13. Golmankhaneh, A. K., Arefi, R., & Baleanu, D. (2015). Synchronization in a nonidentical fractional order of a proposed modified system. *Journal of Vibration and Control*, *21*(6), 1154–1161. doi:10.1177/1077546313494953.
14. Grosu, I. (1997). Robust synchronization. *Physcial Review E*, *56*(3), 3709–3712.
15. Huang, L. L., Zhang, J., & Shi, S. S. (2015). Circuit simulation on control and synchronization of fractional order switching chaotic system. *Mathematics and Computers in Simulation*, *113*, 28–39. doi:10.1016/j.matcom.2015.03.001.
16. Jackson, E. A., & Grosu, I. (1995). An open-plus-closed-loop (OPCL) control of complex dynamic systems. *Physica D: Nonlinear Phenomena*, *85*(1), 1–9.
17. Kapitaniak, M., Czolczynski, K., Perlikowski, P., Stefanski, A., & Kapitaniak, T. (2012). Synchronization of clocks. *Physics Reports*, *517*(1–2), 1–69.
18. Kareem, S. O., Ojo, K. S., & Njah, A. N. (2012). Function projective synchronization of identical and non-identical modified finance and Shimizu Morioka systems, *79*(1), 71–79. doi:10.1007/s12043-012-0281-x.
19. Lorenz, E. N. (1963). Deterministic nonperiodic flow. *Journal of the Atmospheric Sciences*, *20*, 130–141.
20. Mahmoud, G. M., & Mahmoud, E. E. (2011). Modified projective lag synchronization of two nonidentical hyperchaotic complex nonlinear systems. *International Journal of Bifurcation and Chaos*, *21*(08), 2369–2379. doi:10.1142/S0218127411029859.

21. Mahmoud, G. M., Abed-Elhameed, T. M., & Ahmed, M. E. (2016). Generalization of combination—combination synchronization of chaotic n-dimensional fractional-order dynamical systems. *Nonlinear Dynamics*, *83*(4), 1885–1893. doi:10.1007/s11071-015-2453-y.
22. Mathiyalagan, K., Park, J. H., & Sakthivel, R. (2015). Exponential synchronization for fractional-order chaotic systems with mixed uncertainties. *Complexity*, *21*(1), 114–125. doi:10.1002/cplx.21547.
23. Mohadeszadeh, M., & Delavari, H. (2015). Synchronization of fractional-order hyper-chaotic systems based on a new adaptive sliding mode control. *International Journal of Dynamics and Control*, 1–11. doi:10.1007/s40435-015-0177-y.
24. Ogunjo, S. T. (2013). Increased and reduced order synchronization of 2D and 3D dynamical systems. *International Journal of Nonlinear Science*, *16*(2), 105–112.
25. Ogunjo, S. T., Adediji, A. T., & Dada, J. B. (2015). Investigating chaotic features in solar radiation over a tropical station using recurrence quantification analysis. *Theoretical and Applied Climatology*, 1–7. doi:10.1007/s00704-015-1642-4.
26. Ojo, K., Njah, A., Olusola, O., & Omeike, M. (2014). Generalized reduced-order hybrid combination synchronization of three Josephson junctions via backstepping technique. *Nonlinear Dynamics*, *77*(3), 583–595.
27. Ojo, K., Njah, A., Olusola, O., & Omeike, M. (2014). Reduced order projective and hybrid projective combination-combination synchronization of four chaotic Josephson junctions. *Journal of Chaos*.
28. Ojo, K., Njah, A., & Olusola, O. (2015). Compound-combination synchronization of chaos in identical and different orders chaotic systems. *Archives of Control Sciences*, *25*(4), 463–490.
29. Ojo, K., Njah, A., & Olusola, O. (2015). Generalized function projective combination-combination synchronization of chaos in third order chaotic systems. *Chinese Journal of Physics*, *53*(3), 11–16.
30. Ojo, K. S., & Ogunjo, S. T. (2012). Synchronization of 4D Rabinovich hyperchaotic system for secure communication. *Journal of Nigerian Association of Mathematical Physics*, *21*, 35–40.
31. Ojo, K. S., Njah, A., & Ogunjo, S. T. (2013). Comparison of backstepping and modified active control in projective synchronization of chaos in an extended Bonhoffer van der Pol oscillator. *Pramana*, *80*(5), 825–835.
32. Ojo, K. S., Ogunjo, S. T., & Williams, O. (2013). Mixed tracking and projective synchronization of 5D hyperchaotic system using active control. *Cybernetics and Physics*, *2*, 31–36.
33. Ojo, K. S., Njah, A., Ogunjo, S. T., & Olusola, O. I. (2014). Reduced order function projective combination synchronization of three Josephson junctions using backstepping technique. *Nonlinear Dynamics and System Theory*, *14*(2), 119.
34. Ojo, K. S., Njah, A. N. A., Ogunjo, S. T., Olusola, O. I., et al. (2014). Reduced order hybrid function projective combination synchronization of three Josephson junctions. *Archives of Control Sciences*, *24*(1), 99–113.
35. Ojo, K. S., Ogunjo, S. T., Njah, A. N., & Fuwape, I. A. (2014). Increased-order generalized synchronization of chaotic and hyperchaotic systems, *84*(1), 1–13. doi:10.1007/s12043-014-0835-1.
36. Ouannas, A., Azar, A. T., & Vaidyanathan, S. (2016). A robust method for new fractional hybrid chaos synchronization. *Mathematical Methods in the Applied Sciences*, n/a–n/a. doi:10.1002/mma.4099.
37. Pecora, L. M., & Carroll, T. L. (1990). Synchronization in chaotic systems. *Physical Review Letters*, *64*, 821–824.
38. Ramasubramanian, K., & Sriram, M. S. (2000). A comparative study of computation of lyapunov spectra with different algorithms. *Physica D*, *138*(1–2), 72–86.
39. Sivakumar, B. (2004). Chaos theory in geophysics: Past, present and future. *Chaos, Solitons and Fractals*, *19*, 441–462.
40. Strogatz, S. H. (1994). *Nonlinear dynamics and chaos*. Reading: Addison-Wesley.
41. Wang, X., & Song, J. (2009). Synchronization of the fractional order hyperchaos lorenz system usin activstion feedback control. *Communications in Nonlinear Science and Numerical Simulation*, *14*, 3351–3357.

42. Wu, C. W., & Chua, L. O. (1994). A unified framework for synchronization and control of dynamical systems. *International Journal of Bifurcation and Chaos*, *4*(4), 979–998.
43. Xiao, W., Fu, J., Liu, Z., & Wan, W. (2012). Generalized synchronization of typical fractional order chaos system. *Journal of Computers*, *7*(6), 1519–1526. doi:10.4304/jcp.7.6.1519-1526.
44. Zhou, P., Ding, R., & Cao, Y. X. (2012). Multi drive-one response synchronization for fractional-order chaotic systems. *Nonlinear Dynamics*, *70*(2), 1263–1271. doi:10.1007/s11071-012-0531-y.

On New Fractional Inverse Matrix Projective Synchronization Schemes

Adel Ouannas, Ahmad Taher Azar, Toufik Ziar and Sundarapandian Vaidyanathan

Abstract In this study, the problem of inverse matrix projective synchronization (IMPS) between different dimensional fractional order chaotic systems is investigated. Based on fractional order Lyapunov approach and stability theory of fractional order linear systems, new complex schemes are proposed to achieve inverse matrix projective synchronization (IMPS) between n-dimension and m-dimension fractional order chaotic systems. To validate the theoretical results and to verify the effectiveness of the proposed schemes, numerical applications and computer simulations are used.

Keywords Fractional chaos · Inverse matrix projective synchronization · Fractional-order Lyapunov approach · Different dimensional systems

---

A. Ouannas
Laboratory of Mathematics, Informatics and Systems (LAMIS),
University of Larbi Tebessi, 12002 Tebessa, Algeria
e-mail: ouannas.a@yahoo.com

A.T. Azar (✉)
Faculty of Computers and Information, Benha University, Banha, Egypt
e-mail: ahmad.azar@fci.bu.edu.eg; ahmad_t_azar@ieee.org

A.T. Azar
Nanoelectronics Integrated Systems Center (NISC), Nile University, Cairo, Egypt

T. Ziar
Department of Material Sciences, University of Tebessa, 12002 Tébessa, Algeria
e-mail: Toufik1_ziar@yahoo.fr

S. Vaidyanathan
Research and Development Centre, Vel Tech University,
Avadi, Chennai 600062, Tamil Nadu, India
e-mail: sundarvtu@gmail.com

A.T. Azar et al. (eds.), *Fractional Order Control and Synchronization of Chaotic Systems*, Studies in Computational Intelligence 688,
DOI 10.1007/978-3-319-50249-6_17

1 Introduction

Fractional calculus, as generalization of integer order integration and differentiation to its non-integer (fractional) order counterpart, has proved to be valuable tool in the modeling of many physical phenomena [1–5] and engineering problems [6–13]. Fractional derivatives provide an excellent instrument for the description of memory and hereditary properties of various materials and processes [14, 15]. The main reason for using integer-order models was the absence of solution methods for fractional differential equations [16, 17]. The advantages or the real objects of the fractional order systems are that we have more degrees of freedom in the model and that a "memory" is included in the model [18]. One of the very important areas of application of fractional calculus is the chaos theory [19, 20].

Chaos is a very interesting nonlinear phenomenon which has been intensively studied [21–26]. It is found to be useful or has great application potential in many fields such as secure communication [27], data encryption [28], financial systems [29] and biomedical engineering [30]. The research efforts have been devoted to chaos control [31–33] and chaos synchronization [34–40] problems in nonlinear science because of its extensive applications [41–57].

Recently, studying fractional order systems has become an active research area. The chaotic dynamics of fractional order systems began to attract much attention in recent years. It has been shown that the fractional order systems can also behave chaotically, such as the fractional order Chua's system [58], the fractional order Lorenz system [59], the fractional order Chen system [60, 61], the fractional order Rössler system [62], the fractional-order Arneodo's system [63], the fractional order Lü system [64], the fractional-order Genesio-Tesi system [65], the fractional order modified Duffing system [66], the fractional-order financial system [67], the fractional order Newton–Leipnik system [68], the fractional order Lotka-Volterra system [69] and the fractional order Liu system [70]. Moreover, recent studies show that chaotic fractional order systems can also be synchronized [71–78]. Many scientists who are interested in this field have struggled to achieve the synchronization of fractional–order chaotic systems, mainly due to its potential applications in secure communication and cryptography [79–81].

A wide variety of methods and techniques have been used to study the synchronization of the fractional–order chaotic such as sliding mode controller [82–84], active and adaptive control methods [85–87], feedback control method [88, 89], linear and nonlinear control methods [90, 91], scalar signal technique [92, 93]. Many types of synchronization for the fractional-order chaotic systems have been presented [94–127]. Among all types of synchronization, projective synchronization (PS) has been extensively considered. In PS, slave system variables are scaled replicas of the master system variables. A variation of projective synchronization is the so-called matrix projective synchronization (MPS) (or full state hybrid projective synchronization) [128–130]. Also, matrix projective synchronization (MPS) between fractional order chaotic systems has been studied [131–134]. In this type of synchronization the single scaling parameter originally introduced in [135] is replaced by a diagonal scaling matrix [136, 137] or by a full scaling matrix [138]. Recently, an interesting scheme has been introduced [139], in which each master slave system state achieves synchronization with any arbitrary linear combination of slave system states. Since master system states and slave system states are inverted with respect to the MPS, the proposed scheme is called inverse matrix projective synchronization (IMPS). Obviously, the problem of inverse matrix projective synchronization (IMPS) is an attractive idea and more difficult than the problem of matrix projective synchronization (MPS). The complexity of the IMPS scheme can have important effect in applications.

Based on these considerations, this study presents new control schemes for the problem of IMPS in fractional-order chaotic dynamical systems. Based on Laplace transform and fractional Lyapunov stability theory, the study first analyzes a new IMPS scheme between $n-$dimensional commensurate fractional-order master system and $m-$dimensional commensurate fractional-order slave system. Successively, by using some properties of fractional derivatives and stability theory of fractional-order linear systems, IMPS is proved between $n-$dimensional incommensurate fractional-order master system and $m-$dimensional commensurate fractional-order slave system. Finally, several numerical examples are illustrated, with the aim to show the effectiveness of the approaches developed herein.

This study is organized as follow. In Sect. 2, some basic concepts of fractional-order systems are introduced. In Sect. 3, the master and the slave systems are described to formulate the problem of IMPS. In Sect. 4, two control schemes are proposed which enables IMPS to be achieved for commensurate master system and incommensurate master system cases, respectively. In Sect. 5, simulation results are performed to verify the effectiveness and feasibility of the proposed schemes. Finally, concluding remarks end the study.

2 Basic Concepts

In this section, we present some basic concepts of fractional derivatives and stability of fractional systems which are helpful in the proving analysis of the proposed approaches.

2.1 Caputo Fractional Derivative

The idea of fractional integrals and derivatives has been known since the development of the regular calculus, with the first reference probably being associated with Leibniz in 1695. There are several definitions of fractional derivatives [140]. The Caputo derivative [141] is a time domain computation method. In real applications, the Caputo derivative is more popular since the un-homogenous initial conditions are permitted if such conditions are necessary. Furthermore, these initial values are prone to measure since they all have idiographic meanings [142]. The Caputo derivative definition is given by

$$D_t^p f(t) = J^{m-p} f^m(t), \tag{1}$$

where $0 < p \leq 1, m = [p]$, i.e., m is the first integer which is not less than p, f^m is the m-order derivative in the usual sense, and J^q $(q > 0)$ is the q-order Reimann-Liouville integral operator with expression:

$$J^q f(t) = \frac{1}{\Gamma(q)} \int_0^t (t-\tau)^{q-1} f(\tau)\, d\tau, \tag{2}$$

where Γ denotes Gamma function.

Some basic properties and Lemmas of fractional derivatives and integrals used in this study are listed as follows.

Property 1 *For $p = n$, where n is an integer, the operation D_t^p gives the same result as classical integer order n. Particularly, when $p = 1$, the operation D_t^p is the same as the ordinary derivative, i.e., $D_t^1 f(t) = \frac{df(t)}{dt}$; when $p = 0$, the operation $D_t^p f(t)$ is the identity operation: $D_t^0 f(t) = f(t)$.*

Property 2 *Fractional differentiation (fractional integration) is linear operation:*

$$D_t^p \left[af(t) + bg(t) \right] = aD_t^p f(t) + bD_t^p g(t). \tag{3}$$

Property 3 *The fractional differential operator D_t^p is left-inverse (and not right-inverse) to the fractional integral operator J^p, i.e.*

$$D_t^p J^p f(t) = D^0 f(t) = f(t). \tag{4}$$

Lemma 1 [143] *The Laplace transform of the Caputo fractional derivative rule reads*

$$\mathbf{L}\left(D_t^p f(t)\right) = s^p \mathbf{F}(s) - \sum_{k=0}^{n-1} s^{p-k-1} f^{(k)}(0), \quad (p > 0, \; n-1 < p \leq n). \tag{5}$$

Particularly, when $0 < p \leq 1$, we have

$$\mathbf{L}\left(D_t^p f(t)\right) = s^p \mathbf{F}(s) - s^{p-1} f(0). \tag{6}$$

Lemma 2 [144] *The Laplace transform of the Riemann-Liouville fractional integral rule satisfies*

$$\mathbf{L}\left(J^q f(t)\right) = s^{-q} \mathbf{F}(s), \quad (q > 0). \tag{7}$$

Lemma 3 [103] *Suppose $f(t)$ has a continuous kth derivative on $[0, t]$ $(k \in N$, $t > 0)$, and let $p, q > 0$ be such that there exists some $\ell \in N$ with $\ell \leq k$ and p, $p+q \in [\ell - 1, \ell]$. Then*

$$D_t^p D_t^q f(t) = D_t^{p+q} f(t), \tag{8}$$

Remark 1 Note that the condition requiring the existence of the number ℓ with the above restrictions in the property is essential. In this work, we consider the case that $0 < p$, $q \leq 1$, and $0 < p + q \leq 1$. Apparently, under such conditions this property holds.

2.2 Stability of Linear Fractional Systems

Consider the following linear fractional system

$$D_t^{p_i} x_i(t) = \sum_{j=1}^{n} a_{ij} x_j(t), \qquad i = 1, 2, ..., n, \tag{9}$$

where p_i is a rational number between 0 and 1 and $D_t^{p_i}$ is the Caputo fractional derivative of order p_i, for $i = 1, 2, ..., n$. Assume that $p_i = \frac{\alpha_i}{\beta_i}$, $(\alpha_i, \beta_i) = 1$, $\alpha_i, \beta_i \in \mathbb{N}$, for $i = 1, 2, ..., n$. Let d be the least common multiple of the denominators β_i's of p_i's.

Lemma 4 [145] *If p_i's are different rational numbers between 0 and 1, then the system (9) is asymptotically stable if all roots λ of the equation*

$$\det\left(diag\left(\lambda^{dp_1}, \lambda^{dp_2}, ..., \lambda^{dp_n}\right) - A\right) = 0, \tag{10}$$

satisfy $|\arg(\lambda)| > \frac{\pi}{2d}$, where $A = \left(a_{ij}\right)_{n \times n}$.

2.3 Fractional Lyapunov Method

Definition 1 A continuous function γ is said to belong to class-K if it is strictly increasing and $\gamma(0) = 0$.

Theorem 1 [146] *Let $X = 0$ be an equilibrium point for the following fractional order system*

$$D_t^p X(t) = F(X(t)), \tag{11}$$

where $0 < p \leq 1$. Assume that there exists a Lyapunov function $V(X(t))$ and class-K functions γ_i $(i = 1, 2, 3)$ satisfying

$$\gamma_1(\|X\|) \leq V(X(t)) \leq \gamma_2(\|X\|). \tag{12}$$

$$D_t^p V(X(t)) \leq -\gamma_3 (\|X\|). \tag{13}$$

Then the system (11) is asymptotically stable.

Theorem 2 [147] *If there exists a positive definite Lyapunov function $V(X(t))$ such that $D_t^p V(X(t)) < 0$, for all $t > 0$, then the trivial solution of system (11) is asymptotically stable.*

In the following, a new lemma for the Caputo fractional derivative is presented.

Lemma 5 [148] $\forall X(t) \in \mathbf{R}^n$, $\forall p \in \left]0, 1\right]$ *and* $\forall t > 0$

$$\frac{1}{2} D_t^p \left(X^T(t)X(t)\right) \leq X^T(t) D_t^p (X(t)). \tag{14}$$

3 System Description and Problem Formulation

We consider the following fractional chaotic system as the master system

$$D_t^p X(t) = AX(t) + f(X(t)), \tag{15}$$

where $X(t) = \left(x_1(t), x_2(t), ..., x_n(t)\right)^T$ is the state vector of the master system (15), $A = \left(a_{ij}\right)_{n\times n}$ is a constant matrix, $f = \left(f_i\right)_{1\leq i\leq n}$ is a nonlinear function, $D_t^p = \left[D_t^{p_1}, D_t^{p_2}, ..., D_t^{p_n}\right]$ is the Caputo fractional derivative and p_i, $i = 1, 2, ..., n$, are rational numbers between 0 and 1.

Also, consider the slave system as

$$D_t^q Y(t) = g(Y(t)) + U, \tag{16}$$

where $Y(t) = \left(y_1(t), y_2(t), ..., y_m(t)\right)^T$ is the state vector of the slave system (16), $g = \left(g_i\right)_{1\leq i\leq m}$, D_t^q is the Caputo fractional derivative of order q, where q is a rational number between 0 and 1 and $U = \left(u_i\right)_{1\leq i\leq m}$ is a vector controller to be designed.

Before proceeding to the definition of inverse matrix projective synchronization (IMPS) for the coupled fractional chaotic systems (15) and (16), the definition of matrix projective synchronization (MPS) is provided.

Definition 2 The n-dimensional master system $X(t)$ and the m-dimensional slave system $Y(t)$ are said to be matrix projective synchronization (MPS), if there exists a controller $U = \left(u_i\right)_{1\leq i\leq m}$ and a given constant matrix $\Lambda = \left(\Lambda_{ij}\right)_{m\times n}$, such that the synchronization error

$$e(t) = Y(t) - \Lambda \times X(t), \tag{17}$$

satisfies that $\lim_{t\longrightarrow +\infty} \|e(t)\| = 0$.

Definition 3 The n-dimensional master system $X(t)$ and the m-dimensional slave system $Y(t)$ are said to be inverse matrix projective synchronization (IMPS), if there exists a controller $U = \left(u_i\right)_{1\leq i\leq m}$ and a given constant matrix $M = \left(M_{ij}\right)_{n\times m}$, such that the synchronization error

$$e(t) = X(t) - M \times Y(t), \tag{18}$$

satisfies that $\lim_{t\longrightarrow+\infty} \|e(t)\| = 0$.

Remark 2 The problem of inverse matrix projective synchronization in chaotic discrete-time systems have been studied and carried out, for example, in Ref. [149].

4 Fractional IMPS Schemes

In this section, we discuss two schemes of IMPS between the master system (15) and the slave system (16): The first scheme is proposed when the master system is commensurate system and the second one is constructed when the master system is incommensurate system. In this study, we assume that $n < m$.

4.1 *Case 1*

In this case, we assume that $p_1 = p_2 = \ldots = p_n = p$ and $q < p$. The error system of IMPS, in scalar form, between the master system (15) and the slave system (16) is defined by

$$e_i(t) = x_i(t) - \sum_{j=1}^{m} M_{ij} \times y_j(t), \qquad i = 1, 2, ..., n. \tag{19}$$

Suppose that the controllers u_i, $i = 1, 2, ..., m$, can be designed in the following form

$$u_i = -g_i\left(Y(t)\right) + J^{p-q}\left(v_i\right), \qquad i = 1, 2, ..., m, \tag{20}$$

where v_i, $1 \leq i \leq m$, are new controllers to be determined later.

By substituting Eq. (20) into Eq. (16), we can rewrite the slave system as

$$D_t^q y_i\left(t\right) = J^{p-q}\left(v_i\right), \qquad i = 1, 2, ..., m. \tag{21}$$

Now, applying the Laplace transform to (21) and letting

$$\mathbf{L}\left(y_i(t)\right) = \mathbf{F}_i(s), \qquad i = 1, 2, ..., m, \tag{22}$$

we obtain,

$$s^q \mathbf{F}_i(s) - s^{q-1} y_i(0) = s^{q-p} \mathbf{L}\left(v_i\right), \qquad i = 1, 2, ..., m, \tag{23}$$

multiplying both the left-hand and right-hand sides of (23) by s^{p-q}, and again applying the inverse Laplace transform to the result, we obtain

$$D_t^p y_i(t) = v_i, \qquad i = 1, 2, ..., m. \tag{24}$$

Now, the Caputo fractional derivative for order p of the error system (19) can be derived as

$$\begin{aligned} D_t^p e_i(t) &= D_t^p x_i(t) - \sum_{j=1}^{m} M_{ij} \times D_t^p y_j(t) \\ &= \sum_{j=1}^{n} a_{ij} x_j(t) + f_i(X(t)) - \sum_{j=1}^{m} M_{ij} \times v_j, \qquad i = 1, 2, ..., n. \end{aligned} \tag{25}$$

Furthermore, the error system (25) can be written as

$$D_t^p e_i(t) = \sum_{j=1}^{n} \left(a_{ij} - c_{ij}\right) e_j + R_i - \sum_{j=1}^{m} M_{ij} \times v_j, \qquad i = 1, 2, ..., n, \tag{26}$$

where $\left(c_{ij}\right) \in \mathbf{R}^{n \times n}$ are control constants and

$$R_i = \sum_{j=1}^{n} \left(c_{ij} - a_{ij}\right) e_j + \sum_{j=1}^{n} a_{ij} x_j(t) + f_i(X(t)), \qquad i = 1, 2, ..., n. \tag{27}$$

Rewriting the error system (26) in the compact form

$$D_t^p e(t) = (A - C) e(t) + R - M \times V, \tag{28}$$

where $e(t) = \left(e_1(t), e_2(t), ..., e_n(t)\right)^T$, $C = \left(c_{ij}\right)_{n \times n}$ is a control matrix to be selected later, $R = \left(R_1, R_2, ..., R_n\right)^T$ and $V = \left(v_1, v_2, ..., v_n, v_{n+1}, ..., v_m\right)^T$.

Theorem 3 *If the control matrix $C \in \mathbf{R}^{n \times n}$ is chosen such that $P = A - C$ is a negative definite matrix, then the master system (15) and the slave system (16) are globally inverse matrix projective synchronized under the following control law*

$$\left(v_1, v_2, ..., v_n\right)^T = \hat{M}^{-1} \times R, \tag{29}$$

and

$$v_{n+1} = v_{n+2} = ... = v_m = 0, \tag{30}$$

where $\hat{M}^{-1}$ is the inverse matrix of $\hat{M} = \left(M_{ij}\right)_{n\times n}$.

Proof By using (30), the error system (28) can be writtes as

$$D_t^p e(t) = (A - C)\, e(t) + R - \hat{M} \times \left(v_1, v_2, ..., v_n\right)^T, \tag{31}$$

where $\hat{M} = \left(M_{ij}\right)_{n\times n}$. Applying the control law given in Eqs. (29) to (31) yields the resulting error dynamics as follows

$$D_t^p e(t) = (A - C)\, e(t). \tag{32}$$

If a Lyapunov function candidate is chosen as

$$V(e(t)) = \frac{1}{2} e^T(t)\, e(t), \tag{33}$$

then the time Caputo fractional derivative of order p of V along the trajectory of the system (32) is as follows

$$D_t^p V(e(t)) = \frac{1}{2} D_t^p \left(e^T(t)\, e(t)\right), \tag{34}$$

and by using Lemma 5 in Eq. (34) we get

$$\begin{aligned} D_t^p V(e(t)) &\leq e^T(t)\, D_t^p e(t) \\ &= e^T(t)\,(A - C)\, e(t) = e^T(t)\, P e(t) < 0. \end{aligned}$$

Thus, from Theorem 2, it is immediate that is the zero solution of the system (32) is globally asymptotically stable and therefore, systems (15) and (16) are globally inverse matrix projective synchronized.

4.2 Case 2

Now, in this case, we assume that $p_1 \neq p_2 \neq \ldots \neq p_n$ and $q < p_i$ for $i = 1, 2, ..., n$. The vector controller $U = \left(u_i\right)_{1\leq i\leq m}$ can be designed a

$$\begin{bmatrix} u_1 \\ u_2 \\ \vdots \\ u_n \\ u_{n+1} \\ \vdots \\ u_m \end{bmatrix} = \begin{bmatrix} -g_1(Y(t)) + J^{p_1-q}\left(v_1\right) \\ -g_2(Y(t)) + J^{p_2-q}\left(v_2\right) \\ \vdots \\ -g_n(Y(t)) + J^{p_n-q}\left(v_n\right) \\ -g_{n+1}(Y(t)) \\ \vdots \\ -g_m(Y(t)) \end{bmatrix}, \tag{35}$$

where v_i, $i = 1, ..., n$, are new controllers. By substituting Eq. (35) into Eq. (16), we can rewrite the slave system as

$$D_t^q y_i(t) = J^{p_i - q}\left(v_i\right), \qquad i = 1, ..., n, \tag{36}$$

and

$$D_t^q y_i(t) = 0, \qquad i = n+1, n+2, ..., m. \tag{37}$$

By applying the Caputo fractional derivative of order $p_i - q$ to both the left and right sides of Eq. (36) and by using Lemma (3), we obtain

$$\begin{aligned} D_t^{p_i} y_i(t) &= D_t^{p_i - q}\left(D_t^q y_i(t)\right) \\ &= D_t^{p_i - q} J^{p_i - q}\left(v_i\right) \\ &= v_i, \qquad i = 1, ..., n. \end{aligned} \tag{38}$$

In this case, the error system between the master system (15) and the slave system (16) can be derived as

$$\begin{aligned} D_t^{p_i} e_i(t) &= D_t^{p_i} x_i(t) - D_t^{p_i}\left(\sum_{j=1}^{m} M_{ij} y_j(t)\right) \\ &= \sum_{j=1}^{m} a_{ij} x_j(t) + f_i(X(t)) - \sum_{\substack{j=1 \\ j\neq i}}^{m} M_{ij} D_t^{p_i} y_j(t) - M_{ii} v_i, \qquad i = 1, 2, ..., n. \end{aligned} \tag{39}$$

Furthermore, the error system (39) can be written as

$$D_t^{p_i} e_i(t) = \sum_{j=1}^{n} a_{ij} e_j + T_i - M_{ii} v_i, \qquad i = 1, 2, ..., n, \tag{40}$$

where

$$T_i = -\sum_{j=1}^{n} a_{ij} e_j + \sum_{j=1}^{n} a_{ij} x_j(t) + f_i(X(t)) - \sum_{\substack{j=1 \\ j\neq i}}^{m} M_{ij} D_t^{p_i} y_j(t). \tag{41}$$

Rewriting the error system (41) in the compact form

$$D_t^p e(t) = Ae(t) + T - \operatorname{diag}\left(M_{11}, M_{22}, ..., M_{nn}\right) \times V, \tag{42}$$

where $D_t^p = \left[D_t^{p_1}, D_t^{p_2}, ..., D_t^{p_n}\right]$, $e(t) = \left[e_1(t), e_2(t), ..., e_n(t)\right]^T$, $T = \left(T_1, T_2, ..., T_n\right)^T$ and $V = \left(v_1, v_2, ..., v_n\right)^T$.

To achieve IMPS between the master system (15) and the slave system (16), we assume that $M_{ii} \neq 0$, $i = 1, 2, ..., n$. Hence, we have the following result.

Theorem 4 *There exists a feedback gain matrix* $L \in \mathbf{R}^{n \times n}$ *to realize inverse matrix projective synchronization between the master system (15) and the slave system (16) under the following control law*

$$V = diag\left(\frac{1}{M_{11}}, \frac{1}{M_{22}}, ..., \frac{1}{M_{nn}}\right) \times (T + Le(t)). \tag{43}$$

Proof Applying the control law given in Eq. (43) to Eq. (42), the error system can be described as

$$D_t^p e(t) = (A - L) e(t). \tag{44}$$

The feedback gain matrix L is chosen such that all roots λ, of

$$\det\left(\text{diag}\left(\lambda^{dp_1}, \lambda^{dp_2}, ..., \lambda^{dp_n}\right) + L - A\right) = 0, \tag{45}$$

satisfy $|\arg(\lambda)| > \frac{\pi}{2d}$, where d is the least common multiple of the denominators of p_i, $i = 1, 2, ..., n$. According to Lemma 4, we conclude that the zero solution of the error system (44) is globally asymptotically stable and therefore, systems (15) and (16) are IMPS synchronized.

5 Numerical Examples

In this section, two numerical examples are used to show the effectiveness of the derived results.

5.1 Example 1

In this example, we consider the commensurate fractional order Lorenz system as the master system and the controlled hyperchaotic fractional order, proposed by Zhou et al. [151], as the slave system.

The master system is defined as

$$\begin{aligned} D^p x_1 &= \alpha\left(x_3 - x_1\right), \\ D^p x_2 &= \gamma x_1 - x_2 - x_3 x_1, \\ D^p x_3 &= -\beta x_3 + x_2 x_1, \end{aligned} \tag{46}$$

where x_1, x_2 and x_3 are states. For example, chaotic attractors are found in [150], when $(\alpha, \beta, \gamma) = (10, \frac{8}{3}, 28)$ and $p = 0.993$. Different chaotic attractors of the fractional order Lorenz system (46) are shown in Figs. 1 and 2.

Compare system (46) with system (15), one can have

$$A = \begin{pmatrix} -10 & 10 & 0 \\ 28 & -1 & 0 \\ 0 & 0 & -\frac{8}{3} \end{pmatrix}, \qquad f = \begin{pmatrix} 0 \\ -x_1x_3 \\ x_1x_2 \end{pmatrix}.$$

The slave system is described by

$$\begin{aligned} D^q y_1 &= 0.56y_1 - y_2 + u_1, \\ D^q y_2 &= y_1 - 0.1y_2y_3^2 + u_2, \\ D^q y_3 &= 4y_2 - y_3 - 6y_4 + u_3, \\ D^q y_4 &= 0.5y_3 + 0.8y_4 + u_4, \end{aligned} \tag{47}$$

where y_1, y_2, y_3, y_4 are states and u_i, $i = 1, 2, 3, 4$, are synchronization controllers. The uncontrolled fractional hyperchaotic system (47) (i.e. the system (47) with $u_1 = u_2 = u_3 = u_4 = 0$) exhibits hyperchaotic behavior when $q = 0.95$. Attractors in 2-D and 3-D of the uncontrolled fractional hyperchaotic system (47) are shown in Figs. 3 and 4.

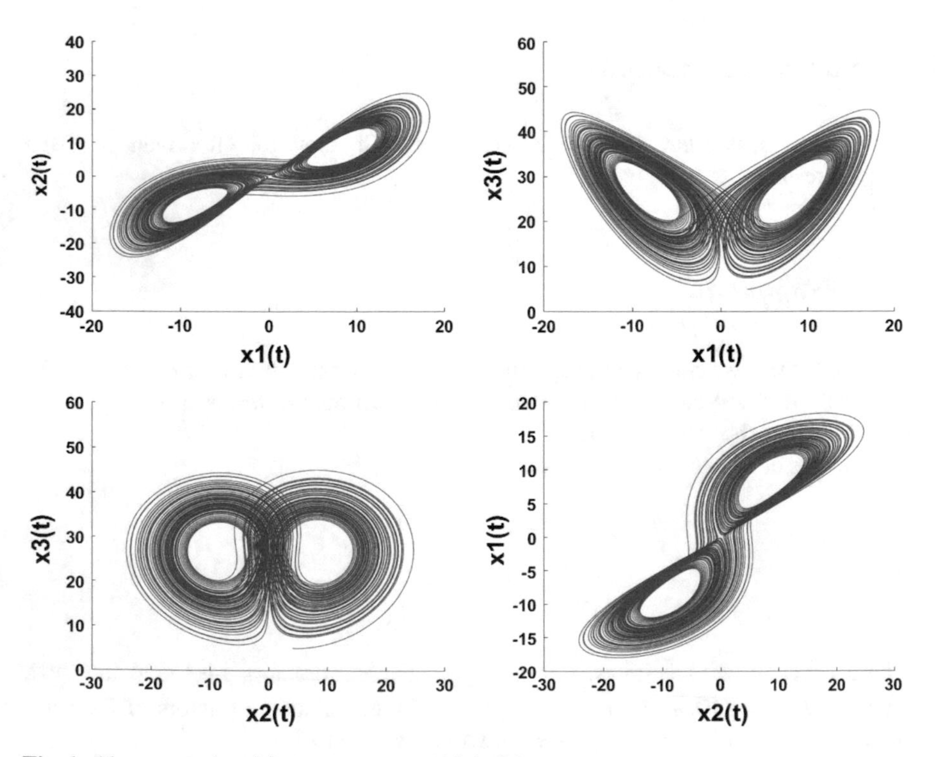

Fig. 1 Phase portraits of the master system (46) in 2-D

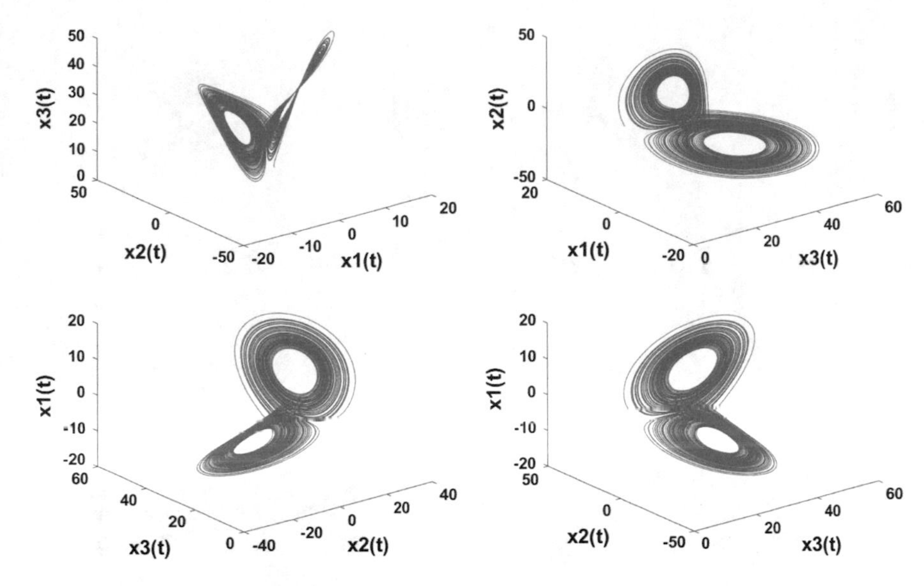

Fig. 2 Phase portraits of the master system (46) in 3-D

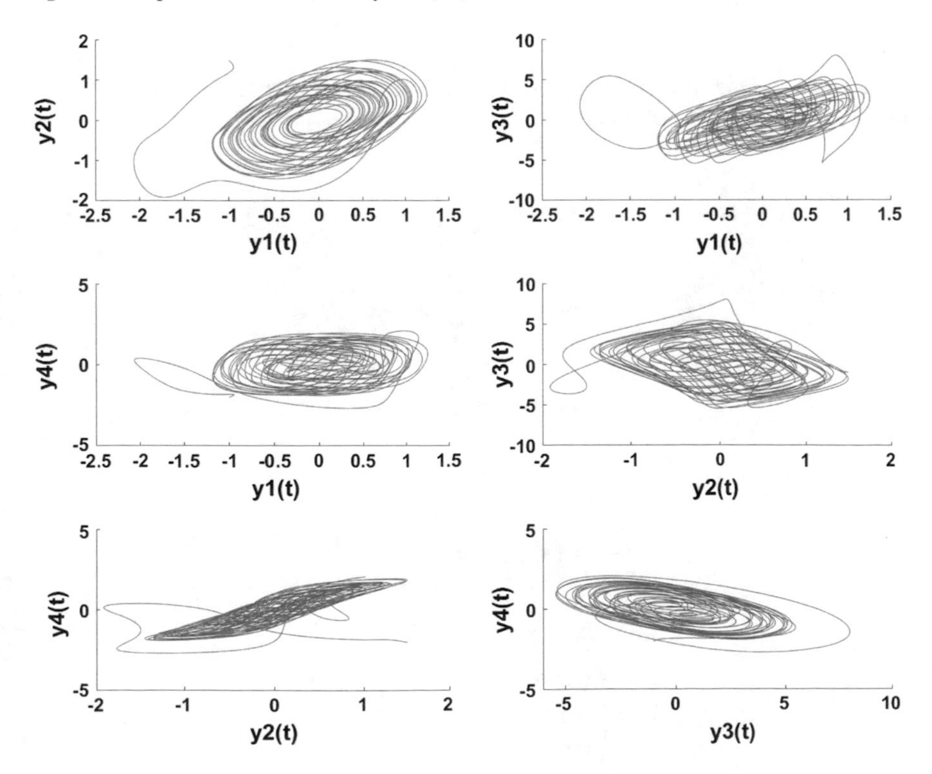

Fig. 3 Phase portraits of the slave system (47) in 2-D

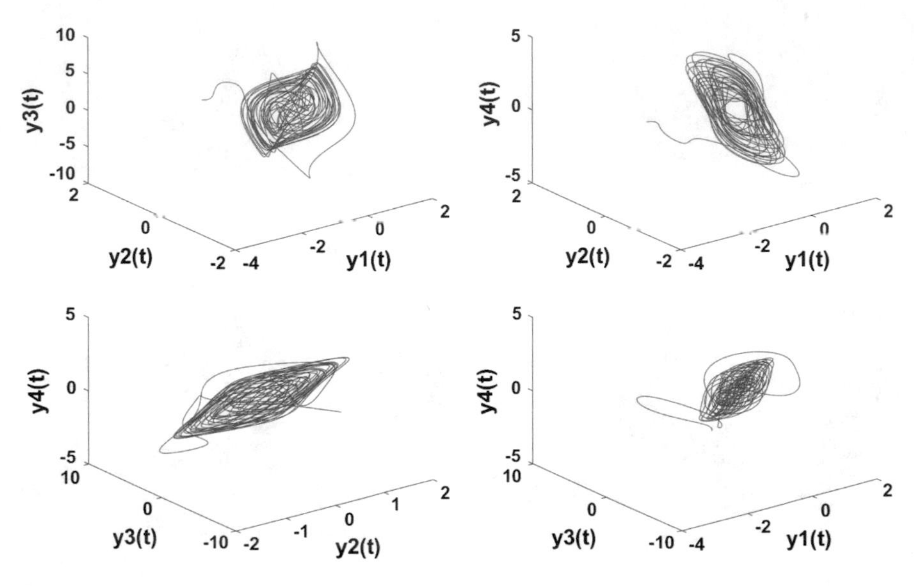

Fig. 4 Phase portraits of the slave system (47) in 3-D

In this example, the error system of IMPS between the master system (46) and the slave system (47) is defined as

$$e_1 = x_1 - \sum_{j=1}^{4} M_{1j} y_j, \tag{48}$$

$$e_2 = x_2 - \sum_{j=1}^{4} M_{2j} y_j,$$

$$e_3 = x_3 - \sum_{j=1}^{4} M_{3j} y_j,$$

where

$$M = \left(M_{ij}\right)_{3\times 4} = \begin{pmatrix} 2 & 0 & 0 & 4 \\ 0 & 1 & 0 & 5 \\ 0 & 0 & 3 & 6 \end{pmatrix}.$$

So,

$$\hat{M} = \begin{pmatrix} 2 & 0 & 0 \\ 0 & 1 & 0 \\ 0 & 0 & 3 \end{pmatrix} \text{ and } \hat{M}^{-1} = \begin{pmatrix} \frac{1}{2} & 0 & 0 \\ 0 & 1 & 0 \\ 0 & 0 & \frac{1}{3} \end{pmatrix}.$$

According to Theorem 3, there exists a control matrix $C \in \mathbf{R}^{3\times3}$ so that systems (46) and (47) realize the IMPS. For example, the control matrix C can be chosen as

$$C = \begin{pmatrix} 0 & 10 & 0 \\ 28 & 0 & 0 \\ 0 & 0 & 0 \end{pmatrix}. \tag{49}$$

It is easy to show that $A - C$ is a negative definite matrix. Then the control functions are designed as

$$\begin{aligned} u_1 &= -0.56y_1 + y_2 + J^{0.043}\left(-10y_1 - 20y_4 + 5x_3\right), \\ u_2 &= -y_1 + 0.1y_2y_3^2 + J^{0.043}\left(-y_2 - 4y_4 + 28x_1 - x_3x_1\right), \\ u_3 &= -4y_2 + y_3 + 6y_4 + J^{0.043}\left(-\frac{8}{3}y_3 - \frac{16}{3}y_4 + \frac{1}{3}x_2x_1\right), \\ u_4 &= -0.5y_3 - 0.8y_4. \end{aligned} \tag{50}$$

Hence, the IMPS between the master system (46) and the slave system (47) is achieved. The error system can be described as follows

$$\begin{aligned} D^{0.993}e_1 &= -10e_1, \\ D^{0.993}e_2 &= -e_2, \\ D^{0.993}e_3 &= -\frac{8}{3}e_3. \end{aligned} \tag{51}$$

For the purpose of numerical simulation, fractional Euler integration method has been used. In addition, simulation time $Tm = 120\,\text{s}$ and time step $h = 0.005s$ have been employed. The initial values of the master system and the slave system are $[x_1(0), x_2(0), x_3(0)] = [3, 4, 5]$ and $[y_1(0), y_2(0), y_3(0), y_4(0)] = [-1, 1.5, -1, -2]$, respectively, and the initial states of the error system are $[e_1(0), e_2(0), e_3(0)] = [13, 12.5, 20]$. Figure 5 displays the time evolution of the errors of IMPS between the master system (46) and the slave system (47).

5.2 *Example 2*

In this example, we assumed that the incommensurate fractional order Liu system is the master system and the incommensurate fractional order hyperchaotic Liu system [153] is the slave system. The master system is defined as

$$\begin{aligned} D^{p_1}x_1 &= a\left(x_2 - x_1\right), \\ D^{p_2}x_2 &= bx_1 - x_1x_3, \\ D^{p_3}x_3 &= -cx_3 + 4x_1^2, \end{aligned} \tag{52}$$

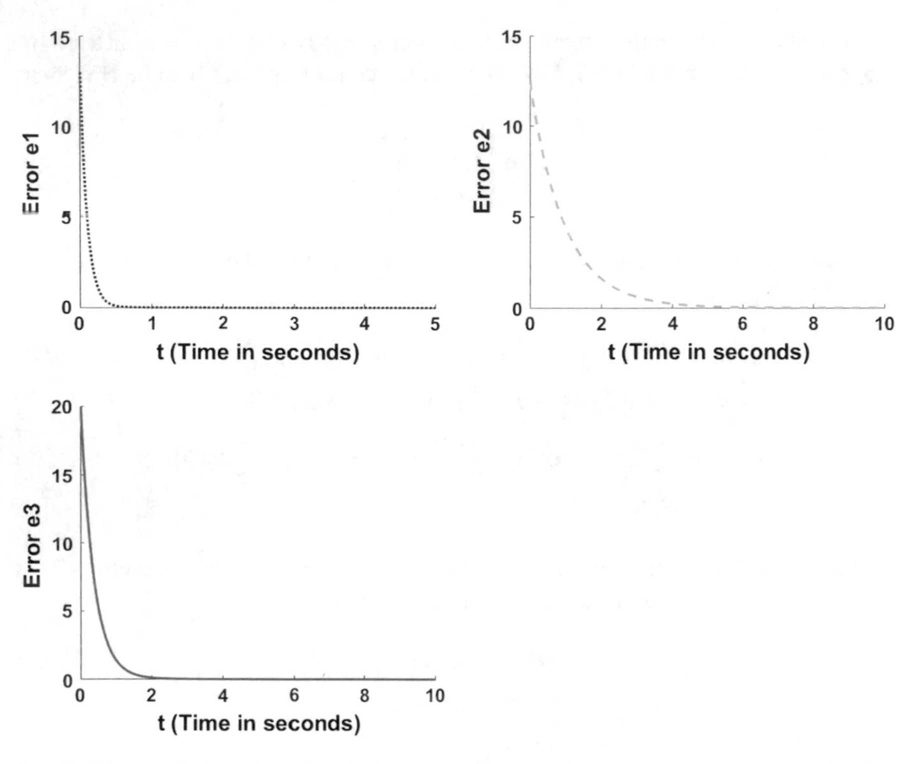

Fig. 5 Time evolution of synchronization errors between the master system (46) and the slave system (47)

where x_1, x_2 and x_3 are states. For example, chaotic attractors are found in [152], when $(p_1, p_2, p_3) = (0.93, 0.94, 0.95)$ and $(a, b, c) = (10, 40, 2.5)$. The Liu chaotic attractors are shown in Figs. 6 and 7.

Compare system (52) with system (15), one can have

$$A = \begin{pmatrix} -10 & 10 & 0 \\ 40 & 0 & 0 \\ 0 & 0 & -2.5 \end{pmatrix}, \qquad f = \begin{pmatrix} 0 \\ -x_1 x_3 \\ 4x_1^2 \end{pmatrix}.$$

The slave system is given by

$$\begin{aligned} D^q y_1 &= 10\left(y_2 - y_1\right) + y_4 + u_1, \\ D^q y_2 &= 40y_1 + 0.5y_4 - y_1 y_3 + u_2, \\ D^q y_3 &= -2.5y_3 + 4y_1^2 - y_4 + u_3, \\ D^q y_4 &= -\left(\frac{10}{15} y_2 + y_4\right) + u_4, \end{aligned} \tag{53}$$

where y_1, y_2, y_3, y_4 are states and u_i, $i = 1, 2, 3, 4$, are synchronization controllers.

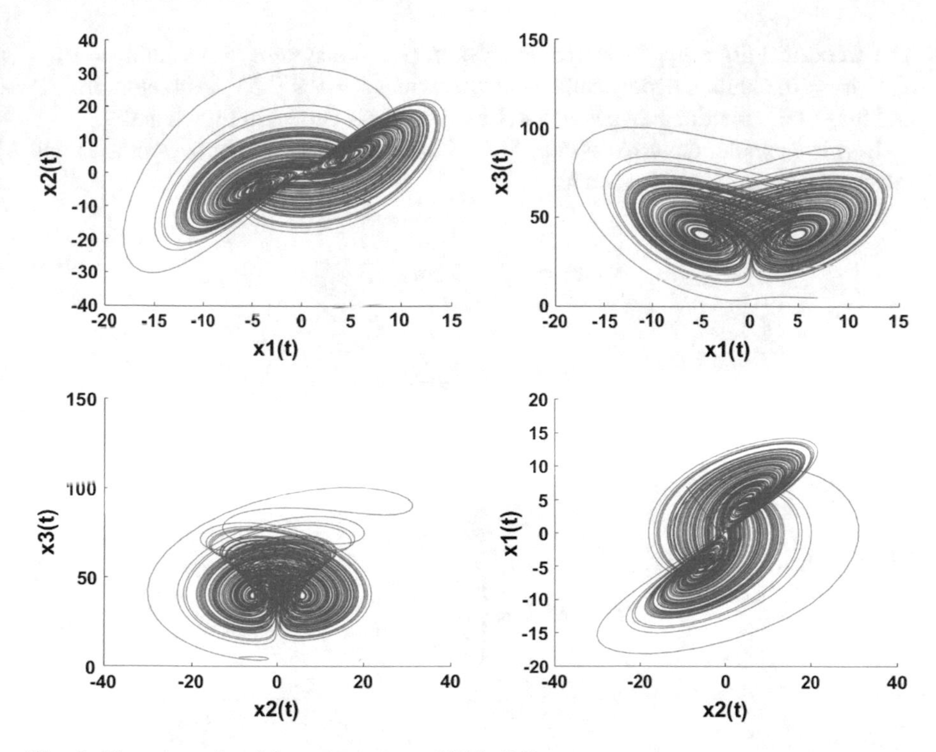

Fig. 6 Phase portraits of the master system (52) in 2-D

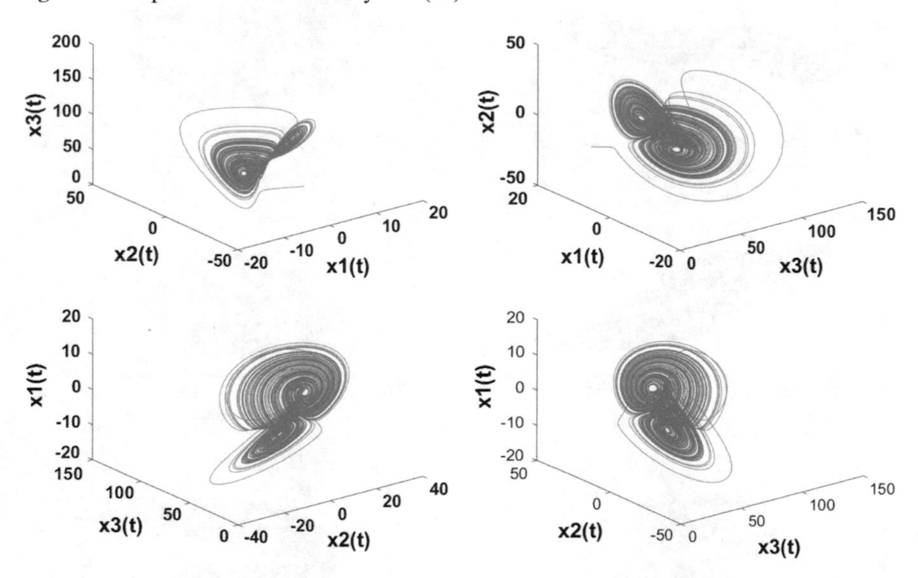

Fig. 7 Phase portraits of the master system (52) in 3-D

The fractional order hyperchaotic Liu system (i.e. the system (53) with $u_1 = u_2 = u_3 = u_4 = 0$) exhibits hyperchaotic behavior when $q = 0.9$ [153]. Attractors in 2-D and 3-D of the fractional hyperchaotic Liu system are shown in Figs. 8 and 9.

In this example, the error system of IMPS between the master system (52) and the slave system (53) is defined as

$$
\begin{aligned}
e_1 &= x_1 - \sum_{j=1}^{4} M_{1j} y_j, \\
e_2 &= x_2 - \sum_{j=1}^{4} M_{2j} y_j, \\
e_3 &= x_3 - \sum_{j=1}^{4} M_{3j} y_j,
\end{aligned} \tag{54}
$$

where

$$
M = \left(M_{ij}\right) = \begin{pmatrix} 6 & 3 & -2 & 4 \\ 0 & -5 & 0 & 5 \\ 2 & 1 & 4 & -1 \end{pmatrix}.
$$

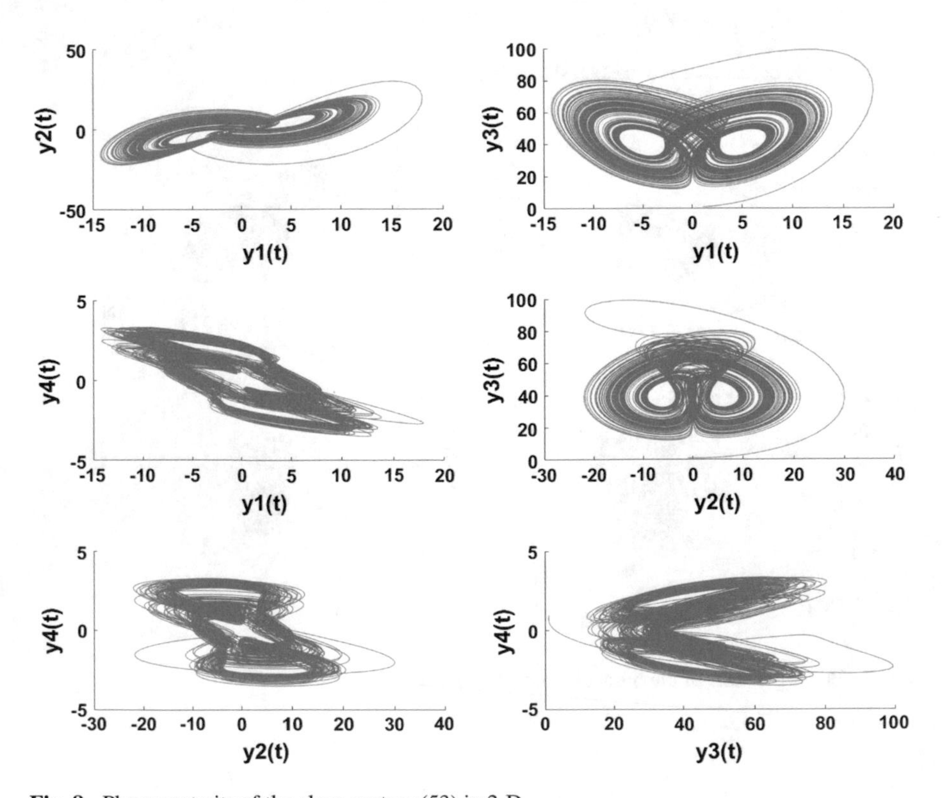

Fig. 8 Phase portraits of the slave system (53) in 2-D

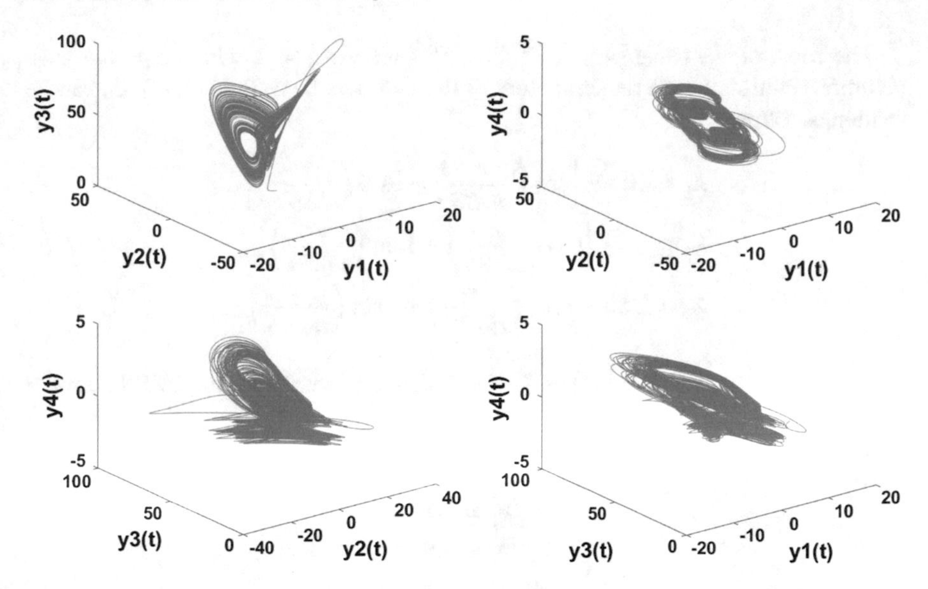

Fig. 9 Phase portraits of the slave system (53) in 3-D

So,

$$\operatorname{diag}\left(M_{11}, M_{22}, M_{33}\right) = \begin{pmatrix} 6 & 0 & 0 \\ 0 & -5 & 0 \\ 0 & 0 & 4 \end{pmatrix}.$$

According to Theorem 4, there exists a feedbak gain matrix $L \in \mathbf{R}^{3\times 3}$ so that systems (52) and (53) realize the IMPS. For example, the feedbak gain matrix L can be selected as

$$L = \begin{pmatrix} 0 & 10 & 0 \\ 40 & 5 & 0 \\ 0 & 0 & 0 \end{pmatrix}.$$

and the control functions are constructed as follows

$$\begin{aligned} u_1 &= 10\left(y_1 - y_2\right) - y_4 + J^{0.03}\frac{1}{6}\left(10e_1 + 10\left(x_2 - x_1\right) - 3D_t^{0.93}y_2\right. \\ &\quad \left. + 2D_t^{0.93}y_3 - 4D_t^{0.093}y_4\right), \qquad (55) \\ u_2 &= -40y_1 - 0.5y_4 + y_1y_3 - J^{0.04}\frac{1}{5}\left(5e_2 + 40x_1 - x_1x_3 - 5D_t^{0.94}y_4\right), \\ u_3 &= 2.5y_3 - 4y_1^2 + J^{0.05}\frac{1}{4}\left(2.5e_3 - 2.5x_3 + 4x_1^2 - 2D_t^{0.95}y_1 - D_t^{0.95}y_2 + D_t^{0.95}y_4\right), \\ u_4 &= \frac{10}{15}y_2 + y_4. \end{aligned}$$

The roots of $det\left(\text{diag}\left(\lambda^{d0.93}, \lambda^{d0.94}, \lambda^{d0.95}\right) + L - A\right) = 0$, where d is the least common multiple of the denominators of the numbers 0.93, 0.94 and 0.95, can be written as follows

$$\begin{aligned}\lambda_1 &= 10^{\frac{1}{d0.93}}\left[\cos\left(\frac{\pi}{d0.93}\right) + \mathbf{i}\sin\left(\frac{\pi}{d0.93}\right)\right],\\ \lambda_2 &= 5^{\frac{1}{d0.94}}\left[\cos\left(\frac{\pi}{d0.94}\right) + \mathbf{i}\sin\left(\frac{\pi}{d0.94}\right)\right],\\ \lambda_3 &= 2.5^{\frac{1}{d0.95}}\left[\cos\left(\frac{\pi}{d0.95}\right) + \mathbf{i}\sin\left(\frac{\pi}{d0.95}\right)\right].\end{aligned}$$

It is easy to see that $\arg\left(\lambda_i\right) > \frac{\pi}{2d}$, $i = 1, 2, 3$, and therefore, the IMPS between systems (52) and (53) is achieved.

The error system can be described as follows

$$\begin{aligned}D^{0.93}e_1 &= -10e_1,\\ D^{0.94}e_2 &= -5e_2,\\ D^{0.95}e_3 &= -2.5e_3.\end{aligned} \tag{56}$$

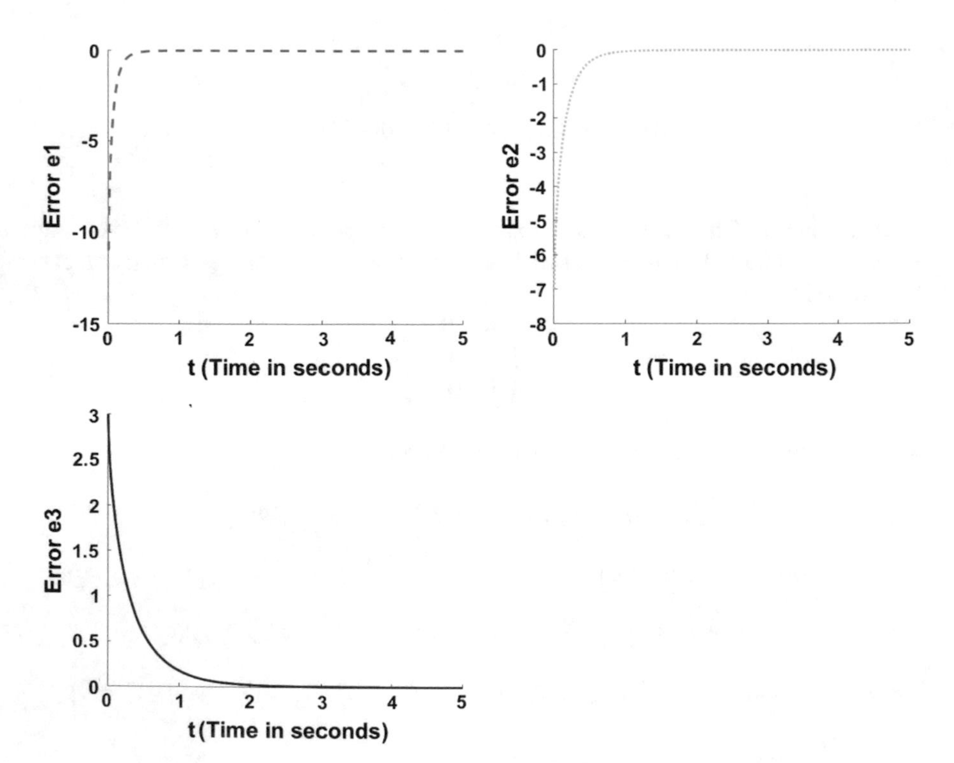

Fig. 10 Time evolution of synchronization errors between the master system (52) and the slave system (53)

For the purpose of numerical simulation, fractional Euler integration method has been used. In addition, simulation time $Tm = 120\,\mathrm{s}$ and time step $h = 0.005s$ have been employed. The initial values of the master system and the slave system are $[x_1(0), x_2(0), x_3(0)] = [0, 3, 9]$ and $[y_1(0), y_2(0), y_3(0), y_4(0)] = [2, -1, 1, 1]$, respectively, and the initial states of the error system are $[e_1(0), e_2(0), e_3(0)] = [-11, -7, 3]$. Figure 10 displays the time evolution of the errors of IMPS between the master system (52) and the slave system (53).

6 Conclusions

In this study, two new complex schemes of the inverse matrix projective synchronization (IMPS) were proposed between a master system of dimension n and a slave system of dimension m. Namely, by exploiting the fractional Lyapunov technique and stability theory of fractional-order linear system, the IMPS is rigorously proved to be achievable including the two cases: commensurate and incommensurate master systems. Finally, the effectiveness of the method has been illustrated by synchronizing a three-dimensional commensurate fractional Lorenz system with four-dimensional commensurate hyperchaotic fractional Zhou-Wei-Cheng system, and a three-dimensional incommensurate fractional order Liu system with four-dimensional commensurate fractional order hyperchaotic Liu system

The proposed approach presents some useful features:

(i) it enables chaotic (hyperchaotic) fractional system with different dimension to be synchronized;
(ii) it is rigorous, being based on theorems;
(iii) it can be applied to a wide class of chaotic (hyperchaotic) fractional systems;
(iv) due to the complexity of the proposed scheme, the fractional IMPS may enhance security in communication and chaotic encryption schemes.

References

1. Jumarie, G. (1992). A Fokker-Planck equation of fractional order with respect to time. *Journal of Mathematical Physics*, *33*, 3536–3541.
2. Metzler, R., Glockle, W. G., & Nonnenmacher, T. F. (1994). Fractional model equation for anomalous diffusion. *Physica A*, *211*, 13–24.
3. Mainardi, F. (1997). Fractional calculus: some basic problems in continuum and statistical mechanics. In A. Carpinteri & F. Mainardi (Eds.), *Fractals and Fractional Calculus in Continuum Mechanics*. Springer.
4. Hilfer, R. (2000). *Applications of fractional calculus in physics*. World Scientific.
5. Parada, F. J. V., Tapia, J. A. O., & Ramirez, J. A. (2007). Effective medium equations for fractional Fick's law in porous media. *Physica A*, *373*, 339–353.
6. Koeller, R. C. (1984). Application of fractional calculus to the theory of viscoelasticity. *Journal of Applied Mechanics*, *51*, 299–307.

7. Axtell, M., & Bise, E. M. (1990). Fractional calculus applications in control systems. In *Proceedings of the IEEE National Aerospace and Electronics Conference New York* (pp. 563–566).
8. Dorčák, L. (1994). *Numerical models for the simulation of the fractional-order control systems, UEF-04-94.*, Institute of Experimental Physics Košice, Slovakia: The Academy of Sciences.
9. Pires, E. J. S., Machado, J. A. T., & de Moura, P. B. (2003). Fractional order dynamics in a GA planner. *Signal Process*, *83*, 2371–2386.
10. Magin, R. L. (2006). *Fractional calculus in bioengineering*. Begell House Publishers.
11. Tseng, C. C. (2007). Design of FIR and IIR fractional order Simpson digital integrators. *Signal Process*, *87*, 1045–1057.
12. da Graca, M. M., Duarte, F. B. M., & Machado, J. A. T. (2008). Fractional dynamics in the trajectory control of redundant manipulators. *Communications in Nonlinear Science and Numerical Simulation*, *13*, 1836–1844.
13. Hedrih, K. S., & Stanojevic, V. N. (2010). A model of gear transmission: fractional order system dynamics. Mathematical Problems in Engineering, 1–23.
14. Nakagava, N., & Sorimachi, K. (1992). Basic characteristic of a fractance device. *IEICE Transactions on Fundamentals of Electronics*, *75*, 1814–1818.
15. Westerlund, S. (2002). Dead Matter Has Memory!. Causal Consulting.
16. Oldham, K. B., & Spanier, J. (1974). *The fractional calculus*. Academic.
17. Miller, K. S., & Ross, B. (1993). *An Introduction to the fractional calculus and fractional differential equations*. Wiley.
18. Petráš, I. (2008). A note on the fractional-order Chua's system. *Chaos, Solitons & Fractals*, *38*, 140–147.
19. West, B. J., Bologna, M., & Grigolini, P. (2002). *Physics of fractal operators*. Springer.
20. Zaslavsky, G.M. (2005). *Hamiltonian chaos and fractional dynamics*. Oxford University Press.
21. Azar, A. T., & Vaidyanathan, S. (2015). *Chaos modeling and control systems design, studies in computational intelligence* (vol. 581). Germany: Springer.
22. Azar, A. T., & Vaidyanathan, S. (2016). *Advances in chaos theory and intelligent control*. Studies in Fuzziness and Soft Computing (vol. 337). Germany: Springer. ISBN 978-3-319-30338-3.
23. Azar, A. T., & Vaidyanathan, S. (2015). *Computational intelligence applications in modeling and control*. Studies in Computational Intelligence (vol. 575). Germany: Springer. ISBN 978-3-319-11016-5.
24. Azar, A. T., & Vaidyanathan, S. (2015). *Handbook of research on advanced intelligent control engineering and automation*. Advances in Computational Intelligence and Robotics (ACIR) Book Series, IGI Global, USA. ISBN 9781466672482.
25. Zhu, Q., & Azar, A. T. (2015). *Complex system modelling and control through intelligent soft computations*. Studies in Fuzziness and Soft Computing (vol. 319). Germany: Springer. ISBN 978-3-319-12882-5.
26. Azar, A. T., & Zhu, Q. (2015). *Advances and applications in sliding mode control systems*. Studies in Computational Intelligence (vol. 576). Germany: Springer. ISBN 978-3-319-11172-8.
27. Filali, R. L., Benrejeb, M., & Borne, P. (2014). On observer-based secure communication design using discrete-time hyperchaotic systems. *Communications in Nonlinear Science and Numerical Simulation*, *19*, 1424–1432.
28. Sheikhan, M., Shahnazi, M., & Garoucy, S. (2013). Hyperchaos synchronization using PSO-optimized RBF-based controllers to improve security of communication systems. *Neural Computing & Applications*, *22*(5), 835–846.
29. Fernando, J. (2011). Applying the theory of chaos and a complex model of health to establish relations among financial indicators. *Procedia Computer Science*, *3*, 982–986.
30. Zsolt, B. (1997). Chaos theory and power spectrum analysis in computerized cardiotocography. *European Journal of Obstetrics & Gynecology and Reproductive Biology*, *71*(2), 163–168.

31. Chen, G., & Dong, X. (1989). *From chaos to order*. World Scientific.
32. Ott, E., Grebogi, C., & Yorke, J. A. (1990). Controling chaos. *Physical Review Letters, 64*, 1196–1199.
33. Boccalettis, C., Grebogi, Y. C., LAI, M. H., & Maza, D. (2000). The control of chaos: theory and application. *Physics Reports, 329*, 103–197.
34. Yamada, T., & Fujisaka, H. (1983). Stability theory of synchroized motion in coupled-oscillator systems. *Progress of Theoretical Physics, 70*, 1240–1248.
35. Pecora, L. M., & Carroll, T. L. (1990). Synchronization in chaotic systems. *Physical Review Letters, 64*, 821–827.
36. Carroll, T. L., & Pecora, L. M. (1991). Synchronizing a chaotic systems. *IEEE Transactions on Circuits and Systems, 38*, 453–456.
37. Pikovsky, A., Rosenblum, M., & Kurths, J. (2001). *Synchronization an universal concept in nonlinear sciences*. Cambridge university press.
38. Boccaletti, S., Kurths, J., Osipov, G., Valladares, D. L., & Zhou, C. S. (2002). The synchronization of chaotic systems. *Physics Reports, 366*, 1–101.
39. Aziz-Alaoui, M. A. (2006). Synchronization of chaos. *Encyclopedia of Mathematical Physics, 5*, 213–226.
40. Luo, A. (2009). A theory for synchronization of dynamical systems. *Communications in Nonlinear Science and Numerical Simulation, 14*, 1901–1951.
41. Vaidyanathan, S., Sampath, S., & Azar, A. T. (2015). Global chaos synchronisation of identical chaotic systems via novel sliding mode control method and its application to Zhu system. *International Journal of Modelling, Identification and Control (IJMIC), 23*(1), 92–100.
42. Vaidyanathan, S., Azar, A. T., Rajagopal, K., & Alexander, P. (2015). Design and SPICE implementation of a 12-term novel hyperchaotic system and its synchronization via active control, (2015). *International Journal of Modelling. Identification and Control (IJMIC), 23*(3), 267–277.
43. Vaidyanathan, S., & Azar, A. T. (2016). Takagi-sugeno fuzzy logic controller for liu-chen four-scroll chaotic system. *International Journal of Intelligent Engineering Informatics, 4*(2), 135–150.
44. Ouannas, A., Azar, A. T., & Abu-Saris, R. (2016). A new type of hybrid synchronization between arbitrary hyperchaotic maps. *International Journal of Machine Learning and Cybernetics*. doi:10.1007/s13042-016-0566-3.
45. Vaidyanathan, S., Azar, A. T. (2015). Anti-Synchronization of identical chaotic systems using sliding mode control and an application to vaidyanathan-madhavan chaotic systems. In A. T. Azar & Q. Zhu (Eds.), *Advances and applications in sliding mode control systems*. Studies in Computational Intelligence book Series (vol. 576, pp. 527–547), GmbH Berlin/Heidelberg: Springer. doi:10.1007/978-3-319-11173-5_19.
46. Vaidyanathan, S., & Azar, A. T. (2015). Hybrid synchronization of identical chaotic systems using sliding mode control and an application to vaidyanathan chaotic systems. In A. T. Azar & Q. Zhu, (Eds.), *Advances and applications in sliding mode control systems*. Studies in Computational Intelligence book Series, (vol. 576, pp. 549–569), GmbH Berlin/Heidelberg: Springer. doi:10.1007/978-3-319-11173-5_20.
47. Vaidyanathan, S., & Azar, A. T. (2015). analysis, control and synchronization of a Nine-Term 3-D novel chaotic system. In A. T. Azar & S. Vaidyanathan (Eds.), *Chaos Modeling and Control Systems Design*, Studies in Computational Intelligence (vol. 581, pp. 3–17), GmbH Berlin/Heidelberg: Springer. doi:10.1007/978-3-319-13132-0_1.
48. Vaidyanathan, S., & Azar, A. T. (2015). Analysis and control of a 4-D novel hyperchaotic system. In A. T. Azar & S. Vaidyanathan (Eds.), *Chaos modeling and control systems design*, Studies in Computational Intelligence (vol. 581, pp. 19–38), GmbH Berlin/Heidelberg: Springer. dpoi:10.1007/978-3-319-13132-0_2.
49. Vaidyanathan, S., Idowu, B. A., & Azar, A. T. (2015). Backstepping controller design for the global chaos synchronization of sprott's jerk systems. In A. T. Azar & S. Vaidyanathan (Eds.), *chaos modeling and control systems design*, Studies in Computational Intelligence, (vol. 581, pp. 39–58), GmbH Berlin/Heidelberg: Springer. doi:10.1007/978-3-319-13132-0_3.

50. Boulkroune, A., Bouzeriba, A., Bouden, T., & Azar, A. T. (2016). Fuzzy adaptive synchronization of uncertain fractional-order chaotic systems. In A. T Azar & S. Vaidyanathan (Eds.), *Advances in chaos theory and intelligent control*. Studies in Fuzziness and Soft Computing (vol. 337). Germany: Springer.
51. Boulkroune, A., Hamel, S., & Azar, A. T. (2016). Fuzzy control-based function synchronization of unknown chaotic systems with dead-zone input. *Advances in chaos theory and intelligent control*. Studies in Fuzziness and Soft Computing (vol. 337), Germany: Springer.
52. Vaidyanathan, S., & Azar, A. T. (2016). Dynamic analysis, adaptive feedback control and synchronization of an Eight-Term 3-D novel chaotic system with three quadratic nonlinearities. *Advances in chaos theory and intelligent control*. Studies in Fuzziness and Soft Computing (vol. 337). Germany: Springer.
53. Vaidyanathan, S., & Azar, A. T. (2016). Qualitative study and adaptive control of a novel 4-D hyperchaotic system with three quadratic nonlinearities. *Advances in chaos theory and intelligent control*. Studies in Fuzziness and Soft Computing (vol. 337). Germany: Springer.
54. Vaidyanathan, S., & Azar, A. T. (2016). A novel 4-D four-wing chaotic system with four quadratic nonlinearities and its synchronization via adaptive control method. *Advances in chaos theory and intelligent control*. Studies in Fuzziness and Soft Computing (vol. 337). Germany: Springer.
55. Vaidyanathan, S., & Azar, A. T. (2016). Adaptive control and synchronization of halvorsen circulant chaotic systems. *Advances in chaos theory and intelligent control*. Studies in Fuzziness and Soft Computing (vol. 337). Germany: Springer.
56. Vaidyanathan, S., & Azar, A. T. (2016). Adaptive backstepping control and synchronization of a novel 3-D jerk system with an exponential nonlinearity. *Advances in chaos theory and intelligent control*. Studies in Fuzziness and Soft Computing (vol. 337). Germany: Springer.
57. Vaidyanathan, S., & Azar, A. T. (2016). Generalized projective synchronization of a novel hyperchaotic Four-Wing system via adaptive control method. *Advances in chaos theory and intelligent control*. Studies in Fuzziness and Soft Computing. (vol. 337). Germany: Springer.
58. Hartley, T., Lorenzo, C., & Qammer, H. (1995). Chaos in a fractional order Chua's system. *IEEE Transactions on Circuits and Systems I Fundamental Theory and Applications*, *42*, 485–490.
59. Grigorenko, I., & Grigorenko, E. (2003). Chaotic dynamics of the fractional Lorenz system. *Physical Review Letters*, *91*, 034101.
60. Li, C., & Chen, G. (2004). Chaos in the fractional order Chen system and its control. *Chaos Solitons Fractals*, *22*, 549–554.
61. Lu, J. G., & Chen, G. (2006). A note on the fractional-order Chen system. *Chaos Solitons Fractals*, *27*, 685–688.
62. Li, C., & Chen, G. (2004). Chaos and hyperchaos in fractional order R össler equations. *Physica A*, *341*, 55–61.
63. Lu, J. G. (2005). Chaotic dynamics and synchronization of fractional-order Arneodo's systems. *Chaos Solitons Fractals*, *26*, 1125–1133.
64. Deng, W. H., & Li, C. P. (2005). Chaos synchronization of the fractional Lü system. *Physica A*, *353*, 61–72.
65. Guo, L. J. (2005). Chaotic dynamics and synchronization of fractional-order Genesio-Tesi systems. *Chinese Physics*, *14*, 1517–1521.
66. Ge, Z. M., & Ou, C. Y. (2007). Chaos in a fractional order modified Duffing system. *Chaos Solitons Fractals*, *34*, 262–291.
67. Chen, W. C. (2008). Nonlinear dynamic and chaos in a fractional-order financial system. *Chaos Solitons Fractals*, *36*, 1305–1314.
68. Sheu, L. J., Chen, H. K., Chen, J. H., Tam, L. M., Chen, W. C., Lin, K. T., et al. (2008). Chaos in the Newton-Leipnik system with fractional order. *Chaos Solitons Fractals*, *36*, 98–103.
69. Ahmed, E., El-Sayed, A. M. A., & El-Saka, H. A. A. (2007). Equilibrium points, stability and numerical solutions of fractional-order predator-prey and rabies models. *Journal of Mathematical Analysis and Applications*, *325*, 542–553.

70. Gejji, V. D., & Bhalekar, S. (2010). Chaos in fractional order Liu system. *Computers & Mathematics with Applications, 59*, 1117–1127.
71. Li, C., & Zhou, T. (2005). Synchronization in fractional-order differential systems. *Physica D, 212*, 111–125.
72. Wang, J., Xiong, X., & Zhang, Y. (2006). Extending synchronization scheme to chaotic fractional-order Chen systems. *Physica A, 370*, 279–285.
73. Li, C. P., Deng, W. H., & Xu, D. (2006). Chaos synchronization of the Chua system with a fractional order. *Physica A, 360*, 171–185.
74. Peng, G. (2007). Synchronization of fractional order chaotic systems. *Physics Letters A, 363*, 426–432.
75. Yan, J., & Li, C. (2007). On chaos synchronization of fractional differential equations. *Chaos Solitons Fractals, 32*, 725–735.
76. Li, C., & Yan, J. (2007). The synchronization of three fractional differential systems. *Chaos Solitons Fractals, 32*, 751–757.
77. Zhou, S., Li, H., Zhu, Z., & Li, C. (2008). Chaos control and synchronization in a fractional neuron network system. *Chaos Solitons Fractals, 36*, 973–984.
78. Zhu, H., Zhou, S., & Zhang, J. (2009). Chaos and synchronization of the fractional-order Chua's system. *Chaos Solitons Fractals, 39*, 1595–1603.
79. Wu, X., Wang, H., & Lu, H. (2012). Modified generalized projective synchronization of a new fractional-order hyperchaotic system and its application to secure communication. *Nonlinear Analysis: Real World Applications, 13*(2012), 1441–1450.
80. Liang, H., Wang, Z., Yue, Z., & Lu, R. (2012). Generalized synchronization and control for incommensurate fractional unified chaotic system and applications in secure communication. *Kybernetika, 48*, 190–205.
81. Muthukumar, P., & Balasubramaniam, P. (2013). Feedback synchronization of the fractional order reverse butterfly-shaped chaotic system and its application to digital cryptography. *Nonlinear Dynamics*, 1169–1181.
82. Chen, L., Wu, R., He, Y., & Chai, Y. (2015). Adaptive sliding-mode control for fractional-order uncertain linear systems with nonlinear disturbances. *Nonlinear Dynamics, 80*, 51–58.
83. Liu, L., Ding, W., Liu, C., Ji, H., & Cao, C. (2014). Hyperchaos synchronization of fractional-order arbitrary dimensional dynamical systems via modified sliding mode control. *Nonlinear Dynamics, 76*, 2059–2071.
84. Zhang, L., & Yan, Y. (2014). Robust synchronization of two different uncertain fractional-order chaotic systems via adaptive sliding mode control. *Nonlinear Dynamics, 76*, 1761–1767.
85. Srivastava, M., Ansari, S. P., Agrawal, S. K., Das, S., & Leung, A. Y. T. (2014). Anti-synchronization between identical and non-identical fractional-order chaotic systems using active control method. *Nonlinear Dynamics, 76*, 905–914.
86. Agrawal, S. K., & Das, S. (2013). A modified adaptive control method for synchronization of some fractional chaotic systems with unknown parameters. *Nonlinear Dynamics, 73*, 907–919.
87. Zhou, P., & Bai, R. (2015). The adaptive synchronization of fractional-order chaotic system with fractional-order $1 < q < 2$ via linear parameter update law. *Nonlinear Dynamics, 80*, 753–765.
88. Odibat, Z. (2010). Adaptive feedback control and synchronization of non-identical chaotic fractional order systems. *Nonlinear Dynamics, 60*, 479–487.
89. Yuan, W. X., & Mei, S. J. (2009). Synchronization of the fractional order hyperchaos Lorenz systems with activation feedback control. *Communications in Nonlinear Science and Numerical Simulation, 14*, 3351–3357.
90. Odibat, Z. M., Corson, N., Aziz-Alaoui, M. A., & Bertelle, C. (2010). Synchronization of chaotic fractional-order systems via linear control. *International Journal of Bifurcation and Chaos, 20*, 81–97.
91. Chen, X. R., & Liu, C. X. (2012). Chaos synchronization of fractional order unified chaotic system via nonlinear control. *International Journal of Modern Physics B, 25*, 407–415.

92. Cafagna, D., & Grassi, G. (2012). Observer-based projective synchronization of fractional systems via a scalar signal: application to hyperchaotic Rössler systems. *Nonlinear Dynamics, 68*, 117–128.
93. Peng, G., & Jiang, Y. (2008). Generalized projective synchronization of a class of fractional-order chaotic systems via a scalar transmitted signal. *Physics Letters A, 372*, 3963–3970.
94. Odibat, Z. M. (2012). A note on phase synchronization in coupled chaotic fractional order systems. *Nonlinear Analysis: Real World Applications, 13*, 779–789.
95. Chen, F., Xia, L., & Li, C. G. (2012). Wavelet phase synchronization of fractional-order chaotic systems. *Chinese Physics Letters, 29*, 6–070501.
96. Razminiaa, A., & Baleanu, D. (2013). Complete synchronization of commensurate fractional order chaotic systems using sliding mode control. *Mechatronics, 23*, 873–879.
97. Pan, L., Zhou, W., Fang, J., & Li, D. (2010). Synchronization and anti-synchronization of new uncertain fractional-order modified unified chaotic systems. *Communications in Nonlinear Science and Numerical Simulation, 15*, 3754–3762.
98. Liu, F. C., Li, J. Y., & Zang, X. F. (2011). Anti-synchronization of different hyperchaotic systems based on adaptive active control and fractional sliding mode control. *Acta Physica Sinica, 60*, 030504.
99. Al-sawalha, M. M., Alomari, A. K., Goh, S. M., & Nooran, M. S. M. (2011). Active anti-synchronization of two identical and different fractional-order chaotic systems. *International Journal of Nonlinear Science, 11*, 267–274.
100. Li, C. G. (2006). Projective synchronization in fractional order chaotic systems and its control. *Progress of Theoretical Physics, 115*, 661–666.
101. Shao, S. Q., Gao, X., & Liu, X. W. (2007). Projective synchronization in coupled fractional order chaotic Rössler system and its control. *Chinese Physics, 16*, 2612–2615.
102. Wang, X. Y., & He, Y. J. (2008). Projective synchronization of fractional order chaotic system based on linear separation. *Physics Letters A, 372*, 435–441.
103. Si, G., Sun, Z., Zhang, Y., & Chen, W. (2012). Projective synchronization of different fractional-order chaotic systems with non-identical orders. *Nonlinear Analysis: Real World Applications, 13*, 1761–1771.
104. Agrawal, S. K., & Das, S. (2014). Projective synchronization between different fractional-order hyperchaotic systems with uncertain parameters using proposed modified adaptive projective synchronization technique. *Mathematical Methods in the Applied Sciences, 37*, 2164–2176.
105. Chang, C. M., & Chen, H. K. (2010). Chaos and hybrid projective synchronization of commensurate and incommensurate fractional-order Chen-Lee systems. *Nonlinear Dynamics, 62*, 851–858.
106. Wang, S., Yu, Y. G., & Diao, M. (2010). Hybrid projective synchronization of chaotic fractional order systems with different dimensions. *Physica A, 389*, 4981–4988.
107. Zhou, P., & Zhu, W. (2011). Function projective synchronization for fractional-order chaotic systems. *Nonlinear Analysis: Real World Applications, 12*, 811–816.
108. Zhou, P., & Cao, Y. X. (2010). Function projective synchronization between fractional-order chaotic systems and integer-order chaotic systems. *Chinese Physics B, 19*, 100507.
109. Xi, H., Li, Y., & Huang, X. (2015). Adaptive function projective combination synchronization of three different fractional-order chaotic systems. *Optik, 126*, 5346–5349.
110. Peng, G. J., Jiang, Y. L., & Chen, F. (2008). Generalized projective synchronization of fractional order chaotic systems. *Physica A, 387*, 3738–3746.
111. Shao, S. Q. (2009). Controlling general projective synchronization of fractional order Rössler systems. *Chaos Solitons Fractals, 39*, 1572–1577.
112. Wu, X. J., & Lu, Y. (2009). Generalized projective synchronization of the fractional-order Chen hyperchaotic system. *Nonlinear Dynamics, 57*, 25–35.
113. Zhou, P., Kuang, F., & Cheng, Y. M. (2010). Generalized projective synchronization for fractional order chaotic systems. *The Chinese Journal of Physics, 48*, 49–56.
114. Deng, W. H. (2007). Generalized synchronization in fractional order systems. *Physical Review E, 75*, 056201.

115. Zhou, P., Cheng, X. F., & Zhang, N. Y. (2008). Generalized synchronization between different fractional-order chaotic systems. *Communications in Theoretical Physics*, *50*, 931–934.
116. Zhang, X. D., Zhao, P. D., & Li, A. H. (2010). Construction of a new fractional chaotic system and generalized synchronization. *Communications in Theoretical Physics*, *53*, 1105–1110.
117. Jun, W. M., & Yuan, W. X. (2011). Generalized synchronization of fractional order chaotic systems. *International Journal of Modern Physics B*, *25*, 1283–1292.
118. Wu, X. J., Lai, D. R., & Lu, H. T. (2012). Generalized synchronization of the fractional-order chaos in weighted complex dynamical networks with nonidentical nodes. *Nonlinear Dynamics*, *69*, 667–683.
119. Xiao, W., Fu, J., Liu, Z., & Wan, W. (2012). Generalized synchronization of typical fractional order chaos system. *Journal of Computer*, *7*, 1519–1526.
120. Martínez-Guerra, R., & Mata-Machuca, J. L. (2014). Fractional generalized synchronization in a class of nonlinear fractional order systems. *Nonlinear Dynamics*, *77*, 1237–1244.
121. Yi, C., Liping, C., Ranchao, W., & Juan, D. (2013). Q-S synchronization of the fractional-order unified system. *Pramana*, *80*, 449–461.
122. Mathiyalagan, K., Park, J. H., & Sakthivel, R. (2015). Exponential synchronization for fractional-order chaotic systems with mixed uncertainties. *Complexity*, *21*, 114–125.
123. Aghababa, M. P. (2012). Finite-time chaos control and synchronization of fractional-order nonautonomous chaotic (hyperchaotic) systems using fractional nonsingular terminal sliding mode technique. *Nonlinear Dynamics*, *69*, 247–261.
124. Li, D., Zhang, X. P., Hu, Y. T., & Yang, Y. Y. (2015). Adaptive impulsive synchronization of fractional order chaotic system with uncertain and unknown parameters. *Neurocomputing*, *167*, 165–171.
125. Xi, H., Yu, S., Zhang, R., & Xu, L. (2014). Adaptive impulsive synchronization for a class of fractional-order chaotic and hyperchaotic systems. *Optik*, *125*, 2036–2040.
126. Ouannas, A., Al-sawalha, M. M., & Ziar, T. (2016). Fractional chaos synchronization schemes for different dimensional systems with non-Identical fractional-orders via two scaling matrices. *Optik*, *127*, 8410–8418.
127. Ouannas, A., Azar, A. T., & Vaidyanathan, S. (2016). A robust method for new fractional hybrid chaos synchronization. *Mathematical Methods in the Applied Sciences*, 1–9.
128. Hu, M., Xu, Z., Zhang, R., & Hu, A. (2007). Adaptive full state hybrid projective synchronization of chaotic systems with the same and different order. *Physics Letters A*, *365*, 315–327.
129. Zhang, Q., & Lu, J. (2008). Full state hybrid lag projective synchronization in chaotic (hyperchaotic) systems. *Physics Letters A*, *372*, 1416–1421.
130. Hu, M., Xu, Z., Zhang, R., & Hu, A. (2007). Parameters identification and adaptive full state hybrid projective synchronization of chaotic (hyperchaotic) systems. *Physics Letters A*, *361*, 231–237.
131. Tang, Y., Fang, J. A., & Chen, L. (2010). Lag full state hybrid projective synchronization in different fractional-order chaotic systems. *International Journal of Modern Physics B*, *24*, 6129–61411.
132. Feng, H., Yang, Y., & Yang, S. P. (2013). A new method for full state hybrid projective synchronization of different fractional order chaotic systems. *Applied Mechanics and Materials*, *385–38*, 919–922.
133. Razminia, A. (2013). Full state hybrid projective synchronization of a novel incommensurate fractional order hyperchaotic system using adaptive mechanism. *Indian Journal of Physics*, *87*, 161–167.
134. Zhang, L., & Liu, T. (2016). Full state hybrid projective synchronization of variable-order fractional chaotic/hyperchaotic systems with nonlinear external disturbances and unknown parameters. *The Journal of Nonlinear Science and Applications*, *9*, 1064–1076.
135. Manieri, R., & Rehacek, J. (1999). Projective synchronization in three-dimensional chaotic systems. *Physical Review Letters*, *82*, 3042–3045.
136. Hu, M., Xu, Z., Zhang, R., & Hu, A. (2008). Full state hybrid projective synchronization in continuous-time chaotic (hyperchaotic) systems. *Communications in Nonlinear Science and Numerical Simulation*, *13*, 456–464.

137. Hu, M., Xu, Z., Zhang, R., & Hu, A. (2008). Full state hybrid projective synchronization of a general class of chaotic maps. *Communications in Nonlinear Science and Numerical Simulation, 13*, 782–789.
138. Grassi, G. (2012). Arbitrary full-state hybrid projective synchronization for chaotic discrete-time systems via a scalar signal. *Chinese Physics B, 21*, 060504.
139. Ouannas, A., & Abu-Saris, R. (2016). On matrix projective synchronization and inverse matrix projective synchronization for different and identical dimensional discrete-time chaotic systems. *Journal of Control Science and Engineering*, 1–7.
140. Gorenflo, R., & Mainardi, F. (1997). Fractional calculus: Integral and differential equations of fractional order. In A. Carpinteri & F. Mainardi (Eds.), *The book Fractals and fractional calculus*, New York.
141. Caputo, M. (1967). Linear models of dissipation whose Q is almost frequency independent-II. *Geophysical Journal of the Royal Astronomical Society, 13*, 529–539.
142. Samko, S. G., Klibas, A. A., & Marichev, O. I. (1993). *Fractional integrals and derivatives: theory and applications*. Gordan and Breach.
143. Podlubny, I. (1999). *Fractional differential equations*. Academic Press.
144. Heymans, N., & Podlubny, I. (2006). Physical interpretation of initial conditions for fractional differential equations with Riemann-Liouville fractional derivatives. *Rheologica Acta, 45*, 765–772.
145. Matignon, D. (1996). Stability results of fractional differential equations with applications to control processing. In: *IMACS, IEEE-SMC*, Lille, France.
146. Li, Y., Chen, Y., & Podlubny, I. (2010). Stability of fractional-order nonlinear dynamic systems: Lyapunov direct method and generalized Mittag Leffler stability. *Computers & Mathematics with Applications, 59*, 21–1810.
147. Chen, D., Zhang, R., Liu, X., & Ma, X. (2014). Fractional order Lyapunov stability theorem and its applications in synchronization of complex dynamical networks. *Communications in Nonlinear Science and Numerical Simulation, 19*, 4105–4121.
148. Aguila-Camacho, N., Duarte-Mermoud, M. A., & Gallegos, J. A. (2014). Lyapunov functions for fractional order systems. *Communications in Nonlinear Science and Numerical Simulation, 19*, 2951–2957.
149. Ouannas, A., & Mahmoud, E. (2014). Inverse matrix projective synchro-nization for discrete chaotic systems with different dimensions. *Intelligence and Electronic Systems, 3*, 188–192.
150. Wang, X.-Y., & Zhang, H. (2013). Bivariate module-phase synchronization of a fractional-order lorenz system in diFFerent dimensions. *Journal of Computational and Nonlinear Dynamics, 8*, 031017.
151. Zhou, P., Wei, L. J., & Cheng, X. F. (2009). A novel fractional-order hyperchaotic system and its synchronization. *Chinese Physics B, 18*, 2674.
152. Liu, C., Liu, T., Liu, L., & Liu, K. (2004). A new chaotic attractor. *Chaos Solitons Fractals, 22*, 1031–1038.
153. Han, Q., Liu, C. X., Sun, L., & Zhu, D. R. (2013). A fractional order hyperchaotic system derived from a Liu system and its circuit realization. *Chinese Physics B, 22*, 6–020502.

Fractional Inverse Generalized Chaos Synchronization Between Different Dimensional Systems

Adel Ouannas, Ahmad Taher Azar, Toufik Ziar and Sundarapandian Vaidyanathan

Abstract In this chapter, new control schemes to achieve inverse generalized synchronization (IGS) between fractional order chaotic (hyperchaotic) systems with different dimensions are presented. Specifically, given a fractional master system with dimension n and a fractional slave system with dimension m, the proposed approach enables each master system state to be synchronized with a functional relationship of slave system states. The method, based on the fractional Lyapunov approach and stability property of integer-order linear differential systems, presents some useful features: (i) it enables synchronization to be achieved for both cases $n < m$ and $n > m$; (ii) it is rigorous, being based on theorems; (iii) it can be readily applied to any chaotic (hyperchaotic) fractional systems. Finally, the capability of the approach is illustrated by synchronization examples.

Keywords Chaos · Inverse generalized synchronization · Fractional systems · Different dimensions · Fractional lyapunov approach

---

A. Ouannas
Laboratory of Mathematics, Informatics and Systems (LAMIS),
University of Larbi Tebessi, 12002 Tebessa, Algeria
e-mail: ouannas.a@yahoo.com

A.T. Azar (✉)
Faculty of Computers and Information, Benha University, Benha, Egypt
e-mail: ahmad.azar@fci.bu.edu.eg; ahmad_t_azar@ieee.org

A.T. Azar
Nanoelectronics Integrated Systems Center (NISC), Nile University, Cairo, Egypt

T. Ziar
Department of Material Sciences, University of Tebessa, 12002 Tebessa, Algeria
e-mail: Toufik1_ziar@yahoo.fr

S. Vaidyanathan
Research and Development Centre, Vel Tech University, Avadi, Chennai 600062, Tamil Nadu, India
e-mail: sundarvtu@gmail.com

A.T. Azar et al. (eds.), *Fractional Order Control and Synchronization of Chaotic Systems*, Studies in Computational Intelligence 688,
DOI 10.1007/978-3-319-50249-6_18

1 Introduction

Because of its deep and natural connections with many fields of applied physics and engineering, in recent decades, researchers from several fields stop to consider fractional calculus as a pure mathematics without real applications [1–8] . It was found that fractional calculus is useful for modeling electromagnetic waves, Mechanic, viscoelasticity, quantum evolution of complex systems, electrical engineering, control systems, robotics, signal processing, chemical mixing, bioengineering and nuclear reactor dynamics [9–30]. That is, the fractional differential systems are more suitable to describe physical phenomena that have memory and genetic characteristics [31–33].

On the other hand, in recent years, many scientists have become aware of the potential use of chaotic dynamics in physics, chemistry, biology, ecology, economics, finance, computer science, engineering, and other areas [34–53]. One of the very important areas that chaos can occur is fractional order systems [54, 55].

Recently, with the development of the fractional-order algorithm, the dynamics of fractional order systems have received much attention. Studying chaos in fractional-order dynamical systems is an interesting topic as well. It is well known that chaos cannot occur in continuous integer order systems of total order less than three. It has been shown that many fractional-order dynamical systems behave chaotically with total order less then three. Similar to nonlinear integer-order differential systems, nonlinear fractional-order differential systems may also have complex dynamics, such as chaos and bifurcation [56–61]. To date, chaotic motions have been found in fractional systems, for example in the fractional versions of the Chua circuit [62, 63], Lorenz system [64], Rössler system [65], Chen system [66, 67], Arneodo system [68], Lü system [69, 70], Duffing system [71, 72], van der Pol system [73, 74], Volta's system [75, 76] and Liu system [77].

Synchronization of chaos has become an active research subject in nonlinear science [78–89]. The current problems of synchronization are very interesting, nontraditional, and indeed very challenging [90–102] . Recently, study on synchronization of chaotic fractional order differential equations has starts to attract increasing attention of many researchers [103–113] , due to its potential applications [114–118].

Until now, a wide variety of fractional techniques have been used to design a synchronization control in fractional–order chaotic systems such as sliding-mode control [119–121], linear control [122], nonlinear control [123], active control [124], adaptive control [125, 126], feedback control [127, 128], scalar signal technique [129–132].

At present, various types of synchronization for the fractional-order chaotic systems have been presented, such as phase synchronization [133, 134], complete synchronization [135], anti-synchronization [136], projective synchronization [137–141], hybrid projective synchronization [142, 143], function projective synchronization [144–146], generalized projective synchronization [147–151], full state hybrid projective synchronization [152], Q-S synchronization [153], exponential synchronization [154], finite-time synchronization [155], impulsive synchronization [156, 157].

Due to the complexities of chaos synchronization, such tasks are always difficult to achieve. In fact, in real physical systems or experimental situations some certain complex systems components cannot supposed to be identical or of the same orders, thus, it is much more attractive and challenging to realize the synchronization of two different chaotic systems with different dimensions or orders. Recently, many approaches have been proposed to study the chaos synchronization between different dimensional systems such as matrix projective synchronization [158], inverse matrix projective synchronization [159], Q-S synchronization [160], $\Lambda - \phi$ generalized synchronization [161, 162], $\Theta - \Phi$ synchronization [163–165] and hybrid synchronization [166, 167].

Among the aforementioned methods, generalized synchronization (GS) is the most effective synchronization method that has been used widely to achieve the chaos synchronization with different dimensions. Generalized synchronization (GS) implies the establishment of functional relation between the master and the slave systems. It has received a great deal of attention for its universality in the recent years. Nowadays, numerous researches of GS in fractional-order chaotic systems have been done theoretically and experimentally [168–174]. Recently, another interesting synchronization type was appeared called inverse generalized synchronization (IGS). The relevant researches for inverse generalized synchronization (IGS) still in an initial stage. However, IGS was applied successfully in continuous and discrete-time chaotic systems with integer-order [175, 176]. The problem of inverse generalized synchronization (IGS) is an attractive idea and more difficult than the problem of generalized synchronization (IGS). The complexity of the inverse generalized synchronization scheme can be used to enhance security in communication and encryption.

Based on these considerations, the aim of this chapter is to present constructive schemes to investigate the problem of inverse generalized synchronization (IGS) between arbitrary fractional chaotic systems with different dimensions. Using a fractional calculus property, Lyapunov stability theory of integer order differential systems, Laplace transform and stability property of integer order linear differential systems, two theorems are proved, which enable synchronization between n-dimensional and m-dimensional fractional chaotic systems to be achieved for the cases: $n < m$ and $n > m$, respectively. Examples of synchronization are given to validate the theoretical results derived in this chapter.

The outline of the rest of this chapter is organized as follows. Section 2 provides a brief review of the fractional order calculus. In Sect. 3, the problem of fractional IGS is formulated. Our main results are presented in Sect. 4. In Sect. 5, illustrative examples are performed to show the effectiveness the proposed analysis. Section 6 is the brief conclusion.

2 Preliminaries

The idea of fractional integrals and derivatives has been known since the development of the regular calculus, with the first reference probably being associated with Leibniz in 1695. There are several definitions of fractional derivatives [177]. The Caputo derivative [178] is a time domain computation method. In real applications, the Caputo derivative is more popular since the un-homogenous initial conditions are permitted if such conditions are necessary. Furthermore, these initial values are prone to measure since they all have idiographic meanings. The Riemann-Liouville fractional integral operator of order $p \geq 0$ of function $f(t)$ is defined as,

$$J^p f(t) = \frac{1}{\Gamma(p)} \int_0^t (t-\tau)^{p-1} f(\tau) d\tau, \quad p > 0, \ t > 0. \tag{1}$$

where Γ denotes Gamma function.
For $p, q \geq 0$ and $\gamma > -1$, we have,

$$J^p J^q f(t) = J^{p+q} f(t), \tag{2}$$

$$J^q t^\gamma = \frac{\Gamma(\gamma+1)}{\Gamma(q+\gamma+1)} t^{q+\gamma}. \tag{3}$$

In this study, Caputo definition is used and the fractional derivative of $f(t)$ is defined as,

$$D_t^p f(t) = J^{m-p} \left(\frac{d^m}{dt^m} f(t) \right) = \frac{1}{\Gamma(m-p)} \int_0^t \frac{f^{(m)}(\tau)}{(t-\tau)^{p-m+1}} d\tau, \tag{4}$$

for $m-1 < p \leq m$, $m \in N$, $t > 0$. The fractional differential operator $D_t^p f(t)$ is left-inverse (and not right-inverse) to the fractional integral operator J^p, i.e. $D_t^p J^p = I$ where I is the identity operator.

Lemma 1 [179] *The Laplace transform of the Caputo fractional derivative rule reads*

$$\mathbf{L}\left(D_t^p f(t)\right) = s^p \mathbf{F}(s) - \sum_{k=0}^{n-1} s^{p-k-1} f^{(k)}(0), \quad (p > 0, \ n-1 < p \leq n). \tag{5}$$

Particularly, when $p \in (0, 1]$, *we have* $\mathbf{L}\left(D_t^p f(t)\right) = s^p \mathbf{F}(s) - s^{p-1} f(0)$.

Lemma 2 [180] *The Laplace transform of the Riemann-Liouville fractional integral rule satisfies*

$$\mathbf{L}\left(J^q f(t)\right) = s^{-q} \mathbf{F}(s), \quad (q > 0). \tag{6}$$

Lemma 3 [140] *Suppose $f(t)$ has a continuous kth derivative on $[0, t]$ $(k \in N, t > 0)$, and let $p, q > 0$ be such that there exists some $\ell \in N$ with $\ell \leq k$ and $p, p + q \in [\ell - 1, \ell]$. Then*

$$D_t^p D_t^q f(t) = D_t^{p+q} f(t) \tag{7}$$

Remark 1 Note that the condition requiring the existence of the number ℓ with the above restrictions in the property is essential. In this paper, we consider the case that $p, q \in (0, 1]$ and $p + q \in (0, 1]$. Apparently, under such conditions this property holds.

3 Problem Statement

The master system is defined by

$$D_t^p X(t) = F(X(t)), \tag{8}$$

where $X(t) = \left(x_1(t), x_2(t), ..., x_n(t)\right)^T$ is the state vector of the master system (8), $F : \mathbf{R}^n \rightarrow \mathbf{R}^n$, p is a rational number between 0 and 1 and D_t^p is the Caputo fractional derivative of order p.

The slave system is described as

$$D_t^q Y(t) = G(Y(t)) + U, \tag{9}$$

where $Y(t) = \left(y_1(t), y_2(t), ..., y_m(t)\right)^T$ is the state vector of the slave system (9), $G : \mathbf{R}^m \rightarrow \mathbf{R}^m$, q is a rational number between 0 and 1, q is the Caputo fractional derivative of order q and $U = \left(u_i\right)_{1 \leq i \leq m}$ is a vector controller to be determined later.

Before proceeding to the definition of inverse generalized synchronization (IGS) between the master system (8) and the slave system (9), the definition of generalized synchronization (IGS) is provided.

Definition 1 The master system (8) and the slave system (9) are said to be generalized synchronized, if there exists a controller $U = \left(u_i\right)_{1 \leq i \leq m}$ and differentiable function $\phi : \mathbf{R}^n \longrightarrow \mathbf{R}^m$ such that the error

$$e(t) = Y(t) - \phi(X(t)), \tag{10}$$

satisfies the $\lim_{t \longrightarrow +\infty} e(t) = 0$.

Definition 2 The master system (8) and the slave system (9) are said to be inverse generalized synchronized, if there exists a controller $U = \left(u_i\right)_{1 \leq i \leq m}$ and differentiable function $\varphi : \mathbf{R}^m \longrightarrow \mathbf{R}^n$ such that the error

$$e(t) = X(t) - \varphi(Y(t)) \tag{11}$$

satisfies the $\lim_{t \longrightarrow +\infty} e(t) = 0$.

Remark 2 Inverse generalized synchronization of integer-order chaotic dynamical systems with different dimensions, based on Lyapunov stability theory, has been studied and carried out, for example, in Refs. [94, 95].

4 Main Results

Now, we consider the master system in the form

$$D_t^p X(t) = AX(t) + f(X(t)), \tag{12}$$

where $X(t) = \left(x_1(t), ..., x_n(t)\right)^T \in \mathbf{R}^n$ is the state vector of the master system (12), $A \in \mathbf{R}^{n\times n}$ and $f : \mathbf{R}^n \to \mathbf{R}^n$ is a nonlinear function, D_t^p is the Caputo fractional derivative of order p and $0 < p \leq 1$.

The slave system is defined by

$$D_t^q Y(t) = g(Y(t)) + U, \tag{13}$$

where $Y(t) = \left(y_1(t), ..., y_m(t)\right)^T \in \mathbf{R}^m$ is the state vector of the slave system (13), $0 < q \leq 1$, D_t^q is the Caputo fractional derivative of order q, $g(Y(t)) = \left(g_i(Y(t))\right)_{1\leq i\leq m}$ and $U = \left(u_i\right)_{1\leq i\leq m}$ is a controller to be designed.

In the following we discuss two kinds of cases: $n < m$ and $n > m$, respectively.

4.1 Case: $n < m$

Theorem 1 *If the control law* $U = \left(u_i\right)_{1\leq i\leq m}$ *is selected as*

$$\left(u_1, u_2, ..., u_n\right)^T = -\left(g_1(Y(t)), ..., g_n(Y(t))\right)^T + J^{1-q}\left[Q^{-1} \times R\right], \tag{14}$$

and

$$\left(u_{n+1}, ..., u_m\right)^T = -\left(g_{n+1}(Y(t)), ..., g_m(Y(t))\right)^T, \tag{15}$$

where Q^{-1} *is the inverse of the of matrix* Q

$$Q = \begin{pmatrix} \frac{\partial \varphi_1}{\partial y_1} & \frac{\partial \varphi_1}{\partial y_2} & \cdots & \frac{\partial \varphi_1}{\partial y_n} \\ \frac{\partial \varphi_2}{\partial y_1} & \frac{\partial \varphi_2}{\partial y_2} & \cdots & \frac{\partial \varphi_2}{\partial y_n} \\ \vdots & \vdots & \ddots & \vdots \\ \frac{\partial \varphi_n}{\partial y_1} & \frac{\partial \varphi_n}{\partial y_2} & \cdots & \frac{\partial \varphi_n}{\partial y_n} \end{pmatrix}, \tag{16}$$

and

$$R = (A - K)\left[X(t) - \varphi(Y(t))\right] + \dot{X}(t), \tag{17}$$

where $K \in \mathbf{R}^{n\times n}$ is a feedback gain matrix. Then the master system (12) and the slave system (13) are globally inverse generalized synchronized.

Proof The error system between the master system (12) and the slave system (13) can be derived as

$$\dot{e}(t) = \dot{X}(t) - \mathbf{D}\varphi(Y(t)) \times \dot{Y}(t), \tag{18}$$

where

$$\mathbf{D}\varphi(Y(t)) = \begin{pmatrix} \frac{\partial\varphi_1}{\partial y_1} & \frac{\partial\varphi_1}{\partial y_2} & \cdots & \frac{\partial\varphi_1}{\partial y_n} \\ \frac{\partial\psi_2}{\partial y_1} & \frac{\partial\psi_2}{\partial y_2} & \cdots & \frac{\partial\varphi_2}{\partial y_n} \\ \vdots & \vdots & \ddots & \vdots \\ \frac{\partial\varphi_n}{\partial y_1} & \frac{\partial\varphi_n}{\partial y_2} & \cdots & \frac{\partial\varphi_n}{\partial y_n} \end{pmatrix}. \tag{19}$$

The error system (18) can be written as

$$\dot{e}(t) = \dot{X}(t) - Q \times \left(\dot{y}_1, \dot{y}_2, ..., \dot{y}_n\right)^T - \hat{Q} \times \left(\dot{y}_{n+1}, \dot{y}_{n+2}, ..., \dot{y}_m\right)^T, \tag{20}$$

where

$$Q = \begin{pmatrix} \frac{\partial\varphi_1}{\partial y_1} & \frac{\partial\varphi_1}{\partial y_2} & \cdots & \frac{\partial\varphi_1}{\partial y_m} \\ \frac{\partial\varphi_2}{\partial y_1} & \frac{\partial\varphi_2}{\partial y_2} & \cdots & \frac{\partial\varphi_2}{\partial y_m} \\ \vdots & \vdots & \ddots & \vdots \\ \frac{\partial\varphi_n}{\partial y_1} & \frac{\partial\varphi_n}{\partial y_2} & \cdots & \frac{\partial\varphi_n}{\partial y_m} \end{pmatrix}, \tag{21}$$

and

$$\hat{Q} = \begin{pmatrix} \frac{\partial\varphi_1}{\partial y_{n+1}} & \frac{\partial\varphi_1}{\partial y_{n+2}} & \cdots & \frac{\partial\varphi_1}{\partial y_m} \\ \frac{\partial\varphi_2}{\partial y_{n+1}} & \frac{\partial\varphi_2}{\partial y_{n+2}} & \cdots & \frac{\partial\varphi_2}{\partial y_m} \\ \vdots & \vdots & \ddots & \vdots \\ \frac{\partial\varphi_n}{\partial y_{n+1}} & \frac{\partial\varphi_n}{\partial y_{n+2}} & \cdots & \frac{\partial\varphi_n}{\partial y_m} \end{pmatrix}. \tag{22}$$

Now, by inserting Eqs. (14) and (15) into Eq. (13), we can rewrite the slave system as follows

$$\left(D_t^q y_1(t), ..., D_t^q y_n(t)\right)^T = J^{1-q}\left[Q^{-1} \times R\right], \tag{23}$$

and

$$\left(D_t^q y_{n+1}(t), ..., D_t^q y_m(t)\right)^T = 0. \tag{24}$$

By applying the fractional derivative of order $1 - q$ to both the left and right sides of Eqs. (23) and (24), we obtain

$$D_t^{1-q}\left(\left(D_t^q y_1(t), ..., D_t^q y_n(t)\right)^T\right) = \left(\dot{y}_1, \dot{y}_2, ..., \dot{y}_n\right)^T$$
$$= D_t^{1-q} J^{1-q}\left(Q^{-1} \times R\right)$$
$$= Q^{-1} \times R, \tag{25}$$

and

$$\left(\dot{y}_{n+1}, \dot{y}_{n+2}, ..., \dot{y}_m\right)^T = 0. \tag{26}$$

By using Eqs. (25) and (26), the error system (20) can be written as follow

$$\dot{e}(t) = (A - K)\left[X(t) - \varphi(Y(t))\right]$$
$$= (A - K)\, e(t). \tag{27}$$

Construct the candidate Lyapunov function in the form

$$V(e(t)) = e^T(t)e(t), \tag{28}$$

we obtain

$$\dot{V}(e(t)) = \dot{e}^T(t)e(t) + e^T(t)\dot{e}(t)$$
$$= e^T(t)(A - K)^T e(t) + e^T(t)(A - K)e(t)$$
$$= e^T(t)\left[(A - K)^T + (A - K)\right] e(t).$$

The control matrix K is selected such that $(A - K)^T + (A - K)$ is a negative definite matrix, then we get $\dot{V}(e(t)) < 0$. Thus, from the Lyapunov stability theory, it is immediate that all solutions of error system (27) go to zero as $t \to \infty$. Therefore, systems (12) and (13) are globally inverse generalized synchronized.

4.2 Case: $m < n$

Now, the error system between the master system (12) and the slave system (13), can be written in scalar form as

$$\dot{e}_i(t) = \dot{x}_i(t) - \sum_{j=1}^{m}\left[\frac{\partial \varphi_i(Y(t))}{\partial y_j(t)} \times \dot{y}_j\right]. \tag{29}$$

In this case, the controllers u_i are given by

$$u_i = -g_i(Y(t)) + J^{1-q}\left(v_i\right), \qquad i = 1, 2, ..., m, \tag{30}$$

where v_i are new controllers. By using (30), the slave system can be written as

$$D_t^q y_i(t) = J^{1-q}\left(v_i\right), \qquad i = 1, 2, ..., m, \tag{31}$$

Applying the Laplace transform to (31) and letting

$$\mathbf{F}_i(s) = \mathbf{L}\left(y_i(t)\right), \qquad i = 1, 2, ..., m, \tag{32}$$

we obtain,

$$s^q \mathbf{F}_i(s) - s^{q-1} y_i(0) = s^{q-1} \mathbf{L}\left(v_i\right), \qquad i = 1, 2, ..., m, \tag{33}$$

multiplying both the left-hand and right-hand sides of (33) by s^{1-q} and applying the inverse Laplace transform to the result, we get the following equation

$$\dot{y}_i(t) = v_i, \qquad i = 1, 2, ..., m, \tag{34}$$

Now, the error system (29), can be described as

$$\begin{aligned}\dot{e}_i(t) &= \dot{x}_i(t) - \sum_{j=1}^{m} \left[\frac{\partial \varphi_i(Y(t))}{\partial y_j(t)} \times v_j\right] \\ &= \sum_{j=1}^{n} \left(a_{ij} - l_{ij}\right) e_j(t) + T_i - \sum_{j=1}^{m} \left[\frac{\partial \varphi_i(Y(t))}{\partial y_j(t)} \times v_j\right], \qquad i = 1, 2, ..., n, \end{aligned} \tag{35}$$

where $\left(l_{ij}\right) \in \mathbf{R}^{n\times n}$ are control constants and

$$T_i = \sum_{j=1}^{n} \left(l_{ij} - a_{ij}\right)\left(x_j(t) - \varphi_j(Y(t))\right) + \dot{x}_i(t), \qquad i = 1, 2, ..., n. \tag{36}$$

To achieve inverse generalized synchonization between the master system (12) and the slave system (13), new controllers are defined as

$$w_i = \sum_{j=1}^{m} \left[\frac{\partial \varphi_i(Y(t))}{\partial y_j(t)} \times v_j\right], \qquad i = 1, 2, ..., n, \tag{37}$$

and we assume that $\frac{\partial \varphi_i(Y(t))}{\partial y_j(t)}$, $1 \le j \le m$, are not all equal zero.

Now, rewriting the error system described in Eq. (35) in the compact form

$$\dot{e}(t) = (A - L)\, e(t) + T - W, \tag{38}$$

where $L = \left(l_{ij}\right)_{n\times n}$ is a control constant matrix to be selected, $T = \left(T_i\right)_{1\le i\le n}$ and $W = \left(w_i\right)_{1\le i\le n}$ is a new vector controller.

Theorem 2 *Inverse generalized synchronization between the master system (12) and the slave system (13), will occur if the following conditions are satisfied:*
(i) $W = T$.
(ii) All eigenvalues of $A - L$ *have negative real part.*

Proof By substituting the control law (i) into Eq. (38), the error system can be described as

$$\dot{e}(t) = (A - L)\, e(t). \tag{39}$$

Thus, by asymptotic stability of linear continuous-time systems, if the control matrix L is selected such that all eigenvalues of $A - L$ have have negative real part, it is immediate that all solutions of error system (39) go to zero as $t \to \infty$. Therefore, systems (12) and (13) are globally inverse generalized synchronized.

5 Illustrative Examples

In this section, in order to show the effectiveness of our approaches, two numerical examples are considered.

5.1 Example 1

In this example, we assume that the fractional order permanent magnet synchronous motor model (PMSM) [181] is the master system and the controlled new fractional order hyperchaotic system, proposed by Wu et al. in [116], is the slave system.

The master system is described by

$$\begin{aligned} D_t^p x_1 &= -x_1 + x_2 x_3, \\ D_t^p x_2 &= -x_2 - x_1 x_3 + a x_3, \\ D_t^p x_3 &= b\left(x_2 - x_3\right), \end{aligned} \tag{40}$$

where $(a, b) = (100, 10)$ and $p = 0.95$. In this case, the matrix $A = \left(a_{ij}\right)_{3\times 3}$ and the nonlinear function f of the master system (40) are given by

$$A = \begin{pmatrix} -1 & 0 & 0 \\ 0 & -1 & 100 \\ 0 & 10 & -10 \end{pmatrix}, \; f = \begin{pmatrix} x_2 x_3 \\ -x_1 x_3 \\ 0 \end{pmatrix}.$$

Figures 1 and 2 show the 2-D and 3-D chaotic attractors of the fractional order PMSM system (40), respectively.

The slave system is given as follows

$$\begin{aligned} D_t^q y_1 &= 10\left(y_2 - y_1\right) + u_1, \\ D_t^q y_2 &= 28 y_1 + y_2 - y_4 - y_1 y_3 + u_2, \end{aligned} \tag{41}$$

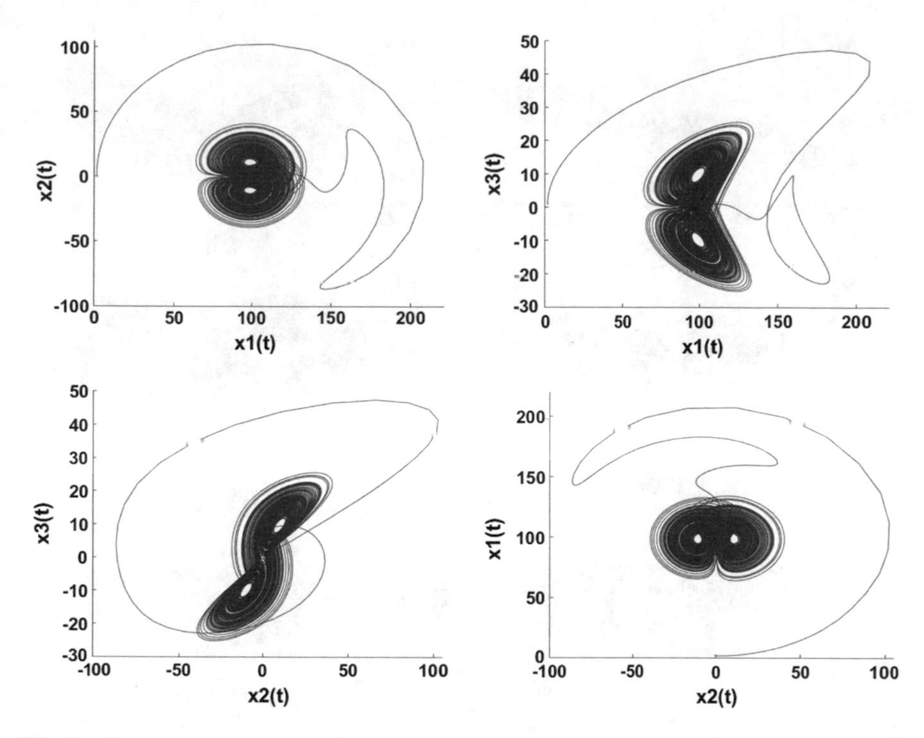

Fig. 1 Chaotic attractors of the fractional order PMSM system (40) in 2-D, when $(a, b) = (100, 10)$ and $p = 0.95$

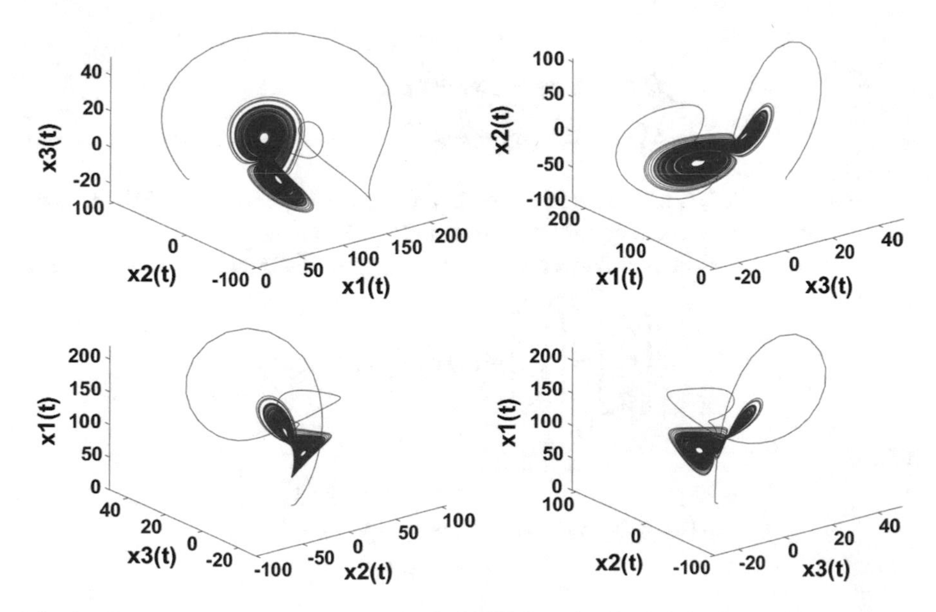

Fig. 2 Chaotic attractors of the fractional order PMSM system (40) in 3-D when $(a, b) = (100, 10)$ and $p = 0.95$

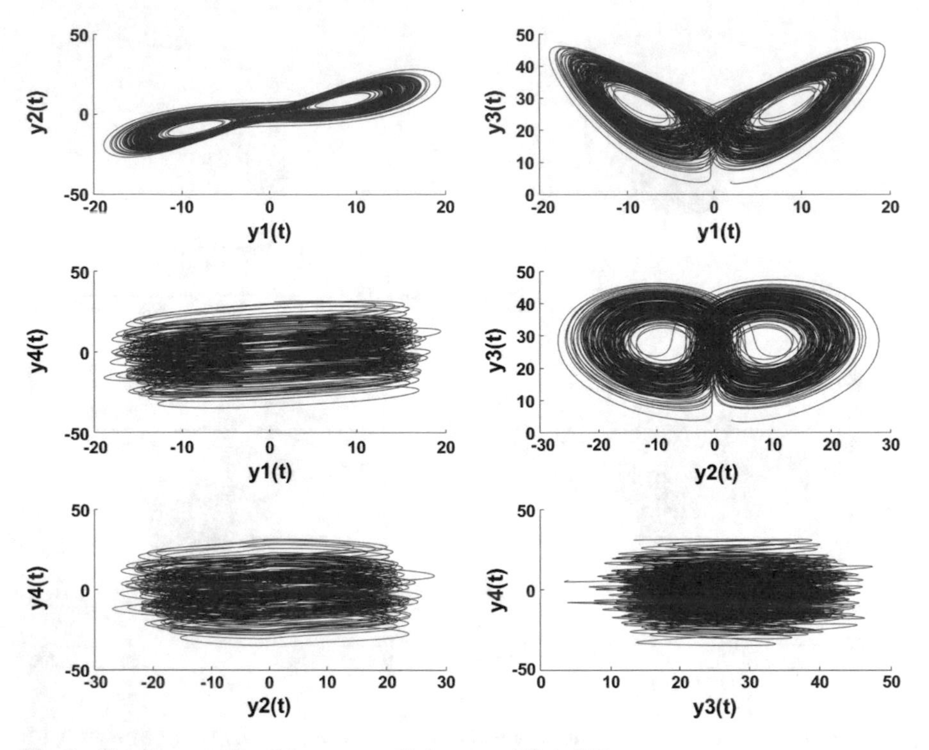

Fig. 3 Chaotic attractors of the uncontrolled system (41) in 2-D

$$D_t^q y_3 = y_1 y_2 - \frac{8}{3} y_3 + u_3,$$
$$D_t^q y_3 = 0.1 y_2 y_3 + u_4,$$

where u_i, $i = 1, 2, 3, 4$, are controllers. System (41), when $q = 0.94$, exhibits hyperchaotic behaviours without control as shown in Figs. 3 and 4.

Define the errors of IGS between the master system (40) and the slave system (41) by:

$$\begin{bmatrix} e_1 \\ e_2 \\ e_3 \end{bmatrix} = \begin{bmatrix} x_1 \\ x_2 \\ x_3 \end{bmatrix} - \varphi\left(y_1, y_2, y_3, y_4\right), \tag{42}$$

where

$$\varphi\left(y_1, y_2, y_3, y_4\right) = \begin{bmatrix} y_1 + y_2 + y_3 + \frac{1}{3} y_4^3 \\ y_1 + y_2 + y_3 + \frac{1}{2} y_4^2 \\ y_1 + y_2 + y_3 + \frac{3}{2} y_4^2 + y_4 \end{bmatrix}.$$

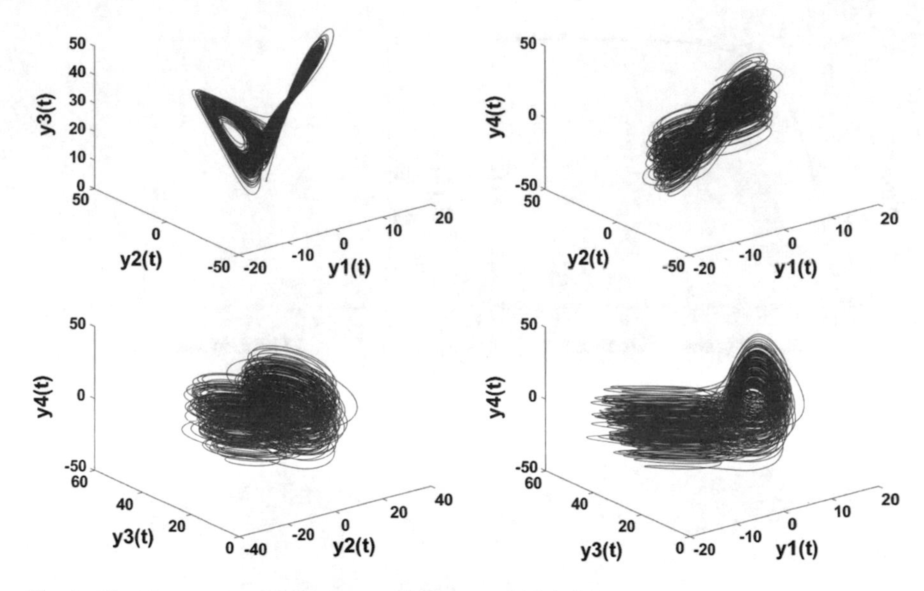

Fig. 4 Chaotic attractors of the uncontrolled system (41) in 3-D

So

$$\mathbf{D}\left(\varphi\left(Y\left(t\right)\right)\right)=\begin{pmatrix}1\,0\,0 & y_4^2\\ 0\,1\,0 & y_4\\ 0\,0\,2 & 3y_4+1\end{pmatrix}.$$

By applying our approach of IGS, described in Sect. 4.1, we get

$$Q=\begin{pmatrix}1\,0\,0\\ 0\,1\,0\\ 0\,0\,2\end{pmatrix},\ Q^{-1}=\begin{pmatrix}1\,0\,0\\ 0\,1\,0\\ 0\,0\,\frac{1}{2}\end{pmatrix},$$

and the feedback gain matrix K is chosen as

$$K=\begin{pmatrix}0 & 0 & 0\\ 0 & 0 & 100\\ 0 & 10 & 0\end{pmatrix}. \tag{43}$$

It is easy to know that $(A-K)^T+(A-K)$ is a negative definite matrix. Then, according to Theorem 1, the master system (40) and the slave system (41) are globally inverse generalized synchronized. In this case, the error system between system (40) and (41) can be written as follow:

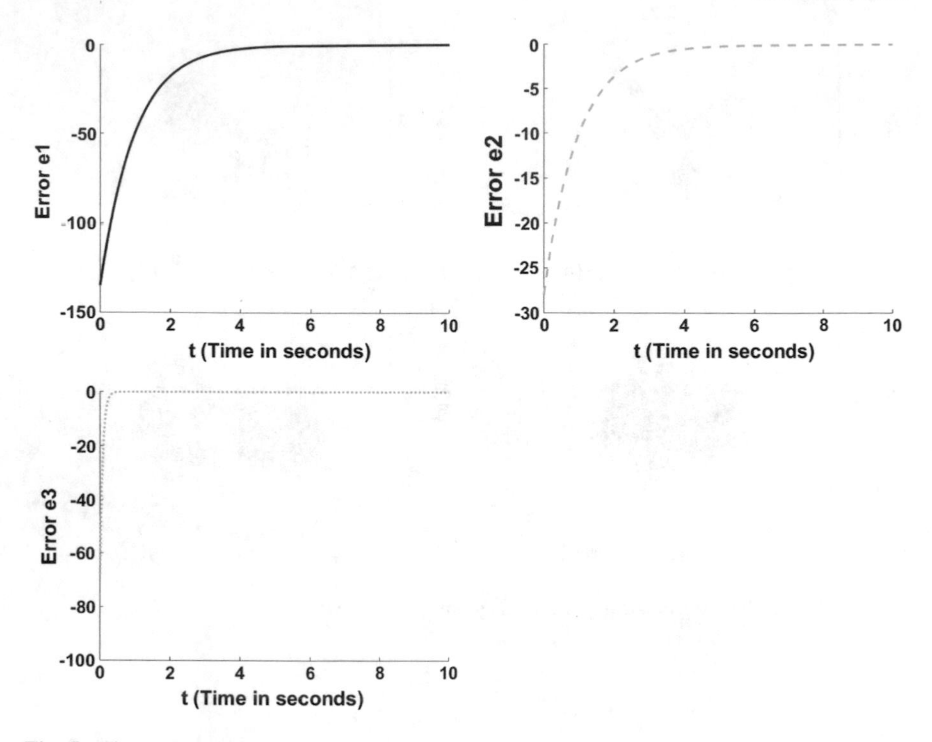

Fig. 5 The evolution of the error functions e_1, e_2 as e_3 as $t \to +\infty$

$$\begin{aligned}\dot{e}_1 &= -e_1, \\ \dot{e}_2 &= -e_2, \\ \dot{e}_3 &= -10e_3.\end{aligned} \tag{44}$$

For the purpose of numerical simulation, fourth order Runge–Kutta integration method method has been used. In addition, simulation time $Tm = 120$ s and time step $h = 0.005$ s have been employed. The initial values of the master system and the slave system are $[x_1(0), x_2(0), x_3(0)] = [2, -1, 1]$ and $[y_1(0), y_2(0), y_3(0), y_4(0)] = [2, 3, 4, 6]$, respectively, and the initial states of the error system are $[e_1(0), e_2(0), e_3(0)] = [-135, -28, -68]$. The error functions evolution, in this case, is shown in Fig. 5.

5.2 *Example 2*

In this example, we choose the fractional order hyperchaotic Liu system [182] as the master system and the controlled fractional order Liu system, presented in [183], as the slave system. The master system is

$$\begin{aligned}
D^p x_1 &= 10\left(x_2 - x_1\right) + x_4, \\
D_t^p x_2 &= 40x_1 - x_1x_3 + 0.5x_4, \\
D_t^p x_3 &= -2.5x_3 + 4x_1^2 - x_4, \\
D_t^p x_4 &= -\frac{10}{15}x_2 - x_4,
\end{aligned} \tag{45}$$

where $p = 0.9$. In this case, the matrix $A = \left(a_{ij}\right)_{3\times 3}$ and the nonlinear function f of thc master system (45) are given by

$$A = \begin{pmatrix} -10 & 10 & 0 & 1 \\ 40 & 0 & 0 & 0.5 \\ 0 & 0 & -2.5 & -1 \\ 0 & -\frac{10}{15} & 0 & -1 \end{pmatrix}, \; f = \begin{pmatrix} 0 \\ -x_1x_3 \\ 4x_1^2 \\ 0 \end{pmatrix}.$$

Figures 6 and 7 show the 2-D and 3-D chaotic attractors of the fractional order hyperchaotic Liu system, respectively.

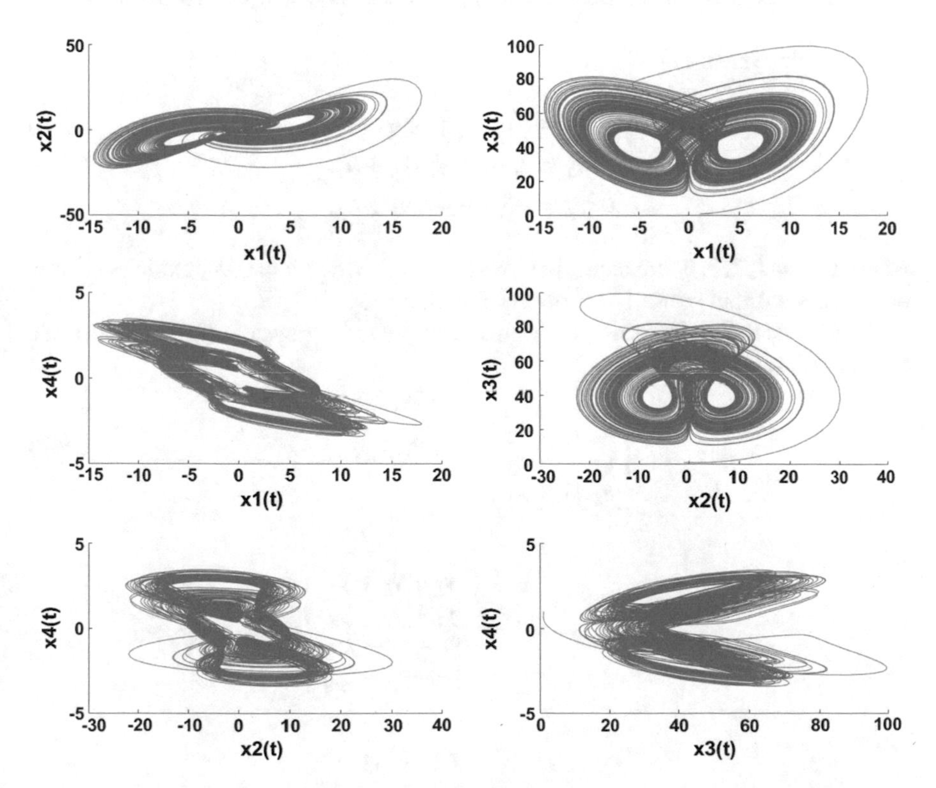

Fig. 6 Chaotic attractors of the fractional order hyperchaotic Liu system (45) in 2-D, when $p = 0.9$

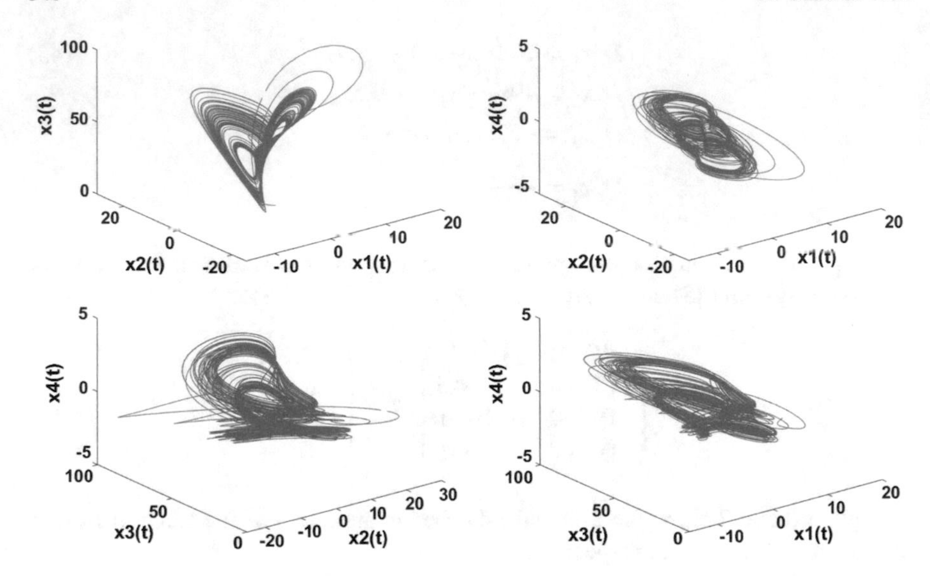

Fig. 7 Chaotic attractors of the fractional order hyperchaotic Liu system (45) in 3-D, when $p = 0.9$

The slave system is

$$\begin{aligned} D^q y_1 &= -y_1 - y_2^2 + u_1, \\ D^q y_2 &= 2.5y_2 - 4y_1y_3 + u_2, \\ D^q y_3 &= -5y_3 + 4y_1y_2 + u_3, \end{aligned} \tag{46}$$

where u_i, $i = 1, 2, 3, 4$, are controllers. System (46), when $q = 0.94$, exhibits chaotic behaviours without control as shown in Figs. 8 and 9.

Now let us define the errors of GS between the master system (26) and the slave system (27) by:

$$\begin{bmatrix} e_1 \\ e_2 \\ e_3 \\ e_4 \end{bmatrix} = \begin{bmatrix} x_1 \\ x_2 \\ x_3 \\ x_4 \end{bmatrix} - \varphi\left(y_1, y_2, y_3\right), \tag{47}$$

where

$$\varphi\left(y_1, y_2, y_3\right) = \begin{bmatrix} y_1 + y_2 + y_3 \\ y_1 + y_2 - y_3 \\ y_1 - y_2 - 2y_3 \\ y_1 - y_2 + 2y_3 \end{bmatrix}.$$

So

$$\mathbf{D}\left(\varphi\left(Y\left(t\right)\right)\right) = \begin{pmatrix} 1 & 1 & 1 \\ 1 & 1 & -1 \\ 1 & -1 & -2 \\ 1 & -1 & 2 \end{pmatrix}.$$

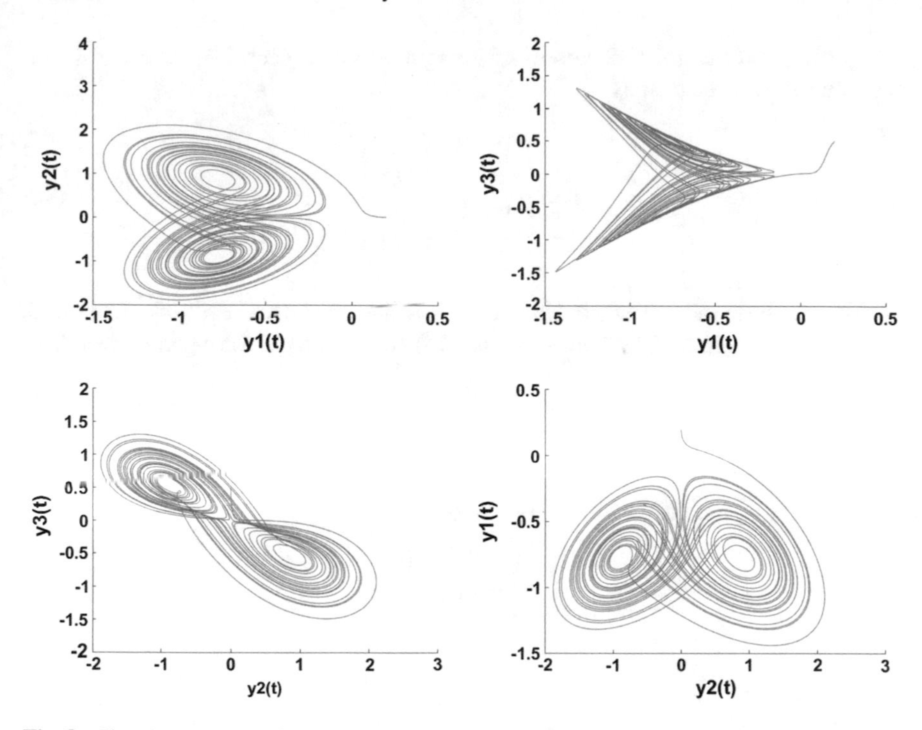

Fig. 8 Chaotic attractors of the uncontrolled system (46) in 2-D

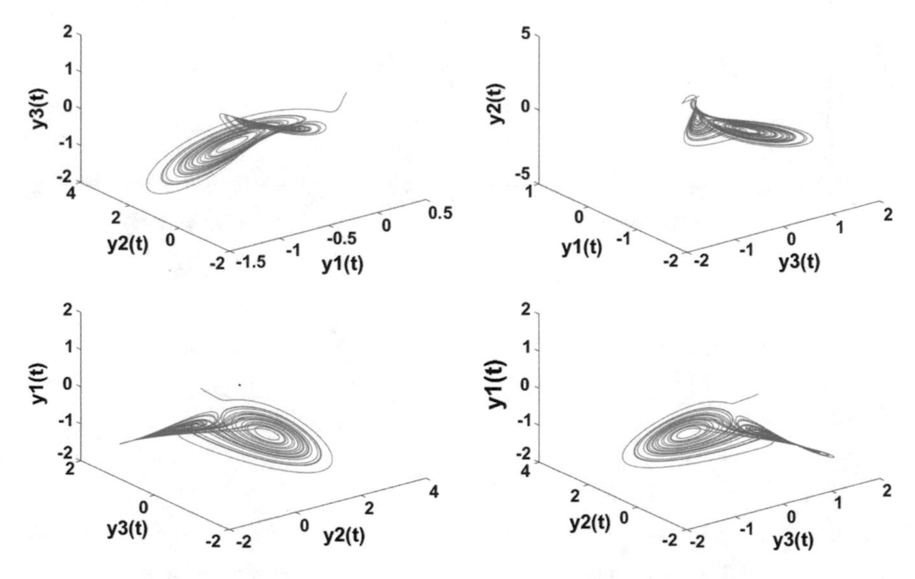

Fig. 9 Chaotic attractors of the uncontrolled system (46) in 3-D

By using our control IGS control scheme presented in Sect. 4.2, the control constant matrix L is selected as

$$L = \begin{pmatrix} 0 & 10 & 0 & 1 \\ 40 & 6 & 0 & 0.5 \\ 0 & 0 & 0 & -1 \\ 0 & -\frac{10}{15} & 0 & -1 \end{pmatrix}. \tag{48}$$

We can show that all eigenvalues of $A - L$ have negative real part. Therefore, according to Theorem 2, systems (45) and (46) are globally inverse generalized synchronized.

In this case, the error system will be

$$\begin{aligned} \dot{e}_1 &= -10e_1, \\ \dot{e}_2 &= -6e_2, \\ \dot{e}_3 &= -2.5e_3, \\ \dot{e}_4 &= -e_4. \end{aligned} \tag{49}$$

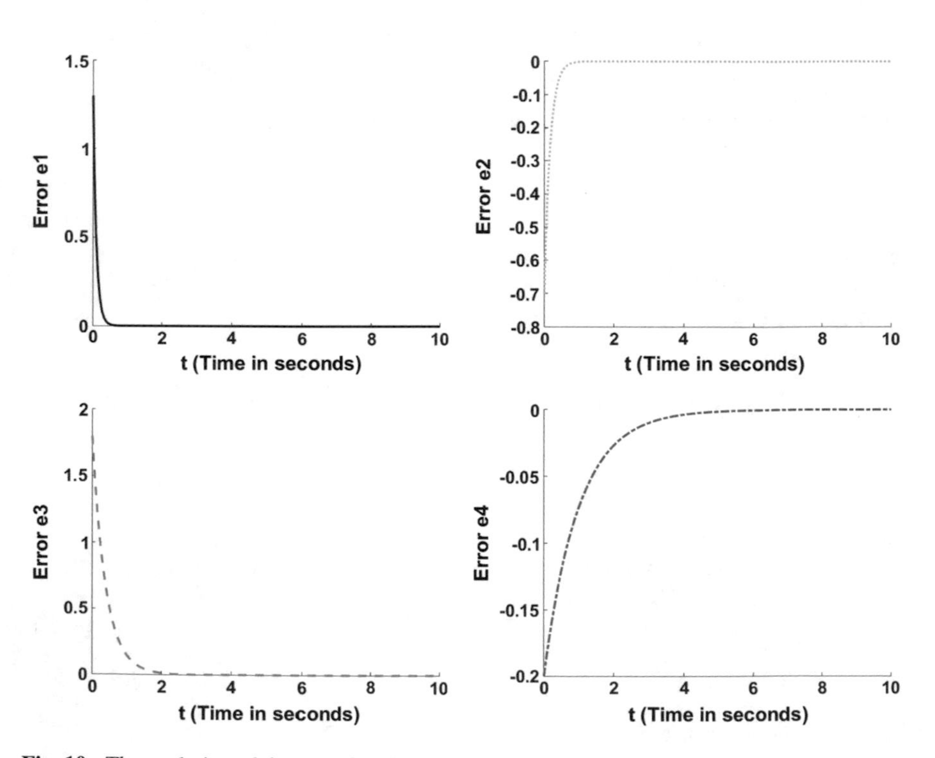

Fig. 10 The evolution of the error functions e_1, e_2, e_3 and e_4 as $t \to +\infty$

For the purpose of numerical simulation, fourth order Runge-Kutta method have been used. In addition, simulation time $Tm = 120$ s and time step $h = 0.005$ s has been employed. The initial values of the master system and the slave system are $[x_1(0), x_2(0), x_3(0), x_4(0)] = [2, -1, 1, 1]$ and $[y_1(0), y_2(0), y_3(0)] = [0.2, 0, 0.5]$, respectively, and the initial states of the error system are $[e_1(0), e_2(0), e_3(0), e_4(0)] = [1.3, -0.7, 1.8, -0.2]$. The error functions evolution, in this case, is shown in Fig. 10.

6 Conclusion

We have illustrated a new scheme of synchronization, called inverse generalized synchronization (IGS), in fractional order chaos (hyperchaos) systems between a master system of dimension n and a slave system of dimension m. The proposed approach, which enables each master system state to be synchronized with a functional relationship of slave system states, has shown some remarkable features. Namely, by exploiting the Lyapunov stability theory and the pole placement technique, inverse generalized synchronization is rigorously proved to be achievable for any fractional order chaotic (hyperchaotic) systems defined to date, including the two cases $n < m$ and $n > m$. Finally, the effectiveness of the method has been illustrated by numerical examples.

References

1. Hilfer, R. (2000). *Applications of fractional calculus in physics*. World Scientific.
2. Kilbas, A. A., Srivastava, H. M., & Trujillo, J. J. (2006). *Theory and applications of fractional differential equations*. Elsevier.
3. Sabatier, J., Agrawal, O. P., & Tenreiro Machado, J. A. (2007). *Advances in fractional calculus: theoretical developments and applications in physics and engineering*. Springer.
4. Tenreiro Machado, J. A., Jesus, I. S., Barbosa, R., Silva, M., & Reis, C. (2011). Application of fractional calculus in engineering. In M. M. Peixoto, A. A. Pinto, & D. A. Rand (Eds.), *Dynamics, games and science I*. Springer.
5. Uchaikin, V. V. (2012). *Fractional derivatives for physicists and engineers*. Higher Education Press.
6. Li, C., Chen, Y. Q., & Kurths, J. (2013). Fractional calculus and its applications. *Philosophical Transctions of the Royal Society A, 371*, 20130037.
7. Varsha, D. G. (2013). *Fractional calculus: Theory and applications*. Narosa Publishing House.
8. Herrmann, R. (2014). *Fractional calculus—an introduction for physicists*. World Scientific.
9. Heaviside, O. (1971). *Electromagnetic theory*. Chelsea.
10. Sugimoto, N. (1991). Burgers equation with a fractional derivative: Hereditary effects on nonlinear acoustic waves. *Journal of Fluid Mechanics, 225*, 631–653.
11. Parada, F. J. V., Tapia, J. A. O., & Ramirez, J. A. (2007). Effective medium equations for fractional Fick's law in porous media. *Physica A, 373*, 339–353.
12. Bagley, R. L., & Torvik, P. J. (1994). On the appearance of the fractional derivative in the behavior of real materials. *Journal of Applied Mechanics, 51*, 294–298.

13. Kulish, V. V., & Lage, J. L. (2002). Application of fractional calculus to fluid mechanics. *Journal of Fluids Engineering*, *124*, 803–806.
14. Atanackovic, T. M., Pilipovic, S., Stankovic, B., & Zorica, D. (2014). *Fractional calculus with applications in mechanics: Vibrations and diffusion processes*. Wiley.
15. Koeller, R. C. (1984). Application of fractional calculus to the theory of viscoelasticity. *Journal of Applied Mechanics*, *51*, 299–307.
16. Bagley, R. L., & Calico, R. A. (1991). Fractional order state equations for the control of viscoelastically damped structures. *Journal of Guidance Control and Dynamics*, *14*, 304–311.
17. Kusnezov, D., Bulgac, A., & Dang, G. D. (1999). Quantum Lévy processes and fractional kinetics. *Physical Review Letters*, *82*, 1136–1139.
18. Arena, P., Caponetto, R., Fortuna, L., & Porto, D. (2000). *Nonlinear noninteger order circuits and systems—an introduction*. World Scientific.
19. Bode, H. W. (1949). *Network analysis and feedback amplifier design*. Tung Hwa Book Company.
20. Carlson, G. E., & Halijak, C. A. (1964). Approximation of fractional capacitors $\left(\frac{1}{s}\right)^{\frac{1}{n}}$ by a regular Newton process. *IEEE Transactions on Circuit Theory*, *11*, 210–213.
21. Nakagava, M., & Sorimachi, K. (1992). Basic characteristics of a fractance device. *IEICE Transactions on Fundamentals E75-A*, 1814–1818.
22. Axtell, M., & Bise, E. M. (1990). Fractional calculus applications in control systems. *Proceedings of IEEE national aerospace electronics conference* (pp. 563–566).
23. Podlubny, I. (1999). Fractional-order systems and $\mathbf{PI}^{\lambda}\mathbf{D}^{\mu}$-controllers. *IEEE Transactions on Automatic Control*, *44*, 208–213.
24. Oustaloup, A. (1995). *La derivation non entiere: Theorie, synthese et applications*. Hermes.
25. da Graca, Marcos M., Duarte, F. B. M., & Machado, J. A. T. (2008). Fractional dynamics in the trajectory control of redundant manipulators. *Communications in Nonlinear Science and Numerical Simulation*, *13*, 1836–1844.
26. Vinagre, B. M., Chen, Y. Q., & Petráš, I. (2003). Two direct Tustin discretization methods for fractional-order differentiator/integrator. *Journal of the Franklin Institute*, *340*, 349–362.
27. Pires, E. J. S., Machado, J. A. T., & de Moura, P. B. (2003). Fractional order dynamics in a GA planner. *Signal Process*, *83*, 2377–2386.
28. Tseng, C. C. (2007). Design of FIR and IIR fractional order Simpson digital integrators. *Signal Process*, *87*, 1045–1057.
29. Magin, R. L. (2006). *Fractional calculus in bioengineering*. Begell House Publishers.
30. Ray, S. S. (2015). *Fractional calculus with applications for nuclear reactor dynamics*. CRC Press.
31. Wang, J. C. (1987). Realizations of generalized Warburg impedance with RC ladder networks and transmission lines. *Journal of the Electrochemical Society*, *134*, 1915–1920.
32. Westerlund, S. (2002). Dead matter has memory!. causal consulting.
33. Petráš, I. (2011). *Fractional-order nonlinear systems: Modeling, analysis and simulation*. Springer.
34. Strogatz, S. H. (2001). *Nonlinear dynamics and chaos: With applications to physics, biology, chemistry, and engineering*. Studies in Nonlinearity: Westview Press.
35. Morbidelli, A. (2001). Chaotic diffusion in celestial mechanics. *Regular and Chaotic Dynamics*, *6*, 339–353.
36. Li, D., Chenga, Y., Wanga, L., Wanga, H., Wanga, L., & Zhou, H. (2011). Prediction method for risks of coal and gas outbursts based on spatial chaos theory using gas desorption index of drill cuttings. *International Journal of Mining Science and Technology*, *21*, 439–443.
37. Li, M., Huanga, X., Liu, H., Liu, B., Wu, Y., Xiong, A., et al. (2013). Prediction of gas solubility in polymers by back propagation artificial neural network based on self-adaptive particle swarm optimization algorithm and chaos theory. *Journal of Fluid Phase Equilibria*, *356*, 11–17.

38. Bozoki, Zsolt. (1997). Chaos theory and power spectrum analysis in computerized cardiotocography. *European Journal of Obstetrics and Gynecology and Reproductive Biology, 71*, 163–168.
39. Sivakumar, B. (2000). Chaos theory in hydrology: Important issues and interpretations. *Journal of Hydrology, 227*, 1–20.
40. Fernando, J. (2011). Applying the theory of chaos and a complex model of health to establish relations among financial indicators. *Procedia Computer Science, 3*, 982–986.
41. Kyrtsou, C., & Labys, W. (2006). Evidence for chaotic dependence between US inflation and commodity prices. *Journal of Macroeconomics, 28*, 256–266.
42. Kyrtsou, C., & Terraza, M. (2003). Is it possible to study chaotic and ARCH behaviour jointly? Application of a noisy Mackey-Glass equation with heteroskedastic errors to the Paris Stock Exchange returns series. *Computational Economics, 21*, 257–276.
43. Wang, X., & Zhao, J. (2012). An improved key agreement protocol based on chaos. *Communications in Nonlinear Science and Numerical Simulation, 15*, 4052–4057.
44. Babaei, M. (2013). A novel text and image encryption method based on chaos theory and DNA computing. *Natural Computing, 12*, 101–107.
45. Nehmzow, U., & Keith, W. (2005). Quantitative description of robot-environment interaction using chaos theory. *Robotics and Autonomous System, 53*, 177–193.
46. Ambarish, G., Benoit, T., & Bernard, E. (1998). A study of the passive gait of a compass-like biped robot: Symmetry and chaos. *International Journal of Robot Research, 17*, 1282–1301.
47. Vaidyanathan, S., & Christos Volos, C. (2016). *Advances and applications in chaotic systems. Studies in computational intelligence*. Springer.
48. Azar, A. T., & Vaidyanathan, S. (2015). *Chaos modeling and control systems design, studies in computational intelligence* (Vol. 581). Germany: Springer.
49. Azar, A. T., Vaidyanathan, S. (2016). *Advances in chaos theory and intelligent control. Studies in fuzziness and soft computing* (Vol. 337). Germany: Springer. ISBN 978-3-319-30338-3.
50. Azar, A. T., & Vaidyanathan, S. (2015). *Computational intelligence applications in modeling and control. Studies in computational intelligence* (Vol. 575). Germany: Springer. ISBN 978-3-319-11016-5.
51. Azar, A. T., & Vaidyanathan, S. (2015). Handbook of research on advanced intelligent control engineering and automation. In *Advances in computational intelligence and robotics (ACIR) book series*, IGI Global, USA. ISBN 9781466672482.
52. Zhu, Q., & Azar, A. T. (2015). Complex system modelling and control through intelligent soft computations. *Studies in fuzziness and soft computing* (Vol. 319). Germany: Springer. ISBN: 978-3-319-12882-5.
53. Azar, A. T., & Zhu, Q. (2015). Advances and Applications in Sliding Mode Control systems. *Studies in computational intelligence* (Vol. 576). Germany: Springer. ISBN: 978-3-319-11172-8.
54. West, B. J., Bologna, M., & Grigolini, P. (2002). *Physics of fractal operators*. Springer.
55. Zaslavsky, G. M. (2005). *Hamiltonian chaos and fractional dynamics*. Oxford University Press.
56. Arena, P., Caponetto, R., Fortuna, L., & Porto, D. (1998). Bifurcation and chaos in noninteger order cellular neural networks. *International Journal of Bifurcation and Chaos, 8*, 1527–1539.
57. Ahmad, W. M., & Sprott, J. C. (2003). Chaos in fractional-order autonomous nonlinear systems. *Chaos Solitons and Fractals, 16*, 339–351.
58. Ahmad, W. M. (2005). Hyperchaos in fractional order nonlinear systems. *Chaos Solitons and Fractals, 26*, 1459–1465.
59. Ahmed, E., El-Sayed, A. M. A., & El-Saka, H. A. A. (2007). Equilibrium points, stability and numerical solutions of fractional-order predator-prey and rabies models. *Journal of Mathematical Analysis and Applictions, 325*, 542–553.
60. Deng, H., Li, T., Wang, Q., & Li, H. (2009). A fractional-order hyperchaotic system and its synchronization. *Chaos Solitons and Fractals, 41*, 962–969.

61. Liu, C., Liu, L., & Liu, T. (2009). A novel three-dimensional autonomous chaos system. *Chaos Solitons and Fractals*, *39*, 1950–1958.
62. Hartley, T., Lorenzo, C., & Qammer, H. (1995). Chaos in a fractional order Chua's system. *IEEE Transactions on Circuit and Systems I: Fundamental Theory and Applications*, *42*, 485–490.
63. Petráš, I. (2008). A note on the fractional-order Chua's system. *Chaos Solitons and Fractals*, *38*, 140–147.
64. Grigorenko, I., & Grigorenko, E. (2003). Chaotic dynamics of the fractional Lorenz system. *Physical Review Letters*, *91*, 034101–39.
65. Li, C., & Chen, G. (2004). Chaos and hyperchaos in fractional order Rössler equations. *Physica A*, *341*, 55–61.
66. Li, C., & Chen, G. (2004). Chaos in the fractional order Chen system and its control. *Chaos Solitons and Fractals*, *22*, 549–554.
67. Lu, J. G., & Chen, G. (2006). A note on the fractional-order Chen system. *Chaos Solitons and Fractals*, *27*, 685–688.
68. Lu, J. G. (2005). Chaotic dynamics and synchronization of fractional-order Arneodo's systems. *Chaos Solitons and Fractals*, *26*, 1125–1133.
69. Deng, W. H., & Li, C. P. (2005). Chaos synchronization of the fractional Lü system. *Physica A*, *353*, 61–72.
70. Lu, J. G. (2006). Chaotic dynamics of the fractional-order Lü system and its synchronization. *Physics Letters A*, *354*, 305–311.
71. Gao, X., & Yu, J. (2005). Chaos in the fractional order periodically forced complex Duffing's oscillators. *Chaos Solitons and Fractals*, *24*, 1097–1104.
72. Ge, Z. M., & Ou, C. Y. (2007). Chaos in a fractional order modified Duffing system. *Chaos Solitons and Fractals*, *34*, 262–291.
73. Ge, Z. M., & Hsu, M. Y. (2007). Chaos in a generalized van der Pol system and in its fractional order system. *Chaos Solitons and Fractals*, *33*, 1711–1745.
74. Barbosa, R. S., Machado, J. A. T., Vinagre, B. M., & Calderón, A. J. (2007). Analysis of the Van der Pol oscillator containing derivatives of fractional order. *Journal of Vibration and Control*, *13*, 1291–1301.
75. Petráš, I. (2009). Chaos in the fractional-order Volta's system: Modeling and simulation. *Nonlinear Dynamics*, *57*, 157–170.
76. Petráš, I. (2010). A note on the fractional-order Volta's system. *Communications in Nonlinear Science and Numerical Simulation*, *15*, 384–393.
77. Gejji, V. D., & Bhalekar, S. (2010). Chaos in fractional ordered Liu system. *Computers and Mathematics with Applications*, *59*, 1117–1127.
78. Vaidyanathan, S., Sampath, S., & Azar, A. T. (2015). Global chaos synchronisation of identical chaotic systems via novel sliding mode control method and its application to Zhu system. *International Journal of Modelling, Identification and Control (IJMIC)*, *23*(1), 92–100.
79. Vaidyanathan, S., Azar, A. T., Rajagopal, K., & Alexander, P. (2015). Design and SPICE implementation of a 12-term novel hyperchaotic system and its synchronization via active control. *International Journal of Modelling, Identification and Control (IJMIC)*, *23*(3), 267–277.
80. Vaidyanathan, S., & Azar, A. T. (2016). Takagi-sugeno fuzzy logic controller for Liu-Chen four-scroll chaotic system. *International Journal of Intelligent Engineering Informatics*, *4*(2), 135–150.
81. Vaidyanathan, S., & Azar, A. T. (2015). Analysis and control of a 4-D novel hyperchaotic system. In A. T. Azar & S. Vaidyanathan (Eds.), *Chaos modeling and control systems design, studies in computational intelligence* (Vol. 581, pp. 19–38). GmbH Berlin/Heidelberg: Springer. doi:10.1007/978-3-319-13132-0_2.
82. Boulkroune, A., Bouzeriba, A., Bouden, T., & Azar, A. T. (2016). Fuzzy adaptive synchronization of uncertain fractional-order chaotic systems. In A. T Azar & S. Vaidyanathan (Eds.), *Advances in chaos theory and intelligent control. Studies in fuzziness and soft computing* (Vol. 337). Germany: Springer.

83. Boulkroune, A., Hamel, S., & Azar, A. T. (2016). Fuzzy control-based function synchronization of unknown chaotic systems with dead-zone input. *Advances in chaos theory and intelligent control. Studies in fuzziness and soft computing* (Vol. 337). Germany: Springer.
84. Vaidyanathan, S., Azar, A. T. (2016). Dynamic analysis, adaptive feedback control and synchronization of an eight-term 3-D novel chaotic system with three quadratic nonlinearities. *Advances in chaos theory and intelligent control. Studies in fuzziness and soft computing* (Vol. 337). Germany: Springer.
85. Vaidyanathan, S., & Azar, A. T. (2016). qualitative study and adaptive control of a novel 4-d hyperchaotic system with three quadratic nonlinearities. *Advances in chaos theory and intelligent control. Studies in fuzziness and soft computing* (Vol. 337). Germany: Springer.
86. Vaidyanathan, S., & Azar, A. T. (2016). A novel 4-d four-wing chaotic system with four quadratic nonlinearities and its synchronization via adaptive control method. *Advances in chaos theory and intelligent control. Studies in fuzziness and soft computing* (Vol. 337). Germany: Springer.
87. Vaidyanathan, S., & Azar, A. T. (2016). Adaptive control and synchronization of halvorsen circulant chaotic systems. *Advances in chaos theory and intelligent control. Studies in fuzziness and soft computing* (Vol. 337). Germany: Springer.
88. Vaidyanathan, S., & Azar, A. T. (2016). Adaptive backstepping control and synchronization of a novel 3-d jerk system with an exponential nonlinearity. *Advances in chaos theory and intelligent control. Studies in fuzziness and soft computing* (Vol. 337). Germany: Springer.
89. Vaidyanathan, S., & Azar, A. T. (2016). Generalized projective synchronization of a novel hyperchaotic four-wing system via adaptive control method. *Advances in chaos theory and intelligent control. Studies in fuzziness and soft computing* (Vol. 337). Germany: Springer.
90. Vaidyanathan, S., & Azar, A. T. (2015). Analysis, control and synchronization of a nine-term 3-D novel chaotic system. In A. T. Azar & S. Vaidyanathan (Eds.), *Chaos modeling and control systems design*. Studies in computational intelligence book series: Springer.
91. Vaidyanathan, S., & Azar, A. T. (2015). Anti-synchronization of identical chaotic systems using sliding mode control and an application to Vaidyanathan-Madhavan chaotic systems. In A. T. Azar & Q. Zhu (Eds.), *Advances and applications in sliding mode control systems*. Studies in computational intelligence book series: Springer.
92. Vaidyanathan, S., & Azar, A. T. (2015). Hybrid synchronization of identical chaotic systems using sliding mode control and an application to Vaidyanathan chaotic systems. In A. T. Azar & Q. Zhu (Eds.), *Advances and applications in sliding mode control systems*. Studies in computational intelligence book series: Springer.
93. Vaidyanathan, S., Idowu, B. A., & Azar, A. T. (2015). Backstepping controller design for the global chaos synchronization of Sprott's jerk systems. In A. T. Azar & S. Vaidyanathan (Eds.), *Chaos modeling and control systems design*. Studies in computational intelligence book series: Springer.
94. Ouannas, A. (2014). Chaos synchronization approach based on new criterion of stability. *Nonlinear Dynamics and System Theory*, *14*, 396–402.
95. Ouannas, A. (2014). On full state hybrid projective synchronization of general discrete chaotic systems. *Journal of Nonlinear Dynamics*, 1–6.
96. Ouannas, A. (2014). Some synchronization criteria for N-dimensional chaotic systems in discrete-time. *Journal of Advanced Research in Applied Mathematics*, *6*, 1–10.
97. Ouannas, A. On inverse full state hybrid projective synchronization of chaotic dynamical systems in discrete-time. *International Journal of Dynamics Control*, 1–7.
98. Ouannas, A. (2015). Synchronization criterion for a class of N-dimensional discrete chaotic systems. *Journal of Advanced Research in Dynamics and Control Systems*, *7*, 82–89.
99. Ouannas, A. (2015). A new synchronization scheme for general 3D quadratic chaotic systems in discrete-time. *Nonlinear Dynamics and Systems Theory*, *15*, 163–170.
100. Ouannas, A., Odibat, Z., Shawagfeh, N. (2016). A new Q–S synchronization results for discrete chaotic systems. *Differential Equations and Dynamical Systems*, 1–10.
101. Ouannas, A. (2016). Co-existence of various synchronization-types in hyperchaotic maps. *Nonlinear Dynamics and Systems Theory*, *16*, 312–321.

102. Ouannas, A., Azar, A. T., & Abu-Saris, R. (2016). A new type of hybrid synchronization between arbitrary hyperchaotic maps. *International Journal of Machine Learning and Cybernetics*, 1–8.
103. Li, C., & Zhou, T. (2005). Synchronization in fractional-order differential systems. *Physica D*, *212*, 111–125.
104. Zhou, S., Li, H., Zhu, Z., & Li, C. (2008). Chaos control and synchronization in a fractional neuron network system. *Chaos Solitons and Fractals*, *36*, 973–984.
105. Peng, G. (2007). Synchronization of fractional order chaotic systems. *Physics Letters A*, *363*, 426–432.
106. Sheu, L. J., Chen, H. K., Chen, J. H., & Tam, L. M. (2007). Chaos in a new system with fractional order. *Chaos Solitons and Fractals*, *31*, 1203–1212.
107. Yan, J., & Li, C. (2007). On chaos synchronization of fractional differential equations. *Chaos Solitons and Fractals*, *32*, 725–735.
108. Li, C., & Yan, J. (2007). The synchronization of three fractional differential systems. *Chaos Solitons and Fractals*, *32*, 751–757.
109. Wang, J., Xiong, X., & Zhang, Y. (2006). Extending synchronization scheme to chaotic fractional-order Chen systems. *Physica A*, *370*, 279–285.
110. Li, C. P., Deng, W. H., & Xu, D. (2006). Chaos synchronization of the Chua system with a fractional order. *Physica A*, *360*, 171–185.
111. Zhu, H., Zhou, S., & Zhang, J. (2009). Chaos and synchronization of the fractional-order Chua's system. *Chaos Solitons and Fractals*, *39*, 1595–1603.
112. Lu, J. G. (2005). Chaotic dynamics and synchronization of fractional-order Arneodo's systems. *Chaos Solitons and Fractals*, *26*, 1125–1133.
113. Ansari, M. A., Arora, D., & Ansari, S. P. (2016). Chaos control and synchronization of fractional order delay-varying computer virus propagation model. *Mathematical Methods in Applied Sciences*, *39*, 1197–1205.
114. Kiani, B. A., Fallahi, K., Pariz, N., & Leung, H. (2009). A chaotic secure communication scheme using fractional chaotic systems based on an extended fractional Kalman filter. *Communications in Nonlinear Science and Numerical Simulation*, *14*, 863–879.
115. Liang, H., Wang, Z., Yue, Z., & Lu, R. (2012). Generalized synchronization and control for incommensurate fractional unified chaotic system and applications in secure communication. *Kybernetika*, *48*, 190–205.
116. Wu, X., Wang, H., & Lu, H. (2012). Modified generalized projective synchronization of a new fractional-order hyperchaotic system and its application to secure communication. *Nonlinear Analysis: Real World Applications*, *13*, 1441–1450.
117. Muthukumar, P., & Balasubramaniam, P. (2013). Feedback synchronization of the fractional order reverse butterfly-shaped chaotic system and its application to digital cryptography. *Nonlinear Dynamics*, *74*, 1169–1181.
118. Muthukumar, P., Balasubramaniam, P., & Ratnavelu, K. (2014). Synchronization of a novel fractional order stretch-twistfold (STF) flow chaotic system and its application to a new authenticated encryption scheme (AES). *Nonlinear Dynamics*, *77*, 1547–1559.
119. Chen, L., Wu, R., He, Y., & Chai, Y. (2015). Adaptive sliding-mode control for fractional-order uncertain linear systems with nonlinear disturbances. *Nonlinear Dynamics*, *80*, 51–58.
120. Liu, L., Ding, W., Liu, C., Ji, H., & Cao, C. (2014). Hyperchaos synchronization of fractional-order arbitrary dimensional dynamical systems via modified sliding mode control. *Nonlinear Dynamics*, *76*, 2059–2071.
121. Zhang, L., & Yan, Y. (2014). Robust synchronization of two different uncertain fractional-order chaotic systems via adaptive sliding mode control. *Nonlinear Dynamics*, *76*, 1761–1767.
122. Odibat, Z., Corson, N., Alaoui, M. A. A., & Bertelle, C. (2010). Synchronization of chaotic fractional-order systems via linear control. *International Journal of Bifurcation and Chaos*, *20*, 81–97.
123. Chen, X. R., & Liu, C. X. (2012). Chaos synchronization of fractional order unified chaotic system via nonlinear control. *International Journal of Modern Physics B*, *25*, 407–415.

124. Srivastava, M., Ansari, S. P., Agrawal, S. K., Das, S., & Leung, A. Y. T. (2014). Anti-synchronization between identical and non-identical fractional-order chaotic systems using active control method. *Nonlinear Dynamics*, *76*, 905–914.
125. Agrawal, S. K., & Das, S. (2013). A modified adaptive control method for synchronization of some fractional chaotic systems with unknown parameters. *Nonlinear Dynamics*, *73*, 907–919.
126. Yuan, W. X., & Mei, S. J. (2009). Synchronization of the fractional order hyperchaos Lorenz systems with activation feedback control. *Communications in Nonlinear Science and Numerical Simulation*, *14*, 3351–3357.
127. Odibat, Z. (2010). Adaptive feedback control and synchronization of non-identical chaotic fractional order systems. *Nonlinear Dynamics*, *60*, 479–487.
128. Zhou, P., & Bai, R. (2015). The adaptive synchronization of fractional-order chaotic system with fractional-order $1<q<2$ via linear parameter update law. *Nonlinear Dynamics*, *80*, 753–765.
129. Peng, G., & Jiang, Y. (2008). Generalized projective synchronization of a class of fractional-order chaotic systems via a scalar transmitted signal. *Physics Letters A*, *372*, 3963–3970.
130. Cafagna, D., & Grassi, G. (2012). Observer-based projective synchronization of fractional systems via a scalar signal: Application to hyperchaotic Rössler systems. *Nonlinear Dynamics*, *68*, 117–128.
131. Li, T., Wang, Y., & Yang, Y. (2014). Designing synchronization schemes for fractional-order chaotic system via a single state fractional-order controller. *Optik*, *125*, 6700–6705.
132. Lai, L. C., Mei, Z., Feng, Z., & Bing, Y. X. (2016). Projective synchronization for a fractional-order chaotic system via single sinusoidal coupling. *Optik*, *127*, 2830–2836.
133. Odibat, Z. (2012). A note on phase synchronization in coupled chaotic fractional order systems. *Nonlinear Analsis: Real World Application*, *13*, 779–789.
134. Chen, F., Xia, L., & Li, C. G. (2012). Wavelet phase synchronization of fractional-order chaotic systems. *Chin Phys Lett*, *29*, 070501–6.
135. Razminia, A., & Baleanu, D. (2013). Complete synchronization of commensurate fractional order chaotic systems using sliding mode control. *Mechatronics*, *23*, 873–879.
136. Al-sawalha, M. M., Alomari, A. K., Goh, S. M., & Nooran, M. S. M. (2011). Active anti-synchronization of two Identical and different fractional-order chaotic systems. *International Journal of Nonlinear Science*, *11*, 267–274.
137. Li, C. G. (2006). Projective synchronization in fractional order chaotic systems and its control. *Progress of Theoretical Physics*, *115*, 661–666.
138. Shao, S. Q., Gao, X., & Liu, X. W. (2007). Projective synchronization in coupled fractional order chaotic Rossler system and its control. *Chinese Physics*, *16*, 2612–2615.
139. Wang, X. Y., & He, Y. J. (2008). Projective synchronization of fractional order chaotic system based on linear separation. *Physics Letters A*, *372*, 435–441.
140. Si, G., Sun, Z., Zhang, Y., & Chen, W. (2012). Projective synchronization of different fractional-order chaotic systems with non-identical orders. *Nonlinear Analysis: Real World Applications*, *13*, 1761–1771.
141. Agrawal, S. K., & Das, S. (2014). Projective synchronization between different fractional-order hyperchaotic systems with uncertain parameters using proposed modified adaptive projective synchronization technique. *Mathematical Methods in the Applied Sciences*, *37*, 2164–2176.
142. Chang, C. M., & Chen, H. K. (2010). Chaos and hybrid projective synchronization of commensurate and incommensurate fractional-order Chen-Lee systems. *Nonlinear Dynamics*, *62*, 851–858.
143. Wang, S., Yu, Y. G., & Diao, M. (2010). Hybrid projective synchronization of chaotic fractional order systems with different dimensions. *Physica A*, *389*, 4981–4988.
144. Zhou, P., & Zhu, W. (2011). Function projective synchronization for fractional-order chaotic systems. *Nonlinear Analysis: Real World Applications*, *12*, 811–816.
145. Zhou, P., & Cao, Y. X. (2010). Function projective synchronization between fractional-order chaotic systems and integer-order chaotic systems. *Chinese Physics B*, *19*, 100507.

146. Xi, H., Li, Y., & Huang, X. (2015). Adaptive function projective combination synchronization of three different fractional-order chaotic systems. *Optik*, *126*, 5346–5349.
147. Chen, H., & Sun, M. (2006). Generalized projective synchronization of the energy resource system. *International Journal of Nonlinear Science*, *2*, 166–170.
148. Peng, G. J., Jiang, Y. L., & Chen, F. (2008). Generalized projective synchronization of fractional order chaotic systems. *Physica A*, *387*, 3738–3746.
149. Shao, S. Q. (2009). Controlling general projective synchronization of fractional order Rössler systems. *Chaos Solitons and Fractals*, *39*, 1572–1577.
150. Wu, X. J., & Lu, Y. (2009). Generalized projective synchronization of the fractional-order Chen hyperchaotic system. *Nonlinear Dynamics*, *57*, 25–35.
151. Zhou, P., Kuang, F., & Cheng, Y. M. (2010). Generalized projective synchronization for fractional order chaotic systems. *Chinese Journal of Physics*, *48*, 49–56.
152. Razminia, A. (2013). Full state hybrid projective synchronization of a novel incommensurate fractional order hyperchaotic system using adaptive mechanism. *Indian Journal of Physics*, *87*(2), 161–167.
153. Yi, C., Liping, C., Ranchao, W., & Juan, D. (2013). Q-S synchronization of the fractional-order unified system. *Pramana*, *80*, 449–461.
154. Mathiyalagan, K., Park, J. H., & Sakthivel, R. (2015). Exponential synchronization for fractional-order chaotic systems with mixed uncertainties. *Complexity*, *21*, 114–125.
155. Aghababa, M. P. (2012). Finite-time chaos control and synchronization of fractional-order nonautonomous chaotic (hyperchaotic) systems using fractional nonsingular terminal sliding mode technique. *Nonlinear Dynamics*, *69*, 247–261.
156. Li, D., Zhang, X. P., Hu, Y. T., & Yang, Y. Y. (2015). Adaptive impulsive synchronization of fractional order chaotic system with uncertain and unknown parameters. *Neurocomputing*, *167*, 165–171.
157. Xi, H., Yu, S., Zhang, R., & Xu, L. (2014). Adaptive impulsive synchronization for a class of fractional-order chaotic and hyperchaotic systems. *Optik*, *125*, 2036–2040.
158. Ouannas, A., & Abu-Saris, R. (2016). On matrix projective synchronization and inverse matrix projective synchronization for different and identical dimensional discrete-time chaotic systems. *Journal of Chaos*, 1–7.
159. Ouannas, A., & Mahmoud, E. (2014). Inverse matrix projective synchronization for discrete chaotic systems with different dimensions. *Intell Electronic System*, *3*, 188–192.
160. Ouannas, A., & Abu-Saris, R. (2015). A robust control method for Q-S synchronization between different dimensional integer-order and fractional-order chaotic systems. *Journal of Control Science and Engineering*, 1–7.
161. Ouannas, A. (2015). A new generalized-type of synchronization for discrete-time chaotic dynamical systems. *Journal of Computational and Nonlinear Dynamics*, *10*, 061019–5.
162. Ouannas, A., Al-sawalha, M. M. (2016). On $\Lambda - \phi$ generalized synchronization of chaotic dynamical systems in continuous-time. *The European Physical Journal Special Topics*, *225*, 187–196.
163. Ouannas, A., & Al-sawalha, M. M. (2015). A new approach to synchronize different dimensional chaotic maps using two scaling matrices. *Nonlinear Dynamics and Systems Theory*, *15*, 400–408.
164. Ouannas, A., & Al-sawalha, M. M. (2016). Synchronization between different dimensional chaotic systems using two scaling matrices. *Optik*, *127*, 959–963.
165. Ouannas, A., Al-sawalha, M. M., & Ziar, T. (2016). Fractional chaos synchronization schemes for different dimensional systems with non-identical fractional-orders via two scaling matrices. *Optik*, *127*, 8410–8418.
166. Ouannas, A., Grassi, G. (2016). A new approach to study co-existence of some synchronization types between chaotic maps with different dimensions. *Nonlinear Dynamics*, 1–10.
167. Ouannas, A., Azar, A. T., & Vaidyanathan, S. (2016). A robust method for new fractional hybrid chaos synchronization. *Mathematical Methods in the Applied Sciences*, 1–9.
168. Deng, W. H. (2007). Generalized synchronization in fractional order systems. *Physical Review E*, *75*, 056201.

169. Zhou, P., Cheng, X. F., & Zhang, N. Y. (2008). Generalized synchronization between different fractional-order chaotic systems. *Communication in Theoretical Physics*, *50*, 931–934.
170. Zhang, X. D., Zhao, P. D., & Li, A. H. (2010). Construction of a new fractional chaotic system and generalized synchronization. *Communication in Theoretical Physics*, *53*, 1105–1110.
171. Jun, W. M., & Yuan, W. X. (2011). Generalized synchronization of fractional order chaotic systems. *International Journal of Modern Physics C*, *25*, 1283–1292.
172. Wu, X. J., Lai, D. R., & Lu, H. T. (2012). Generalized synchronization of the fractional-order chaos in weighted complex dynamical networks with nonidentical nodes. *Nonlinear Dynamics*, *69*, 667–683.
173. Xiao, W., Fu, J., Liu, Z., & Wan, W. (2012). Generalized synchronization of typical fractional order chaos system. *Journal of Computer*, *7*, 1519–1526.
174. Martínez-Guerra, R., & Mata-Machuca, J. L. (2014). Fractional generalized synchronization in a class of nonlinear fractional order systems. *Nonlinear Dynamics*, *77*, 1237–1244.
175. Ouannas, A., & Odibat, Z. (2015). Generalized synchronization of different dimensional chaotic dynamical systems in discrete-time. *Nonlinear Dynamics*, *81*, 7657–71.
176. Ouannas, A. (2016). On inverse generalized synchronization of continuous chaotic dynamical systems. *International Journal of Applied and Computational Mathematics*, *2*, 1–11.
177. Podlubny, I. (1999). *Fractional differential equations*. Academic Press.
178. Caputo, M. (1967). Linear models of dissipation whose Q is almost frequency independent-II. *Geophysical Journal of the Royal Astronomical Society*, *13*, 529–539.
179. Samko, S. G., Klibas, A. A., & Marichev, O. I. (1993). *Fractional integrals and derivatives: Theory and applications*. Gordan and Breach.
180. Gorenflo, R., & Mainardi, F. (1997). Fractional calculus: Integral and differential equations of fractional order. In A. Carpinteri & F. Mainardi (Eds.), *Fractals and fractional calculus in continuum mechanics*. Springer.
181. Xue, W., Li, Y., Cang, S., Jia, H., & Wang, Z. (2015). Chaotic behavior and circuit implementation of a fractional-order permanent magnet synchronous motor model. *Journal of the Franklin Institute*, *352*, 2887–2898.
182. Han, Q., Liu, C. X., Sun, L., & Zhu, D. R. (2013). A fractional order hyperchaotic system derived from a Liu system and its circuit realization. *Chinese Physics B*, *22*, 020502–6.
183. Li, T. Z., Wang, Y., & Luo, M. K. (2014). Control of fractional chaotic and hyperchaotic systems based on a fractional order controller. *Chinese Physics B*, *23*, 080501–11.

Behavioral Modeling of Chaos-Based Applications by Using Verilog-A

J.M. Munoz-Pacheco, V.R. González Díaz, L.C. Gómez-Pavón, S. Romero-Camacho, F. Sánchez-Guzmán, J. Mateo-Juárez, L. Delgado-Toral, J.A. Cocoma-Ortega, A. Luis-Ramos, P. Zaca-Morán and E. Tlelo-Cuautle

Abstract In general, a system can be defined as a collection of interconnected components that transforms a set of inputs received from its environment to a set of outputs. From an engineering point of view, chaos-based applications can be classified as a electronic system where the vast majority of the internal signals used as interconnections are electrical signals. Inputs and outputs are also provided as electrical quantities, or converted from, or to, such signals using sensors or actuators. To gain insight about the overall performance of the particular chaos-based application, the whole system must be characterized and simulated simultaneously. That is not a trivial task because the complexity of each one of the blocks that comprises the system, as well as the intrinsic complex behavior of chaotic generators. In this chapter, a modeling strategy suited to represent chaos-based applications for different control parameters of chaotic systems is presented. Based on behavioral descriptions obtained from a Hardware Description Language (HDL), called Verilog-A, two applications of chaotic systems are analyzed and designed. More specifically, a chaotic sinusoidal pulse width modulator (SPWM) which is useful to develop control algorithms for motor drivers in electric vehicles, and a chaotic pulse position modulator (CPPM) widely used in communication systems are presented as cases under analysis. Those applications are coded in Verilog-A and by using different abstraction levels, the indications of the degree of detail specified on how the function is to be implemented are obtained. Therefore, these behavioral models try to capture as much circuit functionality as possible with far less implementation details than the device-level description of the electronic circuit. Several circuit simulations applying

J.M. Munoz-Pacheco (✉) · V.R. González Díaz · L.C. Gómez-Pavón ·
S. Romero-Camacho · F. Sánchez-Guzmán · J. Mateo-Juárez · L. Delgado-Toral ·
J.A. Cocoma-Ortega · A. Luis-Ramos
Faculty of Electronics Sciencies, Benemérita Universidad Autónoma de Puebla,
Puebla, Pue, Mexico 72000, USA
e-mail: jesusm.pacheco@correo.buap.mx

P. Zaca-Morán
ICUAP, Benemérita Universidad Autónoma de Puebla,
Puebla, Pue, Mexico 72000, USA

E. Tlelo-Cuautle
Department of Electronics, INAOE, Tonantzintla, Pue, Mexico, USA

A.T. Azar et al. (eds.), *Fractional Order Control and Synchronization of Chaotic Systems*, Studies in Computational Intelligence 688,
DOI 10.1007/978-3-319-50249-6_19

H-Spice simulator are presented to demonstrate the usefulness of the proposed models. In this manner, behavioral modeling can be a possible solution for the successful development of robust chaos-based applications due to various types of systems that can be represented and simulated by means of an abstract model.

Keywords Modeling · Verilog · Chaos · Pulse width modulator · Pulse position modulator · Logistic map · Chaotic system

1 Introduction

Nowadays, one of the most studied phenomena is chaos into the nonlinear dynamical systems. Particularly, chaotic systems are mainly characterized by its behavior complex and like random. Their significance has been increased during the last decade because of several applications in diverse fields ranging from living systems, such as synchronization in neurobiology, chemical reactions among pancreatic cells, social events, to non-living systems including robotics, low power high-speed data transceivers for medical applications, chaotic electrochemical oscillators, encrypted communications, control algorithms for motor drivers in electric vehicles, and so on [1–16]. This new and challenging research and development area has become a scientific interdisciplinary, involving system and control engineers, theoretical and experimental physicist, applied mathematicians, physiologists and above all, circuit and devices specialists [1–31].

From engineering point of view, the goal is to obtain a chaos-based application from a mathematical description [32–40]. However, there is a gap between the chaotic system and its physical realization with electronic devices. It means that circuit implementation of chaos-based applications depends on critical design requirements (e.g., dynamic range, bandwidth, slew-rate, voltage supply, appropriate differential amplifiers or a convenient scaling of voltages) that regularly there are no taken into account in a general design flow [3, 10, 13].

Roughly, a chaos-based application can be defined as a system where encompasses various blocks with a chaos generator being the main block. Although some papers have reported chaos-based applications using electronic devices, those are custom designs. This indicates that it is necessary a deeper understanding on chaotic dynamics of the system. In addition, these designs are only valid for a certain range of values for the parameters. If one needs to modify the behavior of the whole system, it should be necessary to evaluate the entire design. Even, physical conditions could limit or prohibit such chaos-based application [8].

Considering the aforementioned issues, a systematic methodology for circuit design of chaos-based applications must be addressed. Therefore, the aim of this chapter is to investigate a modeling approach for determining the design requirements of chaos-based applications using electronic devices whose performance can be analyzed and determined by applying behavioral modeling and simulation. In this manner, the proposed approach is focus on creating a low-level representation (low-level abstraction) of a chaos-based application from a higher-level (more abstract) representation. Where, the obtained representation should have the same function as the higher-level representation.

Verilog-A is used herein to perform the behavioral modeling with aim of the generation of the architecture and the selection of its building blocks and their performance values. Therefore, a design engineer can make decisions at an early stage of the design cycle, thus ensuring correct design.

This chapter contains in Sect. 2 the related work oriented to behavioral modeling, top-down design flows, chaotic systems, and Verilog-A behavioral modeling. An analysis and Verilog-A-based design of a chaotic pulse position modulator (CPPM) is presented in Sect. 3, while Sect. 4 deals with the analysis and Verilog-A-based design of a sinusoidal pulse width modulator based on chaotic amplitude frequency modulator (CAFM-SPWM). Finally, a detailed discussion of the results are given in Sect. 5 and conclusions in Sect. 6.

2 Related Work

2.1 Behavioral Modeling

Engineers analyze and design various types of systems frequently. In general, a system can be defined as a collection of interconnected components that transforms a set of inputs received from its environment to a set of outputs [33–36, 38, 40]. In an electronic system, the vast majority of the internal signals used as interconnections are electrical signals. Actually, behavioral modeling can be a possible solution for the successful development of various types of systems (i.e., chaos-based applications) that can be represented by means of an abstract model where the abstraction levels are indications of the degree of detail specified on how the system is to be implemented [8, 32, 39].

Therefore, behavioral models try to capture as much circuit functionality as possible with far less implementation details than the electronic device-level description of the chaos-based applications. To do that, description and abstraction levels must be described [33, 36, 38, 40].

Description level: A description level is a pair of two sets; a set of elementary elements and a set of interconnection types.

Abstraction level: The abstraction level of a description is the degree to which information about non-ideal effects or structure is neglected compared to the dominant behavior of the entire system.

A system may be described with a certain description level at different levels of abstraction. Although it is clear to consider the functional level at a high abstraction level and the physical level at a low abstraction level, it is not straightforward to compare the abstraction levels of different description levels [33, 36, 38, 40]. Due to the overlap one can easily jump, for instance, from the behavioral to the circuit level.

As consequence, a chaos-based application could be designed by converting the functional specification at the highest abstraction level to a physical realization at the lowest abstraction level via operations between description and abstraction levels [8, 32, 39]. Thereby, automatic synthesis of chaos-based applications can formally be represented as operations between different levels. Four fundamental types of such operations are distinguished in behavioral modeling, which are introduced below.

'Refinement: Translates a system described with a certain description level, into a representation at a lower abstraction level with the same description level.'

'Simplification: Translates a system described with a certain description level, into a representation at a higher abstraction level with the same description level.'

'Translation: Translates a system described with a certain description level, into a representation with another description level preserving the level of abstraction.'

'Transformation: Translates a system described with a certain description level, into a representation at the same abstraction level and the same description level.'

Systematic synthesis implies the application of subsequent refinement operations on the system. Such an operation may introduce more detailed information about the system and therefore the performance of the system can then be re-evaluated. A refinement operation can also be used to introduce a subdivision of the system. Hence, additional knowledge about the structure or building blocks of a system corresponds to descending the hierarchy of descriptions. Such operation also includes the derivation of values for the parameters used in the models of the building blocks for chaos-based applications [33, 36, 38, 40].

2.2 *Top-Down Design Flow for Electronics Systems*

A design process formally consists of the application of a chain of operations defined in the previous subsection from specification (High-level) to implementation (Low-level). The design starts from a description of the functionality of the system, possibly written in some HDL such as Verilog-A, and ends with a layout ready to be fabricated [33, 36]. The four basic operations can be put together in an iterative design process, where if the design fails to meet the specifications, a transformation of either the architecture or its parameters should be applied. Simplification can remove details to make it easier to choose which transformation should be selected. On the other hand, if the specifications are met, but the abstraction level does not correspond to the wishes of the designer, details should be added or removed by applying simplification or refinement operations. Finally, the design within the description level finishes once all specifications are met. Translation will be required to convert the current representation of the system to the next description level in top-down direction [33].

Top-down Methodology [38, 40]: To cope with complex electronic designs starting from functional specifications, a large system is divided into smaller blocks in the top-down design methodology. The design at a high abstraction level of a complex system corresponds to deriving the behavioral models for the building blocks. Simplification operations are not used in this method. From this point of view, the design methodology must indicate the kind and order of the operations to be applied during the design process and include an appropriate modeling strategy to determine how a system is represented. Selecting a good modelling strategy make it easier to execute the synthesis process.

2.3 *Chaotic Systems*

Chaotic systems refer to one type of complex nonlinear dynamical system that possesses some very special features such as extreme sensitivity to tiny variations of initial conditions and parameters, and bounded trajectories in the phase space but with a positive maximum Lyapunov exponent. For deterministic chaos to exist, a nonlinear dynamical system must have a dense set of periodic orbits, be transitive, and sensitive to initial conditions. Density in periodic orbits implies that any periodic trajectory of the orbit visits an arbitrarily small neighborhood of a non-periodic one. Transitivity relates to the existence of points a, b for which a third point c can be found that is arbitrarily close to a and whose orbit passes arbitrarily close to b. Finally, sensitivity to initial conditions is the property to arbitrarily close initial conditions to give rise to orbits that are eventually separated by a finite amount [7].

Over the last two decades, theoretical design and circuit implementation of various chaos generators have been a focal subject of increasing interest due to their promising applications in various real-world chaos-based technologies and information systems [1–16]. In particular, Chua's circuit is considered as a paradigm of chaos and a bridge between electronic circuits and the chaos theory, has been widely studied and used as a platform for engineering applications. This subsection offers an overview of the design methodologies and circuit implementations. Recently, chaotic systems with their variants, i.e., n-scroll chaotic systems, muli-scroll chaotic systems with 1D, 2D and 3D orientations, PWL chaotic systems, chaotic systems with hidden attractors, hyperchaotic systems, fractional order chaotic systems; have been designed by using several electronic devices, such as, OpAmps, CFOAs, OTAs, CCII+, OTRAs, SETs, PICs, FPGAs, FPAAs, micro computers, etc. [3, 7, 8, 10, 12, 13].

Extended from chaotic systems, theoretical design and hardware implementation of different kinds of chaos-based applications have attracted increasing attention, especially for those that can create various complex modulation schemes, for instance PWM, PPM, PAM, and so on. Those modulation schemes are the core of novel engineering applications including navigation techniques for mobile robots, motor drivers of electric vehicles, secure communications and encryption, and power converters.

2.4 Verilog-A: A Hardware Description Language

Synthesis is the process of creating a low-level representation (low-level abstraction) from a higher-level (more abstract) representation. The synthesized representation should have the same function as the higher-level representation. High-level synthesis advantages: continuous and reliable design flow, shorter design cycle, fewer errors, easy and flexible to search the design space, and shared knowledge. In order to support the synthesis process, a Hardware Description Language (HDL) is highly needed because it exists to describe hardware. They differ from traditional programming languages, which generally exist to describe algorithms. To properly describe hardware, one must be able to describe both the behavior of the individual components as well as how they are interconnected [32–40].

In this manner, Verilog-A is used herein to carry out two primary functions: simulation and synthesis of the chaos-based applications [33, 36]. More particularly, Verilog-A can help to find the design requirements for advanced chaos based applications due to it can be used to model components, to create test benches, to accelerate simulation, to support the top-down design process, and to verify mixed-signal systems.

As well known, SPICE provides a limited set of built-in models that limits the design, simulation, and synthesis of chaos based applications. In contrast, Verilog-A provides a very wide variety of features and can be used to efficiently describe a broad range of models since basic components, semiconductor, logics, multidisciplinary until functional blocks [38]. Other advantage by using Verilog-A is the capability to create test benches The term test bench is used to refer to circuitry that is added to the circuit under test so as to provide it with an environment in which the chaos-based application can properly operate. Both aforementioned characteristics enable to accelerate simulation. With designs of chaos-based applications becoming larger and their behavior becoming more complex, it is taking longer and longer to verify them with simulation [36]. Then, the simulation time can be reduced if the non-critical portions of the system are replaced with behavioral models. As a result, the top-down design process is supported, which provides a foundation of designing and verifying the system at an abstract or block diagram level before starting the detailed design of the individual blocks [24]. Also, even if the chaos based application requires a mixed-signal design where Verilog-A is very useful as it allows both digital and analog circuits to be described in a way that is most suitable for each type of circuit [40].

3 Chaotic Pulse Position Modulation, CPPM

In this section is shown the behavioral modeling of a Chaos-based Pulse Position Modulation (CPPM) by using Verilog-A. In the pulse position modulation, we change any parameter of a uniform pulse wave, it could be amplitude, time or pulse. Two cases are recognized: an analog pulse modulation where information is processed in an analog way, but the processing occurs in discrete time; and a digital pulse modulation in which the information signal is discrete both amplitude and time, letting digital transmission as one codified pulse sequence, all with the same amplitude. This kind of transmission has not equivalent in continues wave systems. In latter case, output wave is a binary (logic zeros and ones) flow with same voltage amplitude and time width. Therefore, a pulse stream is generated from a PPM where the elapsed time between pulses is defined by the voltage amplitude of the input signal. Typical input signals are periodic waves such as sinusoidal, ramp, saw-tooth, etc. As consequence, the output bit stream is also periodic. Contrary, when the input is a chaotic signal, the position of pulses is uncertain, i.e., a CPPM [41].

'The Pulse Position Modulation (PPM) is one of the most used methods in satellite communications. When chaotic behavior is added to this kind of modulation, a CPPM is obtained. CPPMs are widely used in secure communications schemes, e.g. in chaos-based optical fiber links [41]'

3.1 Sampling of Chaotic Signals

The sampling process is common in all pulse modulation systems and generally its description is characterized in time domain. By sampling, a continuous time ana-

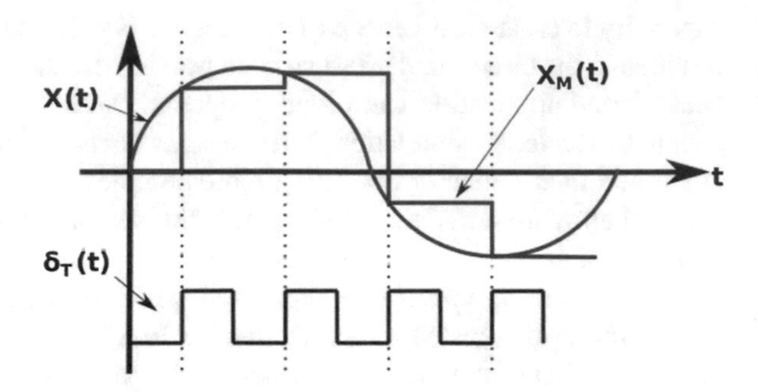

Fig. 1 Ideal sampling of a continuous time signal by satisfying the sample theorem

log signal is transformed in a sequence of discrete signals, with regular intervals. Sampling theorem sets that a finite continuous energy signal and band-limited, without spectral components over a frequency f_{max}, it is complete defined by specifying values for signal in times of $1/2f_{max}$ seconds. The sampled signal can be recovered through with a lowpass filter. The frequency $2f_{max}$ is termed as Nyquist frequency. If a band-limited signal, $x(t)$, is multiplied by a pulse wave with constant interval T, given by; $\delta(t) = \sum_{n=-\infty}^{\infty} \delta(t - nT)$, the resulting signal is $x_M(n) = x(t)\delta_T(t) = x(t - nT)$ Where n represents discrete time intervals every T seconds. The wave $x(t - nT)$ is, therefore, a discrete signal like shown in the Fig. 1 and its sampling points amplitude corresponds to the original signal. Furthermore, the sample and hold is an operation based on sampling method used commonly to hold the sampled value, and it is holded till the next sample.

3.2 Verilog-A-based Behavioral Modeling of a Chaotic PPM

The proposed CPPM is divided in four modules: a sample and hold module, a ramp wave generator, a voltage comparator, and a digital inverter; as shown in Fig. 2.

The basic operation of CPPM is as follows. First, the sample and hold **SH** takes a sample of the input signal. Simultaneously, a ramp function is generated by block ramp wave generator. When the ramp signal equals sample value, the comparator generates a control signal **SH** to reset both **SH** as well as the ramp wave generator. Then, a new sample is obtained and the process is repeated. On the other hand, the pulse width of the control signal **SH** is defined by delay time **dt**, rise time **tr**, and slew-rate **SR** of the voltage comparator. Finally, the position of pulse width is a function of the sample value taken from a chaotic input signal, and also the elapsed time when ramp amplitude reaches the sample value. The parameters **dt**, **tr**, and **SR** are set in the Verilog-A operator **transition**.

The main blocks of CPPM are described in Verilog-A as a structural behavioral model, as given in Listing 1:

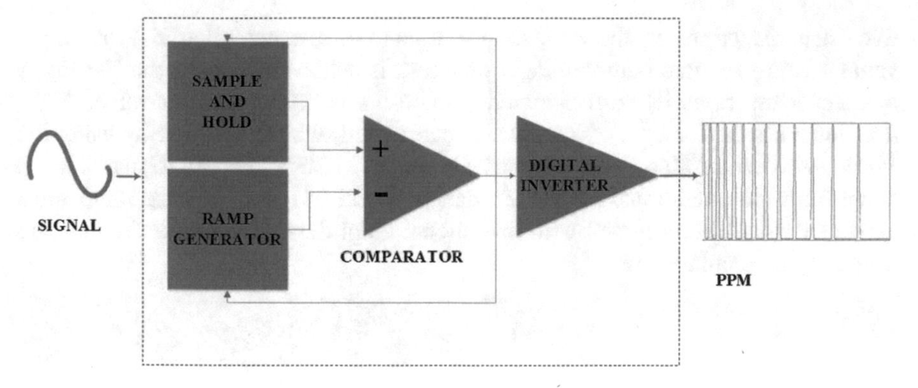

Fig. 2 Main blocks of the proposed CPPM

Listing 1 CPPM behavioral model

```
1  ‘include "disciplines.vams"
2  ‘include "integra2.vams"
3  ‘include "sh1.vams"
4  ‘include "comparador_v1.vams"
5  ‘include "inversor_digital.vams"
6
7  module ppm(in, sal_com, sal_sh, sal_ramp);
8  output sal_com, sal_sh, sal_ramp;
9  input in;
10 electrical in,salida,sal_com,sal_ramp;
11 ground gnd;
12 sh1 #(.dt(0.1u), .tr(0.01u))
13 sample_hold(in, sal_com, sal_sh);
14 integra2 #(.tr(.5), .pendiente(100000), .Vol_Max
      (10.2))
15 gen_rampa(sal_ramp, gnd, sal_com, gnd);
16 comparador_v1 #(.Vmax(5), .Vmin(0), .SR(1.9e7), .dt
      (0.01u), .tr(0.01u))
17 comparador1(sal_sh, sal_ramp, sal_com, gnd);

18 inversor_digital #(.SR(1.9e7), .dt(0.7n), .tr(0.8n),
      .tolerancia_voltaje_maximo(0.8), .
      voltaje_umbral_salida(5))
19 inversordig1(sal_com, sal_ppm);
20 endmodule
```

The ramp wave generator is modeled by integrating a constant value with Verilog-A operator **idtmod** as shown in Listing 2. To reset the ramp signal, it was defined a variable called *reset*, which depends on the value $V(hab1, hab2)$, if it detects a positive value, *reset* transfers the value to the output to be resetted, otherwise the output value is set to the maximum voltage. This task is achieved because the Verilog-A operator **idtmod**, as its third parameter, receives the proposed reset value. Additionally, a non-linear effect, slew-rate, is considered, which is added by using the Verilog-A operator **slew**. This parameter is setting in 1.9×10^7 due to that value is taken from the operational amplifier's data sheet TL081, for the case ramp wave generator would be designed with operational amplifiers. The frequency of ramp frequency is calculated by:

Listing 2 Ramp wave generator behavioral model

```
1  `include "disciplines.vams"
2  module integra2(sal1, sal2, hab1, hab2);
3  parameter real tr = 0;
4  parameter real pendiente = 50000;
5  parameter real Vol_Max = 5;
6  output sal1, sal2;
7  input hab1, hab2;
8  electrical sal1, sal2, hab1, hab2;
9  real val_temp = 0;
10 real val_temp2=0;
11 real reset = 0;
12
13  analog begin
14   val_temp = slew(idtmod(val_temp2, 0, reset, 0)
        ,1.9e7,-1.9e7) - 5;
15   if(V(hab1, hab2) >= tr) begin
16    reset = Vol_Max;


17    val_temp2 = pendiente;
18   end
19   else
20   begin
21    reset = val_temp;
22    val_temp2 = 0;
23    val_temp = -5;//-5 //0
24   end
25   V(sal1,sal2) <+ val_temp;
26  end
27 endmodule
```

$$f = \frac{m}{V_{pp}} \tag{1}$$

where m is the value of the slope; V_{pp} is the value of the peak to peak voltage. In this case, $m = 100000$, $V_{pp} = 10.2$, then $f = 9.8$ KHz. It means is required the input signal frequency (for a chaotic signal the fundamental frequency) is around 10 KHz.

The sample and hold module **SH** is modeled by using Verilog-A operators **cross** and **transition** as shown in Listing 3. The sample is obtained in the falling edge of the input signal *clk*. The response time of the sample and hold is set by *dt* and *tr*.

Listing 3 Sample and hold behavioral model

```
1  `include "disciplines.vams"
2  module sh1(in, clk, salida);
3  parameter dt = 0.1u, tr = 0.01u;
4  output salida;
5  input in, clk;
6  electrical in, clk, salida;
7  real val_temp = 0;
8
9  analog begin
10  @(cross(V(clk)-2.5, -1.0)) begin
11          val_temp = V(in);
12   end
13   V(salida) <+ transition(val_temp, dt, tr);
14   end
15 endmodule
```

The comparator is modeled by using Verilog-A operator **transition**. Slew-rate is included in this comparator to analyze nonlinear effects such as delay and growth rate of the output. The Verilog-A code is given in Listing 4.

Listing 4 Voltage comparator behavioral model

```
1  ‘include "disciplines.vams"
2  module comparador_v1(ps, ns, p,n);
3  parameter real Vmax = 10 from (0:inf);
4  parameter real Vmin = 0 from (-1:inf);
5  parameter real SR = 1.9e9;
6  parameter dt = 0.7u, tr = 0.8u;
7  output p, n;
8  input ps, ns;
9  electrical ps, ns, p, n;
10 real vtemp1 = 0;
11
12 analog begin
13 if(V(ps) >= V(ns))
14         vtemp1 = Vmax;
15 else
16 vtemp1 = Vmin;
17  V(p) <+ slew(transition(vtemp,dt,tr,tr),SR,-1*SR);
18 end
19 endmodule
```

Finally, the behavioral model for the digital inverter is described in Verilog-A as shown in Listing 5. This block is added to CPPM to generate an inverted pulse position modulation, however it does not affect the global response of CPPM.

Listing 5 Digital inverter behavioral model

```
'include "disciplines.vams"
module inversor_digital(entrada,salida);
parameter real SR = 1.9e9;
parameter real
parameter real tolerancia_voltaje_maximo = 0.8;
parameter real voltaje_umbral_salida = 5;
output entrada;
input salida;
electrical entrada, salida;
real lim_inf= 0, lim_sup = 0;
real vtempl = 0;

analog begin

lim_inf=tolerancia_voltaje_maximo*
    voltaje_umbral_salida;
if(lim_inf<=V(entrada)&&V(entrada)<=
    voltaje_umbral_salida)
        vtempl = 0;
else
        vtempl = voltaje_umbral_salida;
V(salida) <+ slew(transition(vtempl, dt, tr, tr),
    SR,-1*SR);
end
endmodule
```

We have analyzed and simulated the resulting CPPM by using the SPICE circuit simulator with the proposed Verilog-A behavioral models. First, the PPM is simulated by considering as its input a sinusoidal signal as shown in Fig. 3. In this figure, we can observe the ramp wave generation, the sample and hold, and the PPM output and its inverted output. The elapsed time between PPM pulses is defined by the voltage amplitude of sinusoidal signal. The behavior of the pulses form PPM is detailed in Fig. 4.

4 Verilog-A-based Behavioral Modeling of a Chaotic SPWM

The chaotic pulse width modulation is a relative recent method used in the power electric generation control because of the high immunity against electric noise. In this section a design of a CAFM-SPWM is developed by employing a chaotic ramp generator and a sinusoidal sign comparator described in Verilog-A.

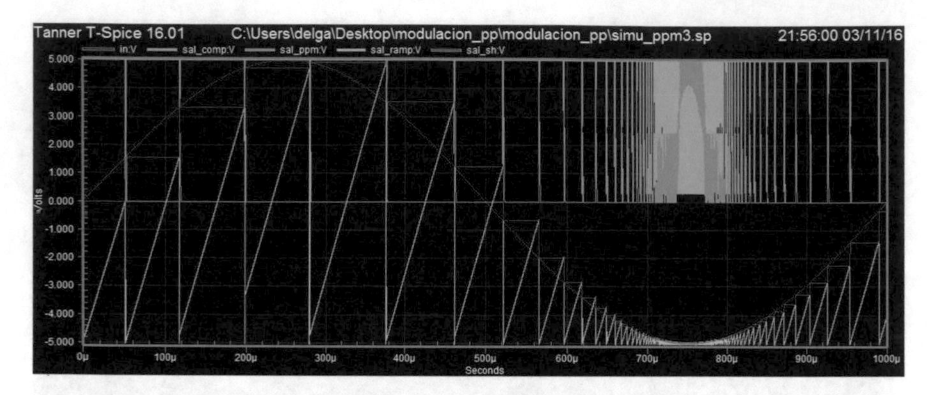

Fig. 3 SPICE simulation results of PPM by a sinusoidal signal as input

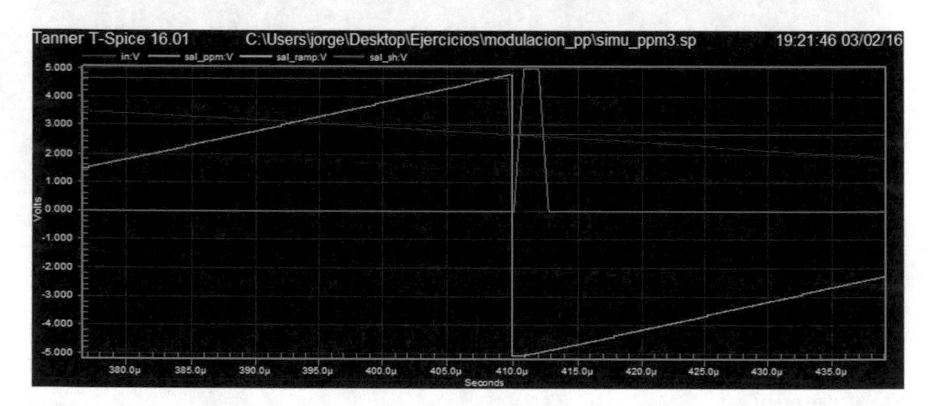

Fig. 4 Zoom on time transitions of PPM to determine the pulse position

Recently there are many modulation techniques based on telecommunications algorithms like pulse wide modulation, this technique are applied in motors or power inverters, control in electric propulsion systems such as hybrid vehicles and full electric vehicles [42]. Each proposed technique try to improve at least one characteristic in the process i.e., commutation loss, conversion deficiencies or harmonic content in the output. This factors are important because of the undesirable effects the harmonic signals. The most relevant undesirable effects are sinusoidal distortion, power factor reduction, increment of leaks in the machinery, over heat, vibrations, torque loss, and machinery life time reduction. Some published papers demonstrate the Sinusoidal Pulse Width Modulation (SPWM) has advantages contrary other approaches because the harmonic content in the resultant signal is minimized. Then, researches have proposed modern PWM methods trying to improve SPWM [42].

Nowadays, chaotic approaches to design PWM have been studied. The basic idea consist on using a chaotic signal as the carrier due to it has a distributed continuous frequency spectrum; this means that the energy or power may be distributed uniformly in the frequency band. As a result, the electromagnetic interference can be suppressed in electric propulsion units when chaotic signals are used [42].

Therefore, the aim of this section is to propose and design a behavioral model of a SPWM. The modulation technique SPWM is a carrier-based signal which in general uses a periodic or like-random triangle signal; and it is classified as a continuous method because each cycle of the carrier a switch is generated [42]. In this method a modulated pulse is obtained by comparing the modulated with the carrier signal at different frequencies, where the carrier signal defines the switching frequency. In this chapter, we study a SPWM based on a chaotic amplitude frequency modulator (CAFM). The logistic map is used as the core to generate the chaotic signal.

Afterwards, the carrier frequency is defined by.

$$f_c = F_C + \xi \Delta_f sin(2\pi f_m t) \tag{2}$$

where F_C is a reference frequency and Δf is the frequency deviation. The f_m is the modulation frequency, in this case a sinusoidal signal. On the other hand, a chaos generator is created by using the logistic map given by.

$$\xi_{i+1} = r\xi_i(1 - \xi) \tag{3}$$

where r is the logistic map constant and ξ is a number within 0 and 1. In order to get the behavioral model, Verilog-A operators *eventcross* and *initialstep* where applied to describe the behavior of sinusoidal and chaotic signal in (2) and (3), respectively. The CAFM-SPWM modulator model is divided in two sections (Figs. 5 and 6).

1. Chaotic triangle signal block.
2. Modulation from the chaotic triangle and sinusoidal signals block.

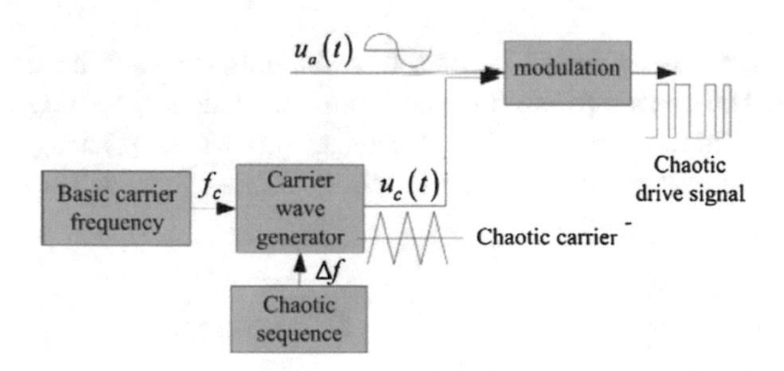

Fig. 5 Block diagram of proposed CAFM-SPWM

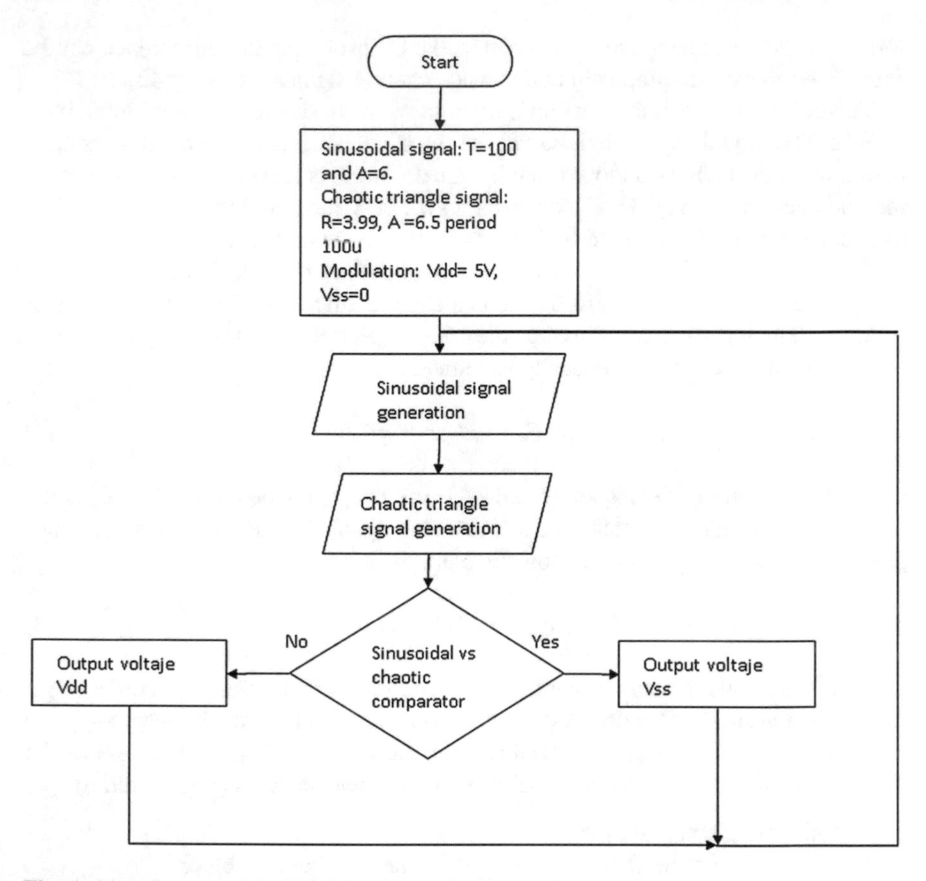

Fig. 6 Flow diagram of CAFM-SPWM described by Verilog-A

These blocks are shown in Fig. 7. Also, flow diagrams of each block are given in Figs. 8 and 9. The chaotic triangle signal block was coded in Verilog-A. This code uses (3) to get different values of ξ who changes the period of triangle signal. This change occurs when a maximum or a minimum value is detected. The code is shown in Listing 6.

Listing 6 Chaotic triangle signal block behavioral model

```
1  'include "disciplines.vams"
2  module V_triangle_generator(out);
3   output out;
4   voltage out;
5   parameter real R=4;
6  real Zhi;
7   parameter real period = 1m from [0:inf),
8                  ampl   = 5;
9  parameter real periodb = 10m from [0:inf);
10
11  integer slope;
12  real offset ,period1;
13  real m;
14
15  analog
16
17  begin
18  @(initial_step)
19         begin
20         Zhi=0.1;
21         Zhi=Zhi*R*(1-Zhi);
22         period1=period+(periodb*Zhi);
23         slope=1;
24         end
25
26    @(cross(V(out)-(ampl),+1))

27         begin
28         Zhi=Zhi*R*(1-Zhi);
29         period1=period+(periodb*Zhi);
30         slope = -1;
31         end
32
33        @(cross(V(out)+(ampl),-1))
34         begin
35         Zhi=Zhi*R*(1-Zhi);
36         period1=period+(periodb*Zhi);
37         slope = +1;
38         end
39
40    @(timer(0,period1/2))
41     begin
```

```
42        offset = \$realtime;
43      end
44
45    V(out) <+ ampl * slope *(4*(\$realtime - offset)
          / period1 - 1) ;
46    end
47  endmodule
```

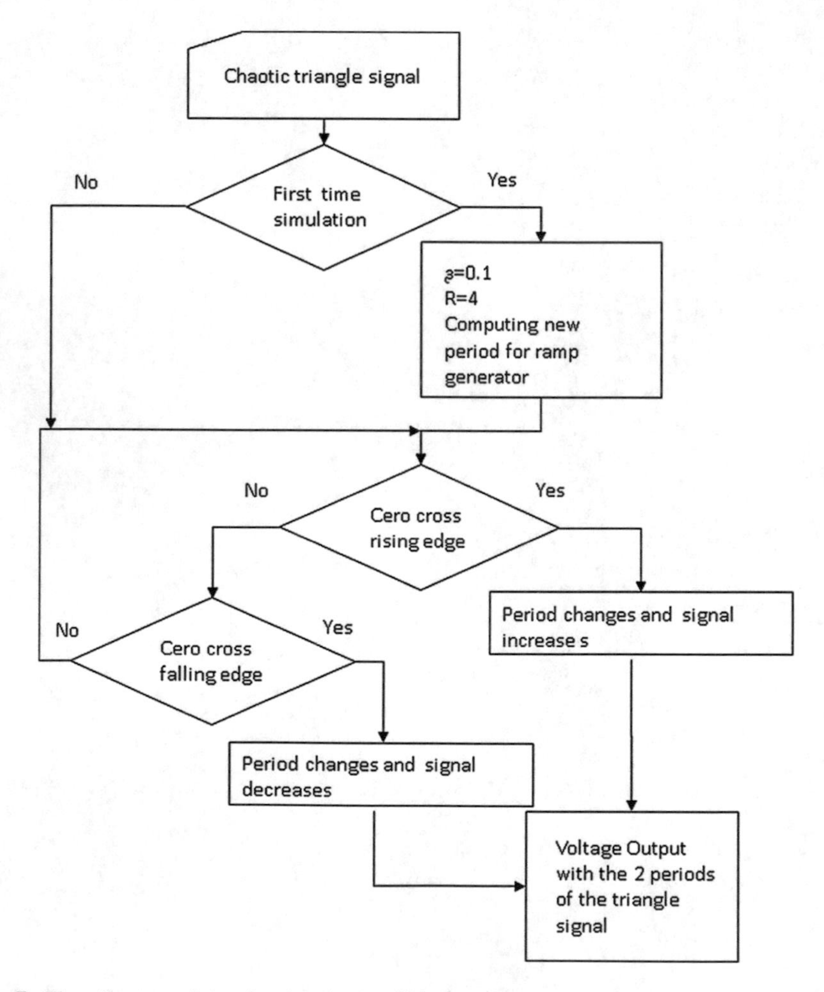

Fig. 7 Flow diagram of chaotic triangle signal block

The other block is the modulator, which realizes a comparison between the chaotic triangle signal and the sinusoidal signal as is described by the Verilog-A code shown in Listing 7. Then, those blocks where simulated with SPICE circuit simulator to get chaotic modulated signal. The code is shown in Listing 8.

Listing 7 CAFM-SPWM behavioral model

```
1  //CAFM–SPWM
2  'include "disciplines.vams"
3
4  module SPWM(mods, chaoss, spwm);
5
6          parameter real  vdd = 5;
7          parameter real  vss = 0;
8          real modsx, chaossx, spwmx1;
9          input mods, chaoss;
10         output spwm;
11         electrical mods, chaoss, spwm;
12
13         analog begin
14         modsx = V(mods);
15         chaossx = V(chaoss);
16         if (modsx >= chaossx)
17                 spwmx1 = vdd;
18         else
19                 spwmx1 = vss;
20         V(spwm) <+ transition(spwmx1,1n,1n);
21 end
22 endmodule
```

Listing 8 SPICE master

```
1  .hdl spwm_code.va
2  .hdl Trianglegen.va
3
4  Vmod mod 0 sin(0,6,100)
5  X1 cha CTWG R=3.99 ampl=6.5 period=100u
6  X2 mod cha spwmo SPWM vdd=5 vss=0
7  .trans 0.05m 40m
8  .print trans V(mod) V(cha) V(spwmo)
9  .end
```

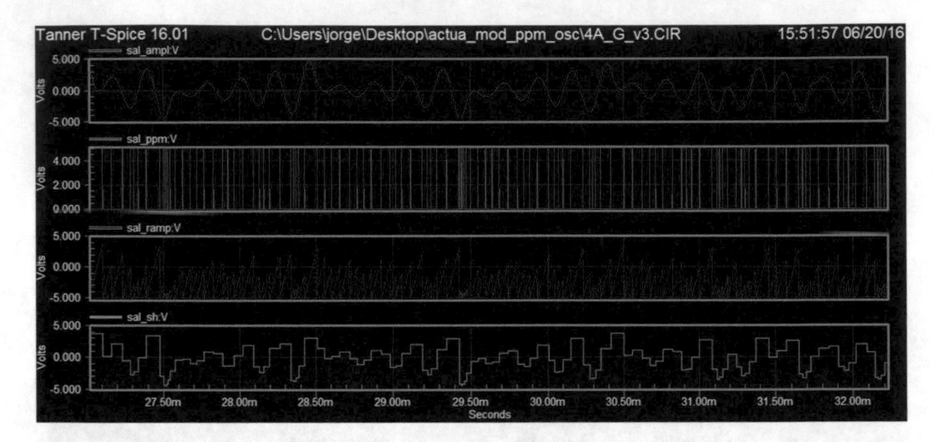

Fig. 8 Chaotic input signal (*red*), pulse position modulation signal (*purple*), ramp generator signal (*red*), sample and hold signal (*blue*)

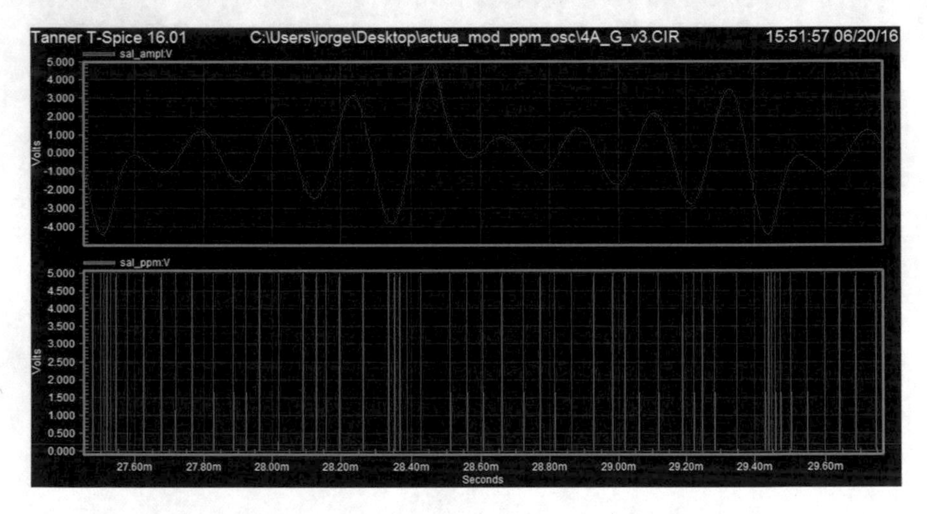

Fig. 9 Inout results of Verilog-A-based CPPM: Chaotic input signal and pulse position modulation signal

SPICE simulations are carried out to demonstrate the chaotic behavior of the triangle signal as a function of the constant mapping in the logistic map, detect the variation rate for each ramp regarding the deviation frequency, and observe CAFM-SPWM resultant signal as shown in next section.

5 Discussion and Analysis

This section is dedicated to analyze and discuss the chaos-based engineering applications by using Verilog-A.

First, we analyze and discuss the CPPM, given in Sect. 3, by considering as input signal to CPPM a chaotic system as shown in Fig. 8. A state variable of a chaotic system is used to feed the PPM. Similarly, the elapsed time between output pulses is set by the amplitude of chaotic signal. The more the amplitude, the bigger the separation between output pulses. Otherwise, the less the amplitude, the smaller the separation between output pulses as demonstrated in Fig. 8. The chaotic ramp wave generation and sample and hold are given in Fig. 8. A close-up of the output in CPPM is sketched in Fig. 9. In simulations is observed that when the ramp reach the sampled value this still is growing due to by the non-ideal effects introduced in the comparator. The trigger of comparator has a delay (line green) plus the delay of digital inverter as shown in Fig. 4. This effect can be reduced if the delays are reduced. As previously mentioned, the digital inverter is used to increase the CPPM performance. Because this CPPM is to transfer signals, it is more efficient the pulse in high than in low level. Herein, the digital outputs as shown in Listing 5.

As second case, we we analyze and discuss the CAFM-SPWM given in Sect. 4. Five simulations were executed as defined below.

1. $F_c = \frac{1}{33.3\,\mu s}, \Delta f = \frac{1}{33.3\,\mu s}, r = 3.6$
2. $F_c = \frac{1}{33.3\,\mu s}, \Delta f = \frac{1}{33.3\,\mu s}, r = 3.89$
3. $F_c = \frac{1}{33.3\,\mu s}, \Delta f = \frac{1}{33.3\,\mu s}, r = 4$
4. $F_c = \frac{1}{33.3\,\mu s}, \Delta f = \frac{1}{66.6\,\mu s}, f_m = 1\,\text{kHz}$, r =4
5. $F_c = \frac{1}{33.3\,\mu s}, \Delta f = \frac{1}{333\,\mu s}, f_m = 273\,\text{Hz}$, r = 4

The amplitudes are set in 4.9V for the sinusoidal signal and 5V for the chaotic triangle signal. The simulation results are given in Figs. 10, 11 and 12.

It is evident that a large variation and randomness in the ramp signal is present regarding the values of r between 3.89 to 4. This behavior depends on the bifurcations of Logistic map. When $r = 4$, the bifurcation reaches a maximum value of x_i from 0 to 1. As r approximates maximum value, the ramp signal is generated with higher variations as shown in Figs. 13, 14 and 15.

In Figs. 14 and 15 is shown the chaotic triangle signals with $r = 4$, however the sudden changes are more evident when $\Delta f = 1/333\,\mu s$, this is because the deviation in the frequency is 10 times the base frequency multiplied with a factor ξ in a interval 0 to 1. The CAFM-SPWM signals generates more pulses with different duty cycles by $\Delta f = 1/333\,\mu s, f_m = 273\,\text{Hz}$, while in $\Delta f = 1/66.6\,\mu s, f_m = 1\,\text{kHz}$ the duty cycle tends to be higher. A window from $15.5ms$ and $18.1ms$ was settled in order to get a improved chaotic triangle signal.

The maximum and minimum period of each ramp is given by the expression $T = T_a + T_b\xi$. Since $Ta = 33.3\,\mu s$ and ξ varies between 0 and 1, the interval of T

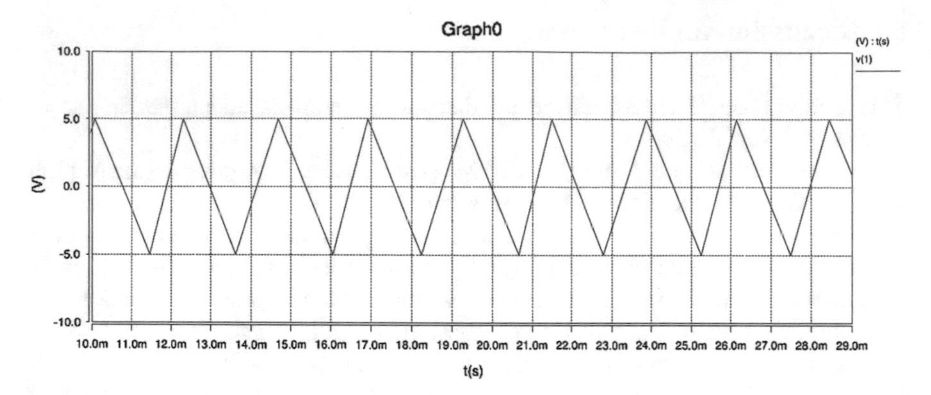

Fig. 10 Resulting chaotic triangle signal with r = 3.6

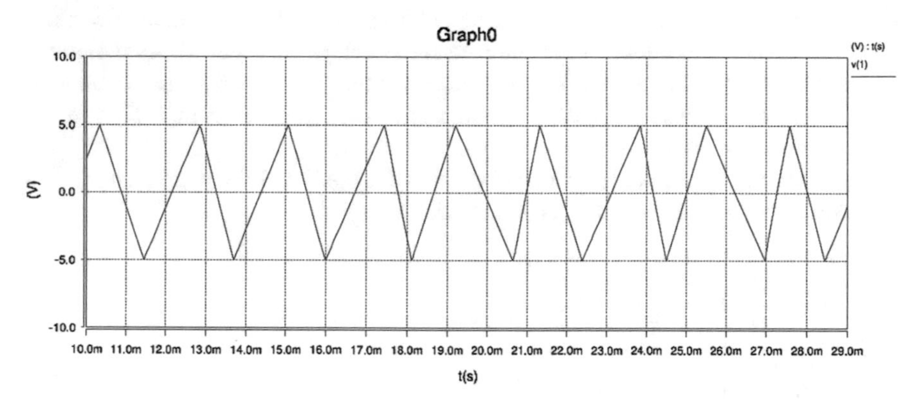

Fig. 11 Resulting chaotic triangle signal with r = 3.89

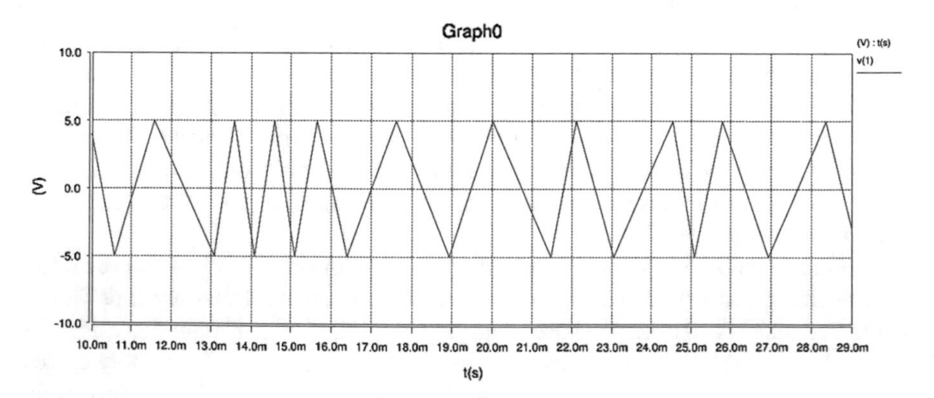

Fig. 12 Resulting chaotic triangle signal with r = 4

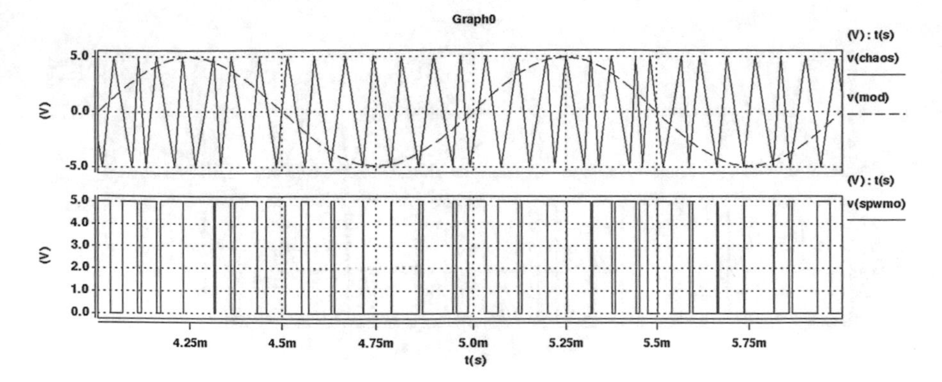

Fig. 13 Chaotic triangle signal, sinusoidal and CAFM-SPWM with simulation parameters case – 4 (Δ_f = 1/66.6 μs, fm = 1 KHz)

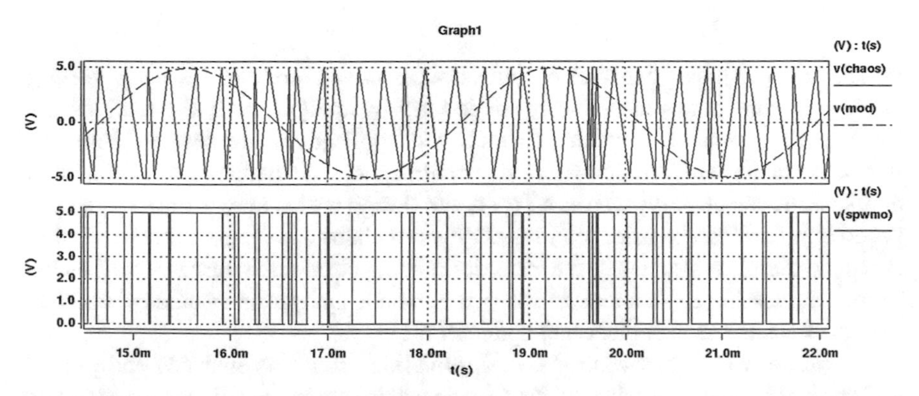

Fig. 14 Chaotic triangle signal, sinusoidal and CAFM-SPWM with simulation parameters number case = 5 (Δ_f = 1/333 μs, fm = 273 Hz)

is from 33.3 s to 99.9 μs. Tt demonstrates that T_b is highly affected from the sudden variations of the ramps in the chaotic triangle signal.

As a conclusion, SPICE simulations of CPPM and CAFM-SPWM agrees with those on literature. However, Verilog-A allows different description levels by incorporating the main effects under test. So, we are able to find the design trade-offs for both chaos-based applications at circuit level.

6 Conclusions

In this chapter, it have been demonstrated the description, analysis and design of modulation schemes based on chaos by using behavioral modeling. In particular, two schemes were analyzed, a chaotic pulse position modulation (CPPM), and a sinusoidal pulse width modulator based on chaotic amplitude frequency modulator

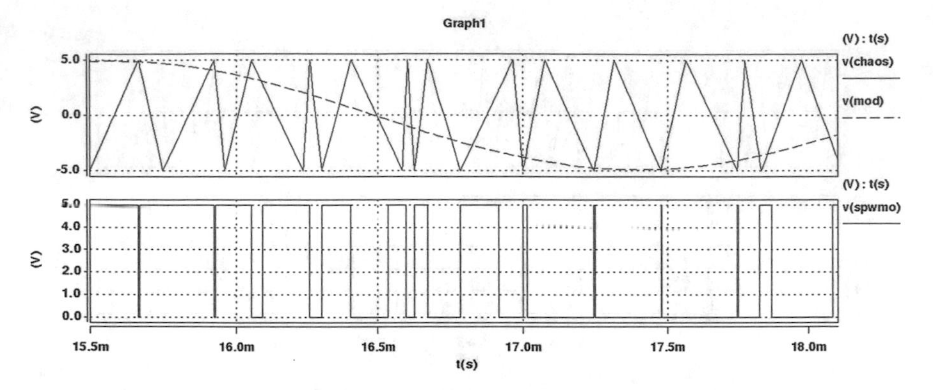

Fig. 15 Zoom of the chaotic triangle signal, sinusoidal signal, and CAFM-SPWM with simulation parameters case = 5 (Δ_f = 1/333 μs, fm = 273 Hz)

(CAFM-SPWM). The behavior of both modulators are codified by using Verilog-A. Several simulations results were carried out by applying SPICE circuit simulator. From the simulation results, chaos-based applications are designed and simulated correctly with Verilog-A. In this manner, circuit synthesis of chaos-based applications can be performed as a Top-down design methodology. Further, the proposed approach can be extended to multiphysics chaos applications due to Verilog-A can co-simulate systems in different domains. Not only this research will enable future engineering applications, but also, will help to analyze and validate theoretical results of chaos-based engineering applications.

As future work, other chaos-based applications could be analyzed and proved, such as synchronization schemes for chaotic secure communications. It is also necessary to consider other second order effects of electronics devices in high-level simulations. It means, moving forward to lower abstraction levels such as: physical implementation of CPPM and CAFM-SPWM with electronic discrete components. A first step forward would be to replace some verilog-a blocks for a macro-model circuit or a transistor level representations in order to perform a mixed simulation. Finally, a integrated circuit design may give insight about precision and robustness of the proposed Verlog-A-based design approach for chaos-based applications.

Acknowledgements This work has been partially supported by the scientific projects: *CONACYT* No. 258880, *PRODEP* Red de Nanociencia y Nanotecnología, *VIEP-BUAP-2016*.

References

1. Arena, P., De Fiore, S., Fortuna, L., Frasca, M., Patané, L., & Vagliasindi, G. (2008). Reactive navigation through multiscroll systems: From theory to real-time implementation. *Autonomous Robots*, *25*(1–2), 123–146. doi:10.1007/s10514-007-9068-1.

2. Cheng, C. J. (2012). Robust synchronization of uncertain unified chaotic systems subject to noise and its application to secure communication. *Applied Mathematics and Computation, 219*(5), 2698–2712. doi:10.1016/j.amc.2012.08.101.
3. Faraji, S., & Tavazoei, M. (2013). The effect of fractionality nature in differences between computer simulation and experimental results of a chaotic circuit. *Central European Journal of Physics, 11*(6), 836–844. doi:10.2478/s11534-013-0255-8.
4. Gotthans, T., & Hrubos, Z. (2013). Multi grid chaotic attractors with discrete jumps. *Journal of Electrical Engineering, 64*(2), 118–122. doi:10.2478/jee-2013-0017.
5. Kanno, T., Miyano, T., Tokuda, I., Galvanovskis, J., & Wakui, M. (2007). Chaotic electrical activity of living β-cells in the mouse pancreatic islet. *Physica D: Nonlinear Phenomena, 226*(2), 107–116. doi:10.1016/j.physd.2006.11.007.
6. Kwon, O., Park, J., & Lee, S. (2011). Secure communication based on chaotic synchronization via interval time-varying delay feedback control. *Nonlinear Dynamics, 63*(1–2), 239–252. doi:10.1007/s11071-010-9800-9.
7. Lu, J., & Chen, G. (2006). Generating multiscroll chaotic attractors: Theories, methods and applications. *International Journal of Bifurcation and Chaos, 16*(4), 775–858. doi:10.1142/S0218127406015179.
8. Munoz-Pacheco, J., & Tlelo-Cuautle, E. (2010). *Electronic design automation of multi-scroll chaos generators*. doi:10.2174/97816080516561100101.
9. Munoz-Pacheco, J., Zambrano-Serrano, E., Felix-Beltran, O., Gomez-Pavon, L., & Luis-Ramos, A. (2012). Synchronization of pwl function-based 2d and 3d multi-scroll chaotic systems. *Nonlinear Dynamics, 70*(2), 1633–1643. doi:10.1007/s11071-012-0562-4.
10. Munoz-Pacheco, J., Tlelo-Cuautle, E., Toxqui-Toxqui, I., Sanchez-Lopez, C., & Trejo-Guerra, R. (2014). Frequency limitations in generating multi-scroll chaotic attractors using cfoas. *International Journal of Electronics, 101*(11), 1559–1569. doi:10.1080/00207217.2014.880999.
11. Pecora, L., & Carroll, T. (1990). Synchronization in chaotic systems. *Physical Review Letters, 64*(8), 821–824. doi:10.1103/PhysRevLett.64.821.
12. Piper, J., & Sprott, J. (2010). Simple autonomous chaotic circuits. *IEEE Transactions on Circuits and Systems II: Express Briefs, 57*(9), 730–734. doi:10.1109/TCSII.2010.2058493.
13. Sanchez-Lopez, C., Munoz-Pacheco, J., Tlelo-Cuautle, E., Carbajal-Gomez, V. & Trejo-Guerra, R. (2011) On the trade-off between the number of scrolls and the operating frequency of the chaotic attractors. In *Proceedings—IEEE International Symposium on Circuits and Systems* (pp. 2950–2953). doi:10.1109/ISCAS.2011.5938210.
14. Sira-Ramirez, H., & Cruz-Hernandez, C. (2001). Synchronization of chaotic systems: A generalized hamiltonian systems approach. *International Journal of Bifurcation and Chaos in Applied Sciences and Engineering, 11*(5), 1381–1395. doi:10.1142/S0218127401002778.
15. Tsuda, I. (2009). Hypotheses on the functional roles of chaotic transitory dynamics. *Chaos, 19*(1). doi:10.1063/1.3076393.
16. Zhang, Z., & Chen, G. (2005). Chaotic motion generation with applications to liquid mixing. *In Proceedings of the 2005 European Conference on Circuit Theory and Design* (Vol.1, pp. 225–228). doi:10.1109/ECCTD.2005.1522951.
17. Azar, A. T., & Vaidyanathan, S. (2014). *Chaos modeling and control systems design*. Incorporated: Springer.
18. Azar, A. T., & Vaidyanathan, S. (2015). *Computational intelligence applications in modeling and control*. Incorporated: Springer.
19. Azar, A. T., & Vaidyanathan, S. (2016). *Advances in chaos theory and intelligent control* (1st ed.). Incorporated: Springer.
20. Boulkroune A, Bouzeriba, A., Bouden, T., & Azar, A. T. (2016) Fuzzy adaptive synchronization of uncertain fractional-order chaotic systems. In T. A. Azar & S. Vaidyanathan (Eds.), *Advances in chaos theory and intelligent control* (pp. 681–697). Springer. doi:10.1007/978-3-319-30340-6_28.
21. Boulkroune, A., Hamel, S., Azar, A. T., & Vaidyanathan, S. (2016) Fuzzy control-based function synchronization of unknown chaotic systems with dead-zone input. In T. A. Azar & S. Vaidyanathan (Eds.) *Advances in Chaos Theory and Intelligent Control, Springer International Publishing* (pp. 699–718) doi:10.1007/978-3-319-30340-6_29.

22. Ouannas, A., Azar, A. T., & Abu-Saris, R. (2016) A new type of hybrid synchronization between arbitrary hyperchaotic maps. *International Journal of Machine Learning and Cybernetics*, 1–8. doi:10.1007/s13042-016-0566-3.
23. Ouannas, A., Azar, A. T., & Vaidyanathan, S. (2016). A robust method for new fractional hybrid chaos synchronization. *Mathematical Methods in the Applied Sciences*. doi:10.1002/mma.4099.
24. Vaidyanathan, S., & Azar, A.T. (2016) Adaptive backstepping control and synchronization of a novel 3-d jerk system with an exponential nonlinearity. In T. A. Azar & S. Vaidyanathan (Eds.), *Advances in chaos theory and intelligent control* (pp. 249–274). Springer. doi:10.1007/978-3-319-30340-6_11.
25. Vaidyanathan, S., & Azar, A.T. (2016) Adaptive control and synchronization of a halvorsen circulant chaotic systems. In T. A. Azar & S. Vaidyanathan (Eds.), *Advances in chaos theory and intelligent control* (pp. 225–247). Springer. doi:10.1007/978-3-319-30340-6_10.
26. Vaidyanathan, S., & Azar, A.T. (2016) Dynamic analysis, adaptive feedback control and synchronization of an eight-term 3-d novel chaotic system with three quadratic nonlinearities. In T.A. Azar & S. Vaidyanathan (Eds.), *Advances in chaos theory and intelligent control* (pp. 155–178). Springer. doi:10.1007/978-3-319-30340-6_7.
27. Vaidyanathan, S., & Azar, A.T. (2016) Generalized projective synchronization of a novel hyperchaotic four-wing system via adaptive control method. In T. A. Azar & S. Vaidyanathan (Eds.), *Advances in chaos theory and intelligent control* (pp. 275–296). Springer. doi:10.1007/978-3-319-30340-6_12.
28. Vaidyanathan, S., & Azar, A.T. (2016) A novel 4-d four-wing chaotic system with four quadratic nonlinearities and its synchronization via adaptive control method. In T. A. Azar & S. Vaidyanathan (Eds.), *Advances in chaos theory and intelligent control* (pp. 203–224) Springer. doi:10.1007/978-3-319-30340-6_9.
29. Vaidyanathan, S., & Azar, A.T. (2016) Qualitative study and adaptive control of a novel 4-d hyperchaotic system with a three quadratic nonlinearities. In T. A. Azar & S. Vaidyanathan (Eds.), *Advances in chaos theory and intelligent control* (pp. 179–202). Springer. doi:10.1007/978-3-319-30340-6_8.
30. Vaidyanathan, S., & Azar, A. T. (2016). Takagi-sugeno fuzzy logic controller for liu-chen four-scroll chaotic system. *Int J Intell Eng Inform*, *4*(2), 135–150. doi:10.1504/IJIEI.2016.076699.
31. Zhu, Q., & Azar, A. T. (Eds.), (2015). *Complex system modelling and control through intelligent soft computations, studies in fuzziness and soft computing*, (Vol. 319). Springer. doi:10.1007/978-3-319-12883-2.
32. Bueno-Ruiz, J., Arriaga-Arriaga, C., Huerta-Barrera, R., Cruz-Dominguez, G., Pimentel-Romero, C., Munoz-Pacheco, J., et al. (2015). *16th Latin-American Test Symposium. LATS, 2015*. doi:10.1109/LATW.2015.7102507.
33. FitzPatrick, D., & Miller, I. (1997). *Analog behavioral modeling with the VERILOG-a language* (1st ed.). Norwell, MA, USA: Kluwer.
34. Gal, G., Fattah, O., & Roberts, G. (2012). A 30–40 ghz fractional-n frequency synthesizer development using a verilog-a high-level design methodology. In *Proceedings of midwest symposium on circuits and systems* (pp. 57–60). doi:10.1109/MWSCAS.2012.6291956.
35. Gonzalez-Diaz, V., Munoz-Pacheco, J., Espinosa-Flores-Verdad, G., & Sanchez-Gaspariano, L. (2016). A verilog-a based fractional frequency synthesizer model for fast and accurate noise assessment. *Radioengineering*, *25*(1), 89–97. doi:10.13164/re.2016.0089.
36. Kundert, K., & Zinke, O. (2013). *The designer's guide to Verilog-AMS*. Incorporated: Springer.
37. Liao, S., & Horowitz, M. (2013). A verilog piecewise-linear analog behavior model for mixed-signal validation. *In Proceedings of the custom integrated circuits conference*. doi:10.1109/CICC.2013.6658461.
38. Martens, E. S. J., & Gielen, G. G. E. (2008). *High-Level Modeling and Synthesis of Analog Integrated Systems* (1st ed.). Incorporated: Springer.
39. Munoz-Pacheco, J., Tlelo-Cuautle, E., Trejo-Guerra, R., & Cruz-Hernandez, C. (2008). Synchronization of n-scrolls chaotic systems synthesized from high-level behavioral modeling. *In Proceedings of the 7th international Caribbean conference on devices, circuits and systems, ICCDCS*. doi:10.1109/ICCDCS.2008.4542634.

40. Rutenbar, R., Gielen, G., & Roychowdhury, J. (2007). Hierarchical modeling, optimization, and synthesis for system-level analog and rf designs. *Proceedings of the IEEE*, *95*(3), 640–669. doi:10.1109/JPROC.2006.889371.
41. Xuan Quyen, N., Van Yem, V., & Manh Hoang, T. (2012). A chaotic pulse-time modulation method for digital communication. *Abstract and Applied Analysis*, *2012*. doi:10.1155/2012/835304.
42. Zhang, Z., Ching, T., Liu, C., & Lee, C. (2012) Comparison of chaotic pwm algorithms for electric vehicle motor drives. In *IECON proceedings (industrial electronics conference)* (pp. 4087–4092). doi:10.1109/IECON.2012.6389236.

A New Method to Synchronize Fractional Chaotic Systems with Different Dimensions

Adel Ouannas, Toufik Ziar, Ahmad Taher Azar and Sundarapandian Vaidyanathan

Abstract By using two scaling function matrices, the synchronization problem of different dimensional fractional order chaotic systems in different dimensions is developed in this chapter. The controller is designed to assure that the synchronization of two different dimensional fractional order chaotic systems is achieved using the Lyapunov direct method. Numerical examples and computer simulations are used to validate numerically the proposed synchronization schemes.

Keywords Fractional order chaotic systems · Synchronization · Hyperchaotic systems · Different dimensional fractional chaotic

A. Ouannas
Laboratory of Mathematics, Informatics and Systems (LAMIS),
University of Larbi Tebessi, 12002 Tébessa, Algeria
e-mail: ouannas.a@yahoo.com

T. Ziar
Department of Material Sciences, University of Tebessa,
12002 Tébessa, Algeria
e-mail: Toufik1_ziar@yahoo.fr

A.T. Azar (✉)
Faculty of Computers and Information, Benha University,
Banha, Egypt
e-mail: ahmad.azar@fci.bu.edu.eg; ahmad_t_azar@ieee.org

A.T. Azar
Nanoelectronics Integrated Systems Center (NISC),
Nile University, Cairo, Egypt

S. Vaidyanathan
Research and Development Centre, Vel Tech University Avadi,
Chennai 600062, Tamil Nadu, India
e-mail: sundarvtu@gmail.com

A.T. Azar et al. (eds.), *Fractional Order Control and Synchronization of Chaotic Systems*, Studies in Computational Intelligence 688,
DOI 10.1007/978-3-319-50249-6_20

1 Introduction

The study of fractional order chaotic systems has become an interdisciplinary field of many scientific disciplines [1–24]. Recently, the applications of fractional chaotic dynamical systems have attracted considerable attention, for example, in secure communication and encryption [25–35].

The synchronization problem means making two systems oscillate in a synchronized manner. The dynamical behavior of two copies of a chaotic system may be identical after a transient when the second system is driven by the first one. Since the synchronization of chaotic system (integer-order differential systems and discrete-time systems) is understood well [36–42] and widely explored [43–67], the synchronization of fractional order chaotic dynamical systems has started to attract increasing attention of many researchers [68–80], due to its potential applications [81–84].

Several control approaches, have already been successfully applied to the problem of synchronization of fractional order chaotic dynamical systems such as sliding-mode control [85–87], linear control [88], nonlinear control [89], active control [90], adaptive control [91, 92], feedback control [93–100] and scalar signal technique [101]. Many types of synchronization of fractional chaotic systems have been studied such as phase synchronization [102, 103], complete synchronization [104], anti-synchronization [105], projective synchronization [106–109], hybrid projective synchronization [110, 111], function projective synchronization [112–114], generalized projective synchronization [115–118], generalized synchronization [119–125], full state hybrid projective synchronization [126], Q-S synchronization [127], exponential synchronization [128], finite-time synchronization [129], impulsive synchronization [130, 131].

To the best of our knowledge most of theoretical results about synchronization of chaos focus on the systems whose models are identical or strictly different systems and systems of different order, especially the systems in biological science and social science. One example is the synchronization that occurs between heart and lung, where one can observe that both circulatory and respiratory systems behave in synchronous way, but their models are essentially different and they have different order. So, the study of synchronization for strictly different dynamical systems and different order dynamical systems is both very important from the perspective of control theory and very necessary from the perspective of practical application. Recently, many effective synchronization approaches have been used widely to achieve synchronization of chaotic systems with different dimensions such as matrix projective synchronization [132], inverse matrix projective synchronization [133], generalized synchronization [134], inverse generalized synchronization [135], $\Lambda - \varphi$ generalized synchronization [136, 137], Q-S synchronization [138] and hybrid synchronization [139, 140].

Not long ago, a new approach to synchronize different dimensional chaotic system by using two scaling matrices has been introduced. The method has been called $\Theta - \Phi$ synchronization. In [141], Ouannas and Al-sawalha used two scaling constant matrices to synchronize different dimensional discrete-time chaotic systems.

The new approach was applied between integer-order differential chaotic systems with different dimension by Ouannas and Al-sawalha in [142]. By using scaling constant matrix and scaling function matrix, the synchronization problem of different dimensional fractional order chaotic systems in different dimensions was developed by Ouannas et al. in [143].

In this chapter, the problem of synchronization for different dimensional fractional chaotic dynamical systems using two scaling function matrices is proposed and experimented. This chapter provides further contribution to the topic of $\Theta - \Phi$ synchronization. The sufficient conditions for achieving $\Theta - \Phi$ synchronization, between n-dimensional fractional chaotic system and m-dimensional fractional, are derived based on Lyapunov stability theory of integer-order differential systems. Analytic expressions of the controllers are shown. The proposed control methods are efficient and easy to implement in practical applications. Illustrative examples of fractional chaotic and hyperchaotic systems are used to show the effectiveness of the proposed approaches.

The rest of the present chapter is organized as follows. In Sect. 2 some theoretical basis are given. In Sect. 3 different schemes for fractional $\Theta - \Phi$ synchronization are proposed. In Sect. 4 the derived criterions and the proposed schemes are applied to some typical different dimensional fractional chaotic and hyperchaotic systems. Finally, the chapter is concluded in Sect. 5.

2 Theoretical Basis

2.1 Fractional Derivative and Integral

Fractional calculus plays an important role in modern science. In this chapter, we use both Reimann Liouville and Caputo fractional operators as our main tools. Caputo fractional derivative is defined as follows [144].

$$D_t^p x(t) = J^{m-p} x^m(t), \qquad 0 < p \leq 1, \tag{1}$$

where $m = [p]$, i.e. m is the first integer which is not less than p, x^m is the m-order derivative in the usual sense, and J^q $(q > 0)$ is the q-order Reimann-Liouville integral operator with expression:

$$J^q y(t) = \frac{1}{\Gamma(q)} \int_0^t (t-\tau)^{q-1} y(\tau)\, d\tau, \tag{2}$$

where Γ denotes Gamma function.

Lemma 1 ([145]) *The Laplace transform of the Caputo fractional derivative rule reads*

$$\mathbf{L}\left\{D_t^p f(t)\right\} = s^p \mathbf{F}(s) - \sum_{k=0}^{n-1} s^{p-k-1} f^{(k)}(0), \quad (p > 0,\ n-1 < p \leq n). \tag{3}$$

Particularly, when $p \in (0, 1]$, *we have* $\mathbf{L}\left\{D_t^p f(t)\right\} = s^p \mathbf{F}(s) - s^{p-1} f(0)$.

Lemma 2 ([146]) *The Laplace transform of the Riemann-Liouville fractional integral rule satisfies*

$$\mathbf{L}\left\{J^q f(t)\right\} = s^{-q} \mathbf{F}(s), \quad (q > 0). \tag{4}$$

Lemma 3 ([147]) *Suppose* $f(t)$ *has a continuous kth derivative on* $[0, t]$ $(k \in N, t > 0)$, *and let* $p, q > 0$ *be such that there exists some* $\ell \in N$ *with* $\ell \leq k$ *and* $p, p+q \in [\ell - 1, \ell]$. *Then*

$$D_t^p D_t^q f(t) = D_t^{p+q} f(t). \tag{5}$$

Remark 1 Note that the condition requiring the existence of the number ℓ with the above restrictions in the property is essential. In this paper, we consider the case that $p, q \in \left]0, 1\right]$ and $p + q \in \left]0, 1\right]$. Apparently, under such conditions this property holds.

2.2 $\Theta - \Phi$ Synchronization

The master and the slave systems are in the following forms

$$D_t^p X(t) = F(X(t)), \tag{6}$$

$$D_t^q Y(t) = G(t)) + U, \tag{7}$$

where $X(t) \in \mathbf{R}^n$, $Y(t) \in \mathbf{R}^m$ are state vectors of the master system and the slave system, respectively, $0 < p, q \leq 1$, D_t^p, D_t^q are the Caputo fractional derivatives of orders p and q, respectively, $F : \mathbf{R}^n \to \mathbf{R}^n$, $G : \mathbf{R}^m \to \mathbf{R}^m$ and $U = \left(u_i\right)_{1 \leq i \leq m}$ is a vector controller.

Definition 1 The master system (6) and the slave system (7) are said to be $\Theta - \Phi$ synchronized in dimension d, if there exists a controller $U = \left(u_i\right)_{1 \leq i \leq m}$ and two function matrices $\Theta(t) = (\Theta(t))_{d \times m}$ and $\Phi(t) = (\Phi(t))_{d \times n}$ such that the synchronization error

$$e(t) = \Theta(t) Y(t) - \Phi(t) X(t), \tag{8}$$

satisfies that $\lim_{t \longrightarrow +\infty} \|e(t)\| = 0$.

3 Theoretical Analysis

In this section, we discuss different schemes of $\Theta - \Phi$ synchronization in different dimensions.

3.1 $\Theta - \Phi$ Synchronization in Dimension m

Consider the following master system

$$D_t^p X(t) = f(X(t)), \tag{9}$$

where $X(t) \in \mathbf{R}^n$ is the state vector of the master system, $0 < p \leq 1$, D_t^p is the Caputo fractional derivative of order p and $f : \mathbf{R}^n \to \mathbf{R}^n$.

As the slave system, we consider the following controlled system

$$D_t^q Y(t) = BY(t) + g(Y(t)) + U, \tag{10}$$

where $Y(t) \in \mathbf{R}^m$ is the state vector of the slave system, respectively, $0 < q \leq 1$, D_t^q is the Caputo fractional derivative of order q, $B \in \mathbf{R}^{m\times m}$, $g : \mathbf{R}^m \to \mathbf{R}^m$ are the linear and the nonlinear parts of the slave system, respectively, and $U = \left(u_i\right)_{1\leq i\leq m}$ is a vector controller.

In this case, the error system between the master system (9) and the slave system (10), is defined as

$$e(t) = \Theta(t)\,Y(t) - \Phi(t)\,X(t), \tag{11}$$

where $\Theta(t) = \left(\Theta_{ij}(t)\right)_{m\times m}$ and $\Phi(t) = \left(\Phi_{ij}(t)\right)_{m\times n}$. Hence, we have the following result.

Theorem 1 *The master system (9) and the slave system (10) are globally $\Theta - \Phi$ synchronized in m-D, if the following conditions are satisfied:*

(i) $\Theta(t)$ is an invertible matrix and $\Theta^{-1}(t)$ its inverse matrix.

(ii) $U = -BY(t) - g(Y(t)) + J^{1-q}\left(-\Theta^{-1}(t) \times R_1\right)$, where

$$R_1 = \left(L_1 - B\right) e(t) + \dot{\Theta}(t)\,Y(t) - \dot{\Phi}(t)\,X(t) - \Phi(t)\,\dot{X}(t), \tag{12}$$

and $L_1 \in \mathbf{R}^{m\times m}$ is a control matrix.

(iii) $\left(L_1 - B\right)^T + \left(L_1 - B\right)$ is a positive definite matrix.

Proof The error system between the master system (9) and the slave system (10) can be derived as

$$\dot{e}(t) = \Theta(t)\,\dot{Y}(t) + \dot{\Theta}(t)\,Y(t) - \dot{\Phi}(t)\,X(t) - \Phi(t)\,\dot{X}(t). \tag{13}$$

Substituting the control law (ii) into Eq. (10), the slave system can be described as

$$D_t^q Y(t) = J^{1-q}\left(-\Theta^{-1}(t) \times R_1\right). \tag{14}$$

Applying the Laplace transform to Eq. (14) and letting

$$\mathbf{F}(s) = \mathbf{L}\left(Y(t)\right), \tag{15}$$

we obtain,

$$s^q \mathbf{F}(s) - s^{q-1} Y(0) = s^{q-1} \mathbf{L}\left(-\Theta^{-1}(t) \times R_1\right), \tag{16}$$

multiplying both the left-hand and right-hand sides of Eq. (16) by s^{1-q} and applying the inverse Laplace transform to the result, we get the following equation

$$\dot{Y}(t) = -\Theta^{-1}(t) \times R_1. \tag{17}$$

The error system (13) can be derived as

$$\dot{e}(t) = \left(B - L_1\right) e(t). \tag{18}$$

Construct the candidate Lyapunov function in the form

$$V(e(t)) = e^T(t) e(t), \tag{19}$$

we obtain,

$$\begin{aligned}\dot{V}(e(t)) &= \dot{e}^T(t) e(t) + e^T(t) \dot{e}(t) \\ &= e^T(t)\left(B - L_1\right)^T e(t) + e^T(t)\left(B - L_1\right) e(t) \\ &= -e^T(t)\left[\left(L_1 - B\right)^T + \left(L_1 - B\right)\right] e(t).\end{aligned}$$

By using (iii), we get $\dot{V}(e(t)) < 0$. Thus, from the Lyapunov stability theory, it is immediate that $\lim_{t \longrightarrow +\infty} \|e(t)\| = 0$. So, the zero solution of the error system (18) is globally asymptotically stable, and therefore, the master system (9) and the slave system (10) are globally $\Theta - \Phi$ synchronized in m-D.

3.2 $\Theta - \Phi$ Synchronization in Dimension n

The master system and the slave system can be described as

$$D_t^p X(t) = AX(t) + f(X(t)), \tag{20}$$

$$D_t^q Y(t) = g(Y(t)) + U, \tag{21}$$

where $X(t) \in \mathbf{R}^n$, $Y(t) \in \mathbf{R}^m$ are state vectors of the master system and the slave system, respectively, $0 < p, q \leq 1$, D_t^p, D_t^q are the Caputo fractional derivatives of orders p and q, respectively, $A \in \mathbf{R}^{n\times n}$, $f : \mathbf{R}^n \to \mathbf{R}^n$ are the linear and the nonlinear parts of the master system, respectively, $g = (g_i)_{1\leq i\leq m} : \mathbf{R}^m \to \mathbf{R}^m$ and $U = (u_i)_{1\leq i\leq m}$ is a vector controller.

In this case, we assume that the synchronization dimension $d = n$, where $n < m$. The error system between the master system (20) and the slave system (21) is considered as

$$e(t) = \Theta(t)Y(t) - \Phi(t)X(t), \tag{22}$$

where $\Theta(t) = (\Theta_{ij}(t))_{n\times m}$ and $\Phi(t) = (\Phi_{ij}(t))_{n\times n}$.

The error system (22) can be described as

$$\dot{e}(t) = (A - L_2)\, e(t) + \Theta(t)\dot{Y}(t) + R_2, \tag{23}$$

where $L_2 \in \mathbf{R}^{n\times n}$ is a feedback gain matrix and

$$R_2 = (L_2 - A)\, e(t) + \dot{\Theta}(t)Y(t) - \dot{\Phi}(t)X(t) - \Phi(t)\dot{X}(t). \tag{24}$$

Hence, we can conclude the following result.

Theorem 2 *The master system (20) and the slave system (21) are globally $\Theta - \Phi$ synchronized in n-D, if the following conditions are satisfied:*

(i) $\hat{\Theta}(t) = (\Theta_{ij}(t))_{n\times n}$ is an invertible matrix and $\hat{\Theta}^{-1}(t)$ its inverse matrix.

(ii) $(u_1, u_2, \ldots, u_n)^T = -\hat{G} - J^{1-q}\left(\hat{\Theta}^{-1}(t) \times R_2\right)$ and $(u_{n+1}, u_{n+2}, \ldots, u_m)^T = -\check{G}(Y(t))$, where $\hat{G} = (g_i)_{1\leq i\leq n}$ and $\check{G} = (g_i)_{n+1\leq i\leq m}$.

(iii) $(A - L_2)^T + (A - L_2)$ is a negative definite matrix.

Proof By inserting the control law (ii) into Eq. (21), we can rewrite the slave system as follows

$$\left(D_t^q y_1(t), \ldots, D_t^q y_n(t)\right)^T = J^{1-q}\left(-\hat{\Theta}^{-1}(t) \times R_2\right), \tag{25}$$

and

$$\left(D_t^q y_{n+1}(t), \ldots, D_t^q y_m(t)\right)^T = 0. \tag{26}$$

By applying the fractional derivative of order $1 - q$ to both the left and right sides of Eqs. (25) and (26), we obtain

$$\begin{aligned} D_t^{1-q}\left(\left(D_t^q y_1(t), \ldots, D_t^q y_n(t)\right)^T\right) &= \left(\dot{y}_1(t), \dot{y}_2(t), \ldots, \dot{y}_n(t)\right)^T \\ &= D_t^{1-q} J^{1-q}\left(\hat{\Theta}^{-1}(t) \times R_2\right) \\ &= -\hat{\Theta}^{-1}(t) \times R_2, \end{aligned} \tag{27}$$

and

$$\left(\dot{y}_{n+1}(t), \dot{y}_{n+2}(t), \ldots, \dot{y}_m(t)\right)^T = 0. \tag{28}$$

The error system (23) can be written as

$$\dot{e}(t) = \left(A - L_2\right) e(t). \tag{29}$$

Construct the candidate Lyapunov function in the form

$$V(e(t)) = e^T(t)e(t), \tag{30}$$

we obtain

$$\begin{aligned}\dot{V}(e(t)) &= \dot{e}^T(t)e(t) + e^T(t)\dot{e}(t)\\ &= e^T(t)\left(A - L_2\right)^T e(t) + e^T(t)\left(A - L_2\right)e(t)\\ &= e^T(t)\left[\left(A - L_2\right)^T + \left(A - L_2\right)\right]e(t),\end{aligned}$$

and by using (iii), we get $\dot{V}(e(t)) < 0$. Thus, from the Lyapunov stability theory, it is immediate that all solutions of error system (29) go to zero as $t \to \infty$. Therefore, the master system (20) and the slave system (21) are globally $\Theta - \Phi$ synchronized in n-D.

3.3 $\Theta - \Phi$ Synchronization in Dimension d

In this case, we assume that $d < m$, $d \neq n$ and the master and the slave systems can be considered in the following forms

$$D_t^p X(t) = f(X(t)), \tag{31}$$

$$D_t^q Y(t) = g(Y(t)) + U, \tag{32}$$

where $X(t) \in \mathbf{R}^n$, $Y(t) \in \mathbf{R}^m$ are state vectors of the master system and the slave system, respectively, $0 < p, q \leq 1$, D_t^p, D_t^q are the Caputo fractional derivatives of orders p and q, respectively, $f : \mathbf{R}^n \to \mathbf{R}^n$, $g : \mathbf{R}^m \to \mathbf{R}^m$ and $U = \left(u_i\right)_{1\leq i\leq m}$ is a vector controller.

The error system between the master system (31) and the slave system (32) is given as

$$e(t) = \Theta(t)Y(t) - \Phi(t)X(t), \tag{33}$$

where $\Theta(t) = \left(\Theta_{ij}(t)\right)_{d\times m}$, $\Phi(t) = \left(\Phi_{ij}(t)\right)_{d\times n}$.

We suppose that the vector controller U can be designed in the following form:

$$U = -g(Y(t)) + J^{1-q}(V), \tag{34}$$

where $V = \left(v_i\right)_{1\leq i\leq m}$ is a new control law.

Now, by using Eqs. (32) and (34), we can get

$$D_t^q Y(t) = J^{1-q}(V). \tag{35}$$

By applying Lemma 3, we obtain

$$\begin{aligned} D_t^{1-q} D_t^q Y(t) &= \dot{Y}(t) \\ &= D_t^{1-q} J^{1-q}(V) \\ &= V. \end{aligned} \tag{36}$$

Then, the error system (33), can be derived as

$$\dot{e}(t) = L_3 e(t) + \Theta(t) V + R_3, \tag{37}$$

where

$$R_3 = -L_3 e(t) + \dot{\Theta}(t) Y(t) - \dot{\Phi}(t) X(t) - \Phi(t) \dot{X}(t), \tag{38}$$

$L_3 = \operatorname{diag}\left(l_1, l_2, \ldots, l_d\right)$ and l_i, $i = 1, 2, \ldots, d$, are unknown control constants.

To achieve $\Theta - \Phi$ synchronization between systems (31) and (32), we choose the controller V as

$$V = \left(v_1, \ldots, v_d, 0, \ldots, 0\right)^T. \tag{39}$$

By substituting Eq. (39) into Eq. (37), the error system can be described as

$$\dot{e}(t) = L_3 e(t) + \check{\Theta}(t) \hat{V} + R_3, \tag{40}$$

where

$$\check{\Theta}(t) = \begin{pmatrix} \Theta_{11}(t) & \cdots & \Theta_{1d}(t) \\ \vdots & \ddots & \vdots \\ \Theta_{d1}(t) & \cdots & \Theta_{dd}(t) \end{pmatrix}, \tag{41}$$

and $\hat{V} = \left(v_1, \ldots, v_d\right)^T$.

Theorem 3 *For an invertible matrix $\check{\Theta}(t)$, the $\Theta - \Phi$ synchronization between the master systems (31) and the slave system (32) will occur in d-D, if the following conditions are satisfied:*

(i) $\hat{V} = -\check{\Theta}^{-1}(t) \times R_3$, where $\check{\Theta}^{-1}(t)$ is the inverse matrix of $\check{\Theta}(t)$.

(ii) The control constants l_i are chosen as $l_i < 0$, $i = 1, 2, \ldots, d$.

Proof By substituting (i) into Eq. (40), the error system can be written as:

$$\dot{e}_i(t) = l_i e_i(t), \quad 1 \leq i \leq d, \tag{42}$$

and let us consider the following quadratic Lyapunov function: $V(e(t)) = \sum_{i=1}^{d} \frac{1}{2}e_i^2(t)$, then we obtain $\dot{V}(e(t)) = \sum_{i=1}^{d} l_i e_i^2(t)$. By using (iii), we get $\dot{V}(e(t)) < 0$. Thus, by Lyapunov stability it is immediate that $\lim_{t\to\infty} e_i(t) = 0$, $(1 \le i \le d)$, then the master system (31) and the slave system (32) are globally $\Theta - \Phi$ synchronized.

4 Numerical Analysis

In this section, we will present some numerical simulations for $\Theta - \Phi$ synchronization to verify and illustrate the effectiveness of the theoretical analysis introduced in Sect. 3.

4.1 $\Theta - \Phi$ Synchronization Between Fractional Order Liu System and Fractional Order Hyperchaotic Lorenz System in 4D

As the master system we consider the fractional order Liu system and the controlled hyperchaotic fractional Lorenz system as the slave system. The master system is described as follows

$$\begin{aligned} D^p x_1 &= \alpha \left(x_2 - x_1\right), \\ D^p x_2 &= \beta x_1 - x_1 x_3, \\ D^p x_3 &= -\gamma x_3 + 4x_1^2, \end{aligned} \tag{43}$$

where x_1, x_2 and x_3 are states. This system, as shown in [148], exhibits chaotic behaviors when $(\alpha, \beta, \gamma) = (10, 40, 2.5)$ and $p = 0.9$. The chaotic attractors of the system (43) are shown in Figs. 1 and 2.

The slave system is defined as

$$\begin{aligned} D^q y_1 &= a\left(y_2 - y_1\right) + y_4 + u_1, \\ D^q y_2 &= c y_1 - y_2 - y_1 y_3 + u_2, \\ D^q y_3 &= y_1 y_2 - b y_3 + u_3, \\ D^q y_4 &= -y_2 y_3 + d y_4 + u_4, \end{aligned} \tag{44}$$

where y_1, y_2, y_3, y_4 are states and $U = (u_1, u_2, u_3, u_4)^T$ is the vector controller. The fractional-order hyperchaotic Lorenz system (i.e. the uncontrolled system (44)) exhibits hyperchaotic behaviors when $q = 0.98$ and $(a, b, c, d) = (10, 28, \frac{8}{3}, -1)$ [149]. Attractors of the fractional order hyperchaotic Lorenz system are shown in Figs. 3 and 4.

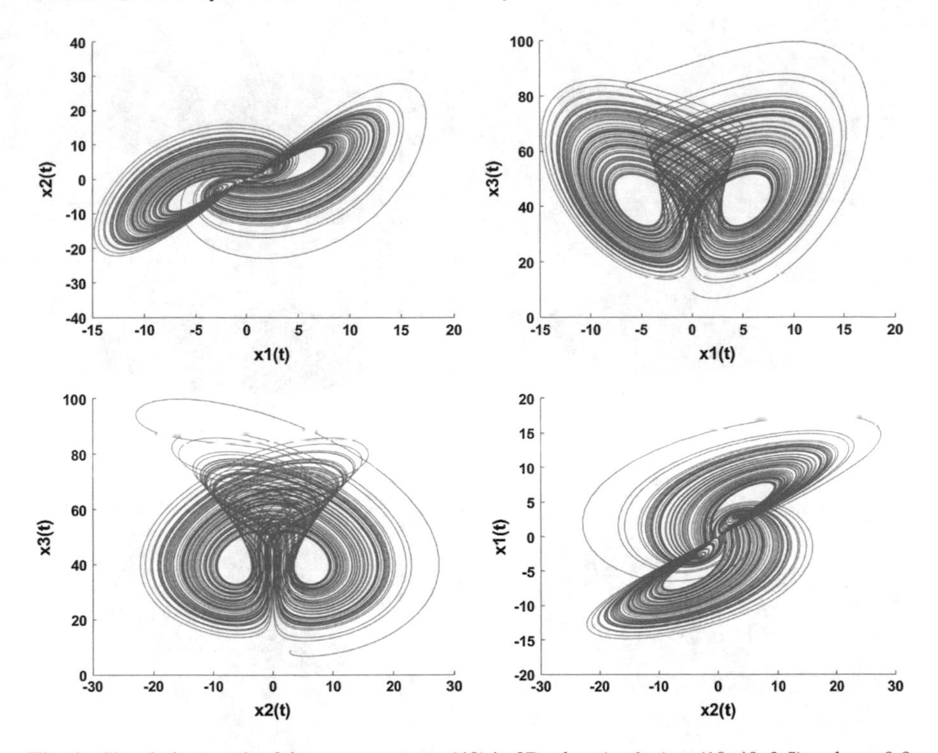

Fig. 1 Simulation result of the master system (43) in 2D when $(\alpha, \beta, \gamma) = (10, 40, 2.5)$ and $p = 0.9$

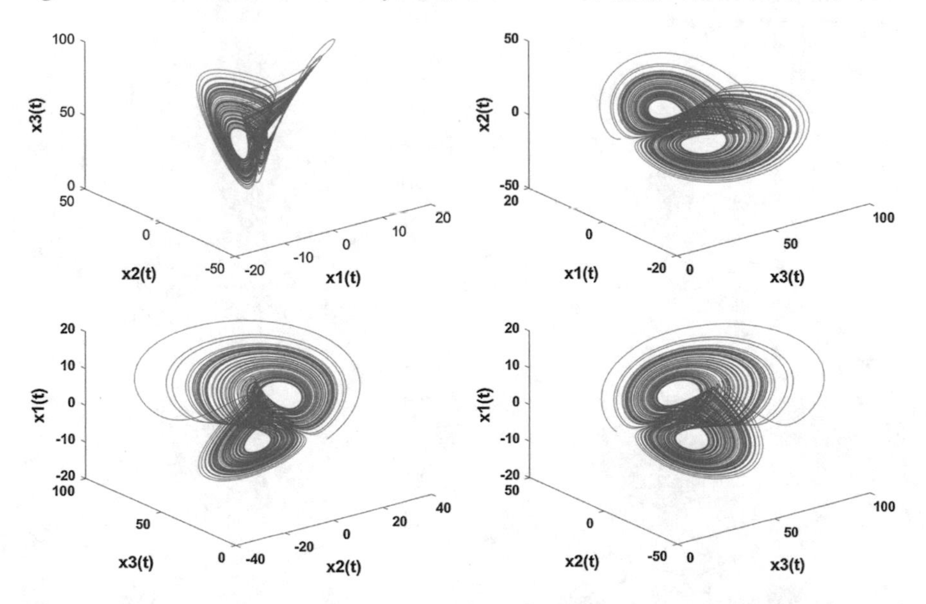

Fig. 2 Simulation result of the master system (43) in 3D when $(\alpha, \beta, \gamma) = (10, 40, 2.5)$ and $p = 0.9$

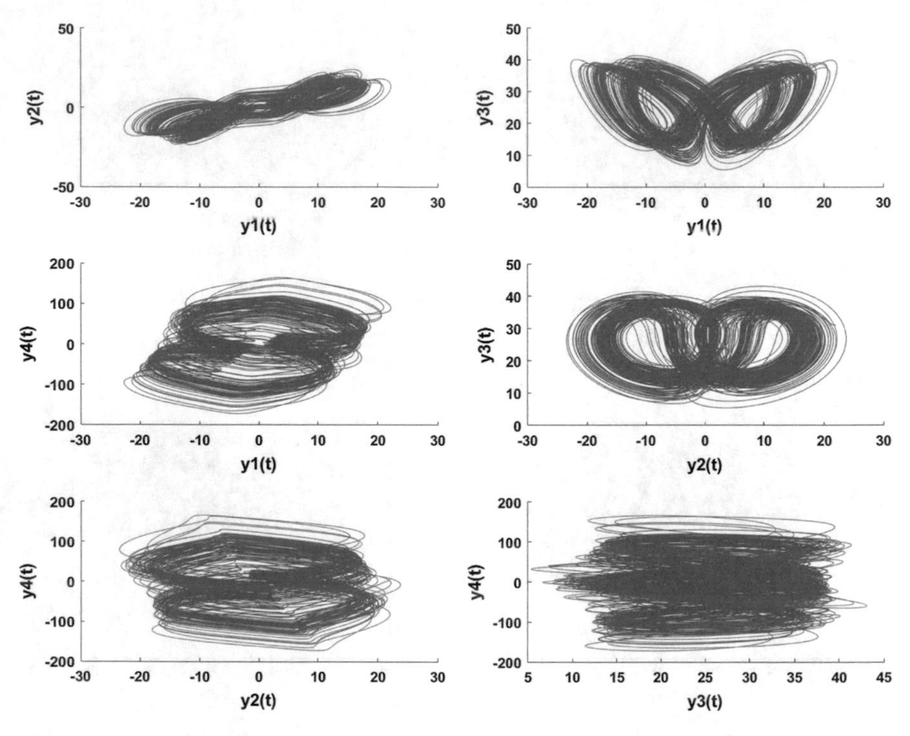

Fig. 3 Simulation result of the slave system (44) in 2D when $(a, b, c, d) = (10, \frac{8}{3}, 28, -1)$, $q = 0.98$ and $u_1 = u_2 = u_3 = u_4 = 0$

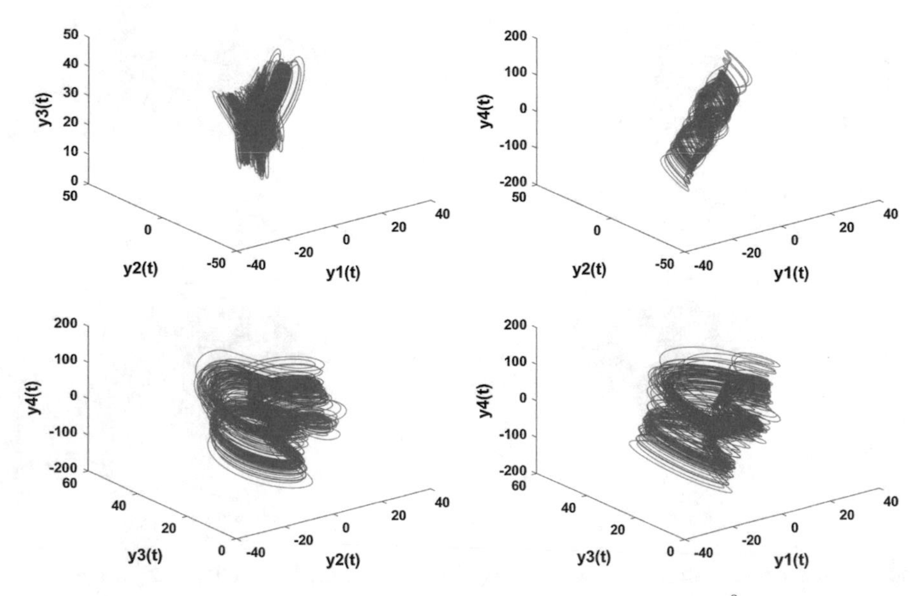

Fig. 4 Simulation result of the slave system (44) in 3D when $(a, b, c, d) = (10, \frac{8}{3}, 28, -1)$, $q = 0.98$ and $u_1 = u_2 = u_3 = u_4 = 0$

Compare the slave system (44) with system (10), one can have

$$B = \begin{pmatrix} -10 & 10 & 0 & 1 \\ 28 & -1 & 0 & 0 \\ 0 & 0 & -\frac{8}{3} & 0 \\ 0 & 0 & 0 & -1 \end{pmatrix},\ g = \begin{pmatrix} 0 \\ -y_1 y_3 \\ y_1 y_2 \\ -y_2 y_3 \end{pmatrix}.$$

The error system of $\Theta - \Phi$ synchronization, between the master system (43) and the slave system (44), is defined as

$$\begin{pmatrix} e_1 \\ e_2 \\ e_3 \\ e_4 \end{pmatrix} = \Theta(t) \times \begin{pmatrix} y_1 \\ y_2 \\ y_3 \\ y_4 \end{pmatrix} - \Phi(t) \times \begin{pmatrix} x_1 \\ x_2 \\ x_3 \end{pmatrix}, \tag{45}$$

where

$$\Theta(t) = \begin{pmatrix} t+1 & 0 & 0 & 0 \\ 0 & t^2+2 & 0 & 0 \\ 0 & 0 & \exp(t) & 0 \\ 0 & 0 & 0 & 4 \end{pmatrix}, \tag{46}$$

and

$$\Phi(t) = \begin{pmatrix} t & 1 & 0 \\ 2 & 0 & t+1 \\ 0 & \exp(t) & 0 \\ 1 & 0 & 1 \end{pmatrix}. \tag{47}$$

So,

$$\Theta^{-1}(t) = \begin{pmatrix} \frac{1}{t+1} & 0 & 0 & 0 \\ 0 & \frac{1}{t^2+2} & 0 & 0 \\ 0 & 0 & \exp(-t) & 0 \\ 0 & 0 & 0 & \frac{1}{4} \end{pmatrix}. \tag{48}$$

Using the notations described in Sect. 3.1, the control matrix L_1 can be chosen as

$$L_1 = \begin{pmatrix} 0 & 10 & 0 & 1 \\ 28 & 0 & 0 & 0 \\ 0 & 0 & 0 & 0 \\ 0 & 0 & 0 & 0 \end{pmatrix}, \tag{49}$$

and the control functions u_1, u_2, u_3 and u_4 are designed as follows

$$u_1 = -10\left(y_2 - y_1\right) - y_4 + J^{0.02}\left[\frac{1}{t+1}\left(-10e_1 - y_1 + x_1 + t\dot{x}_1 + \dot{x}_2\right)\right], \tag{50}$$

$$u_2 = -28y_1 + y_2 + y_1y_3 + J^{0.02}\left[\frac{1}{t^2+2}\left(-e_2 - 2ty_2 + x_3 + 2\dot{x}_1 + (t+1)\dot{x}_3\right)\right],$$
$$u_3 = -y_1y_2 + \frac{8}{3}y_3 + J^{0.02}\left(-\frac{8}{3}\exp(-t)\,e_3 - y_3 + x_2 + \dot{x}_2\right),$$
$$u_4 = y_2y_3 + y_4 + J^{0.02}\left(-\frac{1}{4}e_4 + \frac{1}{4}\dot{x}_1 + \frac{1}{4}\dot{x}_3\right).$$

It is easy to show that $\left(L_1 - B\right)^T + \left(L_1 - B\right)$ is a positive definite matrix. Then the conditions of Theorem 1 are satisfied and the $\Theta - \Phi$ synchronization between systems (43) and (44) is achieved. The error system, in this case, can be described as follows

$$\begin{aligned} \dot{e}_1 &= -10e_1, \\ \dot{e}_2 &= -e_2, \\ \dot{e}_3 &= -\frac{8}{3}e_3, \\ \dot{e}_4 &= -e_4. \end{aligned} \tag{51}$$

For the purpose of numerical simulation, Euler integration method has been used. In addition, simulation time $Tm = 120$ s and time step $h = 0.005$ s have been employed. The initial values of the master system and the slave systems are $[x_1(0), x_2(0), x_3(0)] = [7, -9, 5]$ and $[y_1(0), y_2(0), y_3(0), y_4(0)] = [12, 22, 31, 4]$, respectively, and the initial states of the error system are $[e_1(0), e_2(0), e_3(0), e_4(0)] = [21, 25, 26, 4]$. The error functions evolution, in this case, is shown in Fig. 5.

From Fig. 5, we can conclude that the components of the error system (51), e_1, e_2, e_3 and e_4, decay towards zero as $t \to +\infty$, and so the master system (43) and the slave system (44) are $\Theta - \Phi$ synchronized in 4-D.

4.2 *$\Theta - \Phi$ Synchronization Between Fractional-Order Volta's System and Fractional-Order Modified Hyperchaotic Chen System in 3-D*

In this example, we consider the fractional-order Volta's system as the master system and the controlled hyperchaotic fractional Lorenz system as the slave system. The master system can be described as

$$\begin{aligned} D^p x_1 &= -x_1 - \left(\alpha + x_3\right)x_2, \\ D^p x_2 &= -x_2 - \left(\beta + x_3\right)x_1, \\ D^p x_3 &= \gamma x_3 + x_2x_1 + 1, \end{aligned} \tag{52}$$

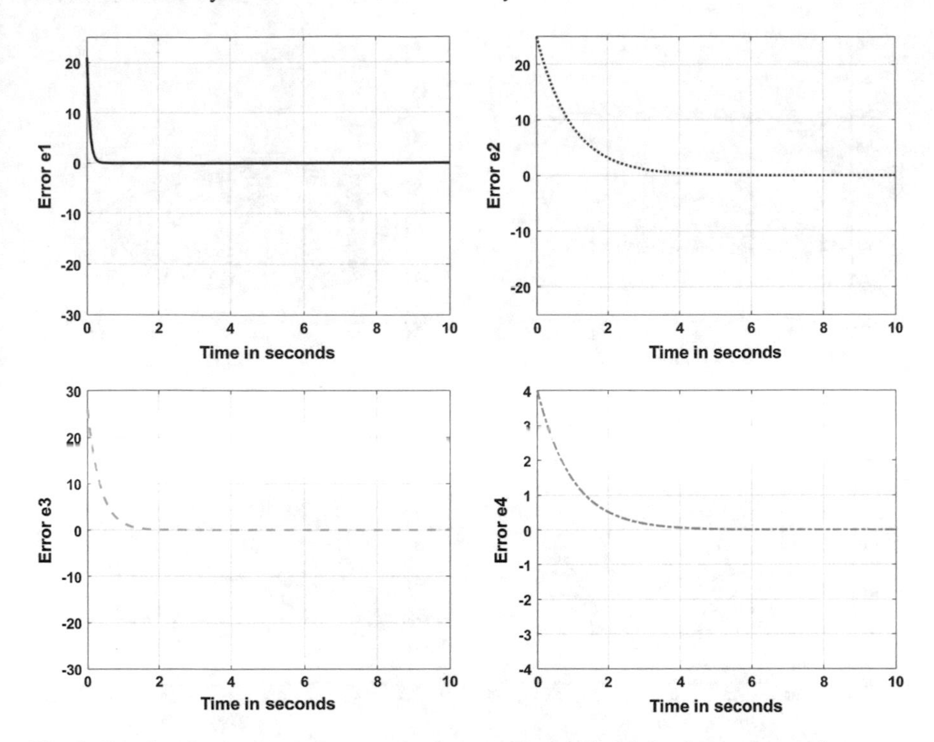

Fig. 5 Synchronization errors between the master system (43) and the slave system (44)

where x_1, x_2 and x_3 are states. This system, as shown in [150], exhibits chaotic behaviors when $(\alpha, \beta, \gamma) = (19, 11, 0.73)$ and $p = 0.98$. The chaotic attractors of the commensurate fractional-order Volta's system (52) are shown in Figs. 6 and 7.

Compare the master system (52) with system (20), one can have

$$A = \begin{pmatrix} -1 & -19 & 0 \\ -11 & -1 & 0 \\ 0 & 0 & 0.73 \end{pmatrix}, f = \begin{pmatrix} -x_2x_3 \\ -x_1x_3 \\ x_2x_1 \end{pmatrix}.$$

The slave is given by

$$\begin{aligned} D^q y_1 &= a\left(y_2 - y_1\right) + u_1, \\ D^q y_2 &= by_1 + cy_2 - y_1y_3 - y_4 + u_2, \\ D^q y_3 &= -dy_3 + y_1y_2 + u_3, \\ D^q y_4 &= y_1 + g + u_4, \end{aligned} \tag{53}$$

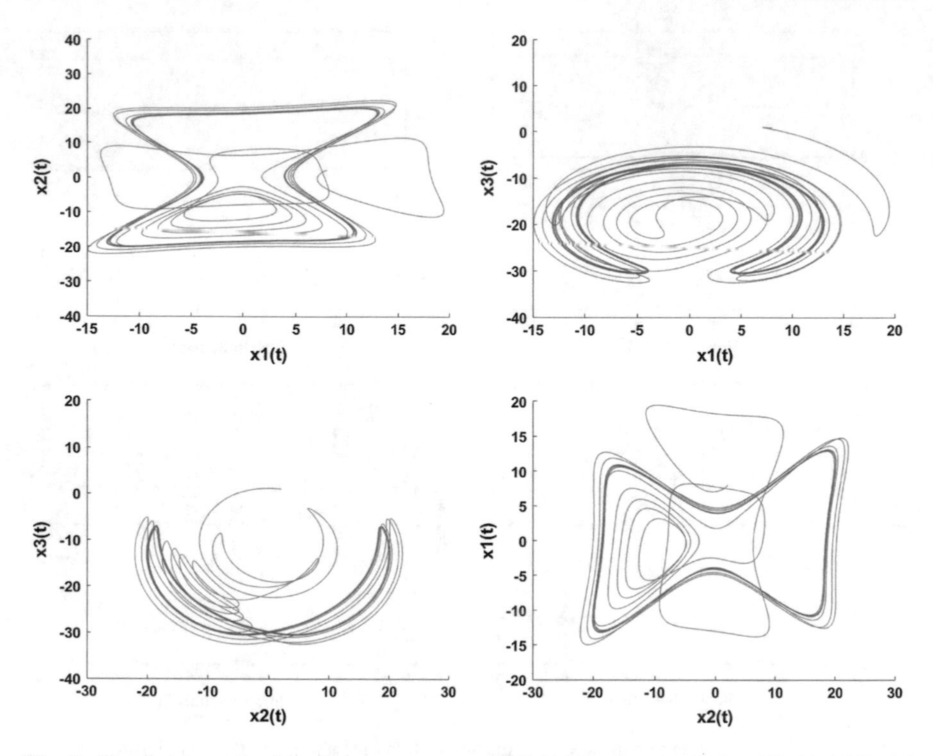

Fig. 6 Simulation result of the slave master system (52) in 2-D when $(\alpha, \beta, \gamma) = (19, 11, 0.73)$ and $p = 0.98$

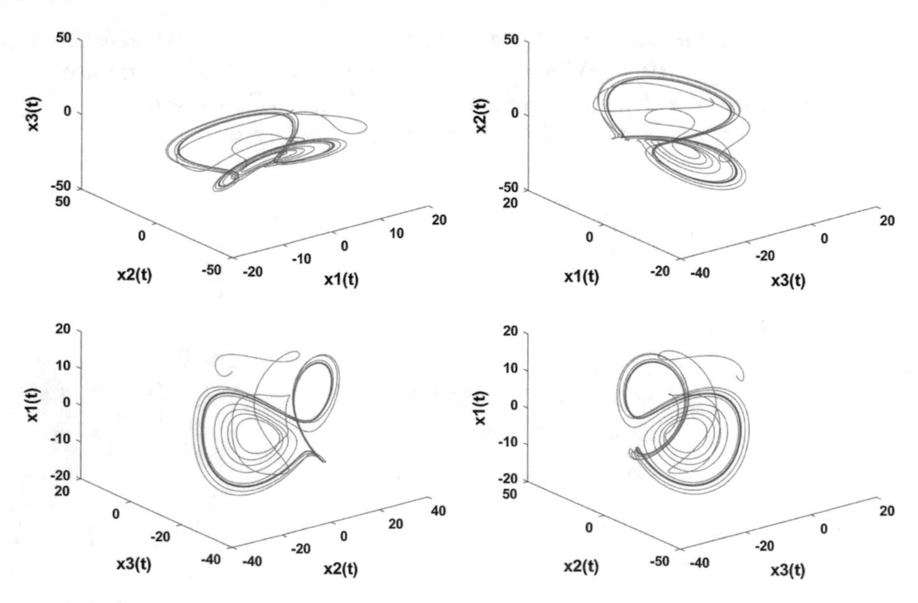

Fig. 7 Simulation result of the slave master system (52) in 3-D when $(\alpha, \beta, \gamma) = (19, 11, 0.73)$ and $p = 0.98$

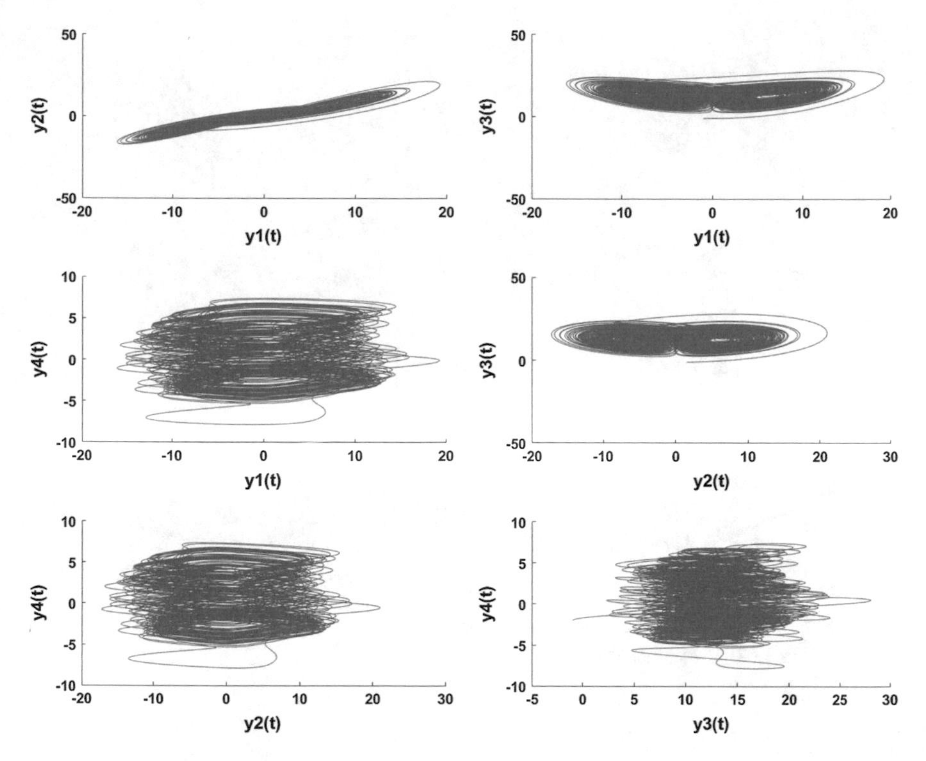

Fig. 8 Simulation result of the slave system (53) in 2D when $(a, b, c, d, g) = (36, -16, 28, 3, 0.5)$, $q = 0.9$ and $u_1 = u_2 = u_3 = u_4 = 0$

where y_1, y_2, y_3, y_4 are states and $U = (u_1, u_2, u_3, u_4)^T$ is the vector controller. The fractional-order modified hyperchaotic Chen system (i.e. the uncontrolled system (53)) has chaotic attractors when $q = 0.9$ and $(a, b, c, d, g) = (36, -16, 28, 3, 0.5)$ [151]. The chaotic attractors of the commensurate fractional order hyperchaotic Lorenz system are shown in Figs. 8 and 9.

In this case, the error system between the master system (52) and the slave system (53) can be defined as

$$\begin{pmatrix} e_1 \\ e_2 \\ e_3 \end{pmatrix} = \Theta(t) \times \begin{pmatrix} y_1 \\ y_2 \\ y_3 \\ y_4 \end{pmatrix} - \Phi(t) \times \begin{pmatrix} x_1 \\ x_2 \\ x_3 \end{pmatrix}, \tag{54}$$

where

$$\Theta(t) = \begin{pmatrix} \frac{1}{t+2} & 0 & 0 & t \\ 0 & \frac{1}{t^2+1} & 0 & \exp(-t) \\ 0 & 0 & t+3 & 1 \end{pmatrix}, \tag{55}$$

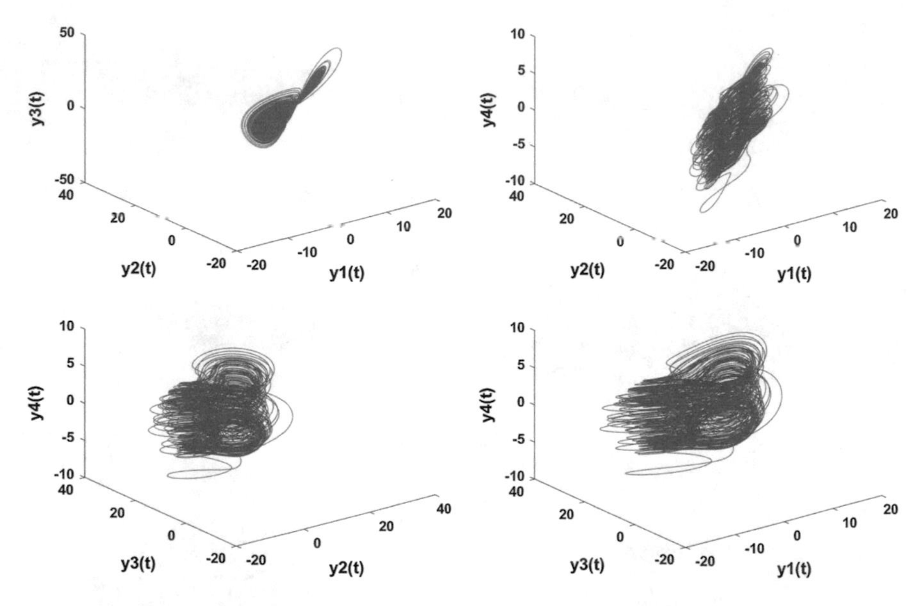

Fig. 9 Simulation result of the slave system (53) in 3D when $(a, b, c, d, g) = (36, -16, 28, 3, 0.5)$, $q = 0.9$ and $u_1 = u_2 = u_3 = u_4 = 0$

and

$$\Phi(t) = \begin{pmatrix} 1 & 0 & 3 \\ 0 & t & 0 \\ 2 & 0 & t^2 + 1 \end{pmatrix}. \tag{56}$$

So,

$$\hat{\Theta}(t) = \begin{pmatrix} \frac{1}{t+2} & 0 & 0 \\ 0 & \frac{1}{t^2+1} & 0 \\ 0 & 0 & t+3 \end{pmatrix} \text{ and } \hat{\Theta}^{-1}(t) = \begin{pmatrix} \frac{1}{t+2} & 0 & 0 \\ 0 & \frac{1}{t^2+1} & 0 \\ 0 & 0 & t+3 \end{pmatrix}.$$

Using the notations presented in Sect. 3.2, the control matrix L_2 is selected as

$$L_2 = \begin{pmatrix} 0 & 0 & 0 \\ 0 & 0 & 0 \\ 0 & 0 & 1 \end{pmatrix}, \tag{57}$$

and the synchronization controllers u_1, u_2, u_3 and u_4 can be constructed as follows

$$
\begin{aligned}
u_1 &= 36\left(y_1 - y_2\right) - J^{0.1}\left[\frac{1}{t+2}\left(e_1 + \frac{-1}{(t+2)^2}y_1 + y_4 - \dot{x}_1 - 3\dot{x}_3\right)\right], \\
u_2 &= 16y_1 - 28y_2 + y_1y_3 + y_4 - J^{0.1}\left[\frac{1}{t+2}\left(e_2 - \frac{2t}{\left(t^2+1\right)^2}y_2 - \exp\left(-t\right)y_4 - x_2 - t\dot{x}_2\right)\right], \\
u_3 &= 3y_3 - y_1y_2 - J^{0.1}\left[\left(t+3\right)\left(0.27e_3 + y_3 - 2tx_3 - 2\dot{x}_1 - \left(t^2+1\right)\dot{x}_3\right)\right], \\
u_4 &= -y_1 - 0.5.
\end{aligned} \tag{58}
$$

We can see $\left(A - L_2\right)^T + \left(A - L_2\right)$ is a negative definite matrix, and the conditions of Theorem 2 are satisfied.

Hence the $\Theta - \Phi$ synchronization between systems (52) and (53) is achieved. The error system, in this case, can be described as follows

$$
\begin{aligned}
\dot{e}_1 &= -e_1, \\
\dot{e}_2 &= -e_2, \\
\dot{e}_3 &= -0.27e_3.
\end{aligned} \tag{59}
$$

For the purpose of numerical simulation, Euler integration method has been used. In addition, simulation time $Tm = 120$ s and time step $h = 0.005$ s have been employed. The initial values of the master system and the slave system are $[x_1(0), x_2(0), x_3(0)] = [8, 2, 1]$ and $[y_1(0), y_2(0), y_3(0), y_4(0)] = [-1, 1.5, -1, -2]$, respectively, and the initial states of the error system are $[e_1(0), e_2(0), e_3(0)] = [-11.5, -0.5, -22]$. The error functions evolution, in this case, is shown in Fig. 10.

From Fig. 10, we can conclude that the components of the error system (59), e_1, e_2, e_3 and e_4, decay towards zero as $t \to +\infty$, and so the master system (52) and the slave system (53) are $\Theta - \Phi$ synchronized in 3D.

4.3 $\Theta - \Phi$ Synchronization Between Fractional Order Lü System and Fractional Order Modified Hyperchaotic Lorenz System in 2D

In this example, the fractional order Lü system is taken as the master system and the controlled fractional-order modified hyperchaotic Lorenz system is taken as the slave system. The master system can be described as

$$
\begin{aligned}
D^{p_1}x_1 &= \alpha\left(x_2 - x_1\right), \\
D^{p_2}x_2 &= \gamma x_2 - x_3x_1, \\
D^{p_3}x_3 &= -\beta x_3 + x_2x_1,
\end{aligned} \tag{60}
$$

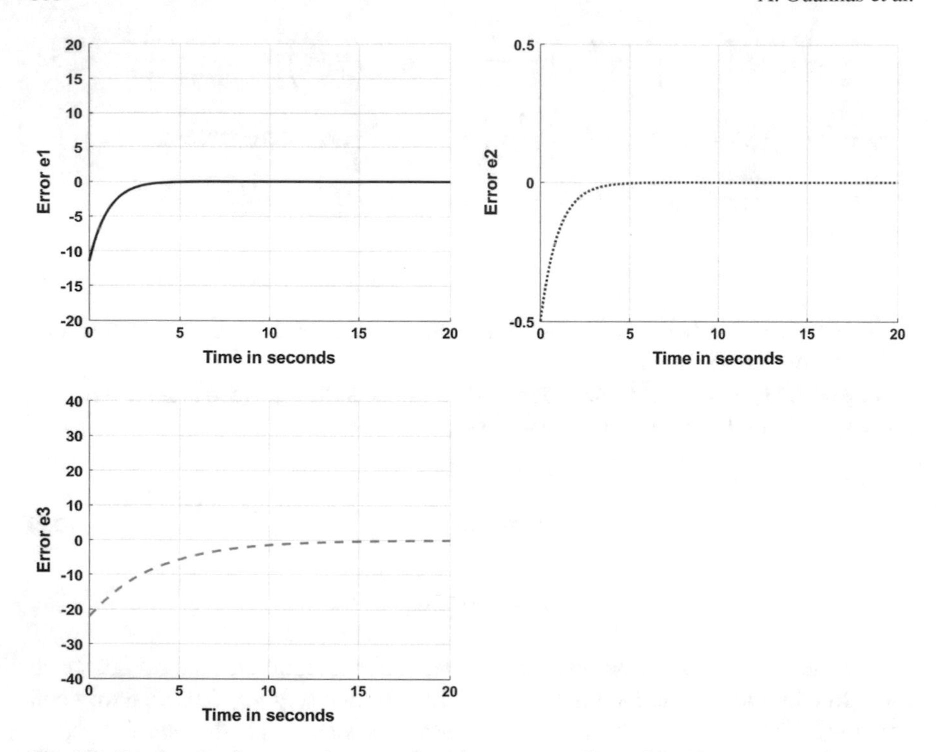

Fig. 10 Synchronization errors between the master system (52) and the slave system (53)

where x_1, x_2 and x_3 are states. This system, as shown in [70], exhibits chaotic behaviors when $(\alpha, \beta, \gamma) = (36, 3, 20)$ and $p = 0.95$. The chaotic attractors of the commensurate fractional order Lü system (60) are shown in Figs. 11 and 12.

The slave system can be described by

$$\begin{aligned} D^q y_1 &= a\left(y_2 - y_1\right) + u_1, \\ D^q y_2 &= b y_1 + y_2 - y_1 y_3 - y_4 + u_2, \\ D^q y_3 &= -c y_3 + y_1 y_2 + u_3, \\ D^q y_4 &= d y_2 y_3 + u_4, \end{aligned} \tag{61}$$

where y_i and u_i, $i = 1, 2, 3, 4$, are states and controllers, respectively. The fractional-order modified hyperchaotic Lorenz system (i.e. the uncontrolled system (61)) has chaotic attractors when $q = 0.94$ and $(a, b, c, d) = \left(10, 28, \frac{8}{3}, 0.1\right)$ [151]. The chaotic attractors of the commensurate fractional order modified hyperchaotic Lorenz system are shown in Figs. 13 and 14.

In this case, the error system between the master system (60) and the slave system (61) is defined as

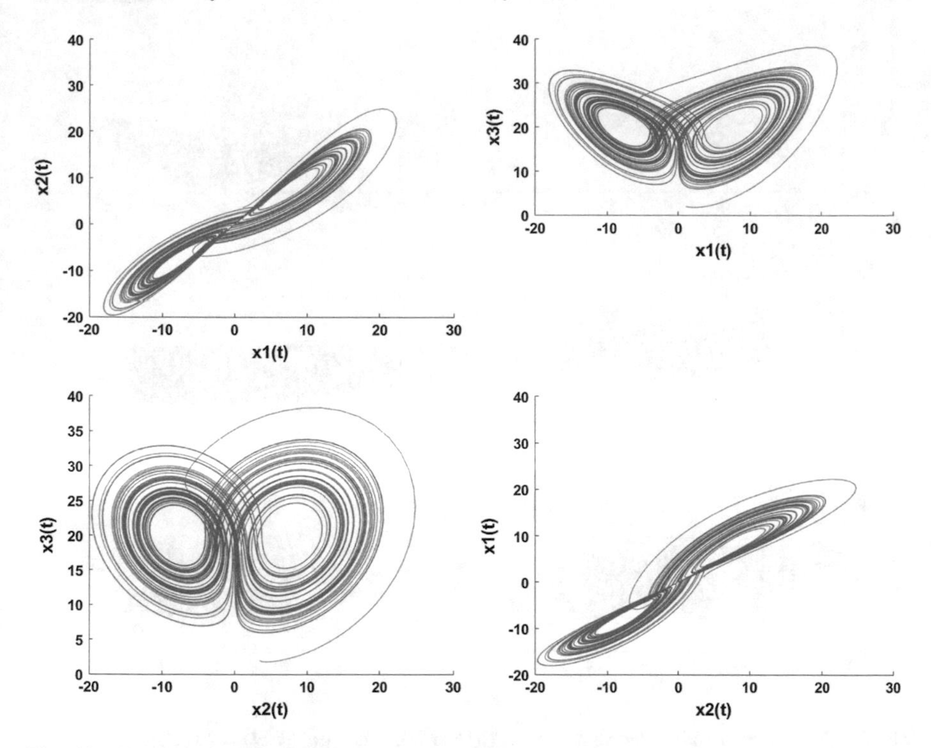

Fig. 11 Simulation result of the master system (60) in 2-D when $(\alpha, \beta, \gamma) = (36, 3, 20)$ and $p = 0.95$

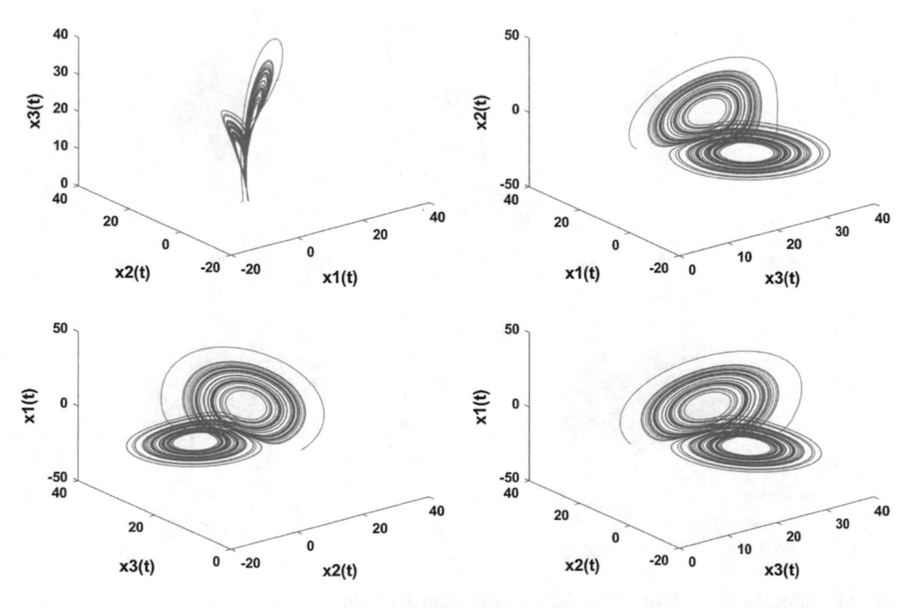

Fig. 12 Simulation result of the master system (60) in 3-D when $(\alpha, \beta, \gamma) = (36, 3, 20)$ and $p = 0.95$

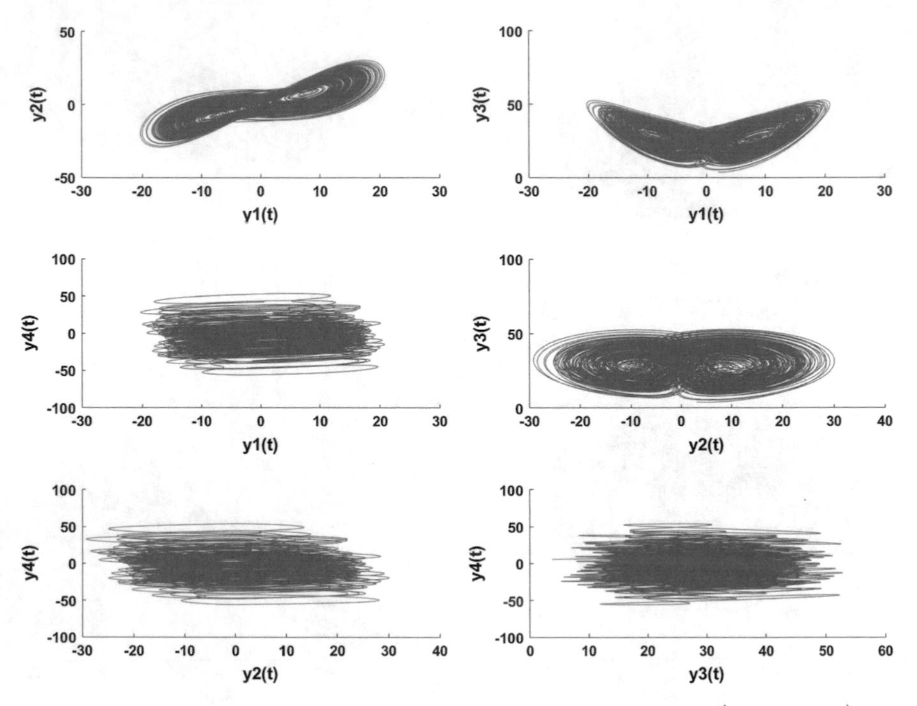

Fig. 13 Simulation result of the slave system (61) in 2-D when $(a, b, c, d) = \left(10, 28, \frac{8}{3}, 0.1\right)$, $q = 0.94$ and $u_1 = u_2 = u_3 = u_4 = 0$

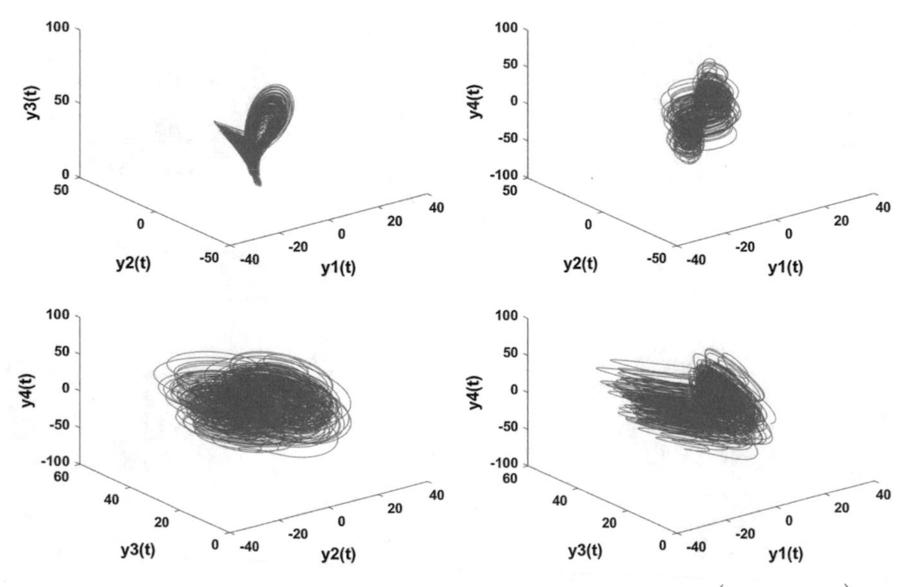

Fig. 14 Simulation result of the slave system (61) in 3-D when $(a, b, c, d) = \left(10, 28, \frac{8}{3}, 0.1\right)$, $q = 0.94$ and $u_1 = u_2 = u_3 = u_4 = 0$

$$\begin{pmatrix} e_1 \\ e_2 \end{pmatrix} = \Theta(t) \times \begin{pmatrix} y_1 \\ y_2 \\ y_3 \\ y_4 \end{pmatrix} - \Phi(t) \times \begin{pmatrix} x_1 \\ x_2 \\ x_3 \end{pmatrix}, \tag{62}$$

where

$$\Theta(t) = \begin{pmatrix} \exp(-t) & 0 & 2t+1 \\ 0 & \exp(t) & 1 & \frac{1}{t^2+1} \end{pmatrix}, \tag{63}$$

and

$$\Phi(t) = \begin{pmatrix} t & 2 & t^2 \\ 0 & t & \exp(t) \end{pmatrix}. \tag{64}$$

So,

$$\breve{\Theta}(t) = \begin{pmatrix} \exp(-t) & 0 \\ 0 & \exp(t) \end{pmatrix} \text{ and } \breve{\Theta}^{-1}(t) = \begin{pmatrix} \exp(t) & 0 \\ 0 & \exp(-t) \end{pmatrix}.$$

Using the same method proposed in Sect. 3.3, we ca find a control matrix L_3 so that systems (30) and (31) realize the $\Theta - \Phi$ synchronization in 2-D. For example, the control matrix L_3 is taken as

$$L_3 = \begin{pmatrix} -1 & 0 \\ 0 & -2 \end{pmatrix}, \tag{65}$$

then the control functions are designed as

$$\begin{aligned} u_1 &= -10(y_2 - y_1) - J^{0.06}\left[\exp(t)\left(-e_1 - \exp(-t)y_1 + y_4 - x_1 - 2tx_3 - t\dot{x}_1 - 2\dot{x}_2 - t^2\dot{x}_3\right)\right], \\ u_2 &= -28y_1 - y_2 + y_1y_3 + y_4 \\ &\quad - J^{0.06}\left(-\exp(-t)e_2 + y_2 - \frac{\exp(-t)2t}{(t^2+1)^2}y_4 - \exp(-t)x_2 - x_3 - \exp(-t)t\dot{x}_2 - \dot{x}_3\right), \\ u_3 &= \frac{8}{3}y_3 - y_1y_2, \\ u_4 &= -0.1y_2y_3. \end{aligned} \tag{66}$$

Simply, we can show that all conditions of Theorem 3 are satisfied. Hence the $\Theta - \Phi$ synchronization between systems (60) and (61) is achieved.

$$\begin{aligned} \dot{e}_1 &= -e_1, \\ \dot{e}_2 &= -2e_2. \end{aligned} \tag{67}$$

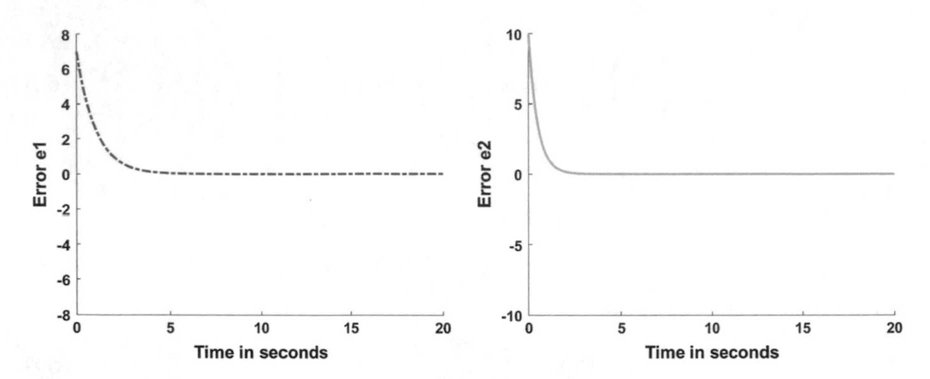

Fig. 15 Synchronization errors between the master system (60) and the slave system (61)

For the purpose of numerical simulation, Euler integration method has been used. In addition, simulation time $Tm = 120\,\text{s}$ and time step $h = 0.005\,\text{s}$ have been employed. The initial values of the master system and the slave system are $[x_1(0), x_2(0), x_3(0)] = [1.1, 3.5, 2]$ and $[y_1(0), y_2(0), y_3(0), y_4(0)] = [2, 3, 4, 6]$, respectively, and the initial states of the error system are $[e_1(0), e_2(0)] = [7, 10]$. The error functions evolution, in this case, is shown in Fig. 15.

From Fig. 15, we can conclude that the components of the error system (67), e_1 and e_2, decay towards zero as $t \to +\infty$, and so the master system (60) and the slave system (61) are $\Theta - \Phi$ synchronized in 2-D.

5 Conclusion

In this chapter, new control approaches were presented to study the problem of $\Theta - \Phi$ synchronization with two scaling function matrices between n-dimensional fractional-order master system and m-dimensional fractional-order slave system. The new criterions derived were proved theoretically using nonlinear fractional controllers and Lyapunov stability theory of integer-order differential systems. Firstly, to achieve the $\Theta - \Phi$ synchronization, when dimension of the synchronization equal m, the synchronization criterion was obtained via controlling the linear part of the slave system. Secondly, to observe $\Theta - \Phi$ synchronization, when the synchronization is equal n, the synchronization scheme was based on the control of the linear part of the master system. Finally, $\Theta - \Phi$ synchronization is guaranteed in dimension d where $d < m$ and $d \neq n$, by a diagonal matrix. Numerical examples and simulations results have been carried out, with the aim to highlight the capabilities of the new synchronization schemes conceived herein.

References

1. Hartley, T., Lorenzo, C., & Qammer, H. (1995). Chaos in a fractional order Chua's system. *IEEE Transactions on Circuits and Systems I: Fundamental Theory and Applications*, *42*, 485–490.
2. Arena, P., Caponetto, R., Fortuna, L., & Porto, D. (1998). Bifurcation and chaos in noninteger order cellular neural networks. *International Journal of Bifurcation and Chaos*, *8*, 1527–1539.
3. Ahmad, W. M., & Sprott, J. C. (2003). Chaos in fractional-order autonomous nonlinear systems. *Chaos Solitons Fractals*, *16*, 339–351.
4. Grigorenko, I., & Grigorenko, E. (2003). Chaotic dynamics of the fractional Lorenz system. *Physical Review Letters*, *91*, 034101.
5. Li, C., & Chen, G. (2004). Chaos and hyperchaos in fractional order Rössler equations. *Physica A*, *341*, 55–61.
6. Li, C., & Chen, G. (2004). Chaos in the fractional order Chen system and its control. *Chaos Solitons Fractals*, *22*, 549–554.
7. Guo, L. J. (2005). Chaotic dynamics and synchronization of fractional-order Genesio-Tesi systems. *Chinese Physics*, *14*, 1517–1521.
8. Lu, J. G. (2005). Chaotic dynamics and synchronization of fractional-order Arneodo's systems. *Chaos Solitons Fractals*, *26*, 1125–1133.
9. Ahmad, W. M. (2005). Hyperchaos in fractional order nonlinear systems. *Chaos Solitons Fractals*, *26*, 1459–1465.
10. Gao, X., & Yu, J. (2005). Chaos in the fractional order periodically forced complex Duffing's oscillators. *Chaos Solitons Fractals*, *24*, 1097–1104.
11. Lu, J. G., & Chen, G. (2006). A note on the fractional-order Chen system. *Chaos Solitons Fractals*, *27*, 685–688.
12. Ge, Z. M., & Hsu, M. Y. (2007). Chaos in a generalized van der Pol system and in its fractional order system. *Chaos Solitons Fractals*, *33*, 1711–1745.
13. Ahmed, E., El-Sayed, A. M. A., & El-Saka, H. A. A. (2007). Equilibrium points, stability and numerical solutions of fractional-order predator-prey and rabies models. *Journal of Mathematical Analysis and Applications*, *325*, 542–553.
14. Li, C., & Yan, J. (2007). The synchronization of three fractional differential systems. *Chaos Solitons Fractals*, *32*, 751–757.
15. Barbosa, R. S., Machado, J. A. T., Vinagre, B. M., & Calderón, A. J. (2007). Analysis of the Van der Pol oscillator containing derivatives of fractional order. *Journal of Vibration and Control*, *13*, 1291–1301.
16. Ge, Z. M., & Ou, C. Y. (2007). Chaos in a fractional order modified Duffing system. *Chaos Solitons Fractals*, *34*, 262–291.
17. Chen, J. H., & Chen, W. C. (2008). Chaotic dynamics of the fractionally damped van der Pol equation. *Chaos Solitons Fractals*, *35*, 188–198.
18. Chen, W. C. (2008). Nonlinear dynamic and chaos in a fractional-order financial system. *Chaos Solitons Fractals*, *36*, 1305–1314.
19. Sheu, L. J., Chen, H. K., Chen, J. H., Tam, L. M., Chen, W. C., Lin, K. T., et al. (2008). Chaos in the Newton-Leipnik system with fractional order. *Chaos Solitons Fractals*, *36*, 98–103.
20. Petráš, I. (2008). A note on the fractional-order Chua's system. *Chaos Solitons Fractals*, *38*, 140–147.
21. Petráš, I. (2009). Chaos in the fractional-order Volta's system: modeling and simulation. *Nonlinear Dynamics*, *57*, 157–170.
22. Petráš, I. (2010). A note on the fractional-order Volta's system. *Communications in Nonlinear Science and Numerical Simulation*, *15*, 384–393.
23. Deng, H., Li, T., Wang, Q., & Li, H. (2009). A fractional-order hyperchaotic system and its synchronization. *Chaos Solitons Fractals*, *41*, 962–969.
24. Gejji, V. D., & Bhalekar, S. (2010). Chaos in fractional ordered Liu system. *Computers & Mathematics with Applications*, *59*, 1117–1127.

25. Deng, W. (2007). Short memory principle anda predictor-corrector approach for fractional differential equations. *Journal of Computational and Applied Mathematics*, *206*, 174–188.
26. Kiani, B. A., Fallahi, K., Pariz, N., & Leung, H. (2009). A chaotic secure communication scheme using fractional chaotic systems based on an extended fractional Kalman filter. *Communications in Nonlinear Science and Numerical Simulation*, *14*, 863–879.
27. Doye, I. N., Zasadzinski, M., Darouach, M., & Radhy, N. (2009). Observer-based control for fractional-order continuous-time systems. In *Proceedings of IEEE Conference on Decision and Control* (1932–1937).
28. Deng, Y. S., & Qin, K. Y. (2010). Fractional order Liu-system synchronization and its application in multimedia security. *ICCCAS*, *23*, 769–772.
29. Sheu, L. J., Chen, W. C., Chen, Y. C., & Weng, W. T. (2010). A two-channel secure communication using fractional chaotic systems. In *International Conference on Computer, Electrical, and Systems Science, and Engineering*, Tokyo.
30. Sheu, L. J. (2011). A speech encryption using fractional chaotic systems. *Nonlinear Dynamics*, *65*, 103–108.
31. Cao, H. F., & Zhang, R. X. (2012). Parameter modulation digital communication and its circuit implementation using fractional-order chaotic system via a single driving variable. *Acta Physica Sinica*, *61*, 123–130.
32. Boroujeni, E. A., & Momeni, H. R. (2012). Observer based control of aclass of nonlinear fractional-order systems using LMI. *International Journal of Science and Engineering Investigations*, *1*, 48–52.
33. Zhao, J. F., Wang, S. H., Chang, Y. X., & Li, X. F. (2015). A novel image encryption scheme based on an improper fractional-order chaotic system. *Nonlinear Dynamics*, *80*, 1721–1729.
34. Chao, L. (2015). Asynchronous error-correcting secure communication scheme based on fractional-order shifting chaotic system. *International Journal of Modern Physics C*, *26*, 1550065-19.
35. Huang, L., Shi, D., & Gao, J. (2016). The design and its application in secure communication and image encryption of a New Lorenz-like system with varying parameter. *Mathematical Problems in Engineering*, 1–11.
36. Yamada, T., & Fujisaka, H. (1983). Stability theory of synchroized motion in coupled-oscillator systems. *Progress of Theoretical Physics*, *70*, 1240–1248.
37. Pecora, L. M., & Carroll, T. L. (1990). Synchronization in chaotic systems. *Physical Review Letters*, *64*, 821–827.
38. Carroll, T. L., & Pecora, L. M. (1991). Synchronizing a chaotic systems. *IEEE Transactions on Circuits and Systems*, *38*, 453–456.
39. Pikovsky, A., Rosenblum, M., & Kurths, J. (2001). *Synchronization an universal concept in nonlinear sciences*. Cambridge University Press.
40. Boccaletti, S., Kurths, J., Osipov, G., Valladares, D. L., & Zhou, C. S. (2002). The synchronization of chaotic systems. *Physics Reports*, *366*, 1–101.
41. Aziz-Alaoui, M. A. (2006). Synchronization of chaos. *Encyclopedia of Mathematical Physics*, *5*, 213–226.
42. Luo, A. (2009). A theory for synchronization of dynamical systems. *Communications in Nonlinear Science and Numerical Simulation*, *14*, 1901–1951.
43. Vaidyanathan, S., & Azar, A. T. (2015). Analysis, control and synchronization of a nine-term 3-D novel chaotic system. In A. T. Azar & S. Vaidyanathan (Eds.), *Chaos modeling and control systems design*. Studies in Computational Intelligence Book Series. Springer.
44. Vaidyanathan, S., & Azar, A. T. (2015). Anti-synchronization of identical chaotic systems using sliding mode control and an application to Vaidyanathan-Madhavan chaotic systems. In A. T. Azar & Q. Zhu (Eds.), *Advances and applications in sliding mode control systems*. Studies in Computational Intelligence Book Series. Springer.
45. Vaidyanathan, S., & Azar, A. T. (2015). Hybrid synchronization of identical chaotic systems using sliding mode control and an application to Vaidyanathan chaotic systems. In A. T. Azar & Q. Zhu (Eds.), *Advances and applications in sliding mode control systems*. Studies in Computational Intelligence Book Series. Springer.

46. Vaidyanathan, S., Idowu, B. A., & Azar, A. T. (2015). Backstepping controller design for the global chaos synchronization of Sprott's Jerk systems. In A. T. Azar & S. Vaidyanathan (Eds.), *Chaos modeling and control systems design*. Studies in Computational Intelligence Book Series. Springer.
47. Vaidyanathan, S., Sampath, S., & Azar, A. T. (2015). Global chaos synchronisation of identical chaotic systems via novel sliding mode control method and its application to Zhu system. *International Journal of Modelling, Identification and Control (IJMIC), 23*(1), 92–100.
48. Vaidyanathan, S., Azar, A. T., Rajagopal, K., Alexander, P., & (2015) Design and SPICE implementation of a 12-term novel hyperchaotic system and its synchronization via active control. *International Journal of Modelling Identification and Control (IJMIC), 23*(3), 267–277.
49. Vaidyanathan, S., & Azar, A. T. (2016). Takagi-Sugeno fuzzy logic controller for Liu-Chen four-scroll chaotic system. *International Journal of Intelligent Engineering Informatics, 4*(2), 135–150.
50. Vaidyanathan, S., & Azar, A. T. (2015). Analysis and control of a 4-D novel hyperchaotic system. In A. T. Azar & S. Vaidyanathan (Eds.), *Chaos modeling and control systems design*. Studies in Computational Intelligence (Vol. 581, pp. 19–38). Berlin/Heidelberg: Springer-Verlag GmbH. doi:10.1007/978-3-319-13132-0_2.
51. Boulkroune, A., Bouzeriba, A., Bouden, T., & Azar A. T. (2016). Fuzzy adaptive synchronization of uncertain fractional-order chaotic systems. In A. T. Azar & S. Vaidyanathan (Eds.), *Advances in chaos theory and intelligent control*. Studies in Fuzziness and Soft Computing (Vol. 337). Germany: Springer.
52. Boulkroune, A., Hamel, S., & Azar, A. T. (2016). Fuzzy control-based function synchronization of unknown chaotic systems with dead-zone input. In *Advances in chaos theory and intelligent control*. Studies in Fuzziness and Soft Computing (Vol. 337). Germany: Springer.
53. Vaidyanathan, S., & Azar, A. T. (2016). Dynamic analysis, adaptive feedback control and synchronization of an eight-term 3-D novel chaotic system with three quadratic nonlinearities. In *Advances in chaos theory and intelligent control*. Studies in Fuzziness and Soft Computing (Vol. 337). Germany: Springer.
54. Vaidyanathan, S., & Azar, A. T. (2016). Qualitative study and adaptive control of a novel 4-d hyperchaotic system with three quadratic nonlinearities. In *Advances in chaos theory and intelligent control*. Studies in Fuzziness and Soft Computing (Vol. 337). Germany: Springer.
55. Vaidyanathan, S., & Azar, A. T. (2016). A novel 4-D four-wing chaotic system with four quadratic nonlinearities and its synchronization via adaptive control method. In *Advances in chaos theory and intelligent control*. Studies in Fuzziness and Soft Computing (Vol. 337). Germany: Springer.
56. Vaidyanathan, S., & Azar, A. T. (2016). Adaptive control and synchronization of Halvorsen circulant chaotic systems. In *Advances in chaos theory and intelligent control*. Studies in Fuzziness and Soft Computing (Vol. 337). Germany: Springer.
57. Vaidyanathan, S., & Azar, A. T. (2016). Adaptive backstepping control and synchronization of a novel 3-D Jerk system with an exponential nonlinearity. In *Advances in chaos theory and intelligent control*. Studies in Fuzziness and Soft Computing (Vol. 337). Germany: Springer.
58. Vaidyanathan, S., & Azar, A. T. (2016). Generalized projective synchronization of a novel hyperchaotic four-wing system via adaptive control method. In *Advances in chaos theory and intelligent control*. Studies in Fuzziness and Soft Computing (Vol. 337). Germany: Springer.
59. Ouannas, A. (2014). Chaos synchronization approach based on new criterion of stability. *Nonlinear Dynamics and Systems Theory, 14*, 396–402.
60. Ouannas, A. (2014). On full state hybrid projective synchronization of general discrete chaotic systems. *Journal of Nonlinear Dynamics*, 1–6.
61. Ouannas, A. (2014). Some synchronization criteria for N-dimensional chaotic systems in discrete-time. *Journal of Advanced Research in Applied Mathematics, 6*, 1–10.
62. Ouannas, A. (2014). On inverse full state hybrid projective synchronization of chaotic dynamical systems in discrete-time. *International Journal of Dynamics and Control*, 1–7.

63. Ouannas, A. (2015). Synchronization criterion for a class of N-dimensional discrete chaotic systems. *Journal of Advanced Research in Dynamical and Control Systems*, *7*, 82–89.
64. Ouannas, A. (2015). A new synchronization scheme for general 3D quadratic chaotic systems in discrete-time. *Nonlinear Dynamics and Systems Theory*, *15*, 163–170.
65. Ouannas, A., Odibat, Z., & Shawagfeh, N. (2016). A new Q–S Synchronization results for discrete chaotic systems. *Differential Equations and Dynamical Systems*, 1–10.
66. Ouannas, A. (2016). Co-existence of various synchronization-types in hyperchaotic maps. *Nonlinear Dynamics and Systems Theory*, *16*, 312–321.
67. Ouannas, A., Azar, A. T., & Abu-Saris, R. (2016). A new type of hybrid synchronization between arbitrary hyperchaotic maps. *International Journal of Machine Learning and Cybernetics*, 1–8.
68. Li, C. G., Liao, X. F., & Yu, J. B. (2003). Synchronization of fractional order chaotic systems. *Physical Review E*, *68*, 067203.
69. Gao, X., & Yu, J. B. (2005). Synchronization of two coupled fractional-order chaotic oscillators. *Chaos Solitons Fractals*, *26*, 141–145.
70. Deng, W. H., & Li, C. P. (2005). Chaos synchronization of the fractional Lü system. *Physica A*, *353*, 61–72.
71. Li, C., & Zhou, T. (2005). Synchronization in fractional-order differential systems. *Physica D*, *212*, 111–125.
72. Zhou, S., Li, H., Zhu, Z., & Li, C. (2008). Chaos control and synchronization in a fractional neuron network system. *Chaos Solitons Fractals*, *36*, 973–984.
73. Peng, G. (2007). Synchronization of fractional order chaotic systems. *Physics Letters A*, *363*, 426–432.
74. Sheu, L. J., Chen, H. K., Chen, J. H., & Tam, L. M. (2007). Chaos in a new system with fractional order. *Chaos Solitons Fractals*, *31*, 1203–1212.
75. Yan, J., & Li, C. (2007). On chaos synchronization of fractional differential equations. *Chaos Solitons Fractals*, *32*, 725–735.
76. Wang, J., Xiong, X., & Zhang, Y. (2006). Extending synchronization scheme to chaotic fractional-order Chen systems. *Physica A*, *370*, 279–285.
77. Li, C. P., Deng, W. H., & Xu, D. (2006). Chaos synchronization of the Chua system with a fractional order. *Physica A*, *360*, 171–185.
78. Zhu, H., Zhou, S., & Zhang, J. (2009). Chaos and synchronization of the fractional-order Chua's system. *Chaos Solitons Fractals*, *39*, 1595–1603.
79. Zhang, F., Chen, G., Li, C., & Kurths, J. (2013). Chaos synchronization in fractional differential systems. *Philosophical Transactions of the Royal Society A*, *371*, 1–26.
80. Ansari, M. A., Arora, D., & Ansari, S. P. (2016). Chaos control and synchronization of fractional order delay-varying computer virus propagation model. *Mathematical Methods in the Applied Sciences*, *39*, 1197–1205.
81. Liang, H., Wang, Z., Yue, Z., & Lu, R. (2012). Generalized synchronization and control for incommensurate fractional unified chaotic system and applications in secure communication. *Kybernetika*, *48*, 190–205.
82. Wu, X., Wang, H., & Lu, H. (2012). Modified generalized projective synchronization of a new fractional-order hyperchaotic system and its application to secure communication. *Nonlinear Analysis: Real World Applications*, *13*, 1441–1450
83. Muthukumar, P., & Balasubramaniam, P. (2013). Feedback synchronization of the fractional order reverse butterfly-shaped chaotic system and its application to digital cryptography. *Nonlinear Dynamics*, *74*, 1169–1181.
84. Muthukumar, P., Balasubramaniam, P., & Ratnavelu, K. (2014). Synchronization of a novel fractional order stretch-twistfold (STF) flow chaotic system and its application to a new authenticated encryption scheme (AES). *Nonlinear Dynamics*, *77*, 1547–1559.
85. Chen, L., Wu, R., He, Y., & Chai, Y. (2015). Adaptive sliding-mode control for fractional-order uncertain linear systems with nonlinear disturbances. *Nonlinear Dynamics*, *80*, 51–58.
86. Liu, L., Ding, W., Liu, C., Ji, H., & Cao, C. (2014). Hyperchaos synchronization of fractional-order arbitrary dimensional dynamical systems via modified sliding mode control. *Nonlinear Dynamics*, *76*, 2059–2071.

87. Zhang, L., & Yan, Y. (2014). Robust synchronization of two different uncertain fractional-order chaotic systems via adaptive sliding mode control. *Nonlinear Dynamics*, *76*, 1761–1767.
88. Odibat, Z., Corson, N., Alaoui, M. A. A., & Bertelle, C. (2010). Synchronization of chaotic fractional-order systems via linear control. *International Journal of Bifurcation and Chaos*, *20*, 81–97.
89. Chen, X. R., & Liu, C. X. (2012). Chaos synchronization of fractional order unified chaotic system via nonlinear control. *International Journal of Modern Physics B*, *25*, 407–415.
90. Srivastava, M., Ansari, S. P., Agrawal, S. K., Das, S., & Leung, A. Y. T. (2014). Anti-synchronization between identical and non-identical fractional-order chaotic systems using active control method. *Nonlinear Dynamics*, *76*, 905–914.
91. Agrawal, S. K., & Das, S. (2012). Amodified adaptive control method for synchronization of some fractional chaotic systems with unknown parameters. *Nonlinear Dynamics*, *73*, 907–919.
92. Yuan, W. X., & Mei, S. J. (2009). Synchronization of the fractional order hyperchaos Lorenz systems with activation feedback control. *Communications in Nonlinear Science and Numerical Simulation*, *14*, 3351–3357.
93. Odibat, Z. (2010). Adaptive feedback control and synchronization of non-identical chaotic fractional order systems. *Nonlinear Dynamics*, *60*, 479–487.
94. Zhou, P., & Bai, R. (2015). The adaptive synchronization of fractional-order chaotic system with fractional-order $1 < q < 2$ via linear parameter update law. *Nonlinear Dynamics*, *80*, 753–765.
95. Azar, A. T., & Vaidyanathan, S. (2015). *Chaos modeling and control systems design*. Studies in Computational Intelligence (Vol. 581). Germany: Springer. ISBN 978-3-319-13131-3.
96. Azar, A. T., & Vaidyanathan, S. (2015). *Advances in chaos theory and intelligent control*. Studies in Fuzziness and Soft Computing (Vol. 337). Germany: Springer. ISBN 978-3-319-30338-3.
97. Azar, A. T., & Vaidyanathan, S. (2015). *Computational intelligence applications in modeling and control*. Studies in Computational Intelligence (Vol. 575). Germany: Springer. ISBN 978-3-319-11016-5.
98. Azar, A. T., & Vaidyanathan, S. (2015). *Handbook of research on advanced intelligent control engineering and automation*. Advances in Computational Intelligence and Robotics (ACIR) Book Series. USA: IGI Global. ISBN 9781466672482.
99. Zhu, Q., & Azar, A. T. (2015). *Complex system modelling and control through intelligent soft computations*. Studies in Fuzziness and Soft Computing (Vol. 319). Germany: Springer. ISBN 978-3-319-12882-5.
100. Azar, A. T., & Zhu, Q. (2015). *Advances and applications in sliding mode control systems*. Studies in Computational Intelligence (Vol. 576). Germany: Springer. ISBN 978-3-319-11172-8.
101. Cafagna, D., & Grassi, G. (2012). Observer-based projective synchronization of fractional systems via a scalar signal: Application to hyperchaotic Rössler systems. *Nonlinear Dynamics*, *68*, 117–128.
102. Odibat, Z. M. (2012). A note on phase synchronization in coupled chaotic fractional order systems. *Nonlinear Analysis: Real World Applications*, *13*, 779–789.
103. Chen, F., Xia, L., & Li, C. G. (2012). Wavelet phase synchronization of fractional-order chaotic systems. *Chinese Physics Letters*, *29*, 070501-6.
104. Razminiaa, A., & Baleanu, D. (2013). Complete synchronization of commensurate fractional order chaotic systems using sliding mode control. *Mechatronics*, *23*, 873–879.
105. Al-sawalha, M. M., Alomari, A. K., Goh, S. M., & Nooran, M. S. M. (2011). Active anti-synchronization of two identical and different fractional-order chaotic systems. *International Journal of Nonlinear Science*, *11*, 267–274.
106. Li, C. G. (2006). Projective synchronization in fractional order chaotic systems and its control. *Progress of Theoretical Physics*, *115*, 661–666.

107. Shao, S. Q., Gao, X., & Liu, X. W. (2007). Projective synchronization in coupled fractional order chaotic Rössler system and its control. *Chinese Physics*, *16*, 2612–2615.
108. Wang, X. Y., & He, Y. J. (2008). Projective synchronization of fractional order chaotic system based on linear separation. *Physics Letters A*, *372*, 435–441.
109. Agrawal, S. K., & Das, S. (2014). Projective synchronization between different fractional-order hyperchaotic systems with uncertain parameters using proposed modified adaptive projective synchronization technique. *Mathematical Methods in the Applied Sciences*, *37*, 2164–2176.
110. Chang, C. M., & Chen, H. K. (2010). Chaos and hybrid projective synchronization of commensurate and incommensurate fractional-order Chen-Lee systems. *Nonlinear Dynamics*, *62*, 851–858.
111. Wang, S., Yu, Y. G., & Diao, M. (2010). Hybrid projective synchronization of chaotic fractional order systems with different dimensions. *Physica A*, *389*, 4981–4988.
112. Zhou, P., & Zhu, W. (2011). Function projective synchronization for fractional-order chaotic systems. *Nonlinear Analysis: Real World Applications*, *12*, 811–816.
113. Zhou, P., & Cao, Y. X. (2010). Function projective synchronization between fractional-order chaotic systems and integer-order chaotic systems. *Chinese Physics B*, *19*, 100507.
114. Xi, H., Li, Y., & Huang, X. (2015). Adaptive function projective combination synchronization of three different fractional-order chaotic systems. *Optik*, *126*, 5346–5349.
115. Peng, G. J., Jiang, Y. L., & Chen, F. (2008). Generalized projective synchronization of fractional order chaotic systems. *Physica A*, *387*, 3738–3746.
116. Shao, S. Q. (2009). Controlling general projective synchronization of fractional order Rössler systems. *Chaos Solitons Fractals*, *39*, 1572–1577.
117. Wu, X. J., & Lu, Y. (2009). Generalized projective synchronization of the fractional-order Chen hyperchaotic system. *Nonlinear Dynamics*, *57*, 25–35.
118. Zhou, P., Kuang, F., & Cheng, Y. M. (2010). Generalized projective synchronization for fractional order chaotic systems. *Chinese Journal of Physics*, *48*, 49–56.
119. Deng, W. H. (2007). Generalized synchronization in fractional order systems. *Physical Review E*, *75*, 056201.
120. Zhou, P., Cheng, X. F., & Zhang, N. Y. (2008). Generalized synchronization between different fractional-order chaotic systems. *Communications in Theoretical Physics*, *50*, 931–934.
121. Zhang, X. D., Zhao, P. D., & Li, A. H. (2010). Construction of a new fractional chaotic system and generalized synchronization. *Communications in Theoretical Physics*, *53*, 1105–1110.
122. Jun, W. M., & Yuan, W. X. (2011). Generalized synchronization of fractional order chaotic systems. *International Journal of Modern Physics B*, *25*, 1283–1292.
123. Wu, X. J., Lai, D. R., & Lu, H. T. (2012). Generalized synchronization of the fractional-order chaos in weighted complex dynamical networks with nonidentical nodes. *Nonlinear Dynamics*, *69*, 667–683.
124. Xiao, W., Fu, J., Liu, Z., & Wan, W. (2012). Generalized synchronization of typical fractional order chaos system. *Journal of Computers*, *7*, 1519–1526.
125. Martínez-Guerra, R., & Mata-Machuca, J. L. (2014). Fractional generalized synchronization in a class of nonlinear fractional order systems. *Nonlinear Dynamics*, *77*, 1237–1244.
126. Razminia, A. (2013). Full state hybrid projective synchronization of a novel incommensurate fractional order hyperchaotic system using adaptive mechanism. *Indian Journal of Physics*, *87*(2), 161–167.
127. Yi, C., Liping, C., Ranchao, W., & Juan, D. (2013). Q-S synchronization of the fractional-order unified system. *Pramana*, *80*, 449–461.
128. Mathiyalagan, K., Park, J. H., & Sakthivel, R. (2015). Exponential synchronization for fractional-order chaotic systems with mixed uncertainties. *Complexity*, *21*, 114–125.
129. Aghababa, M. P. (2012). Finite-time chaos control and synchronization of fractional-order nonautonomous chaotic (hyperchaotic) systems using fractional nonsingular terminal sliding mode technique. *Nonlinear Dynamics*, *69*, 247–261.
130. Li, D., Zhang, X. P., Hu, Y. T., & Yang, Y. Y. (2015). Adaptive impulsive synchronization of fractional order chaotic system with uncertain and unknown parameters. *Neurocomputing*, *167*, 165–171.

131. Xi, H., Yu, S., Zhang, R., & Xu, L. (2014). Adaptive impulsive synchronization for a class of fractional-order chaotic and hyperchaotic systems. *Optik, 125*, 2036–2040.
132. Ouannas, A., & Abu-Saris, R. (2016). On matrix projective synchronization and inverse matrix projective synchronization for different and identical dimensional discrete-time chaotic systems. *Journal of Chaos*, 1–7.
133. Ouannas, A., & Grassi, G. (2016). Inverse full state hybrid projective synchronization for chaotic maps with different dimensions. *Chinese Physics B, 25*, 090503-6.
134. Ouannas, A., & Odibat, Z. (2015). Generalized synchronization of different dimensional chaotic dynamical systems in discrete-time. *Nonlinear Dynamics, 81*, 765–771.
135. Ouannas, A. (2016). On inverse generalized synchronization of continuous chaotic dynamical systems. *International Journal of Applied and Computational Mathematics, 2*, 1–11.
136. Ouannas, A., & Al-sawalha, M. M. (2016). On $\Lambda - \phi$ generalized synchronization of chaotic dynamical systems in continuous-time. *European Physical Journal Special Topics, 225*, 187–196.
137. Ouannas, A. (2015). A new generalized-type of synchronization for discrete-time chaotic dynamical systems. *Journal of Computational and Nonlinear Dynamics, 10*, 061019-5.
138. Ouannas, A., & Abu-Saris, R. (2015). A robust control method for Q-S synchronization between different dimensional integer-order and fractional-order chaotic systems. *Journal of Control Science and Engineering*, 1–7.
139. Ouannas, A., & Grassi, G. (2016). A new approach to study co-existence of some synchronization types between chaotic maps with different dimensions. *Nonlinear Dynamics*.
140. Ouannas, A., Azar, A. T., & Vaidyanathan, S. (2016). A robust method for new fractional hybrid chaos synchronization. *Mathematical Methods in the Applied Sciences*, 1–9.
141. Ouannas, A., & Al-sawalha, M. M. (2015). A new approach to synchronize different dimensional chaotic maps using two scaling matrices. *Nonlinear Dynamics and Systems Theory, 15*, 400–408.
142. Ouannas, A., & Al-sawalha, M. M. (2015). Synchronization between different dimensional chaotic systems using two scaling matrices. *Optik, 127*, 959–963.
143. Ouannas, A., Al-sawalha, M. M., & Ziar, T. (2016). Fractional chaos synchronization schemes for different dimensional systems with non-identical fractional-orders via two scaling matrices. *Optik, 127*, 8410–8418.
144. Caputo, M. (1967). Linear models of dissipation whose Q is almost frequency independent. II. *Geophysical Journal of the Royal Astronomical Society, 13*, 529–539.
145. Samko, S. G., Klibas, A. A., & Marichev, O. I. (1993). *Fractional integrals and derivatives: Theory and applications*. Gordan and Breach.
146. Podlubny, I. (1999). *Fractional differential equations*. Academic Press.
147. Si, G., Sun, Z., Zhang, Y., & Chen, W. (2012). *Projective synchronization of different fractional-order chaotic systems with non-identical orders. Nonlinear Analysis: Real World Applications, 13*, 1761–1771.
148. Liu, C., Liu, T., Liu, L., & Liu, K. (2004). A new chaotic attractor. *Chaos Solitons Fractals, 22*, 1031–1038.
149. Wang, Z., Huang, X., Li, Y.-X., & Song, X. N. (2013). A new image encryption algorithm based on the fractional-order hyperchaotic Lorenz system. *Chinese Physics B, 22*, 010504-7.
150. Petráš, I. (2011). *Fractional-order nonlinear systems: Modeling, analysis and simulation*. Springer.
151. Wang, X.-Y., Zhang, Y.-L., Lin, D., & Zhang, N. (2011). Impulsive synchronisation of a class of fractional-order hyperchaotic systems. *Chinese Physics B, 20*, 030506-7.

A Three-Dimensional Chaotic System with Square Equilibrium and No-Equilibrium

Viet-Thanh Pham, Sundarapandian Vaidyanathan, Christos K. Volos, Sajad Jafari and Tomas Gotthans

Abstract Recently, Leonov and Kuznetsov have introduced a new definition "hidden attractor". Systems with hidden attractors, especially chaotic systems, have attracted significant attention. Some examples of such systems are systems with a line equilibrium, systems without equilibrium or systems with stable equilibria etc. In some interesting new research, systems in which equilibrium points are located on different special curves are reported. This chapter introduces a three-dimensional autonomous system with a square-shaped equilibrium and without equilibrium points. Therefore, such system belongs to a class of systems with hidden attractors. The fundamental dynamics properties of such system are studied through phase portraits, Poincaré map, bifurcation diagram, and Lyapunov exponents. Anti-synchronization scheme for our systems is proposed and confirmed by the Lyapunov stability. Moreover, an electronic circuit is implemented to show the feasibility of the mathematical model. Finally, we introduce the fractional order form of such system.

Keywords Chaos · Hidden attractor · No-equilibrium · Square equilibrium · Lyapunov exponents · Bifurcation · Synchronization · Circuit · SPICE

V.-T. Pham (✉)
School of Electronics and Telecommunications, Hanoi University of Science and Technology, Hanoi, Vietnam
e-mail: pvt3010@gmail.com

S. Vaidyanathan
Research and Development Centre, Vel Tech University, Tamil Nadu, India
e-mail: sundar@veltechuniv.edu.in

C.K. Volos
Physics Department, Aristotle University of Thessaloniki, Thessaloniki, Greece
e-mail: volos@physics.auth.gr

S. Jafari
Biomedical Engineering Department, Amirkabir University of Technology, Tehran, Iran
e-mail: sajadjafari@aut.ac.ir

T. Gotthans
Department of Radio Electronics, Brno University of Technology, Brno, Czech Republic
e-mail: gotthans@feec.vutbr.cz

A.T. Azar et al. (eds.), *Fractional Order Control and Synchronization of Chaotic Systems*, Studies in Computational Intelligence 688,
DOI 10.1007/978-3-319-50249-6_21

1 Introduction

Chaos theory, chaotic systems, and chaos-based applications have been studied in last decades [5–8, 18, 19, 54, 71, 76, 106]. A significant amount of new chaotic systems has been introduced and discovered such as Lorenz [54], Rössler system [66], Arneodo system [4], Chen system [18], Lü system [55], Vaidyanathan system [83], time-delay systems [11], nonlinear finance system [78], four scroll chaotic system [2].

Chaotic systems, that are highly sensitive to initial conditions, were applied in different areas. A new four-scroll chaotic system was used to design a random number generator [2]. Tang et al. implemented image encryption using chaotic coupled map lattices with time-varying delays [79]. Reconfiguration chaotic logic gates based on novel chaotic circuit were discovered in [12]. Chenaghlu et al. introduced a novel keyed parallel hashing scheme based on a new chaotic system [20]. Kajbaf et al. proposed fast synchronization of non-identical chaotic modulation-based secure systems using a modified sliding mode controller [38]. A new hybrid algorithm based on chaotic maps for solving systems of nonlinear equations was presented in [44]. Tacha et al. studied analysis, adaptive control and circuit simulation of a novel nonlinear finance system [78]. Performance improvement of chaotic encryption via energy and frequency location criteria was studied in [70]. Orlando investigated a discrete mathematical model for chaotic dynamics in economics [58].

Recent developments include systems with hidden attractors which are important in engineering applications [34, 35, 48, 61, 85, 110, 112]. Especially, chaotic systems with hidden attractors such as chaotic systems without any equilibrium points, chaotic systems with infinitely many equilibrium points and chaotic systems with stable equilibria have been introduced [34, 35, 43, 56, 99]. Finding new chaotic systems with different families of hidden attractors should be studied further.

In this chapter, we introduce a novel three-dimensional (3D) chaotic system. Especially the new system displays both hidden chaotic attractor with square equilibrium and hidden chaotic attractor without equilibrium. This chapter is organized as follows. The related works are reported in the next section. Section 3 presents the theoretical model of the new system. Dynamics and properties of the new system are investigated in Sect. 4 while the adaptive anti-synchronization scheme for such new system is proposed in Sect. 5. Section 6 presents circuital implementation of the theoretical model. Moreover, fractional-order form of the new no-equilibrium system is described in Sect. 7. Finally, conclusions are drawn in Sect. 8.

2 Related Work

Recently, Leonov and Kuznetsov have proposed a new approach to classify nonlinear systems. They considered dynamical systems with self-excited attractors and dynamical systems with hidden attractors [46, 48, 50, 51]. A self-excited attractor has a basin of attraction that is excited from unstable equilibria. Therefore, self-excited attractors can be localized numerically by using the standard computational procedure. In contrast, hidden attractor cannot be found by using a numerical method in which a trajectory started from a point on the unstable manifold in the neighbourhood of an unstable equilibrium [34, 48]. "Hidden attractor" is important both in nonlinear theory and practical problems [45, 50, 60, 63, 69]. Thus various researches relating hidden attractors have been introduced [16, 36, 68, 73].

Hidden attractors have discovered in a smooth Chua's system [52], in mathematical model of drilling system [49], in a relay system with hysteresis [112], in nonlinear control systems [47], in Van der Pol-Duffing oscillators [16], in a simple four-dimensional system [105], in an impulsive Goodwin oscillator with time delay [111] or in a multilevel DC/DC converter [110]. In addition, hidden chaotic attractors are observed in 3-D chaotic autonomous system with only one stable equilibrium [43], in elementary quadratic chaotic flows with no equilibria [35], in simple chaotic flows with a line equilibrium [34], in a 4-D Rikitake dynamo system [97], in 5-D hyperchaotic Rikitake dynamo system [95], in a 5-D Sprott B system [57], in a chaotic system with an exponential nonlinear term [62] or in a system with memristive devices [10].

It is interesting that chaotic systems with an infinite number of equilibrium points or without equilibrium belong to a class of dynamical systems with "hidden attractor" [35]. A few three-dimensional chaotic systems with infinite equilibria and without equilibrium have been reported. Jafari and Sprott found chaotic flows with a line equilibrium [34]. New class of chaotic systems with circular equilibrium was presented in [26]. Gotthans et al. introduced a 3-D chaotic system with a square equilibrium in [27]. By applying a tiny perturbation into the Sprott D system, Wei obtained a new system with no equilibria [101]. Wang and Chen proposed a no-equilibrium system when constructing a chaotic system with any number of equilibria [100]. Especially, Jafari et al. found a gallery of chaotic flows with no equilibria [35]. However, investigation of new systems which can display both hidden chaotic attractors with infinite equilibria and hidden chaotic attractors without equilibrium is still an attractive research direction.

3 Model of the No-Equilibrium System

Gotthans et al. proposed an interesting three-dimensional chaotic system with a square equilibrium [27]. Gotthans's system is given by

$$\begin{cases} \dot{x} = z \\ \dot{y} = -z\left(ay + b\left|y\right|\right) - x\left|z\right| \\ \dot{z} = |x| + |y| - 1, \end{cases} \tag{1}$$

where x, y, z are state variables, while a, b are two positive parameters. System (1) is the simplest system with a square equilibrium and chaotic behavior. Moreover it is an example of a system with hidden attractor [27].

In this work, we study a new 3-D system based on system (1):

$$\begin{cases} \dot{x} - z \\ \dot{y} = -z\,(ay + b\,|y|) - x\,|z| - c \\ \dot{z} = |x| + |y| - 1, \end{cases} \tag{2}$$

in which x, y, z are state variables and a, b, c are three positive parameters. Dynamics and properties of new nonlinear system (2) are studied in the next section.

4 Dynamics and Properties of the Proposed System

The equilibrium points of system (2) are found by solving $\dot{x} = 0$, $\dot{y} = 0$, and $\dot{z} = 0$. Therefore, we have

$$z = 0, \tag{3}$$

$$-\,z\,(ay + b\,|y|) - x\,|z| - c = 0, \tag{4}$$

$$|x| + |y| - 1 = 0, \tag{5}$$

From (3), (4), we have $z = 0$ and

$$c = 0. \tag{6}$$

Therefore system (2) has an infinite number of equilibrium points when $c = 0$. Moreover equilibrium points are located on a square (5). This case has been studied in [27], so we do not discuss about it. We focus on the case for $c \neq 0$. Obviously, Eq. (6) is inconsistent when $c \neq 0$. On the other word, there is no real equilibrium in system (2). Interestingly, system (2) belongs to a newly introduced class of systems with hidden attractors because its basin of attractor does not contain neighbourhoods of equilibria [48, 50].

We consider the new system (2) for the selected parameters $a = 5, b = 3, c = 0.02$ and the initial conditions are

$$(x\,(0)\,, y\,(0)\,, z\,(0)) = (0, 0.0, 0)\,. \tag{7}$$

Lyapunov exponents, which measure the exponential rates of the divergence and convergence of nearby trajectories in the phase space of the chaotic system [72, 76], are calculated by using the algorithm in [104]. As a result, the Lyapunov exponents of the system (2) are

$$\lambda_1 = 0.1386, \lambda_2 = 0, \lambda_3 = -1.1731. \tag{8}$$

The 2-D and 3-D projections of the chaotic attractors without equilibrium in this case are illustrated in Figs. 1, 2, 3 and 4.

It has been known that the Kaplan–Yorke fractional dimension, which presents the complexity of attractor [23], is given by

$$D_{KY} = j + \frac{1}{\left|\lambda_{j+1}\right|} \sum_{i=1}^{j} \lambda_i, \tag{9}$$

where j is the largest integer satisfying $\sum_{i=1}^{j} \lambda_i > 0$ and $\sum_{i=1}^{j+1} \lambda_i < 0$. Thus, the calculated fractional dimension of no-equilibrium system (2) when $a = 5, b = 3, c = 0.02$ is

$$D_{KY} = 2 + \frac{\lambda_1 + \lambda_2}{|\lambda_3|} = 2.1181. \tag{10}$$

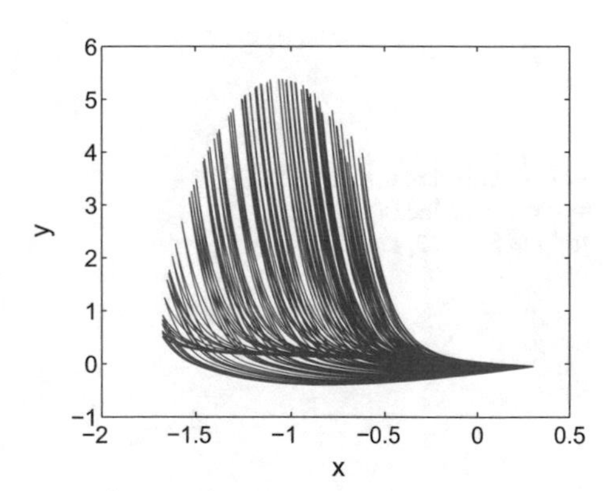

Fig. 1 2-D projection of system (2) in the (x, y)-plane, for $a = 5, b = 3, c = 0.02$

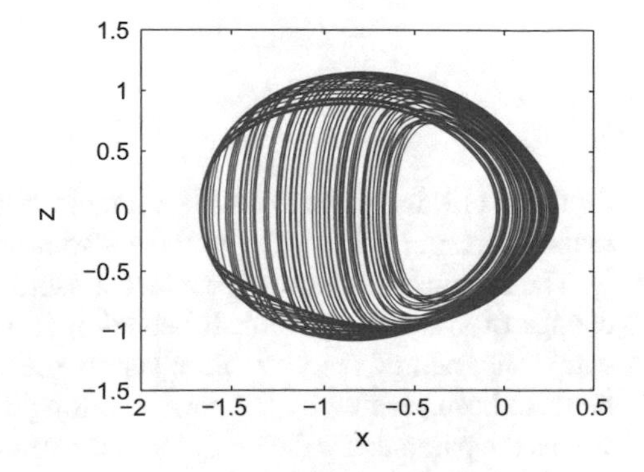

Fig. 2 2-D projection of system (2) in the (x, z)-plane, for $a = 5, b = 3, c = 0.02$

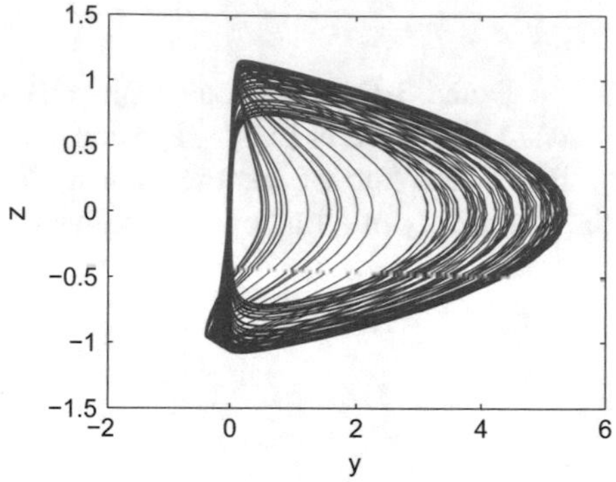

Fig. 3 2-D projection of system (2) in the (y, z)-plane, for $a = 5$, $b = 3$, $c = 0.02$

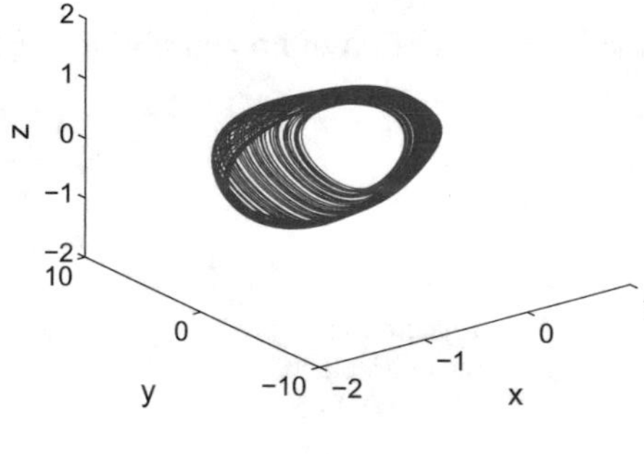

Fig. 4 3-D projection of system (2) in the (x, y, z)-space, for $a = 5$, $b = 3$, $c = 0.02$

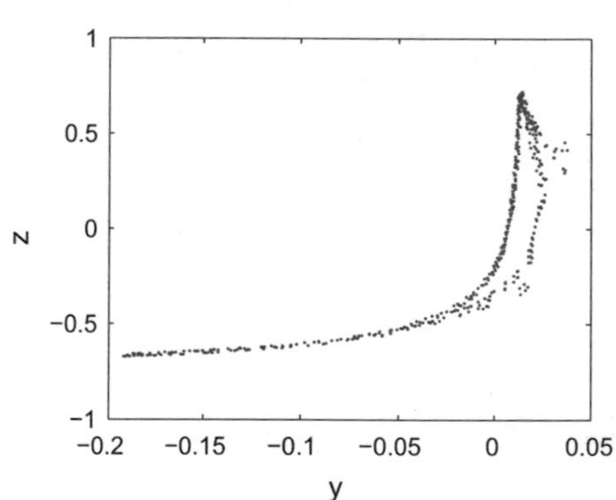

Fig. 5 Poincaré map of system (2) in the (x, y)-plane, for $a = 5$, $b = 3$, $c = 0.02$

Equation (10) indicates a strange attractor. In addition, as seen in Fig. 5, the Poincaré map of system (2) in the (x, y)-plane also illustrates the strange of attractor.

The bifurcation diagram provides a useful tool in nonlinear science. It gives the change of system's dynamical behavior. In more details, Fig. 6 presents the bifurcation diagram of the variable y versus the parameter c. The system's complexity has also been verified by the corresponding diagram of largest Lyapunov exponents versus the parameter c (see Fig. 7). In the regions where the value of the largest Lyapunov exponent is equal to zero the system is in a periodic state, while in the regions where the largest Lyapunov exponent has a positive value the system is in a chaotic

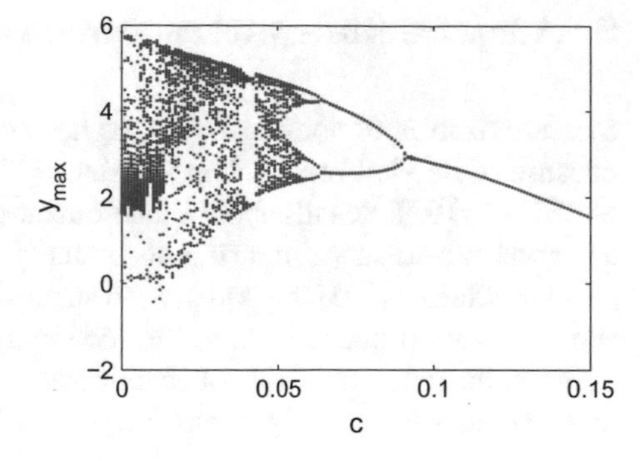

Fig. 6 Bifurcation diagram of system (2) when changing c for $a = 5, b = 3$

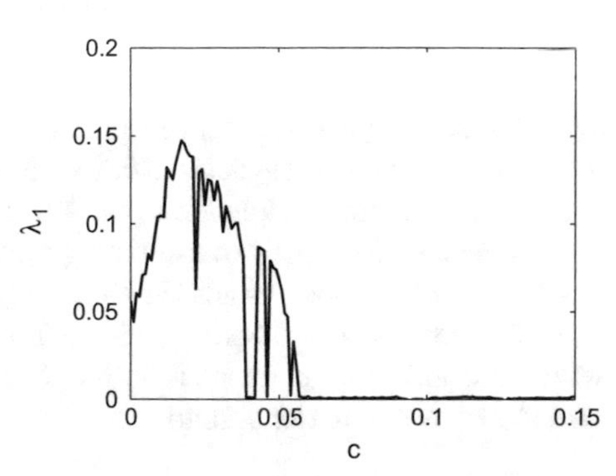

Fig. 7 Largest Lyapunov exponent of system (2) when varying c for $a = 5, b = 3$

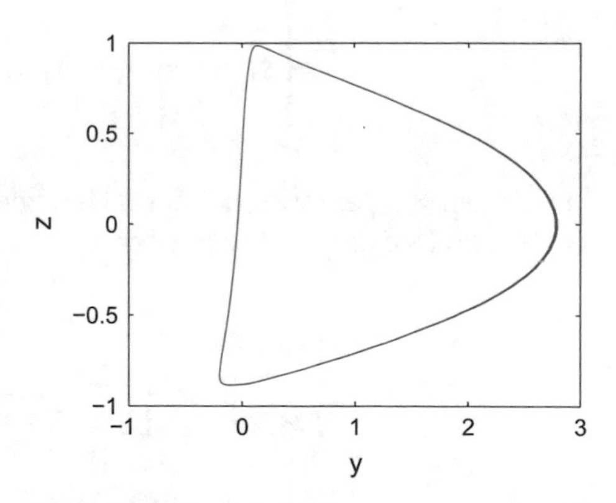

Fig. 8 Limit cycle of system (2) for $a = 5, b = 3$, and $c = 0.1$

state. As seen in Fig. 6, there is a reverse period doubling to chaos when increasing the value of parameter c from 0 to 0.15. When $c < 0.057$ a more complex behavior is emerged. For example, system exhibits chaotic behavior for $c < 0.039$. When $c > 0.057$ the system remains always in periodic states. For instant, system presents periodic behavior for $c = 0.1$ (see Fig. 8).

5 Adaptive Anti-synchronization of the Proposed System

Synchronization of nonlinear systems has been discovered extensively in literature because of its vital practical applications [13, 17, 22, 24, 39, 40, 59, 63, 74, 84, 86, 87, 94, 109]. Results about synchronization of various systems are reported such as synchronized states in a ring of mutually coupled self-sustained nonlinear electrical oscillators [103], ragged synchronizability of coupled oscillators [75], various synchronization phenomena in bidirectionally coupled double-scroll circuits [98], observer for synchronization of chaotic systems with application to secure data transmission was studied in [1], or shape synchronization control [33]. Futhermore different kind of synchronizations have been investigated, for example lag synchronization [65], frequency synchronization [3], projective-anticipating synchronization [31], anti-synchronization [82], adaptive synchronization [88–91, 93, 96], hybrid chaos synchronization [40], generalized projective synchronization [92], fuzzy synchronization [14, 15] or fast synchronization [38] etc. Interestingly, anti-synchronization has received significant attention [32, 42, 82, 108]. Anti-synchronization indicates the relationship between two oscillating systems that have the same absolute values at all times, but opposite signs [32, 42, 108].

In this section, the adaptive anti-synchronization of identical proposed systems with three unknown parameters is proposed. The newly introduced system (2) is considered as the master system:

$$\begin{cases} \dot{x}_1 = z_1 \\ \dot{y}_1 = -ay_1z_1 - b\left|y_1\right|z_1 - x_1\left|z_1\right| - c \\ \dot{z}_1 = \left|x_1\right| + \left|y_1\right| - 1, \end{cases} \tag{11}$$

in which x_1, y_1, z_1 are state variables. The slave system is considered as the controlled system and its dynamics is described by:

$$\begin{cases} \dot{x}_2 = z_2 + u_x \\ \dot{y}_2 = -ay_2z_2 - b\left|y_2\right|z_2 - x_2\left|z_2\right| - c + u_y \\ \dot{z}_2 = \left|x_2\right| + \left|y_2\right| - 1 + u_z, \end{cases} \tag{12}$$

where x_2, y_2, z_2 are the states of the slave system. Here the adaptive controls are u_x, u_y, and u_z. These controls will be designed for the anti-synchronization of the master and slave systems. We used $A(t)$, $B(t)$ and $C(t)$ in order to estimate unknown parameters a, b and c.

The anti-synchronization error between systems (11) and (12) is given by the following relation

$$\begin{cases} e_x = x_1 + x_2 \\ e_y = y_1 + y_2 \\ e_z = z_1 + z_2. \end{cases} \tag{13}$$

As a result, the anti-synchronization error dynamics is described by

$$\begin{cases} \dot{e}_x = e_z + u_x \\ \dot{e}_y = -a\left(y_1 z_1 + y_2 z_2\right) - b\left(\left|y_1\right| z_1 + \left|y_2\right| z_2\right) - \left(x_1\left|z_1\right| + x_2\left|z_2\right|\right) - 2c + u_y \\ \dot{e}_z = \left|x_1\right| + \left|x_2\right| + \left|y_1\right| + \left|y_2\right| - 2 + u_z. \end{cases} \tag{14}$$

Our aim is to construct the appropriate controllers u_x, u_y, u_z to stabilize system (14). Therefore, we propose the following controllers for system (14):

$$\begin{cases} u_x = -e_z - k_x e_x \\ u_y = A(t)\left(y_1 z_1 + y_2 z_2\right) + B(t)\left(\left|y_1\right| z_1 + \left|y_2\right| z_2\right) + \left(x_1\left|z_1\right| + x_2\left|z_2\right|\right) \\ \qquad +2C(t) - k_y e_y \\ u_z = -\left|x_1\right| - \left|x_2\right| - \left|y_1\right| - \left|y_2\right| + 2 - k_z e_z. \end{cases} \tag{15}$$

in which k_x, k_y, k_z are positive gain constants for each controllers and the estimate values for unknown system parameters are $A(t)$, $B(t)$, and $C(t)$. The update laws for the unknown parameters are determined as

$$\begin{cases} \dot{A} = -e_y\left(y_1 z_1 + y_2 z_2\right) \\ \dot{B} = -e_y\left(\left|y_1\right| z_1 + \left|y_2\right| z_2\right) \\ \dot{C} = -2e_y. \end{cases} \tag{16}$$

Then, the main result of this section will be introduced and proved.

Theorem 5.1 *If the adaptive controller (15) and the updating laws of parameter (16) are chosen, the anti-sychronization between the master system (11) and the slave system (12) is achieved.*

Proof It is noting that the parameter estimation errors $e_a(t)$, $e_b(t)$ and $e_c(t)$ are given as

$$\begin{cases} e_a(t) = a - A(t) \\ e_b(t) = b - B(t) \\ e_c(t) = c - C(t). \end{cases} \tag{17}$$

Differentiating (17) with respect to t, we have

$$\begin{cases} \dot{e}_a(t) = -\dot{A}(t) \\ \dot{e}_b(t) = -\dot{B}(t) \\ \dot{e}_c(t) = -\dot{C}(t). \end{cases} \tag{18}$$

Substituting adaptive control law (15) into (14), the closed-loop error dynamics is defined as

$$\begin{cases} \dot{e}_x = -k_x e_x \\ \dot{e}_y = -(a - A(t))\left(y_1 z_1 + y_2 z_2\right) \\ \qquad -(b - B(t))\left(\left|y_1\right| z_1 + \left|y_2\right| z_2\right) - 2(c - C(t)) - k_y e_y \\ \dot{e}_z = -k_z e_z \end{cases} \tag{19}$$

Then substituting (17) into (19), we have

$$\begin{cases} \dot{e}_x = -k_x e_x \\ \dot{e}_y = -e_a(t)\left(y_1 z_1 + y_2 z_2\right) - e_b(t)\left(\left|y_1\right| z_1 + \left|y_2\right| z_2\right) - 2e_c(t) - k_y e_y \\ \dot{e}_z = -k_z e_z. \end{cases} \tag{20}$$

We consider the Lyapunov function given as

$$\begin{aligned} V(t) &= V\left(e_x, e_y, e_z, e_a, e_b, e_c\right) \\ &= \tfrac{1}{2}\left(e_x^2 + e_y^2 + e_z^2 + e_a^2 + e_b^2 + e_c^2\right). \end{aligned} \tag{21}$$

The Lyapunov function (21) is clearly definite positive.

Taking time derivative of (21) along the trajectories of (13) and (17) we have

$$\dot{V}(t) = e_x \dot{e}_x + e_y \dot{e}_y + e_z \dot{e}_z + e_a \dot{e}_a + e_b \dot{e}_b + e_c \dot{e}_c. \tag{22}$$

From (18), (20), and (22) we get

$$\begin{aligned} \dot{V}(t) = & -k_x e_x^2 - k_y e_y^2 - k_z e_z^2 - e_a\left[e_y\left(y_1 z_1 + y_2 z_2\right) + \dot{A}\right] \\ & -e_b\left[e_y\left(\left|y_1\right| z_1 + \left|y_2\right| z_2\right) + \dot{B}\right] - e_c\left(2e_y + \dot{C}\right). \end{aligned} \tag{23}$$

Then by applying the parameter update law (16), Eq. (23) become

$$\dot{V}(t) = -k_x e_x^2 - k_y e_y^2 - k_z e_z^2. \tag{24}$$

Obviously, the time-derivative of the Lyapunov function V is negative semi-definite. According to Barbalat's lemma in the Lyapunov stability theory [41, 67], it follows that $e_x(t) \to 0$, $e_y(t) \to 0$, and $e_z(t) \to 0$, exponentially when $t \to 0$. That is, anti-synchronization between master and slave system exponentially. This completes the proof. □

A numerical example is presented to illustrate the effectiveness of our proposed anti-synchronization scheme. The parameters of the no-equilibrium systems are selected as $a = 5, b = 3, c = 0.02$ and the positive gain constant as $k = 6$. The initial conditions of the master system (11) and the slave system (12) have been chosen as $x_1(0) = 0, y_1(0) = 0, z_1(0) = 0$, and $x_2(0) = 0.5, y_2(0) = 1, z_2(0) = 0.9$, respectively. We assumed that the initial values of the parameter estimates are $A(0) = 10$, $B(0) = 2$, and $C(0) = 0$.

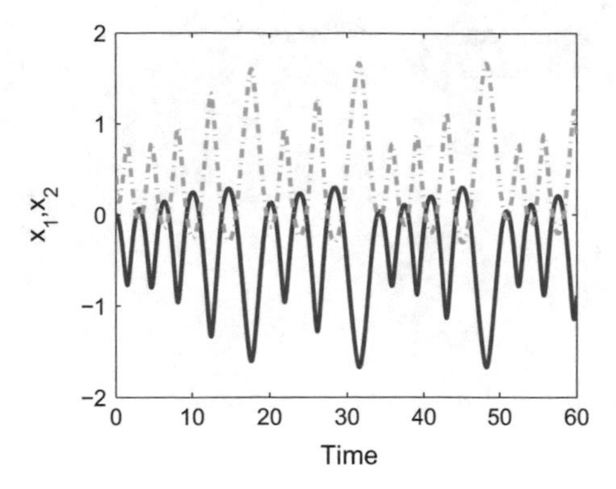

Fig. 9 Anti-synchronization of the states $x_1(t)$ and $x_2(t)$

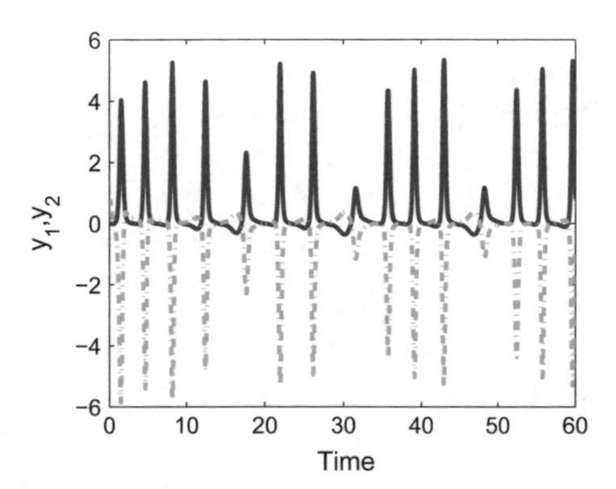

Fig. 10 Anti-synchronization of the states $y_1(t)$ and $y_2(t)$

It is easy to see that when adaptive control law (15) and the update law for the parameter estimates (16) are applied, the anti-synchronization of the master (11) and slave system (12) occurred as illustrated in Figs. 9, 10 and 11. Time series of master states are denoted as blue solid lines while corresponding slave states are plotted as red dash-dot lines in such figures. Moreover, the time-history of the anti-synchronization errors e_x, e_y, and e_z is reported in Fig. 12. The anti-synchronization errors converge to the zero. Therefore the chaos anti-synchronization between the no-equilibrium systems is realized.

6 Electronic Circuit of the Proposed System

Implementation of theoretical chaotic model by electronic circuits is an approach to confirm the feasibility of the theoretical one [2, 64, 78, 97]. In this section, we choose integrator synthesis to synthesize a circuit from the differential equations in system (2) as shown in Fig. 13.

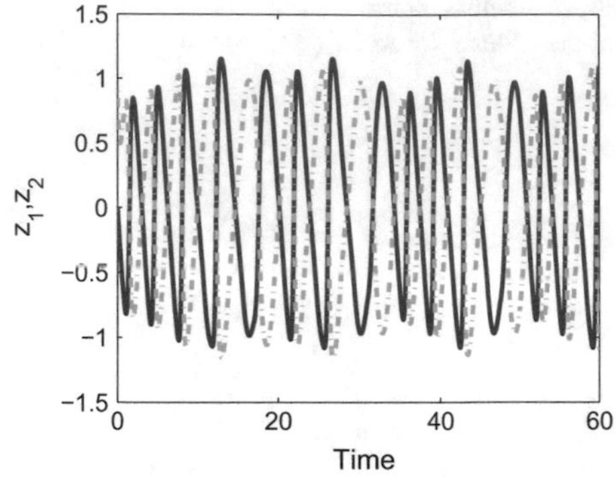

Fig. 11 Anti-synchronization of the states $z_1(t)$ and $z_2(t)$

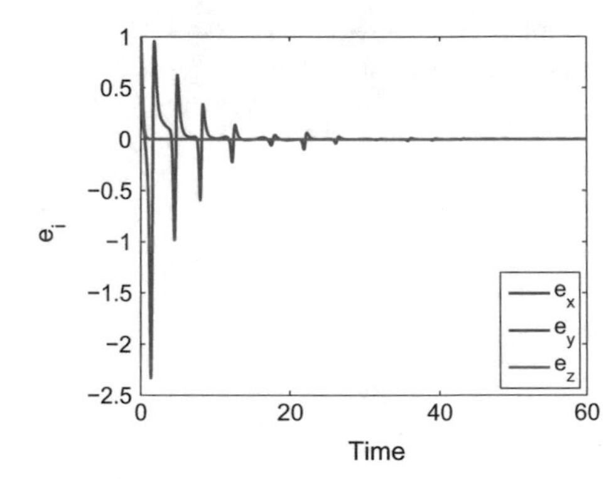

Fig. 12 Time series of the anti-synchronization errors e_x, e_y, and e_z

As seen in Fig. 13, there are only some basic blocks such as integrators, summing amplifiers, multipliers or absolute value blocks. These blocks have been realized easily by electronic components (resistors, capacitors, operational amplifiers, analog multipliers). As a result, the circuit have been implemented in PSpice as illustrated in Fig. 14. Signals in the circuit are measured at the outputs of inverting integrators. Figures 15, 16, 17 present the obtained PSpice results. The designed circuit emulates well the theoretical model.

7 Fractional Order Form of the No-Equilibrium System

As have been known that practical models such as heat conduction, electrode-electrolyte polarization, electronic capacitors, dielectric polarization, viso-elastic systems are more adequately described by the fractional-order different equations [9, 30, 37, 77, 81, 102]. Adams-Bashforth-Mounlton numerical algorithm is often used to investigate fractional-order differential equations [21, 25, 80]. Here we present this algorithm briefly.

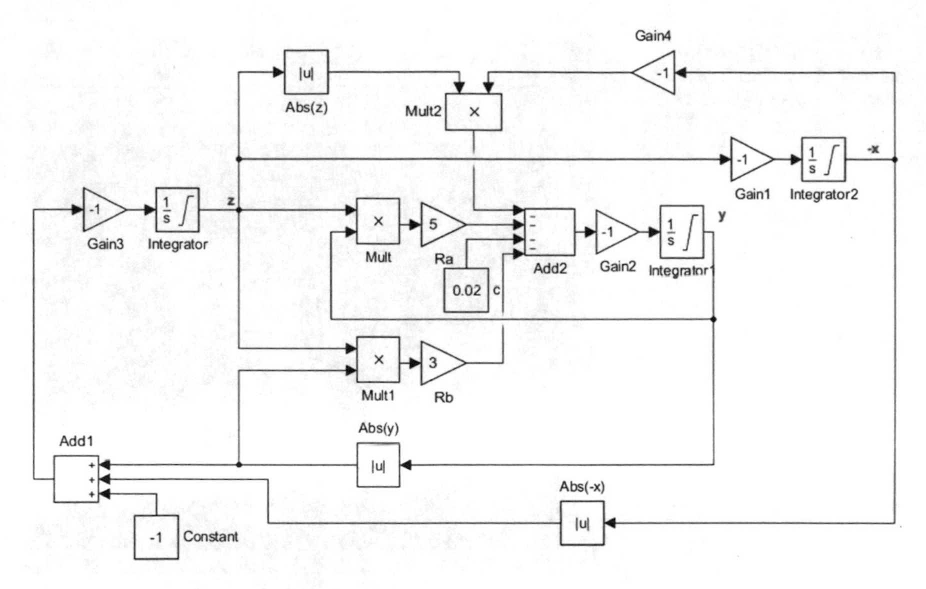

Fig. 13 Block schematic to synthesize the circuit of system (2)

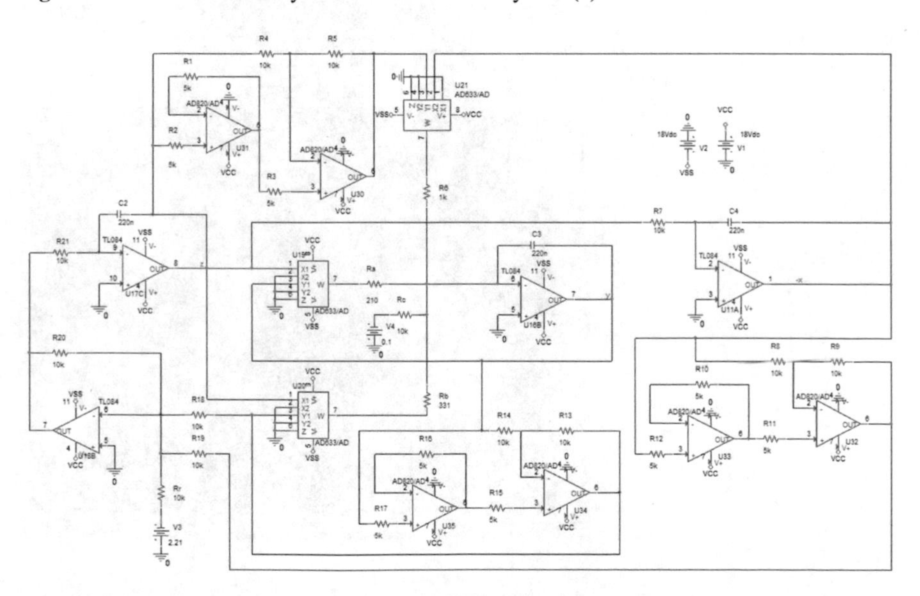

Fig. 14 Observation of the electronic circuit implemented by using PSpice

We consider the fractional-order differential equation as follows:

$$\begin{cases} \frac{d^q x(t)}{dt^q} = f\left(t, x\left(t\right)\right), & 0 \leq t \leq T, \\ x^{(i)}\left(0\right) = x_0^{(i)} & i = 0, 1, ..., m-1, \end{cases} \tag{25}$$

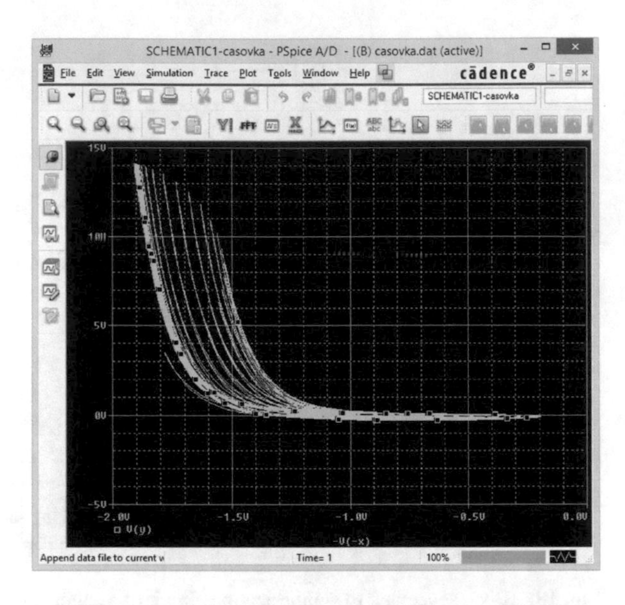

Fig. 15 PSpice phase portrait of the circuit in $-V(-x) - V(y)$ plane

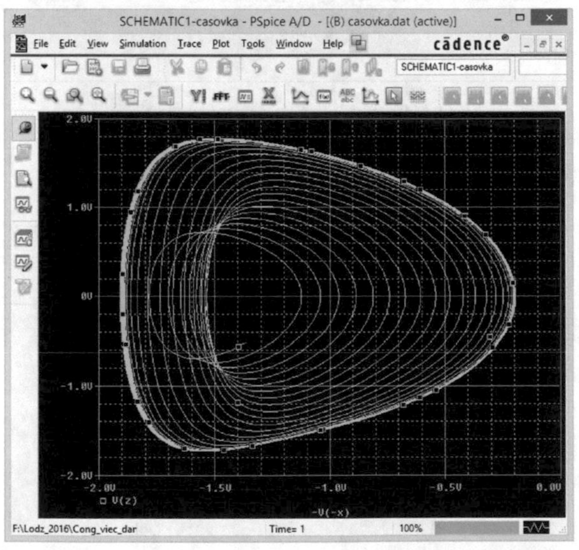

Fig. 16 PSpice phase portrait of the circuit in $-V(-x) - V(z)$ plane

where $m - 1 < q \leq m \in Z^{+}$. Equation (25) is equivalent to the following Volterra integral equation:

$$x(t) = \sum_{i=0}^{m-1} \frac{t^i}{i!} x_0^{(i)} + \frac{1}{\Gamma(q)} \int_0^t (t - \tau)^{q-1} f(\tau, x(\tau)) \, d\tau, \tag{26}$$

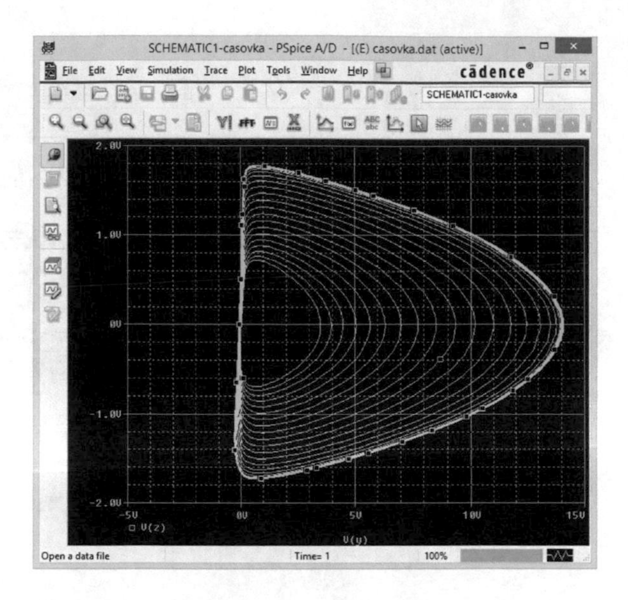

Fig. 17 PSpice phase portrait of the circuit in $V(y) - V(z)$ plane

in which the Gamma function $\Gamma(.)$ is defined as

$$\Gamma(q) = \int_0^\infty e^{-t} t^{q-1} dt. \tag{27}$$

We set $h = \frac{T}{N}, N \in Z^+$, and $t_n = nh \quad (n = 0, 1, ..., N)$. So we can discrete Eq. (26) as follows

$$\begin{aligned} x_h\left(t_{n+1}\right) = &\sum_{i=0}^{m-1} \frac{t_{n+1}^i}{i!} x_0^{(i)} + \frac{h^q}{\Gamma(q+2)} f\left(t_{n+1}, x_h^p\left(t_{n+1}\right)\right) \\ &+ \frac{h^q}{\Gamma(q+2)} \sum_{j=0}^{n} a_{j,n+1} f\left(t_j, x_h\left(t_j\right)\right), \end{aligned} \tag{28}$$

where

$$a_{j,n+1} = \begin{cases} n^{q+1} - (n-q)(n+1)^q, & \text{if } j = 0, \\ (n-j+2)^{q+1} + (n-j)^{q+1} & \\ -2(n-j+1)^{q+1}, & \text{if } 1 \le j \le n, \\ 1, & \text{if } j = n+1. \end{cases} \tag{29}$$

It is noting that the predicted value $x_h^p\left(t_{n+1}\right)$ is calculated as

$$x_h^p\left(t_{n+1}\right) = \sum_{i=0}^{m-1} \frac{t_{n+1}^i}{i!} x_0^{(i)} + \frac{1}{\Gamma(q)} \sum_{j=0}^{n} b_{j,n+1} f\left(t_j, x_h\left(t_j\right)\right), \tag{30}$$

in which

$$b_{j,n+1} = \frac{h^q}{q}\left((n+1-j)^q - (n-j)^q\right), \quad 0 \le j \le n. \tag{31}$$

Here the estimation error e in the method is given by

$$e = \max\left|x\left(t_j\right) - x_h\left(t_j\right)\right| = O\left(h^p\right) \quad (j = 0, 1, ..., N), \tag{32}$$

with $p = \min(2, 1+q)$.

Existence of chaos in fractional-order systems are investigated [28, 29, 53, 107]. In this section, we consider the fractional-order from of the no-equilibrium system which is described as

$$\begin{cases} \frac{d^q x(t)}{dt^q} = z \\ \frac{d^q y(t)}{dt^q} = -z\left(ay + b\left|y\right|\right) - x\left|z\right| - c \\ \frac{d^q z(t)}{dt^q} = \left|x\right| + \left|y\right| - 1, \end{cases} \tag{33}$$

where a, b, c are three positive parameters and $c \neq 0$ for the commensurate order $0 < q \le 1$. Fractional-order system (33) has been studied by applying Adams-Bashforth-Mounlton numerical algorithm [21, 25, 80]. It is interesting that chaos exists in fractional-order system (33). Figures 18, 19, 20 display chaotic attractors generated from fractional-order system (33) for the commensurate order $q = 0.999$, the parameters $a = 5$, $b = 3$, $c = 0.02$ and the initial conditions

$$\left(x(0), y(0), z(0)\right) = (0, 0, 0). \tag{34}$$

However, when decreasing the value of the commensurate order i.e. $q = 0.995$, fractional-order system (33) generates limit cycles as illustrated in Fig. 21.

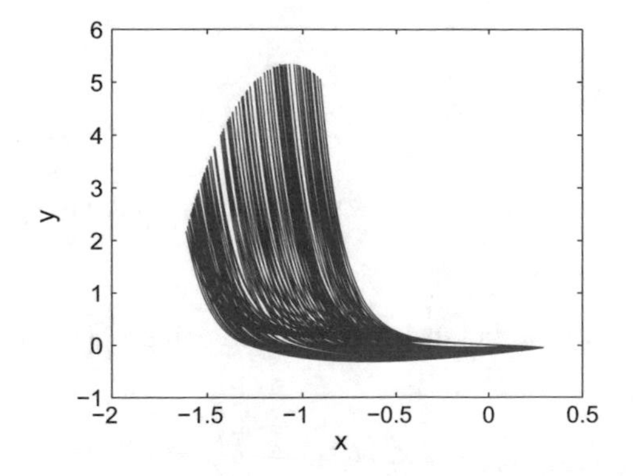

Fig. 18 2-D projection of the fractional-order system (33) in the (x, y)-plane

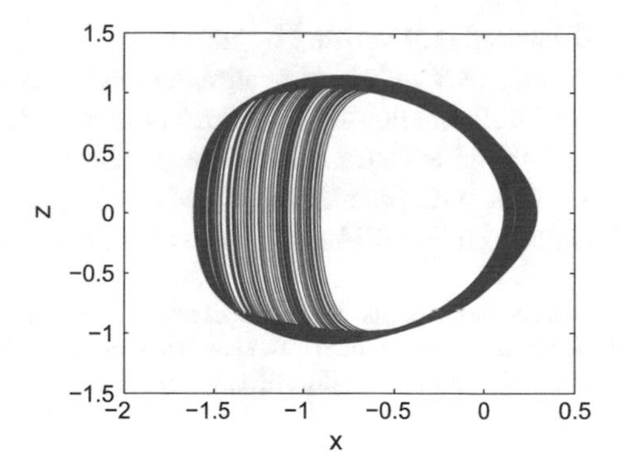

Fig. 19 2-D projection of the fractional-order system (33) in the (x, z)-plane

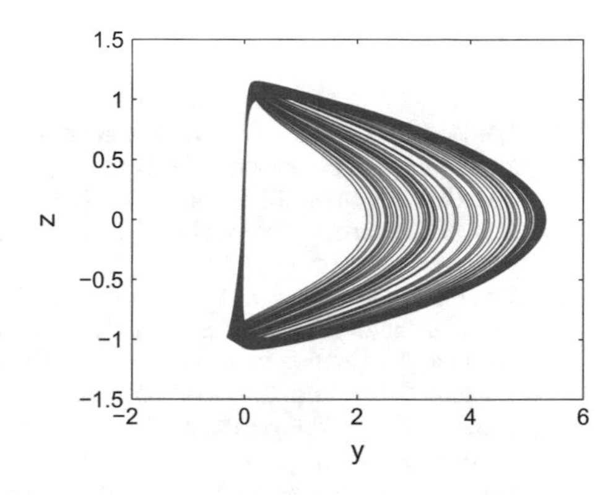

Fig. 20 2-D projection of the fractional-order system (33) in the (y, z)-plane

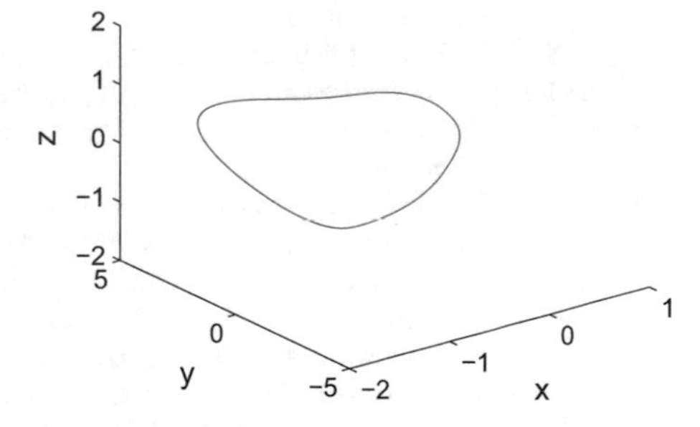

Fig. 21 3-D projection of the fractional-order system (33) in the (x, y, z)-space for the commensurate order $q = 0.995$, the parameters $a = 5$, $b = 3$, $c = 0.02$, and the initial conditions $(x(0), y(0), z(0)) = (0, 0, 0)$

8 Conclusion

A new three-dimensional autonomous system is proposed in this chapter. This system can exhibit chaotic attractors with square equilibrium and without equilibrium. As a result, such system is considered as a system with "hidden attractor". Fundamental dynamical properties of the introduced system are investigated through calculating equilibrium points, phase portraits of chaotic attractors, Poincaré map, bifurcation diagram, largest Lyapunov exponents and Kaplan-Yorke dimension. Moreover, synchronization and electronic implementation of our novel system are discussed and verified by numerical examples. This work is not only to present a new system with hidden attractors but also to extend the knowledge about systems with different families of hidden attractors. Other chaotic systems with different families of hidden attractors will be presented in our next researches. In addition, further studies about potential applications of such system in secure communications and cryptography will be done in our future works.

Acknowledgements Research described in this paper was supported by Czech Ministry of Education in frame of National Sustainability Program under grant GA15-22712S. V.-T. Pham is grateful to Le Thi Van Thu, Philips Electronics—Vietnam, for her help.

References

1. Aguilar-Lopez, R., Martinez-Guerra, R., & Perez-Pinacho, C. (2014). Nonlinear observer for synchronization of chaotic systems with application to secure data transmission. *The European Physical Journal Special Topics*, *223*, 1541–1548.
2. Akgul, A., Moroz, I., Pehlivan, I., & Vaidyanathan, S. (2016). A new four-scroll chaotic attractor and its engineering applications. *Optik*, *127*, 5491–5499.
3. Akopov, A., Astakhov, V., Vadiasova, T., Shabunin, A., & Kapitaniak, T. (2005). Frequency synchronization in clusters in coupled extended systems. *Physics Letters A*, *334*, 169–172.
4. Arneodo, A., Coullet, P., & Tresser, C. (1981). Possible new strange attractors with spiral structure. *Communications in Mathematical Physics*, *79*, 573–579.
5. Azar, A. T., & Vaidyanathan, S. (2015). *Chaos modeling and control systems design*. Germany: Springer.
6. Azar, A. T., & Vaidyanathan, S. (2015). *Computational intelligence applications in modeling and control*. Germany: Springer.
7. Azar, A. T., & Vaidyanathan, S. (2015). *Handbook of research on advanced intelligent control engineering and automation*. USA: IGI Global.
8. Azar, A. T., & Vaidyanathan, S. (2016). *Advances in chaos theory and intelligent control*. Germany: Springer.
9. Bagley, R. L., & Calico, R. A. (1991). Fractional-order state equations for the control of viscoelastically damped structers. *Journal of Guidance, Control, and Dynamics*, *14*, 304–311.
10. Bao, B., Zou, X., Liu, Z., & Hu, F. (2013). Generalized memory element and chaotic memory system. *International Journal of Bifurcation and Chaos*, *23*, 1350135.
11. Barnerjee, T., Biswas, D., & Sarkar, B. C. (2012). Design and analysis of a first order time-delayed chaotic system. *Nonlinear Dynamics*, *70*, 721–734.
12. Behnia, S., Pazhotan, Z., Ezzati, N., & Akhshani, A. (2014). Reconfiguration chaotic logic gates based on novel chaotic circuit. *Chaos, Solitons and Fractals*, *69*, 74–80.
13. Boccaletti, S., Kurths, J., Osipov, G., Valladares, D. L., & Zhou, C. S. (2002). The synchronization of chaotic systems. *Physics Reports*, *366*, 1–101.
14. Boulkroune, A., Bouzeriba, A., Bouden, T., & Azar, A. T. (2016). Fuzzy adaptive synchronization of uncertain fractional-order chaotic systems. In A. T. Azar & S. Vaidyanathan (Eds.), *Advances in chaos theory and intelligent control* (Vol. 337, pp. 681–697). Studies in fuzziness and soft computing. Springer: Germany.
15. Boulkroune, A., Hamel, S., & Azar, A. T. (2016). Fuzzy control-based function synchronization of unknown chaotic systems with dead-zone input. In A. T. Azar & S. Vaidyanathan (Eds.), *Advances in chaos theory and intelligent control* (Vol. 337, pp. 699–718). Studies in fuzziness and soft computing. Springer: Germany.
16. Brezetskyi, S., Dudkowski, D., & Kapitaniak, T. (2015). Rare and hidden attractors in van der pol-duffing oscillators. *The European Physical Journal Special Topics*, *224*, 1459–1467.

17. Buscarino, A., Fortuna, L., & Frasca, M. (2009). Experimental robust synchronization of hyperchaotic circuits. *Physica D, 238*, 1917–1922.
18. Chen, G., & Ueta, T. (1999). Yet another chaotic attractor. *International Journal of Bifurcation and Chaos, 9*, 1465–1466.
19. Chen, G., & Yu, X. (2003). *Chaos control: Theory and applications*. Berlin: Springer.
20. Chenaghlu, M. A., & Khasmakhi, S. J. N. N. (2016). A novel keyed parallel hashing scheme based on a new chaotic system. *Chaos, Solitons and Fractals, 87*, 216–225.
21. Diethelm, K., Ford, N. J., & Freed, A. D. (2002). A predictor-corrector approach for the numerical solution of fractional differential equations. *Nonlinear Dynamics, 29*, 3–22.
22. Fortuna, L., & Frasca, M. (2007). Experimental synchronization of single-transistor-based chaotic circuits. *Chaos, 17*, 043118-1–5.
23. Frederickson, P., Kaplan, J. L., Yorke, E. D., & York, J. (1983). The Lyapunov dimension of strange attractors. *Journal of Differential Equations, 49*, 185–207.
24. Gamez-Guzman, L., Cruz-Hernandez, C., Lopez-Gutierrez, R., & Garcia-Guerrero, E. E. (2009). Synchronization of Chua's circuits with multi scroll attractors: Application to communication. *Communications in Nonlinear Science and Numerical Simulation, 14*, 2765–2775.
25. Gejji, D., & Jafari, H. (2005). A domian decomposition: A tool for solving a system of fractional differential equations. *Journal of Mathematical Analysis and Applications, 301*, 508–518.
26. Gotthans, T., & Petržela, J. (2015). New class of chaotic systems with circular equilibrium. *Nonlinear Dynamics, 73*, 429–436.
27. Gotthans, T., Sportt, J. C., & Petržela, J. (2016). Simple chaotic flow with circle and square equilibrium. *International Journal of Bifurcation and Chaos, 26*, 1650137.
28. Grigorenko, I., & Grigorenko, E. (2003). Chaotic dynamics of the fractional-order lorenz system. *Physical Review Letters, 91*, 034101.
29. Hartley, T. T., Lorenzo, C. F., & Qammer, H. K. (1995). Chaos on a fractional Chua's system. *IEEE Transactions on Circuits and Systems I: Fundamental Theory and Applications, 42*, 485–490.
30. Heaviside, O. (1971). *Electromagnetic theory*. New York, USA: Academic Press.
31. Hoang, T. M., & Nakagawa, M. (2007). Anticipating and projective–anticipating synchronization of coupled multidelay feedback systems. *Physics Letters A, 365*, 407–411.
32. Hu, J., Chen, S., & Chen, L. (2005). Adaptive control for anti-synchronization of Chua's chaotic system. *Physics Letters A, 339*, 455–460.
33. Huang, Y., Wang, Y., Chen, H., & Zhang, S. (2016). Shape synchronization control for three-dimensional chaotic systems. *Chaos, Solitons and Fractals, 87*, 136–145.
34. Jafari, S., & Sprott, J. C. (2013). Simple chaotic flows with a line equilibrium. *Chaos, Solitons and Fractals, 57*, 79–84.
35. Jafari, S., Sprott, J. C., & Golpayegani, S. M. R. H. (2013). Elementary quadratic chaotic flows with no equilibria. *Physics Letters A, 377*, 699–702.
36. Jafari, S., Sprott, J. C., & Nazarimehr, F. (2015). Recent new examples of hidden attractors. *The European Physical Journal Special Topics, 224*, 1469–1476.
37. Jenson, V. G., & Jeffreys, G. V. (1997). *Mathematical methods in chemical engineering*. New York, USA: Academic Press.
38. Kajbaf, A., Akhaee, M. A., & Sheikhan, M. (2016). Fast synchronization of non-identical chaotic modulation-based secure systems using a modified sliding mode controller. *Chaos, Solitons and Fractals, 84*, 49–57.
39. Kapitaniak, T. (1994). Synchronization of chaos using continuous control. *Physical Review E, 50*, 1642–1644.
40. Karthikeyan, R., & Vaidyanathan, S. (2014). Hybrid chaos synchronization of four-scroll systems via active control. *Journal of Electrical Engineering, 65*, 97–103.
41. Khalil, H. (2002). *Nonlinear systems*. New Jersey, USA: Prentice Hall.
42. Kim, C. M., Rim, S., Kye, W. H., Ryu, J. W., & Park, Y. J. (2003). Anti-synchronization of chaotic oscillators. *Physics Letters A, 320*, 39–46.

43. Kingni, S. T., Jafari, S., Simo, H., & Woafo, P. (2014). Three-dimensional chaotic autonomous system with only one stable equilibrium: Analysis, circuit design, parameter estimation, control, synchronization and its fractional-order form. *The European Physical Journal Plus*, *129*, 76.
44. Koupaei, J. A., & Hosseini, S. M. M. (2015). A new hybrid algorithm based on chaotic maps for solving systems of nonlinear equations. *Chaos, Solitons and Fractals*, *81*, 233–245.
45. Kuznetsov, N. V., Leonov, G. A., & Seledzhi, S. M. (2011). Hidden oscillations in nonlinear control systems. *IFAC Proceedings, 18*, 2506–2510.
46. Leonov, G. A., & Kuznetsov, N. V. (2011). Algorithms for searching for hidden oscillations in the Aizerman and Kalman problems. *Doklady Mathematics, 84*, 475–481.
47. Leonov, G. A., & Kuznetsov, N. V. (2011). Analytical–numerical methods for investigation of hidden oscillations in nonlinear control systems. *IFAC Proceedings, 18*, 2494–2505.
48. Leonov, G. A., & Kuznetsov, N. V. (2013). Hidden attractors in dynamical systems: From hidden oscillation in Hilbert-Kolmogorov, Aizerman and Kalman problems to hidden chaotic attractor in Chua circuits. *International Journal of Bifurcation and Chaos*, *23*, 1330002.
49. Leonov, G. A., Kuznetsov, N. V., Kiseleva, M. A., Solovyeva, E. P., & Zaretskiy, A. M. (2014). Hidden oscillations in mathematical model of drilling system actuated by induction motor with a wound rotor. *Nonlinear Dynamics*, *77*, 277–288.
50. Leonov, G. A., Kuznetsov, N. V., Kuznetsova, O. A., Seldedzhi, S. M., & Vagaitsev, V. I. (2011). Hidden oscillations in dynamical systems. *Transmission Systems Control, 6*, 54–67.
51. Leonov, G. A., Kuznetsov, N. V., & Vagaitsev, V. I. (2011). Localization of hidden Chua's attractors. *Physics Letters A, 375*, 2230–2233.
52. Leonov, G. A., Kuznetsov, N. V., and Vagaitsev, V. I. (2012). Hidden attractor in smooth Chua system. *Physica D, 241*, 1482–1486.
53. Li, C. P., & Peng, G. J. (2004). Chaos in Chen's system with a fractional-order. *Chaos, Solitons and Fractals*, *20*, 443–450.
54. Lorenz, E. N. (1963). Deterministic non-periodic flow. *Journal of Atmospheric Science*, *20*, 130–141.
55. Lü, J., & Chen, G. (2002). A new chaotic attractor coined. *International Journal of Bifurcation and Chaos*, *12*, 659–661.
56. Molaei, M., Jafari, S., Sprott, J. C., & Golpayegani, S. (2013). Simple chaotic flows with one stable equilibrium. *International Journal of Bifurcation and Chaos*, *23*, 1350188.
57. Ojoniyi, O. S., & Njah, A. N. (2016). A 5D hyperchaotic Sprott B system with coexisting hidden attractor. *Chaos, Solitons and Fractals*, *87*, 172–181.
58. Orlando, G. (2016). A discrete mathematical model for chaotic dynamics in economics: Kaldor's model on business cycle. *Mathematics and Computers in Simulation*, *125*, 83–98.
59. Pecora, L. M., & Carroll, T. L. (1990). Synchronization in chaotic signals. *Physical Review A*, *64*, 821–824.
60. Pham, V.-T., Jafari, S., Volos, C., Wang, X., & Golpayegani, S. M. R. H. (2014). Is that really hidden? The presence of complex fixed-points in chaotic flows with no equilibria. *International Journal of Bifurcation and Chaos*, *24*, 1450146.
61. Pham, V.-T., Vaidyanathan, S., Volos, C. K., Hoang, T. M., & Yem, V. V. (2016). Dynamics, synchronization and SPICE implementation of a memristive system with hidden hyperchaotic attractor. In A. T. Azar & S. Vaidyanathan (Eds.), *Advances in chaos theory and intelligent control*. Studies in fuzziness and soft computing (Vol. 337, pp. 35–52). Germany: Springer.
62. Pham, V. T., Vaidyanathan, S., Volos, C. K., & Jafari, S. (2015). Hidden attractors in a chaotic system with an exponential nonlinear term. *The European Physical Journal Special Topics*, *224*, 1507–1517.
63. Pham, V.-T., Volos, C. K., Jafari, S., Wei, Z., & Wang, X. (2014). Constructing a novel no-equilibrium chaotic system. *International Journal of Bifurcation and Chaos*, *24*, 1450073.
64. Pham, V. T., Volos, C. K., Vaidyanathan, S., Le, T. P., & Vu, V. Y. (2015). A memristor-based hyperchaotic system with hidden attractors: Dynamics, sychronization and circuital emulating. *Journal of Engineering Science and Technology Review*, *8*, 205–214.

65. Rosenblum, M. G., Pikovsky, A. S., & Kurths, J. (1997). From phase to lag synchronization in coupled chaotic oscillators. *Physical Review Letters, 78*, 4193–4196.
66. Rössler, O. E. (1976). An equation for continuous chaos. *Physics Letters A, 57*, 397–398.
67. Sastry, S. (1999). *Nonlinear systems: Analysis, stability, and control*. USA: Springer.
68. Shahzad, M., Pham, V. T., Ahmad, M. A., Jafari, S., & Hadaeghi, F. (2015). Synchronization and circuit design of a chaotic system with coexisting hidden attractors. *The European Physical Journal Special Topics, 224*, 1637–1652.
69. Sharma, P. R., Shrimali, M. D., Prasad, A., Kuznetsov, N. V., & Leonov, G. A. (2015). Control of multistability in hidden attractors. *The European Physical Journal Special Topics, 224*, 1485–1491.
70. Soriano-Sanchez, A. G., Posadas-Castillo, C., Platas-Garza, M. A., & Diaz-Romero, D. A. (2015). Performance improvement of chaotic encryption via energy and frequency location criteria. *Mathematics and Computers in Simulation, 112*, 14–27.
71. Sprott, J. C. (2003). *Chaos and times-series analysis*. Oxford: Oxford University Press.
72. Sprott, J. C. (2010). *Elegant chaos. Algebraically simple chaotic flows*. Singapore: World Scientific.
73. Sprott, J. C. (2015). Strange attractors with various equilibrium types. *The European Physical Journal Special Topics, 224*, 1409–1419.
74. Srinivasan, K., Senthilkumar, D. V., Murali, K., Lakshmanan, M., & Kurths, J. (2011). Synchronization transitions in coupled time-delay electronic circuits with a threshold nonlinearity. *Chaos, 21*, 023119.
75. Stefanski, A., Perlikowski, P., & Kapitaniak, T. (2007). Ragged synchronizability of coupled oscillators. *Physical Review E, 75*, 016210.
76. Strogatz, S. H. (1994). *Nonlinear dynamics and chaos: With applications to physics, biology, chemistry, and engineering*. Massachusetts: Perseus Books.
77. Sun, H. H., Abdelwahad, A. A., & Onaral, B. (1894). Linear approximation of transfer function with a pole of fractional-order. *IEEE Transactions on Automatic Control, 29*, 441–444.
78. Tacha, O. I., Volos, C. K., Kyprianidis, I. M., Stouboulos, I. N., Vaidyanathan, S., & Pham, V. T. (2016). Analysis, adaptive control and circuit simulation of a novel nonlineaar finance system. *Applied Mathematics and Computation, 276*, 200–217.
79. Tang, Y., Wang, Z., & Fang, J. A. (2010). Image encryption using chaotic coupled map lattices with time-varying delays. *Communications in Nonlinear Science and Numerical Simulation, 15*, 2456–2468.
80. Tavazoei, M. S., & Haeri, M. (2008). Limitations of frequency domain approximation for detecting chaos in fractional-order systems. *Nonlinear Analysis, 69*, 1299–1320.
81. Tavazoei, M. S., & Haeri, M. (2009). A proof for non existence of periodic solutions in time invariant fractional-order systems. *Automatica, 45*, 1886–1890.
82. Vaidyanathan, S. (2012). Anti-synchronization of four-wing chaotic systems via sliding mode control. *International Journal of Automation and Computing, 9*, 274–279.
83. Vaidyanathan, S. (2013). A new six-term 3-D chaotic system with an exponential nonlineariry. *Far East Journal of Mathematical Sciences, 79*, 135–143.
84. Vaidyanathan, S. (2014). Analysis and adaptive synchronization of eight-term novel 3-D chaotic system with three quadratic nonlinearities. *The European Physical Journal Special Topics, 223*, 1519–1529.
85. Vaidyanathan, S. (2016). Analysis, control and synchronization of a novel 4-D highly hyperchaotic system with hidden attractors. In A. T. Azar & S. Vaidyanathan (Eds.), *Advances in chaos theory and intelligent control* (Vol. 337, pp. 529–552). Studies in fuzziness and soft computing. Springer: Germany.
86. Vaidyanathan, S., & Azar, A. T. (2015). Anti-synchronization of identical chaotic systems using sliding mode control and an application to Vaidhyanathan-Madhavan chaotic systems. *Studies in Computational Intelligence, 576*, 527–547.
87. Vaidyanathan, S., & Azar, A. T. (2015). Hybrid synchronization of identical chaotic systems using sliding mode control and an application to Vaidhyanathan chaotic systems. *Studies in Computational Intelligence, 576*, 549–569.

88. Vaidyanathan, S., & Azar, A. T. (2016). A novel 4-D four-wing chaotic system with four quadratic nonlinearities and its synchronization via adaptive control method. In A. T. Azar & S. Vaidyanathan (Eds.), *Advances in chaos theory and intelligent control*. Studies in fuzziness and soft computing (Vol. 337, pp. 203–224). Germany: Springer.
89. Vaidyanathan, S., & Azar, A. T. (2016). Adaptive backstepping control and synchronization of a novel 3-D jerk system with an exponential nonlinearity. In A. T. Azar & S. Vaidyanathan (Eds.), *Advances in chaos theory and intelligent control*. Studies in fuzziness and soft computing (Vol. 337, pp. 249–274). Germany: Springer.
90. Vaidyanathan, S., & Azar, A. T. (2016). Adaptive control and synchronization of Halvorsen circulant chaotic systems. In A. T. Azar & S. Vaidyanathan (Eds.), *Advances in chaos theory and intelligent control*. Studies in fuzziness and soft computing (Vol. 337, pp. 225–247). Germany: Springer.
91. Vaidyanathan, S., & Azar, A. T. (2016). Dynamic analysis, adaptive feedback control and synchronization of an eight-term 3-D novel chaotic system with three quadratic nonlinearities. In A. T. Azar & S. Vaidyanathan (Eds.), *Advances in chaos theory and intelligent control*. Studies in fuzziness and soft computing (Vol. 337, pp. 155–178). Germany: Springer.
92. Vaidyanathan, S., & Azar, A. T. (2016). Generalized projective synchronization of a novel hyperchaotic four-wing system via adaptive control method. In A. T. Azar & S. Vaidyanathan (Eds.), *Advances in chaos theory and intelligent control*. Studies in fuzziness and soft computing (Vol. 337, pp. 275–296). Germany: Springer.
93. Vaidyanathan, S., & Azar, A. T. (2016). Qualitative study and adaptive control of a novel 4-D hyperchaotic system with three quadratic nonlinearities. In A. T. Azar & S. Vaidyanathan (Eds.), *Advances in chaos theory and intelligent control*. Studies in fuzziness and soft computing (Vol. 337, pp. 179–202). Germany: Springer.
94. Vaidyanathan, S., Idowu, B. A., & Azar, A. T. (2015). Backstepping controller design for the global chaos synchronization of Sprott's jerk systems. *Studies in Computational Intelligence, 581*, 39–58.
95. Vaidyanathan, S., Pham, V. T., & Volos, C. K. (2015). A 5-d hyperchaotic rikitake dynamo system with hidden attractors. *The European Physical Journal Special Topics, 224*, 1575–1592.
96. Vaidyanathan, S., Volos, C., Pham, V. T., Madhavan, K., & Idowo, B. A. (2014). Adaptive backstepping control, synchronization and circuit simualtion of a 3-D novel jerk chaotic system with two hyperbolic sinusoidal nonlinearities. *Archives of Control Sciences, 33*, 257–285.
97. Vaidyanathan, S., Volos, C. K., & Pham, V. T. (2015). Analysis, control, synchronization and spice implementation of a novel 4-d hyperchaotic rikitake dynamo system without equilibrium. *Journal of Engineering Science and Technology Review, 8*, 232–244.
98. Volos, C. K., Kyprianidis, I. M., & Stouboulos, I. N. (2011). Various synchronization phenomena in bidirectionally coupled double scroll circuits. *Communications in Nonlinear Science and Numerical Simulation, 71*, 3356–3366.
99. Wang, X., & Chen, G. (2012). A chaotic system with only one stable equilibrium. *Communications in Nonlinear Science and Numerical Simulation, 17*, 1264–1272.
100. Wang, X., & Chen, G. (2013). Constructing a chaotic system with any number of equilibria. *Nonlinear Dynamics, 71*, 429–436.
101. Wei, Z. (2011). Dynamical behaviors of a chaotic system with no equilibria. *Physics Letters A, 376*, 102–108.
102. Westerlund, S., & Ekstam, L. (1994). Capacitor theory. *IEEE Transactions on Dielectrics and Electrical Insulation, 1*, 826–839.
103. Woafo, P., & Kadji, H. G. E. (2004). Synchronized states in a ring of mutually coupled self-sustained electrical oscillators. *Physical Review E, 69*, 046206.
104. Wolf, A., Swift, J. B., Swinney, H. L., & Vastano, J. A. (1985). Determining Lyapunov exponents from a time series. *Physica D, 16*, 285–317.
105. Xiao-Yu, D., Chun-Biao, L., Bo-Cheng, B., & Hua-Gan, W. (2015). Complex transient dynamics of hidden attractors in a simple 4d system. *Chinese Physics B, 24*, 050503.

106. Yalcin, M. E., Suykens, J. A. K., & Vandewalle, J. (2005). *Cellular neural networks: Multi-scroll chaos and synchronization*. Singapore: World Scientific.
107. Yang, Q. G., & Zeng, C. B. (2010). Chaos in fractional conjugate lorenz system and its scaling attractor. *Communications in Nonlinear Science and Numerical Simulation*, *15*, 4041–4051.
108. Zhang, Y., & Sun, J. (2004). Chaotic synchronization and anti-synchronization based on suitable separation. *Physics Letters A*, *330*, 442–447.
109. Zhu, Q., & Azar, A. T. (2015). *Complex system modelling and control through intelligent soft computations*. Germany: Springer.
110. Zhusubaliyev, Z. T., & Mosekilde, E. (2015). Multistability and hidden attractors in a multi-level DC/DC converter. *Mathematics and Computers in Simulation*, *109*, 32–45.
111. Zhusubaliyev, Z. T., Mosekilde, E., Churilov, A., & Medvedev, A. (2015). Multistability and hidden attractors in an impulsive Goodwin oscillator with time delay. *The European Physical Journal Special Topics*, *224*, 1519–1539.
112. Zhusubaliyev, Z. T., Mosekilde, E., Rubanov, V. G., & Nabokov, R. A. (2015). Multistability and hidden attractors in a relay system with hysteresis. *Physica D*, *306*, 6–15.

A Study on Coexistence of Different Types of Synchronization Between Different Dimensional Fractional Chaotic Systems

Adel Ouannas, Ahmad Taher Azar, Toufik Ziar and Ahmed G. Radwan

Abstract In this study, robust approaches are proposed to investigate the problem of the coexistence of various types of synchronization between different dimensional fractional chaotic systems. Based on stability theory of linear fractional order systems, the co-existence of full state hybrid function projective synchronization (FSHFPS), inverse generalized synchronization (IGS), inverse full state hybrid projective synchronization (IFSHPS) and generalized synchronization (GS) is demonstrated. Using integer-order Lyapunov stability theory and fractional Lyapunov method, the co-existence of FSHFPS, inverse full state hybrid function projective synchronization (IFSHFPS), IGS and GS is also proved. Finally, numerical results are reported, with the aim to illustrate the capabilities of the novel schemes proposed herein.

Keywords Fractional chaos · Hybrid synchronization · Inverse full state hybrid projective synchronization · Fractional Lyapunov method

---

A. Ouannas
Laboratory of Mathematics Informatics and Systems (LAMIS),
University of Larbi Tebessi, 12002 Tebessa, Algeria
e-mail: ouannas.a@yahoo.com

A.T. Azar (✉)
Faculty of Computers and Information, Benha University, Banha, Egypt
e-mail: ahmad_t_azar@ieee.org; ahmad.azar@fci.bu.edu.eg

A.T. Azar
Nanoelectronics Integrated Systems Center (NISC), Nile University, Cairo, Egypt

T. Ziar
Department of Material Sciences, University of Tebessa, 12002 Tebessa, Algeria
e-mail: Toufik1_ziar@yahoo.fr

A.G. Radwan
Nanoelectronics Integrated Systems Center (NISC), Nile University, Giza, Egypt
e-mail: agradwan@ieee.org

A.G. Radwan
Engineering Mathematics and Physics Department, Cairo University, Cairo, Egypt

A.T. Azar et al. (eds.), *Fractional Order Control and Synchronization of Chaotic Systems*, Studies in Computational Intelligence 688,
DOI 10.1007/978-3-319-50249-6_22

1 Introduction

Fractional order calculus theory is one of the hottest topics of discussion in recent years and it's importance has been documented extensively [1–5]. However, the applications of fractional calculus [6–34], are just a recent focus of interest. It was found that many nonlinear fractional order differential systems that exhibit chaotic or hyperchaotic behavior are interesting as its study links between the science and nature [35–53]. Scientists who inquire into the laws of nature have struggled to achieve the synchronization of fractional order chaotic and hyperchaotic systems, mainly due to its potential applications especially in information processing and secure communications [54–58].

Recently there are many recent publications that discuss the analog and digital realizations of chaotic systems as well as some hardware applications. Some of them are focused on the generalization of the conventional chaotic maps such as the generalized logistic map [59] where extra degrees of freedom have been added for more control. Other realization are based on transistor-level realization of some famous chaotic generators such as Lorenz and Chua's circuit which are suitable for on-chip fabrication [60–62]. Moreover, the concept of mixed analog-digital system design has been introduced to build different chaotic generators such as [63] where digital counter can control the number of scrolls as well as their locations. In addition due to the extensive need of digital applications, more multi-scroll chaotic generators as well as pseudo random number generator (PRNG) have been introduced totally based on digital designs and FPGA for encryption applications [64, 65].

In most of the chaos synchronization approaches, the master-slave or drive-response formalism is used. If a particular chaotic system is called the master or drive system and another chaotic system is called the slave or response system, then the idea of synchronization is to use the output of the master system to control the slave system so that the output of the slave system tracks the output of the master system asymptotically. Since the seminal work by Pecora and Carroll [66], a variety of impressive approaches and applications have been proposed for the synchronization of the chaotic systems [67–88].

Synchronization of fractional chaotic systems becomes a challenging and interesting problem [89–100]. The research on fractional order chaos synchronization can be classified into two main directions (i) analysis and (ii) synthesis. However, the problem of synchronization analysis consists of understanding and/or giving theoretical description of synchronization, and there exist many types of synchronization such as phase synchronization [101, 102], complete synchronization [103], anti-synchronization [104], projective synchronization [105, 106], hybrid projective synchronization [107, 108], generalized projective synchronization [109, 110], function projective synchronization [111, 112], generalized synchronization [113, 114], Q-S synchronization [115], full state hybrid projective synchronization [116–118], finite-time synchronization [119], impulsive synchronization [120, 121] and exponential synchronization [122]. On the other hand the problem of synchronization synthesis concerns on finding or designing a synchronization control, such that two chaotic systems exhibit different types of synchronization behaviors. Until now, a wide variety of fractional techniques have been used to design a synchronization control, for example, sliding-mode control [123–125], linear control [126], nonlinear control [127], active control [128], adaptive control [129, 130], feedback control [131, 132], scalar signal technique [133, 134].

In full state hybrid projective synchronization (FSHPS), each slave system state achieves synchronization with linear constant combination of master system states [135–138]. When the linear constant combinations in FSHPS are replaced by linear function combinations, an interesting type of synchronization that may occur is the full state hybrid function projective synchronization (FSHFPS) [139]. Recently, an interesting type of synchronization has been introduced, in which each slave system state synchronizes with a linear function combination of slave system states. The proposed scheme is called inverse full state hybrid function projective synchronization (IFSHFPS). A particular case is represented, when the scaling functions are constants, for which the inverse full state hybrid projective synchronization (IFSHPS) is obtained [140].

In generalized synchronization (GS), the master system and the slave system are nonidentical dynamical systems. This type of synchronization is characterized by the existence of a functional relationship ϕ between the state X of the master system and the state Y of the slave system, so that $Y = \phi(X)$ after a transient time. Nowadays numerous researches of GS of fractional chaotic systems have been done theoretically and experimentally [141–143]. Recently, another interesting synchronization type was appeared called inverse generalized synchronization (IGS), where the synchronization condition becomes $X = \varphi(Y)$, where φ is a differentiable function, after a transient time. The relevant researches for IGS in fractional chaotic systems still in an initial stage. However, IGS was applied successfully in continuous-time and discrete-time chaotic systems with integer-order [144, 145].

When studying the synchronization of chaotic systems, an interesting phenomenon that may occur is the co-existence of several synchronization types. Recently, there are many papers studing the problem of co-existence of different types of chaos synchronization. For example, the co-existence of PS, FSHPS and GS between hyperchaotic maps was studied in [146]. A general control scheme was proposed in [147], to study the co-existence of inverse projective synchronization (IPS), IGS and Q-S synchronization between arbitrary 3D hyperchaotic maps. For integer-order chaotic systems, two synchronization schemes of co-existence has been introduced in [148]. The co-existence of FSHPS and IFSHPS between different dimensional incommensurate fractional order chaotic systems was presented in [149]. In [150], a robust method was applied to study coexistence of GS and IGS in fractional chaotic system with different dimensions. The co-existence of synchronization types is very useful in secure communication and chaotic encryption schemes.

Based on these considerations, this work presents new approaches to rigorously study the co-existence of some chaos synchronization types between fractional dynamical systems with different dimensions. By exploiting stability theory of fractional-order differential equations, the study first analyzes the coexistence of FSHFPS, IGS, IFSHPS and GS when the master system is a three-dimensional incommensurate fractional system and the slave system is a four-dimensional incommensurate fractional system. Successively, the co-existence of four different synchronization types is illustrated, i.e., FSHFPS, IFSHFPS, IGS and GS are proved to co-exist between three-dimensional incommensurate fractional master system and four-dimensional commensurate fractional slave system. Numerical examples of co-existence of synchronization types are illustrated, with the aim to show the effectiveness of the novel approaches developed herein.

The rest of this study is organized as follows. In Sect. 2, some preliminaries of fractional-order calculus and stability results of fractional systems are provided. In Sect. 3, the definitions of FSHFPS, IFSHFPS, GS and IGS are presented. Our main synchronization approaches are introduced in Sects. 4 and 5. Finally, Sect. 6 draws some concluding remarks.

2 Preliminaries

2.1 *Fractional Calculus*

There are several definitions of a fractional derivative [1, 151]. The two most commonly used are the Riemann-Liouville and Caputo definitions. Each definition uses Riemann-Liouville fractional integration and derivatives of whole order. The difference between the two definitions is in the order of evaluation. The Riemann-Liouville fractional integral operator of order $p > 0$ of the function $f(t)$ is defined as,

$$J^p f(t) = \frac{1}{\Gamma(p)} \int_0^t (t-\tau)^{p-1} f(\tau) d\tau, \qquad t > 0. \tag{1}$$

where Γ denotes Gamma function. Some properties of the operator J^p can be found, for example, in [2, 4]. We recall only the following, for $p, q > 0$ and $\gamma > -1$, we have,

$$J^p J^q f(t) = J^{p+q} f(t), \tag{2}$$

$$J^p t^\gamma = \frac{\Gamma(\gamma+1)}{\Gamma(p+\gamma+1)} t^{p+\gamma}. \tag{3}$$

In this study, Caputo definition is used and the fractional derivative of $f(t)$ is defined as,

$$D_t^p f(t) = J^{m-p}\left(\frac{d^m}{dt^m} f(t)\right) = \frac{1}{\Gamma(m-p)} \int_0^t \frac{f^{(m)}(\tau)}{(t-\tau)^{p-m+1}} d\tau, \tag{4}$$

for $m - 1 < p \leq m$, $m \in N$, $t > 0$.

The fractional differential operator D_t^p is left-inverse (and not right-inverse) to the fractional integral operator J^p, i.e. $D_t^p J^p = I$ where I is the identity operator.

The Laplace transform of the Caputo fractional derivative rule reads

$$\mathbf{L}\left(D_t^p f(t)\right) = s^p \mathbf{F}(s) - \sum_{k=0}^{n-1} s^{p-k-1} f^{(k)}(0), \quad (p > 0,\ n-1 < p \leq n). \tag{5}$$

Particularly, when $p \in \left]0, 1\right]$, we have $\mathbf{L}\left(D_t^p f(t)\right) = s^p \mathbf{F}(s) - s^{p-1} f(0)$.

The Laplace transform of the Riemann-Liouville fractional integral rule satisfies

$$\mathbf{L}(J^p f(t)) = s^{-p} \mathbf{F}(s), \quad (p > 0). \tag{6}$$

Caputo fractional derivative appears more suitable to be treated by the Laplace transform technique in that it requires the knowledge of the (bounded) initial values of the function and of its integer derivatives of order $k = 1, 2, \ldots, m-1$, in analogy with the case when $p = n$.

Lemma 1 [105] *Suppose $f(t)$ has a continuous kth derivative on $[0, t]$ ($k \in N$, $t > 0$), and let $p, q > 0$ be such that there exists some $\ell \in N$ with $\ell \leq k$ and $p, p+q \in [\ell - 1, \ell]$. Then*

$$D_t^p D_t^q f(t) = D_t^{p+q} f(t). \tag{7}$$

Remark 1 Note that the condition requiring the existence of the number ℓ with the above restrictions in the property is essential. In this study, we consider the case that $p, q \in \left]0, 1\right]$ and $p + q \in \left]0, 1\right]$. Apparently, under such conditions this property holds.

2.2 *Stability of Linear Fractional Systems*

Consider the following linear fractional system

$$D_t^p X(t) = AX(t), \tag{8}$$

where $D_t^p = \left[D_t^{p_1}, D_t^{p_2}, \ldots, D_t^{p_n}\right]$ is the Caputo fractional derivative, p_i are different rational numbers between 0 and 1, for $i = 1, 2, \ldots, n$. Assume that $p_i = \frac{\alpha_i}{\beta_i}$, $\left(\alpha_i, \beta_i\right) = 1$, $\alpha_i, \beta_i \in \mathbb{N}$, for $i = 1, 2, \ldots, n$. Let M be the least common multiple of the denominators β_i's of p_i's.

Lemma 2 [152] *The system (8) is asymptotically stable if all roots λ of the equation*

$$\det\left(diag\left(\lambda^{Mp_1}, \lambda^{Mp_2}, \ldots, \lambda^{Mp_n}\right) - A\right) = 0, \tag{9}$$

satisfy $|\arg(\lambda)| > \frac{\pi}{2M}$.

2.3 Fractional Lyapunov Method

For stability analysis of fractional order systems, a fractional-order extension of Lyapunov direct method has been proposed.

Definition 1 A continuous function γ is said to belong to class-K if it is strictly increasing and $\gamma(0) = 0$.

Theorem 1 [153] *Let $X = 0$ be an equilibrium point for the following fractional order system*

$$D_t^p X(t) = F(X(t)), \tag{10}$$

where $p \in \left]0, 1\right]$. Assume that there exists a Lyapunov function $V(X(t))$ and class-K functions γ_i $(i = 1, 2, 3)$ satisfying

$$\gamma_1(\|X\|) \leq V(X(t)) \leq \gamma_2(\|X\|), \tag{11}$$

$$D_t^p V(X(t)) \leq -\gamma_3(\|X\|). \tag{12}$$

Then the system (10) is asymptotically stable.

From Theorem 1, we can come to the following theorem.

Theorem 2 [154] If there exists a positive definite Lyapunov function $V(X(t))$ such that $D_t^p V(X(t)) < 0$, for all $t > 0$, then the trivial solution of system (10) is asymptotically stable.

Lemma 3 [155] $\forall X(t) \in \mathbf{R}^n$, $\forall p \in \left]0, 1\right]$ *and* $\forall t > 0$

$$\frac{1}{2} D_t^p \left(X^T(t) X(t) \right) \leq X^T(t) D_t^p (X(t)). \tag{13}$$

3 Definitions of FSHFPS, IFSHPS, IFSHFPS, GS and IGS

Consider the following master system

$$D_t^{p_i} x_i(t) = F_i(X(t)), \quad i = 1, 2, ..., n, \tag{14}$$

where $X(t) = \left(x_1(t), x_2(t), ..., x_n(t) \right)^T$ is the state vector of the master system, $F_i : \mathbf{R}^n \to \mathbf{R}$, $0 < p_i \leq 1$, $D_t^{p_i}$ is the Caputo fractional derivative of order p_i.

The slave system is described by the following fractional system

$$D_t^{q_i} y_i(t) = G_i(Y(t)) + u_i, \quad i = 1, 2, ..., m, \tag{15}$$

where $Y(t) = (y_1(t), y_2(t), ..., y_m(t))^T$ is the state vector of the slave system, $G_i : \mathbf{R}^m \to \mathbf{R}$, $0 < q_i \leq 1$, $D_t^{q_i}$ is the Caputo fractional derivative of order q_i and u_i, $i = 1, 2, ..., m$, are synchronization controllers.

Definition 2 The master system (14) and the slave system (15) are said to be full state hybrid function projective synchronized (FSHFPS), if there exist controllers u_i, $i = 1, 2, ..., m$, and given differentiable functions $\alpha_j(t)$, $j = 1, 2, ..., n$, such that the synchronization errors

$$e_i(t) = y_i(t) - \sum_{j=1}^{n} \alpha_j(t) x_j(t), \quad i = 1, 2, ..., m, \tag{16}$$

satisfy that $\lim_{t\to\infty} e_i(t) = 0$.

Definition 3 The master system (14) and the slave system (15) are said to be inverse full state hybrid function projective synchronized (IFSHFPS), if there exist controllers u_i, $i = 1, 2, ..., m$, and given differentiable functions $\beta_j(t)$, $j = 1, 2, ..., m$, such that the synchronization errors

$$e_i(t) = \sum_{j=1}^{m} \beta_j(t) y_i(t) - x_i(t), \quad i = 1, 2, ..., n, \tag{17}$$

satisfy that $\lim_{t\to\infty} e_i(t) = 0$.

Definition 4 The master system (14) and the slave system (15) are said to be inverse full state hybrid projective synchronized (IFSHPS), if there exist controllers u_i, $i = 1, 2, ..., m$, and given real numbers β_i, $i = 1, 2, ..., m$, such that the synchronization errors

$$e_i(t) = \sum_{j=1}^{m} \beta_j y_i(t) - x_i(t), \quad i = 1, 2, ..., n, \tag{18}$$

satisfy that $\lim_{t\to\infty} e_i(t) = 0$.

Definition 5 The master system (14) and the slave system (15) are said to be generalized synchronized (GS), if there exist controllers u_i, $i = 1, 2, ..., m$, and given differentiable functions $\phi_j : \mathbf{R}^n \to \mathbf{R}$, $j = 1, 2, ..., m$, such that the synchronization errors

$$e_i(t) = y_i(t) - \phi_i(X(t)), \quad i = 1, 2, ..., m, \tag{19}$$

satisfy that $\lim_{t\to\infty} e_i(t) = 0$.

Definition 6 The master system (14) and the slave system (15) are said to be inverse generalized synchronized (IGS), if there exist controllers u_i, $i = 1, 2, ..., m$, and given differentiable functions $\varphi_i : \mathbf{R}^m \to \mathbf{R}$, $i = 1, 2, ..., n$, such that the synchronization errors

$$e_i(t) = \varphi_i(Y(t)) - x_i(t), \quad i = 1, 2, ..., n, \tag{20}$$

satisfy that $\lim_{t\to\infty} e_i(t) = 0$.

4 Coexistence of FSHFPS, IGS, IFSHPS and GS

4.1 Systems Description and Problem Statement

Consider the following master system

$$\begin{aligned} D_t^{p_1} x_1(t) &= f_1(X(t)), \\ D_t^{p_2} x_2(t) &= f_2(X(t)), \\ D_t^{p_3} x_3(t) &= f_3(X(t)), \end{aligned} \tag{21}$$

where $X(t) = (x_1(t), x_2(t), x_3(t))^T$ is the state vector of the master system (21), $f_i : \mathbf{R}^3 \to \mathbf{R}$, $i = 1, 2, 3$, $0 < p_i \leq 1$ for $i = 1, 2, 3$, $D_t^{p_i}$ is the Caputo fractional derivative of order p_i.

Suppose that the slave system is in the following form

$$\begin{aligned} D_t^{q_1} y_1(t) &= \sum_{j=1}^{4} b_{1j} y_j(t) + g_1(Y(t)) + u_1, \\ D_t^{q_2} y_2(t) &= \sum_{j=1}^{4} b_{2j} y_j(t) + g_2(Y(t)) + u_2, \\ D_t^{q_3} y_3(t) &= \sum_{j=1}^{4} b_{3j} y_j(t) + g_3(Y(t)) + u_3, \\ D_t^{q_4} y_4(t) &= \sum_{j=1}^{4} b_{4j} y_j(t) + g_4(Y(t)) + u_4, \end{aligned} \tag{22}$$

where $Y(t) = (y_1(t), y_2(t), y_3(t), y_4(t))^T$ is the state vector of the slave system (22), $(b_{ij}) \in \mathbf{R}^{4\times4}$, $g_i : \mathbf{R}^4 \to \mathbf{R}$, $i = 1, 2, 3, 4$, are nonlinear function, $0 < q_i \leq 1$ for $i = 1, 2, 3, 4$, $D_t^{q_i}$ is the Caputo fractional derivative of order q_i and u_i, $i = 1, 2, 3, 4$, are controllers to be determined.

The problem of the coexistence of FSHFPS, IGS, IFSHPS and GS between the master system (21) and the slave system (22) is formulated by the following definition.

Definition 7 We say that FSHFPS, IGS, IFSHPS and GS co-exist in the synchronization of the master system (21) and the slave system (22), if there exists controllers u_i, $i = 1, 2, 3, 4$, differentiable functions $\left(\alpha_j(t)\right)_{1 \le j \le 3}$, differentiable function $\varphi : \mathbf{R}^4 \to \mathbf{R}$, real numbers $\left(\beta_j\right)_{1 \le j \le 3}$ and differentiable function $\phi : \mathbf{R}^3 \to \mathbf{R}$, such that the synchronization errors

$$
\begin{aligned}
e_1(t) &= y_1(t) - \sum_{j=1}^{3} \alpha_j(t) x_j(t), \\
e_2(t) &= \varphi(Y(t)) - x_2(t), \\
e_3(t) &= \sum_{j=1}^{4} \beta_j y_j(t) - x_3, \\
e_4(t) &= y_4(t) - \phi(X(t)),
\end{aligned}
\tag{23}
$$

satisfy that $\lim_{t\to\infty} e_i(t) = 0$, $i = 1, 2, 3, 4$.

4.2 Analytical Results

The error system between the master system (21) and the slave system (22), can be derived as follows

$$
\begin{aligned}
D_t^{q_1} e_1(t) &= D_t^{q_1} y_1(t) - D_t^{q_1}\left(\sum_{j=1}^{3} \alpha_j(t) x_j(t)\right), \\
\dot{e}_2(t) &= \frac{\partial \varphi}{\partial y_2} \dot{y}_2(t) + \sum_{\substack{j=1 \\ j\neq 2}}^{4} \frac{\partial \varphi}{\partial y_j} \dot{y}_j - \dot{x}_2(t), \\
D_t^{q_3} e_3(t) &= \beta_3 D_t^{q_3} y_3(t) + \sum_{\substack{j=1 \\ j\neq 3}}^{4} \beta_j D_t^{q_3} y_j(t) - D_t^{q_3} x_3(t), \\
D_t^{q_4} e_4(t) &= D_t^{q_4} y_4(t) - D_t^{q_4}(\phi(X(t))).
\end{aligned}
\tag{24}
$$

We suppose that the controllers u_i, $i = 1, 2, 3, 4$, can be designed in the following forms

$$
\begin{aligned}
u_1 &= v_1, \\
u_2 &= -\sum_{j=1}^{4} b_{2j} y_j(t) - g_2(Y(t)) + J^{1-q_2}(v_2), \\
u_3 &= v_3, \\
u_4 &= v_4,
\end{aligned}
\tag{25}
$$

where v_i, $1 \le i \le 4$, are new controllers to be determined later.

By substituting (25) into (22), we can rewrite the slave system as

$$D_t^{q_i} y_i(t) = \sum_{j=1}^{4} b_{ij} y_j(t) + g_i(Y(t)) + v_i, \quad i = 1, 3, 4, \tag{26}$$

and

$$D_t^{q_2} y_i(t) = J^{1-q_2}\left(v_2\right). \tag{27}$$

By applying the Caputo fractional derivative of order $1 - q_2$, to both the left and right sides of Eq. (27), we obtain

$$\begin{aligned} \dot{y}_2(t) &= D_t^{1-q_2}\left(D_t^{q_2} y_2(t)\right) \\ &= D_t^{1-q_2} J^{1-q_2}\left(v_2\right) \\ &= v_2. \end{aligned} \tag{28}$$

Note that $1 - q_2$ satisfies $1 - q_2 \in (0, 1]$. According to Lemma 1 the above statement holds.

Furthermore, the error system (24) can be written as

$$\begin{aligned} D_t^{q_1} e_1(t) &= -\left|b_{11}\right| e_1(t) + v_1 + R_1, \\ D_t^{q_3} e_3(t) &= -\left|b_{33}\right| e_3(t) + \beta_3 v_3 + R_3, \\ D_t^{q_4} e_4(t) &= -\left|b_{44}\right| e_4(t) + v_4 + R_4, \end{aligned} \tag{29}$$

and

$$\dot{e}_2(t) = \left(b_{22} - c\right) e_2(t) + \frac{\partial \varphi}{\partial y_2} v_2 + R_2, \tag{30}$$

where c is a control constant and

$$\begin{aligned} R_1 &= \left|b_{11}\right| e_1(t) + \sum_{j=1}^{4} b_{1j} y_j(t) + g_1(Y(t)) - D_t^{q_1}\left(\sum_{j=1}^{3} \alpha_j(t) x_j(t)\right), \\ R_2 &= \left(c - b_{22}\right) e_2(t) + \sum_{\substack{j=1 \\ j \neq 2}}^{4} \frac{\partial \varphi}{\partial y_j} \dot{y}_j - \dot{x}_2(t), \\ R_3 &= \left|b_{33}\right| e_3(t) + \sum_{j=1}^{4} \beta_3 b_{3j} y_j(t) + \beta_3 g_3(Y(t)) + \sum_{\substack{j=1 \\ j \neq 3}}^{4} \beta_j D_t^{q_3} y_j(t) - D_t^{q_3} x_3(t), \\ R_4 &= \left|b_{44}\right| e_4(t) + \sum_{j=1}^{4} b_{4j} y_j(t) + g_4(Y(t)) - D_t^{q_4}(\phi(X(t))). \end{aligned} \tag{31}$$

To achieve synchronization between the master system (21) and the slave system (22), we assume that $\frac{\partial \varphi}{\partial y_2}, \beta_3 \neq 0$. Hence, we have the following result.

Theorem 3 *The co-existence of FSHFPS, IFSHPS, IGS and GS between the master system (21) and the slave system (22) is achieved under the following control laws*

$$\begin{aligned} v_1 &= -R_1, \\ v_3 &= -\frac{R_3}{\beta_3}, \\ v_4 &= -R_4, \end{aligned} \tag{32}$$

and

$$v_2 = -\frac{R_2}{\frac{\partial \varphi}{\partial y_2}}. \tag{33}$$

Proof Firstly, by substituting Eq. (32) into Eq. (29), on can have

$$D_t^q e(t) = K e(t), \tag{34}$$

where $D_t^q e(t) = (D_t^{q1} e_1(t), D_t^{q3} e_3(t), D_t^{q4} e_4(t))$ and

$$K = \begin{pmatrix} -|b_{11}| & 0 & 0 \\ 0 & -|b_{33}| & 0 \\ 0 & 0 & -|b_{44}| \end{pmatrix}. \tag{35}$$

The roots of $det\left(diag\left(\lambda^{Mq_1}, \lambda^{Mq_3}, \lambda^{Mq_4}\right) - K\right) = 0$, where M is the least common multiple of the denominators of q_i's, $i = 1, 3, 4$, can be described as

$$\begin{aligned} \lambda_1 &= |b_{11}|^{\frac{1}{Mq_1}} \left(\cos \frac{\pi}{Mq_1} + \mathbf{i} \sin \frac{\pi}{Mq_1} \right), \\ \lambda_2 &= |b_{33}|^{\frac{1}{Mq_3}} \left(\cos \frac{\pi}{Mq_3} + \mathbf{i} \sin \frac{\pi}{Mq_3} \right), \\ \lambda_3 &= |b_{44}|^{\frac{1}{Mq_4}} \left(\cos \frac{\pi}{Mq_4} + \mathbf{i} \sin \frac{\pi}{Mq_4} \right). \end{aligned}$$

It is easy to see that $\arg\left(\lambda_i\right) = \frac{\pi}{q_i M} > \frac{\pi}{2M}$, $i = 1, 3, 4$. Then, according to Lemma 2, the zero solution of the system (34) is a globally asymptotically stable, i.e.

$$\lim_{t \to \infty} e_1(t) = \lim_{t \to \infty} e_3(t) = \lim_{t \to \infty} e_4(t) = 0. \tag{36}$$

Secondly, by applying the control law described by Eq. (33) to Eq. (30), the dynamics of error $e_2(t)$ can be written as

$$\dot{e}_2(t) = -\left(b_{22} - c\right) e_2(t). \tag{37}$$

If the control constant c is chosen such that $b_{22} - c < 0$, then we get

$$\lim_{t \to \infty} e_2(t) = 0. \tag{38}$$

Finally, from Eqs. (36) and (37), we conclude that the master system (21) and the slave system (22) are globally synchronized.

4.3 Simulation Results

Consider the incommensurate fractional order financial system studied by Chen in [156]

$$\begin{aligned} D^{p_1}x_1 &= x_3 + \left(x_2 - a\right) x_1, \\ D^{p_2}x_2 &= 1 - bx_2 - x_1^2, \\ D^{p_3}x_3 &= -x_1 - cx_3, \end{aligned} \tag{39}$$

as a master system. This system exhibits chaotic behaviors when $\left(p_1, p_2, p_3\right) = (0.97, 0.98, 0.99)$, $(a, b, c) = (1, 0.1, 1)$ and the initial point is $(2, -1, 1)$. Attractors of the master system (39) are shown in Figs. 1 and 2.

The slave system is

$$\begin{aligned} D^{q_1}y_1 &= 0.56y_1 - y_2 + u_1, \\ D^{q_2}y_2 &= y_1 - 0.1y_2y_3^2 + u_2, \\ D^{q_3}y_3 &= 4y_2 - y_3 - 6y_4 + u_3, \\ D^{q_4}y_4 &= 0.5y_3 + 0.8y_4 + u_4, \end{aligned} \tag{40}$$

where y_1, y_2, y_3, y_4 are states and u_1, u_2, u_3, u_4 are synchronization controllers. This system, as shown in [157], exhibits hyperchaotic behavior when $(u_1, u_2, u_3, u_4) = (0, 0, 0, 0)$ and $\left(q_1, q_1, q_1, q_1\right) = (0.98, 0.98, 0.95, 0.95)$. The attractors of the uncontrolled system (40) are shown in Figs. 3 and 4.

Using the notations presented in Sect. 4.1, the errors between the master system (39) and the slave system (40) are defined as follows

$$\begin{aligned} e_1 &= y_1 - \alpha_1(t) x_1 - \alpha_2(t) x_2 - \alpha_3(t) x_3, \\ e_2 &= \varphi\left(y_1, y_2, y_3, y_4\right) - x_2, \\ e_3 &= \beta_1 y_1 + \beta_2 y_2 + \beta_3 y_3 + \beta_4 y_4 - x_3, \\ e_4 &= y_4 - \phi\left(x_1, x_2, x_3\right), \end{aligned} \tag{41}$$

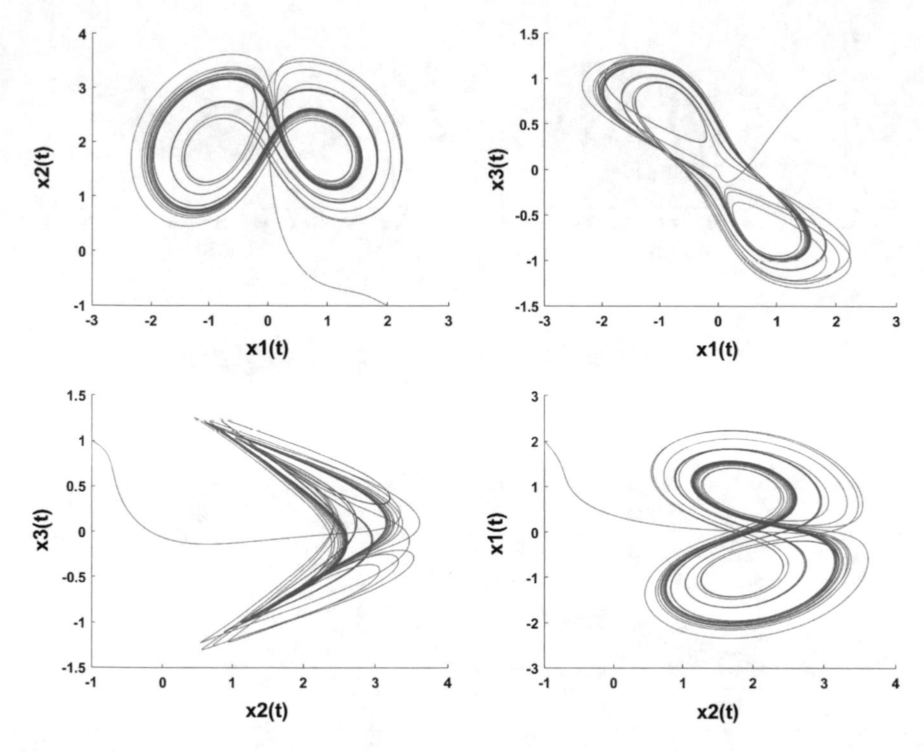

Fig. 1 Chaotic attractors in 2-D of the master system (39)

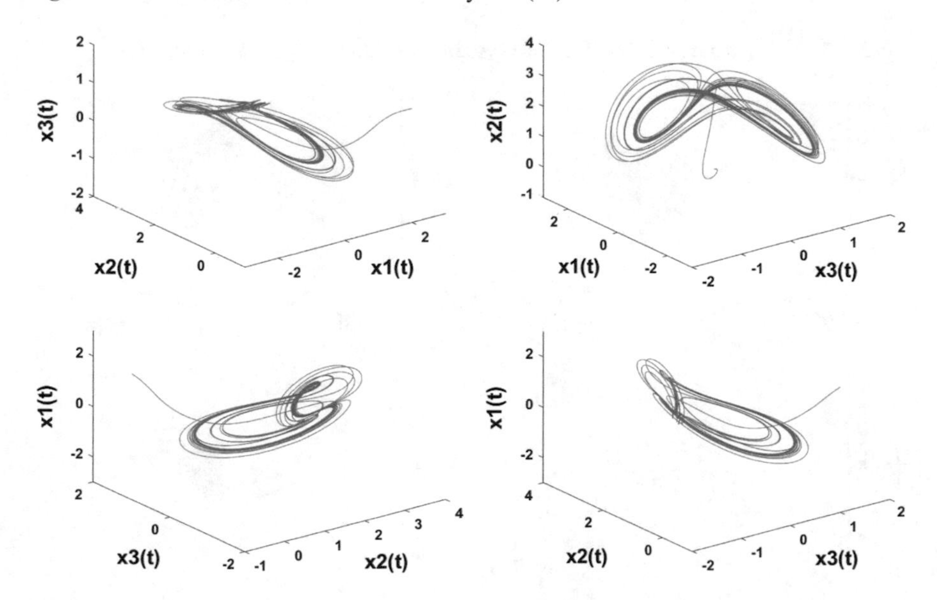

Fig. 2 Chaotic attractors in 3-D of the master system (39)

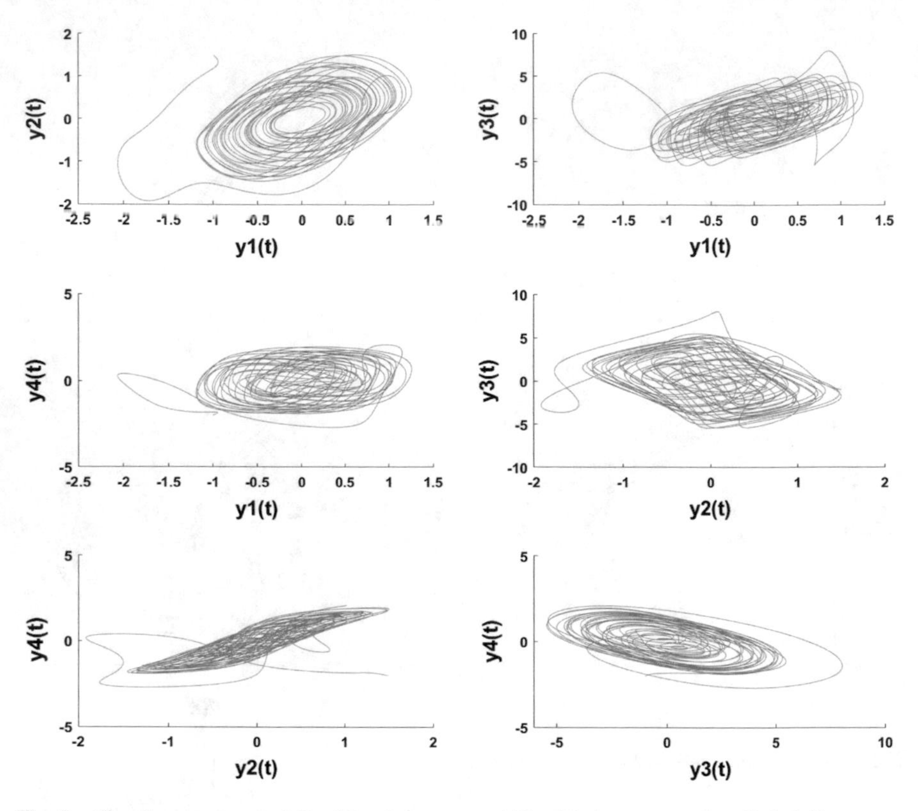

Fig. 3 Chaotic attractors in 2-D of the slave system (40) with $(u_1, u_2, u_3, u_4) = (0, 0, 0, 0)$

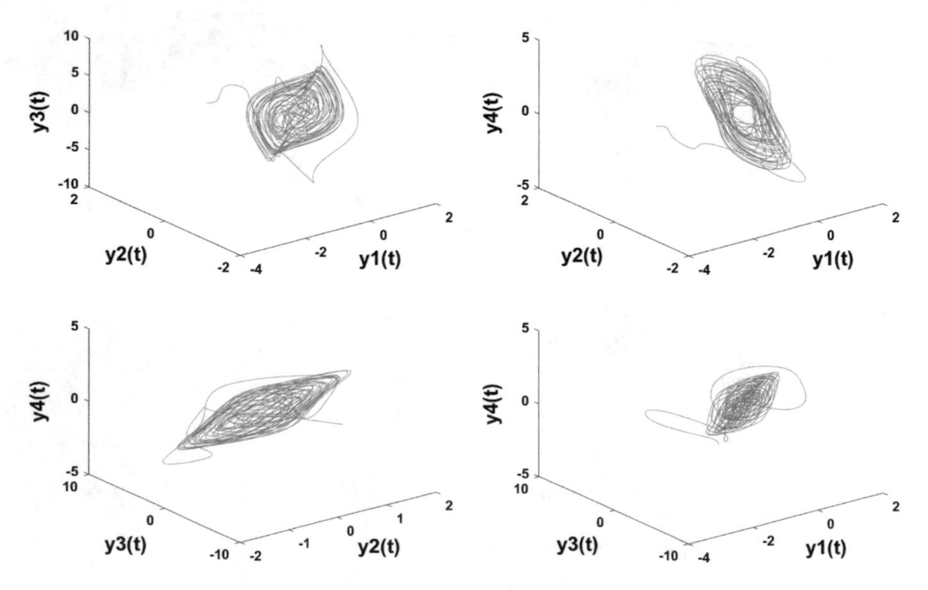

Fig. 4 Chaotic attractors in 3-D of the slave system (40) with $(u_1, u_2, u_3, u_4) = (0, 0, 0, 0)$

where

$$\begin{aligned}
(\alpha_1(t), \alpha_2(t), \alpha_3(t)) &= (t, t+2, t^2), \\
\varphi(y_1, y_2, y_3, y_4) &= y_2 + y_3 y_1 y_4, \\
(\beta_1, \beta_2, \beta_3, \beta_4) &= (1, 2, 3, 4), \\
\phi(x_1, x_2, x_3) &= x_1 x_2 + x_3.
\end{aligned}$$

So,

$$\frac{\partial}{\partial y_2}\varphi(y_1, y_2, y_3, y_4) = 1 \text{ and } \beta_3 = 3.$$

Then according to Theorem 3, the error system can be described as

$$\begin{aligned}
D^{0.98} e_1 &= -56 e_1, \\
D^{0.95} e_3 &= -e_3, \\
D^{0.95} e_4 &= -0.8 e_4,
\end{aligned} \tag{42}$$

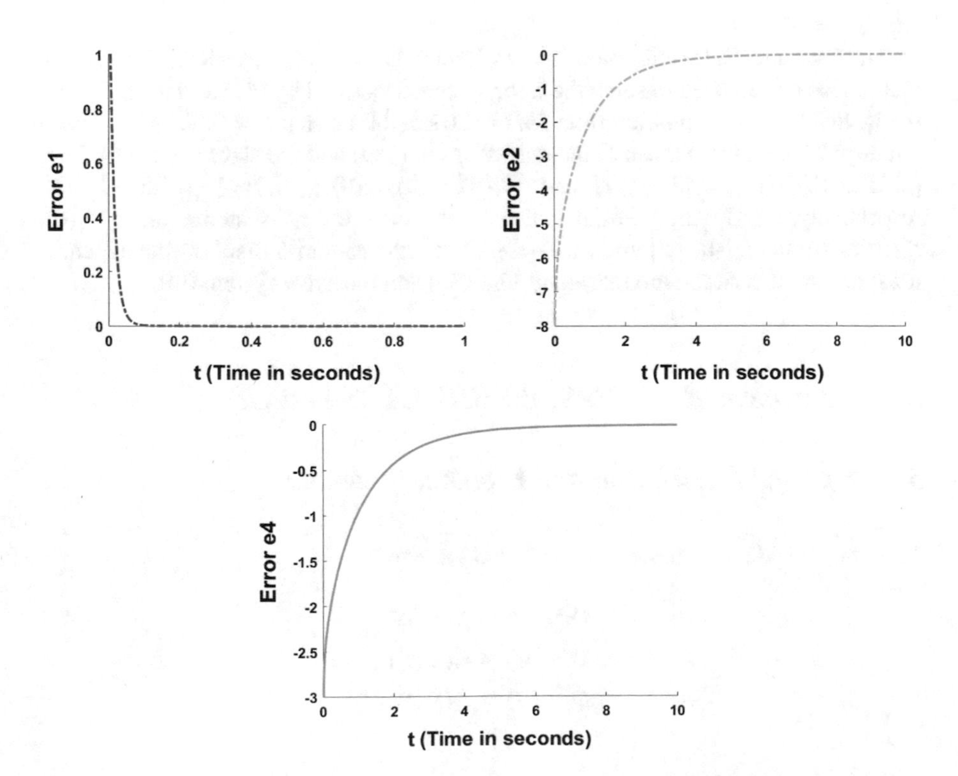

Fig. 5 Time evolution of the errors e_1, e_3 and e_4 the master system (39) and the slave system (40)

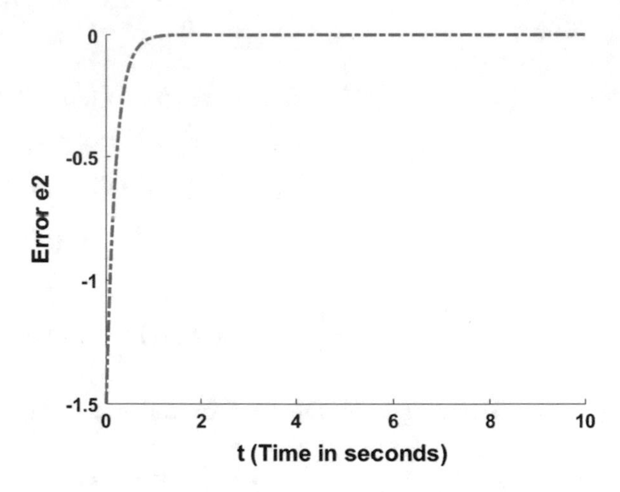

Fig. 6 Time evolution of the error e_2 between the master system (39) and the slave system (40)

and

$$\dot{e}_2 = ce_2, \tag{43}$$

where $c = -5$.

The fractional Euler integration method and fourth order Runge–Kutta integration method have been used to solve the error system described by (42) and (43), respectively. In addition, simulation time $Tm = 120$ s and time step $h = 0.005$ s have been employed. The initial values of the master system (39) and the slave system (40) are $[x_1(0), x_2(0), x_3(0)] = [2, -1, 1]$ and $[y_1(0), y_2(0), y_3(0), y_4(0)] = [-1, 1.5, -1, -2]$, respectively, and the initial states of the error system are $[e_1(0), e_3(0), e_4(0)] = [1, -8, -3]$ and $e_2(0) = -1.5$. Figures 5 and 6 displays the synchronization errors between the master system (39) and the slave system (40).

5 Coexistence of FSHFPS, IFSHFPS, IGS and GS

5.1 Systems Description and Problem Statement

Now, we consider the master and the slave systems as,

$$\begin{aligned} D_t^{p_1} x_1(t) &= f_1(X(t)), \\ D_t^{p_2} x_2(t) &= f_2(X(t)), \\ D_t^{p_3} x_3(t) &= f_3(X(t)), \end{aligned} \tag{44}$$

and

$$D_t^q y_1(t) = \sum_{j=1}^{4} b_{1j} y_j(t) + g_1(Y(t)) + u_1, \tag{45}$$

$$D_t^q y_2(t) = \sum_{j=1}^{4} b_{2j} y_j(t) + g_2(Y(t)) + u_2,$$

$$D_t^q y_3(t) = \sum_{j=1}^{4} b_{3j} y_j(t) + g_3(Y(t)) + u_3,$$

$$D_t^q y_4(t) = \sum_{j=1}^{4} b_{4j} y_j(t) + g_4(Y(t)) + u_4,$$

where $X(t) = \left(x_1(t), x_2(t), x_3(t)\right)^T$, $Y(t) = \left(y_1(t), y_2(t), y_3(t), y_4(t)\right)^T$ are the states vector of the master system (44) and the slave system (45), respectively, $f_i : \mathbf{R}^3 \to \mathbf{R}$, $i = 1, 2, 3$, $\left(b_{ij}\right) \in \mathbf{R}^{4\times 4}$, $g_i : \mathbf{R}^4 \to \mathbf{R}$, $i = 1, 2, 3, 4$, are nonlinear function, $0 < p_i \leq 1$ for $i = 1, 2, 3$, $0 < q \leq 1$, $D_t^{p_i}$, D_t^q are the Caputo fractional derivatives of orders p_i and q, respectively, and u_i, $i = 1, 2, 3, 4$, are controllers to be determined.

The definition of coexistence of FSHFPS, IFSHFPS, IGS and GS between the master system (44) and the slave system (45) is given by

Definition 8 We say that FSHFPS, IFSHFPS, IGS and GS co-exist in the synchronization of the master system (44) and the slave system (45), if there exists controllers u_i, $i = 1, 2, 3, 4$, given differentiable functions $\left(\alpha_j(t)\right)_{1\leq j\leq 3}$, $\left(\beta_j(t)\right)_{1\leq j\leq 4}$, and two differentiable functions $\varphi : \mathbf{R}^4 \to \mathbf{R}$, $\phi : \mathbf{R}^3 \to \mathbf{R}$, such that the synchronization errors

$$e_1(t) = y_1(t) - \sum_{j=1}^{3} \alpha_j(t) x_j(t), \tag{46}$$

$$e_2(t) = \sum_{j=1}^{4} \beta_j(t) y_j(t) - x_2(t),$$

$$e_3(t) = \varphi(Y(t)) - x_3,$$

$$e_4(t) = y_4(t) - \phi(X(t)),$$

satisfy that $\lim_{t\to\infty} e_i(t) = 0$, $i = 1, 2, 3, 4$.

5.2 Analytical Results

The error system (46) can be differentiated as follows:

$$
\begin{aligned}
D_t^q e_1(t) &= D_t^q y_1(t) - D_t^q \left(\sum_{j=1}^{3} \alpha_j(t) x_j(t) \right), \\
\dot{e}_2(t) &= \sum_{j=1}^{4} \dot{\beta}_j(t) y_j(t) + \sum_{j=1}^{4} \beta_j(t) \dot{y}_j(t) - \dot{x}_2(t), \\
\dot{e}_3(t) &= \sum_{j=1}^{4} \frac{\partial \varphi}{\partial y_j} \dot{y}_j - \dot{x}_3, \\
D_t^q e_4(t) &= D_t^q y_4(t) - D_t^q (\phi(X(t))).
\end{aligned}
\tag{47}
$$

In this case, the controllers u_i, $i = 1, 2, 3, 4$, can be constructed as

$$
\begin{aligned}
u_1 &= v_1, \\
u_2 &= -\sum_{j=1}^{4} b_{2j} y_j(t) - g_2(Y(t)) + J^{1-q_2}(v_2), \\
u_3 &= -\sum_{j=1}^{4} b_{3j} y_j(t) - g_3(Y(t)) + J^{1-q_3}(v_3), \\
u_4 &= v_4,
\end{aligned}
\tag{48}
$$

where v_i, $1 \le i \le 4$, are new controllers to be determined later.

From Eqs. (45)–(48), we get

$$
D_t^q y_i(t) = \sum_{j=1}^{4} b_{ij} y_j(t) + g_i(Y(t)) + v_i, \quad i = 1, 4, \tag{49}
$$

and

$$
D_t^q y_i(t) = J^{1-q}(v_i), \quad i = 2, 3. \tag{50}
$$

Now, applying the Laplace transform to Eq. (50), and letting

$$
\mathbf{F}_i(s) = \mathbf{L}(y_i(t)), \quad i = 2, 3, \tag{51}
$$

we obtain

$$
s^q \mathbf{F}_i(s) - s^{q-1} y_i(0) = s^{q-1} \mathbf{L}(v_i), \quad i = 2, 3, \tag{52}
$$

multiplying both the left-hand and right-hand sides of Eq. (52) by s^{1-q} and again applying the inverse Laplace transform to the result, we obtain

$$\dot{y}_i(t) = v_i, \quad i = 2, 3. \tag{53}$$

From Eqs. (47)–(49) and (53), we get the error dynamical system as follows

$$\begin{aligned} D_t^q e_1(t) &= \left(b_{11} - c_{11}\right) e_1(t) + \left(b_{14} - c_{14}\right) e_4(t) + v_1 + R_1, \\ \dot{e}_2(t) &= \left(b_{22} - c_{22}\right) e_2(t) + \left(b_{23} - c_{23}\right) e_3(t) + \beta_2(t) v_2 + \beta_3(t) v_3 + R_2, \\ \dot{e}_3(t) &= \left(b_{32} - c_{32}\right) e_3(t) + \left(b_{33} - c_{33}\right) e_3(t) + \frac{\partial \varphi}{\partial y_2} v_2 + \frac{\partial \varphi}{\partial y_3} v_3 + R_3, \\ D_t^q e_4(t) &= \left(b_{41} - c_{41}\right) e_1(t) + \left(b_{44} - c_{44}\right) e_4(t) + v_4 + R_4, \end{aligned} \tag{54}$$

where c11, c14, c22, c23, c32, c33, c41, c44 are control constants and

$$\begin{aligned} R_1 &= \left(c_{11} - b_{11}\right) e_1(t) + \left(c_{14} - b_{14}\right) e_4(t) + \sum_{j=1}^{4} b_{1j} y_j(t) + g_1(Y(t)) \\ &\quad - D_t^q \left(\sum_{j=1}^{3} \alpha_j(t) x_j(t) \right), \end{aligned} \tag{55}$$

$$\begin{aligned} R_2 &= \left(c_{22} - b_{22}\right) e_2(t) + \left(c_{23} - b_{23}\right) e_3(t) + \sum_{j=1}^{4} \dot{\beta}_j(t) y_j(t) + \beta_1(t) \dot{y}_1(t) + \beta_4(t) \dot{y}_4(t) - \dot{x}_2(t), \\ R_3 &= \left(c_{32} - b_{32}\right) e_2(t) + \left(c_{33} - b_{33}\right) e_3(t) + \frac{\partial \varphi}{\partial y_1} \dot{y}_1(t) + \frac{\partial \varphi}{\partial y_4} \dot{y}_4(t) - \dot{x}_3(t), \\ R_4 &= \left(c_{41} - b_{41}\right) e_1(t) + \left(c_{44} - b_{44}\right) e_4(t) + \sum_{j=1}^{4} b_{4j} y_j(t) + g_4(Y(t)) - D_t^q(\phi(X(t))). \end{aligned}$$

Rewriting the error system described by Eq. (54) in the following compact forms

$$D_t^q e_I(t) = \left(B_I - C_I\right) e_I(t) + V_I + R_I, \tag{56}$$

and

$$\dot{e}(t) = \left(B_{II} - C_{II}\right) e_{II}(t) + \mathbf{M} \times V_{II} + R_{II}, \tag{57}$$

where $D_t^q e_I(t) = \left(D_t^q e_1(t), D_t^q e_4(t)\right)^T$, $\dot{e}_{II}(t) = \left(\dot{e}_2(t), \dot{e}_3(t)\right)^T$, $B_I = \begin{pmatrix} b_{11} & b_{14} \\ b_{41} & b_{44} \end{pmatrix}$, $B_{II} = \begin{pmatrix} b_{22} & b_{23} \\ b_{32} & b_{33} \end{pmatrix}$, $C_I = \begin{pmatrix} c_{11} & c_{12} \\ c_{21} & c_{22} \end{pmatrix}$, $C_{II} = \begin{pmatrix} c_{33} & c_{34} \\ c_{43} & c_{44} \end{pmatrix}$, $\mathbf{M} = \begin{pmatrix} \beta_2(t) & \beta_3(t) \\ \frac{\partial \varphi}{\partial y_2} & \frac{\partial \varphi}{\partial y_3} \end{pmatrix}$, $R_I = \left(R_1, R_4\right)^T$, $R_{II} = \left(R_2, R_3\right)^T$, $V_I = \left(v_1, v_4\right)^T$ and $V_{II} = \left(v_2, v_3\right)^T$.

To achieve synchronization between the master system (44) and the slave system (45), we assume that the matrix **M** is invertible. Hence, we have the following result.

Theorem 4 *If the control matrices C_I and C_{II} are chosen such that $\left(B_I - C_I\right)$ and $(B_I - C_I)^T + (B_I - C_I)$ are negative matrices, then the co-existence of FSHFPS, IFSHFPS, IGS and GS between the master system (44) and the slave system (45)* *is achieved under the following control laws*

$$V_I = -R_I, \tag{58}$$

and

$$V_{II} = -\mathbf{M}^{-1} \times R_{II}, \tag{59}$$

where $\mathbf{M}^{-1}$ *is the inverse matrix of* $\mathbf{M}$.

Proof Firstly, by substituting Eq. (58) into Eq. (56), one can have

$$D_t^q e_I(t) = \left(B_I - C_I\right) e_I(t). \tag{60}$$

If a Lyapunov function candidate is chosen as

$$V\left(e_I(t)\right) = \frac{1}{2} e_I^T(t) e_I(t). \tag{61}$$

Then, the time Caputo fractional derivative of order q of V along the trajectory of the system (60) is as follows

$$D_t^q V(e(t)) = D_t^q \left(\frac{1}{2} e_I^T(t) e_I(t)\right), \tag{62}$$

using Lemma 3 in Eq. (62), we get

$$\begin{aligned} D_t^q V(e(t)) &\leq e_I^T(t) D_t^q e_I(t) \\ &= e_I^T(t) \left(B_I - C_I\right) e_I(t) < 0, \end{aligned}$$

from Theorem 2, the zero solution of the system (60) is a globally asymptotically stable, i.e.

$$\lim_{t\to\infty} e_1(t) = \lim_{t\to\infty} e_4(t) = 0. \tag{63}$$

Secondly, applying the control law described by Eq. (59) to Eq. (57) yields the resulting error dynamics as follows:

$$\dot{e}_{II}(t) = \left(B_{II} - C_{II}\right) e_{II}(t). \tag{64}$$

Construct the candidate Lyapunov function in the form: $V\left(e_{II}(t)\right) = e_{II}^T(t) e_{II}(t)$, we obtain

$$\begin{aligned}\dot{V}(e(t)) &= \dot{e}_{II}^{T}(t)e_{II}(t) + e_{II}^{T}(t)\dot{e}_{II}(t)\\ &= e_{II}^{T}(t)(B_{II} - C_{II})^{T}e_{II}(t) + e_{II}^{T}(t)\left(B_{II} - C_{II}\right)e_{II}(t)\\ &= e_{II}^{T}(t)\left[(B_{II} - C_{II})^{T} + (B_{II} - C_{II})\right]e_{II}(t) < 0.\end{aligned}$$

Thus, from the Lyapunov stability theory of integer-order systems, that is

$$\lim_{t\to\infty} e_2(t) = \lim_{t\to\infty} e_3(t) = 0. \tag{65}$$

Finally, from Eqs. (63) and (65), we conclude that the master system (44) and the slave system (45) are globally synchronized.

5.3 Simulation Results

Consider the fractional version of the modified coupled dynamos system proposed by Wang and Wang [158]

$$\begin{aligned}D^{p_1}x_1 &= -\alpha x_1 + \left(x_3 + \beta\right)x_2,\\ D^{p_2}x_2 &= -\alpha x_2 + \left(x_3 - \beta\right)x_1,\\ D^{p_3}x_3 &= x_3 - x_1x_2,\end{aligned} \tag{66}$$

as a master system. The system (66) exhibits chaotic behaviors when $(p_1, p_2, p_3) = (0.9, 0.93, 0.96)$ and $(\alpha, \beta) = (2, 1)$. Attractors of the master system (66) are shown in Figs. 7 and 8.

As a slave system, we consider the following system

$$\begin{aligned}D^{q}y_1 &= a\left(y_2 - y_1\right) + y_4 + u_1,\\ D^{q}y_2 &= dy_1 + cy_2 - y_1y_3 + u_2,\\ D^{q}y_3 &= -by_3 + y_1y_2 + u_3,\\ D^{q}y_4 &= ry_4 + y_2y_3 + u_4,\end{aligned} \tag{67}$$

where y_1, y_2, y_3, y_4 are states and u_1, u_2, u_3, u_4 are synchronization controllers. This system is the fractional version of the new hyperchaotic Chen proposed by Li et al. [159]. the slave system (67), as shown in [160], exhibits hyperchaotic behavior when $(u_1, u_2, u_3, u_4) = (0, 0, 0, 0)$, $(a, b, c, d, r) = (35, 3, 12, 7, 0.5)$ and $q = 0.96$. The attractors of the uncontrolled system (67) are shown in Figs. 9 and 10.

Using the notations described in Sect. 5.1, the errors system between the master system (66) and the slave system (67) are given as

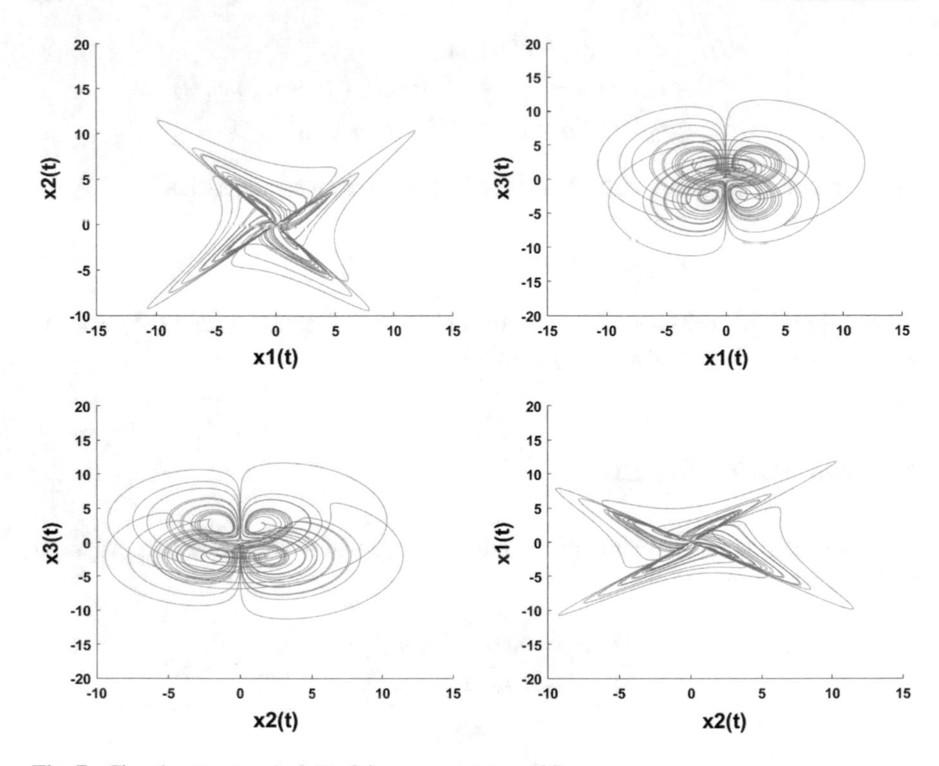

Fig. 7 Chaotic attractors in 2-D of the master system (66)

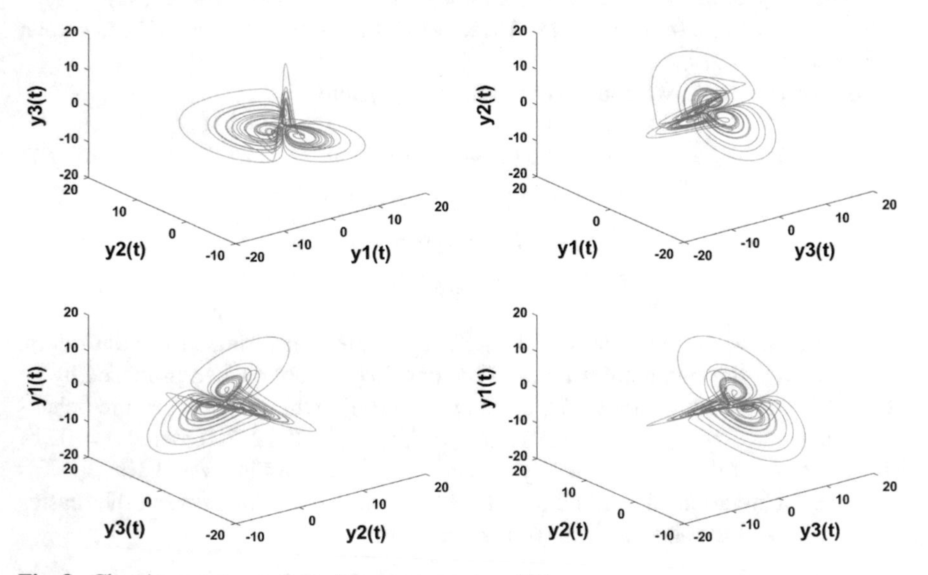

Fig. 8 Chaotic attractors in 3-D of the master system (66)

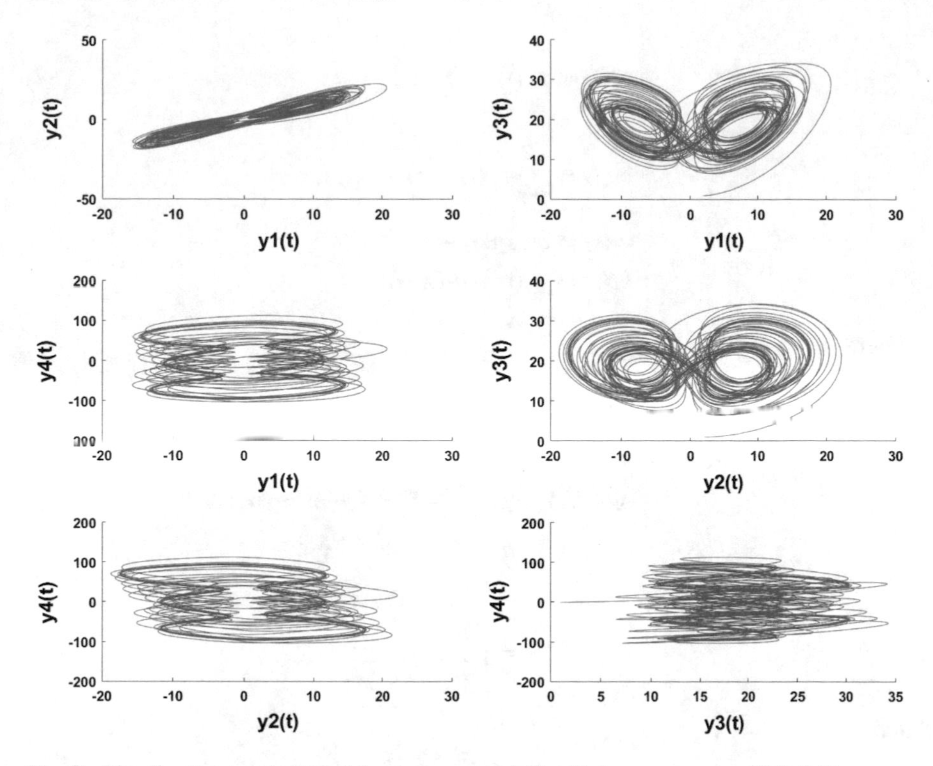

Fig. 9 Chaotic attractors in 2-D of the slave system (67) with $(u_1, u_2, u_3, u_4) = (0, 0, 0, 0)$

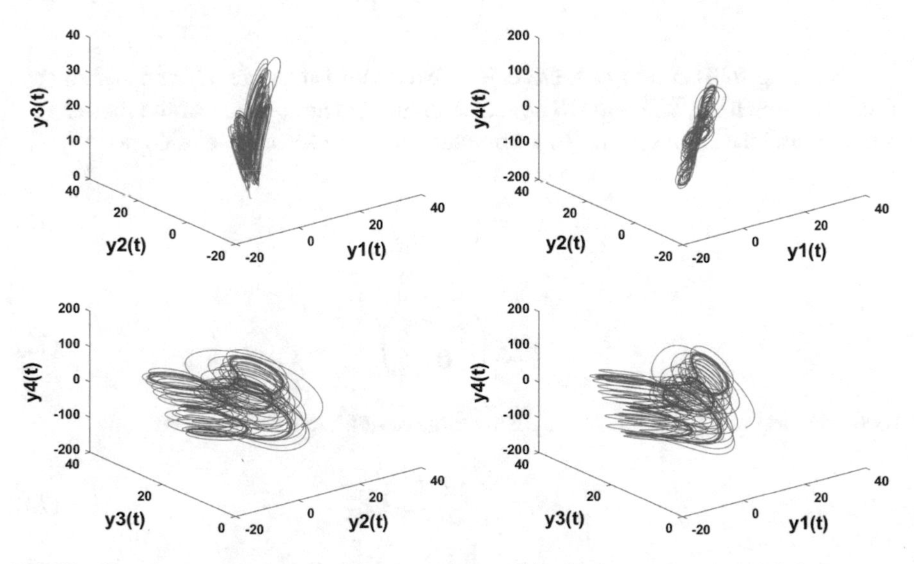

Fig. 10 Chaotic attractors in 3-D of the slave system (67) with $(u_1, u_2, u_3, u_4) = (0, 0, 0, 0)$

$$e_1(t) = y_1(t) - \sum_{j=1}^{3} \alpha_j(t) x_j(t), \tag{68}$$
$$e_2(t) = \sum_{j=1}^{4} \beta_j(t) y_j(t) - x_2(t),$$
$$e_3(t) = \varphi(Y(t)) - x_3,$$
$$e_4(t) = y_4(t) - \phi(X(t)),$$

where

$$\left(\alpha_1(t), \alpha_2(t), \alpha_3(t)\right) = (t, 2, 3t),$$
$$\left(\beta_1(t), \beta_2(t), \beta_3(t), \beta_4(t)\right) = \left(t+1, 2t, 3t, t^2\right),$$
$$\varphi\left(y_1, y_2, y_3, y_4\right) = y_1^2 + 6y_2 + 8y_3 + y_4^2,$$
$$\phi\left(x_1, x_2, x_3\right) = x_1 x_2^2 + x_3^2.$$

So,

$$\mathbf{M} = \begin{pmatrix} 6 & 8 \\ 2 & 2 \end{pmatrix}, \tag{69}$$

and

$$\mathbf{M}^{-1} = \begin{pmatrix} \frac{-1}{2} & 2 \\ \frac{1}{2} & -\frac{3}{2} \end{pmatrix}. \tag{70}$$

According to Theorem 4, there exists two control matrices C_I and C_{II} so that FSHFPS, IFSHFPS, IGS and GS co-exists in the synchronization of the master system (66) and the slave system (67). For example, if we select C_I and C_{II} as

$$C_I = \begin{pmatrix} 14 & 0 \\ 0 & 0 \end{pmatrix}, \tag{71}$$

and

$$C_{II} = \begin{pmatrix} -35 & 1 \\ 0 & 1 \end{pmatrix}, \tag{72}$$

then it is easy to know that $B_I - C_I$ is a negative definite matrix where

$$B_I = \begin{pmatrix} 12 & 0 \\ 0 & -7 \end{pmatrix}. \tag{73}$$

Also we can see that $(B_{II} - C_{II})^T + (B_{II} - C_{II})$ is a negative definite matrix where

$$B_{II} = \begin{pmatrix} -35 & 1 \\ 0 & 0.5 \end{pmatrix}. \tag{74}$$

Hence the synchronization between systems (66) and (67) is achieved and the error system can be described as follows:

$$D^{0.96}e_1 = -2e_1,\\ D^{0.96}e_4 = -7e_4, \tag{75}$$

and

$$\dot{e}_2 = -35e_2,\\ \dot{e}_3 = -0.5e_3. \tag{76}$$

Fractional Euler integration method and fourth order Runge–Kutta integration method have been used to solve the error system described by (75) and (76), respectively. In addition, simulation time $Tm = 120s$ and time step $h = 0.005s$ have been employed. The initial values of the master system (66) and the slave

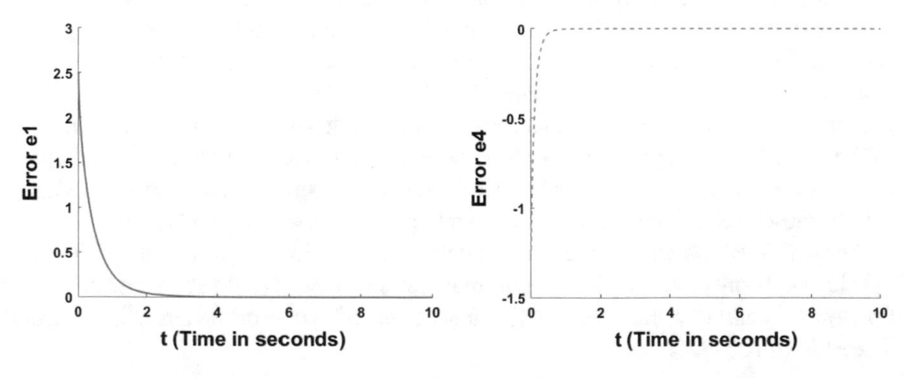

Fig. 11 Time evolution of the errors e_1 and e_4 between the master system (66) and the slave system (67)

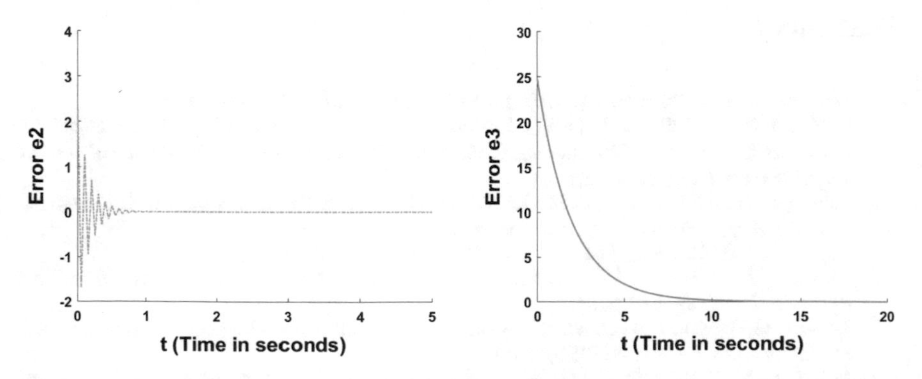

Fig. 12 Time evolution of the errors e_2 and e_3 between the master system (66) and the slave system (67)

system (67) are $[x_1(0), x_2(0), x_3(0)] = [-0.25, -0.25, 0.25]$ and $[y_1(0), y_2(0), y_3(0), y_4(0)] = [2, 2, 1, -1]$, respectively, and the initial states of the error system are $[e_1(0), e_4(0)] = [2.5, -1.2343]$ and $[e_2(0), e_3(0)] = [2.25, 24.75]$. Figures 11 and 12 displays the synchronization errors between the master system (66) and the slave system (67).

6 Conclusion

This study has developed new approaches to study the co-existence of various types of chaos synchronization between different dimensional fractional order chaotic systems. Firstly, Specifically, by using some properties of fractional derivatives, stability theory of linear fractional-order differential equations and stability theory linear integer–order differential equations, the study has analyzed the co-existence of FSHFPS, IGS, IFSHPS and GS between 3D incommensurate fractional order master system and 4D incommensurate fractional order slave system. Secondly, based on Laplace transform, fractional Lyapunov-based approach and integer–order Lyapunov stability theory, the authors have proved the co-existence of four different types of chaos synchronization. More in detail, FSHFPS, IFSHFPS, IGS and GS have been proved to co-exist between 3D incommensurate fractional order master system and 4D commensurate fractional order slave system. Finally, several numerical examples of co-existence of synchronization types have been illustrated. These examples have clearly shown the effectiveness of the novel approaches developed herein.

Further developments and extended analysis related to the application of the new hybrid synchronization in chaotic communication devices and secure communication systems and new complex fractional schemes of synchronization will be investigated in future works.

References

1. Oldham, K. B., & Spanier, J. (1974). *The fractional calculus*. Academic Press.
2. Gorenflo, R., & Mainardi, F. (1997). Fractional calculus: Integral and differential equations of fractional order. In A. Carpinteri & F. Mainardi (Eds.), *Fractals and fractional calculus in continuum mechanics*. Springer.
3. Samko, S. G., Klibas, A. A., & Marichev, O. I. (1993). *Fractional integrals and derivatives: theory and applications*. Gordan and Breach.
4. Podlubny, I. (1999). *Fractional differential equations*. Academic Press.
5. Kilbas, A. A., Srivastava, H. M., & Trujillo, J. J. (2006). *Theory and applications of fractional differential equations*. Elsevier.
6. Jumarie, G. (1992). A Fokker-Planck equation of fractional order with respect to time. *Journal of Mathematica Physics*, *33*, 3536–3541.
7. Miller, K. S., & Ross, B. (1993). *An introduction to the fractional calculus and fractional differential equations*. Wiley.
8. Metzler, R., Glockle, W. G., & Nonnenmacher, T. F. (1994). Fractional model equation for anomalous diffusion. *Physica A*, *211*, 13–24.

9. Mainardi, F. (1997). Fractional calculus: some basic problems in continuum and statistical mechanics. In A. Carpinteri & F. Mainardi (Eds.), *Fractals and fractional calculus in continuum mechanics*. Springer.
10. Bode, H. W. (1949). *Network analysis and feedback amplifier design*. Tung Hwa Book Company.
11. Carlson, G. E., & Halijak, C. A. (1964). Approximation of fractional capacitors $\left(\frac{1}{s}\right)^{\frac{1}{n}}$ by a regular Newton process. *IEEE Transactions on Circuit Theory*, *11*, 210–213.
12. Torvik, P. J., & Bagley, R. L. (1984). On the appearance of the fractional derivative in the behavior of real materials. *Transactions of the ASME*, *51*, 294–298.
13. Nakagava, M., & Sorimachi, K. (1992). Basic characteristics of a fractance device. *IEICE Transactions on Fundamentals*, *E75-A*, 1814–1818.
14. Axtell, M., & Bise, E. M. (1990). Fractional calculus applications in control systems. *Proceedings of the IEEE National Aerospace and Electronics Conference* (pp. 563–566). New York.
15. Koeller, R. C. (1984). Application of fractional calculus to the theory of viscoelasticity. *Journal of Applied Mechanics*, *51*, 299–307.
16. Dorčák, L. (1994). *Numerical models for the simulation of the fractional-order control systems, UEF-04-94, The Academy of Sciences*. Košice, Slovakia: Inst. of Experimental Physic.
17. Bagley, R. L., & Torvik, P. J. (1994). On the appearance of the fractional derivative in the behavior of real materials. *Journal of Applied Mechanics*, *51*, 294–298.
18. Oustaloup A. (1995). La derivation non entiere: theorie, synthese et applications. Hermes.
19. Parada, F. J. V., Tapia, J. A. O., & Ramirez, J. A. (2007). Effective medium equations for fractional Fick's law in porous media. *Physica A*, *373*, 339–353.
20. Podlubny, I. (1999). Fractional-order systems and $\mathbf{PI}^{\lambda}\mathbf{D}^{\mu}$-controllers. *IEEE Transactions on Automatic Control*, *44*, 208–213.
21. Arena, P., Caponetto, R., Fortuna, L., & Porto, D. (2000). Nonlinear noninteger order circuits and systems—An Introduction. World Scientific.
22. Hilfer, R. (2000). Applications of fractional calculus in physics. World Scientific.
23. Westerlund, S. (2002). Dead Matter Has Memory! Causal Consulting.
24. Pires, E. J. S., Machado, J. A. T., & de Moura, P. B. (2003). Fractional order dynamics in a GA planner. *Signal Processing*, *83*, 2377–2386.
25. Vinagre, B. M., Chen, Y. Q., & Petráš, I. (2003). Two direct Tustin discretization methods for fractional-order differentiator/integrator. *Journal of The Franklin Institute*, *340*, 349–362.
26. Magin, R. L. (2006) Fractional calculus in bioengineering. Begell House.
27. Tseng, C. C. (2007). Design of FIR and IIR fractional order Simpson digital integrators. *Signal Processing*, *87*, 1045–1057.
28. Da Graca, Marcos M., Duarte, F. B. M., & Machado, J. A. T. (2008). Fractional dynamics in the trajectory control of redundant manipulators. *Communications in Nonlinear Science and Numerical Simulation*, *13*, 1836–1844.
29. Soltan, A., Radwan, A. G., Soliman, A. M. (2013). Fractional order Butterworth filter: active and passive realizations. *IEEE Journal of Emerging and Selected Topics in Circuits and Systems*, *3:3*, 346–354.
30. Soltan, A., Radwan, A. G., & Soliman, A. M. (2012). Fractional order filter with two fractional elements of dependent orders. *Journal of Microelectronics*, *43*, 818–827.
31. Radwan, A. G., & Fouda, M. E. (2013). Optimization of fractional-order RLC filters. *Journal of Circuits, Systems, and Signal Processing*, *32*, 2097–2118.
32. Moaddy, K., Radwan, A. G., Salama, K. N., Momani, S., & Hashim, I. (2012). The fractional-order modeling and synchronization of electrically coupled neurons system. *Computers and Mathematics with Applications*, *64*, 3329–3339.
33. Radwan, A. G., Moaddy, K., & Momani, S. (2011). Stability and nonstandard finite difference method of the generalized Chua's circuit. *Computers and Mathematics with Applications*, *62*, 961–970.

34. Radwan, A. G., Moaddy, K., Salama, K. N., Momani, S., & Hasim, I. (2014). Control and switching synchronization of fractional order chaotic systems using active control technique. *Journal of Advanced Research*, *5*(1), 125–132.
35. Hartley, T., Lorenzo, C., & Qammer, H. (1995). Chaos in a fractional order Chua's system. *IEEE Transactions on Circuits and Systems I Fundamental Theory and Applications*, *42*, 485–490.
36. Arena, P., Caponetto, R., Fortuna, L., & Porto, D. (1998). Bifurcation and chaos in non-integer order cellular neural networks. *International Journal of Bifurcation and Chaos*, *8*, 1527–1539.
37. Ahmad, W. M., & Sprott, J. C. (2003). Chaos in fractional-order autonomous nonlinear systems. *Chaos Solitons & Fractals*, *16*, 339–351.
38. Grigorenko, I., & Grigorenko, E. (2003). Chaotic dynamics of the fractional Lorenz system. *Physical Review Letters*, *91*, 034101.
39. Li, C., & Chen, G. (2004). Chaos and hyperchaos in fractional order Rössler equations. *Physica A*, *341*, 55–61.
40. Li, C., & Chen, G. (2004). Chaos in the fractional order Chen system and its control. *Chaos Solitons & Fractals*, *22*, 549–554.
41. Ahmad, W. M. (2005). Hyperchaos in fractional order nonlinear systems. *Chaos Solitons & Fractals*, *26*, 1459–1465.
42. Gao, X., & Yu, J. (2005). Chaos in the fractional order periodically forced complex Duffing's oscillators. *Chaos Solitons & Fractals*, *24*, 1097–1104.
43. Lu, J. G., & Chen, G. (2006). A note on the fractional-order Chen system. *Chaos Solitons & Fractals*, *27*, 685–688.
44. Ge, Z. M., & Hsu, M. Y. (2007). Chaos in a generalized van der Pol system and in its fractional order system. *Chaos Solitons & Fractals*, *33*, 1711–1745.
45. Ahmed, E., El-Sayed, A. M. A., & El-Saka, H. A. A. (2007). Equilibrium points, stability and numerical solutions of fractional-order predator-prey and rabies models. *Journal of Mathematical Analysis and Applications*, *325*, 542–553.
46. Barbosa, R. S., Machado, J. A. T., Vinagre, B. M., & Calderón, A. J. (2007). Analysis of the Van der Pol oscillator containing derivatives of fractional order. *Journal of Vibration and Control*, *13*, 1291–1301.
47. Ge, Z. M., & Ou, C. Y. (2007). Chaos in a fractional order modified Duffing system. *Chaos Solitons & Fractals*, *34*, 262–291.
48. Chen, W. C. (2008). Nonlinear dynamics and chaos in a fractional-order financial system. *Chaos Solitons & Fractals*, *36*, 1305–1314.
49. Petráš, I. (2008). A note on the fractional-order Chua's system. *Chaos Solitons & Fractals*, *38*, 140–147.
50. Petráš, I. (2009). Chaos in the fractional-order Volta's system: modeling and simulation. *Nonlinear Dynamics*, *57*, 157–170.
51. Petráš, I. (2010). A note on the fractional-order Volta's system. *Communications in Nonlinear Science and Numerical Simulation*, *15*, 384–393.
52. Deng, H., Li, T., Wang, Q., & Li, H. (2009). A fractional-order hyperchaotic system and its synchronization. *Chaos Solitons & Fractals*, *41*, 962–969.
53. Gejji, V. D., & Bhalekar, S. (2010). Chaos in fractional ordered Liu system. *Computers & Mathematics with Applications*, *59*, 1117–1127.
54. Kiani, B. A., Fallahi, K., Pariz, N., & Leung, H. (2009). A chaotic secure communication scheme using fractional chaotic systems based on an extended fractional Kalman filter. *Communications in Nonlinear Science and Numerical Simulation*, *14*, 863–879.
55. Liang, H., Wang, Z., Yue, Z., & Lu, R. (2012). Generalized synchronization and control for incommensurate fractional unified chaotic system and applications in secure communication. *Kybernetika*, *48*, 190–205.
56. Wu, X., Wang, H., & Lu, H. (2012). Modified generalized projective synchronization of a new fractional-order hyperchaotic system and its application to secure communication. *Nonlinear Analysis: Real World Applications*, *13*, 1441–1450.

57. Muthukumar, P., & Balasubramaniam, P. (2013). Feedback synchronization of the fractional order reverse butterfly-shaped chaotic system and its application to digital cryptography. *Nonlinear Dynamics*, *74*, 1169–1181.
58. Muthukumar, P., Balasubramaniam, P., & Ratnavelu, K. (2014). Synchronization of a novel fractional order stretch-twistfold (STF) flow chaotic system and its application to a new authenticated encryption scheme (AES). *Nonlinear Dynamics*, *77*, 1547–1559.
59. Radwan, A. G. (2013). On some generalized logistic maps with arbitrary power. *Journal of Advanced Research (JAR)*, *4*, 163–171.
60. Radwan, A. G., Soliman, A. M., & EL-Sedeek A. L. (2003). MOS realization of the double scroll-like chaotic equation. *IEEE Circuits and systems-I*, *50*(2), 285–288.
61. Radwan, A. G., Soliman, A. M., & EL-sedeek A. L. (2004). MOS realization of the modified Lorenz chaotic system. *Chaos, Solitons & Fractals*, *21*, 553–561.
62. Radwan, A. G., Soliman, A. M., & EL-sedeek A. L. (2003). An inductorless CMOS realization of Chua's circuit. *Chaos, Solitons and Fractals*, *18*, 149–158.
63. Radwan, A. G., Soliman, A. M., & Elwakil, A. S. (2007). 1-D digitally-controlled multi-scroll chaos generator. *International Journal of Bifurcation and Chaos*, *17*(1), 227–242.
64. Zidan, M. A., Radwan, A. G., & Salama, K. N. (2012). Controllable v-shape multi-scroll butterfly attractor: System and circuit implementation. *Int.International Journal of Bifurcation and Chaos (IJBC)*, *22*, 6.
65. Barakat, M. L., Mansingka, A. S., Radwan, A. G., & Salama, K. N. (2013). Generalized hardware post processing technique for chaos-based pseudo random number generators. *ETRI Journal*, *35*(3), 448–458.
66. Pecora, L. M., & Carroll, T. L. (1990). Synchronization in chaotic systems. *Physical Review Letters*, *64*, 821–824.
67. Azar, A. T., & Vaidyanathan, S. (2015). *Chaos modeling and control systems design, Studies in computational intelligence* (Vol. 581). Germany: Springer.
68. Azar, A. T., & Vaidyanathan, S. (2016). *Advances in chaos theory and intelligent control. Studies in fuzziness and soft computing* (Vol. 337). Germany: Springer. ISBN 978-3-319-30338-3.
69. Azar, A. T., & Vaidyanathan, S. (2015). *Computational intelligence applications in modeling and control*. Studies in computational intelligence (Vol. 575). Germany: Springer. ISBN 978-3-319-11016-5.
70. Azar, A. T., & Vaidyanathan, S. (2015). *Handbook of research on advanced intelligent control engineering and automation*. Advances in Computational Intelligence and Robotics (ACIR) Book Series. USA: IGI Global. ISBN 9781466672482.
71. Zhu, Q., & Azar, A. T. (2015) Complex system modelling and control through intelligent soft computations. Studies in Fuzziness and Soft Computing (Vol. 319). Springer-Verlag, Germany. ISBN: 978-3-319-12882-5.
72. Azar, A. T., & Zhu, Q. (2015). *Advances and applications in sliding mode control systems*. Studies in computational intelligence (Vol. 576). Germany: Springer. ISBN: 978-3-319-11172-8.
73. Vaidyanathan, S., & Azar, A. T. (2015). Analysis, control and synchronization of a nine-term 3-D novel chaotic system. In A. T. Azar & S. Vaidyanathan (Eds.), *Chaos modeling and control systems design*. Studies in computational intelligence book series. Springer.
74. Vaidyanathan, S., & Azar, A. T. (2015). Anti-synchronization of identical chaotic systems using sliding mode control and an application to Vaidyanathan-Madhavan chaotic systems. In A. T. Azar & Q. Zhu (Eds.), *Advances and applications in sliding mode control systems*. Studies in computational intelligence book series. Springer.
75. Vaidyanathan, S., & Azar, A. T. (2015). Hybrid synchronization of identical chaotic systems using sliding mode control and an application to Vaidyanathan chaotic systems. In A. T. Azar & Q. Zhu (Eds.), *Advances and applications in sliding mode control systems*. Studies in computational intelligence book series. Springer.
76. Vaidyanathan, S., Idowu, B. A., & Azar, A. T. (2015). Backstepping controller design for the global chaos synchronization of Sprott's jerk systems. In A. T. Azar & S. Vaidyanathan

(Eds.), *Chaos modeling and control systems design*. Studies in computational intelligence book series. Springer.

77. Vaidyanathan, S., Sampath, S., & Azar, A.T. (2015). Global chaos synchronisation of identical chaotic systems via novel sliding mode control method and its application to Zhu system. *International Journal of Modelling, Identification and Control (IJMIC), 23*(1), 92–100.
78. Vaidyanathan, S., Azar, A. T., Rajagopal, K., & Alexander, P. (2015) Design and SPICE implementation of a 12-term novel hyperchaotic system and its synchronization via active control. International Journal of Modelling. *Identification and Control (IJMIC), 23*(3), 267–277.
79. Vaidyanathan, S., & Azar, A. T. (2016). Takagi-Sugeno fuzzy logic controller for Liu-Chen four-scroll chaotic system. *International Journal of Intelligent Engineering Informatics, 4*(2), 135–150.
80. Vaidyanathan, S., & Azar, A. T. (2015). Analysis and control of a 4-D novel hyperchaotic system. In: A. T. Azar, & S. Vaidyanathan (Eds.), Chaos modeling and control systems design. Studies in computational intelligence (Vol. 581, pp. 19–38). Berlin/Heidelberg: Springer-Verlag GmbH. doi:10.1007/978-3-319-13132-0_2.
81. Boulkroune, A., Bouzeriba, A., Bouden, T., & Azar, A. T. (2016). Fuzzy Adaptive synchronization of uncertain fractional-order chaotic systems. In: A. T. Azar, & S. Vaidyanathan (Eds.), Advances in chaos theory and intelligent control. Studies in fuzziness and soft computing (Vol. 337). Germany: Springer.
82. Boulkroune, A., Hamel, S., & Azar, A. T. (2016). Fuzzy control-based function synchronization of unknown chaotic systems with dead-zone input. In *Advances in chaos theory and intelligent control*. Studies in Fuzziness and Soft Computing (Vol. 337). Germany: Springer.
83. Vaidyanathan, S., & Azar, A. T. (2016). Dynamic analysis, adaptive feedback control and synchronization of an eight-term 3-d novel chaotic system with three quadratic nonlinearities. In *Advances in chaos theory and intelligent control*. Studies in fuzziness and soft computing (Vol. 337). Germany: Springer.
84. Vaidyanathan, S., & Azar, A. T. (2016). Qualitative study and adaptive control of a novel 4-D hyperchaotic system with three quadratic nonlinearities. In *Advances in chaos theory and intelligent control*. Studies in fuzziness and soft computing (Vol. 337). Germany: Springer.
85. Vaidyanathan, S., & Azar, A. T. (2016). A novel 4-d four-wing chaotic system with four quadratic nonlinearities and its synchronization via adaptive control method. In: Advances in chaos theory and intelligent control. Studies in Fuzziness and Soft Computing (Vol. 337). Germany: Springer.
86. Vaidyanathan, S., & Azar, A. T. (2016). Adaptive control and synchronization of Halvorsen circulant chaotic systems. In *Advances in chaos theory and intelligent control*. Studies in Fuzziness and Soft Computing (Vol. 337). Germany: Springer.
87. Vaidyanathan, S., & Azar, A. T. (2016) adaptive backstepping control and synchronization of a novel 3-d jerk system with an exponential nonlinearity. In *Advances in chaos theory and intelligent control*. Studies in Fuzziness and Soft Computing (Vol. 337). Germany: Springer.
88. Vaidyanathan, S., Azar, A. T. (2016). Generalized projective synchronization of a novel hyperchaotic four-wing system via adaptive control method. In *Advances in chaos theory and intelligent control*. Studies in Fuzziness and Soft Computing (Vol. 337). Germany: Springer.
89. Deng, W. H., & Li, C. P. (2005). Chaos synchronization of the fractional Lü system. *Physica A, 353*, 61–72.
90. Li, C., & Zhou, T. (2005). Synchronization in fractional-order differential systems. *Physica D, 212*, 11–125.
91. Lu, J. G. (2005). Chaotic dynamics and synchronization of fractional-order Arneodo's systems. *Chaos Solitons & Fractals, 26*, 1125–1133.
92. Wang, J., Xiong, X., & Zhang, Y. (2006). Extending synchronization scheme to chaotic fractional-order Chen systems. *Physica A, 370*, 279–285.
93. Li, C. P., Deng, W. H., & Xu, D. (2006). Chaos synchronization of the Chua system with a fractional order. *Physica A, 360*, 171–185.

94. Peng, G. (2007). Synchronization of fractional order chaotic systems. *Physics Letters A, 363*, 426–432.
95. Sheu, L. J., Chen, H. K., Chen, J. H., & Tam, L. M. (2007). Chaos in a new system with fractional order. *Chaos Solitons & Fractals, 31*, 1203–1212.
96. Yan, J., & Li, C. (2007). On chaos synchronization of fractional differential equations. *Chaos Solitons & Fractals, 32*, 725–735.
97. Li, C., & Yan, J. (2007). The synchronization of three fractional differential systems. *Chaos Solitons & Fractals, 32*, 751–757.
98. Zhou, S., Li, H., Zhu, Z., & Li, C. (2008). Chaos control and synchronization in a fractional neuron network system. *Chaos Solitons & Fractals, 36*, 973–984.
99. Zhu, H., Zhou, S., & Zhang, J. (2009). Chaos and synchronization of the fractional-order Chua's system. *Chaos Solitons & Fractals, 39*, 1595–1603.
100. Liu, C., Liu, L., & Liu, T. (2009). A novel three-dimensional autonomous chaos system. *Chaos Solitons & Fractals, 39*, 1950–1958.
101. Odibat, Z. (2012). A note on phase synchronization in coupled chaotic fractional order systems. *Nonlinear Analysis: Real World Applications, 13*, 779–789.
102. Chen, F., Xia, L., & Li, C. G. (2012). Wavelet phase synchronization of fractional-order chaotic systems. *Chinese Physics Letters, 29*, 070501–070506.
103. Razminia, A., & Baleanu, D. (2013). Complete synchronization of commensurate fractional order chaotic systems using sliding mode control. *Mechatronics, 23*, 873–879.
104. Al-sawalha, M. M., Alomari, A. K., Goh, S. M., & Nooran, M. S. M. (2011). Active anti-synchronization of two identical and different fractional-order chaotic systems. *International Journal of Nonlinear Science, 11*, 267–274.
105. Si, G., Sun, Z., Zhang, Y., & Chen, W. (2012). Projective synchronization of different fractional-order chaotic systems with non-identical orders. Nonlinear Analytics: Real World Applications *13*, 1761–1771.
106. Agrawal, S. K., & Das, S. (2014). Projective synchronization between different fractional-order hyperchaotic systems with uncertain parameters using proposed modified adaptive projective synchronization technique. *Mathematical Methods in the Applied Sciences, 37*, 2164–2176.
107. Chang, C. M., & Chen, H. K. (2010). Chaos and hybrid projective synchronization of commensurate and incommensurate fractional-order Chen-Lee systems. *Nonlinear Dynamics, 62*, 851–858.
108. Velmurugan, G., & Rakkiyappan, R. (2016). Hybrid projective synchronization of fractional-order memristor-based neural networks with time delays. *Nonlinear Dynamics, 83*, 419–432.
109. Shao, S. Q. (2009). Controlling general projective synchronization of fractional order Rössler systems. *Chaos Solitons & Fractals, 39*, 1572–1577.
110. Zhou, P., Kuang, F., & Cheng, Y. M. (2010). Generalized projective synchronization for fractional order chaotic systems. *Chinese Journal of Physics, 48*, 49–56.
111. Zhou, P., & Zhu, W. (2011). Function projective synchronization for fractional-order chaotic systems. *Nonlinear Analysis: Real World Applications, 12*, 811–816.
112. Xi, H., Li, Y., & Huang, X. (2015). Adaptive function projective combination synchronization of three different fractional-order chaotic systems. *Optik, 126*, 5346–5349.
113. Zhang, X. D., Zhao, P. D., & Li, A. H. (2010). Construction of a new fractional chaotic system and generalized synchronization. *Communications in Theoretical Physics, 53*, 1105–1110.
114. Jun, W. M., & Yuan, W. X. (2011). Generalized synchronization of fractional order chaotic systems. *International Journal of Modern Physics B, 25*, 1283–1292.
115. Yi, C., Liping, C., Ranchao, W., & Juan, D. (2013). Q-S synchronization of the fractional-order unified system. *Pramana, 80*, 449–461.
116. Feng, H., Yang, Y., & Yang, S. P. (2013). A new method for full state hybrid projective synchronization of different fractional order chaotic systems. *Applied Mechanics and Materials, 385–38*, 919–922.
117. Razminia, A. (2013). Full state hybrid projective synchronization of a novel incommensurate fractional order hyperchaotic system using adaptive mechanism. *Indian Journal of Physics, 87*, 161–167.

118. Zhang, L., & Liu, T. (2016). Full state hybrid projective synchronization of variable-order fractional chaotic/hyperchaotic systems with nonlinear external disturbances and unknown parameters. *Journal of Nonlinear Science and Applications*, *9*, 1064–1076.
119. Aghababa, M. P. (2012). Finite-time chaos control and synchronization of fractional-order nonautonomous chaotic (hyperchaotic) systems using fractional nonsingular terminal sliding mode technique. *Nonlinear Dynamics*, *69*, 247–261.
120. Xi, H., Yu, S., Zhang, R., & Xu, L. (2014). Adaptive impulsive synchronization for a class of fractional-order chaotic and hyperchaotic systems. *Optik*, *125*, 2036–2040.
121. Li, D., Zhang, X. P., Hu, Y. T., & Yang, Y. Y. (2015). Adaptive impulsive synchronization of fractional order chaotic system with uncertain and unknown parameters. *Neurocomputing*, *167*, 165–171.
122. Mathiyalagan, K., Park, J. H., & Sakthivel, R. (2015). Exponential synchronization for fractional-order chaotic systems with mixed uncertainties. *Complexity*, *21*, 114–125.
123. Chen, L., Wu, R., He, Y., & Chai, Y. (2015). Adaptive sliding-mode control for fractional-order uncertain linear systems with nonlinear disturbances. *Nonlinear Dynamics*, *80*, 51–58.
124. Liu, L., Ding, W., Liu, C., Ji, H., & Cao, C. (2014). Hyperchaos synchronization of fractional-order arbitrary dimensional dynamical systems via modified sliding mode control. *Nonlinear Dynamics*, *76*, 2059–2071.
125. Zhang, L., & Yan, Y. (2014). Robust synchronization of two different uncertain fractional-order chaotic systems via adaptive sliding mode control. *Nonlinear Dynamics*, *76*, 1761–1767.
126. Odibat, Z., Corson, N., Alaoui, M. A. A., & Bertelle, C. (2010). Synchronization of chaotic fractional-order systems via linear control. *International Journal of Bifurcation and Chaos*, *20*, 81–97.
127. Chen, X. R., & Liu, C. X. (2012). Chaos Synchronization of fractional order unified chaotic system via nonlinear control. *International Journal of Modern Physics B*, *25*, 407–415.
128. Srivastava, M., Ansari, S. P., Agrawal, S. K., Das, S., & Leung, A. Y. T. (2014). Anti-synchronization between identical and non-identical fractional-order chaotic systems using active control method. *Nonlinear Dynamics*, *76*, 905–914.
129. Agrawal, S. K., & Das, S. A. (2013). modified adaptive control method for synchronization of some fractional chaotic systems with unknown parameters. *Nonlinear Dynamics*, *73*, 907–919.
130. Yuan, W. X., & Mei, S. J. (2009). Synchronization of the fractional order hyperchaos Lorenz systems with activation feedback control. *Communications in Nonlinear Science and Numerical Simulation*, *14*, 3351–3357.
131. Odibat, Z. (2010). Adaptive feedback control and synchronization of non-identical chaotic fractional order systems. *Nonlinear Dynamics*, *60*, 479–487.
132. Zhou, P., & Bai, R. (2015). The adaptive synchronization of fractional-order chaotic system with fractional-order $1<q<2$ via linear parameter update law. *Nonlinear Dynamics*, *80*, 753–765.
133. Cafagna, D., & Grassi, G. (2012). Observer-based projective synchronization of fractional systems via a scalar signal: Application to hyperchaotic Rössler systems. *Nonlinear Dynamics*, *68*, 117–128.
134. Peng, G., & Jiang, Y. (2008). Generalized projective synchronization of a class of fractional-order chaotic systems via a scalar transmitted signal. *Physics Letters A*, *372*, 3963–3970.
135. Hu, M., Xu, Z., Zhang, R., & Hu, A. (2007). Adaptive full state hybrid projective synchronization of chaotic systems with the same and different order. *Physics Letters A*, *365*, 315–327.
136. Hu, M., Xu, Z., Zhang, R., & Hu, A. (2007). Parameters identification and adaptive full state hybrid projective synchronization of chaotic (hyperchaotic) systems. *Physics Letters A*, *361*, 231–237.
137. Hu, M., Xu, Z., Zhang, R., & Hu, A. (2008). Full state hybrid projective synchronization in continuous-time chaotic (hyperchaotic) systems. *Communications in Nonlinear Science and Numerical Simulation*, *13*, 456–464.

138. Hu, M., Xu, Z., Zhang, R., & Hu, A. (2008). Full state hybrid projective synchronization of a general class of chaotic maps. *Communications in Nonlinear Science and Numerical Simulation, 13*, 782–789.
139. Cai, G., Yao, L., Hu, P., & Fang, X. (2013). Adaptive full state hybrid function projective synchronization of financial hyperchaotic systems with uncertain parameters. *Discrete and Continuous Dynamical Systems Series B, 18*, 2019–2028.
140. Ouannas, A., & Grassi, G. (2016). Inverse full state hybrid projective synchronization for chaotic maps with different dimensions. *Chinese Physics B, 25*, 090503–090506.
141. Wu, X. J., Lai, D. R., & Lu, H. T. (2012). Generalized synchronization of the fractional-order chaos in weighted complex dynamical networks with nonidentical nodes. *Nonlinear Dynamics, 69*, 667–683.
142. Xiao, W., Fu, J., Liu, Z., & Wan, W. (2012). Generalized synchronization of typical fractional order chaos system. *Computers Journal, 7*, 519–1526.
143. Martínez-Guerra, R., & Mata-Machuca, J. L. (2014). Fractional generalized synchronization in a class of nonlinear fractional order systems. *Nonlinear Dynamics, 77*, 1237–1244.
144. Ouannas, A., & Odibat, Z. (2015). Generalized synchronization of different dimensional chaotic dynamical systems in discrete time. *Nonlinear Dynamics, 81*, 765–771.
145. Ouannas, A., & Odibat, Z. (2016). On inverse generalized synchronization of continuous chaotic dynamical systems. *International Journal of Applied Mathematics and Computation, 2*, 1–11.
146. Ouannas, A. (2016). Co-existence of various synchronization-types in hyperchaotic maps. *Nonlinear Dynamics and Systems Theory, 16*, 312–321.
147. Ouannas, A., Azar, A. T., & Abu-Saris, R. (2016). A new type of hybrid synchronization between arbitrary hyperchaotic maps. *International Journal of Machine Learning and Cybernetics*, 1–8.
148. Ouannas, A., Azar, A. T., & Sundarapandian, V. (2016). New hybrid synchronization schemes based on coexistence of various types of synchronization between master-slave hyperchaotic systems. *International Journal of Computer Applications in Technology* (To be appear).
149. Ouannas, A., Azar, A. T., & Sundarapandian, V. (2016). A new fractional hybrid chaos synchronisation. International Journal of Modelling Identification and Control (To be appear).
150. Ouannas, A., Azar, A. T., & Sundarapandian, V. (2016). A robust method for new fractional hybrid chaos synchronization. *Mathematical Methods in the Applied Sciences*, 1–9.
151. Caputo, M. (1967). Linear models of dissipation whose Q is almost frequency independent.II. *Geophysical Journal of the Royal Astronomical Society, 13*, 529–539.
152. Matignon, D. (1996). Stability results of fractional differential equations with applications to control processing, In *IMACS, IEEE-SMC*, Lille, France.
153. Li, Y., Chen, Y., & Podlubny, I. (2010). Stability of fractional-order nonlinear dynamic systems: Lyapunov direct method and generalized Mittag Leffler stability. *Computers & Mathematics with Applications, 59*, 1810–1821.
154. Chen, D., Zhang, R., Liu, X., & Ma, X. (2014). Fractional order Lyapunov stability theorem and its applications in synchronization of complex dynamical networks. *Communications in Nonlinear Science and Numerical Simulation, 19*, 4105–4121.
155. Aguila-Camacho, N., Duarte-Mermoud, M. A., & Gallegos, J. A. (2014). Lyapunov functions for fractional order systems. *Communications in Nonlinear Science and Numerical Simulation, 19*, 2951–2957.
156. Chen, W. C. (2008). Dynamics and control of a financial system with time-delayed feedbacks. *Chaos Solitons & Fractals, 37*, 1198–1207.
157. Zhou, P., Wei, L. J., & Cheng, X. F. (2009). A novel fractional-order hyperchaotic system and its synchronization. *Chinese Physics B, 18*, 2674.
158. Wang, M. J., & Wang, X. Y. (2010). Dynamic analysis of the fractional order Newton-Leipnik system. *Acta Physica Sinica, 59*, 01583–01587.
159. Li, Y. X., Tang, W. K. S., & Chen, G. R. (2005). Generating hyperchaos via state feedback control. *International Journal of Bifurcation and Chaos, 15*, 3367–3375.
160. Li, T. Z., Wang, Y., & Luo, K. (2014). Control of fractional chaotic and hyperchaotic systems based on a fractional order controller. *Chinese Physics B, 23*, 080501.

Generalized Synchronization of Different Dimensional Integer-Order and Fractional Order Chaotic Systems

Adel Ouannas, Ahmad Taher Azar, Toufik Ziar and Ahmed G. Radwan

Abstract In this work different control schemes are proposed to study the problem of generalized synchronization (GS) between integer-order and fractional-order chaotic systems with different dimensions. Based on Lyapunov stability theory of integer-order differential systems, fractional Lyapunov-based approach and nonlinear controllers, different criterions are derived to achieve generalized synchronization. The effectiveness of the proposed control schemes are verified by numerical examples and computer simulations.

Keywords Chaos · Generalized synchronization · Integer-order system · Fractional-order system · Different dimensions

A. Ouannas
Laboratory of Mathematics, Informatics and Systems (LAMIS),
University of Larbi Tebessi, 12002 Tebessa, Algeria
e-mail: ouannas_adel@yahoo.fr; ouannas.adel@yahoo.fr

A.T. Azar (✉)
Faculty of Computers and Information, Benha University, Benha, Egypt
e-mail: ahmad.azar@fci.bu.edu.eg; ahmad_t_azar@ieee.org

A.T. Azar · A.G. Radwan
Nanoelectronics Integrated Systems Center (NISC), Nile University, Cairo, Egypt
e-mail: agradwan@ieee.org

T. Ziar
Department of Material Sciences, University of Tebessa, 12002 Tebessa, Algeria
e-mail: Toufik1_ziar@yahoo.fr

A.G. Radwan
Engineering Mathematics and Physics Department, Cairo University, Cairo, Egypt

A.T. Azar et al. (eds.), *Fractional Order Control and Synchronization of Chaotic Systems*, Studies in Computational Intelligence 688,
DOI 10.1007/978-3-319-50249-6_23

1 Introduction

Nature is intrinsically nonlinear. So, it is not surprising that most of the systems we encounter in the real world are nonlinear. These nonlinear models of real-life problems generally exhibit chaotic behaviors which possess some special features, such as having bounded trajectories with a positive leading positive Lyapunov exponent of the dynamics of the chaotic system, extreme sensitivity to initial conditions and having noise-like behaviors. This behavior makes chaos undesirable and unwanted in many cases of research as it reduces their predictability over long time scales. But this special attribute may be a valuable advantage in certain areas of research [1–8]. Chaotic dynamics has the ability to amplify small perturbations which improves their utility for reaching specific desired states with very high flexibility and low energy cost. In other words, we could try to control chaos for the benefit of our needs [9–16]. Stability, synchronization of different chaotic or hyperchaotic systems is one of the few main control methods popularly discussed recently [17–36, 36–43]. Many chaotic realizations have been discussed recently from different prospectives based on: generalized chaotic maps [44], analog transistor-level with grounded capacitors [45–47], digital implementation [48, 49], and mixed designs such as in [50].

The idea of chaos synchronization is to use the output of the master system to control the slave system so that the output of the response system follows the output of the master system asymptotically. Recently there has been growing interest in the investigation of various kinds of synchronization in chaotic or hyperchaotic systems [51–59]. This interest is spurred by the possible applications of synchronous chaos particularly in secure communications [60–70].

In the last recent decades, it has been found that nonlinear fractional models are useful for modeling many physical systems [71–96]. It has been demonstrated by scientists that nonlinear fractional order differential systems can display a variety of behaviors including chaos and hyperchaos [97–121]. Chaotic and hyperchaotic fractional order systems can also be synchronized [103, 104, 122–133]. Recently, more and more attentions were paid to the synchronization of different fractional–order chaotic (hyperchaotic) systems [134–151], due to its potential applications [152–156].

The topic of synchronization between fractional order and integer-order chaotic systems is a new domain in the research field of chaos synchronization. However, there are many papers studying the problem of synchronization between integer order chaotic systems and fractional order chaotic system. Various methods and techniques have been proposed to synchronize integer order and fractional-order chaotic systems. For example, general control schemes have been described in [157, 158]. A sliding mode method has been designed in [159–161] and a new fuzzy sliding mode method has been proposed in [162]. A practical method based on circuit simulation has been presented in [163]. The idea of tracking control has been applied in [164, 165]. In [166], a nonlinear feedback control method has been introduced and a robust observer technique has been used in [167]. Also, some synchronization types has been studied between integer order and fractional order chaotic system such as complete synchronization [168], anti-synchronization [169], anticipating synchronization [170], function projective synchronization [171] and Q-S synchronization [172].

Amongst all types of chaos synchronization, generalized synchronization (GS) is one of the most noticeable types. It has received a great deal of attention for its universality in the recent years [173–177]. It has been widely used in the synchronization of chaotic maps [178–181], integer-order chaotic systems [182–184] and fractional-order chaotic systems [186–192]. In generalized synchronization (GS), the master system and the slave system are said to be synchronized if a functional relationship exists between the states of the master and the slave systems.

The main aim of this work is to present constructive schemes to investigate GS between different dimensional integer order and fractional order chaotic systems. Based on the integer order Lyapunov stability and suitable integer order controllers, GS between n-dimensional fractional order master system and m-dimensional integer order slave system is achieved. By using fractional control law and fractional Lyapunov-based approach, GS between integer order master system with dimension n and fractional order slave system with dimension m is observed. The effectiveness and feasibility of the proposed schemes are illustrated by numerical examples.

The outline of the rest of this work is organized as follows. Section 2 provides some preliminaries and basic concepts which are helpful in the proving analysis of the proposed approaches. In Sect. 3, GS between n-D fractional order master system and m-D integer order slave system is studied. In Sect. 4, GS between n-D integer-order master system and m-D fractional order slave system is investigated. Finally, conclusion is drawn in Sect. 5.

2 Preliminaries and Basic Concepts

2.1 Fractional Calculus

Definition 1 [193] The Riemann-Liouville fractional integral operator of order $p > 0$ of the function $f(t)$ is defined as,

$$J^p f(t) = \frac{1}{\Gamma(p)} \int_0^t (t-\tau)^{p-1} f(\tau) d\tau, \quad t > 0. \tag{1}$$

Remark 1 Some properties of the operator J^p can be found, for example, in [194, 195].

Definition 2 [196] The Caputo fractional derivative of $f(t)$ is defined as,

$$D_t^p f(t) = J^{m-p}\left(\frac{d^m}{dt^m} f(t)\right) = \frac{1}{\Gamma(m-p)} \int_0^t \frac{f^{(m)}(\tau)}{(t-\tau)^{p-m+1}} d\tau, \tag{2}$$

for $m-1<p\leq m$, $m\in N$, $t>0$.

Remark 2 The fractional differential operator D_t^p is left-inverse (and not right-inverse) to the fractional integral operator J^p, i.e. $D_t^p J^p = I$ where I is the identity operator [197, 198] and has been used for many applications [199–201].

2.2 Fractional Lyapunov-Based Approach

Definition 3 A continuous function γ is said to belong to class$-K$ if it is strictly increasing and $\gamma(0)=0$.

Theorem 1 *[202] Let $X=0$ be an equilibrium point for the following fractional order system*

$$D_t^p X(t) = F(X(t)), \tag{3}$$

where $p\in\left]0,1\right]$. Assume that there exists a Lyapunov function $V(X(t))$ and class-K functions γ_i $(i=1,2,3)$ satisfying

$$\gamma_1(\|X\|) \leq V(X(t)) \leq \gamma_2(\|X\|), \tag{4}$$

$$D_t^p V(X(t)) \leq -\gamma_3(\|X\|). \tag{5}$$

Then the system (3) is asymptotically stable.

Theorem 2 *[203] If there exists a positive definite Lyapunov function $V(X(t))$ such that $D_t^p V(X(t))<0$, for all $t>0$, then the trivial solution of system (3) is asymptotically stable.*

Lemma 1 *([204])* $\forall X(t)\in\mathbf{R}^n$, $\forall p\in\left]0,1\right]$ *and* $\forall t>0$:

$$\frac{1}{2}D_t^p\left(X^T(t)X(t)\right) \leq X^T(t)D_t^p\left(X(t)\right) \tag{6}$$

2.3 Fractional Numerical Method

Consider the following initial value problem of fractional-order differential equation in the Caputo sense:

$$D_t^p Y(t) = F(t, Y(t)),\ \ 0\leq t\leq T, \tag{7}$$
$$Y^{(k)}(0) = Y_0^{(k)},\ \ k=0,1,2,...,m-1,\ \ (m=\lceil p\rceil).$$

In [205], authors have given a predictor correctors scheme for numerical solution of fractional differential equation, which is a generalization of the classical one-step

Adams–Bashforth–Moulton algorithm for first order ordinary differential equations. The fractional predictor–corrector (PC) algorithm is based on the analytical property. The fractional differential equation (7), is equivalent to the following Volterra integral equation:

$$Y(t) = \sum_{k=0}^{m-1} Y_0^{(k)} \frac{t^k}{k!} + \frac{1}{\Gamma(p)} \int_0^t (t-s)^{p-1} F(s, Y(s))\, ds. \tag{8}$$

Now, set $h = \frac{T}{N}$, $t_n = nh$, $n = 0, 1, 2, ..., N$. Let $Y_h(t_n)$ be approximation to $Y(t_n)$. Assume that we have already calculated approximations $Y_h(t_j)$ and we want to obtain $Y_h(t_{n+1})$ by means of the equation

$$Y_h(t_{n+1}) = \sum_{k=0}^{m-1} C_k \frac{t_{n+1}^k}{k!} + \frac{h^p}{\Gamma(p+2)} F\left(t_{n+1}, Y_h^p(t_{n+1})\right) + \frac{h^p}{\Gamma(p+2)} \sum_{j=0}^{n} a_{j,n+1} F\left(t_j, Y_h(t_j)\right), \tag{9}$$

where

$$a_{j,n+1} = \begin{cases} n^{p+1} - (n-p)(n+1)^p & \text{if } j = 0, \\ (n-j+2)^{p+1} + (n-j)^{p+1} - 2(n-j-1)^{p+1} & \text{if } 1 \le j \le n, \\ 1 & \text{if } j = n+1, \end{cases} \tag{10}$$

and

$$Y_h^q(t_{n+1}) = \sum_{k=0}^{m-1} C_k \frac{t_{n+1}^k}{k!} + \frac{1}{\Gamma(p)} \sum_{j=0}^{n} b_{j,n+1} F\left(t_j, Y_h(t_j)\right), \tag{11}$$

in which $b_{j,n+1} = \frac{h^p}{p}\left((n+1-j)^p - (n-j)^p\right)$. Therefore, the estimation error of this approximation is $\max \left| Y(t_h) - Y_h(t_j) \right| = O(h^p)$, where $q = \min(2, 1+p)$.

3 GS Between *n*-D Fractional Order Master System and *m*-D Integer order Slave System

3.1 Theory

In this case, the master system is considered as follow

$$D_t^p X(t) = f_1(X(t)), \tag{12}$$

where $X(t) \in \mathbf{R}^n$ is the state vector of the master system (12), $f_1 : \mathbf{R}^n \to \mathbf{R}^n$, p is a rational number between 0 and 1 and D_t^p is the Caputo fractional derivative of order p.

The slave system is defined as

$$\dot{Y}(t) = AY(t) + g_1(Y(t)) + U, \tag{13}$$

where $Y(t) \in \mathbf{R}^m$ is the state vector of the slave system (13), $A \in \mathbf{R}^{m\times m}$ is a constant matrix, $g_1 : \mathbf{R}^m \to \mathbf{R}^m$ is a nonlinear function and $U = (u_i)_{1\leq i\leq m}$ is a vector controller to be determined.

The error system of GS between the master system (12) and the slave system (13) is defined as

$$e(t) = Y(t) - \phi_1 (X(t)), \tag{14}$$

where $\phi_1 : \mathbf{R}^n \to \mathbf{R}^m$ is a differentiable function. Then, the aim of GS is to find a controller U such that the error system satisfies

$$\lim_{t\to\infty} \|e(t)\| = 0, \tag{15}$$

where $\|.\|$ is the Euclidian norm.

The integer-order derivative of the error system (14) can be derived as follow

$$\begin{aligned} \dot{e}(t) &= \dot{Y}(t) - \dot{\phi}_1 (X(t)) \\ &= AY(t) + g_1(Y(t)) + U - \mathbf{D}\phi_1 (X(t)) \times \dot{X}(t), \end{aligned} \tag{16}$$

where $\mathbf{D}\phi_1 (X(t))$ is the jacobian matrix of the function ϕ_1.

The error system (16), can be written as

$$\begin{aligned} \dot{e}(t) = &\left(A - L_1\right) e(t) + L_1 Y(t) + \left(A - L_1\right) \phi_1 (X(t)) + g_1(Y(t)) \\ &- \mathbf{D}\phi_1 (X(t)) \times \dot{X}(t) + U, \end{aligned} \tag{17}$$

where $L_1 \in \mathbf{R}^{m\times m}$ is a constant control matrix to be selected.

Theorem 3 *If we select the control matrix L_1 such that $P_1 = \left(A - L_1\right)^T + \left(A - L_1\right)$ is a negative definite matrix, then the master system (12) and the slave system (13) are globally generalized synchronized with respect to ϕ_1, under the following controller*

$$U = -L_1 Y(t) - g_1(Y(t)) - \left(A - L_1\right) \phi_1 (X(t)) + \mathbf{D}\phi_1 (X(t)) \times \dot{X}(t). \tag{18}$$

Proof Applying the control law given in Eq. (18) to Eq. (17) yields the resulting error dynamics as follows

$$\dot{e}(t) = \left(A - L_1\right) e(t). \tag{19}$$

Construct the candidate Lyapunov function in the form

$$V(e(t)) = e^T(t) e(t), \tag{20}$$

we obtain

$$\begin{aligned}\dot{V}(e(t)) &= \dot{e}^T(t)e(t) + e^T(t)\dot{e}(t) \\ &= e^T(t)\left(A - L_1\right)^T e(t) + e^T(t)\left(A - L_1\right)e(t) \\ &= e^T(t)\left[\left(A - L_2\right)^T + \left(A - L_1\right)\right]e(t) = e^T(t)P_1e(t) < 0\end{aligned}$$

Thus, from the Lyapunov stability theory, it is immediate that all solutions of error system (19) go to zero as $t \to \infty$. Therefore, the master system (12) and the slave system (13) are globally generalized synchronized.

3.2 *Application*

As the master system we consider the following 3-D fractional order Lorenz system

$$\begin{aligned}D^p x_1 &= 10\left(x_2 - x_1\right), \\ D^p x_2 &= 28x_1 - x_1x_3 - x_2, \\ D^p x_3 &= x_1x_2 - \frac{8}{3}x_3.\end{aligned} \tag{21}$$

The fractional order Lorenz system (21) has a chaotic attractor, shown in Figs. 1 and 2, for $p = 0.993$ [206].

We consider the controlled integer order hyperchaotic Lorenz system as the slave system. The slave system can be written in the form

$$\dot{Y}(t) = AY(t) + g_1(Y(t)) + U, \tag{22}$$

where $Y(t) = (y_i(t))_{1 \le i \le 4}$,

$$A = \begin{bmatrix} -a_1 & a_1 & 0 & 1 \\ a_2 & -1 & 0 & -1 \\ 0 & 0 & -a_3 & 0 \\ 0 & 0 & 0 & 0 \end{bmatrix}, \quad g_1(Y(t)) = \begin{bmatrix} 0 \\ -y_1y_3 \\ y_1y_2 \\ 0.1y_2y_3 \end{bmatrix},$$

and $U = (u_1, u_2, u_3, u_4)^T$ is a controller to be determined. The integer order hyperchaotic Lorenz system (i.e., the uncontrolled system (22)) has a chaotic behavior when $\left(a_1, a_2, a_3\right) = \left(10, 28, \frac{8}{3}\right)$ [207]. Different views of the attractor of integer-order Lorenz hyperchaotic system are shown in Figs. 3 and 4.

Let us define the errors of GS between the master system (21) and the slave system (22) as

$$\begin{bmatrix} e_1 \\ e_2 \\ e_3 \\ e_4 \end{bmatrix} = \begin{bmatrix} y_1 \\ y_2 \\ y_3 \\ y_4 \end{bmatrix} - \phi_1\left(x_1, x_2, x_3\right), \tag{23}$$

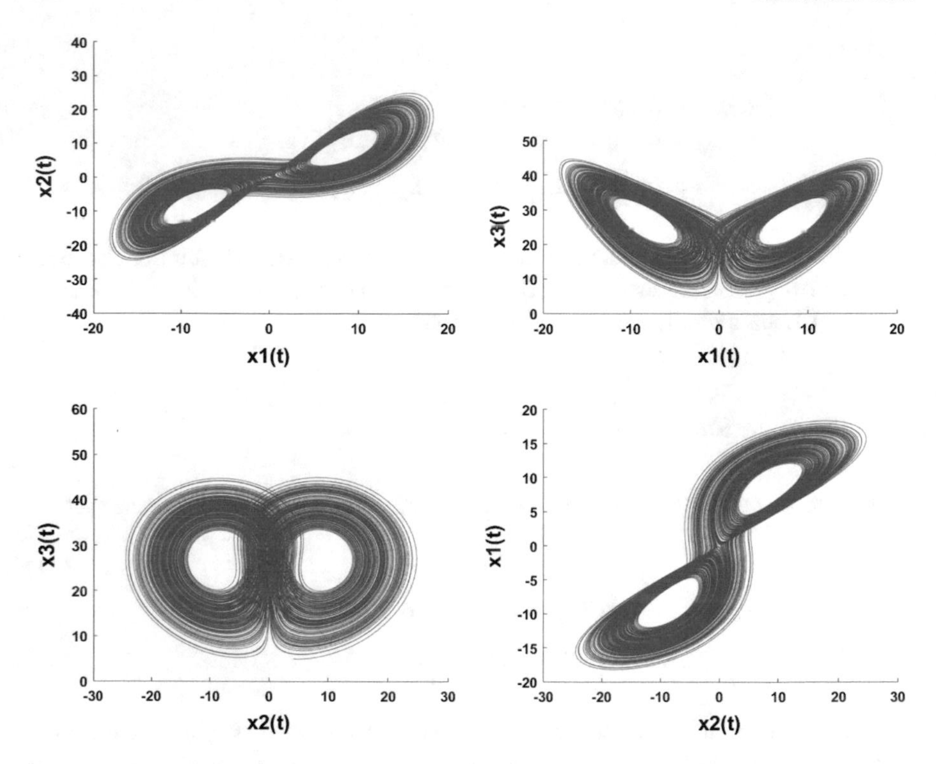

Fig. 1 Different 2-D projections of the fractional-order Lorenz chaotic system (21)

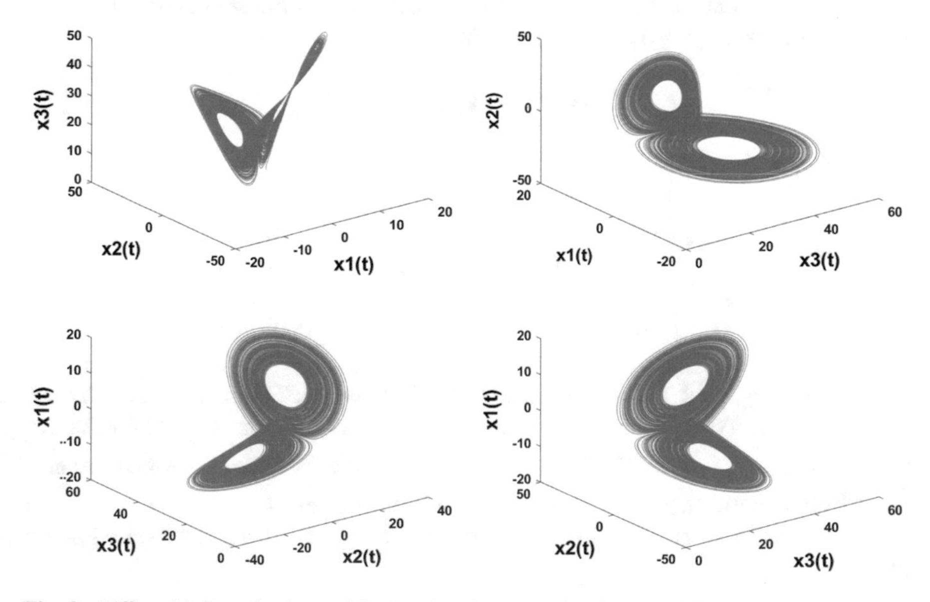

Fig. 2 Different 3-D projections of the fractional Lorenz chaotic system (21)

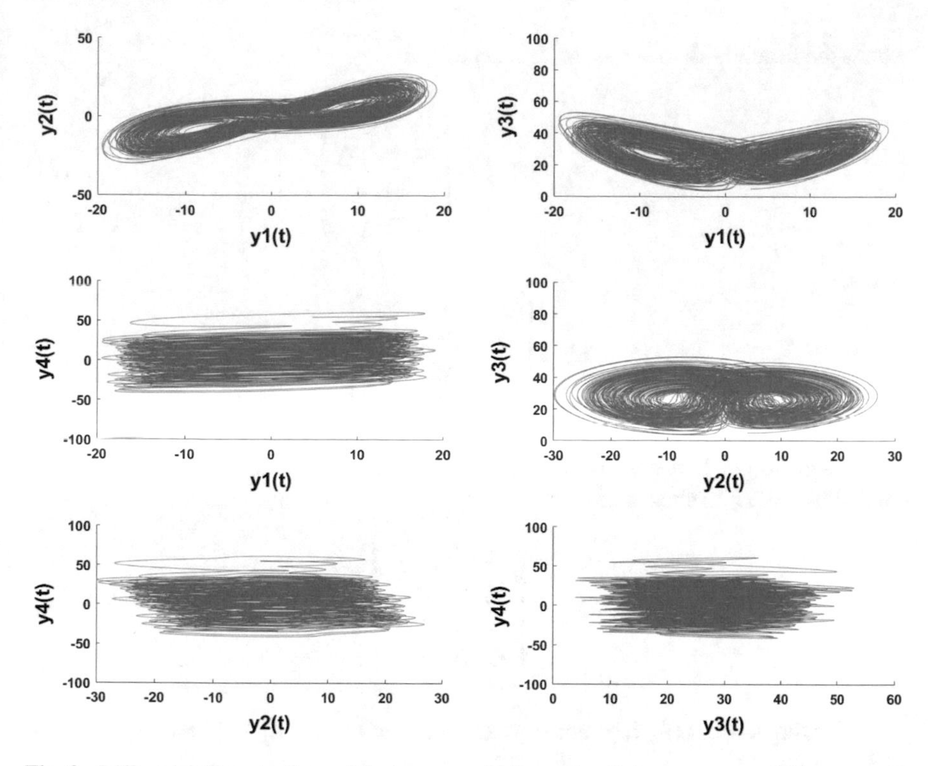

Fig. 3 Different 2-D projections of the integer order hyperchaotic Lorenz system (22)

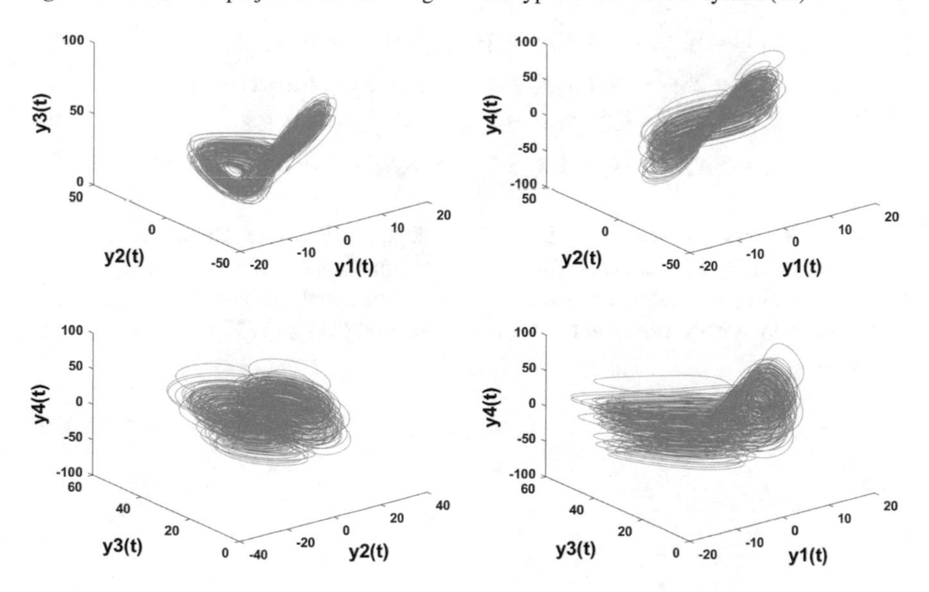

Fig. 4 Different 3-D projections of the integer order Lorenz hyperchaotic system (22)

where the function $\phi_1 : \mathbf{R}^3 \to \mathbf{R}^4$ is selected as

$$\phi_1(x_1, x_2, x_3) = \begin{bmatrix} x_1 + x_2x_3 \\ x_2^2 + x_1x_3 \\ x_3x_2x_1 \\ x_2 + x_1x_3 \end{bmatrix}. \tag{24}$$

So,

$$\mathbf{D}\phi = \begin{bmatrix} 1 & x_3 & x_2 \\ x_3 & 2x_2 & x_1 \\ x_3x_2 & x_1x_3 & x_2x_1 \\ x_3 & 1 & x_1 \end{bmatrix}. \tag{25}$$

To achieve GS between the master system (21) and the slave system (22), the control matrix L_1 is chosen as

$$L_1 = \begin{bmatrix} 0 & 10 & 0 & 1 \\ 28 & 0 & 0 & -1 \\ 0 & 0 & 0 & 0 \\ 0 & 0 & 0 & 1 \end{bmatrix}, \tag{26}$$

According to Eq. (18), the vector controller $U = (u_1, u_2, u_3, u_4)^T$ can be designed as follow

$$\begin{aligned} u_1 &= -10y_2 - y_4 + 10x_1 + 10x_2x_3 + \dot{x}_1 + x_3\dot{x}_2 + x_2\dot{x}_3, \\ u_2 &= -28y_1 + y_4 + y_1y_3 + x_2^2 + x_1x_3 + x_3\dot{x}_1 + 2x_2\dot{x}_2 + x_1\dot{x}_3, \\ u_3 &= -y_1y_2 + 8/3x_3x_2x_1 + x_3x_2\dot{x}_1 + x_1x_3\dot{x}_2 + x_2x_1\dot{x}_3, \\ u_4 &= -y_4 - 0.1y_2y_3 + x_2 + x_1x_3 + x_3\dot{x}_1 + \dot{x}_2 + x_1\dot{x}_3. \end{aligned} \tag{27}$$

We can see that $(A - L_1)^T + (A - L_1)$ is a negative definite matrix. Then, according to Theorem 9, the fractional- order Lorenz chaotic system and the integer order hyperchaotic Lorenz system are globally generalized synchronized. The error system of generalized synchronization between the systems (21) and (22) can be described as follow:

$$\begin{aligned} \dot{e}_1 &= -10e_1, \\ \dot{e}_2 &= -e_2, \\ \dot{e}_3 &= -\frac{8}{3}e_3, \\ \dot{e}_4 &= -e_4. \end{aligned} \tag{28}$$

Fourth order Runge–Kutta integration method has been used. In addition, time step size simulation time $Tm = 120$ s and time step $h = 0 : 005$ s have been employed. The initial values of the master system and the slave systems are $[x_1(0), x_2(0),$

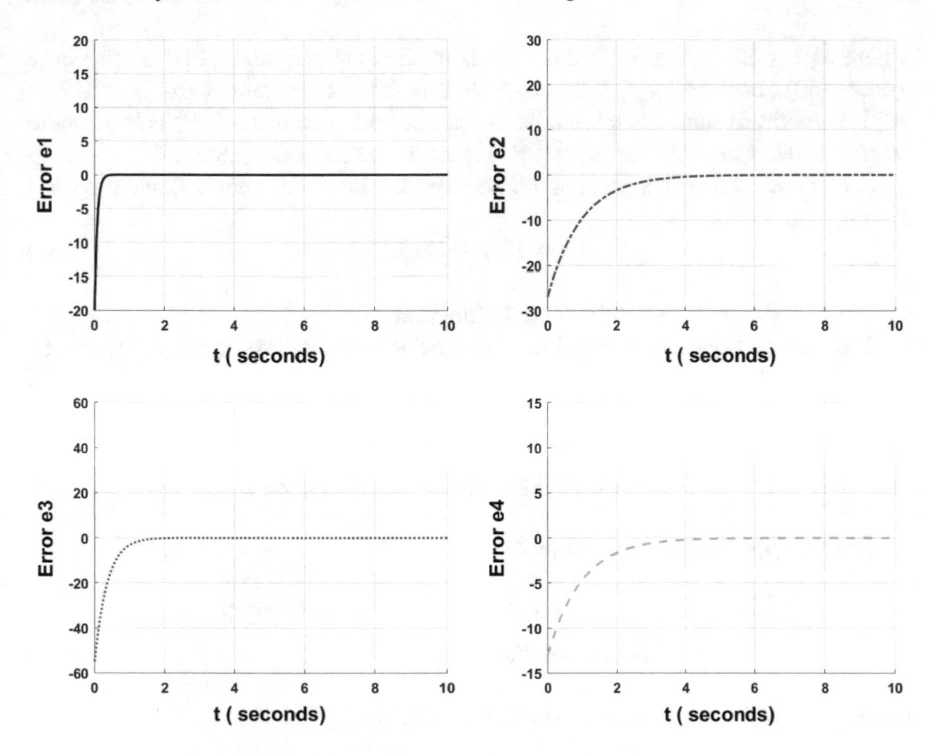

Fig. 5 Time-history of the synchronization errors e_1, e_2, e_3 and e_4

$x_3(0)] = [3, 4, 5]$ and $[y_1(0), y_2(0), y_3(0), y_4(0)] = [3, 4, 5, 6]$, respectively, and the initial states of the error system are $[e_1(0), e_2(0), e_3(0), e_4(0)] = [-20, -27, -55, -13]$. The state variables of the error system (28) versus time are depicted in Fig. 5.

4 GS Between *n*-D Integer-Order Master System and *m*-D fractional-Order Slave System

4.1 Theory

Now, the master and the slave chaotic systems are considered in the following forms

$$\dot{X}(t) = f_2(X(t)), \tag{29}$$

and

$$D_t^q Y(t) = BY(t) + g_2(Y(t)) + V, \tag{30}$$

where $X(t) \in \mathbf{R}^n$, $Y(t) \in \mathbf{R}^m$ are the states of the master system (29) and the slave system (30), respectively, $f_2 : \mathbf{R}^n \to \mathbf{R}^n$, $B \in \mathbf{R}^{m\times m}$ is a constant matrix, $g_2 : \mathbf{R}^m \to \mathbf{R}^m$ is a nonlinear function, q is a rational number between 0 and 1, D_t^q is the Caputo fractional derivative of order q, and V is a controller to be designed.

In this case, the error system of GS between the master system (29) and the slave system (30) is given by

$$e(t) = Y(t) - \phi_2 (X(t)), \tag{31}$$

where $\phi_2 : \mathbf{R}^n \to \mathbf{R}^m$ is a differentiable function.

The Caputo fractional derivative of the order q of the error system (31) can be derived as follow

$$\begin{aligned} D_t^q e(t) &= D_t^q Y(t) - D_t^q \phi (X(t)) \\ &= BY(t) + g_2(Y(t)) + V - D_t^q \phi_2 (X(t)). \end{aligned} \tag{32}$$

The error system can be described as

$$\begin{aligned} D_t^q e(t) = &\left(B - L_2\right) e(t) + L_2 Y(t) + \left(B - L_2\right) \phi_2 (X(t)) + g_2(Y(t)) \\ &- D_t^q \phi_2 (X(t)) + V, \end{aligned} \tag{33}$$

where $L_2 \in \mathbf{R}^{m\times m}$ is a constant control matrix to be chosen.

Theorem 4 *If we select the control matrix $L_2 \in \mathbf{R}^{m\times m}$ such that $P_2 = B - L_2$ is a negative definite matrix, then the master system (29) and the slave system (30) are globally generalized synchronized with respect to ϕ_2, under the following controller*

$$V = -L_2 Y(t) - \left(B - L_2\right) \phi_2 (X(t)) - g_2(Y(t)) + D_t^q \phi_2 (X(t)). \tag{34}$$

Proof From Eqs. (34)–(33), we get the error dynamical system as follows

$$D_t^q e(t) = \left(B - L_2\right) e(t). \tag{35}$$

Constructing the candidate Lyapunov function in the form

$$V(e(t)) = \frac{1}{2} e^T (t) e(t), \tag{36}$$

we obtain

$$D_t^q V(e(t)) = D_t^p \left(\frac{1}{2} e^T (t) e(t)\right), \tag{37}$$

and by using Lemma 8

$$\begin{aligned} D_t^q V(e(t)) &\leq e^T (t) D_t^q e(t) \\ &= e^T(t) \left(B - L_2\right) e(t) = e^T(t) P_2 e(t) < 0. \end{aligned}$$

Thus, from Theorem 7, that is the zero solution of the error system (35) is globally asymptotically stable and therefore the master system (29) and the slave system (30) are globally generalized synchronized.

4.2 Application

We consider the integer order hyperchaotic chen system as the master system. The master system can be described as follows

$$\begin{aligned} \dot{x}_1 &= 35(x_2 - x_1) + x_4, \\ \dot{x}_2 &= 7x_1 - x_1x_3 + 12x_2, \\ \dot{x}_3 &= x_1x_2 - 3x_3, \\ \dot{x}_4 &= x_2x_3 + rx_4, \end{aligned} \tag{38}$$

where r is a parameter. System (38) is hyperchaotic for $r = 0.5$ [208]. Hyperchaotic attractors of the system (38) are given in Figs. 6 and 7.

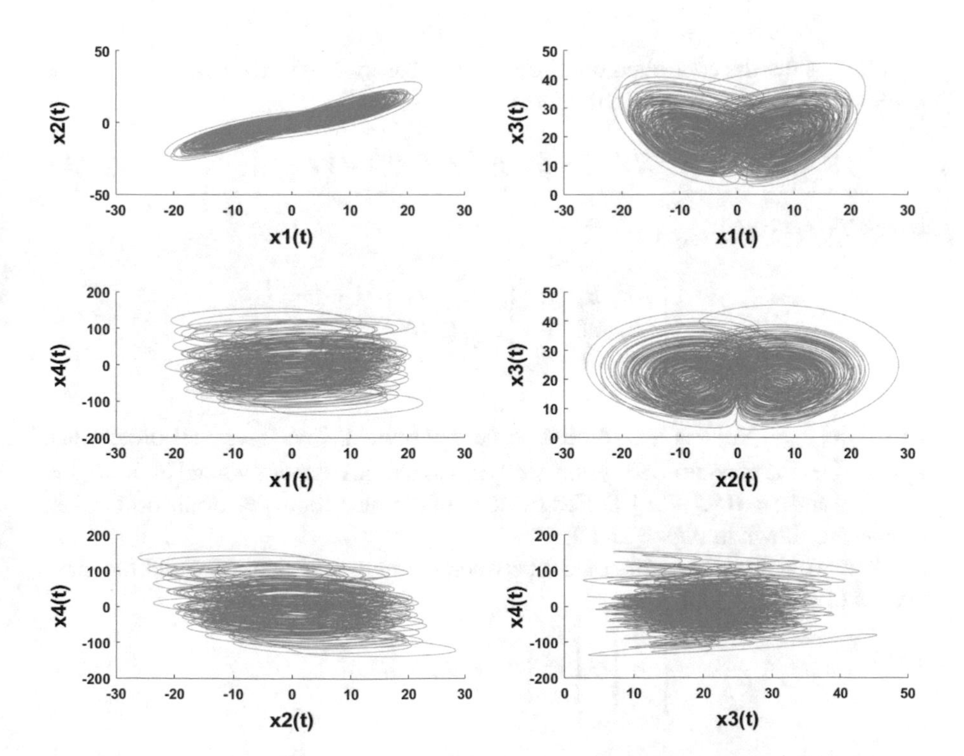

Fig. 6 Different 2-D projections of the integer-order hyperchaotic Chen system (38)

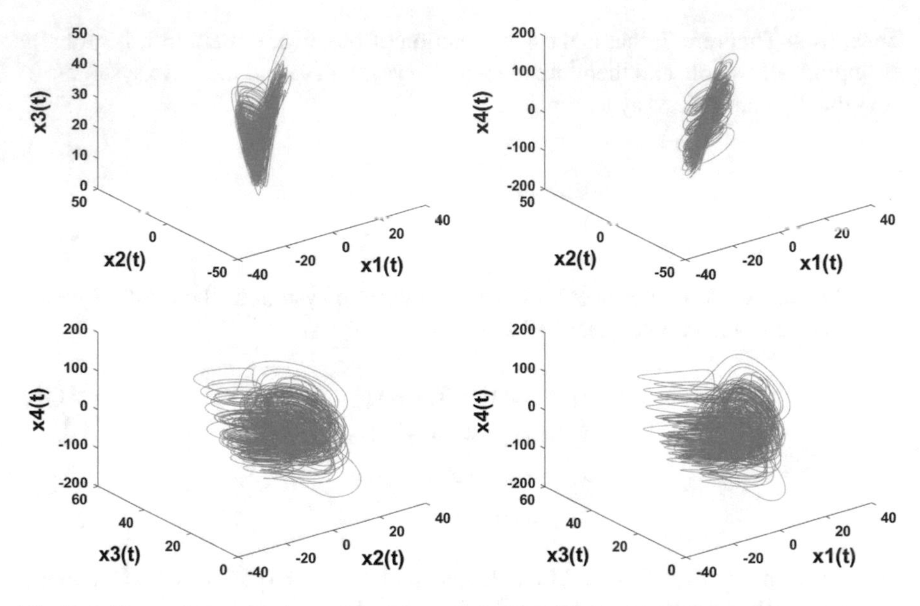

Fig. 7 Different 3-D projections of the integer order hyperchaotic Chen system (38)

We consider the controlled fractional order Chen system as the slave system. The slave system can be written in the form

$$D_t^q Y(t) = BY(t) + g_2(Y(t)) + V, \tag{39}$$

where $Y(t) = (y_i(t))_{1 \leq i \leq 3}$,

$$B = \begin{bmatrix} -b_1 & b_1 & 0 \\ b_2 & 28 & 0 \\ 0 & 0 & -b_3 \end{bmatrix}, \quad g_2(Y(t)) = \begin{bmatrix} 0 \\ -y_1 y_3 \\ y_1 y_2 \end{bmatrix},$$

and $V = \left(v_1, v_2, v_3\right)^T$ is a controller to be determined. The fractional order Chen system (i.e., the uncontrolled system (39)) has a chaotic behavior when $\left(b_1, b_2, b_3\right) = (35, 7, 3)$ and $q = 0.95$ [209]. Different views of the attractor of fractional order Chen system are shown in Figs. 8 and 9.

Now, we define the errors of GS between the master system (38) and the slave system (39) as

$$\begin{bmatrix} e_1 \\ e_2 \\ e_3 \end{bmatrix} = \begin{bmatrix} y_1 \\ y_2 \\ y_3 \end{bmatrix} - \phi_2\left(x_1, x_2, x_3, x_4\right), \tag{40}$$

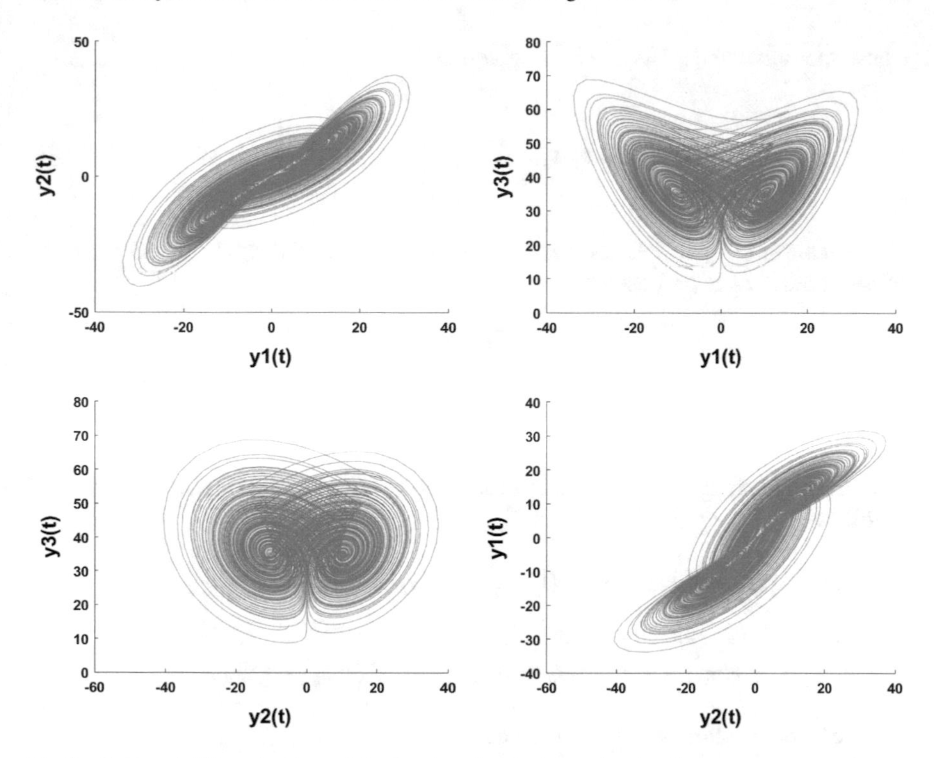

Fig. 8 Different 2-D projections of the fractional order hyperchaotic Chen system

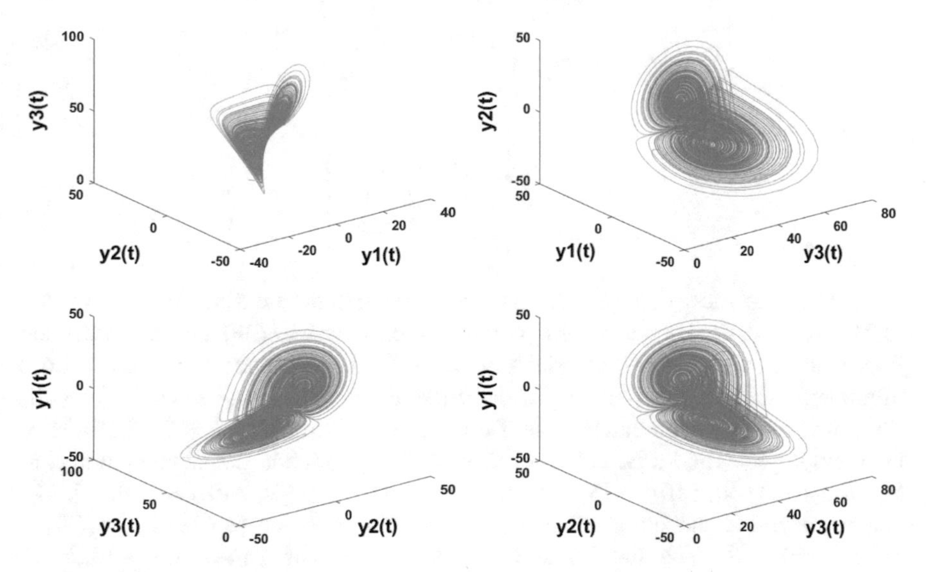

Fig. 9 Different 3-D projections of the fractional order Chen system

where the function $\phi_2 : \mathbf{R}^4 \to \mathbf{R}^3$ is given as

$$\phi_2\left(x_1, x_2, x_3, x_4\right) = \begin{bmatrix} x_1x_2x_3x_4 \\ x_1^2 + x_2^2 + x_3x_4 \\ x_1x_2 + x_3x_4 \end{bmatrix}. \tag{41}$$

To achieve GS between the master system (38) and the slave system (39), the control matrix L_2 is proposed as

$$L_2 = \begin{bmatrix} 0 & 35 & 0 \\ 7 & 56 & 0 \\ 0 & 0 & 19 \end{bmatrix}, \tag{42}$$

According to Eq. (34), the vector controller $V = \left(v_1, v_2, v_3\right)^T$ can be constructed as follow

$$\begin{aligned} v_1 &= -35y_2 + 35x_1x_2x_3x_4 + D_t^{0.95}\left(x_1x_2x_3x_4\right), \\ v_2 &= -7y_1 - 56y_2 + y_1y_3 + 28\left(x_1^2 + x_2^2 + x_3x_4\right) + D^{0.95}\left(x_1^2 + x_2^2 + x_3x_4\right), \\ v_3 &= 19y_3 + -y_1y_2 + 21\left(x_1x_2 + x_3x_4\right) + D^{0.95}\left(x_1x_2 + x_3x_4\right), \end{aligned} \tag{43}$$

and the error functions can be written as

$$\begin{bmatrix} D^{0.95}e_1 \\ D^{0.95}e_2 \\ D^{0.95}e_3 \end{bmatrix} = \left(B - L_2\right)\begin{bmatrix} e_1 \\ e_2 \\ e_3 \end{bmatrix}, \tag{44}$$

where

$$B - L_2 = \begin{bmatrix} -35 & 0 & 0 \\ 0 & -28 & 0 \\ 0 & 0 & -21 \end{bmatrix}. \tag{45}$$

It is easy to show that $\left(B - L_2\right)$ is a negative definite matrix. Then, according to Theorem 10, the integer order hyperchaotic chen system (38) and the controlled fractional Chen system (39) are globally generalized synchronized. For the purpose of numerical simulation, fractional Euler integration method has been used. In addition, time step size simulation time $Tm = 120$ s and time step $h = 0 : 005$ s have been employed. The initial values of the master system and the slave systems are $[x_1(0), x_2(0), x_3(0), x_4(0)] = [3, 4, 5, 6]$ and $[y_1(0), y_2(0), y_3(0)] = [-9, -5, 14]$, respectively, and the initial states of the error system are $[e_1(0), e_2(0), e_3(0)] = [-369, -60, -28]$. The state variables of the error system (44) versus time are depicted in Fig. 10.

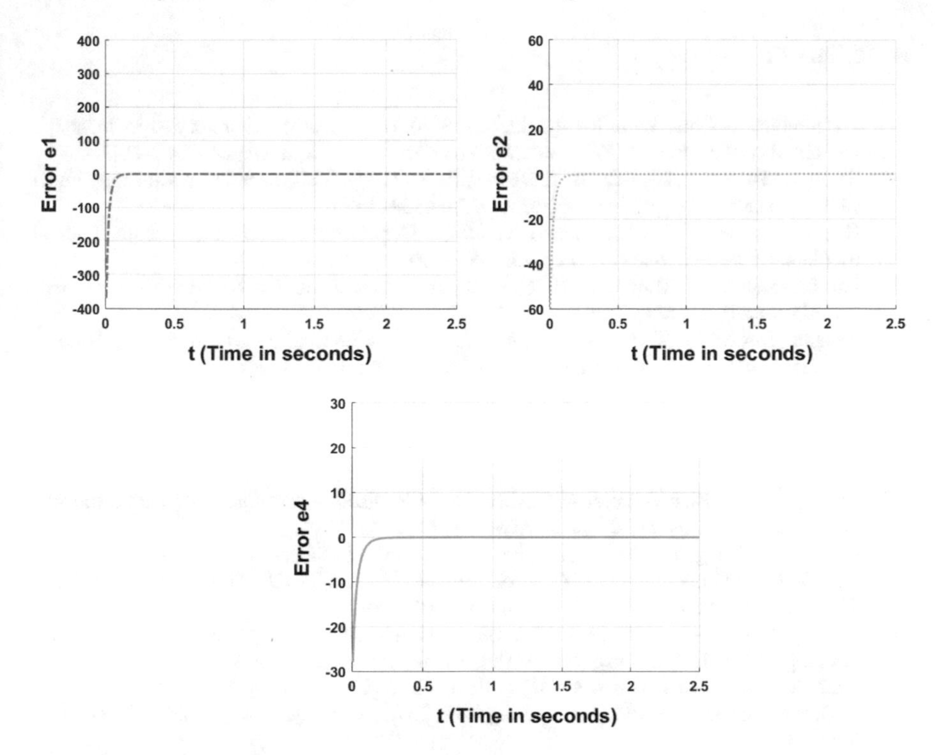

Fig. 10 Time-history of the synchronization errors e_1, e_2 and e_3

5 Conclusion

In this work, new generalized synchronization schemes between integer-order and fractional-order chaotic systems with different dimcnsional were presented. The New control schemes were proved theoretically using integer order Lyapunov stability theory and fractional order Lyapunov-based approach. Firstly, to achieve GS between n-dimensional fractional order master system and m-dimensional integer order slave system, the synchronization criterion was obtained via integer order control of the linear part of the slave system. Secondly, to observe GS behavior between $n-$dimensional integer order master system and m-dimensional fractional order slave system, the synchronization scheme was proposed based on the fractional order control of the linear part of the slave system. Simulations results were used to verify the effectiveness of the proposed control schemes.

References

1. Matsumoto, T., Chua, L., & Kobayashi, K. (1986). Hyperchaos: Laboratory experiment and numerical confirmation. *IEEE Transactions on Circuits and Systems*, *11*, 1143–1147.
2. Stoop, R., Peinke, J., Röhricht, B., & Übener, R. H. (1989). A p-Ge semiconductor experiment showing chaos and hyperchaos. *Physica D*, *35*, 4352–4425.
3. Han, S. K., Kerrer, C., & Kuramoto, Y. (1995). Dephasing and bursting in coupled neural oscillators. *Physical Review Letters*, *75*, 3190–3193.
4. Lakshmanan, M., & Murali, K. (1996). *Chaos in nonlinear oscillators: Controlling and synchronization*. World Scientific.
5. Strogatz, S. (2001). *Nonlinear dynamics and chaos: With applications to physics, biology, chemistry, and engineering*. Studies in Nonlinearity: Westview Press.
6. Blasius, B., & Stone, L. (2000). Chaos and phase synchronization in ecological systems. *International Journal of Bifurcation and Chaos*, *10*, 2361–2380.
7. Zhang, W. (2006). *Discrete dynamical systems, bifurcations, and chaos in economics*. Elsevier.
8. Eduardo, L., & Ruiz-Herrera, A. (2012). Chaos in discrete structured population models. *SIAM Journal on Applied Dynamical Systems*, *11*, 1200–1214.
9. Chen, G., & Dong, X. (1989). *From chaos to order*. World Scientific.
10. Boccaletti, S., Grebogi, C., Lai, Y. C., Mancini, H., & Maza, D. (2001). The control of chaos: Theory and application. *Physical Reports*, *329*(103), 197.
11. Azar, A. T., & Vaidyanathan, S. (2015). *Chaos modeling and control systems design*. Studies in Computational Intelligence (Vol. 581). Germany: Springer. ISBN 978-3-319-13131-3.
12. Azar, A. T., & Vaidyanathan, S. (2016). *Advances in Chaos theory and intelligent control*. Studies in Fuzziness and Soft Computing (Vol. 337). Germany: Springer. ISBN 978-3-319-30338-3.
13. Azar, A. T., & Vaidyanathan, S. (2015). *Computational intelligence applications in modeling and control*. Studies in Computational Intelligence (Vol. 575). Germany: Springer. ISBN 978-3-319-11016-5.
14. Azar, A. T., & Vaidyanathan, S. (2015). *Handbook of research on advanced intelligent control engineering and automation*. Advances in Computational Intelligence and Robotics (ACIR) Book Series, IGI Global, USA. ISBN 9781466672482.
15. Zhu, Q., & Azar, A. T. (2015). *Complex system modelling and control through intelligent soft computations*. Studies in Fuzziness and Soft Computing (Vol. 319). Germany: Springer. ISBN: 978-3-319-12882-5.
16. Azar, A. T., & Zhu, Q. (2015). *Advances and applications in sliding mode control systems*. Studies in Computational Intelligence (Vol. 576). Germany: Springer. ISBN: 978-3-319-11172-8.
17. Yamada, T., & Fujisaka, H. (1983). Stability theory of synchroized motion in coupled-oscillator systems. *Progress of Theoretical Physics*, *70*, 1240–1248.
18. Pecora, L. M., & Carroll, T. L. (1990). Synchronization in chaotic systems. *Physical Review Letters*, *64*, 821–827.
19. Carroll, T. L., & Pecora, L. M. (1991). Synchronizing a chaotic systems. *IEEE Transactions on Circuits and Systems*, *38*, 453–456.
20. Pikovsky, A., Rosenblum, M., & Kurths, J. (2001). *Synchronization an universal concept in nonlinear sciences*. Cambridge University Press.
21. Boccaletti, S., Kurths, J., Osipov, G., Valladares, D. L., & Zhou, C. S. (2002). The synchronization of chaotic systems. *Physical Reports*, *366*, 1–101.
22. Aziz-Alaoui, M. A. (2006). Synchronization of chaos. *Encyclopedia of Mathematical Physics*, *5*, 213–226.
23. Luo, A. (2009). A theory for synchronization of dynamical systems. *Communications in Nonlinear Science and Numerical Simulation*, *14*, 1901–1951.
24. Luo, Albert C. J. (2013). *Dynamical system and synchronization*.

25. Vaidyanathan, S., & Azar, A. T. (2015). Analysis, control and synchronization of a nine-term 3-D novel chaotic system. In A. T. Azar & S. Vaidyanathan (Eds.), *Chaos modeling and control systems design*. Studies in Computational Intelligence Book Series. Springer.
26. Vaidyanathan, S., & Azar, A. T. (2015). Anti-synchronization of identical chaotic systems using sliding mode control and an application to Vaidyanathan-Madhavan chaotic systems. In A. T. Azar & Q. Zhu (Eds.), *Advances and applications in sliding mode control systems*. Studies in Computational Intelligence Book Series: Springer.
27. Vaidyanathan, S., & Azar, A. T. (2015). Hybrid synchronization of identical chaotic systems using sliding mode control and an application to Vaidyanathan chaotic systems. In A. T. Azar & Q. Zhu (Eds.), *Advances and applications in sliding mode control systems*. Studies in Computational Intelligence Book Series: Springer.
28. Vaidyanathan, S., Idowu, B. A., & Azar, A. T. (2015). Backstepping controller design for the global chaos synchronization of Sprott's jerk systems. In A. T. Azar & S. Vaidyanathan (Eds.), *Chaos modeling and control systems design*. Studies In Computational Intelligence Book Series: Springer.
29. Vaidyanathan, S., Sampath, S., & Azar, A. T. (2015). Global chaos synchronisation of identical chaotic systems via novel sliding mode control method and its application to Zhu system. *International Journal of Modelling, Identification and Control (IJMIC), 23*(1), 92–100.
30. Vaidyanathan, S., Azar, A. T., Rajagopal, K., & Alexander, P. (2015). Design and SPICE implementation of a 12-term novel hyperchaotic system and its synchronization via active control,. (2015). *International Journal of Modelling on Identification and Control (IJMIC), 23*(3), 267–277.
31. Vaidyanathan, S., & Azar, A. T. (2016). Takagi-Sugeno fuzzy logic controller for Liu-Chen four-Scroll chaotic system. *International Journal of Intelligent Engineering Informatics, 4*(2), 135–150.
32. Vaidyanathan, S., & Azar, A. T. (2015). Analysis and control of a 4-D novel hyperchaotic system. In A. T. Azar & S. Vaidyanathan (Eds.), *Chaos modeling and control systems design*. Studies in Computational Intelligence (Vol. 581, pp. 19-38). Berlin: Springer GmbH. doi:10.1007/978-3-319-13132-0_2.
33. Vaidyanathan, S., & Azar, A. T. (2016). Dynamic analysis, adaptive feedback control and synchronization of an eight-term 3-D novel chaotic system with three quadratic nonlinearities. *Advances in Chaos theory and intelligent control*. Studies in Fuzziness and Soft Computing (Vol. 337). Germany: Springer.
34. Vaidyanathan, S., & Azar, A. T. (2016). Qualitative study and adaptive control of a novel 4-D hyperchaotic system with three quadratic nonlinearities. *Advances in Chaos theory and intelligent control*. Studies in Fuzziness and Soft Computing (Vol. 337). Germany: Springer.
35. Vaidyanathan, S., & Azar, A. T. (2016). A novel 4-D four-wing chaotic system with four quadratic nonlinearities and its synchronization via adaptive control method. *Advances in Chaos theory and intelligent control*. Studies in Fuzziness and Soft Computing (Vol. 337). Germany: Springer.
36. Vaidyanathan, S., & Azar, A. T. (2016). Adaptive control and synchronization of Halvorsen circulant chaotic systems. *Advances in Chaos theory and intelligent control*. Studies in Fuzziness and Soft Computing (Vol. 337). Germany: Springer.
37. Vaidyanathan, S., & Azar, A. T. (2016). Adaptive backstepping control and synchronization of a novel 3-D jerk system with an exponential nonlinearity. *Advances in Chaos theory and intelligent control*. Studies in Fuzziness and Soft Computing (Vol. 337). Germany: Springer.
38. Vaidyanathan, S., & Azar, A. T. (2016). Generalized projective synchronization of a novel hyperchaotic four-wing system via adaptive control method. *Advances in Chaos theory and intelligent control*. Studies in Fuzziness and Soft Computing (Vol. 337). Germany: Springer.
39. Boulkroune, A., Bouzeriba, A., Bouden, T., & Azar, A. T. (2016). Fuzzy adaptive synchronization of uncertain fractional-order chaotic systems. In A.T Azar & S. Vaidyanathan (Eds.), *Advances in Chaos theory and intelligent control*. Studies in Fuzziness and Soft Computing (Vol. 337). Germany: Springer.

40. Boulkroune, A., Hamel, S., & Azar, A. T. (2016). Fuzzy control-based function synchronization of unknown chaotic systems with dead-zone input. *Advances in Chaos theory and intelligent control*. Studies in Fuzziness and Soft Computing (Vol. 337). Germany: Springer.
41. Moaddy, K., Radwan, A. G., Salama, K. N., Momani, S., & Hashim, I. (2012). The fractional-order modeling and synchronization of electrically coupled neurons system. *Computers and Mathematics with Applications*, *64*, 3329–3339.
42. Radwan, A. G., Moaddy, K., & Momani, S. (2011). Stability and nonstandard finite difference method of the generalized Chua's circuit. *Computers and Mathematics with Applications*, *62*, 961–970.
43. Radwan, A. G., Moaddy, K., Salama, K. N., & Momani, S. (2014). Control and switching synchronization of fractional order chaotic systems using active control technique. *Journal of Advanced Research*, *5*(1), 125–132.
44. Radwan, A. G. (2013). On some generalized logistic maps with arbitrary power. *Journal of Advanced Research (JAR)*, *4*, 163–171.
45. Radwan, A. G., Soliman, A. M., & EL-Sedeek A. L. (2003). MOS realization of the double scroll-like chaotic equation. *IEEE Circuits and Sstems-I*, *50*(2), 285–288.
46. Radwan, A. G., Soliman, A. M., & EL-sedeek A. L. (2004). MOS realization of the modified Lorenz chaotic system. *Chaos, Solitons and Fractals*, *21*, 553–561.
47. Radwan, A. G., Soliman, A. M., & EL-sedeek A. L. (2003). An inductorless CMOS realization of Chua's circuit. *Chaos, Solitons and Fractals*, *18*, 149–158.
48. Zidan, M. A., Radwan, A. G., & Salama, K. N. (2012). Controllable V-shape multi-scroll butterfly attractor: System and circuit implementation. *International Journal of Bifurcation and Chaos (IJBC)*, *22*, 6.
49. Barakat, M. L., Mansingka, A. S., Radwan, A. G., & Salama, K. N. (2013). Generalized hardware post processing technique for chaos-based pseudo random number generators. *ETRI Journal*, *35*(3), 448–458.
50. Radwan, A. G., Soliman, A. M., & Elwakil, A. S. (2007). 1-D digitally-controlled multi-scroll chaos generator. *International Journal of Bifurcation and Chaos*, *17*(1), 227–242.
51. Ouannas, A. (2014). Chaos synchronization approach based on new criterion of stability. *Nonlinear Dynamics and Systems Theory*, *14*, 396–402.
52. Ouannas, A. (2014). On full state hybrid projective synchronization of general discrete chaotic systems. *Journal of Nonlinear Dynamics,* 1–6.
53. Ouannas, A. (2014). Some synchronization criteria for N-dimensional chaotic systems in discrete-time. *Journal of Advanced Research in Applied Mathematics*, *6*, 1–10.
54. Ouannas, A. (2014). On inverse full state hybrid projective synchronization of chaotic dynamical systems in discrete-time. *International Journal of Nonlinear Dynamics and Control,* 1–7.
55. Ouannas, A. (2015). Synchronization criterion for a class of N-dimensional discrete chaotic systems. *Journal of Advanced Research in Dynamical and Control Systems*, *7*, 82–89.
56. Ouannas, A. (2015). A new synchronization scheme for general 3D quadratic chaotic systems in discrete-time. *Nonlinear Dynamics and Systems Theory*, *15*, 163–170.
57. Ouannas, A., Odibat, Z., & Shawagfeh, N. (2016). A new Q–S synchronization results for discrete chaotic systems. *Differential Equations and Dynamical Systems,* 1–10.
58. Ouannas, A. (2016). Co-existence of various synchronization-types in hyperchaotic Maps. *Nonlinear Dynamics and Systems Theory*, *16*, 312–321.
59. Ouannas, A., Azar, A. T., Abu-Saris, R. (2016). A new type of hybrid synchronization between arbitrary hyperchaotic maps. *International Journal of Machine Learning and Cybernetics,* 1–8.
60. Cuomo, K. M., Oppenheim, A. V., & Strogatz, S. H. (1993). Synchronization of Lorenz-based chaotic circuits with applications to communications. *IEEE Transactions on Circuits and Systems I*, *40*, 626–633.
61. Yang, T., & Chua, L. O. (1997). Impulsive stabilization for control and synchronization of chaotic systems theory and application to secure communication. *IEEE Transactions on Circuits and Systems I*, *44*, 976–988.

62. Liao, T. L., & Huang, N. S. (1999). An observer-based approach for chaotic synchronization with applications to secure communication. *IEEE Transactions on Circuits and Systems II, 46*, 1144–1150.
63. Morgul, O., & Feki, M. (1999). A chaotic masking scheme by using synchronized chaotic systems. *Physics Letters A, 251*, 169–176.
64. He, Z., Li, K., Yuang, L., & Sui, Y. (2000). A robust digital structure communications scheme based on sporadic chaos synchronization. *IEEE Transactions on Circuits and Systems I, 47*, 397–403.
65. Lian, K. Y., Chiang, T. S., Chiu, C. S., & Liu, P. (2001). Synthesis of fuzzy model-based designs to synchronization and secure communication for chaotic systems. *IEEE Transactions on Systems, Man, and Cybernetics B, 31*, 66–83.
66. Boutayeb, M., Darouach, M., & Rafaralahy, H. (2002). Generalized state observers for chaotic synchronization and secure communication. *IEEE Transactions on Circuits and Systems I, 49*, 345–349.
67. Feki, M. (2003). An adaptive chaos synchronization scheme applied to secure communication. *Chaos Solitons Fractals, 18*, 141–148.
68. Bowong, S. (2004). Stability analysis for the synchronization of chaotic systems with different order: application to secure communications. *Physics Letters A, 326*, 102–113.
69. Zhu, F. (2009). Observer-based synchronization of uncertain chaotic systems and its application to secure communications. *Chaos Solitons Fractals, 40*, 2384–2391.
70. Aguilar-Bustos, A. Y., & Cruz Hernandez, Y. C. (2009). Synchronization of discrete-time hyperchaotic systems: An application in communications. *Chaos Solitons Fractals, 41*, 1301–1310.
71. Bagley, R. L., & Calico, R. A. (1991). Fractional order state equations for the control of viscoelastically damped structures. *Journal of Guidance Control and Dynamics, 14*, 304–311.
72. Heaviside, O. (1971). *Electromagnetic theory*. Chelsea.
73. Sugimoto, N. (1991). Burgers equation with a fractional derivative: Hereditary effects on nonlinear acoustic waves. *Journal of Fluid Mechanics, 225*, 631–653.
74. Kusnezov, D., Bulgac, A., & Dang, G. D. (1999). Quantum Lévy processes and fractional kinetics. *Physical Review Letters, 82*, 1136–1139.
75. Jiao, Z., Chen, Y. Q., Podlubny, I. (2012). *Distributed-order dynamic systems*. Springer.
76. Parada, F. J. V., Tapia, J. A. O., & Ramirez, J. A. (2007). Effective medium equations for fractional Fick's law in porous media. *Physica A, 373*, 339–353.
77. Torvik, P. J., & Bagley, R. L. (1984). On the appearance of the fractional derivative in the behavior of real materials. *Transactions of ASME, 51*, 294–298.
78. Arena, P., Caponetto, R., Fortuna, L., & Porto, D. (2000). *Nonlinear noninteger order circuits and systems–An Introduction*. World Scientific.
79. Bode, H. W. (1949). *Network analysis and feedback amplifier design*. Tung Hwa Book Company.
80. Carlson, G. E., & Halijak, C. A. (1964). Approximation of fractional capacitors $\left(\frac{1}{s}\right)^{\frac{1}{n}}$ by a regular Newton process. *IEEE Transactions on Circuits and Theory, 11*, 210–213.
81. Nakagava, M., & Sorimachi, K. (1992). Basic characteristics of a fractance device. *IEICE Transactions on Fundamentals of Electronics, 75*, 1814–1818.
82. Westerlund, S. (2002). Dead matter has memory! Causal consulting.
83. Axtell, M., & Bise, E. M. (1990). Fractional calculus applications in control systems. In *Proceedings of the IEEE National Aerospace and Electronics Conference*, New York, pp. 563–566.
84. Dorčák, L. (1994). *Numerical models for the simulation of the fractional-order control systems, UEF-04-94*. Košice, Slovakia: The Academy of Sciences, Institute of Experimental Physics
85. Bagley, R. L., & Torvik, P. J. (1994). On the appearance of the fractional derivative in the behavior of real materials. *Journal of Applied Mechanics, 51*, 294–298.

86. Kilbas, A. A., Srivastava, H. M., & Trujillo, J. J. (2006). *Theory and applications of fractional differential equations*. Elsevier.
87. Hilfer, R. (2000). *Applications of fractional calculus in physics*. World Scientific.
88. Koeller, R. C. (1984). Application of fractional calculus to the theory of viscoelasticity. *Journal of Applied Mechanics*, *51*, 299–307.
89. Pires, E. J. S., Machado, J. A. T., & de Moura, P. B. (2003). Fractional order dynamics in a GA planner. *Signal Processing*, *83*, 2377–2386.
90. Podlubny, I. (1999). Fractional-order systems and $\mathbf{PI}^{\lambda}\mathbf{D}^{\mu}$-controllers. *IEEE Transactions on Automatic Control*, *44*, 208–213.
91. Oustaloup, A. (1995). La derivation non entiere: theorie, synthese et applications. Hermes.
92. Da Graca, Marcos M., Duarte, F. B. M., & Machado, J. A. T. (2008). Fractional dynamics in the trajectory control of redundant manipulators. *Communications in Nonlinear Science and Numerical Simulation*, *13*, 1836–1844.
93. Tseng, C. C. (2007). Design of FIR and IIR fractional order Simpson digital integrators. *Signal Processing*, *87*, 1045–1057.
94. Vinagre, B. M., Chen, Y. Q., & Petráš, I. (2003). Two direct Tustin discretization methods for fractional-order differentiator/integrator. *Journal of The Franklin Institute*, *340*, 349–362.
95. Magin, R. L. (2006). *Fractional calculus in bioengineering*. Begell House Publishers.
96. Wang, J. C. (1987). Realizations of generalized Warburg impedance with RC ladder networks and transmission lines. *Journal of The Electrochemical Society*, *134*, 1915–1920.
97. Arena, P., Caponetto, R., Fortuna, L., & Porto, D. (1998). Bifurcation and chaos in noninteger order cellular neural networks. *International Journal of Bifurcation and Chaos*, *8*, 1527–1539.
98. Hartley, T., Lorenzo, C., & Qammer, H. (1995). Chaos in a fractional order Chua's system. *IEEE Transactions on Circuits and Systems I: Fundamental Theory and Applications*, *42*, 485–490.
99. Ahmad, W. M., & Sprott, J. C. (2003). Chaos in fractional-order autonomous nonlinear systems. *Chaos Solitons, Fractals*, *16*, 339–351.
100. Grigorenko, I., & Grigorenko, E. (2003). Chaotic dynamics of the fractional Lorenz system. *Physical Review Letters*, *91*(034101), 39.
101. Li, C., & Chen, G. (2004). Chaos and hyperchaos in fractional order Rössler equations. *Physica A*, *341*, 55–61.
102. Li, C., & Chen, G. (2004). Chaos in the fractional order Chen system and its control. *Chaos Solitons Fractals*, *22*, 549–554.
103. Lu, J. G. (2005). Chaotic dynamics and synchronization of fractional-order Arneodo's systems. *Chaos Solitons Fractals*, *26*, 1125–1133.
104. Deng, W. H., & Li, C. P. (2005). Chaos synchronization of the fractional Lü system. *Physica A*, *353*, 61–72.
105. Guo, L. J. (2005). Chaotic dynamics and synchronization of fractional-order Genesio-Tesi systems. *Chinese Physics*, *14*, 1517–1521.
106. Ahmad, W. M. (2005). Hyperchaos in fractional order nonlinear systems. *Chaos Solitons Fractals*, *26*, 1459–1465.
107. Gao, X., & Yu, J. (2005). Chaos in the fractional order periodically forced complex Duffing's oscillators. *Chaos Solitons Fractals*, *24*, 1097–1104.
108. Lu, J. G., & Chen, G. (2006). A note on the fractional-order Chen system. *Chaos Solitons Fractals*, *27*, 685–688.
109. Petráš, I. (2006). A Note on the fractional-order cellular neural networks. *Proceedings of the International Journal of Conference Neural Networks,* 1021–1024.
110. Ge, Z. M., & Hsu, M. Y. (2007). Chaos in a generalized van der Pol system and in its fractional order system. *Chaos Solitons Fractals*, *33*, 1711–1745.
111. Ahmed, E., El-Sayed, A. M. A., & El-Saka, H. A. A. (2007). Equilibrium points, stability and numerical solutions of fractional-order predator-prey and rabies models. *Journal of Mathematical Analysis*, *325*, 542–553.

112. Barbosa, R. S., Machado, J. A. T., Vinagre, B. M., & Calderón, A. J. (2007). Analysis of the Van der Pol oscillator containing derivatives of fractional order. *Journal of Vibration and Control, 13*, 1291–1301.
113. Ge, Z. M., & Ou, C. Y. (2007). Chaos in a fractional order modified Duffing system. *Chaos Solitons Fractals, 34*, 262–291.
114. Chen, W. C. (2008). Nonlinear dynamic and chaos in a fractional-order financial system. *Chaos Solitons Fractals, 36*, 1305–1314.
115. Sheu, L. J., Chen, H. K., Chen, J. H., Tam, L. M., Chen, W. C., Lin, K. T., et al. (2008). Chaos in the Newton-Leipnik system with fractional order. *Chaos Solitons Fractals, 36*, 98–103.
116. Petráš, I. (2008). A note on the fractional-order Chua's system. *Chaos Solitons Fractals, 38*, 140–147.
117. Liu, C., Liu, L., & Liu, T. (2009). A novel three-dimensional autonomous chaos system. *Chaos Solitons Fractals, 39*, 1950–1958.
118. Petráš, I. (2009). Chaos in the fractional-order Volta's system: Modeling and simulation. *Nonlinear Dynamics, 57*, 157–170.
119. Petráš, I. (2010). A note on the fractional-order Volta's system. *Communications in Nonlinear Science and Numerical Simulation, 15*, 384–393.
120. Deng, H., Li, T., Wang, Q., & Li, H. (2009). A fractional-order hyperchaotic system and its synchronization. *Chaos Solitons Fractals, 41*, 962–969.
121. Gejji, V. D., & Bhalekar, S. (2010). Chaos in fractional ordered Liu system. *Computers & Mathematics with Applications, 59*, 1117–1127.
122. Li, C. G., Liao, X. F., & Yu, J. B. (2003). Synchronization of fractional order chaotic systems. *Physics Review E, 68*, 067203.
123. Gao, X., & Yu, J. B. (2005). Synchronization of two coupled fractional-order chaotic oscillators. *Chaos Solitons Fractals, 26*, 141–145.
124. Li, C., & Zhou, T. (2005). Synchronization in fractional-order differential systems. *Physica D, 212*, 111–125.
125. Zhou, S., Li, H., Zhu, Z., & Li, C. (2008). Chaos control and synchronization in a fractional neuron network system. *Chaos Solitons Fractals, 36*, 973–984.
126. Peng, G. (2007). Synchronization of fractional order chaotic systems. *Physics Letters A, 363*, 426–432.
127. Sheu, L. J., Chen, H. K., Chen, J. H., & Tam, L. M. (2007). Chaos in a new system with fractional order. *Chaos Solitons Fractals, 31*, 1203–1212.
128. Li, C., & Yan, J. (2007). The synchronization of three fractional differential systems. *Chaos Solitons Fractals, 32*, 751–757.
129. Wang, J., Xiong, X., & Zhang, Y. (2006). Extending synchronization scheme to chaotic fractional-order Chen systems. *Physica A, 370*, 279–285.
130. Li, C. P., Deng, W. H., & Xu, D. (2006). Chaos synchronization of the Chua system with a fractional order. *Physica A, 360*, 171–185.
131. Zhu, H., Zhou, S., & Zhang, J. (2009). Chaos and synchronization of the fractional-order Chua's system. *Chaos Solitons Fractals, 39*, 1595–1603.
132. Zhang, F., Chen, G., Li, C., & Kurths, J. (2013). Chaos synchronization in fractional differential systems. *Philosophical Transactions of the Royal Society A, 371*, 1–26.
133. Ansari, M. A., Arora, D., & Ansari, S. P. (2016). Chaos control and synchronization of fractional order delay-varying computer virus propagation model. *Mathematical Methods in the Applied Sciences, 39*, 1197–1205.
134. Chen, L., Wu, R., He, Y., & Chai, Y. (2015). Adaptive sliding-mode control for fractional-order uncertain linear systems with nonlinear disturbances. *Nonlinear Dynamics, 80*, 51–58.
135. Srivastava, M., Ansari, S. P., Agrawal, S. K., Das, S., & Leung, A. Y. T. (2014). Anti-synchronization between identical and non-identical fractional-order chaotic systems using active control method. *Nonlinear Dynamics, 76*, 905–914.
136. Agrawal, S. K., & Das, S. (2013). A modified adaptive control method for synchronization of some fractional chaotic systems with unknown parameters. *Nonlinear Dynamics, 73*, 907–919.

137. Chen, X. R., & Liu, C. X. (2012). Chaos synchronization of fractional order unified chaotic system via nonlinear control. *International Journal of Modern Physics B*, *25*, 407–415.
138. Cafagna, D., & Grassi, G. (2012). Observer-based projective synchronization of fractional systems via a scalar signal: Application to hyperchaotic Rössler systems. *Nonlinear Dynamics*, *68*, 117–128.
139. Odibat, Z. M. (2012). A note on phase synchronization in coupled chaotic fractional order systems. *Nonlinear Analysis: Real World Applications*, *13*, 779–789.
140. Chen, F., Xia, L., & Li, C. G. (2012). Wavelet phase synchronization of fractional-order chaotic systems. *Chinese Physics Letters*, *29*, 070501–070506.
141. Razminiaa, A., & Baleanu, D. (2013). Complete synchronization of commensurate fractional order chaotic systems using sliding mode control. *Mechatronics*, *23*, 873–879.
142. Agrawal, S. K., & Das, S. (2014). Projective synchronization between different fractional-order hyperchaotic systems with uncertain parameters using proposed modified adaptive projective synchronization technique. *Mathematical Methods in the Applied Sciences*, *37*, 2164–2176.
143. Chang, C. M., & Chen, H. K. (2010). Chaos and hybrid projective synchronization of commensurate and incommensurate fractional-order Chen-Lee systems. *Nonlinear Dynamics*, *62*, 851–858.
144. Xi, H., Li, Y., & Huang, X. (2015). Adaptive function projective combination synchronization of three different fractional-order chaotic systems. *Optik*, *126*, 5346–5349.
145. Zhou, P., Kuang, F., & Cheng, Y. M. (2010). Generalized projective synchronization for fractional order chaotic systems. *Chinese Journal of Physics*, *48*, 49–56.
146. Yi, C., Liping, C., Ranchao, W., & Juan, D. (2013). Q-S synchronization of the fractional-order unified system. *Pramana*, *80*, 449–461.
147. Mathiyalagan, K., Park, J. H., & Sakthivel, R. (2015). Exponential synchronization for fractional-order chaotic systems with mixed uncertainties. *Complexity*, *21*, 114–125.
148. Aghababa, M. P. (2012). Finite-time chaos control and synchronization of fractional-order nonautonomous chaotic (hyperchaotic) systems using fractional nonsingular terminal sliding mode technique. *Nonlinear Dynamics*, *69*, 247–261.
149. Li, D., Zhang, X. P., Hu, Y. T., & Yang, Y. Y. (2015). Adaptive impulsive synchronization of fractional order chaotic system with uncertain and unknown parameters. *Neurocomputing*, *167*, 165–171.
150. Ouannas, A., Al-sawalha, M. M., & Ziar, T. (2016). Fractional chaos synchronization schemes for different dimensional systems with non-Identical fractional-orders via two scaling matrices. *Optik*, *127*, 8410–8418.
151. Ouannas, A., Azar, A. T., & Vaidyanathan, S. (2016). A robust method for new fractional hybrid chaos synchronization. *Mathematical Methods in the Applied Sciences,* 1–9.
152. Kiani, B. A., Fallahi, K., Pariz, N., & Leung, H. (2009). A chaotic secure communication scheme using fractional chaotic systems based on an extended fractional Kalman filter. *Communications in Nonlinear Science and Numerical Simulation*, *14*, 863–879.
153. Liang, H., Wang, Z., Yue, Z., & Lu, R. (2012). Generalized synchronization and control for incommensurate fractional unified chaotic system and applications in secure communication. *Kybernetika*, *48*, 190–205.
154. Wu, X., Wang, H., & Lu, H. (2013). Modified generalized projective synchronization of a new fractional-order hyperchaotic system and its application to secure communication. *Nonlinear Analysis: Real World Applications*, *13*, 1441–1450.
155. Muthukumar, P., & Balasubramaniam, P. (2013). Feedback synchronization of the fractional order reverse butterfly-shaped chaotic system and its application to digital cryptography. *Nonlinear Dynamics*, *74*, 1169–1181.
156. Muthukumar, P., Balasubramaniam, P., & Ratnavelu, K. (2014). Synchronization of a novel fractional order stretch-twistfold (STF) flow chaotic system and its application to a new authenticated encryption scheme (AES). *Nonlinear Dynamics*, *77*, 1547–1559.
157. Si, G. Q., Sun, Z. Y., & Zhang, Y. B. (2011). A general method for synchronizing an integer-order chaotic system and a fractional-order chaotic system. *Chinese Physics B*, *20*, 080505–080507.

158. Wu, Y., & Wang, G. (2014). Synchronization of a class of fractional-order and integer order hyperchaotic systems. *Journal of Vibration and Control*, *20*, 1584–1588.
159. Chen, D., Zhang, R., Ma, X., & Wang, J. (2012). Synchronization between a novel class of fractional-order and integer-order chaotic systems via a sliding mode controller. *Chinese Physics B*, *21*, 120507.
160. Chen, D., Zhang, R., Sprott, J. C., Chen, H., & Ma, X. (2012). Synchronization between integer-order chaotic systems and a class of fractional-order chaotic systems via sliding mode control. *Chaos*, *22*, 023130.
161. Wu, Y. P., & Wang, G. D. (2013) Synchronization between Fractional-Order and Integer-Order Hyperchaotic Systems via Sliding Mode Controller. Journal of Applied Mathematics, 1–5.
162. Chen, D., Zhang, R., Sprott, J. C., & Ma, X. (2012). Synchronization between integer-order chaotic systems and a class of fractional-order chaotic system based on fuzzy sliding mode control. *Nonlinear Dynamics*, *70*, 1549–1561.
163. Chen, D., Wu, C., Iu, H. H. C., & Ma, X. (2013). Circuit simulation for synchronization of a fractional-order and integer-order chaotic system. *Nonlinear Dynamics*, *73*, 1671–1686.
164. Zhou, P., Cheng, Y. M., & Kuang, F. (2010). Synchronization between fractional-order chaotic systems and integer orders chaotic systems (fractional-order chaotic systems). *Chinese Physics B*, *19*, 090503.
165. Yang, L. X., He, W. S., & Liu, X. J. (2011). Synchronization between a fractional-order system and an integer order system. *Computers & Mathematics with Applications*, *62*, 4708–4716.
166. Jia, L. X., Dai, H., & Hui, M. (2010). Nonlinear feedback synchronisation control between fractional-order and integer-order chaotic systems. *Chinese Physics B*, *19*, 110509.
167. El Gammoudi, I., & Feki, M. (2013). Synchronization of integer order and fractional order Chua's systems using robust observer. *Communications in Nonlinear Science and Numerical Simulation*, *18*, 625–638.
168. Khan, A., & Tripathi, P. (2013). Synchronization between a fractional order chaotic system and an integer order chaotic system. *Nonlinear Dynamics and Systems Theory*, *13*, 425–436.
169. Wu, Y., & Wang, G. (2013). Synchronization and between a class of fractional order and integer order chaotic systems with only one controller term. *Journal of Theoretical and Applied Information Technology*, *48*, 145–151.
170. Dong, P., Shang, G., & Liul, J. (2012). Anticiping synchronization of integer order and fractional order hperchaotic Chen system. *International Journal of Modern Physics*, *26*, 1250211–1250215.
171. Zhou, P., & Cao, Y. X. (2010). Function projective synchronization between fractional-order chaotic systems and integer-order chaotic systems. *Chinese Physics B*, *19*, 100507.
172. Ouannas, A., & Abu-Saris, R. (2015). A Robust control method for Q-S synchronization between different dimensional integer-order and fractional-order chaotic systems. *Journal of Control Science and Engineering*, 1–7.
173. Rulkov, N. F., Sushchik, M. M., Tsimring, L. S., & Abarbanel, H. D. (1995). Generalized synchronization of chaos in directionally coupled chaotic systems. *Physical Review E*, *51*, 980–994.
174. Kocarev, L., & Parlitz, U. (1996). Generalized synchronization, predictability, and equivalence of unidirectionally coupled dynamical systems. *Physical Review Letters*, *76*, 1816–1819.
175. Abarbanel, H. D., Rulkov, N. F., & Sushchik, M. M. (1996). Generalized synchronization of chaos: The auxiliary system approach. *Physical Review E*, *53*, 4528–4535.
176. Pyragas, K. (1998). Properties of generalized synchronization of chaos. *Nonlinear Analysis: Modelling and Control*, *3*, 1–29.
177. Ji, Y., Liu, T., & Min, L. Q. (2008). Generalized chaos synchronization theorems for bidirectional differential equations and discrete systems with applications. *Physics Letters A*, *372*, 3645–3652.
178. Ma, Z., Liu, Z., & Zhang, G. (2007). Generalized synchronization of discrete systems. *Applied Mathematics and Mechanics*, *28*, 609–614.

179. Grassi, G. (2012). Generalized synchronization between different chaotic maps via dead-beat control. *Chinese Physics B*, *21*, 050505.
180. Koronovskii, A. A., Moskalenko, O. I., Shurygina, S. A., & Hramov, A. E. (2013). Generalized synchronization in discrete maps. New point of view on weak and strong synchronization. *Chaos Solitons Fractals*, *46*, 12–18.
181. Ouannas, A., & Odibat, Z. (2015). Generalized synchronization of different dimensional chaotic dynamical systems in discrete time. *Nonlinear Dynamics*, *81*, 765–771.
182. Hunt, B. R., Ott, E., & Yorke, J. A. (1997). Differentiable generalized synchronization of chaos. *Physics Review E*, *55*, 4029–4034.
183. Yang, T., & Chua, L. O. (1999). Generalized synchronization of chaos via linear transformations. *International Journal of Bifurcation and Chaos*, *9*, 215–219.
184. Wang, Y., & Guan, Z. (2006). Generalized synchronization of continuous chaotic systems. *Chaos Solitons Fractals*, *27*, 97–101.
185. Zhang, G., Liu, Z., & Ma, Z. (2007). Generalized synchronization of different dimensional chaotic dynamical systems. *Chaos Solitons Fractals*, *32*, 773–779.
186. Deng, W. H. (2007). Generalized synchronization in fractional order systems. *Physics Review E*, *75*, 056201.
187. Zhou, P., Cheng, X. F., & Zhang, N. Y. (2008). Generalized synchronization between different fractional-order chaotic systems. *Communications in Theoretical Physics*, *50*, 931–934.
188. Zhang, X. D., Zhao, P. D., & Li, A. H. (2010). Construction of a new fractional chaotic system and generalized synchronization. *Communications in Theoretical Physics*, *53*, 1105–1110.
189. Jun, W. M., & Yuan, W. N. (2011). Generalized synchronization of fractional order chaotic systems. *International Journal of Modern Physics B*, *25*, 1283–1292.
190. Wu, X. J., Lai, D. R., & Lu, H. T. (2012). Generalized synchronization of the fractional-order chaos in weighted complex dynamical networks with nonidentical nodes. *Nonlinear Dynamics*, *69*, 667–683.
191. Xiao, W., Fu, J., Liu, Z., & Wan, W. (2012). Generalized synchronization of typical fractional order chaos system. *Journal of Computing*, *7*, 1519–1526.
192. Martínez-Guerra, R., & Mata-Machuca, J. L. (2014). Fractional generalized synchronization in a class of nonlinear fractional order systems. *Nonlinear Dynamics*, *77*, 1237–1244.
193. Samko, S. G., Klibas, A. A., & Marichev, O. I. (1993). *Fractional integrals and derivatives: theory and applications*. Gordan and Breach.
194. Podlubny, I. (1999). *Fractional differential equations*. Academic Press.
195. Gorenflo, R., & Mainardi, F. (1997). Fractional calculus: Integral and differential equations of fractional order. In A. Carpinteri & F. Mainardi (Eds.), *Fractals and Fractional Calculus in Continuum Mechanics*. Springer.
196. Caputo, M. (1967). Linear models of dissipation whose Q is almost frequency independent-II. *Geophysical Journal of the Royal Astronomical Society*, *13*, 529–539.
197. Oldham, K. B., & Spanier, J. (1974). *The fractional calculus*. Academic.
198. Miller, K. S., & Ross, B. (1993). *An introduction to the fractional calculus and fractional differential equations*. Wiley.
199. Soltan, A., Radwan, A. G., & Soliman, A. M. (2013). Fractional order Butterworth filter: Active and passive realizations. *IEEE Journal of Emerging and Selected Topics in Circuits and Systems, 3*, 346–354.
200. Soltan, A., Radwan, A. G., & Soliman, A. M. (2012). Fractional order filter with two fractional elements of dependent orders. *Journal of Microelectronics*, *43*, 818–827.
201. Radwan, A. G., & Fouda, M. E. (2013). Optimization of fractional-order RLC filters. *Journal of Circuits, Systems, and Signal Processing*, *32*, 2097–2118.
202. Li, Y., Chen, Y., & Podlubny, I. (2010). Stability of fractional-order nonlinear dynamic systems: Lyapunov direct method and generalized Mittag Leffler stability. *Computers & Mathematics with Applications*, *59*, 1810–1821.
203. Chen, D., Zhang, R., Liu, X., & Ma, X. (2016). Fractional order Lyapunov stability theorem and its applications in synchronization of complex dynamical networks. *Communications in Nonlinear Science and Numerical Simulation, 19*, 4105–4121.

204. Aguila-Camacho, N., Duarte-Mermoud, M. A., & Gallegos, J. A. (2014). Lyapunov functions for fractional order systems. *Communications in Nonlinear Science and Numerical Simulation*, *19*, 2951–2957.
205. Diethelm, K., & Ford, N. J. (2002). Analysis of fractional differential equations. *Journal of Mathematical Analysis and Applications*, *265*, 229–248.
206. Wang, X. Y., & Zhang, H. (2013). Bivariate module-phase synchronization of a fractional-order lorenz system in different dimensions. *Journal of Computational and Nonlinear Dynamics*, *8*, 031017.
207. Qiang, J. (2007). Projective synchronization of a new hyperchaotic Lorenz system chaotic systems. *Physics Letters A*, *370*, 40–45.
208. Li, Y., Tang, W. K., & Chen, G. (2005). Generating hyperchaos via state feedback control. *International Journal of Bifurcation and Chaos*, *15*, 3367–3375.
209. Yan, J., & Li, C. (2007). On chaos synchronization of fractional differential equations. *Chaos Solitons Fractals*, *32*, 725–735.

A New Fractional-Order Jerk System and Its Hybrid Synchronization

Abir Lassoued and Olfa Boubaker

Abstract In this chapter, a new Jerk chaotic system with a piecewise nonlinear (PWNL) function and its fractional-order (FO) generalization are proposed. Both the FO and the PWNL function, serving as chaotic generators, make the proposed system more adopting for electrical engineering applications. The highly complex dynamics of the novel system are investigated by theoretical analysis pointing out its elementary characteristics such as the Lyapunov exponents, the attractor forms and the equilibrium points. To focus on the application values of the novel FO system in multilateral communication, hybrid synchronization (HS) with ring connection is investigated. For such schema, where all systems are coupled on a chain, complete synchronization (CS) and complete anti-synchronization (AS) co-exist where the state variables of the first system couple the *Nth* system and the state variables of the *Nth* system couple the $(N-1)th$ system. Simulations results prove that the synchronization problem is achieved with success for the multiple coupled FO systems.

Keywords Fractional-Order (FO) · Chaotic systems · Jerk equation · Piecewise nonlinear (PWNL) functions · Hybrid synchronization (HS)

1 Introduction

Chaotic systems are defined as nonlinear systems exhibiting unpredictable dynamical trajectories. In this frameworks, several chaotic systems have been discovered these last decades [15] and have been used in many engineering applications [11, 13]. Many researches have attempted to build such dynamical systems with simple algebraic structures highly recommended for electronic implementations. One of

A. Lassoued (✉) · O. Boubaker
National Institute of Applied Sciences and Technology,
Centre Urbain Nord BP.676, 1080 Tunis Cedex, Tunisia
e-mail: lassoued.abir5@gmail.com

O. Boubaker
e-mail: olfa.boubaker@insat.rnu.tn

A.T. Azar et al. (eds.), *Fractional Order Control and Synchronization of Chaotic Systems*, Studies in Computational Intelligence 688,
DOI 10.1007/978-3-319-50249-6_24

the most famous chaotic systems with a simple algebraic structure is the Jerk system. For the general Jerk system [19], some specific nonlinearities are proposed such as the exponential function [29] and the piecewise linear function [2, 25] whereas, in our best knowledge, the piecewise nonlinear (PWNL) function is not used, yet.

On the other hand, the fractional calculus is an old concept discovered 300 years ago. Furthermore, many basic chaotic systems have been generalized into their corresponding fractional order (FO) models generating stranger attractors. Moreover, it has been proved in many research works, that FO systems are more adopting for modeling effectively complex real problems.

On the other side, synchronization of chaotic systems has attracted considerable interest nowadays due to its great potential application in control processing and secure communication. In this context, several synchronization schemas are proposed [18, 21]. The hybrid synchronization (HS) is one of the most interesting schemas. The HS is applied for multiple different or identical chaotic systems [30], with chain connection [16] or with ring connection [3]. The synchronization of several different systems is considered higher useful in engineering applications such as it is more effective to enhance the security in digital communication and leads to more bright future in multilateral communication. Note that the HS with ring connection is not applied for FO systems, yet.

The purpose of this paper is to propose a new FO system with a simple algebraic structure and a very complex dynamical behavior, at the same time. Expecting that the PWNL function gives us more complex chaotic proprieties than the piecewise linear one, a new Jerk chaotic system with PWNL function is proposed. This PWNL function is constructed from an absolute function. To enhance the potential application of the Jerk system, HS with ring connection for multiple FO systems will be achieved. We realize, first, the AS behavior and after we apply the CS behavior under the AS controllers.

The rest of this chapter is organized as follows. In Sect. 2, basic definitions of fractional calculus are presented. In Sect. 3, the new chaotic system based on the general Jerk equation is proposed and analyzed. Then, its FO model is designed and analyzed based on stability theory of fractional calculus. In Sect. 4, HS problem of multiple FO systems with ring connection is achieved.

2 Preliminaries

Despite the inherent complexity of fractional calculus [9], the application of the FO has attracted the attention of many researches in nontraditional fields like chaos theory [6], secure communication [31] and encryption [1]. All definitions of fractional calculus used in this paper are given below.

Definition 1 The Caputo FO definition is given by [23]

$$D^q x(t) = J^{m-q} x^{(m)}(t), q > 0. \tag{1}$$

where q is the FO. $x^{(m)}$ is the m order derivative of x with m the integer part of q. J is the integral operator described as

$$J^{\beta}x(t) = \frac{1}{\Gamma(\beta)} \int_{0}^{\infty} (t-\sigma)^{\beta-1} y(\sigma)\, \mathrm{d}\sigma. \tag{2}$$

where $\Gamma(\beta) = \mathrm{e}^{-t} t^{\beta-1} \mathrm{d}t$ is the Gamma function.

In the rest of this paper, we consider the following FO model [12, 17]

$$\frac{d^q x(t)}{\mathrm{d}t^q} = D^q x(t) = f(t, x(t)). \tag{3}$$

where D^q is the Caputo fractional derivative and q is the corresponding FO such as $q \in]0, 1[$.

The stability analysis of FO systems are different from the integer ones. Thus, we introduce the corresponding definitions.

Definition 2 [7] Consider the following FO system described by

$$\frac{d^q x(t)}{dt^q} = f(x(t)); \tag{4}$$

where $0 < q < 1$ and $x \in \mathbb{R}^n$. The equilibrium points of $f(x(t))$ are locally asymptotically stable if all eigenvalues λ_i of the Jacobian matrix $J = \partial f(x(t))/\partial x(t)$ evaluated at the equilibrium points statisfy $|arg(\lambda_i)| > q\frac{\pi}{2}$. For more details, Fig. 1 illustrates the stable and unstable regions [26].

Definition 3 [26] A three dimensional system has three eigenvalues for each equilibrium as λ_1 a real number and (λ_2, λ_3) a pair of complex conjugate number. An equilibrium is a saddle point if it has at least one eigenvalue in the stable region and one in the unstable region.

Nevertheless, there are two types of saddle points: a saddle point of index 1 and a saddle point of index 2. The first type has one eigenvalue in the unstable region and two eigenvalues in the stable region. However, the saddle point of index 2 has one

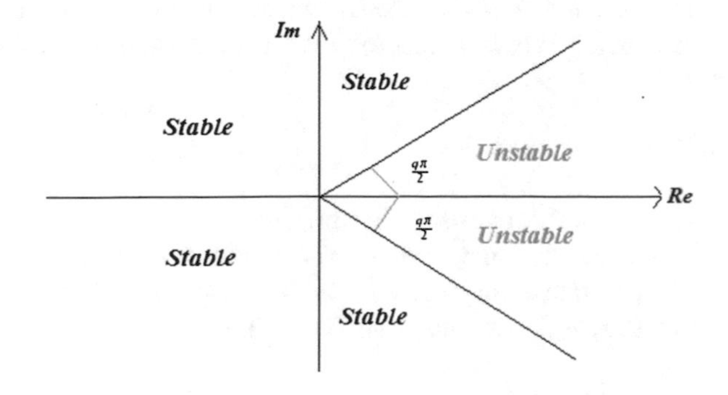

Fig. 1 Stable and unstable regions of the fractional-order system with $0 < q < 1$

eigenvalue in the stable region and two eigenvalues in the unstable region. In strange attractors, scrolls are generated only around the saddle point of index 2.

3 The Fractional-Order Jerk Chaotic System

In this section, a new Jerk chaotic system will be designed and its FO model will be developed. The first Jerk system was discovered by Sprott [24], in 1994, based on the well known third order Jerk equation $\dddot{x} = J(\ddot{x}, \dot{x}, x)$ [8].

3.1 The Jerk Chaotic System

Let consider the following Jerk chaotic system

$$\begin{cases} \dot{x} = y, \\ \dot{y} = z, \\ \dot{z} = -az - by - cx + kx|x|. \end{cases} \tag{5}$$

where (x, y, z) are the state variables and (a, b, c, k) are the system parameters.

To ensure that the Jerk chaotic system can exhibit chaotic behaviors, the general condition of dissipativity should be satisfied. Note that

$$\nabla V = \frac{\partial \dot{x}}{\partial x} + \frac{\partial \dot{y}}{\partial y} + \frac{\partial \dot{z}}{\partial z} = -a < 0$$

with V the volume element of the flow. Thus, system (5) is dissipative only if the parameter a is positive. Hence, Jerk system converges to an exponential rate. This means that its asymptomatic behaviors are fixed on a strange attractor. When the initial conditions are chosen as (1, 1, 1) and the system parameters (a, b, c, k) are equal to (1, 1, −2.625, −0.25), system (5) generates the strange attractor displayed in Fig. 2 and characterized by two scrolls and composite dynamical behaviors. For system (5), it is clear that the PWNL function described by

$$\begin{cases} x^2, \; if \;\; x > 0, \\ -x^2, \; if \;\; x < 0, \end{cases} \tag{6}$$

represents the main chaotic generator.

On the one hand, system (5) is sensitive to initial conditions as a small variation can generate a big change in the final trajectory. Three time series of the state variables (x, y, z) are shown in Fig. 3. They start from the initial conditions (1, 1, 1)

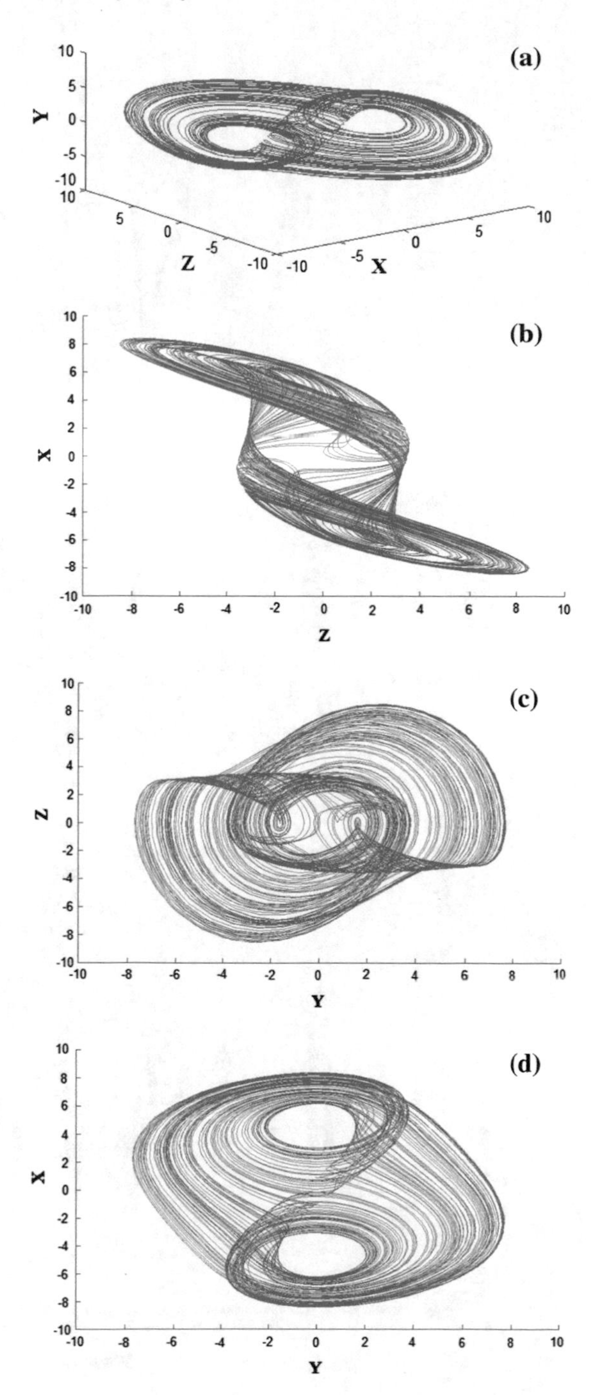

Fig. 2 Phase portraits of system (1) with the parameters $a = 1$, $b = 1$, $c = -2.625$ and $k = -0.25$ and initial values (1, 1, 1): **a** x-y-z; **b** x-y; **c** x-z; **d** y-z

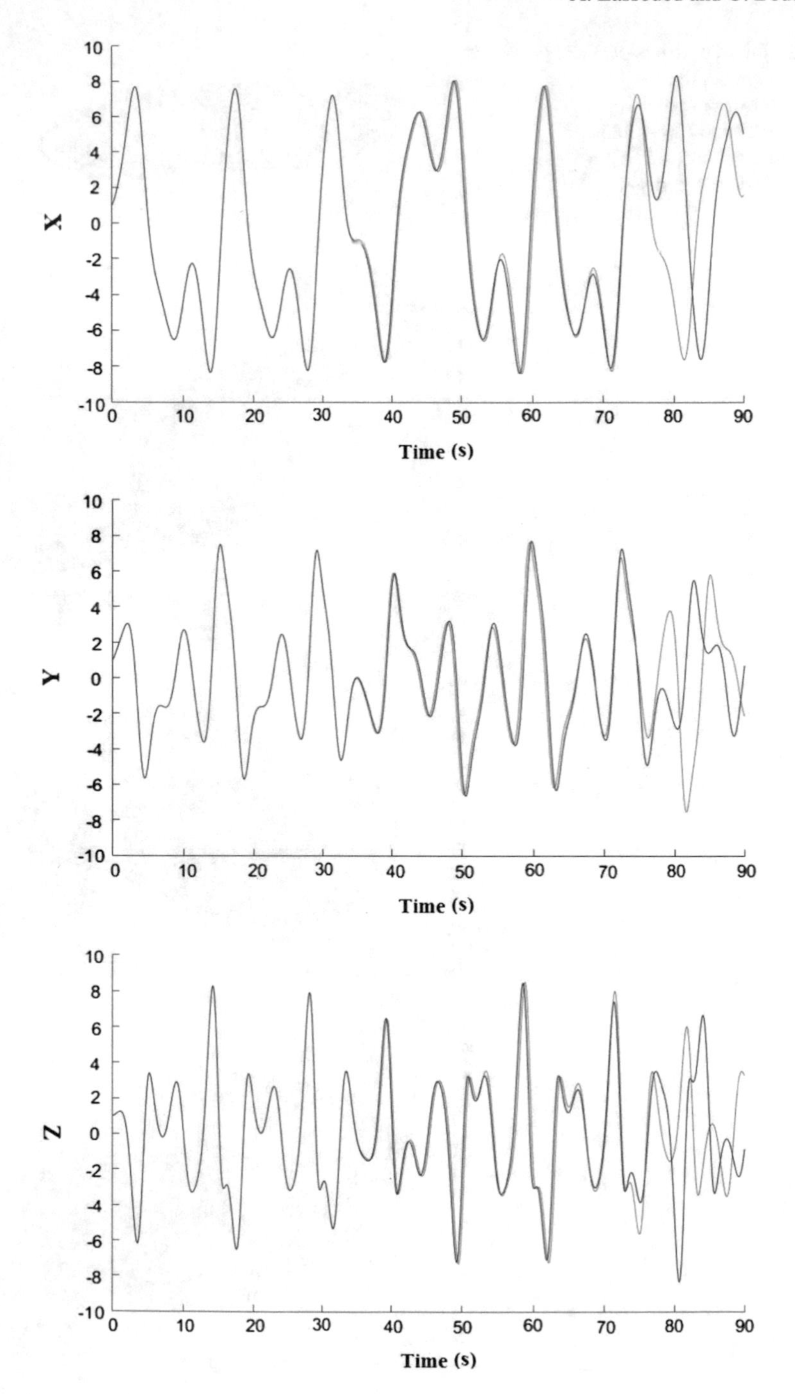

Fig. 3 Sensitive dependence on initial conditions: trajectories of x, y and z, respectively

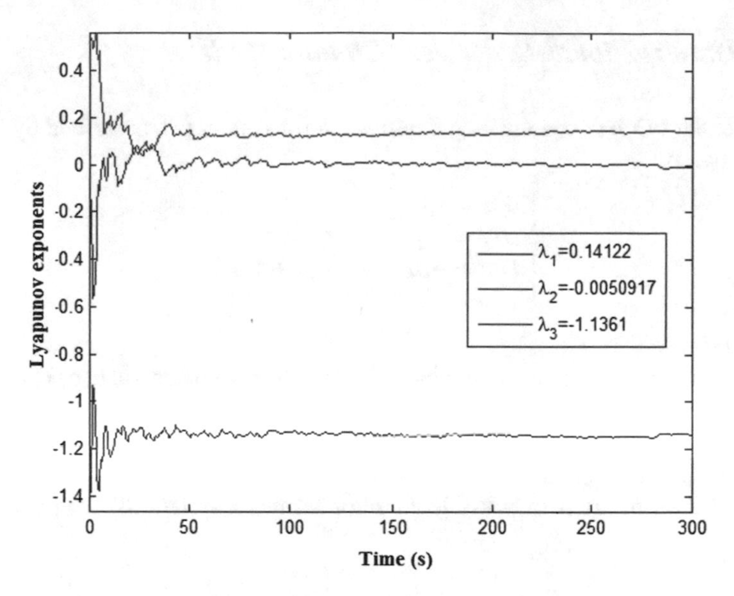

Fig. 4 Lyapunov spectrum of the Jerk chaotic system

(blue) and (1.001, 1, 1) (red), respectively. We can observe that the two trajectories are the same in the beginning but after a short period they diverge completely.

On the other hand, it is known that a chaotic system exhibits three Lyapunov exponents giving the measure of the growth decline rate related to the small perturbation along the dynamical trajectories. In system (5), the largest positive exponent λ_1, which increases the expansion degree of the attractor in the phase space, is equal to 0.14122 where $\lambda_3 = -1.1361$ and $\lambda_2 = 0$. Note that λ_2 and λ_3 increase the contraction degree and the critical nature of the chaotic attractor, respectively. The corresponding Lyapunov exponent spectrum is depicted in Fig. 4. The Lyapunov dimension of system (5) is given by [10]

$$D_L = j + \frac{1}{|\lambda_{j+1}|} \sum_{i=1}^{j} (\lambda_i) = 2 + \frac{\lambda_1 + \lambda_2}{|\lambda_3|} = 2.124. \quad (7)$$

where j is equal to $n - 1$ with n the number of Lyapunov exponents. Thus, system (5) exhibits an attractor with a FO dimension.

Comparing the results obtained for a Jerk system with a piecewise linear function proposed in [2], to those obtained using the system (5), we can conclude that the PWNL function is able to give more complex dynamical behaviors with greater Lyapunov exponent (0.14122 > 0.036).

3.2 *The Fractional-Order Jerk Chaotic Model*

Consider the FO Jerk chaotic model derived from system (5), described by the following system

$$\begin{cases} D^q x = y, \\ D^q y = z, \\ D^q z = -az - by - cx + kx|x|. \end{cases} \quad (8)$$

where $0 < q < 1$.

By maintaining the same numerical values of the parameters (a, b, c, k) and varying the parameter q, different attractor forms are obtained. When $q < 0.85$, system (8) converges to a fixed point. However, when $0.91 < q < 0.98$, system (8) generates periodic orbits. Finally, the FO system (8) is chaotic only for $q \in]0.99..1[$. The different phase portraits of possible attractor forms of system (8) are presented in Fig. 5.

4 Stability Analysis

In this section, we will identify and analyze the equilibrium points of system (8). The equilibrium points are the roots of the equations $\dot{x} = 0$, $\dot{y} = 0$ and $\dot{z} = 0$. Thus, by solving the last equations for $(a, b, c, d) = (1, 1, -2.625, -0.25)$, we obtain

$$\begin{cases} y = 0, \\ z = 0, \\ -cx + kx|x| = 0. \end{cases} \quad (9)$$

Therefore, since the parameters c and k are negative constants, system (8) admits only three equilibrium points. These equilibrium points are the origin $E_1(0, 0, 0)$, $E_2(\frac{c}{k}, 0, 0)$ and $E_3(\frac{-c}{k}, 0, 0)$. We aim, now, to determine the Jacobian matrix J and its corresponding eigenvalues for each equilibrium point. The numerical results are summarized in Table 1.

According to the Definition 2, the FO system can exhibit chaotic behavior if the following condition is satisfied at least for one equilibrium point

$$q > \frac{2}{\pi} arctan \frac{|Im(\lambda)|}{|Re(\lambda)|} \quad (10)$$

where λ is an eigenvalue corresponding to one equilibrium, and $Im(\lambda)$ and $Re(\lambda)$ are the imaginary and real part of λ.

According to the Definition 3, Fig. 2 and referring to Table 1, we find that for the first equilibrium E_1, λ_1 belongs to the unstable region where λ_2 and λ_3 belong to the

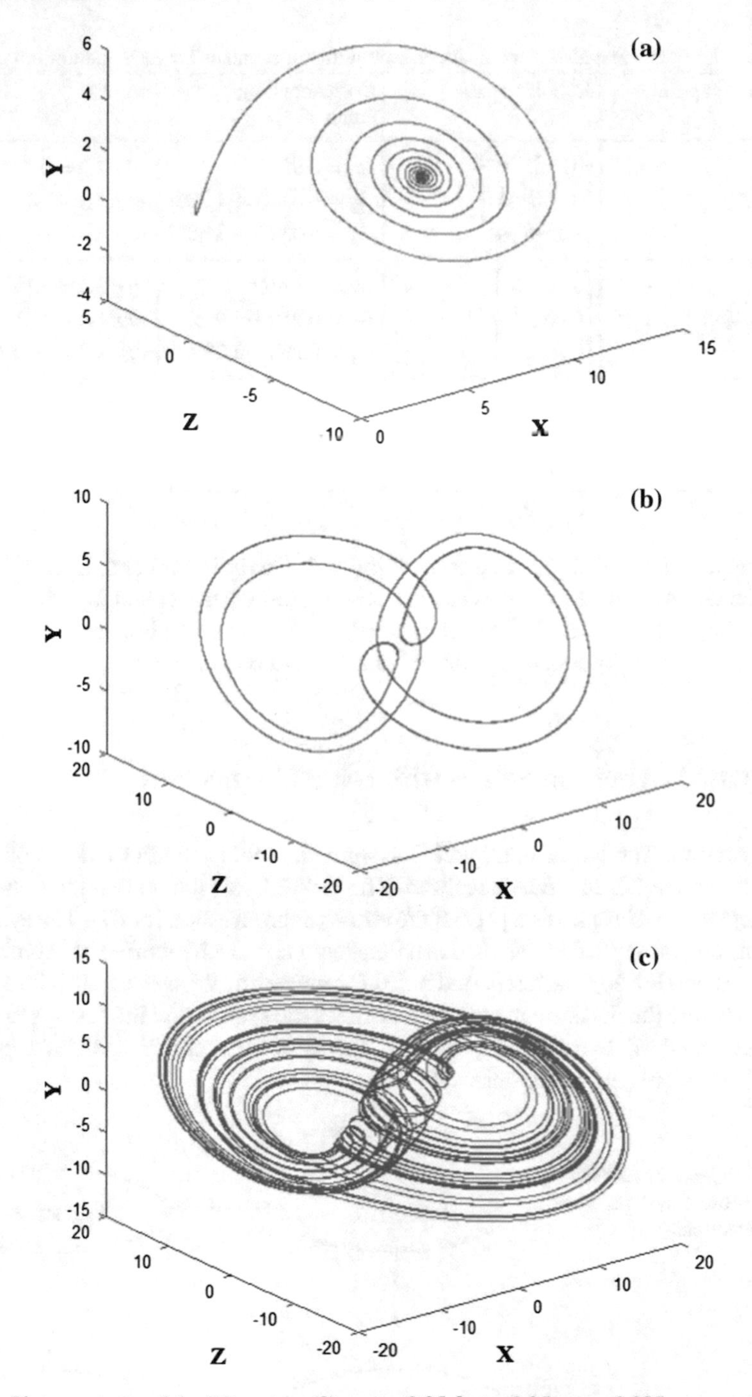

Fig. 5 Phase portraits of the FO system (8): **a** q = 0.85, **b** q = 0.95, **c** q = 0.995

Table 1 The Jacobian matrix J and their corresponding eigenvalues for each equilibrium point

Equilibrium point	Jacobian matrix	Corresponding eigenvalues	$arg(\lambda_i)$
$E_1(0,0,0)$	$\begin{pmatrix} 0 & 1 & 0 \\ 0 & 0 & 1 \\ -c & -b & -a \end{pmatrix}$	$\lambda_1 = 0.934$ $\lambda_2 = -0.967 + 1.386i$ $\lambda_3 = -0.967 - 1.386i$	$arg(\lambda_1) = 0$ $arg(\lambda_2) = 2.186$ $arg(\lambda_3) = -2.186$
$E_2(\frac{c}{k},0,0)$	$\begin{pmatrix} 0 & 1 & 0 \\ 0 & 0 & 1 \\ c & -b & -a \end{pmatrix}$	$\lambda_1 = -1.500$ $\lambda_2 = 0.250 + 1.299i$ $\lambda_3 = 0.250 - 1.299i$	$arg(\lambda_1) = 3.141$ $arg(\lambda_2) = 1.380$ $arg(\lambda_3) = -1.380$
$E_3(\frac{-c}{k},0,0)$	$\begin{pmatrix} 0 & 1 & 0 \\ 0 & 0 & 1 \\ c & -b & -a \end{pmatrix}$	$\lambda_1 = -1.500$ $\lambda_2 = 0.250 + 1.299i$ $\lambda_3 = 0.250 - 1.299i$	$arg(\lambda_1) = 3.141$ $arg(\lambda_2) = 1.380$ $arg(\lambda_3) = -1.380$

stable region. Thus, E_1 is a saddle point of index 1. On the other hand, E_2 and E_3 have the same eigenvalues. λ_1 belongs to the stable region where λ_2 and λ_3 belong to the unstable region. Thus, E_2 and E_3 are saddle points of index 2. Finally, we conclude that system (8) had two saddle points of index 2 and one of index 1.

5 Hybrid Synchronization with Ring Connection

In this section, The HS of multiple FO systems including the FO Jerk chaotic system with the PWNL function is realized. The HS [20, 28] includes at the same time the complete synchronization [4] and the anti-synchronization [5, 27]. The synchronization schema applied in this section extends our approach reported in a conference paper [14] and the approach reported in [3] for integer-order systems. It is illustrated in Fig. 6 where the first system anti-synchronizes the second and the first system synchronizes the third. In additional, the $(N-2)th$ system synchronizes the Nth and the $(N-1)th$ system anti-synchronizes the Nth.

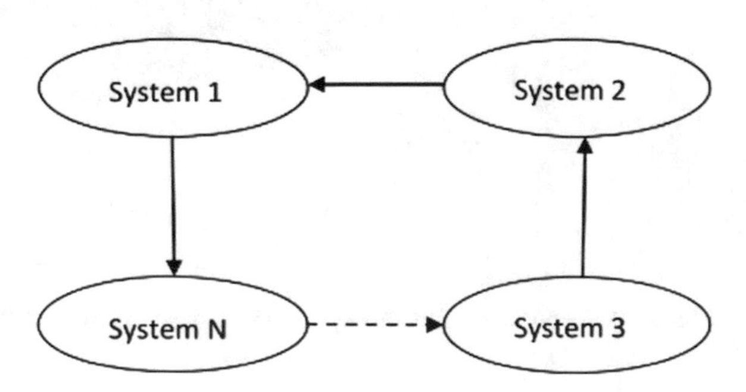

Fig. 6 Diagram of coupled fractional-order systems with ring connections

5.1 Problem Formulation

Let consider a multiple FO system coupled via a ring connection and described as follows

$$\begin{cases} D^q x_1 = A_1 x_1 + g_1(x_1) + D_1(x_N - x_1), \\ D^q x_2 = A_2 x_2 + g_2(x_2) + D_2(x_1 - x_2), \\ \quad \vdots \\ D^q x_N = A_N x_N + g_N(x_N) + D_N(x_{N-1} - x_N), \end{cases} \tag{11}$$

where the state vector x_i, with $(1 \leq i \leq N)$, is defined as $x_i = (x_{i1}, x_{i2}, x_{i3})$ and N is the number of multiple FO system. Each system is divided in two parts, a linear part and a nonlinear one, expressed by the matrices A_i and the g_i functions, respectively. D_i are diagonal matrices to be designed such that the d_{ij}, $i \in [1..N]$ and $j \in [1..3]$, represent the coupled positive parameters of the diagonal matrices.

To achieve the HS, the FO controlled system is given by

$$\begin{cases} D^q x_1 = A_1 x_1 + g_1(x_1) + D_1(x_N - x_1), \\ D^q x_2 = A_2 x_2 + g_2(x_2) + D_2(x_1 - x_2) + u_1, \\ \quad \vdots \\ D^q x_N = A_N x_N + g_N(x_N) + D_N(x_{N-1} - x_N) + u_{N-1}. \end{cases} \tag{12}$$

where u_i, $i \in [1..N-1]$, are the control laws.

Definition 4 The AS error vector $e_A = [e_{A1} \cdots e_{A(N-1)}]$ and the CS error vector $e_C = [e_{C1} \cdots e_{C(N-1)}]$, corresponding to system (12), are depicted as $e_{Ai} = x_i(t) + x_{i+1}(t)$ and $e_{Cj} = x_{j+2}(t) - x_j(t)$, respectively. Also, both parameters i and j are included in the interval $[1..N-1]$. The HS is achieved if the error vectors e_A and e_C satisfy the following conditions

$$\begin{aligned} \lim_{t \to +\infty} \| e_{Ai} \| = 0,\ i \in [1..N-1], \\ \lim_{t \to +\infty} \| e_{Cj} \| = 0,\ j \in [1..N-1]. \end{aligned} \tag{13}$$

Thus, we can conclude that the dynamical errors, corresponding to system (12), are asymptotically stable if the conditions (13) are satisfied.

The purpose of the HS problem is to design the matrices D_i $(i \in [1..N])$ and the controllers laws u_j $(j \in [1..N-1])$ such as the conditions (13) are satisfied.

5.2 Main Results

Theorem 1 *For the system (12), if there exist controller laws $u_i(t)$, $i \in [1 \cdots N-1]$, defined by*

$$\begin{cases} u_1 = v_1 - \begin{pmatrix} (-1-(-1)^N)D_1 + 2D_2 \\ -(A_2 - A_1) \end{pmatrix} x_1 - g_2(x_2) - g_1(x_1), \\ u_2 = v_2 - (-2(D_2 - D_3) - (A_3 - A_2))x_2 - g_3(x_3) - g_2(x_2) - u_1, \\ \vdots \\ u_{N-1} = v_{N-1} - \begin{pmatrix} -2(D_{N-1} - D_N)x_{N-1} - (A_N - A_{N-1})x_{N-1} \\ +g_N(x_N) + g_{N-1}(x_{N-1}) \end{pmatrix} - u_{N-2}. \end{cases} \tag{14}$$

such that the conditions (13) are satisfied, thus, the synchronization error vectors e_A *and* e_C *are asymptotically stable and the HS is correctly achieved. For system (14), the matrices* v_i *are expressed such as* $[v_1..v_{N-1}]^T = H[e_{A1}..e_{A(N-1)}]^T$ *where H is a constant matrix to be computed.*

Proof The AS errors are defined as $e_{Ai}(t) = x_i(t) + x_{i+1}$. Their corresponding FO derivatives can be written as

$$D^q e_{Ai} = D^q x_i + D^q x_{i+1}. \tag{15}$$

We obtain

$$D^q e_A = L(e_A)e_A = (M + H)e_A. \tag{16}$$

where $L(e_A) = L_1(e_A) + L_2$. L_1 and L_2 are two constant matrices to be computed and M is defined as

$$M = \begin{bmatrix} K_1 & K_2 & K_3 & & D_1 \\ D_2 & A_3 - D_3 & 0 & \cdots & 0 \\ 0 & D_3 & A_4 - D_4 & & 0 \\ & \vdots & & \ddots & \vdots \\ 0 & 0 & 0 & & A_N - D_N \end{bmatrix}$$

where $K_1 = A_2 - (-1)^{N-1}D_1 - D_2$, $K_2 = -(-1)^{N-2}D_1$, $K_3 = -(-1)^{N-3}D_1$ and $K_4 = A_N - D_N$.

If L_1 and L_2 satisfy the following condition

$$L_1^T(e_A) = -L_1(e_A) \tag{17}$$

such that $L_2 = diag(l_1, ..., l_N)$ and $l_i < 0$, $i \in [1 \cdots N]$, then the AS error vector e_A is asymptotically stable. Obviously, the AS is correctly applied under the controllers $u_i(t)$, $i \in [1 \cdots N-1]$, defined in Eq. (14).

Therefore, according to Definition 4, we can rewrite the CS error vector e_C by the following equations with $j \in [1 \cdots N-1]$

$$e_{Cj}(t) = x_{j+2}(t) - x_{j+1}(t) + x_{j+1}(t) - x_j(t). \tag{18}$$

Indeed, we obtain

$$\lim_{t \to +\infty} \| x_{j+2} + x_{j+1} \| - \lim_{t \to -\infty} \| x_{j+1} + x_j \| = 0. \tag{19}$$

We can conclude, then, that the CS under the same controllers is achieved. At that point, the HS of multiple coupled FO systems is also realized.

5.3 *Application*

To verify the effectiveness and the feasibility of the proposed synchronization schema, four non-identical FO systems are used. The AS is applied such as the first FO system anti-synchronizes the second system. The second FO system, in turn, anti-synchronizes the third system. Finally, the third system anti-synchronizes the FO fourth system. Also, the CS is applied such as the first system synchronizes the third system and the second system synchronizes the fourth system. The four FO systems include two FO Jerk systems with PWNL functions, a FO Chen system [22] and a Lü system [22], respectively. The four synchronized systems are described, respectively, by

$$\begin{cases} D^q x_{11} = x_{12} + d_{11}(x_{41} - x_{11}), \\ D^q x_{12} = x_{13} + d_{12}(x_{42} - x_{12}), \\ D^q x_{13} = -a x_{13} - b x_{12} - c x_{11} + k x_{11}|x_{11}| + d_{13}(x_{43} - x_{13}). \end{cases} \tag{20}$$

$$\begin{cases} D^q x_{21} = x_{22} + d_{21}(x_{11} - x_{21}) + u_{11}, \\ D^q x_{22} = x_{23} + d_{22}(x_{12} - x_{22}) + u_{12}, \\ D^q x_{23} = -a x_{23} - b x_{22} - c x_{21} + k x_{21}|x_{21}| + d_{23}(x_{13} - x_{23}) + u_{13}. \end{cases} \tag{21}$$

$$\begin{cases} D^q x_{31} = -35 x_{31} + 35 x_{32} + d_{31}(x_{21} - x_{31}) + u_{21}, \\ D^q x_{32} = -7 x_{31} + 28 x_{32} - x_{31} x_{33} + d_{32}(x_{22} - x_{32}) + u_{22}, \\ D^q x_{33} = -3 x_{33} + x_{31} x_{32} + d_{33}(x_{23} - x_{33}) + u_{23}. \end{cases} \tag{22}$$

$$\begin{cases} D^q x_{41} = -36 x_{41} + 36 x_{42} + d_{41}(x_{31} - x_{41}) + u_{31}, \\ D^q x_{42} = 20 x_{42} - x_{41} x_{43} + d_{42}(x_{32} - x_{42}) + u_{32}, \\ D^q x_{43} = -3 x_{43} + x_{41} x_{42} + d_{43}(x_{33} - x_{43}) + u_{33}. \end{cases} \tag{23}$$

where q is a rational number as $0 < q < 1$. u_1, u_2 and u_3 are the control inputs with $u_i = [u_{i1}, u_{i2}, u_{i3}], i = 1...3$. $D_i = diag(d_{i1}, ..., d_{i4}), i = 1...4$, are the coupled matrices.

For each system, the matrices A_i $(i = 1...4)$ are given by

$$A_1 = A_2 = \begin{bmatrix} 0 & 1 & 0 \\ 0 & 0 & 1 \\ -c & -b & -a \end{bmatrix}, A_3 = \begin{bmatrix} -35 & 35 & 0 \\ -7 & 28 & 0 \\ 0 & 0 & -3 \end{bmatrix} \text{ and } A_4 = \begin{bmatrix} -36 & 36 & 0 \\ 0 & 26 & 0 \\ 0 & 0 & -3 \end{bmatrix}.$$

Similarly, the matrices g_i $(i = 1...4)$ are given by

$$g_1(x_1) = \begin{bmatrix} 0 \\ 0 \\ kx_{11}|x_{11}| \end{bmatrix}, g_2(x_2) = \begin{bmatrix} 0 \\ 0 \\ kx_{21}|x_{21}| \end{bmatrix}, g_3(x_3) = \begin{bmatrix} 0 \\ -x_{31}x_{33} \\ -x_{31}x_{32} \end{bmatrix}$$

and $g_4(x_4) = \begin{bmatrix} 0 \\ -x_{31}x_{33} \\ -x_{31}x_{32} \end{bmatrix}$.

The AS is achieved when the AS state errors converge to zero. The error vector e_A is defined as $e_A = [e_{A1}e_{A2}e_{A3}]$ such that $e_{Ai} = x_i(t) + x_{i+1}(t)$, $i \in [1..3]$.

Using Eq. (14), the designed controllers are given by

$$\begin{cases} u_1 = H_1 e_A + 2(D_1 - D_2)x_1 - g_2(x_2) - g_1(x_1), \\ u_2 = H_2 e_A - (-2(D_2 - D_3) - (A_3 - A_2))x_2 - g_3(x_3) - g_2(x_2) - u_1, \\ u_3 = H_3 e_A - (-2(D_3 - D_4) - (A_4 - A_3))x_3 - g_4(x_4) - g_3(x_3) - u_2. \end{cases} \quad (24)$$

where H_1, H_2 and H_3 are constant matrices computed using the relation (17). They are given by

$$H_1 = \begin{bmatrix} 0 & 0 & c & 0 & 0 & 0 & 0 & 0 & 0 \\ -1 & 0 & b & 0 & 0 & 0 & 0 & 0 & 0 \\ 0 & -1 & 0 & 0 & 0 & 0 & 0 & 0 & 0 \end{bmatrix},$$

$$H_2 = \begin{bmatrix} N_1 & 0 & 0 & 0 & 0 & 0 & -d_{31} & 0 & 0 \\ 0 & N_2 & 0 & -28 & 0 & 0 & 0 & -d_{32} & 0 \\ 0 & 0 & N_3 & 0 & 0 & 0 & 0 & 0 & -d_{33} \end{bmatrix}$$

with $N_1 = d_{11} - d_{21}$, $N_2 = d_{12} - d_{22}$ and $N_3 = d_{13} - d_{23}$,

$$H_3 = \begin{bmatrix} -d_{11} & 0 & 0 & 0 & 0 & 0 & 0 & 0 & 0 \\ 0 & -d_{12} & 0 & 0 & 0 & 0 & -36 & 0 & 0 \\ 0 & 0 & -d_{13} & 0 & 0 & 0 & 0 & 0 & 0 \end{bmatrix}.$$

For the four FO systems, the matrices M and H are obtained using the Eq. (16). They are given by

$$M = \begin{bmatrix} A_2 + D_1 - D_2 & -D_2 & D_1 \\ -D_2 & A_3 - D_3 & 0 \\ 0 & D_3 & A_4 - D_4 \end{bmatrix} \text{ and } H = \begin{bmatrix} H_1 \\ H_2 \\ H_3 \end{bmatrix},$$

The final expression of the dynamical errors are written as

$D^q e_A = L(e_A)e_A = \left[L_{ij}\right] e_A$ $(i = 1 \cdots 3$ and $j = 1 \cdots 3)$ where

$$L_{11} = \begin{bmatrix} d_{11} - d_{21} & 1 & c \\ -1 & d_{12} - d_{22} & 1 + b \\ -c & -1 - b & -a + d_{13} - d23 \end{bmatrix},$$

$$L_{13} = L_{21} = -L_{12} = -L_{31} = \begin{bmatrix} d_{11} & 0 & 0 \\ 0 & d_{12} & 0 \\ 0 & 0 & d_{13} \end{bmatrix},$$

$$L_{22} = \begin{bmatrix} -35 - d_{31} & 35 & 0 \\ -35 & 28 - d_{32} & 0 \\ 0 & 0 & -3 - d_{33} \end{bmatrix},$$

$$L_{32} = -L_{23} = \begin{bmatrix} d_{31} & 0 & 0 \\ 0 & d_{32} & 0 \\ 0 & 0 & d_{33} \end{bmatrix},$$

$$L_{33} = \begin{bmatrix} -36 - d_{41} & 36 & 0 \\ -36 & 26 - d_{42} & 0 \\ 0 & 0 & -3 - d_{43} \end{bmatrix}$$

According to Eq. (17), the following conditions must be satisfied in order to achieve correctly the AS.
$d_{11} - d_{21} < 0,$
$d_{12} - d_{22} < 0,$
$-a + d_{13} - d_{23} < 0,$
$35 - d_{31} < 0,$
$28 - d_{32} < 0,$
$-3 - d_{33} < 0,$
$-36 - d_{41} < 0,$
$26 - d_{42} < 0$
and $-3 - d_{43} < 0.$
Relying on the previous conditions, we choose the diagonal parameters such as $(d_{11}, d_{12}, d_{13}, d_{21}, d_{22}, d_{23}, d_{31}, d_{32}, d_{33}, d_{41}, d_{42}, d_{43},)$ are equal to $(0, 0, 0, 20, 60, 80, 60, 80, 100, 80,)$ For numerical simulations, the initial conditions are fixed such as $x_1(0) = (1, 1, 1), x_2(0) = (1, 1, 1), x_3(0) = (4, 5, 4)$ and $x_4(0) = (4, 5, 3)$. The parameter q is fixed to 0.98. All numerical simulations are obtained by using the package MATCONT in MATLAB.

The initial conditions of the AS errors are chosen such as $e_{11}(0) = e_{12}(0) = e_{13}(0) = e_{21}(0) = e_{22}(0) = e_{23}(0) = e_{31}(0) = e_{32}(0) = e_{33}(0) = 10$. Figure 7 illustrates the evolution of the state trajectories of th AS error vector e_A. We can observe that the errors e_{Ai} converge to zero as $t \to +\infty$, which means that the AS and CS are realized. Thus, the HS is achieved under the proposed controller laws.

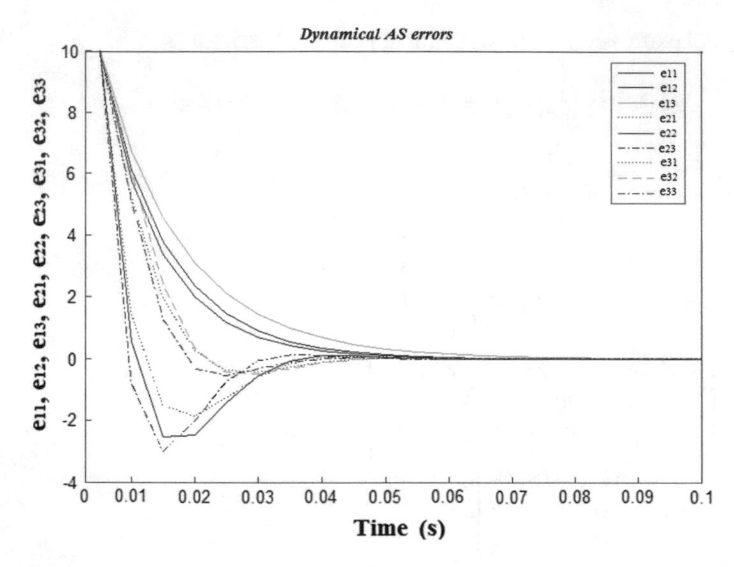

Fig. 7 State trajectories of the dynamical errors $e_A = [e_{ij}]$, $i = 1 \cdots 3$ and $j = 1 \cdots 3$

The state trajectories of the four coupled FO systems are shown in Figs. 8 and 9, under the controllers u_1, u_2 and u_3. In Fig. 8, note that the couples $(x_1(t), x_2(t))$, $(x_2(t), x_3(t))$ and $(x_3(t), x_4(t))$ express the AS where as the couples $(x_1(t), x_3(t))$ and $(x_2(t), x_4(t))$ express the CS. It is also clear in Fig. 9, displaying a zoom on the dynamical modes, that the four state trajectories are different in the beginning but after a short period, they converge, in pairs.

6 Conclusions

In this chapter, a new Jerk chaotic system with a PWNL function is proposed and its FO generalization is developed. For both systems, chaotic behaviors are analyzed. It is shown that the PWNL function gives more complex dynamical proprieties than the piecewise linear one. Finally, HS problem of multiple coupled fractional order systems with ring connections is formulated and solved. This technology provided evidence for the potential application of the proposed system in engineering applications. We have investigated the complete synchronization and the anti-synchronization approaches in order to coexist. Furthermore, numerical simulations have been carried out in order to prove the well achievement of the synchronization problem.

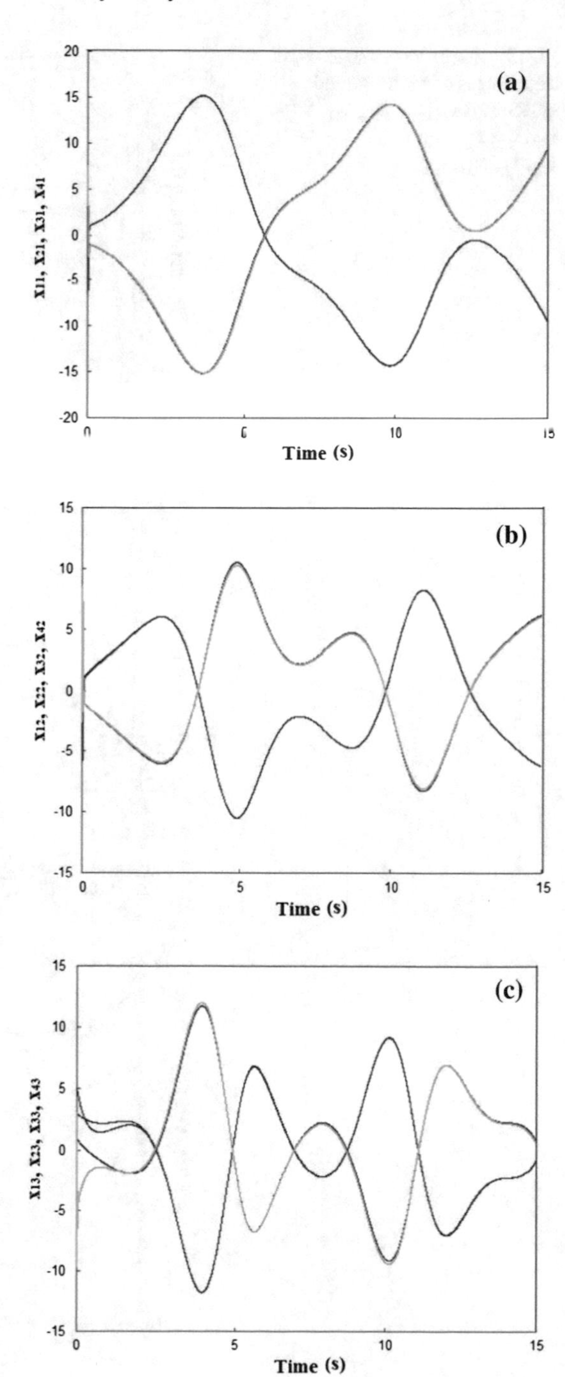

Fig. 8 State trajectories of each coupled systems: **a** $x_{11}, x_{21}, x_{31}, x_{41}$; **b** $x_{12}, x_{22}, x_{32}, x_{42}$; **c** $x_{13}, x_{23}, x_{33}, x_{43}$

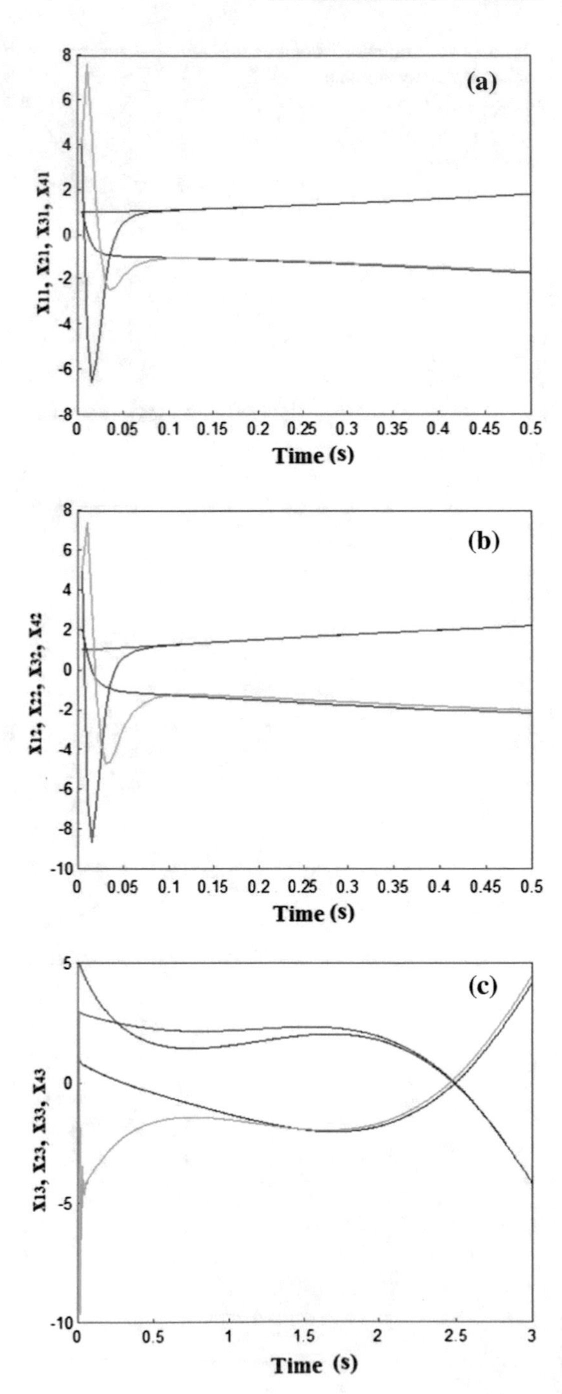

Fig. 9 Zoom on the state trajectories of each coupled systems: **a** $x_{11}, x_{21}, x_{31}, x_{41}$; **b** $x_{12}, x_{22}, x_{32}, x_{42}$; **c** $x_{13}, x_{23}, x_{33}, x_{43}$

References

1. Ahmad, M., Khan, I. R., & Alam, S. (2015). Cryptanalysis of image encryption algorithm based on fractional-order lorenz-like chaotic system. In *'Emerging ICT for Bridging the Future-Proceedings of the 49th Annual Convention of the Computer Society of India CSI'* (Vol. 2, pp. 381–388).
2. Campos-Cantón, E. (2016). Chaotic attractors based on unstable dissipative systems via third-order differential equation. *International Journal of Modern Physics C*, *27*(01), 11.
3. Chen, X., Qiu, J., Cao, J., & He, H. (2016). Hybrid synchronization behavior in an array of coupled chaotic systems with ring connection. *Neurocomputing*, *173*, 1299–1309.
4. Chen, X., Qiu, J., Song, Q., & Zhang, A. (2013). Synchronization of coupled chaotic systems with ring connection based on special antisymmetric structure. *Abstract and Applied Analysis*, *2013*, 7.
5. Chen, X., Wang, C., & Qiu, J. (2014). Synchronization and anti-synchronization of n different coupled chaotic systems with ring connection. *International Journal of Modern Physics C*, *25*(05), 11.
6. Dalir, M., & Bashour, M. (2010). Applications of fractional calculus. *Applied Mathematical Sciences*, *4*(21), 1021–1032.
7. Diethelm, K., & Ford, N. J. (2002). Analysis of fractional differential equations. *Journal of Mathematical Analysis and Applications*, *265*(2), 229–248.
8. Gottlieb, H. (1996). What is the simplest jerk function that gives chaos? *American Journal of Physics*, *64*(5), 525–525.
9. Gutiérrez, R. E., Rosário, J. M., & Tenreiro Machado, J. (2010). Fractional order calculus: Basic concepts and engineering applications. *Mathematical Problems in Engineering*, *2010*, 19.
10. Henry, B., Lovell, N., & Camacho, F. (2012). Nonlinear dynamics time series analysis. *Nonlinear Biomedical Signal Processing: Dynamic Analysis and Modeling*, *2*, 1–39.
11. Hussain, I., Alqahtani, A., & Gondal, M. A. (2015). An efficient method for secure communication of biometric information based on chaos. *3D Research*, *6*(2), 1–7.
12. Jian-Bing, H., & Ling-Dong, Z. (2013). Stability theorem and control of fractional systems. *Acta Physica Sinica*, *62*(24), 7.
13. Lang, J. (2015). Color image encryption based on color blend and chaos permutation in the reality-preserving multiple-parameter fractional fourier transform domain. *Optics Communications*, *338*, 181–192.
14. Lassoued, A., & Boubaker, O. (2016). Hybrid synchronization of multiple fractional-order chaotic systems with ring connection. In *The 8 International Conference On Modelling, Identification and Control*.
15. Lassoued, A., & Boubaker, O. (2016). On new chaotic and hyperchaotic systems: A literature survey. *Nonlinear Analysis: Modelling and Control*, *21*(6), 770–789.
16. Liu, Y., & Lü, L. (2008). Synchronization of n different coupled chaotic systems with ring and chain connections. *Applied Mathematics and Mechanics*, *29*, 1299–1308.
17. Matignon, D. (1996). Stability results for fractional differential equations with applications to control processing. In *Computational engineering in systems applications* (Vol. 2, , pp. 963–968). WSEAS Press.
18. Mkaouar, H., & Boubaker, O. (2012). Chaos synchronization for master slave piecewise linear systems: Application to chuas circuit. *Communications in Nonlinear Science and Numerical Simulation*, *17*(3), 1292–1302.
19. Muthuswamy, B., & Chua, L. O. (2010). Simplest chaotic circuit. *International Journal of Bifurcation and Chaos*, *20*(05), 1567–1580.
20. Ouannas, A., Azar, A. T., & Abu-Saris, R. (2016). A new type of hybrid synchronization between arbitrary hyperchaotic maps. *International Journal of Machine Learning and Cybernetics*, 1–8.
21. Pan, L., Zhou, L., & Li, D. (2013). Synchronization of three-scroll unified chaotic system and its hyper-chaotic system using active pinning control. *Nonlinear Dynamics*, *73*(3), 2059–2071.

22. Petráš, I. (2011). Fractional-order chaotic systems. In *Fractional-order nonlinear systems* (pp. 103–184). Springer.
23. Podlubny, I. (1998). *Fractional differential equations: an introduction to fractional derivatives, fractional differential equations, to methods of their solution and some of their applications* (Vol. 198). New York: Academic press.
24. Sprott, J. (1994). Some simple chaotic flows. *Physical Review E, 50*(2), R647.
25. Sprott, J. C. (2000). A new class of chaotic circuit. *Physics Letters A*, *266*(1), 19–23.
26. Tavazoei, M. S., & Haeri, M. (2007). A necessary condition for double scroll attractor existence in fractional-order systems. *Physics Letters A*, *367*(1), 102–113.
27. Vaidyanathan, S., & Azar, A. T. (2015). Anti-synchronization of identical chaotic systems using sliding mode control and an application to vaidyanathan–madhavan chaotic systems. In *Advances and applications in sliding mode control systems* (pp. 527–547). Springer.
28. Vaidyanathan, S., & Azar, A. T. (2015). Hybrid synchronization of identical chaotic systems using sliding mode control and an application to vaidyanathan chaotic systems. In *Advances and applications in sliding mode control systems* (pp. 549–569). Springer.
29. Vaidyanathan, S., & Azar, A. T. (2016). Adaptive backstepping control and synchronization of a novel 3-d jerk system with an exponential nonlinearity. In *Advances in chaos theory and intelligent control* (pp. 249–274). Springer.
30. Vaidyanathan, S., & Rasappan, S. (2011). Hybrid synchronization of hyperchaotic qi and lü systems by nonlinear control. In *Advances in computer science and information technology* (pp. 585–593). Springer.
31. Wen, T., Fengling, J., Xianqun, L., Xun, L. J., & Feng, W. (2011). Synchronization of fractional-order chaotic system with application to communication. In *Informatics in control, automation and robotics* (pp. 227–234). Springer.

An Eight-Term 3-D Novel Chaotic System with Three Quadratic Nonlinearities, Its Adaptive Feedback Control and Synchronization

Sundarapandian Vaidyanathan, Ahmad Taher Azar and Adel Ouannas

Abstract This research work describes an eight-term 3-D novel polynomial chaotic system consisting of three quadratic nonlinearities. First, this work presents the 3-D dynamics of the novel chaotic system and depicts the phase portraits of the system. Next, the qualitative properties of the novel chaotic system are discussed in detail. The novel chaotic system has four equilibrium points. We show that two equilibrium points are saddle points and the other equilibrium points are saddle-foci. The Lyapunov exponents of the novel chaotic system are obtained as $L_1 = 0.4715, L_2 = 0$ and $L_3 = -2.4728$. The Lyapunov dimension of the novel chaotic system is obtained as $D_L = 2.1907$. Next, we present the design of adaptive feedback controller for globally stabilizing the trajectories of the novel chaotic system with unknown parameters. Furthermore, we present the design of adaptive feedback controller for achieving complete synchronization of the identical novel chaotic systems with unknown parameters. The main adaptive control results are proved using Lyapunov stability theory. MATLAB simulations are depicted to illustrate all the main results derived in this research work for eight-term 3-D novel chaotic system.

Keywords Chaos · Chaotic systems · Adaptive control · Feedback control · Synchronization

S. Vaidyanathan (✉)
Research and Development Centre, Vel Tech University, Avadi, Chennai 600062, Tamil Nadu, India
e-mail: sundarcontrol@gmail.com

A.T. Azar
Faculty of Computers and Information, Benha University, Benha, Egypt
e-mail: ahmad_t_azar@ieee.org; ahmad.azar@fci.bu.edu.eg

A.T. Azar
Nanoelectronics Integrated Systems Center (NISC), Nile University, Cairo, Egypt

A. Ouannas
Laboratory of Mathematics, Informatics and Systems (LAMIS),
University of Larbi Tebessi, 12002 Tébessa, Algeria
e-mail: ouannas_adel@yahoo.fr

A.T. Azar et al. (eds.), *Fractional Order Control and Synchronization of Chaotic Systems*, Studies in Computational Intelligence 688,
DOI 10.1007/978-3-319-50249-6_25

1 Introduction

Chaotic systems are defined as nonlinear dynamical systems which are sensitive to initial conditions, topologically mixing and with dense periodic orbits [10, 11, 13].

Sensitivity to initial conditions of chaotic systems is popularly known as the *butterfly effect*. Small changes in an initial state will make a very large difference in the behavior of the system at future states.

Poincaré [15] suspected chaotic behaviour in the study of three bodies problem at the end of the 19th century, but chaos was experimentally established by Lorenz [41] only a few decades ago in the study of 3-D weather models.

The Lyapunov exponent is a measure of the divergence of phase points that are initially very close and can be used to quantify chaotic systems. It is common to refer to the largest Lyapunov exponent as the *Maximal Lyapunov Exponent* (MLE). A positive maximal Lyapunov exponent and phase space compactness are usually taken as defining conditions for a chaotic system.

In the last five decades, there is significant interest in the literature in discovering new chaotic systems [73]. Some popular chaotic systems are Lorenz system [41], Rössler system [63], Arneodo system [2], Henon-Heiles system [27], Genesio-Tesi system [25], Sprott systems [72], Chen system [19], Lü system [42], Rikitake dynamo system [62], Liu system [40], Shimizu system [71], etc.

In the recent years, many new chaotic systems have been found such as Pandey system [46], Qi system [54], Li system [35], Wei-Yang system [171], Zhou system [178], Zhu system [179], Sundarapandian systems [76, 81], Dadras system [21], Tacha system [84], Vaidyanathan systems [92, 93, 95–98, 101, 112, 126–133, 135–137, 139, 148, 150, 159, 161, 163, 165–167], Vaidyanathan-Azar systems [142, 143, 145–147], Pehlivan system [48], Sampath system [64], Akgul system [1], Pham system [49, 51–53], etc.

Chaos theory and control systems have many important applications in science and engineering [3, 10–12, 14, 180]. Some commonly known applications are oscillators [115, 119, 121–124, 134], lasers [37, 174], chemical reactions [102, 103, 107–109, 111, 113, 114, 117, 118, 120], biology [22, 33, 100, 104–106, 110, 116], ecology [26, 74], encryption [34, 177], cryptosystems [61, 85], mechanical systems [5–9], secure communications [23, 44, 175], robotics [43, 45, 169], cardiology [55, 172], intelligent control [4, 38], neural networks [28, 31, 39], memristors [50, 170], etc.

Synchronization of chaotic systems is a phenomenon that occurs when two or more chaotic systems are coupled or when a chaotic system drives another chaotic system. Because of the butterfly effect which causes exponential divergence of the trajectories of two identical chaotic systems started with nearly the same initial conditions, the synchronization of chaotic systems is a challenging research problem in the chaos literature.

Major works on synchronization of chaotic systems deal with the complete synchronization of a pair of chaotic systems called the *master* and *slave* systems. The design goal of the complete synchronization is to apply the output of the master system to control the slave system so that the output of the slave system tracks the output of the master system asymptotically with time. Active feedback control is used when the system parameters are available for measurement. Adaptive feedback control is used when the system parameters are unknown.

Pecora and Carroll pioneered the research on synchronization of chaotic systems with their seminal papers [18, 47]. The active control method [30, 65, 66, 75, 80, 86, 90, 151, 152, 155] is typically used when the system parameters are available for measurement.

Adaptive control method [67–69, 77–79, 88, 94, 125, 138, 144, 149, 153, 154, 160, 164, 168] is typically used when some or all the system parameters are not available for measurement and estimates for the uncertain parameters of the systems. Adaptive control method has more relevant for many practical situations for systems with unknown parameters. In the literature, adaptive control method is preferred over active control method due to the wide applicability of the adaptive control method.

Intelligent control methods like fuzzy control method [16, 17] are also used for the synchronization of chaotic systems. Intelligent control methods have advantages like robustness, insensitive to small variations in the parameters, etc.

Sampled-data feedback control method [24, 36, 173, 176] and time-delay feedback control method [20, 29, 70] are also used for synchronization of chaotic systems. Backstepping control method [56–60, 83, 156, 162] is also used for the synchronization of chaotic systems, which is a recursive method for stabilizing the origin of a control system in strict-feedback form.

Another popular method for the synchronization of chaotic systems is the sliding mode control method [82, 87, 89, 91, 99, 140, 141, 157, 158], which is a nonlinear control method that alters the dynamics of a nonlinear system by application of a discontinuous control signal that forces the system to "slide" along a cross-section of the system's normal behavior.

In this research work, we describe an eight-term 3-D novel polynomial chaotic system with three quadratic nonlinearities. Section 2 describes the 3-D dynamical model and phase portraits of the novel chaotic system.

Section 3 describes the dynamic analysis of the novel chaotic system. We show that the novel chaotic system has four equilibrium points of which two equilibrium points are saddle points and the other two equilibrium points are saddle-foci.

The Lyapunov exponents of the eight-term novel chaotic system are obtained as $L_1 = 0.4715$, $L_2 = 0$ and $L_3 = -2.4728$. Since the sum of the Lyapunov exponents of the novel chaotic system is negative, this chaotic system is dissipative. Also, the Lyapunov dimension of the novel chaotic system is obtained as $D_L = 2.1907$.

Section 4 describes the adaptive feedback control of the novel chaotic system with unknown parameters. Section 5 describes the adaptive feedback synchronization of the identical novel chaotic systems with unknown parameters. The adaptive feedback control and synchronization results are proved using Lyapunov stability theory [32].

MATLAB simulations are depicted to illustrate all the main results for the 3-D novel chaotic system. Section 6 concludes this work with a summary of the main results.

2 A Novel 3-D Chaotic System

In this research work, we announce an eight-term 3-D chaotic system described by

$$\begin{aligned}\dot{x}_1 &= -ax_2 + x_2x_3 \\ \dot{x}_2 &= px_1 + bx_2 - x_1x_3 \\ \dot{x}_3 &= x_1 - cx_3 + x_1x_2\end{aligned} \tag{1}$$

where x_1, x_2, x_3 are the states and a, b, c, p are constant, positive parameters.

The 3-D system (1) is *chaotic* when the parameter values are taken as

$$a = 2.2, \quad b = 3, \quad c = 5, \quad p = 0.1 \tag{2}$$

For numerical simulations, we take the initial state of the chaotic system (1) as

$$x_1(0) = 0.2, \quad x_2(0) = 0.2, \quad x_3(0) = 0.2 \tag{3}$$

The Lyapunov exponents of the novel chaotic system (1) for the parameter values (2) and the initial values (3) are numerically determined as

$$L_1 = 0.4715, \quad L_2 = 0, \quad L_3 = -2.4728 \tag{4}$$

The Lyapunov dimension of the novel chaotic system (1) is calculated as

$$D_L = 2 + \frac{L_1 + L_2}{|L_3|} = 2.1907, \tag{5}$$

which is fractional.

The presence of a positive Lyapunov exponent in (4) shows that the 3-D novel system (1) is *chaotic* (Fig. 1).

The novel 3-D chaotic system (1) exhibits a strange chaotic attractor. It is interesting to note that the strange chaotic attractor looks like a *trumpet*. Hence, the novel chaotic system (1) can be also called as a *trumpet attractor*.

Figure 2 describes the 2-D projection of the strange chaotic attractor of the novel chaotic system (1) on (x_1, x_2)-plane.

Figure 3 describes the 2-D projection of the strange chaotic attractor of the novel chaotic system (1) on (x_2, x_3)-plane.

Figure 4 describes the 2-D projection of the strange chaotic attractor of the novel chaotic system (1) on (x_1, x_3)-plane.

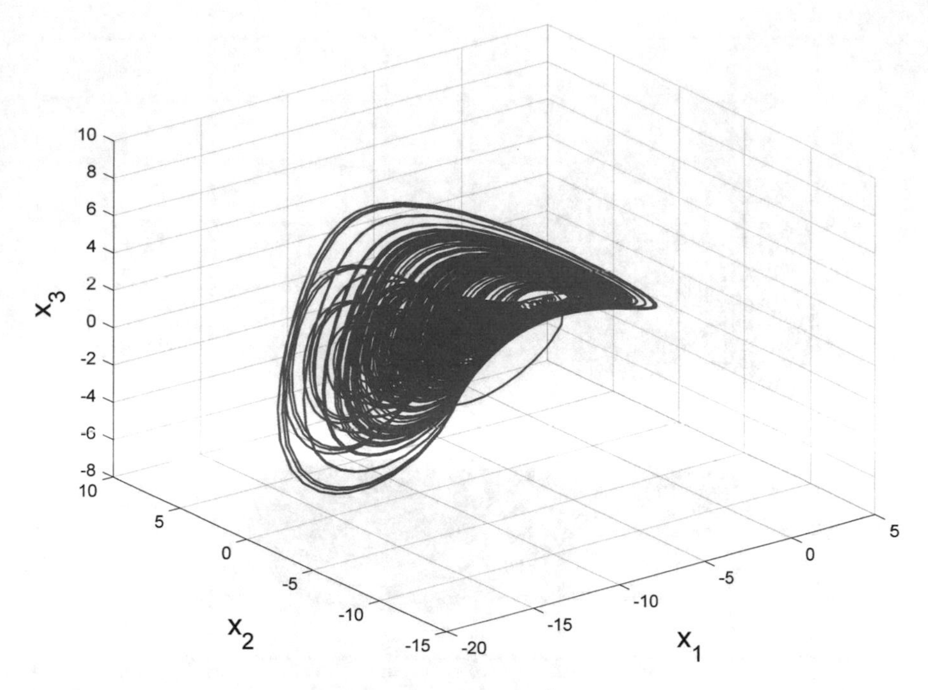

Fig. 1 Strange attractor of the novel chaotic system in $\mathbf{R}^3$

3 Analysis of the 3-D Novel Chaotic System

This section gives the qualitative properties of the novel chaotic system (1).

3.1 Dissipativity

In vector notation, the system (1) can be expressed as

$$\dot{x} = f(x) = \begin{bmatrix} f_1(x) \\ f_2(x) \\ f_3(x) \end{bmatrix}, \tag{6}$$

where

$$\begin{aligned} f_1(x) &= -ax_2 + x_2x_3 \\ f_2(x) &= px_1 + bx_2 - x_1x_3 \\ f_3(x) &= x_1 - cx_3 + x_1x_2 \end{aligned} \tag{7}$$

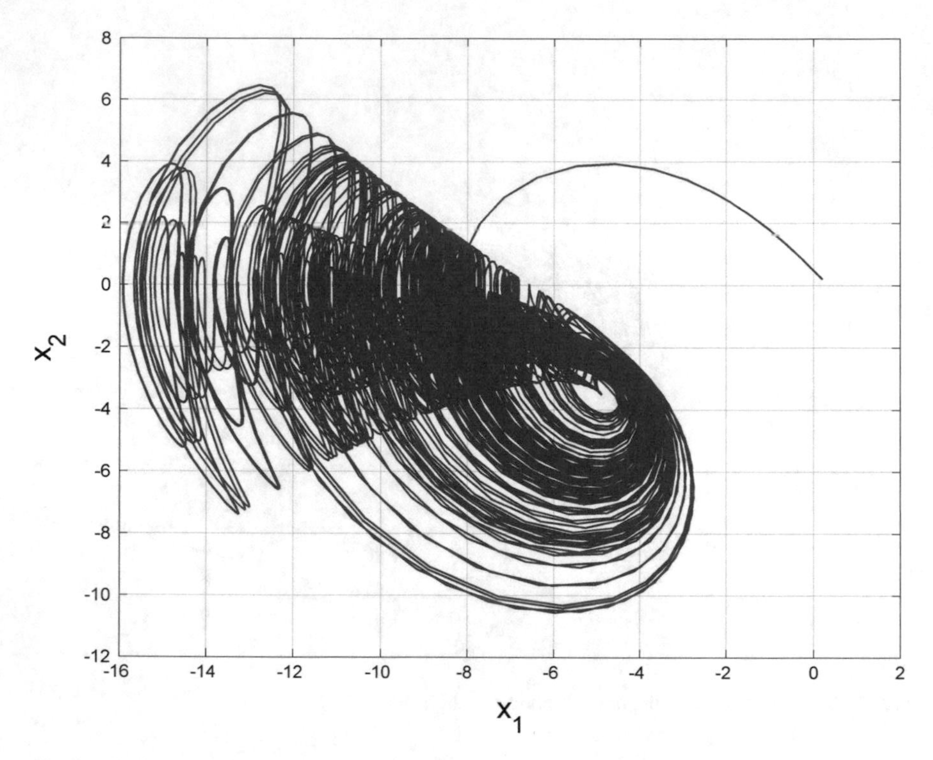

Fig. 2 2-D projection of the novel chaotic system on (x_1, x_2)-plane

We take the parameter values as

$$a = 2.2, \quad b = 3, \quad c = 5, \quad p = 0.1 \tag{8}$$

The divergence of the vector field f on $\mathbb{R}^3$ is obtained as

$$\operatorname{div} f = \frac{\partial f_1(x)}{\partial x_1} + \frac{\partial f_2(x)}{\partial x_2} + \frac{\partial f_3(x)}{\partial x_3} = -(c - b) = -\mu \tag{9}$$

where

$$\mu = c - b = 2 > 0 \tag{10}$$

Let Ω be any region in $\mathbb{R}^3$ with a smooth boundary. Let $\Omega(t) = \Phi_t(\Omega)$, where Φ_t is the flow of the vector field f. Let $V(t)$ denote the volume of $\Omega(t)$.

By Liouville's theorem, it follows that

$$\frac{dV(t)}{dt} = \int\limits_{\Omega(t)} (\operatorname{div} f) dx_1 \, dx_2 \, dx_3 \tag{11}$$

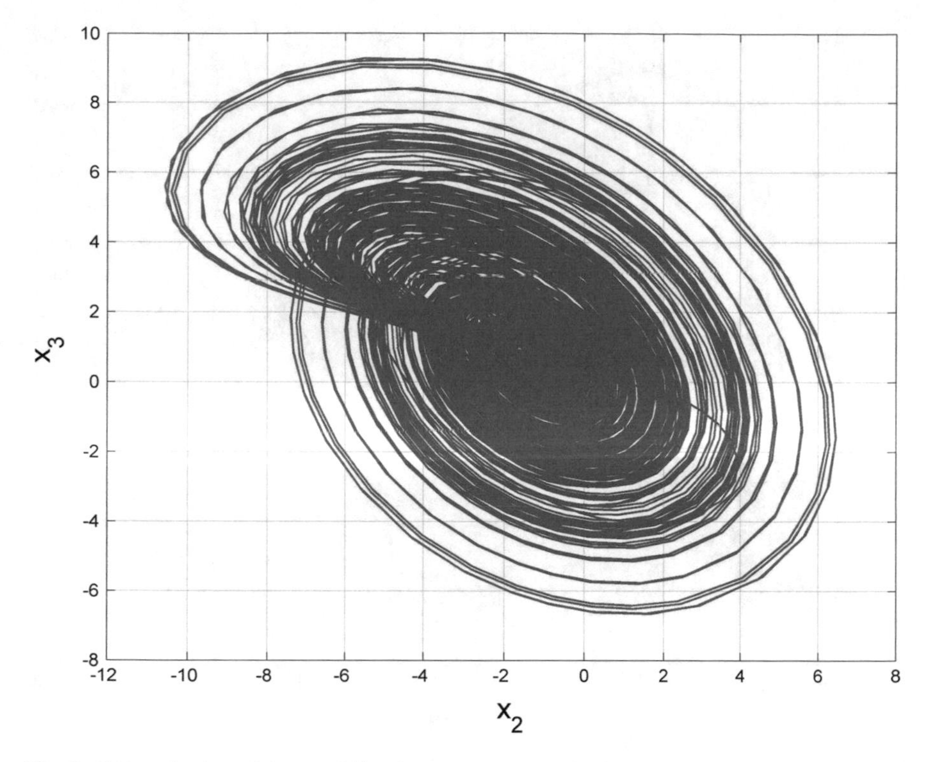

Fig. 3 2-D projection of the novel chaotic system on (x_2, x_3)-plane

Substituting the value of div$f$ in (11) leads to

$$\frac{dV(t)}{dt} = -\mu \int_{\Omega(t)} dx_1\, dx_2\, dx_3 = -\mu V(t) \tag{12}$$

Integrating the linear differential equation (12), $V(t)$ is obtained as

$$V(t) = V(0)\exp(-\mu t) \tag{13}$$

From Eq. (13), it follows that the volume $V(t)$ shrinks to zero exponentially as $t \to \infty$.

Thus, the novel chaotic system (1) is dissipative. Hence, any asymptotic motion of the system (1) settles onto a set of measure zero, i.e. a strange attractor.

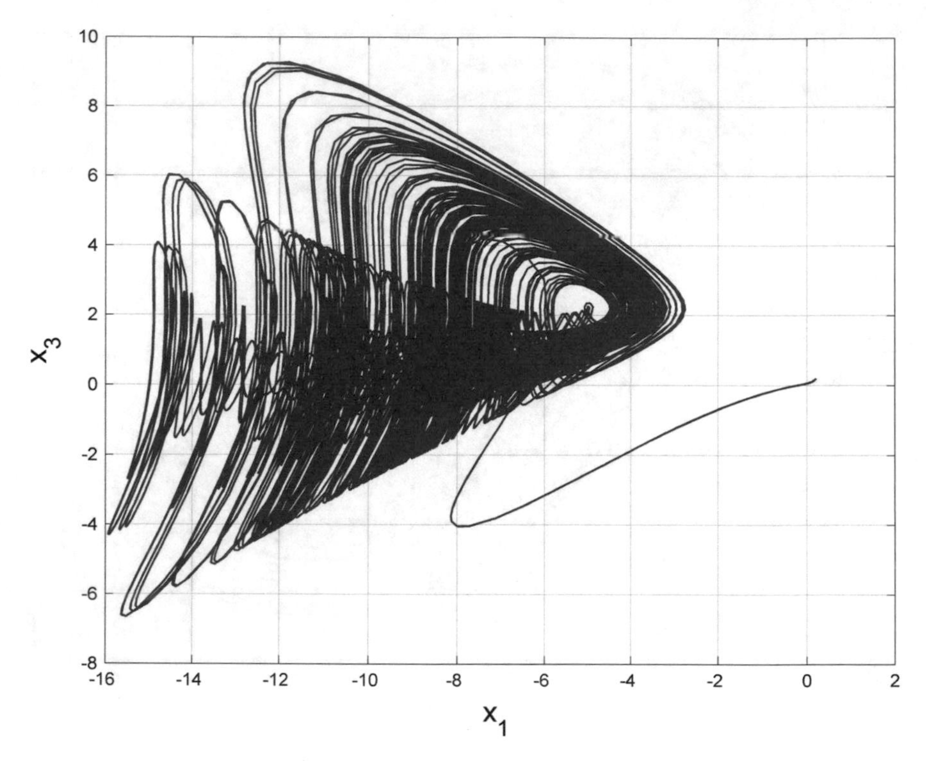

Fig. 4 2-D projection of the novel chaotic system on (x_1, x_3)-plane

3.2 *Invariance*

It is easily seen that the x_3-axis is invariant for the flow of the novel chaotic system (1). The invariant motion along the x_3-axis is characterized by the scalar dynamics

$$\dot{x}_3 = -cx_3, \quad (c > 0) \tag{14}$$

which is globally exponentially stable.

3.3 *Equilibria*

The equilibrium points of the novel chaotic system (1) are obtained by solving the nonlinear equations

$$\begin{aligned} f_1(x) &= -ax_2 + x_2x_3 &&= 0 \\ f_2(x) &= px_1 + bx_2 - x_1x_3 &&= 0 \\ f_3(x) &= x_1 - cx_3 + x_1x_2 &&= 0 \end{aligned} \tag{15}$$

We take the parameter values as in the chaotic case, viz.

$$a = 2.2, \ \ b = 3, \ \ c = 5, \ \ p = 0.1 \tag{16}$$

Solving the nonlinear system (15) with the parameter values (16), we obtain four equilibrium points of the novel chaotic system (1), *viz.*

$$E_0 = \begin{bmatrix} 0 \\ 0 \\ 0 \end{bmatrix}, \ E_1 = \begin{bmatrix} 0.5 \\ 0 \\ 0.1 \end{bmatrix}, E_2 = \begin{bmatrix} -4.7422 \\ -3.3196 \\ 2.2000 \end{bmatrix}, \ E_3 = \begin{bmatrix} 3.3137 \\ 2.3196 \\ 2.2000 \end{bmatrix} \tag{17}$$

The Jacobian matrix of the novel chaotic system (1) at $(x_1^\star, x_2^\star, x_3^\star)$ is obtained as

$$J(x^\star) = \begin{bmatrix} 0 & -a + x_3^\star & x_2^\star \\ p - x_3^\star & b & -x_1^\star \\ 1 + x_2^\star & x_1^\star & -c \end{bmatrix} \tag{18}$$

The matrix $J_0 = J(E_0)$ has the eigenvalues

$$\lambda_1 = -5, \ \ \lambda_2 = 0.0752, \ \ \lambda_3 = -5 \tag{19}$$

This shows that the equilibrium point E_0 is a saddle-point, which is unstable.
The matrix $J_1 = J(E_1)$ has the eigenvalues

$$\lambda_1 = -0.0705, \ \ \lambda_2 = -4.9418, \ \ \lambda_3 = 3.0123 \tag{20}$$

This shows that the equilibrium point E_1 is a saddle point, which is unstable.
The matrix $J_2 = J(E_2)$ has the eigenvalues

$$\lambda_1 = -4.6466, \ \ \lambda_{2,3} = 1.3233 \pm 3.2148i \tag{21}$$

This shows that the equilibrium point E_2 is a saddle-focus, which is unstable.
The matrix $J_3 = J(E_3)$ has the eigenvalues

$$\lambda_1 = -5.4615, \ \ \lambda_{2,3} = 1.7307 \pm 2.0469i \tag{22}$$

This shows that the equilibrium point E_3 is a saddle-focus, which is unstable.

Hence, E_0, E_1, E_2, E_3 are all unstable equilibrium points of the 3-D novel chaotic system (1), where E_0, E_1 are saddle points and E_3, E_4 are saddle-foci.

3.4 Lyapunov Exponents and Lyapunov Dimension

We take the initial values of the novel chaotic system (1) as in (3) and the parameter values of the novel chaotic system (1) as in (2).

Then the Lyapunov exponents of the novel chaotic system (1) are numerically obtained as

$$L_1 = 0.4715, \quad L_2 = 0, \quad L_3 = -2.4728 \tag{23}$$

Since $L_1 + L_2 + L_3 = -2.0013 < 0$, the system (1) is dissipative.

Also, the Lyapunov dimension of the system (1) is obtained as

$$D_L = 2 + \frac{L_1 + L_2}{|L_3|} = 2.1907, \tag{24}$$

which is fractional.

Figure 5 depicts the Lyapunov exponents of the novel chaotic system (1). From this figure, it is seen that the Maximal Lyapunov Exponent (MLE) of the novel chaotic system (1) is $L_1 = 0.4715$.

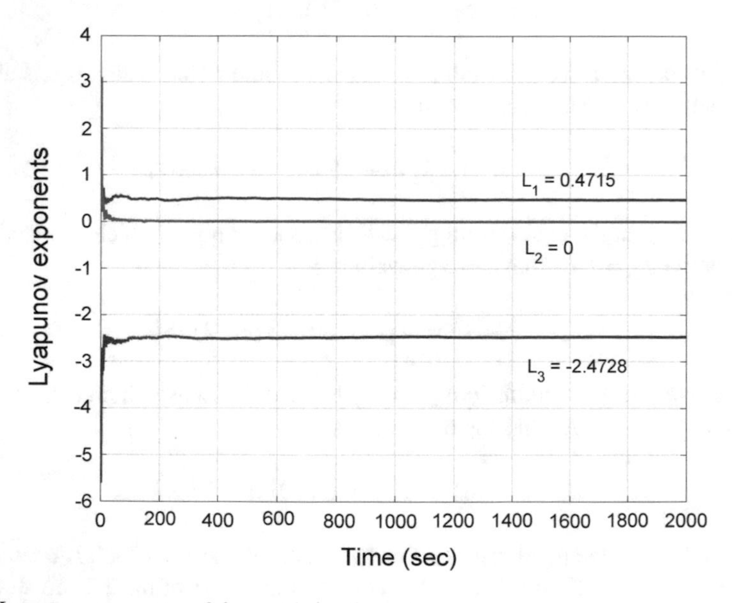

Fig. 5 Lyapunov exponents of the novel chaotic system

4 Adaptive Feedback Control of the 3-D Novel Chaotic System

This section derives new results for adaptive feedback controller design in order to stabilize the unstable novel chaotic system with unknown parameters for all initial conditions.

The controlled novel 3-D chaotic system is given by

$$\begin{aligned} \dot{x}_1 &= -ax_2 + x_2x_3 + u_1 \\ \dot{x}_2 &= px_1 + bx_2 - x_1x_3 + u_2 \\ \dot{x}_3 &= x_1 - cx_3 + x_1x_2 + u_3 \end{aligned} \tag{25}$$

where x_1, x_2, x_3 are state variables, a, b, c, p are constant, unknown, parameters of the system and u_1, u_2, u_3 are adaptive feedback controls to be designed.

An adaptive feedback control law is taken as

$$\begin{aligned} u_1 &= \hat{a}(t)x_2 - x_2x_3 - k_1x_1 \\ u_2 &= -\hat{p}(t)x_1 - \hat{b}(t)x_2 + x_1x_3 - k_2x_2 \\ u_3 &= -x_1 + \hat{c}(t)x_3 - x_1x_2 - k_3x_3 \end{aligned} \tag{26}$$

In (26), $\hat{a}(t), \hat{b}(t), \hat{c}(t), \hat{p}(t)$ are estimates for the unknown parameters a, b, c, p, respectively, and k_1, k_2, k_3 are positive gain constants.

The closed-loop control system is obtained by substituting (26) into (25) as

$$\begin{aligned} \dot{x}_1 &= -[a - \hat{a}(t)]x_2 - k_1x_1 \\ \dot{x}_2 &= [p - \hat{p}(t)]x_1 + [b - \hat{b}(t)]x_2 - k_2x_2 \\ \dot{x}_3 &= -[c - \hat{c}(t)]x_3 - k_3x_3 \end{aligned} \tag{27}$$

To simplify (27), we define the parameter estimation error as

$$\begin{aligned} e_a(t) &= a - \hat{a}(t) \\ e_b(t) &= b - \hat{b}(t) \\ e_c(t) &= c - \hat{c}(t) \\ e_p(t) &= d - \hat{p}(t) \end{aligned} \tag{28}$$

Using (28), the closed-loop system (27) can be simplified as

$$\begin{aligned} \dot{x}_1 &= -e_ax_2 - k_1x_1 \\ \dot{x}_2 &= e_px_1 + e_bx_2 - k_2x_2 \\ \dot{x}_3 &= -e_cx_3 - k_3x_3 \end{aligned} \tag{29}$$

Differentiating the parameter estimation error (28) with respect to t, we get

$$\begin{aligned} \dot{e}_a &= -\dot{\hat{a}} \\ \dot{e}_b &= -\dot{\hat{b}} \\ \dot{e}_c &= -\dot{\hat{c}} \\ \dot{e}_p &= -\dot{\hat{p}} \end{aligned} \tag{30}$$

Next, we find an update law for parameter estimates using Lyapunov stability theory.

Consider the quadratic Lyapunov function defined by

$$V(x_1, x_2, x_3, e_a, e_b, e_c, e_p) = \frac{1}{2}\left(x_1^2 + x_2^2 + x_3^2 + e_a^2 + e_b^2 + e_c^2 + e_p^2\right), \tag{31}$$

which is positive definite on $\mathbf{R}^7$.

Differentiating V along the trajectories of (29) and (30), we get

$$\begin{aligned} \dot{V} = &-k_1x_1^2 - k_2x_2^2 - k_3x_3^2 + e_a[-x_1x_2 - \dot{\hat{a}}] + e_b[x_2^2 - \dot{\hat{b}}] + e_c[-x_3^2 - \dot{\hat{c}}] \\ &+e_p[x_1x_2 - \dot{\hat{p}}] \end{aligned} \tag{32}$$

In view of Eq. (32), an update law for the parameter estimates is taken as

$$\begin{aligned} \dot{\hat{a}} &= -x_1x_2 \\ \dot{\hat{b}} &= x_2^2 \\ \dot{\hat{c}} &= -x_3^2 \\ \dot{\hat{p}} &= x_1x_2 \end{aligned} \tag{33}$$

Theorem 1 *The novel chaotic system (25) with unknown system parameters is globally and exponentially stabilized for all initial conditions $x(0) \in \mathbf{R}^3$ by the adaptive control law (26) and the parameter update law (33), where $k_i, (i = 1, 2, 3)$ are positive constants.*

Proof The result is proved using Lyapunov stability theory [32]. We consider the quadratic Lyapunov function V defined by (31), which is positive definite on $\mathbf{R}^7$.

Substitution of the parameter update law (33) into (32) yields

$$\dot{V} = -k_1x_1^2 - k_2x_2^2 - k_3x_3^2, \tag{34}$$

which is a negative semi-definite function on $\mathbf{R}^7$.

Therefore, it can be concluded that the state vector $x(t)$ and the parameter estimation error are globally bounded, i.e.

$$\left[x_1(t)\ x_2(t)\ x_3(t)\ e_a(t)\ e_b(t)\ e_c(t)\ e_p(t)\right]^T \in \mathbf{L}_\infty. \tag{35}$$

Define

$$k = \min\left\{k_1, k_2, k_3\right\} \tag{36}$$

Then it follows from (34) that

$$\dot{V} \le -k\|\mathbf{x}\|^2 \;\text{ or }\; k\|\mathbf{x}\|^2 \le -\dot{V} \tag{37}$$

Integrating the inequality (37) from 0 to t, we get

$$k\int_0^t \|\mathbf{x}(\tau)\|^2\, d\tau \le -\int_0^t \dot{V}(\tau)\, d\tau = V(0) - V(t) \tag{38}$$

From (38), it follows that $\mathbf{x}(t) \in \mathbf{L}_2$.

Using (29), it can be deduced that $\dot{x}(t) \in \mathbf{L}_\infty$.

Hence, using Barbalat's lemma, we can conclude that $\mathbf{x}(t) \to 0$ exponentially as $t \to \infty$ for all initial conditions $\mathbf{x}(0) \in \mathbb{R}^3$.

This completes the proof. □

For numerical simulations, the parameter values of the novel system (25) are taken as in the chaotic case, *viz.*

$$a = 2.2, \;\; b = 3, \;\; c = 5, \;\; p = 0.1 \tag{39}$$

The gain constants are taken as $k_i = 6, (i = 1, 2, 3)$.

The initial values of the parameter estimates are taken as

$$\hat{a}(0) = 5.4, \;\; \hat{b}(0) = 12.7, \;\; \hat{c}(0) = 21.3, \;\; \hat{p}(0) = 16.2 \tag{40}$$

The initial values of the novel system (25) are taken as

$$x_1(0) = 18.3, \;\; x_2(0) = 11.6, \;\; x_3(0) = 7.9 \tag{41}$$

Figure 6 shows the time-history of the controlled states $x_1(t), x_2(t), x_3(t)$.

Figure 6 depicts the exponential convergence of the controlled states and the efficiency of the adaptive controller defined by (26).

5 Adaptive Synchronization of the Identical 3-D Novel Chaotic Systems

This section derives new results for the adaptive synchronization of the identical novel chaotic systems with unknown parameters.

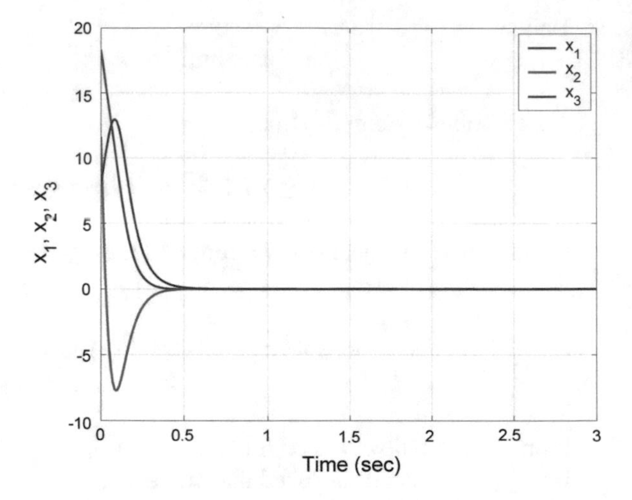

Fig. 6 Time-history of the states $x_1(t), x_2(t), x_3(t)$

The master system is given by the novel chaotic system

$$\begin{aligned}\dot{x}_1 &= -ax_2 + x_2x_3 \\ \dot{x}_2 &= px_1 + bx_2 - x_1x_3 \\ \dot{x}_3 &= x_1 - cx_3 + x_1x_2\end{aligned} \tag{42}$$

where x_1, x_2, x_3 are state variables and a, b, c, p are constant, unknown, parameters of the system.

The slave system is given by the controlled novel chaotic system

$$\begin{aligned}\dot{y}_1 &= -ay_2 + y_2y_3 + u_1 \\ \dot{y}_2 &= py_1 + by_2 - y_1y_3 + u_2 \\ \dot{y}_3 &= y_1 - cy_3 + y_1y_2 + u_3\end{aligned} \tag{43}$$

where y_1, y_2, y_3 are state variables and u_1, u_2, u_3 are adaptive controls to be designed.

The synchronization error is defined as

$$\begin{aligned}e_1 &= y_1 - x_1 \\ e_2 &= y_2 - x_2 \\ e_3 &= y_3 - x_3\end{aligned} \tag{44}$$

The error dynamics is easily obtained as

$$\begin{aligned}\dot{e}_1 &= -ae_2 + y_2y_3 - x_2x_3 + u_1 \\ \dot{e}_2 &= pe_1 + be_2 - y_1y_3 + x_1x_3 + u_2 \\ \dot{e}_3 &= e_1 - ce_3 + y_1y_2 - x_1x_2 + u_3\end{aligned} \tag{45}$$

An adaptive control law is taken as

$$\begin{aligned} u_1 &= \hat{a}(t)e_2 - y_2y_3 + x_2x_3 - k_1e_1 \\ u_2 &= -\hat{p}(t)e_1 - \hat{b}(t)e_2 + y_1y_3 - x_1x_3 - k_2e_2 \\ u_3 &= -e_1 + \hat{c}(t)e_3 - y_1y_2 + x_1x_2 - k_3e_3 \end{aligned} \tag{46}$$

where $\hat{a}(t), \hat{b}(t), \hat{c}(t), \hat{p}(t)$ are estimates for the unknown parameters a, b, c, p, respectively, and k_1, k_2, k_3 are positive gain constants.

The closed-loop control system is obtained by substituting (46) into (45) as

$$\begin{aligned} \dot{e}_1 &= -[a - \hat{a}(t)]e_2 - k_1e_1 \\ \dot{e}_2 &= [p - \hat{p}(t)]e_1 + [b - \hat{b}(t)]e_2 - k_2e_2 \\ \dot{e}_3 &= -[c - \hat{c}(t)]e_3 - k_3e_3 \end{aligned} \tag{47}$$

To simplify (47), we define the parameter estimation error as

$$\begin{aligned} e_a(t) &= a - \hat{a}(t) \\ e_b(t) &= b - \hat{b}(t) \\ e_c(t) &= c - \hat{c}(t) \\ e_p(t) &= p - \hat{p}(t) \end{aligned} \tag{48}$$

Using (48), the closed-loop system (47) can be simplified as

$$\begin{aligned} \dot{e}_1 &= -e_ae_2 - k_1e_1 \\ \dot{e}_2 &= e_pe_1 + e_be_2 - k_2e_2 \\ \dot{e}_3 &= -e_ce_3 - k_3e_3 \end{aligned} \tag{49}$$

Differentiating the parameter estimation error (48) with respect to t, we get

$$\begin{aligned} \dot{e}_a &= -\dot{\hat{a}} \\ \dot{e}_b &= -\dot{\hat{b}} \\ \dot{e}_c &= -\dot{\hat{c}} \\ \dot{e}_p &= -\dot{\hat{p}} \end{aligned} \tag{50}$$

Next, we find an update law for parameter estimates using Lyapunov stability theory.

Consider the quadratic Lyapunov function defined by

$$V(e_1, e_2, e_3, e_a, e_b, e_c, e_p) = \frac{1}{2}\left(e_1^2 + e_2^2 + e_3^2 + e_a^2 + e_b^2 + e_c^2 + e_p^2\right), \tag{51}$$

which is positive definite on $\mathbf{R}^7$.

Differentiating V along the trajectories of (49) and (50), we get

$$\begin{aligned}\dot{V} = & -k_1 e_1^2 - k_2 e_2^2 - k_3 e_3^2 + e_a \left[-e_1 e_2 - \dot{\hat{a}}\right] + e_b \left[e_2^2 - \dot{\hat{b}}\right] \\ & + e_c \left[-e_3^2 - \dot{\hat{c}}\right] + e_p \left[e_1 e_2 - \dot{\hat{p}}\right]\end{aligned} \tag{52}$$

In view of Eq. (52), an update law for the parameter estimates is taken as

$$\begin{aligned}\dot{\hat{a}} &= -e_1 e_2 \\ \dot{\hat{b}} &= e_2^2 \\ \dot{\hat{c}} &= -e_3^2 \\ \dot{\hat{p}} &= e_1 e_2\end{aligned} \tag{53}$$

Theorem 2 *The identical novel chaotic systems (42) and (43) with unknown system parameters are globally and exponentially synchronized for all initial conditions* $x(0), y(0) \in \mathbf{R}^3$ *by the adaptive control law (46) and the parameter update law (53), where* $k_i, (i = 1, 2, 3)$ *are positive constants.*

Proof The result is proved using Lyapunov stability theory [32].

We consider the quadratic Lyapunov function V defined by (51), which is positive definite on $\mathbf{R}^7$.

Substitution of the parameter update law (53) into (52) yields

$$\dot{V} = -k_1 e_1^2 - k_2 e_2^2 - k_3 e_3^2, \tag{54}$$

which is a negative semi-definite function on $\mathbf{R}^7$.

Therefore, it can be concluded that the synchronization error vector $e(t)$ and the parameter estimation error are globally bounded, i.e.

$$\left[e_1(t)\ e_2(t)\ e_3(t)\ e_a(t)\ e_b(t)\ e_c(t)\ e_p(t)\right]^T \in \mathbf{L}_\infty. \tag{55}$$

Define

$$k = \min\left\{k_1, k_2, k_3\right\} \tag{56}$$

Then it follows from (54) that

$$\dot{V} \leq -k\|e\|^2 \ \text{ or } \ k\|\mathbf{e}\|^2 \leq -\dot{V} \tag{57}$$

Integrating the inequality (57) from 0 to t, we get

$$k \int_0^t \|\mathbf{e}(\tau)\|^2 \, d\tau \leq -\int_0^t \dot{V}(\tau)\, d\tau = V(0) - V(t) \tag{58}$$

From (58), it follows that $\mathbf{e}(t) \in \mathbf{L}_2$.

Using (49), it can be deduced that $\dot{\mathbf{e}}(t) \in \mathbf{L}_\infty$.

Hence, using Barbalat's lemma, we can conclude that $\mathbf{e}(t) \to 0$ exponentially as $t \to \infty$ for all initial conditions $\mathbf{e}(0) \in \mathbb{R}^3$.

This completes the proof. □

For numerical simulations, the parameter values of the novel systems (42) and (43) are taken as in the chaotic case, *viz.*

$$a = 2.2, \quad b = 3, \quad c = 5, \quad p = 0.1 \tag{59}$$

The gain constants are taken as $k_i = 6$ for $i = 1, 2, 3$.

The initial values of the parameter estimates are taken as

$$\hat{a}(0) = 6.2, \quad \hat{b}(0) = 12.9, \quad \hat{c}(0) = 28.5, \quad \hat{p}(0) = 17.3 \tag{60}$$

The initial values of the master system (42) are taken as

$$x_1(0) = 5.8, \quad x_2(0) = 18.3, \quad x_3(0) = -12.1 \tag{61}$$

The initial values of the slave system (43) are taken as

$$y_1(0) = 16.4, \quad y_2(0) = 4.5, \quad y_3(0) = -7.8 \tag{62}$$

Figures 7-9 show the complete synchronization of the identical chaotic systems (42) and (43).

Figure 7 shows that the states $x_1(t)$ and $y_1(t)$ are synchronized in one second (MATLAB).

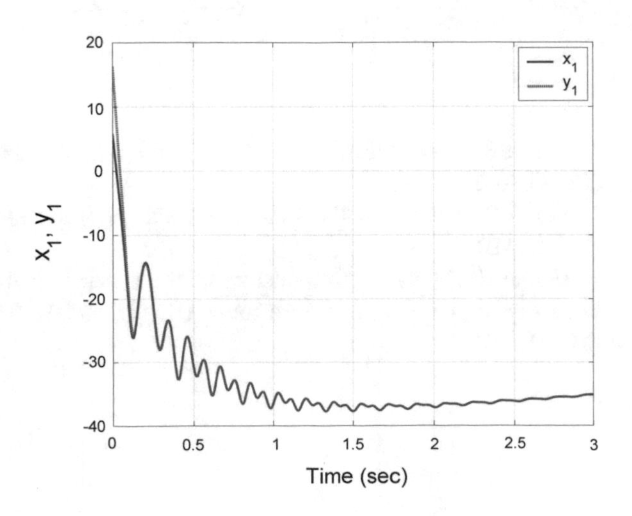

Fig. 7 Synchronization of the states x_1 and y_1

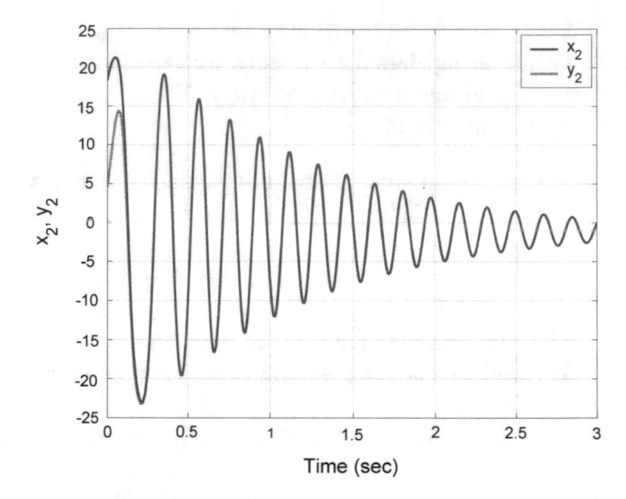

Fig. 8 Synchronization of the states x_2 and y_2

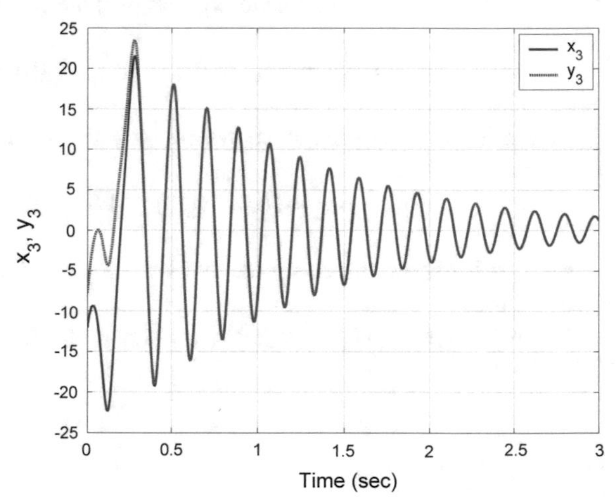

Fig. 9 Synchronization of the states x_3 and y_3

Figure 8 shows that the states $x_2(t)$ and $y_2(t)$ are synchronized in one second (MATLAB).

Figure 9 shows that the states $x_3(t)$ and $y_3(t)$ are synchronized in one second (MATLAB).

Figure 10 shows the time-history of the synchronization errors $e_1(t), e_2(t), e_3(t)$. From Fig. 10, it is seen that the errors $e_1(t), e_2(t)$ and $e_3(t)$ are stabilized in one second (MATLAB).

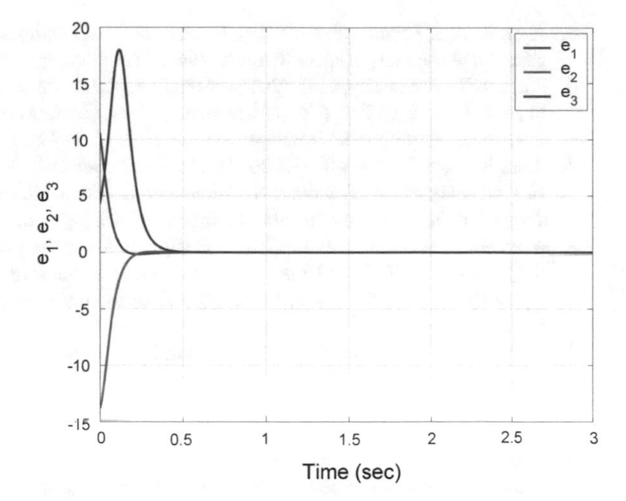

Fig. 10 Time-history of the synchronization errors e_1, e_2, e_3

6 Conclusions

In this work, we described an eight-term 3-D novel polynomial chaotic system consisting of three quadratic nonlinearities. The qualitative properties of the novel chaotic system have been discussed in detail. We showed that the novel chaotic system has four equilibrium points of which two equilibrium points are saddle points and the other equilibrium points are saddle-foci. The Lyapunov exponents of the novel chaotic system were derived as $L_1 = 0.4715, L_2 = 0$ and $L_3 = -2.4728$. The Lyapunov dimension of the novel chaotic system was obtained as $D_L = 2.1907$. Next, we worked on the design of adaptive feedback controller for globally stabilizing the trajectories of the novel chaotic system with unknown parameters. Furthermore, we derived new results for the design of adaptive feedback controller for achieving complete synchronization of the identical novel chaotic systems with unknown parameters. The main adaptive control results were proved using Lyapunov stability theory. MATLAB simulations were displayed to illustrate all the main results presented in this research work.

References

1. Akgul, A., Moroz, I., Pehlivan, I., & Vaidyanathan, S. (2016). A new four-scroll chaotic attractor and its engineering applications. *Optik*, *127*, 5491–5499.
2. Arneodo, A., Coullet, P., & Tresser, C. (1981). Possible new strange attractors with spiral structure. *Communications in Mathematical Physics*, *79*, 573–579.
3. Azar, A. T. (2010). *Fuzzy systems*. Vienna, Austria: IN-TECH.
4. Azar, A. T. (2012). Overview of type-2 fuzzy logic systems. *International Journal of Fuzzy System Applications*, *2*(4), 1–28.

5. Azar, A. T., & Serrano, F. E. (2014). Robust IMC-PID tuning for cascade control systems with gain and phase margin specifications. *Neural Computing and Applications*, *25*(5), 983–995.
6. Azar, A. T., & Serrano, F. E. (2015). Adaptive sliding mode control of the Furuta pendulum. In A. T. Azar & Q. Zhu (Eds.), *Advances and applications in sliding mode control systems*. Studies in computational intelligence (Vol. 576, pp. 1–42). Germany: Springer.
7. Azar, A. T., & Serrano, F. E. (2015). Deadbeat control for multivariable systems with time varying delays. In A. T. Azar & S. Vaidyanathan (Eds.), *Chaos modeling and control systems design*. Studies in computational intelligence (Vol. 581, pp. 97–132). Germany: Springer.
8. Azar, A. T., & Serrano, F. E. (2015). Design and modeling of anti wind up PID controllers. In Q. Zhu & A. T. Azar (Eds.), *Complex system modelling and control through intelligent soft computations*. Studies in fuzziness and soft computing (Vol. 319, pp. 1–44). Germany: Springer.
9. Azar, A. T., & Serrano, F. E. (2015). Stabilizatoin and control of mechanical systems with backlash. In A. T. Azar & S. Vaidyanathan (Eds.), *Handbook of research on advanced intelligent control engineering and automation*. Advances in computational intelligence and robotics (ACIR) (pp. 1–60). USA: IGI-Global.
10. Azar, A. T., & Vaidyanathan, S. (2015). *Chaos modeling and control systems design*. Studies in computational intelligence (Vol. 581). Germany: Springer.
11. Azar, A. T., & Vaidyanathan, S. (2015). *Computational intelligence applications in modeling and control*. Studies in computational intelligence (Vol. 575). Germany: Springer.
12. Azar, A. T., & Vaidyanathan, S. (2015). *Handbook of research on advanced intelligent control engineering and automation*. Advances in computational intelligence and robotics (ACIR). USA: IGI-Global.
13. Azar, A. T., & Vaidyanathan, S. (2016). *Advances in chaos theory and intelligent control*. Studies in fuzziness and soft computing (Vol. 337). Germany: Springer.
14. Azar, A. T., & Zhu, Q. (2015). *Advances and applications in sliding mode control systems*. Studies in computational intelligence (Vol. 576). Germany: Springer.
15. Barrow-Green, J. (1997). *Poincaré and the three body problem*. American Mathematical Society.
16. Boulkroune, A., Bouzeriba, A., Bouden, T., & Azar, A. T. (2016a). Fuzzy adaptive synchronization of uncertain fractional-order chaotic systems. In A. T. Azar & S. Vaidyanathan (Eds.), *Advances in chaos theory and intelligent control*. Studies in fuzziness and soft computing (Vol. 337, pp. 681–697). Germany: Springer.
17. Boulkroune, A., Hamel, S., Azar, A. T., & Vaidyanathan, S. (2016b). Fuzzy control-based function synchronization of unknown chaotic systems with dead-zone input. In A. T. Azar & S. Vaidyanathan (Eds.), *Advances in chaos theory and intelligent control*. Studies in fuzziness and soft computing (Vol. 337, pp. 699–718). Germany: Springer.
18. Carroll, T. L., & Pecora, L. M. (1991). Synchronizing chaotic circuits. *IEEE Transactions on Circuits and Systems*, *38*(4), 453–456.
19. Chen, G., & Ueta, T. (1999). Yet another chaotic attractor. *International Journal of Bifurcation and Chaos*, *9*(7), 1465–1466.
20. Chen, W. H., Wei, D., & Lu, X. (2014). Global exponential synchronization of nonlinear time-delay Lur'e systems via delayed impulsive control. *Communications in Nonlinear Science and Numerical Simulation*, *19*(9), 3298–3312.
21. Dadras, S., & Momeni, H. R. (2009). A novel three-dimensional autonomous chaotic system generating two, three and four-scroll attractors. *Physics Letters A*, *373*, 3637–3642.
22. Das, S., Goswami, D., Chatterjee, S., & Mukherjee, S. (2014). Stability and chaos analysis of a novel swarm dynamics with applications to multi-agent systems. *Engineering Applications of Artificial Intelligence*, *30*, 189–198.
23. Feki, M. (2003). An adaptive chaos synchronization scheme applied to secure communication. *Chaos, Solitons and Fractals*, *18*(1), 141–148.
24. Gan, Q., & Liang, Y. (2012). Synchronization of chaotic neural networks with time delay in the leakage term and parametric uncertainties based on sampled-data control. *Journal of the Franklin Institute*, *349*(6), 1955–1971.

25. Genesio, R., & Tesi, A. (1992). Harmonic balance methods for the analysis of chaotic dynamics in nonlinear systems. *Automatica*, *28*(3), 531–548.
26. Gibson, W. T., & Wilson, W. G. (2013). Individual-based chaos: Extensions of the discrete logistic model. *Journal of Theoretical Biology*, *339*, 84–92.
27. Henon, M., & Heiles, C. (1964). The applicability of the third integral of motion: Some numerical experiments. *The Astrophysical Journal*, *69*, 73–79.
28. Huang, X., Zhao, Z., Wang, Z., & Li, Y. (2012). Chaos and hyperchaos in fractional-order cellular neural networks. *Neurocomputing*, *94*, 13–21.
29. Jiang, G. P., Zheng, W. X., & Chen, G. (2004). Global chaos synchronization with channel time-delay. *Chaos, Solitons & Fractals*, *20*(2), 267–275.
30. Karthikeyan, R., & Sundarapandian, V. (2014). Hybrid chaos synchronization of four-scroll systems via active control. *Journal of Electrical Engineering*, *65*(2), 97–103.
31. Kaslik, E., & Sivasundaram, S. (2012). Nonlinear dynamics and chaos in fractional-order neural networks. *Neural Networks*, *32*, 245–256.
32. Khalil, H. K. (2001). *Nonlinear systems*. New Jersey, USA: Prentice Hall.
33. Kyriazis, M. (1991). Applications of chaos theory to the molecular biology of aging. *Experimental Gerontology*, *26*(6), 569–572.
34. Lang, J. (2015). Color image encryption based on color blend and chaos permutation in the reality-preserving multiple-parameter fractional Fourier transform domain. *Optics Communications*, *338*, 181–192.
35. Li, D. (2008). A three-scroll chaotic attractor. *Physics Letters A*, *372*(4), 387–393.
36. Li, N., Zhang, Y., & Nie, Z. (2011). Synchronization for general complex dynamical networks with sampled-data. *Neurocomputing*, *74*(5), 805–811.
37. Li, N., Pan, W., Yan, L., Luo, B., & Zou, X. (2014). Enhanced chaos synchronization and communication in cascade-coupled semiconductor ring lasers. *Communications in Nonlinear Science and Numerical Simulation*, *19*(6), 1874–1883.
38. Li, Z., & Chen, G. (2006). *Integration of fuzzy logic and chaos theory*. Studies in fuzziness and soft computing (Vol. 187). Germany: Springer.
39. Lian, S., & Chen, X. (2011). Traceable content protection based on chaos and neural networks. *Applied Soft Computing*, *11*(7), 4293–4301.
40. Liu, C., Liu, T., Liu, L., & Liu, K. (2004). A new chaotic attractor. *Chaos, Solitions and Fractals*, *22*(5), 1031–1038.
41. Lorenz, E. N. (1963). Deterministic periodic flow. *Journal of the Atmospheric Sciences*, *20*(2), 130–141.
42. Lü, J., & Chen, G. (2002). A new chaotic attractor coined. *International Journal of Bifurcation and Chaos*, *12*(3), 659–661.
43. Mondal, S., & Mahanta, C. (2014). Adaptive second order terminal sliding mode controller for robotic manipulators. *Journal of the Franklin Institute*, *351*(4), 2356–2377.
44. Murali, K., & Lakshmanan, M. (1998). Secure communication using a compound signal from generalized chaotic systems. *Physics Letters A*, *241*(6), 303–310.
45. Nehmzow, U., & Walker, K. (2005). Quantitative description of robot-environment interaction using chaos theory. *Robotics and Autonomous Systems*, *53*(3–4), 177–193.
46. Pandey, A., Baghel, R. K., & Singh, R. P. (2012). Synchronization analysis of a new autonomous chaotic system with its application in signal masking. *IOSR Journal of Electronics and Communication Engineering*, *1*(5), 16–22.
47. Pecora, L. M., & Carroll, T. L. (1990). Synchronization in chaotic systems. *Physical Review Letters*, *64*(8), 821–824.
48. Pehlivan, I., Moroz, I. M., & Vaidyanathan, S. (2014). Analysis, synchronization and circuit design of a novel butterfly attractor. *Journal of Sound and Vibration*, *333*(20), 5077–5096.
49. Pham, V. T., Vaidyanathan, S., Volos, C. K., & Jafari, S. (2015). Hidden attractors in a chaotic system with an exponential nonlinear term. *European Physical Journal Special Topics*, *224*(8), 1507–1517.
50. Pham, V. T., Volos, C. K., Vaidyanathan, S., Le, T. P., & Vu, V. Y. (2015). A memristor-based hyperchaotic system with hidden attractors: Dynamics, synchronization and circuital emulating. *Journal of Engineering Science and Technology Review*, *8*(2), 205–214.

51. Pham, V. T., Jafari, S., Vaidyanathan, S., Volos, C., & Wang, X. (2016). A novel memristive neural network with hidden attractors and its circuitry implementation. *Science China Technological Sciences*, *59*(3), 358–363.
52. Pham, V. T., Vaidyanathan, S., Volos, C., Jafari, S., & Kingni, S. T. (2016). A no-equilibrium hyperchaotic system with a cubic nonlinear term. *Optik*, *127*(6), 3259–3265.
53. Pham, V. T., Vaidyanathan, S., Volos, C. K., Jafari, S., Kuznetsov, N. V., & Hoang, T. M. (2016). A novel memristive time-delay chaotic system without equilibrium points. *European Physical Journal Special Topics*, *225*(1), 127–136.
54. Qi, G., & Chen, G. (2006). Analysis and circuit implementation of a new 4D chaotic system. *Physics Letters A*, *352*, 386–397.
55. Qu, Z. (2011). Chaos in the genesis and maintenance of cardiac arrhythmias. *Progress in Biophysics and Molecular Biology*, *105*(3), 247–257.
56. Rasappan, S., & Vaidyanathan, S. (2012). Global chaos synchronization of WINDMI and Coullet chaotic systems by backstepping control. *Far East Journal of Mathematical Sciences*, *67*(2), 265–287.
57. Rasappan, S., & Vaidyanathan, S. (2012). Hybrid synchronization of n-scroll Chua and Lur'e chaotic systems via backstepping control with novel feedback. *Archives of Control Sciences*, *22*(3), 343–365.
58. Rasappan, S., & Vaidyanathan, S. (2012). Synchronization of hyperchaotic Liu system via backstepping control with recursive feedback. *Communications in Computer and Information Science*, *305*, 212–221.
59. Rasappan, S., & Vaidyanathan, S. (2013). Hybrid synchronization of *n*-scroll chaotic Chua circuits using adaptive backstepping control design with recursive feedback. *Malaysian Journal of Mathematical Sciences*, *7*(2), 219–246.
60. Rasappan, S., & Vaidyanathan, S. (2014). Global chaos synchronization of WINDMI and Coullet chaotic systems using adaptive backstepping control design. *Kyungpook Mathematical Journal*, *54*(1), 293–320.
61. Rhouma, R., & Belghith, S. (2011). Cryptoanalysis of a chaos based cryptosystem on DSP. *Communications in Nonlinear Science and Numerical Simulation*, *16*(2), 876–884.
62. Rikitake, T. (1958). Oscillations of a system of disk dynamos. *Mathematical Proceedings of the Cambridge Philosophical Society*, *54*(1), 89–105.
63. Rössler, O. E. (1976). An equation for continuous chaos. *Physics Letters A*, *57*(5), 397–398.
64. Sampath, S., Vaidyanathan, S., Volos, C. K., & Pham, V. T. (2015). An eight-term novel four-scroll chaotic system with cubic nonlinearity and its circuit simulation. *Journal of Engineering Science and Technology Review*, *8*(2), 1–6.
65. Sarasu, P., & Sundarapandian, V. (2011). Active controller design for the generalized projective synchronization of four-scroll chaotic systems. *International Journal of Systems Signal Control and Engineering Application*, *4*(2), 26–33.
66. Sarasu, P., & Sundarapandian, V. (2011). The generalized projective synchronization of hyperchaotic Lorenz and hyperchaotic Qi systems via active control. *International Journal of Soft Computing*, *6*(5), 216–223.
67. Sarasu, P., & Sundarapandian, V. (2012). Adaptive controller design for the generalized projective synchronization of 4-scroll systems. *International Journal of Systems Signal Control and Engineering Application*, *5*(2), 21–30.
68. Sarasu, P., & Sundarapandian, V. (2012). Generalized projective synchronization of three-scroll chaotic systems via adaptive control. *European Journal of Scientific Research*, *72*(4), 504–522.
69. Sarasu, P., & Sundarapandian, V. (2012). Generalized projective synchronization of two-scroll systems via adaptive control. *International Journal of Soft Computing*, *7*(4), 146–156.
70. Shahverdiev, E. M., & Shore, K. A. (2009). Impact of modulated multiple optical feedback time delays on laser diode chaos synchronization. *Optics Communications*, *282*(17), 3568–2572.
71. Shimizu, T., & Morioka, N. (1980). On the bifurcation of a symmetric limit cycle to an asymmetric one in a simple model. *Physics Letters A*, *76*(3–4), 201–204.

72. Sprott, J. C. (1994). Some simple chaotic flows. *Physical Review E*, *50*(2), 647–650.
73. Sprott, J. C. (2010). *Elegant chaos*. World Scientific.
74. Suérez, I. (1999). Mastering chaos in ecology. *Ecological Modelling*, *117*(2–3), 305–314.
75. Sundarapandian, V. (2010). Output regulation of the Lorenz attractor. *Far East Journal of Mathematical Sciences*, *42*(2), 289–299.
76. Sundarapandian, V. (2013). Analysis and anti-synchronization of a novel chaotic system via active and adaptive controllers. *Journal of Engineering Science and Technology Review*, *6*(4), 45–52.
77. Sundarapandian, V., & Karthikeyan, R. (2011). Anti-synchronization of hyperchaotic Lorenz and hyperchaotic Chen systems by adaptive control. *International Journal of Systmes Signal Control and Engineering Application*, *4*(2), 18–25.
78. Sundarapandian, V., & Karthikeyan, R. (2011). Anti-synchronization of Lü and Pan chaotic systems by adaptive nonlinear control. *European Journal of Scientific Research*, *64*(1), 94–106.
79. Sundarapandian, V., & Karthikeyan, R. (2012). Adaptive anti-synchronization of uncertain Tigan and Li systems. *Journal of Engineering and Applied Sciences*, *7*(1), 45–52.
80. Sundarapandian, V., & Karthikeyan, R. (2012). Hybrid synchronization of hyperchaotic Lorenz and hyperchaotic Chen systems via active control. *Journal of Engineering and Applied Sciences*, *7*(3), 254–264.
81. Sundarapandian, V., & Pehlivan, I. (2012). Analysis, control, synchronization, and circuit design of a novel chaotic system. *Mathematical and Computer Modelling*, *55*(7–8), 1904–1915.
82. Sundarapandian, V., & Sivaperumal, S. (2011). Sliding controller design of hybrid synchronization of four-wing chaotic systems. *International Journal of Soft Computing*, *6*(5), 224–231.
83. Suresh, R., & Sundarapandian, V. (2013). Global chaos synchronization of a family of *n*-scroll hyperchaotic Chua circuits using backstepping control with recursive feedback. *Far East Journal of Mathematical Sciences*, *73*(1), 73–95.
84. Tacha, O. I., Volos, C. K., Kyprianidis, I. M., Stouboulos, I. N., Vaidyanathan, S., & Pham, V. T. (2016). Analysis, adaptive control and circuit simulation of a novel nonlinear finance system. *Applied Mathematics and Computation*, *276*, 200–217.
85. Usama, M., Khan, M. K., Alghatbar, K., & Lee, C. (2010). Chaos-based secure satellite imagery cryptosystem. *Computers and Mathematics with Applications*, *60*(2), 326–337.
86. Vaidyanathan, S. (2011). Hybrid chaos synchronization of Liu and Lu systems by active nonlinear control. *Communications in Computer and Information Science*, *204*, 1–10.
87. Vaidyanathan, S. (2012). Analysis and synchronization of the hyperchaotic Yujun systems via sliding mode control. *Advances in Intelligent Systems and Computing*, *176*, 329–337.
88. Vaidyanathan, S. (2012). Anti-synchronization of Sprott-L and Sprott-M chaotic systems via adaptive control. *International Journal of Control Theory and Applications*, *5*(1), 41–59.
89. Vaidyanathan, S. (2012). Global chaos control of hyperchaotic Liu system via sliding control method. *International Journal of Control Theory and Applications*, *5*(2), 117–123.
90. Vaidyanathan, S. (2012). Output regulation of the Liu chaotic system. *Applied Mechanics and Materials*, *110–116*, 3982–3989.
91. Vaidyanathan, S. (2012). Sliding mode control based global chaos control of Liu-Liu-Liu-Su chaotic system. *International Journal of Control Theory and Applications*, *5*(1), 15–20.
92. Vaidyanathan, S. (2013). A new six-term 3-D chaotic system with an exponential nonlinearity. *Far East Journal of Mathematical Sciences*, *79*(1), 135–143.
93. Vaidyanathan, S. (2013). Analysis and adaptive synchronization of two novel chaotic systems with hyperbolic sinusoidal and cosinusoidal nonlinearity and unknown parameters. *Journal of Engineering Science and Technology Review*, *6*(4), 53–65.
94. Vaidyanathan, S. (2013). Analysis, control and synchronization of hyperchaotic Zhou system via adaptive control. *Advances in Intelligent Systems and Computing*, *177*, 1–10.
95. Vaidyanathan, S. (2014). A new eight-term 3-D polynomial chaotic system with three quadratic nonlinearities. *Far East Journal of Mathematical Sciences*, *84*(2), 219–226.

96. Vaidyanathan, S. (2014). Analysis and adaptive synchronization of eight-term 3-D polynomial chaotic systems with three quadratic nonlinearities. *European Physical Journal Special Topics, 223*(8), 1519–1529.
97. Vaidyanathan, S. (2014). Analysis, control and synchronisation of a six-term novel chaotic system with three quadratic nonlinearities. *International Journal of Modelling, Identification and Control, 22*(1), 41–53.
98. Vaidyanathan, S. (2014). Generalized projective synchronisation of novel 3-D chaotic systems with an exponential non-linearity via active and adaptive control. *International Journal of Modelling, Identification and Control, 22*(3), 207–217.
99. Vaidyanathan, S. (2014). Global chaos synchronization of identical Li-Wu chaotic systems via sliding mode control. *International Journal of Modelling, Identification and Control, 22*(2), 170–177.
100. Vaidyanathan, S. (2015). 3-cells cellular neural network (CNN) attractor and its adaptive biological control. *International Journal of PharmTech Research, 8*(4), 632–640.
101. Vaidyanathan, S. (2015). A 3-D novel highly chaotic system with four quadratic nonlinearities, its adaptive control and anti-synchronization with unknown parameters. *Journal of Engineering Science and Technology Review, 8*(2), 106–115.
102. Vaidyanathan, S. (2015). A novel chemical chaotic reactor system and its adaptive control. *International Journal of ChemTech Research, 8*(7), 146–158.
103. Vaidyanathan, S. (2015). A novel chemical chaotic reactor system and its output regulation via integral sliding mode control. *International Journal of ChemTech Research, 8*(11), 669–683.
104. Vaidyanathan, S. (2015). Adaptive backstepping control of enzymes-substrates system with ferroelectric behaviour in brain waves. *International Journal of PharmTech Research, 8*(2), 256–261.
105. Vaidyanathan, S. (2015). Adaptive biological control of generalized Lotka-Volterra three-species biological system. *International Journal of PharmTech Research, 8*(4), 622–631.
106. Vaidyanathan, S. (2015). Adaptive chaotic synchronization of enzymes-substrates system with ferroelectric behaviour in brain waves. *International Journal of PharmTech Research, 8*(5), 964–973.
107. Vaidyanathan, S. (2015). Adaptive control design for the anti-synchronization of novel 3-D chemical chaotic reactor systems. *International Journal of ChemTech Research, 8*(11), 654–668.
108. Vaidyanathan, S. (2015). Adaptive control of a chemical chaotic reactor. *International Journal of PharmTech Research, 8*(3), 377–382.
109. Vaidyanathan, S. (2015). Adaptive synchronization of chemical chaotic reactors. *International Journal of ChemTech Research, 8*(2), 612–621.
110. Vaidyanathan, S. (2015). Adaptive synchronization of generalized Lotka-Volterra three-species biological systems. *International Journal of PharmTech Research, 8*(5), 928–937.
111. Vaidyanathan, S. (2015). Adaptive synchronization of novel 3-D chemical chaotic reactor systems. *International Journal of ChemTech Research, 8*(7), 159–171.
112. Vaidyanathan, S. (2015). Analysis, properties and control of an eight-term 3-D chaotic system with an exponential nonlinearity. *International Journal of Modelling, Identification and Control, 23*(2), 164–172.
113. Vaidyanathan, S. (2015). Anti-synchronization of Brusselator chemical reaction systems via adaptive control. *International Journal of ChemTech Research, 8*(6), 759–768.
114. Vaidyanathan, S. (2015). Anti-synchronization of chemical chaotic reactors via adaptive control method. *International Journal of ChemTech Research, 8*(8), 73–85.
115. Vaidyanathan, S. (2015). Anti-synchronization of Mathieu-Van der Pol chaotic systems via adaptive control method. *International Journal of ChemTech Research, 8*(11), 638–653.
116. Vaidyanathan, S. (2015). Chaos in neurons and adaptive control of Birkhoff-Shaw strange chaotic attractor. *International Journal of PharmTech Research, 8*(5), 956–963.
117. Vaidyanathan, S. (2015). Dynamics and control of Brusselator chemical reaction. *International Journal of ChemTech Research, 8*(6), 740–749.

118. Vaidyanathan, S. (2015). Dynamics and control of Tokamak system with symmetric and magnetically confined plasma. *International Journal of ChemTech Research*, *8*(6), 795–803.
119. Vaidyanathan, S. (2015). Global chaos control of Mathieu-Van der pol system via adaptive control method. *International Journal of ChemTech Research*, *8*(9), 406–417.
120. Vaidyanathan, S. (2015). Global chaos synchronization of chemical chaotic reactors via novel sliding mode control method. *International Journal of ChemTech Research*, *8*(7), 209–221.
121. Vaidyanathan, S. (2015). Global chaos synchronization of Duffing double-well chaotic oscillators via integral sliding mode control. *International Journal of ChemTech Research*, *8*(11), 141–151.
122. Vaidyanathan, S. (2015). Global chaos synchronization of Mathieu-Van der Pol chaotic systems via adaptive control method. *International Journal of ChemTech Research*, *8*(10), 148–162.
123. Vaidyanathan, S. (2015). Global chaos synchronization of novel coupled Van der Pol conservative chaotic systems via adaptive control method. *International Journal of ChemTech Research*, *8*(8), 95–111.
124. Vaidyanathan, S. (2015). Global chaos synchronization of the forced Van der Pol chaotic oscillators via adaptive control method. *International Journal of PharmTech Research*, *8*(6), 156–166.
125. Vaidyanathan, S. (2015). Hyperchaos, qualitative analysis, control and synchronisation of a ten-term 4-D hyperchaotic system with an exponential nonlinearity and three quadratic nonlinearities. *International Journal of Modelling, Identification and Control*, *23*(4), 380–392.
126. Vaidyanathan, S. (2016). A novel 2-D chaotic enzymes-substrates reaction system and its adaptive backstepping control. In A. T. Azar & S. Vaidyanathan (Eds.), *Advances in chaos theory and intelligent control*. Studies in fuzziness and soft computing (Vol. 337, pp. 507–528). Germany: Springer.
127. Vaidyanathan, S. (2016). A novel 3-D conservative jerk chaotic system with two quadratic nonlinearities and its adaptive control. In A. T. Azar & S. Vaidyanathan (Eds.), *Advances in chaos theory and intelligent control*. Studies in fuzziness and soft computing (Vol. 337, pp. 349–376). Germany: Springer.
128. Vaidyanathan, S. (2016). A novel 3-D jerk chaotic system with three quadratic nonlinearities and its adaptive control. *Archives of Control Sciences*, *26*(1), 19–47.
129. Vaidyanathan, S. (2016). A novel 4-D hyperchaotic thermal convection system and its adaptive control. In A. T. Azar & S. Vaidyanathan (Eds.), *Advances in chaos theory and intelligent control*. Studies in fuzziness and soft computing (Vol. 337, pp. 75–100). Germany: Springer.
130. Vaidyanathan, S. (2016). A novel double convecton system, its analysis, adaptive control and synchronization. In A. T. Azar & S. Vaidyanathan (Eds.), *Advances in chaos theory and intelligent control*. Studies in fuzziness and soft computing (Vol. 337, pp. 553–579). Germany: Springer.
131. Vaidyanathan, S. (2016). A seven-term novel 3-D jerk chaotic system with two quadratic nonlinearities and its adaptive backstepping control. In A. T. Azar & S. Vaidyanathan (Eds.), *Advances in chaos theory and intelligent control*. Studies in fuzziness and soft computing (Vol. 337, pp. 581–607). Germany: Springer.
132. Vaidyanathan, S. (2016). Analysis, adaptive control and synchronization of a novel 3-D chaotic system with a quartic nonlinearity and two quadratic nonlinearities. In A. T. Azar & S. Vaidyanathan (Eds.), *Advances in chaos theory and intelligent control*. Studies in fuzziness and soft computing (Vol. 337, pp. 429–453). Germany: Springer.
133. Vaidyanathan, S. (2016). Analysis, control and synchronization of a novel 4-D highly hyperchaotic system with hidden attractors. In A. T. Azar & S. Vaidyanathan (Eds.), *Advances in chaos theory and intelligent control*. Studies in fuzziness and soft computing (Vol. 337, pp. 529–552). Germany: Springer.
134. Vaidyanathan, S. (2016). Anti-synchronization of duffing double-well chaotic oscillators via integral sliding mode control. *International Journal of ChemTech Research*, *9*(2), 297–304.
135. Vaidyanathan, S. (2016). Dynamic analysis, adaptive control and synchronization of a novel highly chaotic system with four quadratic nonlinearities. In A. T. Azar & S. Vaidyanathan

(Eds.), *Advances in chaos theory and intelligent control*. Studies in fuzziness and soft computing (Vol. 337, pp. 405–428). Germany: Springer.
136. Vaidyanathan, S. (2016). Global chaos synchronization of a novel 3-D chaotic system with two quadratic nonlinearities via active and adaptive control. In A. T. Azar & S. Vaidyanathan (Eds.), *Advances in chaos theory and intelligent control*. Studies in fuzziness and soft computing (Vol. 337, pp. 481–506). Germany: Springer.
137. Vaidyanathan, S. (2016). Qualitative analysis and properties of a novel 4-D hyperchaotic system with two quadratic nonlinearities and its adaptive control. In A. T. Azar & S. Vaidyanathan (Eds.), *Advances in chaos theory and intelligent control*. Studies in fuzziness and soft computing (Vol. 337, pp. 455–480). Germany: Springer.
138. Vaidyanathan, S., & Azar, A. T. (2015). Analysis and control of a 4-D novel hyperchaotic system. In A. T. Azar & S. Vaidyanathan (Eds.), *Chaos modeling and control systems design*. Studies in computational intelligence (Vol. 581, pp. 19–38). Germany: Springer.
139. Vaidyanathan, S., & Azar, A. T. (2015). Analysis, control and synchronization of a nine-term 3-D novel chaotic system. In A. T. Azar & S. Vaidyanathan (Eds.), *Chaos modelling and control systems design*. Studies in computational intelligence (Vol. 581, pp. 19–38). Germany: Springer.
140. Vaidyanathan, S., & Azar, A. T. (2015). Anti-synchronization of identical chaotic systems using sliding mode control and an application to Vaidhyanathan-Madhavan chaotic systems. *Studies in Computational Intelligence*, *576*, 527–547.
141. Vaidyanathan, S., & Azar, A. T. (2015). Hybrid synchronization of identical chaotic systems using sliding mode control and an application to Vaidhyanathan chaotic systems. *Studies in Computational Intelligence*, *576*, 549–569.
142. Vaidyanathan, S., & Azar, A. T. (2016). A novel 4-D four-wing chaotic system with four quadratic nonlinearities and its synchronization via adaptive control method. In A. T. Azar & S. Vaidyanathan (Eds.), *Advances in chaos theory and intelligent control*. Studies in fuzziness and soft computing (Vol. 337, pp. 203–224). Germany: Springer.
143. Vaidyanathan, S., & Azar, A. T. (2016). Adaptive backstepping control and synchronization of a novel 3-D jerk system with an exponential nonlinearity. In A. T. Azar & S. Vaidyanathan (Eds.), *Advances in chaos theory and intelligent control*. Studies in fuzziness and soft computing (Vol. 337, pp. 249–274). Germany: Springer.
144. Vaidyanathan, S., & Azar, A. T. (2016). Adaptive control and synchronization of Halvorsen circulant chaotic systems. In A. T. Azar & S. Vaidyanathan (Eds.), *Advances in chaos theory and intelligent control*. Studies in fuzziness and soft computing (Vol. 337, pp. 225–247). Germany: Springer.
145. Vaidyanathan, S., & Azar, A. T. (2016). Dynamic analysis, adaptive feedback control and synchronization of an eight-term 3-D novel chaotic system with three quadratic nonlinearities. In A. T. Azar & S. Vaidyanathan (Eds.), *Advances in chaos theory and intelligent control*. Studies in fuzziness and soft computing (Vol. 337, pp. 155–178). Germany: Springer.
146. Vaidyanathan, S., & Azar, A. T. (2016). Generalized projective synchronization of a novel hyperchaotic four-wing system via adaptive control method. In A. T. Azar & S. Vaidyanathan (Eds.), *Advances in chaos theory and intelligent control*. Studies in fuzziness and soft computing (Vol. 337, pp. 275–296). Germany: Springer.
147. Vaidyanathan, S., & Azar, A. T. (2016). Qualitative study and adaptive control of a novel 4-D hyperchaotic system with three quadratic nonlinearities. In A. T. Azar & S. Vaidyanathan (Eds.), *Advances in chaos theory and intelligent control*. Studies in fuzziness and soft computing (Vol. 337, pp. 179–202). Germany: Springer.
148. Vaidyanathan, S., & Madhavan, K. (2013). Analysis, adaptive control and synchronization of a seven-term novel 3-D chaotic system. *International Journal of Control Theory and Applications*, *6*(2), 121–137.
149. Vaidyanathan, S., & Pakiriswamy, S. (2013). Generalized projective synchronization of six-term Sundarapandian chaotic systems by adaptive control. *International Journal of Control Theory and Applications*, *6*(2), 153–163.

150. Vaidyanathan, S., & Pakiriswamy, S. (2015). A 3-D novel conservative chaotic system and its generalized projective synchronization via adaptive control. *Journal of Engineering Science and Technology Review*, *8*(2), 52–60.
151. Vaidyanathan, S., & Rajagopal, K. (2011a). Anti-synchronization of Li and T chaotic systems by active nonlinear control. *Communications in Computer and Information Science*, *198*, 175–184.
152. Vaidyanathan, S., & Rajagopal, K. (2011b). Global chaos synchronization of hyperchaotic Pang and Wang systems by active nonlinear control. *Communications in Computer and Information Science*, *204*, 84–93.
153. Vaidyanathan, S., & Rajagopal, K. (2011c). Global chaos synchronization of Lü and Pan systems by adaptive nonlinear control. *Communications in Computer and Information Science*, *205*, 193–202.
154. Vaidyanathan, S., & Rajagopal, K. (2012). Global chaos synchronization of hyperchaotic Pang and hyperchaotic Wang systems via adaptive control. *International Journal of Soft Computing*, *7*(1), 28–37.
155. Vaidyanathan, S., & Rasappan, S. (2011). Global chaos synchronization of hyperchaotic Bao and Xu systems by active nonlinear control. *Communications in Computer and Information Science*, *198*, 10–17.
156. Vaidyanathan, S., & Rasappan, S. (2014). Global chaos synchronization of *n*-scroll Chua circuit and Lur'e system using backstepping control design with recursive feedback. *Arabian Journal for Science and Engineering*, *39*(4), 3351–3364.
157. Vaidyanathan, S., & Sampath, S. (2011). Global chaos synchronization of hyperchaotic Lorenz systems by sliding mode control. *Communications in Computer and Information Science*, *205*, 156–164.
158. Vaidyanathan, S., & Sampath, S. (2012). Anti-synchronization of four-wing chaotic systems via sliding mode control. *International Journal of Automation and Computing*, *9*(3), 274–279.
159. Vaidyanathan, S., & Volos, C. (2015). Analysis and adaptive control of a novel 3-D conservative no-equilibrium chaotic system. *Archives of Control Sciences*, *25*(3), 333–353.
160. Vaidyanathan, S., Volos, C., & Pham, V. T. (2014). Hyperchaos, adaptive control and synchronization of a novel 5-D hyperchaotic system with three positive Lyapunov exponents and its SPICE implementation. *Archives of Control Sciences*, *24*(4), 409–446.
161. Vaidyanathan, S., Volos, C., Pham, V. T., Madhavan, K., & Idowu, B. A. (2014). Adaptive backstepping control, synchronization and circuit simulation of a 3-D novel jerk chaotic system with two hyperbolic sinusoidal nonlinearities. *Archives of Control Sciences*, *24*(3), 375–403.
162. Vaidyanathan, S., Idowu, B. A., & Azar, A. T. (2015). Backstepping controller design for the global chaos synchronization of Sprott's jerk systems. *Studies in Computational Intelligence*, *581*, 39–58.
163. Vaidyanathan, S., Rajagopal, K., Volos, C. K., Kyprianidis, I. M., & Stouboulos, I. N. (2015). Analysis, adaptive control and synchronization of a seven-term novel 3-D chaotic system with three quadratic nonlinearities and its digital implementation in LabVIEW. *Journal of Engineering Science and Technology Review*, *8*(2), 130–141.
164. Vaidyanathan, S., Volos, C., Pham, V. T., & Madhavan, K. (2015). Analysis, adaptive control and synchronization of a novel 4-D hyperchaotic hyperjerk system and its SPICE implementation. *Archives of Control Sciences*, *25*(1), 5–28.
165. Vaidyanathan, S., Volos, C. K., Kyprianidis, I. M., Stouboulos, I. N., & Pham, V. T. (2015). Analysis, adaptive control and anti-synchronization of a six-term novel jerk chaotic system with two exponential nonlinearities and its circuit simulation. *Journal of Engineering Science and Technology Review*, *8*(2), 24–36.
166. Vaidyanathan, S., Volos, C. K., & Pham, V. T. (2015). Analysis, adaptive control and adaptive synchronization of a nine-term novel 3-D chaotic system with four quadratic nonlinearities and its circuit simulation. *Journal of Engineering Science and Technology Review*, *8*(2), 174–184.

167. Vaidyanathan, S., Volos, C. K., & Pham, V. T. (2015). Global chaos control of a novel nine-term chaotic system via sliding mode control. In A. T. Azar & Q. Zhu (Eds.), *Advances and applications in sliding mode control systems*. Studies in computational intelligence (Vol. 576, pp. 571–590). Germany: Springer.
168. Vaidyanathan, S., Pham, V. T., & Volos, C. K. (2015). A 5-D hyperchaotic Rikitake dynamo system with hidden attractors. *European Physical Journal Special Topics*, *224*(8), 1575–1592.
169. Volos, C. K., Kyprianidis, I. M., & Stouboulos, I. N. (2013). Experimental investigation on coverage performance of a chaotic autonomous mobile robot. *Robotics and Autonomous Systems*, *61*(12), 1314–1322.
170. Volos, C. K., Kyprianidis, I. M., Stouboulos, I. N., Tlelo-Cuautle, E., & Vaidyanathan, S. (2015). Memristor: A new concept in synchronization of coupled neuromorphic circuits. *Journal of Engineering Science and Technology Review*, *8*(2), 157–173.
171. Wei, Z., & Yang, Q. (2010). Anti-control of Hopf bifurcation in the new chaotic system with two stable node-foci. *Applied Mathematics and Computation*, *217*(1), 422–429.
172. Witte, C. L., & Witte, M. H. (1991). Chaos and predicting varix hemorrhage. *Medical Hypotheses*, *36*(4), 312–317.
173. Xiao, X., Zhou, L., & Zhang, Z. (2014). Synchronization of chaotic Lur'e systems with quantized sampled-data controller. *Communications in Nonlinear Science and Numerical Simulation*, *19*(6), 2039–2047.
174. Yuan, G., Zhang, X., & Wang, Z. (2014). Generation and synchronization of feedback-induced chaos in semiconductor ring lasers by injection-locking. *Optik - International Journal for Light and Electron Optics*, *125*(8), 1950–1953.
175. Zaher, A. A., & Abu-Rezq, A. (2011). On the design of chaos-based secure communication systems. *Communications in Nonlinear Systems and Numerical Simulation*, *16*(9), 3721–3727.
176. Zhang, H., & Zhou, J. (2012). Synchronization of sampled-data coupled harmonic oscillators with control inputs missing. *Systems & Control Letters*, *61*(12), 1277–1285.
177. Zhang, X., Zhao, Z., & Wang, J. (2014). Chaotic image encryption based on circular substitution box and key stream buffer. *Signal Processing: Image Communication*, *29*(8), 902–913.
178. Zhou, W., Xu, Y., Lu, H., & Pan, L. (2008). On dynamics analysis of a new chaotic attractor. *Physics Letters A*, *372*(36), 5773–5777.
179. Zhu, C., Liu, Y., & Guo, Y. (2010). Theoretic and numerical study of a new chaotic system. *Intelligent Information Management*, *2*, 104–109.
180. Zhu, Q., & Azar, A. T. (2015). *Complex system modelling and control through intelligent soft computations*. Studies in fuzzines and soft computing (Vol. 319). Germany: Springer.

Dynamics of Fractional Order Complex Uçar System

Sachin Bhalekar

Abstract The fractional order delay differential equations are models with rich dynamical properties. Both fractional order and delay are useful in modelling memory and hereditary properties in the physical system. In this chapter, we have proposed a complex version of fractional order Uçar system with delay. The stability of the numerical methods for solving such equations is discussed. It is observed that a slight modification in the proposed system generates chaotic trajectories. The bifurcation and chaos is studied in the modified system. The delayed feedback method is used to control chaos in the system. Finally, the system is synchronized by using the method of projective synchronization.

Keywords Fractional derivative · Delay differential equation · Uçar system · Chaos · Bifurcation · Synchronization

1 Introduction

For any positive integer n, Leibniz introduced a notation $D^n f(x) = \frac{d^n f(x)}{dx^n}$ for n-th order differentiation of suitable function f. In a letter dated September 30, 1695 L'Hopital asked Leibniz about the meaning of above operation when $n = 1/2$. This moment is popularly believed as origin of Fractional Calculus (FC). The day September 30 is now celebrated as Fractional Calculus Day or the Birthday of Fractional Calculus. When the order of differentiation is any real or complex number or some function of t then it is called as Fractional Derivative (FD). Though this branch of mathematics is as old as conventional calculus, the applications are rather recent.

S. Bhalekar (✉)
Department of Mathematics, Shivaji University, Vidyanagar,
Kolhapur 416004, India
e-mail: sachin.math@yahoo.co.in; sbb_maths@unishivaji.ac.in

A.T. Azar et al. (eds.), *Fractional Order Control and Synchronization of Chaotic Systems*, Studies in Computational Intelligence 688,
DOI 10.1007/978-3-319-50249-6_26

The subject was not popular among scientists because of following weird things related to FD.

(1) Though the fractional integration was given uniquely by Riemann-Liouville (RL) sense, FD had several inequivalent definitions.
(2) Several simple rules involving integer order derivative (ID) take a complicated form during the generalization to FD. For example, the finite sum

$$D^n(fg) = \sum_{k=0}^{n} \binom{n}{k} \left(D^{n-k}f\right)\left(D^k g\right) \tag{1}$$

in a Leibniz rule of ordinary differentiation turned out to be an infinite series [109]

$$^{R}D^{\alpha}(fg) = \sum_{k=0}^{\infty} \frac{\Gamma(\alpha+1)}{\Gamma(k+1)\Gamma(\alpha-k+1)} \left(D^{\alpha-k}f\right)\left(D^k g\right) \tag{2}$$

for Riemann-Liouville (RL) fractional differentiation of order α.
(3) One has to provide initial conditions at the arbitrary ordered derivative of the function while considering the initial value problems (IVP) involving RL fractional derivatives. Such initial conditions are not physically relevant in many cases. (Of course this drawback was overcame by Caputo fractional derivative).
(4) The scientists were unable to provide satisfactory geometrical or physical meaning of any FD unlike an integer order derivative (ID).
(5) The ID of a suitable function f at a point t_0 can be approximated by using values of f at some points in a neighborhood of t_0. However, one has to consider all the history from the initial point while approximating FDs. In other words, the FDs are nonlocal operators in contrast with the local operator ID. Hence the numerical solutions of fractional differential equations (FDE) require a lot of time.

Later on, it was realized that the FDs are able to model nonlocal properties in a system such as *memory* which ID cannot. Further, there are some natural systems which show *intermediate* behaviour (e.g. visco-elasticity [93]) which can be modelled in a more accurate way by using FDs. In result the subject become popular in pure as well as applied scientists.

We now list some applications of fractional calculus in various branches of Science, Engineering and Social Sciences. A well-known fractional diffusion equation is studied by researchers to model numerous phenomena [45, 76–78, 92]. Mainardi in his monograph [93] discussed the application of fractional calculus in linear viscoelasticity. Applications of this branch to bioengineering are discussed by Magin [91]. Few more applications include signal processing [6, 102], image processing [113], image encryption [107, 132, 138], cryptography [130], control theory [108], thermodynamics [97] and nonlinear dynamics [62, 88, 134]. Chaotic fractional order systems can be used to generate reliable cryptographic schemes [99–101, 136].

Various results on existence and uniqueness (EU)of the solutions of fractional differential equations (FDE) are derived by researchers. EU result for linear equations of the form

$$^{R}D^{\alpha}y(x) - \lambda y(x) = f(x), \ (n-1 \leq Re(\alpha) < n), \tag{3}$$

where $^{R}D^{\alpha}$ is Riemann-Lioville fractional derivative is derived by Barret in [21]. Method of successive approximation is used by Al-Bassam [5] to derive EU result for nonlinear problem

$$^{R}D^{\alpha}y(x) = f(x, y(x)), \ (0 < \alpha \leq 1). \tag{4}$$

Schauder's fixed point theorem was utilized by Delbosco and Rodino [49] to derive EU results for nonlinear equations. Cauchy problems of the form (4) with Caputo derivative are discussed by Diethelm and Ford [53]. Existence, uniqueness and stability of solutions of systems of FDEs are proposed by Daftardar-Gejji and Babakhani in [41].

Linear FDEs can be solved exactly by using transform methods and operational method. Laplace transform is used to solve linear FDEs of the form (5) with Caputo fractional derivative by Gorenflo and Mainardi [64]. An operational method is proposed by Luchko and Gorenflo [89] to solve multi-term linear FDEs of Caputo type. Daftardar-Gejji and coworkers [42, 44] used method of separation of variables to solve these equations.

However, the methods described above cannot be used to solve nonlinear FDEs. In this case, one has to use either some numerical method or approximate analytical method (AAM). A popular numerical method is fractional Adam's method developed by Diethelm et al. [54]. An improved method derived by Daftardar-Gejji et al. [39] is more efficient. If the analytical solutions are required instead of numerical approximations then one has to consider some AAM. Few terms of the solution series provided by AAM can be used as a good approximation to the exact solution. However, it should be noted that the AAM solutions are local whereas the numerical solutions are global. Adomian decomposition method (ADM) [1], variational iteration method (VIM) [69], homotopy perturbation method (HPM) [70] and Daftardar-Gejji-Jafari method (DJM) [43] are popular AAMs used by researchers. The chapter is organized as below: Review of the work related to chaos control, synchronization, delay differential equations and the Uçar system is taken in Sect. 2. Basic definitions regarding fractional calculus are listed in Sect. 3. We have described two numerical methods viz. FAM and NPCM in Sect. 4. Section 4.3 deals with the stability of these methods. The complex version of Uçar system is presented in Sect. 5. The modified system is proposed in Sect. 6. Further, the bifurcation analysis, chaos control and synchronization in the modified system is also discussed in this section. Sections 7 and 8 deal with discussion and conclusions respectively.

2 Related Work

2.1 Chaos Control and Synchronization

Nonlinear autonomous systems of ordinary differential equations (ODE) of order three or higher may exhibit aperiodic oscillations which are extremely sensitive to initial conditions. Such bounded orbits are known to be chaotic. The first chaotic system reported in the literature is by a meteorologist and mathematician Lorenz [90]. A large number of examples of dynamical systems which exhibit chaos have been presented in the literature [4, 18]. Chaos has been shown to exist in various branches of science e.g. Chua's circuit in electronics [95], Belousov-Zhabotinsky chemical reaction [2, 60], economics and finance [36, 58, 110], fluid dynamics [133], population dynamics [68], physiology [22, 56], pharmacodynamics [55], artificial intelligence systems [10, 11] and so on.

The Poincare-Bendixson theorem [4] gives necessary condition for the existence of chaos. However, there does not exists any sufficient condition for the chaos. Some tests such as maximum Lyapunov exponent are useful in chaos detection.

Since the chaotic systems are unpredictable, the control of chaos is necessary in various physical applications [12, 139]. However, one can also utilize this unpredictability to generate secure communication schemes [71, 72] by applying the techniques of chaos synchronization. If the trajectory of a chaotic system follow the path of some (same or distinct) chaotic system with different initial conditions after applying a suitable control then the phenomena is called as synchronization [103, 104]. Different control strategies are described in the literature such as nonlinear feedback control [74], adaptive control [33, 87, 122–124, 137], adaptive feedback control [121], adaptive backstepping control [125], active control [19, 20, 116], sliding mode control [9, 13, 15, 98, 117–119, 135], deadbeat control [14], backstepping control [120], function synchronization [34], PID control [17], IMC-PID control [8], backlash control [16] and so on.

Grigorenko and Grigorenko [65] discussed chaos in fractional order system first time in the literature. The system was obtained by replacing IDs in Lorenz system by FDs of order between 0 and 1. Though there was an error in numerical solutions of this system, it gave rise to a new field of research. Till date, there are so many fractional order chaotic dynamical systems (FCDS) are reported in the literature. It is observed in all these systems that the FCDS shows chaos for system order less than that of its integer counterpart. i.e. A three dimensional FCDS can be chaotic for the system order (sum of all the fractional orders of the derivatives in the system) less than three. The bifurcation value of the system order is called as the minimum effective dimension (MED). For a given system, the stable orbits are observed for the system order less than MED. Further increase in system order leads to chaotic trajectories. Fractional order chaotic systems discussed in the literature are Lorenz system [65], Chua system [67], Rossler system [83], Newton-Leipnik system [111], Liu system [46] and so on.

Synchronization of fractional order chaotic systems was first studied by Deng and Li [51, 52] who carried out synchronization in the fractional Lü system. The theory and techniques of fractional order chaos synchronization are summarized in [84] by Li and Deng. Li et al. [85] used the Pecora-Carroll (PC) method, the active-passive decomposition method, the one-way coupling and the bidirectional coupling methods to synchronize two identical fractional order Chua systems. PC and one way coupling methods were used to synchronize unified system by Wang and Zhang [129]. Nonlinear control theory is successfully extended by Wang et al. [128] to fractional order Chen systems to achieve synchronization. The same technique is further used by Jun et al. [79] for chaotic synchronization between fractional order financial system and financial system of integer orders.

The technique of active control (AC) is extended by Bhalekar and Daftardar-Gejji [23, 24] to synchronize and anti-synchronize non-identical commensurate fractional chaotic systems. It is observed that the synchronization is faster as the system order tends to one. Further, Bhalekar utilized AC to synchronize incommensurate order systems [29] and hyperchaotic systems [30] of fractional order.

A new approach named fractional-order dynamics rejection scheme is proposed for fractional-order system based on active disturbance rejection control (ADRC) method in [86]. The ARDC is then used by Gao and Liao [63] to synchronize different fractional order chaotic systems.

2.2 *Delay Differential Equations*

If the rate of change of the current state depends on its values at earlier points then the system can be modelled by using delay differential equation (DDE). The DDEs $\dot{x}(t) = f(x(t), x(t-\tau))$ are particular cases of more general class called functional equations. The term τ in this DDE is known as a delay. The τ may be constant, time dependent or a state dependent. DDEs are proved useful in control systems [61], traffic models [47], epidemiology [38], neuroscience [35], population dynamics [81], chemical kinetics [57], economics [7, 48, 58] etc. It is observed that some nonlinear DDEs of order one can generate chaotic solutions also.

Fractional order delay differential equations (FDDE) are the models containing both non-integer order derivatives and time delays. FDDEs have applications in Control Theory [112], Agriculture [59], Chaos [127], NMR [26] and so on. Daftardar-Gejji and coworkers presented efficient numerical methods [25, 40] for solving FDDEs. A good number of research articles is devoted to analyze the stability of linear time invariant fractional delay systems (LTIFDS). LTIFDS with characteristic equation $(as^\alpha + b) + (cs^\alpha + d)e^{-\rho s} = 0$ are discussed by Hotzel [73]. Lambert function is used to study the stability of $\dot{y}(t) = ay(t-1)$ in [37]. The generalization of this equation is analyzed using Laplace transform by Deng et al. [50]. Hwang and Cheng [75] developed a numerical algorithm to study the BIBO (bounded input and bounded output) stability of LTIFDS. The fractional order $PI^\lambda D^\mu$ controller is utilized in [66] to stabilize LTIFDS. Recently, Bhalekar discussed stability of generalized nonlinear FDDE in [32].

2.3 Uçar System

In [114, 115] Uçar proposed and analyzed a simple scalar delay differential equation exhibiting chaos. This scalar equation consists of a cubic nonlinearity. The modelling equation is

$$\dot{x}(t) = \delta x(t \quad \tau) - \varepsilon \left[x(t-\tau)\right]^3, \tag{5}$$

where δ and ε are positive parameters and τ is time delay. There are three equilibrium points namely $0, \pm\sqrt{\delta/\varepsilon}$ when the delay $\tau = 0$ in (5). It is easy to verify that 0 is unstable and $\pm\sqrt{\delta/\varepsilon}$ are stable equilibriums. The chaotic behaviour is observed in this model for some positive values of delay element τ. Stable orbits or limit cycles are observed for the parameters $\varepsilon = \tau = 1$ and for $\delta < 1$. Multiple limit cycles are observed by the author for $1 < \delta < 1.56$. The system was chaotic when $1.64 < \delta < 1.8$ and for higher values of δ the solutions become unbounded.

Further the author [115] has discussed the effect of τ on the chaotic behaviour of the system for fixed values $\varepsilon = \delta = 1$. The solutions were stable for $0 \leq \tau \leq 0.76$ whereas the limit cycles for $0.8 < \tau \leq 1.25$. The first chaotic region sets around $\tau \cong 1.55$. The multiple bifurcation and the second chaotic region is shown by the system for higher values of τ.

The system was generalized by present author [27] to a fractional order case and discussed the chaotic properties in new model. Further, in [28] he generalized (5) to involve two delays. In both the generalizations, the system was exhibiting one-scroll as well as two-scroll attractors. The stability analysis of (5) is presented in [31].

3 Preliminaries

In this section, we discuss some basic definitions and analytical results regarding fractional calculus.

Definition 3.1 [80, 105] Riemann-Liouville fractional integration of order α is defined as

$$I^{\alpha}f(t) = \frac{1}{\Gamma(\alpha)}\int_0^t (t-y)^{\alpha-1}f(y)\,dy, \quad t > 0. \tag{6}$$

Definition 3.2 [80, 105] Caputo fractional derivative of order α is defined as

$$D^{\alpha}f(t) = I^{m-\alpha}\left(\frac{d^m f(t)}{dt^m}\right), \quad 0 \leq m-1 < \alpha \leq m. \tag{7}$$

Note that for $0 \leq m-1 < \alpha \leq m$, $a \geq 0$ and $\gamma > -1$

$$I^{\alpha}(t-a)^{\gamma} = \frac{\Gamma(\gamma+1)}{\Gamma(\gamma+\alpha+1)}(t-a)^{\gamma+\alpha}, \tag{8}$$

$$(I^{\alpha}D^{\alpha}f)(t) = f(t) - \sum_{k=0}^{m-1} f^{(k)}(0)\frac{t^k}{k!}. \tag{9}$$

Lemma 3.1 ([96]) *The fractional order linear system $D^{\alpha}\mathbf{X} = \mathbf{B}\mathbf{X}$ is asymptotically stable if and only if $|arg(\lambda)| > \alpha\pi/2$, for all eigenvalues λ of matrix $\mathbf{B}$. In this case, the component of the state decay towards 0 like $t^{-\alpha}$.*

4 Numerical Methods for FDDEs

4.1 Fractional Adams Method (FAM)

In this section, we present the modified Adams-Bashforth-Moulton predictor-corrector scheme (FAM) described in [25] to solve delay differential equations of fractional order (FDDE). Consider the following FDDE

$$D^{\alpha}y(t) = f(t, y(t), y(t-\tau)), \quad t \in [0, T], \quad 0 < \alpha \leq 1 \tag{10}$$

$$y(t) = g(t), t \in [-\tau, 0]. \tag{11}$$

We consider a uniform grid $\{t_n = nh : n = -k, -k+1, \ldots, -1, 0, 1, \ldots, N\}$ where k and N are integers satisfying $h = T/N$ and $h = \tau/k$. The corrector formula for this scheme is given by

$$\begin{aligned} y_h(t_{n+1}) = \; & g(0) + \frac{h^{\alpha}}{\Gamma(\alpha+2)} f\left(t_{n+1}, y_h(t_{n+1}), y_h(t_{n+1-k})\right) \\ & + \frac{h^{\alpha}}{\Gamma(\alpha+2)} \sum_{j=0}^{n} a_{j,n+1} f\left(t_j, y_h(t_j), y_h(t_{j-k})\right), \end{aligned} \tag{12}$$

where $a_{j,n+1}$ are given by

$$a_{j,n+1} = \begin{cases} n^{\alpha+1} - (n-\alpha)(n+1)^{\alpha}, & \text{if } j = 0, \\ (n-j+2)^{\alpha+1} + (n-j)^{\alpha+1} - 2(n-j+1)^{\alpha+1}, & \text{if } 1 \leq j \leq n, \\ 1, & \text{if } j = n+1. \end{cases} \tag{13}$$

The unknown term $y_h(t_{n+1})$ on the right hand side of (12) is replaced by an approximation

$$y_h^P(t_{n+1}) = g(0) + \frac{1}{\Gamma(\alpha)} \sum_{j=0}^{n} b_{j,n+1} f\left(t_j, y_h(t_j), y_h(t_{j-k})\right), \tag{14}$$

where

$$b_{j,n+1} = \frac{h^\alpha}{\alpha}\left((n+1-j)^\alpha - (n-j)^\alpha\right). \quad (15)$$

The term y_h^P is termed as a predictor.

4.2 New Predictor-Corrector Method (NPCM)

The FAM described in previous section is improved by Daftardar-Gejji et al. using an iterative method [43]. The new predictor-corrector method (NPCM) [39, 40] is as below:

$$y_{n+1}^p = \sum_{k=0}^{\lceil\alpha\rceil-1} \phi_k \frac{t_{n+1}^k}{k!} + \frac{h^\alpha}{\Gamma(\alpha+2)} \sum_{j=0}^{n} a_{j,n+1} f_1(t_j, y_j, y(t_j-\tau)), \quad (16)$$

$$z_{n+1}^p = \frac{h^\alpha}{\Gamma(\alpha+2)} f_1(t_{n+1}, y_{n+1}^p, y(t_{n+1}-\tau)), \quad (17)$$

$$y_{n+1}^c = y_{n+1}^p + \frac{h^\alpha}{\Gamma(\alpha+2)} f_1(t_{n+1}, y_{n+1}^p + z_{n+1}^p, y(t_{n+1}-\tau)). \quad (18)$$

Here y_{n+1}^p and z_{n+1}^p are called as predictors and y_{n+1}^c is the corrector, and y_j denotes the approximate value of the solution.

4.3 Numerical Stability

Numerical stability is a property which determines the suitability of the scheme for sufficiently large value of step size. A simple linear equation with known stability properties is chosen as a test equation. The difference equation is then obtained by applying numerical scheme to this test equation. Further, the stability properties of these two equations are compared. The stability region is a region in parameter space in which the numerical solution is stable. We consider the following test equation [3]

$$y'(t) = \lambda y(t) + \mu y(t-1). \quad (19)$$

We use the result in [82] to state the stability of (19) when λ and μ are real.

Theorem 1 *The zero solution of (19) is asymptotically stable if one of the following conditions is satisfied:*

(i) $|\mu| < -\lambda$
(ii) $\arccos(-\lambda/\mu) > \sqrt{\mu^2 - \lambda^2}$.

Applying FAM and NPCM to (19) and setting $h = 1/k, k \in \mathbb{N}$ we get the difference equations

$$y_n = \left(1 + \lambda h\left(1 + \frac{\lambda h}{2}\right)\right) y_{n-1} + \frac{\mu h}{2}(1 + \lambda h)\, y_{n-1-k} + \frac{\mu h}{2} y_{n-k} \tag{20}$$

and

$$y_n = \left(1 + \lambda h\left(1 + \frac{\lambda h}{2} + \frac{(\lambda h)^2}{4}\right)\right) y_{n-1} + \mu h\left(\frac{1}{2} + \frac{\lambda h}{4} + \frac{(\lambda h)^2}{4}\right) y_{n-1-k} + \frac{\mu h}{2}\left(1 + \frac{\lambda h}{2}\right) y_{n-k} \tag{21}$$

respectively. The characteristic equations of these equations are given by

$$r^{k+1} - \left(1 + \lambda h\left(1 + \frac{\lambda h}{2}\right)\right) r^k - \frac{\mu h}{2} r - \frac{\mu h}{2}(1 + \lambda h) = 0 \tag{22}$$

and

$$r^{k+1} - \left(1 + \lambda h\left(1 + \frac{\lambda h}{2} + \frac{(\lambda h)^2}{4}\right)\right) r^k - \frac{\mu h}{2}\left(1 + \frac{\lambda h}{2}\right) r - \mu h\left(\frac{1}{2} + \frac{\lambda h}{4} + \frac{(\lambda h)^2}{4}\right) = 0 \tag{23}$$

respectively.

Definition 4.1 The stability region of a numerical method is a collection of points (λ_0, μ_0) in $\lambda\mu$-parameter space for which all the roots s of characteristic equation satisfy $|s| < 1$.

We plot stability regions of FAM and NPCM for $k = 1, 2, 3, 4$ in Figs. 1 and 2 respectively and compare with the stability region of DDE (19) described in Theorem 1.

Now we consider another case $\lambda = 0$ and $\mu = \rho exp(\iota\phi)$ complex number. The stability of test equation (19) is described in the following Theorem.

Theorem 2 *The zero solution of* $y'(t) = \rho exp(\iota\phi)y(t-1)$ *is asymptotically stable if* $\frac{\pi}{2} < \phi < \frac{3\pi}{2}$ *and* $0 < \rho < min\left(\frac{3\pi}{2} - \phi, \phi - \frac{\pi}{2}\right)$.

In this case, the characteristic equations of FAM and NPCM solutions coincide and are given by

$$r^{k+1} - r^k - \rho exp(\iota\phi)\frac{h}{2}(r + 1) = 0. \tag{24}$$

The stability region is shown in Fig. 3.

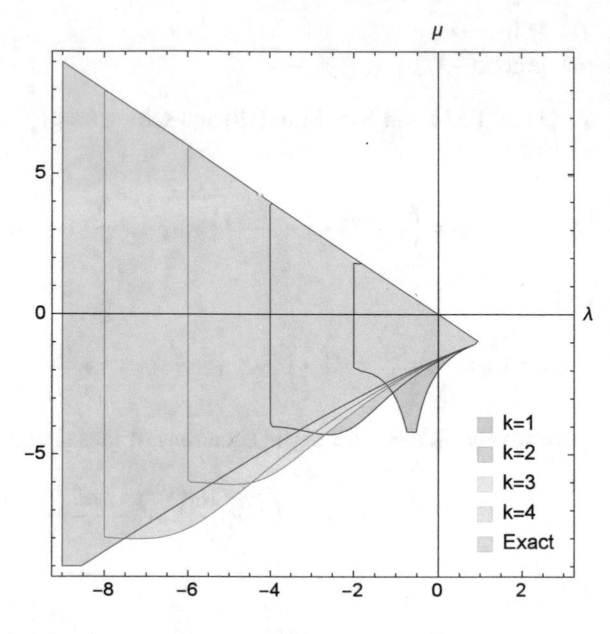

Fig. 1 Stability regions of FAM

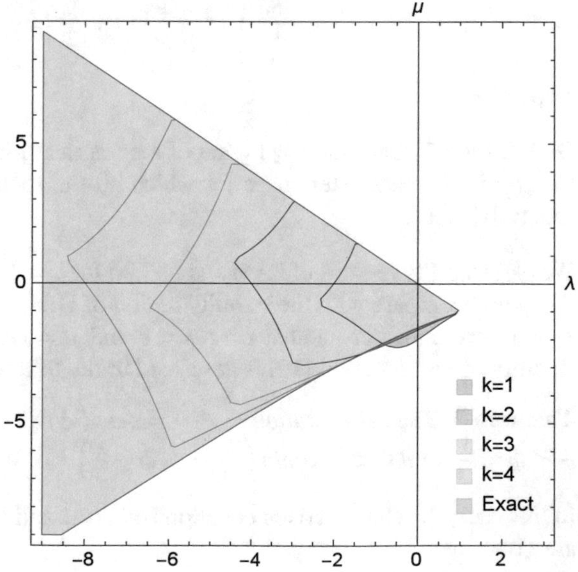

Fig. 2 Stability regions of NPCM

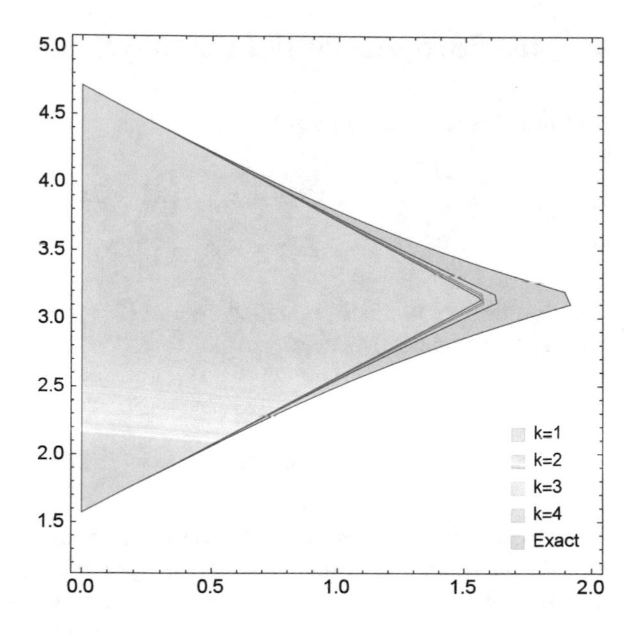

Fig. 3 Stability regions of FAM and NPCM for $\lambda = 0$ and $\mu = \rho exp(\imath\phi)$

5 Complex Uçar System

The complex Uçar system of fractional order can be described as

$$D^{\alpha}z = \delta z_{\tau} - \varepsilon z_{\tau}^{3}, \tag{25}$$

where $z = x + \imath y$, x, y are real functions and $z_{\tau}(t) = z(t - \tau)$. Equivalently, we can write system of real equations as

$$\begin{aligned} D^{\alpha}x &= \delta x_{\tau} - \varepsilon \left(x_{\tau}^{3} - 3x_{\tau}y_{\tau}^{2}\right), \\ D^{\alpha}y &= \delta y_{\tau} - \varepsilon \left(3x_{\tau}^{2}y_{\tau} - y_{\tau}^{3}\right). \end{aligned} \tag{26}$$

There are three equilibrium points viz. $O = (0, 0)$ which is unstable and $E_{\pm} = \left(\pm\sqrt{\delta}/\sqrt{\varepsilon}, 0\right)$ which are asymptotically stable. The system doesn't shows chaotic behavior for a wide range of parameters. However, the modified system described in next section involving incommensurate delays does exhibit chaotic oscillations for some values of delays.

6 Modified System with Incommensurate Delays

Consider the modified system as

$$
\begin{aligned}
D^{\alpha}x &= x_{\tau_1} - \left(x_{\tau_1}^3 + 3x_{\tau_1}y^2\right), \\
D^{\alpha}y &= -y_{\tau_1} - \left(3x^2 y_{\tau_2} + y^3\right).
\end{aligned} \tag{27}
$$

The equilibrium points of modified system (27) are $O_1 = (0, 0)$ (saddle) and $P_{\pm} = (\pm 1, 0)$ (asymptotically stable).

6.1 *Bifurcation and Chaos*

We consider $\alpha = 0.9$ and $\tau_2 = 0.5$. The stable solutions are observed for $0 \leq \tau_1 < 0.2$. At the bifurcation value $\tau_1 = 0.2$ the system starts periodic one-cycle. Period doubling at $\tau_1 = 1.03$ gives rise to chaotic trajectories at $\tau_1 \approx 1.12$. The bifurcation diagrams are shown in Figs. 4 and 5. The limit cycle at $\tau_1 = 1.1$ is presented in Fig. 6. The Fig. 7 shows chaotic attractor observed for $\tau_1 = 1.12$.

We also provide bifurcation diagrams (cf. Figs. 8 and 9) for the system (27) with $\alpha = 0.8$.

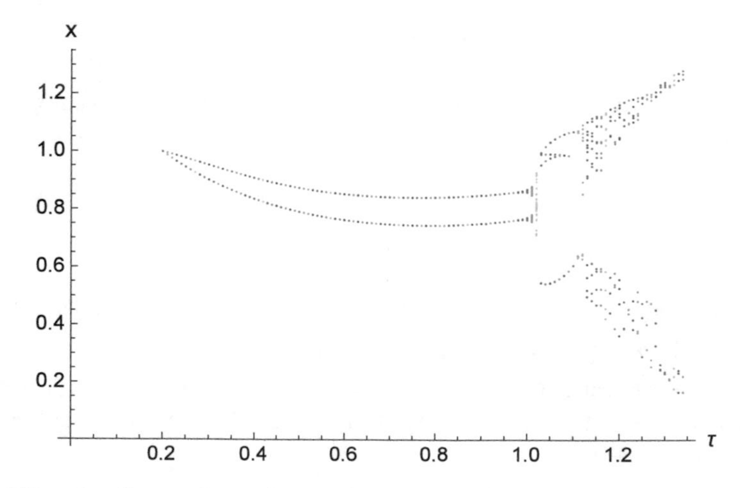

Fig. 4 Bifurcation diagram for x trajectory of system (27) with $\alpha = 0.9$

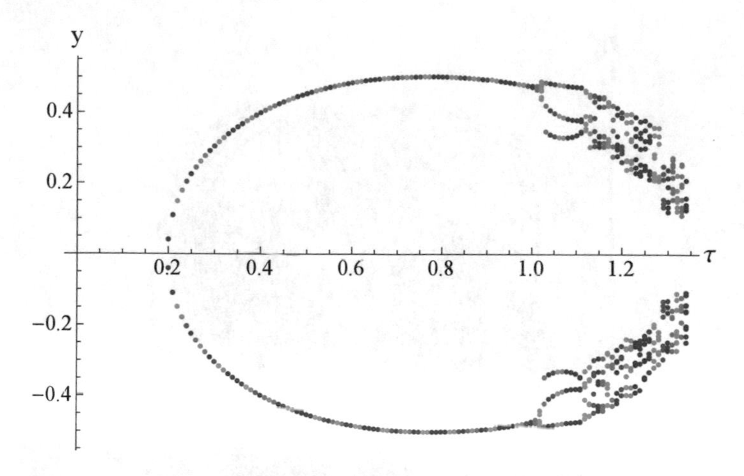

Fig. 5 Bifurcation diagram for y trajectory of system (27) with $\alpha = 0.9$

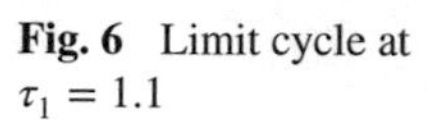

Fig. 6 Limit cycle at $\tau_1 = 1.1$

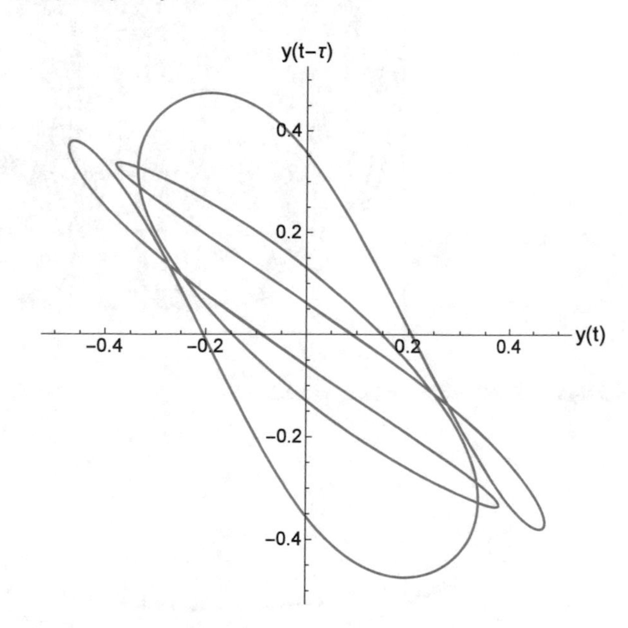

6.2 *Chaos Control*

We use delayed feedback control i.e. Pyragas method [106] to control chaos in the system (27). The controlled system can be described as

$$
\begin{aligned}
D^{\alpha}x &= x_{\tau_1} - \left(x_{\tau_1}^{3} + 3x_{\tau_1}y^{2}\right) + k\left(x - x_{T}\right), \\
D^{\alpha}y &= -y_{\tau_1} - \left(3x^{2}y_{\tau_2} + y^{3}\right).
\end{aligned}
\tag{28}
$$

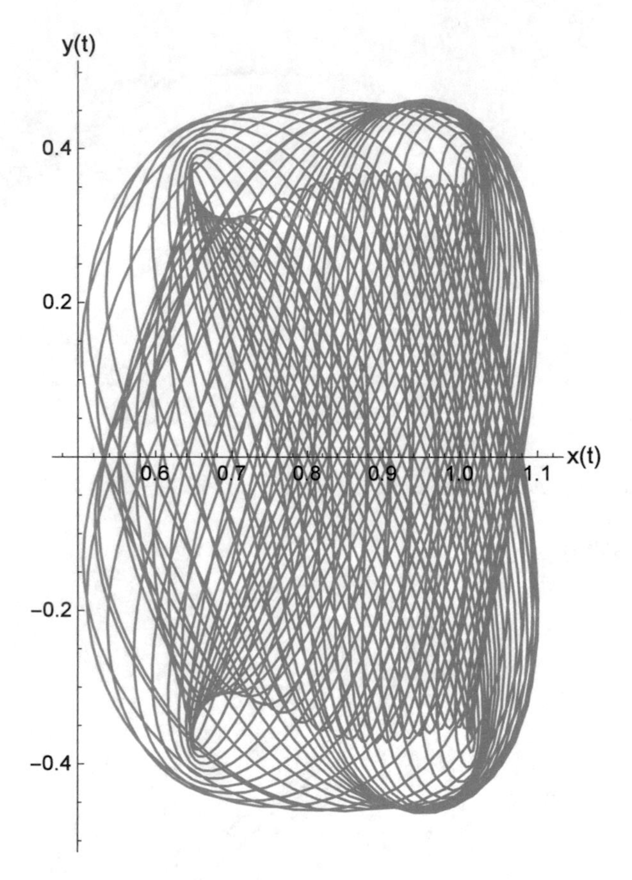

Fig. 7 Chaotic attractor at $\tau_1 = 1.12$

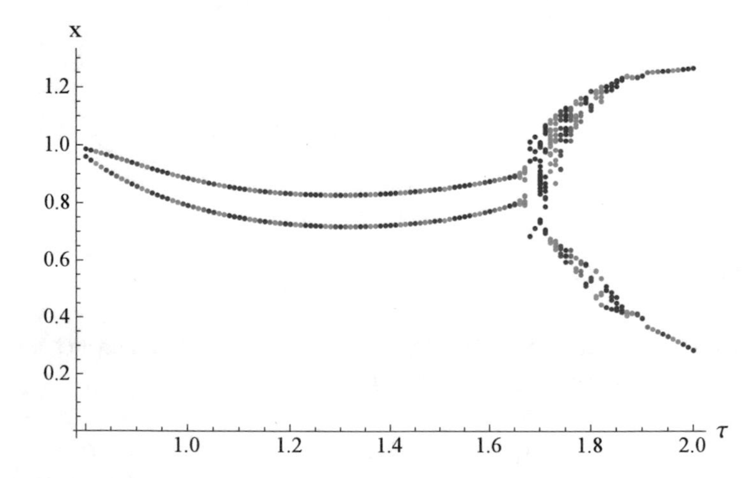

Fig. 8 Bifurcation diagram for x trajectory of system (27) with $\alpha = 0.8$

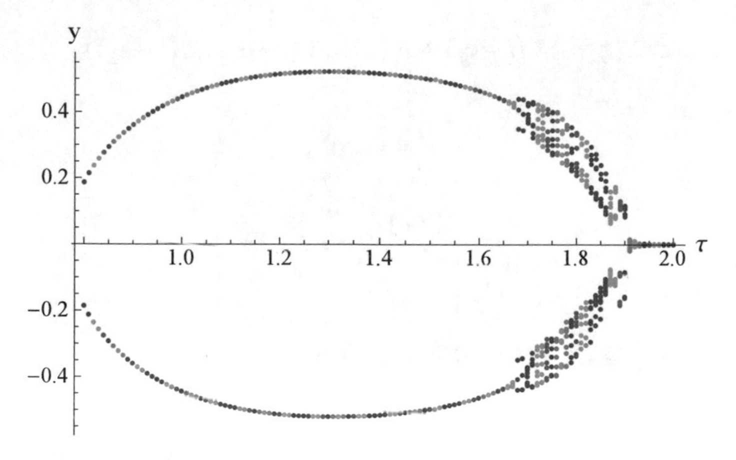

Fig. 9 Bifurcation diagram for y trajectory of system (27) with $\alpha = 0.8$

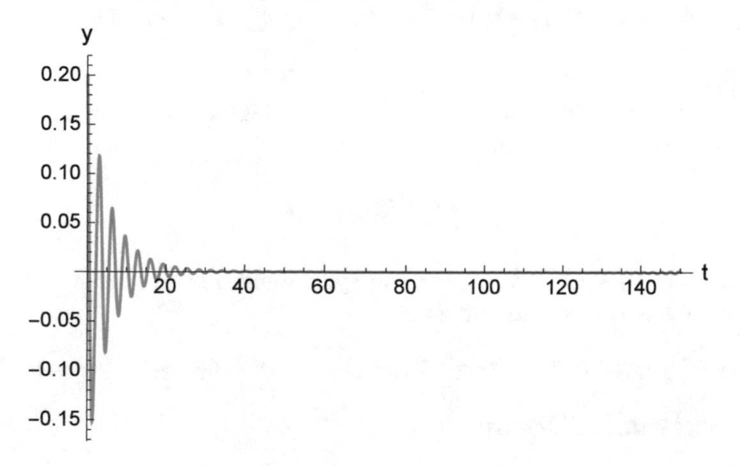

Fig. 10 Controlled trajectory of system (27)

As we have seen, the system (27) is chaotic for the parameter values $\alpha = 0.9$, $\tau_1 = 1.14$ and $\tau_2 = 0.5$. It is observed that the chaos can be controlled for the values $k = -22$ and $T = 3.8$, as shown in Fig. 10.

6.3 *Chaos Synchronization*

In some nonlinear systems, it is not possible to synchronize all the state variables. However, one can synchronize such states up to a scaling factor. This type of strategy is termed as Projective Synchronization [94, 126, 131]. We consider (27) as a master system. We rewrite this system as

$$D^{\alpha}X(t) = AX\left(t-\tau_1\right) + F\left(X(t), X\left(t-\tau_1\right), X\left(t-\tau_2\right)\right), \tag{29}$$

where

$$X = (x, y)^{tr}, \tag{30}$$

$$A = \begin{pmatrix} 1 & 0 \\ 0 & -1 \end{pmatrix} \tag{31}$$

and

$$F\left(X(t), X\left(t-\tau_1\right), X\left(t-\tau_2\right)\right) = \begin{pmatrix} -\left(x_{\tau_1}^3 + 3x_{\tau_1}y^2\right) \\ -\left(3x^2y_{\tau_2} + y^3\right) \end{pmatrix}. \tag{32}$$

The slave system is described as

$$D^{\alpha}Y(t) = AY\left(t-\tau_1\right) + F\left(Y(t), Y\left(t-\tau_1\right), Y\left(t-\tau_2\right)\right) + U, \tag{33}$$

where U is a control term.

We say that projective synchronization (PS) is achieved between systems (29) and (33) if there exists a scaling matrix Λ such that

$$\lim_{t\to\infty} ||E(t)|| = 0, \tag{34}$$

where $E(t) = Y(t) - \Lambda X(t)$ is the error in synchronization. We set $\Lambda = diag(2, 2)$. The error system can now be written as

$$D^{\alpha}E(t) = AE\left(t-\tau_1\right) + F\left(Y(t), Y\left(t-\tau_1\right), Y\left(t-\tau_2\right)\right) - \Lambda F\left(X(t), X\left(t-\tau_1\right), X\left(t-\tau_2\right)\right) + U. \tag{35}$$

If the control term is chosen as

$$U = \Lambda F\left(X(t), X\left(t-\tau_1\right), X\left(t-\tau_2\right)\right) - F\left(Y(t), Y\left(t-\tau_1\right), Y\left(t-\tau_2\right)\right) + BE\left(t-\tau_1\right), \tag{36}$$

where B is a matrix of same dimension as A then (35) becomes

$$D^{\alpha}E(t) = CE\left(t-\tau_1\right), \tag{37}$$

with $C = A + B$.

From a Corollary 3 in [50], the system (37) is globally asymptotically stable if the characteristic equation

$$\lambda^{2\alpha} - (c_{11} + c_{22})\exp(-\lambda\tau_1)\lambda^{\alpha} - c_{12}c_{21}\exp(-2\lambda\tau_1) = 0 \tag{38}$$

has no purely imaginary eigenvalues for any $\tau_1 > 0$. If we set $B = diag(-3, 0)$ then the (38) becomes

$$\lambda^{2\alpha} + 3\exp(-\lambda\tau_1)\lambda^{\alpha} = 0. \tag{39}$$

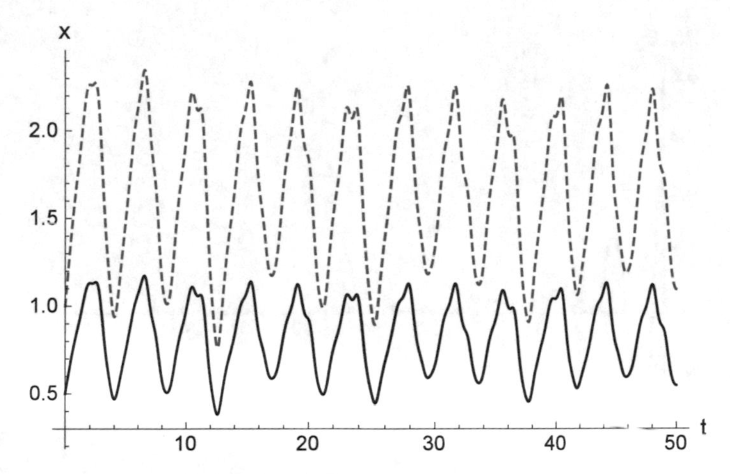

Fig. 11 Chaos synchronization: x trajectories

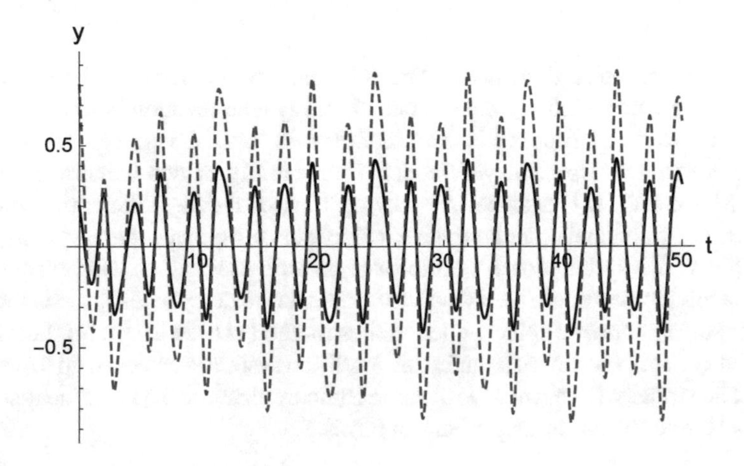

Fig. 12 Chaos synchronization: y trajectories

The condition is now satisfied and the synchronization can be achieved. We show the synchronized orbits in Figs. 11 and 12 and the synchronization errors in Fig. 13. We have taken $\alpha = 0.9$, $\tau_1 = 1.16$ and kept other parameters unchanged. The slave systems and the second component in error system are shown by dashed lines in these figures.

7 Discussion

Fractional order systems are now used to model natural systems in a more realistic way than those involving integer order derivatives. The memory and hereditary properties of such systems cannot be modeled properly unless one uses nonlocal oper-

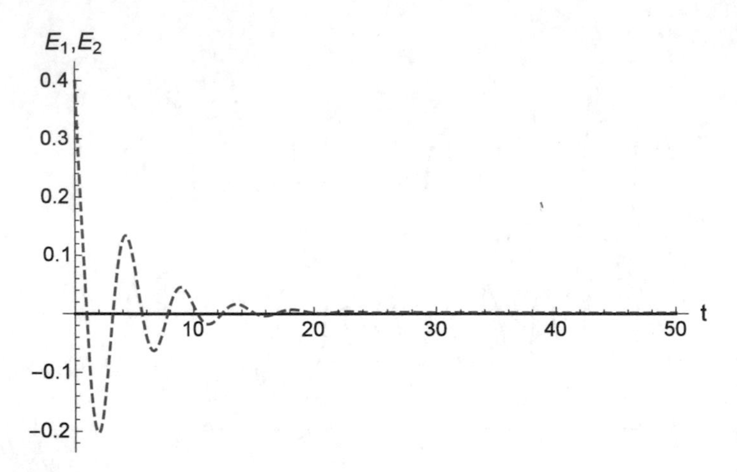

Fig. 13 Synchronization errors

ators such as fractional derivatives. The scientists from diverse field are attracting to the relatively new field viz. chaos control and synchronization in fractional order systems due to its applications in various fields. In this chapter, we have discussed complex version of Uçar delayed model. The stability of two numerical methods viz. FAM and NPCM is analyzed for solving fractional delay differential equations (FDDE). The fractional order complex Uçar system is equivalent to a system of two real FDDEs. The system doesn't exhibit any chaotic oscillations. However, a slight modification in this system involving incommensurate delay leads to a chaotic solutions for some parameter values. The Pyragas method is used to control the chaos in proposed system. Synchronization in this modified system is achieved up to a scaling factor. The strategy is known as Projective Synchronization. The nonlinear stability analysis is used to choose proper control terms.

8 Conclusions

Complex versions of dynamical systems are obtained by replacing real state variables in a system by complex ones. There are very few research articles devoted to fractional order complex dynamical systems (FOCDS). In the present work, we have presented a complex Uçar system of fractional order. Different qualitative structures including chaos are observed in the system by varying the parameter values. The bifurcation diagrams are provided to illustrate the results. We plan in future to analyzed the system in higher dimensions. Further, the system can be generalized to include time-dependent delays. We hope that this chapter will motivate the researchers to work in the field of FOCDS.

Acknowledgements Author acknowledges Council of Scientific and Industrial Research, New Delhi, India for funding through Research Project [25(0245)/15/EMR-II]. The author is grateful to Prof. Ahmad Taher Azar for his encouragement and support.

References

1. Adomian, G. (1994). *Solving frontier problems of physics: The decomposition method*. Dordrecht: Kluwer Academic.
2. Agladze, K. I., Krinsky, V. I., & Pertsov, A. M. (1984). Chaos in the non-stirred Belousov-Zhabotinsky reaction is induced by interaction of waves and stationary dissipative structures. *Nature*, *308*(5962), 834–835.
3. AL-Mutib, A. N. (1984). Stability properties of numerical methods for solving delay differential equations. *Journal of Computational and Applied Mathematics*, *10*(1), 71–79.
4. Alligood, K. T., Sauer, T. D., & Yorke, J. A. (2008). *Chaos: An introduction to dynamical systems*. New York: Springer.
5. Al-Bassam, M. A. (1965). Some existence theorems on differential equations of generalized order. *Journal für die reine und angewandte Mathematik*, *218*, 70–78.
6. Anastasio, T. J. (1994). The fractional-order dynamics of Brainstem Vestibulo-Oculomotor neurons. *Biological Cybernetics*, *72*(1), 69–79.
7. Asea, P., & Zak, P. (1999). Time-to-build and cycles. *Journal of Economic Dynamics and Control*, *23*(8), 1155–1175.
8. Azar, A. T., & Serrano, F. E. (2014). Robust IMC-PID tuning for cascade control systems with gain and phase margin specifications. *Neural Computing and Applications*, *25*(5), 983–995. Springer. doi:10.1007/s00521-014-1560-x.
9. Azar, A. T., & Serrano, F. E. (2015). Adaptive sliding mode control of the Furuta pendulum. In A. T. Azar & Q. Zhu (Eds.), *Advances and applications in sliding mode control systems* (Vol. 576, pp. 1–42). Studies in computational intelligence. Berlin: Springer.
10. Azar, A. T., & Vaidyanathan, S. (2015). *Handbook of research on advanced intelligent control engineering and automation*. Advances in computational intelligence and robotics (ACIR) book series. USA: IGI Global.
11. Azar, A. T., & Vaidyanathan, S. (2015). *Computational intelligence applications in modeling and control* (Vol. 575). Studies in computational intelligence. Germany: Springer. ISBN: 978-3-319-11016-5.
12. Azar, A. T., & Vaidyanathan, S. (2015). *Chaos modeling and control systems design* (Vol. 581)., Studies in computational intelligence Germany: Springer.
13. Azar, A. T., & Zhu, Q. (2015). *Advances and applications in sliding mode control systems* (Vol. 576). Studies in computational intelligence. Germany: Springer. ISBN: 978-3-319-11172-8.
14. Azar, A. T., & Serrano, F. E. (2015). Deadbeat control for multivariable systems with time varying delays. In A. T. Azar & S. Vaidyanathan (Eds.), *Chaos modeling and control systems design* (Vol. 581, pp. 97–132). Studies in computational intelligence. Berlin: Springer.
15. Azar, A. T., & Zhu, Q. (2015). *Advances and applications in sliding mode control systems* (Vol. 576). Studies in computational intelligence. Germany: Springer.
16. Azar, A. T., & Serrano, F. E. (2015). Stabilization and control of mechanical systems with backlash. In A. T. Azar & S. Vaidyanathan (Eds.), *Advanced intelligent control engineering and automation*. Advances in computational intelligence and robotics (ACIR) book series. USA: IGI-Global.
17. Azar, A. T., & Serrano, F. E. (2015). Design and modeling of anti wind up PID controllers. In Q. Zhu & A. T. Azar (Eds.), *Complex system modelling and control through intelligent soft computations* (Vol. 319, pp. 1–44). Studies in fuzziness and soft computing. Germany: Springer. doi:10.1007/978-3-319-12883-2_1.

18. Azar, A. T., & Vaidyanathan, S. (2016). *Advances in chaos theory and intelligent control* (Vol. 337). Studies in fuzziness and soft computing. Germany: Springer. ISBN: 978-3-319-30338-3.
19. Bai, E. W., & Lonngren, K. E. (1997). Synchronization of two Lorenz systems using active control. *Chaos Solitons and Fractals*, *8*(1), 51–58.
20. Bai, E. W., & Lonngren, K. E. (2000). Sequential synchronization of two Lorenz systems using active control. *Chaos Solitons and Fractals*, *11*(7), 1041–1044.
21. Barrett, J. H. (1954). Differential equations of non-integer order. *Canadian Journal of Mathematics*, *6*(4), 529–541.
22. Erol, B. (Ed.). (1990). *Chaos in brain function*. New York: Springer.
23. Bhalekar, S., & Daftardar-Gejji, V. (2010). Synchronization of different fractional order chaotic systems using active control. *Communications in Nonlinear Science and Numerical Simulations*, *15*(11), 3536–3546.
24. Bhalekar, S., & Daftardar-Gejji, V. (2011). Anti-synchronization of non-identical fractional order chaotic systems using active control. *International Journal of Differential Equations*. Article ID 250763.
25. Bhalekar, S., & Daftardar-Gejji, V. (2011). A predictor-corrector scheme for solving non-linear delay differential equations of fractional order. *Journal of Fractional Calculus and Applications*, *1*(5), 1–8.
26. Bhalekar, S., Daftardar-Gejji, V., Baleanu, D., & Magin, R. (2011). Fractional Bloch equation with delay. *Computers and Mathematics with Applications*, *61*(5), 1355–1365.
27. Bhalekar, S. (2012). Dynamical analysis of fractional order Uçar prototype delayed system. *Signal Image Video Processing*, *6*(3), 513–519.
28. Bhalekar, S. (2014). On the Uçar prototype model with incommensurate delays. *Signal Image Video Processing*, *8*(4), 635–639.
29. Bhalekar, S. (2014). Synchronization of incommensurate non-identical fractional order chaotic systems using active control. *The European Physical Journal—Special Topics*, *223*(8), 1495–1508.
30. Bhalekar, S. (2014). Synchronization of non-identical fractional order hyperchaotic systems using active control. *World Journal of Modelling and Simulation*, *10*(1), 60–68.
31. Bhalekar, S. (2016). Stability analysis of Uçar prototype delayed system. *Signal Image Video Processing*, *10*(4), 777–781.
32. Bhalekar, S. (2016). Stability and bifurcation analysis of a generalized scalar delay differential equation. *Chaos: An Interdisciplinary. Journal of Nonlinear Science*, *26*(8), 084306. doi:10.1063/1.4958923.
33. Boulkroune, A., Bouzeriba, A., Bouden, T., & Azar, A. T. (2016). Fuzzy adaptive synchronization of uncertain fractional-order chaotic systems. In *Advances in chaos theory and intelligent control* (Vol. 337). Studies in fuzziness and soft computing. Germany: Springer.
34. Boulkroune. A., Hamel, S., & Azar, A. T. (2016). Fuzzy control-based function synchronization of unknown chaotic systems with dead-zone input. In *Advances in chaos theory and intelligent control* (Vol. 337). Studies in fuzziness and soft computing. Germany: Springer.
35. Campbell, S. A. (2007). Time delays in neural systems. *Handbook of brain connectivity* (pp. 65–90). Berlin: Springer.
36. Chen, W. C. (2008). Nonlinear dynamics and chaos in a fractional-order financial system. *Chaos Solitons and Fractals*, *36*(5), 1305–1314.
37. Chen, Y., & Moore, K. L. (2002). Analytical stability bound for a class of delayed fractional-order dynamic systems. *Nonlinear Dynamics*, *29*(1), 191–200.
38. Cooke, K. L., & Yorke, J. A. (1973). Some equations modelling growth processes and gonorrhea epidemics. *Mathematical Biosciences*, *16*(1), 75–101.
39. Daftardar-Gejji, V., Sukale, Y., & Bhalekar, S. (2014). A new predictor-corrector method for fractional differential equations. *Applied Mathematics and Computations*, *244*(2014), 158–182.
40. Daftardar-Gejji, V., Sukale, Y., & Bhalekar, S. (2015). Solving fractional delay differential equations: A new approach. *Fractional Calculus and Applied Analysis*, *18*(2), 400–418.

41. Daftardar-Gejji, V., & Babakhani, A. (2004). Analysis of a system of fractional differential equations. *Journal of Mathematical Analysis and Applications*, *293*(2), 511–522.
42. Daftardar-Gejji, V., & Jafari, H. (2006). Boundary value problems for fractional diffusion-wave equation. *Australian Journal of Mathematical Analysis and Applications*, *3*(1), 8.
43. Daftardar-Gejji, V., & Jafari, H. (2006). An iterative method for solving non linear functional equations. *Journal of Mathematical Analysis and Applications*, *316*(2), 753–763.
44. Daftardar-Gejji, V., & Bhalekar, S. (2008). Boundary value problems for multi-term fractional differential equations. *Journal of Mathematical Analysis and Applications*, *345*(2), 754–765.
45. Daftardar-Gejji, V., & Bhalekar, S. (2008). Solving multi-term linear and non-linear diffusion-wave equations of fractional order by Adomian method. *Applied Mathematics and Computation*, *202*(1), 113–120.
46. Daftardar-Gejji, V., & Bhalekar, S. (2010). Chaos in fractional ordered Liu system. *Computers and Mathematics with Applications*, *59*(3), 1117–1127.
47. Davis, L. C. (2003). Modification of the optimal velocity traffic model to include delay due to driver reaction time. *Physica A*, *319*, 557–567.
48. De Cesare, L., & Sportelli, M. (2005). A dynamic IS-LM model with delayed taxation revenues. *Chaos Solitons and Fractals*, *25*(1), 233–244.
49. Delbosco, D., & Rodino, L. (1996). Existence and uniqueness for a nonlinear fractional differential equation. *Journal of Mathematical Analysis and Applications*, *204*(2), 609–625.
50. Deng, W., Li, C., & Lü, J. (2007). Stability analysis of linear fractional differential system with multiple time delays. *Nonlinear Dynamics*, *48*(4), 409–416.
51. Deng, W. H., & Li, C. P. (2005). Chaos synchronization of the fractional Lü system. *Physica A*, *353*, 61–72.
52. Deng, W. H., & Li, C. P. (2005). Synchronization of chaotic fractional Chen system. *Journal of the Physical Society of Japan*, *74*(6), 1645–1648.
53. Diethelm, K., & Ford, N. J. (2002). Analysis of fractional differential equations. *Journal of Mathematical Analysis and Applications*, *265*(2), 229–248.
54. Diethelm, K., Ford, N. J., & Freed, A. D. (2002). A predictor-corrector approach for the numerical solution of fractional differential equations. *Nonlinear Dynamics*, *29*(1–4), 3–22.
55. Dokoumetzidis, A., Iliadis, A., & Macheras, P. (2001). Nonlinear dynamics and chaos theory: Concepts and applications relevant to pharmacodynamics. *Pharmaceutical Research*, *18*(4), 415–426.
56. Eidukaitis, A., Varoneckas, G., & Emaitytee, D. (2004). Application of chaos theory in analyzing the cardiac rhythm in healthy subjects at different sleep stages. *Human Physiology*, *30*(5), 551–555.
57. Epstein, I., & Luo, Y. (1991). Differential delay equations in chemical kinetics. Nonlinear models: The cross-shaped phase diagram and the Oregonator. *The Journal of Chemical Physics*, *95*(1), 244–254.
58. Fanti, L., & Manfredi, P. (2007). Chaotic business cycles and fiscal policy: An IS-LM model with distributed tax collection lags. *Chaos Solitons and Fractals*, *32*(2), 736–744.
59. Feliu, V., Rivas, R., & Castillo, F. (2009). Fractional order controller robust to time delay for water distribution in an irrigation main canal pool. *Computers and Electronics in Agriculture*, *69*(2), 185–197.
60. Field, R. J., & Gyorgyi, L. (1993). *Chaos in chemistry and biochemistry*. Singapore: World Scientific.
61. Fridman, E., Fridman, L., & Shustin, E. (2000). Steady modes in relay control systems with time delay and periodic disturbances. *Journal of Dynamic Systems, Measurement, and Control*, *122*(4), 732–737.
62. Fu-Hong, M., Shu-Yi, S., Wen-Di, H., & En-Rong, W. (2015). Circuit implementations, bifurcations and chaos of a novel fractional-order dynamical system. *Chinese Physics Letters*, *32*(3), 030503.
63. Gao, Z., & Liao, X. (2014). Active disturbance rejection control for synchronization of different fractional-order chaotic systems. In *11th World Congress on Intelligent Control and Automation (WCICA). June 29 2014–July 4 2014, Shenyang* (pp. 2699–2704). IEEE.

64. Gorenflo, R., & Mainardi, F. (1996). Fractional oscillations and Mittag-Leffler functions. In *Kuwait University, Department of Mathematics and Computer Science, International Workshop on the Recent Advances in Applied Mathematics. May 4–7, 1996, State of Kuwait* (pp. 193–208).
65. Grigorenko, I., & Grigorenko, E. (2003). Chaotic dynamics of the fractional Lorenz system. *Physical Review Letters*, *91*(3), 034101.
66. Hamamci, S.E. (2007). An algorithm for stabilization of fractional order time delay systems using fractional order PID Controllers. *IEEE Transactions on Automatic Control*, *52*(10), 1964–1969.
67. Hartley, T. T., Lorenzo, C. F., & Qammer, H. K. (1995). Chaos in a fractional order Chua's system. *IEEE Transactions on Circuits and Systems I*, *42*(8), 485–490.
68. Hassell, M. P., Comins, H. N., & May, R. M. (1991). Spatial structure and chaos in insect population dynamics. *Nature*, *353*(6341), 255–258.
69. He, J. H. (1998). Approximate analytical solution for seepage flow with fractional derivatives in porous media. *Computer Methods in Applied Mechanical Engineering*, *167*(1), 57–68.
70. He, J. H. (1999). Homotopy perturbation technique. *Computer Methods in Applied Mechanical Engineering*, *178*(3), 257–262.
71. He, R., & Vaidya, P. G. (1998). Implementation of chaotic cryptography with chaotic synchronization. *Physical Review E*, *57*(2), 1532.
72. Hilfer, R. (Ed.). (2001). *Applications of fractional calculus in physics*. Singapore: World Scientific.
73. Hotzel, R. (1998). Summary: Some stability conditions for fractional delay systems. *Journal of Mathematical Systems Estimation and Control*, *8*, 499–502.
74. Huang, L., Feng, R., & Wang, M. (2004). Synchronization of chaotic systems via nonlinear control. *Physics Letters A*, *320*(4), 271–275.
75. Hwang, C., & Cheng, Y. C. (2006). A numerical algorithm for stability testing of fractional delay systems. *Automatica*, *42*(5), 825–831.
76. Ingo, C., Magin, R. L., & Parrish, T. B. (2014). New insights into the fractional order diffusion equation using entropy and kurtosis. *Entropy*, *16*(11), 5838–5852.
77. Jesus, I. S., & Machado, J. A. T. (2008). Fractional control of heat diffusion systems. *Nonlinear Dynamics*, *54*(3), 263–282.
78. Jesus, I. S., Machado, J. A. T., & Barbosa, R. S. (2010). Control of a heat diffusion system through a fractional order nonlinear algorithm. *Computers and Mathematics with Applications*, *59*(5), 1687–1694.
79. Jun, D., Guangjun, Z., Shaoying, W., & Qiongyao, L. (2014). Chaotic synchronization between fractional-order financial system and financial system of integer orders. In *Control and Decision Conference (2014 CCDC), The 26th Chinese IEEE. May 31, 2014–June 2, 2014, Changsha* (pp. 4924–4928). IEEE.
80. Kilbas, A. A., Srivastava, H. M., & Trujillo, J. J. (2006). *Theory and applications of fractional differential equations*. Amsterdam: Elsevier.
81. Kuang, Y. (1993). *Delay differential equations with applications in population biology*. Boston: Academic Press.
82. Lakshmanan, M., & Senthilkumar, D. V. (2010). *Dynamics of nonlinear time-delay systems*. Heidelberg: Springer.
83. Li, C., & Chen, G. (2004). Chaos and hyperchaos in the fractional order Rossler equations. *Physica A: Statistical Mechanics and its Applications*, *341*, 55–61.
84. Li, C. P., & Deng, W. H. (2006). Chaos synchronization of fractional order differential system. *International Journal of Modern Physics B*, *20*(7), 791–803.
85. Li, C. P., Deng, W. H., & Xu, D. (2006). Chaos synchronization of the Chua system with a fractional order. *Physica A*, *360*(2), 171–185.
86. Li, M., Li, D., Wang, J., & Zhao, C. (2013). Active disturbance rejection control for fractional-order system. *ISA Transactions*, *52*(3), 365–374.
87. Liao, T. L. (1998). Adaptive synchronization of two Lorenz systems. *Chaos Solitons and Fractals*, *9*(9), 1555–1561.

88. Liao, H. (2014). Optimization analysis of Duffing oscillator with fractional derivatives. *Nonlinear Dynamics, 79*(2), 1311–1328.
89. Luchko, Y. F., & Gorenflo, R. (1999). An operational method for solving fractional differential equations with the Caputo derivatives. *Acta Mathematica Vietnamica, 24*(2), 207–233.
90. Lorenz, E. N. (1963). Deterministic nonperiodic flow. *Journal of the Atmospheric Sciences, 20*(2), 130–141.
91. Magin, R. L. (2006). *Fractional calculus in bioengineering*. Redding: Begll House Publishers.
92. Mainardi, F., Luchko, Y., & Pagnini, G. (2001). The fundamental solution of the space-time fractional diffusion equation. *Fractional Calculus and Applied Analysis, 4*(2), 153–192.
93. Mainardi, F. (2010). *Fractional calculus and waves in linear viscoelasticity: An introduction to mathematical models*. London: Imperial College Press.
94. Mainieri, R., & Rehacek, J. (1999). Projective synchronization in three-dimensional chaotic systems. *Physical Review Letters, 82*(15), 3042–3045.
95. Matsumoto, T. (1984). A chaotic attractor from Chua's circuit. *IEEE Transactions on Circuits and Systems, 31*(12), 1055–1058.
96. Matignon, D. (1996). Stability results for fractional differential equations with applications to control processing. In *Computational engineering in systems and application multiconference* (pp. 963–968). Lille: Gerf EC Lille, Villeneuve d'Ascq.
97. Meilanov, R. P., & Magomedov, R. A. (2014). Thermodynamics in fractional calculus. *Journal of Engineering Physics and Thermophysics, 87*(6), 1521–1531.
98. Mekki, H., Boukhetala, D., & Azar, A. T. (2015). Sliding modes for fault tolerant control. In *Advances and applications in sliding mode control systems* (Vol. 576, pp. 407–433). Studies in computational intelligence book series. Berlin/Heidelberg: Springer. doi:10.1007/978-3-319-11173-5_15.
99. Muthukumar, P., & Balasubramaniam, P. (2013). Feedback synchronization of the fractional order reverse butterfly-shaped chaotic system and its application to digital cryptography. *Nonlinear Dynamics, 74*(4), 1169–1181.
100. Muthukumar, P., Balasubramaniam, P., & Ratnavelu, K. (2014). Synchronization and an application of a novel fractional order King Cobra chaotic system. *Chaos, 24*(3), 033105.
101. Muthukumar, P., Balasubramaniam, P., & Ratnavelu, K. (2015). Sliding mode control design for synchronization of fractional order chaotic systems and its application to a new cryptosystem. *International Journal of Dynamics and Control*. In Press. doi:10.1007/s40435-015-0169-y.
102. Ortigueira, M. D., & Machado, J. A. T. (2006). Fractional calculus applications in signals and systems. *Signal Processing, 86*(10), 2503–2504.
103. Pecora, L. M., & Carroll, T. L. (1990). Synchronization in chaotic systems. *Physical Review Letters, 64*(8), 821.
104. Pecora, L. M., & Carroll, T. L. (1991). Driving systems with chaotic signals. *Physical Review A, 44*(4), 2374.
105. Podlubny, I. (1999). *Fractional differential equations*. San Diego: Academic Press.
106. Pyragas, K. (1992). Continuous control of chaos by self-controlling feedback. *Physics letters A, 170*(6), 421–428.
107. Ran, Q., Yuan, L., & Zhao, T. (2015). Image encryption based on nonseparable fractional Fourier transform and chaotic map. *Optics Communications, 348*, 43–49.
108. Sabatier, J., Poullain, S., Latteux, P., Thomas, J., & Oustaloup, A. (2004). Robust speed control of a low damped electromechanical system based on CRONE control: Application to a four mass experimental test bench. *Nonlinear Dynamics, 38*(1–4), 383–400.
109. Samko, S. G., Kilbas, A. A., & Maricev, O. I. (1993). *Fractional integrals and derivatives*. Yverdon: Gordon and Breach.
110. Serletic, A. (1996). Is there chaos in economic series? *Canadian Journal of Economics, 29*, S210–S212.
111. Sheu, L. J., Chen, H. K., Chen, J. H., Tam, L. M., Chen, W. C., Lin, K. T., et al. (2008). Chaos in the Newton-Leipnik system with fractional order. *Chaos Solitons and Fractals, 36*(1), 98–103.

112. Si-Ammour, A., Djennoune, S., & Bettayeb, M. (2009). A sliding mode control for linear fractional systems with input and state delays. *Communications in Nonlinear Science and Numerical Simulation*, *14*(5), 2310–2318.
113. Tseng, C., & Lee, S. L. (2014). Digital image sharpening using Riesz fractional order derivative and discrete hartley transform. In *2014 IEEE Asia Pacific Conference on Circuits and Systems (APCCAS)* (pp. 483–486). Ishigaki: IEEE.
114. Uçar, A. (2002). A prototype model for chaos studies. *International Journal of Engineering Science*, *40*(3), 251–258.
115. Uçar, A. (2003). On the chaotic behaviour of a prototype delayed dynamical system. *Chaos Solitons and Fractals*, *16*(2), 187–194.
116. Vaidyanathan, S., Azar, A. T., Rajagopal, K., & Alexander, P. (2015). Design and SPICE implementation of a 12-term novel hyperchaotic system and its synchronization via active control. *International Journal of Modelling Identification and Control*, *23*(3), 267–277.
117. Vaidyanathan, S., & Azar, A. T. (2015). Anti-synchronization of identical chaotic systems using sliding mode control and an application to Vaidyanathan-Madhavan chaotic systems. In A. T. Azar & Q. Zhu (Eds.), *Advances and applications in sliding mode control systems* (pp. 527–547). Studies in computational intelligence book series. Berlin: Springer.
118. Vaidyanathan, S., & Azar, A. T. (2015). Hybrid synchronization of identical chaotic systems using sliding mode control and an application to Vaidyanathan chaotic systems. In A. T. Azar & Q. Zhu (Eds.), *Advances and applications in sliding mode control systems* (pp. 549–569). Studies in computational intelligence book series. Berlin: Springer.
119. Vaidyanathan, S., Sampath, S., & Azar, A. T. (2015). Global chaos synchronisation of identical chaotic systems via novel sliding mode control method and its application to Zhu system. *International Journal of Modelling, Identification and Control*, *23*(1), 92–100.
120. Vaidyanathan, S., Idowu, B. A., & Azar, A. T. (2015). Backstepping controller design for the global chaos synchronization of Sprott's jerk systems. In A. T. Azar & S. Vaidyanathan (Eds.), *Chaos modeling and control systems design* (pp. 39–58). Berlin: Springer.
121. Vaidyanathan, S., & Azar, A. T. (2016). Dynamic analysis, adaptive feedback control and synchronization of an eight-term 3-D novel chaotic system with three quadratic nonlinearities. In *Advances in chaos theory and intelligent control* (Vol. 337). Studies in fuzziness and soft computing. Germany: Springer.
122. Vaidyanathan, S., & Azar, A. T. (2016). Qualitative study and adaptive control of a novel 4-D hyperchaotic system with three quadratic nonlinearities. In *Advances in chaos theory and intelligent control* (Vol. 337). Studies in fuzziness and soft computing. Germany: Springer.
123. Vaidyanathan, S., & Azar, A. T. (2016). A novel 4-D four-wing chaotic system with four quadratic nonlinearities and its synchronization via adaptive control method. In *Advances in chaos theory and intelligent control* (Vol. 337). Studies in fuzziness and soft computing. Germany: Springer.
124. Vaidyanathan, S., & Azar, A. T. (2016). Adaptive control and synchronization of halvorsen circulant chaotic systems. In *Advances in chaos theory and intelligent control* (Vol. 337). Studies in fuzziness and soft computing. Germany: Springer.
125. Vaidyanathan, S., & Azar, A. T. (2016). Adaptive backstepping control and synchronization of a novel 3-D jerk system with an exponential nonlinearity. In *Advances in chaos theory and intelligent control* (Vol. 337). Studies in fuzziness and soft computing. Germany: Springer.
126. Vaidyanathan, S., & Azar, A. T. (2016). Generalized projective synchronization of a novel hyperchaotic four-wing system via adaptive control method. In *Advances in chaos theory and intelligent control* (Vol. 337). Studies in fuzziness and soft computing. Germany: Springer.
127. Wang, D., & Yu, J. (2008). Chaos in the fractional order logistic delay system. *Journal of Electronic Science and Technology of China*, *6*(3), 225–229.
128. Wang, J., Xionga, X., & Zhang, Y. (2006). Extending synchronization scheme to chaotic fractional-order Chen systems. *Physica A*, *370*(2), 279–285.
129. Wang, J., & Zhang, Y. (2006). Designing synchronization schemes for chaotic fractional-order unified systems. *Chaos Solitons and Fractals*, *30*(5), 1265–1272.

130. Wang, S., Sun, W., Ma, C. Y., Wang, D., & Chen, Z. (2013). Secure communication based on a fractional order chaotic system. *International Journal of Security and Its Applications*, *7*(5), 205–216.
131. Wang, S., Yu, Y., & Wen, G. (2014). Hybrid projective synchronization of time-delayed fractional order chaotic systems. *Nonlinear Analysis: Hybrid Systems*, *11*, 129–138.
132. Wu, G. C., Baleanu, D., & Lin, Z. X. (2016). Image encryption technique based on fractional chaotic time series. *Journal of Vibration and Control*, *22*(8), 2092–2099.
133. Xi, H., & Gunton, J. D. (1995). Spatiotemporal chaos in a model of Rayleigh-Benard convection. *Physical Review E*, *52*(4), 4963–4975.
134. Xu, B., Chen, D., Zhang, H., & Wang, F. (2015). Modeling and stability analysis of a fractional-order Francis hydro-turbine governing system. *Chaos Solitons and Fractals*, *75*, 50–61.
135. Xu, Y., & Wang, H. (2013). Synchronization of fractional-order chaotic systems with Gaussian fluctuation by sliding mode control. *Abstract and Applied Analysis*. Article ID 948782.
136. Xu, Y., Wang, H., Li, Y., & Pei, B. (2014). Image encryption based on synchronization of fractional chaotic systems. *Communications in Nonlinear Science and Numerical Simulation*, *19*(10), 3735–3744.
137. Yassen, M. T. (2001). Adaptive control and synchronization of a modified Chua's circuit system. *Applied Mathematics and Computation*, *135*(1), 113–128.
138. Zhao, J., Wang, S., Chang, Y., & Li, X. (2015). A novel image encryption scheme based on an improper fractional-order chaotic system. *Nonlinear Dynamics*, *80*(4), 1721–1729.
139. Zhu, Q., & Azar, A. T. (2015). Complex system modelling and control through intelligent soft computations. In *Studies in fuzziness and soft computing* (Vol. 319). Germany: Springer. ISBN: 978-3-319-12882-5.

Hyperchaos and Adaptive Control of a Novel Hyperchaotic System with Two Quadratic Nonlinearities

Sundarapandian Vaidyanathan, Ahmad Taher Azar and Adel Ouannas

Abstract Liu-Su-Liu chaotic system (2007) is one of the classical 3-D chaotic systems in the literature. By introducing a feedback control to the Liu-Su-Liu chaotic system,we obtain a novel hyperchaotic system in this work, which has two quadratic nonlinearities. The phase portraits of the novel hyperchaotic system are displayed and the qualitative properties of the novel hyperchaotic system are discussed. We show that the novel hyperchaotic system has a unique equilibrium point at the origin, which is unstable. The Lyapunov exponents of the novel 4-D hyperchaotic system are obtained as $L_1 = 1.1097$, $L_2 = 0.1584$, $L_3 = 0$ and $L_4 = -14.1666$. The maximal Lyapunov exponent (MLE) of the novel hyperchaotic system is obtained as $L_1 = 1.1097$ and Lyapunov dimension as $D_L = 3.0895$. Since the sum of the Lyapunov exponents of the novel hyperchaotic system is negative, it follows that the novel hyperchaotic system is dissipative. Next, we derive new results for the adaptive control design of the novel hyperchaotic system with unknown parameters. We also derive new results for the adaptive synchronization design of identical novel hyperchaotic systems with unknown parameters. The adaptive control results derived in this work for the novel hyperchaotic system are proved using Lyapunov stability theory. Numerical simulations in MATLAB are shown to validate and illustrate all the main results derived in this work.

S. Vaidyanathan (✉)
Research and Development Centre, Vel Tech University,
Avadi, Chennai 600062, Tamil Nadu, India
e-mail: sundarcontrol@gmail.com

A.T. Azar
Faculty of Computers and Information, Benha University, Benha, Egypt
e-mail: ahmad_t_azar@ieee.org; ahmad.azar@fci.bu.edu.eg

A.T. Azar
Nanoelectronics Integrated Systems Center (NISC), Nile University, Cairo, Egypt

A. Ouannas
Laboratory of Mathematics, Informatics and Systems (LAMIS),
University of Larbi Tebessi, 12002 Tebessa, Algeria
e-mail: ouannas_adel@yahoo.fr

A.T. Azar et al. (eds.), *Fractional Order Control and Synchronization of Chaotic Systems*, Studies in Computational Intelligence 688,
DOI 10.1007/978-3-319-50249-6_27

Keywords Chaos · Chaotic systems · Hyperchaos · Hyperchaotic systems · Adaptive control · Feedback control · Synchronization

1 Introduction

Chaotic systems are defined as nonlinear dynamical systems which are sensitive to initial conditions, topologically mixing and with dense periodic orbits [10, 11, 13].

Sensitivity to initial conditions of chaotic systems is popularly known as the *butterfly effect*. Small changes in an initial state will make a very large difference in the behavior of the system at future states.

Poincaré [15] suspected chaotic behaviour in the study of three bodies problem at the end of the 19th century, but chaos was experimentally established by Lorenz [45] only a few decades ago in the study of 3-D weather models.

The Lyapunov exponent is a measure of the divergence of phase points that are initially very close and can be used to quantify chaotic systems. It is common to refer to the largest Lyapunov exponent as the *Maximal Lyapunov Exponent* (MLE). A positive maximal Lyapunov exponent and phase space compactness are usually taken as defining conditions for a chaotic system.

Hyperchaotic systems are chaotic systems with more than one positive Lyapunov exponent. They have important applications in control, cryptography and secure communication. Some recently discovered hyperchaotic systems are hyperchaotic Lorenz system [32], hyperchaotic Chen system [76], hyperchaotic Lü system [19], hyperchaotic Fang system [25], hyperchaotic Qi system [22], etc.

In the last five decades, there is significant interest in the literature in discovering new chaotic systems [78]. Some popular chaotic systems are Lorenz system [45], Rössler system [67], Arneodo system [2], Henon-Heiles system [30], Genesio-Tesi system [28], Sprott systems [77], Chen system [20], Lü system [46], Rikitake dynamo system [66], Liu-Su-Liu system [44], Shimizu system [75], etc.

In the recent years, many new chaotic systems have been found such as Pandey system [50], Qi system [58], Li system [39], Wei-Yang system [176], Zhou system [183], Zhu system [184], Sundarapandian systems [81, 86], Dadras system [23], Tacha system [89], Vaidyanathan systems [97, 98, 100–103, 106, 117, 131–138, 140–142, 144, 153, 155, 164, 166, 168, 171–173], Vaidyanathan-Azar systems [147, 148, 150–152], Pehlivan system [52], Sampath system [68], Akgul system [1], Pham system [53, 55–57], etc.

Chaos theory and control systems have many important applications in science and engineering [3, 10–12, 14, 185]. Some commonly known applications are oscillators [120, 124, 126–129, 139], lasers [41, 179], chemical reactions [107, 108, 112–114, 116, 118, 119, 122, 123, 125], biology [24, 37, 105, 109–111, 115, 121], ecology [29, 79], encryption [38, 182], cryptosystems [65, 90], mechanical systems [5–9], secure communications [26, 48, 180], robotics [47, 49, 174], cardiology [59, 177], intelligent control [4, 42], neural networks [31, 35, 43], memristors [54, 175], etc.

Synchronization of chaotic systems is a phenomenon that occurs when two or more chaotic systems are coupled or when a chaotic system drives another chaotic system. Because of the butterfly effect which causes exponential divergence of the trajectories of two identical chaotic systems started with nearly the same initial conditions, the synchronization of chaotic systems is a challenging research problem in the chaos literature.

Major works on synchronization of chaotic systems deal with the complete synchronization of a pair of chaotic systems called the *master* and *slave* systems. The design goal of the complete synchronization is to apply the output of the master system to control the slave system so that the output of the slave system tracks the output of the master system asymptotically with time. Active feedback control is used when the system parameters are available for measurement. Adaptive feedback control is used when the system parameters are unknown.

Pecora and Carroll pioneered the research on synchronization of chaotic systems with their seminal papers [18, 51]. The active control method [34, 69, 70, 80, 85, 91, 95, 156, 157, 160] is typically used when the system parameters are available for measurement.

Adaptive control method [71–73, 82–84, 93, 99, 130, 143, 149, 154, 158, 159, 165, 169, 170] is typically used when some or all the system parameters are not available for measurement and estimates for the uncertain parameters of the systems. Adaptive control method has more relevant for many practical situations for systems with unknown parameters. In the literature, adaptive control method is preferred over active control method due to the wide applicability of the adaptive control method.

Intelligent control methods like fuzzy control method [16, 17] are also used for the synchronization of chaotic systems. Intelligent control methods have advantages like robustness, insensitive to small variations in the parameters, etc.

Sampled-data feedback control method [27, 40, 178, 181] and time-delay feedback control method [21, 33, 74] are also used for synchronization of chaotic systems. Backstepping control method [60–64, 88, 161, 167] is also used for the synchronization of chaotic systems, which is a recursive method for stabilizing the origin of a control system in strict-feedback form.

Another popular method for the synchronization of chaotic systems is the sliding mode control method [87, 92, 94, 96, 104, 145, 146, 162, 163], which is a nonlinear control method that alters the dynamics of a nonlinear system by application of a discontinuous control signal that forces the system to "slide" along a cross-section of the system's normal behavior.

Liu-Su-Liu chaotic system [44] is one of the classical 3-D chaotic systems in the literature. By introducing a feedback control to the Liu-Su-Liu chaotic system, we obtain a novel hyperchaotic system in this work, which has two quadratic nonlinearities.

Section 2 describes the dynamics and phase portraits of the novel hyperchaotic Liu system. Section 3 describes the dynamic analysis of the novel hyperchaotic system. We show that the novel hyperchaotic system has a unique equilibrium at the origin, which is a saddle point and unstable.

The Lyapunov exponents of the novel 4-D hyperchaotic system are obtained as $L_1 = 1.1097$, $L_2 = 0.1584$, $L_3 = 0$ and $L_4 = -14.1666$. The maximal Lyapunov exponent (MLE) of the novel hyperchaotic system is obtained as $L_1 = 1.1097$ and Lyapunov dimension as $D_L = 3.0895$.

Section 4 describes the adaptive feedback control of the novel hyperchaotic system with unknown parameters. Section 5 describes the adaptive synchronization of the identical novel hyperchaotic systems with unknown parameters. The adaptive feedback control and synchronization results are proved using Lyapunov stability theory [36].

MATLAB simulations are depicted to illustrate all the main results for the 4-D novel hyperchaotic system. Section 6 concludes this work with a summary of the main results.

2 A Novel 4-D Hyperchaotic System

Liu-Su-Liu chaotic system [44] is modelled by the 3-D dynamics

$$\begin{aligned} \dot{x}_1 &= a(x_2 - x_1) \\ \dot{x}_2 &= bx_1 + dx_1x_3 \\ \dot{x}_3 &= -cx_3 - x_1x_2 \end{aligned} \tag{1}$$

where x_1, x_2, x_3 are the states and a, b, c, d are constant, positive parameters of the system.

Liu-Su-Liu system (1) describes a *strange chaotic attractor* for the parameter values

$$a = 10, \quad b = 40, \quad c = 2.5, \quad d = 16 \tag{2}$$

For numerical simulations, we take the initial values of the Liu-Su-Liu system (1) as

$$x_1(0) = 0.2, \quad x_2(0) = 0.2, \quad x_3(0) = 0.2 \tag{3}$$

Figure 1 shows the strange, two-scroll, chaotic attractor of the Liu-Su-Liu system (1). This strange attractor is a reversed butterfly-shaped attractor for the Liu-Su-Liu system (1).

The Lyapunov exponents for the Liu-Su-Liu system (1) for the parameter values (2) and the initial conditions (3) are numerically obtained as

$$L_1 = 1.2047, \quad L_2 = 0, \quad L_3 = -13.6986 \tag{4}$$

Since the sum of the Lyapunov exponents in (4) is negative, it follows that Liu-Su-Liu system (1) is a dissipative chaotic system.

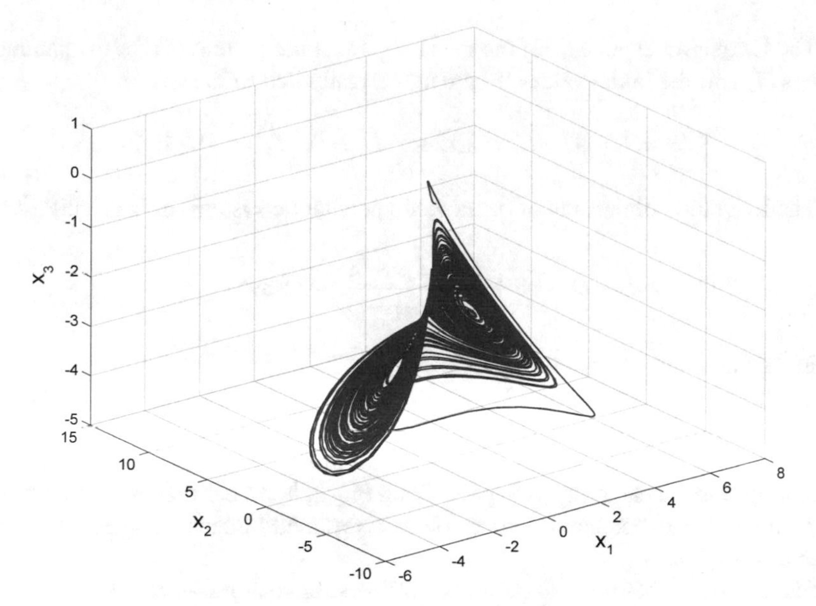

Fig. 1 Strange attractor of the Liu-Su-Liu chaotic system in $\mathbb{R}^3$

The Lyapunov dimension of the Liu-Su-Liu system (1) is numerically obtained as

$$D_L = 2 + \frac{L_1 + L_2}{|L_3|} = 2.0879 \tag{5}$$

In this research work, we obtain a novel 4-D hyperchaotic system by introducing a feedback control to the 3-D Liu-Su-Liu chaotic system (1).

We announce our novel 4-D hyperchaotic system as

$$\begin{aligned} \dot{x}_1 &= a(x_2 - x_1) + x_4 \\ \dot{x}_2 &= bx_1 + dx_1x_3 \\ \dot{x}_3 &= -cx_3 - x_1x_2 \\ \dot{x}_4 &= -x_1 - x_2 + px_4 \end{aligned} \tag{6}$$

where x_1, x_2, x_3, x_4 are the states and a, b, c, d, p are constant, positive parameters.

The 4-D system (6) is *hyperchaotic* when the parameter values are taken as

$$a = 10, \quad b = 40, \quad c = 3, \quad d = 18, \quad p = 0.1 \tag{7}$$

For numerical simulations, we take the initial state of the hyperchaotic system (6) as

$$x_1(0) = 0.2, \quad x_2(0) = 0.2, \quad x_3(0) = 0.2, \quad x_4(0) = 0.2 \tag{8}$$

The Lyapunov exponents of the novel hyperchaotic system (6) for the parameter values (7) and the initial values (8) are numerically determined as

$$L_1 = 1.1097, \ L_2 = 0.1584, \ L_3 = 0, \ L_4 = -14.1666 \tag{9}$$

The Lyapunov dimension of the novel hyperchaotic system (6) is calculated as

$$D_L = 3 + \frac{L_1 + L_2 + L_3}{|L_4|} = 3.0895, \tag{10}$$

which is fractional.

The presence of two positive Lyapunov exponents in (9) shows that the 4-D novel system (6) is *hyperchaotic*.

Figure 2 describes the 3-D projection of the strange attractor of the novel hyperchaotic system (6) on (x_1, x_2, x_3)-space. From Fig. 2, it is clear that the strange attractor of the novel hyperchaotic system (6) is a reversed butterfly-shaped attractor in (x_1, x_2, x_3)-space.

Figure 3 describes the 3-D projection of the strange attractor of the novel hyperchaotic system (6) on (x_1, x_2, x_4)-space.

Figure 4 describes the 3-D projection of the strange attractor of the novel hyperchaotic system (6) on (x_1, x_3, x_4)-space.

Figure 5 describes the 3-D projection of the strange attractor of the novel hyperchaotic system (6) on (x_2, x_3, x_4)-space.

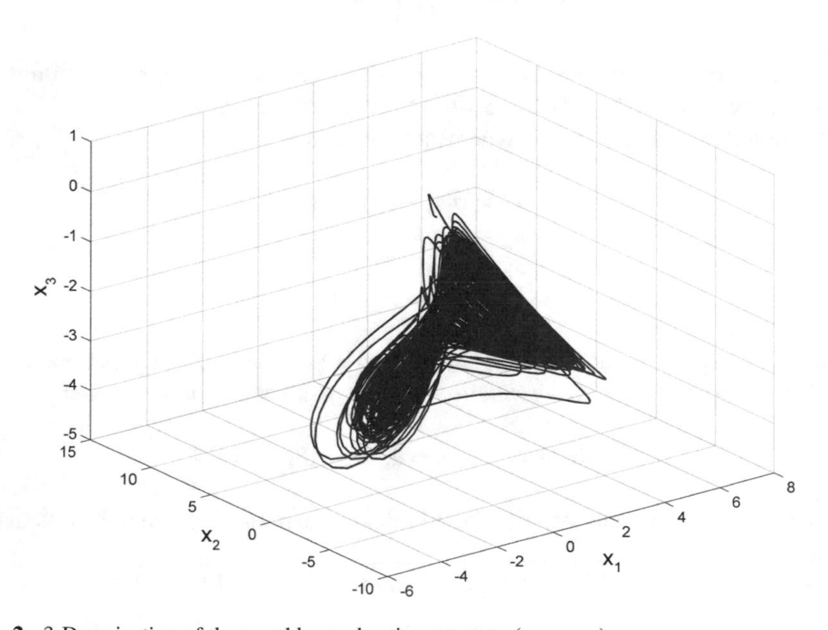

Fig. 2 3-D projection of the novel hyperchaotic system on (x_1, x_2, x_3)-space

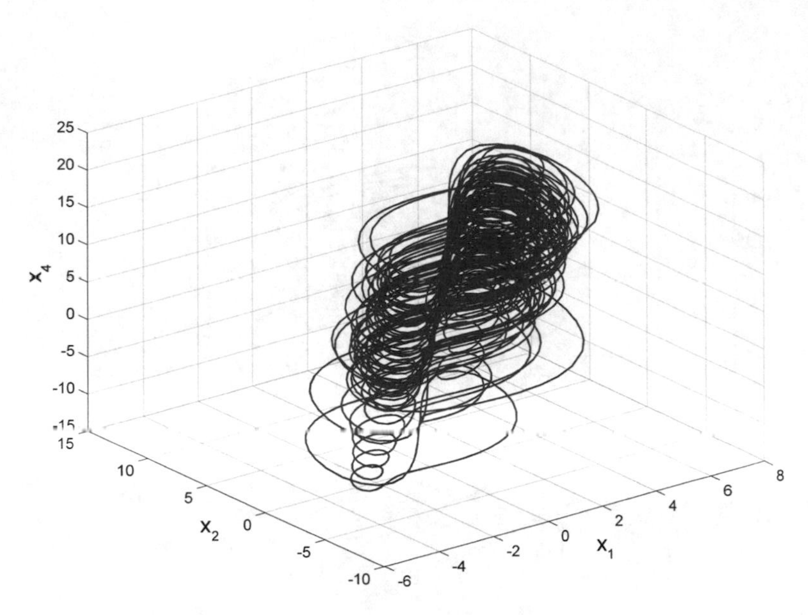

Fig. 3 3-D projection of the novel hyperchaotic system on (x_1, x_2, x_4)-space

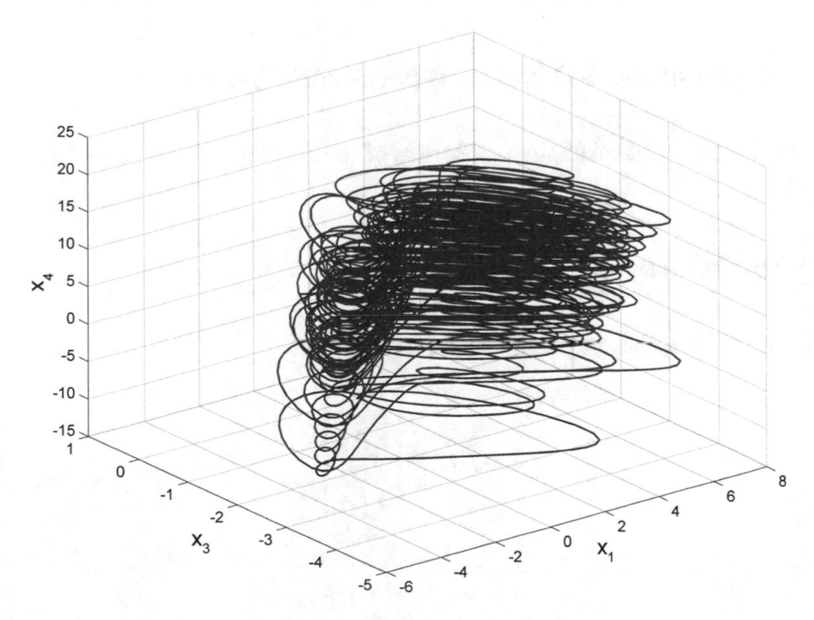

Fig. 4 3-D projection of the novel hyperchaotic system on (x_1, x_3, x_4)-space

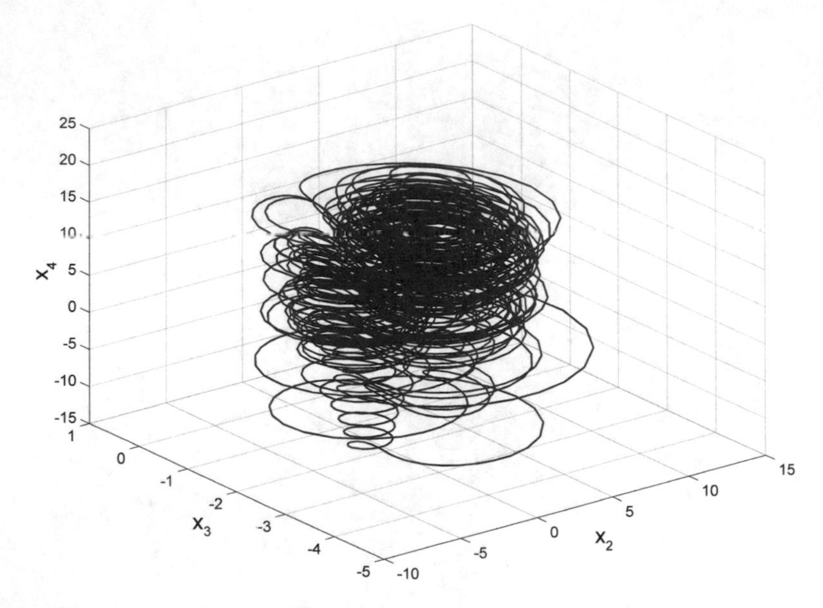

Fig. 5 3-D projection of the novel hyperchaotic system on (x_2, x_3, x_4)-space

3 Analysis of the 4-D Novel Hyperchaotic System

This section gives the qualitative properties of the novel hyperchaotic system (6).

3.1 *Dissipativity*

In vector notation, the system (6) can be expressed as

$$\dot{x} = f(x) = \begin{bmatrix} f_1(x) \\ f_2(x) \\ f_3(x) \\ f_4(x) \end{bmatrix}, \tag{11}$$

where

$$\begin{aligned} f_1(x) &= a(x_2 - x_1) + x_4 \\ f_2(x) &= bx_1 + dx_1x_3 \\ f_3(x) &= -cx_3 - x_1x_2 \\ f_4(x) &= -x_1 - x_2 + px_4 \end{aligned} \tag{12}$$

We take the parameter values as

$$a = 10, \;\; b = 40, \;\; c = 3, \;\; d = 18, \;\; p = 0.1 \tag{13}$$

The divergence of the vector field f on $\mathbb{R}^4$ is obtained as

$$\operatorname{div} f = \frac{\partial f_1(x)}{\partial x_1} + \frac{\partial f_2(x)}{\partial x_2} + \frac{\partial f_3(x)}{\partial x_3} + \frac{\partial f_3(x)}{\partial x_4} = -a - c + p = -\mu \tag{14}$$

where

$$\mu = a + c - p = 12.9 > 0 \tag{15}$$

Let Ω be any region in $\mathbb{R}^4$ with a smooth boundary. Let $\Omega(t) = \Phi_t(\Omega)$, where Φ_t is the flow of the vector field f. Let $V(t)$ denote the hypervolume of $\Omega(t)$.

By Liouville's theorem, it follows that

$$\frac{dV(t)}{dt} = \int\limits_{\Omega(t)} (\operatorname{div} f) dx_1\, dx_2\, dx_3\, dx_4 \tag{16}$$

Substituting the value of div$f$ in (16) leads to

$$\frac{dV(t)}{dt} = -\mu \int\limits_{\Omega(t)} dx_1\, dx_2\, dx_3\, dx_4 = -\mu V(t) \tag{17}$$

Integrating the linear differential equation (17), $V(t)$ is obtained as

$$V(t) = V(0) \exp(-\mu t) \tag{18}$$

From Eq. (18), it follows that the hypervolume $V(t)$ shrinks to zero exponentially as $t \to \infty$.

Thus, the novel hyperchaotic system (6) is dissipative. Hence, any asymptotic motion of the system (6) settles onto a set of measure zero, *i.e.* a strange attractor.

3.2 Symmetry

It is easy to check that novel hyperchaotic system (6) is invariant under the change of coordinates

$$(x_1, x_2, x_3, x_4) \mapsto (-x_1, -x_2, x_3, -x_4) \tag{19}$$

Since the transformation (19) persists for all values of the system parameters, it follows that the novel hyperchaotic system (6) has rotation symmetry about the x_3-axis and any non-trivial trajectory must have a twin trajectory.

3.3 Invariance

It is easily seen that the x_3-axis is invariant for the flow of the novel chaotic system (6). The invariant motion along the x_3-axis is characterized by the scalar dynamics

$$\dot{x}_3 = -cx_3, \quad (c > 0) \tag{20}$$

which is globally exponentially stable.

3.4 Equilibria

The equilibrium points of the novel hyperchaotic system (6) are obtained by solving the nonlinear equations

$$\begin{aligned} f_1(x) &= a(x_2 - x_1) + x_4 = 0 \\ f_2(x) &= bx_1 + dx_1x_3 \quad\;\; = 0 \\ f_3(x) &= -cx_3 - x_1x_2 \quad\; = 0 \\ f_4(x) &= -x_1 - x_2 + px_4 = 0 \end{aligned} \tag{21}$$

We take the parameter values as in the hyperchaotic case (7), viz.

$$a = 10, \quad b = 40, \quad c = 3, \quad d = 18, \quad p = 0.1 \tag{22}$$

Solving the nonlinear system (21) with the parameter values (22), we obtain a unique equilibrium point at the origin, viz.

$$E_0 = \begin{bmatrix} 0 \\ 0 \\ 0 \\ 0 \end{bmatrix} \tag{23}$$

The Jacobian matrix of the novel hyperchaotic system (6) at E_0 is obtained as

$$J_0 = J(E_0) = \begin{bmatrix} -10 & 10 & 0 & 1 \\ 40 & 0 & 0 & 0 \\ 0 & 0 & -3 & 0 \\ -1 & -1 & 0 & 0.1 \end{bmatrix} \tag{24}$$

The matrix J_0 has the eigenvalues

$$\lambda_1 = -3, \quad \lambda_2 = -25.6291, \quad \lambda_3 = 0.2010, \quad \lambda_4 = 15.5281 \tag{25}$$

This shows that the equilibrium point E_0 is a saddle-point, which is unstable.

3.5 *Lyapunov Exponents and Lyapunov Dimension*

We take the initial values of the novel hyperchaotic system (6) as in (8) and the parameter values of the novel hyperchaotic system (6) as in (7).

Then the Lyapunov exponents of the novel hyperchaotic system (6) are numerically obtained as

$$L_1 = 1.1097, \quad L_2 = 0.1584, \quad L_3 = 0, \quad L_4 = -14.1666 \tag{26}$$

Since $L_1 + L_2 + L_3 + L_4 = -12.8985 < 0$, the system (6) is dissipative.

Also, the Lyapunov dimension of the system (6) is obtained as

$$D_L = 3 + \frac{L_1 + L_2 + L_3}{|L_4|} = 3.0895, \tag{27}$$

which is fractional.

Figure 6 depicts the Lyapunov exponents of the novel hyperchaotic system (6). From this figure, it is seen that the Maximal Lyapunov Exponent (MLE) of the novel hyperchaotic system (6) is $L_1 = 1.1097$.

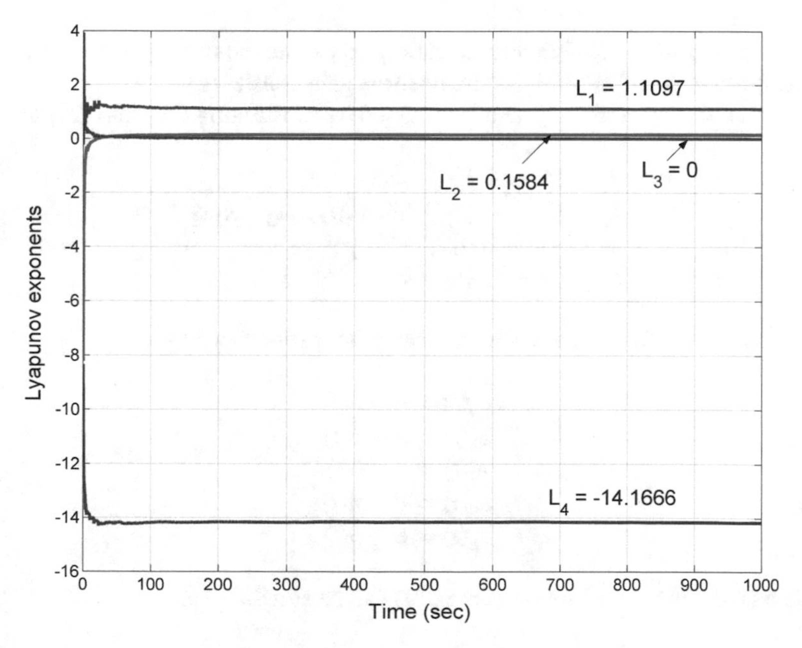

Fig. 6 Lyapunov exponents of the novel hyperchaotic system

4 Adaptive Control of the Novel Hyperchaotic System

This section derives new results for the adaptive controller design in order to stabilize the unstable novel hyperchaotic system with unknown parameters for all initial conditions of the system.

The controlled novel hyperchaotic system is given by

$$\begin{aligned} \dot{x}_1 &= a(x_2 - x_1) + x_4 + u_1 \\ \dot{x}_2 &= bx_1 + dx_1x_3 + u_2 \\ \dot{x}_3 &= -cx_3 - x_1x_2 + u_3 \\ \dot{x}_4 &= -x_1 - x_2 + px_4 + u_4 \end{aligned} \tag{28}$$

where x_1, x_2, x_3, x_4 are state variables, a, b, c, d, p are constant, unknown, parameters of the system and u_1, u_2, u_3, u_4 are adaptive controls to be designed.

An adaptive feedback control law is taken as

$$\begin{aligned} u_1 &= -\hat{a}(t)(x_2 - x_1) - x_4 - k_1x_1 \\ u_2 &= -\hat{b}(t)x_1 - \hat{d}(t)x_1x_3 - k_2x_2 \\ u_3 &= \hat{c}(t)x_3 + x_1x_2 - k_3x_3 \\ u_4 &= x_1 + x_2 - \hat{p}(t)x_4 - k_4x_4 \end{aligned} \tag{29}$$

In (29), $\hat{a}(t), \hat{b}(t), \hat{c}(t), \hat{d}(t), \hat{p}(t)$ are estimates for the unknown parameters a, b, c, d, p, respectively, and k_1, k_2, k_3, k_4 are positive gain constants.

The closed-loop control system is obtained by substituting (29) into (28) as

$$\begin{aligned} \dot{x}_1 &= [a - \hat{a}(t)](x_2 - x_1) - k_1x_1 \\ \dot{x}_2 &= [b - \hat{b}(t)]x_1 + [d - \hat{d}(t)]x_1x_3 - k_2x_2 \\ \dot{x}_3 &= -[c - \hat{c}(t)]x_3 - k_3x_3 \\ \dot{x}_4 &= [p - \hat{p}(t)]x_4 - k_4x_4 \end{aligned} \tag{30}$$

To simplify (30), we define the parameter estimation error as

$$\begin{aligned} e_a(t) &= a - \hat{a}(t) \\ e_b(t) &= b - \hat{b}(t) \\ e_c(t) &= c - \hat{c}(t) \\ e_d(t) &= d - \hat{d}(t) \\ e_p(t) &= p - \hat{p}(t) \end{aligned} \tag{31}$$

Using (31), the closed-loop system (30) can be simplified as

$$\begin{aligned} \dot{x}_1 &= e_a(x_2 - x_1) - k_1x_1 \\ \dot{x}_2 &= e_bx_1 + e_dx_1x_3 - k_2x_2 \\ \dot{x}_3 &= -e_cx_3 - k_3x_3 \\ \dot{x}_4 &= e_px_4 - k_4x_4 \end{aligned} \tag{32}$$

Differentiating the parameter estimation error (31) with respect to t, we get

$$\begin{aligned} \dot{e}_a &= -\dot{\hat{a}} \\ \dot{e}_b &= -\dot{\hat{b}} \\ \dot{e}_c &= -\dot{\hat{c}} \\ \dot{e}_d &= -\dot{\hat{d}} \\ \dot{e}_p &= -\dot{\hat{p}} \end{aligned} \tag{33}$$

Next, we find an update law for parameter estimates using Lyapunov stability theory.

Consider the quadratic Lyapunov function defined by

$$V(\mathbf{x}, e_a, e_b, e_c, e_d, e_p) = \frac{1}{2}\left(x_1^2 + x_2^2 + x_3^2 + x_4^2 + e_a^2 + e_b^2 + e_c^2 + e_d^2 + e_p^2\right), \tag{34}$$

which is positive definite on $\mathbb{R}^9$.

Differentiating V along the trajectories of (32) and (33), we get

$$\begin{aligned} \dot{V} = &-k_1x_1^2 - k_2x_2^2 - k_3x_3^2 - k_4x_4^2 + e_a[x_1(x_2 - x_1) - \dot{\hat{a}}] + e_b[x_1x_2 - \dot{\hat{b}}] \\ &+ e_c[-x_3^2 - \dot{\hat{c}}] + e_d[x_1x_2x_3 - \dot{\hat{d}}] + e_p[x_4^2 - \dot{\hat{p}}] \end{aligned} \tag{35}$$

In view of Eq. (35), an update law for the parameter estimates is taken as

$$\begin{aligned} \dot{\hat{a}} &= x_1(x_2 - x_1) \\ \dot{\hat{b}} &= x_1x_2 \\ \dot{\hat{c}} &= -x_3^2 \\ \dot{\hat{d}} &= x_1x_2x_3 \\ \dot{\hat{p}} &= x_4^2 \end{aligned} \tag{36}$$

Theorem 1 *The novel hyperchaotic system (28) with unknown system parameters is globally and exponentially stabilized for all initial conditions $x(0) \in \mathbb{R}^4$ by the adaptive control law (29) and the parameter update law (36), where k_i, $(i = 1, 2, 3, 4)$ are positive constants.*

Proof The result is proved using Lyapunov stability theory [36]. We consider the quadratic Lyapunov function V defined by (34), which is positive definite on $\mathbb{R}^9$.

Substitution of the parameter update law (36) into (35) yields

$$\dot{V} = -k_1x_1^2 - k_2x_2^2 - k_3x_3^2 - k_4x_4^2, \tag{37}$$

which is a negative semi-definite function on $\mathbb{R}^9$.

Therefore, it can be concluded that the state vector $x(t)$ and the parameter estimation error are globally bounded, *i.e.*

$$\left[x_1(t)\ x_2(t)\ x_3(t)\ x_4(t)\ e_a(t)\ e_b(t)\ e_c(t)\ e_d(t)\ e_p(t)\right]^T \in \mathbf{L}_\infty. \tag{38}$$

Define

$$k = \min\left\{k_1, k_2, k_3, k_4\right\} \tag{39}$$

Then it follows from (37) that

$$\dot{V} \leq -k\|\mathbf{x}\|^2 \ \text{ or } \ k\|\mathbf{x}\|^2 \leq -\dot{V} \tag{40}$$

Integrating the inequality (40) from 0 to t, we get

$$k\int_0^t \|\mathbf{x}(\tau)\|^2\, d\tau \leq -\int_0^t \dot{V}(\tau)\, d\tau = V(0) - V(t) \tag{41}$$

From (41), it follows that $\mathbf{x}(t) \in \mathbf{L}_2$.
Using (32), it can be deduced that $\dot{x}(t) \in \mathbf{L}_\infty$.
Hence, using Barbalat's lemma [36], we can conclude that $\mathbf{x}(t) \to 0$ exponentially as $t \to \infty$ for all initial conditions $\mathbf{x}(0) \in \mathbb{R}^4$.
This completes the proof. □

For numerical simulations, the parameter values of the novel system (28) are taken as in the hyperchaotic case (7), viz.

$$a = 10,\ \ b = 40,\ \ c = 3,\ \ d = 18,\ \ p = 0.1 \tag{42}$$

The gain constants are taken as $k_i = 6, (i = 1, 2, 3, 4)$.
The initial values of the parameter estimates are taken as

$$\hat{a}(0) = 7.2,\ \ \hat{b}(0) = 13.7,\ \ \hat{c}(0) = 18.5,\ \ \hat{d}(0) = 5.4,\ \ \hat{p}(0) = 16.2 \tag{43}$$

The initial values of the novel hyperchaotic system (28) are taken as

$$x_1(0) = 2.3,\ \ x_2(0) = 4.6,\ \ x_3(0) = 6.5,\ \ x_4(0) = 12.8 \tag{44}$$

Figure 7 shows the time-history of the controlled states $x_1(t), x_2(t), x_3(t), x_4(t)$.
Figure 7 depicts the exponential convergence of the controlled states and the efficiency of the adaptive controller defined by (29).

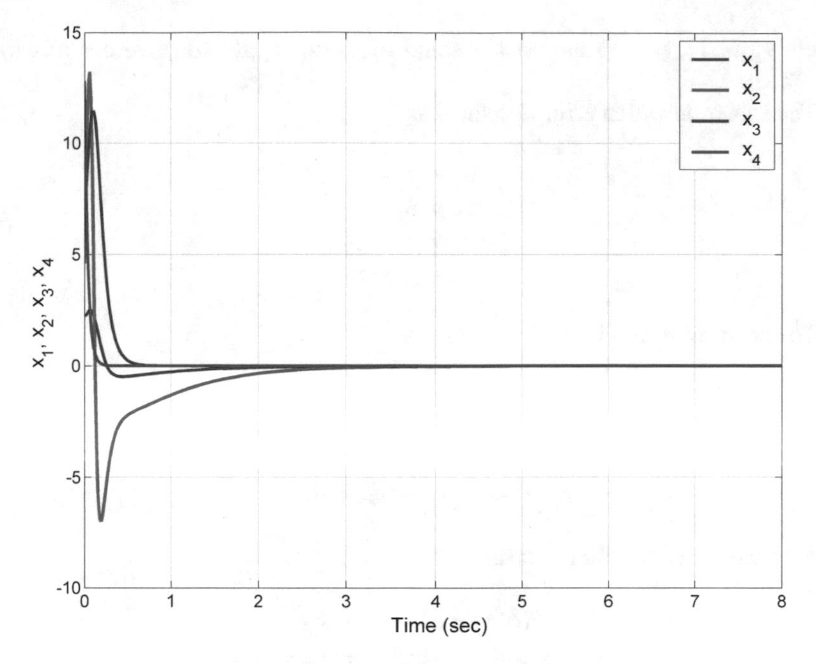

Fig. 7 Time-history of the states $x_1(t), x_2(t), x_3(t), x_4(t)$

5 Adaptive Synchronization of the Identical 4-D Novel Hyperchaotic Systems

This section derives new results for the adaptive synchronization of the identical novel hyperchaotic systems with unknown parameters.

The master system is given by the novel hyperchaotic system

$$\begin{aligned} \dot{x}_1 &= a(x_2 - x_1) + x_4 \\ \dot{x}_2 &= bx_1 + dx_1x_3 \\ \dot{x}_3 &= -cx_3 - x_1x_2 \\ \dot{x}_4 &= -x_1 - x_2 + px_4 \end{aligned} \tag{45}$$

where x_1, x_2, x_3, x_4 are state variables and a, b, c, d, p are constant, unknown, parameters of the system.

The slave system is given by the controlled novel hyperchaotic system

$$\begin{aligned} \dot{y}_1 &= a(y_2 - y_1) + y_4 + u_1 \\ \dot{y}_2 &= by_1 + dy_1y_3 + u_2 \\ \dot{y}_3 &= -cy_3 - y_1y_2 + u_3 \\ \dot{y}_4 &= -y_1 - y_2 + py_4 + u_4 \end{aligned} \tag{46}$$

where y_1, y_2, y_3, y_4 are state variables and u_1, u_2, u_3, u_4 are adaptive controls to be designed.

The synchronization error is defined as

$$\begin{aligned} e_1 &= y_1 - x_1 \\ e_2 &= y_2 - x_2 \\ e_3 &= y_3 - x_3 \\ e_4 &= y_4 - x_4 \end{aligned} \tag{47}$$

The error dynamics is easily obtained as

$$\begin{aligned} \dot{e}_1 &= a(e_2 - e_1) + e_4 + u_1 \\ \dot{e}_2 &= be_1 + d(y_1 y_3 - x_1 x_3) + u_2 \\ \dot{e}_3 &= -ce_3 - y_1 y_2 + x_1 x_2 + u_3 \\ \dot{e}_4 &= -e_1 - e_2 + pe_4 + u_4 \end{aligned} \tag{48}$$

An adaptive control law is taken as

$$\begin{aligned} u_1 &= -\hat{a}(t)(e_2 - e_1) - e_4 - k_1 e_1 \\ u_2 &= -\hat{b}(t)e_1 - \hat{d}(t)(y_1 y_3 - x_1 x_3) - k_2 e_2 \\ u_3 &= \hat{c}(t)e_3 + y_1 y_2 - x_1 x_2 - k_3 e_3 \\ u_4 &= e_1 + e_2 - \hat{p}(t)e_4 - k_4 e_4 \end{aligned} \tag{49}$$

where $\hat{a}(t), \hat{b}(t), \hat{c}(t), \hat{d}(t), \hat{p}(t)$ are estimates for the unknown parameters a, b, c, d, p, respectively, and k_1, k_2, k_3, k_4 are positive gain constants.

The closed-loop control system is obtained by substituting (49) into (48) as

$$\begin{aligned} \dot{e}_1 &= [a - \hat{a}(t)](e_2 - e_1) - k_1 e_1 \\ \dot{e}_2 &= [b - \hat{b}(t)]e_1 + [d - \hat{d}(t)](y_1 y_3 - x_1 x_3) - k_2 e_2 \\ \dot{e}_3 &= -[c - \hat{c}(t)]e_3 - k_3 e_3 \\ \dot{e}_4 &= [p - \hat{p}(t)]e_4 - k_4 e_4 \end{aligned} \tag{50}$$

To simplify (50), we define the parameter estimation error as

$$\begin{aligned} e_a(t) &= a - \hat{a}(t) \\ e_b(t) &= b - \hat{b}(t) \\ e_c(t) &= c - \hat{c}(t) \\ e_d(t) &= d - \hat{d}(t) \\ e_p(t) &= p - \hat{p}(t) \end{aligned} \tag{51}$$

Using (51), the closed-loop system (50) can be simplified as

$$\begin{aligned} \dot{e}_1 &= e_a(e_2 - e_1) - k_1 e_1 \\ \dot{e}_2 &= e_b e_1 + e_d(y_1 y_3 - x_1 x_3) - k_2 e_2 \\ \dot{e}_3 &= -e_c e_3 - k_3 e_3 \\ \dot{e}_4 &= e_p e_4 - k_4 e_4 \end{aligned} \tag{52}$$

Differentiating the parameter estimation error (51) with respect to t, we get

$$\begin{aligned} \dot{e}_a &= -\dot{\hat{a}} \\ \dot{e}_b &= -\dot{\hat{b}} \\ \dot{e}_c &= -\dot{\hat{c}} \\ \dot{e}_d &= -\dot{\hat{d}} \\ \dot{e}_p &= -\dot{\hat{p}} \end{aligned} \tag{53}$$

Next, we find an update law for parameter estimates using Lyapunov stability theory.

Consider the quadratic Lyapunov function defined by

$$V(\mathbf{e}, e_a, e_b, e_c, e_d, e_p) = \frac{1}{2}\left(e_1^2 + e_2^2 + e_3^2 + e_4^2 + e_a^2 + e_b^2 + e_c^2 + e_d^2 + e_p^2\right), \tag{54}$$

which is positive definite on $\mathbb{R}^9$.

Differentiating V along the trajectories of (52) and (53), we get

$$\begin{aligned} \dot{V} = &-k_1 e_1^2 - k_2 e_2^2 - k_3 e_3^2 - k_4 e_4^2 + e_a\left[e_1(e_2 - e_1) - \dot{\hat{a}}\right] + e_b\left[e_1 e_2 - \dot{\hat{b}}\right] \\ &+ e_c\left[-e_3^2 - \dot{\hat{c}}\right] + e_d[e_2(y_1 y_3 - x_1 x_3) - \dot{\hat{d}}] + e_p\left[e_4^2 - \dot{\hat{p}}\right] \end{aligned} \tag{55}$$

In view of Eq. (55), an update law for the parameter estimates is taken as

$$\begin{aligned} \dot{\hat{a}} &= e_1(e_2 - e_1) \\ \dot{\hat{b}} &= e_1 e_2 \\ \dot{\hat{c}} &= -e_3^2 \\ \dot{\hat{d}} &= e_2(y_1 y_3 - x_1 x_3) \\ \dot{\hat{p}} &= e_4^2 \end{aligned} \tag{56}$$

Theorem 2 *The identical novel hyperchaotic systems (45) and (46) with unknown system parameters are globally and exponentially synchronized for all initial conditions* $x(0), y(0) \in \mathbb{R}^4$ *by the adaptive control law (49) and the parameter update law (56), where* $k_i, (i = 1, 2, 3, 4)$ *are positive constants.*

Proof The result is proved using Lyapunov stability theory [36].

We consider the quadratic Lyapunov function V defined by (54), which is positive definite on $\mathbb{R}^9$.

Substitution of the parameter update law (56) into (55) yields

$$\dot{V} = -k_1 e_1^2 - k_2 e_2^2 - k_3 e_3^2 - k_4 e_4^2, \tag{57}$$

which is a negative semi-definite function on $\mathbb{R}^9$.

Therefore, it can be concluded that the synchronization error vector $e(t)$ and the parameter estimation error are globally bounded, *i.e.*

$$\left[e_1(t)\ e_2(t)\ e_3(t)\ e_4(t)\ e_a(t)\ e_b(t)\ e_c(t)\ e_d(t)\ e_p(t) \right]^T \in \mathbf{L}_\infty. \tag{58}$$

Define

$$k = \min\left\{k_1, k_2, k_3, k_4\right\} \tag{59}$$

Then it follows from (57) that

$$\dot{V} \leq -k\|\boldsymbol{e}\|^2 \ \text{ or } \ k\|\mathbf{e}\|^2 \leq -\dot{V} \tag{60}$$

Integrating the inequality (60) from 0 to t, we get

$$k\int_0^t \|\mathbf{e}(\tau)\|^2\, d\tau \ \leq \ -\int_0^t \dot{V}(\tau)\, d\tau = V(0) - V(t) \tag{61}$$

From (61), it follows that $\mathbf{e}(t) \in \mathbf{L}_2$.

Using (52), it can be deduced that $\dot{\mathbf{e}}(t) \in \mathbf{L}_\infty$.

Hence, using Barbalat's lemma, we can conclude that $\mathbf{e}(t) \to 0$ exponentially as $t \to \infty$ for all initial conditions $\mathbf{e}(0) \in \mathbb{R}^4$.

This completes the proof. □

For numerical simulations, the parameter values of the novel systems (45) and (46) are taken as in the hyperchaotic case (7), viz. $a = 10$, $b = 40$, $c = 3$, $d = 18$ and $p = 0.1$.

The gain constants are taken as $k_i = 6$ for $i = 1, 2, 3, 4$.

The initial values of the parameter estimates are taken as

$$\hat{a}(0) = 6.2,\ \ \hat{b}(0) = 5.8,\ \ \hat{c}(0) = 18.2,\ \hat{d}(0) = 7.5,\ \hat{p}(0) = 12.4 \tag{62}$$

The initial values of the master system (45) are taken as

$$x_1(0) = 3.4,\ \ x_2(0) = 12.1,\ \ x_3(0) = 14.7,\ \ x_4(0) = 25.6 \tag{63}$$

The initial values of the slave system (46) are taken as

$$y_1(0) = 12.6,\ \ y_2(0) = 5.3,\ \ y_3(0) = 5.2,\ \ y_4(0) = -12.5 \tag{64}$$

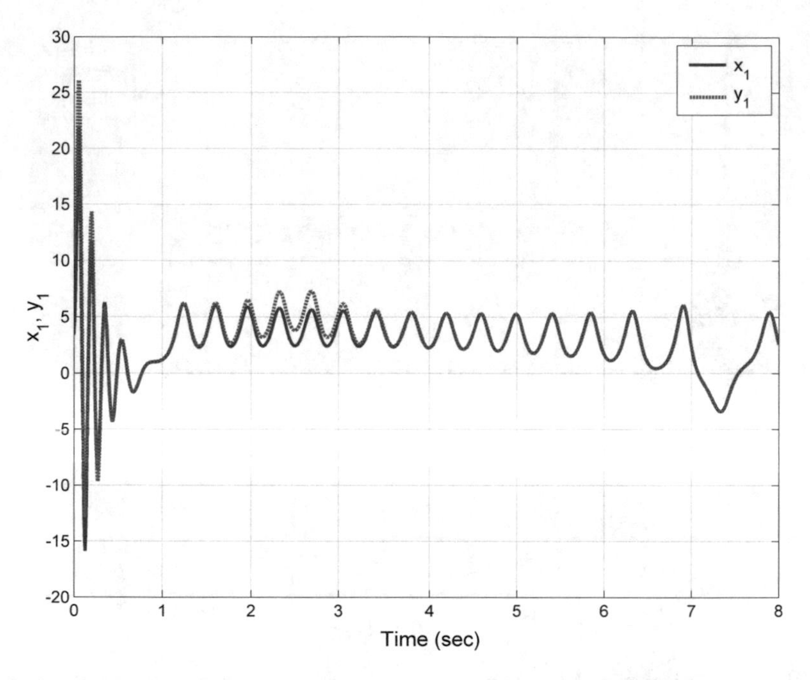

Fig. 8 Synchronization of the states x_1 and y_1

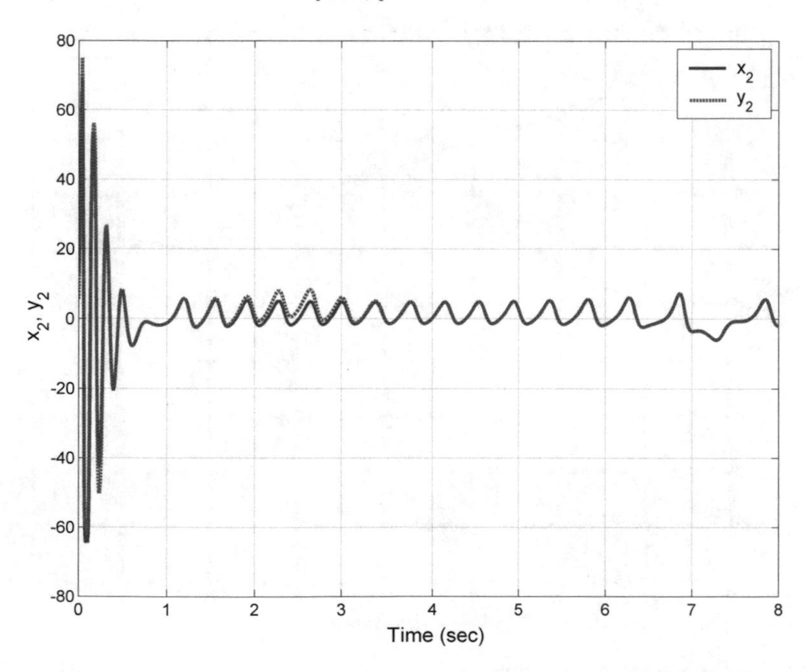

Fig. 9 Synchronization of the states x_2 and y_2

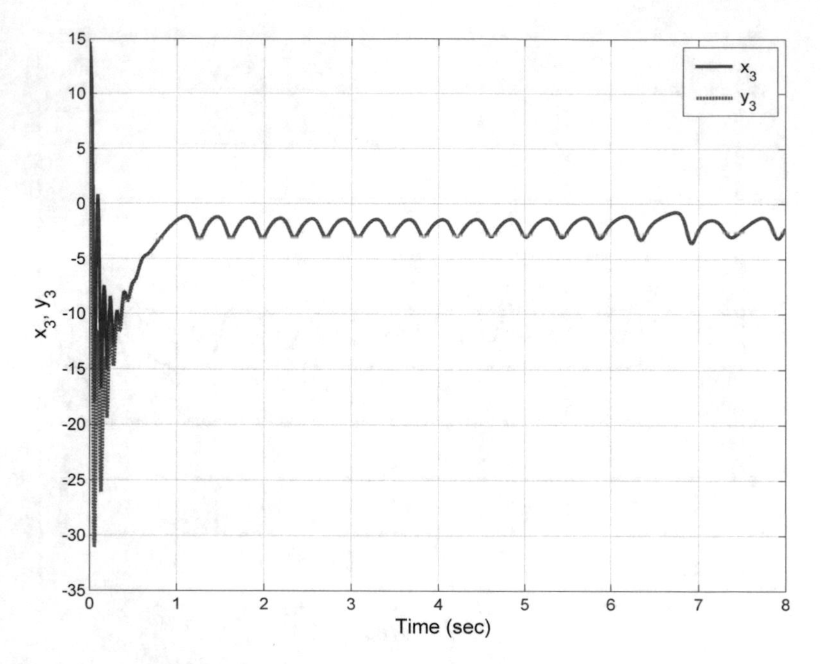

Fig. 10 Synchronization of the states x_3 and y_3

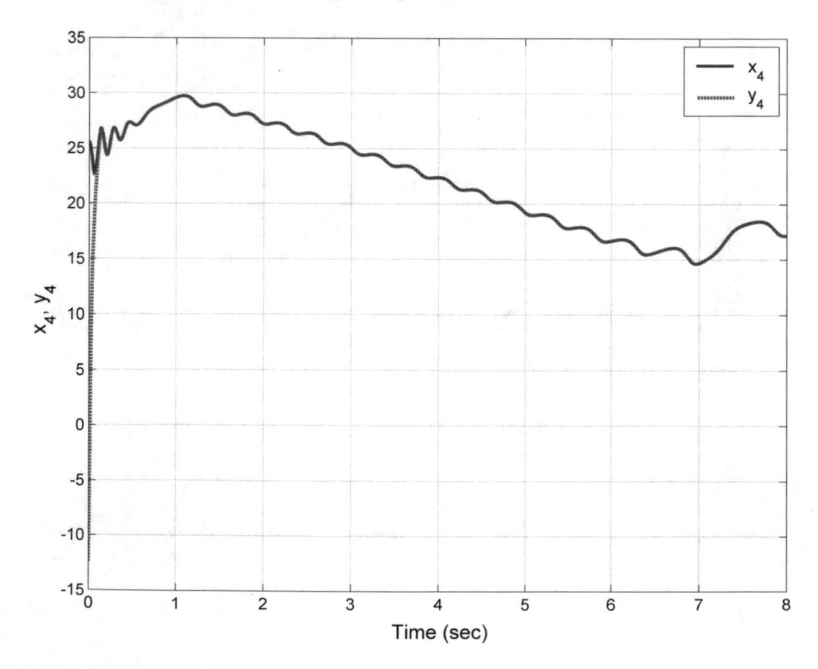

Fig. 11 Synchronization of the states x_4 and y_4

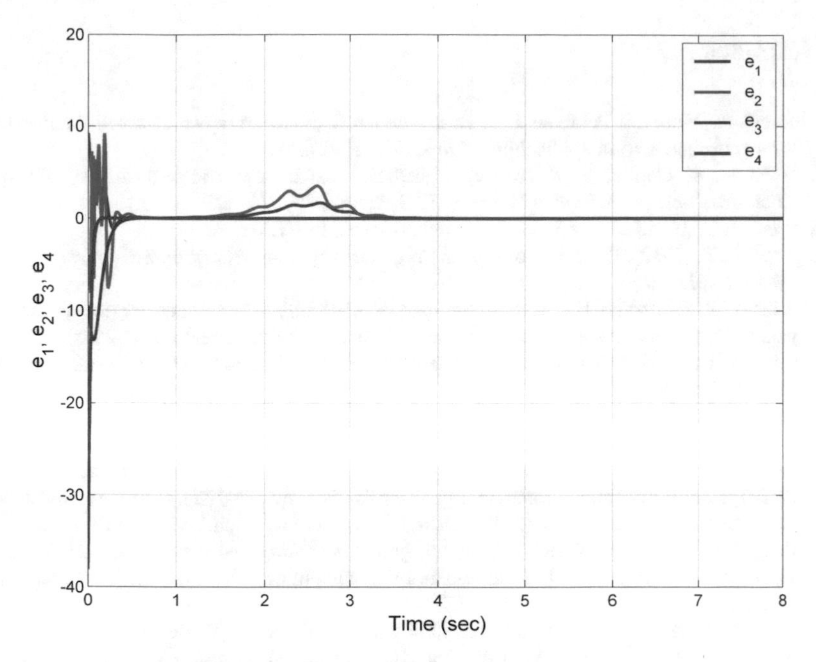

Fig. 12 Time-history of the synchronization errors e_1, e_2, e_3, e_4

Figures 8, 9, 10 and 11 show the complete synchronization of the identical chaotic systems (45) and (46). Figure 12 shows the time-history of the synchronization errors $e_1(t), e_2(t), e_3(t), e_4(t)$.

6 Conclusions

In this work, we derived a novel hyperchaotic system by introducing a feedback control to the Liu-Su-Liu chaotic system (2007). The qualitative properties of the novel hyperchaotic system were discussed. We showed that the novel hyperchaotic system has a unique equilibrium point at the origin, which is unstable. The Lyapunov exponents of the novel 4-D hyperchaotic system were obtained as $L_1 = 1.1097$, $L_2 = 0.1584$, $L_3 = 0$ and $L_4 = -14.1666$. Lyapunov dimension of the novel hyperchaotic system was derived as $D_L = 3.0895$. Next, we derived new results for the adaptive control design of the novel hyperchaotic system with unknown parameters. We also derived new results for the adaptive synchronization design of identical novel hyperchaotic systems with unknown parameters. We established all the main results of this work by using Lyapunov stability theory. Numerical simulations in MATLAB were shown to validate and illustrate all the main results derived in this work. As future work, fractional-order models of the novel hyperchaotic system may be investigated and new controllers may be designed for the adaptive control and synchronization of such systems.

References

1. Akgul, A., Moroz, I., Pehlivan, I., & Vaidyanathan, S. (2016). A new four-scroll chaotic attractor and its engineering applications. *Optik*, *127*, 5491–5499.
2. Arneodo, A., Coullet, P., & Tresser, C. (1981). Possible new strange attractors with spiral structure. *Communications in Mathematical Physics*, *79*, 573–579.
3. Azar, A. T. (2010). *Fuzzy Systems*. Vienna, Austria: IN-TECH.
4. Azar, A. T. (2012). Overview of type-2 fuzzy logic systems. *International Journal of Fuzzy System Applications*, *2*(4), 1–28.
5. Azar, A. T., & Serrano, F. E. (2014). Robust IMC-PID tuning for cascade control systems with gain and phase margin specifications. *Neural Computing and Applications*, *25*(5), 983–995.
6. Azar, A. T., & Serrano, F. E. (2015). Adaptive sliding mode control of the Furuta pendulum. In A. T. Azar & Q. Zhu (Eds.), *Advances and applications in sliding mode control systems*. Studies in computational intelligence (Vol. 576, pp. 1–42). Germany: Springer.
7. Azar, A. T., & Serrano, F. E. (2015). Deadbeat control for multivariable systems with time varying delays. In A. T. Azar & S. Vaidyanathan (Eds.), *Chaos modeling and control systems design*. Studies in computational intelligence (Vol. 581, pp. 97–132). Germany: Springer.
8. Azar, A. T., & Serrano, F. E. (2015). Design and modeling of anti wind up PID controllers. In Q. Zhu & A. T. Azar (Eds.), *Complex system modelling and control through intelligent soft computations*. Studies in fuzziness and soft computing (Vol. 319, pp. 1–44). Germany: Springer.
9. Azar, A. T., & Serrano, F. E. (2015). Stabilizatoin and control of mechanical systems with backlash. In A. T. Azar & S. Vaidyanathan (Eds.), *Handbook of research on advanced intelligent control engineering and automation*. Advances in computational intelligence and robotics (ACIR) (pp. 1–60). USA: IGI-Global.
10. Azar, A. T., & Vaidyanathan, S. (2015). *Chaos modeling and control systems design*. Studies in computational intelligence (Vol. 581). Germany: Springer.
11. Azar, A. T., & Vaidyanathan, S. (2015). *Computational intelligence applications in modeling and control*. Studies in computational intelligence (Vol. 575). Germany: Springer.
12. Azar, A. T., & Vaidyanathan, S. (2015) Handbook of research on advanced intelligent control engineering and automation. *Advances in Computational Intelligence and Robotics (ACIR)*. USA: IGI-Global.
13. Azar, A. T., & Vaidyanathan, S. (2016). *Advances in chaos theory and intelligent control*. Studies in fuzziness and soft computing (Vol. 337). Germany: Springer.
14. Azar, A. T., & Zhu, Q. (2015). *Advances and applications in sliding mode control systems*. Studies in computational intelligence (Vol. 576). Germany: Springer.
15. Barrow-Green, J. (1997). *Poincaré and the three body problem*. American Mathematical Society.
16. Boulkroune, A., Bouzeriba, A., Bouden, T., & Azar, A. T. (2016). Fuzzy adaptive synchronization of uncertain fractional-order chaotic systems. In A. T. Azar & S. Vaidyanathan (Eds.), *Advances in chaos theory and intelligent control*. Studies in fuzziness and soft computing (Vol. 337, pp. 681–697). Germany: Springer.
17. Boulkroune, A., Hamel, S., Azar, A. T., & Vaidyanathan, S. (2016). Fuzzy control-based function synchronization of unknown chaotic systems with dead-zone input. In A. T. Azar & S. Vaidyanathan (Eds.), *Advances in chaos theory and intelligent control*. Studies in fuzziness and soft computing (Vol. 337, pp. 699–718). Germany: Springer.
18. Carroll, T. L., & Pecora, L. M. (1991). Synchronizing chaotic circuits. *IEEE Transactions on Circuits and Systems*, *38*(4), 453–456.
19. Chen, A., Lu, J., Lü, J., & Yu, S. (2006). Generating hyperchaotic Lü attractor via state feedback control. *Physica A*, *364*, 103–110.
20. Chen, G., & Ueta, T. (1999). Yet another chaotic attractor. *International Journal of Bifurcation and Chaos*, *9*(7), 1465–1466.

21. Chen, W. H., Wei, D., & Lu, X. (2014). Global exponential synchronization of nonlinear time-delay Lur'e systems via delayed impulsive control. *Communications in Nonlinear Science and Numerical Simulation*, *19*(9), 3298–3312.
22. Chen, Z., Yang, Y., Qi, G., & Yuan, Z. (2007). A novel hyperchaos system only with one equilibrium. *Physics Letters A*, *360*, 696–701.
23. Dadras, S., & Momeni, H. R. (2009). A novel three-dimensional autonomous chaotic system generating two, three and four-scroll attractors. *Physics Letters A*, *373*, 3637–3642.
24. Das, S., Goswami, D., Chatterjee, S., & Mukherjee, S. (2014). Stability and chaos analysis of a novel swarm dynamics with applications to multi-agent systems. *Engineering Applications of Artificial Intelligence*, *30*, 189–198.
25. Fang, J., Deng, W., Wu, Y., & Ding, G. (2014). A novel hyperchaotic system and its circuit implementation. *Optik*, *125*(20), 6305–6311.
26. Feki, M. (2003). An adaptive chaos synchronization scheme applied to secure communication. *Chaos, Solitons and Fractals*, *18*(1), 141–148.
27. Gan, Q., & Liang, Y. (2012). Synchronization of chaotic neural networks with time delay in the leakage term and parametric uncertainties based on sampled-data control. *Journal of the Franklin Institute*, *349*(6), 1955–1971.
28. Genesio, R., & Tesi, A. (1992). Harmonic balance methods for the analysis of chaotic dynamics in nonlinear systems. *Automatica*, *28*(3), 531–548.
29. Gibson, W. T., & Wilson, W. G. (2013). Individual-based chaos: Extensions of the discrete logistic model. *Journal of Theoretical Biology*, *339*, 84–92.
30. Henon, M., & Heiles, C. (1964). The applicability of the third integral of motion: Some numerical experiments. *The Astrophysical Journal*, *69*, 73–79.
31. Huang, X., Zhao, Z., Wang, Z., & Li, Y. (2012). Chaos and hyperchaos in fractional-order cellular neural networks. *Neurocomputing*, *94*, 13–21.
32. Jia, Q. (2007). Hyperchaos generated from the Lorenz chaotic system and its control. *Physics Letters A*, *366*, 217–222.
33. Jiang, G. P., Zheng, W. X., & Chen, G. (2004). Global chaos synchronization with channel time-delay. *Chaos, Solitons and Fractals*, *20*(2), 267–275.
34. Karthikeyan, R., & Sundarapandian, V. (2014). Hybrid chaos synchronization of four-scroll systems via active control. *Journal of Electrical Engineering*, *65*(2), 97–103.
35. Kaslik, E., & Sivasundaram, S. (2012). Nonlinear dynamics and chaos in fractional-order neural networks. *Neural Networks*, *32*, 245–256.
36. Khalil, H. K. (2001). *Nonlinear systems*. New Jersey, USA: Prentice Hall.
37. Kyriazis, M. (1991). Applications of chaos theory to the molecular biology of aging. *Experimental Gerontology*, *26*(6), 569–572.
38. Lang, J. (2015). Color image encryption based on color blend and chaos permutation in the reality-preserving multiple-parameter fractional Fourier transform domain. *Optics Communications*, *338*, 181–192.
39. Li, D. (2008). A three-scroll chaotic attractor. *Physics Letters A*, *372*(4), 387–393.
40. Li, N., Zhang, Y., & Nie, Z. (2011). Synchronization for general complex dynamical networks with sampled-data. *Neurocomputing*, *74*(5), 805–811.
41. Li, N., Pan, W., Yan, L., Luo, B., & Zou, X. (2014). Enhanced chaos synchronization and communication in cascade-coupled semiconductor ring lasers. *Communications in Nonlinear Science and Numerical Simulation*, *19*(6), 1874–1883.
42. Li, Z., & Chen, G. (2006). *Integration of fuzzy logic and chaos theory*. Studies in fuzziness and soft computing (Vol. 187). Germany: Springer.
43. Lian, S., & Chen, X. (2011). Traceable content protection based on chaos and neural networks. *Applied Soft Computing*, *11*(7), 4293–4301.
44. Liu, L., Su, Y. C., & Liu, C. X. (2007). Experimental confirmation of a new reversed butterfly-shaped attractor. *Chinese Physics B*, *16*, 1897–1900.
45. Lorenz, E. N. (1963). Deterministic periodic flow. *Journal of the Atmospheric Sciences*, *20*(2), 130–141.

46. Lü, J., & Chen, G. (2002). A new chaotic attractor coined. *International Journal of Bifurcation and Chaos*, *12*(3), 659–661.
47. Mondal, S., & Mahanta, C. (2014). Adaptive second order terminal sliding mode controller for robotic manipulators. *Journal of the Franklin Institute*, *351*(4), 2356–2377.
48. Murali, K., & Lakshmanan, M. (1998). Secure communication using a compound signal from generalized chaotic systems. *Physics Letters A*, *241*(6), 303–310.
49. Nehmzow, U., & Walker, K. (2005). Quantitative description of robot-environment interaction using chaos theory. *Robotics and Autonomous Systems*, *53*(3–4), 177–193.
50. Pandey, A., Baghel, R. K., & Singh, R. P. (2012). Synchronization analysis of a new autonomous chaotic system with its application in signal masking. *IOSR Journal of Electronics and Communication Engineering*, *1*(5), 16–22.
51. Pecora, L. M., & Carroll, T. L. (1990). Synchronization in chaotic systems. *Physical Review Letters*, *64*(8), 821–824.
52. Pehlivan, I., Moroz, I. M., & Vaidyanathan, S. (2014). Analysis, synchronization and circuit design of a novel butterfly attractor. *Journal of Sound and Vibration*, *333*(20), 5077–5096.
53. Pham, V. T., Vaidyanathan, S., Volos, C. K., & Jafari, S. (2015). Hidden attractors in a chaotic system with an exponential nonlinear term. *European Physical Journal—Special Topics, 224*(8), 1507–1517.
54. Pham, V. T., Volos, C. K., Vaidyanathan, S., Le, T. P., & Vu, V. Y. (2015). A memristor-based hyperchaotic system with hidden attractors: Dynamics, synchronization and circuital emulating. *Journal of Engineering Science and Technology Review*, *8*(2), 205–214.
55. Pham, V. T., Jafari, S., Vaidyanathan, S., Volos, C., & Wang, X. (2016). A novel memristive neural network with hidden attractors and its circuitry implementation. *Science China Technological Sciences, 59*(3), 358–363.
56. Pham, V. T., Vaidyanathan, S., Volos, C., Jafari, S., & Kingni, S. T. (2016). A no-equilibrium hyperchaotic system with a cubic nonlinear term. *Optik*, *127*(6), 3259–3265.
57. Pham, V. T., Vaidyanathan, S., Volos, C. K., Jafari, S., Kuznetsov, N. V., & Hoang, T. M. (2016). A novel memristive time-delay chaotic system without equilibrium points. *European Physical Journal: Special Topics, 225*(1), 127–136.
58. Qi, G., & Chen, G. (2006). Analysis and circuit implementation of a new 4D chaotic system. *Physics Letters A*, *352*, 386–397.
59. Qu, Z. (2011). Chaos in the genesis and maintenance of cardiac arrhythmias. *Progress in Biophysics and Molecular Biology*, *105*(3), 247–257.
60. Rasappan, S., & Vaidyanathan, S. (2012). Global chaos synchronization of WINDMI and Coullet chaotic systems by backstepping control. *Far East Journal of Mathematical Sciences, 67*(2), 265–287.
61. Rasappan, S., & Vaidyanathan, S. (2012). Hybrid synchronization of n-scroll Chua and Lur'e chaotic systems via backstepping control with novel feedback. *Archives of Control Sciences, 22*(3), 343–365.
62. Rasappan, S., & Vaidyanathan, S. (2012). Synchronization of hyperchaotic Liu system via backstepping control with recursive feedback. *Communications in Computer and Information Science, 305*, 212–221.
63. Rasappan, S., & Vaidyanathan, S. (2013). Hybrid synchronization of *n*-scroll chaotic Chua circuits using adaptive backstepping control design with recursive feedback. *Malaysian Journal of Mathematical Sciences*, *7*(2), 219–246.
64. Rasappan, S., & Vaidyanathan, S. (2014). Global chaos synchronization of WINDMI and Coullet chaotic systems using adaptive backstepping control design. *Kyungpook Mathematical Journal*, *54*(1), 293–320.
65. Rhouma, R., & Belghith, S. (2011). Cryptoanalysis of a chaos based cryptosystem on DSP. *Communications in Nonlinear Science and Numerical Simulation*, *16*(2), 876–884.
66. Rikitake, T. (1958). Oscillations of a system of disk dynamos. *Mathematical Proceedings of the Cambridge Philosophical Society*, *54*(1), 89–105.
67. Rössler, O. E. (1976). An equation for continuous chaos. *Physics Letters A*, *57*(5), 397–398.

68. Sampath, S., Vaidyanathan, S., Volos, C. K., & Pham, V. T. (2015). An eight-term novel four-scroll chaotic system with cubic nonlinearity and its circuit simulation. *Journal of Engineering Science and Technology Review*, *8*(2), 1–6.
69. Sarasu, P., & Sundarapandian, V. (2011). Active controller design for the generalized projective synchronization of four-scroll chaotic systems. *International Journal of Systems Signal Control and Engineering Application, 4*(2), 26–33.
70. Sarasu, P., & Sundarapandian, V. (2011). The generalized projective synchronization of hyperchaotic Lorenz and hyperchaotic Qi systems via active control. *International Journal of Soft Computing*, *6*(5), 216–223.
71. Sarasu, P., & Sundarapandian, V. (2012). Adaptive controller design for the generalized projective synchronization of 4-scroll systems. *International Journal of Systems Signal Control and Engineering Application, 5*(2), 21–30.
72. Sarasu, P., & Sundarapandian, V. (2012). Generalized projective synchronization of three-scroll chaotic systems via adaptive control. *European Journal of Scientific Research, 72*(4), 504–522.
73. Sarasu, P., & Sundarapandian, V. (2012). Generalized projective synchronization of two-scroll systems via adaptive control. *International Journal of Soft Computing, 7*(4), 146–156.
74. Shahverdiev, E. M., & Shore, K. A. (2009). Impact of modulated multiple optical feedback time delays on laser diode chaos synchronization. *Optics Communications*, *282*(17), 3568–2572.
75. Shimizu, T., & Morioka, N. (1980). On the bifurcation of a symmetric limit cycle to an asymmetric one in a simple model. *Physics Letters A*, *76*(3–4), 201–204.
76. Smaoui, N., Karouma, A., & Zribi, M. (2013). Adaptive synchronization of hyperchaotic Chen systems with application to secure communication. *International Journal of Innovative Computing, Information and Control*, *9*(3), 1127–1144.
77. Sprott, J. C. (1994). Some simple chaotic flows. *Physical Review E, 50*(2), 647–650.
78. Sprott, J. C. (2010). *Elegant chaos*. World Scientific.
79. Suérez, I. (1999). Mastering chaos in ecology. *Ecological Modelling*, *117*(2–3), 305–314.
80. Sundarapandian, V. (2010). Output regulation of the Lorenz attractor. *Far East Journal of Mathematical Sciences*, *42*(2), 289–299.
81. Sundarapandian, V. (2013). Analysis and anti-synchronization of a novel chaotic system via active and adaptive controllers. *Journal of Engineering Science and Technology Review*, *6*(4), 45–52.
82. Sundarapandian, V., & Karthikeyan, R. (2011). Anti-synchronization of hyperchaotic Lorenz and hyperchaotic Chen systems by adaptive control. *International Journal of Systems Signal Control and Engineering Application, 4*(2), 18–25.
83. Sundarapandian, V., & Karthikeyan, R. (2011). Anti-synchronization of Lü and Pan chaotic systems by adaptive nonlinear control. *European Journal of Scientific Research, 64*(1), 94–106.
84. Sundarapandian, V., & Karthikeyan, R. (2012). Adaptive anti-synchronization of uncertain Tigan and Li systems. *Journal of Engineering and Applied Sciences, 7*(1), 45–52.
85. Sundarapandian, V., & Karthikeyan, R. (2012). Hybrid synchronization of hyperchaotic Lorenz and hyperchaotic Chen systems via active control. *Journal of Engineering and Applied Sciences, 7*(3), 254–264.
86. Sundarapandian, V., & Pehlivan, I. (2012). Analysis, control, synchronization, and circuit design of a novel chaotic system. *Mathematical and Computer Modelling*, *55*(7–8), 1904–1915.
87. Sundarapandian, V., & Sivaperumal, S. (2011). Sliding controller design of hybrid synchronization of four-wing chaotic systems. *International Journal of Soft Computing*, *6*(5), 224–231.
88. Suresh, R., & Sundarapandian, V. (2013). Global chaos synchronization of a family of *n*-scroll hyperchaotic Chua circuits using backstepping control with recursive feedback. *Far East Journal of Mathematical Sciences*, *73*(1), 73–95.

89. Tacha, O. I., Volos, C. K., Kyprianidis, I. M., Stouboulos, I. N., Vaidyanathan, S., & Pham, V. T. (2016). Analysis, adaptive control and circuit simulation of a novel nonlinear finance system. *Applied Mathematics and Computation*, *276*, 200–217.
90. Usama, M., Khan, M. K., Alghatbar, K., & Lee, C. (2010). Chaos-based secure satellite imagery cryptosystem. *Computers and Mathematics with Applications*, *60*(2), 326–337.
91. Vaidyanathan, S. (2011). Hybrid chaos synchronization of Liu and Lu systems by active nonlinear control. *Communications in Computer and Information Science*, *204*, 1–10.
92. Vaidyanathan, S. (2012). Analysis and synchronization of the hyperchaotic Yujun systems via sliding mode control. *Advances in Intelligent Systems and Computing*, *176*, 329–337.
93. Vaidyanathan, S. (2012). Anti-synchronization of Sprott-L and Sprott-M chaotic systems via adaptive control. *International Journal of Control Theory and Applications*, *5*(1), 41–59.
94. Vaidyanathan, S. (2012). Global chaos control of hyperchaotic Liu system via sliding control method. *International Journal of Control Theory and Applications*, *5*(2), 117–123.
95. Vaidyanathan, S. (2012). Output regulation of the Liu chaotic system. *Applied Mechanics and Materials*, *110–116*, 3982–3989.
96. Vaidyanathan, S. (2012). Sliding mode control based global chaos control of Liu-Liu-Liu-Su chaotic system. *International Journal of Control Theory and Applications*, *5*(1), 15–20.
97. Vaidyanathan, S. (2013). A new six-term 3-D chaotic system with an exponential nonlinearity. *Far East Journal of Mathematical Sciences*, *79*(1), 135–143.
98. Vaidyanathan, S. (2013). Analysis and adaptive synchronization of two novel chaotic systems with hyperbolic sinusoidal and cosinusoidal nonlinearity and unknown parameters. *Journal of Engineering Science and Technology Review*, *6*(4), 53–65.
99. Vaidyanathan, S. (2013). Analysis, control and synchronization of hyperchaotic Zhou system via adaptive control. *Advances in Intelligent Systems and Computing*, *177*, 1–10.
100. Vaidyanathan, S. (2014). A new eight-term 3-D polynomial chaotic system with three quadratic nonlinearities. *Far East Journal of Mathematical Sciences*, *84*(2), 219–226.
101. Vaidyanathan, S. (2014). Analysis and adaptive synchronization of eight-term 3-D polynomial chaotic systems with three quadratic nonlinearities. *European Physical Journal: Special Topics*, *223*(8), 1519–1529.
102. Vaidyanathan, S. (2014). Analysis, control and synchronisation of a six-term novel chaotic system with three quadratic nonlinearities. *International Journal of Modelling, Identification and Control*, *22*(1), 41–53.
103. Vaidyanathan, S. (2014). Generalized projective synchronisation of novel 3-D chaotic systems with an exponential non-linearity via active and adaptive control. *International Journal of Modelling, Identification and Control*, *22*(3), 207–217.
104. Vaidyanathan, S. (2014). Global chaos synchronization of identical Li-Wu chaotic systems via sliding mode control. *International Journal of Modelling, Identification and Control*, *22*(2), 170–177.
105. Vaidyanathan, S. (2015). 3-cells cellular neural network (CNN) attractor and its adaptive biological control. *International Journal of PharmTech Research*, *8*(4), 632–640.
106. Vaidyanathan, S. (2015). A 3-D novel highly chaotic system with four quadratic nonlinearities, its adaptive control and anti-synchronization with unknown parameters. *Journal of Engineering Science and Technology Review*, *8*(2), 106–115.
107. Vaidyanathan, S. (2015). A novel chemical chaotic reactor system and its adaptive control. *International Journal of ChemTech Research*, *8*(7), 146–158.
108. Vaidyanathan, S. (2015). A novel chemical chaotic reactor system and its output regulation via integral sliding mode control. *International Journal of ChemTech Research*, *8*(11), 669–683.
109. Vaidyanathan, S. (2015). Adaptive backstepping control of enzymes-substrates system with ferroelectric behaviour in brain waves. *International Journal of PharmTech Research*, *8*(2), 256–261.
110. Vaidyanathan, S. (2015). Adaptive biological control of generalized Lotka-Volterra three-species biological system. *International Journal of PharmTech Research*, *8*(4), 622–631.
111. Vaidyanathan, S. (2015). Adaptive chaotic synchronization of enzymes-substrates system with ferroelectric behaviour in brain waves. *International Journal of PharmTech Research*, *8*(5), 964–973.

112. Vaidyanathan, S. (2015). Adaptive control design for the anti-synchronization of novel 3-D chemical chaotic reactor systems. *International Journal of ChemTech Research, 8*(11), 654–668.
113. Vaidyanathan, S. (2015). Adaptive control of a chemical chaotic reactor. *International Journal of PharmTech Research, 8*(3), 377–382.
114. Vaidyanathan, S. (2015). Adaptive synchronization of chemical chaotic reactors. *International Journal of ChemTech Research, 8*(2), 612–621.
115. Vaidyanathan, S. (2015). Adaptive synchronization of generalized Lotka-Volterra three-species biological systems. *International Journal of PharmTech Research, 8*(5), 928–937.
116. Vaidyanathan, S. (2015). Adaptive synchronization of novel 3-D chemical chaotic reactor systems. *International Journal of ChemTech Research, 8*(7), 159–171.
117. Vaidyanathan, S. (2015). Analysis, properties and control of an eight-term 3-D chaotic system with an exponential nonlinearity. *International Journal of Modelling, Identification and Control, 23*(2), 164–172.
118. Vaidyanathan, S. (2015). Anti-synchronization of Brusselator chemical reaction systems via adaptive control. *International Journal of ChemTech Research, 8*(6), 759–768.
119. Vaidyanathan, S. (2015). Anti-synchronization of chemical chaotic reactors via adaptive control method. *International Journal of ChemTech Research, 8*(8), 73–85.
120. Vaidyanathan, S. (2015). Anti-synchronization of Mathieu-Van der Pol chaotic systems via adaptive control method. *International Journal of ChemTech Research, 8*(11), 638–653.
121. Vaidyanathan, S. (2015). Chaos in neurons and adaptive control of Birkhoff-Shaw strange chaotic attractor. *International Journal of PharmTech Research, 8*(5), 956–963.
122. Vaidyanathan, S. (2015). Dynamics and control of Brusselator chemical reaction. *International Journal of ChemTech Research, 8*(6), 740–749.
123. Vaidyanathan, S. (2015). Dynamics and control of Tokamak system with symmetric and magnetically confined plasma. *International Journal of ChemTech Research, 8*(6), 795–803.
124. Vaidyanathan, S. (2015). Global chaos control of Mathieu-Van der pol system via adaptive control method. *International Journal of ChemTech Research, 8*(9), 406–417.
125. Vaidyanathan, S. (2015). Global chaos synchronization of chemical chaotic reactors via novel sliding mode control method. *International Journal of ChemTech Research, 8*(7), 209–221.
126. Vaidyanathan, S. (2015). Global chaos synchronization of Duffing double-well chaotic oscillators via integral sliding mode control. *International Journal of ChemTech Research, 8*(11), 141–151.
127. Vaidyanathan, S. (2015). Global chaos synchronization of Mathieu-Van der Pol chaotic systems via adaptive control method. *International Journal of ChemTech Research, 8*(10), 148–162.
128. Vaidyanathan, S. (2015). Global chaos synchronization of novel coupled Van der Pol conservative chaotic systems via adaptive control method. *International Journal of ChemTech Research, 8*(8), 95–111.
129. Vaidyanathan, S. (2015). Global chaos synchronization of the forced Van der Pol chaotic oscillators via adaptive control method. *International Journal of PharmTech Research, 8*(6), 156–166.
130. Vaidyanathan, S. (2015). Hyperchaos, qualitative analysis, control and synchronisation of a ten-term 4-D hyperchaotic system with an exponential nonlinearity and three quadratic nonlinearities. *International Journal of Modelling, Identification and Control, 23*(4), 380–392.
131. Vaidyanathan, S. (2016). A novel 2-D chaotic enzymes-substrates reaction system and its adaptive backstepping control. In A. T. Azar & S. Vaidyanathan (Eds.), *Advances in chaos theory and intelligent control*. Studies in fuzziness and soft computing (Vol. 337, pp. 507–528). Germany: Springer.
132. Vaidyanathan, S. (2016). A novel 3-D conservative jerk chaotic system with two quadratic nonlinearities and its adaptive control. In A. T. Azar & S. Vaidyanathan (Eds.), *Advances in chaos theory and intelligent control*. Studies in fuzziness and soft computing (Vol. 337, pp. 349–376). Germany: Springer.

133. Vaidyanathan, S. (2016). A novel 3-D jerk chaotic system with three quadratic nonlinearities and its adaptive control. *Archives of Control Sciences, 26*(1), 19–47.
134. Vaidyanathan, S. (2016). A novel 4-D hyperchaotic thermal convection system and its adaptive control. In A. T. Azar & S. Vaidyanathan (Eds.), *Advances in chaos theory and intelligent control*. Studies in fuzziness and soft computing (Vol. 337, pp. 75–100). Germany: Springer.
135. Vaidyanathan, S. (2016). A novel double convecton system, its analysis, adaptive control and synchronization. In A. T. Azar & S. Vaidyanathan (Eds.), *Advances in chaos theory and intelligent control*. Studies in fuzziness and soft computing (Vol. 337, pp. 553–579). Germany: Springer.
136. Vaidyanathan, S. (2016). A seven-term novel 3-D jerk chaotic system with two quadratic nonlinearities and its adaptive backstepping control. In A. T. Azar & S. Vaidyanathan (Eds.), *Advances in chaos theory and intelligent control*. Studies in fuzziness and soft computing (Vol. 337, pp. 581–607). Germany: Springer.
137. Vaidyanathan, S. (2016). Analysis, adaptive control and synchronization of a novel 3-D chaotic system with a quadratic nonlinearity and two quadratic nonlinearities. In A. T. Azar & S. Vaidyanathan (Eds.), *Advances in chaos theory and intelligent control*. Studies in fuzziness and soft computing (Vol. 337, pp. 429–453). Germany: Springer.
138. Vaidyanathan, S. (2016). Analysis, control and synchronization of a novel 4-D highly hyperchaotic system with hidden attractors. In A. T. Azar & S. Vaidyanathan (Eds.), *Advances in chaos theory and intelligent control*. Studies in fuzziness and soft computing (Vol. 337, pp. 529–552). Germany: Springer.
139. Vaidyanathan, S. (2016). Anti-synchronization of Duffing double-well chaotic oscillators via integral sliding mode control. *International Journal of ChemTech Research, 9*(2), 297–304.
140. Vaidyanathan, S. (2016). Dynamic analysis, adaptive control and synchronization of a novel highly chaotic system with four quadratic nonlinearities. In A. T. Azar & S. Vaidyanathan (Eds.), *Advances in chaos theory and intelligent control* Studies in fuzziness and soft computing (Vol. 337, pp. 405–428). Germany: Springer.
141. Vaidyanathan, S. (2016). Global chaos synchronization of a novel 3-D chaotic system with two quadratic nonlinearities via active and adaptive control. In A. T. Azar & S. Vaidyanathan (Eds.), *Advances in chaos theory and intelligent control*. Studies in fuzziness and soft computing (Vol. 337, pp. 481–506). Germany: Springer.
142. Vaidyanathan, S. (2016). Qualitative analysis and properties of a novel 4-D hyperchaotic system with two quadratic nonlinearities and its adaptive control. In A. T. Azar & S. Vaidyanathan (Eds.), *Advances in chaos theory and intelligent control*. Studies in fuzziness and soft computing (Vol. 337, pp. 455–480). Germany: Springer.
143. Vaidyanathan, S., & Azar, A. T. (2015). Analysis and control of a 4-D novel hyperchaotic system. In A. T. Azar & S. Vaidyanathan (Eds.), *Chaos modeling and control systems design*. Studies in computational intelligence (Vol. 581, pp. 19–38). Germany: Springer.
144. Vaidyanathan, S., & Azar, A. T. (2015). Analysis, control and synchronization of a nine-term 3-D novel chaotic system. In A. T. Azar & S. Vaidyanathan (Eds.), *Chaos modelling and control systems design*. Studies in computational intelligence (Vol. 581, pp. 19–38). Germany: Springer.
145. Vaidyanathan, S., & Azar, A. T. (2015). Anti-synchronization of identical chaotic systems using sliding mode control and an application to Vaidhyanathan-Madhavan chaotic systems. *Studies in Computational Intelligence, 576*, 527–547.
146. Vaidyanathan, S., & Azar, A. T. (2015). Hybrid synchronization of identical chaotic systems using sliding mode control and an application to Vaidhyanathan chaotic systems. *Studies in Computational Intelligence, 576*, 549–569.
147. Vaidyanathan, S., & Azar, A. T. (2016). A novel 4-D four-wing chaotic system with four quadratic nonlinearities and its synchronization via adaptive control method. In A. T. Azar & S. Vaidyanathan (Eds.), *Advances in chaos theory and intelligent control*. Studies in fuzziness and soft computing (Vol. 337, pp. 203–224). Germany: Springer.
148. Vaidyanathan, S., & Azar, A. T. (2016). Adaptive backstepping control and synchronization of a novel 3-D jerk system with an exponential nonlinearity. In A. T. Azar & S. Vaidyanathan

(Eds.), *Advances in chaos theory and intelligent control*. Studies in fuzziness and soft computing (Vol. 337, pp. 249–274). Germany: Springer.
149. Vaidyanathan, S., & Azar, A. T. (2016). Adaptive control and synchronization of Halvorsen circulant chaotic systems. In A. T. Azar & S. Vaidyanathan (Eds.), *Advances in chaos theory and intelligent control*. Studies in fuzziness and soft computing (Vol. 337, pp. 225–247). Germany: Springer.
150. Vaidyanathan, S., & Azar, A. T. (2016). Dynamic analysis, adaptive feedback control and synchronization of an eight-term 3-D novel chaotic system with three quadratic nonlinearities. In A. T. Azar & S. Vaidyanathan (Eds.), *Advances in chaos theory and intelligent control*. Studies in fuzziness and soft computing (Vol. 337, pp. 155–178). Germany: Springer.
151. Vaidyanathan, S., & Azar, A. T. (2016). Generalized projective synchronization of a novel hyperchaotic four-wing system via adaptive control method. In A. T. Azar & S. Vaidyanathan (Eds.), *Advances in chaos theory and intelligent control*. Studies in fuzziness and soft computing (Vol. 337, pp. 275–296). Germany: Springer.
152. Vaidyanathan, S., & Azar, A. T. (2016). Qualitative study and adaptive control of a novel 4-D hyperchaotic system with three quadratic nonlinearities. In A. T. Azar & S. Vaidyanathan (Eds.), *Advances in chaos theory and intelligent control*. Studies in fuzziness and soft computing (Vol. 337, pp. 179–202). Germany: Springer.
153. Vaidyanathan, S., & Madhavan, K. (2013). Analysis, adaptive control and synchronization of a seven-term novel 3-D chaotic system. *International Journal of Control Theory and Applications*, *6*(2), 121–137.
154. Vaidyanathan, S., & Pakiriswamy, S. (2013). Generalized projective synchronization of six-term Sundarapandian chaotic systems by adaptive control. *International Journal of Control Theory and Applications*, *6*(2), 153–163.
155. Vaidyanathan, S., & Pakiriswamy, S. (2015). A 3-D novel conservative chaotic System and its generalized projective synchronization via adaptive control. *Journal of Engineering Science and Technology Review*, *8*(2), 52–60.
156. Vaidyanathan, S., & Rajagopal, K. (2011). Anti-synchronization of Li and T chaotic systems by active nonlinear control. *Communications in Computer and Information Science, 198*, 175–184.
157. Vaidyanathan, S., & Rajagopal, K. (2011). Global chaos synchronization of hyperchaotic Pang and Wang systems by active nonlinear control. *Communications in Computer and Information Science, 204*, 84–93.
158. Vaidyanathan, S., & Rajagopal, K. (2011). Global chaos synchronization of Lü and Pan systems by adaptive nonlinear control. *Communications in Computer and Information Science, 205*, 193–202.
159. Vaidyanathan, S., & Rajagopal, K. (2012). Global chaos synchronization of hyperchaotic Pang and hyperchaotic Wang systems via adaptive control. *International Journal of Soft Computing*, *7*(1), 28–37.
160. Vaidyanathan, S., & Rasappan, S. (2011). Global chaos synchronization of hyperchaotic Bao and Xu systems by active nonlinear control. *Communications in Computer and Information Science, 198*, 10–17.
161. Vaidyanathan, S., & Rasappan, S. (2014). Global chaos synchronization of *n*-scroll Chua circuit and Lur'e system using backstepping control design with recursive feedback. *Arabian Journal for Science and Engineering*, *39*(4), 3351–3364.
162. Vaidyanathan, S., & Sampath, S. (2011). Global chaos synchronization of hyperchaotic Lorenz systems by sliding mode control. *Communications in Computer and Information Science, 205*, 156–164.
163. Vaidyanathan, S., & Sampath, S. (2012). Anti-synchronization of four-wing chaotic systems via sliding mode control. *International Journal of Automation and Computing*, *9*(3), 274–279.
164. Vaidyanathan, S., & Volos, C. (2015). Analysis and adaptive control of a novel 3-D conservative no-equilibrium chaotic system. *Archives of Control Sciences*, *25*(3), 333–353.

165. Vaidyanathan, S., Volos, C., & Pham, V. T. (2014). Hyperchaos, adaptive control and synchronization of a novel 5-D hyperchaotic system with three positive Lyapunov exponents and its SPICE implementation. *Archives of Control Sciences, 24*(4), 409–446.
166. Vaidyanathan, S., Volos, C., Pham, V. T., Madhavan, K., & Idowu, B. A. (2014). Adaptive backstepping control, synchronization and circuit simulation of a 3-D novel jerk chaotic system with two hyperbolic sinusoidal nonlinearities. *Archives of Control Sciences, 24*(3), 375–403.
167. Vaidyanathan, S., Idowu, B. A., & Azar, A. T. (2015). Backstepping controller design for the global chaos synchronization of Sprott's jerk systems. *Studies in Computational Intelligence, 581*, 39–58.
168. Vaidyanathan, S., Rajagopal, K., Volos, C. K., Kyprianidis, I. M., & Stouboulos, I. N. (2015). Analysis, adaptive control and synchronization of a seven-term novel 3-D chaotic system with three quadratic nonlinearities and its digital implementation in LabVIEW. *Journal of Engineering Science and Technology Review, 8*(2), 130–141.
169. Vaidyanathan, S., Pham, V. T., & Volos, C. K. (2015). A 5-D hyperchaotic Rikitake dynamo system with hidden attractors. *European Physical Journal: Special Topics, 224*(8), 1575–1592.
170. Vaidyanathan, S., Volos, C., Pham, V. T., & Madhavan, K. (2015). Analysis, adaptive control and synchronization of a novel 4-D hyperchaotic hyperjerk system and its SPICE implementation. *Archives of Control Sciences, 25*(1), 5–28.
171. Vaidyanathan, S., Volos, C. K., Kyprianidis, I. M., Stouboulos, I. N., & Pham, V. T. (2015). Analysis, adaptive control and anti-synchronization of a six-term novel jerk chaotic system with two exponential nonlinearities and its circuit simulation. *Journal of Engineering Science and Technology Review, 8*(2), 24–36.
172. Vaidyanathan, S., Volos, C. K., & Pham, V. T. (2015). Analysis, adaptive control and adaptive synchronization of a nine-term novel 3-D chaotic system with four quadratic nonlinearities and its circuit simulation. *Journal of Engineering Science and Technology Review, 8*(2), 174–184.
173. Vaidyanathan, S., Volos, C. K., & Pham, V. T. (2015). Global chaos control of a novel nine-term chaotic system via sliding mode control. In A. T. Azar & Q. Zhu (Eds.), *Advances and applications in sliding mode control systems*. Studies in computational intelligence (Vol. 576, pp. 571–590). Germany: Springer.
174. Volos, C. K., Kyprianidis, I. M., & Stouboulos, I. N. (2013). Experimental investigation on coverage performance of a chaotic autonomous mobile robot. *Robotics and Autonomous Systems*, *61*(12), 1314–1322.
175. Volos, C. K., Kyprianidis, I. M., Stouboulos, I. N., Tlelo-Cuautle, E., & Vaidyanathan, S. (2015). Memristor: A new concept in synchronization of coupled neuromorphic circuits. *Journal of Engineering Science and Technology Review*, *8*(2), 157–173.
176. Wei, Z., & Yang, Q. (2010). Anti-control of Hopf bifurcation in the new chaotic system with two stable node-foci. *Applied Mathematics and Computation*, *217*(1), 422–429.
177. Witte, C. L., & Witte, M. H. (1991). Chaos and predicting varix hemorrhage. *Medical Hypotheses*, *36*(4), 312–317.
178. Xiao, X., Zhou, L., & Zhang, Z. (2014). Synchronization of chaotic Lur'e systems with quantized sampled-data controller. *Communications in Nonlinear Science and Numerical Simulation*, *19*(6), 2039–2047.
179. Yuan, G., Zhang, X., & Wang, Z. (2014). Generation and synchronization of feedback-induced chaos in semiconductor ring lasers by injection-locking. *Optik—International Journal for Light and Electron Optics*, *125*(8), 1950–1953.
180. Zaher, A. A., & Abu-Rezq, A. (2011). On the design of chaos-based secure communication systems. *Communications in Nonlinear Systems and Numerical Simulation*, *16*(9), 3721–3727.
181. Zhang, H., & Zhou, J. (2012). Synchronization of sampled-data coupled harmonic oscillators with control inputs missing. *Systems and Control Letters*, *61*(12), 1277–1285.

182. Zhang, X., Zhao, Z., & Wang, J. (2014). Chaotic image encryption based on circular substitution box and key stream buffer. *Signal Processing: Image Communication*, *29*(8), 902–913.
183. Zhou, W., Xu, Y., Lu, H., & Pan, L. (2008). On dynamics analysis of a new chaotic attractor. *Physics Letters A*, *372*(36), 5773–5777.
184. Zhu, C., Liu, Y., & Guo, Y. (2010). Theoretic and numerical study of a new chaotic system. *Intelligent Information Management*, *2*, 104–109.
185. Zhu, Q., & Azar, A. T. (2015). *Complex system modelling and control through intelligent soft computations*. Studies in fuzziness and soft computing (Vol. 319). Germany: Springer.

Chaotic Planning Paths Generators by Using Performance Surfaces

C.H. Pimentel-Romero, J.M. Munoz-Pacheco, O. Felix-Beltran, L.C. Gomez-Pavon and Ch. K. Volos

Abstract Chaotic systems have been widely used as path planning generators in autonomous mobile robots due to the unpredictability of the generated trajectories and the coverage rate of the robots workplace. In order to obtain a chaotic mobile robot, the chaotic signals are used to generate True RNGs (TRNGs), which, as is known, exploit the nondeterministic nature of chaotic controllers. Then, the bits obtained from TRNGs can be continuously mapped to coordinates (x_n, y_n) for positioning the robot on the terrain. A frequent technique to obtain a chaotic bitstream is to sample analog chaotic signals by using thresholds. However, the performance of chaotic path planning is a function of optimal values for those levels. In this framework, several chaotic systems which are used to obtain TRNGs but by computing a quasi-optimal performance surface for the thresholds is presented. The proposed study is based on sweeping the Poincaré sections to find quasi-optimal values for thresholds where the coverage rate is higher than those obtained by using the equilibrium points as reference values. Various scenarios are evaluated. First, two scroll chaotic systems such as Chua's circuit, saturated function, and Lorenz are used as entropy sources to obtain TRNGS by using its computed performance surface. Afterwards, n-scrolls chaotic systems are evaluated to get chaotic bitstreams with the analyzed performance surface. Another scenario is dedicated to find the performance surface of hybrid chaotic systems, which are composed by three chaotic systems where one chaotic system determines which one of the remaining chaotic signals will be used to obtain the chaotic bitstream. Additionally, TRNGs from two chaotic systems with optimized Lyapunov exponents are studied. Several numerical simulations to compute diverse metrics such as coverage rate against planned points, robot's trajectory evolution, covered terrain, and color map are carried out to analyze the

C.H. Pimentel-Romero · J.M. Munoz-Pacheco (✉) · O. Felix-Beltran ·
L.C. Gomez-Pavon
Fac. de Cs. de la Electronica, Benemerita Universidad Autonoma de Puebla,
Apdo. Postal 542, 72000 Puebla, Pue, Mexico
e-mail: jesusm.pacheco@correo.buap.mx

Ch.K. Volos
Department of Physics, Aristotle University of Thessaloniki,
54124 Thessaloniki, Greece

A.T. Azar et al. (eds.), *Fractional Order Control and Synchronization of Chaotic Systems*, Studies in Computational Intelligence 688,
DOI 10.1007/978-3-319-50249-6_28

resulting TRNGs. This investigation will enable to increase several applications of TRNGs by considering the proposed performance surfaces.

Keywords Chaos · Planning paths · Poincaré map · Robots · *n*-scroll

1 Introduction

First attempt to describe the physical reality in a quantitative way back to the Pythagoreans with their effort to explain the tangible world through integers. From a conceptual point of view, the main Galileo and Newton's legacy is the idea that nature obeys immutable laws that can be formulated in mathematical language, physical events can be predicted with certainty (determinism).

Ironically, the first clear example of what we now know as chaos was found in celestial mechanics. Science of regular and predictable phenomena. Taken into a count the law of gravity, positions and initial velocities of three bodies that interact gravitationally, for example, Sun-Earth-Moon, the equations of mechanics determine the positions and higher speeds. However, despite the deterministic nature of the system, H. Poincaré found that the evolution of these celestial bodies can be chaotic in nature, which means that small disturbances in initial state, as a slight change in the initial position of one body, could lead to dramatic differences in the later stages of the system [10].

The deep implication of these results involves determinism and predictability are different problems. However, Poincaré results not received due attention for a long time. There are probably two main reasons for this delay. First, in the early twentieth century, scientists and philosophers lost interest in classical mechanics because they were attracted primarily by two new revolutionary theories: relativity and quantum mechanics. Second, an important role in the recognition of the importance and ubiquity of *chaos* has been interpreted by the development of the computer, which came long after the contribution of Poincaré. In fact, only thanks to the advent of computer visualization was possible to compute (numerically) and see the complexity of chaotic behavior emerging nonlinear deterministic systems. A widespread opinion holds that the line of scientific inquiry opened by Poincaré was neglected until 1963, when the American meteorologist E. Lorenz rediscovered deterministic chaos while studying the evolution of a simplified model of the atmosphere.

In that framework, chaos behavior has been widely pointed out as a potential solution a different kinds of engineering problems. One of them is focused on finding a method to generate a trajectory for autonomous mobile robots. Recently, the engineering applications for autonomous mobile robots have been increased in different fields, such as industrial, civil and, mainly, military activities (searching for explosives or intruders). The key parameter for the success of those tasks is the *path planning*. It means the positioning of the robot on the terrain and how it moves through the terrain to find an objective. The path planning approaches must try to guarantee and maximize two performance metrics, i.e., the exploration of the whole terrain needs to be increased and the trajectory described by the robot system must be the more erratic or random. The first metric is related to the efficiency of the proposed path planning approaches. The latter, it is required to avoid the anticipation of the trajectory by the possible intruders.

Therefore, scientific community is trying to propose novel methods to improved for path planning. Due to high sensitivity to initial conditions of autonomous chaotic systems, they have been used as the core to generate the trajectories for the autonomous mobile robots. To do that, the chaotic signals are continuously mapped to coordinates (x_n, y_n) by using true random numbers generators (RNGs) for positioning the robot on the terrain. However, a typical approach to obtain path planning is using thresholds. In this manner, the performance of chaotic path planning is a function of the optimal values for those levels.

In this chapter, we analyze several chaotic systems which are used to obtain chaotic trajectories but by computing a quasi-optimal performance surface for the thresholds. The proposed study is based on sweeping the Poincaré sections to find quasi-optimal values for thresholds where the coverage rate is higher than those obtained by using the equilibrium points as reference values. Various scenarios are evaluated by using several chaotic systems, such as; two scroll chaotic systems, *n*-scrolls chaotic systems, hybrid chaotic systems (composed by three chaotic systems), and chaotic systems with optimized Lyapunov exponents. Numerical simulations to compute coverage rate against planned points, robot's trajectory evolution, covered terrain, and color map are carried out to analyze the resulting trajectories.

This chapter contains in Sect. 2 the related work. An overview of three nonlinear systems: Chua's circuit, Lorenz system and the multi-scrolls saturated function-based system is presented in Sect. 3, while Sect. 4 deals with the chaotic attractors with 2D multi-scrolls as well as random number generators. Furthermore, two techniques (thresholds and hybrid signals) for path planning are presented in Sect. 5. Sections 6 and 7 show the results of the proposed RNGs for different nonlinear systems. Section 8 introduces advanced RNGs by considering a dual RNG and a hybrid RNG. Finally, a detailed discussion of the results are given in Sect. 8 and conclusions in Sect. 9.

2 Related Work

During a decade the nonlinear dynamical systems have been researched, including chaotic behavior applied in several fields of applications, such as mathematics, physics, engineering, economics, sociology, etcetera [3, 11, 12, 14, 15, 19, 20, 25, 37, 39, 40].

Chaotic systems have special and attractive characteristics, which have been exploited to solve several problems in science and engineering. These systems are extremely sensitives to tiny variations on the initial conditions. Just one small difference between these one, the future behavior is completely different. Moreover, it is difficult to distinguish from a random system, which make them highly unpredictable systems [7, 13, 19, 26–28].

On the other hand, autonomous robots design is essential to explore narrow and dangerous spaces [17, 38]. There are many applications, which are need to cover large areas, i.e., exploration of tunnels in archaeological excavations, exploration of planets, detection of mines in military missions, or more simple an autonomous robots like vacuum cleaners, lawnmowers or even toys [7, 19, 27, 36].

Autonomous robots design merges to describe optimal planning paths on exploration task, besides maximal performance search through properties of chaotic signal. In robotics, the first mobile robot following a chaotic path was proposed by T. Nakamura and S. Kikuchi using the Arnold equation to generate the desired movements [19]. Chaotic systems have been widely used as path planning generators in autonomous mobile robots due to the unpredictability of the generated trajectories and the coverage rate of the robots workplace. In order to obtain a chaotic mobile robot, the chaotic signals are used to generate True RNGs (TRNGs), which, as is known, exploit the nondeterministic nature of chaotic controllers. Other advanced controllers based on fuzzy, adaptive, or intelligent soft computing control techniques can be also used to control the chaotic patterns in order to obtain chaotic paths [4–6, 8, 9, 29–34, 42].

A simple and frequent technique to obtain a chaotic bitstream is to sample analog chaotic signals by using thresholds as shown in latest reported research [7, 13, 17, 19, 21, 24, 27, 35]. However, the performance of chaotic path planning is a function of optimal values for those levels. Therefore, quasi-optimal thresholds are vital to improved chaos-based path planning. The published papers in [7, 13, 17, 19, 21, 24, 27, 35] have not taken into account this issue. Regularly, the definition of the threshold lies on an heuristic approach. This means that the experience of engineer is used to approximate its best value. In other cases, the equilibrium points, zero-cross, or a point closer to the basin of attraction are defined as reference values for thresholds. In this scenario, the study of techniques for path planning based on chaotic systems is still an open problem.

In this chapter, a quasi-optimal performance surface for the thresholds is presented. The proposed study is based on sweeping the Poincaré sections to find quasi-optimal values for thresholds where the coverage rate is higher than those obtained by using the basic reference values. Thus, this work presents the explorations planning paths strategies search of a mobile robot using chaotic signals. Chua, Lorenz and function based on a saturated system are proposed as nonlinear systems. In particular, last one with two and multiple scrolls on 1-Dimension (1D) and 2-Dimension (2D). Furthermore, through numerical simulations in MATLAB a study of chaotic dynamics in the generation of planning paths is realized. The main goal is achieve high efficiency to cover a determinate area at the same time is highly unpredictable. In addition, the effect and the dependence of maximum Lyapunov exponent (MLE) on planning paths is determined.

3 Overview

Although the chaos is seem in classical mechanics since B.C., Poincaré brought pay attention in the deeply implication of the difference between the determinism and predictively. An important role in the recognition of the importance and ubiquity of chaos has been played by the development of the computer, which came long after the contribution of Poincaré. In fact, only thanks to the advent of computer visualization was possible to compute (numerically) and see the complexity of chaotic behavior emerging nonlinear deterministic systems.

A widespread opinion holds that the line of scientific inquiry opened by Poincaré was neglected until 1963, when the American meteorologist E. Lorenz rediscovered deterministic chaos while studying the evolution of a simplified model of the atmosphere (see Eq. (1)) [16]:

$$\begin{aligned}\dot{x} &= \sigma(y - x),\\ \dot{y} &= \gamma x - y - xz,\\ \dot{z} &= xy - bz.\end{aligned} \tag{1}$$

Lorenz studied the temporal evolution of the signal of a nonlinear system for a parameters values set, and the relation of the integration with an initial condition. Irregular aperiodic oscillation with $t \to \infty$ are presents in the develop of the dynamical. Also, an height sensibility on the initial conditions was observed. This one was plot on a beautiful structure as phase path, z versus x, as you can see in Fig. 1a [28].

3.1 *Chaotic Systems*

Chaos refers to a kind of dynamic behavior with special features [28]: (i) extreme sensitivity to small variations in initial conditions, (ii) defined paths in phase space with an exponent of positive Lyapunov, (iii) an finite entropy of Komogorov-Sinai, (iv) a spectrum of continuous power, and (v) a fractional topological dimension, among other [18]. That is, **chaos** is a long term aperiodic behavior of a deterministic system that exhibits sensitive dependence on initial conditions.

Attractor is a limit set to which all neighboring paths converge when $t \longrightarrow \infty$ [10]. A continuous autonomous system requires more time to display two-dimensional chaos. The behavior of the paths is more complex as asymptotic attraction for neighboring paths keep. These are known as **strange attractors**. Moreover, the paths contained in this kind of attractor may be locally divergent each other within the whole attraction. Such structures are associated with the quasi-random behavior of solutions called chaos [18].

A method for the analysis of dynamical systems oscillations is the phase space representation, which was introduced to the oscillations theory by Andronov et al. [1]. This method has become the standard tool for studying various oscillatory

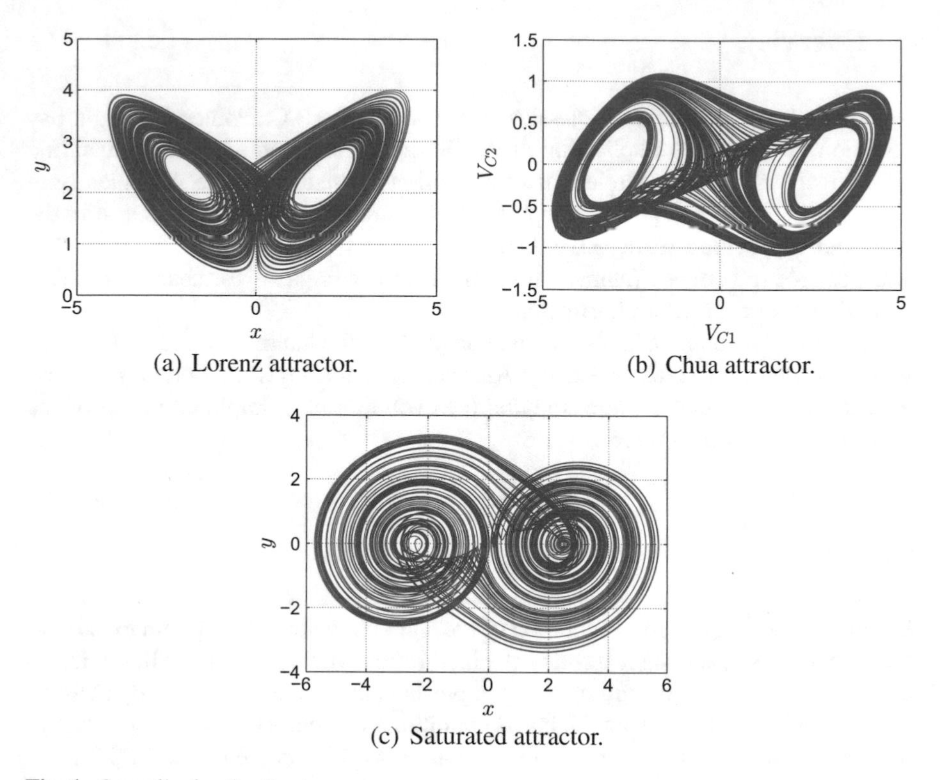

Fig. 1 2-scrolls chaotic attractors

phenomena. When the oscillations of complex shape were discovered, such as dynamic chaos, this method increases in relevance [2].

A nonlinear systems are numerically solved and this behavior is shown through the phase portrait. Then, the phase portrait of the system is obtained directly from the properties of $\mathbf{f}(\mathbf{x})$. So, it can to generate a variety of phase portraits.

Since it is not possible to display paths on $d > 3$, the technique called map of Poincaré is used, construction can be done as follows: (i) To simplify the representation, it is considered an autonomous system of three-dimensional $f(x)$, and focus on one of their trajectories; (ii) to define a plane (generally a surface $(d - 1)$) and consider all points $P - n$ in which the path crossing the plane from the same side.

Poincaré flux map $\mathbf{f}$ is defined as the map $\mathbf{G}$ related with two crossing associated points:

$$\mathbf{P}_{n+1} = \mathbf{G}(\mathbf{P}_n). \tag{2}$$

This results is obtained with the integration of differential ordinal equation into to account n to $(n + 1)$ intersections, which is well defined ever, as well as the inverse function $P_{n-1} = \mathbf{G}^{-}1(\mathbf{P}_n)$. Then, the map (2) is invertible. Poincaré maps allow a space of d-dimensional phase to be reduced to a $(d-1)$-representation, which identifies the path periodicity. Such maps preserve the stability properties of the points and curves [10].

3.1.1 System Based on the Chua's Circuit

Chua's circuit consists of five elements: a linear resistor (R), an inductor (L), two capacitors (C_1 and C_2), and a non-linear resistance known as Chua's diode (N_R). Using Kirchhoff's Laws, the system is represented by state variables as follows:

$$\begin{aligned}
\frac{dV_{C1}}{dt} &= -\frac{V_{C1}}{RC_1} + \frac{V_{C2}}{RC_1} - \frac{I_{NR}}{C_1}, \\
\frac{dV_{C2}}{dt} &= \frac{V_{C1}}{RC_2} - \frac{V_{C2}}{RC_2} - \frac{i_L}{C_2}, \\
\frac{dI_L}{dt} &= -\frac{V_{C2}}{L}
\end{aligned} \tag{3}$$

i_{NR} is a nonlinear function of the Chua's diode, this could be linearized with a PWL function [18]:

$$i_{NR} = \begin{cases} -g2V_{C1} + (g1 - g2)BP_1, & V_{C1} < -BP_1, \\ -g1V_{C1}, & -BP_1 \leq V_{C1} \leq BP_1, \\ -g2V_{C1} + (g2 - g1)BP_1, & V_{C1} > BP_1. \end{cases} \tag{4}$$

where $g1, g2$ and $g3$ are the slopes which behave as negative resistances; $\pm BP1$ and $\pm BP2$ represent the breaking points.

This nonlinear system exhibits chaos if its electronic components take the numerical values [18]: $C1 = 450\,\text{pF}$, $C2 = 1.5\,\text{nF}$, $L = 1\,\text{mH}$, $g1 = 1/1358$, $g2 = 1/2464$, $g3 = 1/1600$, $BP1 = 0.114\,\text{V}$, $BP2 = 0.4\,\text{V}$ and $R = 1625\,\Omega$. Numerical simulation results using MATLAB software are shown in Fig. 1b.

3.1.2 Lorenz System

Lorenz system is a mathematical model describing a meteorological phenomenon known as Rayleigh-Benard convection and reduces it to a set of three ordinary differential equations [10]:

$$\begin{aligned}
\dot{x} &= \sigma(y - x), \\
\dot{y} &= \gamma x - y - xz, \\
\dot{z} &= xy - bz.
\end{aligned} \tag{5}$$

Physical variables are linked to the convection intensity x, upstream-downstream difference temperature y and temperature deviation seen from linear profile z. Constants

σ, γ and b are positive dimensionless parameters related to the physical problem: σ is the Prandtl number, which measures the relationship between fluid viscosity and thermal diffusivity; γ can be regarded as the standard temperature difference imposed (more precisely is the relationship between the value of Rayleigh number and its critical value) and is the main control parameter; finally, b is a geometric factor [10, 28]. However, defining nominal parameters given in literature: $\sigma = 10, \gamma = 24, b = 8/3$, the solution of system is such shown in Fig. 1a.

3.1.3 System Based on a Saturated Function

Chaotic system based on a series of saturated functions is described in Eq. (6), where x, y, z are state variables; the constants $a, b, c, d \in +\mathbb{R}$, and $f(x)$ is defined by (7), where $\pm k$ are called saturated levels, $\pm sp$ are the breaking points and (k/sp) is the saturated slope.

$$\begin{aligned} \dot{x} &= y, \\ \dot{y} &= z, \\ \dot{z} &= -ax - by - cz + df(x). \end{aligned} \tag{6}$$

$$f(x) = \begin{cases} k & x > sp \\ (k/sp)x & -sp \leq x \leq sp \\ -k, & x < -sp. \end{cases} \tag{7}$$

If we define the saturated levels and slope values, we can construct a PWL function to obtain the desired amplitude on the signal $x(t)$. However, in case on one defines a $f(x)$, expressed in (7), considering the numerical values $\pm k = 2.5$, $(k/sp) = 100$ and $a = b = c = d = 0.7$ in Eq. (6), the obtained attractor has the shape as shown in Fig. 1c, where we can see two scrolls around the equilibrium points of the system.

3.1.4 Multi-scroll Chaotic Attractors on 1D

To generate multi-scrolls dimension (1D), one piecewise linear function (PWL) is added $f(x; k, h, p, q)$. Such that is given in Eq. (8) defined with the function (9).

$$\begin{aligned} \dot{x} &= y, \\ \dot{y} &= z, \\ \dot{z} &= -ax - by - cz + df(x; k, h, p, q). \end{aligned} \tag{8}$$

$$f(x; \alpha, k, h, p, q) = \begin{cases} (2q+1)k & x > qh + \alpha \\ k/\alpha(x - ih) + 2ik & |x - ih| \leq \alpha, -p \leq i \leq q \\ (2i+1)k & ih + \alpha < x < (i+1)h - \alpha, -p \leq i \leq q-1 \\ -(2p+1)k & x < -ph - \alpha \end{cases} \tag{9}$$

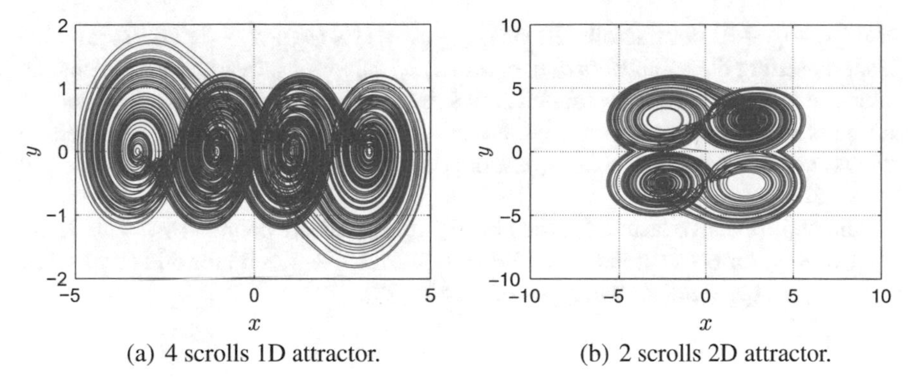

(a) 4 scrolls 1D attractor.

(b) 2 scrolls 2D attractor.

Fig. 2 Multi-scrolls chaotic attractors

Equations system (8) have $2(p + q) + 3$ equilibrium point on the axis x, called saddle points of 1 and 2 indices. These are defines in concordance with the system eigenvalues. The scrolls are generated into around to the saddle points of index 2. Equation (8) is capable of generating scrolls defining the suitable values on a, b, c, d, k and h. Responsible to connect the scrolls are $(p + q + 1)$, saddle points of index 1, forming an attractor. We can say that on the saddle points of index 2 have a saturated time delay just corresponding to a scroll, whereas the saddle points of index 1 have a saturated slope corresponds to ones connection between two scrolls-neighbors.

Saturated plateau on the function of saturated series defined by (9) is: $plateau = \pm nk$ for even scrolls and $plateau = \pm mk$ for odds scrolls. The saturated time delay for the core of the slopes is defined by $h_i = \pm mk$ for even scrolls and $h_i = \pm nk$ for odds scrolls. Multiplicative factors for the previous expressions are defined by $n = 1, 3, \dots, (p + q + 1)$ for odd scrolls and $n = 1, 3, \dots, (p + q - 1)$ for even scrolls; and $m = 2, 4, \dots, (p + q)$ for two kinds of scrolls [18].

If a PWL is constructed with four saturated levels $k = 1.1$, $slope = 100$, $p = 1$, $q = 1$, $h = 2.2$ and the numerical values $a = b = c = d = 0.7$ in Eq. (9), the obtained attractor has the shape as shown in Fig. 2a, where we can see an attractor formed by four scrolls around the equilibrium points of the system in 1D.

4 Chaotic Attractors of Multiple Scrolls on 2D

To generate a chaotic behavior on two-dimensions (2D) is needed to change the chaotic system given by the Eq. (6). This kind of 2D dynamic is shaped by applying approximation through the state variables as (10), with x, y, z as the state variables and the constants a, b, c, d_1 and $d_2 \in +\mathbb{R}$.

A chaotic system for two-dimensional, two series of functions saturated $f(x)$ and $f(y)$ are needed in the system of equations (10) also defined by the function (9), where p_1, p_2, q_1 and q_2 are positive integers. Then the chaotic system can create $(p_1 + q_1 +$

$2) \times (p_2 + q_2 + 2)$ pairs scrolls 2D or $(p_1 + q_1 + 1) \times (p_2 + q_2 + 1)$ odd scrolls 2D properly setting the parameters a, b, c, d_1, d_2, k_1, k_2, h_1 y h_2. In addition, the plateau saturated in the function of saturated series described in Eq. (9) is $meseta = \pm nk$ for 2D pairs scrolls and $meseta = \pm mk$ for odd scrolls 2D. Saturated for outstanding delays centers are defined by $h_i = \pm mk$ mk pairs scrolls for 2D and $h_i = \pm nk$ for odd scrolls 2D.

The multiplicative factors for the above expressions are defined by $n = 1, 3, \ldots,$ $(p_2 + q_2 + 1)$ for even scrolls 2D and $n = 1, 3, \ldots, (p_2 + q_2 - 1)$ for odd scrolls 2D; $m = 2, 4, \ldots, (p_1 + q_1)$ for two types of scrolls [18].

$$\begin{aligned} \dot{x} &= y - \frac{d_2}{b} f(y; k_2, h_2, p_2, q_2), \\ \dot{y} &= z, \\ \dot{z} &= -ax - by - cz + d_1 f(x; k_1, h_1, p_1, q_1) + d_2 f(y; k_2, h_2, p_2, q_2). \end{aligned} \tag{10}$$

If we consider the $f(x)$ PWL function defined in the Sect. 3.1.3 and we use the same values of $f(x)$ to $f(y)$, the attractor obtained has the shape as shown in Fig. 2b, where we can see two scrolls in two dimensions (2D).

4.1 Random Number Generators

A RNG is an unpredictable source of numbers. Mathematically defined as a source of long sequences of symbols independent and identically distributed [23]. There are basically two types of generators used to produce random sequences. RNGs (RNG) and pseudorandom number generators (PRNGs) [22–24].

Generators RNG generally use a non-deterministic source (Entropy) along with some processing functions to produce randomness. It is necessary to have a post-processing to overcome any weakness in the entropy source, do not result in the production of non-random numbers (for example, the appearance of long strings of zeros or ones). The entropy source is typical of any physical quantity, such as noise in an electrical circuit, interrupt processing by the user (for example, keystrokes or mouse movements), quantum effects in a semiconductor or using various combinations of above entries. Outputs a RNG may be used directly as a random number or may be fed in a pseudorandom number generator (PRNG). To be used directly, the output of any RNG must meet strict criteria of randomness measured by statistical tests to determine the physical sources of the RNG entries appear at random [24].

A PRNG uses one or more multiple inputs and generates pseudo-random numbers. At the inputs of a PRNG they are called seeds, which must also be random and unpredictable. Hence, by default, a PRNG should get their seeds from the outputs of a RNG, i.e., a PRNG requires a RNG as partner. The outputs of a PRNG are typically deterministic functions of the seed, that is, all true randomness is confined to the generation of seeds. The deterministic nature of the process leads to the expression "pseudo". Since each element of a pseudorandom sequence is reproducible seed.

Seed needs only be saved if the reproduction or validation of the pseudorandom sequence is required. Ironically, the pseudorandom numbers often seem to have more randomness than those obtained random numbers from physical sources. If a pseudorandom sequence is constructed correctly, each value of the sequence occurs from the previous value through transformations introduce additional randomness. A number of these transformations can eliminate self-statistical correlations between input and output. Therefore, the outputs of a PRNG may have better statistical properties and be faster than RNG.

A sequence of random bits could be interpreted as the result of coin tosses with sides labeled as "0" and "1", with a probability of 0.5 for each side. Moreover, all elements of the sequence must be independent of each other, and the value of the next element in the sequence should not be predictable regardless of the number of items that have already occurred [24].

5 Generation Planning Paths

5.1 Technique 1: Thresholds Levels

A technique to generate random numbers reported in the literature with chaotic systems consists basically of five blocks, as it is shown in Fig. 3 [35, 41]. Each block contain the following informations:

- The block $\mathbf{S_1}$ is responsible to generate the chaotic signal $x(t)$.
- In the block $\mathbf{S_2}$, the bits are obtained when the Poincaré's sections (c_1 and c_2) are crossed by the signal $x(t)$ (Fig. 4). The bits are obtained as follows:

$$S_2 : \begin{array}{l} \sigma^1(x(t)) = \begin{cases} 0, \text{ si } x(t) < c_1 \\ 1, \text{ si } x(t) \geq c_1 \end{cases} \\ \sigma^0(x(t)) = \begin{cases} 0, \text{ si } x(t) > c_2 \\ 1, \text{ si } x(t) \leq c_2 \end{cases} \end{array}$$

- In the third block $\mathbf{S_3}$ a sequence of bits is generated by:

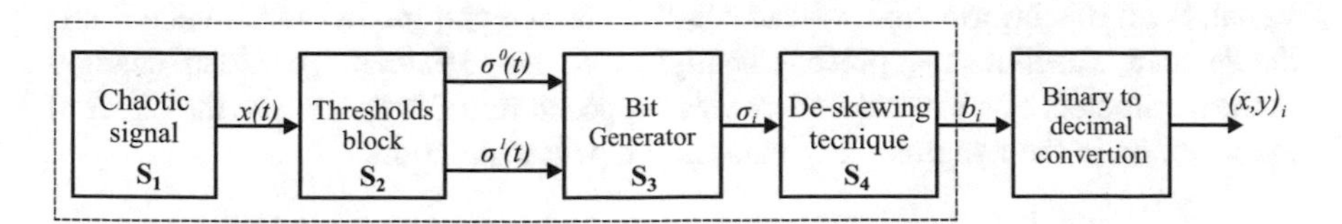

Fig. 3 Basic structure of the RNGs used

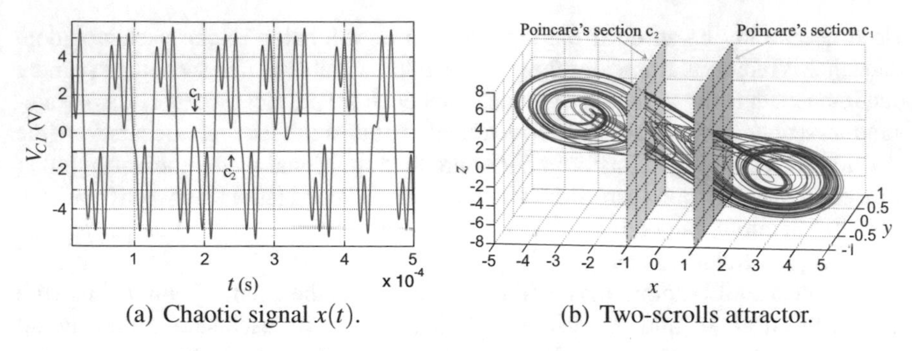

(a) Chaotic signal $x(t)$. (b) Two-scrolls attractor.

Fig. 4 Sampling process of the chaotic signal $x(t)$, using c_1 and c_2 thresholds

$$S_3 : \sigma_i(\sigma^0, \sigma^1) = \begin{cases} 0, \text{ si } \sigma^0 = 0, \sigma^1 : 0 \uparrow^1 \\ 1, \text{ si } \sigma^1 = 0, \sigma^0 : 0 \uparrow^1 \end{cases}$$

where σ^0:0 $\uparrow^1$ is the logical transition '0' to '1' of σ^0 and $i \in \{0, 1, 2, \ldots, n\}$.

- The block $\mathbf{S_4}$ implements the Von Neumann's technique (VN) [41] to reduce the correlation into the bits sequence. The VN post-processing consists in convert the pair of bits "01" in the logic output '0', "10" in the logic output '1' and the pairs "00" y "11" are discarded.
- Finally a binary to decimal conversion is required to obtain the pair of coordinates (x, y).

Then, to generate a pair of coordinates x and y, a total of 10 bits (5 bits for x and 5 bits for y) are needed, discarding the numbers 0 and 31. As result, each RNG designed, has a set of coordinates (x, y), which are the paths of motion of a mobile robot. The coverage rated is quantified by the coordinates that were visited in an area of 30×30 units, so there are 900 different coordinates. In each RNG 3000 random numbers were generated to form 1500 trajectories, this because the coverage rate curve has an exponential behavior, so the probability to get a new coordinate is reduced significantly along the planning of the coordinates, also 1500 paths are sufficient to cover the most of the area proposed and is possible to determine if the location of the Poincaré section is efficient in generating scan paths or not. As a starting point, the Poincaré's sections (c_1 and c_2) were located in the fixed points for this analysis, then they are moved to optimal sections by sweeping along the chaotic signal. With this information we can create a surface graphic in order to determine the Poincaré's sections that present the highest coverage percentages. Also, changes in the parameters of a chaotic system was proposed in order to analyze the effect of the variation in the Lyapunov exponent in the path generation.

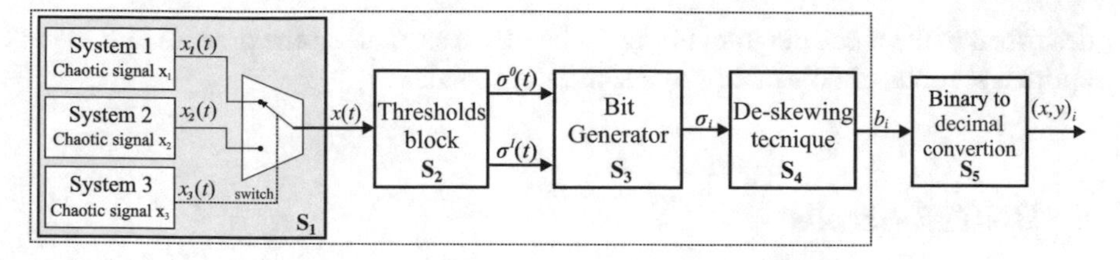

Fig. 5 Blocks diagram describes the hybrids RNGs

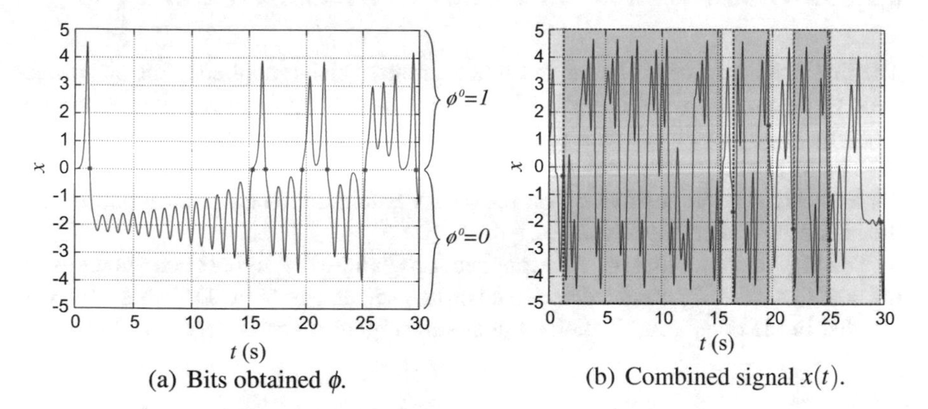

Fig. 6 Construction of the hybrid signal $x(t)$

5.2 *Technique 2: Hybrid Signal*

The generation technique scan paths from hybrid chaotic systems is the generation of routes from RNGs combining the non-linear dynamical systems with chaotic behavior of Chua, Lorenz and saturated function based on previously analyzed system. The main idea is switching between two chaotic systems during the integration process with the goal of creating a more complex dynamic, and especially the sources of entropy not be correlated. Figure 5 shows the blocks that make up this type of RNGs.

In this case, at block $\mathbf{S_1}$ is performed switching between two chaotic systems, which are determined by the bits generated by the signal $x_3(t)$ corresponding to a third system chaotic. That is, if the signal evaluating $x_3(t)$ is in its positive or negative part during the integration process (Fig. 6a). These control bits are obtained as follows:

$$\phi(x_3(t)) = \begin{cases} 0, \text{ si } x_3(t) < 0 \\ 1, \text{ si } x_3(t) \geq 0 \end{cases} \tag{11}$$

Thus, when $\phi = 1$ the system is integrated with the number one chaotic system and when $\phi = 0$ the system is integrated with the chaotic system number two. As result, one obtains a signal $x(t)$ composed of two chaotic systems such as shown in Fig. 6b. In blocks $\mathbf{S_2}$, $\mathbf{S_3}$, $\mathbf{S_4}$ and $\mathbf{S_5}$, the technique for obtaining bits is the same described in the block diagram of Fig. 3. Finally it carried out the process to find the optimal sections of Poincaré by sweeping the signal.

6 RNGs 2-Scrolls

6.1 RNG Based on the Chua's Circuit Without VN Technique

The first RNG proposed is based on Chua's circuit, which is applied to the bits generation technique described in Sect. 3.1.1 regardless of the post-processing technique. The Fig. 7a shows the surface plot of coverage percentages after 1500 simulated scan paths in each pair of Poincaré sections included in the sweep. In the axis x is the Poincaré section positive c_1, on the axis y is found the negative section c_2 and the axis z the coverage percentages.

The Fig. 7a shows that the Poincaré sections, where the highest percentage were ($c_1 = 4.7, c_2 = -4.7$) obtained with a coverage percentage of 66.11%. Also you can see that in the other combinations, the coverage percentage is very low. In Fig. 7b

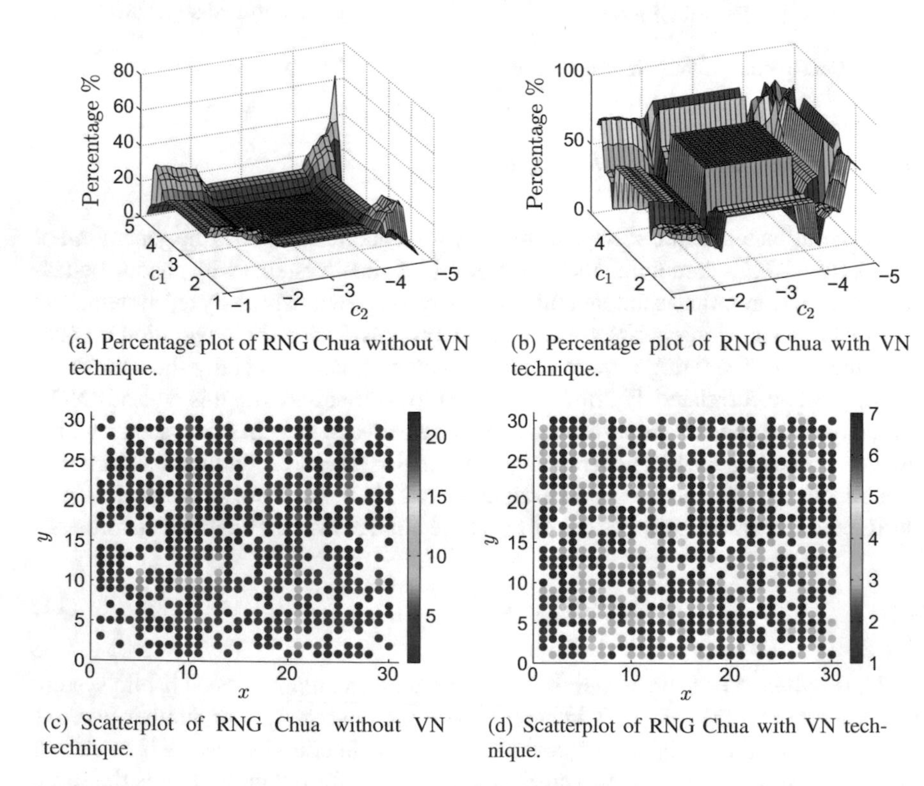

(a) Percentage plot of RNG Chua without VN technique.

(b) Percentage plot of RNG Chua with VN technique.

(c) Scatterplot of RNG Chua without VN technique.

(d) Scatterplot of RNG Chua with VN technique.

Fig. 7 RNG based on Chua circuit with and without VN technique

the evolution of the coverage percentage is observed related to the scanned planning paths in optimal Poincaré sections, where an exponential behavior is observed which it will saturate as the number of paths. In Fig. 7c, 1500 planning paths that would follow a robot in an area of 30×30 units are shown, using the RNG based on the Chua's circuit without post-processing techniques the optimal sections ($c_1 = 4.7$, $c_2 = -4.7$). It can be seen that the area edges are rarely visited and therefore the coverage is not very good, this can check with the scatterplot of Fig. 7d, which shows the number of times that the coordinates were visited in the exploration area.

6.2 RNG Basen on Chua's Circuit with VN Technique

Now, in this section we consider the implementation of a bits post-processing technique, particularly the known Von Neumann Technique. Thus, this RNG is also based on the Chua's circuit, the difference with the previous one lies in the implementation of bits post-processing technique Von Neumann (VN). Figure 7b shows that by implementing the VN technique, coverage percentages increase considerably. Poincaré sections were optimal in this RNG ($c_1 = 4.3$, $c_2 = -4.7$) with a coverage percentage 82.22%, while in Fig. 7d shows the scatterplot of the planning paths.

Table 1 shows the results of RNGs based on Chua's circuit. As you can see, implementing the post-processing VN technique the coverage percentages improved considerably and the repetition in the coordinates decreases. Although, this is also reflected in a much larger number of iterations because the bits discarded by VN technique and therefore a more computationally time. For the purpose of checking

Table 1 Results obtained of RGN based on Chua circuit

Data	RNG Chua	RNG Chua with VN
Integration step width	1×10^{-7}	1×10^{-7}
Coverage percentage in equilibrium points	13.1111	68.5556
Optimal Poincaré sections	$c_1 = 4.7$ $c_2 = -4.7$	$c_1 = 4.3$ $c_2 = -4.7$
M_{1500} on the optimal sections equilibrium points (%)	66.1111	82.2222
Average percent	17.0632	52.6895
Total bits generated in optimal sections	15550	101392
Bits discarded by Von Neumann technique	–	68692
Bits discarded by out of limit ($1 \leq x \leq 30$)	550	1350
MLE	0.0036	0.0036

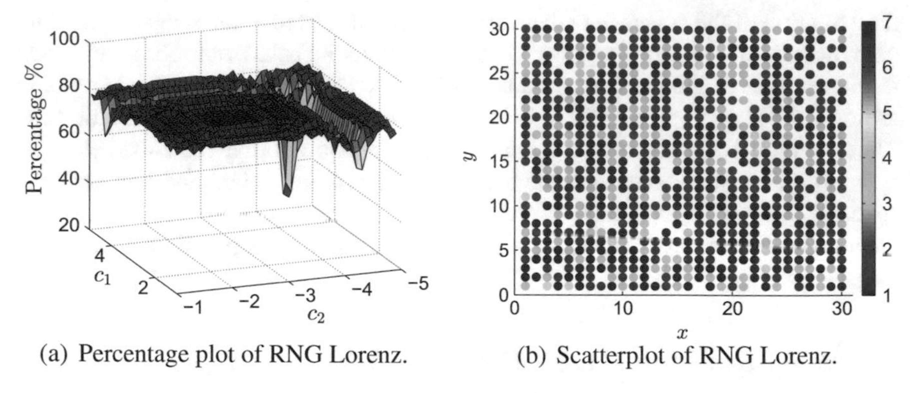

(a) Percentage plot of RNG Lorenz. (b) Scatterplot of RNG Lorenz.

Fig. 8 RNG based on Lorenz system

the relevance of the post-processing technique in generating random numbers, only in this system is designed a generator without the VN technique and other one with this technique.

In the following RNGs, the analysis is carry out with the VN post-processing technique.

6.3 RNG Based on the Lorenz System

In the RNG based on the Lorenz System, the signal $x(t)$ is the entropy source. Like the early RNGs, a sweep was performed on the signal $x(t)$ to determine the optimal Poincaré sections in generating random numbers for planning paths exploration. In Fig. 8a, the coverage percentages evaluated in all sections, where one can see that is larger and more uniform compared to previous RNGs shown. Sections $c_1 = 3.4$ and $c_2 = -2, \ldots, -2.8$ proved to be the optimal coverage percentage with 84.33%.

In Fig. 8 one can observe the increased coverage percentage under the planned trajectories in optimal sections. In Fig. 8a, 1500 trajectories follow a robot in an area of 30×30 units using the RNG based on the Lorenz system in sections optimum cutting are shown, while in Fig. 8b the scatterplot of these planning paths is shown.

6.4 RNG Based on a 2×1 Saturated Function (L1)

Other kind of interesting RNGs are based on saturated functions. In RNG based on a 2×1 saturated function, the signal $x(t)$-based function as a entropy source is used. This system is based on a PWL function with a slope of 100 and an optimal step width as function of the system eigenvalues as well as two saturation levels of 2.5

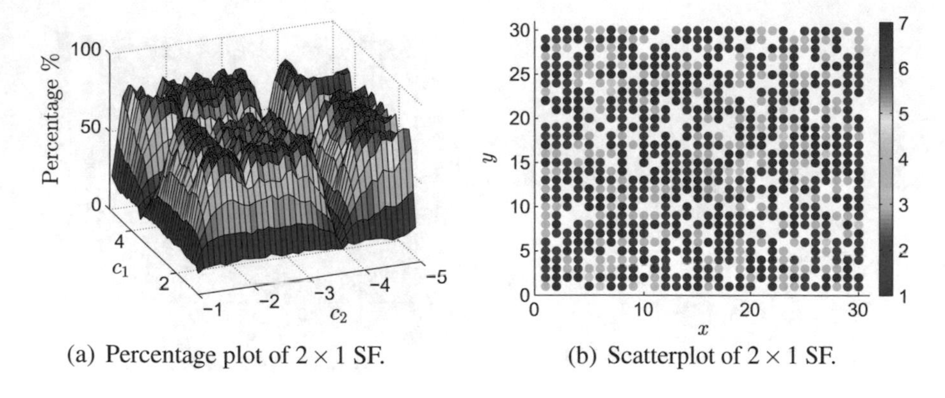

(a) Percentage plot of 2×1 SF.

(b) Scatterplot of 2×1 SF.

Fig. 9 RNG based on a 2×1 SF (L1)

units with the parameters values $a = b = c = d = 0.7$. After performing the scanning for optimum thresholds sections, the coverage percentages of all combinations are collected. In Fig. 9a we can see the coverage percentages of all sections, in which sections are $c_1 = 3.2$, $c_2 = -4.4$. They turn out to be the best, since the highest coverage percentage is obtained with 83.2222%, while in Fig. 9b the scatterplot of these paths is shown. Table 5 the most important data are presented.

7 RNGs Multi-scrolls

7.1 *RNG Based on a* 4×1 *Saturated Function*

Now, in the case of the RNG based on a 4×1 saturated function, this uses the signal $x(t)$-based saturated function in one dimension (4×1) in the RNG system. The Fig. 10 shows the coverage percentages corresponding to all Poincaré sections included in the sweep, where $c_1 = 1.7$ and $c_2 = -1.8$ turn out to be the optimal sections with a coverage percentage of 84.1111%. In Fig. 10a 1500 planning paths are displayed in an area of 30×30 units using the RNG based on a saturated 4×1 function on the optimal Poincaré sections, while Fig. 10b shows the scatterplot of these. In Fig. 10 can see the coverage percentage increases along the 1500 paths on the optimal cutting sections. In Table 5, the most important data are presented.

7.2 *RNG Based on a* 2×2 *Saturated Function*

At the same way, RNG with entropy given by the signal $x(t)$ based on a 2×2 saturated function is now analyzed. For this, the PWL function take into account the

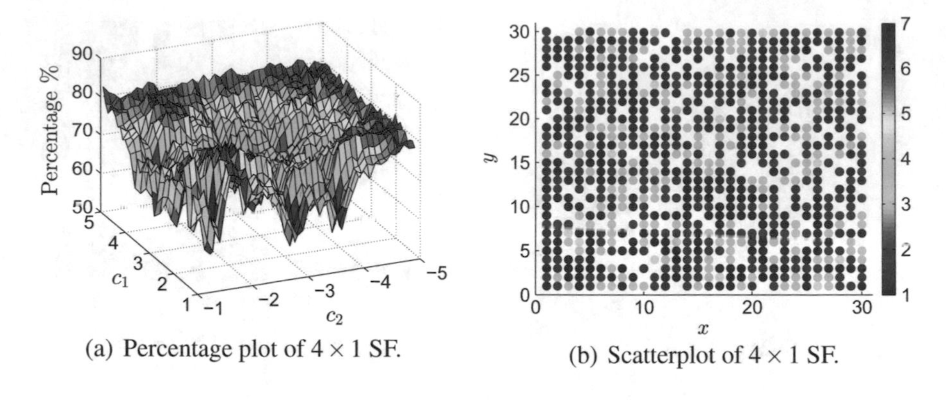

(a) Percentage plot of 4 × 1 SF.

(b) Scatterplot of 4 × 1 SF.

Fig. 10 Planning paths based on RNG-4 × 1 chaotic PWL

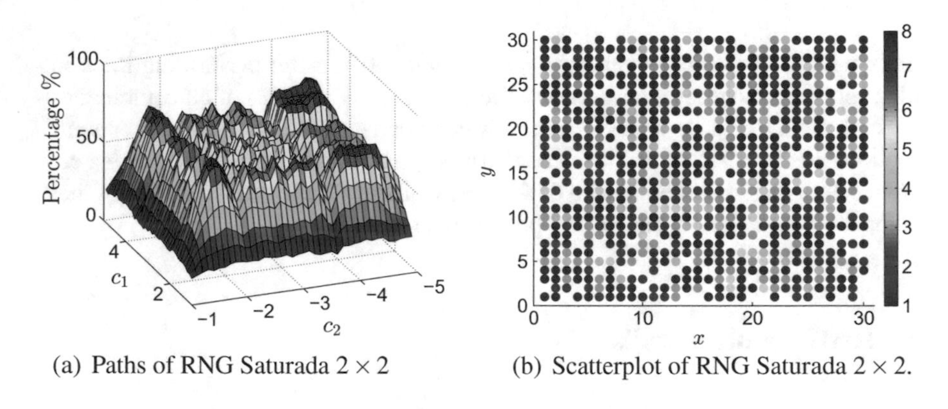

(a) Paths of RNG Saturada 2 × 2

(b) Scatterplot of RNG Saturada 2 × 2.

Fig. 11 Planning paths obtained with RNG based on a chaotic PWL system 2 × 2

slope of 100, a step integration optimum width depending on the system eigenvalues and two saturation levels of 2.5 units. Figure 11 shows the coverage percentages with $c_1 = 1.7$ and $c_2 = -1.6$ turn out to be sections optimal coverage with a percentage of 81.7777%, and the scatterplot of the visited coordinates, (a) and (b) respectively. The results of the aforementioned cases (subsections 6.3, 6.4, 7.1, 7.2) are summarized in Table 2.

8 Advanced RNGs

8.1 RNG Dual

In this RNG, the signals $x(t)$ and $y(t)$-based saturated functions 2 × 2 are used to obtain the simultaneous random number function system. That is, they will have

Table 2 Results obtained from RGN based on one system

Data	Results			
	Lorenz	2×1 SF (L1)	4×1 SF	2×2 SF
Integration step width	0.0037	0.0152	0.1102	0.0152
Coverage percentage in equilibrium points	81.8889	66.1111	71.2222 75.6667 68.1111	58.3333
Optimal Poincaré sections	$c_1 = 3.4$ $c_2 = -2.5$	$c_1 = 3.2$ $c_2 = -4.4$	$c_1 = 1.7$ $c_2 = -1.8$	$c_1 = 1.7$ $c_2 = -1.6$
Coverage percentage on optimal sections equilibrium points	84.3333	83.2222	84.1111	81.7777
Average percent	78.9815	57.4542	73.38	56.0615
Total bits generated in optimal sections equilibrium points	72286	60548	90114	81762
MLE equilibrium points	0.0029	0.0011	0.0156	0.0010

two sources to generate random numbers from a single system. To do this, Poincaré sections corresponding to signal $x(t)$, where the coverage percentage was found, are also taken by the signal $y(t)$.

In Fig. 12 the percentage increase coverage throughout the 1500 coordinates generated from the signal $x(t)$ (blue line) and the signal y (red line), where coverage percentage are 81.444% and 77.8889% respectively. Additionally, in Fig. 13, scatterplots of each used signal are shows, these results are given with 1500 trajectories in an exploration area of 30×30 units. Using the RNG Dual in optimal Poincaré sections, scatterplots show the coordinates distribution the scan. Dual RNG results are shown in Table 3.

8.2 Hybrid RNG Chua-Lorenz-Saturated

We define a hybrid system such that will be in charge of Chua switch between Lorenz system and function based on a saturated two scrolls for generating random numbers (Fig. 14) system. When the $x(t)$ signal system Chua is in its positive part of the integration method solves the Lorenz system and when it is in its negative part based on

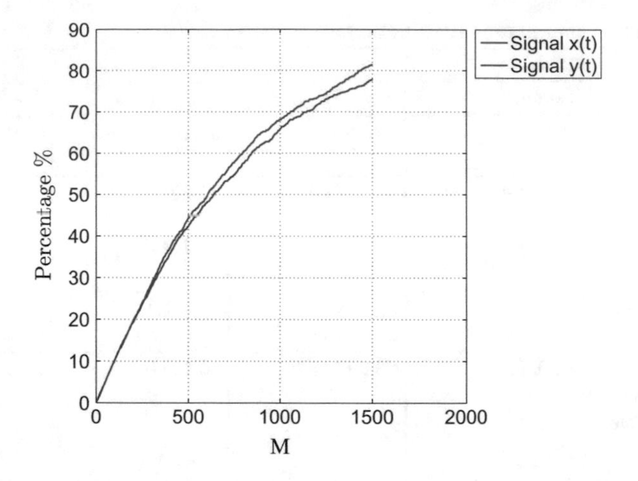

Fig. 12 Coverage percentage plot of the RNG dual with respect to M

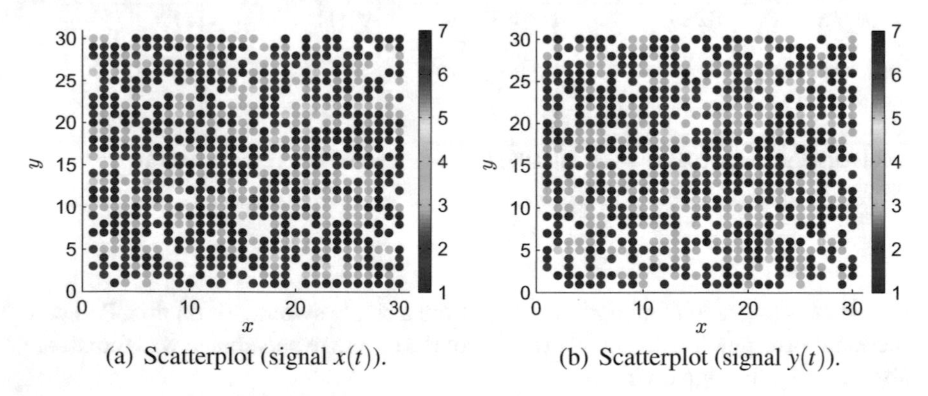

Fig. 13 Scatterplot of planning paths obtained with RNG Dual for each signal: **a** $x(t)$ and **b** $y(t)$

Table 3 Results of RGN Dual

Data	Signal $x(t)$	Signal $y(t)$
Integration step width	0.0152	
Coverage percentage on equilibrium points	58.3333%	63.4444
Optimal sections $x(t)$	($c_1 = 1.7$ $c_2 = -1.6$)	
Coverage percentage in optimal sections	81.7777	78.8889
Total bits generated in optimal sections equilibrium points	81762	168222
Bits discarded by the VN technique	49002	136122
Bits discarded for being off limits	1380	1050

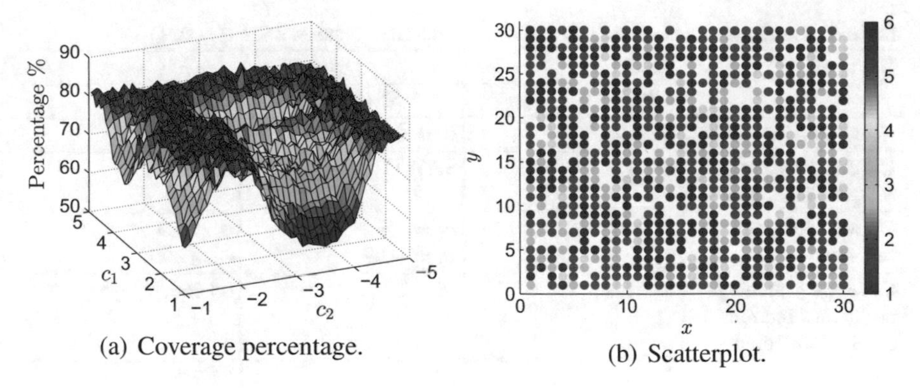

(a) Coverage percentage.

(b) Scatterplot.

Fig. 14 Planning paths based on RNG-hybrid chaotic system

Table 4 Results obtained from RGN-hybrid system

Data	Results
Optimal sections	$c_1 = 3.0$ $c_2 = -4.8$
Coverage percentage in optimal sections equilibrium points	83.8889
Average percent optimal sections equilibrium points	76.3085
Total bits generated in optimal sections equilibrium points	320676
MLE	0.00072478

a saturated function system is solved, forming thus a combined signal by these two systems.

Once you have the combined signal is necessary to perform the scanning process to determine the optimal Poincaré sections according to the coverage percentage. In the graph in Fig. 14a the corresponding percentages are observed at all sections including, where $c_1 = 3$ and $c_2 = -4.8$ turn out to be the optimal sections as the highest coverage percentages is obtained with 83.8889%. In Fig. 14b the dispersion of the planned routes shown. The most important data are presented in Table 4.

8.3 *RNGs Based on Chaotic Systems with Optimized Lyapunov Exponents*

In this section the dependence of the maximum Lyapunov exponent in the generation of random numbers is analyzed. For this, the standard system Chua and based

Table 5 Results obtained from RGN based on a saturated function 2×1 (L3, 2, 1)

Data	Results		
	2×1 SF (L1)	2×1 SF (L2)	2×1 SF (L3)
Integration step width	0.0152	0.0152	0.0152
Coverage percentage in equilibrium points	66.1111	53.1111	27.8889
Optimal Poincaré sections	$c_1 = 3.2$ $c_2 = -4.4$	$c_1 = 1.6$ $c_2 = -3.8$	$c_1 = 3.9$ $c_2 = -2.6$
Coverage percentage on optimal sections equilibrium points	83.2222	82.7778	83.6667
Average percent	57.4542	69.1299	73.2444
Total bits generated in optimal sections equilibrium points	60548	50270	126086
Bits discarded by the VN technique	25208	17430	90886
Bits discarded for being off limits ($1 \leq x \leq 30$)	2670	1420	2600
MLE equilibrium points	0.0011	0.0018	0.0031

on a 2×1 saturated function. The analysis implements the technique of generating random bits. After performing the sweep signal to obtain Poincaré sections, the optimal parameters of system are modified to optimize the maximum Lyapunov exponent computed with an iterative process. For each system parameters are changed twice, this provides us information about the correlation between the coverage percentage and maximum Lyapunov exponent, which is calculated using time-series approaches.

In the first case (L2), the signal $x(t)$ -based saturated function as input is used. Thus, the difference lies in the change in the system parameters to optimize the MLE. To increase the value of the MLE of the system parameters are set as follows: $a = 1, b = 1, c = 0.499, d = 1$. The results show that the highest coverage percentage was obtained after performing the sweep signal $x(t)$ is found in sections ($c_1 = 1.6$, $c_2 = -3.8$) to 82.7778%.

Second case (L3), the signal $x(t)$-based saturated as a entropy source function system is used, in this case the system parameters were modified as follows: $a = 1, b = 0.788, c = 0.643, d = 0.666$. The results indicate that in sections $c_1 = 3.9$ and $c_2 = -2.6$ and higher coverage percentage is 83.6667%.

Table 5 shows RNG results based on a saturated function, which has the optimizations of Lyapunov exponents (L2 and L3) (Fig. 15).

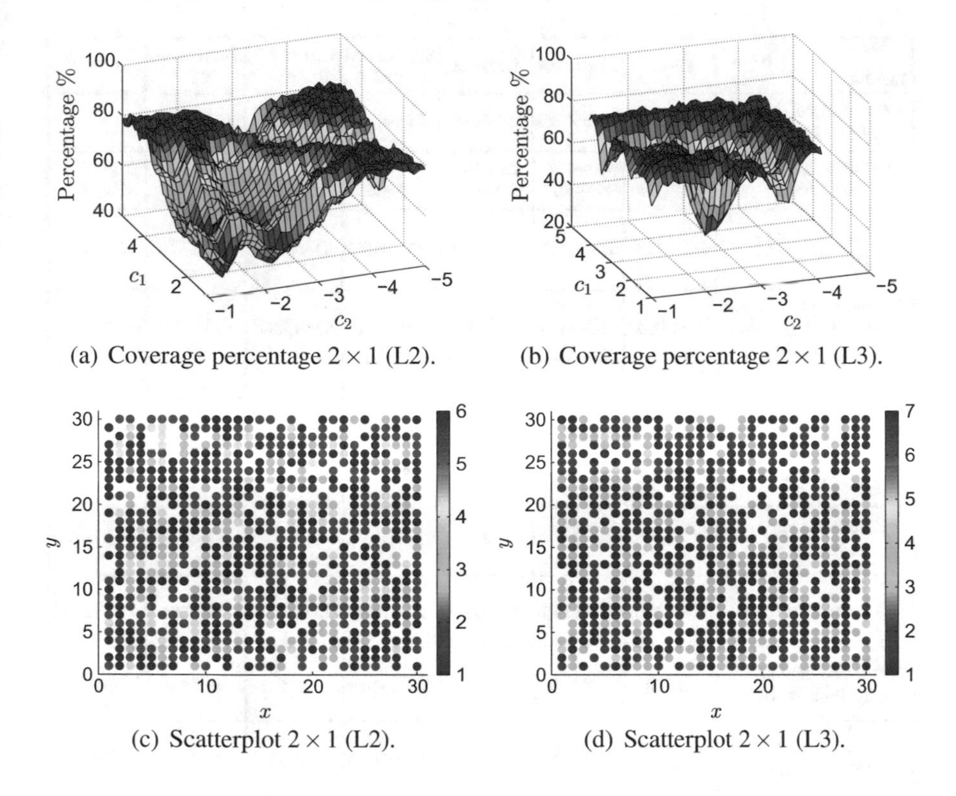

(a) Coverage percentage 2×1 (L2).

(b) Coverage percentage 2×1 (L3).

(c) Scatterplot 2×1 (L2).

(d) Scatterplot 2×1 (L3).

Fig. 15 RNG based on a saturated function 2×1 (L3 2)

9 Discussion

First, we have demonstrated the relevance of post-processing techniques in the RNG's design. Particularly, it is shown that implementing the Von Neumann technique the bits distribution is improved, reducing repetition number generation, and therefore the coverage percentage increases considerably.

However, a series of tests with which, it is concluded that any Poincaré section that is near or above 80% can be considered as an optimal combination for obtaining random bits, were made even more if the sections cuts are located in the middle of the signal. This ensures that fewer bits discarded by the VN technique as shown in comparative Table 1 bits, are obtained faster and there.

Because most of the RNGs give us percentages of ~80%, a way to conclude that a RNG is more efficient than the other would be through the analysis of the average percentages and through the plots of surface, which give us an idea of uniformity that exists in the coverage coverage percentages during the sweep of the Poincaré sections as shown in Fig. 16.

DATA \ RNG	Chua	Chua VN	Lorenz	Saturated 2x1	Saturated 2x2	Saturated 4x1	Hybrid
Coverage Percentage Equilibrium points	13.1111%	68.5556%	81.8889%	54.0000%	58.3333%	75.6667%	--------
Optimal Poincare's sections	c1=4.7 c2=-4.7	c1=4.3 c2=-4.7	c1=3.4 c2=-2.5	c1=4.8 c2=-2.5	c1=1.7 c2=-1.6	c1=1.7 c2=-1.8	c1=3 c2=-4.8
Percentage in optimal sections	66.1111%	82.2222%	84.3333%	83.5555%	81.7777%	84.1111%	82.4444%
Number of iterations to generate 1500 exploration paths	15514580	32443934	15964878	123962019	51445000	15899497	143301074
Percentages above 60%	1	539	1586	881	737	1584	1583
Percentages above 70%	0	167	1548	568	220	1158	1355
Percentages above 80%	0	3	947	181	4	142	516
Average percent	17.0632%	52.6895%	78.9815%	59.4619%	56.0615%	73.38%	76.3085%
Total bits generated in optimal sections	15550	101392	72286	28099528	81762	90114	320676
Bits discarded by VN technique	-----	68692	37566	28065648	49002	55034	285916
Bits discarded by out of limit	550	1350	2360	1940	1380	2540	2380

Fig. 16 Comparative table of RNGs designed

Based on the fact that a RNG has submitted to the higher coverage percentage in the optimal Poincaré sections, other cases not guaranteed to be more efficient. This is because, after performing a series of tests and comparisons with the PRNG MATLAB, it was determined that any section of Poincaré that is near or above 80% can be considered as an optimal combination for obtaining random bits. Thus, the way was determined if a RNG is more efficient than another, was through analysis of the coverage percentages average, uniformity in surface plot, and the number of combinations in Poincaré sections that they found over 80% in the coverage percentage. With this approach, the Lorenz system presented a higher average coverage percentage, more uniformity than others and a greater number of sections with rates of around or above 80% coverage (see Fig. 16). However, from the point of view of hardware implementation, the PWL chaotic system turns out to be the most convenient. Importantly, there is a strong dependence on the location of the Poincaré sections in the efficiency of the planning paths, even for minor variations.

An advantage of these RNGs with respect to those existing in the literature, is that obtaining bits is performed by two cross sections defined by quasi-optimal sections where there is a higher coverage percentage. This proves to be of great importance from the point of view of implementation. Faced with the real situation, one can conclude that the RNGs based on PWL chaotic systems have great advantages over all other RNGs that were designed in this chapter. In this regard, the RNG-Saturated-4×1 it be the most convenient according to the criteria mentioned above (see Fig. 16).

Moreover, a technique for the generation of hybrid systems by combining three different chaotic systems (Chua-Lorenz-Saturated) for the design of a proposed RNG. The results indicate that no significant or radical change in coverage percentages regarding designed with a single system. Moreover, it is difficult to propose algebraic operations for systems not destabilized when switching. Therefore, it is concluded that it is not convenient to combine the dynamics of three systems to achieve similar results to those provided by the RNGs with a single chaotic system as given in Fig. 16.

To determine the dependence of the maximal Lyapunov exponent (MLE), equilibrium points were designed and analyzed based RNGs Chua standardized systems and function based on a saturated system. It was found that there exists a correlation between increased MLE equilibrium points and increased the coverage percentage, which is most noticeable in the RNGs of saturated function based systems as demonstrated in Fig. 16.

10 Conclusions

It have been proposed and analyzed a technique to generate planning paths with RNGs by using Poincaré sections as thresholds to sample the chaotic signals. Various chaotic systems have been considered for the analysis, such as chaotic systems with double scroll attractors (Chua, Lorenz, Saturated function system); multi-scroll attractors in 1D (4×1-scrolls saturated function system); multi-scroll attractors in 2D (2×2-scrolls saturated function system); hybrid systems composed by three different chaotic systems; and double scroll chaotic systems but with an optimized Lyapunov exponent.

In each RNG was designed a comprehensive study focus to determine the optimal Poncaré sections, where the high coverage percentages occurred and the importance of implementing post processing techniques (Von Neumann technique) in the generation of random numbers was found. After comparing the coverage percentages of all RNGs it was observed that the RNG based on the Lorenz system present a more uniform and higher coverage percentage than other coverage distribution. However, from the point of view of hardware implementation, the PWL chaotic system turns out to be the most convenient. Importantly, there is a strong dependence on the location of the Poincaré sections in the efficiency of the planning paths, even for minor variations. Overall, we conclude that the RNGs based on Poincaré sections were designed to prove that they are a suitable tool in generating random paths for autonomous mobile robots.

This investigation will enable to increase several applications of TRNGs by considering the proposed performance surfaces. Also, it is necessary to consider another kind of chaotic oscillators, such as, fractional order chaotic systems, chaotic systems with hidden attractors, hyperchaotic systems, non-equilibrium chaotic systems and so on. In this manner, a general conclusion about the quasi-optimal performance surfaces can be obtained. Additionally, circuit implementation is a must in order to validate the proposed analysis. This physical implementation may require advanced circuit design methodologies that can be adjusted with required accuracy of the value for quasi-optimal thresholds. Otherwise, it would be neccessary to propose novel techniques for generating RNGs but by using zero-cross in order to avoid circuit variations for the quasi-optimal thresholds. Finally, fuzzy, adaptive, or soft computing control techniques could be proved to increase the performance of path planning generators.

Acknowledgements This work has been partially supported by the scientific projects: *CONACYT* No. 258880, *PRODEP* Red de Nanociencia y Nanotecnología, *VIEP-BUAP-2016*. OFB acknowledges the financial support received from *PRODEP (Mexico)*. Also, the authors thankfully acknowledge the computer resources, technical expertise and support provided by the Laboratorio Nacional de Supercómputo del Sureste de México through the grant number O-2016/039.

References

1. Andronov, A., Vitt, E., & Khaikin, S. (1966). *Theory of oscillations*. Springer series in synergetics. Oxford, New York: Addison-Wesley Pub Co.
2. Anishchenko, V. S., Astakhov, V., Vadivasova, T., Neiman, A., & Schimansky-Geier, L. (1995). *Dynamical chaos-models and experiments*. Springer series in synergetics. Singapore: World Scientific.
3. Arita, S. (2014). Relationship between logistic chaos and randomness using a matrix of probabilities and its application to the classification of time series: The line from chaos point to the reversed type of chaos is perpendicular to that from randomness's point to its reversed type. In *15th International Symposium on Soft Computing and Intelligent Systems (SCIS), 2014 Joint 7th International Conference on and Advanced Intelligent Systems (ISIS)* (pp. 1021–1028). doi:10.1109/SCIS-ISIS.2014.7044908.
4. Azar, A. T., & Vaidyanathan, S. (2014). *Chaos modeling and control systems design*. Springer.
5. Azar, A. T., & Vaidyanathan, S. (2015). *Computational intelligence applications in modeling and control*. Springer.
6. Azar, A. T., & Vaidyanathan, S. (2016). *Advances in chaos theory and intelligent control* (1st ed.). Springer.
7. Bae, Y. (2004). Target searching method in the chaotic mobile robot. In *Digital Avionics Systems Conference, 2004, DASC 04* (Vol. 2, pp. 12.D.7–12.1–9).
8. Boulkroune, A., Bouzeriba, A., Bouden, T., & Azar, A. T. (2016). Fuzzy adaptive synchronization of uncertain fractional-order chaotic systems. In T. A. Azar & S. Vaidyanathan (Eds.), *Advances in chaos theory and intelligent control* (pp. 681–697). Springer. doi:10.1007/978-3-319-30340-6_28.
9. Boulkroune, A., Hamel, S., Azar, A. T., & Vaidyanathan, S. (2016). Fuzzy control-based function synchronization of unknown chaotic systems with dead-zone input. In T. A. Azar & S.

Vaidyanathan (Eds.), *Advances in chaos theory and intelligent control* (pp. 699–718). Springer. doi:10.1007/978-3-319-30340-6_29.
10. Cencini, M., Cecconi, F., & Vulpiani, A. (2010). *Chaos from simple models to complex systems* (Vol. 17). World Scientific Publishing.
11. Dettmer, R. (1993). Chaos and engineering. *IEE Review*, *39*(5), 199–203.
12. Jefferies, D. J., Deane, J. H. B., & Johnstone, G. G. (1989). An introduction to chaos. *Electronics Communication Engineering Journal*, *1*(3), 115–123. doi:10.1049/ecej:19890024.
13. Jia, Q., & Wang, X. (2008). Path planning of mobile robots in dynamic environment using chaotic prediction. In *Control and Decision Conference, 2008. CCDC 2008. Chinese* (pp. 925–930).
14. Lawrance, A. J. (2010). Recent theory and new applications in chaos communications. In *In Proceedings of 2010 IEEE International Symposium on Circuits and Systems (ISCAS)* (pp. 2446–2449). doi:10.1109/ISCAS.2010.5537154.
15. Lin, C. F., Shih, S. H., Zhu, J. D., & Lee, S. H. (2012). Implementation of an offline chaos-based EEG encryption software. In *2012 14th International Conference on Advanced Communication Technology (ICACT)* (pp. 430–433).
16. Lorenz, E. N. (1963). Deterministic non periodic flow. *Journal of the Atmospheric Sciences*, *20*(2), 130–141. doi:10.1175/1520-0469(1963)020<0130:DNF>2.0.CO;2.
17. Lu, J., & Chen, G. (2006). A brief overview of multi-scroll chaotic attractors generation. In *2006 IEEE International Symposium on Circuits and Systems (ISCAS). Proceedings* (Vol. 4, pp. 702–705).
18. Muñoz-Pacheco, J., & Tlelo-Cuautle, E. (2010). *Electronic design automation of multi-scroll chaos generators*. Bentham Sciences Publishers.
19. Nakamura, Y., & Sekiguchi, A. (2001). The chaotic mobile robot. *IEEE Transactions on Robotics and Automation*, *17*(6), 898–904. doi:10.1109/70.976022.
20. Nguimdo, R., Lavrov, R., Colet, P., Jacquot, M., Chembo, Y., & Larger, L. (2010). Effect of fiber dispersion on broadband chaos communications implemented by electro-optic nonlinear delay phase dynamics. *Journal of Lightwave Technology*, *28*(18), 2688–2696.
21. Palacin, J., Salse, J. A., Valganon, I., & Clua, X. (2003). Building a mobile robot for a floor-cleaning operation in domestic environments. In *Instrumentation and Measurement Technology Conference, 2003. IMTC '03. Proceedings of the 20th IEEE* (Vol. 2, pp. 1391–1396). doi:10.1109/IMTC.2003.1207979.
22. Pareschi, F., Rovatti, R., & Setti, G. (2007). Second-level nist randomness tests for improving test reliability. In *2007 IEEE International Symposium on Circuits and Systems* (pp. 1437–1440). doi:10.1109/ISCAS.2007.378572.
23. Pareschi, F., Setti, G., & Rovatti, R. (2010). Implementation and testing of high-speed cmos true random number generators based on chaotic systems. *IEEE Transactions on Circuits and Systems I: Regular Papers*, *57*(12), 3124–3137. doi:10.1109/TCSI.2010.2052515.
24. Rukhin, A., & e Soto, J. (2010). *A statistical test suite for random and pseudorandom number generators for cryptographic applications*. Springer series in synergetics. U.S: Department of Commerce, National Institute of Standards and Technology, Gaithersburg, MD.
25. Ryu, H. G., & Lee, J. H. (2013). High security wireless cdsk-based chaos communication with new chaos map. In *Military Communications Conference, MILCOM 2013—2013 IEEE* (pp. 786–790). doi:10.1109/MILCOM.2013.139.
26. Shen, C., Yu, S., Lu, J., & Chen, G. (2014). A systematic methodology for constructing hyperchaotic systems with multiple positive lyapunov exponents and circuit implementation. *IEEE Transactions on Circuits and Systems I: Regular Papers*, *61*(3), 854–864.
27. Sooraska, P., & Klomkarn, K. (2010). "No-cpu" chaotic robots: From classroom to commerce. *IEEE Circuits and Systems Magazine*, *10*(1), 46–53. doi:10.1109/MCAS.2010.935740.
28. Strogatz, S. H. (1994). *Nonlinear dynamics and chaos: With applications to physics, biology, chemistry and engineering*. Perseus Books.
29. Vaidyanathan, S., & Azar, A. T. (2016). Adaptive backstepping control and synchronization of a novel 3-d jerk system with an exponential nonlinearity. In T. A. Azar & S. Vaidyanathan (Eds.), *Advances in chaos theory and intelligent control* (pp. 249–274). Springer. doi:10.1007/978-3-319-30340-6_11.

30. Vaidyanathan, S., & Azar, A. T. (2016). Adaptive control and synchronization of a halvorsen circulant chaotic systems. In T. A. Azar & S. Vaidyanathan (Eds.), *Advances in chaos theory and intelligent control* (pp. 225–247). Springer. doi:10.1007/978-3-319-30340-6_10.
31. Vaidyanathan, S., & Azar, A. T. (2016). Dynamic analysis, adaptive feedback control and synchronization of an eight-term 3-d novel chaotic system with three quadratic nonlinearities. In T. A. Azar & S. Vaidyanathan (Eds.), *Advances in chaos theory and intelligent control* (pp. 155–178). Springer. doi:10.1007/978-3-319-30340-6_7.
32. Vaidyanathan, S., & Azar, A. T. (2016). Generalized projective synchronization of a novel hyperchaotic four-wing system via adaptive control method. In T. A. Azar & S. Vaidyanathan (Eds.), *Advances in chaos theory and intelligent control* (pp. 275–296). Springer. doi:10.1007/978-3-319-30340-6_12.
33. Vaidyanathan, S., & Azar, A. T. (2016). A novel 4-d four-wing chaotic system with four quadratic nonlinearities and its synchronization via adaptive control method. In T. A. Azar & S. Vaidyanathan (Eds.), *Advances in chaos theory and intelligent control* (pp. 203–224). Springer. doi:10.1007/978-3-319-30340-6_9.
34. Vaidyanathan, S., & Azar, A. T. (2016). Qualitative study and adaptive control of a novel 4-d hyperchaotic system with a three quadratic nonlinearities. In T. A. Azar & S. Vaidyanathan (Eds.), *Advances in chaos theory and intelligent control* (pp. 179–202). Springer. doi:10.1007/978-3-319-30340-6_8.
35. Volos, C., Kyprianidis, I., & Stouboulos, I. (2012). A chaotic path planning generator for autonomous mobile robots. *Robotics and Autonomous Systems*, *60*(4), 651–656. doi:10.1016/j.robot.2012.01.001.
36. Wang, L., Xing, X., & Chu, Z. (2008). On definitions of chaos in discrete dynamical system. In *The 9th International Conference for Young Computer Scientists, 2008. ICYCS 2008* (pp. 2874–2878). doi:10.1109/ICYCS.2008.296.
37. Wang, X., Ma, L., & Du, X. (2009). An encryption method based on dual-chaos system. In *Second International Conference on Intelligent Networks and Intelligent Systems, 2009. ICINIS '09* (pp. 217–220). doi:10.1109/ICINIS.2009.63.
38. Xia, F., Tyoan, L., Yang, Z., Uzoije, I., Zhang, G., & Vela, P. (2015). Human-aware mobile robot exploration and motion planner. *SoutheastCon*, *2015*, 1–4. doi:10.1109/SECON.2015.7133021.
39. Xiang, S. Y., Pan, W., Luo, B., Yan, L. S., Zou, X. H., Jiang, N., et al. (2011). Synchronization of unpredictability-enhanced chaos in vcsels with variable-polarization optical feedback. *IEEE Journal of Quantum Electronics*, *47*(10), 1354–1361. doi:10.1109/JQE.2011.2166536.
40. Xu, X., Guo, J., & Leung, H. (2013). Blind equalization for power-line communications using chaos. *IEEE Transactions on Power Delivery*, *99*, 1–8. doi:10.1109/TPWRD.2013.2296834.
41. Yalcin, M. E., Suykens, J. A. K., & Vandewalle, J. (2004). True random bit generation from a double-scroll attractor. *IEEE Transactions on Circuits and Systems I: Regular Papers*, *51*(7), 1395–1404. doi:10.1109/TCSI.2004.830683.
42. Zhu, Q., & Azar, A. T. (Eds.). (2015). *Complex system modelling and control through intelligent soft computations* (Vol. 319). Studies in fuzziness and soft computing. Springer. doi:10.1007/978-3-319-12883-2.

Chaotic System Modelling Using a Neural Network with Optimized Structure

Kheireddine Lamamra, Sundarapandian Vaidyanathan, Ahmad Taher Azar and Chokri Ben Salah

Abstract In this work, the Artificial Neural Networks (ANN) are used to model a chaotic system. A method based on the Non-dominated Sorting Genetic Algorithm II (NSGA-II) is used to determine the best parameters of a Multilayer Perceptron (MLP) artificial neural network. Using NSGA-II, the optimal connection weights between the input layer and the hidden layer are obtained. Using NSGA-II, the connection weights between the hidden layer and the output layer are also obtained. This ensures the necessary learning to the neural network. The optimized functions by NSGA-II are the number of neurons in the hidden layer of MLP and the modelling error between the desired output and the output of the neural model. After the construction and training of the neural model, the selected model is used for the prediction of the chaotic system behaviour. This method is applied to model the chaotic system of Mackey-Glass time series prediction problem. Simulation results are presented to illustrate the proposed methodology.

K. Lamamra (✉)
Department of Electrical Engineering, University of Oum El Bouaghi, Oum El Bouaghi, Algeria
e-mail: l_kheir@yahoo.fr

K. Lamamra
Laboratory of Mastering of Renewable Energies, University of Bejaia, Bejaia, Algeria

S. Vaidyanathan
Research and Development Centre, Vel Tech University, Avadi, Chennai, Tamil Nadu, India
e-mail: sundarvtu@gmail.com

A.T. Azar
Faculty of Computer and Information, Benha University, Benha, Egypt
e-mail: ahmad.azar@fci.bu.edu.eg; ahmad_t_azar@ieee.org

A.T. Azar
Nanoelectronics Integrated Systems Center (NISC), Nile University, Cairo, Egypt

C. Ben Salah
Control and Energy Management Laboratory (CEMLab), Department of Electrical Engineering, National School of Engineers of Sfax, Sfax, Tunisia
e-mail: chokribs@yahoo.fr

A.T. Azar et al. (eds.), *Fractional Order Control and Synchronization of Chaotic Systems*, Studies in Computational Intelligence 688,
DOI 10.1007/978-3-319-50249-6_29

Keywords Modelling • Neural networks • Chaotic system • Multilayer perceptron • Mackey-glass time series • Prediction • NSGA II

1 Introduction

Dynamical systems serve as mathematical models for many exciting real-world problems in many fields of science, engineering and economics [1]. Dynamical systems can be classified into two categories: (1) continuous-time and (2) discrete-time dynamical systems.

A continuous-time dynamical system is generally represented as a system of differential equations given by

$$\frac{dx}{dt} = F(x, t) \tag{1}$$

where F is a continuously differentiable function and $x \in R^n$ is the state vector.

A discrete-time dynamical system is generally described by

$$x(k+1) = G(x(k), k) \tag{2}$$

where G is a continuous function and $x \in R^n$ is the state vector.

A dynamical system is called *chaotic* if it is very sensitive to initial conditions, topologically mixing and exhibits dense periodic orbits [2–4, 5, 6].

Chaos modeling with nonlinear dynamical systems is an active area of research [2–4, 7, 8].

Chaos theory has applications in different fields such as medicine [9, 10], chemical reactions [11–14], biology [15], Tokamak systems [16, 17], dynamos [18, 19], population biology systems [20, 21], oscillators [22, 23], etc.

Several methods have been designed to control a chaotic system about its unstable equilibrium such as active control [24]; Pehlivan et al. [25], adaptive control [26–31], backstepping control [32], sliding mode control [33, 34], fuzzy control [35, 36], artificial neural networks [37, 38], etc.

Artificial Neural Network (ANNs) are networks inspired by biological neuron networks. In machine learning and cognitive science, Artificial Neural Networks (ANNs) are used to estimate functions that can depend on a large number of inputs.

Usually, research in the field of Artificial Neural Network (ANN) is focused on architectures by which neurons are combined and the methodologies by which the weight of the interconnections are calculated or adjusted [39, 40].

The use of neural networks is very large because of its advantages such as the ability to adapt to difficult environments and that change its behaviour, etc. [41]. The Artificial Neural Networks (ANN) are used to model and control both linear and nonlinear dynamical systems [42, 43].

The research on artificial neural networks gives great importance on the construction of an appropriate network, how to combine neurons together, how to calculate and adjust the connection weights and how to find other parameters of the Artificial Neural Networks [44, 45].

Currently some biologists, physicists and psychologists are carrying out research to develop a neural model that is able to simulate the behaviour of the brain by improving the accuracy and the precision. Some engineers are interested in the neural network structure and how to improve their powerful computing capabilities [46, 47].

Recently, a Multi-Layer Perceptron neural network with a fast learning algorithms [48] was developed for the prediction of the concentration of the post-dialysis blood urea. Eight different learning algorithms are used to study their capabilities and compared their performances. Artificial neural networks are used for breast cancer classification [49]. Artificial Neural Networks (ANNs) are used to model a photovoltaic power generation system, by applying the Levenberg–Marquardt algorithm adopted into back propagation learning algorithm for training a feed-forward neural network [50].

In this chapter, the ANNs are used to model a nonlinear chaotic system of Mackey Glass and the structure of the neural model is optimized by Non-dominated Sorting Genetic Algorithm II (NSGA-II), which ensures the obtaining of an optimal neural structure and its learning. We use this neuronal model of Mackey Glass chaotic system to predict the time series of this system and also to test the effectiveness of this technique for the construction of optimal structures for chaotic systems. The resulting model can be also used for the control and synchronization of chaotic systems.

This chapter is organized as follows. Section 2 details the related work in modelling chaotic systems with neural networks. Section 3 details the chaotic systems and the prediction in chaotic systems. Section 4 outlines the modelling of systems using neural networks. Section 5 describes the operating principle of the NSGA-II multi-objective genetic algorithm. Section 6 details the construction and the learning of the MLP neural model using NSGA-II algorithm. As an application, Sect. 7 discusses the Mackey–Glass chaotic time series prediction. Section 8 details the simulation results. Section 9 contains a discussion of the main results. Section 10 contains the conclusions of this work and suggests future research directions.

2 Related Work

Neural networks have many applications in chaos theory. Many researchers have used neural networks in their work for modelling or control of chaotic systems.

Recently, Cellular Neural Networks (CNN) have been applied for the modelling and control of chaotic systems [51–55]. Cellular Neural Networks are similar to neural networks with the difference that communication is allowed between neighbouring units only.

Chen [56] presented a work in which he proposed the application of neural-network based fuzzy logic control to a nonlinear time-delay chaotic system.

Zhang and Shen [57] have studied memristor-based chaotic neural networks with both time-varying delays and general activation functions. Pham et al. [58] proposed a novel simple neural network having a memristive synaptic weight.

Wen et al. [59] have studied the problem of exponential lag synchronization control of memristive neural networks via the fuzzy method and applications in pseudorandom number generators.

Kyprianidis and Makri [60] have presented a study of a complex dynamics of a system of two nonlinear neuronal cells, coupled by a gap junction, which is modelled as a linear variable resistor and the coupled cells of the FitzHugh-Nagumo oscillators systems.

He et al. [61] proposed a pinning control method focused on the chaotic neural network, and they demonstrated that the chaos in the chaotic neural network can be controlled with this method and the states of the network can converge in one of its stored patterns if the control strength and the pinning density are chosen in a suitable manner.

Ramesh and Narayanan [62] discussed the chaos control of Bonhoeffer–van der Pol oscillator using neural networks. Ren et al. [63] have proposed a dynamic control method using a neural network for unknown continuous nonlinear systems through an online identification and adaptive control.

Neural networks have been applied for the synchronization of chaotic systems. Many control methods have been used for the synchronization of chaotic systems such as sampled-data control [64], fuzzy neural control [65], adaptive control [66], chaotic time-series method [67], simulated annealing method [68], sliding mode control [69], time-delay control [70], etc.

In this chapter, we use the Multi-Layer Perceptron (MLP) neural networks optimized by multi-objective genetic algorithms of NSGA-II type for the modelling of chaotic systems and making the prediction.

In this approach, the training of the resulting neural network is provided by the NSGA-II algorithm and the structure is considered optimal, since the NSGA-II algorithm evolves over many generations. Each generation of the NSGA-II algorithm is composed of several individuals who are neural network models proposed to model the chaotic system. At the end of the evolution of the last generation, we obtain a set of solutions (called Pareto front) that are a set of models of the chaotic system, and we choose from this set, the model that suits us most. The details of the NSGA-II algorithm are presented in Sect. 5.

This optimization technique of the neural network structure has been used in further work and has proven its efficiency. We cite for example; the use of radial basis function (RBF) neural networks optimized by the NSGA-II multi-objective genetic algorithm for modelling of a nonlinear systems of Box and Jenkins which is a gas-fired boiler. For this model, the input is the gas at the inlet and the output is the concentration of released CO2. In the neural networks model for the gas-fired boiler, the NSGA-II algorithm is used to find the best number of neurons in the hidden layer of the RBF neural network and provide the best connection weights between neurons in the hidden layer and the output layer, and also find the parameters of the radial function of neurons hidden layer. A Gaussian form of the radial functions is used in all the neurons. The NSGA-II algorithm finds the best centers and the best widths sigma for the Gaussian functions. The chromosome then contains the number of neurons in the hidden layer, Gaussian functions centers and widths of the hidden layer neurons, and the weights of connections between the hidden layer and the output layer [71].

3 Chaotic Systems and Prediction

3.1 Chaotic Systems

Chaotic systems are nonlinear dynamical systems which are very sensitive to initial conditions [38]. Chaotic behaviour is observed in many branches of science and engineering [2–4]. The behaviour of chaotic systems can be studied with chaotic mathematical models [2, 3, 72]. Chaos theory has applications in several areas such as computer science, engineering, physics, biology, meteorology, sociology, economics, etc. [4, 8, 72].

3.2 Prediction in Chaotic Systems

The prediction in chaotic systems has important applications in science and engineering [73, 74]. Chaotic systems like Lorenz system have important applications in weather models [4]. Several errors of weather forecasts are caused by the use of overly simplified models and the lack of accurate measurement of various parameters such as pressure, temperature, wind speed, etc. [74].

The existence of chaos was first introduced by Poincaré at the end of the nineteenth century [4]. Poincaré discovered chaos when he investigated the stability of the three-body model (e.g. Earth-Moon-Sun). Poincaré tested the stability of the three-body model by comparing the trajectories followed by one of the bodies from two very close initial positions. These phase trajectories remain close to each other in the short term and can therefore predict eclipses, but they become completely different in the long term. This is due to the chaotic nature of the three-body problem (e.g. Earth-Moon-Sun).

Thus, there is a need to build a good mathematical model that can make good predictions of the trajectories of the chaotic system. Neural networks serve as a good mathematical model for building a chaotic system and making predictions of the trajectories of the system. In this work, a Multi-Layer Perceptron (MLP) neural network is used to model a chaotic system and make prediction thereafter starting from this model.

4 Modelling Using a Neural Network

Artificial neural networks have the approximation property which may be stated as follows: Every bounded function sufficiently regular can be approximated with arbitrary precision in a finite area of space of its variables, by a neural network comprising a layer of hidden neurons finite in number, all possessing the same activation function and a linear output neuron [75, 76].

Usually we do not use neural networks to make approximations of known functions. The system identification problem is typically to find a relationship between a set of outputs of a given process, and all the corresponding entries, through the measurements. It is assumed that this relationship exists despite the measures are finite in number, which are often blemished by noise [77]. Also, all the variables that determine the outcome of the process may not be measurable. Thus, neural networks are considered as important models for system identification using available outputs of the system [3, 78–80].

Generally, a neural network allows to make better use of available measures than the conventional linear approximation methods [77, 81]. This advantage is especially important when the process to be modelled depends on several variables such as in the case of shaping processes where it intervenes several types of non-linearity and multiple hardware and technological parameters. Thus, the concept of classification can be conducted to a problem of approximation of nonlinear regression functions. For this reason, neural networks are frequently used as classifiers or discriminators [82].

When the data related to the system are well controlled and with significant number and where the boundaries between each collected data class are not overly complex, neural networks are used to perform sorting and pattern recognition of surfaces, characters, or symbols [83–85].

5 The NSGA-II Multi-objective Genetic Algorithm

The construction and learning of the Multi-Layer Perceptron (MLP) neural model for a chaotic system are performed by the Multi-Objective Genetic Algorithm of NSGA-II type (Non-dominated Sorting Genetic Algorithm). NSGA-II algorithm has important applications in multi-objective optimization problems [86].

NSGA-II algorithm is a popular genetic algorithm in multi-objective function optimization theory [86]. NSGA-II algorithm has been used in many research works in multi-objective optimization problems [87–91].

NSGA-II algorithm is an improved version of NSGA algorithm for multi-objective function optimization [92]. NSGA-II algorithm has helped to solve the problems encountered in the NSGA algorithm such as the complexity, non-elitist and the use of sharing.

NSGA-II algorithm makes the dominance relationships between individuals and provides a fast sorting method of chromosomes [86, 93]. It uses a selection operator based on the measure of crowding around individuals to ensure diversity in the population.

NSGA-II algorithm is an elitist algorithm. In order to manage elitism, NSGA-II algorithm evolves so that at each new generation, the best individuals encountered are retained. The operating principle of NSGA-II algorithm is shown in Fig. 1 [94].

The working principle of NSGA-II algorithm is described as follows: At first, an initial population is randomly created. Then a sorting operation is performed using the non-domination concept. For each solution, we assign a rank equal to the level of non-dominance, viz. the rank 1 for best, 2 for the next level, etc. Then, a tournament of selection of parents is performed during the reproduction process.

Once two individuals of the population are randomly chosen, the tournament is performed on a comparison of the domination with constraints of the two individuals.

For a given generation t, after creating a children population Q_t from the previous population P_t (generated from the parents via the genetic operators, crossover and mutation), a population R_t is created that includes the parents population P_t and the children population Q_t such that $R_t = P_t \cup Q_t$. This ensures the elite nature of the NSGA-II algorithm. Then the population R_t contains twice the size of the previous population, i.e. R_t has $2N$ individuals consisting of N parents and N children).

Next, the concept of non-dominance of Preto is applied to sort the population R_t. Then the individuals of R_t will be grouped in successive fronts $(F_1, F_2, \ldots)$ where F_1 represents individuals of rank 1, F_2 represents individuals of rank 2, etc.

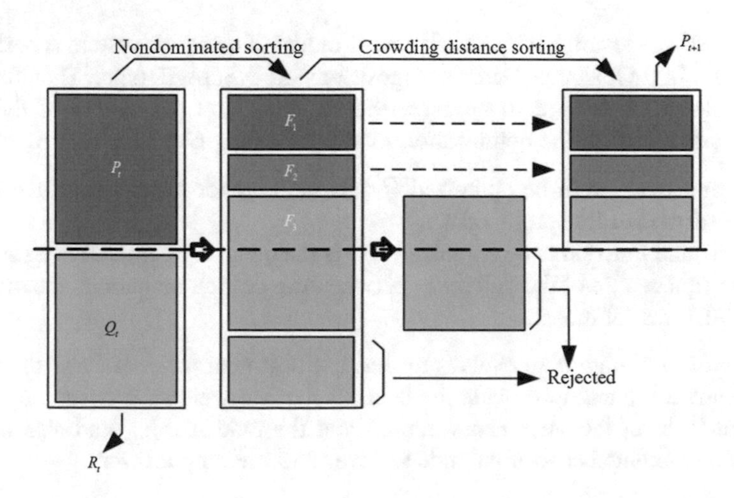

Fig. 1 Operating principle of NSGA-II algorithm

After the sorting, the size of R_t should be reduced to N individuals in order to form the next population P_{t+1}. Thus, N individuals from R_t must be excluded for forming the next population. If the size of the front F_1 is less than N, then all its individuals are preserved and the same procedure is followed for the other fronts while the number of the preserved individuals does not exceed the size N.

In the example illustrated by Fig. 1, both fronts F_1 and F_2 are fully preserved. However, keeping the front F_3 will result in exceeding the size N of the population P_{t+1}. Thus, it is necessary to make a selection to determine which individuals to keep. For this purpose, NSGA-II performs a mechanism for preserving the diversity in the population based on the evaluation of the density of the individuals around each solution across a calculating procedure of the "distance proximity". Thereby, a low value of the proximity distance for an individual is an individual "well surrounded".

NSGA-II algorithm then proceeds with a descending sorting according to this proximity distance to preserve the necessary number of individuals of F_3 front and remove some individuals from the densest areas. In this manner, the population P_{t+1} is made up to N individuals and diversity is ensured. The individuals with extreme values of the criteria are maintained by this selection mechanism, which keeps the external bounds of the Pareto front.

At the end of this phase, the population P_{t+1} is created. Then a new population Q_{t+1} is generated from P_{t+1} by the reproduction operators.

The above procedure is repeated by ensuring elitism and diversity until the stopping criteria defined beforehand is reached.

6 Construction and Learning of the MLP Neural Model by NSGA-II Algorithm

In this work, the learning of the MLP neural model of a chaotic system is performed by using Multi-Objective Genetic Algorithms of NSGA-II type. Therefore this genetic algorithm is used to optimize the structure and parameters of the MLP neural model through the optimization of the following objective functions:

- The first function to be optimized (f_1) is the number of neuron of the hidden layer of the MLP neural model.
- The second function to be optimized (f_2) is the quadratic cumulative error which is the square sum of the difference between the desired output and the output of the MLP neural model.

The NSGA-II algorithm evolves to find the best neurons number of the hidden layer denoted N_{hl} and to provide the best connection weights between neurons of the input layer of the MLP neural model and the hidden layer, and also the best connection weights between the hidden layer and the output layer.

Then the chromosome contains the number of neurons in the hidden layer, and the connection widths connecting the different layers of the MLP neural model.

The MLP neural network used here to model a chaotic system is composed of three layers:

- The Input layer, which is composed of two neurons. The first neuron corresponds to the input data of the chaotic system, whereas the second neuron corresponds to the modelling instantaneous error.
- The hidden layer, which is composed of an undetermined number of neurons (N_{hl}). In fact, the number N_{hl} varies according to the optimized neural network model. The multi-objective genetic algorithm NSGA-II seeks to find the best number that ensures the best possible structure with the smallest model error in order to guarantee good modelling of the chaotic system.
- The output layer, which is composed of one neuron. This neuron corresponds to the output of the neural model of the chaotic system. Our objective is to keep this output the nearest possible to the desired output of the chaotic system to ensure thereafter (after the construction of the neural model) a good prediction of the future states of the chaotic system.

The structure of the neural model used is shown in Fig. 2.

In Fig. 2, x_1 corresponds to the chaotic system input data, x_2 corresponds to the modelling instantaneous error and y corresponds to the output of the neural model of the chaotic system.

W is the matrix of the connection weights of the input layer neurons and the neurons of the hidden layer. In our case, this matrix is composed of W_{11} to W_{1Nhl} which are the x_1 input connection weights with the neurons of the hidden layer N_{hl} and W_{21} to W_{2Nhl} that are the x_2 input connection weights with the neurons of the hidden layer N_{hl}. Therefore, the size of the matrix W is $2 \times N_{hl}$.

Z is a vector composed of Z_1 to Z_{Nhl} which are the connection weights between the neurons of the hidden layer N_{hl} and the neuron of the output layer y. Therefore, the size of the matrix Z is $1 \times N_{hl}$.

The chromosome of the multi-objective algorithm NSGA-II generic therefore takes the following form:

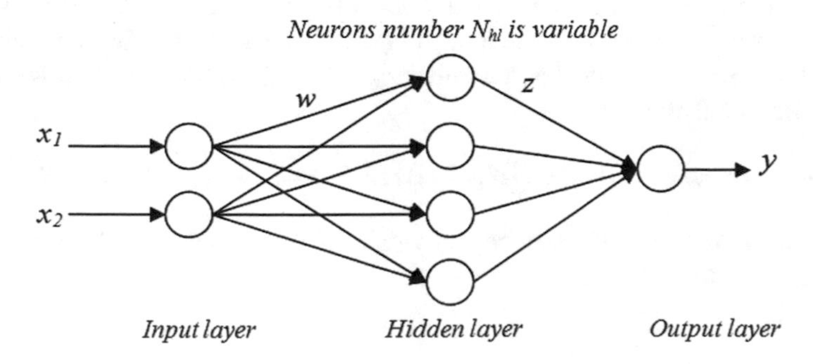

Fig. 2 Structure of the MLP neural model

$$Chromosome = [N_{hl} \quad W_{11} \quad \cdots \quad W_{1Nh1} \quad W_{21} \quad \cdots \quad W_{2Nh1} \quad Z_1 \quad \cdots \quad Z_{Nh1}]$$

The chromosome length (Lc) depends on the number of neurons in the hidden layer N_{hl} and it is given by the formula:

$$Lc = (N_{hl} \times 3) + 1.$$

For example, for a neural model with 4 neurons in the hidden layer (i.e. $N_{hl} = 4$), the chromosome will have a length equal to 13 (i.e. $Lc = 13$) and it will have the following form:

$$Chromosome = [4 \quad W_{11} \quad W_{12} \quad W_{13} \quad W_{14} \quad W_{21} \quad W_{22} \quad W_{23} \quad W_{24} \quad Z_1 \quad Z_2 \quad Z_3 \quad Z_4]$$

We have chosen a variation range of the number of neurons in the hidden layer N_{hl} between 2 and 20.

Then the population size matrix (Spm) will have the following size.

$$\mathrm{Size}(Spm) = N \times Lc_{\max} = N \times [(N_{hl\max} \times 3) + 1]$$

where N is the number of individuals in the population, and $N_{hl\max}$ is the maximum number of neurons of the hidden layer of the current population. This number is not necessarily equal to 20, because the number of neurons in the hidden layer of each individual (or neural model) is randomly initialized (such as the initial connection weights) during the creation of the initial population at the beginning of the evolution of the multi-objective genetic algorithm NSGA-II.

Therefore, the size of the population in columns is not fixed and it takes every time a corresponding size to the greatest number of neurons in the hidden layer.

It is noted that this value (which cannot exceed 20 neurons, according to the interval that we have chosen) may change over the generations following the process of crossover and mutation.

Also, since we have to keep the same number of columns of the population matrix for all individuals, we have proceeded to put zeros in columns so that their N_{hl} is less than N_{hlmax}.

For example, we consider two individuals of the same population i, having a different number of neurons in the hidden layer. Suppose that the first individual has a $N_{hl1}(i) = 4$ and the second has a number $N_{hl2}(i) = N_{hlmax}(i) = 16$. In this case, the size of the population is:

$$Spm(i) = N \times Lc_{max}(i) = N \times (N_{hlmax}(i) \times 3 + 1) = N \times (16 \times 3 + 1) = N \times 49;$$

The chromosome of the second individual of this population (i) will have the following structure:

$$Chromosome_2(i) = [16\, W_{11}\, W_{12}\, \ldots\, W_{116}\, W_{21}\, W_{22}\, \ldots\, W_{216}\, Z_1\, Z_2\, \ldots\, Z_{N16}]$$

whereas the chromosome of the first individual of this population will have the following structure:

$$Chromosome_1(i) = [4\, W_{11}\, W_{12}\, \ldots\, W_{14}\, W_{21}\, W_{22}\, \ldots\, W_{24}\, Z_1\, Z_2\, \ldots\, Z_{N4}\, 0\, 0\, \ldots\, 0]$$

Thus, the lengths of the vectors of these two individuals are equal and they have the value of $L_{c1}(i) = L_{c2}(i) = N_{hlmax}(i) \times 3 + 1 = 49$. This is the length of all individuals within the population i. However, the chromosome of the first individual will include 36 zeros to reach the same size as its counterparts individuals within the population i.

It is quite obvious that the evolution process of the multi-objective genetic algorithm takes much time to provide in the end the non-dominated solutions (often they are the optimal solutions), but it does not hamper our technique. that This is because the NSGA-II algorithm is used to evolve off-line and at the end we choose the model which gives the smallest modelling error among those of the Pareto front to make the prediction without the intervention of NSGA-II algorithm.

7 Application to the Chaotic System: Mackey–Glass Chaotic Time Series Prediction

Our method is applied to model the chaotic system of MacKey–Glass Time Series with 1000 data for predicting the time series [95].

A time series is a sequence of observations on a variable measured at successive points in time or over successive periods of time. The measurements may be taken every hour, day, week, month, or year, or at any other regular interval. The pattern of the data is an important factor in understanding how the time series has behaved in the past. If such behaviour can be expected to continue in the future, we can use the past pattern to guide us in selecting an appropriate forecasting method.

To identify the underlying pattern in the data, a useful first step is to construct a time series plot. A time series plot is a graphical presentation of the relationship between time and the time series variable; time is on the horizontal axis and the time series values are shown on the vertical axis. Let us review some of the common types of data patterns that can be identified when examining a time series plot [96].

The chaotic system of MacKey–Glass Time Series is the first delay chaos discovered in 1977 from a physiological model. It is generated by the following differential equation:

$$\dot{x}(t) = \beta x(t) + \frac{\alpha x(t-\tau)}{1 + x^{10}(t-\tau)}$$

where $\alpha = 0.2, \beta = -0.1$ and $\tau = 17$. It is well-known that the MacKey-Glass system is chaotic with the fractal dimension 2.1 for these parameters [97].

For this chaotic system, when $x(0) = 1.2$ and $\tau = 17$, there is a non-periodic series and non-convergent time series which are highly sensitive to initial conditions.

The training phase of the neural model is done on 200 data and the validation on the 800 remaining data. The NSGA-II algorithm is applied to minimize simultaneously the following fitness functions:

1. The number of neurons in the hidden layer (N_{hl})
2. The quadratic cumulative error (Ec) that is the difference between the reference (or desired output) and neural model output.

$$Ec(W) = \sum_{i=1}^{m} (y_r(i) - y_d(i))^2$$

where:

- W is the matrix of connection weights.
- m is the number of *data of* input vector (equal to the number of data vector of desired outputs y_d); m is equal to 1000; (200 *data* for the training phase of the neural model; and 800 data for the validation phase*)*.
- $y_r(i)$ is the ith output value of the MLP neural model.
- $y_d(i)$ is the ith output of the desired data.

8 Results of Simulation

The multi-objective genetic algorithm NSGA-II evolves along the generations with the aim to build, train and provide in the end a neural model with optimal structure and connection weights ensuring a good training (construction and training phase). The parameters of the NSGA-II algorithm used are the following:

- Search intervals:
  - Connexion weights: $W \in [-30, 30]$; $Z \in [-30, 30]$;
  - Neural number in the hidden layer $N_{hl} \in [2, 20]$
- The population size (number of individuals): $N = 100$
- Number of generations = 300
- Crossover Probability = 0.9
- Mutation Probability = 0.08
- Selection type: Stochastic selection

Table 1 Pareto front of the MLP NN Model

N° of individual	Hidden layer neurons number N_{hl}	Training error E_{ct}	Prediction error E_{cp}	Global error E_g
1	2	93.2314	399.7845	493.0159
2	3	15.3047	71.6695	86.9742
3	4	10.4758	35.8125	46.2883
4	5	4.2154	16.8541	21.0695
5	6	1.6245	7.5231	9.1476
6	7	1.1687	5.0054	6.1741
7	8	0.8876	3.3251	4.2127
8	9	0.8340	2.4975	3.3315

At the end of the evolution of the latest generation of the NSGA-II algorithm, a set of neural models is provided in the Pareto front (set of non-dominated solutions).

When the construction and training of the neural model are finished (construction and training phase), the selected model (among the Pareto front) is placed separately (without the intervention of the NSGA-II) for the prediction of the future of the Mackey Glass chaotic system.

The neural models of the Pareto front of the latest generation are given in the Table 1.

The training phase of the MLP neural model is carried out on 200 data, and the prediction phase is done on the 800 data.

E_{ct} is the training quadratic cumulative error; E_{cp} is the prediction quadratic cumulative error and E_g is the global quadratic cumulative error.

All the individuals (MLP neural models) of the Table 1 are non-dominated solutions of the Pareto front. Nonetheless, the individual N° 6 is chosen as the best solution because it provides the smallest error and also there is a great difference between this value and that of other individuals.

The choice of this individual is done through the first fitness function (which is the error) which has a higher importance than the second fitness function which is the number of neurons in the hidden layer of the neural model. This individual is a neural model, with nine neurons in the hidden layer ($N_{hl} = 9$) and offers a global modelling error equal to 3.3315.

Figure 3 shows the Mackey Glass chaotic time series considered as the desired output y_d. Figs. 4 and 5 show the results of the training phase. Figures 5 and 6 show the results of the prediction phase. Figures 8 and 9 depict the representation of the overall data of the two phases together. Figure 10 shows the non-dominated individuals of the Pareto front of the latest generation of NSGA-II algorithm.

The Mackey Galss time series

See Fig. 3.

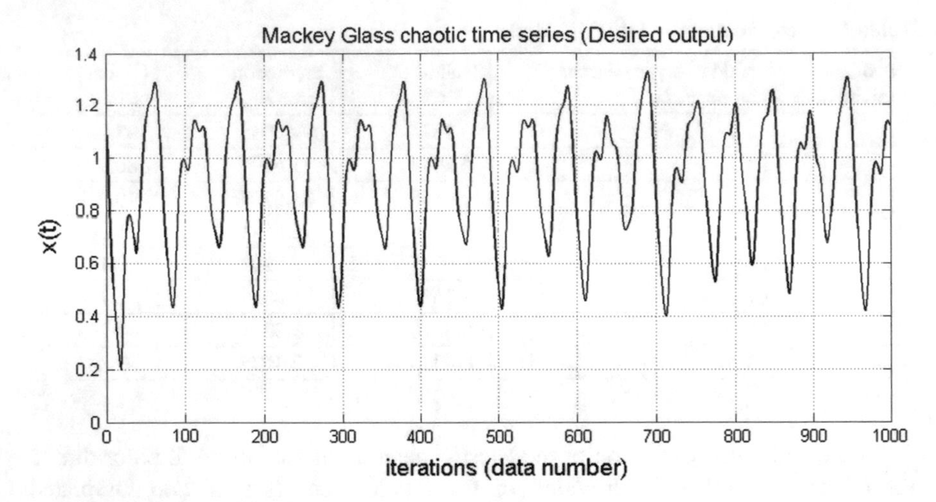

Fig. 3 The Mackey glass chaotic system time series

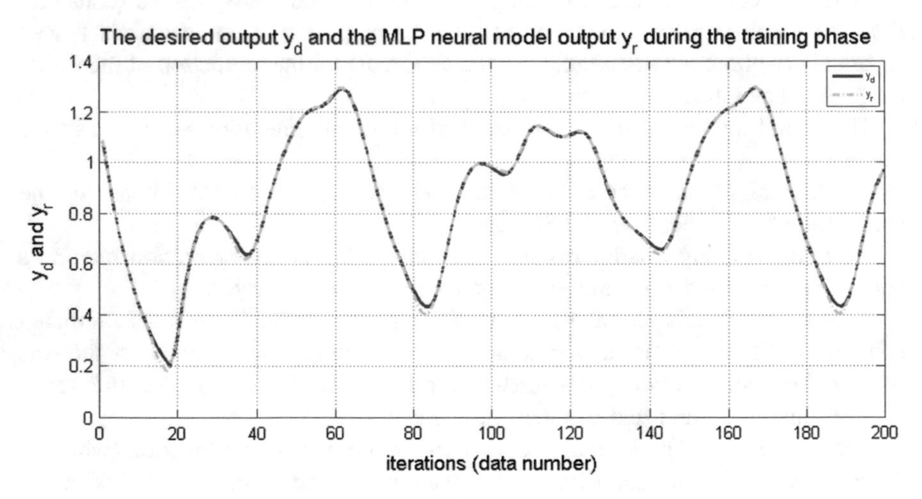

Fig. 4 The Mackey glass chaotic system time series and the output of MLP neural model during the training phase

The training phase:

The Mackey Glass chaotic time series data which are the desired output y_d and the output of MLP neural model y_r of the training phase are shown in Fig. 4 and the global instantaneous quadratic training error is represented in the Fig. 5.

The prediction phase:

The desired output y_d and the output of MLP neural model y_r of the prediction phase are shown in Fig. 6 and the global instantaneous quadratic prediction error is represented in Fig. 7.

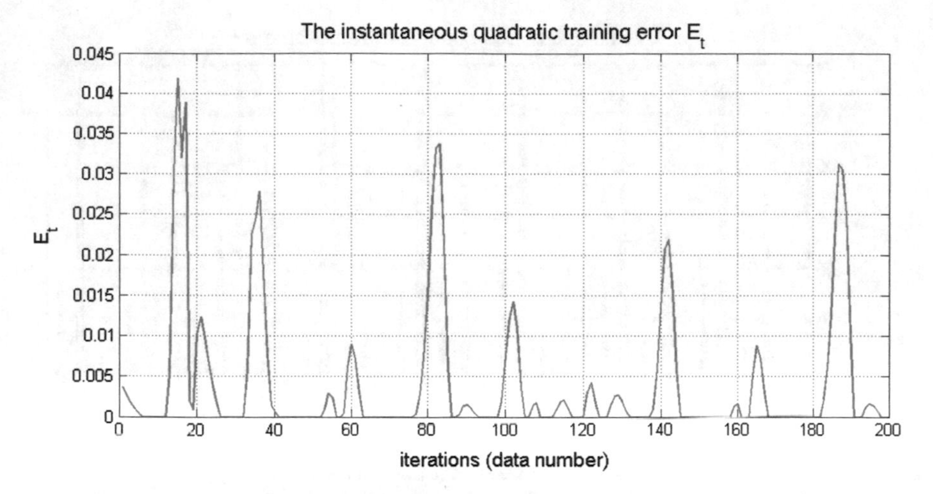

Fig. 5 The instantaneous quadratic training error

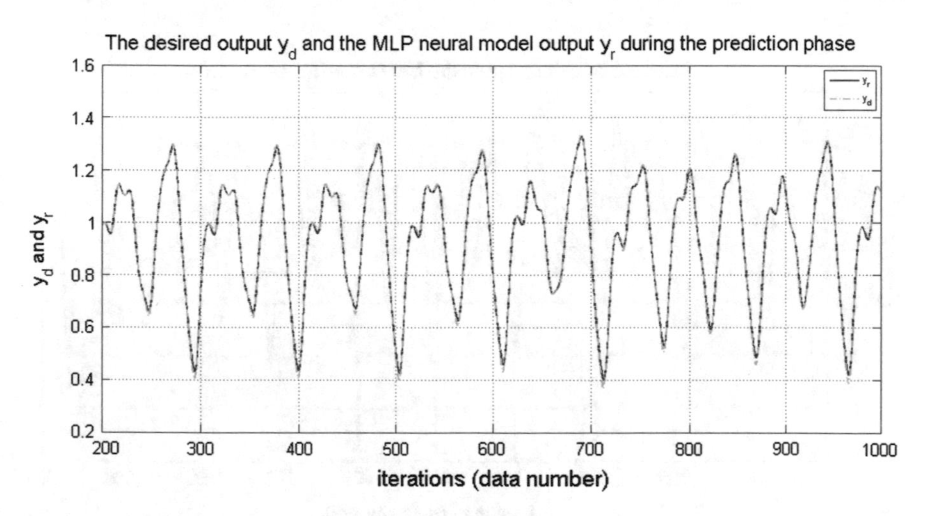

Fig. 6 The Mackey glass chaotic system time series and the output of MLP neural model during the prediction phase

The global data:

The global data of the desired output y_d and the output of MLP neural model y_r are shown in Fig. 8 and the global instantaneous quadratic error is represented in the Fig. 9.

The Pareto front:

Pareto front which is composed of non-dominated individuals (that are the MLP neural models) of the last generation of the multi-objective genetic algorithm NSGA-II is shown in Fig. 10.

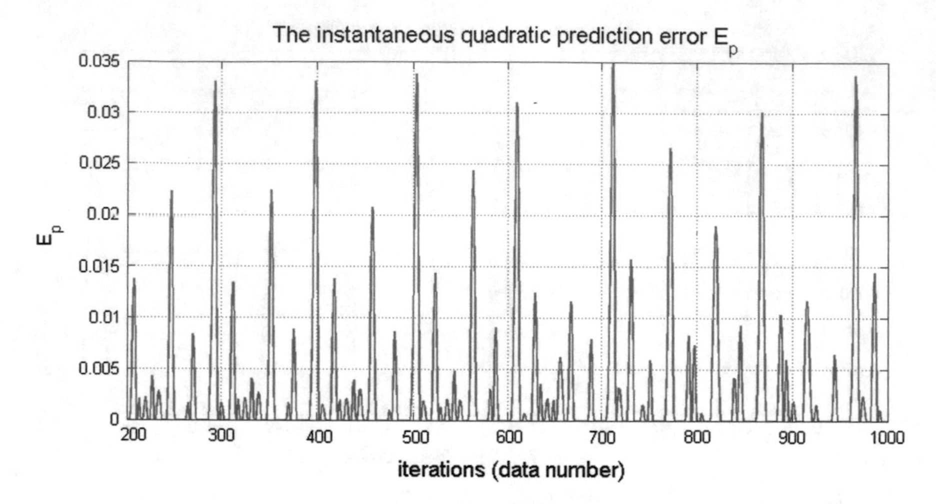

Fig. 7 The instantaneous quadratic prediction error

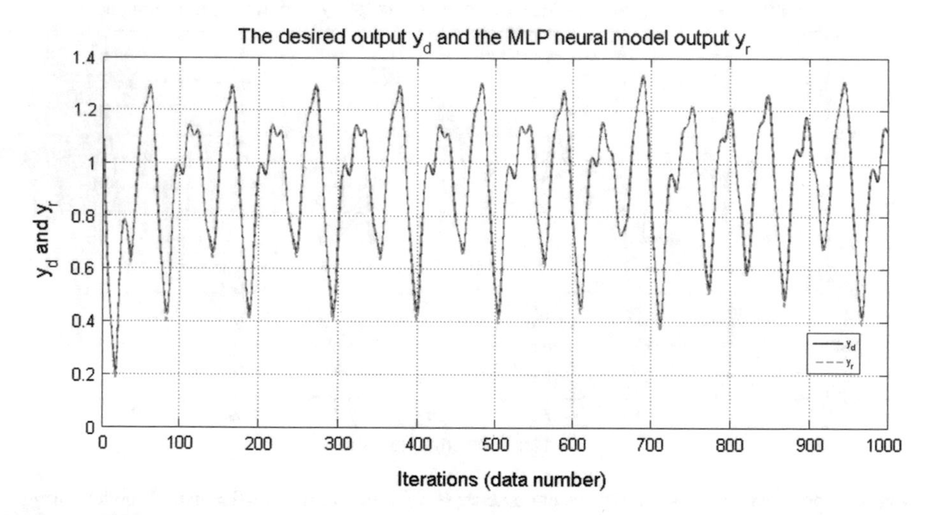

Fig. 8 The Mackey Glass chaotic system time series and the MLP neural model output

9 Discussion of Results

The figures shown in Sect. 8 represent the results of the individual which gives a global quadratic error equal to $E_g = 3.3315$ for 1000 data. This gives rise to an average modelling error $E_{gm} = 3.3315 \times 10^{-3}$. The training quadratic cumulative error is $E_{ct} = 0.8340$ for 200 data and the predictive quadratic cumulative error is $E_{cp} = 2.4975$ for 800 data. Thus, the average training error is $E_{ctm} = 4.17 \times 10^{-3}$

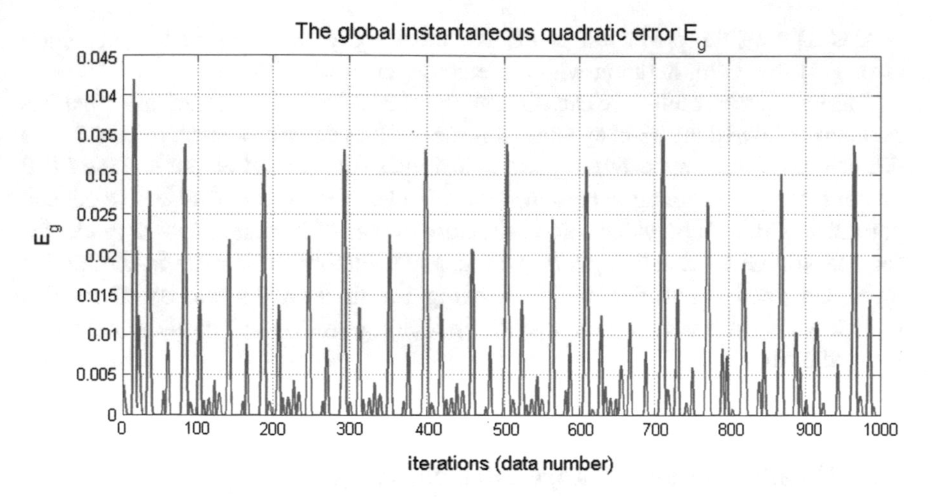

Fig. 9 The global instantaneous quadratic error

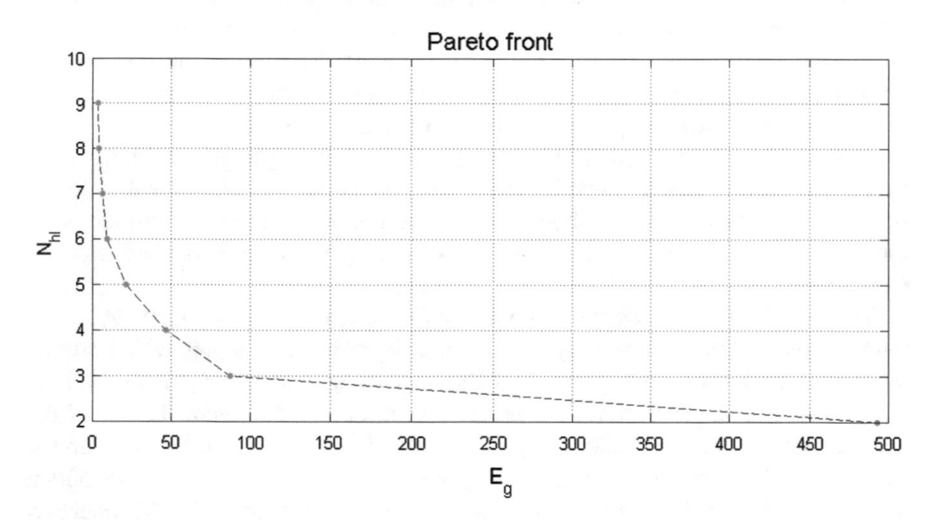

Fig. 10 The Pareto front

and the average prediction error is $E_{cpm} = 3.1218 \times 10^{-3}$. These errors are very small for a chaotic system modelling.

It can be noted that the average training error is bigger than the average prediction error. This is caused from the fact that during the training phase, the neural network needs to adapt to the changes coming to the chaotic system. This is a very important property of a neural network (it has the ability to adapt to the conditions imposed by any environment). During the training phase, the biggest peak in the training error (Fig. 4) appears at 15th iteration with a training error equal to 0.04192, and this error is the greatest instantaneous error for all data of the two phases. During the prediction phase, the most significant peak in the prediction error is at the 712th iteration with a predictive error of 0.03486.

Through these results, we can observe that the output of MLP neural model has perfectly followed the Mackey Glass time series (the desired output). In Fig. 7, it is difficult to differentiate between these two signals (the desired output and the MLP neural model). According to these results, it is clear that the multi-objective genetic algorithm NSGA-II provided a good structure of the MLP neural network model for the Mackey Glass chaotic system, with a good number of neurons in the hidden layer and good connection weights linking the three layers that constitute this model, which brings us to note that the modelling process using this technique is very efficient.

10 Conclusion and Future Directions

Chaos theory has wide applications in several fields of physics, biology, chemistry, robotics, communication theory, networks, medicine, economics, etc.

Chaotic systems are very sensitive to initial conditions. Thus, the evolution of phase trajectories of chaotic systems follows a very complex pattern and it is a very difficult task to predict long-term behaviour of chaotic systems

In this work we presented the modelling of chaotic systems by Multi-Layer Perceptron (MLP) neural networks. Since the construction of an optimal structure of neural networks is usually difficult, we proceeded to optimize their structure by multi-objective genetic algorithm of NSGA-II type (Non-dominated Sorting Genetic Algorithm).

The proposed neural model is composed of three layers: an input layer, the output layer and the hidden lawyer. The input layer has two inputs which are the input data of the chaotic system and the modelling error. The output layer is composed of a single neuron that represents the output the neural model of the chaotic system. Also, the hidden layer consists of a variable number of neurons. The NSGA-II algorithm determines the optimum number of neurons in the hidden layer. Also, the NSGA-II algorithm is intended to provide the optimal connection weights between the three layers of the neural model of the chaotic system and thereby ensuring its training. Thus it operates to minimize the following two fitness functions: the number of neurons in the hidden layer and the modelling error. This method is applied in two phases. The first phase is the construction and training phase of the neural model with the NSGA-II algorithm. The second phase is the application of the neural model for predicting the future of the modelled chaotic system. Thus, once the construction and training phase of the neural model are finished, the model chosen (among the Pareto front of the latest generation of the NSGA-II) is used alone (without the NSGA-II) in the prediction phase.

This technique is applied to model the Mackey Glass chaotic system. The objective is to make from this neuronal model, the predicting of the Mackey Glass time series and test whether this technique is effective for the construction of optimal structures for chaotic systems. The resulting model can be also used for the study, control and synchronization of the chaotic systems.

The obtained results show that the two signals of the desired data and of the neural model data are almost completely identical whether during the training phase or into the prediction phase. These results are very satisfying and encouraging and show that this method provides a good model of the chaotic system.

In future work, we plan to improve this technique for the neural networks type of Multi-Layer Perceptron (MLP) and for those of Radial basis function type, by the application of NSGA-II algorithm and even the use of others algorithms such that the Particle Swarm Optimization (PSO). These techniques will also be applied in different areas, such as the modelling and the control in the field of renewable energy mainly for the maximum power point tracking for dynamics photovoltaic systems, and also for the control of highly complex nonlinear systems.

References

1. Sundarapandian, V. (2012). *Ordinary and partial differential equations*. New Delhi, India: McGraw Hill Education.
2. Azar, A. T., & Vaidyanathan, S. (2015). *Chaos modeling and control systems design, Studies in computational intelligence* (Vol. 581). Germany: Springer. ISBN 978-3-319-13131-3.
3. Azar, A. T., & Vaidyanathan, S. (2015). *Computational intelligence applications in modeling and control. Studies in computational intelligence* (Vol. 575). Germany: Springer. ISBN 978-3-319-11016-5.
4. Azar, A. T., & Vaidyanathan, S. (2016). *Advances in chaos theory and intelligent control. Studies in fuzziness and soft computing* (Vol. 337). Germany: Springer. ISBN 978-3-319-30338-3.
5. Marek, M., & Schreiber, I. (1995). *Chaotic behaviour of deterministic dissipative systems* (Vol. 1). Cambridge University Press. ISBN: 0-521-43830-6.
6. Schuster, H. G., & Just, W. (2006). *Deterministic chaos: An introduction.* Wiley. ISBN 978-3-527-40415-5.
7. Iooss, G., Helleman, R. H., & Stora, R. (1983). *Chaotic behaviour of deterministic systems* (Vol. 16). Les Houches. ISBN 0-444-86542-X.
8. Thompson, J. M. T., & Stewart, H. B. (2002). *Nonlinear dynamics and chaos.* Wiley. ISBN 0-471-87645-3.
9. Vaidyanathan, S. (2016). Global chaos control of the FitzHugh-Nagumo chaotic neuron model via integral sliding mode control. *International Journal of PharmTech Research, 9*(4), 413–425.
10. Vaidyanathan, S. (2016). Anti-synchronization of enzymes-substrates biological systems via adaptive backstepping control. *International Journal of PharmTech Research, 9*(2), 193–205.
11. Vaidyanathan, S. (2015). Anti-synchronization of chemical chaotic reactors via adaptive control method. *International Journal of ChemTech Research, 8*(8), 73–85.
12. Vaidyanathan, S. (2015). Global chaos synchronization of chemical chaotic reactors via novel sliding mode control method. *International Journal of ChemTech Research, 8*(7), 209–221.
13. Vaidyanathan, S. (2015). A novel chemical chaotic reactor system and its output regulation via integral sliding mode control. *International Journal of ChemTech Research, 8*(11), 669–683.
14. Vaidyanathan, S. (2015). Integral sliding mode control design for the global chaos synchronization of identical novel chemical chaotic reactor systems. *International Journal of ChemTech Research, 8*(11), 684–699.

15. Meng, X., Xiang, W., Jia, L., & Xu, J. (2015). Train flow chaos analysis based on an improved automata model. *Chaos, Solitons and Fractals, 81*, 43–51.
16. Vaidyanathan, S. (2015). Dynamics and control of Tokamak system with symmetric and magnetically confined plasma. *International Journal of ChemTech Research, 8*(6), 795–803.
17. Vaidyanathan, S. (2015). Synchronization of Tokamak systems with symmetric and magnetically confined plasma via adaptive control. *International Journal of ChemTech Research, 8*(6), 818–827.
18. Vaidyanathan, S. (2015). Adaptive synchronization of Rikitake two-disk dynamo chaotic systems. *International Journal of ChemTech Research, 8*(8), 100–111.
19. Vaidyanathan, S. (2015). Hybrid chaos synchronization of Rikitake two-disk dynamo chaotic systems via adaptive control method. *International Journal of ChemTech Research, 8*(11), 12–25.
20. Vaidyanathan, S. (2015). Active control design for the hybrid chaos synchronization of Lotka-Volterra biological systems with four competitive species. *International Journal of PharmTech Research, 8*(8), 30–42.
21. Vaidyanathan, S. (2015). Adaptive biological control of generalized Lotka-Volterra three-species biological system. *International Journal of PharmTech Research, 8*(4), 622–631.
22. Vaidyanathan, S. (2015). Sliding controller design for the global chaos synchronization of forced Van der Pol chaotic oscillators. *International Journal of PharmTech Research, 8*(7), 100–111.
23. Vaidyanathan, S. (2015). Output regulation of the forced Van der Pol chaotic oscillator via adaptive control method. *International Journal of PharmTech Research, 8*(6), 106–116.
24. Karthikeyan, R., & Sundarapandian, V. (2014). Hybrid chaos synchronization of four-scroll systems via active control. *Journal of Electrical Engineering, 65*(2), 97–103.
25. Pehlivan, I., Moroz, I. M., & Vaidyanathan, S. (2014). Analysis, synchronization and circuit design of a novel butterfly attractor. *Journal of Sound and Vibration, 333*(20), 5077–5096.
26. Akgul, A., Moroz, I., Pehlivan, I., & Vaidyanathan, S. (2016). A new four-scroll chaotic attractor and its engineering applications. *Optik, 127*(13), 5491–5499.
27. Vaidyanathan, S., & Azar, A. T. (2016). A novel 4-D four-wing chaotic system with four quadratic nonlinearities and its synchronization via adaptive control method. In *Advances in chaos theory and intelligent control. Studies in fuzziness and soft computing* (Vol. 337, pp. 203–224). Berlin, Germany: Springer.
28. Vaidyanathan, S., & Azar, A. T. (2016). Adaptive control and synchronization of Halvorsen circulant chaotic systems. In *Advances in chaos theory and intelligent control. Studies in fuzziness and soft computing* (Vol. 337, pp. 225–247). Berlin, Germany: Springer.
29. Vaidyanathan, S., & Azar, A. T. (2016). Dynamic analysis, adaptive feedback control and synchronization of an eight-term 3-D novel chaotic system with three quadratic nonlinearities. In *Advances in chaos theory and intelligent control. Studies in fuzziness and soft computing* (Vol. 337, pp. 155–178). Berlin, Germany: Springer.
30. Vaidyanathan, S., & Azar, A. T. (2016). Generalized projective synchronization of a novel hyperchaotic four-wing system via adaptive control method. In *Advances in chaos theory and intelligent control. Studies in fuzziness and soft computing* (Vol. 337, pp. 275–296). Berlin, Germany: Springer.
31. Vaidyanathan, S., & Azar, A. T. (2016). Qualitative study and adaptive control of a novel 4-D hyperchaotic system with three quadratic nonlinearities. In *Advances in chaos theory and intelligent control. Studies* in *fuzziness and soft computing* (Vol. 337, pp. 179–202). Berlin, Germany: Springer.
32. Vaidyanathan, S., & Azar, A. T. (2016). Adaptive backstepping control and synchronization of a novel 3-D jerk system with an exponential nonlinearity. In: *Advances in chaos theory and intelligent control. Studies in fuzziness and soft computing* (Vol. 337, pp. 249–274). Berlin, Germany: Springer.

33. Sampath, S., & Vaidyanathan, S. (2016). Hybrid synchronization of identical chaotic systems via novel sliding control method with application to Sampath four-scroll chaotic system. *International Journal of Control Theory and Applications, 9*(1), 221–235.
34. Vaidyanathan, S., Sampath, S., & Azar, A. T. (2015). Global chaos synchronisation of identical chaotic systems via novel sliding mode control method. *International Journal of Modelling, Identification and Control, 23*(1), 92–100.
35. Boulkroune, A., Bouzeriba, A., Bouden, T., & Azar, A. T. (2016). *fuzzy adaptive synchronization of uncertain fractional-order chaotic systems, advances in chaos theory and intelligent control, studies in fuzziness and soft computing* (Vol. 337, pp. 681–697). Germany: Springer.
36. Vaidyanathan, S., & Azar, A. T. (2016). Takagi-Sugeno fuzzy logic controller for Liu-Chen four-scroll chaotic system. *International Journal of Intelligent Engineering Informatics, 4*(2), 135–150.
37. Adachi, M., & Aihara, K. (1997). Associative dynamics in a chaotic neural network. *Neural Networks, 10*(1), 83–98.
38. Hoppensteadt, F. C. (2013). *Analysis and simulation of chaotic systems, applied mathematical sciences* (Vol. 94). Berlin, Germany: Springer.
39. Hajihassani, M., Armaghani, D. J., Marto, A., & Mohamad, E. T. (2015). Ground vibration prediction in quarry blasting through an artificial neural network optimized by imperialist competitive algorithm. *Bulletin of Engineering Geology and the Environment, 74*(3), 873–886.
40. Wasserman, P. D. (1993). *Advanced methods in neural computing*. New Jersey, USA: Wiley.
41. Amiri, A., Niaki, S. T. A., & Moghadam, A. T. (2015). A probabilistic artificial neural network-based procedure for variance change point estimation. *Soft Computing, 19*(3), 691–700.
42. Maind, S. B., & Wankar, P. (2014). Research paper on basic of artificial neural network. *International Journal on Recent and Innovation Trends in Computing and Communication, 2* (1), 96–100.
43. Mekki, H., Mellit, A., & Salhi, H. (2016). Artificial neural network-based modelling and fault detection of partial shaded photovoltaic modules. *Simulation Modelling Practice and Theory, 67*, 1–13.
44. Dou, J., Yamagishi, H., Pourghasemi, H. R., Yunus, A. P., Song, X., Xu, Y., et al. (2015). An integrated artificial neural network model for the landslide susceptibility assessment of Osado Island. *Japan Natural Hazards, 78*(3), 1749–1776.
45. Wang, T., Gao, H., & Qiu, J. (2016). A combined adaptive neural network and nonlinear model predictive control for multirate networked industrial process control. *IEEE Transactions on Neural Networks and Learning Systems, 27*(2), 416–425.
46. Grézl, F., Karafiát, M., & Veselý, K. (2014). Adaptation of multilingual stacked bottle-neck neural network structure for new language. In *2014 IEEE International Conference on Acoustics, Speech and Signal Processing (ICASSP)* (pp. 7654–7658). IEEE.
47. Nickl-Jockschat, T., Rottschy, C., Thommes, J., Schneider, F., Laird, A. R., Fox, P. T., et al. (2015). Neural networks related to dysfunctional face processing in autism spectrum disorder. *Brain Structure and Function, 220*(4), 2355–2371.
48. Azar, A. T. (2013). Fast neural network learning algorithms for medical applications. *Neural Computing and Applications, 23*(3–4), 1019–1034.
49. Azar, A. T., & El-Said, S. A. (2013). Probabilistic neural network for breast cancer classification. *Neural Computing and Applications, 23*(6), 1737–1751.
50. Hsu, C.-T., Korimara, R., Tsai, L.-J., & Cheng, T.-J. (2016). Photovoltaic power generation system modeling using an artificial neural network. In J. Juang (Ed.), *Proceedings of the 3rd International Conference on Intelligent Technologies and Engineering Systems (ICITES2014)* (pp. 365–371). Cham: Springer.
51. Huang, X., Zhao, Z., Wang, Z., & Li, Y. (2012). Chaos and hyperchaos in fractional-order cellular neural networks. *Neurocomputing, 94*, 13–21.
52. Vaidyanathan, S. (2015). 3-cells cellular neural network (CNN) attractor and its adaptive biological control. *International Journal of PharmTech Research, 8*(4), 632–640.

53. Vaidyanathan, S. (2015). Synchronization of 3-cells cellular neural network (CNN) attractors via adaptive control method. *International Journal of PharmTech Research, 8*(5), 946–955.
54. Vaidyanathan, S. (2015). Anti-synchronization of 3-cells cellular neural network attractors via adaptive control method. *International Journal of PharmTech Research, 8*(7), 26–38.
55. Vaidyanathan, S. (2016). Anti-synchronization of 3-cells cellular neural network attractors via integral sliding mode control. *International Journal of PharmTech Research, 9*(1), 193–205.
56. Chen, C.-W. (2014). Applications of neural-network-based fuzzy logic control to a nonlinear time-delay chaotic system. *Journal of Vibration and Control, 20*(4), 589–605.
57. Zhang, G., & Shen, Y. (2014). Exponential synchronization of delayed memristor-based chaotic neural networks via periodically intermittent control. *Neural Networks, 55*, 1–10.
58. Pham, V. T., Volos, C., Jafari, S., Wang, X., & Vaidyanathan, S. (2014). Hidden hyperchaotic attractor in a novel simple memristive neural network. *Optoelectronics and Advanced Materials, Rapid Communications, 8*(11–12), 1157–1163.
59. Wen, S., Zeng, Z., Huang, T., & Zhang, Y. (2014). Exponential adaptive lag synchronization of memristive neural networks via fuzzy method and applications in pseudorandom number generators. *IEEE Transactions on Fuzzy Systems, 22*(6), 1704–1713.
60. Kyprianidis, I. M., & Makri, A. T. (2013). Complex dynamics of FitzHugh-Nagumo type neurons coupled with gap junction under external voltage stimulation. *Journal of Engineering Science and Technology Review, 6*(4), 104–114.
61. He, G., Cao, Z., Zhu, P., & Ogura, H. (2003). Controlling chaos in a chaotic neural network. *Neural Networks, 16*(8), 1195–1200.
62. Ramesh, M., & Narayanan, S. (2001). Chaos control of Bonhoeffer–van der Pol oscillator using neural networks. *Chaos, Solitons and Fractals, 12*(13), 2395–2405.
63. Ren, X. M., Rad, A. B., Chan, P. T., & Lo, W. L. (2003). Identification and control of continuous-time nonlinear systems via dynamic neural networks. *IEEE Transactions on Industrial Electronics, 50*(3), 478–486.
64. Gan, Q., & Liang, Y. (2012). Synchronization of chaotic neural networks with time delay in the leakage term and parametric uncertainties based on sampled-data control. *Journal of the Franklin Institute, 349*(6), 1955–1971.
65. Lin, D., & Wang, X. (2011). Self-organizing adaptive fuzzy neural control for the synchronization of uncertain chaotic systems with random-varying parameters. *Neurocomputing, 74*(12), 2241–2249.
66. Zhang, H., Ma, T., Huang, G.-B., & Wang, Z. (2010). Robust global exponential synchronization of uncertain chaotic delayed neural networks via dual-stage impulsive control. *IEEE Transactions on Systems, Man, and Cybernetics, Part B (Cybernetics), 40*(3), 831–844.
67. Ardalani-Farsa, M., & Zolfaghari, S. (2010). Chaotic time series prediction with residual analysis method using hybrid Elman–NARX neural networks. *Neurocomputing, 73*(13), 2540–2553.
68. Chen, L., & Aihara, K. (1995). Chaotic simulated annealing by a neural network model with transient chaos. *Neural Networks, 8*(6), 915–930.
69. Mou, C., Jiang, C., Bin, J., & Wu, Q. (2009). Sliding mode synchronization controller design with neural network for uncertain chaotic systems. *Chaos, Solitons and Fractals, 39*(4), 1856–1863.
70. Chen, M., & Chen, W. (2009). Robust adaptive neural network synchronization controller design for a class of time delay uncertain chaotic systems. *Chaos, Solitons and Fractals, 41* (5), 2716–2724.
71. Lamamra, K., Belarbi, K., & Boukhtini, S. (2015). Box and Jenkins nonlinear system modelling using RBF neural networks designed by NSGAII. In *Computational Intelligence Applications in Modeling and Control* (pp. 229–254). Springer.
72. Sellnow, T. L., Seeger, M. W., & Ulmer, R. R. (2002). Chaos theory, informational needs, and natural disasters. *Journal of Applied Communication Research, 30*(4), 269–292.
73. Ata, N. A. A., & Schmandt, R. (2016). Systemic and systematic Risk. *The tyranny of uncertainty* (pp. 57–62). Berlin Heidelberg: Springer.

74. Elhadj, Z. (2006). *Etude de quelques types de systemes chaotiques : Generalisation d'un modele issu du modele de chen*. Doctoral thesis, University of Constantine.
75. Hecht-Nielsen, R. (1989). Theory of the backpropagation neural network. In *IEEE International Joint Conference on Neural Networks (IJCNN-1989)* (pp. 593–605). IEEE.
76. Yadav, N., Yadav, A., & Kumar, M. (2015). *An Introduction to Neural Network Methods for Differential Equations*. Berlin, Germany: Springer.
77. Lamamra, K., & Belarbi, K. (2011). Comparison of neural networks and fuzzy logic control designed by multi-objective genetic algorithm. *IJACT: International Journal of Advancements in Computing Technology, 3*(4), 137–143.
78. Sajikumar, N., & Thandaveswara, B. S. (1999). A non-linear rainfall–runoff model using an artificial neural network. *Journal of Hydrology, 216*(1), 32–55.
79. Suykens, J. A. (2001). Support vector machines: A nonlinear modelling and control perspective. *European Journal of Control, 7*(2), 311–327.
80. Zhu, Q., & Azar A. T. (2015). *Complex system modelling and control through intelligent soft computations. Studies in fuzziness and soft computing* (Vol. 319). Berlin, Germany: Springer. ISBN: 978-3-319-12883-2.
81. Graupe, D. (2013). *Principles of artificial neural networks* (Vol. 7). World Scientific. ISBN 978-981-4522-73-1.
82. Dong, L., Wesseloo, J., Potvin, Y., & Li, X. (2016). Discrimination of mine seismic events and blasts using the Fisher classifier, Naive Bayesian classifier and logistic regression. *Rock Mechanics and Rock Engineering, 49*(1), 183–211.
83. Greenhalgh, J., & Mirmehdi, M. (2015). Automatic detection and recognition of symbols and text on the road surface. In *International Conference on Pattern Recognition Applications and Methods* (pp. 124–140). Springer.
84. Greenhalgh, J., & Mirmehdi, M. (2015). Detection and recognition of painted road surface markings. In *Proceedings of 4th International Conference on Pattern Recognition Applications and Methods* (pp. 130–138).
85. Lamamra, K., Belarbi, K., & Mokhtari, F. (2006). Optimization of the structure of a neural networks by multi-objective genetic algorithms. *Proceedings of the ICGST International Journal on Automation, Robotics and Autonomous Systems*, *6*(1), 1–4.
86. Deb, K. (2001). *Multi-objective optimization using evolutionary algorithms* (Vol. 16). Wiley. ISBN 0-471-87339-X.
87. Lamamra, K., Belarbi, K., Belhani, A., & Boukhtini, S. (2014). NSGA2 based of multi-criteria decision analysis for multi-objective optimization of fuzzy logic controller for non linear system. *Journal of Next Generation Information Technology, 5*(1), 57–64.
88. Rao, R. V., Rai, D. P., & Balic, J. (2016). Multi-objective optimization of machining and micro-machining processes using non-dominated sorting teaching–learning-based optimization algorithm. *Journal of Intelligent Manufacturing*, 1–23.
89. Sardou, I. G., & Ameli, M. T. (2016). A fuzzy-based non-dominated sorting genetic algorithm-II for joint energy and reserves market clearing. *Soft Computing, 20*(3), 1161–1177.
90. Wong, J. Y., Sharma, S., & Rangaiah, G. P. (2016). Design of shell-and-tube heat exchangers for multiple objectives using elitist non-dominated sorting genetic algorithm with termination criteria. *Applied Thermal Engineering, 93*, 888–899.
91. Yang, M.-D., Lin, M.-D., Lin, Y.-H., & Tsai, K.-T. (2016). Multiobjective optimization design of green building envelope material using a non-dominated sorting genetic algorithm. *Applied Thermal Engineering*. In Press.
92. Srinvas, N., & Deb, K. (1994). Multi-objective function optimization using non-dominated sorting genetic algorithms. *Evolutionary Computation, 2*(3), 221–248.
93. Deb, K., Agrawal, S., Pratap, A., & Meyarivan, T. (2000). A fast elitist non-dominated sorting genetic algorithm for multi-objective optimization: NSGA-II. In *International Conference on Parallel Problem Solving From Nature* (pp. 849–858). Springer.
94. Versèle, C., Deblecker, O., & Lobry, J. (2011). Multiobjective optimal design of an inverter fed axial flux permanent magnet in-wheel motor for electric vehicles. *Electric Vehicles-Modelling And Simulations, 287*, 300–321.

95. Mackey, M. C., & Glass, L. (1977). Oscillation and chaos in physiological control systems. *Science, 197*(4300), 287–289.
96. Anderson, D., Sweeney, D., Williams, T., Camm, J., & Martin, R. (2011). *an introduction to management science: Quantitative approaches to decision making*. Revised: Cengage Learning.
97. Martinetz, T. M., Berkovich, S. G., & Schulten, K. J. (1993). Neural-gas' network for vector quantization and its application to time-series prediction. *IEEE Transactions on Neural Networks, 4*(4), 558–569.

A New Fractional-Order Predator-Prey System with Allee Effect

Afef Ben Saad and Olfa Boubaker

Abstract In this chapter, a new Fractional-order (FO) predator-prey system with Allee Effect is proposed and its dynamical analysis is investigated. The two case studies of weak and strong Allee Effects are considered to bring out the consequence of such extra factors on the FO system's dynamics. Not only it will be proven, via analytic and numerical results, that the system's stability is governed by the type of the Allee Effect but also it will be shown that such extra factor is a destabilizing force. Finally, simulation results reveal that rich dynamic behaviors of the (FO) predator-prey model are exhibited and dependent on the order value of the FO system.

Keywords Fractional-order (FO) · Predator-prey model · Weak Allee effect · Strong Allee effect · Bifurcation analysis

1 Introduction

In recent times, the protection of biological and ecological systems is central to a huge range of scientific areas. Therefore, analyzing and controlling the complex dynamics of these systems are a great challenge for researchers [27].

The complexity of such systems is primarily introduced by the interaction between populations [4, 5]. In this framework, Predator-Prey models are the most popular models for interacting populations [6, 29]. The interaction between the predator and prey as well as extra factors effects have long been an important topic in mathematical ecology. Among the most famous types of interaction, there are the predation [4] and the competition species [3]. The competition occurs either when both populations are independent of each other or when the populations interact and each population

A. Ben Saad (✉) · O. Boubaker
National Institute of Applied Sciences Technology, Centre Urbain Nord BP. 676, 1080 Tunis Cedex, Tunisia
e-mail: afef.ben.saad88@gmail.com

O. Boubaker
e-mail: olfa.boubaker@insat.rnu.tn

A.T. Azar et al. (eds.), *Fractional Order Control and Synchronization of Chaotic Systems*, Studies in Computational Intelligence 688,
DOI 10.1007/978-3-319-50249-6_30

exerts downward pressure on the other. These interactions are generally described by nonlinear polynomial models. Lotka Volterra model [29] is the most famous one. It is based on polynomial differential equations inducing mostly a functional response which describes the relationship between individuals rate of consumption and food density. This functional response can be prey dependent [24], predator dependent [32] or ratio dependent [1, 31]. Prey dependent is when the predators rate of prey consumption depends on the densities of both prey and predator. Predator dependent is when predators have to search, share or compete for food [15]. However, ratio-dependent is a combination between the two previous functional responses. In the same context, there are several types of polynomial functional responses such as the Michaelis Menten type [31], the Beddington DeAngelis type [26] and the Holling type functional response [13]. Among the known predator-prey systems introducing the holling type I functional response, there are the BB-model proposed by Bazykin and Berezovskay [4] and the Lesli Gower model proposed by Leslie and Gower [21]. Modeling of such systems by integer order polynomials permits to describe a limited set of system features. Polynomial models can not fit all the system's data [25]. However, modeling of predator-prey systems by FO is recently used to fit nonlinear functions and offer an extended family of curves. Fractional order calculus (FOC) is an old mathematical field [16] which has attracted, in recent years, many attentions. The FOC advantages over classical mathematics are investigated with studies on real world process. It has been applied to several areas in science and engineering [12, 19, 20, 25, 30] such as biological systems, neural systems, diffusion processes, etc. Stability analysis of systems based on fractional order polynomial models permits to describe more features of dynamics [18, 25] and creates more and more complex dynamic behaviors. Its power term are restricted to a small set of integer and non integer values. Despite, research in this area is still in its early stages due to the complexity of the FOC understanding. Hence, such research papers are rare which motivate the present work.

Thus, in this chapter, based on the predator-prey BB-model inducing an Allee Effect extra factor, a new FO predator-prey BB-model with strong and weak Allee Effects is proposed. The model is, first, designed with FO prey's growth rate and a FO predator's functional response. Then, a full stability and bifurcation analysis is investigated for the two case studies of Allee Effects.

This chapter is organized in three sections: Sect. 2 recalls the stability and bifurcation analysis of the basic integer polynomial order BB-model with Allee Effect. In Sect. 3, an accurate stability and bifurcation analysis of the new BB-model with a FO prey's growth rate and a FO predator's functional response is investigated. Finally, the theoretical and numerical results of the proposed Fractional models are discussed in Sect. 4.

2 Related Work

In a previous work [7], a full dynamical behavior analysis of the polynomial BB-model was presented. It introduces one of the most important extra factors which is the Allee Effect factor [8, 11, 17, 22]. It is defined as positive relationships between any component of fitness of species and eithers numbers or densities of conspecific, meaning that an individual of species that is subject to an Allee Effects will suffer a decrease in some aspect of its fitness when conspecific density is low [11, 22]. The stability and bifurcation analysis proves that the incorporation of Allee Effect factor destabilizes the predator-prey system and creates a set of complex dynamic behavior according to the type of Allee Effect which can be strong [10, 28] or weak [23].

Consider the general Predator-prey model described by Abrams and Ginzburg [2], Berryman [9], Van Voorn et al. [28]

$$\begin{cases} \frac{dx_1}{dt} = ax_1 - f(x_1)x_2 \\ \frac{dx_2}{dt} = cx_1x_2 - dx_2 \end{cases} \tag{1}$$

where

x_1 is the size of the prey population,
x_2 is the size of the predator population,
a is the prey's growth rate in absence of the predator,
$f(x_1)$ is the functional response of the predator to prey density
c is the predator's conversion efficiency
d is the mortality rate of the predator depending on the predator's efficiency.

Let $a = (x_1 - l)(k - x_1), f(x_1) = x_1$ and $d = c\,m$, where l, k, m are the Allee Effect threshold, the carrying capacity and the predator's mortality rate, respectively. System (1) can be then written as:

$$\begin{cases} \frac{dx_1}{dt} = x_1(x_1 - l)(1 - x_1) - x_1x_2 \\ \frac{dx_2}{dt} = c(x_1 - m)x_2 \end{cases} \tag{2}$$

System (2) describes the BB-model predator prey system [4, 28]. The prey's growth rate is modeled by a nonlinear polynomial introducing the Allee Effect extra factor [8, 11, 17, 22]. However, the predation is modeled by a polynomial prey-dependent functional response. A strong Allee Effect is obtained for $l \in [0\ 1]$ whereas the weak Allee Effect is obtained for $l \in [-1\ \ 0]$.

Stability analysis of system (2) is investigated for the two case studies of Allee Effects by linearizing the system (2) and using the following Jacobian matrix

$$J = \begin{pmatrix} -3x_1^2 + 2(l+1)x_1 - (l + x_2) & -x_1 \\ x_2 & (x_1 - m) \end{pmatrix}$$

Table 1 Stability analysis of system (2) with strong Allee Effect for $m \in [0, l]$ [7]

$m \in [0, l]$		
Equilibrium	Eigenvalues	Singularity
$E_0(0, 0)$	$\lambda_1 = -l$ $\lambda_2 = -m$	Stable node 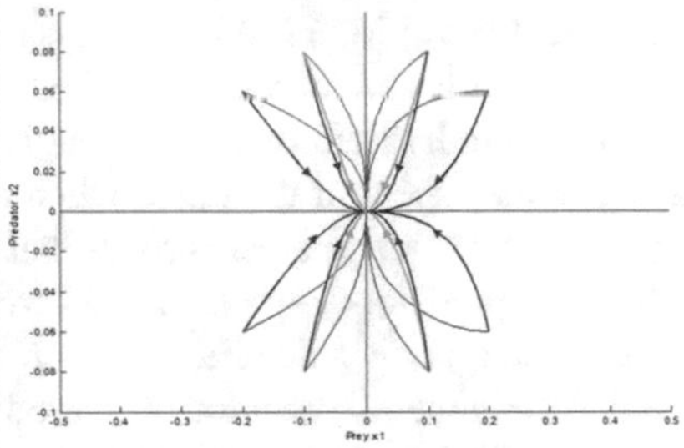
$E_1(l, 0)$	$\lambda_1 = l(1-l)$ $\lambda_2 = l - m$	Unstable node 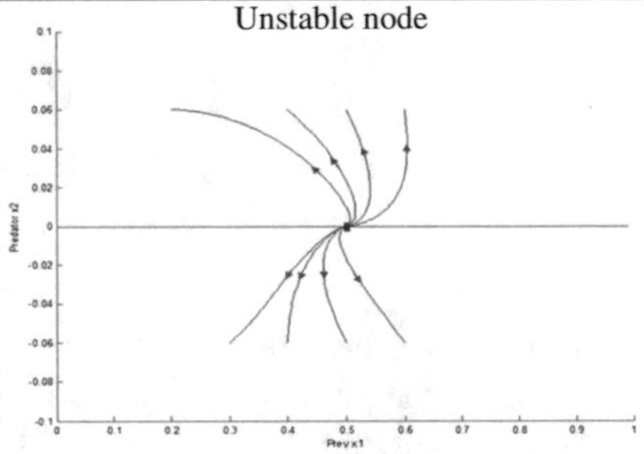
$E_2(1, 0)$	$\lambda_1 = l - 1$ $\lambda_2 = l - m$	Saddle point 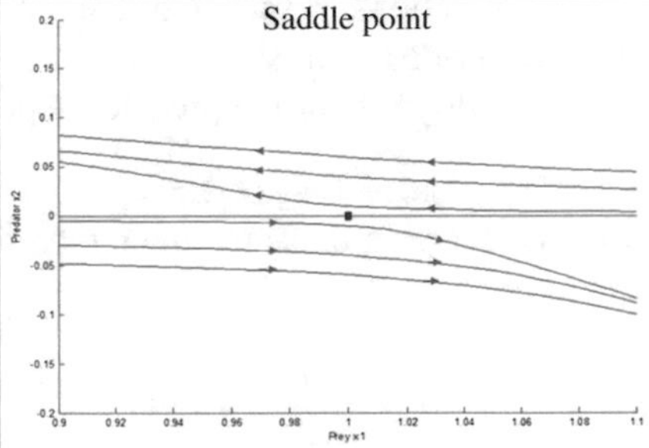

As shown in Tables 1, 2 and 3, system (2) with strong Allee Effect admits three equilibrium points when $m \in [0\ 1]$ but, when $m \in [l\ 1]$ it admits four equilibrium points. Furthermore, the system (2) with weak Allee Effect admits only three equilibrium points $\forall\, m \in [0\ 1]$ as shown in Table 3.

Calculus of the Jacobian matrix J at the equilibrium point $E_3(m, (m-l)(1-m))$ gives two complex eigenvalues as shown in Tables 2 and 3 with δ, $tr(J_3)$ and $det(J_3)$ are defined as follow:

Table 2 Stability analysis of system (2) with strong Allee Effect for $m \in [l, 1]$ [7]

$m \in [l, 1]$		
Equilibrium	Eigenvalues	Singularity
$E_0(0, 0)$	$\lambda_1 = -l$ $\lambda_2 = -m$	Stable node
$E_1(l, 0)$	$\lambda_1 = l(1 - l)$ $\lambda_2 = l - m$	Saddle point
$E_2(1, 0)$	$\lambda_1 = l - 1$ $\lambda_2 = l - m$	Saddle point

(continued)

Table 2 (continued)

$m \in [l, 1]$		
Equilibrium	Eigenvalues	Singularity
$E_3(m, (m-l)(1-m))$	$\lambda_1 = \frac{tr(J_3)-i\sqrt{\delta}}{2}$ $\lambda_2 = \frac{tr(J_3)+i\sqrt{\delta}}{2}$	**$m = 0.75$** Center 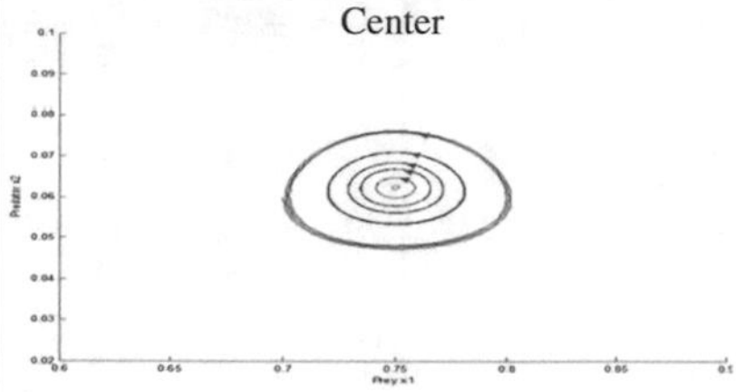 **$m < 0.75$** Unstable Focus 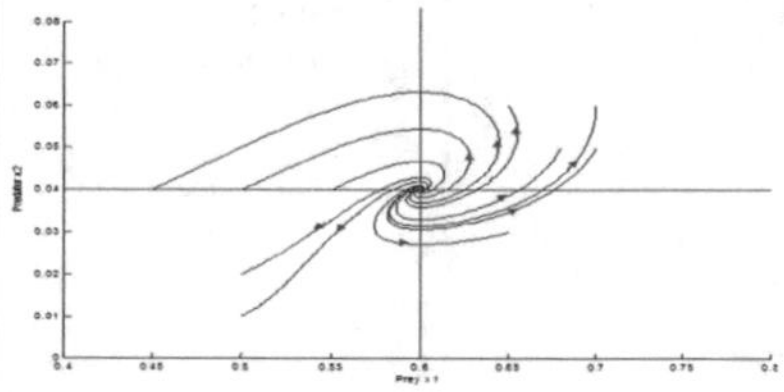 **$m > 0.75$** Stable Focus 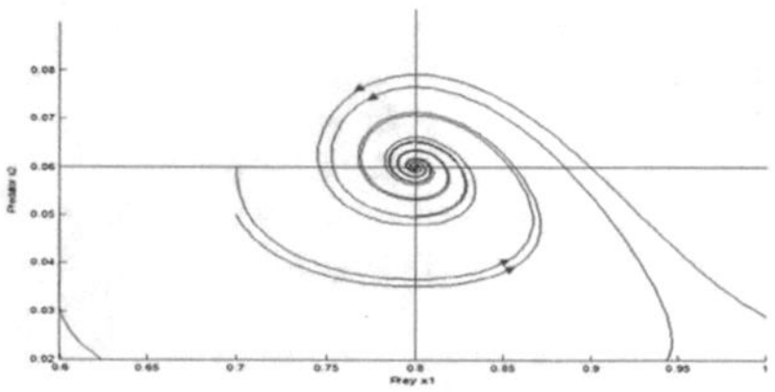

$$tr(J_3) = m(l + 1 - 2m)$$
$$\delta = tr^2(J_3) - 4det(J_3)$$
$$det(J_3) = m(m - l)(1 - m)$$

The previous theoretical and numerical results show that the number as well as the stability of equilibrium points depend on the value of the parameters m and l. For the weak Allee Effect, the extinction equilibrium E_0 is unstable whereas for the strong Allee Effect the extinction equilibrium becomes stable. Thus, system (2) admits the extinction of the two species at low density if and only if the Allee Effect is strong.

As shown by Fig. 1, bifurcation analysis of the system (2) is with respect to the bifurcation parameter m for both case studies of Allee Effects.

Table 3 Stability analysis of system (2) with weak Allee Effect for $m \in [0, 1]$ [7]

$m \in [l, 1]$

Equilibrium	Eigenvalues	Singularity
$E_0(0,0)$	$\lambda_1 = -l$ $\lambda_2 = -m$	Saddle point 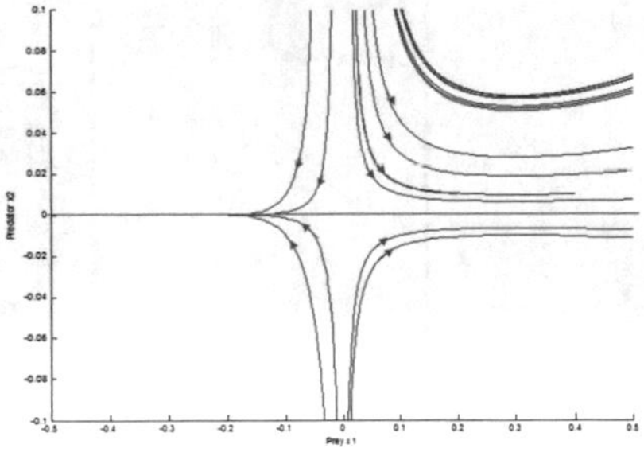
$E_2(1,0)$	$\lambda_1 = (l-1)$ $\lambda_2 = l - m$	Saddle point 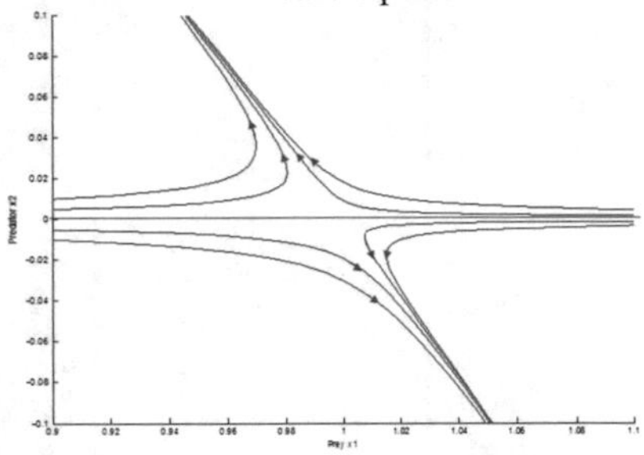

(continued)

Table 3 (continued)

$m \in [l, 1]$		
Equilibrium	Eigenvalues	Singularity
$E_3(m, (m-l)(1-m))$	$\lambda_1 = \frac{tr(J_3) - i\sqrt{\delta}}{2}$ $\lambda_2 = \frac{tr(J_3) + i\sqrt{\delta}}{2}$	***m* = 0.4** Center 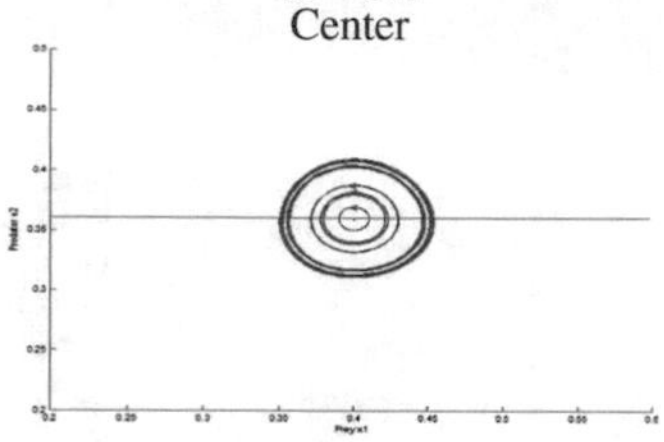 ***m* < 0.4** Unstable Focus 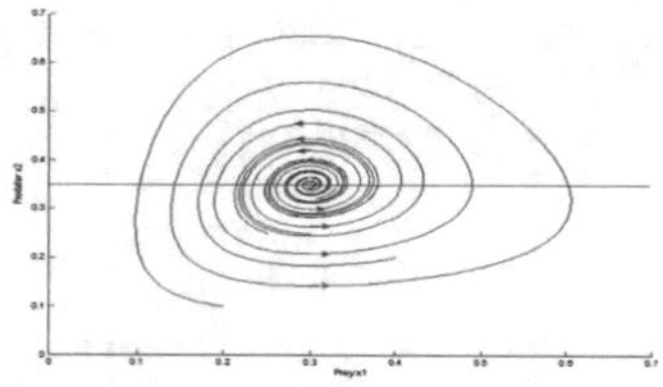 ***m* > 0.4** Stable Focus 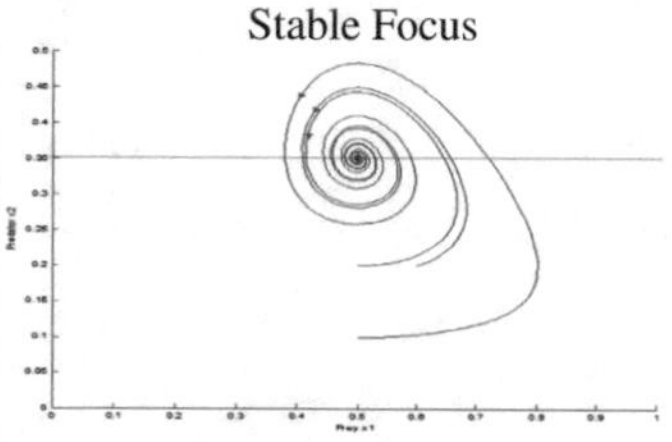

Bifurcation analysis of system (2) proves that the predator prey BB-model admits three bifurcation points: one Hopf point bifurcation and two branch points bifurcations. Moreover, the population growth rate of the predator is sensitive to the value of the mortality rate of the prey. In Fig. 1, for the strong Allee Effect, the population growth rate of the predator decreases below a critical Allee threshold and becomes negative, while for the weak Allee Effect, it decreases but never becomes negative which confirms that the two species never go to extinct, expect in the case of the strong Allee Effect. As a conclusion, the strong Allee Effect is a critical Allee Effect whereas the weak Allee Effect is not.

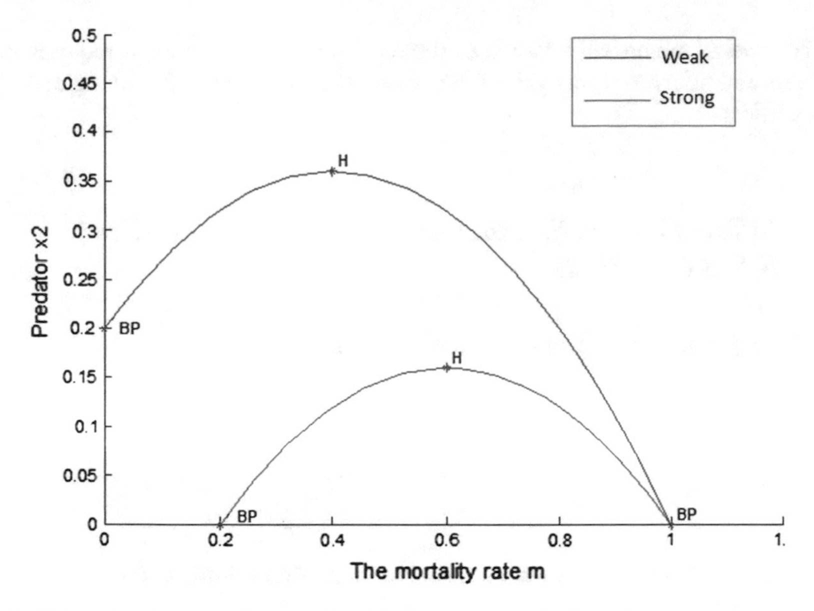

Fig. 1 Bifurcation diagram of system (1) with the two types of Allee Effect

3 The New Predator-Prey BB Model with a FO Growth Rate of the Prey and a FO Predator's Functional Response

In this section, the dynamical behavior of the new BB-model with a FO prey's growth rate and a FO predator's functional response is proposed and analyzed.

3.1 *The FO Model*

Let for system (2):

$$a = x_1^{q_1 - 1}(x_1 - l)(1 - x_1),$$
$$f(x_1) = x_1^{q_2}$$

where q_1 and q_2 are two fractional order elements (FOE) pertaining to the interval $]0, 1[\cup]1, 2[$, so:

$$\begin{cases} \frac{dx_1}{dt} = x_1^{q_1}(x_1 - l)(1 - x_1) - x_1^{q_2} x_2 \\ \frac{dx_2}{dt} = c(x_1 - m)x_2 \end{cases} \tag{3}$$

The overall dynamical analysis of the new FO model BB-model requires a full stability and bifurcation analysis of the system (3) with each of the strong and weak Allee Effects.

3.2 Bifurcation and Stability Analysis: The Strong Allee Effect Case Study

For $0 \leq l \leq 1$, the equilibrium solutions are given by:

$$\begin{array}{ll} E_0: & x_1^{q_1} = 0; x_2 = 0 \\ E_1: & x_1 = l; x_2 = 0 \\ E_2: & x_1 = 1; x_2 = 0 \\ E_3: & x_1 = m; x_2 = \frac{(m-l)(1-m)}{m^{q_2-q_1}} \end{array}$$

- For $m \in [0, l]$, the system admits the same three equilibriums as those of the original system (2):

  –The extinction equilibrium $E_0(0, 0)$,
  –The positive equilibrium $E_1(l, 0)$,
  –The boundary equilibrium $E_2(1, 0)$.

- For $m \in [l, 1]$, the system (3) has four equilibriums:

  –The zero equilibrium $E_0(0, 0)$,
  –The equilibrium $E_1(l, 0)$,
  –The boundary equilibrium $E_2(1, 0)$,
  –The non-isolated equilibrium $E_3(m, \dfrac{(m-l)(1-m)}{m^{q_2-q_1}})$.

The Jacobian matrix of system (3) is given by:

$$J = \begin{pmatrix} a_{11} & a_{12} \\ a_{21} & a_{22} \end{pmatrix}$$

$$tr(J_3) = a_{11} + a_{22}$$
$$\delta = tr^2(J_3) - 4det(J_3)$$
$$det(J_3) = a_{11}a_{22} - a_{12}a_{21}$$

where

$$
\begin{aligned}
a_{11} &= -(2+q_1)x_1^{q_1+1} - (1+q_1)(l+1)x_1^{q1} - lq_1x_1^{q_1-1} - q_2x_1^{q2-1}x_2 \\
a_{12} &= -x_1^{q_2} \\
a_{21} &= x_2 \\
a_{22} &= (x_1 - m)
\end{aligned}
$$

Theoretical and numerical analysis in the neighborhood of all equilibrium points is investigated by determining their eigenvalues as well as their singularities as presented in the Tables 4 and 5 using the graphical MATLAB package MATCONT [14].

As shown in Tables 4 and 5 the FO system (3) admits the same singularities as those of the original system (2) at equilibiriums $E_1(l, 0)$ and $E_2(1, 0)$. However, at equilibriums E_0 and $E_3(m, \frac{(m-l)(1-m)}{m^{q_2-q_1}})$, new types of singularities are occurred:

- For the extinction equilibrium E_0, a collision of saddle point and stable node singularities according to the value of parameters m, q_1 and q_2 is obtained. Thus, the system may undergo a saddle-node bifurcation. In this bifurcation the fixed point can appear and disappear depending on q_1 and q_2.
- For the non-isolated equilibrium E_3, the same singularities as those of the original system (2) at this equilibrium are obtained if and only if $q_1 = q_2$. However, when $q_1 \neq q_2$ and $m = 0.75$, the equilibrium is destabilized. Its singularity is switched from a stable focus to an unstable focus while the value of the FO parameter q_1 increases and becomes higher than the FO parameter q_2. In other word, at a special value of mortality rate equal to 0.75, the FO system (3) is destabilized when the FO prey's growth rate becomes higher than FO predator's functional response to prey density and may undergo a Hopf bifurcation.

Stability of system (3) is sensitive to the variation of the parameters m, q_1 and q_2.

Thus, a numerical bifurcation analysis of the system (3) with strong Allee Effect according to the bifurcation parameter m is proposed for a set of q_1 and q_2 values.

Bifurcation diagram of system (3) with strong Allee Effect confirms the previous stability analysis and proves that the system keeps the same bifurcation types as those of system (2) a Hopf codimension 0 bifurcation point and two Branch points bifurcation as it is shown in Fig. 2. Existence of the Hopf bifurcation indicates the change of the system (3) behavior from a stationary dynamic to an oscillatory one. For each value of (q_1, q_2), a new Hopf point with a new first lyapunov coefficient is created corresponding to a new value of mortality rate m.

Table 4 Stability analysis of system (3) with strong Allee Effect for $m \in [0, l]$

$m \in [0, l]$		
Equilibrium	Eigenvalues	Singularity
$E_0(0, 0)$	$\lambda_1 = 0$ $\lambda_2 = -m$	Stable node 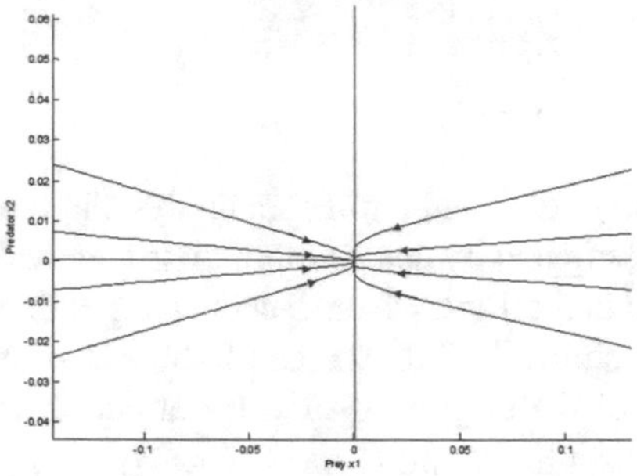
$E_1(l, 0)$	$\lambda_1 = l^{q_1}(1 - l)$ $\lambda_2 = l - m$	Unstable node 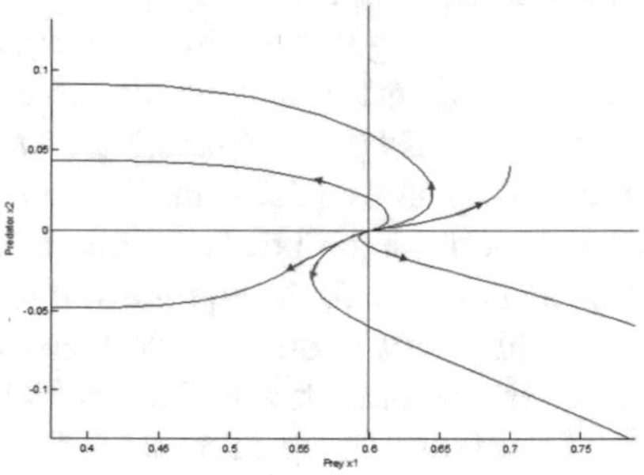
$E_2(1, 0)$	$\lambda_1 = l - 1$ $\lambda_2 = l - m$	Saddle point 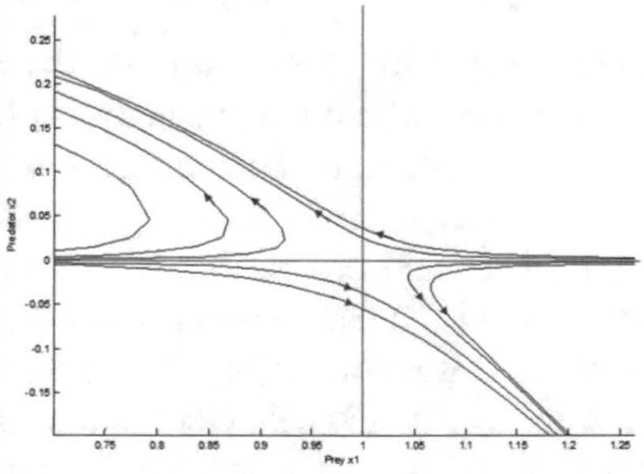

Table 5 Stability analysis of system (3) with strong Allee Effect for $m \in [l, 1]$

$m \in [l, 1]$		
Equilibrium	Eigenvalues	Singularity
$E_0(0, 0)$	$\lambda_1 = 0$ $\lambda_2 = -m$	Saddle point-Stable node 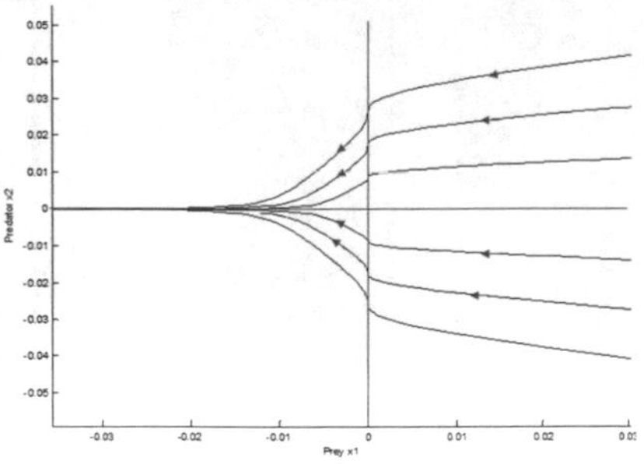
$E_1(l, 0)$	$\lambda_1 = l^{q_1}(1 - l)$ $\lambda_2 = l - m$	Saddle point 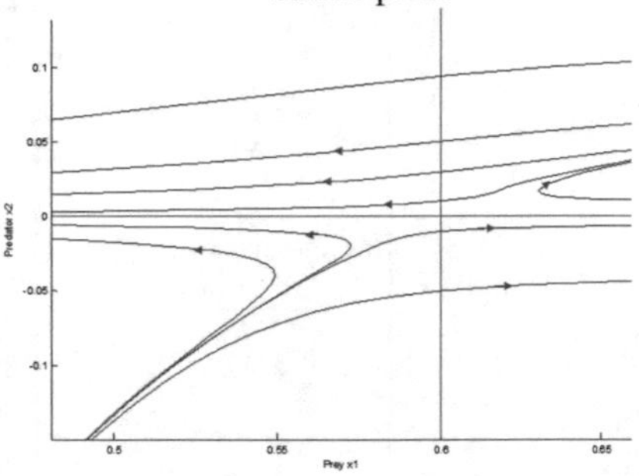
$E_2(1, 0)$	$\lambda_1 = l - 1$ $\lambda_2 = l - m$	Saddle point 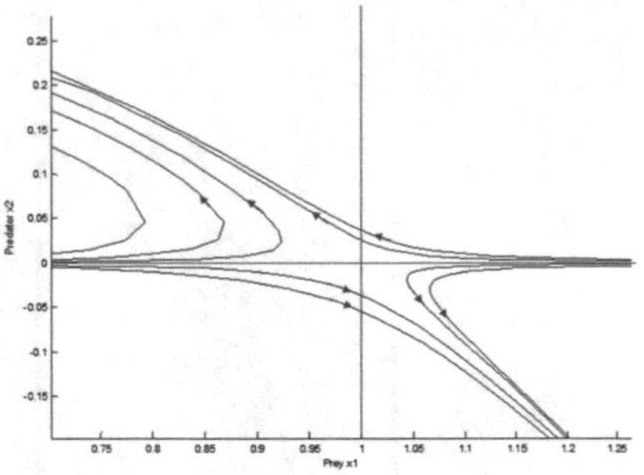

(continued)

Table 5 (continued)

$m \in [l, 1]$		
Equilibrium	Eigenvalues	Singularity
$E_3(m, \frac{(m-l)(1-m)}{m^{q_2-q_1}})$	$\lambda_1 = \frac{tr(J_3)-i\sqrt{\delta}}{2}$ $\lambda_2 = \frac{tr(J_3)+i\sqrt{\delta}}{2}$	***m* = 0.75** If $q_1 = q_2$ Center 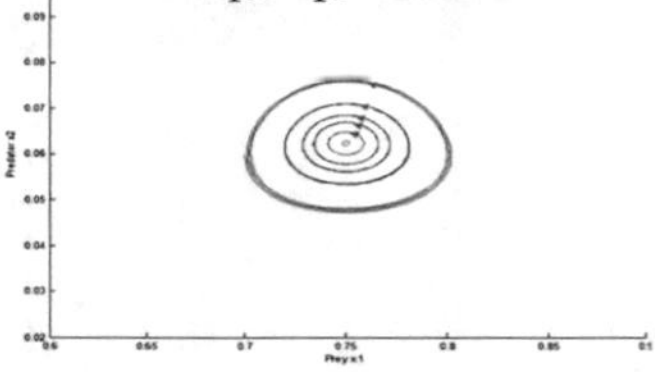 If $q_1 < q_2$ Stable Focus 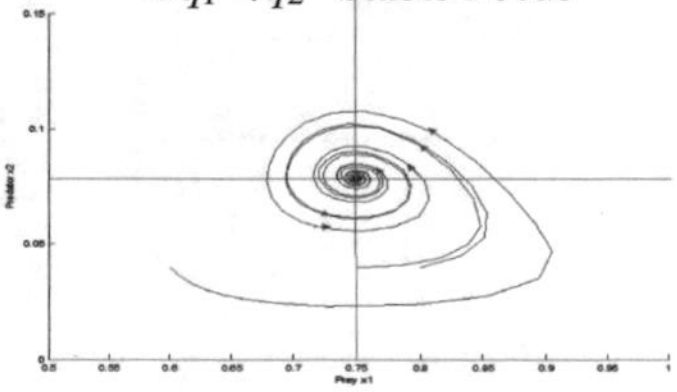 If $q_1 > q_2$ UnStable Focus 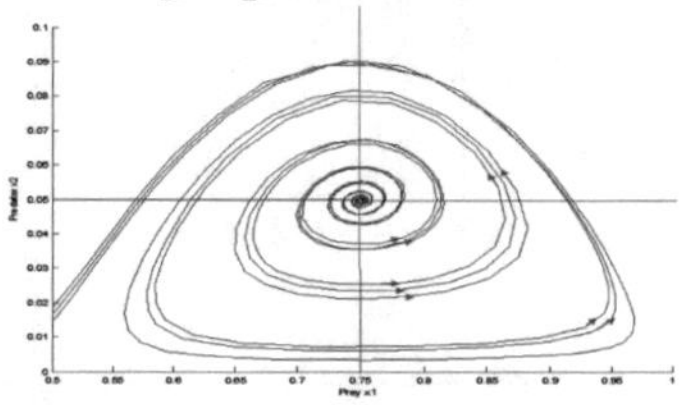 ***m* < 0.75** Unstable Focus 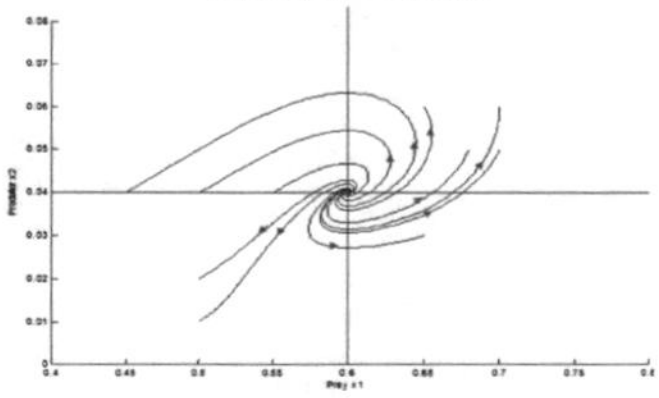 ***m* > 0.75** Stable Focus 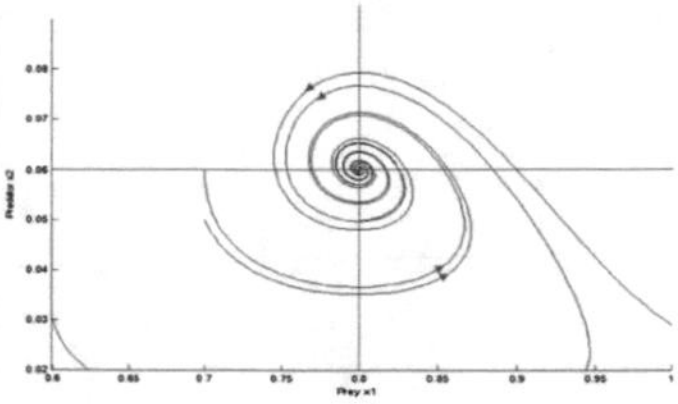

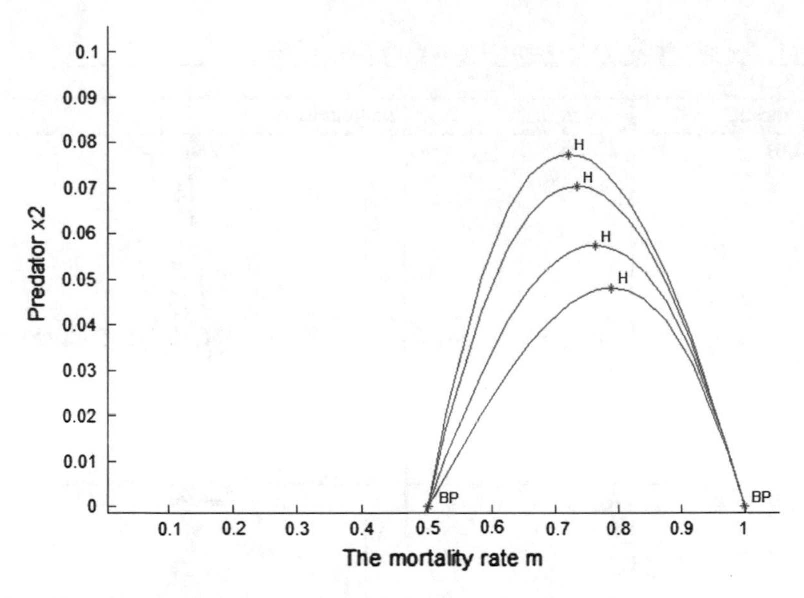

Fig. 2 Bifurcation diagram of system (3): Strong Allee Effect case study

3.3 Bifurcation and Stability Analysis: The Weak Allee Effect Case Study

For $-1 \leq l \leq 0$, the equilibrium solutions are given by:

$$\begin{array}{ll} E_0: & x_1^{q_1} = 0\ ; x_2 = 0 \\ E_1: & x_1 = l\ \ ; x_2 = 0 \\ E_2: & x_1 = 1\ \ ; x_2 = 0 \\ E_3: & x_1 = m\ ; x_2 = \frac{(m-l)(1-m)}{m^{q_2-q_1}} \end{array}$$

By considering the biological meaningful condition, $x_1 \geq 0$, system (3) admits only this three equilibrium points for all parameter $m \in [0\ 1]$:

–The zero equilibrium $E_0(0, 0)$,
–The trivial equilibrium $E_2(1, 0)$,
–The non-isolated equilibrium $E_3(m, \dfrac{(m-l)(1-m)}{m^{q_2-q_1}})$.

Using the same Jacobian matrix of system (3), the corresponding eigenvalues as well as the corresponding singularities are determined for each equilibrium as shown in Table 6.

Table 6 Stability analysis of system (3) with weak Allee Effect

$m \in [0, 1]$		
Equilibrium	Eigenvalues	Singularity
$E_0(0, 0)$	$\lambda_1 = 0$ $\lambda_2 = -m$	Saddle point 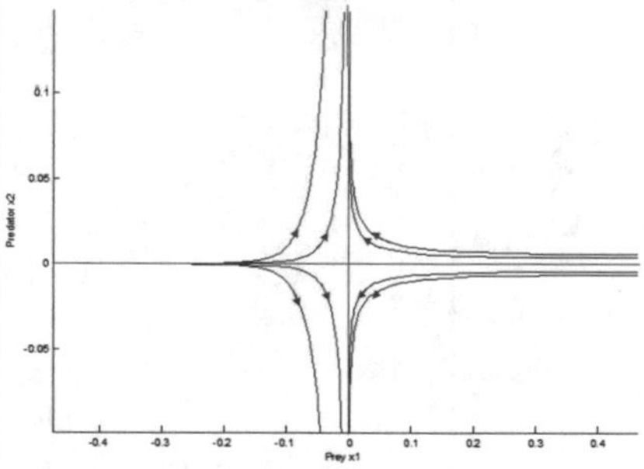
$E_2(1, 0)$	$\lambda_1 = l - 1$ $\lambda_2 = l - m$	Saddle point 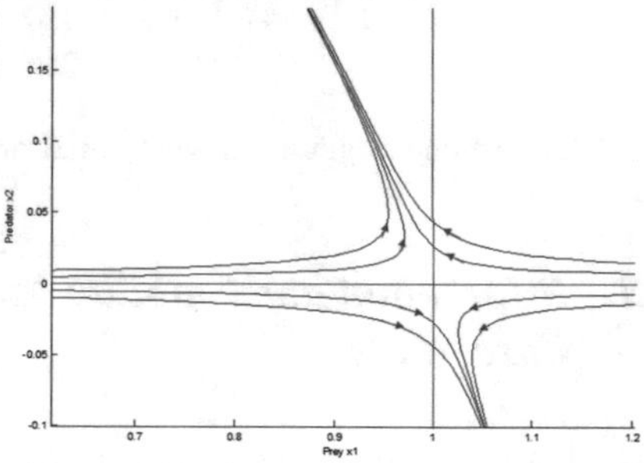

(continued)

Table 6 (continued)

$m \in [0, 1]$		
Equilibrium	Eigenvalues	Singularity
$E_3(m, \frac{(m-l)(1-m)}{m^{q_2-q_1}})$	$\lambda_1 = \frac{tr(J)-i\sqrt{\delta}}{2}$ $\lambda_2 = \frac{tr(J)+i\sqrt{\delta}}{2}$	$\boldsymbol{m \leq 0.4}$ If $q_1 < q_2$ Stable Focus If $q_1 > q_2$ Unstable Focus 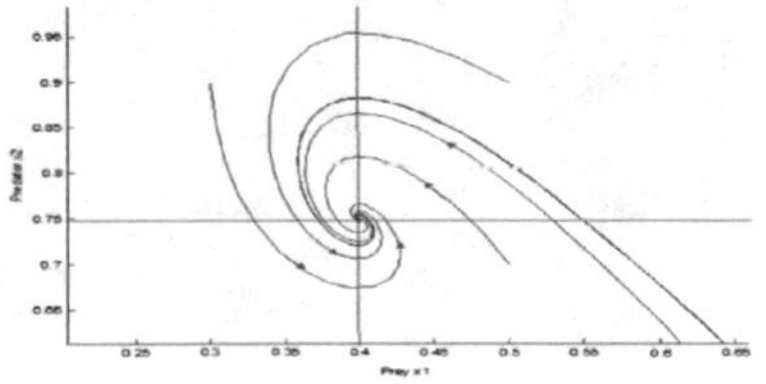 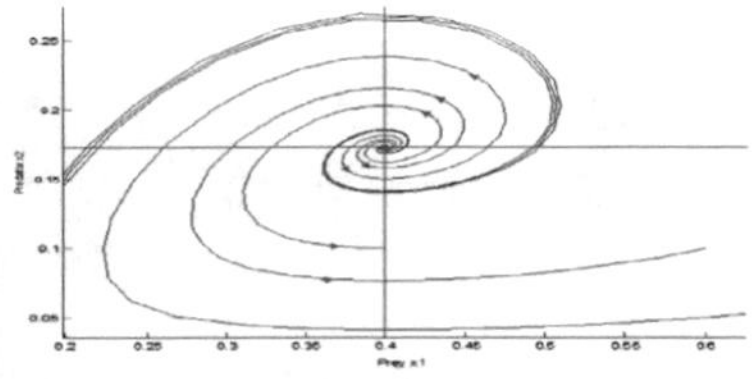 $\boldsymbol{m > 0.4}$ Stable Focus $\forall q_1 and q_2$ 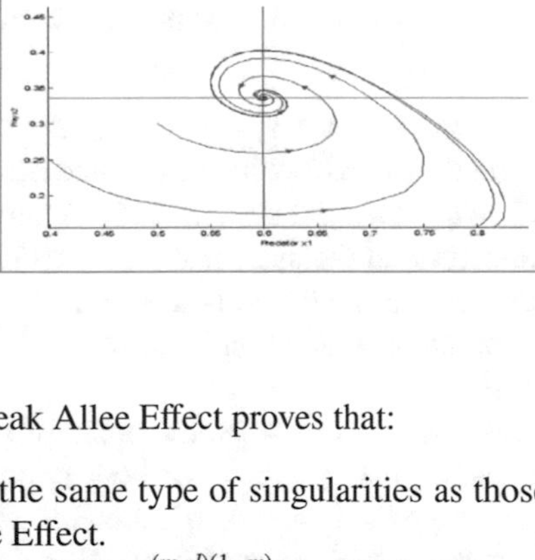

Stability analysis of system (3) with weak Allee Effect proves that:

- The equilibrium point E_0 and E_2 keep the same type of singularities as those of the original system (2) with weak Allee Effect.
- However, the singularity of the equilibrium $E_3(m, \frac{(m-l)(1-m)}{m^{q_2-q_1}})$ is switched from stable focus to an unstable focus when the value of the FOE q_1 increases under the value of FOE q_2 and the mortality rate m is lower or equal to 0.4. When q_1 is small, the system (3) admits a stable focus singularity for all m values. It means biologically, that when the FO growth rate is smaller than the FO predator's functional response, the effect of the mortality rate variation on the dynamical behavior of the system (3) at E_3 will be omitted.

For that, numerical bifurcation analysis according to m and a set of FOE q_1 and q_2 values is realized and is shown in Fig. 3.

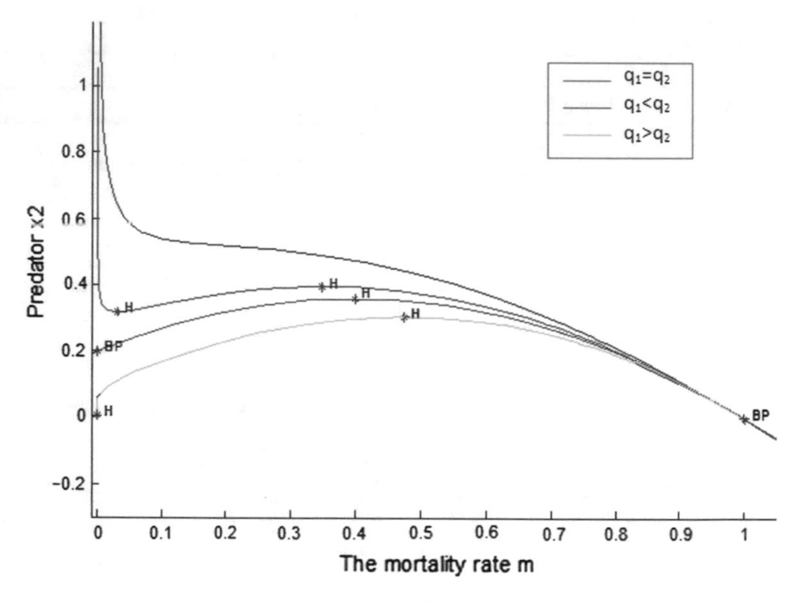

Fig. 3 Bifurcation diagram of system (3): Weak Allee Effect case study

Bifurcation diagram of system (3) with weak Allee Effect illustrates a set of bifurcation points

- When $q_1 = q_2$, the system (3) admits the same bifurcation points as those of the original system (2) with weak Allee Effect: one Hopf point and two Branch points.
- When $q_1 < q_2$, the system exhibits a new Hopf point in the right of the Fig. 3. It indicates that the system starts to oscillate around the equilibrium E_3. When the FO prey's growth rate is less than the FO predator's functional response value, the predator population increases and can never go to the extinction at low prey density.
- However, when $q_1 > q_2$, the predator growth rate decreases and can be negative indicating the extinction of both populations. Thus, the FO prey's growth rate aggravates the influence of Allee Effect on the dynamical behavior of the system (3) and can lead the weak Allee Effect to be a critical Allee Effect.

4 Comparative Analysis

The dynamical behavior analysis of the system (3) with both types of Allee Effects proves that FO prey's growth rate and the predator's functional response to prey density destabilizes the system (3) especially at E_3.

- For the strong Allee Effect case study, the system (3) is destabilized at the zero equilibrium E_0 and the non isolated equilibrium E_3 when the FO prey's growth rate is very high. In other hand, the system (3) exhibits new types of bifurcation which are the saddle-node bifurcation and the Hopf point bifurcation. This latter can be supercritical Hopf bifurcation or a subcritical hopf bifurcation according to the first lyapunov coefficient (FLC) sign. It is dependent on the value of the FO prey's growth rate and the FO predator's functional response.
- For the weak Allee Effect, the system (3) is destabilized only at the non isolated equilibrium E_3 especially when the mortality rate is below or equal the value 0.4 and the FO prey's growth rate is very high. The system (3) exhibits too new Hopf bifurcation points when the FO prey's growth rate is unequal to the FO predator's functional response. In addition, when the FO prey's growth rate is highly increasing, the influence of Allee Effect on the system becomes more important. As a conclusion, modeling the relationship between the predator's prey consumption rate and prey density by a FO prey dependent functional response and the prey's growth rate by a FO polynomial permit to describe more features of system (2) and to obtain more complex dynamic behaviors.

5 Conclusion

In this chapter, a new FO predator-Prey BB-model is proposed for both case studies of Allee Effects. The model is proposed for analyzing the dynamical behavior of the BB-model with a FO prey's growth rate and a FO predator's functional response to prey density. Bifurcation and stability analysis of the new FO system prove that FO modeling of the prey's growth rate and predator's functional response creates more complex dynamical behaviors according to the FO values but never converge to chaos.

In future works, a FO modeling of predator's efficiency will be proposed and its analysis will permit to obtain a period doubling bifurcation. Such bifurcation gives a new behavior by doubling the period of the original system. Its existence can even indicate the onset of a chaotic behavior at particular value.

References

1. Abrams, P. A. (1994). The fallacies of "ratio-dependent" predation. *Ecology, 75*(6), 1842–1850.
2. Abrams, P. A., & Ginzburg, L. R. (2000). The nature of predation: Prey dependent, ratio dependent or neither? *Trends in Ecology and Evolution, 15*(8), 337–341.
3. Ayala, F. J., Gilpin, M. E., & Ehrenfeld, J. G. (1973). Competition between species: Theoretical models and experimental tests. *Theoretical Population Biology, 4*, 331–356.
4. Bazykin, A. (1998). *Nonlinear dynamics of interacting populations* (Vol. 11). World Scientific.

5. Bazykin, A. (2008). *Encyclopedia of ecology*. Elsevier.
6. Bazykin, A., Berezovskaya, F., Denisov, G. A., & Kuznetzov, Y. A. (1981). The influence of predator saturation effect and competition among predators on predator-prey system dynamics. *Ecological Modelling*, *14*(1–2), 39–57.
7. Ben Saad, A., & Boubaker, O. (2015). On bifurcation analysis of the predator-prey bb-model with weak allee effect. In *16th International Conference on Sciences and Techniques of Automatic Control and Computer Engineering (STA)*, IEEE (pp. 19–23).
8. Berezovskaya, F., Wirkus, S., Song, B., & Castillo-Chavez, C. (2011). Dynamics of population communities with prey migrations and Allee effects: A bifurcation approach. *Mathematical Medicine and Biology*, *28*(2), 129–152.
9. Berryman, A. A. (1992). The origins and evolution of predator-prey theory. *Ecology*, *73*(5), 1530–1535.
10. Cai, L., Chen, G., & Xiao, D. (2013). Multiparametric bifurcations of an epidemiological model with strong allee effect. *Mathematical Medicine and Biology*, *67*(2), 185–215.
11. Courchamp, F., Berec, L., & Gascoigne, J. (2008). *Allee effects in ecology and conservation*. Oxford University Press.
12. Das, S. (2008). *Functional fractional calculus for system identification and controls*. Springer.
13. Dawes, J. H. P., & Souza, M. (2013). A derivation of holling's type I, II and III functional responses in predator prey systems. *Journal of Theoretical Biology*, *327*, 11–22.
14. Dhooge, A., Govaerts, W., Kuznetsov, Y. A., Meijer, H. G. E., & Sautois, B. (2007). New features of the software matcont for bifurcation analysis of dynamical systems. *Mathematical and Computer Modelling of Dynamical Systems*, *14*(2), 147–175.
15. Dubey, B., & Upadhyay, R. K. (2004). Persistence and extinction of one-prey and two-predators system. *Nonlinear Analysis: Modelling and Control*, *9*, 307–329.
16. Dumitru, B., Guvenc, Z. B., & Machado, J. T. (2010). *New trends in nanotechnology and fractional calculus applications*. Springer.
17. Elaydixz, S. N., & Sacker, R. J. (2008). Population models with Allee effect: A new model. *Journal of Biological Dynamics*, *00*(00), 1–11.
18. Gilmour, S. G., & Trinca, L. A. (2005). Fractional polynomial response surface models. *Journal of Agricultural, Biological, and Environmental Statistics*, *10*(50), 200–203.
19. Kaczorek, T. (2011). *Selected problems of fractional systems theory*. Springer.
20. Kilbas, A. A., Srivastava, H. M., & Trujillo, J. (2006). *Theory and applications of fractional differential equations* (Vol. 204). Elsevier.
21. Leslie, P. H., & Gower, J. C. (1960). The properties of a stochastic model for the predator-prey type of interaction between two species. *Biometrika*, *47*(3–4), 219–234.
22. Lidicker, W. Z., Jr. (2010). The allee effect: Its history and future importance. *The Open Ecology Journal*, *3*, 71–82.
23. Lin, R., Liu, S., & Lai, X. (2013). Bifurcation of a predator-prey model system with weak Allee effects. *Journal of the Korean Mathematical Society*, *50*(4), 695–713.
24. Mingxin, W. (2004). Stationary patterns for a prey predator model with prey-dependent and ratio-dependent functional responses and diffusion. *Physica D*, *196*(1–2), 172–192.
25. Royston, P., & Altman, D. G. (1994). Regression using fractional polynomials of continuous covariates: Parsimonious parametric modelling. *Journal of the Royal Statistical Society*, *43*(3), 429–467.
26. Tzy-Wei, H. (2003). Global analysis of the predator prey system with beddington deangelis functional response. *Journal of Mathamtical Analysis and Applications*, *281*(1), 395–401.
27. Upadhyay, R. K., Iyengar, S. R. K., & Vikas, R. (2000). Stability and complexity in ecological systems. *Chaos, Solitons and Fractals*, *11*(4), 533–542.
28. Van Voorn, G. A., Hemerik, L., Boer, M. P., & Kooi, B. W. (2007). Heteroclinic orbits indicate overexploitation in predatorprey systems with a strong allee effect. *Mathematical Biosciences*, *209*(5), 451–469.
29. Volterra, V. (1928). Variations and fluctuations of the number of individuals in animal species living together. *ICES Journal of Marine Science*, *3*(1), 3–51.

30. Xiong, L., Zhao, Y., & Jiang, T. (2011). Stability analysis of linear fractional order neutral system with multiple delays by algebraic approach. *World Academy of Science, Engineering and Technology*, *52*, 983–986.
31. Yuan-Ming, W. (2009). Numerical solutions of a michaelis menten-type ratio-dependent predatorprey system with diffusion. *Applied Numerical Mathematics*, *59*(5), 1075–1093.
32. Zimmermann, B., Sand, H., Wabakken, P., Liberg, O., & Andreassen, H. P. (2014). Predator-dependent functional response in wolves: From food limitation to surplus killing. *Journal of Animal Ecology*, *84*(1), 102–112.